Lecture Notes in Computer Science 4247

Commenced Publication in 1973
Founding and Former Series Editors:
Gerhard Goos, Juris Hartmanis, and Jan van Leeuwen

Tzai-Der Wang Xiaodong Li
Shu-Heng Chen Xufa Wang
Hussein Abbass Hitoshi Iba
Guoliang Chen Xin Yao (Eds.)

Simulated Evolution and Learning

6th International Conference, SEAL 2006
Hefei, China, October 15-18, 2006
Proceedings

Springer

Volume Editors

Tzai-Der Wang
Cheng Shiu University, Taiwan, E-mail: dougwang@csu.edu.tw

Xiaodong Li
RMIT, Australia, E-mail: xiaodong@cs.rmit.edu.au

Shu-Heng Chen
AI-ECON Research Center, Taiwan, E-mail: chchen@nccu.edu.tw

Xufa Wang
University of Science and Technology of China, E-mail: xfwang@ustc.edu.cn

Hussein Abbass
University of New South Wales, Australia, E-mail: abbass@itee.adfa.edu.au

Hitoshi Iba
University of Tokyo, Japan, E-mail: iba@iba.k.u-tokyo.ac.jp

Guoliang Chen
University of Science and Technology of China, E-mail: glchen@ustc.edu.cn

Xin Yao
University of Birmingham, UK, E-mail: x.yao@cs.bham.ac.uk

Library of Congress Control Number: Applied for

CR Subject Classification (1998): F.1.1, I.2.6, I.6, G.1.6, D.2.2, J.3-4

LNCS Sublibrary: SL 1 – Theoretical Computer Science and General Issues

ISSN 0302-9743
ISBN-10 3-540-47331-2 Springer Berlin Heidelberg New York
ISBN-13 978-3-540-47331-2 Springer Berlin Heidelberg New York

Springer is a part of Springer Science+Business Media

springer.com

Typesetting: Camera-ready by author, data conversion by Scientific Publishing Services, Chennai, India
Printed on acid-free paper SPIN: 11903697 06/3142 5 4 3 2 1 0

Preface

We are very delighted to present this LNCS volume, the proceedings of the Sixth International Conference on Simulated Evolution And Learning (SEAL 2006). SEAL is a prestigious international conference series in evolutionary computation and learning. This biennial event was first held in Seoul, Korea, in 1996, and then in Canberra, Australia (1998), Nagoya, Japan (2000), Singapore (2002), and Busan, Korea (2004).

SEAL 2006 received a record 420 paper submissions this year. After an extensive peer review process involving more than 1100 reviews, the best 117 papers were selected by the programme committee to be presented at the conference and included in this volume, resulting in an acceptance rate of less than 30%.

The papers included in this volume cover a wide range of topics in simulated evolution and learning: from evolutionary learning to evolutionary optimisation, from hybrid systems to adaptive systems, from theoretical issues to real-world applications. They represent some of the latest and best research in simulated evolution and learning in the world.

The conference featured four distinguished keynote speakers: Karl Sigmund, Zbigniew Michalewicz, Han La Poutré and Gary Yen. Karl Sigmund's talk was on "The Evolution of Cooperation in Groups." Zbigniew Michalewicz's talk was on "Adaptive Business Intelligence." Han La Poutré's talk was on "Learning Agents in Socio-economic Games." Gary G. Yen's talk was on "Adaptive Critics for Fault Tolerant Control". We were very fortunate to have such distinguished speakers giving talks at SEAL 2006 despite their busy schedules. Their presence at the conference was yet another indicator of the importance of SEAL on the international research map.

SEAL 2006 also included five tutorials, which were free to all conference participants. The five tutorials covered some of the hottest topics in evolutionary computation and its applications, i.e., Evolutionary Multiobjective Optimization and its Applications (Gary Yen), Evolutionary Computation for Real-World Problems (Zbigniew Michalewicz), Automatic Decomposition in Evolutionary Computation for Optimization and Learning (Hussein Abbass), Particle Swarm Optimization (Xiaodong Li) and Recent Advances in Real Parameter Optimization (P.N. Suganthan). They provided an excellent start to the four-day event.

Furthermore, Jim Kennedy, a pioneer in particle swarm optimisation, also gave a brilliant plenary speech at SEAL 2006.

The success of a conference depends on its authors, reviewers and organisers. SEAL 2006 was no exception. We were very grateful to all the authors for their paper submissions and to all the reviewers for their outstanding work in refereeing the papers within a very tight schedule. We relied heavily upon a team of volunteers to keep the SEAL 2006 wheel turning. They were true heros working behind the scene. In particular, Wenjian Luo from USTC in Hefei, China, played a crucial role in organising the conference. We are most grateful to all the volunteers for their great efforts and contributions.

August 2006

Xin Yao
Tzai-Der Wang
Xiaodong Li

Organisation

The Sixth International Conference on Simulated Evolution And Learning (SEAL 2006) was organised and hosted by the University of Science and Technology of China, Hefei, China.

SEAL 2006 Conference Committee

General Chair	Guo-Liang Chen (China)
Programme Chairs	Xin Yao (UK) Xufa Wang (China)
Technical Co-chairs	Shu-Heng Chen (Taiwan) Tzai-Der Wang (Taiwan) Hussein Abbass (Australia) Hitoshi Iba (Japan) Zengqi Sun (China) Bob McKay (South Korea)
Tutorials and Special Sessions Chair	Xiaodong Li (Australia)
Organising Committee Co-chairs	Xufa Wang (China) Liusheng Huang (China) Naijie Gu (China)
Organising Committee Members	Haoran Zheng, Xianbin Cao, Enhong Chen, Houning Wang, Bin Li, Wenjian Luo, Jinlong Li, Shangfei Wang, Jianhui Ma, Sihai Zhang, Kaiming Chen, Hai Qian

SEAL 2006 Steering Committee

Takeshi Furuhashi	Japan
Jong-Hwan Kim	South Korea
Bob McKay	South Korea
Xin Yao	UK

SEAL 2006 Tutorials

Evolutionary Multiobjective Optimization and Its Applications
Gary G. Yen

Evolutionary Computation for Real-World Problems
Zbigniew Michalewicz

Automatic Decomposition in Evolutionary Computation for Optimization and Learning
Hussein Abbass

Particle Swarm Optimization
Xiaodong Li

Recent Advances in Real Parameter Optimization
P.N. Suganthan

SEAL 2006 Programme Committee

Abbass, Hussein
Alam, Sameer
Bonabeau, Eric
Bui, Lam Thu
Cao, Wenming
Chandra, Arjun
Chen, Enhong
Chen, Guoqing
Chen, Shu-Heng
Chen, Shyi-Ming
Cheng, Xiaochun
Cho, Sung-Bae
Chong, Siang Yew
Chu, Chao-Hsien
Chuang, Shang-Jen
Coello, Carlos A. Coello
Dam, Helen
Darwen, Paul
de Carvalho, André
Engelbrecht, AP
Fyfe, Colin
Gallegati, Mauro
Gao, Jun
Handa, Hisashi
He, Jun
Heymann, Daniel
Hiot, Lim-Meng
Hiroshi, Furutani
Iba, Hitoshi
Ishibuchi, Hisao
Izumi, Kiyoshi
Jain, Lakhmi
Jiao, Licheng
Jin, Yaochu
Jo, Jun
Kahraman, Cengiz
Kaizoji, Taisei
Kendall, Graham
Klos, Tomas
Kubota, Naoyuki
Laing, Yiwen
Lajbcygier, Paul
Lee, JangMyung
Lee, Yuh-Jye
Leung, Kwong-Sak
Li, Bin
Li, Jin
Li, Tong
Li, Xiaodong
Lin, Ping-Chen
Luo, Wenjian
McKay, Bob
McMullan, Paul
Michalewicz, Zbigniew
Miyamoto, Sadaaki
Mo, Hongwei
Nakao, Zensho
Navet, Nicolas
Nishizaki, Ichiro
North, Michael
Ohkura, Kazuhiro
Pei, Jihong
Pichl, Lukas
Riechmann, Thomas
Sasaki, Yuya
Sato, Yuji
Schnier, Thorsten
Shi, Zhongzhi
Soon, Ong Yew
Suganthan, Ponnuthurai Nagaratnam
Sun, Jianyong
Sun, Zengqi
Sun, Zhaohao
Szeto, Kwokyip

Takadama, Keiki
Tan, Kay Chen
Tang, Maolin
Terano, Takao
Tohme, Fernando
Tsaih, Rua-huan
Tsao, Chueh-Yung
Tseng, Chiu-Che
Velez-Langs, Oswaldo
Verma, Brijesh
Vila, Xavier
Vriend, Nicholas J.
Wang, Binghong
Wang, Han
Wang, Lipo
Wang, Shinn-Wen
Wang, Tzai-Der
Wang, Xufa
Wang, Zengfu
Westerhoff, Frank
Whigham, Peter
Xu, Yong
Xue, Weimin
Yang, Shengxiang
Yao, Xin
Yen, Gary
Yu, Xinghuo
Zhang, Byoung-Tak
Zhang, Ling
Zhao, Mingsheng
Zhao, Qiangfu
Zheng, Zijian
Zhou, Zhihua

Sponsoring Institutions

National Natural Science Foundation of China

Chinese Academy of Sciences

Anhui Computer Federation

University of Science and Technology of China

The Centre of Excellence for Research in Computational Intelligence and Applications (Cercia), The University of Birmingham, UK

Anhui Province Key Laboratory for Computing and Communication Software Engineering

Table of Contents

Evolutionary Learning

Hybrid Learning

Adaptive Systems

Theoretical Issue in Evolutionary Computation

Real-World Applications of Evolutionary Computation Techniques

Evolutionary Dynamics on Graphs: The Moran Process

P.A. Whigham and G. Dick

Spatial Information Research Centre,
Information Science Dept., Univ. of Otago,
Dunedin, New Zealand
Ph: +64 3 4797391
pwhigham@infoscience.otago.ac.nz

Abstract. Evolutionary dynamics on graphs for the Moran process have been previously examined within the context of fixation behaviour for introduced mutants, where it was demonstrated that certain spatial structures act as amplifiers of selection. This paper will revisit the assumptions for this spatial Moran process and show that the assumption of proportional global fitness, introduced as part of the Moran process, is necessary for the amplification of selection to occur. Under the circumstances of local proportional fitness selection the amplification property no longer holds, which supports the original results from population genetics that spatial structure does not alter fixation probability for fixed population sizes with constant migration.

1 Introduction

The probability of fixation of an introduced allele in a spatially-structured population is a fundamental concept in population genetics, evolutionary biology and evolutionary computation [1]. The question of how the organisation of subpopulations alters the fate of alleles in terms of their probability of fixation and time to fixation has relevance in understanding the dynamics of the evolutionary process.

The evolution of a population is characterised by two forces: genetic drift (the stochastic effect of sampling) and selection. In addition, how these forces interact can be altered by the spatial arrangement of the population, so that subpopulations, or demes, of local interaction are formed. As would be expected, forming subpopulations which are distantly connected to each other affect genetic drift by increasing the mean time to fixation (loss of all variation) of the population [2,3,4,5]. Further work on structured populations in the field of population genetics showed that, when migration is conservative (i.e. fixed local population sizes) the probability of fixation of an introduced mutant was equal to that of a well-mixed (panmictic) population [6,7]. This has lead to the assumption that the probability of fixation of an introduced beneficial alelle was invariant to the neighbourhood structure of a population.

In the field of Evolutionary Computation the effect of spatially-structured populations on the dynamics of evolution have been mainly considered from the standpoint of takeover times [8,9,10]. Although these approaches have shown that the takeover times vary as a measure of selection pressure, the models use an elitist strategy so that nonextinctive selection occurs. Hence the concept of probability of fixation is meaningless, and therefore these approaches will not be considered in this paper.

T.-D. Wang et al. (Eds.): SEAL 2006, LNCS 4247, pp. 1–8, 2006.

2.1 Selection Amplifiers

When W is not isothermal the fixation probability is no longer described by Eqn. 1. Lieberman et al. (2005) showed that for a variety of spatial structures that selection was amplified. This paper will examine just one of these structures, a star structure, which was shown to have fixation probability, ρ_K, with an amplification of $K = 2$:

$$\rho_K = \frac{1 - 1/r^K}{1 - 1/r^{KN}} \tag{2}$$

The star structure has a central node that is connected to all other vertices. For example, a star with $N = 5$ has the form and stochastic matrix W shown in Fig. 2, assuming the central vertex is individual I_1. The original interpretation of an entry w_{ij} for the Moran process was that it defined the probability of j being the location of the offspring from parent I_i. Here we change the interpretation of W so that each row W_i defines the deme (subpopulation) associated with I_i, and therefore each individual is part of its own deme ($\forall i w_{ii} > 0$). The Moran process does not allow the offspring to be placed at the location of the parent and therefore this interpretation does not alter the behaviour of the model. However, it will be found to be a more appropriate description when local Moran processes are considered in §3. Empirical results for this property are shown in

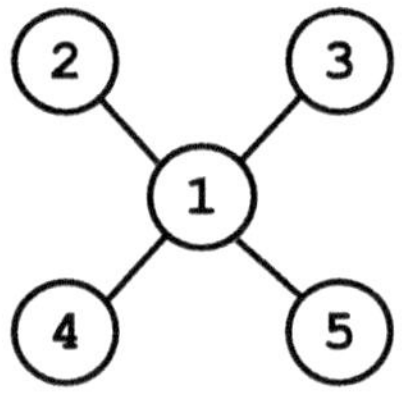

$$W = \begin{bmatrix} 0.2 & 0.2 & 0.2 & 0.2 & 0.2 \\ 0.5 & 0.5 & 0 & 0 & 0 \\ 0.5 & 0 & 0.5 & 0 & 0 \\ 0.5 & 0 & 0 & 0.5 & 0 \\ 0.5 & 0 & 0 & 0 & 0.5 \end{bmatrix}$$

Fig. 2. The star population (N=5) and associated W matrix

Fig. 3, where the probability of fixation for a range of mutant relative fitness r, and the mean time to fixation for a star structure with $N = 100$ averaged over $10,000$ runs, is shown.These results confirm Eqn. 2 for the star structure, and stand in contrast to the population genetics results which imply spatial structure has no effect on fixation probability. Note that although the probability of fixation is amplified, the mean time to fixation has increased by two orders of magnitude over a panmictic neighbourhood structure that satisfies ρ_1. Hence although the balance between drift and selection has tilted in favour of selection, the increase in time to fixation indicates that drift is interacting to combat this selective advantage. When the mutant fitness $r = 1$ the only force acting on fixation is drift. In this situation (Fig. 3) the star structure increases the time to fixation from approximately $10,000$ for regular structures to $1,000,000$ time steps. Although the homogenising effect of drift is slowed for regular spatial structures [12], for spatial structures such as a star drift acts to rapidly reduce the diversity in the population [13]. Hence for a star structure, if fixation of the mutant is to occur the action of drift needs to be overcome, and therefore the mean time to fixation increases.

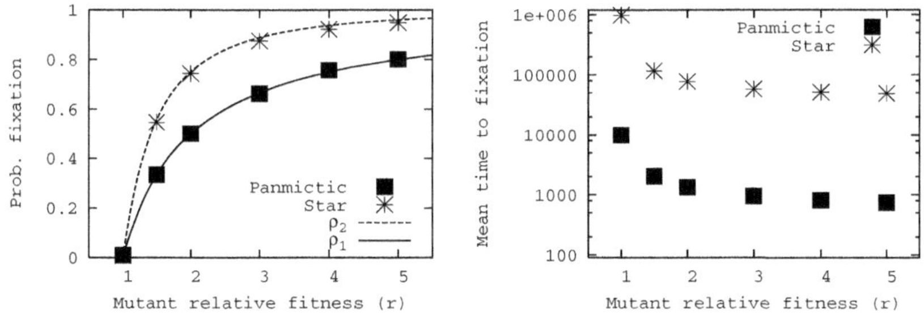

Fig. 3. The Moran process for a star population (N=100)

3 The Local Moran Process

The previous section has demonstrated that certain spatial structures act as amplifiers for the Moran process. However, the Moran process described in Algorithm 1 used a global fitness proportionate selection (step 1) to select the parent at each time step. A spatial structure implies that subpopulations are only distantly connected to one another, the use of a global selection scheme seems inappropriate. Since replacement acts locally, the deme structures defined by W should also be used to select parents. A local spatial version of the Moran process is shown as Algorithm 2. For this process, a deme is

Algorithm 2. The local Moran process for a structured population

input : Population P_0 at time zero, Connectivity Matrix W
output: Fixated Population P_t after t time steps

1 **while** P_t *not fixed* **do**
2 $t \leftarrow t + 1$;
3 I_i = `Random(`I`);`
4 *fitness* ← `SumFitness(`W_i`);`
5 I_p ← `SelectParent(`*fitness*, W_i`);`
6 j ← `OffspringLoc(`I_p, W_i`);`
7 $I_j \leftarrow I_p$;
8 **end**
9 **return** P_t;

selected (line 2) by randomly picking a location I_i. The sum of the fitness for the deme associated with I_i is calculated (line 2), and a parent from the deme is then selected with probability proportional to fitness of the individuals in the deme (line 2). The offspring location is then determined from W_i as for Algorithm 1. Hence the local Moran process differs from the global algorithm only in the mechanism for selecting the parent: the probability of selection is proportional to the relative fitness of the deme, rather than the relative fitness compared with the entire population. The results for the local Moran process using the star structure with $N = 100$, averaged over $10,000$ runs, are shown in Fig. 4. The redefinition of the parent selection to a local deme has resulted in the probability and time to fixation returning to the original ρ_1 behaviour for a panmictic population. Hence, for the local Moran process, no amplification of selection or adjustment in the time to fixation occurs, in agreement with the original findings from population genetics [6,7].

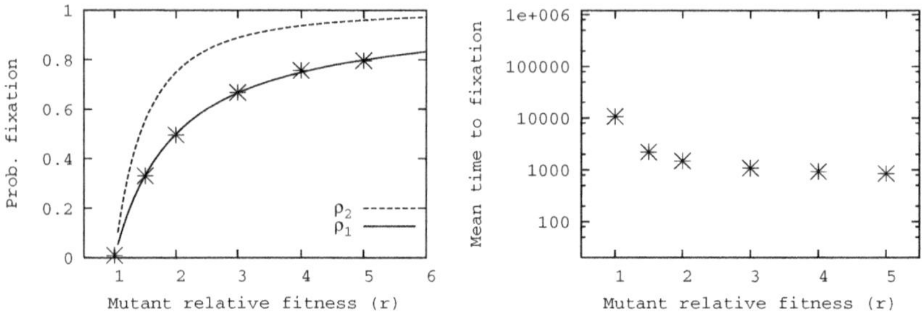

Fig. 4. The Local Moran process for a star population (N=100)

4 Discussion

What are the differences between the local and global Moran process on a star structure, and why do these change the probability of fixation? Selection of the parent for the global Moran process is independent of spatial structure. Hence it is just the influence of drift on the selection of the offspring that varies with different neighbourhood arrangements. Since the global selection of a parent is greater than the corresponding probability of selecting a mutant parent under local selection (Fig. 5(a)), this allows the global Moran process to propagate mutants more rapidly than the local Moran on a star. Each time a non-central mutant parent is selected the central node of the star becomes a mutant, and therefore this increases the chance of a mutant being moved to a new non-central node. Since the star structure accelerates the homogenising effect of drift, the global selection of the parent increases the probability of a mutant creation at the star's centre, which can then produce an offspring on the star edge. Although drift reduces the variation in the population, global selection amplifies the relative fitness advantage of the introduced mutant for the star structure. The local Moran process interacts with the star structure in a different way. Since the probability of picking a parent depends on the initial deme selection (which is random), there is a constant probability that, for a particular deme, the parent will be a mutant. This is lower than the corresponding global selection probability. However, it is the influence of the spatial structure that ultimately determines the fixation probability - spatial structures where a small number of nodes are highly connected, and therefore act as hubs, become amplifiers for the global Moran process. This is shown in Fig 5(b), where the probability of fixation for spatial structures defined by a set of Kawachi-style networks [14] are shown. As the Kawachi parameter increases towards one the network structure transitions from regular towards scale-free, with a subsequent increase in the dominance of a few highly connected nodes. The probability of fixation increases for the global Moran process, whereas the local process remains constant.

Generalized Embedded Landscape and Its Decomposed Representation*

Shude Zhou[1], Robert B. Heckendorn[2], and Zengqi Sun[1]

[1] Department of Computer Science and Technology, Tsinghua University, Beijing, China
zsd03@mails.tsinghua.edu.cn,
szq-dcs@tsinghua.edu.cn
[2] Department of Computer Science, University of Idaho, Moscow, Idaho, USA
heckendo@uidaho.edu

Abstract. In this paper, embedded landscapes are extended to a non-binary discrete domain. Generalized embedded landscapes (GEL) are a class of additive decomposable problems where the representation can be expressed as a simple sum of subfunctions over subsets of the representation domain. The paper proposes a Generalized Embedding Theorem that reveals the close relationship between the underlying structure and the Walsh coefficients. Theoretical inductions show that the Walsh coefficients of any GEL with bounded difficulty can be calculated with a polynomial number of function evaluations. A deterministic algorithm is proposed to construct the decomposed representation of GEL. It offers an efficient way to detect the decomposable structure of the search space.

1 Introduction

Genetic algorithms have been found to be not very effective at solving optimization problems where there is a large amount of interactions between variables [1]. It has been realized that competent and scalable performance can only be obtained when the information of building block linkage is already in the problem coding or is identified before or during the search process of genetic algorithm [2]. Due to the above fact, during the last decade the technology of learning the linkage information has become an intensively studied research topic in the field of evolutionary computation. Various attempts have been made to learn the linkage information between variables. In GA literature, such technologies mainly include the messy GA [3], the gene expression messy GA [4], linkage learning GA [5], estimation of distribution algorithms [6,7] and linkage detection algorithms [8,9,10]. This paper will focus the discussion on linkage detection algorithms based on Walsh transformation.

In GA field, embedded landscapes [8] (also named additively decomposed functions [7]) are a class of typical optimization problems that have been widely studied to help design competent genetic algorithms. Embedded landscape is valid for any binary problem where the representation can be expressed as a sum of subfunctions over subsets of the representation domain [8]. It has been theoretically shown that

* Shude Zhou and Zengqi Sun are funded by the National Key Project for Basic Research of China (G2002cb312205).

T.-D. Wang et al. (Eds.): SEAL 2006, LNCS 4247, pp. 9 – 17, 2006.

there is a close relationship between Walsh coefficients of the function and the underlying decomposed structure. Efficient detection algorithms to compute the decomposed representation of embedded landscape have been proposed [8,9,10]. One of the shortcomings of the existent research is that the current research is limited to binary domain. In order to design efficient genetic algorithm for more general problems, detection of linkage information in non-binary domain is desirable.

This paper studied generalized embedded landscapes (GEL) whose domain is a non-binary discretization. We adopt generalized Walsh transformation as mathematical tool to analyze the underlying structure of GEL. The relationship between the GEL underlying structure and the Walsh coefficients is revealed in the Generalized Embedding Theorem. Theoretical inductions will show that the Walsh coefficients of any GEL with bounded difficulty can be calculated with a polynomial number of function evaluations. A deterministic algorithm is constructed to obtain the decomposed representation of GEL. It offers an efficient way to detect the decomposed structure of the search space.

2 Basic Notation and Definition

This section introduces some basic notations and definitions related to generalized embedded landscapes.

Rather than look at a simple alphabet of 0 and 1 we look at an alphabet from the set $\tilde{M} = \{0,1,\cdots,M-1\}$. The space $\tilde{M}^L$ is the space of L dimensional vectors with each component selected from $\tilde{M}$. When $\tilde{M} = \{0,1\}$ then $\tilde{M}^L = B^L$. The size of $\tilde{M}$ is M.

$\|x\|$, where $x \in \tilde{M}^L$ denotes the number of positions with non-zero value. For example, $x = (0,2,3,0,1,2)$ and $\|x\| = 4$.

The *supplement* of x is denoted by $\bar{x}$, $\forall k = 1 \cdots L, \bar{x}_k = \text{mod}_M (M - x_k)$.

We define a *template set* as follows: let $S(x)$ be defined over $x \in \tilde{M}^L$ such that:

$$S(x) = \{y \mid y \in \tilde{M}^L \;\&\; y_k = 0 \text{ if } x_k = 0\}.$$

For example, a 6-length string $x = (0,2,3,0,0,1) \in \{0,\cdots 3\}^6$, then $S(x) = \{y \mid y_1 = y_4 = y_5 = 0, y_2 \in \{0,\cdots,3\}, y_3 \in \{0,\cdots,3\}, y_6 \in \{0,\cdots,3\}\}$ and $|S(x)| = 4^3$. For any $x \in \tilde{M}^L, \|x\| \neq 0$, there are $M^{\|x\|}$ members in $S(x)$; if $\|x\| = 0$, then $S(x) = \phi$.

$[b]_M$ denotes the remainder when non-negative integer b is divided by M.

For $i, j \in \tilde{M}^L$: $i \subseteq j$ iff $\forall i_k \; i_i = j_k$ if $j_k \neq 0$.

A function $pack(x,y)$ is defined as $pack$: $\tilde{M}^L \times B^L \mapsto \tilde{M}^H$, where $H \leq L$ is the number of nonzero elements in y. $pack(x,y)$ is the vector composed of elements of x that are in the same positions as the elements of y that are nonzero. Order is preserved. For example, $pack((0,2,3,0,1,2),(0,1,0,1,1,0)) = (2,0,1)$.

Generalized embedded landscape (GEL) is defined as

$$f(x)=\sum_{i=1}^{P} g_i\left(pack(x,m_i)\right) \tag{1}$$

where P is the number of subfunctions, $g_i\left(pack(x,m_i)\right)$ is the i-th subfunction whose domain is $\tilde{M}^{\|m_i\|}$, $x\in\tilde{M}^L$ and $m_i\in B^L$. This is an extension of embedded landscapes which is usually restricted to binary domain in GA literature [8]. A GEL is defined of *order-k*, if $\max_{1\le i\le P}\|m_i\|=k$.

Example: $f(x_1,x_2,x_3,x_4)=g_1(x_1,x_2)+g_2(x_3,x_4)$, where $(x_1,x_2,x_3,x_4)\in\{0,1,2\}^4$, $m_1=(1,1,0,0)$, $m_2=(0,0,1,1)$.

3 Generalized Walsh Transform

Walsh transform is a useful mathematical tool to analyze the information of building block linkage in binary-coded genetic algorithms [8,9,10]. In order to study the characters of generalized embedded landscapes, Walsh transform has to extend to more generalized non-binary discrete domain [11].

The generalized Walsh function ψ can be defined as:

$$\psi_j^{(M)}(x)=\frac{1}{\sqrt{M^L}}e^{\frac{2\pi\sqrt{-1}}{M}(x\cdot j)} \tag{2}$$

where $x,j\in\{0,1,\cdots,M-1\}^L$. And $\bar{\psi}$ indicates the complex conjugate of ψ. There are total M^L orthogonal Walsh basis functions. Any function $f:\tilde{M}^L\mapsto R$ can be represented as the linear sum of the orthogonal Walsh basis functions:

$$f(x)=\sum_j \omega_j\psi_j^{(M)}(x) \tag{3}$$

where ω_j is the Walsh coefficient corresponding to the j-th Walsh basis function.

The Walsh transform creates an invertible linear transformation between an enumeration of the M^L function values ordered by counting in $\tilde{M}^L$ space expressed as a vector and a vector of Walsh coefficients. The Walsh coefficients can be calculated by:

$$\omega_j=\sum_x f(x)\bar{\psi}_j^{(M)}(x) \tag{4}$$

4 Generalized Embedding Theorem

The important characteristic of GEL is its underlying decomposable structure. In this section, a generalized Walsh transform will be used to study the embedding feature of GEL. It will be proved that the decomposable structure (also named epistatic linkage [8]) has close relation with Walsh coefficients.

First of all, we show how embedding a single function in a higher dimensional space imposes a relation between the Walsh coefficients of the two functions. Let $g:\tilde{M}^H \mapsto R$ and $f:\tilde{M}^L \mapsto R$ with $H \le L$. Function g is said to be *embedded* in f if there is a mask m, $m \in B^L$ and $\|m\| = H$ such that $f(x) = g(pack(x,m))$, $\forall x \in \tilde{M}^L$. The following theorem shows the Walsh coefficients of f only depend on its embedding function g. Embedding a lower dimensional function in a higher dimensional space will not increase the number of nonzero Walsh coefficients.

Theorem 1 (Generalized Embedding Theorem): *Let $g(x): \tilde{M}^H \mapsto R$ and $f(x): \tilde{M}^L \mapsto R$ and g is embedded in f, then Walsh coefficients can be calculated:*

$$\omega_i^f = \begin{cases} \sqrt{M^{L-H}}\,\omega^g_{pack(i,m)} & \text{if } i \in S(m) \\ 0 & \text{otherwise} \end{cases} \qquad (5)$$

where ω^g and ω^f are Walsh coefficients for the g and f function, respectively, and $i \in \tilde{M}^L$, $m \in B^L$ and $\|m\| = H$.

Proof:

$$\begin{aligned}
\omega_i^f &= \sum_{x=0}^{M^L-1} f(x)\overline{\psi}_x^{(M)}(i) \\
&= \sum_{x=0}^{M^L-1} g(pack(x,m))\overline{\psi}_x^{(M)}(i) \\
&= \sum_{j=0}^{M^H-1} \sum_{x:pack(x,m)=j} g(pack(x,m))\overline{\psi}_x^{(M)}(i) \\
&= \sum_{j=0}^{M^H-1} \sum_{x:pack(x,m)=j} g(j)\overline{\psi}_x^{(M)}(i) \\
&= \sum_{j=0}^{M^H-1} g(j) \sum_{x:pack(x,m)=j} \overline{\psi}_x^{(M)}(i)
\end{aligned}$$

We consider $\sum_{x:pack(x,m)=j} \overline{\psi}_x^{(M)}(i)$ in two cases: $i \notin S(m)$ and $i \in S(m)$.

Case 1: Assume $i \notin S(m)$:

$$\begin{aligned}
\sum_{x:pack(x,m)=j} \overline{\psi}_x^{(M)}(i) &= \sum_{x:pack(x,m)=j} \frac{1}{\sqrt{M^L}} e^{\frac{2\pi\sqrt{-1}}{M}(-x\cdot i)} \\
&= \sum_{x:pack(x,m)=j} \frac{1}{\sqrt{M^L}} e^{\frac{2\pi\sqrt{-1}}{M}\left[-\sum_{k=1}^{L} x_k i_k\right]_M}
\end{aligned}$$

$x \in S(m), i \notin S(m)$, according to characters of Modular Algebraic:

$= 0$

So:

$$\begin{aligned}\omega_i^f &= \sum_{j=0}^{M^H-1} g(j) \sum_{x:pack(x,m)=j} \overline{\psi}_x^{(M)}(i)\\ &= \sum_{j=0}^{M^H-1} g(j)0\\ &=0\end{aligned}$$

Case 2: Assume $i \in S(m)$:

$$\begin{aligned}\sum_{x:pack(x,m)=j} \overline{\psi}_x^{(M)}(i) &= \sum_{x:pack(x,m)=j} \frac{1}{\sqrt{M^L}} e^{\frac{2\pi\sqrt{-1}}{M}\left(-\sum_{k=1}^{L} x_k i_k\right)}\\ &= \sum_{x:pack(x,m)=j} \frac{1}{\sqrt{M^L}} e^{\frac{2\pi\sqrt{-1}}{M}\left(-\sum_{k=1}^{H}((pack(i,m))_k j_k\right)}\\ &= M^{L-H}\left(\frac{1}{\sqrt{M^{L-H}}}\frac{1}{\sqrt{M^H}} e^{\frac{2\pi\sqrt{-1}}{M}(-pack(i,m)\cdot j)}\right)\\ &= \sqrt{M^{L-H}}\,\overline{\psi}_{pack(i,m)}^{(M)}(j)\end{aligned}$$

So:

$$\begin{aligned}\omega_i^f &= \sum_{j=0}^{M^H-1} g(j) \sum_{x:pack(x,m)=j} \overline{\psi}_x^{(M)}(i)\\ &= \sum_{j=0}^{M^H-1} g(j)\sqrt{M^{L-H}}\,\overline{\psi}_{pack(i,m)}^{(M)}(j)\\ &= \sqrt{M^{L-H}}\,\omega_{pack(i,m)}^g\end{aligned}$$

It is obvious that if $i \in S(m)$, the Walsh coefficient ω_i is equal to $\sqrt{M^{L-H}}\,\omega_{pack(i,m)}^g$, otherwise, the Walsh coefficient is zero. □

Generalized embedding theorem shows that the relation between f and g can be exhibited by the Walsh coefficients. If we restrict $\tilde{M} = \{0,1\}$, the theorem is reduced to embedding theorem proposed by R. B. Heckendorn [8, 10].

Next, we extend GEL f to more general function formulation which comprises many subfunctions. Because the GEL is a sum of subfunctions and the Walsh transform is a linear transformation, then by Generalized Embedding Theorem we have Theorem 2 that reveals the relationship between its underlying structure and the Walsh coefficients. Detailed proof is omitted due to the page limit.

Theorem 2. $f(x)=\sum_{i=1}^{P} g_i(pack(x,m_i))$ *is a generalized embedded landscape, then its Walsh coefficients can be calculated:*

$$\omega_j^f = \begin{cases}\alpha & \text{if } \exists i \in \{1\cdots P\}, j \in S(m_i)\\ 0 & \text{otherwise}\end{cases}, \tag{6}$$

First of all, we consider Walsh coefficient ω_i, where $\|i\| = k-1$. According to Equation (7) in which $\forall j, j \subseteq i$ and $\|j\| \geq k-1$, ω_j has been computed before, we can compute ω_i by

$$\omega_i = \frac{\sum_{x \in S(i)} f(x)\overline{\psi}_i^{(M)}(x) - \sum_{j \subseteq i \wedge \|j\| > \|i\|} \omega_j \sum_{x \in S(i)} \psi_j^{(M)}(x)\overline{\psi}_i^{(M)}(x)}{\sum_{x \subseteq S(i)} \psi_i^{(M)}(x)\overline{\psi}_i^{(M)}(x)}, \qquad (9)$$

in which the values of $f(x)$, $x \in S(i)$ has been calculated in Case 2. Using the Equation (9), we can continue iteratively to calculate Walsh coefficients of $(k-2), (k-3), \cdots$, 1-order partitions.

It should be noted that, the computation of order-*k* GEL's Walsh coefficients requires only $M^k \binom{L}{k}$ function evaluations that happens in Case 2. The bounded decomposition property can therefore be exploited using only a number of function evaluations $M^k \binom{L}{k}$ compared to the usual M^L evaluations needed using regular approach. Based on the above analyses, a simple, deterministic algorithm to compute the Walsh coefficients of order-*k* GEL can be summarized in Fig. 1.

Computing the WCs of order-k GEL

Setp 1. *Set* $k' = k$.

Setp 2. *Compute the Walsh coefficients* ω_i, *where* $\|i\| = k'$, *using Equation (8).* $M^k \binom{L}{k}$ *fitness evaluations are required.*

Setp 3. $k' = k' - 1$. *Compute the Walsh coefficients* ω_i, *where* $\|i\| = k'$, *using Equation (9). No fitness evaluation is needed.*

Setp 4. *If* $k' = 0$, *then all the WCs have been computed; otherwise go to Setp 3.*

Fig. 1. Deterministic algorithm to computer Walsh coefficients of order-*k* GEL

Given an order-*k* GEL, where *k* is fixed independently of dimension *L*, this proposed algorithm can find all the nonzero Walsh coefficients using only polynomial number of function evaluations $M^k \binom{L}{k}$. According to Generalized Embedding Theorem, once the Walsh coefficients are calculated, the decomposed structure can be obtained by its decomposed Walsh representation.

6 Conclusions

This paper extends embedded landscapes to a non-binary discrete domain. Generalized embedded landscapes (GEL) are a class of additive decomposable problems where the representation can be expressed as a simple sum of subfunctions over subsets of the representation domain.

There are mainly two contributions in theoretical aspect:

First, the paper proposes Generalized Embedding Theorem that is the rigorous mathematical foundation for Walsh analyses of GEL. This theorem shows that the close relationship between the underlying decomposable structure of GEL and its Walsh coefficients.

Second, theoretical inductions show that the Walsh coefficients of any GEL with bounded epistatic order can be calculated with a polynomial number of function evaluations. A deterministic algorithm is constructed to calculate the decomposed representation of GEL. It offers an efficient way to detect the decomposed structure of the search space.

The paper is our first attempt to consider the decomposition algorithm of GEL based on generalized Walsh transform. Future work will include extending embedded landscapes to continuous domains and designing efficient decomposition algorithms.

References

1. Kargupta, H., Bandyopadhyay, S.: A Perspective on the Foundation and Evolution of the Linkage Learning Genetic Algorithms. Journal of Computer Methods in Applied Mechanics and Engineering, Vol. 186. (2000) 266–294
2. Thierens D.: Scalability Problems of Simple Genetic Algorithms. Evolutionary Computation, Vol. 7.4. (1999) 331–352
3. Goldberg, D.E., Deb, K., Kargupta, H., Harik, G.: Rapid, Accurate Optimization of Difficult Optimization Problems Using Fast Messy Genetic Agorithms. In: Proceedings of the Fifth International Conference on Genetic Algorithms, San Mateo, USA (1993)
4. Kargupta, H.: The Gene Expression Messy Genetic Algorithm. In: Proceedings of the IEEE international conference on Evolutionary Computation, Nogoya, Japan (1996)
5. Harik, G.R.: Learning Gene Linkage to Efficiently Solve Problems of Bounded Difficulty Using Genetic Algorithms. Ph.D. Thesis, University of Michigan, Ann Arbor, MI (1997)
6. Larrañaga, P., Lozano, J.A.: Estimation of Distribution Algorithms: A new Tool for Evolutionary Computation. Kluwer Academic Publishers (2002)
7. Pelikan, M.: Hierarchical Bayesian Optimization Algorithm: Toward a New Generation of Evolutionary algorithms. Springer Publication (2005)
8. Heckendorn, R.B.: Embedded Landscapes. Evolutionary Computation, Vol. 10.4. (2002) 345–369
9. Kargupta, H., Park, B.: Gene Expression and Fast Construction of Distributed Evolutionary Representation. Evolutionary Computation, Vol. 9.1. (2001) 43–69
10. Hechendorn, R.B., Wright A.H.: Efficient Linkage Discovery by Limited Probing. Evolutionary Computation, Vol. 12.4. (2004) 517–545
11. Oei, C. K.: Walsh Function Analysis of Genetic Algorithms of Nonbinary Strings. Unpublished Master's Thesis, UIUC (1992)
12. Song, Y. Y.: Number Theory for Computing. Springer-Verlag (2002)
13. Bak, J., Newman, D. J.: Complex Analysis. Springer-Verlag (1996)

Representative Selection for Cooperative Co-evolutionary Genetic Algorithms

Sun Xiao-yan, Gong Dun-wei, and Hao Guo-sheng

School of Information and Electrical Engineering,
China University of Mining & Technology Xuzhou,
Jiangsu, 221008, China
Xysun78@126.com, Dwgong@vip.163.com, Hgs0754@163.com

Abstract. The performance of cooperative co-evolutionary genetic algorithms is highly affected by the representative selection strategy. But rational method is absent now. Oriented to the shortage, the representative selection strategy is studied based on the parallel implementation of cooperative co-evolutionary genetic algorithms in LAN. Firstly, the active cooperation ideology for representative selection and the dynamical determinate method on cooperation pool size are put forward. The methods for determining cooperation pool size, selecting cooperators and permuting cooperations are presented based on the evolutionary ability of sub-population and distributive performance of the individuals. Thirdly, the implementation steps are given. Lastly, the results of benchmark functions optimization show the validation of the method.

1 Introduction

In cooperative co-evolutionary genetic algorithms (CCGAs), the variable space is decomposed firstly, and then the decompositions are encoded to form sub-population. To perform the fitness evaluation, each sub-population submits representatives to form cooperators and these cooperators are applied to objective function to obtain fitness [1]. Because of its good parallelism and high efficiency in processing short coding chromosomes, it has been widely used in complicated function optimization [2], multi-objective function optimization [3], robots behavior study [4], neural network optimization [5], agent control system learning [6] and so on.

In CCGAs, the selection strategy for representatives is critical to the fitness evaluation. Three main factors such as selection pressure, credit assignment and size of cooperation pool make great influence on the representatives selection [7]. Among the three factors, the size of cooperation pool makes severe impact on the performance of the algorithm. Increasing the size of cooperation pool can largely improve the optimization performance of CCGAs and of course the computational complexity is also obviously increased. In work [7], it only gave qualitative discussions on the affect of the factors but no practicable ones, especially about how to decide the size of cooperation pool. So it is very necessary and important to make study on representative selection.

Oriented to the representative selection shortage, the performance of cooperative co-evolutionary genetic algorithms is highly impacted on by the representative selection strategy. But rational method is absent now. The representative selection strategy

T.-D. Wang et al. (Eds.): SEAL 2006, LNCS 4247, pp. 18 – 25, 2006.

is studied based on the parallel implementation of cooperative co-evolutionary genetic algorithms in LAN. Firstly, the active cooperation ideology for representative selection and cooperation pool size dynamical determination is presented. The methods for determining cooperation pool size, selecting cooperators and permuting cooperations are presented based on the evolutionary ability of sub-population and their individuals' distributive performance.

2 The Strategy for Representative Selection

In traditional CCGAs, when fitness is evaluated, each sub-population is asked to present representatives for other sub-population. That is to say each sub-population submits their representatives passively without consideration to its self-performance which here is called a passive cooperative method. In this paper the traditional manner is broken and a positive cooperative method is presented. The main idea is that each sub-population submits their representatives to the database positively based on its self-evolutionary performance and the representatives are selected from the database based on the self-evolutionary performance of the cooperative sub-population. In this case, the cooperators have much autonomy and diversity to improve the performance of CCGAs.

The representative selection strategy given in this paper includes the determination of representatives' number, the selection of representatives, the permutation of cooperator groups, and so on. The number of submitted representatives is very critical and the evolutionary competence [8] and the distribution performance of individuals in each sub-population are considered.

2.1 The Evolutionary Competence of Sub-population

Definition 1: For sub-population *k*, time *t* and $t-\tau$, the fitness of the optimum individual is $f_k^b(t)$, $f_k^b(t-\tau)$, the evolutionary competence is defined as follows:

$$E_k^\tau(t) = \frac{f_k^b(t) - f_k^b(t-\tau)}{\tau} \tag{1}$$

If $E_k^\tau(t) > 0$ and it has a rising tendency, which indicates that the evolutionary competence of this sub-population is very high, it will submit more representatives to other sub-population to increase the cooperative competence. If $E_k^\tau(t)$ is a constant for a long time, it indicates that the evolutionary competence of this sub-population is weak and it is a premature population. It will submit fewer representatives to other sub-population to reduce the computational cost.

2.2 Distribution Performance of Sub-population

The standard variance of a sub-population is used to measure its distribution. The number of sub-population is M and the size of each sub-population is S_i, the number of optimized variance is V_i, $i=1,2\ldots$M。Generally, each sub-population is encoded

3 The Implementation Steps

For pages limitation, the parallel implementation of CCGAs is abbreviated. Only the steps of CCGAs realized in parallel based on LAN are given as follows:

Step1. Decomposing variables, determining the parameters such as population size, evolutionary generation, crossover probability and mutation probability for each sub-population.
Step2. Based on the performance of each client, assigning the decomposed variables; $t=0$, generating sub-populations randomly, and each client randomly submitting a representative to server.
Step3. Server and each sub-population interacting to obtain the representatives information for fitness evaluation and genetic evolution continued.
Step4. $t=t+1$, each client computing the evolutionary performance and distribution of each sub-population. The number of representatives submitted is determined by expression (4) and the submitted representatives are determined.
Step5. Each sub-population submitting its representatives to the server and the information that each sub-population submit is received by the server, including representatives.
Step6. Each sub-population submitting the information of representatives' requirements, the evolving process information of the representatives and the identifying code to the server.
Step7. Each sub-population obtaining the synchronized representatives and implementing genetic operators.
Step8. If termination condition is satisfied, outputting the optimal cooperator information, otherwise turning back to step4.

The genetic operators of each sub-population are as follows: real coding, tournament selection, arithmetic crossover and Guess mutation.

4 Experiments

The optimization of benchmark numerical functions is used to examine the rationality of the representative selection strategy. They are given as follows.

$$\min \quad f_1(x) = 418.9829n + \sum_{i=1}^{n} x_i \sin(\sqrt{|x_i|}) \quad x_i \in [-500,500], n=10$$

$$\min \quad f_2(x) = \sum_{i=1,3}^{j=2,4} 100(x_j - x_i^2)^2 + (1-x_i^2)^2, x_i, x_j \in [-10,10]$$

f_1 has a global minimal value 0 at $x=(-420.9687,-420.9687,\cdots)$ and many local optima, which are far away from the global one. The local optima are easy trapped into if normal algorithms are used. f_2 represents a sort of incompletely decomposable function whose variables of (x_1 , x_2) are linked tightly and (x_3, x_4) linked tightly. The global minimal value 0 is acquired at x= (1, 1, 1...). The method of breaking linkage is adopted to examine the rationality of representative selection method given in this paper.

In order to illustrate the rationality of the method for selecting the representatives given in this paper, a comparison is made between this method and the method of an elitist individual taken as representative in CCGA. The algorithm of this paper is denoted as RCCGA and the traditional algorithm is denoted as CCGA.

Genetic algorithm is used for each sub-population evolution. The genetic operators used have also given in section 3.The values of the parameters are shown in Table 1. The evolutionary time T is used to compare the quality of solutions .The solution precision ε is used to compare the evolutionary time of the two algorithms.

Table 1. The values of parameters

	Crossover Probability	Mutation Probability	Population Size	Evolutionary Time T	Precision ε of Solutions
f_1	0.98	0.2	600	120	0.02
f_2	0.95	0.3	500	120	0.01

Two aspects of these two algorithms are compared. Firstly, the optimization time, namely the iterative times and the running time of software are compared and the results are given in table 2.

Table 2. Results of RCCGA and CCGA based on LAN parallel implementation

	The optimal solution in the same evolutionary time		The iterative times needed		The running time	
	CCGA	RCCGA	CCGA	RCCGA	CCGA	RCCGA
f_1	0. 12	0.021	23.4	9.3	6.13	10.25
f_2	0.42	0.031	50.2	19.4	23.13	22.36

There is a contradiction between the optimal performance and the computational complexity of the algorithms. The computational complexity generally increases if the optimal performance increases. Also, the optimal performance of algorithms may be sacrificed if small computational complexity is pursued. The external computation time increases for the method of determining the size of cooperation pool intuitively, but it is seen from table 2 that the running time is not increased dramatically when the same optimal solution is found. The computational cost can be accepted completely. But the new method gains an obviously better solution than the traditional one. The main reason is that the rate of losing effective cooperative group is reduced so that there are more effective cooperators to be reserved and exploited. Chance of CCGA escaping from the "Nash balance" is increased so that the algorithm have more chances to get a better solution, especially for function f_2. That is to say the computational complexity of RCCGA is not increased severely. So the method of this paper is effective and feasible. It is a guidance for representative selection of CCGAs.

In order to know the self-adaptable variation of the number of representatives directly, figure 1 gives the variation of representatives of a sub-population in the evolutionary process of function f_1. The figure shows that sub-population evolve 10 generation. The number of the representatives is(1,35,1,14,1,27,11,11,1,1).The number of representatives for submitting of the first generation is 1 illustrates that the sub-population selects an individual as representative randomly .In the progress, the parameter τ of the first five generations is 3 and of course the value of τ is subjective ,which should be selected technically. The R_p value is 1 twice when the generation is 9 and 10 which illustrates that the sub-population is convergent at this time. During the middle of evolution, the value of R_p is 1 just illustrating that the optimal individual is not improved between the two generations, and not meaning the whole sub-population is convergent.

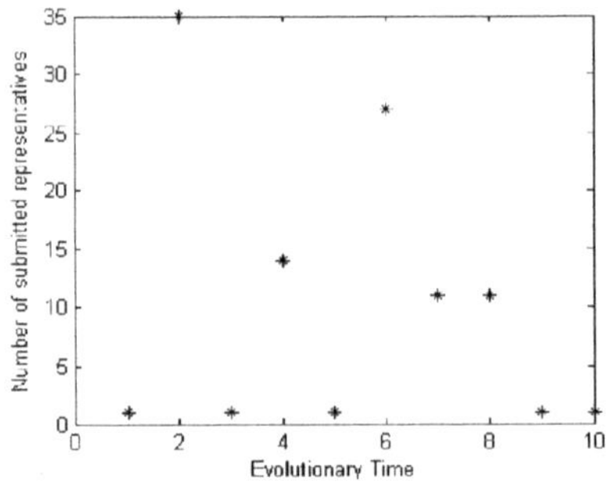

Fig. 1. Figure of representatives' number

5 Conclusions

The performance of cooperative co-evolutionary genetic algorithm is highly influenced by the strategy of representative selection. But effective method is absent. Aimed at the shortages, based on the parallel implementation of cooperative co-evolutionary genetic algorithm in LAN, the representative selection strategy is studied. The main idea is that the representatives are submitted actively against passively by each sub-population and the size of cooperation pool is varied dynamically. Based on the evolutionary competence and the distribution performance of individuals in each population, the size of cooperation pool, the method for cooperators selection and cooperator permutation, and so on are presented. The rationality and validity of the representative selection strategy is validated by benchmark functions. The determination of the evolving competence value τ and the other methods of selecting and cooperating are further research issues.

Acknowledgments

Project *Research on Key Issues of Co-interactive Evolutionary Computation and Their Implementation based on Internet (60304016)* supported by National Natural Science Foundation of China

References

1. Potter, M.A.:The Design and Analysis of a Copmutational Model of Cooperative Coevolution.Doctorate Degree Dissertation of George Mason University.Fairfax(1997)
2. Mitchell, A.,Potter,K.A.,De Jong,.A.: Coorperative Coevolutionary Approach to Function Optimization.Proceedings of The Third Conference of Parallel Problem Solving From Performance.(1994)249–257
3. Nattavut Keerativuttitumrong,Nachol Chaiyaratana,Vara Varavithya.Multi-Objective Cooperative Coevolutionary Genetic Algorithm. Genetic Algorithms in Engineering Systems: Innovations and Applications (2001)69–74
4. Westra,R.,Paredis,J.: Coevolutionary Computation for Path Planning. 5th European Conference on Intelligent Techniques and Soft Computing (EUFIT), Aachen Germany (1997) 394–399
5. Pedrajas,N.G., Hervas-Martinez, C., Munoz-perez. J.: Multi-objective Cooperative Coevolution of Artificial Neural Networks. Neural Networks(2002)1259–1278
6. Chern, H. Y., ,Miikkulainen,R.: Cooperative Coevolution of Multi-Agent Systems. Technical Report AI01–287(2000)
7. Wiegand, R.P., Liles, W.C., De Jong. K.A.: An Empirical Analysis of Collaboration Methods In Cooperative Coevolutionary Algorithms.Proceedings of the Genetic and Evolutionary Computation Conference,Morgan Kaufmann Publishers(2001)1235–1245
8. Li, Z.Y., Tong, T.S.: Research on ANN Evolutionary Design Method Based on Populations Evolution Niche Genetic Algorithm. Control and Decision,Vol.18(5) (2003)607–610
9. Wiegand, R.P.: An Analysis of Cooperative Coevolutionary Algorithms. Doctor Degree Dissertation of Computer Science of George Mason University(2003)

Kernel Matching Pursuit Based on Immune Clonal Algorithm for Image Recognition

Shuiping Gou, Licheng Jiao, Yangyang Li, and Qing Li

National Key Laboratoty for Radar Signal Processing and Institute of Intelligent Information Processing, Xidian University, Xi'an, 710071, China
shpgou@mail.xidian.edu.cn

Abstract. A method for object recognition of Kernel matching pursuits (KMP) [1] based on Immune Clonal algorithm (ICA) [2] is presented. Using the immune clonal select algorithm, which combines the global optimal searching ability and the locally quickly searching ability in search basic function data in function dictionary, this method can reduces computational complexity of basic matching pursuits algorithm. As compared with kernel matching pursuits the method has higher accurate recognition rate.

1 Introduction

S.Mallat[1] and S.Qian[3] presented respectively matching pursuit(MP) algorithm on the basis of projection pursuit [4,5] in the early of 1990s. MP was introduced in the signal-processing community as an algorithm "that decomposes any signal into a linear expansion of waveforms that are selected from a redundant dictionary of function". It is a general, greedy sparse-approximation scheme. The MP algorithm is adaptive signal decomposing method, and any signal is decomposed step by step into a linear expansion of waveforms that are selected from a redundant dictionary of functions. These waveforms are chosen in order to best match the signal structure. So the MP algorithm can describe best inherent character of the signal and decompose best the signal. The KMP algorithm [1] is a matching pursuit method using kernel function sets to optimize. Its basic idea is reflection for feature of kernel function from low-dimension to high-dimension, which transforms low-dimension space non-linear into high-dimension linear problem. KMP is a relatively new learning algorithm utilizing Mercer kernels to produce non-linear version of conventional supervised and unsupervised learning algorithm. Theoretically MP algorithm is an excellent method but its implement is a greedy algorithm. It is well known that it is NP-hard [6] to optimally approximate a function with a linear expansion over a redundant dictionary of waveforms. The greedy MP algorithm and its orthogonal variant produce suboptimal function expansions by iteratively choosing dictionary waveforms that best match the function's structures. Then computing time can be increased greatly with searching spaces increasing, which may be more serious when the dictionary has two or few function.

However the essential parallelism of evolutionary computing algorithm make it can reduce the number of the solution sets and communicate information mutually, which

T.-D. Wang et al. (Eds.): SEAL 2006, LNCS 4247, pp. 26–33, 2006.

obtains the large and valid solution knowledge by smaller computing. Immunity Clonal Algorithm (ICA) [2], a new artificial intelligence method, is proposed by Du, H.F. and jiao, L.C. etc. On the basis of the evolutionary algorithms, the method introduces avidity, clonal and immune genetic mechanism while adopts corresponding operator to make it convergence quickly to the global optimal value. So the MP based on evolutionary computing algorithm has been successfully applied in many areas such as video coding, pattern recognition and image processing. In the year of 2003, A.R.Silva implemented matching pursuit using genetic algorithm, named, atomic decomposition with evolutionary pursuit, and he proposed a method for atomic decomposing of several dictionary [7]. Hence, an object recognition algorithm of Kernel matching pursuits based on Immune Clonal is given by us, which overcomes large computational number of the basic matching pursuit algorithm. The presented algorithm is effective on the image recognition.

The organization of this paper is as follows. In section 2, we briefly describe the basic principle of kernel matching pursuit algorithm and immune clonal algorithm. In section 3, operating steps of object recognition algorithm of Kernel matching pursuits based on Immune Clonal Algorithm is discussed. Simulation results and its analysis are given in section 4. Finally, a conclusion is drawn in section 5.

2 Review of Matching Pursuit

2.1 Basic Matching Pursuit Algorithm

Given a objection function $f \in h$, l noisy observations $\{y_1, \ldots, y_l\}$ and a finite dictionary D of functions in Hilbert space H, the aim of the algorithm to find sparse approximations of $\{y_1, \ldots, y_l\}$ that is the expansion of the form $\tilde{f}_n = \sum_{k=1}^{n} \alpha_k g_k$ that approximate f, where n means the maximum of the basis functions g_k, α_k is weight coefficient , $(\alpha_1 \cdots \alpha_n) \in R^n$, $(g_1, \ldots, g_n) \subset D$. It is a general rule by $\min(\|R_n\|^2 = \|f - \tilde{f}_n\|^2)$, R_n named residue, to search the sets of $(\alpha_1 \cdots \alpha_n)$ and $(g_1 \cdots g_n)$ .The above searching process is named matching pursuit process.

Basic MP Algorithm is presented as follows: we first find out α_1 and g_1 on the rule $\min\|R_n\|^2$, obtain α_n and g_n by iterative repetition, and then $\tilde{f}_n$ is gotten. Based on the equation (1) and minimize the equation (2) can get α_{n+1}, g_{n+1}. Three equations are as follows:

$$\tilde{f}_{n+1} = \tilde{f}_n + \alpha_{n+1} g_{n+1} \quad (1)$$

$$\|R_{n+1}\|^2 = \|R_n - \alpha_{n+1} g_{n+1}\|^2 . \quad (2)$$

$$(\alpha_{n+1}, g_{n+1}) = \arg\min_{(g\in D,\alpha\in IR)} \left\| (\sum_{k=1}^{n} \alpha_k g_k) + \alpha g - f \right\|^2 . \quad (3)$$

A preferable alternation of the matching pursuit is back-fitting matching pursuit. In basic algorithm, when appending $\alpha_{i+1} g_{i+1}$ in ith iterative, the expansion may not be optimal, so doing back-fitting is to recompute the optimal set of coefficients $\alpha_1 \cdots \alpha_{n+1}$ at each step instead of only the last α_{i+1} to approximate the objective function more accurately. While this can be quite time-consuming, one trade-off method is to do back-fitting algorithm in every few steps instead of every step [8, 9].

2.2 Kernel Matching Pursuit Algorithm

Kernel matching pursuit (KMP) is simply the idea of applying the Matching Pursuit (MP) family of algorithms to problem in machine learning, using a kernel-based dictionary [8,10]. KMP uses Mercer kernel to map the data in input space to a high dimensional feature space in which we can process a problem in linear form. Given a kernel function *K*, we construct the basis dictionary of MP by the kernel centered on the training data: $D = \{d_i = K(x, x_i) | i = 1 \cdots l\}$. Method of kernel is enlightened in great part to the success of the Support Vector Machine (SVM), and there exist a lot of commonly used Mercer kernels, such as polynomial kernel with the form of $K(x, x_i) = [(x, x_i) + 1]^d$ and RBF kernel with the form of $K(x, x_i) = \exp(-\|x - x_i\| / 2p)$. Running matching pursuit algorithm to get the approximation functions in regression or the decision function in classification.

$$f_N(x) = \sum_{k=1}^{N} \beta_k g_k(x) = \sum_{k=1}^{N} \beta_k K(x, x_k) . \quad (4)$$

$$f_N(x) = \mathrm{sgn}\left(\sum_{k=1}^{N} \beta_k g_k(x) \right) = \mathrm{sgn}\left(\sum_{k=1}^{N} \beta_k K(x, x_k) \right). \quad (5)$$

where $\{x_k | k = 1 \cdots N\} \in \{x_1, \cdots, x_l\}$ is support point. There is two ways to stop the algorithm. One is that the basis functions reach the maximum *N*, the other is that the error goes below a predefined given threshold. More details about KMP can be found in [8, 2].

3 Immunity Clonal Algorithm

In order to enhance the diversity of the population in GA and avoid prematurity, immunity clonal algorithm is proposed. The clonal operator is an antibody random map induced by the affinity. The state transfer of antibody population is denoted as follows:

$$C_{MA}: A(k) \xrightarrow{clone} A'(k) \xrightarrow{mutation} A''(k) \xrightarrow{selection} A(k+1)$$

According to the affinity function $f(*)$, a point $a_i(k) \in A(k)$ in the solution space will be divided into q_i different points $a_i'(k) \in A'(k)$, by using clonal operator, a new antibody population is attained after performing clonal mutation and clonal selection. In fact, the process includes three operating steps: clone, clonal mutation, and clonal selection. Here antibody, antigen, the avidity between antibody and antigen are similar to the definition of the objective function and restrictive condition, the possible solution, match between solution and the fitting function in Artificial Intelligent Computation respectively [11]. The clonal operation can be embodied as follows:

Clonal: If the antibody $a \in B^l$, and $B^l = \{0,1\}^l$, the antibody population is $A = \{a_1, a_2 \cdots a_n\}$, the clonal operator is defined as:

$$\Theta(A) = [\Theta(a_1) \quad \Theta(a_2) \quad \cdots \quad \Theta(a_n)]^T . \tag{6}$$

Where $\Theta(a_i) = I_i \times a_i \quad i = 1,2 \cdots n$, and I_i is q_i dimension row vector.

$$q_i = g(N, aff(a_i)) . \tag{7}$$

Generally, q_i is given by:

$$q_i = N * \frac{aff(a_i)}{\sum_{j=1}^{n} aff(a_j)} \qquad i = 1,2 \cdots n . \tag{8}$$

Let N>n be a given integer relating to the clonal size. After the clone step, the population becomes:

$$A' = \{A, A_1', A_2', \cdots, A_n'\} . \tag{9}$$

Where:

$$A_i' = \{a_{i1}, a_{i2}, \cdots, a_{iq_i - 1}\} , \quad a_{ij} = a_i , \quad j = 1,2, \cdots, q_i - 1 . \tag{10}$$

Clonal mutation T_m^c: In order to save the information of the aboriginal population, clonal mutation is not performed on original antibody population A, and we apply the Gaussian mutation to the cloned population, namely:

$$p(a_i \to a_i') = \begin{cases} p_m^{H(a_i, a_i')} (1 - p_m)^{l - H(a_i, a_i')} & a_i \in A_j' \\ 1 & a_i \in A \end{cases} . \tag{11}$$

clonal selection T_s^c : $\forall i = 1,2, \cdots n$, if $b = \max\{ f(a_{ij}) \mid j = 2,3, \cdots q_i - 1\}$, and

$$f(a_i) < f(b) \qquad a_i \in A . \tag{12}$$

Then b replaces the antibody a_i in the aboriginal population. So the antibody population is updated, and the information exchanging among the antibody population is realized.

4 Object Recognition Algorithm of KMP Based on ICA

MP algorithm requires every step of searching process be global optimal searching in the redundant dictionary of function in order to select best matching signal structure, from which the large amounts of computing time has often not suffered. So the KMP of wide use is seriously held up. As an optimal method, immune clonal selection algorithm solve problem by transforming low dimention space into high dimention space and the result is projected to low dimention space, which gets full-scale understanding of the problem. It can be realized that the essence of immune clone, which is that clonal operator can produce a variation population around the optimal of parents according to their affinity, and then the searching area is enlarged which is helpful for avoiding prematurity in the course of the evolution. The optimal process of the KMP based on ICS can reduce to the scale of the optimization problem. Then, object recognition algorithm is as follows:

Step1. Given the data sets of training classifier $S=\{(x_1, y_1),\ldots,(x_l, y_l)\}$, kernel function K, here we adopt for RBF kernel and Set a positive number p as kernel parameter, the size of clonal population, popsize. Set a small positive number, ε , the number of iteration L , then compute each training data, $g_i(x) = K(x, x_i)$, i=1...l and suppose generation t is 1

Step2. *Based* on the rule $\min_{\omega_i}\|y-\omega_i g_i(x)\|$ obtain weight coefficient $\omega_i = g_i^T(x)\cdot y \Big/ \|g_i(x)\|^2$. Set affinity f_t (ω_j, x_j) $=\sum_{t=1}^{L}\omega_j^t g_j^t(x)$, j=1...*popsize*. Select the anterior minimal residue corresponding vectors $\bar{x}_j$ and ω_j from kernel function sets and get the initial basic function individual with the size of popsize.

Step3. clonal weight coefficient:

$$(\omega_j^t)' = \Theta(\omega_j^t) = [\Theta(\omega_{j1}^t) \quad \Theta(\omega_{j2}^t) \quad \cdots \quad \Theta(\omega_{jN}^t)]^T \quad j=1...popsize$$

where N is clonal scale.

Step4. Clonal mutation: $(\omega_j^t)'' = T_m^c(\omega_j^t)'$,j=1...*popsize*

Step5. Calculate the affinity:

$$f_t((\omega_j^t)'',\ \bar{x}_j) = \sum_{t=1}^{L} w_{jk}^t g^t(x,\bar{x}_j),\ k=1...N,\ j=1...popsize$$

Step6. clonal selection: $w_j^t = \max(f_t(w_j^t)'', \bar{x}_j)$, j=1...*popsize*

Step7. Let $y = y - f_t$, if $y >= \varepsilon$ go to step2 and solve the next basic function and weight coefficient for each individual.

Step8. Select the best individual with better affinity from population and get f_t , namely training classifier, subsequently object is recognized by equation (5).

5 Experiment Results and Analysis

In this section we will describe a number of experiments that have been performed in order to assess the performance of the algorithms presented and to study their behavior on various datasets.

In the first case, we used datasets for UCI repository from the Web page http://www.ics.uci.edu/~mlearn/MLRepository.html. On the Pima and Cancer datasets we used the same kernel function parameters *P* setting as the one in [10] and the Waveform dataset we used to test the performance of the pretended algorithm for the large size dataset.

We randomly select one third as training data and the rest as test data. In this experiment, the parameters of the method presented are defined as follows, the size of initial population, 10, mutation probability, 0.1 and the clonal scale, 3. The termination condition is specified the recognition accuracy rate of training samples be 100%, or the residue *R* be 0.01. Simultaneity, the comparisons of the algorithm of KMP is drawn. The average error recognition results of 50 times are shown in Table 1, in which between parentheses is indicated the number of basis functions.

Table 1. Comparison of error recognition results on 4 UCI datasets

	KMP[10]	KMP on ICA	KMP on ICA
Iris	N/A	2.5% (6)	1.67% (7)
Wisc.Cancer	3.40%(7)	2.13% (5)	2.13% (7)
Pima Indians	23.9% (7)	24.53% (7)	23.23% (10)
Waveform	N/A	8.02%(6)	6.48% (7)

Firstly, the results of table 1 are all based on limited iteration number of the KMP algorithm. From Table 1, it is obvious that the KMP based on ICA has much higher recognition accuracy rate than the KMP. Secondly, recognition accuracy rate of the algorithm may increase with the number of basic function but improve not remarkable while computing time is increased when the number of basic function is added a certain extent. So we must trade off time and accuracy rate in fact.

The second set of tests was carried out on a number of datasets from Brodatz texture repository, in which Image collection is composed of 16 texture images, as shown in Fig.1. Each of the images is cropped into 25 non-overlap sub-images with size of 128×128. Then the 400 images are decomposed into 2-level and 3 levels by brushlet transform [12], which gets 32-dimension feature vectors.

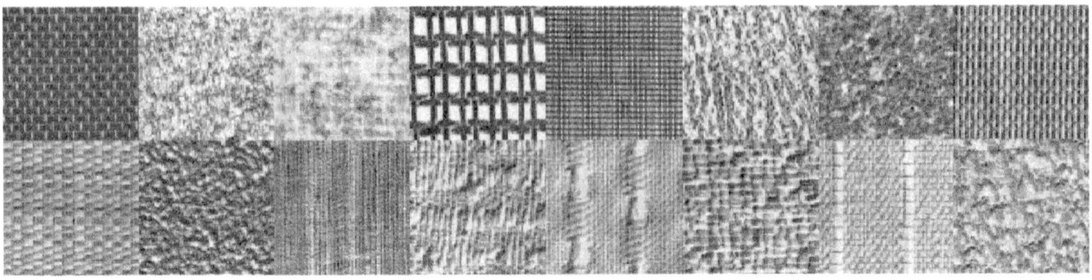

Fig. 1. 16 texture images from Brodatz texture library

We select 10 training data and 15 test data from each class respectively, the kernel function parameter, P is 0.22. In addition, the comparisons of the algorithm of KMP based on GA is drawn, in which the size of population is 10, mutation probability 0.1, crossover probability 0.8, and terminative iteration number 100. Then the images collection discrimination is conducted using the same method as experiment 1. The corresponding 50 times recognition results are shown in Table2.

Table 2. Comparison of recognition results of texture images

	KMP	KMP on GA	KMP on ICA
training Time (s)	1.781	1.052	0.986
number of basic function	6	7	7
ARR of test sets (%)	97.1	97.51	98.75

From Table 2, we can see that the proposed algorithm in this paper may recognize effectually image object and the related recognition accuracy rate is improved in comparative with experiment 1. This owes to the fact that the KMP is a good classification method and 2-D brushlet decomposition can provide orientation information of image, which is very important to classify image object. In other words, the kernel matching pursuit algorithm depends on the data a certain extent. Secondly, as much the number of basic function, recognition accuracy rate of KMP based on GA is decreased slightly while KMP based on ICA is higher than the KMP. The reason is that GA has not always found out the global optimal solution, and ICA makes use of the global optimal search ability and the local and quick search ability to get the global optimal solution by finite repeat.

6 Conclusions and Discussion

In this paper, we have presented an approach to optimize searching process in KMP based on immune clonal algorithm, by which UCI database and texture image recognition are implemented and then recognition result is satisfied. In addition, we argue that the existing KMP algorithm do not provide better multi-class recognition performance. Further, convergence decision and selection of the parameters of the kernel function in KMP deserve more research.

References

1. Mallat, S., Zhang, Z.: Matching Pursuits with Time-frequency Dictionaries. IEEE Trans. Signal Process. 12 (41) (1993) 3397-3415
2. Du, H.F., Jiao, L.C.: Clonal Operator and Antibody Clone Algorithms. Proceedings of the First International Conference on Machine Learning and Cybernetics, Beijing, China: (2002)506–510
3. Qian, S., Chen, D.: Joint Time-Frequecy Analysis Methods and Applications. Pub. Prentice-Hall, (1996) ISBN 0-13-254384-2
4. Morton, S.C.: Interpretable exploratory projection pursuit. Proc. Symp.Interface. Berlin: Springer-Verlag (1990) 470-474
5. Friedman, J.H.: Exploratory projection pursuit. J. Am. Statist. Ass.,82(1987)249-266
6. Davis, G.M., Mallat, S., Avelanedo, M.: Greedy adaptive approximations. J. Constr. Approx. 13 (1997) 57-98
7. Silva, A.R.: Approximations with evolutionary pursuit. Signal Processing, Vol. 83. 3(2003) 465 – 481
8. Engel,Y., Mannor, S., Meir, R.: The kernel recursive least-squares algorithm. IEEE Trans. Signal Processing, vol. 52: 8, (2004)2275-2285
9. Davis, G., Mallat, S., Zhang, Z.: Adaptive Time-frequency Decompositions. Optical Engineering. 33(7) 2183-2191
10. Pascal, V., Yoshua,B.: Kernel Matching Pursuit. Machine Learning. 48(2002)165--187
11. Liu, R.C., Du, H.F., Jiao, L.C.: Immunity Clonal Strategies, In Proceedings of Fifth International Conference on Computational Intelligence and Multimedia Applications, Xi' an,China:IEEE(2003)290–295
12. Francois, G.M., Ronald, R.C.: Brushlets: A Tool for Directional Image Analysis and Image Compression, Applied and Computational Harmonic analysis. 4(1997) 147–187

Power Quality Disturbance Detection and Classification Using Chirplet Transforms

Guo-Sheng Hu[1,3], Feng-Feng Zhu[2], and Yong-Jun Tu[3]

[1] Electric Power College, South China University of Tech., Guangzhou, P.R.C., 510640
jam_hu@21cn.com
[2] Mathematics Science School, South China University of Tech., Guangzhou, P.R.C., 510640
zhuffeng@163.com
[3] Guangdong Vocational College of Science & Tech., Guangzhou, P.R.C., 510640
yongjtou@tom.com

Abstract. In this paper, a new approach is presented for the detection and classification of PQ disturbance in power system by Chirplet transforms(CT), which is the generalized forms of Fourier transform(FT), short-time Fourier transform(STFT) and wavelet transform(WT). WT and wavelet ridge are very useful tools to analyze PQ disturbance signals, but invalid for nonlinear time-varying harmonic signals. CT can detect and identify voltage quality and frequency quality visually, i.e., according to the contour of CT matrix of PQ harmonic signals, the harmonics can be detect and identify to fixed, linear time-varying and nonlinear time-varying visually. It is helpful to choose appropriate WT to analyze harmonics. Simulations show the contours of CT can effectively detect harmonic disturbance occurrence time and duration. Finally, it is validated that the harmonics of the stator current fault signal of the bar-broken electric machine is nonlinear time-varying, and tend to stable status in a short time.

1 Introduction

Power Qualities(PQ) have been studying for a long time, In the early development of electric power system, power loads consist of linear equipments, such as synchronous motors, inductive motors and illuminate apparatus. So the indices of PQ are very simple. From 80's of 20th century, non-linear power electrical apparatus and devices are used extensively in modern industry with power electric developing. Meanwhile, in order to solve the problems of electric power system himself, DC transmissions and FACTS are in used in engineering. Speed-modulating motors in operation bring on waveform aberrance of voltage and current in power networks. In other ways, impulsion and fluctuate loads, such as arc furnaces, large rolling mills and electric trans, make a large numbers of harmonics, voltage fluctuate, flick and three-phase unbalance. However, because of many kinds of complicated, precise, advanced monitoring advices used widely, high power qualities are demand in our life. They are why PQ are always interesting topic recently(see Ref.[1-6]).

The chirplet transform(CT) is a time-frequency representation of a time series. It uniquely combines a frequency- dependent resolution that simultaneously localizes the real and imaginary spectra. The basis functions for the CT are linear frequency

T.-D. Wang et al. (Eds.): SEAL 2006, LNCS 4247, pp. 34 – 41, 2006.

modulation Gaussian. With the advantage of fast lossless invertibility from time, to time-frequency, and back to the time domain, the usage of the CT is very analogous to the FT. In the case of PQ disturbances, the CT visually illustrates voltage interruption, swell and detect harmonic types by the contours of CT matrix. It is helpful and necessary to choose WT, wavelet ridge or other tools to detect PQ disturbances. It can avoid blindness, because WT and wavelet ridge are invalid to process nonlinear time-varying harmonics.

The paper is organized as follows. Section 2 presents Gaussian Chirplet transform theory and its algorithm. The detection and identification of voltage and disturbance signals are presented in Section 3 and 4. The conclusions are presented in Section 5.

2 Gaussian Chirplet Transform

In signal processing, for the sake of reducing signal energy letting out, a window-added method is usually used to process signal transform. Gaussian function is most popular window showed as follow

$$g_0(t)=\frac{1}{\sqrt{2\pi\sigma}}\exp\left(-\frac{(t-b)^2}{2\sigma 2}\right) \tag{1}$$

In fact, Gaussian window function is a special case of the following function[5,6]

$$g(t)=\frac{1}{\sqrt{2\pi\sigma}}\exp\left[-\frac{1}{2}\left(\frac{t-b}{\sigma}\right)^2+i\phi\right]\exp(i\omega t) \tag{2}$$

Function (2) is called to Gaussian wavelet packet $g(t)$ has better localization and high resolution feature. Where phase shift $\phi\in R$ is a constant, ω is modulated angle frequency ($\omega=2\pi f$) , scale parameter σ controls waveform width.

Then we define non-orthogonal and abundant Gaussian chirplet packet

$$g(t)=\frac{1}{\sqrt{2\pi\sigma}}\exp\left[-\frac{1}{2}\left(\frac{t-b}{\sigma}\right)^2\right]\exp\left(i\left(\omega t+qt^2\right)\right) \tag{3}$$

FT, STFT and WT are special cases of CT(see Ref.[9,10]). Imitating the expression of inner product, CT is naturally defined as follow(see Ref.[5-8])

$CT(\sigma,b)=\langle h(t),g(t)\rangle=$

$$\int_{-\infty}^{+\infty} f(t)\times\frac{1}{\sqrt{2\pi\sigma}}\exp\left[-\frac{1}{2}\left(\frac{t-b}{\sigma}\right)^2\right]\exp\left(-i\left(\omega t+qt^2\right)\right)dt \tag{4}$$

The scalar σ controls the width of window function $g(t,\sigma)$.we set $\sigma=1/f$.

Formula (4) can be re-written as follow

$$CT(f,b)=\int_{-\infty}^{+\infty} f(t)\times\frac{|f|}{\sqrt{2\pi}}\exp\left[-\frac{1}{2}\left(\frac{b-t}{\sigma}\right)^2\right]\exp\left(-i\left(2\pi t+qt^2\right)\right)dt \quad (5)$$

Discrete Chirplet transforms of time series $f[pT]$ can be calculated as

$$CT\left[\frac{n}{NT}, jT\right]=\sum_{m=0}^{N-1} F\left[\frac{m+n}{NT}\right]\exp\left(-\frac{2\pi^2 m^2}{n^2}\right)\exp\left(i\frac{2\pi mj}{N}\right) \quad (6)$$

when $n=0$

$$C[0, jT]=\frac{1}{N}\sum_{m=0}^{N-1} F\left[\frac{m}{NT}\right] \quad (7)$$

Integer indexes j, m, $n=0,1,\cdots,N-1$.

From formula (6), it is easy to calculate Chirplet transforms using convolution theory and fast Fourier transforms(FFT).

Chirplet transform is a matrix with dimension $n\times m$, the rows represent frequencies, the columns represent time. Each element represents "local spectrum".

3 Voltage Distortion Detection and Identification

3.1 Voltage Interruption Signal

Voltage interruption presents a phase or multi phase lost voltage amplitude(lower than 10% rating voltage amplitude) in a short time. Let T to be a waveform period of voltage(T=1/50 second).When voltage normalized amplitude is lower than 0.8 in certain period, the momentary interruption is called broken voltage. According to duration time, it can be classified instantaneously broken voltage(0.5T~ 3s), transiently broken voltage (3s~ 60s) , and continuously broken voltage (more than 60s) . Fig. 1 (a)

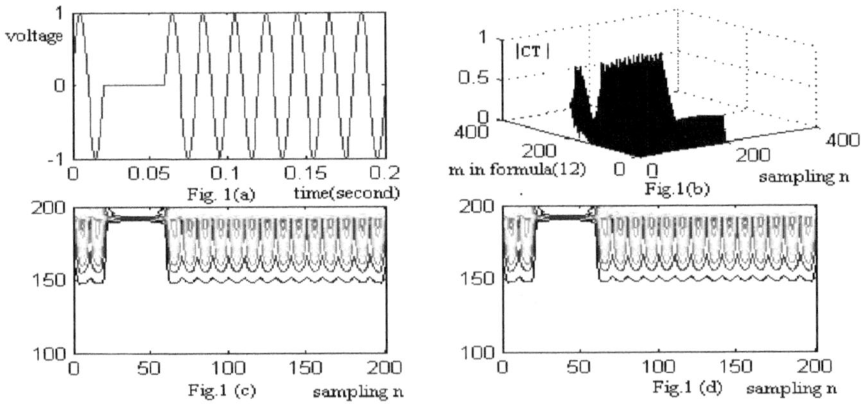

Fig. 1. (a) a momentary interruption voltage signal, (b) 3-dimension plot of CT, (c) contours with parameter q=0.01, (d) contours with parameter q=0.05

shows simulating waveform of instantaneously voltage interruption. Interruption duration time is from 20ms to 60ms(sampling frequency is always 1000 in this paper, interruption occur from 20th to 60th sampling points). Interruption last 40ms(40 sampling points). From 3-dimension figure of CT, we can observe the voltage signal change, where *X*-horizontal coordinate represents sampling points, *Y*-vertical coordinate denotes column in formula (6), *Z*-coordinate shows the elements of matrix. Fig.1 (c) and (d) illustrate the contours of the voltage interruption signal in Fig.1(a), they are visual figures. According to them, we can easily to find out the interruption occurrence time and duration time.

3.2 Voltage Swell Signal

Voltage swell means voltage amplitude exceeding 10 to 90 percent of rating voltage, which last $0.5T^{\sim}$ 1min. Fig.2 shows a voltage swell signal, 3-dimension plot and its contours with different parameter q. swell duration last from $0.5T$ to 1min (start at 20ms, end at 40ms), normalized amplitude is 1.4. From Fig.2(c) and (d), we can find out start time and duration of swell visually.

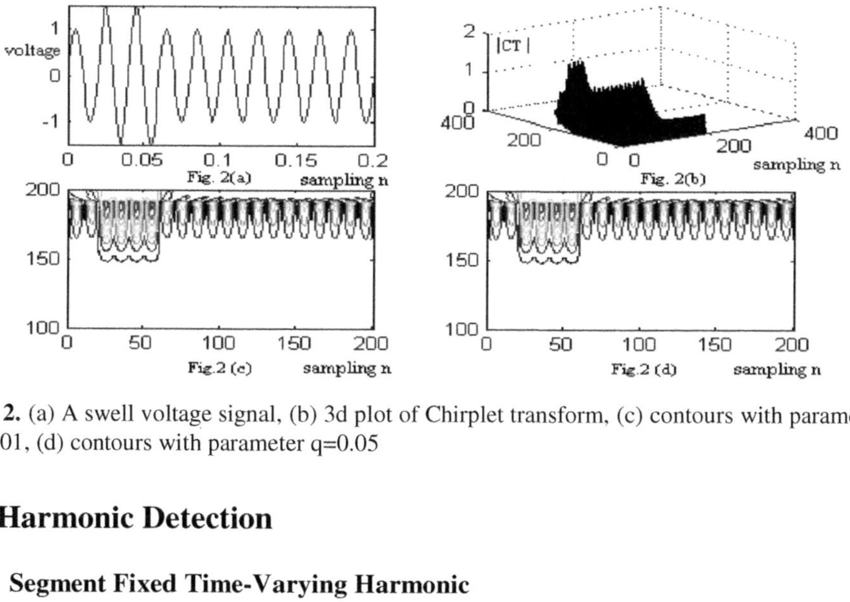

Fig. 2. (a) A swell voltage signal, (b) 3d plot of Chirplet transform, (c) contours with parameter q=0.01, (d) contours with parameter q=0.05

4 Harmonic Detection

4.1 Segment Fixed Time-Varying Harmonic

Set time-varying harmonic simulating signal to be

$$s(t)=\begin{cases}\sin(2\pi\times 50t) & (0s\le t\le 0.10s)\\ 2\sin(2\pi\times 50t)+1.6\sin(2\pi\times 150t)+ & \\ \quad 1.2\sin(2\pi\times 350t) & (0.10s<t\le 0.20s)\\ \sin(2\pi\times 50t)+0.8\sin(2\pi\times 150t)+ & \\ \quad 0.6\sin(2\pi\times 350t) & (0.20s<t\le 0.30s)\end{cases}$$

Fig.3 shows waveform of the simulating signal and its contour of CT matrix.

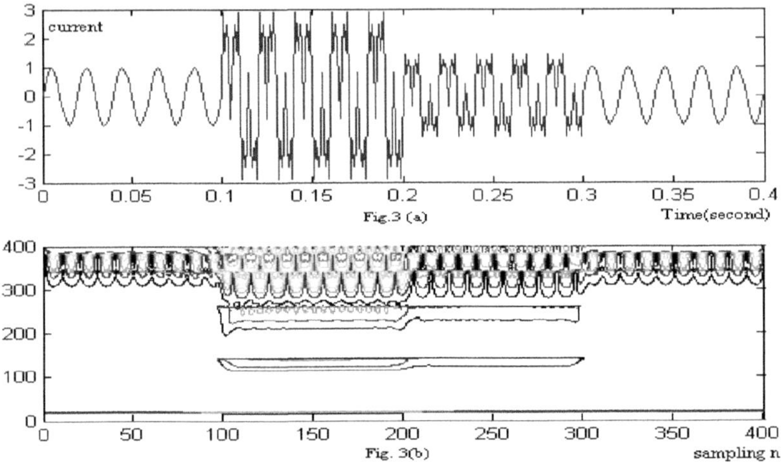

Fig. 3. (a) a time varying ladder harmonic signal and (b) the contour of its CT matrix

Watching Fig.3, we can get two results, firstly, the simulating time-varying harmonic signal contains two disturbance harmonics. Secondly, 150Hz harmonic and 350Hz harmonic occurrence time and duration time are nearly as seem as actual case. All these can be seen from contour of Fig. 3(b).

4.2 Linear Time-Varying Harmonic Signal

The linear time-varying harmonic simulating signal is

$$s(t)=\begin{cases}\sin(2\pi\times 50t) & (0s\le t\le 0.06s)\\ \sin(2\pi\times(50+0.5t)t) & (0.06s<t\le 0.14s)\\ \sin(2\pi\times 50t) & (0.14s<t\le 0.2s)\end{cases}$$

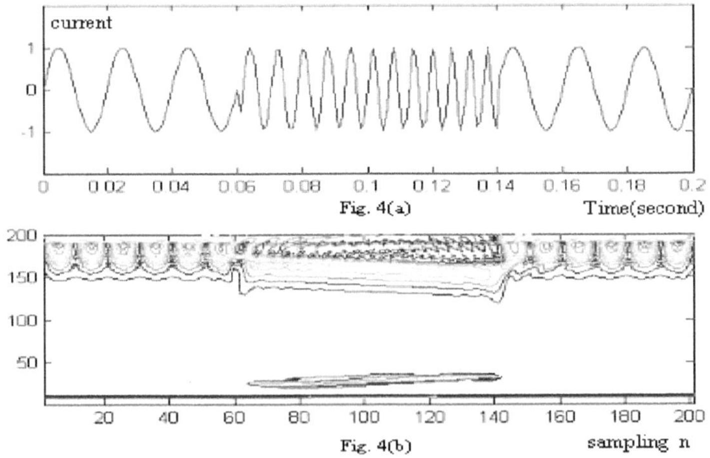

Fig. 4. (a) a frequency linear time-varying harmonic signal and (b) the contour of its CT matrix

When t belongs to $0.06s \le t \le 0.14s$, the signal contains linear frequency modulation component, the frequencies are changing by linear time expression of $f = 50 + 0.5t$. The waveform and contour of CT matrix are illustrated in Fig.4(a) and (b). From Fig.4(b), we can visually see the frequencies are changing linear about time t. Occurrence and end time of harmonic disturbances are almost identical with actual hypothesis.

4.3 Stator Current Signal

Figure 5(a) and (b) are a actual stator current signal waveform of two bar-broken electric machine and the contour of CT matrix. Figure 5(b) illustrates the stator current signal contains three kinds of time-varying harmonics tending a stable status in a short time(1.2s). However, it is very difficult to know what the harmonics are.

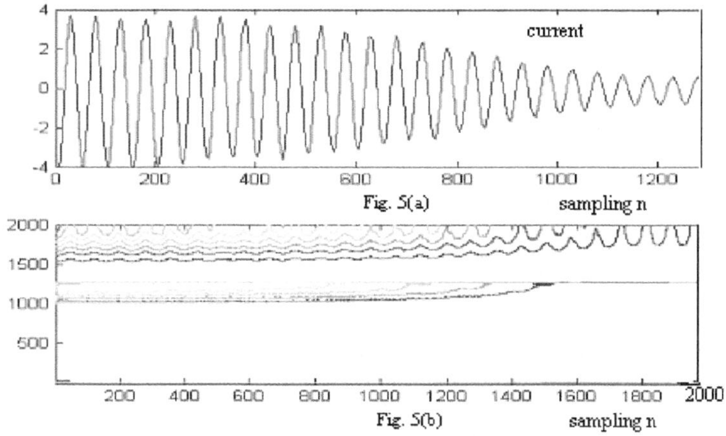

Fig. 5. (a) a fault current signal of the bar broken electric machine and (b) the contour of its CT matrix

WTs have been using to analyze PQ disturbances(see Ref. [3-5,10,12-15]). However, WTs are not always effective. According to the contours of CT matrix, we can firstly detect and identify PQ harmonic types visually, then choosing WTs or wavelet ridges([16,17]). Generally, to process voltage interruption, sag, swell, fixed harmonic, WTs is very effective. Meanwhile, wavelet ridges are better than WTs to process linear time-varying harmonic signals,. However, both of them are feeble to detect non-linear time-varying harmonics. So before processing PQ harmonic disturbances, it is very important to know PQ harmonic disturbance types, CTs are an useful tool to process them previously. If the instantaneous harmonic is nonlinear time- varying, such as the stator current signal of bar-broken electric machine in figure 10, it needs do hard work to find out a new tool to process the disturbance signals. We will discuss this issue in another paper.

5 Conclusions

Chirplet transforms has been proposed in this paper for detecting and classifying PQ disturbances. Fourier transform, short-time Fourier transform and wavelet transform are special cases of Chirplet transforms. Using contours of Chirplet transform matrix, we can detect and identify voltage interruption, voltage sag, voltage swell, momentary impulse, fixed harmonic, linear time-varying harmonic, nonlinear time-varying harmonic and so on. We can visually identify voltage interruption, voltage sag, voltage swell and impulse and detect how many harmonics the PQ disturbance signal contains and what kind the harmonics are. It is very helpful to choose signal processing tool for analyzing PQ disturbance and avoid blindness.

Acknowledgement

The authors would like to thank the GDNSF #033044 for financial assistance.

References

1. Gaouda A M, Kanoun S H, Alama M A et al. Pattern recognition applications for power system disturbance[J]. IEEE Trans. on Power Delivery, 2002, 17(3): 677-683.
2. Gaouda A M, Salama M M, Sultan M R et al. Application of multiresolution signal decomposition for monitoring short-duration variations in distribution systems[J]. IEEE Trans. on Power Delivery, 2000, 15(2): 478-485.
3. Santoso S, Powers E J, Grady W M and Hoffman P. Power quality assessment via wavelet transform analysis[J]. IEEE Trans. Power Delivery, 1996, vol. 11: 920-930.
4. Pillay P and Bhattacharijee A. Application of wavelets to model short term power system disturbances[J]. IEEE Trans. Power Syst., 1996, vol. 11:2031-2037.
5. Gouda M, Salama M A, Sultan M R, Chikhani A Y. Power quality detection and classification using wavelet multiresolution signal decomposition[J]. IEEE Trans. Power Delivery, 1999, vol. 14: 1469-1476.
6. Borras D. Wavelet and neural structure: A new tool for diagnostic of power system disturbances[J] IEEE Trans. Ind. Applicat., 2001, vol. 14:1469-1476.
7. Mann S, Haykin S. "Chirplets and wavelets": Novel time-frequency methods. Electronic Letters[J] 1992,28:114-----116.
8. Mann S, Haykin S. The chirplet transform: Physical considerations[J]. IEEE Trans. Signal Processing,1995,43(11):2745—2761.
9. Hu G S. Motor Fault Signals Denosing Based Chirplet Transform[J]. Trans. of china electrotechnical society, 2002, 17(3):59-62.
10. Hu G S. Location of slight fault in electric machine using trigonometric spline chirplet transforms. Power System Technology, 2003, 27(3): 28-31.
11. IEEE Std 1159-1995, IEEE recommended practice for monitoring electric power quality. New York: IEEE; 1995.
12. Xue H, Yang R G. Power quality disturbance detection method using wavelet packet transform based de-nosing scheme, Proceedings of the CSEE, 2004, 24(3):85-90.

13. Liu A D, Xiao X Y, Deng W J. Detection and analysis of power quality disturbance signal based on discrete cosine transform and wavelet transform. Power System Technology, 2005, 29(10): 70-74.
14. Zhu F F, Ren Z, Huang W Y. A new method to quantitatively detect power quality disturbances based on two real wavelet transforms. Power System Technology, 2003, 27(11): 52-54.
15. Wang C S, Wang J D. Classification method of poer quality disturbance based on wavelet packet decomposition. Power System Technology, 2004, 28(15): 78-82.
16. Zhang Z P. A novel method of motor faulted rotor detection based on wavelet ridge. Proceedings of the CSEE, 2003, 23(1): 97-81.
17. Mallat S. A wavelet tour of signal processing. Second edition, Academic Press, 2001, USA.

Ensemble Learning Classifier System and Compact Ruleset

Yang Gao[1], Lei Wu[1], and Joshua Zhexue Huang[2]

[1] State Key Laboratory for Novel Software Technology, Nanjing University, China
gaoy@nju.edu.cn
[2] E-business Technology Institute, The University of Hong Kong, China
jhuang@eti.hku.hk

Abstract. The aim of this paper is twofold, to improve the generalization ability, and to improve the readability of learning classifier system. Firstly, an ensemble architecture of LCS (LCSE) is described in order to improve the generalization ability of the original LCS. Secondly, an algorithm is presented for compacting the final classifier population set in order to improve the readability of LCSE, which is an amendatory version of CRA brought by Wilson. Some test experiments are conducted based on the benchmark data sets of UCI repository. The experimental results show that LCSE has better generalization ability than single LCS, decision tree, neural network and their bagging methods. Comparing with the original population rulesets, compact rulesets have readily interpretable knowledge like decision tree, whereas decrease the prediction precision lightly.

1 Introduction

The learning classifier system (LCS) is brought by Holland et al. in 1989, which consists of two important components[4]. One is genetic algorithm module, with which LCS creates new classifier rules; another is reinforcement learning module, with which LCS receives payoff from environment, distributes the message and adjusts the classifier's strength. Since 1994, Wilson made some pivotal changes from original LCS and brought forward ZCS and XCS[5][6]. A lot of researches show that the XCS can obtain the most optimal performance in many ideal learning problems, such as BOOLEAN multiplex. However, when face the noise data(incorrect measure value) and the missing data in the practical data, XCS would over-fit these data and may not classify unseen data correctly. So, Gao et al. presented the ensemble architecture of XCS in order to improve the generalization ability[3]. In former papers, LCSE is only applied in single data set of UCI repository. So, we investigate the learning performance in multiply data sets further in this paper.

Since the population set of LCSE consists of thousands of classifier rules, it is not easy to determine which rules play the important role in classifying new cases. Especially when need the explicit knowledge to help users to understand system in practice, lacking of readability of the population set becomes a vital shortcoming.

T.-D. Wang et al. (Eds.): SEAL 2006, LNCS 4247, pp. 42–49, 2006.

So we must face up to improving the readability of LCSE. Wilson and Dixon et al. have already presented some compact ruleset algorithms, such as CRA, CRA2 and XCSQ, for reducing the size of evolved classifier populations[8][2]. But, because CRA must run off-line after learn the whole dataset and it has high time complexity, we revise the compact ruleset algorithm in order to get the reducing rulesets with the training stage to meet the requirement of readability.

The paper is organized as follows. In Section 2, a two-level learning classifier system ensemble and its learning process are discussed. Then we design a compact ruleset algorithm based on CRA to reduce the size of final population set and improve the readability performance in Section 3. We conduct some test experiments on three data sets and investigate performance of the respective approaches in Section 4. Finally, we draw some conclusions and outline future works.

2 Ensemble Learning Classifier System

In 2005, Gao et al. firstly designed the architecture of learning classifier system ensemble, which is abbreviated to LCSE[3]. LCSE bases on XCSR proposed by Wilson, which is broadly applied into the problems in continuous-valued environments[7]. LCSE combines multiple XCSRs with ensemble learning so that improve the generality capability of whole system. Besides several single sub-LCSs, the bootstrap module and voting module are added into LCSE. The bootstrap module is used to distribute the input into different sub-LCSs. And the voting module is used to combine all classification results of the respective sub-LCS in order to generate the global system's output. In Fig. 1, the architecture of LCSE is described.

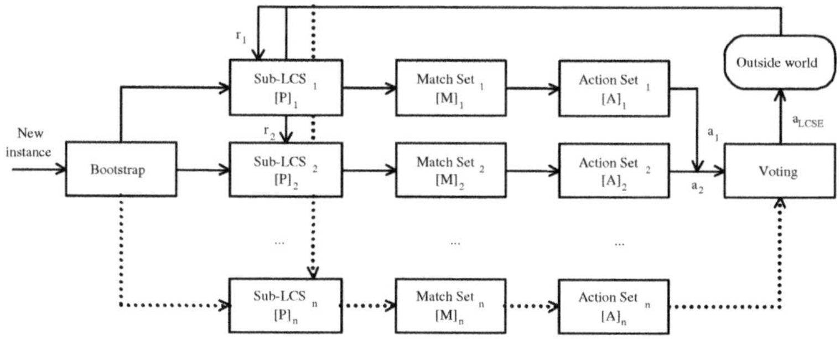

Fig. 1. The Architecture of Learning Classifier System Ensemble

When initialize the system, all sub-LCSs generate their population sets respectively. In each episode, the bootstrap module inputs the new coming instance to every sub-LCSs randomly with the sampling probability λ, which is set as 0.632 in experiments. If the ith sub-LCS is activated and receives this input, then constructs the match set $[M]_i$. Whereafter, using the wheel selecting algorithm to get the action set $[A]_i$. Finally, the ith sub-LCS output its classification result a_i. The voting module ensembles all activated sub-LCSs' outputs according to the basic plurality voting method in order to get the final system's classification result a_{LCSE}. It must be emphasized that each activated sub-LCSs receives respective payoff r_i from the outside according to itself output a_i. Then, every activated sub-LCS processes reinforcement learning of the rule's strength and rule's discovery basic on itself payoff. In Fig. 1, if the ith sub-LCS is activated then its running process are described using the solid line. Else, the dot line means that the respective sub-LCS couldn't be activated in current learning episode. After finish learning current instance, all activated sub-LCSs are modified to be inactive. Essentially, the bootstrap module in LCSE aims at developing multiple sub-LCSs with different classification performance by means of learning different samples, initialing different population set, undertaking different rule's learning and discovery processes. So, the learning classifier system can achieve stronger generality performance when combing multiple different sub-LCSs using ensemble learning.

3 Compact Ruleset Algorithm

After learn from training data, the population set consists of thousands of classifier rules which include interesting patterns of dataset. However, the size of population set is too large to determine which rules are most important during the classification process. For example, in Tab.2 the population set's size of each sub-LCS is 800. In addition, when need the explicit knowledge to help users to understand system in practice, lacking of readability of the population set becomes a vital shortcoming of LCS. So, we need reduce the size of population set and hope to generate some available, intelligible rules while maintain performance on test sets.

Wilson's compact ruleset algorithm, which is abbreviated to CRA, is the elementary work for producing more understandable rules[8]. The algorithm has three steps and run after finish all test tasks. Firstly, ordering the classifier based on a selected rule's parameter and form the smallest subset which has highest performance. It is a heuristic knowledge that 'better' rules should have greater value of the selected parameter. Then, CRA algorithm removes the classifiers which do not advance performance from the smallest subset. Finally, the algorithm compares the marginal contribution of each rule with respect to data set. And removing the classifiers which have weak relative importance. Using CRA, the slightly smaller subset is produced and has same performance as the original population set of LCS.

CRA has very high time complexity, $O(r^2 a)$, where r is the final reduced set size and a is the time to evaluate a single rule against whole data set[2]. In addition, CRA must run off-line after completely learn on the dataset (number of iteration is about millions) so that try to achieve 100% performance on the same data set. In practice, it is almost unfeasible during on-line, incremental learning process. So, we modify and simplify the compact ruleset algorithm as in Tab.1.

Table 1. The process of amendatory compact ruleset algorithm

On-line part
1. remove the ith classifier from the population set which doesn't meet $e_i < e_1$ and $p_i < p_1$;
2. order the remaining classifier based on the predictive reward.
Off-line part
3. add the classifier rule with the highest marginal contribution to final compacted ruleset{
  3.1 compute the number of inputs matched with respect to the train data set;
  3.2 add the rule with the highest value to final compacted ruleset.
  3.3 remove the matched instances from the train data set.
  3.4 until the train data set is empty. }

In step 1 in Tab.1, p_1 is a threshold of predictive reward and e_1 is a threshold of experience of a classifier. We simplify the step 1 and step 2 to select satisfied useful rule in the amendatory compact ruleset algorithm, unlike finding the best smallest subset that has 100% performance in Wilson's compact ruleset algorithm. The step 3 is similar with CRA which is used to measure their relative importance. And the former two steps could be done on-line whereas the step 3 must run after the learning process. Obliviously, the amendatory compact ruleset algorithm has a lower time complexity of $O(ca)$, where c is the reduced set size after step 2 and a is the time to evaluate a single rule against whole data set.

4 Experimental Results and Analysis

Firstly, LCSE, LCS and other learning methods are compared on three data sets from the UCI machine learning repository[1]. Ten runs of 10-fold cross validation are performed on each data set, and the average results are reported. The parameters of LCSE and their description are listed in Tab. 2 as below.

In Tab. 3, we compare the average fraction correct of LCSE which has multiple sub-LCSs with other learning methods. Among these methods, we use the decision tree and neural network method of Weka toolbox designed by the university of Waikato in New Zealand [1]. In this first experiment, the number of total training steps is 2000. The experiment results are different from our former reports after we optimal the program of LCSE[3]. In Tab. 3, the numbers previous '±' are the average predictive precision and the numbers following '±' are the standard deviations. Known from Tab. 3, all LCSE have the better performance than XCSR finally. And the LCSE with 9 sub-LCSs has the best average predictive precision of all LCSE listed in Tab. 3 and has better performance than C4.5, Neural Network and their bagging methods. It also can be seen from Tab. 3 that the average predictive precision become better when increasing in number of sub-LCS of LCSE. But, we argue that the performance improving becomes more difficult when continuing to increase the sub-LCS's number. It's because

Table 2. The description and values of parameters in LCSE

Para.	Description	Value
N	number of sub-LCSs	3, 5, 7 or 9
P	population set's size of each sub-LCS	800
T	total training steps	2000-20000
α	learning rate of reinforcement learning	0.2
χ	probability of crossover operator	0.8
μ	probability of mutation operator	0.04
θ_{GA}	the number steps when activating GA process	12
$\theta_{Covering}$	the threshold for covering process	5
ϵ_0	a threshold of computing error of a classifier	10
s_0	a parameter of create random scope of a condition in classifier	1.0
p_1	a threshold of predictive reward of a classifier	50
exp_1	a threshold of experience of a classifier	10

of there are some missing data and inconsistent data in the data set so that the predictive accuracy couldn't reach at 100% regardless of using any learning method. However, the number of sub-LCSs should be justified by rigorous theoretical analysis and far more experiments.

Table 3. Comparing LCSE with XCSR, Decision Tree, Neural Network and their bagging learning methods

Learning Method	Diabetes	Live-disorders	Wine
LCSE with 9 sub-LCSs	0.7969±0.02915	0.7261±0.07799	0.9983±0.00500
LCSE with 7 sub-LCSs	0.7891±0.02898	0.7229±0.08006	0.9961±0.00661
LCSE with 5 sub-LCSs	0.7865±0.02608	0.7249±0.06026	0.9899±0.00989
LCSE with 3 sub-LCSs	0.7852±0.02702	0.6874±0.06550	0.9758±0.01544
XCSR	0.7617±0.02025	0.6754±0.05242	0.9433±0.02281
C4.5	0.7155±0.01161	0.6116±0.03370	0.8989±0.02390
C4.5 bagging	0.7474±0.01082	0.6959±0.01601	0.9545±0.01079
Neural Network	0.7318±0.01443	0.6362±0.03346	0.9635±0.00628
Neural Network bagging	0.7647±0.00359	0.7099±0.00854	0.9753±0.00275

Known from the first experiment, we think that the LCSE is not enough stable since their standard deviations are greater than XCSR. But, we think that the reason might be the total training steps are not enough large, especially in former two data sets. So, we conduct the second experiment in order to know whether the number of training steps would influence the predictive precision and standard deviation of LCSE. In second experiment, we increase the training steps of LCSE and experimental results are reported in the table 4. But, we could not get the expected results in this experiment. In Diabetes and Live disorders data sets, the average predictive precision is obviously improved with increasing the training steps. However the standard deviation changes lightly and could not exhibition any rules. So, we have doubts that the learning capacity of LCSE and XCSR are instability since they depend on their populations.

Table 4. Comparing the predictive precisions and standard deviations of LCSE with different training steps

Learning Method	Diabetes				Live-disorders			
	2000	5000	10000	20000	2000	5000	10000	20000
LCSE with 9 sub-LCSs	0.7969 ±0.02915	0.8173 ±0.02808	0.8379 ±0.02357	0.8474 ±0.02638	0.7261 ±0.07799	0.7049 ±0.10521	0.7681 ±0.10518	0.7507 ±0.07079
LCSE with 7 sub-LCSs	0.7891 ±0.02898	0.8137 ±0.03330	0.8374 ±0.02571	0.8436 ±0.02570	0.7229 ±0.08006	0.6945 ±0.10773	0.7568 ±0.09936	0.7400 ±0.07098
LCSE with 5 sub-LCSs	0.7865 ±0.02608	0.8112 ±0.03008	0.8340 ±0.02678	0.8378 ±0.02524	0.7249 ±0.06026	0.7072 ±0.10080	0.7754 ±0.08480	0.7307 ±0.07213
LCSE with 3 sub-LCSs	0.7852 ±0.02702	0.8010 ±0.02881	0.8151 ±0.02574	0.8324 ±0.02537	0.6874 ±0.06550	0.6974 ±0.07777	0.7591 ±0.07241	0.7249 ±0.07148
XCSR	0.7617 ±0.02025	0.7855 ±0.03087	0.8104 ±0.02463	0.8107 ±0.02281	0.6754 ±0.05242	0.6704 ±0.08947	0.7252 ±0.08518	0.7107 ±0.06677

The third experiment is conducted to investigate the performance of the amendatory compact ruleset algorithm. This experiment only is test in the Pima Indians Diabetes data set. We also use the stratified tenfold cross-validation method in the LCSE with 7 sub-LCSs. Table 5 shows the results. The second column refers to the size of final compacted ruleset and the predictive performance is listed in the third column using the compacted ruleset. The predictive performance of LCSE also be listed in the forth column. Known from the table 5, the mean prediction precision of compacted(about 73%) ruleset is less than LCSE with 7 sub-LCSs(about 79%) whereas is almost equal to performance of decision tree(about 73%). In our opinion, discarding some classifiers in step 1 of table 1 is the reason for decreasing performance. In addition, it is also observed that the size of compacted ruleset isn't stationary and ranges from 24 to 40.

Table 6 shows some pieces of the final compacted ruleset. We can link the value interval of all attributes in each row to construct a classifier rule. In order to understand these rules well, we compare the classifier rules with the results of decision tree algorithm. The results of decision tree are described in table 7 and consist of 8 rules. Observed from table 6, some important thresholds of attributes occur in the compacted ruleset of LCSE, such as 127 of attribute 2 and 30 of attribute 6. It means that the compacted ruleset expresses readily, interpretable and actual knowledge about the dataset. Of course, the compacted ruleset does not produce same rules as decision tree algorithm because some information are hidden in other pieces. So we will develop other merge technique to combine these rules in order to generate more general rules in the future.

Table 5. The result of amendatory compact ruleset algorithm in Diabetes data set

Fold	1	2	3	4	5	6	7	8	9	10
Size of compacted ruleset	37	38	24	36	31	34	38	40	33	36
Prediction performance of compacted ruleset	0.7368	0.7368	0.6812	0.7237	0.7368	0.7403	0.7403	0.7368	0.7237	0.7403
Average predictive precision ± standard deviation	0.7267 ± 0.01747									

Table 6. Some pieces of final classifier using amendatory compact ruleset algorithm

attr.1	attr.2	attr.3	attr.4	attr.5	attr.6	attr.7	attr.8	Action	prediction	fitness	experience
[2, 17]	[44, **127**]	[24, 78]	[7, 95]	[14, 478]	[18, 29]	[0, 2]	[52, 67]	0	100	0.04	26
[0, 17]	[44, 150]	[24, 122]	[16, 23]	[14, 846]	[18, 29]	[0, 1]	[49, 81]	0	100	0.05	13
[6, 17]	[77, 106]	[24, 122]	[7, 53]	[14, 837]	[20, **30**]	[0, 1]	[21, 81]	0	100	0.06	49
[0, 17]	[44, **127**]	[24, 122]	[7, 95]	[14, 478]	[18, 29]	[0, 2]	[52, 67]	0	100	0.04	24
[12, 13]	[142, 199]	[38, 69]	[7, 99]	[14, 846]	[22, 49]	[0, 2]	[21, 63]	1	95	0.1	18

Table 7. The result of decision tree algorithm applied in Diabetes data set

rule 1. (a2 $\leq$ 127):0;
rule 2. (a2 $>$ 164) and (a6 $\leq$ 29.9) and (a3 $\leq$ 78):1;
rule 3. (a2 $>$ 164) and (a6 $\leq$ 29.9) and (a3 $>$ 78):0;
rule 4. (127 $<$ a2 $<$ 164) and (a6 $\leq$ 29.9):0;
rule 5. (a2 $>$ 127) and (a6 $>$ 29.9) and (a3 $\leq$ 61):1;
rule 6. (a2 $>$ 127) and (a6 $>$ 29.9) and (a3 $>$ 61) and (a2 $>$ 157):1;
rule 7. (157 $\geq$ a2 $>$ 127) and (a6 $>$ 29.9) and (a3 $>$ 61) and (a8 $\leq$ 30):0;
rule 8. (157 $\geq$ a2 $>$ 127) and (a6 $>$ 29.9) and (a3 $>$ 61) and (a8 $>$ 30):1;

5 Conclusion and Future Work

The aim of this paper has twofold. Firstly, we demonstrated that LCSE, Learning Classifier System Ensemble, combine learning classifier system with ensemble learning in order to improve the generality ability than single learning classifier system. Known from the conducted experiments, the LCSE with much more sub-LCSs has better generality ability than other LCSE with less sub-LCSs, LCS and other supervised learning methods such as decision tree and neural network. In addition, with increasing the total training steps, LCSE can improve the average predictive precision further. Secondly, we brought forward an amendatory compact rulesets algorithms in order to reduce the size of final population ruleset and improve the readability. In the future, we will try to combine the rules in compact rulesets into actual knowledge and make an effort to apply LCSE into other data mining domains.

Acknowledgement

The paper is supported by the Natural Science Foundation of China (No.60475026), the National Outstanding Youth Foundation of China (No.60325207) and the National Grand Fundamental Research 973 Program of China (No.2002CB312002).

References

1. Http://www.cs.waikato.ac.nz/ ml/weka/. Last visit at 23, July, 2006.
2. Phillip William Dixon, David Corne, and Martin J. Oates. A ruleset reduction algorithm for the xcs learning classifier system. In W. Stolzmann P. L. Lanzi and S. W. Wilson, editors, *Learning Classifier System. The 5th International Workshop of Learning Classifier System. LNAI 2661*, pages 20–29, Berlin, 2003. Springer-Verlag.
3. Yang Gao, Joshua Zhexue Huang, Hongqing Rong, and Daqian Gu. Learning classifier system ensemble for data mining. In *Genetic and Evolutionary Computation Conference (GECCO2005) workshop program*, pages 63–66, Washington, D.C., USA, 2005. ACM Press.
4. Booker L., Goldberg D.E., and Holland J.H. Classifier systems and genetic algorithms. *Artificial Intelligence*, 40(1-3):235–282, 1989.
5. Stewart W. Wilson. Zcs: A zeroth level classifier system. *Evolutionary Computation*, 2(1):1–18, 1994.
6. Stewart W. Wilson. Classifier fitness based on accuracy. *Evolutionary Computation*, 3(2):149–175, 1995.
7. Stewart W. Wilson. Get real! xcs with continous-valued inputs. In W. Stolzmann P. L. Lanzi and S. W., editors, *Learning Classifier Systems. From Foundations to Applications. LNAI 1813*, pages 209–219, Berlin, 2000. Springer-Verlag.
8. Stewart W. Wilson. Compact rulesets from xcsi. In W. Stolzmann P. L. Lanzi and S. W. Wilson, editors, *Advances in Learning Classifier Systems. LNCS 2321*, pages 192–210, Berlin, 2001. Springer-Verlag.

The Role of Early Stopping and Population Size in XCS for Intrusion Detection

Kamran Shafi, Hussein A. Abbass, and Weiping Zhu

Defence and Security Applications Research Centre
Univ. of New South Wales @ ADFA
Canberra ACT 2600, Australia
{k.shafi, h.abbass, w.zhu}@adfa.edu.au

Abstract. Evolutionary Learning Classifier Systems (LCSs) are rule based systems that have been used effectively in concept learning. XCS is a prominent LCS that uses genetic algorithms and reinforcement learning techniques. In traditional machine learning (ML), early stopping has been investigated extensively to the extent that it is now a default mechanism in many systems. However, there has been a belief that EC methods are more resilient to overfitting. Therefore, this topic is under-investigated in the evolutionary computation literature and has not been investigated in LCS. In this paper, we show that it is necessary to stop evolution in LCS using a stopping criteria other than a maximum number of generations and that evolution may suffer from overfitting similar to other ML methods.

1 Introduction

XCS, a state of the art LCS, is a population based system, where the population consists of if-then form of rules called *classifiers*. The early work with XCS focused on reinforcement learning problems and used synthetic environments like the binary multiplexer and maze problems as the test beds. Later on Wilson extended XCS for continuous variables, called XCSR [8], which has expanded the XCS' applications domain from artificial problems to real world problems. Several studies have been carried out since then to explore the use of XCS as a generic data miner in concept learning tasks [4,6,1,2].

Offline intrusion detection (ID) is a typical concept learning task where a training dataset with normal and attack traffic is presented to the learning system which then has to classify correctly the unseen instances in the test set. Training sets are normally generated from network traffic dumps and event logs and are thus typically very large. The test set normally has many dissimilar or completely unseen cases caused, for example, by the attacker's activity to evade firewalls and other computer security mechanisms. Thus, learning systems need to generalize well in order to classify accurately attack patterns in the test set. However, the generalization is achieved as the system learns over time going through the training data multiple times. Given the large training sets for ID and other similar domains it can become quite time consuming, especially for a

T.-D. Wang et al. (Eds.): SEAL 2006, LNCS 4247, pp. 50–57, 2006.

population based system like XCS. Over training can also lower the test accuracy if the learning algorithm is susceptible to overfitting. It is thus essential to identify when to stop training in LCS.

Wilson used the *performance* criteria [10] to evaluate an XCS, which is the fraction of correctly classified instances calculated over a window of n exploit trials (usually 50). Other researchers have shown that *performance* might not correctly represent the classifier's ability to learn maximally general solutions, they instead used the percentage of optimal population size (%[O]) as another measure of generalization [7]. This measure however, has its own limitations and can only be applied in domains where the size of the population size for optimal solutions is known in advance. Nevertheless, the only stopping criterion used for the algorithm to terminate, for either measures, is reaching the user defined maximum number of exploit trails. The system is run for a pre-specified number of exploit trials sufficient enough to reach the best performance. The number of exploit trials are increased as the problem complexity increases e.g. 10000 trails for a 6-multiplexer problem and 20000 trails for an 11-multiplexer problem.

In this paper we compare three stopping criteria for XCSR on a subset of a benchmark ID dataset. We show that using an early stopping gives better generalization than stopping at the end of a pre-determined fixed number of generations. We also show the average overall (over 5 classes) XCSR performance using 9 different population sizes on validation and training sets. The results suggest that a smaller population size can be traded off for some loss of accuracy.

The rest of this paper is organized as follows. In section 2 we detail the XCSR modifications, performed in order to address certain test situations. Section 3 gives a description of the datasets and the methodology used in this work. Section 3.3 provides the detail of the stopping critera experiments followed by a discussion of their results. We conclude in section 4 by summarizing our contributions in this work and the future work.

2 System Modifications

The core algorithm remains the same in both XCS and XCSR. In XCSR, a classifier's condition consists of a conjunction of interval predicates represented in the form of (l_i, u_i) corresponding to upper and lower bounds of each interval, and a predicted action. A classifier matches an input data instance if all of its attribute values lies between the corresponding predicate intervals. GA operators and covering mechanism are also modified to suit the real representation. Due to space limitations, we only provide the modifications done in the XCSR system for the current work. Interested readers can refer to [10][8] for a complete description of XCS and XCSR.

In XCS[1], if no classifier from the current population matches an input instance, n covering classifiers are created, where n is usually the total number of classes, and their parameters are initialized to their default values. During

[1] We shall use XCS instead of XCSR when refereing to the main algorithm.

training, one of these n classifiers is chosen randomly and its parameters are updated based on the reward received from the environment. During testing, the best action is chosen based on its fitness weighted prediction. However there can be situations arising during testing when all classifiers in a *matchset* have the same prediction values. This can happen for example during initial runs when there are many classifiers with no experience and zero predictions or when the *matchset* is empty and n new classifiers are created as a result with the same initial prediction values. In these cases, a random decision on the class value may not be the best choice. We introduced two methods to deal with this situation. First, in case of an empty *matchset* during testing i.e. upon encountering a test instance for which there is no matching classifier in the current population we do not invoke covering, instead we either (i) choose a default class based on the majority class in the training dataset or (ii) choose the nearest matching classifier based on the shortest distance D from the input calculated as:

$$d_i = max\left\{0, (l_i - x_i), (x_i - u_i)\right\}$$

$$D = \sum_{i=1}^{n} \quad d_i$$

where n is the number of input attributes in the input vector, x_i is the ith attribute of the input instance, and u_i and l_i are the upper and lower bounds of the ith interval of a classifier respectively. If an input attribute value lies between the upper and lower bound of the corresponding classifier interval, it is considered a match. If a value falls outside the interval then the distance is measured between this value and the closest bound of the interval.

Secondly, we deal with the tie situation during testing i.e. when the *matchset* is not empty but there is more than one best actions available. In this case, we choose the action with the least cost according to the given cost matrix for the KDD dataset [3] i.e.

$$C_c = min_i \left\{C_i | i \epsilon Tie\right\}$$

where C_c is the chosen class among the set of equally best actions represented above as *Tie*.

3 Experimental Setup

3.1 Datasets

1999 KDD cup ID datasets [5] have been extensively used in ID research. The datasets contain four categories of attacks among the normal traffic:

- Probe - e.g. IP sweep, Port scans
- Denial Of Service (DOS) - e.g. SYN flood, Smurf
- User to Root (U2R) - e.g. buffer overflow
- Remote to Local (R2L) - e.g. password guessing

Table 1. Class distribution in FTP-only Dataset

Class	Training Instances	Test Instances
Normal	373	122
Probe	5	2
DOS	104	57
U2R	3	15
R2L	313	641
Total	798	837

We extracted a small subset consisting of all FTP control records (port 21 only) from the distinct training and test datasets. We call it FTP-only dataset. Also some of the irrelevant features (features whose values remain constant over FTP connection records) were excluded from this dataset which reduced the feature space from 41 to 29 for the FTP-only dataset. All 29 features have mixed values - symbolic, continuous and discrete. For this study we mapped all the symbolic features including the class (attack) categories to integer values and then all features are normalized between 0 and 1 to suit the real representation of the XCSR. The class distribution of FTP-only dataset is presented in table 1.

3.2 Methodology

In these experiments we use a batch mode testing, i.e. the system first explores complete training dataset and then the resulting population at the end of each training pass is used to predict the test instances. The overall and individual class accuracy is recorded at the end of each test pass and the cycle is repeated 100 times. The batch mode testing can perform better than the traditional online mode testing, at least in the initial runs of the algorithm, as the system does not have to classify any test cases until it has trained through the complete training dataset.

We used the same parameter settings as used by Wilson for XCSR [8] and both GA and *actionset* subsumption techniques in these experiments.

3.3 Experiments

We investigated three different stopping criteria for XCSR in this work i.e. the best overall test accuracy achieved using (i) best validation accuracy (ii) best training accuracy and (iii) the last generation accuracy. Each stopping criterion is tested for the two testing techniques i.e. using default class and distance metric.

For validation set experiments, FTP-only training set is divided into 10 subsets in a stratified sample. A 10-fold cross validation is used with a 9:1 train/validation proportion. The system is trained on each training set in batch mode and then tested on validation set and subsequently on a stand alone FTP-only test set. Each experiment was run 30 times with different seeds and folds. In case of no match, random covering is used during the explore phase as well as on the validation set during the exploit phase. For test set the two options discussed above, i.e. the default class and the distance metric techniques, are used. The corresponding test accuracy for the best overall validation accuracy is recorded for each run and then averaged over all 30 runs. All experiments are repeated for different population sizes (100, 500, 1000, 1500, 2000, 3000, 5000, 8000, 10000).

For training set experiments, the same setup is used except that the training set itself is used as the validation set and test performance is recorded corresponding to the best overall training accuracy. Each experiment is repeated with 30 different seeds to make it equivalent to the validation runs.

3.4 Results

Table 2 presents the results of three different stopping criteria. The highest accuracy achieved using a stopping criteria is given in bold face. The results which show 95% significant (one tailed) improvement over the last generation are italicized except where mentioned for 90% significance.

Table 2. The performance comparison of two early stopping criteria with the last generation results on FTP-only test set

Method	Max on Validation	Max on Training	Last Generation
Default Class			
100	***0.312(0.148)***	*0.261(0.166)*	0.158(0.110)
500	**0.313(0.080), 90%**	0.308(0.099)	0.280(0.098)
1000	***0.354(0.087)***	*0.348(0.111)*	0.307(0.066)
1500	***0.378(0.072)***	0.335(0.085), 90%	0.314(0.078)
2000	0.358(0.099)	***0.381(0.101)***	0.339(0.090)
3000	0.342(0.072)	**0.353(0.077)**	0.329(0.082)
5000	0.365(0.067)	**0.369(0.069)**	0.350(0.077)
8000	***0.383(0.087)***	*0.371(0.072)*	0.331(0.075)
10000	**0.394(0.095), 90%**	0.393(0.080), 90%	0.362(0.086)
Distance Metric			
100	***0.335(0.142)***	*0.230(0.140)*	0.170(0.084)
500	***0.359(0.116)***	0.305(0.064)	0.307(0.083)
1000	***0.376(0.104)***	0.366(0.104)	0.327(0.082)
1500	***0.377(0.088)***	*0.349(0.086)*	0.301(0.068)
2000	0.350(0.075), 90%	***0.386(0.103)***	0.320(0.100)
3000	***0.382(0.123)***	0.357(0.086)	0.334(0.081)
5000	***0.398(0.079)***	0.395(0.091), 90%	0.350(0.090)
8000	***0.399(0.092)***	*0.376(0.078)*	0.353(0.075)
10000	**0.401(0.094), 90%**	0.395(0.068), 90%	0.365(0.077)

The results show that the two early stopping criteria using validation and training always outperform the last generation results. This also suggests that XCS, like other traditional machine learning techniques, is susceptible to overfitting. Also note that stopping on validation yields better generalization most of the time than stopping on the training alone. Furthermore, a better overall accuracy is achieved when using the distance metric technique for unmatched instances during testing, which suggests its superiority over the default class technique.

Table 3. Best test accuracy for individual classes using distance metric based testing and validation based stopping for different Population sizes. The number of test instances for each class is given in the parenthesis in the header.

Pop. Size	Normal(122)	Probe(2)	DOS(57)	U2R(15)	R2L(641)	Overall(837)
100	0.655(0.10)	0.233(0.29)	0.295(0.25)	0.261(0.24)	0.280(0.19)	0.335(0.14)
500	0.921(0.02)	0.317(0.36)	0.930(0.08)	0.273(0.23)	0.204(0.15)	0.359(0.12)
1000	0.932(0.03)	0.467(0.37)	0.965(0.04)	0.341(0.28)	0.217(0.13)	0.376(0.10)
1500	0.942(0.02)	0.583(0.30)	0.962(0.04)	0.407(0.24)	0.217(0.12)	0.377(0.09)
2000	0.945(0.02)	0.650(0.37)	0.974(0.05)	0.479(0.26)	0.178(0.10)	0.350(0.07)
3000	0.948(0.02)	0.667(0.27)	0.978(0.04)	0.565(0.24)	0.215(0.16)	0.382(0.12)
5000	0.946(0.02)	0.783(0.28)	0.990(0.02)	0.500(0.23)	0.238(0.10)	0.398(0.08)
8000	0.946(0.02)	0.817(0.25)	0.987(0.03)	0.521(0.27)	0.238(0.12)	0.399(0.09)
10000	0.951(0.02)	0.867(0.26)	0.993(0.02)	0.517(0.26)	0.239(0.12)	0.401(0.09)

We can see that the overall accuracy on testing is not particularly high. This is happening partly because of the different class distribution in the training and test sets. By looking at the class distribution (table 1) in the FTP-only test set, it can be seen that R2L class constitutes bulk of the test set, which is a difficult class to predict [3], whereas Normal is the majority class in the training set. Table 3 shows the average individual class accuracies (along with their standard deviations) using distance metric technique for unmatched instances during testing and validation based early stopping. It can be seen that although the overall accuracies are low, the individual class accuracies other than R2L are quite high especially for the larger population sizes. Nevertheless, accuracy improvement is not the focus of this paper.

Population Sizing. The maximum population size is an important factor in determining the learning complexity of the LCSs. Difficult problems with high dimensions and oblique classification boundaries need higher number of rules to cover the input space. However, increasing the population size increases the learning time exponentially and may still not result in a significant improvement in accuracy. Figure 1 shows the average overall performance for various population sizes over 100 training passes on validation and training sets using distance metric based testing (refer to [9] for default class results).

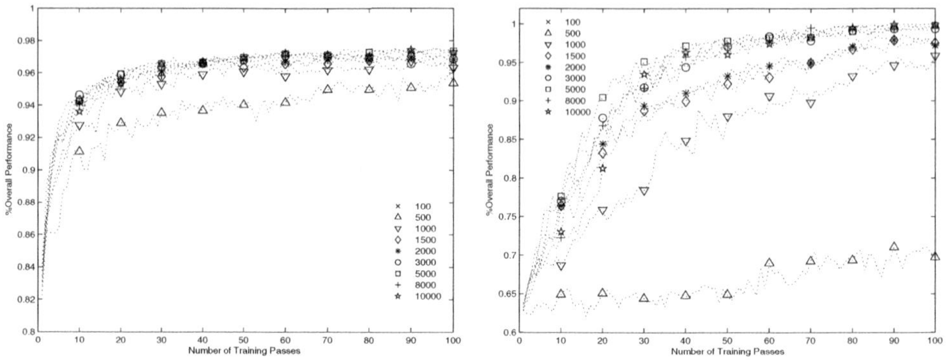

(a) Validation Set Performance (b) Training Set Performance

Fig. 1. Average overall performance with different population sizes using Distance Metric

It can be seen that higher population sizes i.e. 2000 and above reach almost 100% accuracy on the training set and around 95% on the validation set. However as the stopping criteria results show that better generalization is achieved on the test set using validation stopping, a population size decision based on validation stopping might be better than training alone. Also it can be seen that accuracy achieved using a population size of 2000 and above remains within a few percents of each other. This suggests that a tradeoff can be made between the optimal population size and the computational cost without a significant amount of loss in the accuracy.

4 Conclusions and Future work

This work investigated the effect of using an early stopping criterion in the XCSR classifier system, an XCS with real numbers representation, and showed that higher generalization can be achieved using an early stopping. The results also show that XCS, similar to other machine learning techniques, is susceptible to overfitting. We also looked at the population sizing issue and showed that a smaller population size and thus a much less computational cost can be traded off for a small amount of loss in the accuracy. Both of these issues i.e. optimal generalization and speed are critical for the intrusion detection domain which is characterized by highly unbalanced class distributions and very large training datasets.

In this work we have seen that increasing the population size does not necessarily increase the accuracy significantly, whilst an exponential rise in the computational cost. Currently, we are investigating other heuristics that can improve the XCS accuracy for the intrusion detection domain with smaller population sizes. We are also looking at mechanisms to reduce the computational time of XCS so that it can be deployed in a real time networking environment.

Acknowledgment

This work is funded by University College Postgraduate Research Scholarship (UCPRS). Most of these experiments are run on Australian Center for Advanced Computing (AC3) super computing facilities.

References

1. J. Bacardit and M. V. Butz. *Data Mining in Learning Classifier Systems: Comparing XCS with GAssist.* Illinois Genetic Algorithms Laboratory, University of Illinois at Urbana-Champaign, June 2004. IlliGAL Report No. 2004030.
2. E. Bernadó, X. Llorà, and J. M. Garrell. XCS and GALE: a comparative study of two learning classifier systems with six other learning algorithms on classification tasks. In *Proceedings of the 4th International Workshop on Learning Classifier Systems (IWLCS-2001)*, pages 337–341, 2001.
3. C. Elkan. Results of the kdd'99 classifier learning. *SIGKDD Explor. Newsl.*, 1(2):63–64, 2000.
4. A. Greenyer. CoIL Challenge 2000. The use of a learning classifier system JXCS. Technical Report LIACS Technical Report 2000-09, Sentient Machine Research, Amsterdam and Leiden Institute of Advanced Computer Science, Leiden, June 22, 2000.
5. S. Hettich and S. D. Bay. The UCI KDD Archive. http://kdd.ics.uci.edu/databases/kddcup99/kddcup99.html, 1999.
6. N. S. Inc. Online. available at http://www.nutechsolutions.com/, 2003.
7. T. Kovacs and M. Kerber. High classification accuracy does not imply effective genetic search. In *Genetic and Evolutionary Computation Conference Seattle - GECCO 2004*, Seattle,WA, USA, 6 2004.
8. S. W. Wilson. Get Real! XCS with Continuous-Valued Inputs. In P. Lanzi, W. Stolzmann, and S. Wilson, editors, *Learning Classifier Systems, From Foundations to Applications, LNAI-1813*, pages 209–219, Berlin, 2000.
9. K. Shafi, H. Abbass, and W. Zhu. The role of early stopping and population size in XCS for intrusion detection. Technical Report TR-ALAR-200604006, Defence and Security Applications Research Centre, University of New South Wales @ ADFA, Canberra, Australia, 2006.
10. S. W. Wilson. Classifier fitness based on accuracy. *Evolutionary Computation*, 3(2):149–175, 1995.

Improving Radial Basis Function Networks for Human Face Recognition Using a Soft Computing Approach

Wanida Pensuwon[1,2], Rod Adams[2], Neil Davey[2], and Wiroj Taweepworadej[1]

[1] Department of Computer Engineering, Khon Kaen University, Khon Kaen, 40002, Thailand
{wanida, wirtaw}@kku.ac.th
[2] Department of Computer Science, University of Hertfordshire, Hatfield, Herts, AL10 9AB, United Kingdom
{w.pensuwon, r.g.adams, n.davey}@herts.ac.uk

Abstract. In this paper, a new efficient method is proposed based on the radial basis function neural networks (RBFNs) architecture for human face recognition system using a soft computing approach. The performance of the present method has been evaluated using the BioID Face Database and compared with traditional radial basis function neural networks. The new approach produces successful results and shows significant recognition error reduction and learning efficiency relative to existing technique.

1 Introduction

Recently radial basis function neural networks (RBFNs) have been found to be very attractive for many real world problems. An important property of the RBFNs is that they form a unifying link among many different research fields such as function approximation, regularisation, noisy interpolation and pattern recognition [5]. In addition, RBFNs can provide a fast, linear algorithm capable of representing complex non-linear mappings [13] and can approximate any regular function [9].

As one of the most popular neural network models, RBFNs attracts lots of attentions on the improvement of its approximate ability as well as the construction of its architecture [7,8]. A support vector machine (SVM) was used to calculate support vectors and then used these support vectors as radial basis function centres. Their experimental results showed that the support-vector-based RBF outperforms conventional RBFNs [12]. The Expectation Maximization (EM) algorithm was introduced to optimise the cluster centres with two steps: obtaining initial centres by clustering and optimisation of the basis functions [2]. The extend model for mixture of experts to estimate basis functions, output neurons and the number of basis functions all together was later introduced [14]. A supervised fuzzy clustering for the RBFN training has been proposed in [10,11].

The problem of RBFN learning remains rather complex for large practical applications, and finding global search training algorithms is the subject of interest. In this work, a new technique to improve the performances of RBFNs utilises the radial basis function (RBF) architecture is presented. The method is based on the soft computing [15], which is used in order to allocate the input data to different clusters stochastically.

T.-D. Wang et al. (Eds.): SEAL 2006, LNCS 4247, pp. 58–63, 2006.

The paper is organised as follows: Section 2 gives a brief description of the RBFN classifier. In Section 3, the improving RBFN is presented. The face recognition method and the design of experiment are described in Section 4. Experimental results are presented in Section 5 and the conclusions are attained in Section 6.

2 Radial Basis Function Neural Networks

In this section, the basic characteristics of the RBF neural network architecture are presented. An RBFN can be considered as a special three-layered network depicted in Figure 1.

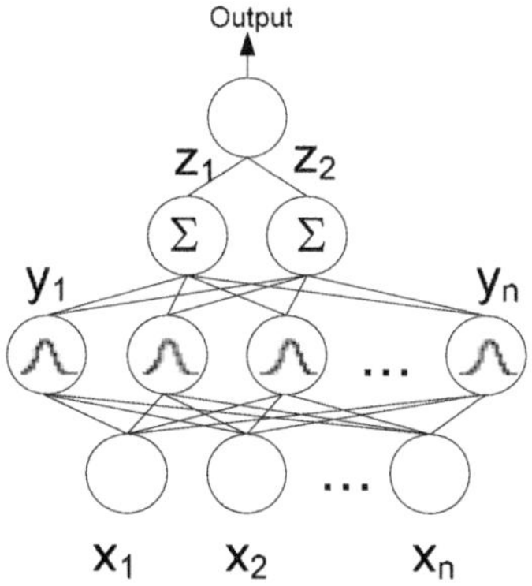

Fig. 1. An RBFN architecture

The input nodes pass the input values to the internal nodes that formulate the hidden layer. The input units are fully connected to the hidden layer. Connections between the input and the hidden layer have unit weights and, as a result, do not have to be trained. In this structure the hidden units are referred to as the RBF units. The goal of the RBF units is to cluster the data and reduce its dimensionality with a nonlinear transformation and to map the input data to a new space. The RBF units are also fully connected to the output layer. The output layer implements a linear combination on this new space.

In the pattern recognition's point of view, the main idea is to divide the input space into subclasses, and to assign a prototype vector for every subclass in the centre of it. Then the membership of every input vector in each subclass will be measured by a function of its distance from the prototype, that is:

$$f_n(x) = f(\|x - p_n\|) \tag{1}$$

This membership function should attend the maximum value in the centre (zero distance), and has considerable value in the close neighborhood of centre. The RBFN in Fig. 1 is capable of performing all the operations, and is called the RBF network. The neurons in the hidden layer of network have a Gaussian activity function and their input–output relationship is:

$$y_n = f_n(x) = \exp\left(-\frac{\|x - p_n\|^2}{2\sigma_n^2}\right) \tag{2}$$

where p_n is the prototype vector or the centre of the nth subclass
σ_n is the spread parameter

After obtaining the membership values of input vector in the subclasses, the results should be combined to obtain the membership degrees in every class.

3 Improving RBFN Network Using Soft Computing Approach

Intelligent techniques such as neural computing, fuzzy reasoning, soft computing and evolutionary computing for data analysis and interpretation are an increasingly powerful tool for making breakthroughs in the science and engineering fields by transforming the data into information and information into knowledge [6]. Soft Computing is a recently coined term describing the symbiotic use of many emerging computing disciplines [16]. The guiding principle of soft computing is: Exploit the tolerance for imprecision, uncertainty, partial truth, and approximation to achieve tractability, robustness and low solution cost. In effect, the role model for soft computing is the human mind.

In the improved RBF method, the calculation of the outcome of the RBF units is done by adding stochasticity into it. The improved RBF in Fig.2 on the right illustrates that each centre value of the RBF units is calculated using the sigmoid function, that is:

$$y_n^{'} = 0.5 * (1 + (\tanh(0.5 * y_n))) \tag{3}$$

where $y_n^{'}$ is the stochastic value of the n centre values of RBF units.

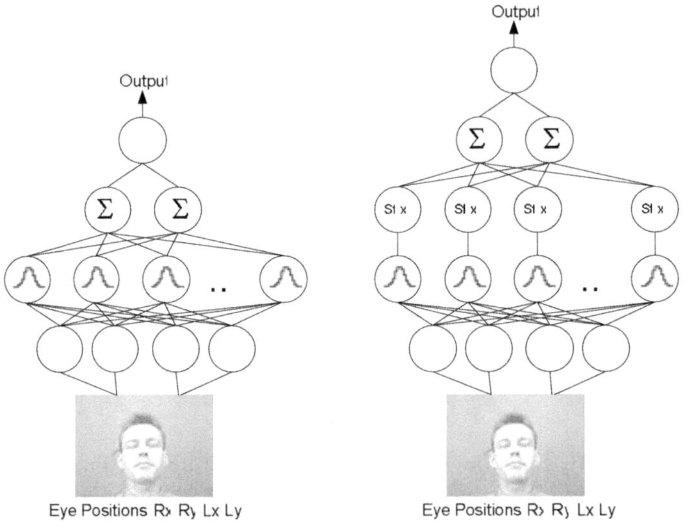

(a) Traditional RBFN model (b) Improving RBFN

Fig. 2. Improving RBFN by introducing a stochastic mean clustering algorithm for the output

Then, the new centre value of RBF units obtained by the sigmoid function is compared to a random number between 0 and 1, and if larger, the new centre value of RBF units remains unchanged. Otherwise, the new centre value will be set to the original value of RBF units. In this way, values of centre variable less than the original threshold can lead to correct classification, and value greater than then precise one can lead to negative classification.

The reason for adding stochasticity to the RBF units is that it may be useful for the network to tentative produce better classification results. The stochasticity softens the strong outcome from the RBF units and allows the possibility of more correct classification.

4 Experiments

The human face recognition system has been used as a benchmark in this study. A complete conventional human face recognition system should include two stages. The first stage requires extraction of pertinent features from the facial images and the creation of the feature vectors. The second stage involves classification of facial images based on the derived feature vector obtained in the first stage. The designed RBFNN with the proposed learning algorithm has been used as a classifier in the second stage of the human face recognition system [3,4].

4.1 Datasets

The human face database used in this study was obtained from the BioID face database at http://www.bioid.com/downloads/facedb/index.php[1]. The BioID face dataset consists of 1,521 sets of eye positions (Rx, Ry, Lx, Ly) of 23 different persons. Note that the Rx and the Ry coordinate of the left eye and the Lx and the Ly coordinate of the right eye. Samples of eye position datasets are depicted in Fig.3.

Fig. 3. Samples of eye positions data sets from BioID face database (http://www.bioid.com/downloads/facedb/index.php)

5 Results and Discussion

The experimental study conducted in this paper evaluates the effect of 2 different learning algorithms on the 2 different sets of data, training speed, sensitivity of the learning algorithms to the training and testing set and the overall recognition rate. Also, the effect of the irrelevant data on the overall recognition rate is studied. Finally the proposed human face recognition system with the soft computing approach is compared with the traditional RBFN networks presented in Table 1. The improving RBFN results were calculated from the average values of 6 times running.

In the context of this paper, learning processes in the context of Soft Computing are only considered. Therefore, structural and parametric learning, which are the counterpart of system identification and parameter estimation in classical system theory are not discussed.

Table 1. Recognition rates in % of RBFN and improving RBFN using soft computing approach

Models	No. of Data Train:Test	Recognition Rate (%)	Training Time(s)	Testing Time(s)
RBF	1217: 304	87.03	4.119	0.944
	1369: 152	90.15	4.993	0.662
RBF+Soft Computing	1217: 304	87.28	1.372	0.746
	1369: 152	94.48	1.602	0.566

From the results reported in Tables 1, one can observe that, the ratio of the size of the training and testing data are 1217:304, there is slightly improvement of the proposed model in terms of recognition rate. However, surprisingly it spends less time in both training and testing processes. Another observation one can make is that the improving RBFN clearly increases the recognition rate when the ratio of the size of the training and testing data are 1369: 152. In addition, the times in both training and testing process of the improving RBFN are obviously less than the traditional RBFN. As Table 1 shows, human face data classification based on the proposed improving RBFN model performs overall better than the conventional RBFN in terms of recognition rate and time in training and testing processes.

6 Conclusion

The conventional approach to constructing an RBFN network is to search for the optimal cluster centers among the training examples. In this paper the soft computing approach improves the performance of RBFN networks substantially on the given BioID face dataset, in terms of convergence speed and recognition rate. A strong advantage of the new improvement algorithm is its ability to gradually change and adapt radial basis functions within the learning procedure which includes alternative refinement of the radial basis functions. The successful results indicated that the proposed method can be used as a reliable technique for developing artificial classifiers for human face recognition.

References

1. BioID face database: http://www.bioid.com/downloads/facedb/index.php
2. Bishop, C. M.: Improving the generalization properties of radialbasis function neural networks. Neural Computation 3(4) (1991) 579–581
3. Howell, A.J., Buxton, H.: Face recognition using radial basis function neural networks, in: R.B. Fisher, E. Trucco (Eds.), Proc. British Machine Vision Conf., BMVA Press, Edinburgh (1996) 455-464
4. Lawrence, S., Giles, C.L., Tsoi, A.C., Back, A.D.: Face recognition: a convolutional neural network approach, IEEE Trans. Neural Networks 8 (1997) 98-113
5. Leonard, J.A., Kramer, M.A.: Radial basis function networks for classifying process faults. IEEE Control System 11 (1991) 31–8
6. Lin, S.H., Kung, S.Y., Lin, L. J.: Face recognition/detection by probabilistic decision-based neural network, IEEE Trans. Neural Networks 8 (1997) 114-132
7. Looney, C.G.: Pattern Recognition Using Neural Networks, Oxford University Press, New York (1997)
8. Mao, K. Z.: RBF neural network centre selection based on fisher ratio class separability measure. IEEE Transactions on Neural Networks 13(5) (2002) 1211–1217.
9. Park, J., Sandberg, I. W.:Approximation and radial basis function networks. Neural Computation 5 (1993) 305–316
10. Ryoo, Y.J., Lim, W.C., Kim, K.H.: Classification of materials using temperature response curve fitting and fuzzy neural network. Sensor Actuators A: Physics 94(1–2) (2001) 11–18
11. Sarimveis H, Alexandridis A, Tsekouras G, Bafas G.: A fast and efficient algorithm for training radial basis function neural networks based on a fuzzy partition of the input space. Ind Eng Chem Research 41 (2002) 751–9
12. Scholkopf, B., Sung, K. K., Burges, C. J. C., Girosi, F., Niyogi, P., Poggio, T., et al. Comparing support vector machines with Gaussian kernels to radial basis function classifiers. IEEE Transactions on Signal Processing 45(11) (1997) 2758–2765.
13. Walczak, B., Massart, D.L.: Local modeling with radial basis function networks, Chemometr. Intelligent Laboratory System 50 (2000) 179–198
14. Xu, L., Krzyzak, A., Yuille, A.: On radial basis function nets and kernel regression: statistical consistency, convergence rates, and receptive field size. Neural Networks 7(4) (1994) 609–628
15. Zadeh, L. A.: Fuzzy logic, neural networks and soft computing. One page course announcement of CS 294-4, Spring 1993, the University of California at Berkeley, November (1992)
16. Zadeh, L. A.: Fuzzy logic computing with words. IEEE Transactions on Fuzzy Systems 4 (1996) 103–111

Solving Traveling Salesman Problems by Artificial Immune Response

Maoguo Gong, Licheng Jiao, and Lining Zhang

Institute of Intelligent Information Processing, Xidian University,
710071, Xi'an, China
Maoguo_gong@hotmail.com

Abstract. This paper introduces a computational model simulating the dynamic process of human immune response for solving Traveling Salesman Problems (TSPs). The new model is a quaternion (***G***, ***I***, ***R***, ***Al***), where ***G*** denotes exterior stimulus or antigen, ***I*** denotes the set of valid antibodies, ***R*** denotes the set of reaction rules describing the interactions between antibodies, and ***Al*** denotes the dynamic algorithm describing how the reaction rules are applied to antibody population. The set of immunodominance rules, the set of clonal selection rules, and a dynamic algorithm TSP-PAISA are designed. The immunodominance rules construct an immunodominance set based on the prior knowledge of the problem. The antibodies can gain the immunodominance from the set. The clonal selection rules strengthen these superior antibodies. The experiments indicate that TSP-PAISA is efficient in solving TSPs and outperforms a known TSP algorithm, the evolved integrated self-organizing map.

1 Introduction

The traveling salesman problem (TSP) is one of the typical and most widely-studied combinatorial optimization problems.[1] It can be stated as this: given a finite number of cities along with the cost of travel between each pair of them, find the cheapest way of visiting all the cities and returning to the starting point. TSPs raise important issues because various real-world problems can be formulated as TSPs.[2][3]

The human immune system (HIS) is a highly evolved, parallel and distributed adaptive system. The information processing abilities of HIS provide important aspects in the field of computation. This emerging field is referring to as the Immunological Computation, Immunocomputing or Artificial Immune Systems (AIS).[4] In reference [5], we proposed a novel immune response computational model-the quaternion model of immune response, and analyzed its convergent conditions based on Lyapunov's Theorem. Then we applied it to solving optimal approximation of linear systems successfully.[6]

In this study, we introduced the quaternion model of immune response to solve TSPs. The set of immunodominance rules, the set of clonal selection rules, and a dynamic algorithm, named TSP-PAISA, are designed. The rest of the paper is organized as follows: section 2 describes the quaternion model of immune response. Section 3 describes the set of immunodominance rules, the set of clonal selection

T.-D. Wang et al. (Eds.): SEAL 2006, LNCS 4247, pp. 64 – 71, 2006.

rules, and the dynamic algorithm TSP-PAISA. Section 4 describes the experimental studies on some benchmark TSPs. Finally, concluding remarks are presented in section 5.

2 Quaternion Model of Immune Response

The dynamic process of human immune response can be modeled as a quaternion (G, I, R, Al) [5][6], where G denotes exterior stimulus or antigen, I denotes the set of valid antibodies, R denotes the set of reaction rules describing the interactions between antibodies, Al denotes the dynamic algorithm describing how the reaction rules are applied to antibodies. Among the four elements of the model (G, I, R, Al), antibody space I and dynamic algorithm Al depend on the antigen G, and the practical design of reaction rules in set R depend on the antigen G and the representation method of antibodies.

2.1 Antigen G

In immunology, an antigen is any substance that causes immune system to produce antibodies against it. In (G, I, R, Al), antigens refer to the pending problems. Taking optimization problem (*P*) for example

$$(P)\begin{cases}\text{minimize} & f(\boldsymbol{x}) = f(x_1, x_2, \ldots x_n) \\ \text{subject to} & g_i(\boldsymbol{x}) < 0 \quad i = 1, 2, \cdots, p \\ & h_j(\boldsymbol{x}) = 0 \quad j = p+1, p+2, \cdots, q\end{cases} \tag{1}$$

where $\boldsymbol{x} = (x_1, x_2, \ldots x_n)$, antigen is the function of objective function $f(\boldsymbol{x})$, i.e. $G(\boldsymbol{x}) = g(f(\boldsymbol{x}))$.

2.2 Antibody Space I

In (G, I, R, Al), B cells, T cells and antigen-specific lymphocytes are generally called antibodies. An antibody represents a search point in the space of potential solutions. The antibody $\boldsymbol{a} = a_1 a_2 \cdots a_l$ is the coding of variable $\boldsymbol{x}$, and $\boldsymbol{x}$ is called the decoding of antibody $\boldsymbol{a}$.

The space of potential solutions is called antibody space denoted by I, where $\boldsymbol{a} \in I$. An antibody population $A = (\boldsymbol{a}_1, \boldsymbol{a}_2, \cdots, \boldsymbol{a}_n)$, $\boldsymbol{a}_k \in I$, $1 \le k \le n$, is an n-dimensional group of antibody $\boldsymbol{a}$, where the positive integer n is the size of antibody population A.

2.3 The Set of Reaction Rules R

The set R describes all the possible interactions between antibodies in antibody space I. For antibody population $A = (\boldsymbol{a}_1, \boldsymbol{a}_2, \cdots, \boldsymbol{a}_n)$, a rule $R \in \boldsymbol{R}$ can be expressed as

$$R(\boldsymbol{A}) = R(\boldsymbol{a}_1 + \boldsymbol{a}_2 + \cdots + \boldsymbol{a}_n) = \boldsymbol{a}_1' + \boldsymbol{a}_2' + \cdots + \boldsymbol{a}_m' \tag{2}$$

where n, m are positive integers, the value of m depends on the rule R, and the representation '$+$' is not the arithmetical operator, but only separates the antibodies on either side in Equation (2). Equation (2) shows that the n antibodies of $\boldsymbol{A}$ evolve into m antibodies on the right-hand side by the effect of reaction rule R.

2.4 Dynamic Algorithm *Al*

Al is the algorithm simulating the process of antibody evolution and dominating interactions among antibodies during artificial immune response, including the format of the set $\boldsymbol{R}$ acting on antibody space $\boldsymbol{I}$, the computing of antibody-antigen affinity, the judgment of halt conditions in artificial immune response, and so on.

3 Reaction Rules and Dynamic Algorithm for TSPs

In order to solving TSPs, the reaction rules $\boldsymbol{R}$ and dynamic algorithm $\boldsymbol{Al}$ in the model $(\boldsymbol{G}, \boldsymbol{I}, \boldsymbol{R}, \boldsymbol{Al})$ are designed as follows.

3.1 Reaction Rules for TSPs

In this paper, we design the set $\boldsymbol{R}$ composed of two subsets, i.e. the set of immunodominance rules $\boldsymbol{R}_{\mathrm{D}}$ and the set of clonal selection rules $\boldsymbol{R}_{\mathrm{CS}}$.

3.1.1 The Set of Immunodominance Rules $\boldsymbol{R}_{\mathrm{D}}$

By the theory of immunology[7], there are many epistasises on an antibody, but only one epistasis works when the immune response takes place. This phenomenon is called immunodominance.

For TSPs, an antibody denotes a candidate tour:

$$\boldsymbol{a} = \{a_1, a_2, \cdots, a_l\} \tag{3}$$

Where l is the number of cities, $a_i \in \{1,2,3,\cdots,l\}$, and $\forall i, j \in \{1,2,3,\cdots,l\} \wedge i \neq j$, $a_i \neq a_j$. So the antibody population $\boldsymbol{A}$ is

$$\boldsymbol{A} = \{\boldsymbol{a}_1, \boldsymbol{a}_2, \cdots, \boldsymbol{a}_m\} \tag{4}$$

Where m is the size of antibody population.

For TSPs, an immunodominance set $\boldsymbol{id} = ido(e_1, e_2)$, where $\boldsymbol{e}_1 \subset \boldsymbol{a}$, $\boldsymbol{e}_2 \subset \boldsymbol{a}$, are the two subsets of $\boldsymbol{a}$, and $\boldsymbol{e}_1 \cap \boldsymbol{e}_2 = \Phi$, $ido(e_1, e_2)$ satisfies the requirement

$$\{a_{e1}, a_{e2}\} = ido(\boldsymbol{e}_1, \boldsymbol{e}_2) = \arg \min_{a_i \in \boldsymbol{e}_1, a_j \in \boldsymbol{e}_2} \left(dis(a_i, a_j) \right) \tag{5}$$

where $dis(a_i, a_j)$ denotes the distance between a_i and a_j.

For the antibody $\boldsymbol{a}=\{a_1,a_2,\cdots,a_l\}\in \boldsymbol{A}$, if the kth position is selected to be changed, $\boldsymbol{e}_1=\{a_1,a_2,\cdots,a_k\}$, $\boldsymbol{e}_2=\{a_{k+1},a_{k+2},\cdots,a_l\}$. Then, the immunodominance rules $\boldsymbol{R}_{\mathrm{D}}$ are described as follows.

```
 Begin
{
      while( e1 ≠ Φ )do
            {
              Get the immunodominance set id = {a_e1, a_e2};
              Insert a_e1 into e2 behind a_e2 •
              Delete a_e1 from e1 •
            }
}
```

3.1.2 The Set of Clonal Selection Rules R_{CS}

The clonal selection theory[8] is used in immunology to describe the basic features of an immune response.[9] Biological clonal occurs to the degree that a B-cell's antibodies match antigen. A strong match causes a B-cell to be cloned many times, and a weak match results in few clones.[10] Inspired by the clonal selection theory, the set of clonal selection rules $\boldsymbol{R}_{\mathrm{CS}}$ include Clonal Proliferation Rule ($R_{\mathrm{P}}^{\mathrm{C}}$), Affinity Maturation Rule ($R_{\mathrm{M}}^{\mathrm{A}}$) and Clonal Selection Rule ($R_{\mathrm{S}}^{\mathrm{C}}$) on the antibody population $\boldsymbol{A}(k)$, where the antibody population at time k is represented by the time-dependent variable matrix $\boldsymbol{A}(k)=\{\boldsymbol{a}_1(k),\boldsymbol{a}_2(k),\cdots,\boldsymbol{a}_n(k)\}$. The evolution process can be described as:

$$\boldsymbol{A}(k)\xrightarrow{R_{\mathrm{P}}^{\mathrm{C}}}\boldsymbol{Y}(k)\xrightarrow{R_{\mathrm{M}}^{\mathrm{A}}}\boldsymbol{Z}(k)\cup\boldsymbol{A}(k)\xrightarrow{R_{\mathrm{S}}^{\mathrm{C}}}\boldsymbol{A}(k+1) \tag{6}$$

Clonal Proliferation Rule $R_{\mathrm{P}}^{\mathrm{C}}$: Define

$$\boldsymbol{Y}(k)=R_{\mathrm{P}}^{\mathrm{C}}(\boldsymbol{A}(k))=[R_{\mathrm{P}}^{\mathrm{C}}(\boldsymbol{a}_1(k)),\,R_{\mathrm{P}}^{\mathrm{C}}(\boldsymbol{a}_2(k)),\cdots,R_{\mathrm{P}}^{\mathrm{C}}(\boldsymbol{a}_n(k))]^{\mathrm{T}} \tag{7}$$

where $\boldsymbol{Y}_i(k)=R_{\mathrm{P}}^{\mathrm{C}}(\boldsymbol{a}_i(k))=\boldsymbol{I}_i\times\boldsymbol{a}_i(k)$, $i=1,2,\cdots,n$, $\boldsymbol{I}_i$ is a q_i-dimensional unit column vector. $q_i\in[1,n_{\mathrm{c}}]$ is an self-adaptive parameter, or set as a constant, n_{c} is a given value related to the upper limit of clone scale.

After Clonal Proliferation, the population becomes:

$$\boldsymbol{Y}(k)=\{\boldsymbol{Y}_1(k),\boldsymbol{Y}_2(k),\cdots,\boldsymbol{Y}_n(k)\} \tag{8}$$

where

$$\begin{gathered}\boldsymbol{Y}_i(k)=\left\{\boldsymbol{y}_{ij}(k)\right\}=\left\{\boldsymbol{y}_{i1}(k),\boldsymbol{y}_{i2}(k),\cdots,\boldsymbol{y}_{iq_i}(k)\right\}\\ \text{and }\boldsymbol{y}_{ij}(k)=\boldsymbol{a}_i(k),\ j=1,2,\cdots,q_i\ \ i=1,2,\cdots,n\end{gathered} \tag{9}$$

Affinity Maturation Rule R_M^A: The Affinity Maturation Operation R_M^A is diversified basically by hypermutation[11]. Random changes are introduced into the genes, i.e. mutation, such changes may lead to an increase in the affinity of the clonal antibody occasionally.

After Affinity Maturation Operation, the population becomes:

$$\boldsymbol{Z}(k) = \{\boldsymbol{Z}_1(k), \boldsymbol{Z}_2(k), \cdots, \boldsymbol{Z}_n(k)\} \tag{10}$$

where

$$\begin{aligned} &\boldsymbol{Z}_i(k) = \left\{\boldsymbol{z}_{ij}(k)\right\} = \left\{\boldsymbol{z}_{i1}(k), \boldsymbol{z}_{i2}(k), \cdots, \boldsymbol{z}_{iq_i}(k)\right\} \\ &\text{and } \boldsymbol{z}_{ij}(k) = R_M^A\left(\boldsymbol{y}_{ij}(k)\right),\ j = 1, 2, \cdots, q_i \quad i = 1, 2, \cdots, n \end{aligned} \tag{11}$$

Clonal Selection Rule R_S^C: Define $\forall\ i = 1, 2 \cdots n$, if $\boldsymbol{b}_i(k) \in \boldsymbol{Z}_i(k)$ is the best antibody (the antibody with the highest antibody-antigen affinity) in $\boldsymbol{Z}_i(k)$, then

$$\boldsymbol{a}_i(k+1) = R_S^C\left(\boldsymbol{Z}_i(k) \cup \boldsymbol{a}_i(k)\right) = \begin{cases} \boldsymbol{b}_i(k) & \text{if } \boldsymbol{b}_i(k) \text{ is better than } \boldsymbol{a}_i(k) \\ \boldsymbol{a}_i(k) & \text{else} \end{cases} \tag{12}$$

The newcome population is

$$\begin{aligned} &\boldsymbol{A}(k+1) = R_S^C\left(\boldsymbol{Z}(k) \cup \boldsymbol{A}(k)\right) \\ &= \left[R_S^C\left(\boldsymbol{Z}_1(k) \cup \boldsymbol{a}_1(k)\right),\ R_S^C\left(\boldsymbol{Z}_2(k) \cup \boldsymbol{a}_2(k)\right),\ \cdots,\ R_S^C\left(\boldsymbol{Z}_n(k) \cup \boldsymbol{a}_n(k)\right)\right]^T \\ &= \left\{\boldsymbol{a}_1(k+1), \boldsymbol{a}_2(k+1), \cdots, \boldsymbol{a}_n(k+1)\right\} \end{aligned} \tag{13}$$

where $\boldsymbol{a}_i(k+1) = R_S^C\left(\boldsymbol{Z}_i(k) \cup \boldsymbol{a}_i(k)\right) \quad i = 1, 2 \cdots n$.

3.2 Dynamic Algorithm Driving the Population Evolution

Dynamic algorithm $\boldsymbol{Al}$ is the algorithm dominating the interactions between antibodies and driving the antibody population evolution, including the format of the set $\boldsymbol{R}$ acting on antibody populations, the affinity assessment, the judgment of halt conditions, and so on. ALGORITHM 1 describes the details of the population-based artificial immune system algorithm for TSPs (TSP-PAISA).

ALGORITHM 1. Population-based AIS Algorithm for TSPs (TSP-PAISA)

Step 1) Initialization: Give the termination criterion. Randomly generate the initial antibody population:

$$\boldsymbol{A}(0) = \left\{\boldsymbol{a}_1(0), \boldsymbol{a}_2(0), \cdots \boldsymbol{a}_n(0)\right\}.$$

Calculate the antibody-antigen affinities of all antibodies of $\boldsymbol{A}(0)$, k=0.

Step 2) Perform $\boldsymbol{R}_{\mathrm{D}}$: Update $A(k)$ by applying $\boldsymbol{R}_{\mathrm{D}}$ to $A(k)$.
Step 3) Perform $\boldsymbol{R}_{\mathrm{CS}}$:
    Step 3.1) Clonal Proliferation: Get $\boldsymbol{Y}(k)$ by applying $R_{\mathrm{P}}^{\mathrm{C}}$ to $\boldsymbol{A}(k)$.
    Step 3.2) Affinity Maturation: Get $\boldsymbol{Z}(k)$ by applying $R_{\mathrm{M}}^{\mathrm{A}}$ to $\boldsymbol{Y}(k)$.
    Step 3.3) Evaluation: Calculate the antibody-antigen affinities of all antibodies of $\boldsymbol{Z}(k)$.
    Step 3.4) Clonal Selection: Get $\boldsymbol{A}(k+1)$ by applying $R_{\mathrm{S}}^{\mathrm{C}}$ to $\boldsymbol{Z}(k)$ and $\boldsymbol{A}(k)$.
Step 4) Termination test: If a stopping condition is satisfied, stop the algorithm. Otherwise, $k=k+1$, go to Step 2).

The antibody-antigen affinity is defined as the length of the corresponding tour for an antibody. The termination criterion of TSP-PAISA is a maximum number of generations being reached.

4 Experimental Studies

In this study, the problems in standard test set[12] will be tested by TSP-PAISA and the evolved integrated self-organizing map (eISOM)[1] for TSPs. The experiments are carried on a PC with PIV 3.2 GHz CPU and 2G RAM. All the problems are tested with programming language VC++6.0.

In TSP-PAISA, the size of population size is 25, clonal scale is 4. The maximum number of generations is 1000. In eISOM, the population size is 100, the crossover and the mutation probabilities are 0.99 and 0.01, respectively. The maximum number of generations is 6000. Under this condition, the computational cost of TSP-PAISA is much less than that of eISOM.

The experimental results of TSP-PAISA and eISOM over 30 independent runs are shown in Table 1, where l is the number of cities, S_0 denotes the known minimal tour length, σ denotes the relative difference[1] which is defined as Equation (14). In Table 1, eISOM1 denotes the eISOM without local improvement heuristic, eISOM2 denotes the eISOM improved by local improvement heuristic. The bold-faced text indicates the best solution among all the algorithms for a TSP.

$$\sigma\% = \left(\sum_{i=1}^{T} (S_i - S_0) \right) / (T \times S_0) \times 100\% \tag{14}$$

Where S_i denotes the minimal tour length obtained from the i-th run, T is the number of independent runs.

Table 1. Performance comparison between TSP-PAISA and eISOM

TSPs	l	S_0	σ (%)		
			eISOM1	eISOM2	TSP-PAISA
Eil51	51	429	2.56	1.97	**0.78**
Eil101	101	629	3.59	2.92	**1.06**
KroA150	150	26524	1.83	1.26	**1.23**
KroA200	200	29368	1.64	1.21	**1.09**
Lk318	318	42029	2.05	1.93	**1.80**
Pcb442	442	50779	6.11	5.67	**3.87**
Att532	532	87550	3.35	2.39	**2.21**
Tk1002	1002	259045	4.82	4.01	**3.63**
Tk2392	2392	378032	6.44	**5.83**	6.80
Average			3.60	3.02	**2.50**

The experimental studies indicate that TSP-PAISA is efficient in most of the selected TSPs and outperforms eISOM clearly, but we also find that its performance is not very good in some special problems, especially in the large scale TSP. We can enlarge the size of population to improve the performance of TSP-PAISA. However, it will take a very long time to getting a satisfying solution.

5 Concluding Remarks

In this paper, we introduced an immune inspired computational model (***G***, ***I***, ***R***, ***Al***), where ***G*** denotes exterior stimulus or antigen, ***I*** denotes the set of valid antibodies, ***R*** denotes the set of reaction rules describing the interactions between antibodies, and ***Al*** denotes the dynamic algorithm describing how the reaction rules are applied to antibody population. A specific ***Al***, TSP-PAISA, based on the set of clonal selection rules and the set of immunodominance rules was proposed for solving TSPs. TSP-PAISA was tested on 32 benchmark TSPs with 16 to 2392 cities and compared with a known TSP algorithm eISOM. The experiments indicated that TSP-PAISA outperforms the eISOM clearly. The results indicated that TSP-PAISA can obtain high quality solutions for most selected TSPs.

Acknowledgements. This work was supported by the National Natural Science Foundation of China (No. 60372045) and the Graduate Innovation Fund of Xidian University (No. 05004).

References

1. Jin, H. D., Leung, K. S., Wong, M. L., Xu, Z. B.: An Efficient Self-Organizing Map Designed by Genetic Algorithms for the Traveling Salesman Problem. IEEE Transactions on Systems, Man, and Cybernetics-Part B. Vol. 33, No. 6 (2003) 877–888
2. Durbin, R., Willshaw, D.: An analogue approach to the traveling salesman problem. Nature, vol. 326 (1987) 689–691
3. Reinelt, G.: The Traveling Salesman: Computational Solutions for TSP Applications. New York: Springer-Verlag (1994)

4. Garrett, S. M.: How Do We Evaluate Artificial Immune Systems. Evolutionary Computation, Vol. 13, No. 2 (2005) 145–178
5. Gong, M. G., Jiao, L. C., Liu, F., Du, H. F.: The Quaternion Model of Artificial Immune Response. In: The Proceedings of 4th International Conference on Artificial Immune Systems, ICARIS 2005, Banff, Canada. August 14-17, 2005, Lecture Notes in Computer Science, Vol. 3627 (2005) 207–219
6. Gong, M. G., Du, H. F., Jiao, L. C.: Optimal approximation of linear systems by artificial immune response. Science in China: Series F Information Sciences. Science in China Press, co-published with Springer-Verlag GmbH. Vol. 49, No.1 (2006) 63–79
7. Yewdell, J.W., Bennink, J. R.: Immunodominance in major histocompatibility complex class I-restricted T lymphocyte responses. Annual Review of Immunology. Vol. 17 (1999) 51–88
8. Burnet, F. M.: The Clonal Selection Theory of Acquired Immunity. Cambridge University Press (1959)
9. Burnet, F. M.: Clonal selection and after. Theoretical Immunology. New York: Marcel Dekker (1978) 63–85
10. Garrett, S. M.:. Parameter-free, Adaptive Clonal Selection. In: The Proceedings of IEEE Congress on Evolutionary Computing (CEC2004), Portland Oregon, June (2004) 1052–1058
11. de Castro, L. N., Von Zuben, F. J.: Learning and Optimization Using the Clonal Selection Principle. IEEE Transactions on Evolutionary Computation, Vol.6, No. 3 (2002) 239–251
12. Reinelt, G.: TSPLIB—A traveling salesman problem library. ORSA Journal of Computing. Vol. 3, No.4 (1991) 376–384

A Strategy of Mutation History Learning in Immune Clonal Selection Algorithm*

Yutao Qi, Xiaoying Pan, Fang Liu, and Licheng Jiao

Institute of Intelligent Information Processing and National Key Lab of Radar Signal Processing, Xidian University, Xi'an, China, 710071
{qi_yutao, xiaoying_pan, f63liu}@163.com
lchjiao@mail.xidian.edu.cn

Abstract. A novel strategy termed as mutation history learning strategy (MHLS) is proposed in this paper. In MHLS, a vector called mutation memory is introduced for each antibody and a new type of mutation operation based on mutation memory is also designed. The vector of mutation memory is learned from a certain antibody's iteration history and used as guidance for its further evolution. The learning and usage of history information, which is absent from immune clonal selection algorithm (CSA), is shown to be an efficient measure to guide the direction of the evolution and accelerate algorithm's converging speed. Experimental results show that MHLS improves the performance of CSA greatly in dealing with the function optimization problems.

1 Introduction

The famous antibody clonal selection theory was put forward by Burnet in 1958. It establishes the idea that the antibodies can selectively react to the antigens. The clonal selection is a dynamic process of the immune system that self-adapting antigen stimulation. The cells are selected when they recognize the antigens and then proliferate. These biological characteristics can be used in the artificial immune system.

On the basis of the antibody clonal selection theory, De Castro pioneered the Clonal Selection Algorithm (CSA) [1] in 2000. He applied it to many problems, such as binary character recognition, multi-modal optimization and traveling salesman problem (TSP). With the aim of solving the network intrusion detection problem, Kim constructed Dynamic Clonal Selection Algorithm (DynamiCS) [2] in 2002. DynamiCS endeavors to reduce the number of parameters of the algorithm and adjust them adaptively, which makes the algorithm much more robust. Licheng Jiao and Haifeng Du proposed a polyclonal strategy [3] in 2003. In this strategy, the recombination operation which dramatically speeds up the algorithm's convergence is introduced. In recent years, researchers have shown an increasing interest in using the immune system as a powerful metaphor for the development of novel computational intelligence paradigms. Artificial immune systems become a new research hot spot

* Supported by the National Natural Science Foundation of China under Grant Nos. 60372045, 60133010 and the National Grand Fundamental Research 973 Program of China under Grant No.2001CB309403.

T.-D. Wang et al. (Eds.): SEAL 2006, LNCS 4247, pp. 72–79, 2006.

after the neural network, fuzzy logic and evolutionary algorithm [4]. The concept clone has been extensively applied to many fields such as learning and optimization [5], computer programming [6,7], system control [8], and interactive parallel simulation [9] and so on.

Of all these clonal selection algorithms mentioned above, there is a common drawback. Among all the clone offsprings of a certain antibody, only the best one will be kept in the next generation and replace the old one, while the others which also embody some useful information about the optimization destination are discarded. It is obviously a waste of computing resources. The main idea of this study is to explore a method which can obtain more information from antibodies' clonal mutation offsprings and use it as guidance for their further evolution. So, the mutation history learning strategy (MHLS) is proposed. In order to see weather MHLS helps to reinforce CSA's searching capability, the improve CSA based on MHLS (MHLS_CSA) is used to solve the function optimization problems.

2 Descriptions of Clonal Selection Algorithm

Clonal selection algorithm was straightforward by Leandro Nunes de Castro in 2000. The algorithm works as in Fig.1, after each six steps we have one cell generation: [1].

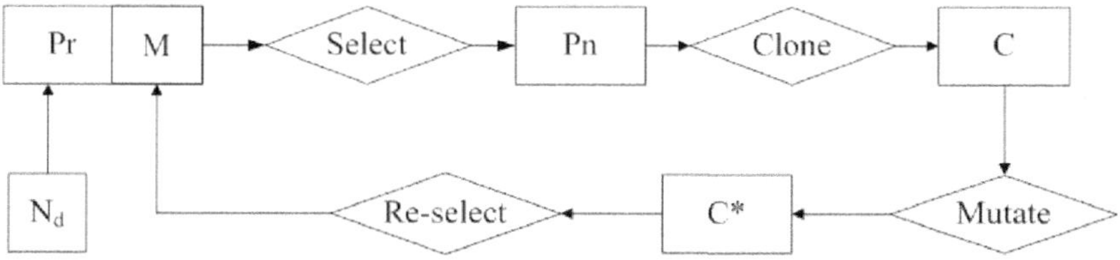

Fig. 1. Block diagram of the clonal selection algorithm

Step1: Generate a set (P) of candidate solutions, composed of the subset of memory cells (M) added to the remaining (P_r) population ($P = P_r + M$);

Step2: Determine (Select) the n best individuals of the population (P_n), based on an affinity measure;

Step3: Reproduce (Clone) these n best individuals of the population, giving rise to a temporary population of clones (C). The clone size is an increasing function of the affinity with the antigen;

Step4: Submit the population of clones to a hypermutation scheme, where the hypermutation is proportional to the affinity of the antibody with the antigen. A maturated antibody population is generated (C^*);

Step5: Re-select the improved individuals from C^* to compose the memory set M. Some members of P can be replaced by other improved members of C^*;

Step6: Replace d antibodies by novel ones (diversity introduction). The lower affinity cells have higher probabilities of being replaced.

3 MHLS Based CSA for Function Optimization

The mathematical model of function optimization is:

$$\text{minimize } f(\mathbf{x}), \quad \mathbf{x} = (x_1, x_2, ..., x_m) \in S$$

in which, $f(\mathbf{x})$ is the object function. $\mathbf{S} \subseteq R^m$ is the m-dimensional searching space with border $\underline{x_i} \le x_i \le \overline{x_i}, i = 1, 2, ..., m$. And it can be described as following:

$$\mathbf{S} = [\underline{x}, \overline{x}], \underline{\mathbf{x}} = (\underline{x_1}, \underline{x_2}, \cdots, \underline{x_m}), \overline{\mathbf{x}} = (\overline{x_1}, \overline{x_2}, ..., \overline{x_m}).$$

As the affinity of an antibody must be a positive value, we construct a negative real function $g(\mathbf{x})$ in our algorithm. $g(\mathbf{x})$ is consistent with $f(\mathbf{x})$, in other words, for any two variables $\mathbf{X}_1, \mathbf{X}_2 \in S$, if $g(\mathbf{X}_1) > g(\mathbf{X}_2)$ then $f(\mathbf{X}_1) > f(\mathbf{X}_2)$. So the original optimization problem becomes $\text{minimize}\{g(e^{-1}(\mathbf{A})) \mid \mathbf{A} \in \mathbf{I}\}$. Note $\mathbf{X} = (x_1, x_2, ..., x_m)$ as a variable of the new optimization problem, and the limited-length real number string $\mathbf{A} = g_1, g_2, ..., g_m$, $g_i \in [0,1]$ $i = 1, 2, ..., m$ is the antibody coding of the variable $\mathbf{X}$. Then $\mathbf{A}$ can be described by $\mathbf{A} = e(\mathbf{X})$ .We call $\mathbf{X}$ is the decoding of antibody $\mathbf{A}$. And it can be described by $\mathbf{X} = e^{-1}(\mathbf{A})$. Then we can get:

$$x_i = \underline{x_i} + (\overline{x_i} - \underline{x_i}) \times g_i, \; i = 1, 2, ..., m \tag{1}$$

The set of all antibodies $\mathbf{I}$ is called antibody space. *affinity*($\mathbf{A}$) is a positive real function on the set $\mathbf{I}$, and it is called the antibody-antigen affinity function, defined as following:

$$affinity(\mathbf{A}) = -g(e^{-1}(\mathbf{A})) \tag{2}$$

The antibody space is:

$$\mathbf{I}^n = \{\vec{\mathbf{A}} : \vec{\mathbf{A}} = (\mathbf{A}_1, \mathbf{A}_2, \cdots, \mathbf{A}_n), \;\; \mathbf{A}_k \in \mathbf{I}, \;\; 1 \le k \le n\} \tag{3}$$

In which the positive integer n is the size of antibody population, and the antibody population $\vec{A} = \{A_1, A_2, \cdots, A_n\}$ is a set of n antibodies.

3.1 Mutation Memory

In order to make full use of information which is achieved during iterations, an m-dimensional vector termed as mutation memory is introduced for each antibody to act as the carrier of history information. And the mutation memory updating operation is added in to CSA after the modified mutation operation to from the MHLS_CSA.

Note $\mathbf{A}(T)$ as any antibody in current antibody population that has been iterated for T generations. $\mathbf{Q}(T) = \{\mathbf{A}_1(T), \mathbf{A}_2(T), ..., \mathbf{A}_q(T)\}$ as q clones of $\mathbf{A}(T)$. Then submit the cloned antibody set $\mathbf{Q}(T)$ to the following described mutation operation, and note

$\mathbf{Q}'(T)=\{\mathbf{A}'_1(T),\mathbf{A}'_2(T),...,\mathbf{A}'_q(T)\}$ as the set of mutated offsprings. Evaluate antibodies in $\mathbf{Q}'(T)$, and compare them with $\mathbf{A}(T)$. Then reselect the improved and degenerate antibodies to compose two populations $\mathbf{P}_1(T)=\{\mathbf{A}''_1(T),\mathbf{A}''_2(T),...,\mathbf{A}''_j(T)\}$ and $\mathbf{P}_2(T)=\{\mathbf{A}'''_1(T),\mathbf{A}'''_2(T),...,\mathbf{A}'''_k(T)\}$ respectively. The sizes of the improved antibody population $\mathbf{P}_1(T)$ and the degenerated antibody population $\mathbf{P}_2(T)$ satisfy the following inequation $j+k\le q$.

Definition 1. The mutation memory $\mathbf{M}(T)$ for any antibody $\mathbf{A}(T)$ in the population can be described by Equation (4) (5) and (6).

$$\mathbf{S}_i=\sum_{i=1}^{j}(\mathbf{A}''_i(T)-\mathbf{A}(T)) \tag{4}$$

$$\mathbf{S}_d=\sum_{i=1}^{k}(\mathbf{A}'''_i(T)-\mathbf{A}(T)) \tag{5}$$

$$\mathbf{M}(T)=\begin{cases}\vec{0} & T=0\\ \mathbf{S}_i-\mathbf{S}_d & T=1\\ \alpha\times\mathbf{M}(T-1)+(1-\alpha)\times(\mathbf{S}_i-\mathbf{S}_d) & T>1\end{cases} \tag{6}$$

In Equation (6), $\vec{0}$ is an m-dimensional zero vector, the symbols "+","$\sum$" and"–" are linear operators on vectors. The parameter α which is termed as inertia factor is a real number parameter between 0 and 1. It represents the degree of retaining or forgetting of mutation history information. The operation of updating mutation memory will be taken after the mutation operation in MHLS_CSA.

Seen from Equation (6), a part of the old mutation history information is fade from the history memory vector and the new mutation information provide by the latest iteration is introduced. The mutation memory updating operation picks up useful information from both advantage and disadvantage mutations in each iteration.

3.2 Mutation Operation Using Mutation Memory

The second part of MHLS is the modification of the CSA's mutation operation to make use of useful information provided by mutation memories. The mutation operation straightforward can be divided in to two steps: first apply random hypermutation on the target antibody, and then regulate the maturated antibody using the mutation memory information.

Let $\mathbf{A}(T)$ is any antibody in current antibody population. $\mathbf{M}(T)$ is corresponding mutation memory vector. $\mathbf{Q}(T)=\{\mathbf{A}_1(T),\mathbf{A}_2(T),...,\mathbf{A}_q(T)\}$ is q clones of $\mathbf{A}(T)$. For each antibody $\mathbf{A}_i(T)$ (i = 1,2,…,q) in $\mathbf{Q}(T)$, apply a random hypermutation on it, giving rise to a temporary antibody $\tilde{\mathbf{A}}_i(T)$. Then $\tilde{\mathbf{A}}_i(T)$ will be regulated by the following approach. Note $\mathbf{A}'_i(T)$ as the regulated antibody, the regulation approach is defined as Equation (7) and (8):

$$\mathbf{D} = \tilde{\mathbf{A}}_i(T) - \mathbf{A}_i(T) \tag{7}$$

$$\mathbf{A}'_i(T) = \mathbf{A}_i(T) + \beta \times \mathbf{M}(T) + (1-\beta) \times \mathbf{D} \tag{8}$$

This regulation above may lead to some invalid antibodies. If a certain antibody's gene goes beyond the upper limit 1, this gene will be coded as 1. On the contrary, if the coding of a gene is smaller than the lower limit 0, it will be replaced by 0.

To see the efficiency of MHLS, MHLS_CSA keeps the main operations of CSA unchanged except the modification of mutation operation and the affiliating of the mutation memory updating operation after it.

4 Simulation Experiments

In order to validate our improved strategy, MHLS_CSA is executed to solve the following test functions.

$$f_1(x) = \sum_{i=1}^{N} x_i^2, \quad -100 \le x_i \le 100 \quad f_{\min} = 0$$

$$f_2(x) = \sum_{i=1}^{N} |x_i| + \prod_{i=1}^{N} |x_i|, \quad -10 \le x_i \le 10 \quad f_{\min} = 0$$

$$f_3(x) = \sum_{i=1}^{N} i x_i^4 + random[0,1), \quad -1.28 \le x_i \le 1.28 \quad f_{\min} = 0$$

$$f_4(x) = \sum_{i=1}^{N} (x_i^2 - 10\cos(2\pi x_i) + 10), \quad -5.12 \le x_i \le 5.12 \quad f_{\min} = 0$$

$$f_5(x) = \sum_{i=1}^{N} \frac{x_i^2}{4000} - \prod_{i=1}^{N} \cos(\frac{x_i}{\sqrt{i}}) + 1, \quad -600 \le x_i \le 600 \quad f_{\min} = 0$$

$$f_6(x) = -20\exp\left(-0.2\sqrt{\frac{1}{N}\sum_{i}^{N} x_i^2}\right) - \exp\left(\frac{1}{n}\sum_{i=1}^{N}\cos(2\pi x_i)\right) + 20 + e, -30 \le x_i \le 30, \quad f_{\min} = 0$$

These six standard test functions contain diversity of types. $f_1 \sim f_2$ are single-peak functions. f_3 is a four times function with noise. $f_4 \sim f_6$ are multi-peak functions and there local optimal values increase with the dimension of variables [10].

4.1 Experimental Results

The performances of CSA and MHLS_CSA in the task of function optimization are compared in the following figures. The average function evaluation curves in the following figures represent computing cost the algorithm spend to solve the function optimization problems.

For both CAS and MHLS_CSA, the population size is set as 20, the number of cloned antibodies n is set as 15, and the diversity introduction scale d is set as 3. For MHLS_CSA, there are two special parameters, the inertia factor α is set as 0.2 and the reviewing factor β is set as 0.1.

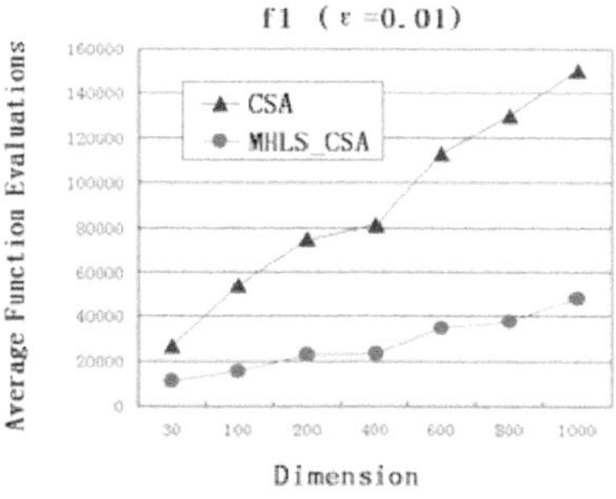

Fig. 2. Performance comparisons for f1

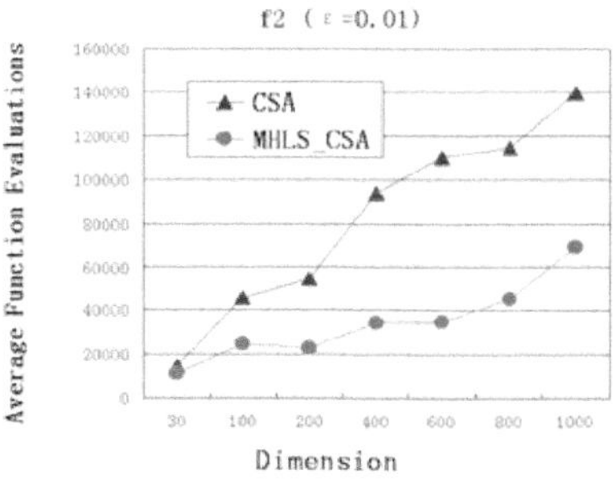

Fig. 3. Performance comparisons for f2

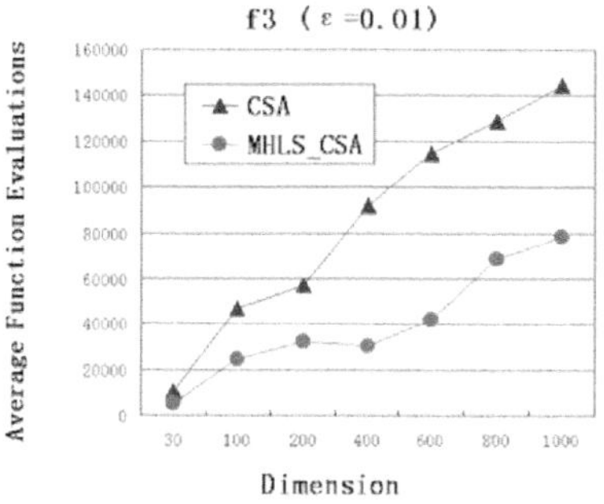

Fig. 4. Performance comparisons for f3

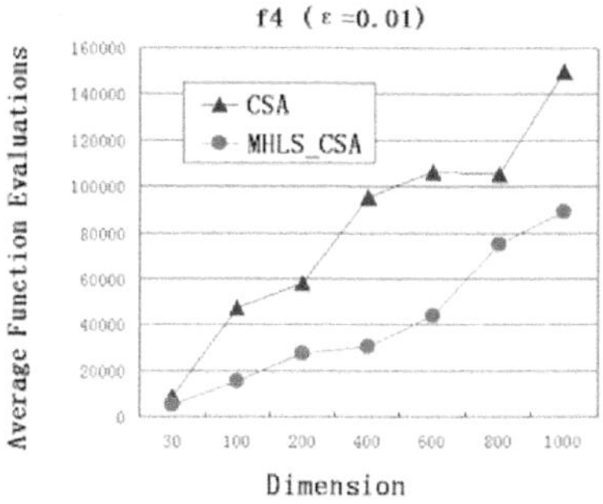

Fig. 5. Performance comparisons for f4

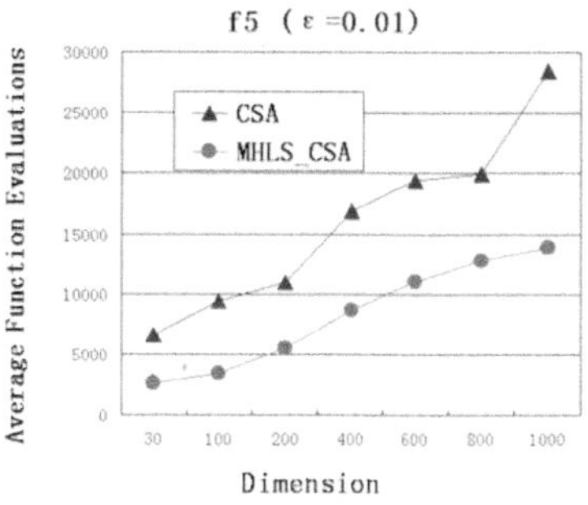

Fig. 6. Performance comparisons for f5

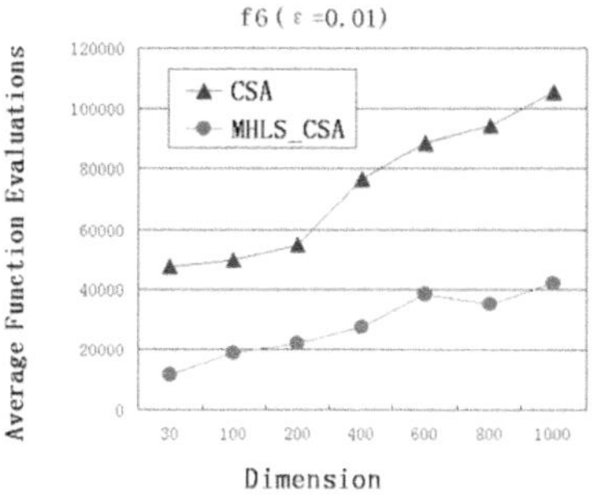

Fig. 7. Performance comparisons for f6

The algorithm will be finished when the generation time reaches its upper limit 10000 or the precision demand $\varepsilon = 0.01$ has been satisfied. The experimental data in Fig.2 to Fig.7 are obtained from 50 times of independent running.

As shown in Fig.2 to Fig.7, MHLS_CSA solves function optimization problem with less function evaluation cost than CSA. On the basis of these experimental data, MHLS_CSA converges much faster than CSA. In addition, comparing with CSA, the average function evaluation curves of MHLS_CSA climb much slower as the variable dimension increases. The strategy of mutation history learning proposed in this study is shown to be effective for improving CSA's searching performance.

4.2 Parameter Selection

There are two parameters special in MHLS: the inertia factor α and the reviewing factor β. The values of these two parameters are both real number between 0 and 1. Experiments have been done to investigate the performance for different parameters. Four typical test functions f1, f3, f4 and f6 are taken into account.

Fig.8 to Fig.11 are the graphs of the variation of MHLS_CSA's performance with the two parameters. Experimental data are obtained from 50 times of random running.

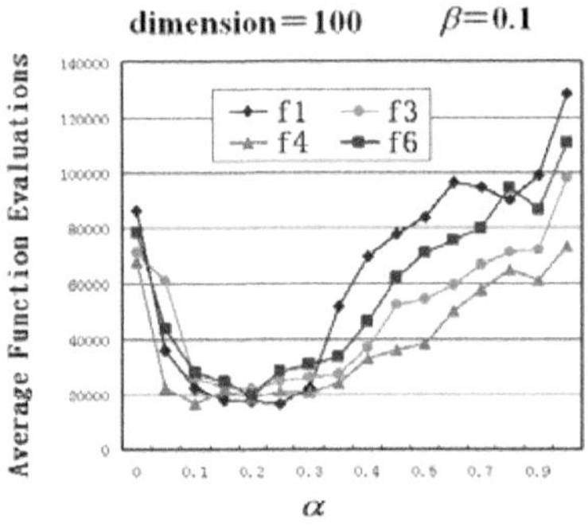

Fig. 8. Affection of α with dimension 100

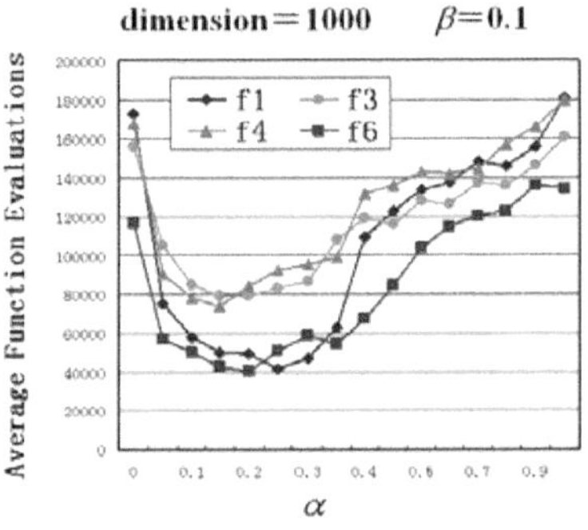

Fig. 9. Affection of α with dimension 1000

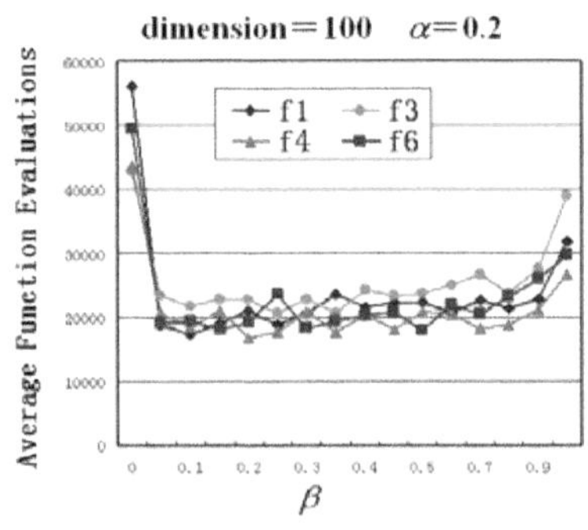

Fig. 10. Affection of β with dimension 100

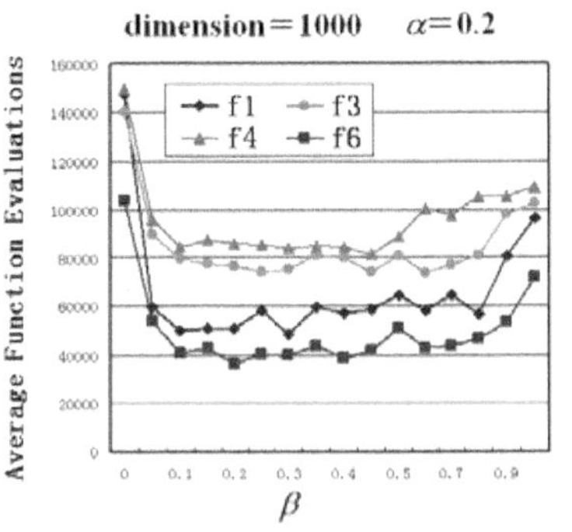

Fig. 11. Affection of β with dimension 1000

As shown in Fig.8 and Fig.9, these curves obtain their minimum vales when the value of inertia factor α is between 0.05 and 0.3. Seen from these curves in Fig.10 and Fig.11, the algorithm is not sensitive to the setting of β.

5 Concluding Remarks

In this paper, a strategy which is termed as mutation history learning strategy (MHLS) was proposed. Based on MHLS an improved CSA (MHLS_CSA) was also put forward. To validate the effectiveness of MHLS, the MHLS_CSA was executed to solve six standard test functions with different types. By comparing the performance of MHLS_CSA with that of standard CSA, we can find that MHLS is capable of making full use of the history information which is achieved from the assuming and testing iterations and accelerating CSA's convergent speed. Finally, the selection of two special parameters of MHLS is investigated. Experiment results indicate that the proposed algorithm performs stably at a large range of parameter values, which proves the algorithm to be quite robust and easy to use.

It is necessary to note that MHLS was designed for real number coding CSA. The development of suitable mutation history learning strategy for other coding modes suggests a natural direction for the future work.

References

1. L. N. De Castro, F. J. Von Zuben. The Clonal Selection Algorithm with Engineering Applications[C], In Proceedings of GECCO'00, Workshop on Artificial Immune Systems and Their Applications, 2000: 36-37.
2. Kim, J. and Bentley, P. J. (2000) Towards an Artificial Immune System for Network Intrusion Detection: An Investigation of Dynamic Clonal Selection, Proceedings of Congress on Evolutionary Computation, 2002: 1015-1020.
3. Ruochen Liu, Haifeng Du, Licheng Jiao, Immunity Clonal Strategies. Fifth International Conference on Computational Intelligence and Multimedia Applications (ICCIMA'03) 2003: 290.
4. Dasgupta D, et al. Artificial immune systems in industrial applications. In: IPMM '99. Proceedings of the Second International Conference on Intelligent Processing and Manufacturing of Materials. IEEE press, 1999. 257-267.
5. de Castro, L. N., Von Zuben, F. J. Learning and Optimization Using the Clonal Selection Principle. IEEE Transactions on Evolutionary Computation, Special Issue on Artificial Immune Systems, 6(3), 2002: 239-251.
6. Cooper K D, et al. Procedure Cloning. In: Proceedings of the 1992 International Conference on Computer Languages, 1992. 96-105.
7. Balazinska M, et al. Advanced clone-analysis to support object-oriented system refactoring. In: Proceedings: Seventh Working Conference on Reverse Engineering, 2000. 98~107.
8. Esmaili N, et al. Behavioural cloning in control of a dynamic system. In: IEEE International Conference on Systems, Man and Cybernetics Intelligent Systems for the 21st Century. 1995, 3. 2904-2909.
9. Hybinette M, et al. Cloning: A Novel Method for Interactive Parallel Simulation. In: Proceedings of the 1997 Winter Simulation Conference, 1997. 444-451.
10. Minqiang LI, Jisong KOU, Dan LIN, Shuquan LI .Principles and Applications of Genetic Algorithm [M]. Science press.2002: 399-403.

Quantum-Inspired Immune Clonal Algorithm for Multiuser Detection in DS-CDMA Systems

Yangyang Li, Licheng Jiao, and Shuiping Gou

Institute of Intelligent Information Processing, Xidian University, Xi'an 710071, China
lyy_791@yahoo.com.cn, {lchjiao, shpgou}@mail.xidian.edu.cn

Abstract. This paper proposes a new immune clonal algorithm, called a quantum-inspired immune clonal algorithm (QICA), which is based on the concept and principles of quantum computing, such as a quantum bit and superposition of states. Like other evolutionary algorithms, QICA is also characterized by the representation of the antibody (individual), the evaluation function, and the population dynamics. However, in QICA, an antibody is proliferated and divided into a subpopulation. Antibodies in a subpopulation are represented by multi-state gene quantum bits. For the novel representation, we put forward the quantum mutation operator which is used at the inner subpopulation to accelerate the convergence. Finally, QICA is applied to a practical case, the multiuser detection in DS-CDMA systems, with a satisfactory result.

1 Introduction

Immune clonal algorithm (ICA) [1] is principally a stochastic search and optimization method based on the clonal selection principle in the artificial immune system (AIS). Compared to traditional optimization methods, such as calculus-based and enumerative strategies, ICA are robust, global, and may be applied generally without recourse to domain-specific heuristics.

In [2], quantum-inspired computing was proposed. Unlike other research areas, there has been relatively little work done in applying quantum computing to AIS. We firstly proposed a quantum-inspired immune clonal algorithm (QICA) for solving the high dimensional function optimization problems in [3]. It should be noted that although QICA is based on the concept of quantum computing, QICA is not a quantum algorithm, but a novel optimization algorithm for a classical computer. In this paper, we apply QICA to the multiuser detection in asynchronous DS-CDMA systems.

The paper is organized as follows. Section 2 describes QICA. Section 3 presents an application example with QICA, and summarizes the experimental results. Concluding remarks follow in Section 4.

2 QICA

2.1 Representation

QICA use a new representation, called on a quantum bit or qubit, for the probabilistic representation that is based on the concept of qubits, and a qubit antibody as a string of qubits, which are defined below.

T.-D. Wang et al. (Eds.): SEAL 2006, LNCS 4247, pp. 80 – 87, 2006.

Definition 1. The probability amplitude of one qubit is defined with a pair of numbers (α, β) as

$$\begin{pmatrix} \alpha \\ \beta \end{pmatrix}, \tag{1}$$

satisfying

$$|\alpha|^2 + |\beta|^2 = 1, \tag{2}$$

where $|\alpha|^2$ gives the probability that the qubit will be found in the '0' state and $|\beta|^2$ gives the probability that the qubit will be found in the '1' state.

Definition 2. A qubit antibody as a string of m qubits is defined as:

$$\begin{pmatrix} \alpha_1 & \alpha_2 & \cdots & \alpha_m \\ \beta_1 & \beta_2 & \cdots & \beta_m \end{pmatrix}, \tag{3}$$

where $|\alpha_l|^2 + |\beta_l|^2 = 1, (l = 1, 2, \ldots, m)$.

2.2 Immune Clonal Algorithm

Clonal Selection Theory is put forward by Burnet [4]. It is used in the immune system to describe the basic features of an immune response to an antigenic stimulus. It establishes the idea that only those cells that recognize the antigens proliferate, thus being selected against those which do not. Based on the clonal selection theory, Immune Clonal Algorithm (ICA) is proposed. The ICA used in this paper is an antibody random map induced by the affinity including three steps [5]: clone operator, immune genetic operator and clonal selection operator. The state transfer of antibody population is denoted as follows:

$$A(t) \xrightarrow{\text{clone operator}} A'(t) \xrightarrow{\text{immune genetic operator}} A''(t) \xrightarrow{\text{selection operator}} A(t+1).$$

Here $A(t)$ is antibody population based on classical bit at t-th generation. Antibody, antigen, affinity between antibody and antigen are similar to the definitions of the possible solution, the objective function (and restrictive condition), the fitness between solution and the objective function, respectively. It can be found out that ICA obtains good local searching ability at a cost of adding the scale of the population by clone operation. As a result, we adopt qubit representation which has powerful parallel and the corresponding immune genetic operator in order to speed up the convergence. We present the proposed QICA in the following section.

2.3 The Quantum-Inspired Immune Clonal Algorithm (QICA)

In Fig 1, we describe its critical steps in detail. Where $Q(t)$, $P(t)$, $D(*)$ and $B(t)$ mean the antibody population based on qubit at the t-th generation, the antibody population based on classical bit at the t-th generation, the avidity function, and the best solutions in the t-th generation's subpopulation, respectively.

Algorithm1. The quantum-inspired immune clonal algorithm
Step1. Initialize $Q(t)$ and $B(t)$, t=0.
Step2. Generate $Q'(t)$ from $Q(t)$ by the clonal operator Θ.
Step3. Update $Q'(t)$ by quantum mutation.
Step4. Produce $P'(t)$ by observing the updated $Q'(t)$.
Step5. Evaluate the affinity of $P'(t)$, by clonal selection operator produce $B(t)$ and record the corresponding qubit to generate next generation $Q(t+1)$.
Step6. Store the best solutions among $B(t)$ to b and judge the termination condition, if it is satisfied, then output the best solution, and else go to *step2*.

Fig. 1. The quantum-inspired immune clonal algorithm

The major elements of QICA are presented as follows.

- **The Quantum Population**

QICA maintains a quantum population $Q(t)=\{q_1^t, q_2^t, \cdots q_n^t\}$ at the t-th generation where n is the size of population, and m is the length of the qubit antibody q'_i which is defined as: $q_i^t = \begin{Bmatrix} \alpha'_1 & \alpha'_2 & \cdots & \alpha'_m \\ \beta'_1 & \beta'_2 & \cdots & \beta'_m \end{Bmatrix}, i=1,2,...,n$. In *step1* all α_l^t and β_l^t of q_i^t ($l=1,2,...,m$; $t=0$) are randomly generated between -1 and 1 and satisfying $|\alpha_l^t|^2 + |\beta_l^t|^2 = 1, (l=1,2,...,m)$.

- **Clonal Operator**

The clonal operator Θ is defined as:

$$\Theta(Q(t)) = [\Theta(q_1) \quad \Theta(q_2) \quad \cdots \quad \Theta(q_n)]^T, \tag{4}$$

where $\Theta(q_i) = I_i \times q_i, i=1,2\cdots n$, and I_i is C_i dimension row vectors. Generally, C_i is given by:

$$C_i = g(N_c, D(q_i)) \qquad i=1,2\cdots n, \tag{5}$$

which can be adjusted self-adaptively by the affinity $D(*)$. N_c is a given value relating to the clone scale. After clone, the population becomes:

$$Q'(t) = \{Q(t), q'_{1,} \cdots q'_n\}, \tag{6}$$

where:

$$q'_i(t) = \{q_{i1}(t), q_{i2}(t), \cdots, q_{iC_i-1}(t)\}, q_{ij}(t) = q_i(t) \; j=1,2,\cdots,C_i-1. \tag{7}$$

- **Immune Gene Operator**

In the following, we give a simple mutation method (namely quantum mutation) to evolve the qubit antibody. It deduces a probability distribution in terms of the current best antibody. It is also much simpler, whose process is: define a guide qubit antibody from the current best antibody which is stored in $B(t)$ and spread the mutated qubit subpopulation with this guide qubit antibody being the center. Quantum mutation can be written as:

$$Q_{guide}(t) = a \times B_{currentbest}(t) + (1-a) \times (1 - B_{currentbest}(t)), \tag{8}$$

$$q_i''(t) = Q_{guide}(t) + I_i' * b \times N(0,1), \quad (i = 1,2,...,n), \tag{9}$$

Where $B_{currentbest}(t)$, Q_{guide} and $N(0,1)$ are the current best antibody based on classical bit in subpopulation, the guide qubit antibody at the t-th generation and a random number, chosen from a normal distribution with mean zero, variance one and standard deviation one, respectively. $q_i''(t)$ is the mutated qubit subpopulation. a is the guide factor of Q_{guide} and b is the spread variance. I_i' is $(C_i - 1)$ dimension unit row vectors (see equation 5), namely the quantum mutation is unused to $Q(t) \in Q'(t)$. For easy to comprehend, an example is given. Obviously, we only need to let q'' = (0 0 1 1 0) to get B = (1 1 0 0 1) with probability 1, i.e., $q'' = \overline{B}$. If P is the optimum, the probability of getting the optimum becomes larger with a becoming smaller. When a=0, $Q_{guide} = \overline{B}$, one will get B with probability 1 after observing Q_{guide}. Often we let $a \in [0.1, 0.5]$, $b \in [0.05, 0.15]$.

- **Observing Operator**

In *Step4*, in the act of observing a quantum state, it collapses to a single state (namely classical representation). For binary coding problem, we observe the updated $Q'(t)$ (namely $Q''(t)$) and produce binary stings population $P'(t) = \{x_1^t, x_2^t, \cdots, x_n^t\}$ and x_i^t $(i = 1,2,...,n)$ is a numeric string of length m which derives from $\alpha_l''^t$ or $\beta_l''^t$ (l=1,..,m). The process is: generate a random number $p \in [0,1]$. If it is larger than $|\alpha_l''^t|^2$, the corresponding bit in $P'(t)$ takes '1', else takes '0'.

- **Clonal Selection Operator**

The operation as follows, if we will search maximal value of object function: for $\forall i = 1,2,\cdots n$, if there is the mutated and observed classical antibody b_i and $D(b_i) = \max\{D(x_{ij}) \mid j = 2,3,\cdots C_i\}$, namely: $D(b_i) > D(x'_i)$, then b_i replaces the antibody x'_i in the original population. And record the corresponding qubit of b_i as the next generation population $Q(t+1)$ at the same time. The antibody population is updated, and the information exchanging among the antibody population is realized.

- **The Termination Condition**

The termination condition is the given number of generation.

We have proved that QICA is convergent with probability of 1 based on Markov Chain. [3]

3 Simulations

In recent years, DS-CDMA systems have emerged as one of prime multiple-access solutions for 3G. In the DS-CDMA framework, multiple-access interference (MAI) existing at the received signal creates "near-far" effects. Multiuser detection (MUD) techniques can efficiently suppress MAI and substantially increase the capacity of CDMA systems, so it has gained significant research interest since the Optimal MUD (OMD) was proposed by Verdu [6].

In this section, we apply QICA to solve MUD, and named by QICAD. We present some simulation results and comparisons that demonstrate the advantage of our algorithm. The performance of the QICAD is evaluated via computer simulations and compared with that of Standard ICA (see section 2.2) (ICAD) and Optimal Multiuser Detector (OMUD) as well as with that of conventional Matched Filters Detector (MFD) in asynchronous DS-CDMA systems.

3.1 Problem Statements

Consider a base-band digital DS-CDMA network with K active users operating with a coherent BPSK modulation format. The signal received at the output of the sensor is:

$$r(t)=\sum_{i=0}^{M-1}\sum_{k=1}^{K}A_k b_k(i)s_k(t-iT_b)+n(t)=S(t,b)+n(t)\,, \tag{10}$$

here $n(t)$ is the additive white noise vector whose standard deviation is σ, T_b is the symbol interval, M is the packet length, A_k is the signal's amplitude of the k-th user, $b_k(m)$ is the m-th coded modulated symbol of the k-th user and $b_k(m)\in\{\pm 1\}$, $s_k(t)$ is the k-th user's signature sequence.

The matched filter output corresponding to the m-th bit of the k-th user is given by:

$$y_k(m)=\int_{-\infty}^{\infty} r(t)s_k(t-mT_b-\tau_k)dt\,. \tag{11}$$

If set $y(m)=[y_1(m),y_2(m),...,y_K(m)]^T$, $b(m)=[b_1(m),b_2(m),...,b_K(m)]^T$, $A(m)=\text{diag}\,(A_1,A_2,...,A_K)$, $n(m)=[n_1(m),n_2(m),...,n_K(m)]^T$ and $R(q)=(\rho_{kl}(q))_{K\times K}$, where $\rho_{kl}(q)=\int_{\tau_k}^{T+\tau_k} s_k(t-\tau_k)s_l(t+qT-\tau_l)dt$ then

$$y=RAb+n\,, \tag{12}$$

where $y=[y(m),y(m+1),...,y(m+M-1)]^T$, $b=[b(m),b(m+1),...,b(m+M-1)]^T$, $A=\text{diag}\,(A(m),A(m+1),...,A(m+M-1))$, $n=[n(m),n(m+1),...,n(m+M-1)]^T$.

The OMD produces an estimate for the information vector transmitted at the discrete-time instant m. In the asynchronous systems it holds that

$$\hat{b}_{\text{optimal}}=\arg\max_{\substack{b_k(m)\in\{\pm1\}\\ 1\le k\le K,1\le m\le M}}\left\{2b^T Ay-b^T ARAb\right\}. \tag{13}$$

3.2 QICA for Multiuser Detection

Assume that K active users share the same channel and the packet length is M, then definition (13) can be described as a combination optimization problem as

$$\max\left\{f(b)=2b^T Ay-b^T ARAb, b\in I\right\}, \tag{14}$$

where $b=\left\{[b_1^{(1)},b_2^{(1)},...,b_K^{(1)}],...,[b_1^{(M)},b_2^{(M)},...,b_K^{(M)}]\right\}, b_k^{(m)}\in\{\pm1\}$ is the variants to be optimized, I denotes the antibody space. The value of affinity $D(b)$ is equal to the value of the objective function $f(b)$, and set I^n denotes the antibody population space as

$$I^n=\{B:B=(b_1,b_2,...,b_n),b_k\in I,1\le k\le n\}, \tag{15}$$

in which $B=(b_1,b_2,...,b_n)$ is the antibody population, n is the size of the antibody population, and antibody $b_i=\left\{[b_{1i}^{(1)},b_{2i}^{(1)},...,b_{Ki}^{(1)}],...,[b_{1i}^{(M)},b_{2i}^{(M)},...,b_{Ki}^{(M)}]\right\}$. In this experiment, the main operators of QICA are the same as those given in section 2.3, but the observing operator is: generate a random number $p\in[0,1]$. If it is larger than $\left|\alpha_i^t\right|^2$, the corresponding bit in $P'(t)$ takes '1', else takes '-1'. For Standard ICA, it does not apply qubit antibody design and quantum mutation operator. Gaussian mutation is used on the classical numeric representation.

It is assumed that the number of users is K and the packet length is M, Gold sequences of length 31 are used as code sequences. The signal to noise ratio of the k-th user is $\text{SNR}_k=A_k^2/2*\sigma^2$ where $\sigma=1$. For QICAD and ICAD, we will terminate the search at the Y-th generation where $Y=2\times K\times M$. In two algorithms above, the size of initial population is 5, the clonal sizes is 10, and p_m=0.5. We take all the experiments based on 10000 bits signals. Our performance metric is the average Bit Error Ratio (BER).

A. In order to gain the results of the OMUD, we assumed that K=3, M=3 and SNR=10 dB. The first user is the desired user while other users are disturbing users and all users have the same power. The ratio of power between disturbing users and desired user denotes the ratio of 'near-far'. The performances in 'near-far' resistance of mentioned receivers are shown in Fig. 2(a).

B. It is assumed that K=10 and M=10. All users have the same power, changing the value of SNR from -2 dB to 10 dB. The performances in eliminating noise's disturbing of mentioned receivers are shown in Fig. 2(b).

C. It is assumed that M=10 and SNR=10 dB, the number of users K is changed from 5 to 30, all users have the same power. The performances in accommodating users of mentioned receivers are shown in Fig. 2(c).

D. It is assumed that SNR=10 dB, K=10, the packet length is changed from 5 to 30, all users have the same power. The performances in accommodating packet length of mentioned receivers are shown in Fig 2(d).

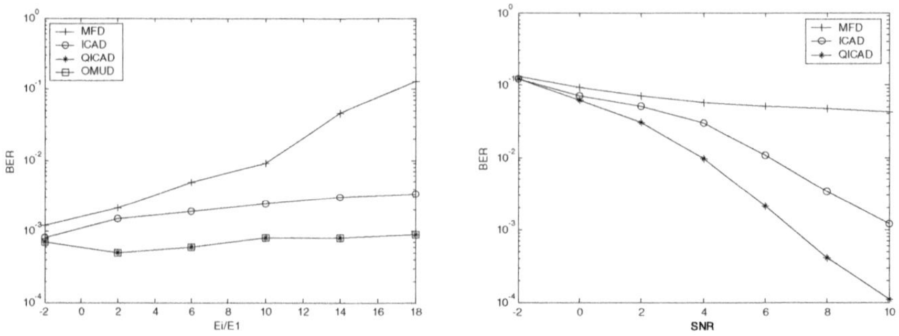

(a) The performances in 'near-far' resistance (b) The performances in eliminating noise's disturbing

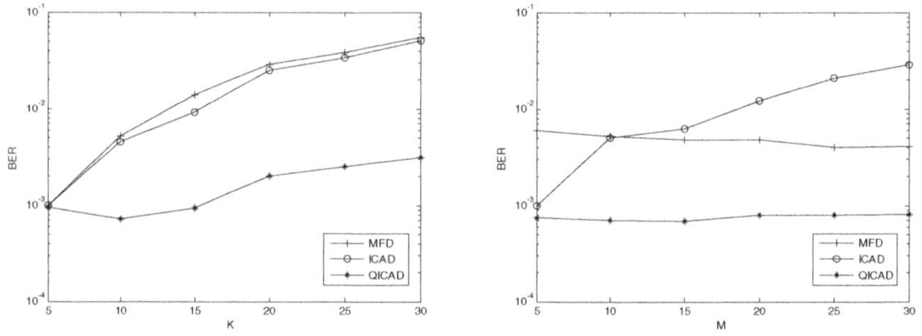

(c) The performances in accommodating users (d) The performances in accommodating packet length

Fig. 2. The simulation results

As we can see from Fig. 2(a), the conventional detector produces the receivable estimate only when powers of the users are close to each other. The QICAD and ICAD are better than conventional detector. But ICAD's performance is unacceptable when powers of disturbing users are much larger than that of desired user. As we expect, QICAD exhibits the best performance and seldom fails to produce the correct estimate for the transmitted symbols, so its performance is almost the same good as the OMD. When the cumulative BER is evaluated versus the value of the SNR of all the users, from Fig. 2(b) we can see that QICAD receiver achieves acceptable performance, whereas the performances of conventional detector and ICAD are very poor. When the number of users or the transmitted packet length is relatively large, the advantage of QICAD can be seen in Fig 2(c) and Fig 2(d). The simulations suggest that, QICAD detector still performs quite well when K and M are relatively large. These results indicate that quantum representation design and quantum mutation operator can effectively improve the standard ICA.

4 Conclusions

This paper proposed a novel immune clonal algorithm-QICA, inspired by the concept of quantum computing. The qubit representation has a better characteristic of population diversity than other representations. Due to the novel representation, we put forward the quantum mutation operator which is used at the inner subpopulation to accelerate the convergence. The application in the CDMA proves its superiority to its counterpart. The application of QICA to other problems such as the multiobjective optimization problem deserves our further research.

References

1. De Castro, L. N., Von Zuben, F. J.: Artificial Immune Systems: Part II—A Survey of Applications. FEEC/Univ. Campinas, Campinas, Brazil. [Online]. Available: http://www.dca.fee.unicamp.br/~lnunes/immune.html (2000)
2. Moore, M., Narayanan, A.: Quantum-Inspired Computing. Dept. Comput. Sci., Univ. Exeter, Exeter, U.K., (1995)
3. Li, Y.Y., Jiao, L.C.: Quantum-Inspired Immune Clonal Algorithm. in Proceedings of the 4th International Conference on Artificial Immune Systems, Christian Jacob, Marcin L. Pilat, Peter J. Bentley, et al, Eds. Banff, Alberta, Canada, Aug (2005) 304 – 317
4. Burnet, F. M.: Clonal Selection and After .In Theoretical Immunology, Bell, G. I., Perelson, A. S., pimbley Jr, g. H.(eds.) Marcel Dekker Inc., (1978) 63-85
5. Du, H. F., Jiao, L.C., Wang, S.A.: Clonal Operator and Antibody Clone Algorithms. in Proceedings of the First International Conference on Machine Learning and Cybernetics. Shichao, Z., Qiang, Y. and Chengqi, Z., Eds. IEEE, Beijing, (2002) 506–510
6. Sergio, V.: Optimum Multiuser Asymptotic Efficiency. IEEE Trans. Commun., Vol. 34, No. 9, (1986) 890-897

Innate and Adaptive Principles for an Artificial Immune System

M. Middlemiss and P.A. Whigham

Information Science Department,
University of Otago,
Dunedin, N.Z.
MMiddlemiss@infoscience.otago.ac.nz

Abstract. This paper summarises the current literature on immune system function and behaviour, including pattern recognition receptors, danger theory, central and peripheral tolerance, and memory cells. An artificial immune system framework is then presented based on the analogies of these natural system components and a rule and feature-based problem representation. A data set for intrusion detection is used to highlight the principles of the framework.

1 Introduction

The vertebrate immune system provides a multilayered form of defence capable of identifying and responding to harmful stimulus. This paper presents a framework for an artificial immune system (AIS) that incorporates innate and adaptive concepts capable of recognising dangerous behaviour within an intrusion detection data set (KDD'99 Cup Data set [1]). The reader is referred to [2] for a review of current AIS literature. The paper is structured as follows: §2 presents properties of the innate immune system; §3 presents the basic operations of the adaptive immune systems; §4 describes the analogies between the immune system and the components of the AIS framework; and §5 outlines the framework itself.

2 Innate Immune System

The innate immune system consists of germ-line encoded specialised cells that are responsible for the initial immune response [3]. There are two theories for how these cells determine the need to initiate an immune response. According to Janeway [4] the antigen presenting cells (APCs) of the innate immune system (such as macrophages) have pattern recognition receptors (PRRs) on their cell surface. These PRRs recognise pathogen associated molecular patterns (PAMPs) which are found on the cell surface of invading pathogens and never on the host (refer Figure 1.1.a). These molecular patterns have been conserved through evolution and are common to many pathogens. For example, the CD14 receptor binds to a particular molecule (*lipopolysaccharide-LPS*) which is

T.-D. Wang et al. (Eds.): SEAL 2006, LNCS 4247, pp. 88–95, 2006.

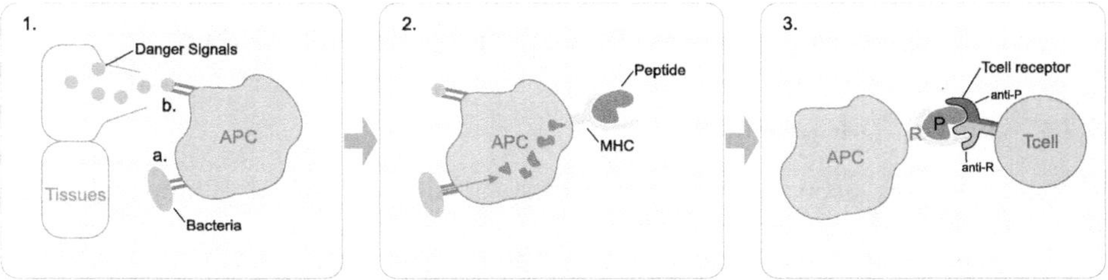

Fig. 1. Simplified model of immune system function

found only in the cell wall of Gram-negative bacteria, e.g. *E-coli.*, *Neisseria*, and *Salmonella* [5].

An alternative view is that of the danger theory proposed by Matzinger [6,7]. This theory proposes that APCs, in particular the dendritic cell (DC), have danger signal receptors (DSRs) which recognise alarm signals sent out by distressed or damaged cells. It is these signals that inform the immune system to initiate an immune response [7] (refer Figure 1.1.b). Evidence for these signals has been found in the form of molecules such as heat shock proteins and mitochondrial products which are released from necrotic cells and stimulate DCs.

Although the theories behind PRRs and DSRs are different, they do share some commonalities. In particular they both share the idea that APCs require a signal to initiate an immune response. The danger signals released by necrotic cells during death are mitochondria and as such are essentially intracellular bacteria. This means that the same mechanisms may be employed to detected foreign pathogens as well as to detect the death of host cells [8].

3 Adaptive Immune System

When an adaptive immune response is required the APC internalises the pathogen (antigen) and breaks it down into peptides which are presented on the surface in a complex with a special self-MHC (Major Histocompatibility Complex) molecule (refer Figure 1.2). MHC is a highly polymorphic molecule, with an estimated 10^{13} possible combinations [5]. Humans inherit genes for about six different MHC molecules, promoting diversity within the population [3]. The specialised cells responsible for this functionality are called lymphocytes, of which there are two types: T and B cells[1]. Lymphocytes circulate through the blood and lymphatic systems waiting to encounter antigens. Each antigen has a particular shape that is recognised by the receptor present on the lymphocyte cell surface. The ability of the immune system to recognise and respond to the millions of different antigens that it encounters comes from the large lymphocyte receptor repertoire. Lymphocyte development produces the ability to control the response against harmful

[1] For the purposes of the simplified immune model presented here, we only discuss the involvement of T cells.

pathogens while ignoring harmless pathogens and self-tissues. This ability is developed through the mechanisms of central and peripheral tolerance.

3.1 Central Tolerance

Central tolerance occurs in the thymus during the development of immature lymphocytes through the processes of negative and positive selection. For T cells to perform effectively they must be able to bind to the MHC:peptide complex presented to them by an APC. According to the differential signalling hypothesis [4], two different structures combine to form the T cell receptor: a germline encoded structure (anti-R) which provides allele-specific recognition of the MHC-encoded restricting element (R); and a somatically encoded structure (anti-P) which provides specific recognition of the peptide (P) bound to the MHC [9]. Through positive selection, only T cells whose anti-R structure is able to recognise the MHC restricting element (R) are retained. Additionally through negative selection, T cells whose anti-P structure binds with self peptides (self-P) are removed. This central tolerance reduces the potential of T cells to enter the body and cause an auto-immune reaction against harmless self cells.

While central tolerance removes self-reactive T cells, if it were highly restrictive the available repertoire of T cells would be limited and the immune systems would be compromised [8]. For this reason, central tolerance deletes the most self-reactive T cells and allows some T cells with limited self-reactivity to leave the thymus.

3.2 Peripheral Tolerance

Peripheral tolerance is the process of controlling self-reactive T cells to minimise the potential of auto-immune responses [8]. When a T cell does bind to an antigen, if it does not succumb to peripheral tolerance, it must have recognised the MHC:peptide complex presented by the antigen presenting cell (refer Figure 1.3). This requires the anti-R structure of the T cell receptor to bind to the self-MHC (R) of the antigen presenting cell, and the anti-P structure to bind to the peptide (P) that is from the bacteria or pathogen causing harm. When these bindings occur, an immune response is initiated.

3.3 Memory

When something harmful enters the body only a small number of cells have receptors capable of initiating an immune response. This leads to clonal selection where those lymphocytes that have been activated proliferate (generate clones of the original lymphocyte) and differentiate (become either effector cells or memory cells). As most lymphocyte clones eventually die and effector cells have a short lifespan, the immune system uses memory cells to prevent the loss of all lymphocytes which would leave the individual susceptible to infection and disease [3]. Memory cells provide the immune system with a method of lasting protection [4].

4 Analogy Between HIS and AIS

The main aspects of the model presented are:

- The role of antigen presenting cells in the innate immune system and their ability to recognise conserved PAMPs and danger signals with pattern recognition receptors (PRRs) or danger signal receptors (DSRs);
- The role of antigen presenting cells in the adaptive immune system and their ability to break down and present peptides to T cells in a MHC:peptide complex;
- The development of T cells through the mechanisms of central tolerance;
- The control of auto-immune responses through peripheral tolerance;
- The initiation of immune responses by T cells that recognise the MHC:peptide complex presented by antigen presenting cells; and
- Adaptation in the system provided by the development of memory cells.

Table 1 outlines the components of this model and proposes the component of the artificial immune system model used in this framework.

Table 1. Comparison of components from real and artificial immune system models

Immune System Model	**Artificial Immune System Model**
Antigen Presenting Cell -PRR / DSR -MHC	Antigen Presenting Cell -Rule extracted from background knowledge -Feature mask from local MHC set
MHC	Feature mask
T cell -Anti-P -Anti-R	T cell -Rule negatively selected -Feature mask positively selected
Central Tolerance	Negative and Positive selection of T cells during development
Peripheral Tolerance	Removal of T cells that respond to normal data examples
Memory cells	T cells formed through cloning and mutation after immune response

In the immune system model an APC is a detector that uses its receptor to recognise conserved patterns that are known to be foreign or dangerous. In our AIS model background knowledge of computer attacks are used to extract rules that can describe these attacks.

When an APC detects a foreign or dangerous pathogen it internalises and breaks it down into smaller peptides which are presented by the MHC molecule to a T cell. This process is analogous to the MHC performing a feature selection where the peptide is presented with particular features, that the T cell can recognise. For this reason, MHC is represented in our AIS model as a feature mask.

As described in §2, MHC promotes diversity of recognition and response within a population of individuals. Previous work [10] has used random permutation masks applied to T cells to model the concept of diversity that MHC provides a population. Rather than using a random mask, in our model each invidual has a local MHC set that is derived from background knowledge of normal and dangerous behaviour. This data is used to develop a set of feature masks that represent important features in the data. Within a population, each individual may have a different set of background knowledge and thus a slightly different set of MHC masks. This leads to an artificial APC which comprises a rule and an associated feature mask. The feature mask is selected from the local MHC set. For example, if the rule described a Denial of Service (DOS) attack, the mask selected may describe features specific to Ping-of-Death.

During development, T cells are subjected to central tolerance. In our AIS model a T cell is a randomly initialised rule which represents the anti-P part of the T cell receptor, and an associated feature mask representing the anti-R part of the T cell receptor. During central tolerance, artificial T cells are presented with a subset of background knowledge of normal system behaviour and any cells with a rule that match this data are removed. A threshold is used in this matching, analogous to the immune system where negative selection removes only those cells that bind the strongest and provide the biggest risk of initiating auto-immune responses. The remaining artificial T cells are then presented with the local MHC mask set and only cells that have a mask which matches (above a threshold) of one or more of these MHC masks are retained.

Peripheral tolerance is modelled as follows: periodically during the system operation known normal data examples are presented to the system. It is likely that one or more artificial APC will match these data examples, in a similar manner to how self-proteins are continuously being sampled by APCs and presented to T cells in the immune system. However, in this case if any artificial T cells match the peptide presented by the artificial APC they are removed to reduce the auto-immune, or false positive, responses of the system.

In the AIS memory cells are modelled as a special form of T cell. When a response is given by the AIS the T cell or cells involved in that response are cloned. Each clone is mutated in order to produce a T cell that would provide a better match on subsequent presentations of the data example causing the response.

5 AIS Framework

For this framework the assumption is made that data is classified as either anomalous or normal. It is further assumed that anomalous data can then be classified into one or more classes, with each class having one or more subclasses. The KDD'99 Data Set fits these assumptions and is used to highlight principles of the framework.

5.1 Module 1: Initialisation

Initialisation involves offline generation of the local MHC set, APC set and initial T cell set. The local MHC set is generated using background knowledge of both normal and dangerous (not normal, or anomalous) data. For each subclass within the anomalous data, a feature mask is extracted in the form of a binary string. The method presented here uses an Information Gain (IG) ranking to select features with an IG ratio above a set threshold. This leads to a set of masks with varying numbers of features, which are highly specific to known types of anomalous behaviour. This type of MHC representation differs slightly from previous papers [11,10]. Our approach is similar to that of Hofmeyr, except that their masks are randomly generated whereas ours represent features related to anomalous data.

Generation of the APC set begins with extracting rules describing anomalous data from the background knowledge. A set of rules is extracted for each class within the anomalous data, and then each rule is paired with a mask from the local MHC set. This pairing randomly matches a rule for a particular class of anomalous data with a MHC mask relating to a specific attack within that class. For example a rule describing the DOS class may be paired with a Ping-of-death mask or with a SYNflood mask, both of which are specific types of DOS attacks. This leads to a set of artificial APCs which are each represented by a rule and a mask.

Finally, the initial T cell set, which will evolve over time, is generated. Each T cell consists of a random feature mask (anti-R) and a random rule (anti-P). For each T cell, positive selection is performed where the T cell mask is compared with masks in the local MHC set. Only T cells that match above a given threshold are retained to ensure they are able to recognise the MHC mask of the artificial APCs.

T cells that survive positive selection are subjected to negative selection as a model of central tolerance. The feature mask is applied to the T cell rule and this masked rule is compared with a random sample of the background knowledge of normal data. Negative selection has been widely researched in the area of artificial immune systems [12]. Kim and Bentley [13] have shown that with an increased definition of 'self', or normal, it becomes computationally inefficient to find a set of detectors that adequately covers the "non-self", or dangerous, space while minimising the false negative error. However, by using a generalised representation of background knowledge via rules these scaling problems are largely overcome.

5.2 Module 2: Adaptation

Each T cell has a given life span which decreases with every example presented to the system. When one cell reaches the end of its life span it is removed and a new T cell is created. Whenever a T cell is involved in a response it is cloned and mutated (proportional to strength of the match) in order to find a better matching T cell. This new T cell is then added to the memory T cell set.

5.3 Module 3: Evaluation

An example (refer Fig. 2) is presented to the system and the cells of the artificial immune system work together to determine if a response should be produced.

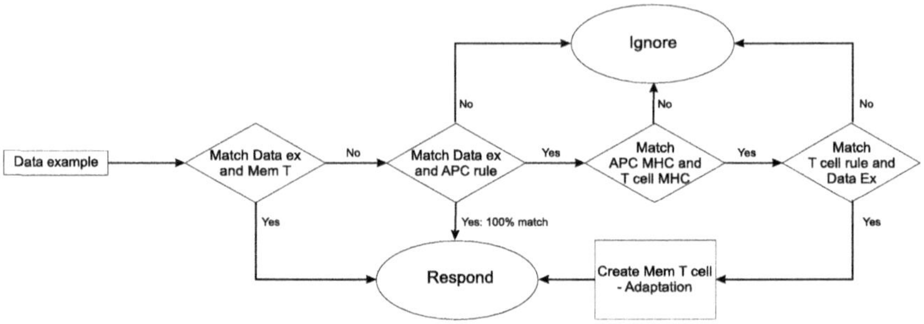

Fig. 2. Flow chart of evaluation module

Initially the example is matched with any T cells in the memory T cell set. Theses cells have been formed through the adaptation module. If there are no memory T cells, or none that match the example, the data is passed to the APC set. The example is evaluated by the antigen presenting cells which use their knowledge of common patterns in anomalous data (their PRR/DSR rules) to assess the example. A 100% match would indicate the example is likely to be anomalous and, without signalling the adaptive immune system, an alert is produced. If there are no APCs with a rule that match the example, it is assumed to be normal or harmless and the evaluation of that data example is complete.

Alternatively, if the match is not exact, but is above a set threshold, the APC signals the T cells by internalising the data example. This is achieved by applying the feature mask to the data example so that features specific to the class of data the APC has detected are presented. T cells are now signalled and any which have a MHC mask that is similar to the APC, or are able to recognise the self-MHC molecule in the MHC:peptide complex presented by the APC, are selected. The T cells must also be able to recognise the peptide presented in the complex. This evaluation is performed by masking the T cell rule with the T cell mask and evaluating the data example with this rule. Note that it is possible that multiple APCs can be activated by a data example, but not be co-stimulated by a T cell. If a response is produced from the system by T cells interacting with the APCs, the adaptation phase is initiated (§5.2).

6 Conclusion

This paper has described a simplified view of the immune system and characterised some of the known properties of this system in terms of a rule and feature-based model. The framework demonstrates how the immune system principles can be mapped to an innate and adaptive set of principles incorporating concepts of MHC, positive and negative selection in a manner that has not previously been considered.

References

1. The UCI KDD Archive: Kdd cup 1999 data, Last Modified: 28 Oct 1999, `http://kdd.ics.uci.edu/databases/kddcup99/kddcup99.html` (1999)
2. Hart, E., Timmis, J.: Application areas of ais: The past, the present and the future. In Jacob, C., Pilat, M., Bentley, P., Timmis, J., eds.: 4th International Conference on Artificial Immune Systems (ICARIS 2005). Volume 3627. (2005) 126–138
3. Playfair, J., G., B.: Infection and immunity. 2nd edn. Oxford University Press, Oxford; New York (2004)
4. Janeway, C., Travers, P.: Immunobiology : the immune system in health and disease. 5th edn. Current Biology ; Garland Pub., London ; San Francisco New York (2001)
5. Lydyard, P.M., Whelan, A., Fanger, M.W.: Immunology. 2nd edn. Instant Notes Series. London: Bios Scientific (2004)
6. Matzinger, P.: Tolerance, danger, and the extended family. Annual Reviews of Immunology **12** (1994) 991–1045
7. Matzinger, P.: The danger model: A renewed sense of self. Science **296** (2002) 301
8. Walker, L., Abbas, A.: The enemy within: keeping self-reactive t cells at bay in the periphery. Nature Reviews Immunology **2** (2002) 11–19
9. Cohn, M.: An alternative to current thinking about positive selection, negative selection and activation of t cells. Immunology **111** (2004) 375–380
10. Hofmeyr, S., Forrest, S.: Immunity by design: An artificial immune system. In Banzhaf, W., et. al., eds.: Genetic and Evolutionary Computation Conference (GECCO). (1999) 1289–1296
11. Kim, J., Ong, A., Overill, R.E.: Design of an artificial immune system as a novel anomaly detector for combating financial fraud in the retail sector. In Sarker, R., et. al., eds.: Congress on Evolutionary Computation, Canberra, Australia, IEEE Press (2003) 405–412
12. Forrest, S., Perelson, A.S., Allen, L., Cherukuri, R.: Self-nonself discrimination in a computer. In: IEEE Symposium on Research in Security and Privacy, IEEE Computer Society Press (1994) 202–212
13. Kim, J., Bentley, P.: Towards an artificial immune system for network intrusion detection: An investigation of clonal selection with a negative selection operator. In: Proceedings of the 2001 Congress on Evolutionary Computation CEC2001. (2001) 1244–1252

Immune-Based Dynamic Intrusion Response Model

SunJun Liu, Tao Li, Kui Zhao, Jin Yang, Xun Gong, and JianHua Zhang

School of Computer Science, Sichuan Univ., Chengdu, 610065, China
liusunjun@163.com, litao@scu.edu.cn

Abstract. Inspired by the immunity theory, a new immune-based dynamic intrusion response model, referred to as IDIR, is presented. An intrusion detection mechanism based on self-tolerance, clone selection, and immune surveillance, is established. The method, which uses antibody concentration to quantitatively describe the degree of intrusion danger, is demonstrated. And quantitative calculations of response cost and benefit are achieved. Then, the response decision-making mechanism of maximum response benefit is developed, and a dynamic intrusion response system which is self-adaptation is set up. The experiment results show that the proposed model is a good solution to intrusion response in the network.

1 Introduction

Intrusion response system, referred to as IRS, can take an active defending measure, according to the secure affair report of an intrusion detection system (IDS). It could effectively prevent intrusions, reduce the system loss, and make up the limitations of passive defending in IDS. So it can guarantee the network security in deep degree.

However, many intrusion response models have some limitations at present. Fisch [1] presented the concept of intrusion response classification for the first time, according to the attacked targets and the time when intrusions are detected. But this method was so rough that it could not be used for response decision-making. Carver [2] created a multidimensional events classification model to solve the problem of classifying intrusion affairs, which lacked evaluations of the response cost. And Thomas Toth [3] constructed dependent relationship trees between the network entities to get the minimum loss, which is taken as the basis of decision-making. But this model didn't consider about the response operation cost and validity.

There are many similarities between computer network security and biological immune system (BIS). Both have to maintain stability in a changing environment. BIS can produce antibodies to resist pathogens through B cells distributing all over the human body. T cells can regulate the antibody concentration. In normal state, this concentration hardly changes except that pathogens are intruding. Therefore, the intrusion intensity of pathogens can be evaluated by variation of the antibody concentration. The artificial immune system (AIS) [4-5] [7-10] based on the theory of BIS has been considered as an important direction in the research of network security.

Inspired by the biological and human immune systems, an immune-based dynamic intrusion response model (IDIR) is presented. The definitions of self, nonself and antibody are developed. The intrusion detection procedure is achieved in this model,

T.-D. Wang et al. (Eds.): SEAL 2006, LNCS 4247, pp. 96 – 103, 2006.

which makes use of some immune mechanisms including self-tolerance [4], clone selection [5], and immune surveillance. Using a quantitative computation of the system danger degree, it eliminates many redundant detected events and avoids a lot of unneeded reduplicate responses. Quantitative computation models of the response operation cost and response negative cost are established, which ameliorate the limitations of the statically analyzing response cost and benefit presented by WenkeLess [6]. Then, the response decision-making mechanism of maximum response benefit is demonstrated, and a dynamic intrusion detection response mechanism, which is self-adaptation, is established. The experiment results prove that this model is a good solution to intrusion response in the network.

The rest of the paper is organized as follows. Section 2 introduce IDIR model. Section 3 introduces the simulation experimental. Section 4 contains our conclusions.

2 IDIR Model

An immune-based dynamic intrusion response system consists of two parts, whose architecture is shown in Fig. 1. The IDoC is an intrusion detection cell, which is used to extract the antigens from IP packets. After the detection of IDS, it can quantitatively describe the intrusion danger degree of a system, according to the concentration changes and send it to the dynamic response cell (DRoC). When DRoC received the quantitative description, it will calculate the response benefit and cost, in order to make an optimal response strategy that producing maximum response benefit. Then the strategy will be executed and the detection of network attacks and the procedure of dynamic response are achieved.

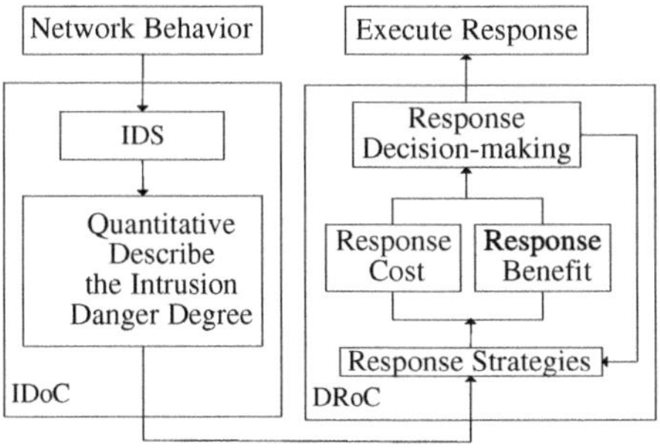

Fig. 1. Architecture of IDIR

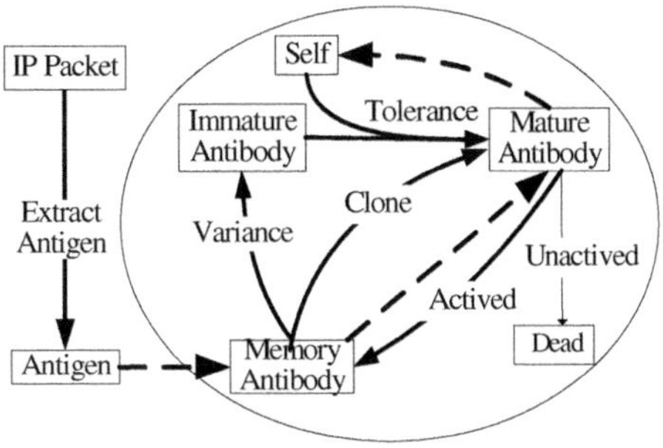

Fig. 2. Immune-based Intrusion Detection

2.1 The Definitions of Immune Components

Antigens (*Ag*) [9] [10]in our approach are binary strings extracted from the IP packets, including IP address, port number, protocol type, etc, which are given by:

$$Ag = \{< a,b >| a \in D \wedge b \in \Psi \wedge |a| = l \wedge a = APCs(b)\}, D = \{0,1\}^l \quad (1)$$

The self set (*Self*) are normal network transactions, nonself set (Non*self*) represent attacks from network, $Self \cup Nonself = Ag$, $Self \cap Nonself = \phi$.

Immunocytes $B=\{<d,p,age,count>|d\in D,\ p\in R,\ age,count\in N\}$, where d is an antibody, p is the antibody concentration, *age* represents the cell age, and *count* is the antigen number matched by antibody d. Immunocytes contains two subsets: mature immunocytes T_b and memory immunocytes M_b, $B=M_b\cup T_b$ and $M_b\cap T_b=\Phi$. When the amount of matched antigens arrives to a certain threshold, mature immuocytes would be activated and evolve into memory ones, such that $M_b=\{x|x\in B, x.count>\beta\wedge x.age<\lambda\}$, where β is the active threshold, and λ is a lifecycle. Immature immunocyte set is $I_b=\{<d,age>|d\in D, age\in N\}$, which will evolve into mature one T_b through the self-tolerance.

Regulate cells $T=\{signal\,|\,signal=signal_h-signal_s, signal_h, signal_s\in R\}$. And T cells are divided into help T cells T_h [8] and restrained T cells T_s [8], where $T=T_h\cup T_s$ and $T_h\cap T_s=\Phi$. $signal_h$ is the signal excreted by T_h, which would stimulate antibody concentration to increase, and $signal_s$ is the signal excreted by T_s, which would restrain the increase of concentration.

The affinity is used to evaluate the match degree between x and y. It is given by:

$$f_{match}(x,y)=\begin{cases}1, & f_{h\_dis}(x,y)\geq\theta\\ 0, & otherwise\end{cases} \tag{2}$$

Where l is the length of x and y, $f_{h\_dis}(x,y)=\sqrt{\sum_{i=1}^{l}(x_i-y_i)^2}$, and θ represents the threshold proportion .

2.2 Immune-Based Intrusion Detection

The immune-based intrusion detection consists of the evolution of immunocytes and antigens detection, both is concomitant. The immune evolution has mechanisms including self-tolerance, clone, variation, etc, whose process is shown in Fig.2 with real lines, while the process of antigen detection is shown in it with dashed lines.

The detection antigens process is fulfilled by memory immunocytes and mature immunocytes. The memory immunocytes will match the antigens to antibodies at first and eliminate nonself antigens. The left antigens will be submitted to mature immunocytes for detection and those nonself antigens would be eliminated. After the above detections, the left would be added to the set of self on behalf of maintaining the dynamic updating. During detections, a single kind of antibodies always can distinguish the similar attacks because of the diversity of antibodies. And they will be considered as a same kind. Then, the problem of classifying attacks is solved.

The process of self-tolerance can make the immature immunocytes evolve into mature ones, through the negative selection. And this process avoids taking self for nonself. The self-tolerance process is defined by:

$$f_{tolerance}(I_b)=I_b-\{d\,|\,d\in I_b\wedge\exists y\in Self\wedge Match(d,y)=1\} \tag{3}$$

At the same time, the mature immunocytes that are not activated and that are much older will be killed. And the mature ones whose matching count arrives to the activation threshold β will be activated and evolve into the memory ones.

The mechanism of clone and variation consists of clone selection and variation. When memory antibodies match to antigens, they will clone themselves and respond to those recognized antigens. Clone selection gives priority to the antibodies owning high affinity. The clone process is accompanied with the variation, which will select $(l_b - f_{h\_dis}(Ab, Ag))$ bits at random in antibody Ab of the identifying antigen Ag. The values of these bits can be randomly 0 or 1, l_b is the length of Ab. The cells that suffered from the variation will be taken as immature ones and start a new self-tolerance. The clone variation not only make the antibodies can recognize the known attacks, but also resolve the problems that recognize the unknown attack type.

2.3 Immune-Based Intrusion Danger Evaluation

During the immune-based intrusion detection process, the antibody concentration can quantitatively describe the danger degrees. The intensity of the regulate signals sent by T cells is decided by the affinity. Suppose the immune system has m kinds of antibody $Ab_i, i = 1 \cdots m$ at time t-1, whose concentrations are $Ab_i \cdot p$. And suppose there are n kinds of antigen $Ag_i, i = 1 \cdots n$, whose concentrations are $Ag_i \cdot p$. Then the signal of Ab_i at time t-1 is defined by:

$$signal = \kappa_1 \cdot \sum_{j=1}^{m} a_{ij} Ag_j \cdot p + \kappa_2 \cdot \sum_{\substack{k=1 \\ k \neq i}}^{n} b_{ik} Ab_k \cdot p - \kappa_3 \cdot \sum_{\substack{k=1 \\ k \neq i}}^{n} c_{ki} Ab_k \cdot p - \sigma_i \cdot Ag_i \cdot p \tag{4}$$

Where κ_1 is the increase velocity of Ab_i stimulated by Ag, κ_2 is the increase velocity of Ab_i paratopes stimulated by other antibody's epitopes, κ_3 is the velocity of Ab_i epitopes restrained by other antibodiy's paratopes. To simplify the calculation, all of these velocities are set to be same, whose values are all equal to κ. a_{ij} is the affinity between Ab_i and Ab_j, b_{ik} represents the affinity between the paratope of Ab_i and the epitope of Ab_k, while c_{ik} is the affinity between the epitope of Ab_i and the paratope of the antibody Ab_k, σ_i is the nature mortality.

The first two items in the right of the formula (4) represent the stimulating signal $signal_h$ sent by assistant T cells, which would increase the concentration. And the next two show the restrained $signal_s$ sent by restrained T cells, which would reduce the concentration. When antibody Ab_i receives the stimulating signal $signal(t-1)$ sent by T cells at time t-1, the concentration $Ab_i \cdot p$ at time t will change. And this calculation formula is as follow:

$$Ab_i.p(t) = Ab_i.p(t-1) \cdot \left(\frac{2}{1 + e^{-signal(t-1)}}\right) \tag{5}$$

Suppose $\varphi_i (0 \le \varphi_i \le 1)$ is the danger coefficient of attack type *i*. $n_j(t)$ is the amount of antibodies on computer *j* at time *t*. And $c_{ij}(t)$ is the amount of antibodies of attack type *i* which is detected on computer *j* in normal state. Then the intrusion danger coefficient of the computer *j* suffering from attack *i* is $dv_{ij}(t)$ ($0 \le dv_{ij}(t) \le 1$) at time *t*, which is defined by:

$$dv_{ij}(t) = 1 - \frac{1}{1 + \ln(\varphi_i (n_j \cdot Ab_i.p(t) - c_{ij}) + 1)} \tag{6}$$

2.4 Response Cost and Benefit Based Response Decision-Making

When the DRoC received the intrusion danger coefficient $dv_{ij}(t)$ from IDoC, the response decision-making mechanism will start and it will calculate the response cost and benefit for each strategy. Then it will choose the optimal strategy.

Suppose $\phi(i,r)(0 \le \phi(i,r) \le 1)$ is the response benefit coefficient, which represents the degree that response strategy *r* eliminates the attack type *i*. Therefore, when the danger coefficient of the computer *j* that suffering from attack type *i* is $dv_{ij}(t)$ at time *t*, the response gross benefit (RGB) of response strategy *r* is defined by:

$$RGB_{ij}(r,t) = dv_{ij}(t) \cdot \phi(i,r) \cdot \omega_j \tag{7}$$

Response cost includes response operation cost (ROC) and response negative cost (RNC). ROC is the system cost resulting from consuming the system resources when the response decision-making is running. RNC is the system cost when eliminating the damage of attacks. Both of these costs consume system resource, such as utilization of CPU and memory, whose results are the same as that of *DoS* attack. So they can be transformed into calculating the system loss of *DoS* attack. ROC is related with the utilization of system resources. The more the utilization is, the higher the ROC is. Suppose there is a direct ration between ROC and the utilization of CPU, such that $RNC(t,r) = r_{cpu}(t) \cdot k_{rnc}(r) \cdot \omega_j \cdot \varphi_{dos}$, where $r_{cpu}(t)$ is the CPU utilization, $k_{roc}(r)$ is ROC influencing coefficient, φ_{dos} is the danger coefficient of *DoS* attack. RNC is related with strategies, such that $ROC(t,r) = k_{roc}(r) \cdot \varphi_{dos}$, $k_{rnc}(r)$ is RNC influencing coefficient decided by the variation range of the strategy.

The response net benefit (RNB) resulting from computer *j* which suffers from attacking at time *t* can be defined by:

$$RNB_{ij}(r,t) = RGB_{ij}(r,t) - ROC(r,t) - RNC(r,t) \tag{8}$$

suppose the response system can support *n* kinds of response strategies $r_1, r_2, \cdots, r_n$. The empty response that means that it cannot respond to any attack is also taken as a strategy, referred to as r_0. The system would select $\hat{r}$ as the optimal response strategy and execute it, which satisfies the equation $RNB(\hat{r},t) = \underset{r \in R}{Max}(RNB(r_i,t))$.

3 Simulations and Analysis of Experiment Results

The following experiments were carried out in the Laboratory and a total of 40 computers were under surveillance. The importance coefficients of the services including *www, ftp* in server *A* were set to 0.4. The importance coefficient of server *A* is set to be 0.8. The network was attacked by many types of attacks, e.g., *synflood, ssping, port_scan,* etc. The corresponding danger coefficients were 0.3, 0.3, and 0.2, respectively. Antigen was defined as a fixed length (l=172) binary string composed of the source/destination IP, port number, protocol type, etc. Suppose the amount of immuocytes is not more than 200, the antibody increase velocity $\kappa = 1$.

Table 1. Response Strategy Parameters

	Null	Alarm	Block_sip	Isolate_service	Shutdown_host
ϕ	0	0.1	1	1	1
$k_{roc}(r)$	0	0.005	0.2	0.1	0.2
$k_{rnc}(r)$	0	0	$rate_{block} \times 1.2$	0.4	0.8

Table 1 shows some corresponding response strategies and parameter for *syn flood* attack. $k_{rnc}(r)$ is related with the meanings of the strategies. In the experiments, $k_{rnc}(r)$ of the strategy *Block_sip* should consider about the services and the ratio ($rate_{block}$) of the amount of attacking source computers to that of whole computers under the surveillance. E.g., there are five computers that are isolated, and then $k_{rnc}(r) = 3 \cdot 0.3 \cdot 5/40$. *Isolate_service* considers about the influence on network when closing FTP. And *Shutdown_host* concern with the influence on networking while closing both *ftp* and *www* at the same time, at which $\varphi_{dos} = 0.3$.

Table 2. Response Strategy Decision-making Examples

		Null	Alarm	Block_sip	Isolate_service	Shutdown_host
RNB	t_1	0	0.004112	0.00224	-0.07984	-0.20176
	t_2	0	0.026048	0.15752	0.13616	0.00752
	t_3	0	0.06736	0.4576	0.5416	0.3976

Table 2 shows server *A* uses different strategy according to RNB. When the attack server *A* suffers attacks from only 4 computers at time t_1, $dv = 0.0526$, $r_{cpu} = 8\%$, the strategy Alarm is used. When the attacks coming from 10 computers at t_2, $dv = 0.331$, $r_{cpu} = 36\%$, it needs to adjust firewall to prevent the attack, so the strategy Block_sip is an optimal selection. When the amount rises to 20 at t_3, $dv = 0.857$, $r_{cpu} = 100\%$, it needs to close *ftp* access. Therefore, it should give the priority to Isolate_service.

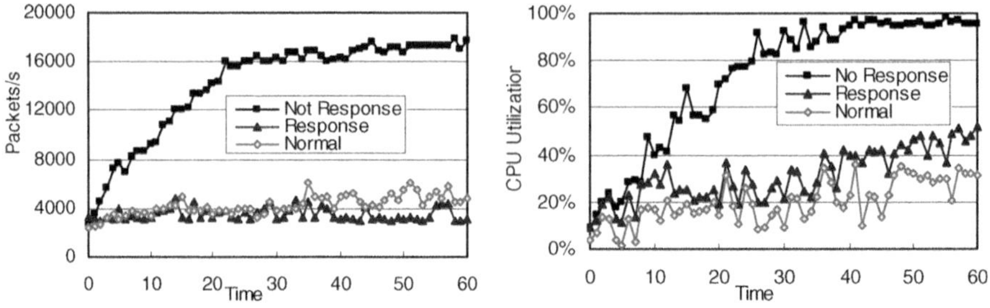

Fig. 3. The contrast of network traffic and the CPU utilization in three states

Fig. 3 compares the network traffic and the CPU utilization on server *A* in three states: normal, no response when suffering from *syn flood* attacks, and respond using IDIR. It proves when IDIR isn't used, with the increment of attack-intensity, the network traffic sharply rises, the CPU utilization is nearly 100%, and system cannot run. when using IDIR, the network traffic will descend, because response strategy is carried out, such as locking attacking origin, closing *FTP*, etc. At this time, the normal packets to server *A* are also to be refused, which results in the amount of actual received packets is lower than the case of normal state. Meanwhile, CPU utilization descend, which is more 10% than that of the normal state. The excess is occupied by IDIR. The reduction of network traffic to server *A* and the CPU utilization demonstrates that IDIR is rational, efficient and real-time.

In order to validate the capability of IDIR, Table 3 shows the synthetic comparison between IDIR model and some typical intrusion response models .

Table 3. The synthetic comparison between IDIR and other intrusion response model

	Affairs classification	Response cost	Self adaptation	Explanations of the intrusion response
Fisch[1]	√			The classification is rough, so it can't calculate response cost.
Carver[2]	√			The events classification can't be practiced in applications.
Thomas Toth[3]		√		It doesn't consider the response operation cost and availability.
IDIR	√	√	√	It could adjust the response strategy according the response cost and benefit.

4 Conclusion

The proposed immune-based dynamic intrusion response model IDIR, depending on the self-learning and diversity of a human immune system, can recognize unknown attacks and classify them. It eliminates the influence of redundant affairs and quantitatively describes the danger facing by the system with the antibody concentration. The quantitative calculations of response cost and benefit are demonstrated. A dynamic response decision-making mechanism is also established, which can dynamically adjust the defending strategies according to the changing environment and use the minimum cost to guarantee the safe of a system. The experiment proves that this model has self-adaptation, rationality, quantitative calculation. This model provides a new method for the intrusion response system.

Acknowledgement

This work was supported by the National Natural Science Foundation of China under Grant No.60373110, No.60573130, NO.60502011, and the National Research Foundation for the Doctoral Program of Higher Education of China under Grant No.20030610003.

References

1. Fisch, E.A.: Intrusion Damage Control and Assessment: A Taxonomy and Implementation of Automated Responses to Intrusive Behavior. Ph.D. Dissertation, Texas A&M University, College Station TX, 1996
2. Carver, C.A. and Pooch, U.W.: An Intrusion Response Taxonomy and its Role in Automatic Intrusion Response; Proceedings of the 2000 IEEE Workshop on Information Assurance and Security, New York: West Point, 2000: 129-135
3. Toth, T.: Evaluating the Impact of Automated Intrusion Response Mechanisms. In 18th Annual Computer Security Applications Conference (ACSAC'02), 2002
4. Forrest, S., Perelson, A. and Cherukuri, R.: Self-Nonself Discrimination in a Computer. Proceedings of IEEE Symposium on Re-search in Security and Privacy, Oakland, (1994)
5. Kim, J., Bentley, P. J.: Immune Memory in the Dynamic Clonal Selection Algorithm. 1st International Conference on Artificial Immune Systems (ICARIS-2002), University of Kent at Canterbury, UK, 2002.9
6. Lee, W., Fan, W., Miller M.: Toward Cost-sensitive Modeling for Intrusion Detection and Response[C]. 1st ACM Workshop on Intrusion Detection Systems, 2000
7. Chao, D.L., Davenport, M.P., Forrest, S., and Perelson, A.: A Stochastic Model of Cytotoxic Tcell Responses. Journal of Theoretical Biology, vol. 228(2) (2004) 227-240.
8. Varela, F.J., Stewart, J.: Dynamic of a Class of Immune Network. Global Stability of Idiotype Interactions .J. Theoretical Biology, 1990(144):93-101
9. Li, T.: An immune based dynamic intrusion detection model. Chinese Science Bulletin, vol. 50, (2005) 2650-2657
10. Li, T.: An immunity based network security risk estimation. Science in China Ser. F. Information Sciences. vol.48 (2005) 557 - 578

Immune Multi-agent Active Defense Model for Network Intrusion

SunJun Liu, Tao Li, DianGang Wang, Kui Zhao, Xun Gong, XiaoQing Hu, Chun Xu, and Gang Liang

School of Computer Science, Sichuan Univ., Chengdu 610065, China
liusunjun@163.com, litao@scu.edu.cn

Abstract. Inspired by the immune theory and multi-agent systems, an immune multi-agent active defense model for network intrusion is established. The concept of immune agent is introduced. While its logical structure and running mechanism are established. The method which uses antibody concentration to quantitatively describe the degree of intrusion danger is presented. The proposed model implements a multi-layer and distributed active defense mechanism for network intrusion, and it is a new way to the network security.

1 Introduction

With the rapid development of information technology and Internet, many network intrusion technologies have continuously changed, which result in more and more serious damages. The network security has been the focus of the public attention. But the present technologies, e.g. accessing controlling, certification authorization, firewall, etc. can only provide a passive defense, which do little to the dynamically changed Internet. Intrusion Detection System, referred to as IDS, has been an indispensable component of the network information security defense system. And it is the focus of the network security researches.

However, these intrusion detection technologies [1], including statistics analysis, feature analysis, data mining, etc., have some limitations, lack self-adaptation and can only detect the known attacks. The system also lacks robustness, and each part of the system is isolated with each other, so that there is not an effective communication, and the early warning and response mechanism cannot be established. In this case, there is an urgent need to build an active defense system, which has the features of self-adaptation, early warning and robustness.

An immune multi-agent active defense model for network intrusion is proposed here, referred to as IMAAD, which makes use of self-adaptation, diversity, memory ability in artificial algorithm [2-5], and combines the robustness and distribution in multi-agent system's architecture [6]. The concept of immune agent (IMA) and its logical structure model are presented. The method, which uses antibody concentration to quantitatively describe the degree of intrusion danger, is developed. The concept of vaccine [7] in biological system is also introduced. It implements the multi-layer and distributed active defense mechanism for network intrusion.

The rest of the paper is organized as follows. In section 2, we discuss the related work. In section 3, the model of IMAAD is provided, section 4 is the conclusions.

T.-D. Wang et al. (Eds.): SEAL 2006, LNCS 4247, pp. 104 – 111, 2006.

2 Related Work

With many different types of lymphocytes distributing all over the body, the biological immune system (BIS) [7] can identify and kill the intrusion antigens. It is seen as a distributed system. Artificial immune system (AIS) [3-5] [7-10] derives from BIS and has become an important research direction in the realm of network security. In 1974, Jerne [7] presented the first mathematic model of the immune system. Farmer J.D. [11] demonstrated the mathematical description of immune system for the firs time in 1986. In 1994, the negative selection algorithm and the concept of computer immune system were proposed by Forrest [9]. In 2002, Castro and Timmis [12] brought the variation into the immune system. Kim and Bentley [3] proposed dynamic clone selection algorithm in the same year, which was used in the network intrusion detection..

Agent [6] is an entity that has the ability of consciousness, solving problems and communication. It can adjust by itself according to the internal reasoning mechanism. With the cooperation during the isolated agents in Multi-agent system (MAS) [10], the problems in the complex environment can be solved.

3 The Theory of IMAAD

On the whole, an immune system is an autonomous one which owns features of distribution and multi-agent. AIS have the abilities of self-learning and memory. Immunocytes are distributed and independent. In this case, a new concept of agent that is immune agent, referred to as IMA, is introduced here. Beside the common features inherited from the general agent, IMA [5] [8] has the desirable features of evolution, identification diversity, memory, tolerance, active, defense, etc.. The mapping relationship between BIS and active defense model is shown in Table 1.

Table 1. The Mapping Relationship between the BIS and an active defense system

Biological immune system	The active defense system
Organism	The whole net work
Organ	The network segment
Cells	hosts
Vaccine	The intrusion information
Antigen	Binary strings extracted from TCP/IP header
B cell, T cell	Antibodies expressed by binary strings

3.1 The Architecture of Active Defense Model

The IMAAD constructs a multi-layer intelligent network security model, using multi-agent technologies and AIS. Its architecture is shown in Fig.1. IMA is the security state of computers surveilled. Local Monitor Agent (LMA) analyzes the state of the local area network. Central Monitor Agent (CMA) surveils the whole network.

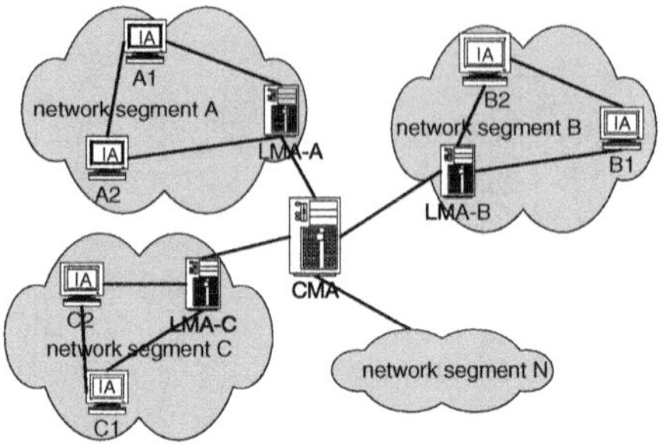

Fig. 1. The Architecture of IMAAD

IMA is the kernel of IMAAD, distributing in every node, identifying the intrusion affairs. It also quantitatively evaluates the risk state facing by the node and sends it to its own network segment's LMA. Meanwhile, it sends the vaccine of the new attack to each node in the same network segment, improves the defense active ability.

LMA takes charge of surveilling the network segment, mixing data from IMA information, evaluating the intrusion risk of network segment, and sending it to Central Monitor Agent (CMA). The segments without intrusion receive the vaccine from the intruded segment to implement the active defense of the whole network.

CMA takes charge of computer interaction. It receives the risk of network segment and shows the secure status through client interface. According to this, administrator takes response measures to control and protect the whole network.

3.2 Active Defending Mechanism of the IMA

The architecture of IMA consists of self, immature antibodies, mature antibodies and memory antibodies, etc. The work flow includes two kinds of circulation: the circulation of immune antibodies' detecting intrusion antigen and the circulation of immune antibodies evolution. Both mutually affect and run at the same time.

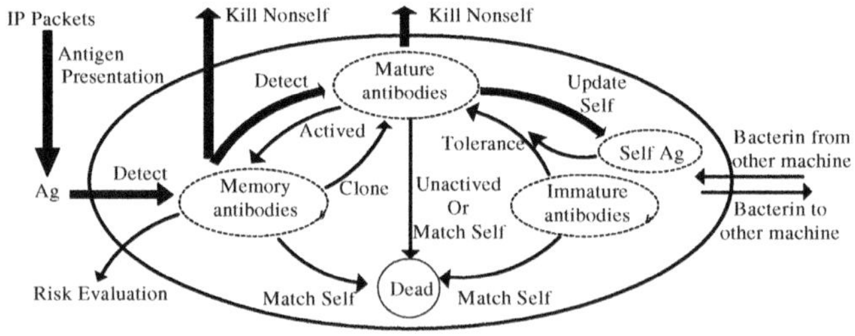

Fig. 2. The Architecture and Work Flow of IMA

3.2.1 The Definition of Immune Elements

Antigens [2] [4] [8]in our approach are binary strings extracted from the IP packets, including IP address, port number, protocol type, etc. The network behavior can be divided into self set (*Self*) that are the normal network behaviors and nonself set (*Nonself*) that is the abnormal network behaviors.

$$Self \cup Nonself = Ag, Self \cap Nonself = \Phi \tag{1}$$

Antibodies are binary strings having the similar features with antigens. Where s is the antibody binary strings, *age* is the cell age, count is the antigen number matched by antibody, *ag* is the antigen detected by the antibody, N is the set of natural number. There are three types of antibodies: memory, mature and immature.

$$B = \{ Ab \mid Ab =< s, age, count, ag >, s, ag \in U \wedge age, count \in N \} \tag{2}$$

The computation of affinity is shown in formula (3) and formula (4) use affinity to judge the match between antibody and antigens, where $x \in Ag$, $y \in B$, x_i and y_i are the corresponding chart of x and y, l is the length of string, θ is the threshold value.

$$f_{h\_dis}(x, y) = \sqrt{\sum_{i=1}^{l} (x_i - y_i)^2} \tag{3}$$

$$f_{match}(x, y) = \begin{cases} 1, & (f_{h\_dis}(x,y)/l) \geq \theta \\ 0, & otherwise \end{cases} \tag{4}$$

3.2.2 The Variation Process of Immature Antibodies

Set I contains immature antibodies, the dynamic variation equation is:

$$I(t + \Delta t) = I(t) + I_{new} \cdot \Delta t - (\frac{\partial I_{mature}}{\partial x_{mature}} \cdot \Delta t + \frac{\partial I_{dead}}{\partial x_{dead}} \cdot \Delta t) \tag{5}$$

It shows the change of I_{Ab} can be divided into two processes: inflow and outflow. In inflow process, new immature antibodies add to the Set I, I_{new} is the immature antibody's producing velocity. In outflow process, immature antibodies are eliminated from Set I through two directions: Some evolve into mature antibodies after self-tolerance, $\frac{\partial I_{mature}}{\partial x_{mature}} \cdot \Delta t$ is the reduced immature ones' number in Set I after a successful tolerance and the immature antibodies in Set I suffer a failed self-tolerance are also can be eliminated, the number is defined as $\frac{\partial I_{dead}}{\partial x_{dead}} \cdot \Delta t$.

In order to prevent antibodies from matching to self, the new immature antibodies can match to antibodies only through the process of self-tolerance, where 1 means passing the self-tolerance and 0 means not, $ab \in I_{Ab}$:

$$f_{tolerate}(ab) = \begin{cases} 0 & iff\ \exists y \in \boldsymbol{Self} \wedge f_{match}(ab.s, y) = 1 \\ 1 & otherwise \end{cases} \tag{6}$$

3.2.3 The Variation Process of Mature Antibodies

Set T contains mature antibodies, the dynamic variation equation is:

$$T(t+\Delta t)=T(t)+(\frac{\partial T_{tolerate}}{\partial x_{tolerate}}\cdot\Delta t+\frac{\partial T_{clone}}{\partial x_{clone}}\cdot\Delta t)-(\frac{\partial T_{active}}{\partial x_{active}}+\frac{\partial T_{dead}}{\partial x_{dead}})\cdot\Delta t \tag{7}$$

It shows the change of mature antibodies can divided into two processes: inflow and outflow. In inflow process, mature antibodies added to the Set T have two ways: Some are evolved from immature antibodies, the corresponsive increased number is $\frac{\partial T_{tolerate}}{\partial x_{tolerate}}\cdot\Delta t$. The others are evolved from memory set, $\frac{\partial T_{clone}}{\partial x_{clone}}\cdot\Delta t$ is the increased number. In outflow process, mature antibodies are eliminated from Set T, which contains two directions: Some are activated into memory set, the number is $\frac{\partial T_{active}}{\partial x_{active}}\cdot\Delta t$. The others activated fail go to death, $\frac{\partial T_{dead}}{\partial x_{dead}}\cdot\Delta t$ is the corresponsive reduced number.

The set of mature antibodies that are activated into memory ones is shown in formula (8), and the set of mature ones that suffer failing activation is shown in formula (9), where β is the activated threshold, and λ is the lifecycle:

$$T_{active} := \{x \mid x\in T_{Ab} \wedge x.count \geq \beta \wedge x.age \leq \lambda\} \tag{8}$$

$$T_{dead} := \{x \mid x\in T_{Ab} \wedge x.count < \beta \wedge x.age > \lambda\} \tag{9}$$

3.2.4 The Variation Process of Memory Antibodies

Set M contains memory antibodies, the dynamic variation equation is:

$$M(t+\Delta t)=M(t)+\frac{\partial M_{active}}{\partial x_{active}}\cdot\Delta t+\frac{\partial M_{bacterin}}{\partial x_{bacterin}}\cdot\Delta t \tag{10}$$

Because of the infinite lifecycle in memory antibodies, the change of M_{Ab} only has inflow process without outflow process. The inflow has two ways: Some are transformed by the activated mature antibodies T_{active}, $\frac{\partial M_{active}}{\partial x_{active}}\cdot\Delta t=\frac{\partial T_{active}}{\partial x_{active}}\cdot\Delta t$, others are obtained from the antibody vaccine sent by other IMA. Once the memory one match to the antigen, it will clone itself, and has a rapid response to the antigens that have been identified before. This is the second response in BIS.

3.2.5 Immune Surveillance

The detection of IMA on network makes use of mature and memory cells to detect the antigens, and filters the mature ones which do well in identifying antigens to evolve into memory ones. These would make them distinguish the nonself antigens from others more quickly and effectively. The detailed steps are as follow [13] [14]:

(1)Antigen extraction: Extract the information from TCP/IP header and Construct a binary string as the antigen, including IP address, port number and protocols, etc.

(2)Memory antibody detects: Use memory antibody to detect Ag, eliminate detected nonself. If memory antibody detects the self, it will be eliminated from Set *M*.

(3)Mature antibody detects: Use mature antibody to detect antigens, eliminate detected nonself. If the mature antibody detect enough antigens, it will be activated and evolve into memory one. Otherwise the mature antibody will go to death.

(4) Self update: The left antigens after the above detection will be added to the self set in order to maintain updating self set. It will endure a self-tolerance process with immature set *I* to keep the dynamic circulation of antibodies evolution.

3.2.6 The Risk Evaluation of IMA

IMA simulates the cells in BIS to apperceive the surrounding. When the host isn't under attack, the antibody amount keeps invariable. When it encounters attack, IMA proliferate antibody to resist the fierce attack. When the attack disappears, IMA restrain the antibody. The types and amount of antibody in IMA reflect the attack type and intensity suffered by host. Therefore, the risk of hosts can be quantitatively computed according to the antibody amount change.

Set $\varphi_j\,(0 \le \varphi_j \le 1)$ is the danger of attack j, $n_{ij}(t)$ is the antibody amount which detects attack j on hosts i, $c_{ij}(t)$ is the antibody amount which detects attack j on host i in normal state. The risk on host i suffering attack j at time t is $d_{node\_ij}(t)$, $0 \le d_{node\_ij}(t) \le 1$ and the synthetical risk of host i is $d_{node\_i}(t)$, $0 \le d_{node\_i}(t) \le 1$.

$$d_{node\_ij}(t) = 1 - \frac{1}{1 + \ln(\varphi_i (n_{ij} - c_{ij}) + 1)} \tag{11}$$

$$d_{node\_i}(t) = 1 - \frac{1}{1 + \ln(\sum_j (\varphi_i (n_{ij} - c_{ij})) + 1)} \tag{12}$$

3.3 Active Defense Mechanism of Network Layers

The LMA in the network takes charge of receiving the intrusion information sent by IMA, mixing data, evaluating risk of a network segment, and sending it to other LMA. Meanwhile, it takes charge of receiving early warning information from other local monitor agents, analyzing this information and simulating vaccination. It will update the antibody set of IMA using the new received antibodies and start the response mechanism to prevent the potential attack in the network segment.

Set $\alpha_i (0 \le \alpha_i \le 1)$ is the importance of host *i* in the network. The risk of network segment *k* suffering attack *j* at time *t* is $d_{LAN\_kj}(t)$, $0 \le d_{LAN\_kj}(t) \le 1$. The synthetical risk of network segment *k* is $d_{LAN\_k}(t)$, $0 \le d_{LAN\_k}(t) \le 1$.

$$d_{LAN\_kj}(t) = 1 - \frac{1}{1 + \ln(\sum_i (\alpha_i \varphi_j (n_{ij} - c_{ij})) + 1)} \tag{13}$$

$$d_{LAN\_k}(t) = 1 - \frac{1}{1 + \ln(\sum_j (\sum_i (\alpha_i \varphi_j (n_{ij} - c_{ij}))) + 1)} \tag{14}$$

3.4 Communication Mechanism of IMAAD

The concept of vaccine is introduced in to the active defense model. And the message mechanism [6] makes the communication during the agent come true. The structure of the messages is defined as:

$$Message ::= < Sender >< Receiver >< SendTime >< ValidTime >< Content > \tag{15}$$

Where *Sender* and *Receiver* can be a transverse communication during the agents in the same level and also can be the longitudinal communication during the agents in the different levels. *Receiver* judges whether to receive this message depending on *Sendtime* and *Validtime*. *Content* contains information about new antibodies and the risk evaluation. Distributing vaccines strengthens the contact during the agents. Sharing the effective antibodies in each IMA improves the response ability of the nodes. And switching intrusion information during local monitor agents guarantees the early warning mechanism. Then the capability of resist attack is highly improve.

4 Conclusion

An immune multi-agent active defense model for network intrusion is proposed. The logical structure of immune agent is established and a multi-layer and distributed architecture is achieved. This model has the following advantages: (1) Self-learning. Memory antibody mechanism make the model can detect both the known and unknown attacks. (2) Multi-layer. The concept of vaccine is introduced, and the connection during the network nodes and segments are strengthened. The active defense is achieved in different layers. (3) Real-time. This model can quantitatively evaluate the risk status facing by the network. (4) Robust. This model uses a distributed architecture, so that the attacks on a single node cannot influent the others, it avoids of invalidation of a single node. This model changes the isolated and passive status in the traditional network security models. It is a good solution to establishing active defense for the network security.

Acknowledgement

This work was supported by the National Natural Science Foundation of China under Grant No.60373110, No.60573130, NO.60502011, and the National Research Foundation for the Doctoral Program of Higher Education of China under Grant No.20030610003, the New Century Excellent Expert Program of Ministry of Education of China under Grant No NCET-04-0870.

References

1. Bai, Y., Kobayashi, H.: Intrusion Detection Systems: technology and development. IEEE Advanced Information Networking and Applications, (2003) 710 – 715
2. Li, T.: An immune based dynamic intrusion detection model. Chinese Science Bulletin, vol. 50, (2005) 2650-2657
3. Kim, J., Bentley, P.: The Artificial Immune Model for Network Intrusion Detection. 7th European Congress on Intelligent Techniques and Soft Computing (EUFIT '99), Aachen, Germany, 1999
4. Li, T.: An immunity based network security risk estimation. Science in China Ser. F Information Sciences, vol.48 (2005) 557 - 578
5. Hegazy, I.M., Faheem, H.M., Al-Arif, T. and Ahmed, T.: Evaluating how well agent-based IDS perform. Potentials, Digital Object Identifier, IEEE, Vol.24 (2005) 27-30
6. Shi, Z.Z.: Intelligent agent and their application [M]. Beijing: Science Press, 2000
7. Jerne, N.K.: Towards a Network Theory of the Immune System. Annnual Immunology, vol.125c, 1974
8. Li, T.: Computer Immunology: Publishing House of Electronics Industry, Beijing (2004).
9. Forrest, S., Perelson, A.S.: Self-Nonself Discrimination in a Computer. Proceedings of IEEE Symposium on Research in Security and Privacy, Oakland (1994)
10. Dasgupta, D.:An Artificial Immune System as a Multi-Agent Decision Support System, Proc of the IEEE International Conference on SMC, San Diego(1998)
11. Farmer, J.D., Packard,N.H., Perelson, A. S.: The Immune System, Adaption, and Machine Learning, Physica, vol.22d (1986)
12. Ayara, Timmis.: Negative Selection: How to Generate Detectors, Proc of 1st International Conference on Artificial Immune Systems, University of Kent Canterbury, (2002)

An Immune Mobile Agent Based Grid Intrusion Detection Model

Xun Gong, Tao Li, Tiefang Wang, Jin Yang, Sunjun Liu, and Gang Liang

School of Computer Science, Sichuan Univ., Chengdu 610065, China
gongxunphd@163.com, orchid_xun@tom.com

Abstract. This paper proposes a novel immune mobile agent based grid intrusion detection (*IMGID*) model, and gives the concepts and formal definitions of *self, nonself, antibody, antigen, agent* and *match algorithm* in the grid security domain. Then, the mathematical models of *mature MoA* (monitoring agent) and *dynamic memory MoA survival* are improved. Besides, effects of the important parameter *P* in the models of dynamic memory MoA survival on system performance are showed. Our theoretical analyses and the experiment results show the model that enhances detection efficiency and assures steady performance is a good solution to grid intrusion detection.

1 Introduction

Current research on grid intrusion detection is still in its infancy, and a few references can be used [1] [2]. According to the distinctive characteristics of grid environments [3], we propose a novel immune mobile agent based grid intrusion detection (*IMGID*) model, and apply mobile agent [4] [5] technology as support for intrusion detection. The model has the features of artificial immune system, such as dynamicity, self-adaptation and diversity [6] - [10] that just meet the constraints derived from the characteristics of the grid environment.

The memory lymphocytes have an unlimited lifecycle except they match the newly added selfs, in the current immunity based intrusion detection systems [11]. Obviously, a considerable number of memory lymphocytes will be generated in the end. Lymphocytes consist of mature and memory cells. As the increasing of memory cells, the cells number will increase, at the same time, the intrusion detection time will increase. When the number of cell reaches or exceeds a certain value, the intrusion detection system will either become a bottleneck or ineffective as some packets are skipped. However, *IMGID* introduces a new idea of *dynamic memory MoAs survival* to overcome this problem: a given number of the least recently used memory MoAs that simulate memory lymphocytes will degrade into mature MoAs and be given a new age and affinity (the number of antigens matched by antibody), if the number of MoAs reaches a given maximum. The method assures the number of memory cells does not exceed a given value, so that the total number of cells does not exceed a given value. So it does not take too long time to detect intrusions. The set of degraded memory MoAs number is very important, which affects the system performances and diversity and is shown in *Theorem 1*.

T.-D. Wang et al. (Eds.): SEAL 2006, LNCS 4247, pp. 112 – 119, 2006.

2 The Proposed Grid Intrusion Detection Model (*IMGID*)

IMGID has two sub-models. The first is the model of the generation of MoAs. The second is the intrusion detection model.

As B-lymphocytes consist of mature lymphocytes and memory ones, MoAs are divided into mature and memory MoAs that simulate, respectively, mature and memory cells. The generation of MoAs is as follows: The random generated immature MoAs evolve into mature MoAs, if they tolerate to selfs. The mature MoAs evolves into memory MoAs if they are tolerant to self and match enough antigens in their lifecycle. Others will die. The least recently used memory MoAs degrade into mature ones if the number of memory MoAs reaches or exceeds a certain maximum.

The intrusion detection model is as follows: We define three kinds of agents: monitoring agents (*MoA*), communicator agents (*CoA*) and beating off agents (*BoA*). MoAs simulate B-lymphocytes to monitoring four level parameters of grid nodes simultaneously and detect intrusions. CoAs are responsible for message transmission among agents. BoAs simulate T killer cells and deal with intrusive activities. Once MoAs detect intrusions, they will stimulate CoAs and present the features of intrusions to BoAs. CoAs will activate BoAs. BoAs will move to the suspected place to counterattack intrusions. MoAs can clone themselves if the intrusion is detected. Some cloned MoAs are left here to detect more intrusions, and others are moved to other grid nodes to detect the similar intrusions.

The correlative concepts and definitions are given as follows:

We define antigens (*Ag*) to be the features of grid services and accesses:

$$Ag = \{ag \mid ag \in D\},$$
$$D = \{d \mid d = <uslevel, syslevel, prolevel, paclevel>\}, D = \{0,1\}^l. \quad (1)$$

uslevel, *syslevel*, *prolevel* and *paclevel* present the parameters, which are monitored by monitoring agents at user, system, process, and packet level respectively. They are also called, respectively, the field of an antigen. l is a natural number (constant). The length of string ag is l. For the convenience using the fields of a antigen x, a subscript operator "." is used to extract a specified field of x, where

$$x.fieldname = \text{the value of field } fieldname \text{ of } x. \quad (2)$$

We define *Self* to be the set of normal grid services and accesses. Similarly, *Nonself* is a set of abnormal services and accesses. *Ag* contains two subsets, *Self* and *Nonself*, where $Self \subset Ag$ and $Nonself \subset Ag$ such that

$$Self \cup Nonself = Ag, \quad Self \cap Nonself = \Phi. \quad (3)$$

All the agents form a set (*Agent*). *Agent* contains three elements, *MoA*, *CoA* and *BoA*. Monitoring agents, communicator agents and beating off agents form, respectively, the sets *MoA*, *CoA* and *BoA*. A monitoring agent is used as a detector to recognize nonself antigens (intrusions). Thus, we have:

$$\begin{aligned}&Agent = \{MoA, CoA, BoA\},\\&CoA = \{c \mid c \text{ is a communicator agent which transmits message among agents}\},\\&BoA = \{b \mid b \text{ is a beating off agent which counterattacks intrusions}\\&\qquad \text{according to the antigen features presented by MoAs}\},\\&MoA = \{< d, age, count > \mid d \in D, age \in N, count \in N\}.\end{aligned} \tag{4}$$

d is the lymphocyte antibody that is used to match an antigen. Each monitoring agent carries several antibodies. *age* is the agent age, *count* (affinity) is the antigen number matched by antibody *d*, and *N* is the set of natural numbers. *d*, *age* and *count* are also called, respectively, field *d*, *age* and *count* of an agent.

Mature and memory MoAs form, respectively, the set MA_{MoA} and ME_{MoA}. Therefore, we have:

$$\begin{aligned}&MoA = MA_{MoA} \cup ME_{MoA}, \qquad MA_{MoA} \cap ME_{MoA} = \Phi,\\&MA_{MoA} = \{x \mid x \in MoA, \forall y \in Self\ (< x.d, y > \notin Match \wedge x.count < \beta)\},\\&ME_{MoA} = \{x \mid x \in MoA, \forall y \in Self\ (< x.d, y > \notin Match \wedge x.count \geq \beta)\}.\end{aligned} \tag{5}$$

Match is a match relation in *D* defined by

$$\begin{aligned}&Match = \{< x, y > \mid (x, y \in D), (f_{match}(x.uslevel, y.uslevel) = 1\\&\vee f_{match}(x.syslevel, y.syslevel) = 1 \vee f_{match}(x.prolevel, y.prolevel) = 1\\&\vee f_{match}(x.paclevel, y.paclevel) = 1)\}.\end{aligned} \tag{6}$$

$f_{match}(x, y)$ is based on the affinity between *x* and *y*: if the affinity greater than a specified threshold, then *1* is returned, otherwise, *0* is returned.

In the following sections, the mathematical models of *mature MoA* and *dynamic memory MoA survival* are established.

2.1 Mature MoAs Model

$$MA_{MoA}(t) = \begin{cases}\Phi, & t = 0\\ MA_{retain}(t) \cup MA_{new}(t) \cup MA_{cycle}(t), & t \geq 1\end{cases}. \tag{7}$$

$$MA_{retain}(t) = MA_{no\,detection}(t) \cup \{x \mid x \in MA_{detection}(t), x.count < \beta\} \tag{8}$$

$$\begin{aligned}MA_{no\,detection}(t) = \{&y \mid y \in MoA, x \in MA_{MoA}(t-1), x.age < \lambda, \forall s \in Self < x.d, s > \notin Match,\\&\forall z \in Ag < x.d, z > \notin Match, y.d = x.d, y.age = x.age + 1, y.count = x.count\}.\end{aligned} \tag{9}$$

$$\begin{aligned}MA_{detection}(t) = \{&y \mid y \in MoA, x \in MA_{MoA}(t-1), x.age < \lambda, \forall s \in Self < x.d, s > \notin Match,\\&\exists z \in Ag < x.d, z > \in Match, y.d = x.d, y.age = x.age + 1, y.count = x.count + 1\}.\end{aligned} \tag{10}$$

$$MA_{cycle}(t)=\{y|y\in MoA, x\in ME_{degradation}(t), 0<T<\lambda, y.d=x.d, y.age=T, y.count=\beta-1\} \tag{11}$$

$$MA_{new}(t)=\{y \mid y\in MoA, x\in I_{new}(t), \forall s\in Self <x,s>\notin Match, y.d=x.d, y.count=0, y.age=0\}. \tag{12}$$

$$I_{new}(t)=\begin{cases}\Phi, & |MA_{MoA}(t-1)\cup MA_{cycle}(t)|\geq\Gamma \\ \{y_1,y_2,\ldots,y_{\Gamma-|MA_{MoA}(t-1)\cup MA_{cycle}(t)|}\}, y_i\in D, & otherwise \\ 1\leq i\leq\Gamma-|MA_{MoA}(t-1)\cup MA_{cycle}(t)|, & \end{cases}. \tag{13}$$

Equation (7) depicts the evolvement of mature cells. Equations (8) - (10) simulate the process that the mature cells evolve into the next generation ones, where the cells do not tolerate to those newly added self elements or have not match enough antigens in lifecycle λ, will be eliminated. Equation (12) depicts the generation of new mature MoAs. Γ (>0) is the max number of mature MoAs in *IMGID*. Equation (11) is the set of the least recently used memory MoAs which degrade into mature MoAs and be given a new age T and count $\beta-1$ ($\beta\geq1$). Once a mature MoA with count $\beta-1$ matches an antigen in lifecycle λ, the count of it will be set to β and it will be activated again. Because the degraded memory MoA has better detection capability than mature MoAs, it is endowed with a priority in evolvement into memory MoAs. When the same antigens arrive again, they will be detected immediately by the re-evolved memory MoA. The method enhances detection efficiency. The value of T cannot be set too large or too small. If the value of T is too large, the survival time of the degraded memory MoAs will be short, and the degraded memory MoA will die soon because it cannot match enough antigens in lifecycle λ. Contrarily, if the value of T is too small, a few immature MoAs and even none of them will be generated. So the diversity of *IMGID* cannot be assured.

2.2 Dynamic Memory MoAs Survival Model

$$ME_{MoA}(t)=\begin{cases}\Phi, & t=0 \\ ME_{retain}(t)\cup ME_{new}(t)-ME_{degradation}(t), & t\geq1\end{cases}. \tag{14}$$

$$ME_{retain}(t)=ME_{nodetection}(t)\cup ME_{nodetection}(t). \tag{15}$$

$$ME_{nodetection}(t)=\{y \mid y\in MoA, x\in ME_{MoA}(t-1), \forall s\in Self <x.d.s>\notin Match, \forall z\in Ag <x.d,z>\notin Match, y.d=x.d, y.age=x.age+1, y.count=x.count\}. \tag{16}$$

$$ME_{detection}(t) = \{y \mid y \in \boldsymbol{MoA}, x \in ME_{MoA}(t-1), \forall s \in \boldsymbol{Self} < x.d.s > \notin Match, \\ \exists z \in Ag < x.d, z > \in Match, \boldsymbol{y.d} = \boldsymbol{x.d}, \boldsymbol{y.age} = \boldsymbol{x.age} + 1, \boldsymbol{y.count} = \boldsymbol{x.count} + 1\}. \quad (17)$$

$$ME_{new}(t) = \{x \mid x \in ME_{MoA}, y \in MA_{detection}(t), y.count \geq \beta, \\ x.d = \boldsymbol{y.d}, \boldsymbol{x.age} = 0, \boldsymbol{x.count} = \boldsymbol{y.count}\}. \quad (18)$$

$$ME_{degradation}(t) = \begin{cases} \{x_1, x_2, \cdots, x_P\}, 1 \leq i \leq P, x_i \in ME_{degradation}(t), & \\ \forall y \in ME_{retain}(t) - ME_{degradation}(t) \boldsymbol{y.age} \leq \boldsymbol{x_i.age}, & |ME_{MoA}(t-1)| \geq \kappa \\ \Phi, & otherwise \end{cases} . \quad (19)$$

Equation (14) simulates the evolvement of memory MoAs. Equations (15) - (17) depict the process that memory MoAs evolve into the next generation ones, where the memory MoAs that match the newly added selfs will die and those matched by an antigen will be activated immediately. Equation (18) is the set of new memory MoAs evolved from mature ones. Equation (19) is the set of memory MoAs that are not activated by antigens lately and are selected randomly to degrade into mature MoAs when the number of memory MoAs reaches or exceeds a given maximum κ. P (>0) is the number of selected memory MoAs. Obviously, the number of memory MoAs does not exceed a certain maximum.

Theorem 1. $|MA_{new}(t+1)|$ is in inverse proportion to *P*. The value of *P* cannot be set too large, otherwise too large value of *P* will affect the diversity of *IMGID*.

Proof: According to equation (7), $|MA_{MoA}(t)|$ is in direct proportion to $|MA_{cycle}(t)|$. The number $|I_{new}(t+1)|$ of new immature MoAs is in inverse proportion to $|MA_{MoA}(t)|$ and in inverse proportion to $|MA_{cycle}(t)|$ according to equation (13). $|MA_{new}(t+1)|$ is in direct proportion to the number $|I_{new}(t+1)|$ according to equation (12). So $|MA_{new}(t+1)|$ is in inverse proportion to $|MA_{cycle}(t)|$. Obviously, $|MA_{cycle}(t)|$ is equal to $|ME_{degradation}(t)|$ according to equation (11) and $|ME_{degradation}(t)|$ is in direct proportion to *P* according to equation (19). So $|MA_{new}(t+1)|$ is in inverse proportion to P.

So the value of *P* cannot be set too large. If the value of *P* is too large, the number of mature MoAs will be large. The larger the number $|MA_{MoA}(t)|$ of mature MoAs is, the less new immature MoAs $I_{new}(t+1)$ and even new mature MoAs $MA_{new}(t+1)$ will be generated. When the number of mature MoAs reaches or exceeds the maximum Γ, immature MoAs will not be generated. Therefore, too large value of *P* will affect the diversity of *IMGID*.

Theorem 2. Suppose the least recently used memory MoAs degrade into mature MoAs at the time point t_1, t_2, …, t_n, …, respectively. To reach $|ME_{MoA}(t_n)| < \kappa$ immediately, we are sure to have $P > |ME_{MoA}(t_n - 1)| + |ME_{new}(t_n)| - \kappa$.

Proof: Because P memory MoAs degrade into mature MoAs at the time point t_1, to reach $|ME_{MoA}(t_n)| < \kappa$, we have according to equations (14) – (19):

$$|ME_{MoA}(t_1)| = |ME_{retain}(t_1)| + |ME_{new}(t_1)| - |ME_{\deg radation}(t_1)|. \tag{20}$$

$$|ME_{MoA}(t_1)| \le |ME_{MoA}(t_1 - 1)| + |ME_{new}(t_1)| - P. \tag{21}$$

$$|ME_{MoA}(t_1 - 1)| + |ME_{new}(t_1)| - P < \kappa. \tag{22}$$

So we have:

$$|ME_{MoA}(t_1 - 1)| < \kappa + P - |ME_{new}(t_1)|. \tag{23}$$

$$P > |ME_{MoA}(t_1 - 1)| + |ME_{new}(t_1)| - \kappa. \tag{24}$$

Similarly, we have $P > |ME_{MoA}(t_n - 1)| + |ME_{new}(t_n)| - \kappa$.

So the value of P cannot be set too small. If the set of P dose not satisfy *Theorem 2*, We can not reach $|ME_{MoA}(t_n)| < \kappa$ immediately. Furthermore, If the value of P is too small, the memory MoAs will reach or exceed the maximum κ again soon due to increasing memory MoAs.

3 Simulations and Experiment Results

We developed abundant experiments to prove the effect of parameter P on steady performance and diversity in *IMGID*. We developed a grid simulation toolkit based on GridSim [12] to satisfy our needs. The simulation environment simulated users with different behaviors, resources and agents. A total of 40 computers in a grid were under surveillance. The length of selfs was 1 K. The affinity threshold was 0.7. Suppose all the mature MoAs, including the degraded memory MoAs had tolerated to those newly added selfs, and the degraded memory MoAs in evolvement into memory MoAs were prior to other mature MoAs. Every MoA carried an antibody. The mature MoAs maximum Γ was set to 256 and lifecycle λ was set to 20. 16 immature MoAs and 16 mature MoAs were evolved at a time. The initial number of memory MoAs was 192. The memory MoAs maximum κ was set to 256. T was set to 5. To simplify environments, we define a function $f_{generation}(t)$ of the generation of new immature MoAs. If new immature MoAs are generated the value of $f_{generation}(t)$ is 1, 2, 3, respectively, when P is 48, 80, and 96. Otherwise, the value of $f_{generation}(t)$ remains 0.

The effects of P on the number of mature MoAs are shown in Fig.1. Given $P = 48$, 48 memory MoAs degraded into mature ones when the number of memory MoAs reached the given maximum 256 at the fifth, eighth, eleventh, fourteenth, seventeenth, twentieth, twenty-third time point, respectively. Given $P = 80$, 80 memory MoAs degraded into mature ones at the fifth, tenth, fifteenth, twentieth time point, respectively. Given $P = 96$, 96 memory MoAs degraded into mature ones at the fifth, fourteenth, twenty-third time point.

The generation of new immature MoAs is shown in Fig.2. Before the fifth time point, there had been new immature MoAs to be generated. Given $P = 48$, new immature MoAs were generated again at the eighth, eleventh, fourteenth, seventeenth, twentieth, twenty-third time point, respectively. Given $P = 80$, new immature MoAs were generated again at the tenth, fifteenth, twentieth time point, respectively. Given $P = 96$, new immature MoAs were generated again only at the fourteenth, twenty-third time point, respectively.

Obviously, the larger P was, the slower rate at which memory MoAs reached the maximum again. So the system performance was steadier. However, the larger P was, the less new generated immature MoAs were. Because larger P caused the mature MoAs to remain in a saturated condition longer, there was less probability of generation of the new immature MoAs. Therefore, too large P will affect the diversity of *IMGID*.

Therefore, the experiments results show that the appropriate value of P assures steady performance and diversity in *IMGID*.

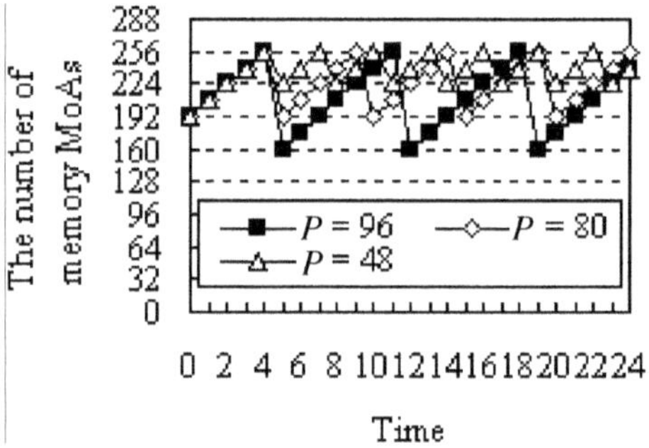

Fig. 1. Effect of P on the number of memory MoAs

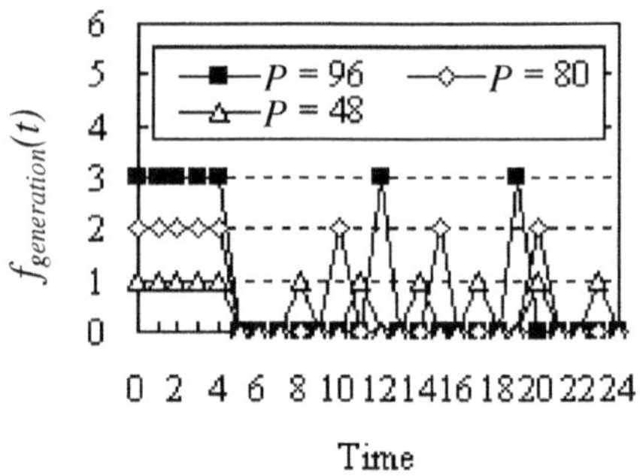

Fig. 2. Effect of P on the generation of new immature MoAs

4 Conclusions

We propose a novel immune mobile agent based grid intrusion detection (*IMGID*) model, and give the correlative concepts, formal definitions and mathematical models. Besides, we introduce a new idea of *dynamic memory MoAs survival*. Furthermore, the theoretical analyses and the experiment results are given to show that the proposed method assures steady performance and diversity.

5 Future Researches

The effects of mature MoAs number and memory MoAs number on steady performance of the system are very important. The models of mature MoA and dynamic memory MoA survival can assure steady MoAs number. The steadiness of MoAs number should be proved theoretically.

Acknowledgment

This work is partially supported by the National Natural Science Foundation of China under Grant No. 60373110, 60573130 and 60502011, the National Research Foundation for the Doctoral Program of Higher Education of China under Grant No.20030610003, and the New Century Excellent Expert Program of Ministry of Education of China under Grant No. NCET-04-0870.

References

1. Fang-Yie Leu, Jia-Chun Lin, Ming-Chang Li, Chao-Tung Yang: A Performance-Based Grid Intrusion Detection System. In Proc. of International Computer Software and Applications, Vol.1 (2005) 525-530
2. M. Tolba, M. Abdel-Wahab, I. Taha, and A. Al-Shishtawy: GIDA: Toward Enabling Grid Intrusion Detection Systems. In Proc. of the Second International Intelligent Computing and Information Systems Conference (2005)
3. Foster, Carl Kesselman, Gene Tsudik and Steven Tuecke: A Security Architecture for Computational Grids. In Proc. of Computer and Communication Security (1998)
4. S. Forrest, S. Hofmeyr, and A. Somayaji: Computer Immunology. In Proc. of Communications of the ACM, Vol. 40 (1997) 88-96
5. A. Somayaji, S. A. Hofmeyr, and S. Forrest: Principles of a Computer Immune System. In Proc. of the New Security Paradigms'97 (1997) 75-82
6. D. Dasgupta: Immunity-based intrusion detection system: A general framework. In Proc. of 22nd National Information Systems Security Conference (1999) 147-160
7. R. B. Machado, A. Boukerche, J. B. M. Sobral, K. R. L. Juca and M. S. M. A. Notare: A Hybird Artificial Immune and Mobile Agent Intrusion Detection Base Model for Computer Network Operations. In Proc. of the 19th IEEE International Parallel and Distributed Processing (2005)
8. T. Li: Compute immunology. Publishing House of Electronics Industry, Beijing (2004)
9. T. Li: An immune based dynamic intrusion detection model. Chinese Science Bulletin (2005) 2650 – 2657
10. T. Li: An immunity based network security risk estimation. Science in China Ser. F Information Sciences (2005) 557 – 578
11. Paul K. Harmer, Paul D. Williams, Gregg H. Gunsch, and Gary B. Lamont: An Artificial Immune System Architecture for Computer Security Applications. In Proc. of IEEE Transactions on Evolutionary Computation, Vol. 6 (2002)
12. M. Murshed, R. Buyya, D. Abramson: GridSim: A Grid Simulation Toolkit for Resource Management and Scheduling in Large-Scale Grid Computing Environments. In Proc. of the 17th IEEE International Symposium on Parallel and Distributed (2002)

Solving Optimization Problem Using Multi-agent Model Based on Belief Interaction

Guo Dongwei, Liu Yanbin, Zhang Na, and Wang Kangping

College of Computer Science & Technology, Jilin University, Changchun130012, China
guodw@jlu.edu.cn, aimd@email.jlu.edu.cn,
eileenz@email.jlu.edu.cn, wangkp@jlu.edu.cn

Abstract. Multi-Agent model based on belief interaction is used to solve the function optimization problems in this paper. "Belief" is led into the Agent, being the parameter of learning machine to decide the searching direction and intensity of the Agent in the environment. It is also the interaction information between Agents. Agent has the ability to evaluate its path in the past. In this way, Agent can find optimization object rapidly and avoid partial extremum at the same time. Finally, several benchmark problems are considered to evaluate the performance of this model. The experimental results prove the efficiency of this model when solving optimization problems.

Keywords: Multi-Agent model, Belief, Interaction, Optimization problem.

1 Introduction

The theory and technology of Agent develop fast as a research area of Distributed Artificial Intelligence since they appeared in 1970s [1]. Multi-Agent system has been a hotspot of research in recent years. Generally speaking, an autonomous agent is a system situated within and a part of an environment that senses that environment and acts on it, over time, in pursuit of its own agenda and so as to effect what it senses in the future [2]. Agent has the knowledge, target and abilities. Many scholars carried on research of Multi-Agent system and its applications, for example, in 1999, Han Jing, etc, put forward a Multi-Agent system model based on artificial beings to solve the N-Queen problem, showing that the Agent had a huge potential ability to solve actual problems [3]. Researchers also made some progress in solving multi-objective optimization problems with Multi-Agent system [4].

In order to set up a valid way solving function optimization problems, Multi-Agent system is taken as the platform to organize reasonable system structure and Agent's cognition ability of interaction in this paper.

2 Multi-agent Model

Agent is an individual that has intelligent behavior, and the individual is given target dissimilarly based on different concrete problems. Agent will change its own

T.-D. Wang et al. (Eds.): SEAL 2006, LNCS 4247, pp. 120 – 125, 2006.

attribute according to its belief in the process of iteration to attain the purpose of tending target [5] [6].

What this paper resolve is the function optimization problem, so we choose the domain of definition of the function as the existence environment of the Agents, and choose the return value as the target of the Agent. Each Agent is given a belief randomly in the beginning, in the period of each iteration of the system, all individuals decide the directions of moving on and the size of step according to its own belief by using the learning machine, then adjust the belief according to the changes of the return value after moving ahead. At last, Agents interact with others in the same scope.

Belief is the attribute of Agent. It is produced to choose path to attain the target through Agent's learning history. Agent decides the direction and size of next step according to its belief. Belief is basically stable to Agent, and is adjusted according to feedback of itself and other Agents. It contains two parts: one is direction, a direction vector naming b_i ; the other is intensity, which will affect the size of the Agent in a specific way.

Agent delivers its belief in the form of parameter to the learning machine, then the learning machine return the distance and direction of the ambulation that Agent will go forward this time. If this ambulation outran the boundary of the environment, a new random direction is given to the Agent in order to carry on new searching.

The adjustment of belief makes Agent can have a simple evaluation of its path, so Agent can vary its belief at any time to make it tend the target soon.

After walking over one step, Agent compare the function value before and after walk according to the target function, if what this step walks is closer to the best solution, strengthen the intensity of the belief. Otherwise, weaken it. If belief is small enough to arrive at a certain value, a new random direction and intensity will be given to make it continue to iterate [7].

Adjusting belief intensity makes the searching of Agent is similar to the process of local climbing, which can tend best solution quickly, and avoid the Agent to sink into local peak because of the variety of the intensity, either.

Interaction is an important function of Agent, also an important part of Multi-Agent model.

After a period of iteration, all Agents interact with other individuals in the sight of itself. What is affected after the interaction is the direction vector of Agent's belief.

Suppose Agent i is meeting with Agent j, the belief vector of Agent i is adjusted as follows:

$$b_i = \lambda b_i + (1-\lambda) b_j \tag{1}$$

According to its own belief, the existence of the learning machine makes Agent able to choose a reasonable decision. It also has the randomicity of stochastic algorithm.

Workflow of the learning machine is as follows: Take belief's direction as the center and produce a random direction to go forward this time, and the probability of choosing each direction should satisfy Gauss distribution ($f(x) = \frac{1}{\sigma\sqrt{2\pi}} e^{-\frac{(x-\mu)^2}{2\sigma^2}}$). The size of the step will be decided by the intensity of Agent's belief, also by the deviation between moving direction and belief direction. If the angle is small enough, the step is set to be a little big, otherwise a little small.

3 Algorithm Workflow and Interpretation

3.1 Workflow

The system constructs the environment in the beginning, and Agent constructs its belief randomly, then system starts iteration. In this process, Agent changes its belief and positions gradually, and records the best solution it finds out. Concrete workflow is as follows:

- Step 1:
  m Agents each with a random belief are distributed randomly in the environment.
- Step 2:
  (2.1) Agent delivers its belief in the form of parameter to the learning machine, getting direction and size of the step of its movement
  (2.2) Record the current return value of the target function, move one more time according to the plan in 2.1.
- Step 3:
  (3.1) Get current return value of target function, and compare it with the value before moving, if it is better than original value, strengthen the belief intensity. Otherwise weaken it.
  (3.2) If the belief intensity is small enough to arrive at a certain value, a new random belief is given to the Agent.
- Step 4:
  (4.1) Agent gets other individuals' believes in the sight of itself
  (4.2) Agent adjusts its own belief according to believes of other individuals. Concrete angle adjustment formula has been given before. Here λ is taken as 0.7.
- Step 5:
  If arrive at designated times of iteration then stop. Otherwise, turn to step 2.

3.2 Algorithm Interpretation

We guaranteed the randomicity of the algorithm by following several aspects: beginning position of the Agent; beginning attribute of the Agent; change of direction when Agent arrives at environment's boundary; and the Gauss distribution which learning machine uses when deciding Agent's direction also has randomicity.

The individual has belief, and every movement of Agent is decided by learning machine reasonably. It is mainly that rule decides the action. It is based on the return value that Agent valuates position of itself.

Agent can simply judge whether its path is good or not, change its own belief, as well as judge whether its position is good or not.

When individual marching forward, it inclines to walk on in the direction which is stronger in its belief. Since the change of belief intensity is gradual, the algorithm is not only a process of climbing, but also can avoid sinking into partial best solution because before Agent's belief is weakened to a certain value it may have crossed the minimum already.

In order to get a converging global best solution, two mechanisms are led to: First, the intelligence behavior of Agent can enlarge its step size gradually when its path is better and better; Secondly, the interaction between individuals is good for Agents to get global best solution quickly.

4 Results of the Experiment

In order to have an overall test of performance of the algorithm in this paper, we adopt three standard testing functions which are usually used in function optimization. The three functions and their attributes are given in Table1.

Table 1. Three testing functions adopted in this experiment

Testing function	Search space	Best solution in theory
$f_1 = -x_1^2 - x_2^2$	$x_i \in [-5.12, 5.12]$	0
$f_2 = -100(x_1^2 - x_2)^2 - (1 - x_1)^2$	$x_i \in [-2.048, 2.048]$	0
$f_3 = -x_1 * \sin(4\pi x_1) - x_2 * \sin(4\pi x_2) - 1$	$x_i \in [-1, 1]$	0.758

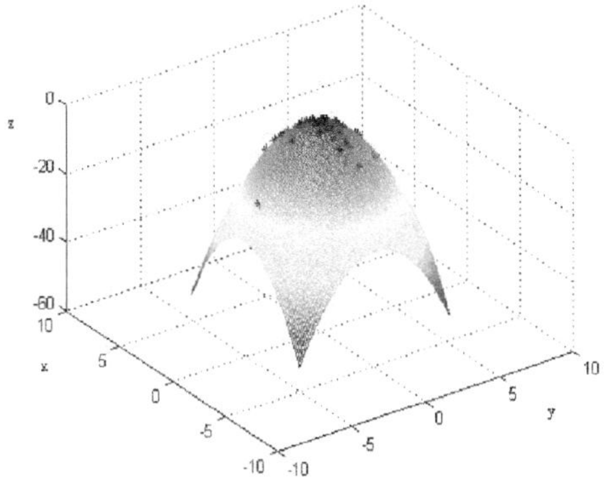

Fig. 1. After optimization, the distribution of each Agent's solution in f1

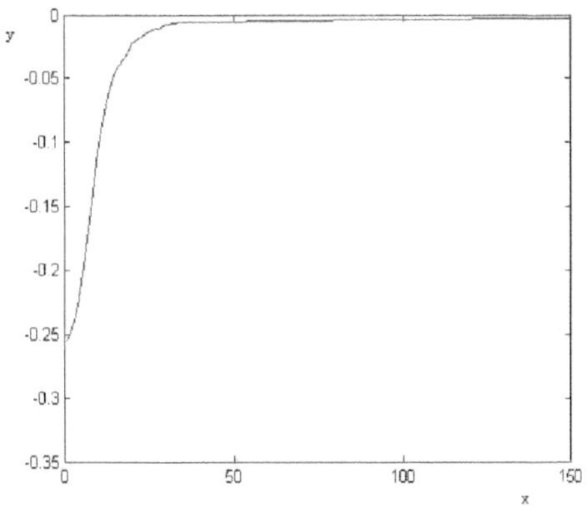

Fig. 2. Global best solution and iteration times

The results of the test of the function can be showed in two ways: distribution of the Agents in the function's domain of definition and the curves of whole Agent colony's solution with times of iteration changing. Here the curve is an average value of 50 times processes, from which we can see the comparison with the global best solutions. In the beginning, the number of Agents is 100, the time of iteration is 100, and the size of the district is 1000 ×1000.

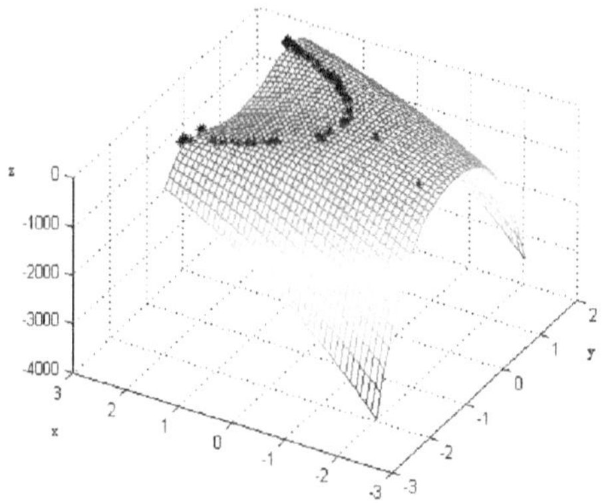

Fig. 3. After optimization, the distribution of each Agent's solution in f2

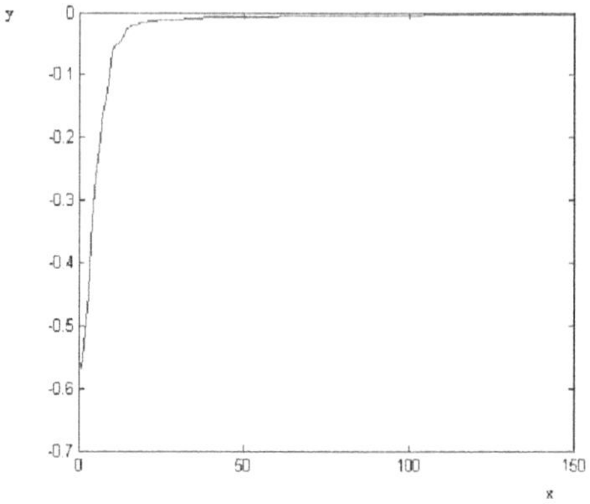

Fig. 4. Global best solution and iteration times

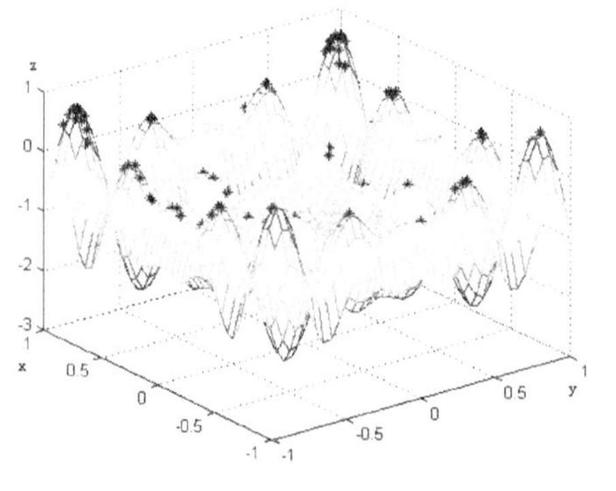

Fig. 5. After optimization, the distribution of each Agent's solution in f3

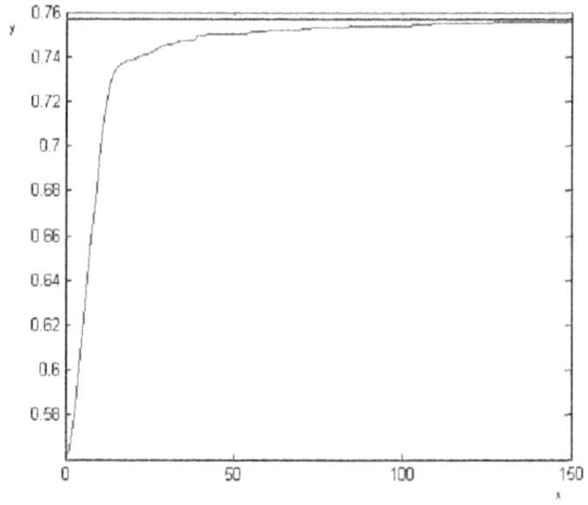

Fig. 6. Global best solution and iteration times

The results of the experiment show that Multi-Agent model based on belief interaction can solve typical function optimization problems very well because of leading into concept "belief" and mechanism of belief adjusting. Figure1 shows the optimizing result of Single-Peak Function; Figure3 shows the optimizing result of seriate Multi-Peak Function; Figure5 shows the optimizing result of Multi-Peak Function with many minimums; Figure2, Figure4 and Figure6 are average curves of colony's best solution with iteration times changing. As long as system has settled time of iteration and number of individuals, we can find out the global best solution quickly, thus prove the validity of the algorithm.

Table 2. Optimizing results of the function

Testing function	Best solution in theory	Best solution of running	Percentage of the individuals which reach the best
$f_1 = -x_1^2 - x_2^2$	0	0	44%
$f_2 = -100(x_1^2 - x_2)^2 - (1 - x_1)^2$	0	0	69%
$f_3 = -x_1 * \sin(4\pi x_1) - x_2 * \sin(4\pi x_2) - 1$	0.758	0.758	55%

5 Conclusions

In this paper, Multi-Agent model is taken to make use of Agent's characteristics of autonomy, intelligence and interaction etc. Leading concept "belief" and a set of belief adjustment mechanisms to Agent take advantages of Multi-Agent model to solve problems. The efficiency of this model is proved by results of experiment.

References

1. Gerhard Weiss. Multi-agent systems, A modern approach to distributed artificial intelligence. Cam-bridge, The MIT Press. 1999, ISBN 0-262-23203-0
2. Petrie, C.J. Agent-based engineering, the Web, and intelligence. Expert, IEEE [see also IEEE Intelligent Systems and Their Applications], Dec 1996, Vol11, Issue: 6, pages 24-29, ISSN: 0885-9000
3. Han Jing, Zhang Hong-jiang, Cai Qing-sheng. Prediction for Visiting Path on Web. Journal of Software, 2001, 1000-9825/2002/13(06)1040-10, Vol.13, No.6
4. John D. Siirola, Steinar Hauan, Arthur W. Westerberg. Computing Pareto fronts using distributed agents. Computers and Chemical Engineering, Vol29 (2004), Pages 113–126.
5. Rao A. S., Georgeff M. P. BDI Agents: From Theory to Practice. In the Proceedings of the First International Conference on Multi-Agent-System (ICMAS), San Francisco 1995.
6. Rao A. S., Georgeff M. P. Modeling rational agents within a BDI-architecture. In Proceedings of the 2rd International Conference on Principles of Knowledge Representation and Reasoning. San Mateo, CA: Morgan Kaufmann Publishers, Inc., 1991. 473~484.
7. Malcolm J.A. Strens. Learning Multi-agent Search Strategies. Adaptive Agents and Multi-Agent Systems. Vol11 (2005), Pages 245-259

Continuous Function Optimization Using Hybrid Ant Colony Approach with Orthogonal Design Scheme*

Jun Zhang[1,**], Wei-neng Chen[1], Jing-hui Zhong[1], Xuan Tan[1], and Yun Li[2]

[1] Department of Computer Science, Sun Yat-sen University, P.R. China
junzhang@ieee.org
[2] Department of Electronics and Electrical Engineering, University of Glasgow, UK

Abstract. A hybrid Orthogonal Scheme Ant Colony Optimization (OSACO) algorithm for continuous function optimization (CFO) is presented in this paper. The methodology integrates the advantages of Ant Colony Optimization (ACO) and Orthogonal Design Scheme (ODS). OSACO is based on the following principles: a) each independent variable space (IVS) of CFO is dispersed into a number of random and movable nodes; b) the carriers of pheromone of ACO are shifted to the nodes; c) solution path can be obtained by choosing one appropriate node from each IVS by ant; d) with the ODS, the best solved path is further improved. The proposed algorithm has been successfully applied to 10 benchmark test functions. The performance and a comparison with CACO and FEP have been studied.

1 Introduction

Ant Colony Optimization (ACO) was first proposed by Marco Dorigo in the early 1990s in the light of how ants manage to establish the shortest path from their nest to food sources [1]. By now, the idea of ACO has been used in a large number of intractable combinatorial problems and become one of the best approaches to TSP [2], quadratic assignment problem [3], data mining [4], and network routing [5].

In spite of its great success in the field of discrete space problems, the uses of ACO in continuous problems are not significant. Bilchev *et al* [6] first introduced an ACO metaphor for continuous problems in 1995 but the mechanism of ACO was only used in the local search procedure. Later, Wodrich *et al* [7] introduced an effective bi-level search procedure using the idea of ants. This algorithm, which was referred to as CACO, also employed some ideas of GA. The algorithm was further extended by Mathur *et al.* [8] and the performances were significantly improved. Based on some other behaviors of ants, two algorithms called API [9] and CIAC [10] were proposed, but they did not follow the framework of ACO strictly, and poor performances are observed in high-dimension problems. Overall, the use of ACO in continuous space optimization problems is not significant.

* This work was supported in part by NSF of China Project No.60573066 and NSF of Guangdong Project No. 5003346.
** Corresponding author.

T.-D. Wang et al. (Eds.): SEAL 2006, LNCS 4247, pp. 126 – 133, 2006.

This paper proposed a hybrid ant colony algorithm with orthogonal scheme (OSACO) for continuous function optimization problems. The methodology integrates the advantages of Ant Colony Optimization (ACO) and Orthogonal Design Scheme (ODS). The proposed algorithm has been successfully applied to 10 benchmark test functions. The performance and a comparison with CACO and FEP have been studied.

2 Background

2.1 ACO

The idea underlying ACO is to simulate the autocatalytic and positive feedback process of the forging behavior of real ants. Once an ant finds a path successfully, pheromone is deposited to the path. By sensing the pheromone ants can follow the path discovered by other ants. This collective pheromone-laying and pheromone-following behavior of ants has become the inspiring source of ACO.

2.2 Orthogonal Experimental Design

The goal of orthogonal design is to perform a minimum number of tests but acquire the most valuable information of the considered problem [11][12]. It performs by judiciously selecting a subset of level combinations using a particular type of array called the orthogonal array (OA). As a result, well-balanced subsets of level combinations will be chosen.

3 The Orthogonal Search ACO Algorithm (OSACO)

The characteristics of OSACO are mainly in the following aspects: a) each independent variable space (IVS) of CFO is dispersed into a number of random nodes; b) the carriers of pheromone of ACO are shifted to the nodes; c) SP can be obtained by choosing one appropriate node from each IVS by ant; d) with the ODS, the best SP is further improved.

Informally, its procedural steps are summarized as follows. ***Step 1)*** *Initialization*: nodes and the pheromone values of nodes are initialized; ***Step 2)*** *Solution Construction*: Ants follow the mechanism of ACO to select nodes separately using pheromone values and form new SPs; ***Step 3)*** *Sorting and Orthogonal Search*: SPs in this iteration are sorted and the orthogonal search procedure is applied to the global best SP; ***Step 4)*** *Pheromone Updating*: Pheromone values on all nodes of all SPs are updated using the pheromone depositing rule and the pheromone evaporating rule; ***Step 5)*** *SP Reconstruction*: A number of the worst SPs are regenerated; ***Step 6)*** *Termination Test*: If the test is passed, stop; otherwise go to ***step 2)***.

To facilitate understanding and explanation of the proposed algorithm, we take the optimization work as minimizing a D-dimension function $f(X)$, $X=(x_1,x_2,\dots,x_D)$. The lower and upper bounds of variable x_i are $lowBound_i$ and $upBound_i$. Nevertheless, without loss of generality, this scheme can also be applied to other continuous space optimization problems.

3.1 Definition of Data Structure

We first define the structure of a Solution Path (SP):

```
structure SP
begin
    real node [D];        % the nodes of the SP;
    real r [D]            % the search radiuses for each node of the SP;
    real t [D]            % the pheromone values for each node of the SP;
    real value;           % the function value of the SP;
end
```

Fig. 1. The structure of a "SP"

In the above definition, each SP includes four attributes: the nodes of the SP in all IVS, the search radiuses for each node which are used during the orthogonal search procedure, the pheromone values for each node which are used in the solution construction procedure, and the function value of the SP.

Assume that the number of SPs (ants) in the algorithm is *SPNUM*. In the following text, we denote the four attributes of the sp_k ($1 \le k \le SPNUM$) as $SP_k.NODE(SP_k.node_1, SP_k.node_2,\ldots,SP_k.node_D)$, $SP_k.R(SP_k.r_1, SP_k.r_2,\ldots, SP_k.r_D)$, $SP_k.T(SP_k.\tau_1, SP_k.\tau_2,\ldots, SP_k.\tau_D)$ and $SP_k.value$.

3.2 Initialization

In the beginning, *SPNUM* nodes are created randomly in each IVS and form *SPNUM* SPs. Pheromone values of all nodes are set to $\tau_{initial}$. ($\tau_{initial}$ is also the unitage of pheromone values and we set $\tau_{initial}$ =1 for computational convenience purpose.) Function values of all SPs are calculated and sorted in ascending order. Pheromone values are updated using the following formula:

$$SP_i.\tau_j \leftarrow SP_i.\tau_j + \alpha \cdot (GOODNUM - rank_i) \cdot \tau_{initial}, \text{ if } GOODNUM - rank_i > 0 \tag{1}$$

$SP_i.\tau_j$ is the pheromone value of the jth node of SP_i. *GOODNUM* and α are two parameters. *GOODNUM* ($1 \le GOODNUM \le SPNUM$) represents the number of SPs that can obtain additional pheromone and $\alpha \in [0,1]$ determines the amount of pheromone deposited to the SPs. $rank_i$ represents the rank of SP_i. Search radiuses of all nodes are set to $(upBound_i - lowBound_i)/SPNUM$. Moreover, the best SP will be preserved additionally and is denoted as $SP_{SPNUM+1}$.

3.3 Solution Construction

All ants build their SPs to the problem incrementally in this phase. The new SPs created in this phase are denoted as *antSP*. The procedure for ant *k* ($1 \le k \le SPNUM$) to build its solution $antSP_k$ is as follows:

$$\begin{cases} antSP_k \leftarrow SP_k, \text{ if } q > q0 \\ antSP_k \text{ is created using pheromone information, otherwise} \end{cases}, \quad 1 \le k \le SPNUM; \tag{2}$$

At first, a random number $q \in (0,1)$ is created and is compared with a parameter $q0$ ($0 \le q0 \le 1$). If $q > q0$, the solution created by ant k is the same as SP_k. All attributes of SP_k are reserved by $antSP_k$. Otherwise, a new solution is built by ant k in terms of the pheromone information using the roulette wheel scheme given by equation (3). The node in each IVS has to be selected separately based on the pheromone values of all nodes in that IVS. The probability of selecting $SP_j.node_i$ ($1 \le j \le SPNUM+1$) as the ith node of SP constructed by ant k is in direct proportion to the pheromone value of the ith node in SP_j. It is important to note that the attributes of the best SP $SP_{SPNUM+1}$ can also be selected by ants.

$$p(antSP_k.node_i = SP_j.node_i) = \frac{SP_j.\tau_i}{\sum_{l=1}^{SPNUM+1} SP_l.\tau_i}, \tag{3}$$

$$(1 \le j \le SPNUM+1, 1 \le k \le SPNUM, 1 \le i \le D)$$

Suppose $SP_j.node_i$ is selected to be the ith node of the SP constructed by ant k, the pheromone value of the ith node on SP_j will also be inherited to $antSP_k$, that is, $antSP_k.\tau_i = SP_j.\tau_i$. After all nodes have been selected by ant k, the complete SP is evaluated, that is, $antSP_k.value = f(antSP_k.NODE)$. Search radiuses of all nodes on $antSP_k$ will also be regenerated, that is, $antSP_k.r_i = (upBound_i - lowBound_i)/SPNUM$ ($1 \le i \le D$).

3.4 Sorting and Orthogonal Search

All new SPs created by ants ($antSP_k$ ($1 \le k \le SPNUM$)) are sorted in ascending order based on their function values. If the function value of the best SP created by ants is smaller than $SP_{SPNUM+1}.value$, the best SP is preserved in the place of $antSP_{SPNUM+1}$, otherwise $antSP_{SPNUM+1} = SP_{SPNUM+1}$. Then, the orthogonal search procedure will be implemented to the global best SP $antSP_{SPNUM+1}$.

1) Orthogonal Search

In order to introduce the orthogonal design technique to this case, we assume that each IVS corresponds to a single factor of the experiment. Also, we divide the search range in each IVS of a SP into l-1 segments to obtain l dividing values. These l dividing values are accordingly considered as the l levels of the experiment. Hence, if we are optimizing an D-dimension object function $f(x_1, x_2, \ldots x_D)$, we can simply take the n variables $(x_1, x_2, \ldots x_D)$ as the D factors. When acquiring the l dividing values of the ith IVS from a SP S located at $S.NODE(S.node_1, S.node_2, \ldots, S.node_D)$, we first compute the search range of each IVS $r_i = S.r_i \cdot ran$, where $S.r_i$ is the search radius in the ith IVS of that SP and ran is a random number distributed in [0,1]. Then we could get $S.node_i - r_i$ as the lower bound of this IVS and $2r_i$ as its search diameter. Thus, $S.node_i - r_i + 2jr_i/(l-1)$, ($0 \le j \le l-1$) are just the l dividing values we need. With that, the SP search problem is formulated as a D-factor experiment with all factors having l levels and an OA can be applied to search the SP. We call the combinations of factor levels generated by OA as the orthogonal points. In a 3-dimension SP, we obtain two levels of each IVS simply by using the two ends of each edge. Then the OA $L_4(2^3)$ is applied and the four points are finally selected.

2) SP Moving
Soon after all orthogonal points have been selected, they are evaluated in the function $f(x_1,x_2,\ldots x_D)$. Once the old SP is worse than the best one of all these orthogonal points, it will be replaced by the best one. That is, all nodes of the old SP are replaced by the nodes of the best orthogonal point.

3) Radius Adapting
The characteristic of a SP is adaptive, that is, the radiuses of all nodes ($S.r_1$, $S.r_2,\ldots$, $S.r_D$) would adjust themselves during the algorithm by applying (4), where $\theta(0\leqslant\theta\leqslant1)$ is a parameter.

$$S.r_i \leftarrow \begin{cases} S.r_i / \theta, \text{ if SP } S \text{ is replaced by a new one} \\ S.r_i * \theta, \text{ otherwise} \end{cases} \quad (4)$$

This can effectively help us to improve the SP by deciding whether to move it faster or to let it shrink. If the SP is not substituted, it is probably that the best solution area is inside the search range of the SP so that we decrease its radius to obtain a solution in higher precision. Otherwise, the best solution of the test function may not lie in the search range of this SP. In this case, we enlarge the search range of the SP to make it move faster to a better area.

3.5 Pheromone Updating

GOODNUM (0≤*GOODNUM*≤*SPNUM*) is a parameter which represents the number of SPs that can obtain additional pheromone, that is, pheromone will only be deposited to the best *GOODNUM* SPs. Additionally, pheromone on all nodes of all SPs will be evaporated. The pheromone updating procedure is executed as follows:

$$\begin{cases} antSP_k.\tau_i \leftarrow \rho \cdot antSP_k.\tau_i + \alpha \cdot (GOODNUM - rank_k) \cdot \tau_{initial}, \text{ if } rank_k < GOODNUM \\ antSP_k.\tau_i \leftarrow \rho \cdot antSP_k.\tau_i, \text{ otherwise} \end{cases}, \quad (5)$$
$$(1 \le k \le SPNUM, 1 \le i \le D)$$

$\alpha\in(0,1)$ and $\rho\in(0,1)$ are two parameters. α determines the amount of pheromone that is deposited to the SPs. ρ determines the evaporating rate of the pheromone. $rank_i$ represents the rank of $antSP_i$.

3.6 SP Reconstruction

At the end of each iteration, a number of the worst SPs will be forgotten by ants and will be regenerated randomly. (The number of the deserted SPs is denoted as *DUMPNUM*, which is a parameter (0≤*DUMPNUM*≤*SPNUM*).) Then, all old SPs are replaced by new generation of SPs constructed by ants, that is, $SP_k=antSP_k$ (1≤k≤*SPNUM*+1).

4 Computational Results and Discussing

To demonstrate the effectiveness of the proposed algorithm, 10 test functions in Table 1 are selected from [13] where we can obtain more information about these functions.

f_1-f_5 are unimodal functions used to test the convergence rate of an algorithm and to evaluate how much precise an algorithm can obtain. f_6-f_{10} are multimodal functions with local optima. Moreover, f_1-f_9 are 30-dimension functions. A good performance on such functions is always taken as a proof of an algorithm's effectiveness.

Table 1. List of 10 Test Functions

Test functions	Search Domain	Test functions	Search Domain
$f_1(x)=\sum_{i=1}^{D} x_i^2$	$[-100, 100]^D$	$f_6(x)=\sum_{i=1}^{D} -x_i \sin(\sqrt{x_i})$	$[-500, 500]^D$
$f_2(x)=\sum_{i=1}^{D}(\sum_{j=1}^{i} x_j)^2$	$[-100, 100]^D$	$f_7(x)=\sum_{i=1}^{D}[x_i^2-10\cos(2\pi x_i)+10]$	$[-5.12, 5.12]^D$
$f_3(x)=\max_i(\lvert x_i\rvert, 1\le i\le D)$	$[-100, 100]^D$	$f_8(x)=\frac{1}{4000}\sum_{i=1}^{D} x_i^2-\prod_{i=1}^{D}\cos(\frac{x_i}{\sqrt{i}})+1$	$[-600, 600]^D$
$f_4(x)=\sum_{i=1}^{D-1}[100(x_{i+1}-x_i^2)^2+(x_i-1)^2]$	$[-30, 30]^D$	$f_9(x)=\frac{\pi}{D}\{10\sin^2(3\pi x_1)+\sum_{i=1}^{D-1}(x_i-1)^2[1+\sin^2(3\pi x_{i+1})]+(x_n-1)^2[1+\sin^2(2\pi x_n)]\}+\sum_{i=1}^{D}u(x_i,10,100,4)$	$[-50, 50]^D$
$f_5(x)=\sum_{i=1}^{D}(\lfloor x_i+0.5\rfloor)^2$	$[-100, 100]^D$	$f_{10}(x)=\sum_{i=1}^{11}[a_i-\frac{x_1(b_i^2+b_i x_2)}{b_i^2+b_i x_3+x_4}]^2$	$[-5,5]^D$

We first set the parameters of the proposed algorithm. Parameters are configured carefully and the ones we use are given below. It is important to note that different test functions may result in different good parameter settings, but the settings we use here are able to balance the performances in all test functions.

Table 2. Comparison of optimization results and computational effort between OSACO, CACO, and FEP (All results are averaged over 100 runs)

	OSACO		CACO		FEP	
	Computational Effort	Mean best (Variance)	Computational Effort	Mean best (Variance)	Computational Effort	Mean best (Variance)
f_1	150000	**1.669e-34** 1.553e-33	150000	3.30e-21 1.21e-20	150000	5.7e-4 1.3e-4
f_2	500000	**1.31e-71** 5.21e-71	500000	0.1173 0.0819	500000	0.016 0.014
f_3	500000	**1.30e-37** 3.89e-37	500000	0.365 0.702	500000	0.30 0.50
f_4	2000000	**0.3596** 1.0443	2000000	38.003 25.715	2000000	5.06 5.87
f_5	50000	0 0	50000	0 0	150000	0 0
f_6	500000	**-12569.49** 1.28e-11	900000	-12446.71 133.928	900000	-12554.5 52.6
f_7	500000	**7.71e-10** 7.71e-9	500000	2.577 1.715	500000	0.046 0.012
f_8	200000	0.01078 0.01136	200000	**0.00826** 0.01391	200000	0.016 0.022
f_9	800000	**1.570e-32** 2.751e-47	1000000	0.00311 0.01778	150000	9.2e-6 3.6e-6
f_{10}	400000	**4.294e-4** 3.460e-4	400000	5.873e-4 1.150e-4	400000	5.0e-4 3.2e-4

Parameters are set as follows: the number of SPs or ants *SPNUM*=20, pheromone depositing rate α=0.9, pheromone evaporating rate ρ=0.8, the probability of selecting a new SP $q0$=0.9, the changing rate of the radiuses θ=0.8, the number of SPs that receive additional pheromone *GOODNUM*=*SPNUM*×0.1=2, and the number of deserted SPs *DUMPNUM*=*SPNUM*×0.05=1. Also, we use the OA $L_{81}(3^{40})$ to satisfy the need of optimizing 30-dimension functions in f_1-f_9, and use $L_9(3^4)$ in f_{10}, which is only a 4-dimension test function.

The proposed algorithm is compared with two other algorithms – CACO [8] and FEP [13]. CACO is by now one of the top algorithms that use the idea of ACO for continuous optimization problems. FEP is one of the state-of-the-art approaches to continuous function optimization problems. Parameters of these two algorithms are set in terms of paper [8] and [13] respectively.

The computational results are shown in Table 2. Obviously, OSACO performs better than CACO and FEP in most cases. In unimodal functions, OSCAO obtains higher precision than CACO and FEP in f_1-f_4, and gets the global best solutions of f_5 much faster than FEP. These prove that the use of the orthogonal search scheme can significantly improve the search precision of the algorithm. In multimodal functions, OSACO obtains the best final results of f_6, f_7, f_9, and f_{10}, and the result of f_8 obtained by OSACO is only slightly worse than CACO, but better than FEP. In f_9, though OSACO seems slower than FEP, but manages to get much higher precision than FEP.

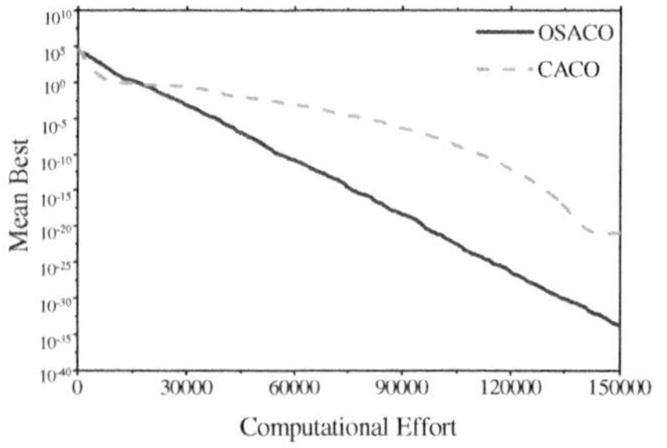

Fig. 2. Convergent speed of OSACO and CACO in f_1

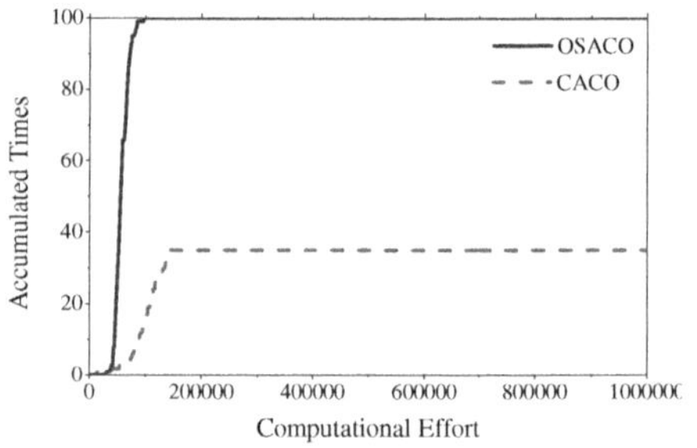

Fig. 3. Accumulative times of errors smaller than 1.0 in f_6

Additionally, Fig. 2 illustrates the comparison of the convergent speed between OSACO and CACO in unimodal function f_1. It is apparent that OSACO is able to obtain higher precision with fewer computational efforts. Fig. 3 reveals the accumulative times of errors smaller than 1.0 in multimodal function f_6 within 100 runs. OSACO successes in all times with at most 100,000 times of function evaluations, while CACO only successes for 35 times after 1,000,000 times of function evaluations. These demonstrated the effectiveness of OSACO.

5 Conclusion

The hybrid orthogonal scheme ant colony optimization (OSACO) algorithm has been proposed. The general idea underlying this algorithm is to use the orthogonal design scheme to improve the performance of ACO in the filed of continuous space optimization problems. Experiments on 10 diverse test functions presented the effectiveness of the algorithm compared with CACO and FEP.

References

[1] M. Dorigo, V. Maniezzo and A. Colorni, "Ant system: optimization by a colony of cooperating agents," *IEEE Trans. on systems man, and cybernetics - part B: cybernetics,* vol. 26, 1996, pp. 29-41.
[2] M. Dorigo and L. M. Gambardella, "Ant colony system: A cooperative learning approach to TSP", *IEEE Trans. Evol. Comput.*, vol. 1, 1997, pp. 53–66.
[3] M. Dorigo, G. Di Caro, and L. M. Gambardella, "Ant algorithms for discrete optimization," *Artificial Life*, 1999,5(2), pp. 137-172.
[4] R. S. Parpinelli, H. S. Lopes, and A. A. Freitas, "Data mining with an ant colony optimization algorithm," *IEEE Transactions on Evolutionary Computation,* vol. 4, 2002, pp. 321-332.
[5] K. M. Sim and W. H. Sun, "Ant colony optimization for routing and load-balancing: survey and new directions," *IEEE Trans. on systems man, and cybernetics - part A: system and humans,* vol. 33, 2003, pp. 560-572.
[6] G. Bilchev and I. C. Parmee, "The ant colony metaphor for searching continuous design spaces," *AISB Workshop on evolutionary computation*, 1995.
[7] M. Wodrich and G. Bilchev, "Corporative distributed search: the ants' way", *Control Cybernetics*, 1997, 26, 413
[8] M. Mathur, S. B. Karale, S. Priye, V. K. Jayaraman, and B. D. Kulkarni, "Ant colony approach to continuous function optimization," *Ind. Eng. Chem. Res.* 2000, pp3814-3822.
[9] N. Monmarché, G. Venturini, and M. Slimane, "On how *Pachycondyla apicalis* ants suggest a new search algorithm," *Future Generation Computer Systems*, 16:937-946, 2000.
[10] J. Dréo and P. Siarry, "Continues interacting ant colony algorithm based on based on dense heterarchy," *Future Generation Computer Systems*, 20(5):841-856, 2004.
[11] K. T. Fang and Y. Wang, *Number-Theoretic Methods in Statistics*, New York: Chapman & Hall, 1994.
[12] A. S. Hedayat, N. J. A. Solane, and John Stufken, *Orthogonal Arrays: Theory and Applications*, New York: Springer-Verlag, 1999.
[13] X. Yao, Y. Liu and G. Lin, "Evolutionary programming made faster", *IEEE Transactions on Evolutionary Computation,* vol. 8, 2004, pp. 456-470.

Niching for Dynamic Environments Using Particle Swarm Optimization

Isabella Schoeman[1] and Andries Engelbrecht[2]

[1] Department of Computer Science, University of Pretoria, Pretoria 0002, South Africa
bschoeman@onsweb.co.za
[2] Department of Computer Science, University of Pretoria, Pretoria 0002, South Africa
engel@cs.up.ac.za

Abstract. Adapting a niching algorithm for dynamic environments is described. The Vector-Based Particle Swarm Optimizer locates multiple optima by identifying niches and optimizing them in parallel. To track optima effectively, information from previous results should be utilized in order to find optima after an environment change, with less effort than complete re-optimization would entail. The Vector-Based PSO was adapted for this purpose. Several scenarios were set up using a test problem generator, in order to assess the behaviour of the algorithm in various environments. Results showed that the algorithm could track multiple optima with a varying success rate and that results were to a large extent problem-dependent.

1 Introduction

Currently Particle Swarm Optimization can be considered to be a well-established population-based optimization technique. Since its inception by Kennedy and Eberhart [1], various developments resulted in an algorithm that proved effective, especially in the case of complex optimization problems. The initial purpose of the strategy was to find the overall best value of a function in a search space, the algorithm being designed in such a way that the population of particles would be discouraged from settling on a suboptimal solution. A balance had to be struck between exploration and exploitation of the search space, an issue addressed elegantly by incorporating both a personal and a social component when updating particle positions.

The kind of difficult and complex optimization problems for which PSO proved to be effective, often contains many suboptimal solutions. In some cases it can be advantageous to find the positions of all the optima in the search space. It might be necessary to know the 'second best' solution, or a parameter not included in the objective function might necessitate a choice between several good solutions. Research has therefore also been directed towards developing PSO algorithms to locate multiple optima. Such algorithms are also called niching algorithms. Various approaches have been followed and promising results were reported [2],[3],[4],[5],[6],[7],[8],[9].

Although PSO has been successfully applied to static problems, most real-world optimization has to be carried out in a dynamic environment, that is, the objective

T.-D. Wang et al. (Eds.): SEAL 2006, LNCS 4247, pp. 134 – 141, 2006.

function changes over time. Locations and values of optima may change while some optimal solutions disappear altogether and others appear in new positions. To locate an optimum in PSO, particles have to converge on a single point. Keeping track of dynamically changing optima therefore requires frequent re-optimization. The best results would be obtained if a good solution has been found before the next objective function change. This implies that particles have already converged on the optimal solution and diversity should therefore have to be increased in order to find the optimum that has moved away. For optimal results a balance must be found between exploitation and exploration. PSO in dynamic environments proved to be very effective when locating and tracking a single optimal value [10],[11]. In a highly multimodal environment, however, temporal changes may result in a suboptimal solution taking on the best overall value. In such circumstances it would be advantageous if an algorithm would locate and track multiple optima in parallel.

In this paper an initial and explorative study of the ability of a niching PSO method to track multiple dynamically changing optima is undertaken. The vector-based PSO [7],[8],[9] is adapted to locate and track those optima in a changing environment. A limited number of functions is set up and used to test this approach.

The paper is organized as follows: section 2 describes PSO techniques to locate multiple optima in a static environment, while research on PSO models to locate a single optimum in a dynamic environment is explored in section 3. Section 4 addresses attempts to locate and track multiple dynamic optima. The experimental setup is described and results reported in section 5 while section 6 contains the conclusion.

2 Niching Algorithms for Multimodal Optimization

Several niching algorithms have been developed to find multiple optima in a static environment. Examples are Kennedy's PSO using k-means clustering [2], the "stretching" technique of Parsopoulos and Vrahatis to transform a function once an optimum has been found in order to set the stage for locating subsequent optima sequentially [3],[4], as well as the NichePSO of Brits *et al* [5]. In all these approaches the initial identification of candidate solutions proved to be a difficult and problematic issue. In his species-based PSO (SPSO) Li devised an ingenious strategy to overcome this challenge [6]. Initial neighbourhood best values – the so-called species seeds – were determined by first sorting all particles in decreasing order of fitness. Particles are checked in turn from best to least-fit against species seeds found so far and assigned to the seed if it falls within a pre-specified radius. The particle becomes a new seed if it does not fall within the radius of any of the seeds identified so far. Particles are now adjusted over a number of iterations to converge on the best position in each particle's neighbourhood. The optimization process therefore takes place in parallel.

While some of these algorithms yielded very good results, it remained a challenge to devise a strategy where multiple optima could be located and optimized with no prior knowledge of the objective function landscape, the number of optima to be found in a designated search area and the niche radii.

Another objective was to find a solution that would be simple, but powerful and where the principles driving PSO would be utilized to its fullest extent. Concepts like the tendency to move towards personal best as well as global best are implemented using vectors, as well as the concept of a velocity associated with a particle. These vectors are manipulated to find a new position. If these vectors could also be manipulated to facilitate niching, the result would be an elegant as well as a powerful solution. Schoeman and Engelbrecht developed sequential and parallel vector-based particle swarm optimizers implementing these principles [7],[8],[9].

In the original PSO, the velocity vector associated with a particle is updated during each iteration. If the position vector towards a particle's personal best position points roughly in the same direction as the position vector towards the best position found so far in the entire neighbourhood, it means that the particle's position will be adjusted towards the swarm's best position, and unless some other obstacle in the function's landscape is encountered, the particle is probably moving towards an optimal solution. If the two vectors are pointing roughly in opposite directions, it is an indication that, without the influence of the current neighbourhood best position, the particle would be adjusted towards another optimal solution. Thus, when identifying niches, this knowledge can be used to identify particles that are not in the niche surrounding the current neighbourhood best position. Not all particles where both vectors point in the same direction would of course be moving towards the current best position, as there may be other optimal solutions between those particles and the current neigh-bourhood best.

Vector addition is used extensively in the original PSO paradigm. In the vector-based PSO, another vector operation is used, namely the *dot product.* In general the dot product of two vectors will be positive when they move in the same direction and negative when moving in opposite directions. Although the stochastic nature of the PSO algorithm is such that the above conjecture will not always hold, it can be used to identify the niche to which a particle belongs. Niches are identified sequentially by first finding the particle with the overall best fitness – the *neighbourhood best* value - of the first niche and calculating the dot product of all particles. A niche radius is then assigned to that niche by calculating the distance between the neighbourhood best particle and the nearest particle with a negative dot product, that is, a particle which will probably converge on a neighbouring niche. A niche identification number starting at one is also assigned to each particle where the dot product is positive and the distance between the particle and its neighbourhood best is smaller than the niche radius. The process is then repeated for particles where niche identification numbers have not yet been assigned, until all particles have been identified. This strategy will yield a number of niches, each with its own neighbourhood best and unique niche radius. Each particle will also be numbered to show to which niche it belongs. Unlike most other niching algorithms developed so far, the niche radius need not be specified in advance. This approach also facilitates niching in functions with irregular landscapes as niche radii are calculated individually and differ from one another. Additional niches may be identified as the landscape is not necessarily symmetrical.

Once all niches have been identified, the particles belonging to each niche are treated as subswarms to be optimized. Optimization takes place in parallel. During this process niches can be merged if they are observed to be converging on the same optimum. A parameter indicating the distance between optima that will trigger merging, is set in advance. This parameter, a small value, is known as the *granularity*. The same value is used to find an initial personal best value when spawning new particles. One of the advantages of this PSO model is therefore that very few parameters have to be set in advance.

3 Particle Swarm Models for Tracking a Single Optimum in a Dynamic Environment

One of the first investigations into the modification of the Particle Swarm Optimizer to locate changing extrema, was by Carlisle and Dozier [10]. The use of a sentry particle was proposed to detect any change in the environment. To prevent a return to outdated positions after an environment change, resetting of the personal best positions to the current position of each particle, was proposed. It should, however, only be reset if the current position is better than the previous personal best position. It must be noted that such a strategy would only be effective when the swarm has not yet converged to a solution. If it has converged, the velocity updates would also be very small. Therefore partial re-initialization of the swarm could be combined with the resetting of personal best positions to increase diversity.

Eberhart and Shi experimented with tracking and optimizing a single optimum in a dynamic system [11]. Successful tracking of a 10-dimensional parabolic function with a severity of up to 1.0 was demonstrated. PSO seemed to be naturally suited to such problems and was perceived to perform better than genetic algorithms on similar problems. Dynamic environments can, however, vary considerably. It was reasoned that searching from the current position works well for small environmental changes, or when the swarm has not yet reached an equilibrium state. For severe environmental changes, reinitialization of the entire swarm could be more effective as previous optimal positions would contribute very little or nothing at all to the effort. A combination of these two approaches, namely retaining the global best position and reinitializing a percentage of the particle positions, was suggested in order to retain potentially good positions as well as increase diversity.

Blackwell and Branke [12] developed a new variant of PSO that would work well in dynamic environments. They constructed interacting multi-swarms by extending the single population PSO as well as the charged particle swarm optimization method. In charged particle swarm optimization a roaming swarm of "charged" particles, that is, particles that are repelled by a subswarm converging on a peak, is maintained to detect new peaks in the search space. Although the purpose of the multiswarms that had to be maintained on different peaks, was to track changing optima, the creation of subswarms *per se* is relevant to multimodal optimization. Two forms of swarm interaction were proposed: exclusion and anti-convergence. Exclusion prevents swarms from settling on the same peak, by re-initializing the lesser swarm if two swarms move too close to one another. Anti-conversion re-initializes the worst swarm in order to track down any new peak that may appear. The latter is of course more relevant in a dynamic environment.

4 Particle Swarm Models for Multiple Dynamic Optima

Most techniques using PSO to track moving optima have been developed for single optimum functions. The SPSO described in section 2 has, however, been modified for operation in dynamic environments [13]. The fact that multiple optima are found in parallel provides a natural mechanism for extending the algorithm. To track optima, each particle's personal best fitness value is re-evaluated before being compared with its current fitness. A strategy for preventing crowding at known optima entails introducing a maximum species population parameter. Only the best candidate members are allocated as members of a species. Redundant particles are reinitialized at random positions in the solution space.

The SPSO was extensively tested on functions generated by Morrison and De Jong's dynamic test function generator [14]. Results indicated that the algorithm can successfully track optima in a dynamic two-dimensional environment. The testing, however, focused mainly on the effect of different values of the population maximum and the species radius on the performance of the algorithm.

The proposed dynamic vector-based PSO uses the parallel vector-based PSO to find initial multiple optima. For a strategy to be effective, all optima should be found after each change with less effort than re-optimization would entail. The only data that would be useful after the objective function has been modified, would be the position where the previous optima were located, and only if the changes were not too severe. For severe changes it could be argued that no benefit can be derived from previous optimal positions and that complete re-optimization would be preferable. The proposed algorithm was designed to retain only those values and then spawn a few particles in the immediate vicinity. Only those particles are then optimized. During this process, niches may be merged if they converge on the same position. New optima, may, however, appear when the objective function changes. To locate them will not be so simple, as the initial strategy to find niches must basically be repeated to find new candidate solutions. The algorithm does, however, deactivate particles falling within the niche radii of the existing optima by marking them not to be updated. The remaining particles are then grouped around candidate solutions and optimized. Some of these niches will form around existing niches and be absorbed during the optimization process. Some will, however, converge on new optima.

The vector-based PSO algorithm for multiple dynamic optima can now be presented:

1. Find niches and optimize by using the parallel vector-based PSO. See section 2.
2. Repeat:

   If function has been modified, then

   Stage 1:

   1. Retain particle with the best value in each niche.
   2. Create a small number of new particles in vicinity of previous best particles. 3 additional particles at random positions within 80% of the niche radius from the neighbourhood best particle in each niche, works well.
   3. Optimize to find new best value in each niche.
   4. Merge niches as described in section 2.

   Stage 2:

   5. Clear vector of particles but retain best particles
   6. Create additional particles over entire search space
   7. Consolidate particles in existing niches as follows (see section 2):
      - Find niche radii

- Set identification numbers of particles inside each niche radius.

8. Optimize remaining particles if any
   - Repeat until all particles have been included in a niche (see section 2):
     - Find best value (neighbourhood best for that niche)
     - Set neighbourhood best of particles to best value
     - Set particles with positive dot product to next niche
     - Optimize niches in parallel and merge if necessary

5 Experimental Setup and Results

To test the above algorithm exhaustively for all possible moving optima is a major undertaking. For this study Morrison and De Jong's test problem generator was used [14]. The test problem generator was devised to be used in the study of EA performance in changing environments. An environment in two dimensions consisting of a number of cone shapes is generated by the following equation:

$$f(X,Y) = \max_{i=1,N}\left(H_i - R_i \cdot \sqrt{(X - X_i)^2 + (Y - Y_i)^2}\right). \quad (1)$$

where the height of each cone is given by H, the slope by R and the position by (X_i, Y_i). The number of optima, their positions, shapes and heights can be specified. An environment with three cones is illustrated in Figure 1. When the environment changes, optima may be obscured and appear again at a later stage.

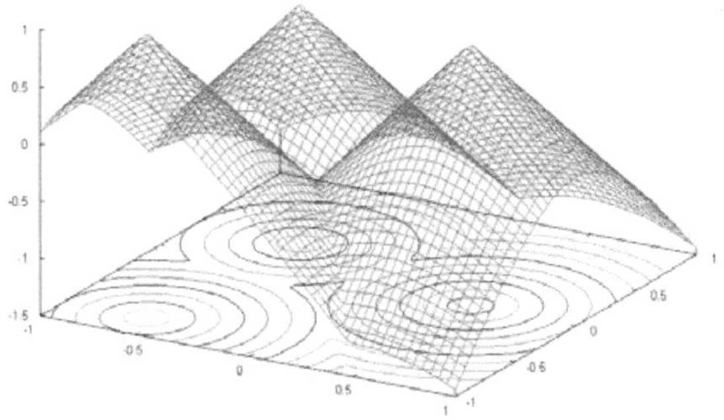

Fig. 1. An environment with three peaks, generated by Morrison and De Jong's test problem generator

Six separate scenarios were implemented; in each case 30 runs were performed in a two-dimensional environment with range [-1.0, 1.0]. The initial number of particles was set to 30 to find the optima for stage 1 of the algorithm. For stage 2 it was increased to 60. The granularity parameter was set to 0.05 for all the experiments. When updating, parameter **w** of the equation given in [1], is set to 0.8, while $\mathbf{c_1}$ and $\mathbf{c_2}$ are both set to 1. $\mathbf{V_{max}}$ was not used in this case [7],[8],[9].

Settings for the six experiments that have been carried out, as well as the results obtained, are summarized in Table 1.

Table 1. Experimental setup and results

Initial environment	Modification	Result
Three peaks	Temporal changes over six steps	All three solutions found after each step
Five peaks	Temporal changes over six steps	All five solutions found after each step
Three peaks	Spatial changes over six steps	All three solutions found after each step
Five peaks	Spatial changes over six steps	All three solutions found after each step
Three peaks	Spatial changes over six steps. One peak is obscured after three steps	All three solutions found after three steps and two solutions after next three steps
Three peaks	Spatial changes over eight steps. One peak is obscured after three steps and appears again after another three steps	All three solutions found after three steps and two solutions after next three steps. In 23 of 30 experiments three optima were located in last two steps, in 5 experiments two optima and in 2 experiments only one. All optima were located in 73% of experiments

6 Conclusion

In this paper it was demonstrated that the parallel vector-based particle swarm optimizer could be modified to track multiple moving optima in a few selected dynamic environments. Good results were found when existing optima were tracked and the spatial severity was low. Results are, however, highly problem-dependent, and locating new peaks is not always successful. It is also computationally more expensive to search the entire problem space for these peaks, and in some cases re-optimization of the search space might have to be considered.

Future research holds many challenges. More complicated test functions will have to be devised to facilitate thorough testing in a variety of environments. The technique should also be refined in order to obtain more conclusive results.

References

1. Kennedy, J., Eberhart, R.C.: Particle Swarm Optimization. In: Proceedings of the IEEE Int. Conf. On Neural Networks. Piscataway N.J. (1995) 1942–1948
2. Kennedy, J.: Stereotyping: Improving Particle Swarm Performance with Cluster Analysis. In: Proceedings of the 2000 Congress on Computational Intelligence. Piscataway N.J. (2000) 1507–1512

3. Parsopoulos, K.E., Plagianakos, V.P., Magoulas, G.D., Vrahatis, M.N.: Stretching Techniques for Obtaining Global Minimizers through Particle Swarm Optimization. In: Proceedings of the Particle Swarm Optimization Workshop. Indianapolis USA (2001) 22-29
4. Parsopoulos, K.E., Vrahatis, M.N.: Modification of the Particle Swarm Optimizer for Locating all the Global Minima. In: Kurkova, V., Steele, N.C., Neruda, R., Karny, M. (eds.): Artificial Neural Networks and Genetic Algorithms, Springer (2001) 324-327
5. Brits, R., Engelbrecht, A.P., van den Bergh, F.: A Niching Particle Swarm Optimizer. In: Proceedings of the 4th Asia-Pacific Conference on Simulated Evolution and Learning (SEAL 2002). Singapore (2002) 692-696
6. Li, X.: Adaptively Choosing Neighbourhood Bests using Species in a Particle Swarm Optimizer for Multimodal Function Optimization. In: Proceedings of the Genetic and Evolutionary Computation Conference 2004 (GECCO 2004). 105-116
7. Schoeman, I.L., Engelbrecht, A.P.: Using Vector Operations to Identify Niches for Particle Swarm Optimization. In: Proceedings of the Conference on Cybernetics and Intelligent Systems. Singapore (2004)
8. Schoeman, I.L., Engelbrecht, A.P.: A Parallel Vector-Based Particle Swarm Optimizer. In: Proceedings of the International Conference on Artificial Neural Networks and Genetic Algorithms. Coimbra Portugal (2005)
9. Schoeman, I.L., Engelbrecht, A.P.: Containing Particles inside Niches when Optimizing Multimodal Functions. In: Proceedings of SAICSIT2005. White River South Africa (2005) 78-85
10. Carlisle, A., Dozier, G.: Adapting Particle Swarm Optimization to Dynamic Environments. In: Proceedings of the International Conference on Artificial Intelligence. Las Vegas Nevada USA (2000) 429-434
11. Eberhart, R.C, Shi, Y.: Tracking and Optimizing Dynamic Systems with Particle Swarms. In: Proceedings of the 2001 Congress on Evolutionary Computation (CEC2001). 94-100
12. Blackwell, T., Branke, J.: Multi-Swarm Optimization in Dynamic Environments. In: Raidl, G.R. (ed.): Applications of Evolutionary Computation, Vol. 1281. Springer (2004) 489-500
13. Parrott, D., Li, X.: A Particle Swarm Model for Tracking Multiple Peaks in a Dynamic Environment using Speciation. In: Proceedings on the 2004 Congress of Evolutionary Computation (CEC2004). 98-103
14. de Jong, K.A., Morrison, R.W.: A Test Problem Generator for Non-Stationary Environments. In: Proceedings of the Congress on Evolutionary Computation. IEEE Press. (1999) 2047-2053

A New Ant Colony Optimization Applied for the Multidimensional Knapsack Problem

Min Kong and Peng Tian

Shanghai Jiaotong University, Shanghai, China, 200052
kongmin@sjtu.edu.cn, ptian@sjtu.edu.cn

Abstract. This paper proposes a Binary Ant System (BAS), a new Ant Colony Optimization applied to multidimensional knapsack problem (MKP). In BAS, artificial ants construct the solutions by selecting either 0 or 1 at every bit stochastically biased by the pheromone level. For ease of implementation, the pheromone is designed specially to directly represent the probability of selection. Experimental results show the advantage of BAS over other ACO based algorithms. The ability of BAS in finding the optimal solutions of various benchmarks indicates its potential in dealing with large size MKP instances.

Keywords: Ant Colony Optimization, Binary Ant System, Combinatorial Optimization, Multidimensional Knapsack Problem.

1 Introduction

The multidimensional knapsack problem (MKP) is a well-known NP-hard combinatorial optimization problem, which can be formulated as:

$$\text{maximize } f(x) = \sum_{j=1}^{n} p_j x_j \tag{1}$$

$$\text{subject to } \sum_{j=1}^{n} r_{ij} x_j \leq b_i, \quad i = 1, ..., m, \tag{2}$$

$$x_j \in \{0, 1\}, \quad j = 1, ..., n. \tag{3}$$

Each of the m constraints described in condition (2) is called a knapsack constraint, so the MKP is also called the m-dimensional knapsack problem. Let $I = \{1, 2, ..., m\}$ and $J = \{1, 2, ..., n\}$, with $b_i > 0$ for all $i \in I$ and $r_{ij} \geq 0$ for all $i \in I, j \in J$, a well-stated MKP assumes that $p_j > 0$ and $r_{ij} \leq b_i < \sum_{j=1}^{n} r_{ij}$ for all $i \in I, j \in J$.

MKP can be regarded as a resource allocation problem, where we have m resources and n objects. Each resource $i \in I$ has a burget b_i, each object $j \in J$ has a profit p_j and consumes r_{ij} of resource i. The problem is to maximize the profit within a limited budget.

Ant Colony Optimization (ACO) is a recently developed, population-based meta-heuristic[3,7], which has been successfully applied to several NP-hard combinatorial optimization problems, such as traveling salesman problem[5,6], vehicle routing problem[9], and quadratic assignment problem[10,12].

T.-D. Wang et al. (Eds.): SEAL 2006, LNCS 4247, pp. 142–149, 2006.

This paper presents a Binary Ant System (BAS), an extension of the hyper-cube frame for Ant Colony Optimization[2] to the constrained combinatorial optimization problem: the MKP. The main idea of BAS is similar to those ACO algorithms dealing with assignment problems. Instead of assigning objects or items to variables, BAS assigns 0 or 1 to every bit of the binary solution string. Furthermore, to reduce the computational complexity and make BAS easy to use, we do not use a fully connected weighed graph as the ACO algorithms usually do, instead, we use a simple graph in which every bit is independently connected to node 0 and node 1. Pheromone are associated to the selection of 0 and 1 for every bit, and its value can be directly regarded as the probability of selection. Because of the independence of the selection of each bit, artificial ants in BAS can construct solutions in a kind of parallel method, rather than the normal sequential solution construction method in ACO. The experimental results over various benchmark problems show the potential of BAS for solving MKP of different characteristics.

The paper is organized as follows. In the next section, we present the description of BAS applied to MKP, together with a short discussion about its difference from other ACO based algorithms available for the MKP. Computational results in section 3 show the performance of BAS in comparison with other ACO based algorithms. Finally, we end with some concluding remarks and discussions in Section 4.

2 Application of BAS to MKP

2.1 Solution Representation and Construction

BAS is designed specially for the problems with binary solution structure, as for MKP, a bit string $x = \{x_1, \cdots, x_n\} \in \{0,1\}^n$ represents a potential solution in BAS. In which, $x_j = 0$ means that the object j is not selected, while $x_j = 1$ means that the object j is selected. By this solution representation, we can see that a solution might not be feasible. An infeasible solution is one for which at least one of the knapsack constraints is violated, i.e., $\sum_{j=1}^{n} r_{ij}x_j > b_i$ for some $i \in I$.

In a solution construction cycle, every ant walks through all the object nodes to construct the solution. At every object node j, an ant selects node 0 or node 1 to go stochastically guided by the pheromone distributed on the selection $j \to 0$ and $j \to 1$:

$$p_{js}^{k}(t) = \frac{\tau_{js}(t)}{\sum_{l \in \{0,1\}} \tau_{jl}(t)}, s \in \{0,1\} \qquad (4)$$

Since all the bits are independent with each other, every ant can construct its solution by making selections for all the bits simultaneously, thus can make the algorithm run in a kind of parallel way to cut down the computing time.

After all the ants have completed their tours, all the solutions generated during the current iteration are checked to see whether they are feasible. Infeasible solutions are repaired by the repair operator, then they are evaluated and compared by the objective function $f(x)$ as described in (1).

BAS incorporates different pheromone update strategy depending on the convergence status of the algorithm. The convergence status of the algorithm is monitored through a convergence factor, which is defined as:

$$cf = \frac{\sum_{j=1}^{n} |\tau_{j0} - \tau_{j1}|}{n} \tag{5}$$

According to the definition of cf, at the beginning of the algorithm, all the pheromone have the value of $\tau_0 = 0.5$, such that $cf = 0$; as the algorithm goes on, the difference between τ_{j0} and τ_{j1} becomes larger and larger, and to be close to 1 when the algorithm falls into a local optima, such that $cf \to 1$. So, the convergence procedure of BAS to a local optima is equivalent to the procedure of cf changes from 0 to 1.

Pheromone update procedure in BAS consists of two parts: the first is the pheromone evaporation procedure, in which pheromone on all the paths evaporate according to the following equation:

$$\tau_{js}(t+1) \leftarrow (1-\rho)\tau_{js}(t), \qquad j = 1, \cdots, n, \ \ s \in \{0,1\} \tag{6}$$

where ρ is the evaporation parameter, and the second part is the pheromone intensification procedure:

$$\tau_{js}(t+1) \leftarrow \rho \sum_{x \in S_{upd} | js \in x} w_x, \qquad j = 1, \cdots, n, \ \ s \in \{0,1\} \tag{7}$$

where S_{upd} is the set of solutions to be intensified in the pheromone update procedure, $w_x \in (0,1)$ is the intensification weight for each solution x, which satisfies $\sum_{x \in Supd} w_x = 1$, and $js \in x$ means that path js is selected by solution x during the solution construction procedure.

S_{upd} consists of three components, which are:

- the global best solution S^{gb}: the best solution generated since the start of the algorithm.
- the iteration best solution S^{ib}: the best solution generated in the current iteration by all the ants.
- the restart best solution S^{rb}: the best solution generated since the last re-initialization of the pheromone.

Depending on the value of cf, BAS uses different combinations of the pheromone intensification weight w_x, and decides whether a pheromone re-initialization procedure should be performed. The details of the pheromone intensification strategy is described in Table 1, in which five different pheromone update strategy are used in different status of cf. w_{ib}, w_{rb}, and w_{gb} are intensification weights for solution S^{ib}, S^{rb}, and S^{gb} respectively. $cf_i (i = 1, \cdots, 5)$ are the threshold parameters for the division of the five stages.

When $cf \geq cf_5$, BAS performs the pheromone re-initialization procedure, in which all the pheromone are set to the initial value τ_0, and followed directly by

Table 1. Pheromone Intensification Strategy for BAS. According to the value of cf, BAS uses different combination of S^{ib}, S^{rb} and S^{gb} to intensify the pheromone belonging to the solutions. While $cf \geq cf_5$, the pheromone re-initialization is performed.

	$cf < cf_1$	$cf \in [cf_1, cf_2)$	$cf \in [cf_2, cf_3)$	$cf \in [cf_3, cf_4)$	$cf \in [cf_4, cf_5)$
w_{ib}	1	2/3	1/3	0	0
w_{rb}	0	1/3	2/3	1	0
w_{gb}	0	0	0	0	1

a pheromone update procedure using S^{gb} as the only intensification solution in order to make the algorithm keep search around the most promising areas in the following iterations.

2.2 The Pseudo Utility

At the initialization step, BAS sorts and renumbers variables according to the decreasing order of their pseudo utility, u_j, which were introduced by the surrogate duality approach of Pirkul [14]. The general idea of this approach is described very briefly as follows.

The surrogate relaxation problem of the MKP can be defined as:

$$\text{maximize} \sum_{j=1}^{n} p_j x_j \tag{8}$$

$$\text{subject to} \sum_{j=1}^{n} \left(\sum_{i=1}^{m} \omega_i r_{ij}\right) x_j \leq \sum_{i=1}^{m} \omega_i b_i \tag{9}$$

$$x_j \in \{0,1\}, \quad j = 1, 2, ..., n \tag{10}$$

where $\omega = \{\omega_1, ..., \omega_m\}$ is a set of surrogate multipliers (or weights) of some positive real numbers. We obtain these weights by a simple method suggested by Pirkul [14], in which we solve the LP relaxation of the original MKP and use the values of the dual variables as the weights. The weight ω_i can be seen as the shadow price of the ith constraint in the LP relaxation of the MKP.

The pseudo-utility ratio for each variable, based on the surrogate constraint coefficient, is defined as:

$$u_j = \frac{p_j}{\sum_{i=1}^{m} \omega_i r_{ij}} \tag{11}$$

2.3 The Repair Operator

In BAS, a repair operator is incorporated into *SolutionRepair* to guarantee feasible solutions generated by ants at *SolutionConstruction* step. The idea comes from Chu and Beasley [4]. The repair operator consists of two phases. The first phase (called DROP) examines each bit of the solution string in increasing order of u_j and changes the bit from one to zero if feasibility is violated. The second phase (called ADD) reverses the process by examining each bit in decreasing order of u_j and changes the bit from zero to one as long as feasibility is not violated.

2.4 Local Search

Local search has been verified effective by most of the previous studies[3,7]. In BAS_MKP, we design a simple random 4-flip method as the local search, that is, randomly select 4 variables from the solution, flip their values from 1 to 0 or 0 to 1, repair the new solution if necessary. If the new generated solution is better, it replace the original solution; otherwise, the original solution is kept for the following flips. The local search method is performed 1000 times at every iteration for the iteration best solution S^{ib} and the global best solution S^{gb}.

2.5 The Difference to Other ACO Based Algorithms

There are several ACO algorithms available to solve MKP problems [1,8,11]. The main differences between BAS and these algorithms are summarized as follows:

- BAS uses a different way of pheromone laying, which is much simple and more general for the binary solution structure $\{0,1\}^n$.
- Contrary to the dynamic heuristic value calculation used in other ACO algorithms, BAS does not use the complicated local heuristic value, but utilize the problem specific information in the repair operator.
- Infeasible solutions during the solution construction step are allowed in BAS. Instead of calculating constraint information to guide the ant tour during the iteration, BAS repair the infeasible solutions at the end of every cycle.

Since BAS bypasses the time-consuming calculation of the dynamic heuristic value and the constraint violation consideration during the tour construction iteration, it is much faster comparing to other ACO based algorithms in each cycle.

3 Comparison Results with Other ACO Approaches

BAS has been tested on benchmarks of MKP from OR-Library. The tested instances are 5.100 with 100 objects and 5 constraints, and 10.100 with 100 objects and 10 constraints. The general parameter settings for all the tests are: $m = 100$, $\tau_0 = 0.5$, $\Delta\tau = 1$, $\rho = 0.3$ and $cf = [0.3, 0.5, 0.7, 0.9, 0.95]$.

We compare the results of BAS_MKP with other three previous ACO algorithms applied to MKP, they are from Leguizamon and Michalewicz [11], Fidanova [8], and Alaya et.al. [1].

Table 2 displays the comparison results of 5.100 from OR library. On these instances, BAS_MKP outperforms all the other three algorithms in the results of the average solution found. Actually, BAS_MKP finds 29 best solutions out of the 30 instances tested.

Table 2. The results of BAS_MKP on 5.100 instances. For each instance, the table reports the best known solutions from OR-library, the best and average solutions found by Leguizamon and Michalewicz[11], the best solution found by Fidanova [8], the best and average solutions found by Alaya et.al. [1], and the results from BAS_MKP, including the best and average solutions over 30 runs for each instance.

$N°$	Best Known	L.&M.		Fidanova	Alaya et.al.		BAS_MKP	
		Best	Avg.	Best	Best	Avg.	Best	Avg.
00	24381	24381	24331	23984	24381	24342	24381	24380.7
01	24274	24274	24245	24145	24274	24247	24274	24270.7
02	23551	23551	23527	23523	23551	23529	23551	23539.7
03	23534	23527	23463	22874	23534	23462	23534	23524.1
04	23991	23991	23949	23751	23991	23946	23991	23978.5
05	24613	24613	24563	24601	24613	24587	24613	24613
06	25591	25591	25504	25293	25591	25512	25591	25591
07	23410	23410	23361	23204	23410	23371	23410	23410
08	24216	24204	24173	23762	24216	24172	24216	24205.4
09	24411	24411	24326	24255	24411	24356	24411	24405.5
10	42757			42705	42757	42704	42757	42736.2
11	42545			42445	42510	42456	42545	42498.9
12	41968			41581	41967	41934	41968	41966.5
13	45090			44911	45071	45056	45090	42074.8
14	42218			42025	42218	42194	42198	42198
15	42927			42671	42927	42911	42927	42927
16	42009			41776	42009	41977	42009	42009
17	45020			44671	45010	44971	45020	45016.5
18	43441			43122	43441	43356	43441	43408.8
19	44554			44471	44554	44506	44554	44554
20	59822			59798	59822	59821	59822	59822
21	62081			61821	62081	62010	62081	62010.4
22	59802			59694	59802	59759	59802	59772.7
23	60479			60479	60479	60428	60479	60471.8
24	61091			60954	61091	61072	61091	61074.2
25	58959			58695	58959	58945	58959	58959
26	61538			61406	61538	61514	61538	61522.5
27	61520			61520	61520	61492	61520	61505.2
28	59453			59121	59453	59436	59453	59453
29	59965			59864	59965	59958	59965	59961.7

Table 3 displays the results for 30 instances of 10.100. On these instances, BAS_MKP also obtains overall better results comparing to other two ACO algorithms. BAS_MKP outperforms all the other algorithms in average solutions found, and finds 28 best solutions out of 30 instances. The instances we did not find the best solutions are 10.100.13 and 10.100.27, in which we found the best value as 45598 and 59366 instead of the best-known value 45624 and 59391.

Table 3. The results of BAS_MKP on 10.100 instances. For each instance, the table reports the best known solutions from OR-library, the best and average solutions found by Leguizamon and Michalewicz[11], the best and average solutions found by Alaya et.al. [1], and the results from BAS_MKP, including the best and average solutions over 30 runs for each instance.

$N°$	Best Known	L.&M.		Alaya et.al.		BAS_MKP	
		Best	Avg.	Best	Avg.	Best	Avg.
00	23064	23057	22996	23064	23016	23064	23058.87
01	22801	22801	22672	22801	22714	22801	22776.33
02	22131	22131	21980	22131	22034	22131	22116.17
03	22772	22772	22631	22717	22634	22772	22766.6
04	22751	22654	22578	22654	22547	22751	22707.8
05	22777	22652	22565	22716	22602	22777	22733.53
06	21875	21875	21758	21875	21777	21875	21861.93
07	22635	22551	22519	22551	22453	22635	22635
08	22511	22418	22292	22511	22351	22511	22433.83
09	22702	22702	22588	22702	22591	22702	22702
10	41395			41395	41329	41395	41387.33
11	42344			42344	42214	42344	42282.83
12	42401			42401	42300	42401	42367
13	45624			45624	45461	45598	45561.73
14	41884			41884	41739	41884	41846.5
15	42995			42995	42909	42995	42991.4
16	43559			43553	43464	43559	43535.8
17	42970			42970	42903	42970	42962.8
18	42212			42212	42146	42212	42212
19	41207			41207	41067	41207	41172
20	57375			57375	57318	57375	57361.43
21	58978			58978	58889	58978	58970.67
22	58391			58391	58333	58391	58382.37
23	61966			61966	61885	61966	61905.67
24	60803			60803	60798	60803	60803
25	61437			61437	61293	61437	61377.2
26	56377			56377	56324	56377	56352
27	59391			59391	59339	59366	59366
28	60205			60205	60146	60205	60171.5
29	60633			60633	60605	60633	60633

4 Conclusions

In this paper we have presented BAS, a HCF based ACO heuristic algorithm for solving the multidimensional knapsack problem. Our approach differs from previous ACO based algorithms in that the way for pheromone laying is specially designed for binary solution structure, in which pheromone can be directly represented as the probability of selection during the solution construction procedure.

We omit the local heuristic information to cut down the time complexity of each iteration, but use a problem-specific repair operator to guarantee the feasibility of solutions. The computational results show that the BAS outperforms other previous ACO algorithms in solution quality, indicating the potential of BAS in dealing with large size MKP problems of different characteristics.

The solution structure and pheromone laying method incorporated by BAS implies that BAS can be a general algorithm for 0-1 integer-programming problems and function optimization problems with binary solution coding. Further works will be on these fields with special focus on how to apply the problem-specific information.

References

1. I. Alaya, C. Solnon, and K. Ghéira: Ant algorithm for the multi-dimensional knapsack problem. International Conference on Bioinspired Optimization Methods and their Applications. (BIOMA 2004) - October 2004 - pages 63-72
2. C. Blum and M. Dorigo. The Hyper-Cube Framework for Ant Colony Optimization. IEEE Transactions on Man, Systems and Cybernetics — Part B, 34(2), 1161–1172, 2004.
3. E. Bonabeau, M. Dorigo, G. Theraulaz. Swarm Intelligence: From Natural to Artificial Systems. Oxford University Press, New York Oxford, 1999.
4. P.C. Chu and J.E. Beasley. A genetic algorithm for the multidimentional knapsack problem. Journal of heuristic, 4: 63-86, 1998
5. M.Dorigo and L.M. Gambardella. Ant colony system: A cooperative learning approach to the traveling salesman problem. IEEE Trans. Evol. Comput., vol.1, pp. 53-66, Apr. 1997
6. Marco Dorigo, Gianni Di Caro, Luca M. Gambardella. Ant Algorithms for Discrete Optimization. Artificial Life, Vol.5, No.3. pp. 137-172, 1999.
7. M. Dorigo and T. Stützle. Ant Colony Optimization. The MIT Press, Cambridge, Massachusetts, London, England 2004
8. S. Fidanova. Evolutionary Algorithm for Multidimensional Knapsack Problem. PPSNVII-Workshop 2002.
9. L.M. Gambardella, E.D. Taillard, and G. Agazzi, MACS-VRPTW: A multiple ant colony system for vehicle routing problems with time windows. in New Ideas in Optimization, D. Corne, M. Dorigo, and F. Glover, Eds. London, U.K.: McGraw-Hill, 1999, pp. 63-76
10. L. Gambardella, E. Taillard, and M. Dorigo. Ant colonies for the quadratic assignment problem. J. Oper. Res. Soc., vol. 50, no. 2, pp. 167-176, Feb. 1999
11. G. Leguizamon and Z. Michalewicz. A new version of Ant System for Subset Problem, Congress on Evolutionary Computation pp1459-1464, 1999.
12. V. Maniezzo and A. Colorni. The ant system applied to the quadratic assignment problem. IEEE Trans. Data Knowl. Eng., vol. 11, pp. 769-778, Sept./Oct. 1999
13. S. Martello, P. Toth. Knapsack Problems: Algorithms and Computer Implementations. John Wiley & Sons. 1990
14. Pirkul, H. A Heuristic Solution Procedure for the Multiconstraint Zero-One Knapsack Problem. Naval Research Logistics 34, 161-172, 1987
15. T.Stützle and H.Hoos(2000) MAX-MIN Ant System. Future Generation Computer Systems Journal. 16(8):889–914, 2000.

Numerical Optimization Using Organizational Particle Swarm Algorithm*

Lin Cong, Yuheng Sha, and Licheng Jiao

Institute of Intelligent Information Processing, Xidian University, Xi'an 710071, China
{conglinsyh, shayuheng}@163.com
lchjiao@mail.xidian.edu.cn

Abstract. The classical particle swarm optimization (PSO) has its own disadvantages, such as low convergence speed and prematurity. All of these make solutions have probability to convergence to local optimizations. In order to overcome the disadvantages of PSO, an organizational particle swarm algorithm (OPSA) is presented in this paper. In OPSA, the initial organization is a set of particles. By competition and cooperation between organizations in every generation, particles can adapt the environment better, and the algorithm can converge to global optimizations. In experiments, OPSA is tested on 6 unconstrained benchmark problems, and the experiment results are compared with PSO_TVIW, MPSO_TVAC, HPSO_TVAC and FEP. The results indicate that OPSA performs much better than other algorithms both in quality of solutions and in computational complexity. Finally, the relationship between parameters and success ratio are analyzed.

1 Introduction

Particle Swarm Optimization (PSO) is an algorithm proposed by James Kennedy and R. C. Eberhart in 1995[1] motivated by social behavior of organisms such as bird flocking and fish schooling. According to the initial definition of PSO, Shi yuhui and R. C. Eberhart [2] introduced the concept of inertia weights in PSO to control the relationship between exploration and exploitation. At the same time, combining PSO with other methods also brings people's interests. A fuzzy adaptive particle swarm optimization method is proposed [3] to adjust the inertia weights. Angeline [4] proposed a particle swarm algorithm using the concept of selection.

In this paper, an organizational particle swarm algorithm (OPSA) is presented. Based on reference [5], the initial organization is a set of particles. By competition and cooperation between organizations in every generation, particles can adapt the environment better, and the algorithm can converge to global optimizations. In order to study the performance of OPSO well, we tested 6 unconstrained benchmark problems, and compared with PSO_TVIW, MPSO_TVAC, HPSO_TVAC [6], and FEP [7]. Finally, the effect of parameters to the algorithm performance is analyzed. The results and analysis demonstrate OPSO has stable performance and high success ration.

---

* This work was supported by the National Natural Science Foundation of China under Grant 60133010 and 60372045, the "863" Project under Grant 2002AA135080, and "973" Project under Grant 2001CB309403.

T.-D. Wang et al. (Eds.): SEAL 2006, LNCS 4247, pp. 150 – 157, 2006.

2 Particle Swarm Algorithm

In particle swarm algorithm, the trajectory of each individual in search space is adjusted by dynamically altering velocity of each particle, according to its own flying experience. The position vector and the velocity vector of the ith particle in the d-dimensional search space can be represented as $X_i=(x_{i,1},x_{i,2},\dots,x_{i,n})$ and $V_i=(v_{i,1},v_{i,2},\dots,v_{i,n})$. According to users defined fitness function, let us say the best position of each particle is $y_i(t)=(y_{i,1},y_{i,2},\cdots,y_{i,n})$, and the fittest particle found so far at time t is $\hat{y}_i(t)=(\hat{y}_{i,1}(t),\hat{y}_{i,2}(t),\cdots,\hat{y}_{i,n}(t))$. Then the new velocities and the positions of particles for the next generation are calculated using the following equations:

$$\begin{cases} v_{i,j}(t+1)=v_{i,j}(t)+c_1\times r_{1,j}(t)(y_{i,j}(t)-x_{i,j}(t))+c_2\times r_{2,j}(t)(\hat{y}_j(t)-x_{i,j}(t)) \\ x_{i,j}(t+1)=x_{i,j}(t)+v_{i,j}(t+1) \end{cases} \quad (1)$$

where c_1 and c_2 are constants known as acceleration coefficients, and $r_{1,j}$ and $r_{2,j}$ are generated uniformly distributed random numbers in range [0,1], separately.

3 Organizational Particle Swarm Algorithm

The model of organization learning was first introduced to GA-based classifier by Wilcox in [8]. And its emphasis puts on making uses of the interaction between organizations. In OEA [5], the conception of the organization defined as a set of individuals. After forming organizations, they will compete and cooperate with each other to enlarge their size and reinforce their strength. In this paper an organizational particle swarm algorithm (OPSA) is presented. Here some definition of organization is given.

An organization, *org*, is a set of individuals, $x=(x_1,x_2,\dots,x_n)$, and the individuals belonging to *org* are called members. The member with best fitness in every organization is called Leader. And the fitness of the ith organization, $Fitness(org_i)$, is equal to the ith organizational leader's fitness value, viz. $Fitness(leader_i)$. Here we consider global minimum numerical optimization as following objective function.

$$\min f(x), x=(x_1,x_2,\cdots,x_n)\in S, \underline{x}_i \le x_i \le \overline{x}_i, i=1,2,\cdots,n \quad (2)$$

where $S \subseteq R^n$ defines the search space of an n-dimensional bounded by parametric constraints.

3.1 Operators of Organizational Particle Swarm Algorithm

Given two parent organizations, $org_{i1}=\{x_1,x_2,\dots,x_{m1}\}$ and $org_{i2}=\{y_1,y_2,\dots,y_{m2}\}$ with $Fitness(org_{i1})\le Fitness(org_{i2})$, then the new organization *org* is generated. Here the operators are defined as follows.

Cooperative Combination Operator. Let the new organization is Q.

$$Q=\{q_1,q_2,\cdots,q_{m1+m2}\}=\{x_1,\cdots,x_{p-1},\overline{x}_p,x_{p+1},\cdots,x_{m1},y_1,\cdots,y_{q-1},\overline{y}_q,y_{q+1},\cdots,y_{m2}\} \quad (3)$$

Here x_p and y_q are the leaders of org_{i1} and org_{i2}, respectively. In order to promote the intercommunion between org_{i1} and org_{i2}, we use leaders to cooperate with each other. If $U(0,1)<CS$, two new individuals $x_p^{'}$ and $y_q^{'}$ are generated by formula (4) in cooperative strategy1 and by formula (5) in cooperative strategy 2, where $CS\in(0,1)$ is a predefined parameter.

$$\begin{cases} x_{p,k}^{'} = \beta_k \times x_{p,k} + (1-\beta_k)\times y_{q,k} \\ y_{q,k}^{'} = (1-\beta_k)\times x_{p,k} + \beta_k \times y_{q,k} \end{cases} \quad k=1,2,\cdots,n \tag{4}$$

$$\begin{cases} x_{p,k}^{'} = (x_{p,1}, x_{p,2}, \cdots, x_{p,i1-1}, y_{q,i1}, y_{q,i1+1}, \cdots, y_{q,i2}, x_{p,i2+1}, x_{p,i2+2}, \cdots, x_{p,n}) \\ y_{q,k}^{'} = (y_{q,1}, y_{q,2}, \cdots, y_{q,i1-1}, x_{p,i1}, x_{p,i1+1}, \cdots, x_{p,i2}, y_{q,i2+1}, y_{q,i2+2}, \cdots, y_{q,n}) \end{cases} \tag{5}$$

where $\beta_k = U_k(0,1)$, $1<i_1<n$, $1<i_2<n$, and $i_1<i_2$.

$$\overline{x}_p = \begin{cases} x_p^{'} & Fitness(x_p^{'}) \le Fitness(x_p) \\ x_p & otherwise \end{cases} \tag{6}$$

$$\overline{y}_q = \begin{cases} y_q^{'} & Fitness(y_q^{'}) \le Fitness(y_q) \\ y_q & otherwise \end{cases} \tag{7}$$

Finally, the organization Q is generated, org_{i1} and org_{i2} are deleted from current generation and org_c is added to the next generation.

Self-study Operator. Self-study operator makes full uses of information of leaders and can find better solutions around the organization. Using self-study operator on Q, then the new organization $Z=\{z_1, z_2, \cdots, z_{m1+m2}\}$ is generated.

As q_p and q_{m1+q} are the individuals obtained from cooperation of leaders org_{i1} and org_{i2}, we keep them in organization Z for preserving the cooperative information between leaders. In organization Z, if $U_j(0,1)<AS$, then z_j, $j=1,2,\ldots,m_1+m_2$ $j\neq p$ and $j\neq m_1+q$, is determined by formula (8) in self-study strategy 1, otherwise it is determined by formula (9) in self-study strategy 2, where $z_p=q_p$ and $z_{m1+q}=q_{m1+q}$, and the subscript in $U_j(0,1)$ indicates that the random number is generated anew for each value of j, $AS\in(0,1)$ is a predefined parameter.

$$z_{j,k} = \begin{cases} \underline{x}_k & r_{j,k} < \underline{x}_k \\ \overline{x}_k & r_{j,k} > \overline{x}_k \\ r_{j,k} & otherwize \end{cases} \quad k=1,2,\cdots,n \tag{8}$$

where $r_{j,k}=x_{p,k}+U_k(0,1)\times(x_{p,k}-q_{j,k})$, $j\neq p$ and $j\neq m_1+q$.

$$z_{j,k} = \begin{cases} \underline{x}_k + \mathrm{U}(0,1)\times(\overline{x}_k - \underline{x}_k) & U_k(0,1) < 1/n \\ x_{p,k} & otherwize \end{cases} \quad k=1,2,\cdots,n \tag{9}$$

where $\underline{x}_k$ and $\overline{x}_k$ are the lower and upper bounds of the search space. After z_j is calculated, Z is determined by formula (10).

$$z_j = \begin{cases} z_j & Fitness(z_j) \geq Fitness(q_j) \\ z_j & Fitness(z_j) < Fitness(q_j) \text{ and } \{U_j(0,1) < \exp(Fitness(z_j) - Fitness(q_j))\} \\ q_j & otherwize \end{cases} \quad (10)$$

As can be seen, if z_j is better than q_j, z_j gets into org_c, otherwise, z_j gets into org_c with probability. The more z_j close to q_j, the greater the probability is.

Splitting Operator. Splitting condition is given by (11):

$$| org | > Max_{os} \quad (11)$$

Where $|org|$ stands for the number of individuals in organization org, Max_{os} ($<N$) is the maximum number of organizations, N denotes the individual number of all organizations. If a parent organization, org_i, satisfies formula (11), it will be split into two child organizations, org_{c1} and org_{c2}, as follows: randomly select $|org_i|/3$~$2|org_i|/3$ members from org_i to create org_{c1}, and the rest members of org_i form org_{c2}. Finally, org_i is deleted from current generation and org_{c1} and org_{c2} are added to next generation.

Swarm Velocity Updating Operator. Most of previous empirical developments of PSO are based on different weight factor method. However, Shi and Eberhart [9] suggest that for complex multimodal functions, the control of diversity of population with a linearly varying inertia weight may lead particles to converge to a local optimum prematurely.

Here we keep the previous velocity term at zero. Then the new velocities and the positions of particles for the next generation are calculated using the following equations (12).

$$\begin{cases} v_{i,j}(t+1) = c_1 \times r_{1,j}(t)(y_{i,j}(t) - x_{i,j}(t)) + c_2 \times r_{2,j}(t)(\hat{y}_i(t) - x_{i,j}(t)) \\ x_{i,j}(t+1) = x_{i,j}(t) + v_{i,j}(t+1) \end{cases} \quad (12)$$

Here $y_i(t) = (y_{i,1}, y_{i,2}, \cdots, y_{i,n})$ is the leader of each organization, and $\hat{y}_i(t) = (\hat{y}_{i,1}(t), \hat{y}_{i,2}(t), \cdots, \hat{y}_{i,n}(t))$ is the fittest leader found so far at that time t among organizations.

3.2 The Procedure of Organizational Particle Swarm Algorithm

The procedure is as follows.

Step1: Initialize population P_t with N organizations, and each organization with org$N$ member, $t \leftarrow 0$.

Step2: Reaching termination criteria, output results and stop; otherwise go to Step 3.

Step3: Updating velocity and position of individual in every organization according to formula (12).

Step4: Randomly select two organizations, org_{i1} and org_{i2} from P_t, then perform cooperating combination operator on them to create organization Q.
Step5: To organization Q, perform self-study operator on it to cerate organization Z.
Step6: To Z, if satisfying splitting conditions, formula (11), the splitting operator is performed on it.
Step7: If the number of organizations in P_t is greater than 1, go to Step4, otherwise go to Step8.
Step8: If only one organization left in P_t, judging splitting condition and performing splitting operator on organization, otherwise go to step9.
Step9: Move the organization left in P_t to P_{t+1}, $t \leftarrow t+1$, go to step2.

4 Experimental Studies on Global Numerical Optimization

In order to test the validity of OPSA, 6 benchmark functions are adopted, and the results are compared with PSO_TVIW, MPSO_TVAC, HPSO_TVAC and FEP.

4.1 Unconstrained Multidimensional Functions

$$f_{01}(\boldsymbol{x}) = \sum_{i=1}^{n} x_i^2, S= [-100, 100]^n, \boldsymbol{x}^*=(0, 0, \cdots, 0), f(\boldsymbol{x}^*)=0$$

$$f_{02}(\boldsymbol{x}) = \sum_{i=1}^{n} |x_i| + \prod_{i=1}^{n} |x_i|, S=[-10, 10]^n, \boldsymbol{x}^*=(0, 0, \cdots, 0), f(\boldsymbol{x}^*)=0$$

$$f_{03}(\boldsymbol{x}) = \sum_{i=1}^{n} \left(\sum_{j=1}^{i} x_j\right)^2, S= [-100, 100]^n, \boldsymbol{x}^*=(0, 0, \cdots, 0), f(\boldsymbol{x}^*)=0$$

$$f_{04}(\boldsymbol{x}) = \sum_{i=1}^{n-1} \left[100(x_{i+1} - x_i^2)^2 + (x_i - 1)^2\right], S= [-30, 30]^n, \boldsymbol{x}^*=(1, 1, \cdots, 1), f(\boldsymbol{x}^*)=0$$

$$f_{05}(\boldsymbol{x}) = \sum_{i=1}^{n} \left[x_i^2 - 10\cos(2\pi x_i) + 10\right], S= [-5.12, 5.12]^n, \boldsymbol{x}^*=(0, 0, \cdots, 0), f(\boldsymbol{x}^*)=0$$

$$f_{06}(\boldsymbol{x}) = \frac{\pi}{n}\left\{10\sin^2(\pi y_1) + \sum_{i=1}^{n-1}(y_i - 1)^2\left[1 + 10\sin^2(\pi y_{i+1})\right] + (y_n - 1)^2\right\}$$
$$+\sum_{i=1}^{n} u(x_i, 10, 100, 4) \quad S= [-50, 50]^n, \boldsymbol{x}^*=(1, 1, \cdots, 1), f(\boldsymbol{x}^*)=0.$$

$$where, u(x_i, a, k, m) = \begin{cases} k(x_i - a)^m & x_i > a \\ 0 & -a \le x_i \le a \\ k(-x_i - a)^m & x_i < -a \end{cases} \qquad y_i = 1 + \tfrac{1}{4}(x_i + 1)$$

4.2 The Experiment Results

The experiment tests the performance of OPSA on functions with 30 dimensions. The termination criterion of OPSA is one of the objectives, $|f_{best} - f_{min}| < \varepsilon$, is achieved, where f_{best} and f_{min} represent the best solution found until the current generation and the global optimum, respectively. To consistent, $\varepsilon = 10^{-4}$ and the maximum generation being 3000 are used for all functions. In the following experiments, the parameter settings are: N=10, $orgN$=10, c_1=c_2=2, Max_{os}=30, AS=0.8, CS=0.6. The results are compared with PSO_TVIW, MPSO_TVAC, HPSO_TVAC, and FEP.

Table 1. Comparisons of PSO_TVIW, MPSO_TVAC, HPSO_TVAC, FEP and OPSA

f	f_{min}	Mean value / standard deviation / mean function values				
		PSO_TVIW	MPSO_TVAC	HPSO_TVAC	FEP	OPSA
f_{01}	0	9.114×10^{-5} (7.811×10^{-6}) (232 208)	9.343×10^{-5} **(5.304×10^{-6})** (125 004)	5.108×10^{-4} (3.0×10^{-3}) (51 274)	5.7×10^{-4} (1.3×10^{-4}) (150 000)	**8.729×10^{-5}** (1.602×10^{-5}) **(24 817)**
f_{02}	0	9.383×10^{-5} **(5.195×10^{-6})** (233 746)	9.276×10^{-5} (1.167×10^{-5}) (136 845)	1.440 (2.242) (290 844)	8.1×10^{-3} (7.7×10^{-4}) (200 000)	**8.285×10^{-5}** (1.911×10^{-5}) **(33 030)**
f_{03}	0	299.0044 (218.5204) (300 000)	2.6×10^{-3} (4.6×10^{-3}) (295 140)	720.726 (342.040) (300 000)	1.6×10^{-2} (1.4×10^{-2}) (500 000)	**9.781×10^{-5}** **(1.681×10^{-6})** **(125 090)**
f_{04}	0	42.3739 (35.3131) (300 000)	40.499 (30.902) (303 000)	47.429 (42.062) **(300 000)**	5.06 (5.87) (2 000 000)	**1.288×10^{-4}** **(1.518×10^{-4})** (427 724)
f_{05}	0	32.4158 (7.0615) (300 000)	56.454 (13.962) (303 000)	23.387 (10.355) (300 000)	4.6×10^{-2} (1.2×10^{-2}) (500 000)	**8.648×10^{-5}** **(3.601×10^{-5})** **(44 231)**
f_{06}	0	3.73×10^{-2} (1.064×10^{-1}) (272 926)	1.265×10^{-1} (2.298×10^{-1}) (201 219)	6.2×10^{-3} (2.49×10^{-2}) (140 088)	9.2×10^{-6} (3.6×10^{-6}) (150 000)	**8.283×10^{-7}** **(2.030×10^{-7})** **(26 850)**

As can be seen from table 1, PSO_TVIW、MPSO_TVAC and HPSO_TVAC can only find satisfied solution to several functions. So the algorithms have no robust. And FEP needs more computational cost than OPSO. To sum up, OPSO has better performance on numerical optimization problems. It can not only obtain the better solution but also need less mean number of function values with little standard deviation.

5 Analyses of OPSA with Different Parameters

In this subsection, the impact of different parameters, viz. Maxos, AS and OS, on the performance of OPSA is analyzed. With the restricted length of paper, we only tested on function 6 and the condition of the experiment is the same to above.

5.1 The Analysis of OPSA with Different Maxos

Parameter Maxos is used in splitting operator. In order to analysis the effect of parameter Maxos to the algorithm performance, it is sampled from [5,100] in step 5, so 20 parameter values are obtained for function 6. 50 trials are carried out at each parameter value. Here we use average success ratio to depict the effect of Maxos to the algorithm.

The definition of average succession ratio, labeled as *Ras* is given in definition 1:

Definition 1. If a trial satisfies (13), and then the trial is called success, otherwise failure:

$$\text{If} \quad f^* \neq 0 \text{ then} \left|f^* - f_{best}\right| < \varepsilon \cdot \left|f^*\right|, \text{ else } \left|f_{best}\right| < \varepsilon \tag{13}$$

Where f_{best} stands for the best solution found by trial, and f^* is the global optimum of the function. Suppose *Ms* out of *M* trials succeed, then average success ratio is given as *Ras=Ms/M*.

Experimental result is shown in figure 1. It illuminate that OPSA have good performance with different Maxos. On the whole, although the best value of Maxos is different for varied functions, OPSA can achieve high average success ratio and perform stably at a large range of parameter values.

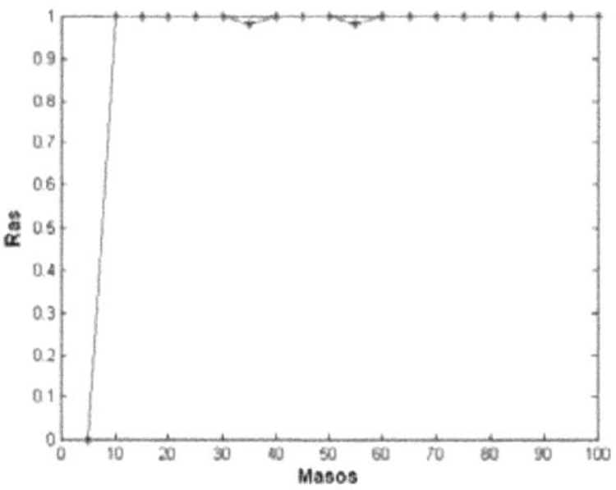

Fig. 1. Relationship between Maxos and *Ras*

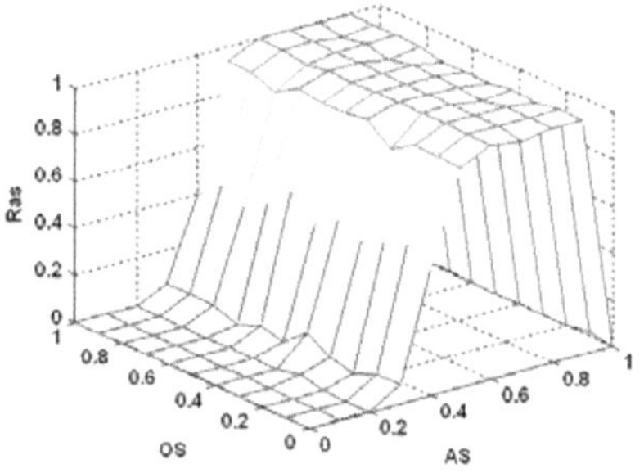

Fig. 2. Relationship between *AS*, *CS* and *Ras*

5.2 Analysis of OPSA with Different *AS* and *CS*

Parameters *AS* and *CS* control the probabilities of using self-study and cooperative combination operators. In experiment, *AS* and *CS* are sampled from [0, 1] in step of 0.1, so 11×11=121 groups of parameter values are obtained for each function. 50 trials are carried out at every group of parameters, and the relationship between *AS*, *CS* and *Ras* is depicted in Fig.2.

From the result we can see that the effects of varied AS to the success ration is larger than *CS*. OPSA performs better of combining two strategies together. On the whole, although the best values of *AS* and *CS* are different for different functions, OPSA can achieve to a high average success ratio and perform stably at a large range of parameter values.

6 Conclusion

OPSA is proposed to solve numerical optimization problems in this paper. By competition and cooperation of organizations in every generation, particles can adapt the environment better, and the algorithm can converge to global optimizations. In experiments, OPSA is tested on 6 unconstrained benchmark problems, and the results are compared with PSO_TVIW, MPSO_TVAC, HPSO_TVAC and FEP. The experimental results show that OPSA can stably find optimal or satisfied solutions. Finally, the affection of random parameters to the algorithm performance is analyzed. The results indicate that the proposed method not only has high stability and fast convergence rate but also is not sensitive to parameters.

References

1. Kennedy J. and Eberhart R.C. Particle swarm optimization. In Proceedings of IEEE International Conference on Neural Networks, volume IV, Perth, Australia, IEEE Service Center Piscataway, NJ. 1995, 1942-1948.
2. Shi Y. and Eberhart R.C. A modified particle swarm optimizer. In IEEE International Conference of Evolutionary Computation. Anchorage, Alaska, 1998, 69-73.
3. Shi Y. and Eberhart R.C. Fuzzy adaptive particle swarm optimization. In Proceedings of the Congress on Evolutionary Computation, Seoul, Korea, 2001.
4. Angeline P.J. Using selection to improve particle swarm optimization. Proc. IEEE Int. Conf. on Evolutionary Computation. Anchorage, 1998, 84-89.
5. Liu Jing, Zhong Weicai, Liu Fang, Jiao Licheng. Numerical optimal using organizational evolutionary algorithm. In: the Proceedings of the Fifth International Conference on Computational Intelligence and Multimedia Applications. IEEE Computer Society Publisher, Xi'an, China, Sept. 2003: 284-289.
6. Ratnaweera, A. Halgamuge, S.K. Watson, H.C.Self-organizing hierarchical particle swarm optimizer with time-varying acceleration. Evolutionary Computation, IEEE Transactions on Volume 8, Issue 3, June 2004 Page(s):240-255.
7. Xin Yao,Yong Liu, Guangming Lin. Evolutionary programming made faster. IEEE Trans. Evolutionary Computation. 1999, 3(2):82-102.
8. J.R. Wilcox, Organizational learning with a learning Classifier System, IlliGAL Report NO.95003, 1995.
9. Shi Y. and Eberhart R.C. Parameter selection in particle swarm optimization. Lecture Notes in Computer Science-Evolutionary Programming VII, vol. 1447. Proc. 7th Int. Conf. Evolutionary Programming, Mar. 1998, pp. 591-600.

A Hybrid Discrete Particle Swarm Algorithm for Open-Shop Problems

Qingyun Yang[1,2], Jigui Sun[1,2,3], Juyang Zhang[1,2], and Chunjie Wang[4]

[1] College of Computer Science and Technology, Jilin University, Changchun, 130012, China
[2] Key Laboratory for Symbolic Computation and Knowledge, Engineering of Ministry of Education, Jilin University, Changchun, 130012, China
[3] Open Laboratory for Intelligence Information Processing, Fudan University, Shanghai, 200433, China
[4] Basic Sciences of ChangChun University of Technology, ChangChun University of Technology, Changchun, 130012, China
qyyang_jlu@yahoo.com.cn, jgsun@jlu.edu.cn, sl_zj@vip.sina.com, chunjie_wang@yahoo.com.cn

Abstract. A hybrid discrete particle swarm algorithm is presented in this paper to solve open-shop problems. The operations are redefined in the discrete particle swarm algorithm. To improve the performance the simulated annealing algorithm is combined with discrete particle swarm. We use SA to enhance the results of local best positions instead of current positions. The experimental results show that our hybrid discrete particle swarm algorithm is effective and efficient to solve open-shop problems.

1 Introduction

Particle Swarm Optimization (PSO) is inspired by the behavior of bird flocking and fish schooling, which is a stochastic search technique based on the simulation of social behavior metaphor. PSO exploits a population of potential solutions to probe the search space. It initializes the population with random candidate solutions, called particles. Each particle is assigned a randomized velocity and is iteratively moved through the search space. It benefits from the experience of its own and that of the other members of the population.

Since its developed by Kennedy and Eberhart[1], PSO is widely used as optimizer for continuous nonlinear functions. Many experimental results denote that PSO is a excellent stochastic search method on continuous numeric optimization problems. Recently PSO is used to solve scheduling problems[2],[3],[4],[5],[6], [7] and is proved its efficiency.

The classical open shop problem (OSP) is a well-known scheduling problem. It involves a collection of n jobs $J_1, J_2, \ldots, J_n$ and m machines $M_1, M_2, \ldots, M_m$. Each job J_i comprises of a collection of m operations (sometimes called tasks) $O_{i1}, O_{i2}, \ldots, O_{im}$ where O_{ij} has a proceeding time of $p_{ij} > 0$ and has to be processed on machine M_j. Furthermore the operations may not be interrupted,

T.-D. Wang et al. (Eds.): SEAL 2006, LNCS 4247, pp. 158–165, 2006.

i.e. preemption is not allowed in their processing time. Each machine can process at most one operation at a time and each job can be processed by at most one machine at any given time. The operations of a job can be processed in any order. The objective of OSP is to obtain a feasible combination of the machine and job orders, i.e. a schedule with the minimum makespan. OSP is NP-hard for $m \geq 3$[8]. Some branch and bound methods as well as some hybrid branch and bound methods are published to solve that problem[9],[10],[11]. Incomplete algorithms are also developed to solve OSPs[12],[13],[14],[15].

In this paper, a discrete particle swarm algorithm combining with simulated annealing algorithm has been proposed for solving a set of benchmark OSPs. The remainder of this paper is organized as follows:Section 2 provides the definition of operators in discrete particle swarm algorithm and the brief overview of simulated annealing algorithm. Then the hybrid algorithm is presented. Experimental results are given in Section 3. Section 4 is the conclusion.

2 Hybrid Discrete Particle Swarm for Open-Shop Problem

Particle swarm is a general heuristic exploration technique which maintains a swarm of candidate solutions, referred to as *particle*. Particle swarm performs effective exploration through memory and feedback. Particles flow through hyper-dimensional search space, attracting towards the best position found by the neighborhood particle and the best historic position by themselves.PSO was fist applied to optimize continuous nonlinear function. Many applications are published in recent years. Later, PSO has been adapted to solve discrete optimization problems[16]. A permutation discrete PSO has been proposed by Clerc[17] and its applications are reported[7],[18].

2.1 Discrete Particle Swarm

Supposing that the search space is N-dimensional, then the i-th particle in the swarm could be presented with a N-dimensional vector, $X_i = (x_{i1}, x_{i2}, \ldots, x_{iN})$. The velocity of this particle can be presented as $V_i = (v_{i1}, v_{i2}, \ldots, v_{iN})$. The best historic position visited by the i-th particle (i.e. local best position) can be denoted as $P_i = (p_{i1}, p_{i2}, \ldots, p_{iN})$. g is defined as the index of the best neighborhood. For the popular PSO, the main manipulation according to the following two equations:

$$v_{ij}^{t+1} = \omega v_{ij}^{t} + c_1 r_{1j}(p_{ij} - x_{ij}^{t}) + c_2 r_{2j}(p_{gj} - x_{ij}^{t}) \quad (1)$$

$$x_{ij}^{t+1} = x_{ij}^{t} + v_{ij}^{t+1} \quad (2)$$

where i = 1, 2, ..., S, and S is the size of the swarm; j = 1, 2, ..., N; t = 1, 2, ..., is the number of iteration; ω is the inertia weight; c_1 and c_2 are acceleration coefficients; $r_{1j}, r_{2j} \sim U(0,1)$.

New velocity of a particle at time $(t+1)$ is determined by three components: current velocity v_i^t; the historic position p_i and the global best position p_g. After the update of velocity, the position at time $(t+1)$ is modified with equation (2). In continuous PSO, current velocity v_i^t is a real value vector but in the discrete particle swarm algorithm we use transpositions as a substitute for a real value vector. Just as described in [17].

$$v_i = \{(x_{ij} \rightarrow x'_{ij})\}; j = 1, \ldots, L \tag{3}$$

where x_{ij} and x'_{ij} are operations of OSP position values, and L represents the length of the list or the maximum number of possible permutation which is randomly generated between 1 and dimension size N.

Here v_i refers to exchange of operation positions $(x_1 \rightarrow x'_1)$, then $(x_2 \rightarrow x'_2)$, etc. In discrete particle swarm algorithm, v_i prefers to generate randomly instead of reserved in last iteration. After generating the velocity index v_i, we can compute the velocity at time $(t+1)$ with formulation (1). The operators presented in equation (1) are redefined below:

Subtraction (position - position) operator: the subtraction of two positions produces a velocity. Suppose x_1 and x_2 are positions, then $x_1 - x_2$ yields a velocity $v = \{(x_1 \rightarrow x_2)\}$. The arrow represents the exchange of position, i.e. when v is applied to a particle. We exchange the same values in x_2 and in this particle position with the corresponding value in x_1.

Addition (velocity + velocity) operator: the addition of two velocity vectors results in a new velocity. Suppose $v_1 = \{(x_1 \rightarrow x'_1)\}$ and $v_2 = \{(x_2 \rightarrow x'_2)\}$, then the addition of v_1 and v_2 we get $v = v_1 + v_2 = \{(x_1 \rightarrow x'_1), (x_2 \rightarrow x'_2)\}$.

Addition (position + velocity) operator: the addition of a position with a velocity results in a position. Let x be the position and $v = \{(x_1 \rightarrow x_2)\}$ be the velocity. New position x' is produced by applying the permutation of v to x, i.e. values in x who equal to x_2 are replaced by x_1.

Multiplication (coefficient × velocity) operator: the multiplication results in a velocity. In our algorithm the acceleration coefficient are set to 1, and inertia weight is set to 1 too. And the r_1, r_2 are still sampled from $U(0,1)$, but now we do not time them with velocity directly, we treat them as probabilistic selection. At the update of velocity we select $(p_i - x_i^t)$ with probability r_1, and $(p_g - x_i^t)$ with probability r_2 respectively.

2.2 Combining Simulated Annealing with DPS

Simulated annealing is an advanced local search method which, introduced by Kirkpatrick, Gelatt and Vecchi[19], inspires the physical annealing process studied in statistical mechanics. Simulated annealing has been applied to many combinatorial optimization problems successfully. It can be viewed as an iterative improvement approach to combinatorial optimization problems. A SA approach repeats iterative small local alterations that look for better solutions while offering the possibility of accepting in a controlled manner worse solutions to escape from local optima.

At each iteration, starting from an initial solution s, a new solution s' in the neighborhood of s is generated randomly. Then a decision is taken to decide whether s' will replace s based on the calculation of $\triangle = f(s') - f(s)$. For a minimization problem, if $\triangle \leq 0$, we move from s to s', otherwise, we move to s' with the probability $e^{-\frac{\triangle}{t}}$, where t is a control parameter called the temperature (higher temperatures lead to higher accepting probabilities and vice versa). Usually SA algorithm starts from a high temperature and decreases gradually. The algorithm will stop until the termination criterion is reached.

In SA the temperature is controlled by a cooling schedule specifying how the temperature should be gradually decreased. The cooling schedule is specified by several parameters generally, namely the initial temperature T0, the epoch length L, the rule of reducing the temperature, and the termination criterion. T0 should be high enough so that all possible configurations have the equal chance to be visited. To prevent too long from running, the temperature should be reduced in a certain way.The method specifies the temperature with $t_k = b \times t_{k-1}$, during the kth epoch ($k = 1, 2, \ldots$), where b is a parameter called decreasing rate, with a value less than 1. The termination condition in our algorithm is near zero, in which a termination temperature t_{end} is set. When current temperature $t_k < t_{end}$ the algorithm will be terminated.

The performance of the SA depends on the choice of the neighborhood mainly. Variable neighborhood search (VNS) method presented in [20] is used in our algorithm, which employs two neighborhood structures: remove the operations at the uth dimension and insert it in the vth dimension (*insert*); exchange two operations between the uth and vth dimension (*exchange* or *interchange*). The interchange method is the same as the exchange in DPS algorithm and we can apply the exchange method to SA directly. In our algorithm, SA is used to search the particles' historic positions (local best positions) instead of the current positions.

The pseudocode of the entire hybrid algorithm is as follows:

```
sa(Particle p) {
  /*T0 is the initial temperature, tend is the end condition
  temperature, b is the decreasing rate*/
  t = T0
  kcount=0
  max_method=2
  while(t > tend) {
    for i = 0 to N {
      kcount =0
      while (kcount < max_method) {
        u = rand() % N
        v=rand()%N
        if(kcount == 0) {
          p' = insert(p, u, v)
        }
        if(kcount == 1) {
```

```
        p' = interchange(p, u, v)
      }
      if(fitness(p') <= fitness(p)) {
        p moves to p'
        kcount=0
      }
      else {
        with probability of exp(-(fitness(p')-fitness(p))/t)
          p moves to p' and kcount=0
        else kcount++
      }
    }
  }
  t = b * t
 }
}

main procedure
 initialize positions with inequality each other
 while(unsatisfy termination criterion) {
   compute the makespan of each particle
   determine gbest and pbest
   for i=0 to popSize {
     sa(particle i's pbest position)
   }
   update velocity
   update position
 }
```

3 Experimental Results

We test the algorithm on a large set of benchmark problem instances have been generated by Taillard[21][1]. All the test instances are quadratic of size n jobs and n machines, with n ranging from 4 to 20.

In our study the parameters of hybrid DPS are as follows: the population size is set to 30, the probability selection of v_i is set to $3/N$, In our test the algorithm stops when it has found a solution or when it has performed 10^5 evaluations in each case. 10^5 evaluations are considered no more solution can be found and are excluded from experimental results. 50 runs are performed for each instance. The parameter setting for SA including: the initial temperature $T0 = 2$, the terminating criterion temperature $t_{end} = 0.02$, the epoch length L is set to N (where N is the number of operations), and decreasing rate $b = 0.95$.

[1] http://ina2.eivd.ch/Collaborateurs/etd/problemes.dir/ordonnancement.dir/ordonnancement.html

Table 1. Partial test results for Taillard benchmark problems

Name	Problem Instances			Hybrid DPS	
	Size	Current best	Lower Bound	Best	Max
tai4_41	4×4	193	186	193	193
tai4_42	4×4	236	229	236	236
tai4_43	4×4	271	262	271	271
tai4_44	4×4	250	245	250	250
tai4_45	4×4	295	287	295	295
tai5_51	5×5	300	295	300	300
tai5_52	5×5	262	255	262	262
tai5_53	5×5	328	321	**323**	328
tai5_54	5×5	310	306	310	310
tai5_55	5×5	329	321	**326**	329
tai7_71	7×7	438	435	**435**	438
tai7_72	7×7	449	443	**446**	449
tai7_73	7×7	479	468	**473**	479
tai7_74	7×7	467	463	**463**	467
tai7_75	7×7	419	416	**416**	419
tai10_101	10×10	652	637	**649**	652
tai10_102	10×10	596	588	**594**	596
tai10_103	10×10	617	598	**614**	617
tai10_104	10×10	581	577	**578**	581
tai10_105	10×10	657	640	**652**	657
tai15_151	15×15	956	937	**952**	956
tai15_152	15×15	957	918	**953**	957
tai15_153	15×15	899	871	**897**	899
tai15_154	15×15	946	934	**942**	946
tai15_155	15×15	992	946	**989**	992
tai20_201	20×20	1215	1155	**1211**	1215
tai20_202	20×20	1332	1241	**1329**	1332
tai20_203	20×20	1294	1257	**1292**	1294
tai20_204	20×20	1310	1248	**1307**	1310
tai20_205	20×20	1301	1256	**1299**	1301

Table 1 reports the partial results of the hybrid DPS algorithm on Taillard benchmarks. As can be seen in Table 1, our hybrid DPS algorithm obtained the current best solutions for all Taillard benchmarks problems (the Max column in Table 1 denotes the maximal makespan of each instance). The Best column in Table 1 presents the minimal makespan found by our hybrid DPS algorithm. Only for all 4×4 benchmark problems and three 5×5 benchmark problems the Best solutions are equal to the Current best solutions, and for the left benchmark problems the Best solutions found by our hybrid DPS algorithm are superior to the Current best solutions.

4 Conclusion

The results presented in this paper are very encouraging and promising for the application of the hybrid DPS algorithm to Open-shop scheduling problems. Our new hybrid algorithm is very effective and efficient. It can find optima for most test instances, and has less running time. Computational experiments on some famous test sets of benchmark problem instances taken from literature demonstrate the efficiency of this approach. It may be an attractive alternative for solving Open-shop scheduling problems and other optimization problems.

Acknowledgments

This work was supported in part by the National Natural Science Foundation of China under grant no. 60273080 and no. 60473003. This work also was supported in part by the Outstanding Youth Foundation of Jilin province of China under grant no. 20030107.

References

1. Kennedy, J, Eberhart, R. C: Particle swarm optimization. In: IEEE Int. Conf. on Neural Networks, Perth, Australia. (1995) 1942–1948
2. M. Fatih Tasgetiren, Mehmet Sevkli, Yun-Chia Liang, and Gunes Gencyilmaz: Particle Swarm Optimization Algorithm for Permutation Flowshop Sequencing Problem. In: Proc. of the 4th International Workshop on Ant Colony Optimization and Swarm Intelligence (ANTS 2004), LNCS 3172, September 5-8, Brussels, Belgium, (2004) 382–390
3. M. Fatih Tasgetiren, Mehmet Sevkli, Yun-Chia Liang, and Gunes Gencyilmaz: Particle Swarm Optimization Algorithm for Single Machine Total Weighted Tardiness Problem. In: Proc. of the 2004 Congress on Evolutionary Computation (CEC'04), June 20-23, Portland, Oregon, (2004) 1412–1419
4. M. Fatih Tasgetiren, Yun-Chia Liang, Mehmet Sevkli, and Gunes Gencyilmaz: Particle Swarm Optimization Algorithm for Makespan and Maximum Lateness Minimization in Permutation Flowshop Sequencing Problem. In: Proc. of the 4th International Symposium on Intelligent Manufacturing Systems (IMS 2004), September 6-8, Sakarya, Turkey, (2004) 431–441
5. Weijun Xia, Zhiming Wu: An effective hybrid optimization approach for multi-objective flexible job-shop scheduling problems. Computers & Industrial Engineering 48 (2005) 409–425
6. Chen Ai-ling, Yang Gen-ke, Wu Zhi-ming: Hybrid discrete particle swarm optimization algorithm for capacitated vehicle routing problem. Journal of Zhejiang University SCIENCE A, 7(4), (2006) 607–614
7. K.Rameshkumar, R.K. Suresh, and K.M.Mohanasundaram: Discrete Particle Swarm Optimization (DPSO) Algorithm for Permutation Flowshop Scheduling to Minimize Makspan. In: Proc. ICNC 2005, LNCS 3612, (2005) 572–581
8. T.Gonzalez and S.Sahni: Open-shop scheduling to minimize finish time. Journal of the Association for Computing Machinery, VOL. 23(4), (1976) 665–679

9. Christelle Guret, Narendra Jussien: Combining AI/OR techniques for solving Open Shop problems. Workshop on Integration of Operations Research and Artifical Intelligence techniques in Constraint Programming (CP-AI-OR), 1999
10. Christelle Guret, N. Jussien and C.Prins: Using intelligent backtracking to improve branch-and bound methods: An application to Open-Shop problems. European Journal of Operational Research, 127 (2000) 344–354
11. U. Dorndorf, E.Pesch and T.Phan-Huy: Constraint Propagation in Open Shop Scheduling. In: Proc. of the 7th International Workshop on Project Management and Scheduling, (2000) 111–113
12. D. Alcaide, J. Sicilia and D.Vigo: A Tabu Search Algorithm for the Open Shop Problem. TOP (Trabajos de Investigacin Operativa), VOL. 5(2), (1997) 283–296
13. S. Khuri and S. R. Miryala: Genetic algorithms for solving open shop scheduling problems. Lecture Notes in Computer Science, 1070, (2001) 10–18
14. Hsiao-Lan Fang, Peter Ross, and Dave Corne: A promising genetic algorithm approach to job-shop scheduling, re-scheduling, and open-shop scheduling problems. In: Proc. of the 5th International Conference on Genetic Algorithms, San Mateo, CA, USA, (1993) 375–382
15. Xu, Z. and Louis, S. J: Genetic algorithms for open shop scheduling and rescheduling. In: Proc. of the ISCA 11th International Conference on Computers and Their Applications, Raleigh, NC, USA, (1996) 99–102
16. J.Kennedy and R.C.Eberhart: A discrete binary version of the particle swarm algorithm. In: International Conference on Systems, Man, and Cybernetics, 1997
17. M.Clerc: Discrete Particle Swarm Optimization: A Fuzzy Combinatorial Black Box. In: http://clerc.maurice.free.fr/pso/Fuzzy_Discrete_PSO/Fuzzy_DPSO.htm
18. M.Clerc: Discrete particle swarm optimization, illustrated by the Traveling Salesman Problem. In: New Optimization Techniques in Engineering. Heidelberg, Germany, (2004) 219–239
19. S.Kirkpatrick, C.D.Gelatt and M.P.Vecchi: Optimization by simulated annealing. Science, Vol. 220(13), (1983) 671–680
20. N.Mladenovic and P.Hansen: Variable Neighborhood Search. Computer and Operations Research, VOL. 24, (1997) 1097–1100
21. E.Taillard: Benchmarks for Basic Scheduling Problems. European Journal of Operations Research, 64, (1993) 278–285

Particle Swarms Cooperative Optimization for Coalition Generation Problem

Guofu Zhang, Jianguo Jiang, Na Xia, and Zhaopin Su

School of Computer and Information,
Hefei University of Technology,
Hefei, 230009 PR China
zhang197933@163.com

Abstract. In this paper, a Particle Swarms Cooperative Optimization is proposed to solve Coalition Generation Problem in parallel manner with each Agent taking part in several different coalitions and each coalition turning its hand to several different tasks. With a novel two-dimensional binary encoding approach, the algorithm performs well on coalition parallel generation. An adaptive disturbance factor is adopted to force swarms getting out of local optimums quickly. Introduced an active-feedback based on island models, the algorithm has a good cooperative searching characteristic. The effectiveness of the proposed algorithm is proved by experiments.

1 Introduction

Coalition is an aggregation in which Agents with certain ability are interdependent and cooperative to accomplish tasks. Forming a coalition, Agents can enhance their ability to solve problems and obtain more income. Accordingly, Coalition is an important manner of Agents coordination and cooperation [1].

Coalition generation, especially multi-task coalition parallel generation, is a hard topic in Multi-Agent System (MAS). It mainly researches how to generate several optimal task-oriented coalitions in dynamic and parallel manner. In order to find the optimal coalition for each task, we must think of all or the mass of possible coalitions. Obviously, it is a very complicated combinatorial optimization problem. To solve the problem, researchers have already put forward a lot of relevant algorithms, such as [2], [3] and [4]. But the condition that each Agent can only take part in a coalition and each coalition can only turn its hand to a task is restricted in existing researches above, which have produced a big waste of resource and ability in MAS. In this paper, a Particle Swarms Cooperative Optimization applied to solving Coalition Generation Problem in parallel manner with each Agent taking part in several different coalitions and each coalition turning its hand to several different tasks is presented. Using the remnant ability of Agents, the proposed algorithm gains more income and partly reduces the waste phenomena. The effectiveness of the proposed algorithm is proved by experiments.

T.-D. Wang et al. (Eds.): SEAL 2006, LNCS 4247, pp. 166–173, 2006.

The rest of the paper is organized as follows: The details of the Coalition Generation Problem are described in section 2. In section 3, the basic Discrete Particle Swarm Optimization and its variants are reviewed. A Particle Swarms Cooperative Optimization is proposed for solving coalition parallel generation in section 4. Experiments are made to illustrate the performance of the proposed algorithm in section 5. And a conclusion is drawn in section 6.

2 Coalition Generation Problem

The Coalition Generation Problem can be described in formalization as follows:

The MAS contains n Agents ($A_1...A_n$). Each Agent has certain ability which can be expressed as an ability vector. For example, the ability vector of A_i is $B_A^i=[b_1^i...b_r^i]$, which shows its r kinds of ability. At the same time, the task t_j has the ability demanded vector $D_t^j=[d_1^j...d_r^j]$. A group of Agents in MAS can form a coalition $C_k=\{A_k^i...A_k^j\}$ through necessary negotiation to finish tasks together. The coalition's ability vector is the sum of the ability vectors of all its members, that is $B_C^k=\sum_{A_i\in C_k} B_A^i=[b_1^k...b_r^k]$. The essential condition for the coalition C_k to finish the task t_j is $d_i^j\leq b_i^k$ ($1\leq i\leq r$). After finishing the task collectively, the coalition members obtain their income:

$$\nu_k = \mu(t_k) - \tau(C_k) - \varepsilon(C_k) \tag{1}$$

Where $\mu(t_k)$ is the guerdon paid for finishing the task t_k, $\tau(C_k)$ is the cost of all members' ability, $\varepsilon(C_k)$ is the communication spending.

To realize coalition parallel generation, m coalitions must be searched parallel according to m tasks with the purpose of maximizing the income of the whole system, that is $\nu_{MAS}=\max\sum_{k=1}^{m}\nu_k$. The number of the possible coalitions in MAS which contains n Agents is 2^n. It is very hard to search the whole space to get each optimal coalition. Therefore, we propose a Particle Swarm Cooperative Optimization which can form some task-oriented coalitions parallel with the maximal or approximately maximal income under the restriction conditions above and finishing within an acceptable time.

3 Discrete Particle Swarm Optimization

The Discrete Particle Swarm Optimization (DPSO) was first proposed by Kennedy and Eberhart in 1997 [5] based on the standard Particle Swarm Optimization (PSO) [6] [7], which is a novel nature-inspired evolutionary computing algorithm. In the description of DPSO, the swarm is made up of a certain number of particles. At each iteration, all the particles move in the problem space to find the global optima. The velocity and position of each particle is adjusted by the following formulas:

$$v_{id}^{t+1} = w\cdot v_{id}^t + c_1\cdot r_1\cdot(P_{id} - x_{id}^t) + c_2\cdot r_2\cdot(P_{gd} - x_{id}^t) \tag{2}$$

$$x_{id}^{t+1} = \begin{cases} 1, \text{ if } \rho_{id}^{t+1} < s(v_{id}^{t+1}), \\ 0, \text{ otherwise.} \end{cases} \tag{3}$$

Where variable i denotes the i-th particle in the swarm, t represents the current iteration number, d is the d-th dimensional value of the vector, v_i is the velocity vector of the i-th particle, x_i is the position vector of the i-th particle, and P_i is the local best position that the i-th particle had reached, P_g is the global best position that all the particles had reached, w is called the inertia weight, c_1 and c_2 are two constant numbers, which are often called the cognitive confidence coefficients, r_1 and r_2 are two random numbers between 0 and 1, ρ_i is a quasi-random number selected from a uniform distribution in [0.0, 1.0], s(v) is a sigmoid limiting transformation function (s(v)=$1/(1+e^{-v})$).

Generally, $v_{\max}$ is retained, that is $|v_{id}^{t+1}| < v_{\max}$, which simply limits the ultimate probability that bit x_{id}^{t+1} will take on a binary value. A smaller $v_{\max}$ will allow a higher mutation rate.

4 Particle Swarms Cooperative Optimization

Memory is an important characteristic in the DPSO, which has a different mechanism of sharing information contrast to the Genetic Algorithm (GA). In the GA, chromosomes share information with each other so that the whole swarm can move to the optimal area symmetrically. But in the DPSO, it is only the global best particle that can broadcast information to the other particles. Quickly the other particles converge at local minimum area in the course of following the global best particle, so it is easy to get the local least value. To improve the global searching ability, we propose a modified DPSO called "Particle Swarms Cooperative Optimization" (PSCO) based on island models. The PSCO contains $S(S \geq 2)$ unaided swarms with each swarm containing M particles.

4.1 Two-Dimensional Binary Encoding

In this paper we design a novel two-dimensional binary encoding approach applied to the PSCO. For example, if there are m tasks and n Agents, the dimension of each particle is $m \times n$. Each row represents a task and each column represents an Agent. Fig.1 shows a matrix representation of a particle x_i in the swarm. If $a_{ij} = 1$, the Agent A_j will take part in the coalition C_i with responsibility for the task t_i.

Here an Agent can take part in several different coalitions, so it is possible that $\sum_{i=0}^{m-1} a_{ij} > 1$ and "lawless particles" which are unfeasible combinations come into being in the course of particles moving continuously, that is an Agent A_j with its narrow ability may not support several coalitions synchronously with responsibility for the corresponding tasks. Then those "lawless particles" must be adjusted as follows:

- if $a_{pj} = 1$ and $a_{qj} = 1$, then B_{rest}^{p}=D_t^p-$\sum_{a_{pk}=1}^{k\neq j} B_A^k$, B_{rest}^{q}=D_t^q-$\sum_{a_{qk}=1}^{k\neq j} B_A^k$,
- if $B_{rest}^{p} \leq 0$ and $B_{rest}^{q} \leq 0$, then $a_{pj} = 0$ and $a_{qj} = 0$,
- if $B_{rest}^{p} + B_{rest}^{q} \leq B_A^j$, then do nothing,
- if $B_{rest}^{p} > B_{rest}^{q}$, then $a_{pj} = 1$ and $a_{qj} = 0$,
- if $B_{rest}^{q} > B_{rest}^{p}$, then $a_{qj} = 1$ and $a_{pj} = 0$,
- if $B_{rest}^{q} = B_{rest}^{p}$, then one selected optionally is 1 and another is 0.

Coalition \ Agent	0	1	...	i	...	n-1
0	0	0	...	0	...	1
1	1	0	...	1	...	0
...	...	...	...	...	...	...
m-1	0	1	...	0	...	0

Fig. 1. Particle x_i

4.2 Cooperative Searching Based on Active-Feedback

Enlightened by the Ant Colony Optimization (ACO), we introduce an active-feedback in the PSCO in order to improve the cooperative searching characteristic of the PSCO. Each swarm searches separately for the optimal position and modifies its particles with the best position P_g that it had reached. Then the global best position P_g^* can be fined in the S swarms and be used to replace the worst position in each swarm. But the P_g each swarm itself had reached will not be modified by the P_g^*. Using the P_g^* after each iteration to have an impact on each swarms evolution, but not changing blindly the moving direction of particles, the PSCO can urge particles to approach to the global optimal area gradually.

4.3 Adaptive Disturbance Factor

Suitable selection of inertia weight w in equation (2) provides a balance between global and local exploration and exploitation, and on average results in less iterations required to find a sufficiently optimal solution. As originally developed, w often decreases linearly from about 0.9 to 0.4 during a run. In general, the inertia weight w is set according to the following equation [4]:

$$w = w_{\max} - t \cdot \frac{w_{\max} - w_{\min}}{t_{\max}} \tag{4}$$

Here $t_{\max}$ is the maximum iteration number. To avoid the algorithm getting in the local minimum area easily, an adaptive disturbance factor "g" is added to force swarms getting out of local optimums quickly. It acts as follows:

$$w = w_{\max} - t \cdot (1 - \frac{g}{2g_{\max}})^g \cdot \frac{w_{\max} - w_{\min}}{t_{\max}} \tag{5}$$

Here "g" is adjusted according to the current optimal solution at each iteration (see equation (6)), $g_{\max}$is a constant number.

$$x_{k,i,j}^{t+1} = \begin{cases} 0, & \text{if } P_g \text{ is still evolving,} \\ g+1, & \text{if } P_g \text{ has't improved for N iterations and } (g+1) \leq g_{\max}, \\ g_{\max}, & \text{otherwise.} \end{cases} \quad (6)$$

Obviously, in the initial stage of the algorithm running, the solution gained by the algorithm is still evolving, so the w decreases linearly according to the equation (4). When the solution hasn't improved for N iterations and the local minimum may occur, the disturbance factor "g" will act. It will remarkably reduce the importance of t in the equation (5) and accordingly improve the importance of $w_{\max}$, thereby, particles will explore new wider space, and the new solutions will likely be detected. If the solution hasn't improved yet, the "g" will act much more, increase the disturbing effect and make the solution jump out of the local minimum area more easily. At the same time, in order to ensure the constringency speed of the algorithm, the constant number $g_{\max}$is introduced to control the intensity of the disturbance. Once the solution jumps out of the local minimum area, and the solution begins to evolve again, g=0, the disturbance factor will not work again and come into "latent period".

4.4 PSCO for Multi-task Coalition Parallel Generation

The fitness function is mainly used to provide a measure of how the particles perform in the problem domain. Therefore, in the PSCO, we define the fitness function shown in equation (1), where ν_k is the function of particle x_i. Using the PSCO to solve the multi-task coalition parallel generation in Coalition Generation Problem, the main steps are described as follows:

- Step1. Generate randomly a population of dissimilar and feasible particles for S swarms. Set g=0.
- Step2. Calculate each particle's fitness value, and compare each particle's fitness value with its P_j. The particle who owns the best fitness value among P_j is set to be P_g.
- Step3. Modify each particle's velocity according to equation (7), which is a three-dimension mathematical equation:

$$v_{k,i,j}^{t+1} = w \cdot v_{k,i,j}^{t} + c_1 \cdot r_1 \cdot (P_{k,i,j} - x_{k,i,j}^{t}) + c_2 \cdot r_2 \cdot (P_{g,i,j} - x_{k,i,j}^{t}) \quad (7)$$

- Step4. If $v_{k,i,j}^{t+1} \geq v_{\max}$, then $v_{k,i,j}^{t+1} = v_{\max}$, if $v_{k,i,j}^{t+1} < -v_{\max}$, then $v_{k,i,j}^{t+1} = -v_{\max}$.
- Step5. Modify each particle's position according to equation (8):

$$x_{k,i,j}^{t+1} = \begin{cases} 1, \text{ if } \rho_{k,i,j}^{t+1} < s(v_{k,i,j}^{t+1}), \\ 0, \text{ otherwise.} \end{cases} \quad (8)$$

- Step6. Test whether every new particle is a feasible combination. If it is a "lawless particle", adjust it according to the approach presented in section 4.1. If its new fitness value of last particle is better than the previous P_j, then the current particle is set to be the new P_j. Subsequently, if the best P_j is better than P_g, the new P_j is set to be the new P_g.
- Step7. Compare swarms' P_g with each other and find the global best particle P_g^* in the S swarms. Use P_g^* to replace the worst particle in each swarm. Adjust g according to equation (6).
- Step8. If the maximum iteration number is reached, then output the maximal income and each optimal coalition with responsibility for each task and end the whole program. Otherwise, go to Step3.

5 Experimental Results

The proposed algorithm is experiment on 20 Agents and 4 tasks contrast to the basic DPSO and GA. In the PSCO algorithm, the number swarms are 4 with 20 particles in each swarm, c_1 and c_2 are 1.8, $g_{\max}$ is 10, $t_{\max}$ is 500.

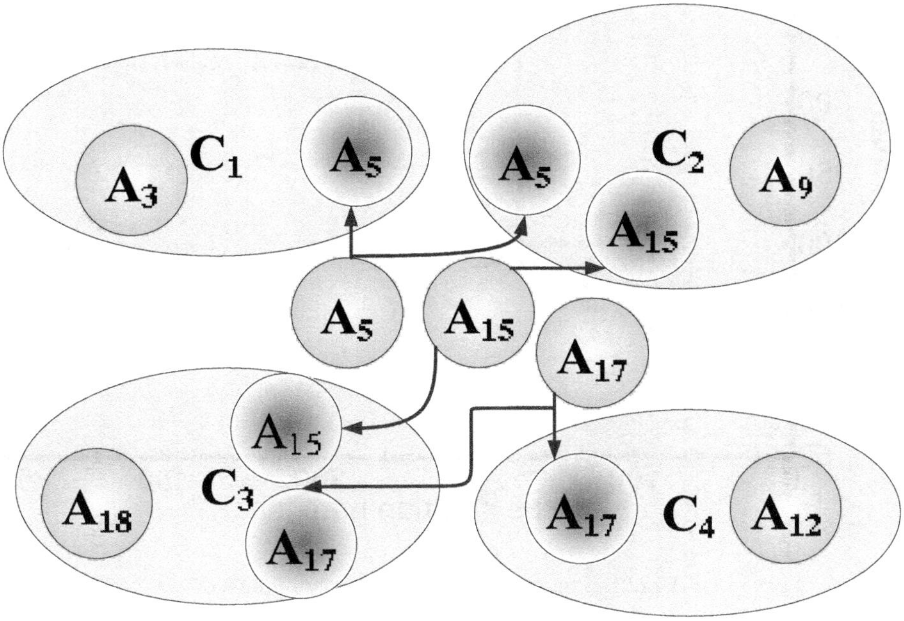

Fig. 2. Coalitions generated by the PSCO

As can be seen, the income of the whole system obtained using the PSCO is superior to the basic DPSO and GA (see Table 1). Using the residual abilities of Agents, the PSCO gains more income and partly reduces the waste phenomena(see Fig.2): Agent A_5 takes part in coalitions C_1 and C_2, Agent A_{15} takes

Table 1. The performance comparison of three algorithms

Algorithm	ν_1	ν_2	ν_3	ν_4	ν_{MAS}
GA	124	75	56	72	327
DPSO	96	59	47	54	256
PSCO	131	94	85	112	422

part in coalitions C_2 and C_3, Agent A_{17} takes part in coalitions C_3 and C_4. Simultaneously, the PSCO gets out of local optimums more quickly and easily than the standard DPSO (see Fig.3), thus verifying that the PSCO has better quality of solution and convergence characteristic.

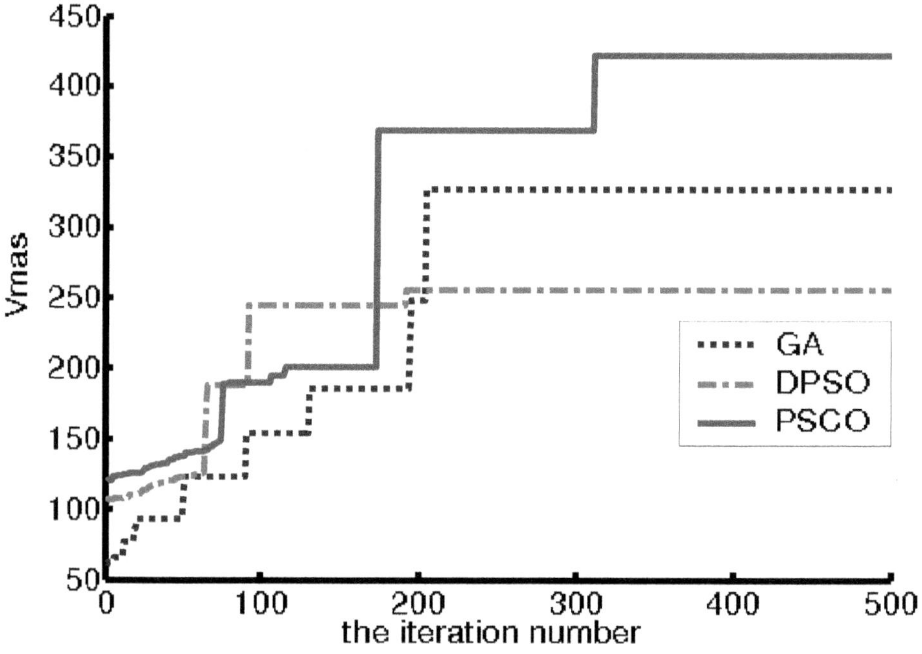

Fig. 3. The optimal solution evolving curve

6 Conclusions and Future Works

This paper has proposed a PSCO based on the DPSO for multi-task coalition parallel generation in Coalition Generation Problem with each Agent taking part in several different coalitions and each coalition turning its hand to several different tasks. Firstly, a novel two-dimensional binary encoding approach is adopted to represent the position and velocity of the particles in DPSO and the operators in the original DPSO formulas are redefined. Then an adaptive disturbance factor and an active-feedback based on island models are used to modify the algorithm. The experimental results demonstrate that the algorithm performs well on Coalition Generation Problem due to its robustness.

In future works, we need to select a better improved method in our PSCO to reduce the computational time.

Acknowledgment

This work is supported by the grants from the NSFC (No.60474035).

References

1. Shehory, O., Kraus, S.: Task allocation via coalition formation among autonomous agents. In Proceedings of the Fourteenth International Joint Conf. on artificial Intelligence, Mont-real, Canada, Morgan Kaufman (1995) 655–661
2. Sen, S., Dutta, P.: Searching for optimal coalition structures. In Proceedings of the 4th ICMAS, Boston, MA, IEEE Press (2000) 287–292
3. Na, X., Jianguo, J., Yalin, H.: Solution to agent coalition problem using improved ant colony optimization algorithm. Intelligent Agent Technology, Beijing (2004) 475–478
4. Zhenghu, L.: Research on key issues of mobile Agent system. PhD Thesis, department of Computer and Information Science, Hefei University of Technology, China (2002) 81–98
5. Kennedy, J., Eberhart, R.: A discrete binary version of the particle swarm optimization algorithm. In Proceedings of Conf. On Systems, Man, and Cybernetics, Piscataway (1997) 4104–4109
6. Kennedy, J., Eberhart, R.: Particle swarm optimization. In Proceedings of Conference on Neural Networks (1995) 1942–1948
7. Eberhart, R., Kennedy, J.: A new optimizer using particle swarm theory. In Proceedings of the 6th International Symposium on Micro Machine and Human Science (1995) 39–43
8. Yuhui, S., Eberhart, R.: Parameter selection in particle swarm optimization. In Proceed-ings of the 7th Annual Conf. on Evolutionary Programming (1998) 591–600

A Novel Multi-objective PSO Algorithm for Constrained Optimization Problems*

Jingxuan Wei[1] and Yuping Wang[2]

[1] School of Science, Xidian University, Xi' an 710071, China
wjxjingxuan@yahoo.com.cn
[2] School of Computer Science and Technology, Xidian University, Xi' an 710071, China
ywang@xidian.edu.cn

Abstract. A new approach is presented to handle constrained optimization by using PSO algorithm. It neither uses any penalty function in the proposed PSO algorithms. The new technique treats constrained optimization as a two-objective optimization, one objective is original objective function, and the other is the degree violation of constraints. As we prefer the second objective, a new crossover operator is designed based on the three-parent crossover operator, which will lead the degree violation of constraints to zero. Then, in order to keep the diversity of the swarm and escape from the local optimum easily, we design a dynamically changing inertia weight. The simulation results indicate the proposed algorithm is effective.

1 Introduction

Particle swarm optimization (PSO) originally developed by Kennedy and Eberhart ([1], [2]) is a population based algorithm. PSO is initialized with a population of candidate solution, each candidate solution in PSO is called particle and moves through the search space by two velocity vectors, one of which is associated with its best solution and the other is associated with the best solution of all particles . The PSO has been found to be fast in solving nonlinear, non-differentiable multimodal optimization problems.

Lately, significant effort has been reported in the literature as researchers figure out ways to enhance the PSO algorithm with a constraint handling mechanism. Coath and Halgamuge[3] proposed the feasible solutions method (FSA) and the penalty function method (PFA) to handle constraint in PSO. But, both of these methods have great disadvantages. FSA requires all particles initialized must be inside the feasible region, and PFA requires careful fine tuning of the penalty function parameters.

Recently, some researchers have suggested the use of multi-objective optimization concepts to handle constraints (e.g., [4] [5] [6]). The main idea is to transform

* This work was supported by the National Natural Science Foundations of China (60374063).

T.-D. Wang et al. (Eds.): SEAL 2006, LNCS 4247, pp. 174–180, 2006.

each constraint as an objective function, and transform the constrained optimization problem as a multi-objective optimization problems with (m+1) objectives, where m is the number of constraints. Then pareto dominance is used as a selection criterion, or pareto ranking is used to assign fitness in such a way that nondominated individuals are assigned the highest fitness value. This means all objectives are seen of the same importance. However, there is a serious drawback if all objectives are seen of the same importance. For example, for an infeasible solution which is far away from the feasible region and has a very small objective value for the constrained problem. This solution will have a great fitness and be seen as a good solution according to pareto dominance or pareto ranking. However it is actually a bad solution, and should not survive in the following generations.

In this paper, we proposed a new constraint handling approach that transforms constrained optimization problem of any number of constraints into a two objective optimization problem. In order to overcome the drawbacks mentioned above, a three-parent crossover operator is designed, and in order to escape from the local optimum easily, a dynamically changing inertia weight is designed. The simulation results show the efficiency of the proposed algorithm.

2 Transformation of Constrained Optimization Problem

We consider the following constrained optimization problem

$$\begin{cases} \min & f(x) \\ \text{s.t.} & g_i(x) \leq 0, i = 1, \cdots, m \end{cases} \tag{1}$$

If a point satisfies all constraints of 1, then it is called a feasible solution. Define functions

$$f_1(x) = f(x),$$

$$f_2(x) = \max\{0, g_i(x), i = 1, 2, \cdots, m\}.$$

We have $f_2(x) \geq 0$, and$f_2(x) = 0$, only if x satisfies all constraints. Therefore, problem 1 can be transformed into the following two objective optimization problems.

$$\min\{f_1(x), f_2(x)\} \tag{2}$$

3 A Novel Multi-objective PSO Algorithm for Constrained Optimization Problems

3.1 PSO Algorithm

PSO initialized the flock of birds randomly over the search space, every birds is called a particle. At each generation, each particle adjusts its velocity, based on its best solution p_{best}^t and the best solution of its neighbors g_{best}^t. The original PSO formulae are:

$$V^{t+1} = \omega V^t + c_1 r_1 (p_{best}^t - p_{present}^t) + c_2 r_2 (g_{best}^t - p_{present}^t) \tag{3}$$

$$p_{present}^{t+1} = p_{present}^t + V^{t+1} \tag{4}$$

Where V is the velocity vector, $p_{present}^t$ is the location vector in the t generation, ω is the inertia weight in the range $[0.1, 0.9]$, r_1 and r_2 are the random numbers in $[0, 1]$, c_1 and c_2 are positive constants.

3.2 Selection Strategy

The selection in the proposed algorithm doesn't use the pareto ranking or the concept of pareto dominance. We use the following selection scheme for any two particles.
1. If the second objective values of two particles are equal to zero, we prefer to select the one with the smaller first objective value.
2. If the second objective values of two particles are both nonzero, we prefer to select the one with the smaller second objective value.
3. If the second objective value of one particle is zero, and that of the other is nonzero, we prefer to choose the one with the zero second objective value.

3.3 Three-Parent Crossover Operator

In order to make the produced solutions move toward the feasible region, we design a three-parent crossover operator. This crossover operator is more flexible than the traditional two-parent crossover operator. The detail is as follows:
1. For three particles x^1, x^2, x^3 chosen to undergo the crossover find the center C of the three particles, $C = \frac{x^1+x^2+x^3}{3}$.
2. For every points $x^i (i = 1, 2, 3)$ find its reflecting points, $o^i = (x^i - C) \times (1 + \varepsilon)(i = 1, 2, 3)$, where $\varepsilon > 0$ is called the expand rate.
3. Generate the offspring:
$o = k_1 \times o^1 + k_2 \times o^2 + k_3 \times o^3$ where k_1, k_2, k_3 are random numbers in (0, 1), and $k_1 + k_2 + k_3 = 1$.

The most important advantage of three-parent crossover operator is that the offspring will have greater probability locating near the feasible region and the constraint violation will be smaller than its parents.

3.4 Improved Inertia Weight

From formula 3, we know that when ω is big, the PSO algorithm is of the ability of global exploration and when ω is small, the PSO algorithm is of the ability of local exploitation. The inertia weight is a very important parameters to balance the global and local search. To evaluate the diversity of the swarm, a measure is defined by

$$s = \frac{1}{NL} \cdot \sum_{i=1}^{N} \sqrt{\sum_{d=1}^{n} (p_{id} - \bar{p_d})^2} \tag{5}$$

where N is the swarm size, n is the number of variables, L is the maximum length between the points in the search space, p_{id} is the dth coordinate of the i-th particle, $\bar{p_{id}}$ is the average value of all particles in the coordinate. The bigger the value of s, the better the diversity of the swarm.The smaller the value of s, the more crowded the swarm is. When the swarm is crowded, it becomes difficult for the algorithm to jump out from the local optimum.

Based on these, inertia weight ω can be defined by

$$\omega = 1 - s\omega_s \tag{6}$$

where $\omega_s \in (0.1, 0.2)$ is a randomly generated number. It can be seen from the definition of ω that when particles are crowded, namely s is small, ω will be big in order to increase the ability of global exploration.

4 Proposed Algorithm

Algorithm 1. (A novel multi-objective PSO algorithm for constrained optimization problems)

1. (Initialization) Given swarm size N. Generate initial swarm $P(t)$ randomly, set $t = 0$.

2. (Evolution)For each particle, find its personal best position and the global best position in the generation t, according to the selection strategy in 3.2. Then update each particle's velocity and position according to the formulae 3 to 6. The new swarm is defined as $P(t+1)$.

3. (Crossover)Find the g_{best}^{t+1} in $P(t+1)$ and then calculate the crossover probability $P_c = \frac{1}{1+e^{\alpha\theta}}$, where $\theta = \| g_{best}^{t+1} - g_{best}^{t} \|$, $\alpha > 0$ is a constant. Let $x^1 = g_{best}^{t+1}$, $x^2 = g_{best}^{t}$, and randomly choose x^3 from $P(t+1)$, then generate an offspring o by crossover operator. If o is better than g_{best}^{t+1} , let $g_{best}^{t+1} = o$, set $t = t+1$.

4. (Stop criterion) If stop criterion is satisfied, then stop, g_{best}^{t} is seen as the global optimum; otherwise, go to step2.

5 Simulation Results

Five widely used test functions are chosen from ([7], [8]) and listed below.

F1 [7]:

$$\begin{aligned}
\min: f(x) &= (x_1-10)^2+x_3^4+3(x_4-11)^2+10x_5^6+7x_6^2+x_7^4-4x_6x_7-10x_6 \\
&\quad -8x_7 \\
\text{s. t.}: g_1(x) &= -127+2x_1^2+3x_2^4+x_3+4x_4^2+5x_5 \le 0 \\
g_2(x) &= -282+7x_1+3x_2+x_3+10x_3^2+x_4-x_5 \le 0 \\
g_3(x) &= -196+23x_1+x_2^2+6x_6^2-8x_7 \le 0 \\
g_4(x) &= 4x_1^2+x_2^2-3x_1x_2+2x_3^2+5x_6-11x_7 \le 0 \\
&-10 \le x_i \le 10 (i=1,\cdots,7)
\end{aligned}$$

The known global optimum is $x^* = (2.330499, 1.951372, -0.4775414, 4.365726, -0.6244870, 1.038131, 1.594227)$ and the corresponding value is 680.6300573. The best solution obtained by the proposed algorithm is $x = (2.330489, 1.951392, -0.4775384, 4.365731, -0.6244830, 1.038101, 1.594228)$, and the corresponding value is 680.6281, the constraint violation is 0.0018.

F2 [8]:

$$\begin{aligned}
\min: f(x) &= 100(x_2-x_1^2)^2+(1-x_1)^2 \\
\text{s. t.}: g_1(x) &= -x_1-x_2^2 \le 0 \\
g_2(x) &= -x_1^2-x_2 \le 0 \\
&-0.5 \le x_1 \le 0.5,\ x_2 \le 1.0
\end{aligned}$$

The known global optimum is $x^* = (0.5, 0.25)$, and the corresponding value is 0.25. The best solution obtained by the proposed algorithm is $x = (0.5, 0.2503)$, the corresponding value is 0.25, and the constraint violation is 0.

F3 [8]:

$$\begin{aligned}
\min: f(x) &= -x_1-x_2 \\
\text{s. t.}: g_1(x) &= x_2-2x_1^4+8x_1^3-8x_1^2-2 \le 0 \\
g_2(x) &= x_2-4x_1^4+32x_1^3-88x_1^2+96x_1-36 \le 0 \\
&0 \le x_1 \le 3,\ 0 \le x_2 \le 4
\end{aligned}$$

The known global optimum is $x^* = (2.3295, 3.1783)$, and the corresponding function value is -5.5079. The best solution obtained by the proposed algorithm is $x = (2.3295, 3.1785)$, the corresponding function value is -5.5080, and the constraint violation is 10^{-4}.

F4 [8]:

$$\begin{aligned} \min: & \ f(x) = 0.01x_1^2 + x_2 \\ \text{s. t.}: & \ g_1(x) = -x_1x_2 + 25 \le 0 \\ & \ g_2(x) = -x_1^2 - x_2^2 + 25 \le 0 \\ & \ 2 \le x_1 \le 50,\ 0 \le x_2 \le 50 \end{aligned}$$

The known global optimum is $x^* = (\sqrt{250}, \sqrt{2.5})$, and the corresponding function value is 5.0. The best solution obtained by the proposed algorithm is $x = (11.3637, 2.9825)$, the corresponding function value is 4.2738, and the constraint violation is 0.

For all test functions, we take the following parameters: $N = 100, \alpha = 1, \varepsilon \in (0.4, 0.6), c_1 = c_2 = 2$. In simulation, the maximum generation number is taken to 3000 when $n > 5$, and 60 when $n < 5$. We execute the proposed algorithm (MO-PSO) 10 independent runs for each problem and record the following data: Best objective value (Best), Mean best objective value (Mean) and the worst objective value (worst) on 10 runs. If the constraint violation is no more than 10^{-3}, we said the solution is a feasible solution. We compare our results with the existing ones obtained in [7][8][9]in table 1.

Table 1. Comparison of best, mean, and worst solutions among MO-PSO and other algorithms in [7][8][9]. "NA" represents the corresponding result is not available.

Problem	Optimal	Methods	Best	Mean	Worst
F1	680.630	MO-PSO	680.628	680.6508	680.9644
		RY [7]	680.630	680.671	680.772
		JH [9]	680.632	680.648	680.761
		Algorithm1[8]	NA	NA	NA
		Algorithm3[8]	NA	NA	NA
F2	0.25	MO-PSO	0.25	0.2503	0.2507
		RY [7]	NA	NA	NA
		JH [9]	NA	NA	NA
		Algorithm1[8]	0.25	0.25	0.25
		Algorithm3[8]	0.25	0.25	0.25
F3	-5.5079	MO-PSO	-5.5080	-5.5080	-5.5080
		RY [7]	NA	NA	NA
		JH [9]	NA	NA	NA
		Algorithm1[8]	-5.5080	-5.5074	-5.50679
		Algorithm3[8]	-5.5080	-5.50679	-4.41998
F4	5	MO-PSO	4.2738	4.5616	4.8579
		RY [7]	NA	NA	NA
		JH [9]	NA	NA	NA
		Algorithm1[8]	5.0000	5.00003	5.00048
		Algorithm3[8]	5.0000	5.00002	5.00005

It can be seen from table 1 that, for F1, F3 and F4, the best solutions found are better than the optimal one, but they violate the constraints a little bit, and the constraint violation of them is less than 10^{-3}. Moreover, for F3 and F4, the solutions found in each run by MO-PSO are better than or equal to these found by algorithm1[8] and algorithm3[8]. For F2, the performance of MO-PSO is almost the same as algorithm1. These results indicate MO-PSO is effective.

6 Conclusions

A novel multi-objective PSO for constrained optimization problem is proposed in this paper. The main ideas are as follows: Firstly, in order to decrease the constraint violation of g_{best} , a new crossover operator is designed. Secondly, in order to keep the diversity of swarm and make it easy to go away from the local optimum, a new scheme for inertia weight is proposed. The simulation results indicate that the proposed algorithm is superior to other compared algorithms.

References

1. J. Kennedy, and R. Eberhart: Particle Swarm Optimization. In Proceedings of the IEEE International Conference on Neural Networks.IEEE Service Center, Piscataway, NJ, IV (1995) 1941-1948.
2. J. Kennedy, R. Eberhart, and Y. Shi: Swarm Intelligence, SanFrancisco: Morgan Kaufmann Publishers (2001).
3. A. Hernandez-Aguirre, C. Coello. Passss: An Implementation of A Novel Diversity Strategy to Handle Constraints. In Proceeding of the 2004 Congress on Evolutionary Computation CEC-2004, IEEE press. Vol.1. June (2004) 403-410.
4. A. H. Aguirre, S.B. Rionda, Carlos A. Coello Coello: Handling Constraints Using Multiobjective Optimization Concepts. International J. for Numerical Methods in Engineering Vol. 59. (2004)1989-2017.
5. Carlos A. Coello Coello: Constraint Handling Using an Evolutionary Multiobjective Opti-mization Technique. Civil Engineering Systems Vol. 17. (2000)319-346.
6. Yuren Zhou, Yuanxiang Li, etc: A Pareto Strength Evolutionary Algorithm for Constrained Optimization. Journal of Software. Vol.14. (2003)1243-1249.
7. Thomas P. Runarsson and Xin Yao: Stochastic Ranking for Constrained Evolutionary Op-timization. IEEE Trans. On Evolutionary Computation, Vol.4, No3, Sep. (2000)284-292.
8. Jong-Hwan Kim and Hyun Myung, Evolutionary Programming Techniques for Constrained Optimization, IEEE Transactions on Evolutionary Computation, April (1997)129-139.
9. Joines and Houck C: On the Use of Non-Stationary Penalty Function to Solve Nonlinear Constrained Optimization Problems with Gas. In Proceedings of the First IEEE Interna-tional Conference on Evolutionary Computation, Piscataway, NJ: IEEE press (1994)579-584.

A Hybrid Discrete Particle Swarm Optimization for the Traveling Salesman Problem

Xiangyong Li, Peng Tian, Jing Hua, and Ning Zhong

College of Economics & Management, Shanghai Jiaotong University,
Shanghai 200052, P.R. China
lixiangyong@163.com, ptian@sjtu.edu.cn, hwa_jing@sjtu.edu.cn,
zhongning@sjtu.edu.cn

Abstract. This paper presents a hybrid discrete particle swarm optimization (HDPSO) for solving the traveling salesman problem (TSP). The HDPSO combines a new discrete particle swarm optimization (DPSO) with a local search. DPSO is an approach designed for the TSP based on the binary version of particle swarm optimization. Unlike in general versions of particle swarm optimization, DPSO redefines the particle's position and velocity, and then updates its state by using a tour construction. The embedded local search is implemented to improve the solutions generated by DPSO. The experimental results on some instances are reported and indicate HDPSO can be used to solve TSPs.

1 Introduction

The Traveling Salesman Problem (TSP) is a classical combinatorial optimization problem. It plays an important role in the combinatorial optimization and has become an early proving ground for many approaches designed for combinatorial optimization. It has proved to be $\mathcal{NP}$-hard and attracted many researchers from different fields. Now TSP has been studied intensively and many algorithms have been developed. These algorithms include construction heuristics, iteratively improvement algorithms, and exact methods such branch & bound or branch & cut, and many metaheuristics. The metaheuristic mainly consists of SA, tabu search, guide local search, evolutionary algorithm, and ACO algorithms etc.

Particle Swarm Optimization (PSO) is a population-based swarm intelligence algorithm. It was originally proposed by Kennedy and Eberhart as a simulation of the social behavior of social organisms such as bird flocking and fish schooling[1]. Because of its easy implementation and inexpensive computation, the PSO has proved to be an effective and competitive algorithm for the optimization problem in continuous space. But most applications of PSO have concentrated on the optimization in continuous space. Recently, some work has been done to the discrete optimization problem, including flowshop scheduling problem[2]. The published work related to the application of PSO to the traveling salesman problem is relatively few. Clerc[3] first proposed a discrete particle swarm optimization for solving the TSP. In this algorithm, he first redefined the particle's position and velocity, where the position is defined as city sequence

T.-D. Wang et al. (Eds.): SEAL 2006, LNCS 4247, pp. 181–188, 2006.

and the velocity is defined as city exchange. The velocity applied to the position can produced a new position. Clerc then defined some operators, i.e. "opposite of velocity ", "position plus velocity", "position minus position", "velocity plus velocity", and "coefficient times velocity". For detailed depiction, the reader can refer to[3]. But the results in[3] show the proposed algorithm is certainly not as powerful as some specific algorithms. We also applied Clerc's algorithm to solve the Multiple TSP(MTSP)[4]. Although the computation result shows that Clerc's PSO can solve the problems, the computation burden is very expensive. But such an idea of Clerc's discrete PSO have been modified and applied for many discrete or combinatorial problems such as flowshop scheduling.

In this paper, we extend the application of PSO to TSP and propose a hybrid discrete PSO inspired from[2]. In the proposed algorithm, the particle's position is redefined by a relation matrix, which denotes the relative position of cities in TSPs tour. The velocity is redefined as the particle's predisposition choosing one city as the predecessor of another city. Unlike the implementation in other versions of PSO, the updating of the particle's position is achieved by a tour construction. The local search method coupled with two speed-up techniques is incorporated with the proposed algorithm to improve the solutions. Finally, the performance of the proposed algorithm is investigated on some testing problems.

2 Brief Overview of Particle Swarm Optimization

PSO belongs to the class of swarm intelligence algorithms, which are inspired by social dynamics and emergent behavior in socially organized colonies. Since its introduction, PSO has gained rapid popularity and proved to be a competitive and effective optimization algorithm in comparison with other metaheuristics such as GA and SA. The position of each individual (called particle) is represented by a n-dimensional vector in problem space $x_i = (x_{i1}, x_{i2}, \ldots, x_{in})$, $i = 1, 2, \ldots, N$ (N is the population size), and its performance is evaluated on the predefined fitness function. The velocity of the *i-th* particle $v_i = (v_{i1}, v_{i2}, \ldots, v_{in})$ is defined as the change of its position. The flying direction of each particle is the dynamical interaction of individual and social flying experience. In the pioneering work of Kennedy and Eberhart[1], the particle swarm is manipulated as follows:

$$v_i(t+1) = v_i(t) + c_1 \cdot rand1 \cdot (p_i - x_i(t)) + c_2 \cdot rand2 \cdot (p_g - x_i(t)) \quad (1)$$

$$x_i(t+1) = x_i(t) + v_i(t+1) \quad (2)$$

where, $p_i = (p_{i1}, \ldots, p_{in})$ is the best position encountered by *i-th* particle so far; p_g represents the best position found by any member in the neighborhood of *i-th* particle; t is iteration counter; c_1 and c_2 are acceleration coefficients; $rand1$ and $rand2$ are two random numbers in $[0, 1]$.

The basic PSO and its variants have successfully operated for continuous optimization functions. In order to extending the application to discrete space, Kennedy and Eberhart proposed a discrete binary version of PSO[5] where the particles' trajectories are defined as the changes in the probability and v_i is a measure of individual's current probability of taking 1. If the velocity is higher, it is more likely to choose 1,and lower values favor choosing 0. A sigmoid function is applied to transform the velocity from real number space to probability space:

$$s(v_{id}) = \frac{1}{1+\exp(-v_{id})} \quad (3)$$

In the binary version of PSO, the velocities and positions of particles are updated as the following formulas[5]:

$$v_{id}(t+1) = v_{id}(t) + c_1 \cdot rand1 \cdot (p_{id} - x_{id}(t)) + c_2 \cdot rand2 \cdot (p_{gd} - x_{id}(t)) \quad (4)$$

$$x_{id}(t+1) = \begin{cases} 1 & \text{if } rand3 < s(v_{id}), \\ 0 & \text{if } rand3 \geq s(v_{id}) \end{cases} \quad (5)$$

where x_{id} is the valued of the d-th dimension of particle x_i, and $x_{id} \in \{0,1\}$; v_{id} is the corresponding velocity; $s(v_{id})$ is calculated according to the equation (3). $rand3$ is a random number distributed in $[0,1]$. As in basic PSO, a parameter V_{max} is incorporated to limit the v_{id} so that $s(v_{id})$ does not approach too closely 0 or 1[6]. Such implementation can ensure that the bit can transfer between 1 and 0 with a positive probability. In practice, V_{max} is often set at ± 4[6].

3 Proposed Particle Swarm Optimization for the TSP

3.1 Discrete Particle Swarm Optimization (DPSO) for the TSP

(1)*Particle Representation*

One of the key issues in designing a successful PSO for TSP is to find a suitable mapping between TSP solutions and particles in PSO. Many different representation have been proposed by solving TSP by Genetic Algorithm, i.e. binary representation, path representation, adjacency representation, ordinal representation and matrix representation. To extend binary PSO to solve TSP, the positions and velocities of the particle are redefined. We uses a 0-1 relation matrix to represent the positions of a particle in discrete space of TSP. Suppose that considered TSP contains n cities. The position of particle i can be represented by a $n \times n$ matrix $x_i = (x^i_{jk})_{n \times n}$, $x^i_{jk} \in \{0,1\}$, and $x^i_{jk} \neq x^i_{kj}$,$(j = 1,\ldots,n; k = 1,\ldots,n)$. The element x^i_{jk} indicates the relative positions of cities j and k in TSP solution. If city k is immediately visited after city j, x^i_{jk} is equal to 1, otherwise 0. The relation matrix x_i defines the visiting sequence of n cities in the TSP tour. For

$$x_i = \begin{pmatrix} 0\,0\,1\,0 \\ 1\,0\,0\,0 \\ 0\,0\,0\,1 \\ 0\,1\,0\,0 \end{pmatrix}$$

This implies that a legal position of the particle is represented by a matrix of which each row and each column contains precisely one 1. In the proposed algorithm, the velocity of particles i is also denoted by a $n \times n$ matrix $v_i = (v^i_{jk})_{n \times n}$,

($j = 1, \ldots, n$; $k = 1, \ldots, n$). $v_{jk}^i \in R$. As in binary version of PSO, here the velocity $v_{jk}^i \in R$ also means the probabilistic, namely the probability of particle choosing to go from city j to city k. A high value indicates that it is more likely to choose city k as the immediate successor of city j in TSP tour.

(2) ***Updating Equations of Particles***

The velocity of particle i is updated according to equation (6)

$$v_i(t+1) = w \cdot v_i(t) + c_1 \cdot rand1 \cdot (p_i - x_i(t)) + c_2 \cdot rand2 \cdot (p_g - x_i(t)) \quad (6)$$

Unlike the updating equations in binary version of PSO[5], the proposed algorithm is established based on standard PSO, namely basic PSO with inertia weight developed by Shi and Eberhart in[7]. w is inertia weight. The inertia weight controls the impact of previous histories of velocities on current velocity, which is often used as a parameter to control the trade-off between exploration and exploitation. The particle adjusts its trajectory based on information about its previous best performance and the best performance of its neighbors. In the binary version of PSO, the trajectories of particles, velocities, etc., are defined in terms of the changes of probabilities. So in the proposed discrete PSO, the velocity also need to be transformed from real number space to probability space, namely the interval $[0, 1]$. It is accomplished by a logistic transformation function:

$$s(v_{jk}^i) = \frac{1}{1 + \exp(-v_{jk}^i)} \quad (7)$$

where, $s(v_{jk}^i)$ represents the probability of x_{jk} taking the value 1. In the context of TSP, it means that the salesman, currently located at city j, chooses to go to city k with probability $s(v_{jk}^i)$. Inspired by the concept of pheromone trail in ant colony optimization, here, we define $s(v_{jk}^i)$ as probability trail and the corresponding matrix of $(s(v_{jk}^i))_{n \times n}$ as probability trail matrix. As in basic PSO, the velocity of the proposed algorithm is also constrained, that is, $|v_{jk}^i| \leq v_{max}$.

In binary version of PSO, the value of each dimension of particle takes the value 1 or 0 according to equation(5). In the proposed PSO, the position of each particle is updated based on probability trail, namely, each particle's position is updated by tour construction defined on probability trail. The position of particle i is updated as follows: the position $x_i = (x_{jk}^i)_{n \times n}$ is initialized to a zero matrix, i.e. the corresponding TSP tour is a null sequence. A random city j is first selected as the first city in TSP tour. The proposed PSO then uses a so-called pseudorandom proportional rule to construct the tour of each particle(the rule is inspired by the one in ACO) and update the position. Next city k to visit is chosen according the following equation:

$$k = \begin{cases} \arg\max_{l \in N_j} s(v_{jl}^i), & \text{if } q \leq q_0 \\ J, & \text{if } q > q_0 \end{cases} \quad (8)$$

where, $s(v_{jk}^i)$ is probability trail of particle i; N_j is the feasible set of unvisited cities, and the corresponding probability trail used to construct tour is the elements of the j-th row of matrix $(s(v_{jk}^i))_{n \times n}$. q is a random variable in $[0, 1]$; q_0

is a control parameter and $0 \leq q_0 \leq 1$; J is a randomly selected city according to the probability distribution as follows:

$$p_{jk}^i = \frac{s(v_{jk}^i)}{\Sigma_{l \in N_j} s(v_{jl}^i)}, \qquad \text{if } k \in N_j \tag{9}$$

By the probability defined in Equation(9), J is selected by the roulette. With the above-mentioned rule, next city k is chosen and the corresponding element in position of particle i is set to 1, that is, $x_{jk}^i = 1$. Then, city k is chosen as current city and particle i determines next city according to equation (8) and set the corresponding element of its position x_i to 1. Such a construction is repeated until all n cities have been inserted into TSP tour. By such a tour construction, the positions of all the particles are updated based at probability trail. In the proposed discrete PSO, we use the *candidate list* N_j, when updating the positions of particles by equation(8). The candidate list contains for each city i those cities j that are at a small distance. At each iteration, each particle update its position and construct TSP tour by choosing next city among the candidate list of current city. Only when all the members of the candidate list of current city have been visited, another remaining city out of N_j is chosen.

As an additional technique to prevent the algorithm being trapped in local area, we reset the algorithm occasionally. When the best found solution is not improved for a consecutive iterations, e.g. 80, the velocity and position of the particle are re-initialized randomly.

3.2 Hybrid Discrete Particle Swarm Optimization (HDPSO)

The hybrid discrete particle swarm optimization (HDPSO) is the DPSO plus local search. The literature on metaheuristics indicates that a hybrid algorithm coupling metaheuristic with local search can get high-quality solution. In this section, we also consider local search embedded in DPSO. Among the simple local search algorithms for TSP, the most widely known algorithms are 2-Opt and 3-Opt, which are widely studied in the literature and have been shown empirically to yield good solution. The 2-Opt and 3-Opt delete two or three edges, thus breaking the tour into two or three paths, and then reconnect those paths in other possible ways. These methods proceed by systematically testing whether the current tour can be improved by replacing two or at most three arcs,respectively. In DPSO, we incorporate 3-Opt.

For 3-Opt, the straightforward implementation would examine $O(n^3)$ potential exchanges. But it is certainly infeasible when trying to apply it to all particles during all the iterations, or deal with larger TSPs in reasonable time. Here two speedup techniques are used to reduce the computation cost. One is the use of nearest-neighbor lists of fixed length k, typically $k = 20$[8]. That is to restrict the examination of exchanges of each city in its k nearest-neighbors. The implementation of fixed k nearest-neighbor for 3-Opt can reduce the computation time for finding an improving move from $O(n^3)$ to $O(n)$.

The computation cost may be still relatively expensive when we applying local search to all particles during all iterations. Another speed-up is to perform

Table 1. Computational results of DPSO over 10 trials

Problem	n	Opt	DPSO				
			Best	Worst	Average	Std.	Time
att48.tsp	48	10628	11038	11038	11038	0.00	36.12
eil51.tsp	51	426	447	448	447.20	0.45	48.15
st70.tsp	70	675	700	700	700	0.00	129.34
eil76.tsp	76	538	552	577	557.0	11.18	84.53
gr96.tsp	96	55209	60982	63009	61486.2	877.74	138.42
kroA100.tsp	100	21282	23075	23075	23075.0	0.0	300.23
ftv38.atsp	39	1530	1591	1595	1591.80	1.79	29.58
p43.atsp	43	5620	5639	5646	5641.4	2.63	50.95
ftv44.atsp	45	1613	1679	1679	1679	0	39.35
ftv64.atsp	65	1839	1969	1969	1969	0	88.77

3-Opt for limited particles. Here, we gradually increase the number of particles applying local search, to obtain a tradeoff between solution quality and computation cost. During the implementation, the proposed algorithm first allows only the iteration-best particle to apply local search. Then the number m of particles applying local search is increased by one every *num_iter* iterations. When triggering local search, the particles are first ranked in terms of their fitness, and then only m best-ranked particles can be allowed to apply local search. Additionally local search is trigger every constant iterations.

4 Experiments and Computational Results

In this section we investigate the performance of HDPSO on some TSP instances in TSPLIB (available at http://www.iwr.uni-heidelberg.de/iwr/comopt/soft/TSPLIB95/TSPLIB.html). All the experiments were implemented on a Pentium IV 2.8G/256MB PC and the program was coded in C language.

HDPSO need a set of parameters to be tuned in order to provide best possible solutions. By preliminary experiments, we set the parameters as: swarm size $N = 25$; acceleration coefficients were set to $c_1 = 2$, and $c_2 = 2$; Inertia weight was set to $w = 1$. The size of nearest-neighbor list of 3-Opt was set to 15. In equation (8), the size of candidate list is set to $\frac{n}{6}$ if $n < 70$, and $\frac{n}{10}$ if $n \geq 70$. The preliminary experiments showed that the value of $num\_iter = 150$, can give a reasonable tradeoff between the solution quality and computation time.

We first investigated the performance of DPSO for some considered symmetric TSPs(x.tsp) and asymmetric TSPs(x.atsp). The stopping criterion is the maximum allowed number of tour construction, $2500 \cdot n$ (n is the number of cities). The reported statistical values are: Opt: the know optimal solution in TSPLIB; Best: best-found solution; Average: average solution quality over 10 runs. Worst: the worst generated solution. Std.: standard deviation of the optima over 10 runs. Time: average time used to find best solutions. Table 1 summarizes the computational results of DPSO. The computational results show that the DPSO is

Table 2. Summary of results of HDPSO

Problem	n	Opt	HDPSO				
			Best	Worst	Average	Std.	Time
att48.tsp	48	10628	10842	11070	10926	11.15	34.36
eil51.tsp	51	426	442	442	442.0	0.00	44.51
st70.tsp	70	675	694	704	696.0	4.47	130.71
eil76.tsp	76	538	550	568	555.60	8.17	85.30
gr96.tsp	96	55209	60263	63087	61323.20	1171.22	164.39
kroA100.tsp	100	21282	22829	23867	23036.6	464.21	251.52
ftv38.atsp	39	1530	1571	1571	1571.0	0.0	29.88
p43.atsp	43	5620	5636	5650	5644	4.06	59.04
ftv44.atsp	45	1613	1613	1708	1648.2	42.56	35.29
ftv64.atsp	65	1839	1951	1951	1951	0	88.78

Table 3. Comparison results of HDPSO and other algorithms

Problem	Opt	$\mathcal{MMAS}$[1]	ACS[2]	AS_{rank}[1]	AS[1]	AS_e[1]	HDPSO
eil51.tsp	426	427.6	428.1	434.5	428.3	437.3	435.6
kroA100.tsp	21282	21320.3	21420	21746	21522.8	22471.4	22523.1
d198.tsp	15780	15972.5	16054	16199.1	16205	16702.1	16200.1
ry48p.atsp	14422	14553.2	14565.4	14511.4	14685.2	15296.4	14675.3
ft70.atsp	38673	39040.2	39099	39410.1	39261.8	39596.3	39173.4
kro124p.atsp	36230	36773.5	36857	36973.5	37510.2	38733.1	37513.2
ftv170.atsp	2755	2828.8	2826.5	2854.2	2952.4	3154.5	2959.4

[1] $\mathcal{MMAS}$: the $\mathcal{MAX}$-$\mathcal{MIN}$ Ant System. AS: the ant system. AS_e: the ant system with elitist strategy. AS_{rank}: the rank-based version of Ant System. The results of $\mathcal{MMAS}$, AS, AS_e, and AS_{rank} are taken from[9].

[2] ACS: Ant Colony System[10].

able to find very high-quality solutions for all instances. Compared to the best-known results in the literatures, the proposed algorithm achieves a good average solutions quality on most TSP instances. Additionally, the values of standard deviation are very small for most instances, which show that the HDPSO is an approach with high robustness.

We now attempt to incorporate local search in DPSO and observe the performance of hybrid algorithm, HDPSO. Table 2 shows the computation results under 10 trials. The results indicate that HDPSO outperformed DPSO. By adding local search, HDPSO found enhanced solutions.

Table 3 compare solutions of HDPSO and the ones found by some ant algorithms. To be fair, the comparison is done based on the same allowed number of tour constructions for all considered algorithms. As in[9], this number is chosen as $k \cdot n \cdot 10000$, where $k = 1$ for symmetric TSPs and $k = 2$ for asymmetric TSPs. HDPSO was implemented for 10 runs. For each instance the average solution qualities are reported for all algorithms. Since only one instance (gr17.tsp) was implemented in Clerc's PSO[3], we didn't compared the performance between HDPSO and Clerc's PSO. The comparison showed that HDPSO also produced good solutions compared to ant algorithms, which now is a more competitive method for TSPs in the literatures.

5 Conclusions and Future Work

This paper proposes a hybrid PSO (HDPSO) to solve TSPs. By redefinition the positions and velocities of particles, we first develop a discrete PSO(DPSO). Unlike general implementation, DPSO updates the position of each particle by using TSP tour construction based on probability trail. Local search is then embedded to improve solution quality. The experimental results indicate HDPSO is a viable approach for TSPs. As we know that many algorithms have been established to solve TSPs, here, we do not claim that our algorithm can provide competitive results compared with some specialized TSP algorithms. Nevertheless, the results indicate the proposed PSO can be used to solve TSPs. Moreover, the performance of HDPSO can be improved further by integrating other more powerful local search algorithms like tabu search, and Lin-Kernighan algorithm[8] etc.

Future work would concentrate on the improvement of computation speed, especially on the application of HDPSO to large scale TSPs.

References

1. Kennedy, J., Eberhart, R.: Particle swarm optimization. In: Proceedings of 1995 IEEE International Conference on Neural Networks. Volume 4. (1995) 1942–1948
2. Liao, C.J., Tseng, C.T., Luarn, P.: A discrete version of particle swarm optimzation for flowshop scheduling problems. Computers and Operations Research (In press)
3. Clerc, M.: Discrete particle swarm optimization, illustrated by the traveling salesman problem. Availabel at http://www.mauriceclerc.net (2000)
4. Li, X., Tian, P., Kong, M.: A modified particle swarm optimization for multiple traveling salesmen problem. In: Proceedings of 2005 International Conference on Management Science & Engineering, Incheon, South Korea, Harbin Institute Technology Publisher (2005) 475–480
5. Kennedy, J., Eberhart, R.: A discrete binary version of the particle swarm algorithm. In: Proceedings of 1997 IEEE International Conference on Systems, Man, and Cybernetics. Volume 5., Orlando, FL (1997) 4104–4108
6. Kennedy, J., Eberhart, R., Shi, Y.: Swarm intelligence. Morgan Kaufmann Publisher, San Francisco (2001)
7. Shi, Y., Eberhart, R.: A modified particle swarm optimizer. In: Proceedings of 1998 IEEE World Congress on Computational Intelligence. (1998) 69–73
8. Johnson, D.S., McGeoch, L.A.: The travelling salesman problem: A case study in local optimization. In Aarts, E.&Lenstra, J., ed.: Local Search in Combinatorial Optimization. Chichester, UK, John Wiley & Sons (1997) 215–310
9. Stützle, T., Hoos, H.H.: $\mathcal{MAX}$-$\mathcal{MIN}$ ant system. Future Generation of Computer Systems **16** (2000) 889–914
10. Gambardella, L., Dorigo, M.: Solving symmetric and asymmetric TSPs by ant colonies. In: Proceedings of IEEE International Conference on Evolutionary Computation (ICEC'96), Piscataway, USA, IEEE Press (1996) 622–627

Incremental Clustering Based on Swarm Intelligence

Bo Liu[1], Jiuhui Pan[1], and R I (Bob) McKay[2]

[1] Department of Computer Science, Jinan University, Guangzhou, China, 510632
lbxldd@sohu.com, jhpan@jnu.edu.cn
[2] Dept of Computer Science and Engineering, Seoul National University, Seoul 151744 Korea
rim@cse.snu.ac.kr
http://sc.snu.ac.kr

Abstract. We propose methods for incrementally constructing a knowledge model for a dynamically changing database, using a swarm of special agents (ie an ant colony) and imitating their natural cluster-forming behavior. We use information-theoretic metrics to overcome some inherent problems of ant-based clustering, obtaining faster and more accurate results. Entropy governs the pick-up and drop behaviors, while movement is guided by pheromones. The primary benefits are fast clustering, and a reduced parameter set. We compared the method both with static clustering (repeatedly applied), and with the previous dynamic approaches of other authors. It generated clusters of similar quality to the static method, at significantly reduced computational cost, so that it can be used in dynamic situations where the static method is infeasible. It gave better results than previous dynamic approaches, with a much-reduced tuning parameter set. It is simple to use, and applicable to continuously- and batch-updated databases.

1 Introduction

Ant Colony Optimization (ACO) is a relatively new and expanding branch of intelligent systems, specifically a form of swarm intelligence [1]. One of its most successful applications has been to clustering. Ants, and the algorithms based on them, have the ability to create clusters of objects without any initial partitioning, and without knowing ahead of time how many clusters will be necessary. One of the first studies of ant-based clustering is found in [2], where a population of ant-like agents randomly moving on a 2D grid are allowed to pick up and drop objects in such a way as to cluster them. Lumer and Faieta [3] developed the basic model and applied it in exploratory data analysis.

To the best of our knowledge, of the rich body of research which has emerged in ant mining, only Ramos et al have tackled the problem of dynamic data, in [4]. One common approach to learning from changing data is to repeatedly apply a traditional learner to the most recent examples. However the computational cost of re-applying a learner may be prohibitive. To meet these challenges, Ramos et al

T.-D. Wang et al. (Eds.): SEAL 2006, LNCS 4247, pp. 189–196, 2006.

presented an online classifier using swarm intelligence, which allowed incremental clustering, avoiding any type of re-training. It was based on the observation that an ant colony can collectively respond to a perturbation, through individuals exhibiting a single simple behavior. Our method is also based on ants' natural behaviors, but uses new measures to guide the agents in moving, and picking up or dropping an item.In this paper, dynamic data means data which changes, in this context, through insertions into and deletions from a database (an update can be viewed as an insertion followed by a deletion).

2 Incremental Ant-Based Clustering

In our dynamic system, the cluster generator consists of a group of agents. Each agent computes the information entropy or pheromone concentration of the area surrounding it, and clusters objects by picking them up, dropping them, and by moving. The incremental clustering algorithm incorporates the following modules: cluster initialization, dynamic cluster modification, and cluster model maintenance (in a changing grid).

2.1 Cluster Intialization

In the initial state ($t = t_0$), the algorithm selects N_0 objects from the current database, and distributes the N_0 objects uniformly randomly on the $Z \times Z$ grid. It initializes all agents to be unladen.

A subspace with clustered data has lower entropy than a subspace with unclustered data [5]. Inspired by this, we introduce information entropy into the clustering algorithm, to incorporate information about the local level of clustering. Each agent lies in an $s \times s$ region. Assuming independence of attributes, the entropy of the $s \times s$ area covering a set of objects is defined by equation 1, where $p(x)$ is defined by equation 2, num_{obj} is the total number of objects in region $s \times s$; and num_x is the number of objects whose attribute X_i has value x.

$$E(s^2) = -\sum_{i=1}^{n} \sum_{xinX_i} p(x)logp(x) \tag{1}$$

$$p(x) = \frac{num_x}{num_{obj}} \tag{2}$$

Both distance-based and density-based clustering metrics suffer some disadvantages. Distance-based methods are intrinsically biased toward hyper-elliptical cluster shapes, while density-based methods require the setting of an arbitrary density threshold parameter. Entropy-based metrics avoid both these disadvantages, requiring no thresholds, and being unbiased regarding cluster shape.

Taking entropy as the criterion for an agent to pick up or drop items, we propose the initial clustering algorithm shown in table 1.

In the algorithm, when all agents stop moving, it indicates that the initial clusters have been formed. For the next incremental clustering, for each pixel

Table 1. Swarm Cluster Algorithm

```
Initialize parameters: Z, s, t_max, N_a
For every object o_i do
   place o_i randomly on the Z × Z plane
For n=1 to N_a do //N_a is the number of agents
   place agent at randomly selected site in the Z × Z plane
For t=1 to t_max do //t_max is maximal times that an agent moves
  For n=1 to N_a do
    If ((agent unladen) and (site occupied by object o_i )) then
       Compute entropy E_1, E_2
       If (E_1 > E_2) then pick up o_i //picking up rule
    Else If ((agent carrying object o_i) and (site empty)) then
       Compute entropy E_1, E_2
       If (E_1 > E_2) then drop o_i //dropping rule
    End if
    Move to randomly selected neighboring site not occupied by an agent
  End For
End For
For each site (x, y) in the Z × Z plane do
  Compute entropy of the surrounding area s × s area
  Compute pheromone τ(x, y)
End For
```

where E_1 and E_2 are the entropies before and after performing the relevant action.

(x, y) in the $Z \times Z$ plane, we compute the entropy of the surrounding $s \times s$ area according to equation 1. If the $s \times s$ area is empty, we set the entropy of the area to the maximum value. We then compute the pheromone concentration $\tau(x, y)$ of the surrounding $s \times s$ area according to equation 3.

$$\tau(x, y) = \frac{num_{obj}}{1 + (s \times s)} \tag{3}$$

where num_{obj} is the number of objects in the surrounding $s \times s$ area.

2.2 Dynamic Cluster Modification

After the initial clusters form, the cluster model needs to be updated periodically as the database changes through insertions. Suppose a monitor transfers new data periodically, placing the inserted data objects randomly at empty sites in the $Z \times Z$ plane. The plane maintains a record of the entropy and pheromone, and this information can guide newly-laden agents to move toward the previously generated clusters. Although the basic behavior for incremental clustering is the same as in the initial clustering, an agent moves according to the algorithm in table 2 instead of moving randomly.

In the movement algorithm, ϕ is a random number. For an agent carrying a new object, the greater $\tau(x, y)$ is, the higher the probability that $\tau(x, y) > \phi$, or

Table 2. Agent Movement Algorithm

```
If (agent carrying a new object)
  If τ(x,y) > φ
    Move to the pixel (x,y), is an empty site within the s × s area
  Else
    Move to a randomly selected site not occupied by another agent
If (agent unladen)
  If τ(x,y) < φ
    Move to the site of the nearest new object
  Else
    Move to a randomly selected site not occupied by another agent
```

in other words, the more likely it is to move to one of the clusters. Conversely, for an unladen agent, the smaller $\tau(x, y)$ is, the higher the probability that $\tau(x, y) < \phi$, or in other words, the more likely it is to move to the nearest new object, pick up a new object, and move it to one of the existing clusters.

After all agents stop moving, at each pixel (x, y) in the $Z \times Z$ plane, the algorithm modifies the entropy of the surrounding $s \times s$ area according to equation 1, and modifies the pheromone $\tau(x, y)$ of the surrounding $s \times s$ area according to equation 4

$$\tau(x, y) = \frac{(1 - \lambda)num_{obj} + \Delta num_{obj}}{1 + (s \times s)} \tag{4}$$

where

- num_{obj} is the number of previous objects in $s \times s$
- Δnum_{obj} is the number of new objects in $s \times s$
- λ is the pheromone evaporation rate

2.3 Cluster Model Maintenance

The clustering model changes with the arrival of dynamically inserted or deleted data. A database is used to save the position of each pixel on the $Z \times Z$ plane. In the initial state, each pixel on the $Z \times Z$ plane is set to a null value. The schema of the database is $(x, y, t, entropy, pheromone)$, where (x, y) is the coordinate of each pixel on the $Z \times Z$ plane, t is the arrival time of the occupying object, *entropy* is the value $E(s^2)$ of the pixel, and the pheromone is the $\tau(x, y)$ value.

3 Features of the Clustering Algorithm

By comparison with non-ant-based incremental methods, our method avoids the need to pre-specify the number of clusters, and can find clusters of arbitrary shape, without requiring a predefined bias. In addition, it has some other important features:

1. The factor influencing an agent's picking up or dropping action is entropy. Each action of picking up or dropping by an agent can reduce the entropy of the previous patch, and thus speed up clustering. Compared with the similarity measure or response thresholds used in [3] and [6], the experiments in section 4 show that entropy is a more effective measure.
2. The number of parameters we need to specify for constructing a clustering model is small. In similar work, more are required. In the LF algorithm [3], seven parameters are required $(k_1, k_2, \alpha, Z, s, t_{max}, N_a)$, and in the ACLUSTER algorithm [6], eleven are needed $(k_1, k_2, k, \eta, \alpha, \beta, \gamma, Z, s, t_{max}, N_a)$. Our method requires only five $(Z, s, t_{max}, N_a, \lambda)$.
3. After initial or incremental clustering, we compute and save the entropy and pheromone for each pixel in the $Z \times Z$ plane for subsequent clustering . When each batch of data arrives, agent movements are guided by the pheromone to locate new objects. We use a simple pheromone update method.
4. The time complexity for locating objects in the initial clustering is $O(t_{max} \times N_a)$, and for computing the entropy and pheromone is $O(Z \times Z)$. Both are independent of the number of objects. This is one important reason why swarm intelligence is especially applicable to dynamic environments.

4 Experiments

Our first experiments compared incremental and non-incremental clustering. We conducted experiments on two data sets from the UCI repository [7]: Tic-tac-toe endgame, and Balance Scale. We used the settings $Z \times Z = 50 \times 50$, $s \times s = 5 \times 5$, $N_a = 50$, $\lambda = 0.1$. For static clustering, we set $t_{max} = 25000$, and for incremental clustering $t_{max} = 15000$.

The Tic-tac-toe endgame database contains 958 instances with two classes. The Balance Scale database contains 625 instances with three classes. To convert these static problems into dynamic problems, the instances in each database were divided into six groups. For the Tic-tac-toe endgame database, the first group G_0 consisted of 358 objects for the initial clustering; the rest of the data were divided into five subsets, G_1 to G_5, each with 120 objects, for the subsequent incremental clustering. For the Balance Scale database, the first group G_0 consisted of 225 objects, the rest being divided into five subsets, G_1 to G_5, each with 80 objects, for subsequent incremental clustering.

The scenario consisted of an initial dataset G_0, then five rounds of insertion, with G_1 to G_5 being in turn added to the dataset. We can more formally describe the datasets S_i by $S_0 = G_0$; $S_1 = (S_0 \cup G_1)$; $S_2 = (S_1 \cup G_2)$; and so on. We replicated our experiments 10 times, with independent sampling of $G_0 \ldots G_5$.

To evaluate the quality of the clusters, we used Ramos' [6] metric (equation 6) For a given location x containing an object of class A,

$$e_x = \frac{\sum_{y \in s \times s} \delta(x, y)}{s^2 - 1} \tag{5}$$

where
$\delta(x, y) = 0$ if location y also contains an element of class A
$\delta(x, y) = 1$ otherwise

$$E_A = \frac{\sum_{x \in A} e_x}{\mid A \mid} \tag{6}$$

For comparison purposes, we also conducted experiments using the different incremental approaches, and obtained the following results and conclusions.

5 Results

5.1 Incremental Vs Non-incremental Approaches

The first experiment compared our incremental approach with a non-incremental approach using the initial clustering algorithm (a typical ant clustering algorithm) over the whole database. The purpose of this comparison was to confirm that there actually was a benefit form incremental clustering – that is, that by using the previous clustering as a starting point, incremental clustering could achieve similar quality clusterings to those given by static clustering, but with a reduced computational cost. In our experiments, we provided almost twice the computational time resources to the static clustering as we did to the incremental clustering (t_{max}). The average quality of the clusters produced by the two methods (incremental vs non-incremental) are shown in table 3. The table records the mean and standard deviation (over the ten trials) for each statistic, together with the p value from Student's two-tailed heteroscedastic T-test. Despite the increased resources provided to the static method, the quality of the two clusterings are statistically indistinguishable. This is not surprising. Population-based methods frequently offer good performance, even for static problems, but at relatively high computational cost; for dynamic problems, where the population can assist in effectively maintaining solution state, the computational cost disadvantages are dramatically reduced.

Table 3. Cluster Quality (Incremental vs Static)

System	Tic-Tac-Toe End-Game	Balance Scale
Static	0.722 ± 0.032	1.485 ± 0.047
Dynamic	0.739 ± 0.024	1.461 ± 0.038
T-test p value	0.22	0.22

5.2 Entropy vs Non-entropy Approaches

The primary purpose of this paper is to introduce our entropy-based approach, and compare its performance with previously-studied dynamic ant-based clustering algorithms. Hence we compared our method with Lumer and Faieta's response function LF [3], and with Ramos and Abraham's RA [6]. The results are shown in table 4. This table records the mean and standard deviation of the final clustering (ie after five rounds of data update), together with T-test p values, for each of the treatments. The clustering performance over time (ie after each of the epochs of updating) are shown in figure 1.

Table 4. Cluster Quality (Different Incremental Methods)

System	Tic-Tac-Toe End-Game	Balance Scale
Entropy	0.739 ± 0.025	1.461 ± 0.037
LF	1.368 ± 0.014	2.219 ± 0.043
p value (LF)	1×10^{-19}	3×10^{-19}
RA	1.671 ± 0.012	2.737 ± 0.019
p value (RA)	8×10^{-21}	4×10^{-20}

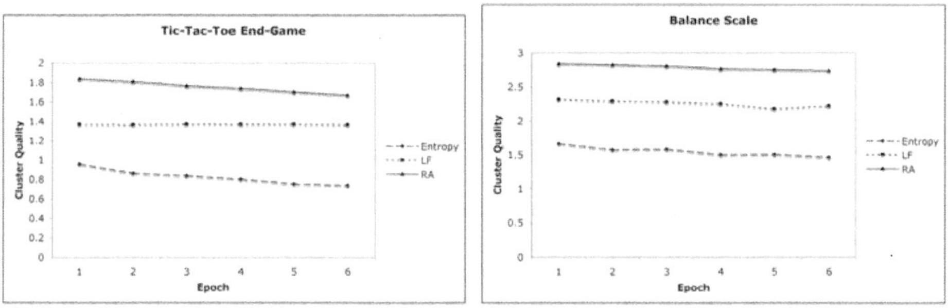

Fig. 1. Dynamic Clustering on Tic-Tac-Toe End-Game and Balance Scale

It is clear from these results that the entropy-based measure systematically out-performs both the LF and RA approaches, yielding better cluster quality for essentially the same algorithmic cost. The more informed approach is better able to rapidly form clusters. What is more, it appears that despite the dynamicity of the problem, the entropy-based method is able to continue to improve the clusters from epoch to epoch.

6 Conclusions and Further Work

Our results suggest that dynamic clustering, for a realistic problem scenario, can provide significant benefits over static clustering, achieving comparable clustering results at lower computational cost. This provides further confirmation of the arguments of Lumer and Faieta, and of Ramos and Abrahams.

Equally important, we have shown that the performance of a dynamic ant-based clustering algorithm can be significantly improved by using an entropy-based metric to guide the clustering, and that such a metric is able to incorporate useful information about the local structure. The performance of the algorithm, with this metric, on two realistic dynamic problems, was substantially better than either of the two earlier metrics. This was achieved despite a substantially reduced number of algorithm parameters.

Future work will concentrate on validating the approach on a wider range of dynamic problem scenarios, and on characterising the robustness of the parameter settings of the algorithm.

References

1. Bonabeau, E., Dorigo, M., and Theraulaz, G., Swarm Intelligence: From Natural to Artificial Systems. New York: Oxford University Press,1999
2. Deneubourg, J.-L.,S.Goss, N.Franks, C.Detrain, and L.Chretien, The Dynamics of Collective Sorting: Robot-Like Ant and Ant-Like Robot, Proceedings First Conference on Simulation of Adaptive Behavior: From Animals to Animats, edited by J.A.Meyer and S.W.Wilson, 356-365, Cambridge, MA:MIT Press, 1991
3. Lumer, E., and Faieta, B., Diversity and Adaptation in Populations of Clustering Ants, Proceedings Third International Conference on Simulation of Adaptive Behavior: From Animal to Animats 3, 499-508,, Cambridge, MA: MIT Press, 1994
4. Ramos, V. and Abraham, A. Swarms on Continuous Data, Proceedings of 2003 IEEE Congress on Evolutionary Computation, 1370-1375, 2003
5. Barbar, D., Couto, J. and Li, Y., COOLCAT: An Entropy-based Algorithm for Categorical Clustering, Proceedings of the Eleventh International Conference on Information and Knowledge management, 582-589,2002
6. Ramos, V. and Merelo, J. J., Self-organized Stigmergic Document Maps: Environment as A Mechanism for Context Learning, in E.Alba, F.Herrera, J.J.Merelo et al. (Eds.),Procs.of AEB02-1st Spanish Conference on Evolutionary and Bio-inspired Algorithms, Spain, 284-293, 2002
7. Hettich, S. and Bay, S.D. The UCI KDD Archive [DB/OL] , http://kdd.ics.uci.edu. , 1999.

Variable Neighborhood Particle Swarm Optimization for Multi-objective Flexible Job-Shop Scheduling Problems

Hongbo Liu[1,2], Ajith Abraham[3,1], Okkyung Choi[3,4], and Seong Hwan Moon[4]

[1] School of Computer Science, Dalian Maritime University, 116024, Dalian, China
[2] Department of Computer, Dalian University of Technology, 116023, Dalian, China
[3] School of Computer Science and Engineering, Chung-Ang University, Seoul 156-756, Korea
[4] Department of Science and Technology, Education for Life, Seoul National University of Education, Seocho-dong, Seocho-gu, Seoul 137-742, Korea
lhb@dlut.edu.cn, ajith.abraham@ieee.org, ok1124@freechal.com, hmoon@snue.ac.kr

Abstract. This paper introduces a hybrid metaheuristic, the Variable Neighborhood Particle Swarm Optimization (VNPSO), consisting of a combination of the Variable Neighborhood Search (VNS) and Particle Swarm Optimization(PSO). The proposed VNPSO method is used for solving the multi-objective Flexible Job-shop Scheduling Problems (FJSP). The details of implementation for the multi-objective FJSP and the corresponding computational experiments are reported. The results indicate that the proposed algorithm is an efficient approach for the multi-objective FJSP, especially for large scale problems.

1 Introduction

Flexible Job-shop Scheduling Problems (FJSP) is an extension of the classical JSP which allows an operation to be processed by any machine from a given set. It incorporates all the difficulties and complexities of its predecessor JSP and is more complex than JSP because of the additional need to determine the assignment of operations to the machines. The scheduling problem of a FJSP consists of a routing sub-problem, that is, assigning each operation to a machine out of a set of capable machines and the scheduling sub-problem, which consists of sequencing the assigned operations on all machines in order to obtain a feasible schedule minimizing a predefined objective function. It is quite difficult to achieve an optimal solution with traditional optimization approaches owing to the high computational complexity. In the literature, different approaches have been proposed to solve this problem. Mastrolilli and Gambardella [1] proposed some neighborhood functions for metaheuristics. Kacem et al. [2,3] studied on modeling genetic algorithms for FJSP. Ong et al. [4] applied the clonal selection principle of the human immune system to solve FJSP with re-circulation. By hybridizing particle swarm optimization and simulated annealing, Xia and

T.-D. Wang et al. (Eds.): SEAL 2006, LNCS 4247, pp. 197–204, 2006.

Wu [5] developed an hybrid approach for the multi-objective flexible job-shop scheduling problem (FJSP). Because of the intractable nature of the problem and its importance in both fields of production management and combinatorial optimization, it is desirable to explore other avenues for developing good heuristic algorithms for the problem.

Particle Swarm Optimization (PSO) incorporates swarming behaviors observed in flocks of birds, schools of fish, or swarms of bees, and even human social behavior, from which the intelligence is emerged [6]. It has become the new focus of research recently. However, its performance deteriorates as the dimensionality of the search space increases, especially for the multi-objective FJSP involving large scale. PSO often demonstrates faster convergence speed in the first phase of the search, and then slows down or even stops as the number of generations is increased. Once the algorithm slows down, it is difficult to achieve better scheduling solutions. To avoid termination at a local minimum, we introduce a novel hybrid meta-heuristic, the Variable Neighborhood Particle Swarm Optimization (VNPSO) for the multi-objective FJSP. The basic idea is to drive those particles by a shaking strategy and get them to explore variable neighborhood spaces for the better scheduling solutions.

2 Problem Formulation

The classical FJSP considers in general the assignment of a set of machines $M = \{M_1, \cdots, M_m\}$ to a set of jobs $J = \{J_1, \cdots, J_n\}$, each of which consists of a set of operations $J_j = \{O_{j,1}, \cdots, O_{j,p}\}$. There are several constraints on the jobs and machines, such as (1) each machine can process only one operation at a time; (2) operations cannot be interrupted; (3) there are no precedence constraints among operations of different jobs; (4) setup times for the operations are sequence-independent and included in the processing times; (5) there is only one of each type of machine; (6) machines are available at any time.

To formulate the objective, define $C_{i,j,k}$ ($i \in \{1, 2, \cdots, m\}$, $j \in \{1, 2, \cdots, n\}$, $k \in \{1, 2, \cdots, p\}$) as the completion time that the machine M_i finishes the operation $O_{j,k}$; $\sum C_i$ represents the time that the machine M_i completes the processing of all the jobs. Define $C_{max} = max\{\sum C_i\}$ as the makespan, and $C_{sum} = \sum_{i=1}^{m}(\sum C_i)$ as the flowtime. The problem is thus to both determine an assignment and a sequence of the operations on all machines that minimize some criteria. Most important optimality criteria are to be minimized: (1) the maximum completion time (makespan): C_{max}; (2) the sum of the completion times (flowtime): C_{sum}.

Minimizing C_{sum} asks the average job finishes quickly, at the expense of the largest job taking a long time, whereas minimizing C_{max}, asks that no job takes too long, at the expense of most jobs taking a long time. Minimization of C_{max} would result in maximization of C_{sum}. The weighted aggregation is the most common approach to the problems. According to this approach, the objectives, $F_1 = min\{C_{max}\}$ and $F_2 = min\{C_{sum}\}$, are summed to a weighted combination:

$$F = w_1 min\{F_1\} + w_2 min\{F_2\} \tag{1}$$

where w_1 and w_2 are non-negative weights, and $w_1 + w_2 = 1$. These weights can be either fixed or adapt dynamically during the optimization. The dynamic weighted aggregation [7] was used in the paper. Alternatively, the weights can be changed gradually according to the Eqs. (2) and (3). The variation for different values of w_1 and w_2 ($R = 200$) are illustrated in Fig. 1.

$$w_1(t) = |sin(2\pi t/R)| \tag{2}$$

$$w_2(t) = 1 - w_1(t) \tag{3}$$

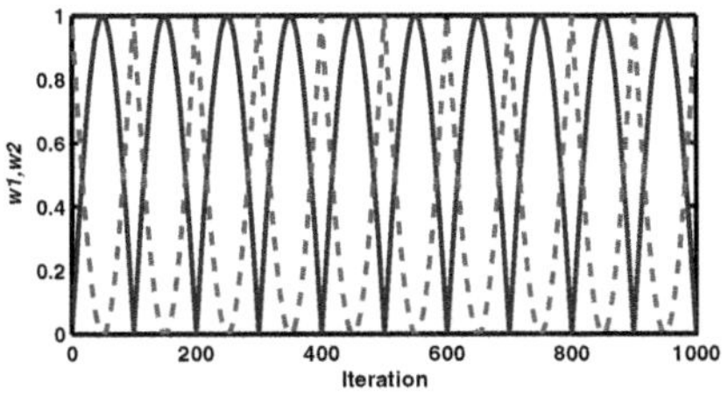

Fig. 1. Dynamic weight variation

3 The VNPSO Heuristic for FJSP

The classical PSO model consists of a swarm of particles, which are initialized with a population of random candidate solutions. They move iteratively through the d-dimensional problem space to search the new solutions, where the fitness, f, can be calculated as the certain qualities measure. Each particle has a position represented by a position-vector $\boldsymbol{x}_i$ (i is the index of the particle), and a velocity represented by a velocity-vector $\boldsymbol{v}_i$. Each particle remembers its own best position so far in a vector $\boldsymbol{x}_i^{\#}$, and its j-th dimensional value is $x_{ij}^{\#}$. The best position-vector among the swarm so far is then stored in a vector $\boldsymbol{x}^*$, and its j-th dimensional value is x_j^*. During the iteration time t, the update of the velocity from the previous velocity to the new velocity is determined by Eq.(4). The new position is then determined by the sum of the previous position and the new velocity by Eq.(5).

$$v_{ij}(t) = wv_{ij}(t-1)+c_1r_1(x_{ij}^{\#}(t-1)-x_{ij}(t-1))+c_2r_2(x_j^*(t-1)-x_{ij}(t-1)) \tag{4}$$

$$x_{ij}(t) = x_{ij}(t-1) + v_{ij}(t) \tag{5}$$

The particle swarm algorithm can be described generally as a population of vectors whose trajectories oscillate around a region which is defined by each individual's previous best success and the success of some other particle. The best particle acts as an attractor, pulling its neighborhood particles towards it. Some previous studies has been shown that the trajectories of the particles oscillate in different sinusoidal waves and converge quickly [8]. During the iteration, the particle is attracted towards the location of the best fitness achieved so far by the particle itself and by the location of the best fitness achieved so far across the swarm. The algorithm has faster convergence. But very often for multi-modal problems involving high dimensions it tends to suffer from premature convergence.

Variable Neighborhood Search (VNS) is a relatively recent metaheuristic which relies on iteratively exploring neighborhoods of growing size to identify better local optima with shaking strategies [9,10]. More precisely, VNS escapes from the current local minimum x^* by initiating other local searches from starting points sampled from a neighborhood of x^*, which increases its size iteratively until a local minimum is better than the current one is found. These steps are repeated until a given termination condition is met. The metaheuristic method we propose, the VNPSO, was originally inspired by VNS. In PSO, if a particle's velocity decreases to a threshold v_c, a new velocity is assigned using Eq.(6):

$$v_{ij}(t) = w\hat{v} + c_1 r_1 (x^{\#}_{ij}(t-1) - x_{ij}(t-1)) + c_2 r_2 (x^*_j(t-1) - x_{ij}(t-1)) \quad (6)$$

Algorithm 1. Variable Neighborhood Particle Swarm Optimization Algorithm

Initialize the size of the particle swarm n, and other parameters.
Initialize the positions and the velocities for all the particles randomly.
Set the flag of iterations without improvement $Nohope = 0$.
While (the end criterion is not met) do
    $t = t + 1$;
    Calculate the fitness value of each particle;
    $\boldsymbol{x}^* = argmin_{i=1}^{n}(f(\boldsymbol{x}^*(t-1)), f(\boldsymbol{x}_1(t)), f(\boldsymbol{x}_2(t)), \cdots, f(\boldsymbol{x}_i(t)), \cdots, f(\boldsymbol{x}_n(t)))$;
    If $\boldsymbol{x}^*$ is improved then $Nohope = 0$, else $Nohope = Nohope + 1$.
        For i= 1 to n
        $\boldsymbol{x}^{\#}_i(t) = argmin_{i=1}^{n}(f(\boldsymbol{x}^{\#}_i(t-1)), f(\boldsymbol{x}_i(t))$;
        For $j = 1$ to d
            If $Nohope < 10$ then
                Update the j-th dimension value of $\boldsymbol{x}_i$ and $\boldsymbol{v}_i$
                according to Eqs.(4),(5)
            else
                Update the j-th dimension value of $\boldsymbol{x}_i$ and $\boldsymbol{v}_i$
                according to Eqs.(7),(6).
        Next j
    Next i
End While.

$$\hat{v} = \begin{cases} v_{ij} & \text{if } |v_{ij}| \geq v_c \\ u(-1,1)v_{max}/\rho & \text{if } |v_{ij}| < v_c \end{cases} \tag{7}$$

Our algorithm scheme is summarized as Algorithm 1. The performance of the algorithm is directly correlated to two parameter values, v_c and ρ. A large v_c shortens the oscillation period, and it provides a great probability for the particles to leap over local minima using the same number of iterations. But a large v_c compels the particles in the quick "flying" state, which leads them not to search the solution and forcing them not to refine the search. The value of ρ changes directly the variable search neighborhoods for the particles. It is to be noted that the algorithm is different from the multi-start technique and the turbulence strategy [11]. We also implemented the Multi-Start PSO (MSPSO) and Velocity Turbulent PSO (VTPSO) to compare their performances.

For applying PSO successfully for the FJSP problem, we setup a search space of O dimension for a $(m - Machines, n - Jobs, O - Operations)$ FJSP problem. Each dimension was limited to $[1, m+1)$. Each dimension of the particle's position maps one operation, and the value of the position indicates the machine number to which this task is assigned to during the course of PSO. The particle's position may appear real values such as 1.4, etc. We usually round off the real optimum value to its nearest integer number.

4 Experiment Settings and Results

To illustrate the effectiveness and performance of the proposed algorithm, three representative instances based on practical data have been selected. Three

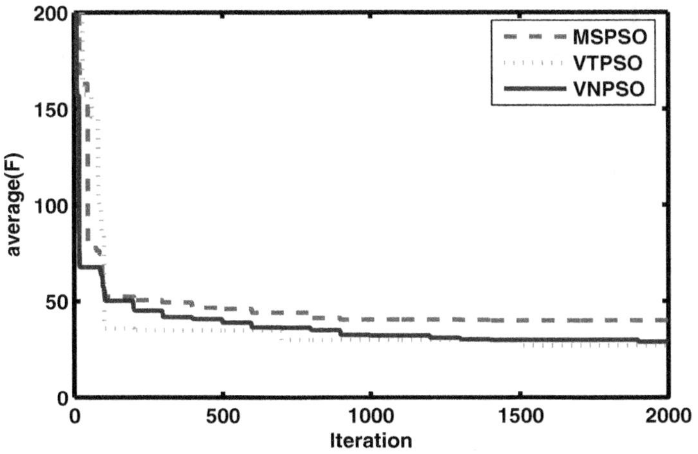

Fig. 2. The performance of the algoriths for $(J8, O27, M8)$ FJSP

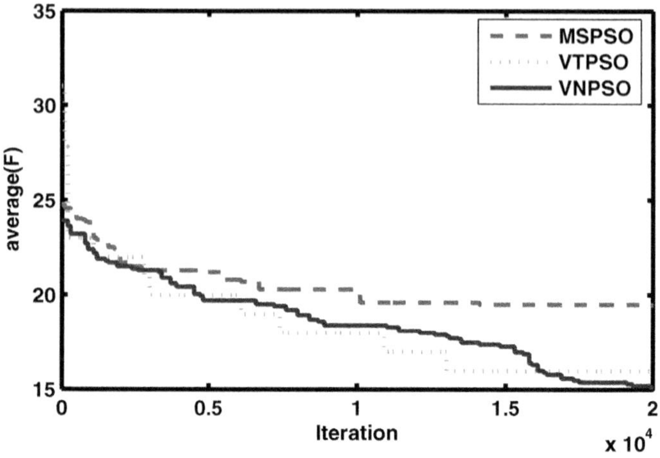

Fig. 3. The performance of the algoriths for $(J10, O30, M10)$ FJSP

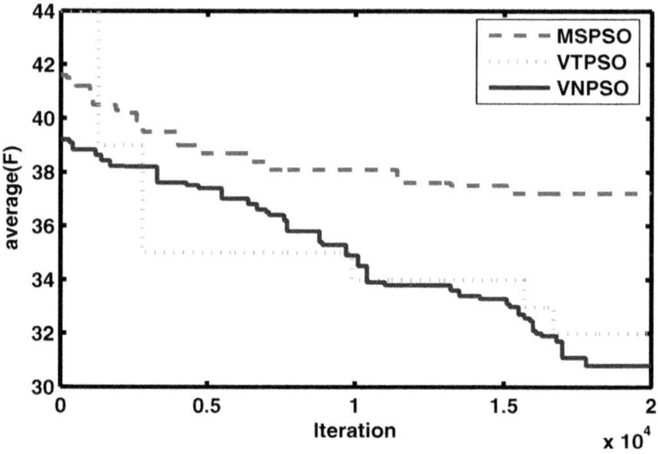

Fig. 4. The performance of the algoriths for $(J15, O56, M10)$ FJSP

problem instances $((J8, O27, M8)$, $(J10, O30, M10)$ and $(J15, O56, M10)$ are taken from Kacem et al. [2,3]. In our experiments, the algorithms used for comparison were MSPSO (Multi-start PSO), VTPSO (Velocity Turbulent PSO) and VNPSO (Variable Neighborhood PSO). The parameters c_1 and c_2 were set to 1.49 for all the PSO algorithms. Inertia weight w was decreased linearly from 0.9 to 0.1. In VTPSO and VNPSO, ρ and v_c were set to 2 and 1e-7 before 15,000 iterations, while they were set to 5 and 1e-10 after 15,000 iterations. The swarm size in all the algorithms were set to 20. The average fitness values of the best solutions throughout the optimization run were recorded. The averages (F) and

Table 1. Comparing the results for FJSPs

Instance	*Items*	MSPSO	VTPSO	VNPSO
	Best makespan	30	26	24
	Best flowtime	168	155	152
$(8, 27, 8)$	average	39.9087	28.3853	28.8000
	std	±5.7140	±2.3146	±3.8239
	time	184.0620	181.2500	181.2970
	Best makespan	19	13	11
	Best flowtime	96	92	75
$(10, 30, 10)$	average	19.4612	15.2000	15.0000
	std	±1.8096	±1.3166	±1.8257
	time	1.7145e+003	1.5891e+003	1.5908e+003
	Best makespan	36	30	29
	Best flowtime	231	241	220
$(15, 56, 10)$	average	37.2000	31.9000	30.8000
	std	±1.0328	±1.2867	±1.7512
	time	2.0497e+003	12.0816e+003	2.0703e+003

the standard deviations (std) were calculated from the 10 different trials. The standard deviation indicates the differences in the results during the 10 different trials. Usually another emphasis will be to generate the schedules at a minimal amount of time. So the completion time for 10 trials were used as one of the criteria to improve their performance. Figs. 2, 3 and 4 illustrate the performance for the three algorithms during the search processes for the three FJSPs. Empirical results are illustrated in Table 1. In general, VNPSO performs better than the other two approaches, although its computational time is worse than VTPSO for the low dimension problem, $(J8, O27, M8)$. VNPSO could be an ideal approach for solving the large scale problems when other algorithms failed to give a better solution.

5 Conclusions

In this paper, we introduce a hybrid metaheuristic, the Variable Neighborhood Particle Swarm Optimization (VNPSO), consisting of a combination of the Variable Neighborhood Search (VNS) and Particle Swarm Optimization(PSO), and considered its application for solving the multi-objective Flexible Job-shop Scheduling Problems (FJSP). The details of implementation for the multi-objective FJSP are provided and its performance was compared using computational experiments. The empirical results have shown that the proposed algorithm is an available and effective approach for the multi-objective FJSP, especially for large scale problems.

Acknowledgments. This work was supported by the Korea Research Foundation Grant funded by the Korean Government (MOEHRD, Basic Research Promotion Fund) (KRF-2006-353-B00059).

References

1. Mastrolilli M. and Gambardella L. M.: Effective neighborhood functions for the flexible job shop problem. Journal of Scheduling, (2002) 3(1):3-20.
2. Kacem I., Hammadi S., and Borne P.: Approach by localization and multiobjective evolutionary optimization for flexible job-shop scheduling problems. IEEE Transactions on Systems, Man and Cybernetics, (2002) 32(1):1-13.
3. Kacem I., Hammadi S. and Borne P.: Pareto-optimality approach for flexible job-shop scheduling problems: hybridization of evolutionary algorithms and fuzzy logic. Mathematics and Computers in Simulation, (2002) 60:245-276.
4. Ong Z. X., Tay J. C. and Kwoh C. K.: Applying the Clonal Selection Principle to Find Flexible Job-Shop Schedules. in: Jacob C. et al. (eds.), ICARIS2005, LNCS3627, 2005, 442-455.
5. Xia W. and Wu Z.: An effective hybrid optimization approach for multi-objective flexible job-shop scheduling problems. Computers and Industrial Engineering, (2005) 48:409-425.
6. Kennedy J. and Eberhart R.: Swarm Intelligence. Morgan Kaufmann (2001).
7. Parsopoulos K. E. and Vrahatis M. N.: Recent Approaches to Global Optimization Problems through Particle Swarm Optimization, Natural Computing, (2002) 1: 235-306.
8. Cristian T. I.: The particle swarm optimization algorithm: convergence analysis and parameter selection. Information Processing Letters, (2003) 85(6):317-325.
9. Hansen P. and Mladenović N.: Variable neighbourhood search: Principles and applications. European Journal of Operations Research, (2001) 130:449-467.
10. Hansen P. and Mladenović N.: Variable neighbourhood search. In: Glover F. W. and Kochenberger G. A. (eds.), Handbook of Metaheuristics, Dordrecht, Kluwer Academic Publishers (2003).
11. Liu H. and Abraham A.: Fuzzy Adaptive Turbulent Particle Swarm Optimization, in: Proceedings of the Fifth International conference on Hybrid Intelligent Systems, (2005) 445-450.

Optimal Designing of Multi-channel WDM Filter Using Intelligent Particle Swarm Optimization Algorithm

Yumin Liu[1,2] and Zhongyuan Yu[1,2]

[1] School of Science, Beijing University of Posts and Telecommunications, 100876, Haidian District Beijing, China
[2] Key Laboratory of Optical Communication and Lightwave Technologies, Ministry of Education 100876, Haidian District Beijing, China
liuyuminhqy@263.net, yuzhongyuan30@hotmail.com

Abstract. An innovation long period grating (LPG) synthesis method based on intelligent particle swarm optimization (PSO) algorithm is demonstrated to be very effective for designing flat band LPG filters. A flatten 3dB loss spectrum LPG with bandwidth over 100nm is designed as an example to show the effectiveness of the PSO algorithm. To improve the capability of optimization algorithm, we use the improved LPSO version proposed recently. This 3dB LPG is a key component of multi-channel filter in optical communications and optical sensors. The results showed that the intelligent PSO algorithm is very powerful and can be used for complex optimization problem.

1 Introduction

Fiber gratings have been evolved into one of the key optical components and found a multitude of applications in optical communications and optical sensor fields [1,2]. LPG is one of the most popular transmission-type grating devices, which couples the fundamental guided mode to one or more the phase-matched cladding modes. The LPG filters have proved to be very useful in band rejection filter [3], high sensitive temperature and strain sensors [4], and EDFA gain flattening [5]. Generally specking, the uniform LPG has little applications because of the limited transmission loss spectrum. In recent years, lots of synthesis methods are developed to design the index modulation profile that can produce the specific spectrum applications. These synthesis methods can be roughly divided into two groups: the fist group is inverse scattering based algorithm including the layer-peeling method, which can be used for both the LPG and fiber Bragg grating (FBG) designing [6,7]. The other kind of synthesis methods is the stochastic optimization approaches. Because the FBG or LPG filters synthesis is generally a complex optimization problem, the used optimization algorithms are not common sense's optimization algorithms. The evolutionary programming (EP) and genetic algorithm (GA) are important branches of the evolution algorithms, which are probabilistic search algorithms gleaned from the organic evolution process. EP is relative simple compared with the GA) and only use the mutation process of continuous variables and does not use the coding and crossover process. However, for both these methods, the controllable parameters are

T.-D. Wang et al. (Eds.): SEAL 2006, LNCS 4247, pp. 205 – 212, 2006.

problem sensitive and not easy selective properly. Recently, PSO has been proposed as a simple and alternative tool for solving optimization problems. PSO has been shown to be effective in optimization for multidimensional discontinuous problems. PSO has also been successfully applied in electromagnetic design problems. In this article, we have successfully utilized this intelligent optimization algorithm to design a LPG with flatten 3dB loss spectrum within the bandwidth about 100nm.

2 The Principle of PSO

The PSO algorithm, like GA, is population-based optimisation tool and can be applied to virtually any problem that can be expressed in terms of an objective function for which extremum must be found [8]. It emulates some aspects of social behaviour of a flock of birds and a school of fish. The PSO algorithm is iterative and involves initializing a number of vectors (particles) randomly within the search space. The collective of the particles is known as the swarm. Each particle presents a potential solution to the problem of the target function. At initial stage, each particle is also randomly initializing a vector called particle speed. During each time step (generation), the objective function is evaluated to establish the fitness of each particle using the particle position itself as the input parameter. Then the particle will fly through the search space being attracted to both their personal best position as well as the best position found by the swarm so far.

We make the position of particle i expressed as $X_i = (x_{i1,}, x_{i2} \ldots, x_{id})$ and the best position of that particle so far expressed $p_i = (p_{i1}, p_{i2} \ldots, p_{id})$. The best position of the whole swarm can be expressed as $p_g = (p_{g1}, p_{g2}, \ldots, p_{gd})$. Then the particle position update can be expressed as $x_{id} = x_{id} + v_{id}$ where the v_{id} denote the speed of the d dimension component of particle i and expressed as:

$$v_{id} = wv_{id} + \phi_1 rand()(p_{id} - x_{id}) + \phi_2 rand()(p_{gd} - x_{gd}) \tag{1}$$

where w is the inertia weight determining how much of the particle's previous speed is preserved, ϕ_1, ϕ_2 are two acceleration constants present the cognition part and the social part respectively, ***rand***() is uniform random sequences from $\{0,1\}$.

It should be noted that, each particle has the memory ability and could remember the best position it has been gone through. What's more, there is information flow among the particles, by which they could communicate with the global particles or with the special particles based on some topological neighbourhood structure. Each particle updates its speed and position based on its memory and communion information of the swarm. So we called the PSO algorithm as intelligent PSO algorithm.

The absolute value of the current speed determines the searching depth and expandability in solve spacing. The direction of the speed determines the path by which the particle could quickly or slowly go to the best solution. So the speed update is very important in the PSO algorithm. Fro the speed update, there are two basic versions: one employ the weight factor W, A large weight facilities the global exploration, while a small exploration facilities the local exploration; the other one employ the K factor, which has proved to be better than the first one in success probability and searching speed . Thus, in later sections, the position vectors and speed vectors in our paper are mainly based on the constriction factor factor K technique, which describes in detailed as follows:

$$v_{id} = \begin{cases} K\left(v_{id} + \phi_1 rand()\left(p_{id} - x_{id}\right)\right. \\ \quad \left. + \phi_2 rand()\left(p_{gd} - x_{id}\right)\right), \ x_{\min} < x_{id} < x_{\max} \\ v_{id} = 0, \ \ otherwise \end{cases} \tag{2}$$

$$x_{id} = \begin{cases} x_{id} + v_{id}, & x_{\min} < x_{id} < x_{\max} \\ x_{\max}, & x_{id} \geq x_{\max} \\ x_{\min}, & x_{id} \leq x_{\min} \end{cases} \tag{3}$$

where K is computed as: $K = 2\left|2 - \varphi - \sqrt{\varphi^2 - 4\varphi}\right|^{-1}$, $\varphi = \phi_1 + \phi_2$, $\varphi > 4$. ϕ_1, ϕ_2 are constants represents the social learning and self recognition components and set 1.1 and 3.0 respectively in our algorithm. Just as can be seen in (2), we used the absorber boundary conditions when the one components of the particle vector overcome the boundary. In other PSO literatures, boundary selection problem have been studied and show that reflection and trap boundary generally can produce better performance.

The iterative process will continue using the formula (2) until the extremum has been found or the number of iteration reached the maximum value. The algorithm in pseudo-code follows:[9]

Intialize population
Do
For $i = 1$ to population swarm size
If $f(\vec{x}_i) < f(\vec{p}_i)$ then $\vec{p}_i = \vec{x}_i$
$\vec{p}_g = \min\left(\vec{p}_{neighors}\right)$
  for $d = 1$ to Dimension
$v_{i,d} = wv_{i,d} + \varphi_1 rand()\left(p_{i,d} - x_{i,d}\right)$
  $+ \varphi_2 rand()\left(p_{g,d} - x_{i,d}\right)$
  $v_{i,d} = sign\left(v_{i,d}\right)\min\left(abs(v_{i,d}), v_{\max}\right)$
  $x_{i,d} = x_{i,d} + v_{i,d}$
 Next d
Next i
Until termination criterion is met

The ability of avoiding being trapped into the sub optimal is an important factor to evaluate the performance of an algorithm. Fro the PSO algorithm, it provides the selective topological structures to solve this problem. The two most common used structures are known as a global neighbor and a local neighbor. In the global neighborhood structure, the trajectory of each particle's search is influenced by the best position found so far and the other particles of the entire swarm. The local neighborhood structure allows each individual particle to be influenced by only some small number of adjacent members. A kind of lore has evolved regarding these sociometric structures. It has been thought that the global structure converges rapidly on problem solutions but has a weakness of being trapped into local optima, while the local structure can flow around local optima and avoids being trapped into local optima. In this paper we use two possible structures just as in reference [7] Generally, there are two kinds of topological neighbourhood structures, global neighbourhood structure, corresponding to the global version of PSO (GPSO), and local neighbourhood structure, corresponding to the local version of the PSO (LPSO). In the GPSO, each particle's search is influenced by the best position found by any member of the entire population. In contrast, in LPSO, the search is influenced only by parts of the adjacent members. It is widely believed that GPSO converges quickly to an optimum but has the weakness of being trapped in local optima occasionally, while the LPSO is able to "flow around" local optima, because the sup-swarm explore different regions in hyperspace. To improve the search ability, for the simple applications, we can use the GPSO algorithm and get the optimised solution in short searching time. For the complex applications, we used the LPSO algorithm to improve the probability of finding the best global solution. The two possible topologies for LPSO are shown in Fig.1 (a) [10].

3 Analysis and Optimization of LPG

The coupled mode equations describing the coupling in long period fibre gratins are given by [11]

$$dA/dz = -ikB - i\delta A$$

$$dB/dz = -ikA + i\delta B \tag{4}$$

where A B are respectively the amplitude of the guide mode and cladding mode, k is the couple strength, $\delta = \pi\Delta_{neff}/\lambda - \pi/\Lambda$ represents the detuning and $\Delta_{neff} = n_{cor} - n_{cla}$, where n_{cor} and n_{cla} are the effective refractive index, and Λ is the period of the LPG. For the uniform LPG there exist analytic solution for the equation (2). If only one cladding mode is excitated, the un-uniform LPG of the equation can be solved by the transfer-matrix method. The LPG is assumed to be divided into N sections and each section is treat as a uniform LPG in index modulation and period, and then the transmission characteristics of the whole grating can be expressed as $T = \prod_{i=1}^{N} T_i$, where T_i is the transmission matrix of the section i and can be seen in general grating literature, we not listed here. In our problem, we first define an error function to evaluate the fitness of the particles (each particle is a set of the index modulations). The spectral window are divided into m discrete wavelength, the sum of the weighted errors is normally used as cost function.

$$Err(Particle_i) = \frac{1}{m}\sum_{l=1}^{m}\left(\frac{T_{target,l} - T_{i,l}}{\sigma_l}\right)^2, i = 1,2\ldots N \tag{5}$$

where the $Err(Particle_i)$ is the deviation of the calculated spectrum of particle i from the target spectrum, $T_{target,l}$ is the target spectrum component at the sampling point l, $T_{i,l}$ is the calculated spectrum component of particle i at the sampling point l, and σ_l is the uncertainty at the sampling point l. In the simulation, we have to specify the upper and lower bounds of the grating physical parameters that should be optimized. The N particles are randomly initialized and each particle is a set of the grating index modulation parameters. In the upper and lower bonds the parameters values are continuous. The parameters of PSO algorithm is set as: ϕ_1=1.1, $\phi_2 = 3.0$, $N = 20$, $l = 500$, and w is tuneable parameter expressed as:

$$w(iter) = w_{\max} - (iter / iter_{\max})(w_{\max} - w_{\min}) \tag{6}$$

where $iter$ and $iter_{\max}$ are the current and the maximum iteration number respectively. The advantage of tuneable inertia weight are as follows: At the beginning stage, the inertia weight can be set a large value, so we can expand the searching space, when the potential particle has been approached to the best solution, the inertia weight must be very small, which make the PSO can implement elaborate searching around the best solution and avoiding escaping from the best solution to bad solutions. To improve the convergence solution to be the best results, we used local PSO (LPSO) where a special neighbour topology is introduced [10]. This can be seen in Fig.2, which shows two possible topologies used in our simulation.

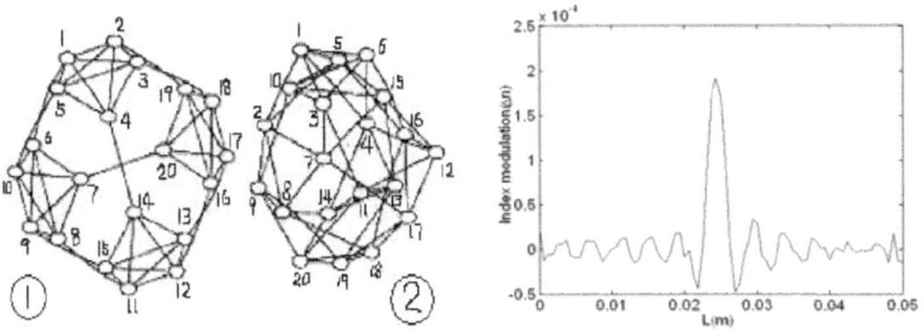

Fig. 1. Two possible particles swarm local topological structure (a), the optimized index modulation envelop for 3dB LPG transmission spectrum (b)

4 Numerical Results and Discussions

Mini Das proposed an improved version to realize wavelength multiplexing isolation filter using concatenated chirped LPG [12]. Although the introducing of chirp increased the useful channels by extending the loss window of the single LPG, however, to get high isolation between the channels, the key factor is to design the flat loss spectrum of the single LPG approach to 50%. The loss spectrum of the chirped LPG is approximately to be parabolic curve, which limited the useful channels, otherwise, in the chirp-type cascaded filter, the isolation undulation among the channels are very large. To overcome these problems, the best method is to design a wide bandwidth 3dss LPG filter within the bandwidth of 100nm, and the characteristics of the LPG pair interferometer made of two identical designed LPGB LPG filter. In the first example, we designed a 3dB los sandwiched a standard fiber are also discussed.

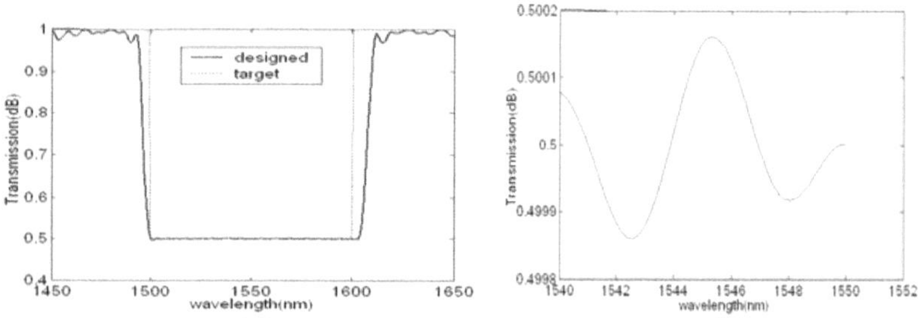

(a) (b)

Fig. 2. The designed (solid line) and target (dotted line) transmission spectrum for the cladding mode (a), the transmission spectrum ripples of the designed LPG

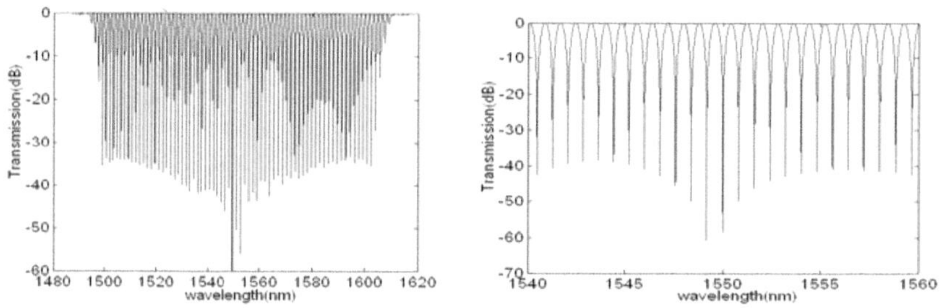

(a) (b)

Fig. 3. (a) the interference spectrum of the LPGs pair with the inserted single mode fibre is 20cm, (b) part of the transmission spectrum of (a)

The target spectrum loss in the region of 1500nm to1600nm should be 3dB, and outer the bandwidth the spectrum loss is 0. In our problem we mainly focus on the 3db loss bandwidth range. Fig.1 (b) shows the optimized index modulations envelop using the improved LPSO method. In our simulation, we divided the fibre grating into 80 sections of the same length and assumed the index modulation in each section is unchanged; the whole length of the LPG is 5cm, the center resonance wavelength is 1550nm, the grating period is uniform along the whole LPG, the effective index difference between the core mode and cladding mode is 0.01. As can be seen, the designed index modulation profile is just like a sinc function, and the envelope is approximately symmetry. The maximum index modulation position is located in the center of the LPG, and the index modulation amplitude is less than 2×10^{-4}, so it is easy to realize such device for the low index modulation. Fig.2(a) shows the output spectrum of the core mode of the single stage LPG calculated by the couple mode equation using the transfer matrix method. As can be seen, within the 100nm bandwidth that starts from 1500nm to 1600nm, the designed spectrums agree well with the target spectrums. There is a little divergence around the side band where the target transmission loss spectrum is a steep step, and break the continuity fortunately, however, those parts are not useful to our problem, and we are not concern about those region. Fig.2 (b) shows the divergence degree of the designed spectrum from the target spectrum within part of the useful bandwidth, we can see the divergence ripple is less than ±0.0002 , the influence of the divergence on the transmission to the LPG pair interferometer can be ignored.

Fig.3 (a) shows the interference spectrum of the two identical optimized LPGs pair interferometer that made by cascading two LPGs together, and length of the sandwiched fibre between them is 20cm. We can see, in the bandwidth scope of more than100nm, the isolation degree at the stop bands is larger than 30db. What's more, the isolation undulation among channels is very small, and almost all the channels are in equal operation in the 100nm bandwidth region. This is because of the little divergence between the designed spectrum and that of the idea 3dB transmission loos filter. The channel spacing is about 1nm, so there are more than 100 useful equal spaced channels produced by the LPGs pair. To give a clear picture, Fig.3 (b) shows part transmission spectrum of Fig.3 (a), in the central part, the isolation degree is about 40dB. The channel spacing can be tuned by changing the length of sandwiched fibre. with increasing the length of sandwiched fibre, the channel spacing will become narrower and we will get more useful channels.

5 Conclusions

In conclusion, we proposed a novel synthesized method based on the improved intelligent LPSO algorithm. To make the algorithm convergence to a global optimal or a better sub-optimal, a special topology is used in our problem. An example to design a flatten loss spectrum in a large bandwidth that approach 50% is demonstrated to test the effectiveness of the algorithm. From the optimized LPG's sinc-type index modulation profile, we obtained a flat transmission spectrum in bandwidth of 100nm with the transmission loss approximately to be 50%. A numerical simulation showed the good performance of the cascaded isolation filter consisted of two identical LPGs. Based on the studied results, we believe that the LPSO algorithm is an effective inveiglement algorithm for optimally designing complicated LPG and other fibre grating filters. Since the LPSO is a stochastic search approach in nature, the required computation time cannot be precisely predicted. However, the advantage computer hardware can remedy this disadvantage. Otherwise, we may change some PSO control parameters to accelerate the convergence speed.

Acknowledgments

This work was supported by the National "973" Basic Research and Development Program of China (No.2003CB314901), the National "863" High Technology Project of China (No. No.2003AA311070).

References

1. Xuewen Shu, Lin Zhang and Ian Bennion, "Sensitivity characteristics of long period fiber gratings," IEEE J. Lightwave Technology, vol.20,no.2, pp.255-266.
2. K. W. Yang, A. G. Liu, Chih-Chun Cheng and Yu-Lung Lo, "Topology and shape optimizations of substrates for chirp fiber bragg grating spectrum tuning," IEEE J. Lightwave Technology, vol.20, no.7, pp.1182-1186, 2002.
3. Ashiish M, Vengsarkar, Paul J. Lemaire, Justin B. Judkins, Vikram Bhatia, Turan Erdogan, and John E. Sipe, "Long period fiber gratings as band-rejection filters," IEEE J. Lithrwave Technology, vol.14, no.1, pp.58-64, 1996
4. Donghui Zhao, Xuewen Shu, Yicheng Lai and Ian Bennion, "Fiber Bragg grating sensor interrogation using chirped fiber grating-based sagnac loop," IEEE J.Sensors, vol.3, no.6, pp734-738.
5. Bai-ou Guang, A-Ping Zhang, Hwa-Yaw Tam, Helen L.W.Chan, Chung-LoongChoy, "Step changed long-period fiber gratings," IEEE Photonics Technology Lett., vol.14, no.5, pp.657-659, 2002
6. L. Wang and T. Erdogan, "Layer peeling algorithm for reconstruction of long- period fiber gratings," IEEE Electronics Lett., vol.37, no.3, pp.154-156, 2001
7. Johannes Skaar, Ligang Wang and Turan Erdogan, "On the synthesis of fiber Bragg gratings by layer peeling," IEEE Lightwave Technology , vol.37, no.2, pp.165-173, 2001
8. Carlos A. Coello, Gregorio Toscano Pulido and Maximino Salazar Lechuga, "Handing Multiple objectives with particle swarm optimization," IEEE Transcations on Evolutionary computations, vol.8, no.3, pp.256-279, 2004
9. Yumin Liu, Zhongyuan Yu, Hongbo Yang, Na Zhang, Qiang Feng and Xiaoguang Zhang, "Numerical optimization and simulation to wavelength-division multiplexing isolation filter consisted of two identical long period fiber grating," Optics Communications 246, pp.367-372, 2005
10. Rui Mendes, James Kennedy and Jose Neves, "The Fully informed particle swarm: simpler, maybe better", IEEE Transactions On Evolutionary computation, vol.8, no.3, pp. 204-210, 2004
11. Yumin Liu, Zhongyuan Yu, Jianzhong Zhang, Bojun Yang, and Xiaoguang Zhang, "Optimization design of flat-band long-period grating," Chinese Optics Letters, vol.2, no.3, pp: 200-202, 2004
12. Mini Das and Krishna Thyagarajan, "Wavelength -division multiplexing isolation filter using concatenated chirped long period gratings," J. Optics Communications, 197, pp. 67-71, 2001

Adaptive Comprehensive Learning Particle Swarm Optimizer with History Learning

J.J. Liang and P.N. Suganthan

School of Electrical and Electronic Engineering,
Nanyang Technological University, Singapore 639798
liangjing@pmail.ntu.edu.sg, epnsugan@ntu.edu.sg

Abstract. This paper proposes an Adaptive Comprehensive Learning Particle Swarm Optimizer with History Learning (AH-CLPSO) based on the previous proposed Learning Particle Swarm Optimizer (CLPSO) [1], which is good at multimodal problems but converges slow on single modal problems. A self-adaptation technique is introduced to adjust the learning probability adaptively in the search process and the historical information is used in the velocity update equation to search more effectively. The experiment results show that the history learning strategy and the adaptation technique improves the performance of CLPSO on problems which need fast convergence and achieve comparable results on the problems requiring slow convergence.

1 Introduction

The Particle Swarm Optimizer (PSO) [2, 3] is introduced by Eberhart and Kennedy as a new optimization technique in 1995. Different from other evolutionary computation techniques, the standard PSO does not use evolution operators such as crossover and mutation. It emulates the swarm behavior of insects, birds flocking and fish schooling and each member in the swarm adapts its search direction by learning from its own experience and other members' experiences.

In PSO, a member in the swarm, called a particle, represents a potential solution which is a point in the search space. Each particle has a fitness value and a velocity to adjust its flying direction and the particles fly in the D dimensional problem space by learning from the historical information of all the particles. Using the useful information collected in the search process, the particles have a tendency to fly towards better search area over the course of search process. The velocity V_i^d and position X_i^d updates of dth dimension of the ith particle are presented below:

$$V_i^d = w * V_i^d + c_1 * rand1_i^d * (pbest_i^d - X_i^d) + c_2 * rand2_i^d * (gbest^d - x_i^d) \tag{1}$$

$$X_i^d = X_i^d + V_i^d \tag{2}$$

where c_1 and c_2 are the acceleration constants, $rand1_i^d$ and $rand2_i^d$ are two uniformly distributed random numbers in the range [0,1]. $\mathbf{X}_i = (X_i^1, X_i^2, ..., X_i^D)$ is the position of the ith particle; $\mathbf{pbest}_i = (pbest_i^1, pbest_i^2, ..., pbest_i^D)$ is the best previous

T.-D. Wang et al. (Eds.): SEAL 2006, LNCS 4247, pp. 213 – 220, 2006.

position yielding the best fitness value $pbest_i$ for the ith particle; $\mathbf{gbest} = (gbest^1, gbest^2, ..., gbest^D)$ is the best position discovered by the whole population; $\mathbf{V}_i = (v_i^1, v_i^2, ..., v_i^D)$ represents the rate of the position change (velocity) for particle *i*. *w* is the inertia weight used to balance between the global and local search abilities [4]. Except the inertia weight, by analyzing the convergence behavior of the PSO, a PSO variant with a constriction factor was introduced by Clerc and Kennedy [5]. Constriction factor guarantees the convergence and improves the convergence velocity.

In the PSO domain, there are two main variants: global PSO and local PSO. In the local version of PSO, each particle's velocity is adjusted according to its personal best and the best performance achieved so far within its neighborhood instead of learning from the personal best and the best position achieved so far by the whole population as in the global version. The velocity updating equation becomes:

$$V_i^d = w * V_i^d + c_1 * rand1_i^d * (pbest_i^d - X_i^d) + c_2 * rand2_i^d * (lbest_i^d - x_i^d) \quad (3)$$

where $\mathbf{lbest}_i = (lbest_i^1, lbest_i^2, ..., lbest_i^D)$ is the best position achieved within its neighborhood. Kennedy [6] claimed that PSO with a small neighborhood might perform better on complex problems while PSO with a large neighborhood would perform better on simple problems.

In the 10 years of development, PSO has attracted a high level of interest and many different interesting and efficient versions are proposed. For example, Parsopoulos and Vrahatis combined the global version and local version together to construct a Unified Particle Swarm Optimizer (UPSO) [7]. Mendes and Kennedy introduced a fully informed PSO in [8] in which instead of using the **pbest** and **gbest** positions in the standard algorithm, all the neighbors of the particle are weighted based on its fitness value and the neighborhood size to update the velocity. Veeramachaneni *et al.* developed the Fitness-Distance-Ratio based PSO (FDR-PSO), with near neighbor interactions [9]. When updating each velocity dimension, the FDR-PSO algorithm selects one other particle, *nbest*, which has a higher fitness value and is nearer to the particle being updated. A Cooperative Particle Swarm Optimizer (CPSO-H) was proposed in [10]. Although CPSO-H uses 1-*D* swarms to search each dimension separately, the results of these searches are integrated by a global swarm to significantly improve the performance of the original PSO on multimodal problems.

Comprehensive Learning Particle Swarm Optimizer proposed in [1] is also one of those variants. It employs a comprehensive learning strategy where other particles' previous best positions are exemplars to be learned from by any particle and each dimension of a particle can potentially learn from a different exemplar. With the comprehensive strategy the particles have more exemplars to learn from, a larger potential space to fly and can make use of the information in swarm more effectively to generate better quality solutions frequently when compared to some other PSO variants. The CLPSO is not the best choice for solving unimodal problems due to its slow convergence. Thus in this paper we propose a new adaptive CLPSO with history learning (AH-CLPSO) to improve the performance of the original CLPSO.

In Section 2, the CLPSO algorithm will be introduced briefly, the history learning strategy and the employed adaptation method is analyzed, then the whole algorithm is summarized in the same Section. The experiments results are presented and discussed in Section 3 and the conclusion is given in Section 4.

2 Adaptive CLPSO with History Learning

2.1 Comprehensive Learning Particle Swarm Optimizer

Before we introduce the adaptive CLPSO with History Learning, some brief description of the CLPSO algorithm is necessary. Though much better results are achieved, CLPSO is very simple, the only difference between the original PSO and CLPSO is the velocity updating equation:

$$V_i^d \leftarrow w * V_i^d + c * rand_i^d * (pbest_{fi(d)}^d - X_i^d) \tag{4}$$

where $f_i = [f_i(1), f_i(2), ..., f_i(D)]$ defines which particle's pbest particle i should follow. $pbest_{fi(d)}^d$ can be the corresponding dimension of any particle's pbest including its own pbest, and the decision depends on probability *Pc*, referred to as the learning probability. For each dimension of particle *i*, we generate a random number. If this random number is larger than Pc_i, this dimension will learn from its own pbest, otherwise it will learn from another particle's pbest. When choosing an exemplar for one dimension of a particle, first randomly choose two particles out of the population which excludes the particle whose velocity is updated, then compare the fitness values of these two particles' **pbests** and select the better one as the exemplar to learn from for that dimension. If all exemplars of a particle are its own **pbest**, we will randomly choose at least one dimension to learn from another particle's **pbest**'s corresponding dimension. New exemplars are chosen for a particle when it fails to improve itself, say finds better position, for *m* generations. Here *m* is called refreshing gap. The flowchart of CLPSO is given in Fig.1.

Step 1: Initialize the particles (position **X** and velocity **V**), calculate *f*(**X**) for each particle. Set **pbest=X**. Define Pc_i and find exemplars for each particle *i*.
Step 2: For each particle, update the velocity $\mathbf{V}_i$ and position $\mathbf{X}_i$ as eq. (4) and eq. (2). Update $\mathbf{pbest}_i$ if better position is found by the particle.
Step 3: Re-choose exemplars for the particle which has not be improved for *m* generations.
Step 4: If no stop criterion is satisfied, go to Step 2.

Fig. 1. The flowchart of CLPSO

2.2 History Learning Strategy

In the original PSO or in CLPSO, only the best positions achieved so far in the search process are directly made use of and no historical search direction information is considered in the updating equation. Velocity seems keeps some historical information, but in fact its main function is giving the particle inertia and adding the diversity of the particles to prevent the premature convergence. If $pbest_i^t$ represents the best position found by particle *i* in *t* generations, and $pbest_i^{t-T}$ represents **pbest** of particle *i* of the $(t\text{-}T)^{th}$ generation, then the vector $\boldsymbol{\Delta}_i = \mathbf{pbest}_i^t - \mathbf{pbest}_i^{t-T}$ represents the improving direction of the passed *T* generations. Illustration of $\boldsymbol{\Delta}_i$ is shown in Fig.2. Obviously $\boldsymbol{\Delta}_i$ is a promising search direction of particle *i*. If $\mathbf{pbest}_i^t = \mathbf{pbest}_i^{t-T}$, $\boldsymbol{\Delta}_i$=0. It means this particle has stop improving for *T* generations, so it's no meaning to continue the search on this direction. Under such situation, this history learning will lose its function and wait for the next good direction.

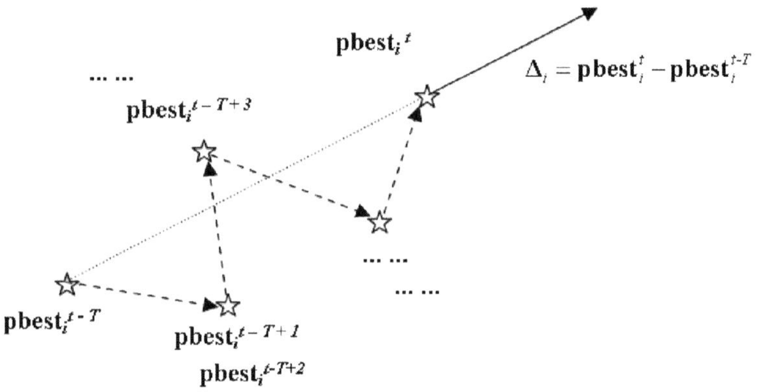

Fig. 2. Illustration of the improving direction $\Delta_i = \mathbf{pbest}_i^t - \mathbf{pbest}_i^{t-T}$

In order to push the particles to fly to the promising correct direction and speed up the search, combining $\boldsymbol{\Delta}_i$ into the velocity updating equation is a reasonable choice. The new updating equation is:

$$\mathbf{V}_i \leftarrow w * \mathbf{V}_i + c_1 * \mathbf{rand}_i * (\mathbf{pbest}_{\mathbf{fi}} - \mathbf{X}_i) + c_2 * \boldsymbol{\Delta}_i \tag{5}$$

From the authors' observation, the history learning strategy works and can achieve better results than the old CLPSO with a proper *Pc* setting. But it becomes sensitive to the parameter *Pc*. Thus making *Pc* self-adaptive is necessary.

2.3 Self-adaptive *Pc*

In the previous study, it is observed that the CLPSO requires *Pc* values in order to solve different problems effectively. In order to deal with this problem, an adaptation technique is employed to adjust *Pc*. In the initialization stage, $\overline{Pc}$ is predefined to take different random values in the range [0,1] with normal distribution of mean $\overline{Pc}$ and standard deviation 0.1. Every T generations, $f(\mathbf{pbest}_i^{t-T}) - f(\mathbf{pbest}_i^t)$ is sorted to identify the best Pc_i achieving the biggest improvement and the best Pc_i chosen as the new $\overline{Pc}$ and a new set of Pc_i is generated using the new $\overline{Pc}$. Hence, $\overline{Pc}$ will learn a proper value for different problems and during different stages of evolution. In order to have a larger diversity in the initial stages, a small initial $\overline{Pc}$ is recommended.

To test the efficiency of the adaptive *Pc*, CLPSO with history learning is tested on 10 dimensional Rosenbrock's Function and Rastrigin's Function with different *Pc* values and with the proposed adaptive *Pc* where the initial $\overline{Pc}$ is set at 1/D. The mean results of 10 runs are shown in Fig. 3. From the results, it is clear that these two problems show two extreme situations: Rosenbrock's Function performs better with a larger *Pc* and the best result is achieved when *Pc*=1; While Rastrigin's Function yields better results with smaller *Pc* and the best *Pc* value is 0. As expected, the adaptive *Pc* performs well on both test functions though it does not achieve the best results in comparison to a well tuned fixed Pc value.

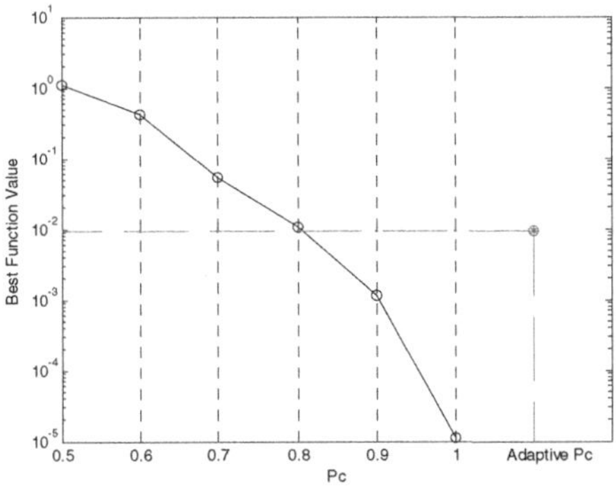

(a) Rosenbrock's Function

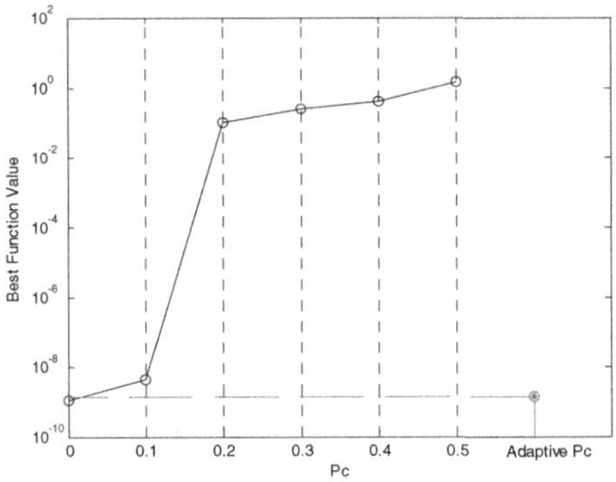

(b) Rastrigin's Function

Fig. 3. Comparison of Results of CLPSO with history learning with different *Pc* and the adaptive *Pc*

2.4 Adaptive CLPSO with History Learning

The entire flowchart of the adaptive CLPSO with history learning is concluded below:

Step 1: Initialize the particles (position **X** and velocity **V**), calculate $f(\mathbf{X})$ for each particle. Set **pbest=X** and fill pbest history archive **pbest_hist** with **pbest.** Generate Pc according to the predefined $\overline{Pc}$ and find exemplars for each particle i as described in Section 2.1.
Step 2: For each particle, update the velocity $\mathbf{V}_i$ and position $\mathbf{X}_i$ using eq. (5) and eq. (2). Update $\mathbf{pbest}_i$ if better position is found by the particle. Add $\mathbf{pbest}_i$ into $\mathbf{pbest\_hist}_i$ and remove the member in $\mathbf{pbest\_hist}_i$ which is older than T.
Step 3: Re-choose exemplars for particles which stop improving for m generations
Step 4: Every T generations, check the improvement of each particle and move $\overline{Pc}$ to Pc_i whose corresponding particle has the biggest improvement.
Step 5: If no stop criterion is satisfied, go to Step 2.

Fig. 4. The flowchart of the AH-CLPSO

4 Experiments

Experiments were conducted to compare ten PSO algorithms including the CLPSO algorithm with the new proposed AH-CLPSO on 4 10 dimensional test problems. The first two problems favour the algorithm which converges fast and has good local search ability and the other two can be solved by having big diversity and has good global search ability. The equations of the four functions are listed below and the global optimum, search ranges and initialization ranges of the test functions are given in Table 1.

1) Sphere function
$$f_1(x) = \sum_{i=1}^{D} x_i^2 \tag{6}$$

2) Rosenbrock's function
$$f_2(x) = \sum_{i=1}^{D-1} (100(x_i^2 - x_{i+1})^2 + (x_i - 1)^2) \tag{7}$$

3) Rastrigin's function
$$f_3(x) = \sum_{i=1}^{D} (x_i^2 - 10\cos(2\pi x_i) + 10) \tag{8}$$

4) Schwefel's function
$$f_4(x) = 418.9829 \times D - \sum_{i=1}^{D} x_i \sin(|x_i|^{1/2}) \tag{9}$$

Table 1. Global optimum, search ranges and initialization ranges of the test functions

f	x^*	$f(x^*)$	Search Range	Initialization Range
f_1	[0,0,…,0]	0	$[-100, 100]^D$	$[-100, 50]^D$
f_2	[1,1,…,1]	0	$[-2.048, 2.048]^D$	$[-2.048, 2.048]^D$
f_3	[0,0,…,0]	0	$[-0.5, 0.5]^D$	$[-0.5, 0.2]^D$
f_4	[420.96, 420.96,…420.96]	0	$[-500, 500]^D$	$[-500, 500]^D$

The algorithms in the experiment include:

- ✧ PSO with inertia weight (PSO-w) [4]
- ✧ PSO with constriction factor (PSO-cf) [5]
- ✧ Local Version of PSO with inertia weight (PSO-w-local)
- ✧ Local Version of PSO with constriction factor (PSO-cf-local) [6]
- ✧ Unified Particle Swarm Optimization (UPSO) [7]
- ✧ Fully informed particle swarm (FIPS) [8]
- ✧ Fitness-distance-ratio based particle swarm optimization (FDR-PSO) [9]
- ✧ Cooperative particle swarm optimization (CPSO-H) [10]
- ✧ Comprehensive Learning particle swarm optimizer (CLPSO) [1]
- ✧ Adaptive CLPSO with history learning (AH-CLPSO)

Among these PSO local versions, PSO_w_local and PSO_cf_local were chosen as these versions yielded the best results [6] with von Neumann neighborhoods where neighbors above, below, and one each side on a two-dimensional lattice were connected. Fully informed particle swarm (FIPS) with U-Ring topology that achieved the highest success rate [8] is used. The population size is set at 10 and the maximum fitness evaluations (FEs) is set at 30,000. The default parameters are used for the other algorithms and c_1=1, c_2=0.2, T=20, m=7 are set for AH-CLPSO. All experiments were run 30 times. When $(f(\mathbf{x})- f(\mathbf{x}^*))<=1e\text{-}10$, the final errors are reported as 0, denoting the algorithm has found the global optimum. The mean values and standard deviation of the errors are presented in Table 2. From the results, we can observe that the proposed adaptive CLPSO with history learning (AH-CLPSO) gives the best performance when compared to the other PSOs on different types of problems.

Table 2. Results achieved by varrious PSO varriants

PSOs \ Func	f_1	f_2	f_3	f_4
PSO-w	**0** **± 0**	3.08e+000 ± 7.69e-001	5.82e+000 ± 2.96e+000	3.20e+002 ± 1.85e+002
PSO-cf	**0** **± 0**	6.98e-001 ± 1.46e+000	1.25e+001 ± 5.17e+000	9.87e+002 ± 2.76e+002
PSO-w-local	**0** **± 0**	3.92e+000 ± 1.19e+000	3.88e+000 ± 2.30e+000	3.26e+002 ± 1.32e+002
PSO-cf-local	**0** **± 0**	8.60e-001 ± 1.56e+000	9.05e+000 ± 3.48e+000	8.78e+002 ± 2.93e+002
UPSO	**0** **± 0**	1.40e+000 ± 1.88e+000	1.17e+001 ± 6.11e+000	1.08e+003 ± 2.68e+002
FDR	**0** **± 0**	8.67e-001 ± 1.63e+000	7.51e+000 ± 3.05e+000	8.51e+002 ± 2.76e+002
FIPS	**0** **± 0**	2.78e+000 ± 2.26e-001	2.12e+000 ± 1.33e+000	7.10e+001 ± 1.50e+002
CPSO-H	**0** **± 0**	1.53e+000 ± 1.70e+000	**0** **± 0**	2.13e+002 ± 1.41e+002
CLPSO	**0** **± 0**	2.46e+000 ± 1.70e+000	**0** **± 0**	**0** **± 0**
AH-CLPSO	**0** **± 0**	**1.14e-002** **± 2.44e-002**	**0** **± 0**	**0** **± 0**

5 Conclusion

A novel adaptive comprehensive learning particle swarm optimizer with history learning algorithm (AH-CLPSO) is proposed in this paper. Through the analysis of the search behavior, a history learning strategy and an adaptive technique are combined in the old comprehensive learning particle swarm optimizer (CLPSO) to improve the performance of the CLPSO on certain types of the problems. The experiments show that the self-adaptation of the learning probability *Pc* is successful and the new algorithm can yield good results on problems requiring faster convergence and good local search ability or problems requiring slower convergence and good global search ability. Compared to the other variants of PSO and the old CLPSO, the improved CLPSO version, AH-CLPSO performs better. In the future work, we will study the history learning strategy's search behavior further and attempt to make more parameters self-adaptive.

References

1. J. J. Liang, P. N. Suganthan A. K. Qin and S. Baskar, "Comprehensive Learning Particle Swarm Optimizer for Global Optimization of Multimodal Functions", accepted by IEEE Transactions on Evolutionary Computation, to appear in June 2006.
2. R. C. Eberhart and J. Kennedy, "A new optimizer using particle swarm theory," *Proc. 6th Int. Symposium on Micromachine and Human Science,* Nagoya, Japan, pp. 39-43, 1995.
3. J. Kennedy and R. C. Eberhart, "Particle swarm optimization," *Proc. IEEE International Conference on Neural Networks,* pp. 1942-1948, 1995.
4. Y. Shi and R. C. Eberhart, "A modified particle swarm optimizer," *Proc. IEEE Congress on Evolutionary Computation*, pp. 69-73, 1998.
5. M. Clerc and J. Kennedy, "The particle swarm-explosion, stability, and convergence in a multidimensional complex space," *IEEE Trans on Evolutionary Computation*, 6(1):58-73, 2002.
6. J. Kennedy and R. Mendes, "Population structure and particle swarm performance," *P. IEEE Congress on Evolutionary Computation (CEC 2002),* Honolulu, Hawaii USA, 2002
7. K. E. Parsopoulos and M. N. Vrahatis, "UPSO- A Unified Particle Swarm Optimization Scheme," Lecture Series on Computational Sciences, pp. 868-873, 2004
8. R. Mendes, J. Kennedy, and J. Neves, "The fully informed particle swarm: simpler, maybe better," *IEEE Transactions on Evolutionary Computation*, 8(3):204 - 210, June 2004.
9. T. Peram, K. Veeramachaneni, and C. K. Mohan, "Fitness-distance-ratio based particle swarm optimization," *P. Swarm Intelligence Symposium*, Indiana, USA., pp. 174-181, 2003.
10. F. van den Bergh and A. P. Engelbrecht, "A cooperative approach to particle swarm optimization," *IEEE Transactions on Evolutionary Computation*, 8(3):225 - 239, June 2004.

Optimal Designing of EDFA Gain Flattening Long Period Fiber Grating by Intelligent Particle Swarm Optimization Algorithm

Yumin Liu[1,2] and Zhongyuan Yu[1,2]

[1] School of science, Beijing University of Posts and Telecommunications, 100876, Beijing China
[2] Key Laboratory of Optical Communication and Lightwave Technologies, Ministry of Education 100876, Haidian District Beijing, China
liuyuminhqy@263.net, yuzhongyuan30@hotmail.com

Abstract. An innovative long-period fiber grating (LPG) used for erbium-doped fiber amplifier (EDFA) gain flattening synthesized by the particle swarm optimization (PSO) algorithm is demonstrated. In our problem, we used the topological neighborhood local PSO algorithm to improve the performance, in addition, we used the damp boundary conditions to avoid the particles escaping out of the solve space. The simulated results are in good coincidence with design targets, and proved the capability and effectiveness of the algorithm. In addition, this algorithm is general and can be used for other similar synthesis problems of fiber Bragg gratings (FBGs).

1 Introduction

Fiber Bragg gratings have evolved into critical components for a multitude of applications in the optics fiber communication systems [1]. Among the various members of FBGs family, LPGs are exhibiting great promise as spectral-selective filters, due to unique features such as low insertion loss, low back-reflection, and excellent polarization-insensitivity [2]. It has proved that the LPGs are very useful devices in applications like band-reject filter, high sensitivity sensors, mode converters and ideal candidates for gain-flattening filter of EDFA etc. LPG is one type of transmission grating device based on the principle of coupling the guide fundamental mode to one or several forward propagating cladding modes. The advantage of the LPGs to FBGs is that the LPGs have a much greater periodicity and can be easily fabricated. In recent years, a lot of FBGs synthesis or inverse methods have been proposed such as Layer-peeling method, the Fourier transform technology, genetic algorithm and the Gel'fand-Levitan-Marchenko method [3][4][5][6]. Although extremely powerful, they require complete information on both the amplitude and the phase of the reflection or transmission coefficient. In this paper, another synthesis method for fiber grating based on PSO algorithm is proposed. This method is population-based evolutionary mechanism. To verify the effectiveness of the approach, we use PSO algorithm together with the transfer-matrix method based on the coupled mode theory to design a single LPG transmission spectrum used for EDFA gain

T.-D. Wang et al. (Eds.): SEAL 2006, LNCS 4247, pp. 221 – 227, 2006.

flattening. Because the phase information is unrelated to the EDFA gain flattening, we ignore the phase information in designing such a device.

2 Synthesis of LPG Using PSO

The PSO was first presented by Kenedy and Eberhart in 1995, which original from the social modal. For its simplex and powerful, PSO algorithm has achieved great successes and become one of special topics by 'CEC' [7][8]. The PSO algorithm can be applied to virtually any problem that can be expressed in terms of an objective function for which extremum must be found. The PSO algorithm is iterative and involves initializing a number of vectors (particles) randomly within the search space. In our problems the vectors of each particle represent the couple information along the grating. The collective of the particles is known as the swarm. Each particle presents a potential solution to the problem of the target function. Each particle also randomly initializes a vector called particle speed. During each time step, the objective function is evaluated to establish the fitness of each particle according the particle itself as the input parameter. Then the particle will fly through the search space being attracted to both their personal best position as well as the best position found by the swarm so far. The flow chart of the PSO algorithm can be seen in Fig.1 (a).

The position of particle i is expressed as $X_i = (x_{i1,}, x_{i2} \ldots, x_{id})$ and the best position of that particle so far is expressed $p_i = (p_{i1}, p_{i2} \ldots, p_{id})$. The best position of the swarm can be expressed as $p_g = (p_{g1}, p_{g2}, \ldots, p_{gd})$. Then the particle position update can be expressed as $x_{id} = x_{id} + v_{id}$ where the v_{id} denotes the speed of the d dimension component of particle i and is expressed as [9]:

$$v_{id} = wv_{id} + \phi_1 rand()(p_{id} - x_{id}) + \phi_2 rand()(p_{gd} - x_{gd}) \tag{1}$$

where the w is the inertia weight determining how much of the particle's previous speed is preserved, ϕ_1, ϕ_2 are two acceleration constants present the cognition part and the social part respectively, $rand()$ is uniform random sequences from $\{0,1\}$. In the stand PSO algorithm (1) only the inertia weight is introduced. To improve the performance, we introduce another parameter $K = 2 \cdot \left| 2 - \varphi - \sqrt{\varphi^2 - 4\varphi} \right|^{-1}$, where $\varphi = \varphi_1 + \varphi_2$, $\varphi > 4$, so the speed update can be expressed as: K

$$v_{id} = K\left(wv_{id} + \varphi_1 rand()\left(p_{id} - x_{id} \right) + \varphi_2 rand()\left(p_{gd} - x_{id} \right)\right) \tag{2}$$

The iterative process will continue using the formula (2) until the extremum has been found or the number of iteration has accessed to the maximum value. In the designing of our problem, the LPG for EDFA gain flattening, we select the parameter $\phi_1 = 1.1, \phi_2 = 3.0$ and w is tunable parameter expressed as:

$$w(iter) = w_{max} - (iter / iter_{max})(w_{max} - w_{min}) \tag{3}$$

where *iter* and $iter_{\max}$ are the current and the maximum iteration number respectively. In order to calculating the fitness, we divide the spectrum window into m discrete sampling wavelength, and try to design the transmission spectrum approach the target spectrum in according to the EDFA gain spectrum. We define the error function as:

$$Err(Particle_i) = \frac{1}{N}\sum_{l=1}^{m}\left(\frac{T_{target,l} - T_{i,l}}{\sigma_l}\right)^2, i = 1, 2 \ldots N \quad (4)$$

where $T_{i,j}$ is the transmission coefficient at the sampling point l of the particle i and $T_{target,l}$ is the target value. N is the number of the particle swarm and σ_i is the uncertainty with which this value is defined. Uncertainties σ_i should be smaller for the parts of the spectrum where it should be fitted better. We divided grating into equal M sections then the whole grating spectrum properties can be calculated by the transfer matrix method.

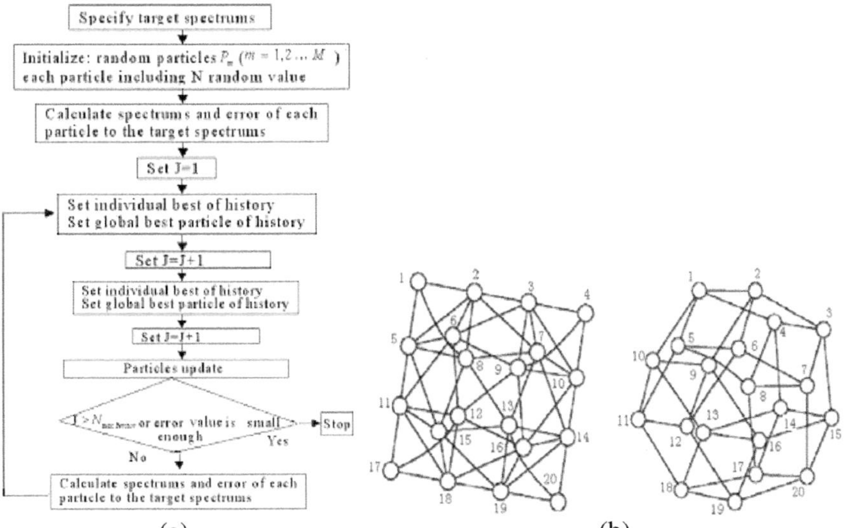

Fig. 1. The flow chart of the PSO for synthesizing the FBG's parameters (a), the possible structures used for the fully informed LPSO algorithm (b)

Although the PSO algorithm is easy to implement and has been empirically shown to perform well on many optimization problems. However, it may easily get trapped into a local optimum in solving complex problems. In order to avoiding such cases, we use the improved fully informed PSO algorithm to solve our problem. The two most common used structures are known as a global neighbor and a local neighbor. In the global neighborhood structure, the trajectory of each particle's search is influenced by the best position found so far and the other particles of the entire swarm. The local neighborhood structure allows each individual particle to be influenced by only some small number of adjacent members. A kind of lore has evolved regarding these social structures. It has been thought that the global structure converges rapidly on problem solutions but has a weakness of being rapidly on problem solutions but has a weakness of being trapped into local optima, while the local structure can flow around local optima and avoids being trapped into local optima. In this paper we use two possible structures just as in reference [7,9] Generally, there are two kinds of topological neighborhood structures, global neighborhood structure, corresponding to the global version of PSO (GPSO), and local neighborhood structure, corresponding to the local version of the PSO (LPSO). For the global neighborhood structure the whole swarm is considered as the neighborhood, while for the local neighborhood structure some smaller number of adjacent members in sub-swarm is taken as the neighborhood. In the GPSO, each particle's search is influenced by the best position found by any member of the entire population. In contrast, in LPSO, the search is influenced only by parts of the adjacent members. It is widely believed that GPSO converges quickly to an optimum but has the weakness of being trapped in local optima occasionally, while the LPSO is able to "flow around" local optima, because the sup-swarm explore different regions in hyperspace. To improve the search ability, for the simple applications, we can use the GPSO algorithm and get the optimized solution in short searching time. For the complex applications, we used the LPSO algorithm to improve the probability of finding the best global solution. The two possible topologies for LPSO are shown in Fig.1 (b) [12].

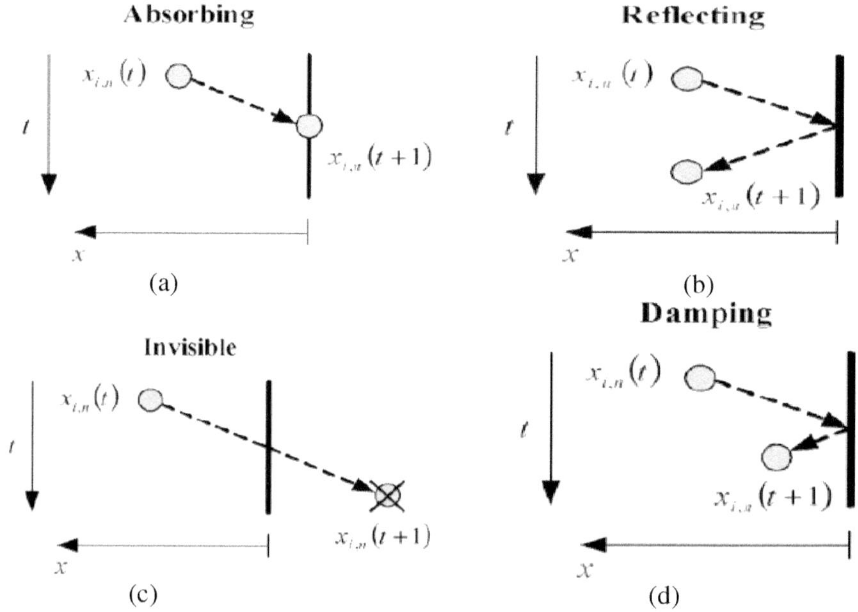

Fig. 2. Four boundaries operations for particles across the boundary of the solve space

In most cases, a parameter V_{max} acts as an upper limit for the achievable velocity of the particles to search and should be confined within the problem solve space. However, in some conditions the position update may lead the next position go through the boundary and hence induce an invalid solution. To solve this problem, one can define several methods. First, if the one dimension of the particle across the boundary in the next update, we make an absorbing boundary, so we make the boundary value as the update value of that dimension of the particle. Second, we can make a reflecting boundary, by doing this operation, the update value will reflected into the solve space again. Third, the damping boundary, by doing this operation, part of the velocity is absorbed by the boundary, during the collision of the particle and the boundary, then the particle if reflected into the solve spacing with a less velocity. Another boundary condition is invisible boundary, in which the particle is allowed to escape the solve space, but the fitness evaluator is ignored. The feature of each boundary condition is illustrated in Fig. 2 (a) (b) (c) (d). In many practical optimization problems, it is desirable to have a single boundary condition that can offer a robust and consistent performance for the PSO technique regardless of the problem dimensionality and the location of the global optimum. In this paper, we used the damping boundary to avoiding the particles escaping out of the space, this boundary condition combine the advantage of the reflective and absorbed boundary conditions and has been prove to be very powerful [12].

3 Numerical Example

To demonstrate the effectiveness of the PSO based synthesis algorithm, a practice design example is given in this section. We designed a LPG used for EDFA gain flattening. In literature, EDFA gain flattening technology is proposed by blaze grating or cascade of one Phase-shifted LPG and one normal LPG [10][11]. In this article we have succeed in designing a single stage LPG for EDFA gain -flattening of the entire C-band by the PSO algorithm proposed above. The grating length is 10cm, and we divide the grating into 20 uniform sections, and the spectrum points are chosen to be $m = 501$, and the center grating period is $1550nm$. The effective core and cladding mode index are assumed to be 1.45 and 1.46. Fig.3 (a) gives the designed and target transmission spectrum. Fig.3 (b) shows the EDFA gain curve before and after flattened. The curve can be flattened to be less $\pm 0.4db$ within the

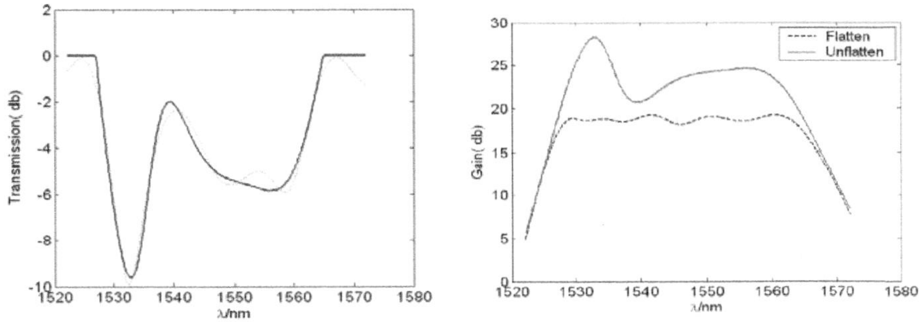

(a) (b)

Fig. 3. (a) The designed spectrum (dotted line) and the target spectrum (solid line), (b) Flattened gain profile (dashed line) and original gain profile (solid line) of a typical EDFA

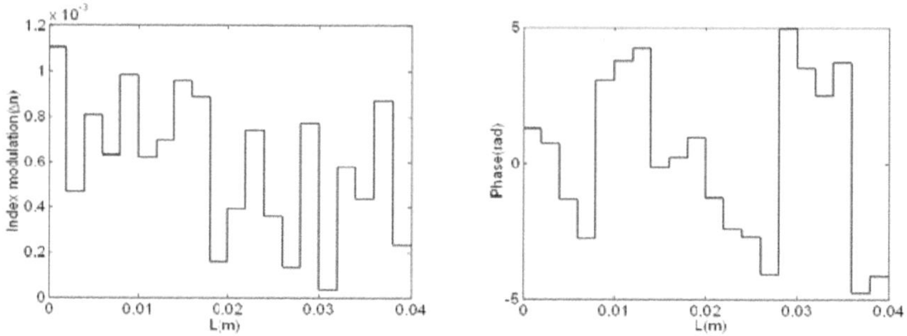

(a) (b)

Fig. 4. The index amplitude modulation (a) and phase modulation (b) distribution along the optimised LPG for EDFA gain flattening

band of 35nm. Fig.4(a) and (b) give the couple coefficient of the whole grating including the amplitude and phase information. From Fig.4, we can see the designed parameters are not difficult to be used for fabricating by the side-writing technology. As the knowledge about author, it is the first time that the PSO algorithm is successfully used for designing only one stage of LPG for EDFA gain flattening in the whole C-band.

4 Conclusions

In conclusion, for the first time, we have presented a novel fiber grating synthesis method based on the PSO algorithm. We used an example to demonstrate the effectiveness and capability in solving such problems. The advantage of the PSO is that this method is not sensitive to the dimensions of the particle. So this method is especially useful for synthesis of fiber grating problems with many of parameters. But when the number of parameters is large, the probability of the algorithm trapped into the local best will increase. To improve the performance one has to use the improved PSO algorithm such as local PSO (LPSO) algorithm, which use the topology neighbor and divide the swarm of particles into many sub-groups and sub-group transfer the information by the interval particles between the sub-groups, can avoid to be trapped into the local best in the early stage and we will not discuss in detail here. In addition the uncertainty σ_i is a critical parameter to increase the convergence speed of the PSO, one should set smaller σ_i value at the parts of spectrum not fitted well.

Acknowledgments

This work was supported by the National "973" Basic Research and Development Program of China (No.2003CB314901), the National "863" High Technology Project of China (No. No.2003AA311070).

References

1. Po Dong, Jose Azna, Andrew G.Kirk, "Synthesis of fiber Bragg grating parameters from reflectivity by means of a simulated annealing algorithm", Optics Communications, (228), pp.303-308, 2003
2. Byeong Ha Lee, Jaehee Cheong, and Un-Chul Paek, "Spectral polarization-dependent loss of cascaded long-period fiber gratings," *Optics Lett.,*vol.27, pp.1096-1098, 2002.
3. Johannes Skaar, Ligang wang, and Euran Erdogan, "On the synthesis of fiber Bragg gratings by layer peeling," vol.37, pp.165-173, 2001.
4. R.C.Alferness, and P.S.Cross, "Filter characteristics of codirectional coupled waveguides with weighted coupling," *IEEE J.Quantum Electon.,* vol.14, pp.843-847, 1978
5. E.Peral,J.Capmany, and J.Marti, "Iterative solution to the Gel'Fand-Levitan-Marchenko coupled equations and application to the synthesis of fiber gratings," IEEE J.Quantum *Electon.,* vol.32, pp. 1526-1528, 1998.
6. Johanes Skaar and knut Magne Risvik, "A genetic algorithm for the inverse problem in synthesis of fiber Gratings," *J.Lightwave Technol.,* vol.16, pp.1928-1932, 1998.
7. K. E. Parsopoulos and M. N. Vrahatis, "Particle swarm optimizer in noisy and continuously changing environments", Artifical Intelligence and Soft Computing, IASTED/ACTA Press, pp.289-294, 2001.
8. J. Kennedy and R. C. Eberhart, "Particle swarm optimization ", Proc. Of IEEE International Conference on Neural Networks, Piscataway, NJ, USA, pp1942-1948, 1995.
9. J. Kennedy and R. C. Eberhart, "Particle swarm optimization ", Proc. Of IEEE International Conference on Neural Networks, Piscataway, NJ, USA, pp1942-1948, 1995.
10. M.K.Pandit et al.,"Tunable long-period fiber gratings for EDFA gain and ASE equalization," Microwave and Optical Tech. Lett., vol.25, 2000.
11. Yin Zhu, Ping Shum, Chao Lu, "EDFA gain flattening using phase-shifted long-period grating," Microwave and Optical Techl. Lett., Vol.37, pp.153-157,2003.
12. Tony Huang, Ananda Sanagavarapu Mohan, "A hybrid boundary condition for robust particle swarm optimization", IEEE antennas and wireless propagation letters. Vol.4, 2005.

Research of Undirected Network Capacity Expansion Based on the Spanning-Tree

Yuhua Liu[1], Kaihua Xu[2], Hao Huang[1], and Wei Teng[1]

[1] Department of Computer Science, Central China, Normal University, Wuhan 430079, China
yhliu@mail.ccnu.edu.cn
[2] Research Center of the Digital Space Technology, Central China, Normal University,Wuhan 430079, China

Abstract. This paper gives a mathematical description of the undirected network capacity definitions and theorems of computing capacities towards the network. Prove of the theorem correctness is also given. According to them, the paper proposes an algorithm of computing the undirected network capacities based on spanning tree and even an advanced optimization algorithm. In addition, it discusses network capacity expansion problems with constraints. And then, the optimization algorithm in the scenario of undirected network with limited costs is outlined, whose feasibility and process are illustrated via examples.

1 Introduction

Various problems of computer network theories and technologies have close connections with network optimization. Researches on related optimization problems mean much to the field of computer science. The importance penetrates through network design and implementation process, especially in allocation of network traffics and capacities [1].

Network optimization is an important branch of operation research. Some theories of operation research, such as combination optimization, etc. apply to network optimization widely. The paper employs some theories of network traffic of combination optimization, combined with the graph theory, to solve expansion problems at network bottlenecks [2]. We begin with expanding edge capacities to obtain a desired maximum traffic, and the expansion goes on till bottleneck problems are solved. In the mean time, we will also take minimizing cost problem into consideration [3].

2 Definitions and Problem Description

This paper researches computing undirected network capacities and expansion problems based on spanning tree. The definition of undirected network is as follows. Assume a network $G = (V, E, C)$, a vertex set $V = \{v_1, v_2, \cdots, v_n\}$ and a edge set

$$E \subseteq \{(v_i, v_j) \mid i = 1, 2, \cdots, n; j = 1, 2, \cdots, n\} ,$$

T.-D. Wang et al. (Eds.): SEAL 2006, LNCS 4247, pp. 228–235, 2006.

where n is the vertex number. C is a non-negative real number function, whose component $c(v_i, v_j)$ denotes the capacity of the edge (v_i, v_j).

Definition 1. *In a network G, if the vertex $(v_i, v_j) \in A$, then $(v_j, v_i) \in A$. These two vertices are called symmetrical if $c(v_i, v_j) = c(v_j, v_i)$. A network is undirected when this condition matches all the vertices in the network; otherwise, it is directed.*

The following definition is the one of network capacity. In this paper, it is the maximum value of all spanning tree capacities which is the minimum capacity of edges on the tree.

Definition 2. *Given the network $G = (V, E, C)$ which consists of a set V of vertices, a set E of edges, and a function C: $E \rightarrow R^+$ of capacity function on edges. Then the capacity of the network*

$$c(G) = \max\{c(T) \mid T \subseteq G,\ T \text{ is the spanning tree of } G\},$$

and the capacity of the spanning tree $c(T) = \min\{c(v_i,\ v_j) \mid (v_i,\ v_j) \in E(T)\}$.

3 Spanning-Tree Based Undirected Network Capacity Expansion

This section proposes the theorem and its corollary. According to the computing process of undirected network capacity, it gives a spanning-tree based capacity computing algorithm and even an advanced one. Then it discusses the capacity expansion problem of undirected networks and an expansion algorithm as well as its pseudocode that could be applicable in computers.

3.1 Computing Undirected Network Capacities and Algorithms

Assume an undirected network $G = (V, E, C)$which consists of a set V of vertices, a set E of edges, and a capacity function C, is connected.

Theorem 1. *For any edge $(x, y) \in E$ in a connected network G, it is always in the spanning tree of G.*

Proof (of theorem). For any give edge $(x, y) \in E$, choose a spanning tree T randomly. The theorem is true if $(x, y) \in E(T)$. Otherwise, as the spanning tree definition, there must be a path between x and y in T. Suppose T denotes the path, then $P + (x, y)$ forms a circle Q. Choose a edge $(u, v) \in E(Q)$, where $(u, v) \neq (x, y)$. Then $T + (x, y) - (u, v) = T_1$ is also a spanning tree of G, where $(x, y) \in E(T_1)$.

Corollary 1. *Suppose a network $G(V, E, C)$. On the basis of not destroying connectivity of G and while $c(T) = \min\{c(x, y)|(x, y) \in T\}$, erase the edges (x, y) with the minimum capacity one by one till someone is inerasable. The generated sub-graph is then the spanning tree T whose capacity is the minimum capacity of edges, say $c(T) = \min\{c(x, y)|(x, y) \in T\}$.*

According to above theorem and corollary, we can come to a methodology of computing undirected network capacity: Order the elements of set E by capacity from small to large and form a sequence L, whose component l_{ij}^k denotes the kth element in L which is also the component (v_i, v_j) of the set E; Let $k = 1 \cdots m$; If G is connected after erasing l_{ij}^k(erase (v_i, v_j) in E), then erase it, otherwise, keep it; The process continues till all edges are inerasable.

Then a spanning tree $T = (V, E^T, C^T)$ is generated, where E^T is the set of retained edges. The minimum capacity r of edges in E^T is the spanning tree capacity as well as network capacity. The spanning tree of a network consists of $(n-1)$ edges. Hence, if there are edges, the connected graph after erasing edges is the spanning tree.

Algorithm description is:

Step 1. Let $u = 0$, $k = 1$.
Step 2. Order the elements of set E by capacity from small to large and form a sequence L, whose component l_{ij}^k denotes the kth element in L which is also the component (v_i, v_j) of the set E.
Step 3. If G is connected after erasing l_{ij}^k(erase (v_i, v_j) in E), then erase it and $u = u + 1$. Otherwise, keep it.
Step 4. Go to step 6 if $u = m - (n-1)$.
Step 5. $k = k + 1$ and go to step 3.
Step 6. The spanning tree $T = (V, E^T, C^T)$ is generated, where E^T is the set of retained edges. The minimum capacity r of edges in E^T is the network capacity. Output T and r.

Implementing the algorithm needs a 2-array $Graph[3][m]$ to store undirected network G. $Graph[0][i]$ stores the capacity of the ith edge, while $Graph[1][i]$ and $Graph[2][i]$ store two vertices of the ith edge separately. m is the number of edges, and n is that of vertices $(1 \le i \le m)$. The algorithm suppose: the capacity of the ith edge sets to -1 if erased, say $Graph[0][i] = -1$. Now introduce a parameter *count* to be the number of erased edges, which is initialized to 0.

The pseudocode is as follows:

Step 1. Initialize $Graph[3][m]$, and order its elements by capacity from small to large. Let $i = 1$, $count = 0$.
Step 2. Let v_j and v_k denote two vertices. $v_j = Graph[1][i]$ and $v_k = Graph[2][i]$. Verify the graph connectivity. If G is connected after erasing (v_i, v_k), then erase it. Let $Graph[0][i] = -1$ and $count = count + 1$.
Step 3. $i = i + 1$. Go to step 4 if $count = m - (n-1)$. Otherwise, go to step 2.
Step 4. Scan the array *Graph* by capacity from small to large. That is, let $i = 0$, and set $i = i + 1$ if $Graph[0][i] = -1$ until $Graph[0][i] \neq -1$. Then output $Graph[0][i]$ which is the final result.

On the basis of maintaining the graph connectivity, the algorithm finds the first inerasable edge is the edge with minimum capacity in the spanning tree while erasing edges with minimum capacity. Since edges are ordered by capacity from small to large, that first inerasable one is the capacity of the spanning tree, say, the capacity of the network G.

3.2 The Advanced Algorithm of Computing

It needs $m-(n-1)$ times of verifying the connectivity for the algorithm in section 3.1 to cut the network G be a minimum connected graph, a spanning tree T. Hence, if the scale parameter m and n are very large, so is the total verifying time. Now we can consider the problem reversely. Suppose the vertices are individual connected components initially, which means the network G is a null graph at that time. And then generate its spanning tree. The capacity could be computed on the basis of the tree. The *Kruskal* algorithm, which is used to compute the minimum spanning tree of a network, is applied to this problem [4]. However, some modifications need to be made.

The *Kruskal* based algorithm to compute network capacity is:

Initially, the network is an unconnected graph $G'=(V,E',C')$ which consists of n vertices but without edges, and $E'=\emptyset$. Each vertex is a connected component. Order the elements of set E by capacity from small to large and form a sequence L, whose components l_{ij}^k denote the kth element in L which is also the component (v_i,v_j) of the set E. Let $k=1\cdots m$. If the attaching vertex v_i and v_j on l_{ij}^k are on various connected components of G', then add l_{ij}^k to E'; otherwise, take no action. This process continues till all the vertices of G' are on the same connected components. The minimum capacity of edges in E' is then the network capacity.

Then, the advance computing algorithm is:

Step 1. Order the elements of set E by capacity from small to large and form a sequence L, whose components l_{ij}^k denote the kth element in L which is also the component (v_i,v_j) of the set E.

Step 2. Let $k=1$, $E'=\emptyset$.

Step 3. If the attaching vertex v_i and v_j on l_{ij}^k are on various connected components of G', then add l_{ij}^k to E', say $E'=E'\cup\{(v_i,v_j)\}$; otherwise, take no action.

Step 4. Turn to next step if the number of edges in E' is $n-1$; otherwise, $k=k+1$ and turn to step 3.

Step 5. A spanning tree $T=(V,E^T,C^T)$ is generated, where $E^T=E'$ and $C^T=\{c(v_i,v_j)|\,(v_i,v_j)\in E^T\}$. The minimum capacity r of edges in E^T is the network capacity. Output T and r.

Implementing the algorithm needs a 2-array $Graph[3][m]$ to store undirected network G. $Graph[0][i]$ stores the capacity of the ith edge, while $Graph[1][i]$ and $Graph[2][i]$ store two vertices of ith edge, separately. m is the number of edges, and n is that of vertices. Then introduce two auxiliary 1-arrays to achieve a greater running efficiency. $EdgeDeasc[m]$ stores subscripts of vertices after ordered by capacity in a descending order. For example, if $EdgeDeasc[k]=t$, the sequence number of ordered tth vertex in the array $Graph$ is k. Because of the introduction of $EdgeDeasc$, classical ordering program needs some little modifications when using it to order vertices. The variables to be compared is $Graph[0][EdgeDeasc[i]]$, and the one to be moved is $EdgeDeasc[i]$. Then initialize the array $EdgeDeasc$. Let $EdgeDeasc[i]=i$, $0\le i\le m-1$.

And another array $VertexIndex[n]$ stores index of connected component of each vertex in order to verify whether two vertices are in the same connected component or not. $VertexIndex[0]$ stores the first vertex index, and so forth. If two indices equal to each other, then they are in the same connected component. The initial values of each index are their own sequence number. Obviously, equality of all indices means all vertices are connected. And if a connected component consists of some vertices, we use the minimum sequence number of these vertices to signify the component.

Combining two components is to make their indices equal. We choose the minimum sequence number as the index of the combined components. And the vertex index in the new combined one needs modifications. The modification process is: if indices of two pre-combined ones are a and b, the new index is then a(suppose a is smaller than b); The algorithm scans all the indices. Change the index to a if one is b; Scan the sequences. If the ith edge could not be added to it (or needs to be erased), then let $EdgeDeasc[i] = -1$.

According to the above analysis, the pseudocode is as follows:

Step 1. Initialize $EdgeDeasc$. Let $EdgeDeasc[i] = i$, $1 \leq i \leq m$. Order vertices by their capacities in an ascending order. $EdgeDeasc$ stores sequence of ordered edges. Namely, let $Graph[EdgeDeasc[i]][0] \geq Graph[EdgeDeasc[i+1][0]]$.

Step 2. Initialize $VertexIndex$. Let $VertexIndex[i] = i$, $1 \leq i \leq n$, and $j = 0$.

Step 3. Scan $VertexIndex$. Go to step 6 if all elements in the array are equal.

Step 4. Let

$$t_1 = VertexIndex[Graph[1][EdgeDeasc[j]]],$$

and

$$t_2 = VertexIndex[Graph[2][EdgeDeasc[j]]].$$

Let $t_3 = t_1$ and $t_4 = t_2$ if $t_1 \neq t_2$(suppose $t_1 < t_2$, $t_3 = t_2$, $t_4 = t_1$). And let $k = 0 \cdots n-1$. If $VertexIndex[k] = t_4$, then $VertexIndex[k] = t_3$. Otherwise, $EdgeDeasc[j] = -1$.

Step 5. $j = j + 1$, and turn to step 3.

Step 6. Let $k = 0, \cdots, n-1$. If $EdgeDeasc[j] = -1$, then $k = k + 1$ until $EdgeDeasc[k] \neq -1$. Then $Graph[EdgeDeasc[k]][0]$ is the network capacity.

The average time complexity is $O(n \log n)$ if the algorithm in section 3.2 uses quick sorting to order vertices [5]. And if not implement ordering, namely, the capacities are in ascending order before inputting the array *Graph*, then it becomes $O(n)$. The need for auxiliary storage space is $m + n$ integer space. The introduction of *EdgeDesc* is to decrease times of exchanging data while sorting. 3 pairs of data needs to be exchanged in a round of exchange only with the *Graph*, but 1 pair after introducing the *EdgeDesc*. This advanced algorithm avoids verifying the graph connectivity via traversing the whole graph. It is highly improved in efficiency that the more complex the networks are, the more obvious advancement would be obtained. Hence, the algorithm has better adaptability in complex networks.

3.3 The Algorithm of Undirected Network Capacity Expansion

Assume $G = (V, E, C, W, D)$ is an undirected network. The definition of V, E, C is the same as above ones, while W is a non-negative real number function on E. Component $w(v_i, v_j)$ of W denotes cost of expanding unit capacity of the edge (v_i, v_j). D denotes a given investment cost and r denotes network capacity.

The algorithm description is as follows.

Step 1. Invoke the algorithm in section 3.2. Compute the spanning tree $T = (V, E^T, C^T)$ and capacity r.
Step 2. Order the elements of set E^T by capacity from small to large and form a sequence L, whose component l_{ij}^k denotes the kth element in L which is also the component e_{ij} of the set E.
Step 3. Let $k = 1$, $R = \emptyset$, and $d = 0$.
Step 4. Turn to step 6 if $c(l_{ij}^k) \neq r$.
Step 5. $R = R \cup \{(v_i, v_j)\}$, $k = k + 1$, and turn to step 4.
Step 6. If $d + \sum_{(v_i,v_j)\in R} w_{ij} < D$, then $d = d + \sum_{(v_i,v_j)\in R} w_{ij}$. For all $(v_i, v_j) \in R$, $c(v_i, v_j) = c(v_i, v_j) + 1$, $r = r + 1$ and turn to step 4. Otherwise, go to next step.
Step 7. Output r, which is the maximum capacity under given investment D.

4 Examples

Example 1. Given a network $G = (V, E, C, W)$. Its capacity distribution is shown as Fig.1. Compute capacity of this network.

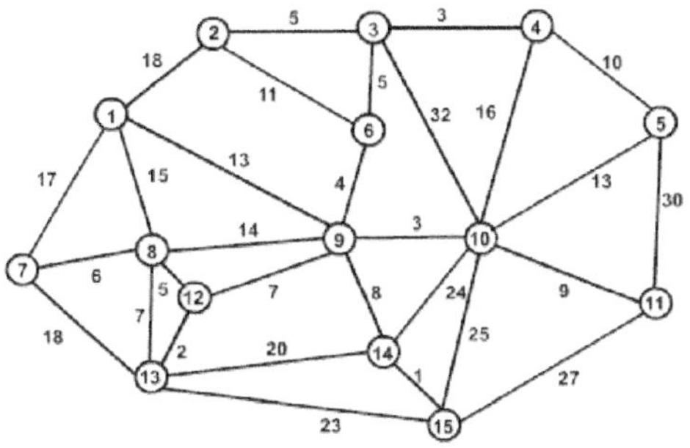

Fig. 1. Topology of the network G

This network consists of 15 vertex and 30 edges. The edge set ordered by their capacities is shown as Table 1. After computing capacities according to the algorithm in section 3.2, edges with sequence number 1,2,3,4,5,6,7,8,9,10,12,13,14,16, 17,24 are to be erased (marked by grey blocks), while the remains construct the spanning tree which is shown via solid lines in Fig.2. The computed network capacity is the minimum capacity of the retained edges, which is $c(v_9, v_{12}) = 7$.

Table 1. Correspondence table of spanning tree vertices and capacities of example 1

SequenceNumber	1	2	3	4	5	6	7	8	9	10	11	12	13	14	15
Capacity	1	2	3	3	4	5	5	5	6	7	7	8	9	10	11
Vertex	14	12	3	9	6	2	3	8	7	8	9	9	10	4	2
Vertex	15	13	4	10	9	3	6	12	8	13	12	14	11	5	6

SequenceNumber	16	17	18	19	20	21	22	23	24	25	26	27	28	29	30
Capacity	13	13	14	15	16	17	18	18	20	23	24	25	27	30	32
Vertex	1	5	8	1	4	1	1	7	13	13	10	10	11	5	3
Vertex	9	10	9	8	10	7	2	13	14	15	14	15	15	11	10

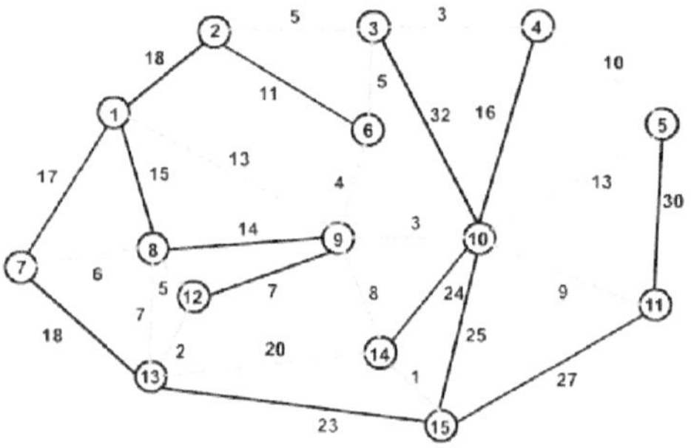

Fig. 2. Network G and its spanning tree

Table 2. Correspondence table of spanning tree vertices and capacities of example 2

SequenceNumber	1	2	3	4	5	6	7	8	9	10	11	12	13	14
Capacity	19	19	19	19	19	19	19	19	23	24	25	27	30	32
Vertex	9	2	8	1	4	1	1	7	13	10	10	11	5	3
Vertex	12	6	9	8	10	7	2	13	15	14	15	15	11	10

Example 2. On the basis of computing capacity in the example 1, give an investment 36. Assume components of W is $w_{ij} = 1$, such that unit cost of capacity expansion of each edge is 1. Now expand capacities of G according to the algorithm of section 3.3. The edges which have been expanded are marked by grey blocks, and the conclusion that the total capacity is expanded to 19 could be drawn. The capacity distribution of edges after expansion is shown as Fig.3 (origin capacities on left side of plus signs, while expanded ones on right side).

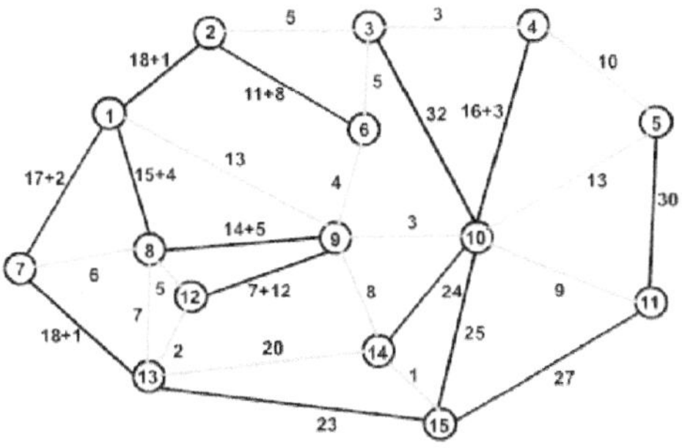

Fig. 3. Capacity distribution of edges after expansion

5 Conclusions

Comparing to the capacity expansion algorithms of existing researches, the greatest innovation of this paper is: it computes network capacity and carries out expansion based on spanning tree capacity. Besides, the expansion algorithm in [6] uses a fixed root vertex of a spanning tree. Expand in this scenario results in a tree that meets capacity needs.

However, the advanced expansion algorithm in this paper uses every single vertex to be the root in a spanning tree. This could achieve the same expansion goals, which matters more in practical applications.

Acknowledgments. The work was jointly supported by the National Key Basic Research Project (973 Project Grant No. 2003CB314804) and the Natural Science Foundation of China (Grant No. 60473085).

References

[1] Ahuja R.K., Magnanti T.L., Orlin J.B.: Network Flows: Theory, Algorithms and Applications. Prentice-Hall, New York (1993)

[2] Zhang, J., Yang, C., Lin, Y.: A class of bottleneck expansion problems. Comput Oper. Res. **28** (2001) 505-519

[3] Yang, C., Zhang J.Z.: A constrained capacity expansion problem on network. **70** (1998) 19-33

[4] Gondran, Minoux, M., Vajda, M.: Graphs and Algorithms. John Wiley and Sons, New Jersey (1984)

[5] Goldberg, A.V.: Recent developments in maximum flow algorithms(Invited Lecture). In: S Arnborg, L Ivansson (eds.): Proc of the Sixth Scandinavian Workshop on Algorith Theory. Lecture Notes in Computer Science, Vol. 1004. Springer-Verlag, Berlin Heidelberg New York (1998) 1-10

[6] Wang H.G., Ma S.H.: Capacity expansion problem of undirected networks, Journal of Shandong University. **28** (2001) 418-425

A New Approach to Solving Dynamic Traveling Salesman Problems

Changhe Li, Ming Yang, and Lishan Kang

China University of Geosciences(Wuhan) School of Computer,
430074 Wuhan, P.R. China
lchwfx@yahoo.com, yangming0702@gmail.com,
kang_whu@yahoo.com

Abstract. The Traveling Salesman Problem (TSP) is one of the classic NP hard optimization problems. The Dynamic TSP (DTSP) is arguably even more difficult than general static TSP. Moreover the DTSP is widely applicable to real-world applications, arguably even more than its static equivalent. However its investigation is only in the preliminary stages. There are many open questions to be investigated. This paper proposes an effective algorithm to solve DTSP. Experiments showed that this algorithm is effective, as it can find very high quality solutions using only a very short time step.

1 Introduction

With the development of computer science and communication technology, from highly centralized computing through distributed computing to today's highly mobile computing, computing environments have changed a great deal. Key research challenges they face in common are the optimization of dynamic networks, arising from network planning and designing, load-balance routing and traffic management. However, guaranteeing that these systems run effectively and reliably is a difficult problem. It leads to a very important theoretical mathematical model: the Dynamic TSP (DTSP).

Because of the characteristics of DTSP itself, the solutions to general static TSPs are usually unsuitable for DTSP. Most cost too much time to gain good solutions, so the general algorithms are inefficient. Though a number of authors have researched [1][2][3][4] the DTSP, since it was proposed by Psaraftis[5], exploration of the DTSP is still in the preliminary stages, and many open questions need to be discussed. The ultimate (but unobtainable) goal is to find an optimum solution at every moment, as time progresses. In fact, if you want to get better solutions, the efficiency will be lower, and conversely, if you require quick solutions, their quality will reduce that is to say the two goals (solution quality and time response)are in conflict. Since we can't get an optimum solution at every instant, we can solve the problem in discrete time steps, finding good solutions in a time step as short as possible.

In this paper, we will introduce an improved Inver-Over[6] algorithm(GSInver-Over) based on a gene pool. Generally, we find that using a gene pool, which reduces the search space sharply, gives solutions much more rapidly, without

T.-D. Wang et al. (Eds.): SEAL 2006, LNCS 4247, pp. 236 – 243, 2006.

degradation of the solution quality. Thus it dramatically improves the system performance in the combined objectives. The GSInver-Over algorithms can improve the individuals using information either from other individuals or from gene pools. We augment this with elastic relaxation as a local search method, which permits the rapid evolution of variants of individuals which were successful in previous situations. Our experiments show that these operators can provide highly satisfactory results.

In the remainder of this paper, there are five sections: (1) description of DTSP; (2) design of the gene pool; (3) the detailed algorithms; (4) analysis of the results; (5) summary and conclusions.

2 Description of DTSP

Definition 1
A dynamic TSP(DTSP) is a TSP determined by a dynamic cost (distance) matrix as follows:

$$D(t) = \{d_{ij}(t)\}_{n(t)\times n(t)} \tag{1}$$

where $d_{ij}(t)$ is the cost from city(node) c_i to city c_j, and t is the real-world time. In this definition, the number of cities $n(t)$ and the cost matrix are time-dependent. The Dynamic Traveling Salesman Problem is to find a minimum-cost route containing the all the $n(t)$ nodes.

Definition 2 DTSP
Given all $n(t)$ $(P_1, P_2, ..., P_{n(t)})$ points, and the corresponding cost matrix $D = \{d_{ij}(t)\}, i, j = 1, 2, ..., n(t)$, find a minimum-cost route containing all the $n(t)$ points, where t stands for the moment of time t; $d_{ij}(t)$ stands for the distance between the objective point P_i and the objective point P_j, and $d_{ij}(t) = d_{ji}(t)$.

For example:

$$Min(d(T(t))) = \min(\sum_{i=1}^{n(t)} d_{T_i, T_{i+1}}(t)) \tag{2}$$

where $T \in \{1, 2, ..., n(t)\}$ if $i \neq j$, then $T_i \neq T_j$, $T_{n(t)+1} = T_1$

In definition 1, we deem the change of a DTSP's cost matrix with time as a continuous process. Practically, we discretize this change process. Thus, A DTSP becomes a series of optimization problems:

$$D(t_k) = \{d_{ij}(t_k)\}_{n(t_k)\times n(t_k)} \tag{3}$$

$k = 0, 1, 2, ..., m-1$, with time windows $[t_k, t_{k+1}]$, where $\{t_k\}_{i=0}^{m}$ is a sequence of real world time sampling points.

3 Design of Gene Pool

Heuristic rules can dramatically affect the efficiency of DTSP algorithms. The TSP is an NP-hard problem, and adding only one node, will increase the candidate search space by $n \times n!$. Thus it is impossible to search all the candidate individuals. One useful heuristic derives from the fact that most of the edges in a minimum-cost route will join nearby vertices. So it is generally desirable for the elements in the gene pool to select from edges which go to nearby vertices. Unfortunately, this heuristic is often violated: for hard TSPs, a small proportion of the edges in optimal routes will need to connect distant vertices. If the heuristic is too rigidly applied, some of the edges in the optimal route won't exist in the gene pool. So a heuristic method based on minimising local distances seems reasonable, but in fact, this kind of restriction usually results in bad performances. We describe an alternative heuristic for the design of the gene pool. It is based on the concept of α-nearness [7], which derives from Minimum Spanning Trees (MSTs). The α-length α(i,j) of an edge $<v_i,v_j>$ can be defined as the difference in length between the true MST, and the length of the 1-tree which is constrained to contain $<v_i, v_j>$.

$$\alpha(i,j) = L(T^+(i,j)) - L(T) \tag{4}$$

where T is an any given MST of length L(T),$T^+(i,j)$ is a 1-tree that contain the edge $< v_i, v_j >$,that is, given a MST of length L(T),α(i,j) is the increase length of a 1-tree required to contain the edge $< v_i, v_j >$.

It is easy to see that: α(i,j)≥0 and α(i,j)=0 when the edge $< v_i, v_j >$ belong to T. It can compute α(i,j) in the following rules:

(1) if the edge $< v_i, v_j >$ belong to T, then α(i,j)=0.
(2) Otherwise, insert the edge $< v_i, v_j >$ into T, this will create a circle containing the edge $< v_i, v_j >$,then α(i,j) is the length difference between the longest edge of the circle and the edge $< v_i, v_j >$.

The gene pool is a candidate set of some most promising edges. The candidate set may, for example, contain k α-nearness edges incident to each node. Generally speaking, the experience value of k is 6 to 9.we set k=8 in CHN144 problem[8].

Through experiments, it can be shown experimentally that 50%-80% (i.e the experiment result of table 1 of instances in TSPLIB) of the connections in an optimal TSP solution are also in the minimum spanning tree. This is a far larger proportion than the proportion of the n shortest edges. Thus we expect that a gene pool constructed based on α-nearness may perform better than a gene pool constructed on the distance. It should be possible to use a smaller gene pool while maintaining the solution quality. Taking the CHN144 problem for example, 76.3% of the connections in the best known solution are also edges in the minimum spanning tree. If we bias the gene pool based on the α-nearness, we expect that it will better match the optimal solution. That is to say, we expect that a gene pool that probabilistically includes elements close to members of the MST will also include elements close to members of the optimum TSP circuit. The gene pool based on the α-nearness has another remarkable advantage that it is independent of instance scale, it means ,when the instance scale increases ,the size of gene pool won't increase. This character of gene pool is much suitable to DTSP.

Table 1. Instances in TSPLIB

CHN144	a280	pr439	u574	u724	rl1323
76.3%	75.7%	79.1%	75.6%	75.2%	89.2%

4 Introducing the Algorithms

Operator design is the key to solving TSPs. A vast range of operators have already been proposed (for example: λ-Opt[9], LK[10], Inver-Over), and we anticipate that this trend will continue. Based on performance, many of these local-search inspired operators are superior to the traditional mutation, crossover and inversion operators. In this paper, we adopt a form of improved Inver-Over operator. We propose a highly efficient dynamic evolution algorithm based on elastic.

4.1 GSInver-Over Operator

Inver-Over is a high-efficiency search operator based on inversion, and having a recombination aspect. It can fully utilize information from other individuals in the population to constantly renew itself. This gives it better search ability than many other operators (within a certain range of problems), yet the complexity of algorithm is low. We can say Inver-Over is a highly adaptable operator which has very effective selection pressure.

However Inver-Over operator has its own constrains, experiments show that reducing the inversion times can sharply increase the convergence, this is favourable to DTSP. We set the maximal inversion times Max_time in the GSInver-Over operator, when the inversion times surpass Max_time, end the algorithm. The search environment has increased and it can get useful heuristic information not only from other individuals, but also from the gene pool. The choice of matching individuals is not random, but biased toward better individuals in the population. This reduces the probability of incorporating bad information that damages the individual. The inversion operator is more rationally designed, incorporating knowledge about the directionality of the route, further improving the performance of the algorithm. Based on the above improvements, the GSInver-Over performed substantially better than the original algorithm. the algorithm for our GSInver-Over operator is as follows:
GSInver-Over Operator:

1. Select a gene g from individual S randomly and set S′=S;
2. If the number p generated randomly is less than p1, then select the gene g′ from the gene pool of g;
3. Else if p<p2 select an individual randomly from some best individuals and g′ is the gene that is next to g in the selected individual;
4. Else select next gene g′ randomly from other genes;
5. If the next gene or the previous gene of g in S′ is g′, then go to step 9;
6. Inverse the section from the next gene of g to g′ in S′;
7. counter++, if counter > Max_time, then go to step 9;
8. g= g′ and go to step 2;
9. If the fitness of S′ is better than S, then replace S with S′;

4.2 The Dynamic Elastic Operator

The changes of a node include three cases: the node disappears, the node appears and the position of the node changes. If the node disappears, it will be deleted directly, then link the two adjacent nodes of the deleted node. if the node appear, find the nearest node to the appearing node, then insert it to the tour that minimize the total length. if the node position changes, it can be seen as the combination of the two former cases. The dynamic-elastic operator is very simple in concept, but we find it is an effective local search operator.

Dynamic Elastic Operator

1. Delete the node c and link the cities adjacent to c;
2. Find the nearest node c* to c in the current individual;
3. Insert c next to C*, on the side that minimize the total length;

4.3 Main Program Loop

In the main program loop, we use a difference list Dlist to store the information of changed nodes. Note that ΔT is the time step.

1. Initialize the population;
2. If Dlist is not empty goto 3, else goto 5;
3. Update gene pool;
4. Dynamic-elactic ();
5. For each individual in the population,do
   GSInver-Over ();
   Optimizing();
6. If the ΔT>0 goto 5;
7. If not termination condition goto 2;

When some nodes change at time t, it needs to update the gene pool, that is to say, create a new MST of the new nodes topology of time t, then construct a gene pool according to α-nearness.

5 Experiments with CHN146+3

We tested our algorithm in a relatively difficult dynamic environment, adapted from the well-known CHN145[11] static TSP benchmark. The problem has 146 static locations (145 Chinese cities, plus a geo-stationary satellite) and three mobile locations, two satellites in polar orbit and one in equatorial orbit (fig. 1).

In dynamic optimisation experiments, since the results represent a trade-off between solution quality, computational cost and problem dynamics, it is important to specify the computational environment in which experiments were conducted. Our experimental environment consisted of the following: CPU: Intel C4 1.7GHz, RAM:256MB. We measure the offline error $\bar{e}$ and μ as our quality metric:

$$\bar{e}(t_k) = d(\pi(t_k)) - d(\bar{\pi}(t_k)) \tag{5}$$

$$\mu(t_k) = \bar{e}(t_k) / d(\bar{\pi}(t_k)) \tag{6}$$

where $d(\bar{\pi}(t_k))$ is the best tour obtained by a TSP solver (which is assumed to be good enough to find an optimal solution for the static TSP) $d(\bar{\pi}(t_k))$ is the best tour obtained by our DTSP solver. Together with:
Maximum error:

$$e_m = \max_{k=0,\cdots,m} \{\bar{e}(t_k)\} \qquad \mu_m = \max_{k=0,\cdots,m} \{\mu(t_k)\} \tag{7}$$

Minimum error:

$$e_r = \min_{k=0,\cdots,m} \{\bar{e}(t_k)\} \qquad \mu_r = \min_{k=0,\cdots,m} \{\mu(t_k)\} \tag{8}$$

Average error:

$$e_a = \frac{1}{m+1} \sum_{t=0}^{m} (\bar{e}(t_k)) \qquad \mu_a = \frac{1}{m+1} \sum_{t=0}^{m} (\mu(t_k)) \tag{9}$$

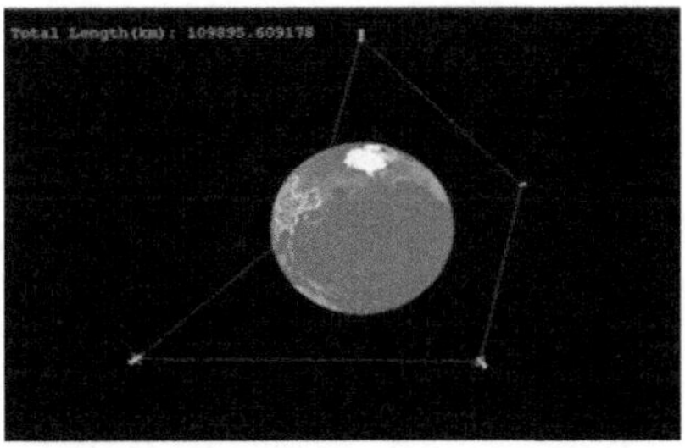

Fig. 1. Experiments with CHN146+3

120 sample time-points in the period of the satellites were performed for the experiments. The results are given in table 2 with ΔT ranging from 0.059s to 1.3s. Fig. 2 to fig. 5 are error curves respectively for ΔT=0.059s, ΔT=0.326s, ΔT=0.579s and ΔT=0.982s.

From the experiments, we can see that, when ΔT is very small, the result is relatively poor. As ΔT increases, the maximal and average errors decrease, and the solution quality improves showing the stability of the algorithm, and its rapid convergence. The experiments also demonstrate the conflict between the two DTSP goals, requiring a compromise through the choice of ΔT. With the exception of the shortest time interval, the data in table 2 are generally acceptable, with the average errors being under 1%, and the maximal errors less than 2%.

Table 2. Error of experiments

ΔT(s)	e_m(km)	μ_m(%)	e_r (km)	μ_r(%)	e_a(km)	μ_a(%)
0.059	1742	1.487	206	0.199	796	0.727
0.222	2020	1.722	773	0.062	406	0.372
0.326	1310	1.122	0	0	289	0.264
0.481	1014	0.866	0	0	283	0.257
0.579	1038	0.884	0	0	203	0.187
0.982	899	0.770	0	0	240	0.218
1.300	1514	1.297	0	0	188	0.170

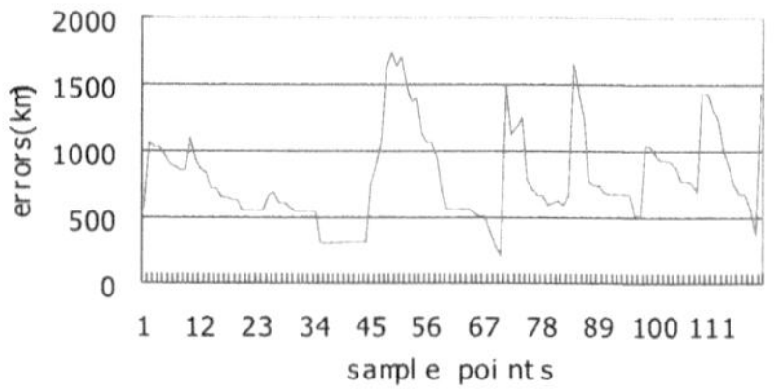

Fig. 2. Error Curve for ΔT = 0.059s

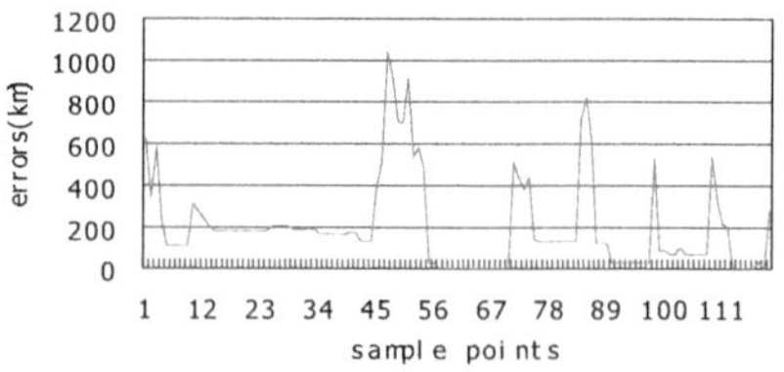

Fig. 4. Error Curve for ΔT = 0.579s

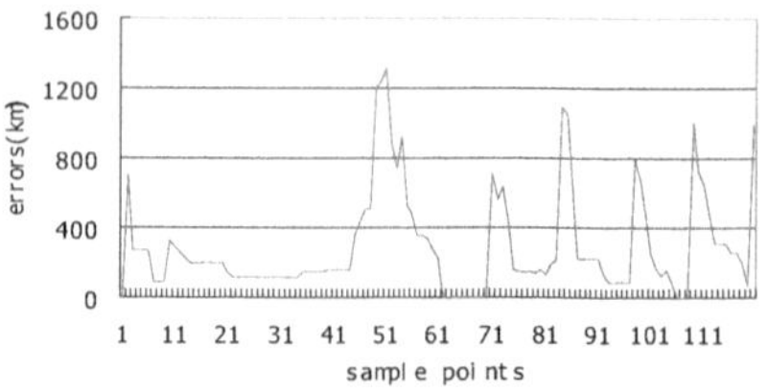

Fig. 3. Error Curve for ΔT = 0.326s

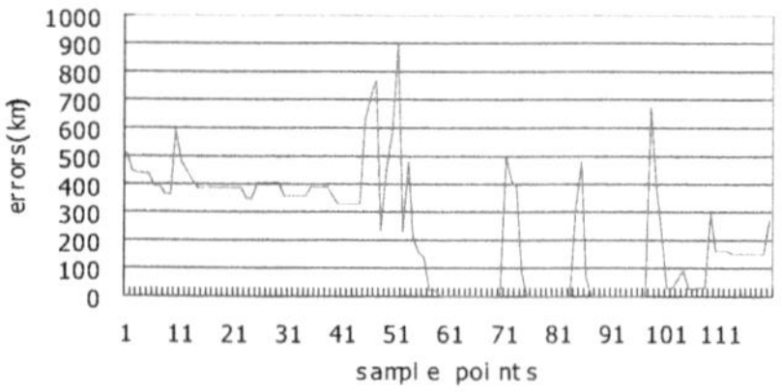

Fig. 5. Error Curve for ΔT = 0.982s

6 Conclusions

In this paper, we analyze the DTSP which can be seen as a two-objective problem by trading off the quality of the result and the reaction time. We propose a solution based on a gene pool, which greatly reduces the search space without degradation of the solution quality. By adding the improved GSInver-Over Operator, we were able to significantly improve the efficiency of the algorithm. Adding the elastic relaxation method as a local search operator improves the system's real-time reaction ability.

Acknowledgement

This work was supported by National Natural Science Foundation of China (NO. 60473081). We would like to thank Bob McKay, of Seoul National University, Korea, for contributions to the presentation of this work.

References

1. C.J.Eyckelhof and M.Snoek. Ant Systems for a Dynamic TSP -Ants caught in a traffic jam. In3rd International Workshop on Ant Algorithms(2002)
2. Z.C.Huang,X.L.Hu and S.D.Chen. Dynamic Traveling Salesman Problem based on Evolutionary Computation. In Congress on Evolutionary Computation(CEC'01),IEEE Press (2001)1283–1288
3. Allan Larsen. The Dynamic Vehicle Routing Problem. Ph.D theis,Department of MathematicalModelling (IMM) at the Technical University of Denmark(DTU)(2001)
4. A.M.Zhou, L.S.Kang and Z.Y.Yan. Solving Dynamic TSP with Evolutionary Approach in Real Time. Proceedings of the Congress on Evolutionary Computation, Canberra, Austrilia, 8-12, December 2003, IEEE Press(2003) 951–957
5. H.N.Psaraftis. Dynamic vehicle routing problems. In Vehicles Routing: Methods and Studies,B.L.Golden and A.A.Assad(eds),Elsevier Science Publishers(1988)
6. T.Guo and Z.Michalewize. Inver-Over operator for the TSP. In Parallel Problem Sovling from Nature(1998)
7. Keld Helsgaun,An Effective Implementation of the Lin-Kernighan Traveling Salesman Heuristic, Department of Computer Science Roskilde University.
8. L.S.Kang, Y.Xie,S.Y.You,etc. Nonnumerical Parallel Algorithms:Simulated Annealing Algorithm. Being:Science Press(1997)
9. S. Lin,"Computer Solutions of the Traveling Salesman Problem",Bell System Tech. J., 44, (1965) 2245–2269
10. S. Lin & B. W. Kernighan,An Effective Heuristic Algorithm for the Traveling-Salesman Problem(1973)
11. Lishan Kang, Aimin Zhou, Bob McKay,Yan Li Zhuo Kang ,Benchmarking Algorithms for Dynamic Travelling Salesman Problems, Benchmarking Algorithms for Dynamic Travelling Salesman Problems, Proceedings, Congress on Evolutionary Computation, Portland, Oregon(2004)

Shannon Wavelet Chaotic Neural Networks

Yao-qun Xu[1,2], Ming Sun[1], and Ji-hong Shen[2]

[1] Institute of System Engineering, Harbin University of Commerce, 150028 Harbin, China
Xuyq@hrbcu.edu.cn, Snogisun@tom.com
[2] Department of Mathematics, Harbin Engineering University, 150001 Harbin, China
Shenjihong@hrbeu.edu.cn

Abstract. Chaotic neural networks have been proved to be strong tools to solve the optimization problems. In order to escape the local minima, a new chaotic neural network model called Shannon wavelet chaotic neural network was presented. The activation function of the new model is non-monotonous, which is composed of sigmoid and Shannon wavelet. First, the figures of the reversed bifurcation and the maximal Lyapunov exponents of single neural unit were given. Second, the new model is applied to solve several function optimizations. Finally, 10-city traveling salesman problem is given and the effects of the non-monotonous degree in the model on solving 10-city traveling salesman problem are discussed. The new model can solve the optimization problems more effectively because of the Shannon wavelet being a kind of basic function. Seen from the simulation results, the new model is powerful.

1 Introduction

Neural networks have been shown to be powerful tools for solving optimization problems. The Hopfield network, proposed by Hopfield and Tank [1, 2], has been extensively applied to many fields in the past years. The Hopfield neural network converges to a stable equilibrium point due to its gradient decent dynamics; however, it causes sever local-minimum problems whenever it is applied to optimization problems. Several chaotic neural networks with non-monotonous activation functions have been proved to be more powerful than Chen's chaotic neural network in solving optimization problems, especially in searching global minima of continuous function and traveling salesman problems [3, 8~9]. The reference [4] has pointed out that the single neural unit can easily behave chaotic motion if its activation function is non-monotonous. And the reference [5] has presented that the effective activation function may adopt kinds of different forms, and should embody non-monotonous nature. In this paper, a new chaotic neural network model is presented to improve the ability to escape the local minima so that it can effectively solve optimization problems. The chaotic mechanism of this new model is introduced by the self-feedback connection weight. The activation function of the new chaotic neural network model is composed of Sigmoid and Shannon Wavelet, therefore the activation function is non-monotonous. And because Shannon wavelet function is a kind of basic function, the model can solve optimization problems more effectively. Finally, the new model is applied to solve both function optimizations and

T.-D. Wang et al. (Eds.): SEAL 2006, LNCS 4247, pp. 244 – 251, 2006.

combinational optimizations and the effects of the non-monotonous degree in the model on solving 10-city TSP are discussed. The simulation results in solving 10-city TSP show that the new model is valid in solving optimization problems.

For any function $f(x) \in L_2(R)$ and any wavelet Ψ which is a basic function, the known formula can be described as follows:

$$f(x) = \sum_{j,k=-\infty}^{\infty} c_{j,k} \Psi_{j,k}(x) \tag{1}$$

2 Shannon Wavelet Chaotic Neural Network (SWCNN)

Shannon wavelet chaotic neural network is described as follows:

$$x_i(\mathrm{t}) = f(y_i(\mathrm{t})(1+\eta_i(\mathrm{t}))) \tag{2}$$

$$y_i(\mathrm{t}+1) = ky_i(\mathrm{t}) + \alpha\left[\sum_{\substack{j=1 \\ j\neq i}}^{n} w_{ij}x_j(\mathrm{t}) + I_i\right] - z_i(\mathrm{t})(x_i(\mathrm{t}) - I_0) \tag{3}$$

$$z_i(\mathrm{t}+1) = (1-\beta)z_i(\mathrm{t}) \tag{4}$$

$$\eta_i(\mathrm{t}+1) = \frac{\eta_i(\mathrm{t})}{\ln(\exp(1) + \lambda(1-\eta_i(\mathrm{t})))} \tag{5}$$

$$f(\mu) = \mathrm{Sigmoid}(\mu, \varepsilon_1) + coef \cdot \mathrm{Shannon}(\mu, \varepsilon_2) \tag{6}$$

$$\mathrm{Sigmoid}(\mu, \varepsilon_1) = \frac{1}{1+\exp(\mu/\varepsilon_1)} \tag{7}$$

$$\mathrm{Shannon}(\mu, \varepsilon_2) = \frac{\sin\pi(\mu/\varepsilon_2 - \frac{1}{2}) - \sin2\pi(\mu/\varepsilon_2 - \frac{1}{2})}{\pi(\mu/\varepsilon_2 - \frac{1}{2})} \tag{8}$$

Where i is the index of neurons and n is the number of neurons, $x_i(t)$ the output of neuron i, $y_i(t)$ the internal state for neuron i, W_{ij} the connection weight from neuron j to neuron i, I_i the input bias of neuron i, α the positive scaling parameter for inputs, k the damping factor of the nerve membrane ($0 \le k \le 1$), $z_i(t)$ the self-feedback connection weight, $\varepsilon_1, \varepsilon_2$ the steepness parameters of the activation function, β the simulated annealing parameter of the self-feedback connection weight $z_i(t)$, $\eta_i(t)$ the other simulated annealing parameter of the activation, I_0 a positive parameter and $coef$ the non-monotonous degree ($0 \le coef \le 1$).

In this model, the variable $z_i(t)$ corresponds to the temperature in the usual stochastic annealing process and the equation (4) is an exponential cooling schedule for the annealing as well as the equation (5). The chaotic mechanism is introduced by the self-feedback connection weight as the value of $z_i(t)$ becomes small step by step. The chaotic behavior plays a global search role in the beginning. When the value of $z_i(t)$ decreases to a certain value, the network functions in a fashion similar to the Hopfield network which functions in gradient descent dynamic behavior. Finally, the neurons arrive at a stable equilibrium state. The reference [6] shows that both the parameter β governed the bifurcation speed of the transient chaos and the parameter α could affect the neuron dynamics; in other words, the influence of the energy function was too strong to generate transient chaos when α was too large, and the energy function could not be sufficiently reflected in the neuron dynamics when α was too small. So in order for the network to have rich dynamics initially, the simulated annealing parameter β must be set to a small value, and α must be set to a suitable value, too.

In this model, the parameter *coef* presents the non-monotonous degree of the activation function. Seen from the equation (6), it is concluded that the equation (6) is similar to the function of Sigmoid alone in form in the circumstance of the value of *coef* being between 0 and 1 without consideration of the monotonous nature. So the parameter *coef* presents a local non-monotonous phenomenon of the activation function. In other words, if the parameter *coef* borders on 1, the non-monotonous phenomenon of the activation function is very apparent; otherwise, if the parameter *coef* borders on 0, the non-monotonous phenomenon of the activation function is very weak.

In order to gain insight into the evolution progress of the single neural unit, the research was made as follows.

3 Research on Single Neural Unit

In this section, we make an analysis of the neural unit of the Shannon Wavelet chaotic neural networks.

The single neural unit can be described as (9) ~ (12) together with (6) ~ (8):

$$x(t) = f(y(t)) \tag{9}$$

$$y(t+1) = ky(t) - z(t)(x(t) - I_0) \tag{10}$$

$$z(t+1) = (1-\beta)z(t) \tag{11}$$

$$\eta(t+1) = \frac{\eta(t)}{\ln(\exp(1) + \lambda(1-\eta(t)))} \tag{12}$$

In order to make the neuron behave transient chaotic behavior, the parameters are set as follows:

$$\varepsilon_1 = 0.004, \varepsilon_2 = 1.25, y(1) = 0.283, z(1) = 0.1, k = 1, \eta(1) = 0.8, \lambda = 0.5, I_0 = 0.5$$

The state bifurcation figures and the time evolution figures of the maximal Lyapunov exponent are respectively shown as Fig.1~Fig.3 when $\beta = 0.004$ and $\beta = 0.002$.

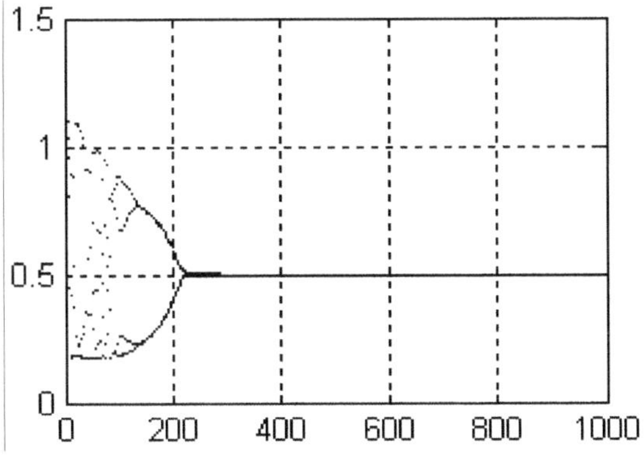

Fig. 1. State bifurcation figure of the neuron when $\beta = 0.004$

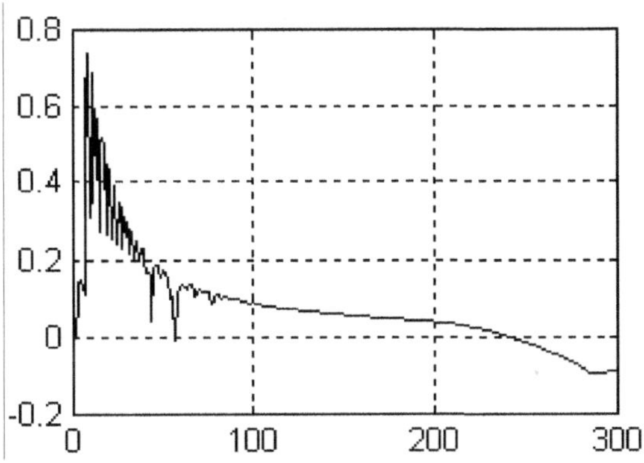

Fig. 2. Time evolution figure of the maximal Lyapunov exponent of the neuron when $\beta = 0.004$

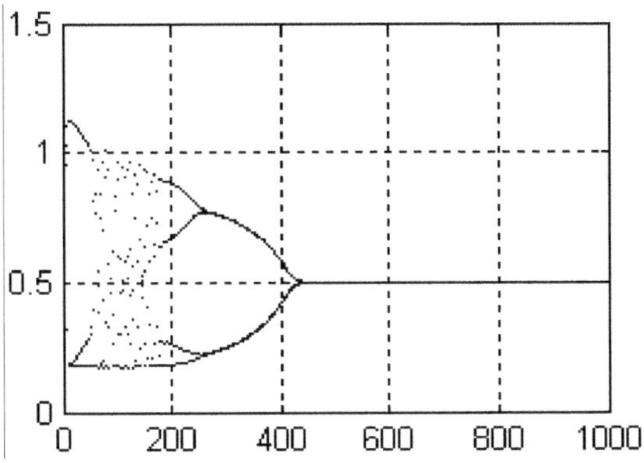

Fig. 3. State bifurcation figure of the neuron when $\beta = 0.002$

Seen from the above state bifurcation figures, the neuron behaves a transient chaotic dynamic behavior. The single neural unit first behaves the global chaotic search, and with the decrease of the value of $z(0,0)$, the reversed bifurcation gradually converges to a stable equilibrium state. After the chaotic dynamic behavior disappears, the dynamic behavior of the single neural unit is controlled by the gradient descent dynamics. When the behavior of the single neural unit is similar to that of Hopfield, the network tends to converge to a stable equilibrium point. The simulated annealing parameter β affects the length of the reversed bifurcation, that is, the lager value of β prolongs the reversed bifurcation.

4 Application to Continuous Function Optimization Problems

In this section, we apply the Shannon wavelet chaotic neural network to search global minima of the following function.

The function is described as follows [7]:

$$f_2(x_1, x_2) = (x_1 - 0.7)^2[(x_2 + 0.6)^2 + 0.1] + (x_2 - 0.5)^2[(x_1 + 0.4)^2 + 0.15] \qquad (13)$$

The minimum value of (13) is 0 and its responding point is (0.7, 0.5).

The parameters are set as follows:

$\varepsilon_1 = 0.05$, $\varepsilon_2 = 10$, $\alpha = 0.08$, $k = 1$, $I_0 = 0.5$, $coef = 1/4$, β =0.002, $z(0,0)$ =[0.8, 0.8], $y(0,0)$ =[0.283, 0.283], $\eta(0,0)$ =[0.8, 0.8], $\lambda(0,0)$ =[0.01, 0.01].

The time evolution figure of the energy function of SWCNN in solving the function is shown as Fig.4.

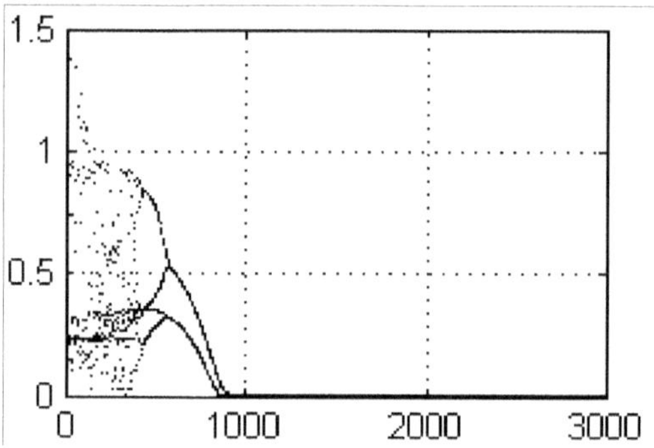

Fig. 4. Time evolution figure of energy function

The global minimum and its responding point of the simulation are respectively 2.1448e-015 and (0.7, 0.5).

This section indicates that SWCNN has a good performance to solve function optimization problems. In order to testify the performance of SWCNN, the new model is applied to solve 10-city traveling salesman problems.

5 Application to 10-City TSP

A solution of TSP with N cities is represented by N×N-permutation matrix, where each entry corresponds to output of a neuron in a network with N×N lattice structure. Assume v_{xi} to be the neuron output which represents city x in visiting order i . A computational energy function which is to minimize the total tour length while simultaneously satisfying all constrains takes the follow form [1]:

$$E = \frac{W_1}{2}\left\{\sum_{i=1}^{n}\left[\sum_{j=1}^{n} x_{ij} - 1\right]^2 + \sum_{j=1}^{n}\left[\sum_{i=1}^{n} x_{ij} - 1\right]^2\right\} + \frac{W_2}{2}\sum_{i=1}^{n}\sum_{j=1}^{n}\sum_{k=1}^{n}(x_{k,j+1} + x_{k,j-1})x_{ij}d_{ik} \quad (14)$$

Where $x_{i0} = x_{in}$ and $x_{i,n+1} = x_{i1}$. W_1 and W_2 are the coupling parameters corresponding to the constrains and the cost function of the tour length, respectively. d_{xy} is the distance between city x and city y .

This paper adopts the following 10-city unitary coordinates:

(0.4, 0.4439),(0.2439, 0.1463),(0.1707, 0.2293),(0.2293, 0.716),(0.5171,0.9414), (0.8732,0.6536),(0.6878,0.5219),(0.8488, 0.3609),(0.6683, 0.2536),(0.6195, 0.2634). The shortest distance of the 10-city is 2.6776.

The reference [6] has presented that the effective activation function may adopt kinds of different forms, and should behave non-monotonous behavior. In this paper, *coef* that represents the non-monotonous degree is analyzed in order to simply ascertain the effect of the non-monotonous degree to SWCNN in solving 10-city TSP. Therefore, the models with different values of *coef* in solving 10-city TSP are analyzed as follows:

The parameters of the network are set as follows:

$W_1 = 1$, $W_2 = 0.8$, $z_i(1) = 0.2$, $\alpha = 0.5$, $k = 1$, $\eta_i(1) = 0.8$, $I_0 = 0.5$, $\lambda = 0.008$, $\varepsilon_1 = 0.004$, $\varepsilon_2 = 2.5$.

2000 different initial conditions of y_{ij} are generated randomly in the region [0, 1] for different β . The results are summarized in Table1, the column 'NL', 'NG', 'LR' and 'GR' respectively represents the number of legal route, the number of global optimal route, the rate of legal route, the rate of global optimal route.

The lager value of the simulated annealing parameter β is regarded stronger if the network can all converge to the global minimum in 2000 different random initial conditions.

Seen from table 1, the follow observations can be drawn according to numerical simulation test:

First, the model with smaller *coef* s such as 0, 1/4, 1/5, 1/6, 1/7, 1/8, 1/9 and 1/10 in solving 10-city TSP can all converge to the global minimum. But, it is not true that the smaller the parameter *coef* is, the more powerful the ability to solve 10-city is. Because, for example, the parameter $coef = 1/10$ can all converge to the global minimum as $\beta = 0.0005$ while the parameter $coef = 1/9$ can all converge to the global minimum as $\beta = 0.0007$.

Table 1. Results of 2000 different initial conditions for each value β on 10-city TSP

coef	β	NL	NG	LR	GR
$coef = 0$ (common network)	0.0003	2000	1923	100%	95.65%
	0.001	1998	1998	99.9%	99.9%
	0.0008	2000	2000	100%	100%
$coef = 1$	0.003	1837	620	91.85%	31%
	0.001	1891	1146	94.55%	57.3%
	0.0008	1904	1075	95.2%	53.75%
$coef = 1/2$	0.003	1962	1791	98.1%	89.55%
	0.0008	1925	1858	96.25%	92.9%
	0.0005	1842	1672	92.1%	83.6%
$coef = 1/4$	0.003	1975	1811	98.75%	90.55%
	0.0008	2000	1997	100%	99.85%
	0.00046	2000	2000	100%	100%
$coef = 1/5$	0.003	1979	1797	98.95%	89.85%
	0.001	2000	1999	100%	99.95%
	0.0009	2000	2000	100%	100%
$coef = 1/6$	0.003	1987	1819	99.35	90.95%
	0.001	2000	2000	100%	100%
$coef = 1/7$	0.003	1989	1806	99.45%	90.3%
	0.001	1999	1999	99.95%	99.95%
	0.0008	2000	2000	100%	100%
$coef = 1/8$	0.003	1990	1713	99.5%	85.65%
	0.0008	1999	1999	99.95%	99.95%
	0.0006	2000	2000	100%	100%
$coef = 1/9$	0.003	1993	1713	99.65%	85.65%
	0.0008	1999	1999	99.95%	99.95%
	0.0007	2000	2000	100%	100%
$coef = 1/10$	0.003	1998	1799	99.9%	89.95%
	0.0008	1999	1998	99.95	99.9%
	0.0005	2000	2000	100%	100%

Second, with the decrease of the value of $coef$, the value of 'NL' becomes large gradually from 1837 ($coef = 1$) to 2000 ($coef = 0$) as $\beta = 0.003$ .In other word, with the decrease of the value of $coef$, the ability to get legal route becomes strong.

Third, when the parameter $coef = 1/5$ and $coef = 1/6$, the ability to all converge to the global minimum is more powerful than that of $coef = 0$, that is, the non-monotonous degree of the activation function has a positive effect on the solution of 10-city TSP.

However, as is analyzed in second, the ability in reaching 'NL' when the parameter $coef = 1/5$ and $coef = 1/6$ is weaker than that of $coef = 0$. So, which model is needed is connected with the concrete request. However, in order to get the tradeoff effect, the value of $coef = 1/6$ may be chose. As the test result is not based on the theoretical analysis, the relationship between $coef$ and the performance need to be studied further.

6 Conclusion

The presented chaotic neural network called SWCNN is proved to be effective in solving optimization problems, and in the section of application to 10-city TSP, the model with different *coef* is analyzed and made a comparison. As a result, the simple rule of the model is disclosed. However, there are a lot of points in the model needed to be studied.

Acknowledgement

This work is supported by the Youth Science Foundation of Harbin 2005AFQXJ040.

References

1. Hopfield J J., Tank D W.: Neural Computation of Decision in Optimization Problems.Biol. Cybern.Vol.52. (1985)141-152
2. Hopfield, J.: Neural Networks and Physical Systems with Emergent Collective Computational Abilities. Vol. 79.In: Proc Natl Acad Sci, (1982) 2554-2558
3. Yao-qun Xu, Ming Sun, Guang-ren Duan. Wavelet Chaotic Neural Networks and Their Application to Optimization Problems. Lecture Notes in Computer Science, Vol. 3791. Springer (2006) 379-384
4. A Potapove, M Kali. Robust chaos in neural networks. Physics Letters A, vol277,no.6, (2000)310-322
5. Shuai J W, Chen Z X, Liu R T, et al. Self-evolution Neural Model. Physics Letters A, vol.221,no.5,(1996)311-316
6. Chen L, Aihara K. Chaotic Simulated Annealing by a Neural Network Model with Transient Chaos. Neural Networks. Vol.8,no.6,(1995)915-930.
7. Wang Ling. Intelligence optimization algorithm and its application. Press of TUP. (2001)
8. Y.-q. Xu and M. Sun. Gauss-Morlet-Sigmoid Chaotic Neural Networks. ICIC 2006, LNCS vol.4113,Springer-Verlag Berlin Heidelberg (2006) 115-125
9. Y.-q. Xu, M. Sun, and J.-h. Shen. Gauss Chaotic Neural Networks. PRICAI 2006, LNAI vol.4099,Springer-Verlag Berlin Heidelberg (2006) 319 – 328

Deterministic Divide-and-Conquer Algorithm for Decomposable Black-Box Optimization Problems with Bounded Difficulty*

Shude Zhou and Zengqi Sun

Department of Computer Science and Technology,
Tsinghua University, Beijing, China
zsd03@mails.tsinghua.edu.cn, szq-dcs@mail.tsinghua.edu.cn

Abstract. There is a class of GA-hard problems for which classical genetic algorithms often fail to obtain optimal solutions. In this paper, we focus on a class of very typical GA-hard problems that we call decomposable black-box optimization problems (DBBOP). Different from random methods in GA literature, two "deterministic" divide-and-conquer algorithms DA1 and DA2 are proposed respectively for non-overlapping and overlapping DBBOP, in which there are no classical genetic operations and even no random operations. Given any DBBOP with dimension l and bounded order k, our algorithms can always reliably and accurately obtain the optimal solutions in deterministic way using $O(l^k)$ function evaluations.

1 Introduction

Genetic algorithms (GAs) are successful stochastic optimization algorithms that guide the exploration of the search space by application of selection and evolutionary operators. However, on a class GA-deceptive problems which can be additively decomposed into subfunctions of bounded order, the simple GAs often experience building block disruption and exhibit very poor performance [9]. That is why there has been a growing interest in the study of designing effective methods to solve the class of GA-deceptive problems. In GA literature, various attempts have been made to solve such problems. One class of approaches refers to manipulating the representation of solutions to make the interacting components of partial solutions less likely to be disrupted. This work includes the messy GA [2], the gene expression messy GA [12], and the linkage learning GA [13]. Another way to deal with interactions between variables is estimation of distribution algorithms (EDAs) [9,10,11,14], that use probabilistic models of promising solutions found so far to guide further exploration of the search space.

Different from the methods in GA literature, this paper develops "deterministic" divide-and-conquer algorithms for the decomposable GA-hard problems. We call this class of problems decomposable black-box optimization problems (DBBOPs), and mathematic description and some related theorem are given. Based on the fact that the

* This work is funded by National Key Project for Basic Research of China (G2002cb312205).

T.-D. Wang et al. (Eds.): SEAL 2006, LNCS 4247, pp. 252 – 260, 2006.

decomposable underlying structure of DBBOP can be explored by its Walsh representation, two efficient algorithms DA1 and DA2 are proposed respectively for non-overlapping and overlapping problems. The advantages of our algorithms include determinacy, reliability, accurateness and scalability. Given any DBBOP with dimension l and bounded order k, our algorithms can always reliably obtain the optimal solutions in deterministic way. Both DA1 and DA2 are developed based rigorous theoretical induction. It can be theoretically guaranteed that the solution obtained by DA1 and DA2 is globally optimal. Our algorithms exhibit good scalability and require polynomial number $O(l^k)$ of function evaluations.

Section 2 gives mathematic description of DBBOP; in Section 3, the close relationship between the underlying structure of DBBOP and Walsh coefficients is described, based on which Walsh decomposition algorithm is given; Section 4 proposes deterministic algorithms DA1 and DA2; the paper is concluded in Section 5.

2 DBBOP

This section gives general mathematic formulation of a class of GA-hard problems, which have the characters of additive decomposition.

The black-box optimization problem with the prior knowledge "decomposable" is called decomposable black-box optimization problems (DBBOP) and formulated as:

$$x = \arg\min_{x \in D} f(x)$$

where f can be represented as $f(x) = \sum_{i=1}^{m} (g_i(s_i))$, in which $s_i \subseteq X$, $X = \{x_1, x_2, \cdots x_l\}$ indicates the parameters set, $|s_i|$ indicates the size of s_i, $g_i(s_i)$ is black-box sub-function that can be also written as $g_i(s_i) = g_i\left(x_{i_1}, x_{i_2}, \cdots, x_{i_{|s_i|}}\right)$, and m is the number of sub-functions. If $\forall i, j \in \{1, \cdots, m\}$, $i \neq j$, $s_i \cap s_j = \phi$, then it is a *non-overlapping* DBBOP. If $\exists i, j \in \{1, \cdots, m\}$, $i \neq j$, we have $s_i \cap s_j \neq \phi$, then it is *overlapping* DBBOP. $k = \arg\max_{1 \le i \le m} |s_i|$ is named the *order* of DBBOP.

In this paper, we will restrict DBBOP to binary domain $D = \{0,1\}^l$. It has been found that the simple GAs often experience building block disruption and exhibit very poor performance for high-dimension DBBOP[2,9,11,13]. However, the decomposable character of DBBOP makes divide-and-conquer strategy available if problem structure can be obtained.

Theorem 1. *For non-overlapping DBBOP* $x = \arg\min_{x \in D} f(x)$ *where* $D = \{0,1\}^l$ *and* $k << l$, *the time-complexity of solving it is no more than* $O(A) + O(l)$, *where* $O(A)$ *indicates the time complexity of decomposition algorithm, k is the order and l is the dimension.*

Proof
Suppose that the DBBOP has been decomposed by algorithm *A*. The parameter set $X=\{x_1,x_2,\cdots x_n\}$ is clustered as $s_1,s_2,\cdots,s_i,\cdots,s_m$ where $s_i\subseteq X$ and *f* can be represented as $f(x)=\sum_{i=1}^{m}(g_i(s_i))$. Then we will attempt obtain the optimal solution by divide-and-conquer strategy. Because of the black-box nature of g_i, we can not solve g_i directly. For each g_i, we randomly set the values of parameters that are not in s_i, and then we form a new optimization problem $x_{s_i}=\arg\min f(s_i,\bar{s}_i)$ where $\bar{s}_i$ is the complementary set of s_i. The parameters in $\bar{s}_i$ has been set randomly, so we can get the optimal settings s_i by enumerating all possible settings for variables in s_i. The time complexity is $O(2^{|s_i|})$. So the total time complexity of solving f is:

$$O(f)=O(A)+\sum_{i=1}^{m}O(g_i(s_i))\le O(A)+2^kO(l) \tag{1}$$

The time-complexity to solve non-overlapping DBBOP is $O(A)+O(l)$. □

The above theorem gives us theoretical foundation to design deterministic algorithm for non-overlapping DBBOP. For overlapping DBBOP, the situation becomes much more complex, that will be discussed in Section 4.

3 The Connection Between Walsh Transformation and DBBOP Structure

In this section, the connection between Walsh transformation and problem structure is introduced, that will provide a clue to implement the "divide" process in the divide-and-conquer strategy for DBBOP.

Before going on, we give the definition of *partition*. Given a DBBOP, $X=\{x_1,x_2,\cdots x_l\}$ is the parameters set, $s\subseteq X$ is an arbitrary subset of *X*, and $\rho(X)$ is the power set of *X*. *partition* is defined as $b=pa(s)$, where $pa:\rho(X)\mapsto\{0.1\}^l$ is a function:

$$b_i=\begin{cases}1 & x_i\in s\\ 0 & x_i\notin s\end{cases}$$

The function *pa* is a bijective function. For example, $X=\{x_1,x_2,x_3,x_4\}$, $s=\{x_1,x_3\}$, the partition corresponding to s can be written as $b=(1010)$. For any partitions $a,b\in\{0,1\}^l$, we say that $a\subseteq b$, if $s_a\subseteq s_b$, where $a=pa(s_a)$ and $b=pa(s_b)$.

It has been proved that the values of Walsh coefficients can reflect the partition structure of DBBOP [6].

Theorem 2. *If a binary-code function f can be written as* $f(x)=\sum_{i=1}^{m}(g_i(s_i))$, *then*

$$\omega_{pa(s)}=\begin{cases}\alpha & \exists \mathrm{i},1\le \mathrm{i}\le \mathrm{m},\mathrm{s}\subseteq \mathrm{s_i}\\ 0 & \text{otherwise}\end{cases} \tag{2}$$

in which $\omega_{pa(s)}$ is Walsh coefficient, $s \in \rho(X)$, pa is the bijective function that map $\rho(X)$ to the partitions, and α is a value depending on the Walsh coefficients of $g_i(s_i)$ where $s \subseteq s_i$ [6].

This theorem explicitly demonstrates that the underlying structure of the given problem can be captured by the partitions with non-zero Walsh coefficients. However, computation of a single Walsh coefficient usually requires all the 2^l values of *f*. In the literature, several attempts has been made to calculate the Walsh coefficients of DBBOP [1,6,15]. Here, we restrict ourselves to only the deterministic Walsh decomposition algorithm proposed by Hillol Kargupta et.al.[1].

According to the Walsh representation $f(x)=\sum_{i=0}^{2^l-1}\omega_i\psi_i(x)$, for any partition *i*, we have:

$$f(x)\psi_i(x)=\sum_{j=0}^{2^l-1}\omega_j\psi_j(x)\psi_i(x) \tag{3}$$

It can be further written as:

$$\begin{aligned}\sum_{x\subseteq i} f(x)\psi_i(x)&=\sum_{j=0}^{2^l-1}\omega_j\sum_{x\subseteq i}\psi_j(x)\psi_i(x)\\&=\sum_{j\supseteq i}\omega_j\sum_{x\subseteq i}\psi_j(x)\psi_i(x)+\sum_{j\not\supseteq i}\omega_j\sum_{x\subseteq i}\psi_j(x)\psi_i(x)\end{aligned} \tag{4}$$

It can be easily proven $\sum_{j\not\supseteq i}\omega_j\sum_{x\subseteq i}\psi_j(x)\psi_i(x)=0$, thus equation (4) can be written as:

$$\sum_{j\supseteq i}\omega_j\sum_{x\subseteq i}\psi_j(x)\psi_i(x)=\sum_{x\subseteq i} f(x)\psi_i(x) \tag{5}$$

According to the characters of Walsh functions, we know that, if $j \supset i$, then $\forall x\subseteq i, \psi_j(x)\psi_i(x)=1$. So Equation (5) can be rewritten as $\sum_{j\supseteq i}\omega_j|s_i|=\sum_{x\subseteq i} f(x)\psi_i(x)$, where s_i is the variables set corresponding to partition *i*, $i=pa(s_i)$. Thus, we have:

$$\sum_{j\supseteq i}\omega_j=\frac{1}{|s_i|}\sum_{x\subseteq i} f(x)\psi_i(x). \tag{6}$$

Now, let's consider the calculation of Walsh coefficients. Given a DBBOP with dimension *l* and order *k*, the computing method of ω_i can be considered in 3 cases.

Case 1: $|s_i|>k$

According to Theorem 2, it is obvious that $\forall s_i\ |s_i|>k$, $\omega_i=0$, where $i=pa(s_i)$.

Case 2: $|s_i|=k$

In Equation (6), we set $|s_i|=k$, then $\forall j, j\supset i$ we have $\omega_j=0$. Thus, for any partition *i* with order *k*, we can calculate its corresponding Walsh coefficient by equation:

$$\omega_i = \frac{1}{|s_i|}\sum_{x \subseteq i} f(x)\psi_i(x). \tag{7}$$

For each order-*k* Walsh coefficient, 2^k function evaluations are needed. So, computing all the order-*k* Walsh coefficients requires $2^k C_l^k$ function evaluations.

Case 3: $|s_i| < k$

First of all, we set $i = k-1$. According to Equation (5) in which $\forall j, j \supset i$, ω_j has been computed before, we can obtain ω_i by

$$\omega_i = \frac{\sum_{x \subseteq i} f(x)\psi_i(x) - \sum_{j \supset i} \omega_j \sum_{x \subseteq i} \psi_j(x)\psi_i(x)}{\sum_{x \subseteq i} \psi_i(x)\psi_i(x)}, \tag{8}$$

where the values of $f(x)$, $x \subseteq i$ has been calculated in Case 2. Using the Equation (8), we can continue iteratively to calculate Walsh coefficients of $(k-2), (k-3), \cdots$, 1-order partitions.

Till now, all the Walsh coefficients have been calculated. According to Theorem 2, once the Walsh coefficients is computed, the underlying decomposable structure is obtained by the partitions with non-zero Walsh coefficients.

Note that, for an order-*k* DBBOP, no additional function evaluation is needed after the order-*k* Walsh coefficients have been computed. The number of function evaluations required is $2^k \binom{l}{k}$. So, the time complexity is polynomial $O(l^k)$ in *l* for DBBOP with fixed bounded order *k*.

4 Deterministic Divide-and-Conquer Algorithms

In this section, we propose two deterministic divide-and-conquer algorithms DA1 and DA2 for non-overlapping and overlapping DBBOPs respectively.

4.1 DA1: Deterministic Algorithm for Non-overlapping DBBOP

Theorem 1 shows that non-overlapping DBBOP can be optimized by divider-and-conquer method with time complexity $O(A) + O(l)$, where $O(A)$ indicates the time complexity of decomposition algorithm. In the proposed algorithm DA1, the optimization process is implemented during the decomposition process.

Given a non-overlapping DBBOP with dimension *l* and order *k*, DA1 is implemented as follows:

1. X indicates the variables set $X = (x_1, x_2, \cdots x_l)$. $X_0 = \phi$. x_0 stores the optimal solution and is initiated as $x_0 = 111\cdots11$. $k' = k$.
2. Select k' variables from $X - X_0$ and there are $C_{l-|X_0|}^{k'}$ possible selections.

3. For each selection $s_i \subseteq X - X_0$, $|s_i| = k'$, $i = pa(s_i)$, calculate the Walsh coefficients: $\omega_i = \begin{cases} Equation\ (7) & k' = k \\ Equation\ (8) & k' \neq k \end{cases}$

   If $\omega_i \neq 0$, $X_0 = X_0 \cup s_i$, $x_0 = x_0 \wedge x^*$, where $x^* = \arg\min_{x \subseteq i} f(x)$.
4. $k' = k' - 1$. If $k' = 0$ or $X - X_0 = \phi$, then x_0 is the optimal solution and exit; else go to step 2.

In step 2, we consider two cases, $k' = k$ and $k' < k$. In case $k' = k$, if the Walsh coefficient corresponding to s_i is non-zero, then we obtain $x^* = \arg\max_{x \subseteq i} f(x)$ by enumeration. $\forall x, x \subseteq i$, $f(x)$ has been calculated during the process of calculating ω_i, so no additional function evaluation is required. In case $k' < k$ and $|s_i| = k'$, if the Walsh coefficient corresponding to s_i is non-zero, then we obtain $x^* = \arg\min_{x \subseteq i} f(x)$ by enumeration, in which all the $f(x)$ have been calculated during the process of calculating ω_j where $i \subseteq j$ and $j = k$. So, In case of $k' < k$, no additional function evaluation is needed. DA1 always finds the optimum using $2^k \binom{l}{k}$ function evaluations.

For each sub-problem, the optimal solution is obtained by enumeration, so it is guaranteed that the solution obtained by DA1 is always global optimal. The reliability and accuracy can be guaranteed by Theorem 1 and Theorem 2.

4.2 DA2: Deterministic Algorithm for Overlapping DBBOP

This subsection presents a deterministic algorithm to solve overlapping DBBOP. Because of the overlapping structure, it gets much more complex to directly divide and conquer DBBOP.

Firstly, *Overlapping variable set* is defined as a set of variables in which each variable appears in at least two sub-problems. If we know f can be represented as $f(x) = \sum_{i=1}^{m} (g_i(s_i))$, the overlapping variable set s^* can be formulated as

$$s^* = \bigcup_{i,j \in \{1,\cdots,m\}, i \neq j} (s_i \cap s_j)$$

For example, in Fig.1., the problem with dimension 7 is partitioned into 3 sub-problems. The shadow part is s^* and the white part indicates $s_i - s^*$, $i = 1,2,3$.

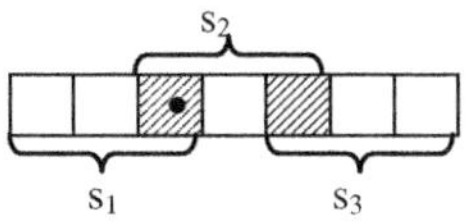

Fig. 1. Partition of variables set. The white part indicates $s_i - s^*$, $i = 1,2,3$, and the shadow part is s^*.

Because the interactions between overlapping partitions, the overlapping variable set makes it impossible to directly divide and conquer the overlapping DBBOP. After the decomposable structure has been captured by algorithm in section 3, we can solve DBBOP in the way as illustrated in Figure 2. First of all, the overlapping variable set s^* can be obtained from the decomposable structure. And then, we set binary values to the variables in s^*. For each setting, we can divide the problem into sub-problems and solve them separately. Thus, we can obtain the optimal settings of other variables in the condition of current s^* setting. There are totally $2^{|s^*|}$ possible settings, so we can get $2^{|s^*|}$ conditional optimal solutions, of which the best one is the global optimum.

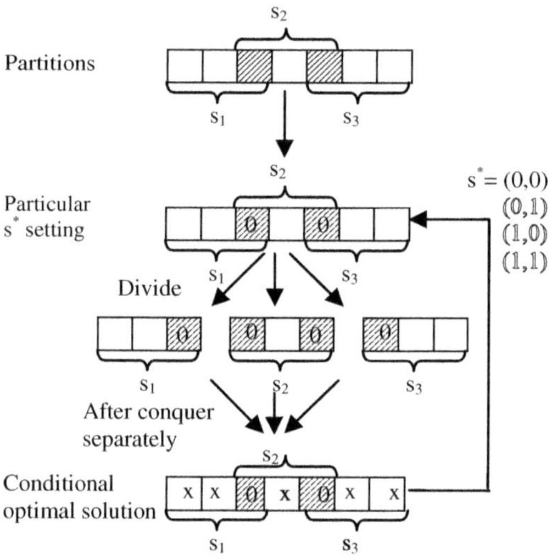

Fig. 2. Divide and conquer process for solving overlapping optimization problem

Given an overlapping DBBOP with dimension l and order k, the algorithm DA2 can obtain its global minimal solution by the procedure described as follows.

1. X indicates the variables set $X=(x_1,x_2,\cdots x_l)$. $X_0=\phi$, $k'=k$.
2. After decomposition algorithm described in section 3, the structure of the problem is represented as $S=\{s_1,s_2,\cdots,s_m\}$.
3. The overlapping variable set is $s^*=\bigcup_{i,j\in\{1,\cdots,m\},i\neq j}(s_i\cap s_j)$ and $i^*=pa(s^*)$. $\overline{i}^*=111\cdots11-i^*$.
4. For every $a\subseteq i^*$,

   $f=0$, $f^*=+\infty$

   $x_0=111\cdots11$ used to store the optimal solution in condition of a.

   For every s_j, $1\le j\le m$

   $x_j=\arg\min_{x\subseteq(i\wedge(a\vee\overline{i}^*))} f(x)$, where $i=pa(s_j)$

   (Note that, for $x\subseteq i$, $f(x)$ has been calculated during the decomposition process. So, no additional fitness evaluation is needed.)

$$x_0 = x_0 \wedge x_j$$
$$f = f + f_j$$

if $f^* > f$

$$f^* = f$$
$$x^* = x_0$$

5. x^* is the global optimal solutions that make DBBOP globally minimal.

Step 4 is used to calculate the conditional optimal solution in condition of detailed s^* setting. In the sixth line of Step 4, $x_j = \arg \min_{x \subseteq (i \wedge (a \vee \bar{i}^*))} f(x)$ is used to compute the best parameter setting of s_j in the condition that the detailed setting of s^* is embedded in a. In the above equation, for every $x \subseteq (i \wedge (a \vee \bar{i}^*))$, $f(x)$ has been calculated during the decomposition process, so no additional fitness evaluation is needed.

In DA2, only Step 2 needs to calculate the function values. So the implementation of DA2 requires totally $2^k \binom{l}{k}$ function evaluations. For any overlapping DBBOP, the proposed DA2 can reliably and accurately obtain optimal solution in a deterministic way using only polynomial number of function evaluations. However, it should note that it does NOT mean that the overlapping DBBOP can be solved in polynomial time (overlapping DBBOP is NP-complete problem). Although only $2^k \binom{l}{k}$ function evaluations are needed, the computational time or space may scale exponentially if the s^* is very large. In practical application, it very often happens that calculating the objective function one time is quite expensive either in computation or money, so we can obtain its optimal solutions by only polynomial number of function evaluations using DA2 plus other technology to enhance the computation speed.

4.3 Experimental Verifications

The purpose of this section is to verify the performance of DA1 and DA2. Because of the determinacy of the algorithms, no parameter setting is needed. What we care about is whether DA1&DA2 can obtain solutions as the previous sections described.

DA1 and DA2 are implemented on Matlab 6.5 platform. Some typical GA-deceptive problems are used as test problems, which include order-3 deceptive function [9] (dimension ranged: 30, 60, 90, 120, 150), oder-5 trap function [9] (dimension: 50, 100, 150), Whitley's order-4 deceptive function [8] (dimension: 40, 80, 120), Overlapping order-3 deceptive function [9] (with dimension 30, 60) and overlapping order-4 deceptive function [9] (with dimension 20, 40, 60).

Because the determinacy of the algorithm, the description of the empirical results is very simple: Experimental results demonstrate that DA1 and DA2 can always reliably obtain optimum in deterministic way. The number of function evaluations in DA1 and DA2 is $2^k \binom{l}{k}$, where l is the dimension and k the bounded order. In addition, experiments also show that DA1 behaves polynomial time complexity, while DA2 behaves exponential time complexity because of the overlapping variable set.

5 Conclusion

The main contribution of the paper is that two deterministic algorithms DA1 and DA2 are proposed to solve DBBOP with bound difficulty that is usually very hard for simple GA. Divide-and-conquer strategy is the spirit of the two algorithms. The advantages of our algorithms include determinacy, reliability, accurateness and scalability. Given any DBBOP with dimension l and bounded order k, the algorithms can always reliably obtain the optimal solutions in deterministic way. DA1 and DA2 are developed based on rigorous theoretical induction. It can be theoretically guaranteed that the solution obtained is globally optimal. For a DBBOP with dimension l and order k, the algorithms can find the optimal solution using $O(l^k)$ function evaluations.

However, it should be noted that DA2 is far from efficient enough to solve overlapping DBBOP, even though it need only polynomial number of function evaluations. Future work will focus on how to handle overlapping variable set intelligently.

References

1. Kargupta, H., Park, B.: Gene Expression and Fast Construction of Distributed Evolutionary Representation. Evolutionary Computation, Vol. 9. (2001) 43–69
2. Goldberg, D. E., Deb, K., Kargupta, H., Harik, G.: Rapid, Accurate Optimization of difficult Optimization Problems Using Fast Messy Genetic Algorithms. In: Proceedings of the fifth international conference on Genetic Algorithm, San Mateo, California (1993) 56–64
3. Goldberg, D. E.: Genetic Algorithms and Walsh Functions: Part I, a Gentle Introduction. Complex Systems, Vol. 3. (1989) 129–152
4. Goldberg, D. E.: Genetic Algorithms and Walsh Functions: Part II, Deception and its Analysis. Complex Systems, Vol. 3. (1989) 153–171
5. David, H. W., William, G. M.: No Free Lunch Theorems for Optimization. IEEE Trans. Evolutionary Computation, Vol. 1. (1997) 67–80
6. Heckendorn, R.B.: Embedded Landscapes. Evolutionary Computation, Vol. 10. (2002)
7. Sami, K.: Transform Methods: Walsh Analysis, Walsh Transforms. In: Handbook of Evolutionary Computation, IOP Publishing Ltd and Oxford University Press (1995)
8. Whitley, D.: Fundamental Principles of Deception in Genetic Search. In: Rawlins, G. J. E. (eds.): Foundations of Genetic Algorithms, Morgan Kaufmann, San Mateo, CA (1991)
9. Pelikan, M., Goldberg, D.E., Cantú-Paz, E.: Linkage Problem, Distribution Estimation, and Bayesian Networks. Evolutionary Computation, Vol. 8(3) (2000) 311–340
10. Pelikan M., Goldberg, D.E., Lobo, F.G.: A survey of optimization by building and using probabilistic models. Computational Optimization and Applications, Vol. 21. (2002) 5–12
11. Larrañaga, P., Lozano, J. A.: Estimation of Distribution Algorithms: A new Tool for Evolutionary Computation. Kluwer Academic Publishers (2002)
12. Kargupta, H.: The gene expression messy genetic algorithm. In: Proceedings of the IEEE international conference on Evolutionary Computation, Nogoya, Japan (1996)
13. Harik, G. R.: Learning Gene Linkage to Efficiently Solve Problems of Bounded Difficulty using Genetic Algorithms. Ph.D. Thesis, University of Michigan, Ann Arbor, MI (1997)
14. Mühlenbein, H., Mahnig, T.: FDA–Scalable Evolutionary Algorithm for the Optimization of Additively Decomposed Functions. Evolutionary Computation Vol.7. (1999) 353–376
15. Hechendorn, R.B., Wright A.H.: Efficient Linkage Discovery by Limited Probing. Evolutionary Computation, Vol.12. (2004) 517–545

QPSO-Based QoS Multicast Routing Algorithm

Jun Sun, Jing Liu, and Wenbo Xu

Center of Intelligent and High Performance Computing,
School of Information Technology, Southern Yangtze University,
No. 1800, Lihudadao Road, Wuxi 214122, Jiangsu, China
sunjun_wx@hotmail.com

Abstract. QoS multicast routing in networks is a very important research issues in the areas of networks and distributed systems. Because of its NP-completeness, many heuristics such as Genetic Algorithms (GAs) are employ solve the QoS routing problem. Base on the previously proposed Quantum-behaved Particle Swarm Optimization (QPSO), this paper proposes a QPSO-based QoS multicast routing algorithm. The proposed method converts the QoS multicast routing problem into an integer programming problem and then solve the problem by QPSO. We test QPSO-base routing algorithm on a network model. For performance comparison, we also test Particle Swarm Optimization (PSO) algorithm and GA. The experiment results show the availability and efficiency of QPSO on the problem and its superiority to PSO and GA.

1 Introduction

Multicast routing including Quality-of-Service (QoS) multicast routing has continued to be a very important research issue in the areas of networks and distributed systems. An efficient allocation of network resources to satisfy the different QoS requirements is the primary goal of QoS-based multicast routing. However the inter-dependency and confliction among multiple QoS parameters makes the problem difficult. It has been demonstrated that it is NP-Complete to find a feasible multicast tree with two independent additive path constraints. Generally, heuristics are employed to solve this NP-complete problem. Some Genetic Algorithms (GAs) has been used to solve the problem from different aspects ([5] etc.).

Particle Swarm Optimization (PSO) is a new efficient evolutionary optimization method originally proposed by J. Kennedy and R.C. Eberhart ([6]). Recently, a novel variant of PSO, called Quantum-behaved Particle Swarm Optimization (QPSO), has been proposed in order to improve the global search performance of the original PSO ([10], [11], [12]). The goal of this paper is to solve QoS multicast routing problem with QPSO, which is the first attempt to explore the applicability of QPSO to Combinatorial Optimization Problems. The rest of the paper is structured as follows. In Section 2, the network model of QoS multicast routing problem is introduced. A brief introduction of PSO and QPSO is given in Section 3. Section 4 is our proposed QPSO-based QoS multicast routing algorithm. The experiment results are presented in Section 5 and the paper is concluded in Section 6.

T.-D. Wang et al. (Eds.): SEAL 2006, LNCS 4247, pp. 261 – 268, 2006.

2 Network Model

A network is usually represented as a weighted digraph $G = (V, E)$, where V denotes the set of nodes and E denotes the set of communication links connecting the nodes. |V| and |E| denote the number of nodes and links in the network, respectively. Without loss of generality, only digraphs are considered in which there exists at most one link between a pair of ordered nodes.

Let $s \in V$ be source node of a multicast tree, and $M \subseteq \{V - \{s\}\}$ be a set of end nodes of the multicast tree. Let R be the positive weight and R^+ be the nonnegative weight. For any link $e \in |E|$, we can define the some QoS metrics: delay function delay (e): $E \to R$, cost function cost (e): $E \to R$, bandwidth function bandwidth (e): $E \to R$; and delay jitter function delay-jitter (e): $E \to R^+$. Similarly, for any node $n \in V$, one can also define some metrics: delay function delay (n): $V \to R$, cost function cost (n): $V \to R$, delay jitter function delay-jitter (n): $V \to R^+$ and packet loss function packet-loss (n): $V \to R^+$. We also use $T(s,M)$ to denote a multicast tree, which has the following relations:

$$delay\,(p(s,T)) = \sum_{e \in p(s,T)} delay\,(e) + \sum_{n \in p(s,T)} delay\,(n) \tag{1}$$

$$\cos t(T(s,M)) = \sum_{e \in p(s,M)} \cos t(e) + \sum_{n \in p(s,M)} \cos t(n) \tag{2}$$

$$bandwidth(p(s,T)) = \min(bandwidth(e)), e \in p(s,T) \tag{3}$$

$$delay-jitter(p(s,T)) = \sum_{e \in p(s,T)} delay-jitter(e) + \sum_{n \in p(s,T)} delay-jitter(n) \tag{4}$$

$$packet-loss(p(s,T)) = 1 - \prod_{n \in p(s,T)} (1 - packet-loss(n)) \tag{5}$$

where $p(s,T)$ denotes the path from source s to end node t to T(s, M). With QoS requirements, the multicast routing problem, which was proved to be NP-complete, can be represented as finding a multicast tree T(s, M) satisfying the following constraints: Delay Constraint: $delay(p(s,T)) \leq D$; Bandwidth Constraint: $bandwidth(p(s,T)) \geq B$; Delay-jitter Constraint: $delay-jitter(p(s,T)) \leq J$; Packet-loss Constraint: $packet-loss(p(s,T)) \leq L$;

3 Quantum-Behaved Particle Swarm Optimization

Particle Swarm Optimization (PSO) was originally proposed by J. Kennedy and R. Eberhart [9]. The underlying motivation for the development of PSO algorithm was social behavior of animals such as bird flocking, fish schooling, and swarm theory. In the Standard PSO with S individuals, each individual is treated as a volume-less particle in the D-dimensional space, with the position and velocity of ith particle represented as $X_i = (X_{i1}, X_{i2}, \cdots, X_{iD})$ and $V_i = (V_{i1}, V_{i2}, \cdots, V_{iD})$. The particles move according to the following equation:

$$V_{id} = w \cdot V_{id} + c_1 \cdot rand(\cdot) \cdot (P_{id} - X_{id}) + c_2 \cdot Rand(\cdot) \cdot (P_g - X_{id}) \tag{6}$$

$$X_{id} = X_{id} + V_{id} \tag{7}$$

where c_1 and c_2 are positive constant and rand() and Rand() are two uniform random functions within [0,1]. Parameter w is the inertia weight introduced to accelerate the convergence speed of the PSO. Vector $P_i = (P_{i1}, P_{i2}, \cdots, P_{iD})$ is the best previous position (the position giving the best fitness value) of particle i called ***personal bet position***, and vector $P_g = (P_{g1}, P_{g2}, \cdots, P_{gD})$ is the position of the best particle among all the particles in the population and called ***global best position***.

Many improved versions of PSO have been proposed ([1], [3], [7] etc.). In the previous work presented in [10], [11] and [12], we proposed a novel variant form of PSO, Quantum-behaved Particle Swarm Optimization algorithm (QPSO), which was inspired by Quantum Mechanics and seems to a promising optimization problem solver. In QPSO, the particle moves according to the following equation:

$$X_{id} = p_{id} \pm \gamma |mbest_d - X_{id}| \ln(1/u), \quad u = Rand() \tag{8}$$

where *mbest* is the mean of personal best positions among the particles, that is

$$mbest = \frac{1}{S}\sum_{i=1}^{M} P_i = \left(\frac{1}{S}\sum_{i=1}^{M} P_{i1}, \ \frac{1}{S}\sum_{i=1}^{M} P_{i2}, \ \cdots, \ \frac{1}{S}\sum_{i=1}^{M} P_{id} \right) \tag{9}$$

and p_{id} is determined by $p_{id} = \varphi \cdot P_{id} + (1-\varphi) \cdot P_{gd}$, $\varphi = rand()$. φ is a random umber distributed uniformly on [0,1] and u is another uniformly-distributed random number within [0,1] and γ is a parameter of QPSO called Contraction-Expansion Coefficient. The stochastic evolution equation (8) comes from a quantum δ potential well model proposed in [10]. One may refer to references [10], [11] and [12] for the origin and development of QSPO.

4 QPSO-Based QoS Routing Algorithm

The first step of solving QoS routing problem by QPSO involves encoding a path serial into a feasible solution (or a position) in search space of the particle. In our proposed coding scheme, the number of paths (no loop) reaching each end node $t \in M$ is worked out first. With the number of end nodes denoted as $|M|$, the number of paths to end node i is represented as an integer $n_i (1 \le i \le |M|)$. Thus each path to end node i can be numbered by an integer variable $t_i (1 \le i \le |M|)$, where $t_i \in [1, n_i] (1 \le i \le |M|)$. Consequently, we can obtain a $|M|$-dimensional integral vector

$(t_1,t_2,\cdots,t_{|M|})$ denoting a possible path serial with each component t_i varying within $[1,n_i]$. In the QPSO for QoS Multicasting routing problem, such an integral vector represents the position of the particle and the QoS routing problem is reduced to a $|M|$-dimensional integral programming.

In our proposed method, the fitness unction is defined as:

$$\max \quad f(x)=\frac{\omega_1}{\cos t(T(s,M))}(\omega_2\cdot f(d)+\omega_3\cdot f(j)+\omega_4\cdot f(p)) \tag{10}$$

where ω_1, ω_2, ω_3 and ω_4 is the weight of cost, delay, delay-jitter and packet loss, respectively; $f(d)$, $f(j)$ and $f(p)$ are defined as

$$f(d)=\prod_{t\in M}F_d(delay(p(s,t))-D),\quad F_d(delay(p(s,t))-D)=\begin{cases}1, delay(p(s,t))<D\\ \alpha, delay(p(s,t))\geq D\end{cases} \tag{11}$$

$$f(j)=\prod_{t\in M}F_j(delay\_jitter(p(s,t))-J),\quad F_j(delay\_jitter(p(s,t))-J)=\begin{cases}1, delay\_jitter(p(s,t))<J\\ \beta, delay\_jitter(p(s,t))\geq J\end{cases} \tag{12}$$

$$f(p)=\prod_{t\in M}F_p(packet\_loss(p(s,t))-L),\quad F_p(packet\_loss(p(s,t))-L)=\begin{cases}1, packet\_loss(p(s,t))<L\\ \sigma, packet\_loss(p(s,t))\geq L\end{cases} \tag{13}$$

where $F_d(x)$, $F_j(x)$ and $F_p(x)$ are penalty functions for delay, delay-jitter and packet loss, respectively, and α, β and σ are positive numbers smaller than 1. With above specification, QPSO-based QoS multicast routing algorithm is described as follows.

QPSO-Based QoS Multicast Routing Algorithm

Input: The dimension of the particles' positions (equal to the number of end nodes); Population size; Parameters of the network model.

Output: The best fitness value after QPSO executes for MAXITER iterations; optimal multicast tree.

Procedure:

1. Initialize the population;
2. **for** t=1 to MAXITER
3. Compute the fitness value of each particle according to (10);
4. Update the personal best position P_i;
5. Update the global best position P_g;
6. Compute the mbest of the population by (9) and the value of γ;
7. **for** each particle in the population
8. Update each component of the particle's position by (8) and adjust the component t_i as an integer in $[1,n_i]$; **endfor**
9. **endfor**

Implementation of QPSO on the problem yields a multicast tree denoted by a path serial. The path serial is a $|M|$-dimensional integral vector with each component being the path number of a path from the source code to corresponding end node. To make the multicast tree a feasible solution, we must delete loops existing in the tree.

5 Experiments

We use the network model in Figure 1 as our testing problem. In the experiments, it is assumed that all the end nodes of multicast satisfy the same set of QoS constraints without regard to the characteristics of the nodes. The characteristics of the edges described by a quaternion (d, j, b, c) with the components representing delay, delay-jitter, bandwidth and cost, respectively. For performance comparison, we also used Particle Swarm Optimization (PSO) and Genetic Algorithm (GA) to test the problem.

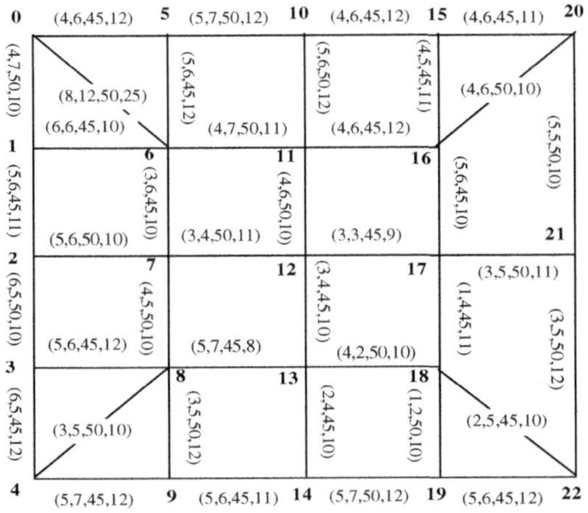

Fig. 1. A network model as the testing paradigm in our experiments

The experiment configuration is as follows. The population size for PSO and QPSO is 50 and maximum number of iterations is 200 for all three algorithms and the number of the end nodes is 5. The fitness function is formula (11) with $\omega_1=1$, $\omega_2=0.5$, $\omega_3=0.5$, $\omega_4=0.3$, $\alpha=0.5$, $\beta=0.5$, $\sigma=0.5$. There are 23 nodes in the network model (Figure 1), and we assume node 0 to be the source node; the set of end nodes to be M={4,9,14,19,22}. For QPSO, the value of γ varies from 1.0 to 0.5 over the course of running. The inertia weight w in PSO decreases linearly from 0.9 to 0.4 over a running and acceleration coefficients c_1 and c_2 are fixed at 2.0. For GA, the individual is real coded, the population size is 100 and binary tournament selection is used. The probability of crossover operation is 0.8 and that of mutation operation is 0.2. We adopt two sets of constraints in the experiments:

1. When delay constraint D=20, delay-jitter constraint J=30, bandwidth constraint B=40 and packet loss constraint L=0.002, the three algorithms executed for 50 runs respectively and the multicast trees corresponding to best fitness value at the end of 200 iterations out of 50 runs are shown in Figure 2(a), Figure 3(a) and Figure 4(a).
2. When delay constraint D=25, delay-jitter constraint J=35 and bandwidth constraint B=40 and packet loss constraint L=0.002, the three algorithms executed for 50 runs respectively and the multicast trees corresponding to best fitness value at the end of 200 iterations out of 50 runs are shown in Figure 2(b), Figure 3(b) and Figure 4(b).

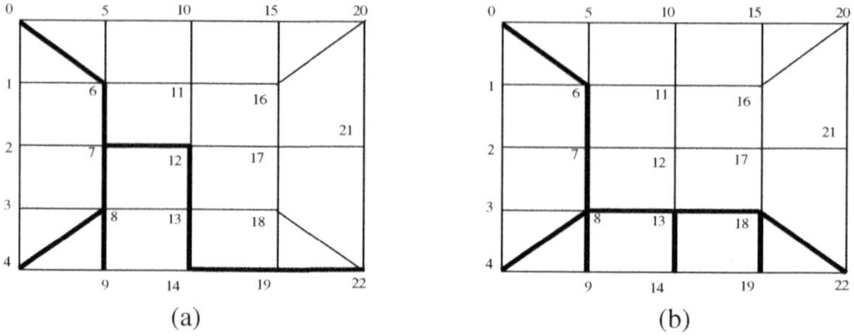

Fig. 2. The best Multicast trees (broad-brush) generated by Genetic Algorithm. (a). D=20, J=30, B=40 and L=0.002; (b). D=25, J=35, B=40 and L=0.002.

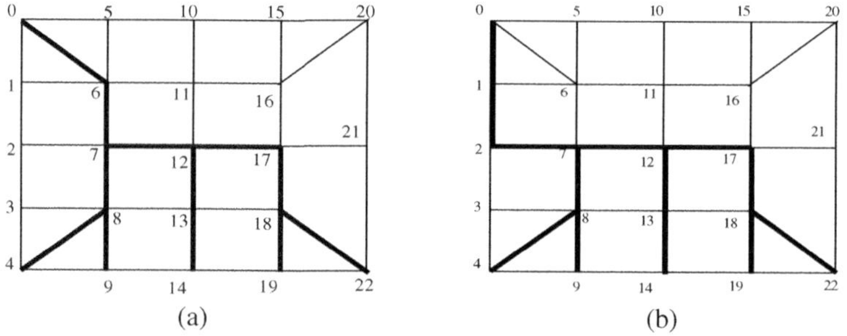

Fig. 3. The best Multicast trees (broad-brush) generated by PSO Algorithm. (a). D=20, J=30, B=40 and L=0.002; (b). D=25, J=35, B=40 and L=0.002.

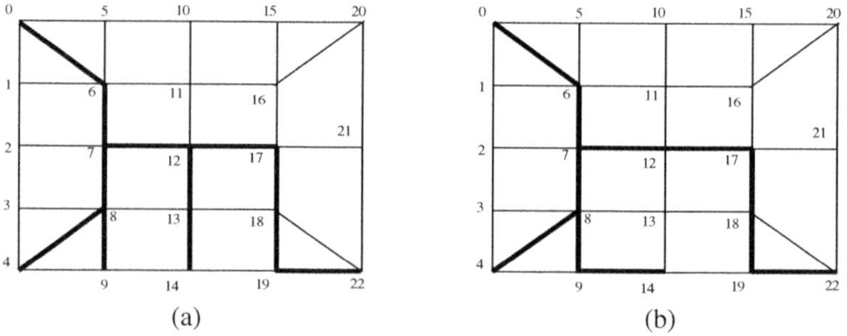

Fig. 4. The best Multicast trees (broad-brush) generated by QPSO Algorithm. (a). D=20, J=30, B=40 and L=0.002; (b). D=25, J=35, B=40 and L=0.002.

Table 1. Average fitness values and standard deviation over 50 runs

Iteration	GA	PSO	QPSO
Average Fitness	0.11312714	0.19680293	0.20772234
Standard Deviation	0.00736464	0.04100095	0.03485647

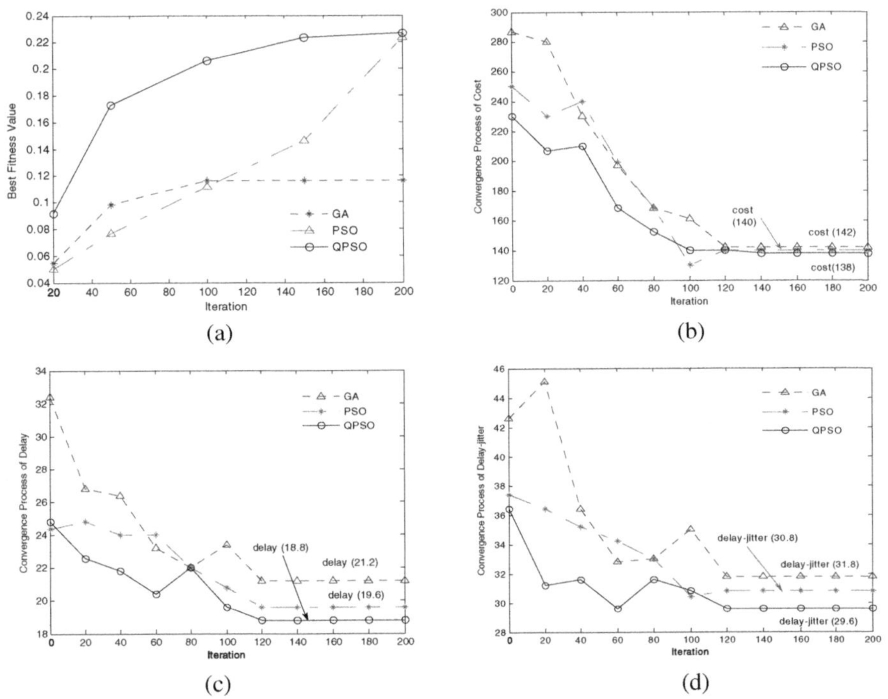

Fig. 5. The figure shows the convergence process of (a). Best fitness value; (b). Cost, (c). Delay and (d). Delay-jitter, with the development of iteration for three algorithms

It can be seen that QPSO could generate better multicast tree than other two algorithms. For example, when constraints is that D=25, J=35, B=40 and L=0.002, we recorded in Table 1 the average best fitness over 50 runs and standard deviations for the three algorithms. It is shown that QPSO has best performance generally. Figure 5 is the visualization of convergence process of fitness value, cost, delay and delay-jitter over 200 iterations corresponding to the runs that lead to best fitness value by the algorithms out of 50 runs under the above constraints that D=25, J=35, B=40 and L=0.002. For QPSO under these constraints, the fitness value after 200 iterations is 0.22644928, the corresponding cost is 138, delay is 18.8, and delay-jitter is 29.6. These results correspond to the multicast tree in Figure 4(b). The results generated by GA and PSO correspond to the multicast tree in figure 2(b) and figure 3(b). It can be seen that convergence speed of QPSO is most rapid. It means that among the three algorithms, QPSO has the strongest global search ability.

6 Conclusion

This paper has presented a QPSO-based multicast routing policy for Internet, mobile network or other high-performance networks. This algorithm provides QoS-sensitive paths in a scalable and flexible way in the networks environment. It can also optimize the network resources such as bandwidth and delay, and can converge to the optimal on near-optimal solution within fewer iterations. The incremental rate of computational cost can close to polynomial and is less than exponential rate. The availability and efficiency of QPSO on the problem have been verified by experiments. We also test the other two heuristics, PSO and GA, for performance comparison, and the experiment results show that QPSO outperforms PSO and GA on QoS the tested multicast routing problem. Our future work will focus on using QPSO to solve QoS multicast routing in network environment with uncertain parameter.

References

1. Angeline, P.J.: Using Selection to Improve Particle Swarm Optimization. Proc. 1998 IEEE International Conference on Evolutionary Computation. Piscataway, NJ (1998) 84-89
2. Charikar, M., Naor, J., Schieber B.: Resource Optimization in QoS Multicast Routing of Real-time Multimedia. Proc. of the 19th Annual IEEE INFOCOM (2000) 1518-1527
3. Clerc, M.: The Swarm and Queen: Towards a Deterministic and Adaptive Particle Swarm Optimization. Proc. 1999 Congress on Evolutionary Computation, Piscataway, NJ (1999) 1951-1957
4. Guerin, R.A., Orda, A.: QoS Routing in Networks with Inaccurate Information: Theory and algorithms. IEEE/ACM. Transactions On Networking, No.3, Vol.7, NJ (1999) 350-363
5. Roy, A., Das, S. K.: QM^{2}RP: A QoS-based Mobile Multicast Routing Protocol Using Multi-Objective Genetic Algorithm. Wireless Networks, Vol. 10, No.3, Kluwer Academic Publishers, Netherlands (2004) 271-286
6. Kennedy, J., Eberhart, R.C.: Particle Swarm Optimization. Proc. IEEE 1995 International Conference on Neural Networks, IV. Piscataway, NJ (1995) 1942-1948
7. Kennedy, J.: Sereotyping: Improving Particle Swarm Performance with Cluster Analysis. Proc. 2000 Congress on Evolutionary Computation, Piscataway, NJ (2000) 1507-1512
8. Kennedy, J.: Small worlds and Mega-minds: Effects of Neighborhood Topology on Particle Swarm Performance. Proc. 1999 Congress on Evolutionary Computation. Piscataway, NJ (1999) 1931-1938
9. Li, L.-Y.: The Routing Protocol for Dynamic and Large Computer Networks. Journal of Computers, Vol.11, No.2, (1998) 137-144
10. Sun, J., Feng, B., Xu, W.-B.: Particle Swarm Optimization with Particles Having Quantum Behavior. Proc. 2004 Congress on Evolutionary Computation, Piscataway, NJ (2004) 325-331
11. Sun, J., Xu, W.-B., Feng, B.: A Global Search Strategy of Quantum-behaved Particle Swarm Optimization. Proc. 2004 IEEE Conference on Cybernetics and Intelligent Systems, Singapore (2004) 111-115
12. Sun, J., Xu, W.-B., Feng, B.: Adaptive Parameter Control for Quantum-behaved Particle Swarm Optimization on Individual Level. Proc. 2005 IEEE International Conference on Systems, Man and Cybernetics. Piscataway, NJ (2005) 3049-3054
13. Shi, Y., Eberhart, R.C.: A Modified Particle Swarm. Proc. 1998 IEEE International Conference on Evolutionary Computation. Piscataway, NJ (1998) 69-73
14. Tan, K.C., Lim, M.H., Yao, X., Wang, L.P. (Eds.): Recent Advances in Simulated Evolution And Learning. World Scientific, Singapore (2004)
15. Yao, X.: Evolutionary Computation: Theory and Applications. World Scientific, Singapore (1999)

ANDYMARK: An Analytical Method to Establish Dynamically the Length of the Markov Chain in Simulated Annealing for the Satisfiability Problem

Juan Frausto-Solís[1], Héctor Sanvicente-Sánchez[2], and Froilán Imperial-Valenzuela[3]

[1] ITESM, Cuernavaca Campus, Paseo de la Reforma 182-A, Col. Lomas de Cuernavaca, C.P. 62589, Temixco Morelos, México
juan.frausto@itesm.mx
[2] IMTA, Paseo Cuauhnáhuac 8532, Col. Progreso, C.P. 62550, Jiutepec Morelos, México
hsanvice@tlaloc.imta.mx
[3] UVM, Querétaro Campus, Blvd. Villas del Mesón 1000, Col. Provincia Juriquilla, C.P. 76230, Querétaro Querétaro, México
froilan.imperial@queretaro.uvmnet.edu

Abstract. Because the efficiency and efficacy in Simulated Annealing (SA) algorithms is determined by their cooling scheme, several methods to set it have been proposed. In this paper an analytical method (ANDYMARK) to tune the parameters of the cooling scheme in SA for the Satisfiability (SAT) problem is presented. This method is based on a relation between the Markov chains length and the cooling scheme. We compared ANDYMARK versus a classical SA algorithm that uses the same constant Markov chain. Experimentation with SAT instances shows that SA using this method obtains similar quality solutions with less effort than the classical one.

Keywords: Satisfiability, Simulated Annealing, Combinatorial Optimization, NP-Hard Problems, Optimization, Heuristics.

1 Introduction

Simulated Annealing algorithm (SA) proposed by Kirkpatrick et al. [1] and Cerny [2] is considered as an extension of the Metropolis algorithm [3] used for the simulation of the physical annealing process. SA is especially applied to solve combinatorial optimization problems where it is very difficult to find the optimum or even near optimal solutions with a fast convergence rate.

Efficiency and efficacy are given to SA by the cooling scheme which consists of initial (c_1) and final (c_f) temperatures, the cooling function ($f(c_k)$) and the length of the Markov chain (L_k) established by the Metropolis algorithm. For each value of the control parameter c_k (temperature), SA accomplishes a certain

T.-D. Wang et al. (Eds.): SEAL 2006, LNCS 4247, pp. 269–276, 2006.

number of Metropolis decisions. In this sense, in order to get a better performance of SA a relation between the temperature and Metropolis cycle may be enacted.

Several methods to tune the parameters of the cooling scheme have been proposed. These methods usually are based on a) a tuning experimental process or b) with an adaptive criteria (mean and standard deviation). Case a) produces a lot of tuning time, while case b) may stop SA prematurely because the stochastic equilibrium can not always be identify correctly by measuring dynamically the mean and standard deviation of several solutions.

In this paper an analytical method to tune the parameters of the cooling scheme for the SA algorithm is presented. It uses a modified Markov chain building cycle in order to explore the solution space that each temperature requires.

We implemented two algorithms: 1) Simulated Annealing with Constant Markov chains (SACM) that uses the same Markov chain length at each temperature and 2) Simulated Annealing with Dynamic Markov chains (SADM) using ANDYMARK. Experimentation with some SAT instances shows that our method reduces the process time maintaining a similar quality of the final solution.

2 Background

In this section the mathematical background of the ANDYMARK method is presented.

2.1 Neighborhood

In [4] is established that SA requires a well defined neighborhood structure and the value of the maximum and the minimum cost increment in the objective function in order to calculate c_1 and c_f temperatures. So following the analysis made in [4] we give the basis of this method. Let $P_A(S_j)$ be the accepting probability of one proposed solution (S_j) generated from a current solution (S_i), and $P_R(S_j)$ the rejecting probability. The probability of rejecting S_j can be established in terms of $P_A(S_j)$ as follows:

$$P_R(S_j) = 1 - P_A(S_j) \ . \tag{1}$$

Accepting or rejecting S_j only depends on the cost deterioration size of the current solution that this change will produce, that means:

$$P_A(S_j) = g\Big(Z(S_i) - Z(S_j)\Big) = g(\Delta Z_{ij}) \tag{2}$$

$Z(S_i)$ and $Z(S_j)$ are the cost associated to S_i and S_j respectively, and $g(\Delta Z_{ij})$ is the probability to accept the cost difference $\Delta Z_{ij} = Z(S_i) - Z(S_j)$.

The solution selected from S_i may be any solution S_j defined by the next neighborhood scheme:

Definition 1. *Let* $\{\forall\ S_i \in S,\ \exists$ *a set* $V_{S_i} \subset S | V_{S_i} = V{:}S \rightarrow S\}$ *be the neighborhood of a solution* S_i*, where* V_{S_i} *is the neighborhood set of* S_i*,* $V{:}S \rightarrow S$ *is a mapping and* S *is the solution space of the problem being solved.*

From the above definition it can be seen that neighbors of a solution S_i only depend on the neighborhood structure V established for a specific problem. Once V is defined, the maximum and minimum cost deteriorations can be written as:

$$\Delta Z_{V\max} = Max\Big\{Z(S_j) - Z(S_i)\Big\} \qquad \forall\ S_j \in V_{S_i}, \forall\ S_i \in S \tag{3}$$

$$\Delta Z_{V\min} = Min\Big\{Z(S_j) - Z(S_i)\Big\} \qquad \forall\ S_j \in V_{S_i}, \forall\ S_i \in S \tag{4}$$

where $\Delta Z_{V\max}$ and $\Delta Z_{V\min}$ are the maximum and minimum cost deteriorations of the objective function through V respectively.

2.2 Markov Chains and Cooling Function

The SA algorithm can be seen like a sequence of homogeneous Markov chains, where each Markov chain is constructed for descending values of the control parameter $c_k > 0$ [5]. The control parameter is set by a cooling function like:

$$c_{k+1} = f(c_k) \tag{5}$$

and c_k must satisfy the next property:

$$\lim_{k \to \infty} c_k = 0 \tag{6}$$

$$c_k \geq c_{k+1} \qquad \forall k \geq 1$$

At the beginning of the process c_k has a high value and the probability to accept one proposed solution is high. When c_k decreases this probability also decreases and only good solutions are accepted at the end of the process. In this sense every Markov chain makes a stochastic walk in the solution space until the stationary distribution is reached. Then a strong relation between the Markov chain length and the cooling speed of SA exists: when $c_k \rightarrow \infty$, $L_k \rightarrow 0$ and when $c_k \rightarrow 0$, $L_k \rightarrow \infty$. Because the Markov chains are built through a neighborhood sampling method, the maximum number of different solutions rejected at c_f when the current solution S_i is the optimal one, is the neighborhood size $|V_{S_i}|$. In this sense the maximum Markov chain length is a function of $|V_{S_i}|$. In general L_k can be established as:

$$L_k \leq L_{\max} = g\big(|V_{S_i}|\big) \tag{7}$$

where $L_{\max}$ is the Markov chain length when $c_k = c_f$, and $g\big(|V_{S_i}|\big)$ is a function that gives the maximum number of samples that must be taken from the neighborhood V_{S_i} in order to evaluate an expected fraction of different solutions at c_f. The value of $L_{\max}$ only depends on the number of elements of V_{S_i} that will be explored at c_f.

Usually a SA algorithm uses an uniform probability distribution function $G(c_k)$ given by a random replacement sampling method to explore V_{S_i} at any temperature c_k, where $G(c_k)$ is established as follows:

$$G(c_k) = G = \begin{cases} 1/|V_{S_i}|, & \forall S_j \in V_{S_i}; \\ 0, & \forall S_j \notin V_{S_i}. \end{cases} \tag{8}$$

In this sense, the probability to get the solution S_j in N samples is:

$$P(S_j) = 1 - \exp\Big(- (N/|V_{S_i}|)\Big) \tag{9}$$

notice that $P(S_j)$ may be understood as the expected fraction of different solutions obtained when N samples are taken. From (9) N can be obtained as:

$$N = -\ln\Big(1 - P(S_j)\Big)|V_{S_i}| = C|V_{S_i}| \tag{10}$$

where C establishes the level of exploration to be done, $C = -\ln\Big(1 - P(S_j)\Big) = \ln P_R(S_j)$. In this way different levels of exploration can be applied. For example if a 99% of the solution space is going to be explored, the rejection probability will be $P_R(S_j) = 0.01$, so $C = 4.6$. Therefore $N = 4.6|V_{S_i}|$; in the same way if 86% of the solution space will be explored, $N = 2|V_{S_i}|$.

Then in any SA algorithm the maximum Markov chain length (when $c_k = c_f$) may be set as:

$$L_{max} = N = C|V_{S_i}| \tag{11}$$

because a high percentage of the solution space should be explored C varies from $1 \leq C \leq 4.6$ which guaranties a good level of exploration of the neighborhood at c_f [4].

3 ANDYMARK Method

As shown before, a strong relation between the cooling function and the length of the Markov chain exists. For the SA algorithm, the stationary distribution for each Markov chain is given by the Boltzmann probability distribution, which is a family of curves that vary from an uniform distribution to a pulse function. At the very beginning of the process (with $c_k = c_1$), SA has an uniform distribution, henceforth any guess would be accepted as a solution. Besides any neighbor of the current solution is also accepted as a new solution. In this way when SA is just at the beginning the Markov chain length is really small ($L_k = L_1 \approx 1$). By the time $k \to \infty$ the value of c_k is decremented by the next cooling function until the final temperature is reached ($c_k = c_f$):

$$c_{k+1} = \alpha c_k \tag{12}$$

where α is normally in the range of $0.7 \leq \alpha \leq 0.99$ [5].

In this sense the length of each Markov chain must be incremented at any temperature cycle in a similar but in inverse way that c_k is decremented. This means that L_k must be incremented until L_{max} is reached at c_f by applying an increment Markov chain factor (β). The cooling function given by (12) is applied many times until the final temperature c_f is reached. Because Metropolis cycle is finished when the stochastic equilibrium is reached [5], it can be also modeled as a Markov chain as follows:

$$L_{k+1} = \beta L_k \tag{13}$$

L_k represents the length of the current Markov chain at a given temperature, that means the number of iterations of the Metropolis cycle for a k temperature. So L_{k+1} represents the length of the next Markov chain. In this Markov Model, β represents an increment of the number of iterations in the next Metropolis cycle.

If the cooling function given by (12) is applied over and over again until $c_k = c_f$, the next geometrical function is easily gotten:

$$c_f = \alpha^n c_1 \tag{14}$$

knowing the initial (c_1) and the final (c_f) temperature and the cooling coefficient (α), the number of times that the Metropolis cycle is executed (i.e. the number of temperatures) can be calculated as:

$$n = \frac{\ln c_f - \ln c_1}{\ln \alpha} \ . \tag{15}$$

Applying systematically (13), another geometrical function is gotten:

$$L_{max} = \beta^n L_1 \ . \tag{16}$$

Once n is known, the value of the increment coefficient (β) is calculated as:

$$\beta = \exp\left(\frac{\ln L_{max} - \ln L_1}{n}\right) \ . \tag{17}$$

Once L_{max}, L_1 and β are known, the length of each Markov chain for each temperature cycle can be calculated using (13). In this way L_k is computed dynamically from $L_1 = 1$ for c_1 until L_{max} at c_f.

In the next section is shown that if the cooling scheme parameters are established with ANDYMARK, then the same quality of the final solution is gotten with less effort than using the same Markov chain for each temperature cycle.

4 Computational Results

To measure the efficiency of ANDYMARK, two SA algorithms were implemented. One using the same Markov chain length for any temperature cycle (i.e. $L_k = L_{max} = L$) or SACM algorithm, and other using ANDYMARK named

Simulated Annealing with Dynamic Markov chains (SADM). Both SACM and SADM were tested and compared using the same SAT instances obtained from [6]. Efficiency and efficacy were measured respectively according to the next parameters: a) executing process time and b) quality of the final solution.

The processing time was measured in seconds. Quality means how many clauses where made true for each instance. Table 1 shows the SAT instances solved for both algorithms. Table 2 shows the characteristics of each SAT instance and the

Table 1. SAT instances tested

Category	Name	Id
Random	balancedhidden-k3-n918-pos5-neg5-03-S166036896.cnf	r1
Random	glassyb-v399-s500582891.cnf	r2
Random	unif-r4.25-v600-c2550-03-S1158627995.cnf	r3
Random	gencnf-k10-r720-v50-c36000-03-S999535890.cnf	r4
Random	hgen6-4-20-n390-01-S682927285.cnf	r5
Handmade	bqwh.60.1080.cnf	h1
Handmade	color-18-4.cnf	h2
Handmade	genurq30Sat.cnf	h3
Planning	huge.cnf	pp1
Planning	medium.cnf	pp2
Planning	bw-large.a.cnf	pp3
Circuit fault analysis	ssa7552-038.cnf	c1

corresponding value of c_1, c_{f} and β used. These parameters were obtained with ANDYMARK as follows:

$$c_1 = \frac{-\Delta Z_{\mathrm{Vmax}}}{\ln\left(P_A(\Delta Z_{\mathrm{Vmax}})\right)} \text{ , and} \tag{18}$$

$$c_{\mathrm{f}} = \frac{-\Delta Z_{\mathrm{Vmin}}}{\ln\left(P_A(\Delta Z_{\mathrm{Vmin}})\right)} \ . \tag{19}$$

At the beginning of the annealing process, the probability to accept any proposed solution as the new solution is high, but this probability is reduced while the temperature is reduced. Thus, the accepting probability for high temperatures $P_A(\Delta Z_{\mathrm{Vmax}})$ must be high (i.e. 0.90, 0.95 or 0.99) and the rejecting probability for low temperatures $P_A(\Delta Z_{\mathrm{Vmin}})$ must be set too low values (i.e. 0.10, 0.05 or 0.01) to accept anly good solutions at the end of the process [5]. Next we show the parameters setting used to test the algorithms:

- From the cooling function $c_{\mathrm{k}+1} = \alpha c_{\mathrm{k}}$ two values of the cooling coefficient were used: $\alpha = 0.85$ and $\alpha = 0.95$.
- The value of L_{max} was set to enable SA to explore around 86% of the neighborhood at the final temperature (c_{f}). This is when $L_{\mathrm{max}} = N = 2|V_{S_{\mathrm{i}}}|$. For SACM algorithm it was set to $L_{\mathrm{k}} = L_{\mathrm{max}} = 2|V_{S_{\mathrm{i}}}|$ for each temperature cycle.

Table 2. Characteristics of the instances and values of c_1, c_f and β

Id	Variables	Clauses	SACM and SADM		β (SADM)	
			c_1	c_f	α=0.85	α=0.95
r1	918	3054	194.957	1.737	1.295	1.085
r2	399	1862	272.94	3.040	1.273	1.079
r3	600	2550	506.889	0.869	1.198	1.058
r4	50	36000	143332.576	1535.665	1.179	1.053
r5	390	1638	292.436	2.606	1.257	1.075
h1	6283	53810	604.367	1.737	1.299	1.086
h2	1296	95904	5712.248	63.624	1.328	1.093
h3	3622	17076	623.863	1.737	1.278	1.080
pp1	459	7054	1267.222	1.520	1.179	1.053
pp2	116	953	721.342	1.520	1.154	1.046
pp3	459	4675	111.256	1.52	1.183	1.054
c1	1501	3575	5302.837	0.217	1.137	1.041

- To obtain c_1 and c_f temperatures, the value of $P_A(\Delta Z_{\mathrm{Vmax}})$ was 0.95 and $P_A(\Delta Z_{\mathrm{Vmin}})$ was set to 0.01.

Both algorithms were implemented in a HP Laptop with 512 MB of Ram memory with a Pentium 4 processor running at 1.8 GHz. From Table 3, notice that the quality of the solution obtained from SACM and SADM algorithms was maintained almost at the same value for both cooling coefficients ($\alpha = 0.85$ and $\alpha = 0.95$). Note that setting the length of L_k for each temperature as is established in Section 3, the processing time of SA is outperformed.

For the instances tested the mean reduction of the processing time that SADM algorithm produces is 90.049% for $\alpha = 0.85$ and 86.157% for $\alpha = 0.95$. For

Table 3. Performance of SACM and SADM with $\alpha = 0.85$, $\alpha = 0.95$ and $C = 2$

Id	Quality (%)				Time (secs)			
	SACM		SADM		SACM		SADM	
	$\alpha = 0.85$	$\alpha = 0.95$	$\alpha = 0.85$	$\alpha = 0.95$	$\alpha = 0.85$	$\alpha = 0.95$	$\alpha = 0.85$	$\alpha = 0.95$
r1	91.650	91.650	91.781	91.912	25	38	1	5
r2	92.213	92.642	92.266	92.589	5	8	< 1	1
r3	95.961	95.882	95.686	96.392	15	23	1	3
r4	99.947	99.961	99.944	99.961	17	28	1	6
r5	92.063	92.125	92.369	92.491	4	7	1	1
h1	96.432	96.486	96.402	96.540	3007	7858	252	841
h2	97.089	97.440	97.431	97.538	782	1865	68	298
h3	96.410	96.422	96.410	96.422	526	1674	47	184
pp1	98.696	98.993	98.625	98.852	18	67	2	9
pp2	97.587	97.901	96.747	97.482	< 1	< 1	< 1	< 1
pp3	97.134	97.647	97.241	97.668	12	41	1	6
c1	98.713	99.413	98.462	99.217	48	198	6	24

example, for the instance h1 (the biggest tested) the processing time was reduced from 3007 to 252 secs (91.619%) when $\alpha = 0.85$ and from 7858 to 841 secs (89.297%) when $\alpha = 0.95$.

5 Conclusions

In this paper a new analytical method to tune the parameters of the cooling scheme in Simulated Annealing for the Satisfiability problem is presented. This method models the temperature and Metropolis cycles with Markov models. The paper shows that using a geometrical cooling scheme, both initial and final temperature can be determined in an analytic way. Once initial and final temperature are derived, the length of the Markov chains for every temperature can be determined dynamically with a first order Markov model.

Satisfiability instances are solved with Simulated Annealing algorithms using two implementations: 1) ANDYMARK, where the length of the Markov chains are changed during the execution of the algorithm, changing in this way the number of iterations of the Metropolis cycle, and 2) Simulated annealing with constant Markov chains (i.e. a constant number of iterations) in the Metropolis cycle. The results presented show that using ANDYMARK to tune the cooling scheme's parameters in SA algorithms produces a mean reduction time of 90.049% and 86.157% for $\alpha = 0.85$ and 0.95 respectively.

An important feature of ANDYMARK is that it enables SADM to produce solutions with similar quality in less time than SACM using the same values of the decrement temperature's parameter.

References

1. Kirkpatrick, S., Gelatt, C.D., Vecchi, M.P.: Optimization by Simulated Annealing. Science, Number 4598, 13 May 1983 **220, 4598** (1983) 671–680
2. Cerny, V.: Thermodynamical Approach to the Traveling Salesman Problem: an Efficient Simulation Algorithm. Journal of Optimization Theory and Applications **45** (1985) 41–51
3. Metropolis, N., Rosenbluth, A.W., Rosenbluth, M.N., Teller, A.H., Teller, E.: Equation of State Calculations by Fast Computing Machines. Chem. Phys. **21** (1953) 1087–1092
4. Sanvicente-Sánchez, H., Frausto-Solís, J.: A Method to Establish the Cooling Scheme in Simulated Annealing Like Algorithms. In: ICCSA 2004, Assisi, Italy, May 14-17. (2004) 755–763
5. Aarts, E., Korst, J.: Simulated Annealing and Boltzmann Machines: A Stochastic Approach to Combinatorial Optimization and Neural Computing. John Wiley & Sons, Inc., New York, NY, USA (1989) pp. 272.
6. Hoos, H.H., Stützle, T.: SATLIB: An Online Resource for Research on SAT. In: SAT 2000. IOS Press (2000) 283–292

An Accelerated Micro Genetic Algorithm for Numerical Optimization

Linsong Sun and Weihua Zhang

College of Hydraulic Science & Engineering, Yangzhou University,
Yangzhou 225009, PR China
sunlinsong@sohu.com, zhanghu6077@sina.com

Abstract. In this paper, we present an accelerated micro genetic algorithm for numerical optimization. It is implemented by incorporating the conventional micro genetic algorithm with a local optimizer based on heuristic pattern move and Aitken Δ^2 acceleration method. Performance tests with three benchmarking functions indicate that the presented algorithm has excellent convergence performance for multimodal optimization problems. The number of objective function evaluations required to obtain global optima is only 5.4-11.9% of that required by using conventional micro genetic algorithm.

1 Introduction

Genetic Algorithms (GAs), initially developed by Holland [1] in 1960s, remain the most recognized and practiced form of Evolutionary Algorithms which are based on stochastic search strategy. These algorithms are able to reach global optima without the dependence on initial guesses, and only using the objective function value, requiring no derivatives or other auxiliary information. However, they are sometimes very poor in terms of convergence performance. To improve the efficiency of GAs, some hybrid genetic algorithms are developed by combining genetic algorithms with heuristic search strategies based on gradient, such as conjugate gradient algorithm [2]. These hybrid algorithms are demonstrated to have good performance by using GAs for global search and gradient-based algorithm for local search. However, the use of gradient-based search strategies limits the applicable area of these algorithms. Micro genetic algorithm (micro-GA) proposed by Krishnakumar [3] uses very small population size (typically 5), the population evolves in normal GA fashion and converges in a few generations, at this point, a new random population is chosen while keeping the best individual from last converged generation and the evolution restart. The micro-GA can avoid premature convergence and converge faster to the near-optimal region than does a simple GA. With the incorporation of micro-GA with a local optimizer based on the heuristic pattern move, an effective hybrid GA (hGA) was proposed by Xu et al [4, 5].

In this study, we proposed an accelerated micro genetic algorithm by incorporating micro-GA with local optimizer based on the heuristic pattern move and Aitken Δ^2 accelerating scheme.

T.-D. Wang et al. (Eds.): SEAL 2006, LNCS 4247, pp. 277 – 283, 2006.

The paper is organized as follows: Section 2 presents the detail of proposed algorithm and Section 3 gives the performance test result with some benchmarking function. Finally some general conclusions are drawn in Section 4.

2 Accelerated Micro Genetic Algorithm

In the proposed algorithm, two local optimization operators are combined with micro genetic algorithm to accelerate the convergence of evolution. The Aitken Δ^2 acceleration scheme is used to update current generation population. The local optimizer based on heuristic pattern move is used to update offspring generated by genetic operators.

2.1 Aitken Δ^2 Acceleration Scheme

It is well known that Aitken Δ^2 scheme has excellent performance in accelerating the convergence of sequence [6]. It uses three points, b_1, b_2 and b_3 to constitute a new point b_m.

$$b_m = b_1 - \frac{(b_2 - b_1)^2}{b_3 - 2b_2 + b_1} \tag{1}$$

In our algorithm, b_1, b_2 and b_3 are three best individuals obtained successively at previous generations in the evolution process. The new individual b_m is evaluated and compared to the worst individual at current generation, if b_m is better then the worst individual is replaced by it.

2.2 Local Optimizer Based on Pattern Move

The local optimization operator is used to update offspring in the proposed algorithm. The best individual c_1 and the second best individual c_2 in the offspring are selected to constitute a pattern move. Three new individuals are obtained by using extrapolation and interpolation operations based on this pattern move. Then, the best individual among these three new individuals is selected to replace the worst individual in the offspring obtained with genetic operations.

The local optimizer can be described as follows:

(1) Constitute the pattern move of the best individual:

$$d = c_1 - c_2 \tag{2}$$

(2) Generate three new individuals:

$$c_1' = c_1 + \alpha d \tag{3a}$$

$$c_2' = c_2 + \beta d \tag{3b}$$

$$c_3' = c_2 - \gamma d \tag{3c}$$

Where α, β and γ are control parameters which are recommended to be 0.3, 0.5 and 0.3, respectively.

(3) Select the best one (c_m) among three individuals, c_1', c_2' and c_3':

$$f(c_m) = \max\{f(c_1'), f(c_2'), f(c_3')\} \quad c_m \in \{c_1', c_2', c_3'\} \tag{4}$$

(4) Replace the worst individual in offspring with c_m.

2.3 Implementation of Proposed Algorithm

The new algorithm can be carried out as follows:

(1) Initialize the optimization process, including:
  a) Select the operation parameters of population size $N_{popsize}$, crossover possibility p_c, random seed i_d, local operator control parameter α, β and γ.
  b) Set generation index i_g=1 and acceleration flag i_a=1.
  c) Generate the initial population $P(i_g)$ by random method
  d) Evaluate each individual use fitness function and select the best one to be $b(i_a)$.

(2) Check if the termination criterion is fulfilled. If yes, the optimization process ends. Otherwise, select the best individual and check if it is different from $b(i_a)$. If yes, let the best individual be $b(i_a+1)$ and set $i_a=i_a+1$.

(3) Check if i_a=3. if yes, apply acceleration by using equation (1) to obtain a new individual b_m and set $b(1)=b(3)$, i_a=1. Check if b_m is better than the worst individual in $P(i_g)$, if yes, replace the worst individual with b_m.

(4) Generate the offspring $C(i_g)$ by using genetic operators: selection, crossover and elitism.

(5) Use the local optimization operator described in previous section to update the obtained offspring $C(i_g)$.

(6) Check if nominal convergence is reached in offspring. If yes, use the restart strategy to obtain a new offspring $C(i_g)$.

(7) Set $i_g= i_g+1$ and go back to (2)

3 Numerical Experiments

3.1 Test Functions

Three typical benchmarking functions, f_1, f_2 and f_3, are used to test the performance of the proposed algorithm.

$$f_1 = \prod_{i=1}^{2} \sin^{80}(5.1\pi x_i + 0.5) \times e^{-4\log 2 \times (x_i - 0.0667)^2 / 0.64} \quad (0 \le x_i \le 1) \tag{5}$$

$$f_2 = 0.5 + \frac{\sin\sqrt{x_1^2 + x_2^2} - 0.5}{[1.0 + 0.001(x_1^2 + x_2^2)]^2} \quad (-100 \le x_1, x_2 \le 100) \tag{6}$$

$$f_3 = \sum_{i=1}^{20} \left(x_i^2 - 10\cos(2\pi x_i) + 10 \right) \qquad (-5.12 \le x_i \le 5.12) \tag{7}$$

These benchmarking functions are all multimode functions. Function f_1 has 25 local optima and the global maximum $f_{1\max} \approx 1.0$ at $x_1 = x_2 \approx 0.066832$. Fig. 1 shows the search space of function f_1.

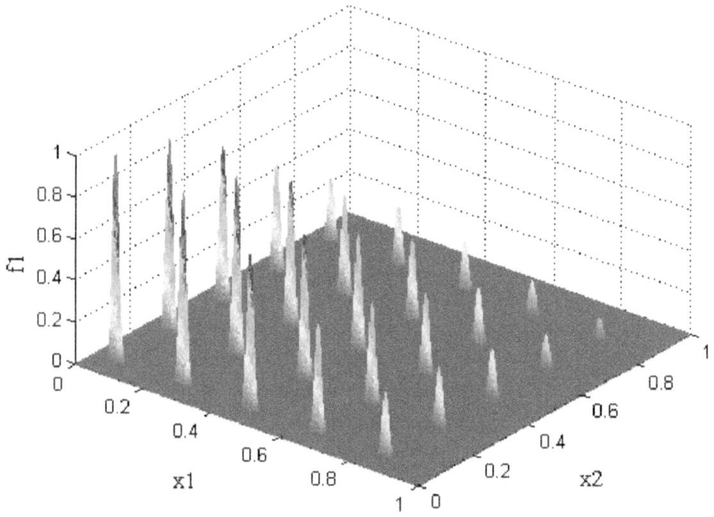

Fig. 1. The solution space of function f_1

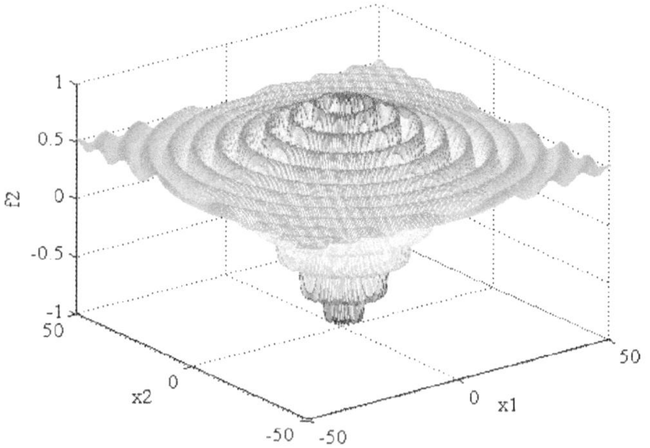

Fig. 2. The solution space of function f_2

Function f_2 is symmetric to grid origin and reaches its global maximum $f_{2\max} \approx 0.997542$ at the points with distance of 1.570796 to grid origin. The solution of f_2 varies smarter with the closing to global optimum. Figure 2 is a part of its solution space.

Function f_3 was first proposed by Rastrigin [7] as a 2-dimensional function and has been generalized by Mühlenbein et al in [8]. This function is a fairly difficult problem due to its large search space and its large number of local minima. The local minima are located at a rectangular grid with size 1. The global minimum is at $x_i = 0$, $i = 1, \cdots, n$, giving $f_{3\min} = 0$. Grid points with $x_i = 0$ except one coordinate, where $x_i = 1$, give $f_3 = 1$, the second best minimum. With increasing distance to the global minimum the local minima become littler.

3.2 Experiment Results

Using the method described in the previous section, the genetic operations and control parameters are set as follows: population size of five, real-number coding, tournament selection, and uniform arithmetic crossover of p_c=0.5 with one child. The local optimizer parameters are α=0.3, β=0.5 and γ=0.3. The number of objective function evaluations that is required to obtain the global optimum (with in ±0.001) are listed in table 1, where n_a, n_h and n_m are evaluation numbers using proposed algorithm with Aitken Δ^2 scheme and local optimizer (A-hGA), proposed algorithm with local

Table 1. Comparision of the number of function evaluations to convergence

No.	n_a	n_h	n_m	n_a/n_m (%)	n_h/n_m (%)
f_1	815	3960	15020	5.4	26.4
f_2	1317	2200	11045	11.9	19.9
f_3	69689	96000	/	/	/

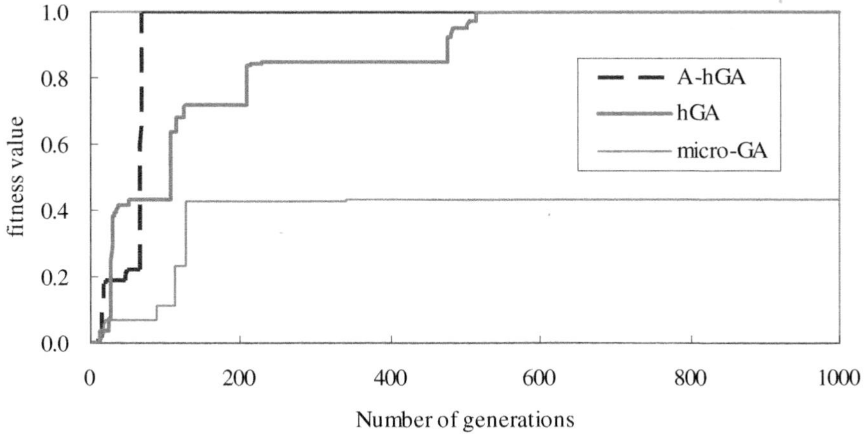

Fig. 3. Convergence processes of function f_1 for different algorithms

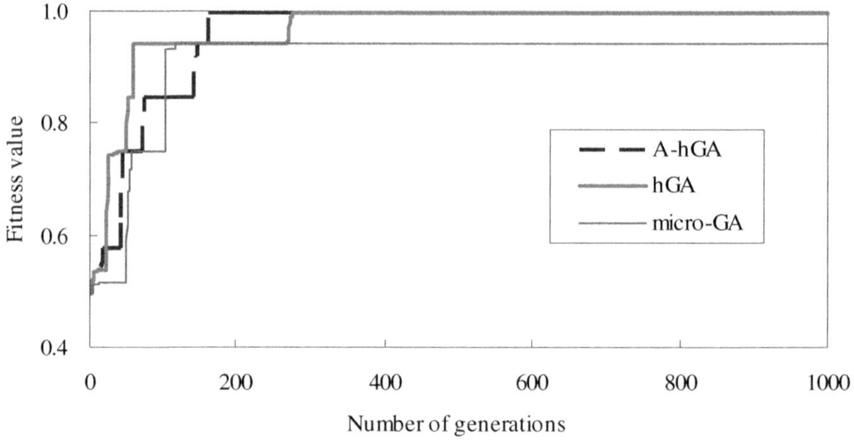

Fig. 4. Convergence processes of function f_2 for different algorithms

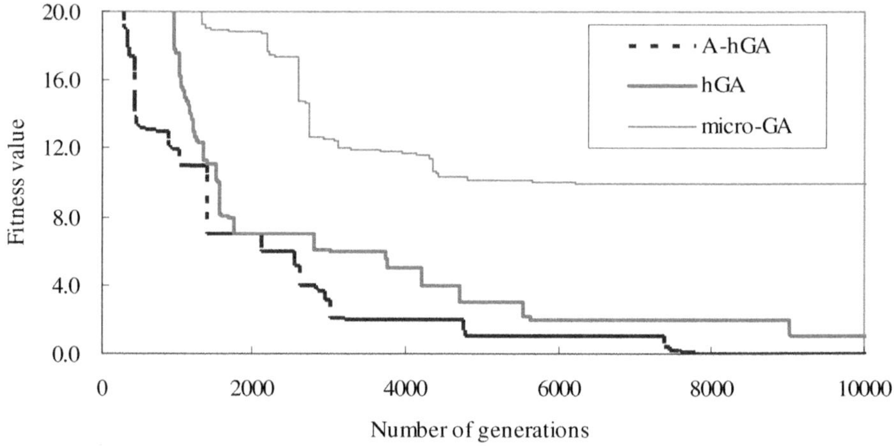

Fig. 5. Convergence processes of function f_3 for different algorithms

optimizer alone (hGA) and conventional micro-GA, respectively. The A-hGA and hGA require only 5.4-11.9% and 19.9-26.4%, respectively, of the number of function evaluations required by micro-GA. It should be noted that, for function f_3, micro-GA can't converge to global optimum. Figures 3 to 5 show the convergence processes of function f_1, f_2, and f_3. It can be seen that the proposed algorithm converges much faster than conventional micro-GA does.

4 Conclusions

This study proposed an accelerated micro genetic algorithm for numerical optimization. Performance tests using three typical benchmarking functions show that the new algorithm has superior performance than conventional micro-GA.

The proposed algorithm only requires objective function value in optimization process. This property makes it highly convenient and universal.

Acknowledgement

The support of National Nature Science Foundation of PR China (NSFC) under grant No. 90410011 is gratefully acknowledged.

References

1. Holland, J.H.: Outline for a Logical Theory of Adaptive Systems. J. Assoc. Comput. Mach. 9 (1962) 297-314
2. Gudla, Pradeep Kumar, Ganguli, Ranjan: An automated Hybrid Genetic-Conjugate Gradient Algorithm for Multimodal Optimization Problems. Applied Mathematics and Computation 167 (2005) 1457-1474
3. Krishnakumar, K.: Micro-Genetic Algorithms for Stationary and Non-Stationary Function Optimization. SPIE Intelligent Control and Adaptive Systems. 1196 (1989) 289-296
4. Xu, Y.G., Liu, G.R., Wu, Z.P.: A Novel Hybrid Genetic Algorithm Using Local Optimizer Based on Heuristic Pattern Move. Appl. Artif. Intell. Int. J. 15 (2001) 601-631
5. Fawaz, Z., Xu, Y.G., Behdinan, K.: Hybrid Evolutionary Algorithm and Application to Structural Optimization. Struct. Multidisc. Optim. 30 (2005) 219-226
6. Mathews, John H.: Numerical Methods for Mathematics, Science, and Engineering. 2nd edn. Prentice-Hall, Englewood Cliffs, New Jersey. (1992)
7. Törn, A., Zilinskas, A. (eds): Global Optimization. Lecture Notes in Computer Science, vol 350, Springer-Verlag, Berlin (1989)
8. Mühlenbein, H., Schomisch, D., Born, J.: The Parallel Genetic Algorithm as Function Optimizer. Parallel Computing, 17 (1991) 619-632

Optimal Ordering and Recovery Policy for Reusable Items with Shortages

H.M. Wee[1], Jonas C.P. Yu[2], Ling-Huey Su[1], and Te-Chin Wu[1]

[1] Department of Industrial Engineering Chung Yuan Christian University
No. 200, Chung Pei Rd.Chung Li City, Taiwan
weehm@cycu.edu.tw
[2] Logistics Management Department, Takming College
No. 56, Huan Shan Rd., Sec. 1, Ney Hwu Taipei City, Taiwan

Abstract. This study considers inventory of reusable items with shortages and develops a replenishing and production policy to satisfy customer need, and to promote the smart use of resources. Distinct from former researches, our study considers reusable items with shortages to derive the best replenishment and production strategy. A hybrid of numerical analysis and search method is used to derive the minimum total cost of the mathematical model. The result is compared with the case when no shortage is allowed.

1 Introduction

The process of returning and reusing materials is not a new concept. In the last several decades, it has been a common practice to reuse metal, paper and glass. This is due to the increase in environmental concerns as well as the economy of recycling [1].

There are four synonyms of reuse according to Thierry et al. [2]. They are: direct reuse, repair, recycling and remanufacturing. In the US, remanufacturing is a fifty billion dollar per year business (Corbett and Kleindorfer, 2001). Schrady [3] was the first author to consider reuse in a deterministic model. He assumed a constant demand; constant return rates and fixed lead times for external orders and recovery. Two costs were considered: fixed setup costs for orders and recovery process, and the linear holding costs for serviceable and recoverable items. In his model, Schrady [3] proposed a control policy with fixed lot sizes. Demand is satisfied as far as possible from the recovered products. Recently, Mabini et al. [4] extended Schrady's model to consider stock out service level constraints, and a multi-item system sharing at the same repair facility. Richter [5] developed a model with a different control policy. In the policy, expressions for the optimal control parameter values were obtained. Koh [6] developed a joint EOQ and EPQ model in which the stationary demand could be satisfied by recycled products and newly purchased products. The recycled products were assumed to be a fixed proportion of the used products collected from the customers. Shortage due to stock-out occasionally occurs in the real world. In order to reduce loss of sales or goodwill, the supplier can reduce shortage with reusable items.

In this study, we extend Koh's model [6] by assuming stationary demand and shortage backorders. To the best of our knowledge, research on recovery inventory with allowable shortage and finite replenishment has not been done previously. The

T.-D. Wang et al. (Eds.): SEAL 2006, LNCS 4247, pp. 284–291, 2006.

demand in our inventory system is met either by external ordered new products or recovered items. The objective of this inventory management system is to minimize the fixed and the variable costs. This study developed a model to analyze an inventory system where the stationary demand can be satisfied by recovered products and newly purchased products. Shortage back order is allowed. The objective is to minimize the total cost and obtain the optimum inventory level to initiate the recovery process for recoverable items. Because a closed form solution is infeasible, we propose a search procedure to solve the problem. When p is equal to d, the optimal quantity for the newly procured items is the same for models with and without shortage. However, issues in recyclable items with shortages are complex, and our understanding of the issues could significantly influence managerial decisions and environmental impacts.

2 Notation and Assumptions

The mathematical model in this paper is based on the following assumptions:

1. The demand rate is deterministic and known.
2. The return rate is known.
3. The quality for reuse items is the same as new ones.
4. The repair capacity is constant and known.
5. The cost parameter is constant and known.
6. The shortage cost is constant and known.
7. The procurement lead-time is constant and known.
8. The repair lead-time is constant and known.
9. The repair costs are less than purchasing costs.
10. The recovery rate is greater than the returning rate ($p>r$).
11. The demand rate is greater than the recovery rate ($d>r$).

The following notation is used:

Input parameters

r	collected rate (units/time)
p	recovery process rate (units/times)
d	demand rate (units/time)
C_{h1}	per unit holding cost for recoverable items
C_{h2}	per unit holding cost for serviceable items
C_J	per unit shortage cost
C_o	ordering cost for the new item
C_s	setup cost for recovery process

Decision variables

J	shortage quantity
R	inventory level of recoverable items to start recovery process
Q	order quantity for newly procured items
n	number of orders for one setup in the recovery shop
T	cycle time of the model
t	idle time length of the recovery facility

3 Mathematical Modeling and Analysis

Three cases are analyzed in accordance to the three parameter conditions: $p>d$, $p=d$ and $p<d$.

3.1 Case 1: $p > d$

From the Fig. 1, one can see that

$$t_a = \frac{1}{n}\left[t - \frac{(p-d)(T-t)-J}{d} - n\frac{J}{d}\right] \tag{1}$$

$$t_{\beta 1} = \frac{(p-d)(T-t)-J}{p-d} \tag{2}$$

$$t_{\beta 2} = \frac{(p-d)(T-t)-J}{d} \tag{3}$$

$$t_r = \frac{J}{p-d} \tag{4}$$

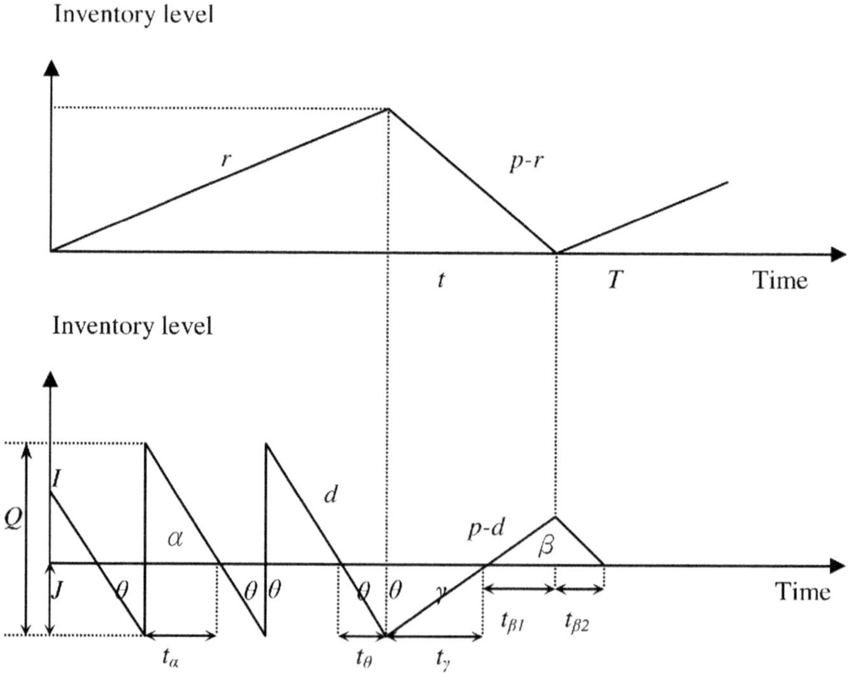

Fig. 1. Recovery inventory system ($p>d>r$)

$$t_{\theta} = \frac{J}{d} \tag{5}$$

From Fig.1, one can see that

$$T = \frac{pR}{r(p-r)} \text{ and } t = \frac{R}{r} \tag{6}$$

The total cost = recoverable inventory cost +ordering cost+ serviceable inventory cost. The total cost per unit time for this case is as follows:

$$\begin{aligned} TVC(n,R,J) &= \frac{r(p-r)(C_s + nC_o)}{pR} + \frac{C_{h1}R}{2} + \frac{C_{h2}dr(p-r)}{2npR}\left[R\frac{p(d-r)}{dr(p-r)} - J\frac{(1+n)}{d}\right]^2 \\ &+ \frac{C_{h2}r(p-r)}{2dR(p-d)}\left[R\frac{p-d}{p-r} - J\right]^2 + \frac{r(p-r)C_J J^2}{2pR(p-d)} + \frac{nr(p-r)C_J J^2}{2dpR} \end{aligned} \tag{7}$$

For a constant n, the optimal solution, R^* and J^*, can be found by differentiating the above equation with respect to R and J and equate them to zero. One has

3.2 Case 2: $p = d$

From Fig. 2, one can easily see that

$$t_{\alpha} = \frac{t}{n} - \frac{J}{d} \tag{8}$$

$$t_{\beta} = \frac{J}{d} \tag{9}$$

Inventory holding cost for n triangles (triangle α in Fig. 2) is

$$\frac{dnC_{h2}}{2}\left(\frac{t}{n} - \frac{J}{d}\right)^2 \tag{10}$$

Shortage cost for n triangles(triangle β in Fig. 2) is

$$n\frac{C_J J^2}{2d} \tag{11}$$

From Fig. 2, one can see that

$$T = \frac{pR}{r(p-r)} \text{ and } t = \frac{R}{r} \tag{12}$$

The total cost per unit time for this case is as follows:

$$TVC(n,R,J)=\frac{r(p-r)(C_s+nC_o)}{pR}+\frac{C_{h1}R}{2}+\frac{dnr(p-r)C_{h2}}{pR}\left(\frac{R}{nr}-\frac{J}{d}\right)^2 + \frac{nr(p-r)C_J J^2}{2dpR} \tag{13}$$

For a fixed n, the optimal solution, R^* and J^*, can be found by differentiating the above equation with respect to R and J and equate them to zero.

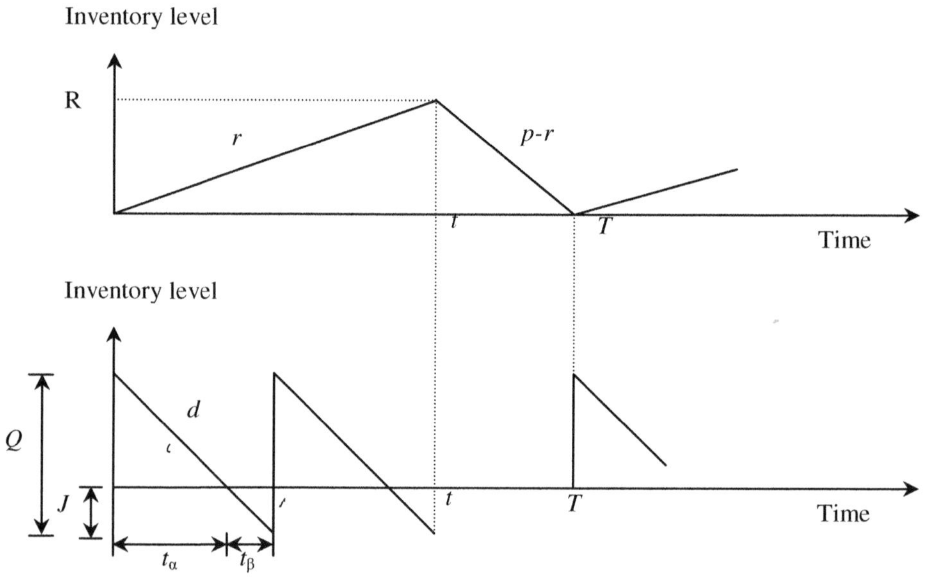

Fig. 2. Recovery inventory system ($p=d$)

3.3 Case 3: $p < d$

From Fig. 3, one can easily see that

$$t_\alpha = \frac{1}{n+1}\left(t+\frac{(d-p)(T-t)}{d}\right)-\frac{nJ}{(n+1)d} \tag{14}$$

$$t_\beta = \frac{J}{d} \tag{15}$$

$$t_\gamma = T - t \tag{16}$$

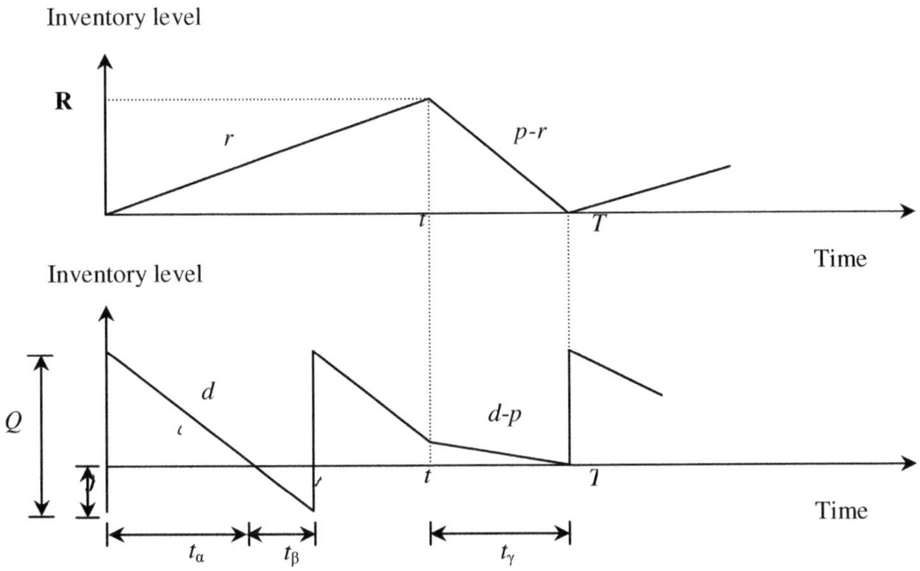

Fig. 3. Recovery inventory system (*p*<*d*)

The cost for recoverable inventory per cycle is

$$C_s + \frac{C_{h1}}{2} RT \tag{17}$$

Inventory holding cost for *n* triangles (triangle α in Fig. 3) is

$$\frac{dnC_{h2}}{2(n+1)^2}\left(t + \frac{(d-p)(T-t)}{d} - \frac{nJ}{d}\right)^2 \tag{18}$$

Shortage cost for *n* triangles(triangle β in Fig. 3) is

$$n\frac{C_J J^2}{2d} \tag{19}$$

From Fig. 3, one can see that

$$T = \frac{pR}{r(p-r)} \text{ and } t = \frac{R}{r} \tag{20}$$

The total cost per unit time for this case is as follows:

$$TVC(n,R,J) = \frac{r(p-r)(C_s + nC_o)}{pR} + \frac{C_{h1}R}{2} + \frac{dnr(p-r)C_{h2}}{2pR(n+1)^2}\left(\frac{R}{r} + \frac{(d-p)R}{d(p-r)} - \frac{nJ}{d}\right)^2 \tag{21}$$

$$+\frac{dr(p-r)C_{h2}}{2pR(n+1)^2}\left(\frac{R}{r}+\frac{(d-p)R}{d(p-r)}-\frac{nJ}{d}\right)^2-C_{h2}\frac{r(d-p)^2R}{2dp(p-r)}$$
$$+\frac{C_{h2}r(d-p)}{2p}\frac{R}{p-r}+\frac{nr(p-r)C_JJ^2}{2dpR}$$

Take the first derivative of TVC with respect to R and J, and then equate the first derivative to zero to derive the optimal solution, R^* and J^*.

3.4 Solution Procedure

From the given parameters and n, it is easily seen that the optimal values of R, J and Q can be obtained. The closed form optimal values of the integer variables is not feasible. Therefore the following computer search procedure is suggested to derive the optimal decision variables.

Step 1 Determine the values of the system parameters r, p, d, C_o, C_s, C_{h1}, C_{h2} and C_J.
Step 2 Calculate $TVC(n=1)$ for $n=1$.
Step 3 Calculate $TVC(n=2)$ for $n=2$.
Step 4 If $TVC(n=1) > TVC(n=2)$, set $n = 1$, else, set $n = n+1$ and calculate the total cost till $TVC(n+1) > TVC(n)$.

4 Conclusion

This study developed a mathematical model to analyze an inventory system where the stationary demand can be satisfied by recovered products and newly purchased products. Shortage backorder is allowed. The objective is to minimize the total cost and obtain the optimum inventory level for the recoverable items. We propose a hybrid of numerical analysis and search procedure to solve the problem since a closed form solution is infeasible. Table 1 compares Koh's model (no shortage) with our model. When p is equal to d, the optimal quantity for the newly procured items is the same for models with shortage and without shortage. Our model considering shortage backorder has shown a significant improvement in cost as compared with Koh's model.

Table 1. Comparing the result of Koh's model with our model

	Koh's model (no shortage)	Our model (shortage allowable)
$p>d$	$Q^*=\frac{p(d-r)}{nr(p-r)}R_1^*$	$Q^*=\frac{p(d-r)}{nr(p-r)}R_2^*-\frac{J_2^*}{n}$
$p=d$	$Q^*=\frac{dR_1^*}{nr}$	$Q^*=\frac{dR_2^*}{nr}$
$p<d$	$Q^*=\left(\frac{d-p}{p-r}\right)\frac{R_1^*}{n}$	$Q^*=\frac{1}{n+1}\left(\frac{dR_2^*}{r}+\frac{r(d-p)(p-r)}{PR_2^*}+J_2^*\right)$

References

1. Fleischmann, M., Bloemhof-Ruwaard, J.M., Dekker, R., Van der Laan, E., Van Numen, J.A., Van Wassenhove, L.N.: Quantitative models for reverse logistics: A review. European Journal of Operation Research 103 (1) (1997)1-17
2. Thierry, M.C., Salomon, M., Van Numen, J., Van Wassenhove, L.: Strategy issues in product recovery management. California Management Review 37 (2) (1995) 114-135
3. Schrady, D.A.: A deterministic inventory model for repairable items. Naval Research Logistics Quarterly 14 (1967) 391-398
4. Mabini, M.C., Pintelon, L.M., Gelders, L.F.: EOQ type formulation for controlling repairable inventories. International Journal of Production Economics, 28 (1992) 21-33
5. Richter, K.: The extended EOQ repair and waste disposal. International Journal of Production Economics 45 (1996) 443-447
6. Koh, S.G., Hwang, H., Sohn, K.I., Ko, C.S.: An optimal ordering and recovery policy for reusable items. Computers & Industrial Engineering 43 (2002) 59-73

A New Algorithm of Automatic Programming: GEGEP

Xin Du[1,2], Yueqiao Li[1], Datong Xie[1], and Lishan Kang[1,3]

[1] Department of Computer Science and Technology,
China University of Geosciences, Wuhan, China
xindu79@126.com
[2] Department of Information and engineering,
Shijiazhuang University of Economics, Shijiazhuang, China
xindu79@126.com
[3] State Key Laboratory of Software Engineering,
Wuhan University, Wuhan, China
kang_whu@yahoo.com

Abstract. Gene Expression Programming (GEP) has wide searching ability, simple representation, powerful genetic operators and the creation of high levels of complexity. However, it has some shortcomings, such as blind searching and when dealing with complex problems, its genotype under Karva notation does not allow hierarchical composition of the solution, which impairs the efficiency of the algorithm. So a new automatic programming method is proposed: Gene Estimated Gene Expression Programming(GEGEP) which combines the advantages of Estimation of Distribution Algorithm (EDA) and basic GEP. Compared with basic GEP, it mainly has the following characteristics: First, improve the gene expression structure, the head of gene is divided into a head and a body, which can be used to introduce learning mechanism. Second, the homeotic gene which is also composed of a head, a body and a tail is used which can increase its searching ability. Third, the idea of EDA is introduced, which can enhance its learning ability and accelerate convergence rate. The results of experiments show that GEGEP has better fitting and predicted precision, faster convergence speed than basic GEP and traditional GP.

Keywords: Gene Expression Programming; Genetic Programming; Gene Estimated Gene Expression Programming,; Estimation of Distribution Algorithm.

1 Introduction

Gene Expression Programming(GEP) was first proposed by Candida Ferreira [1]. In this new syste- m, the complex computer programs (the phenotype) evolved by GEP are totally encoded in simple strings of fixed length (the genotype) .

GEP is the inheritance and development of GA and GP, it synthesizes the merits of basic GA and traditional GP and has stronger ability of solving problems [2][3]. However, the learning procedure of GEP can be improved upon when dealing with complex problems with respect to both time efficiency and solution quality. The biological evolutionary process has revealed the principle of evolving from a self-contained functional

T.-D. Wang et al. (Eds.): SEAL 2006, LNCS 4247, pp. 292 – 301, 2006.

single cell to a well- developed entity with numerous specialized components [4]. We are naturally inspired to assume that solutions to complex problems might be built up incremen- tally from simpler elements. Although the phenotype of expression trees in GEP has retained the struct- ural representation from GP, the linear representation of the genotype conforms to Karva notation, under which the genotype-phenotype mapping mechanism does not guarantee that the levels of functional complexity in the phenotype are also directly reflected in the genotype. Since it is the genotype that is subject to the different genetic operations, it is difficult to follow the approach of incrementally forming solutions with the original GEP. Moreover, an evolved good functional structure is very likely destroyed in the subsequent generations not only by mutations but also by recombination and transpositions, which may require much additional computation to recover before an optimal solution is found. In view of this weakness, we propose a Gene Estimated Gene Expression Programming (GEGEP), it combines EDA with GEP . Compared with GEP, it mainly has the following characteristics: First, improve the gene expression structure, original gene structure is divided the head of chromosomes into a head and a body, which can be used to introduce the learning mechanism. Second, the homeotic gene which is also composed of a head, a body and a tail is used which can increase its searching ability. Third, the idea of EDA is introduced, which can enhance its learning ability and accelerate the convergence rate. We believe GEGEP benefits the evolution in terms of the convergence of a good functional structure.

The other parts of this article is organized as follows. Section2 explains GEGEP algorithm; Section3 proves the validity of the GEGEP algorithm through experiments; Section 4 presents some conclusions and ideas for future work.

2 Gene Estimated Gene Expression Programming

2.1 The Improvement of GEGEP

2.1.1 Improve the Structure of Gene Expression Programming

In GEGEP, gene head in GEP is divided into a head and a body. The head contains only functions, the body like the head of GEP contains symbols representing both functions and terminals, whereas the length of the tail is a function of head ,body and the number of arguments of the function with more arg- uments n(also called maximum arity) and is evaluated by the equation: tail =(head+body)×(n-1) + 1.

2.1.2 Improve the Structure of Homeotic Gene

Homeotic genes have exactly the same kind of structure as conventional genes of GEGEP and are built using an identical process. They are also composed of a head,a body and a tail.The heads only contain linking functions.The bodys contain linking functions, a special class of terminals:genic terminals representing conventional gene and random numerical constants.The tails contain obviously random numerical constants and genic terminals.

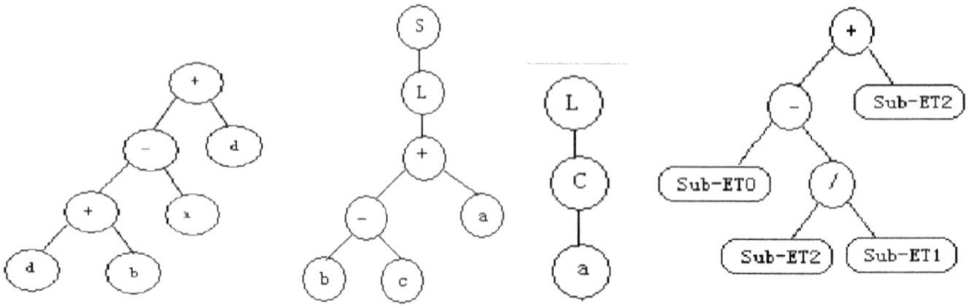

Sub-ET0 Sub-ET1 Sub-ET2

a)The sub-ETs codified by each conventional gene b)The final program

Fig. 1. Expression of chromosomes

The head, body and tail of homeotic gene are respectively represented as homo_head, homo_body, homo_tail.The method of computation of homeotic gene is same as conventional genes. So the length of gene individual is evaluated by the equation:

length=(head+body+tail) *n_gene+ homo_head +homo_body + homo_tail, where n_gene represents the number of genes. For example, if we choose head=2, body=4,n=2,then tail = (head+body)*(2-1)+1=7,the length of conventional genes is 2+4+7 =13.In this case ,if n_gene=3, homo_head =1,homo_body=4,then the length of homeotic gene is equal to 11, the length of gene individual is equal to 50.

One such gene individual is shown below:

0123456789012 0 123456789012 0123456789012 01234567890

+-d+ adbc?d?c c SL+- abc?aa?bc LCa*b- abd? cca +-20 / 21?20?

Where ? represents random numerical constant, 0、 1、 2 respectively represent first、 second、 third conventional gene, L、 C and S respectively represent ln、 cos and sin function. It codes for three conventional genes and one homeotic gene. The conventional genes code as usual for three different sub-ETs. The homeotic gene controls the interactions between the different sub-ETs . The homeotic gene and conventional genes have same functional set and terminal set. The corresponding phenotypes are shown in Figure1.

2.1.3 EDA

EDA[5] is put forward from the viewpoint of statistics by MüHlenbein and Paaß, which is evolutionary algorithm based on probability analysis. EDA is similar to traditional genetic algorithm, the difference between them is that EDA selects evaluation and learning distribution to produce new offspring. New offspring are produced by Bayes network. The Bayes network is a directed acycline graph, which considers specific interactions of the problem among the variables of chromosomes. It can make full use of realm knowledge and information of sampling datum with Bayes statistics. So it has strong learning ability and uses to model on discretional or continual polynomial in common.EDA contains three kinds of models: Univariate EDA Model, Bivariate EDA Model and Multivariate EDA Model which are shown in Figure2. Univariate EDA Model supposes that there have no interdependencies among variables. Bivariate EDA Model supposes that there have dependencies between pairs of variables. Multivariate EDA Model supposes that there have dependencies among variables greater than two.

For the convenience of computation, we only use Univariate EDA Model. The Model is shown in Figure 2(a). This basic thought of GEGEP is to carry on the statistics to the head and body department of each above operator of the conventional genes and homeotic gene , extracts its statistical probability, then carries on the mutation according to its statistical probability.

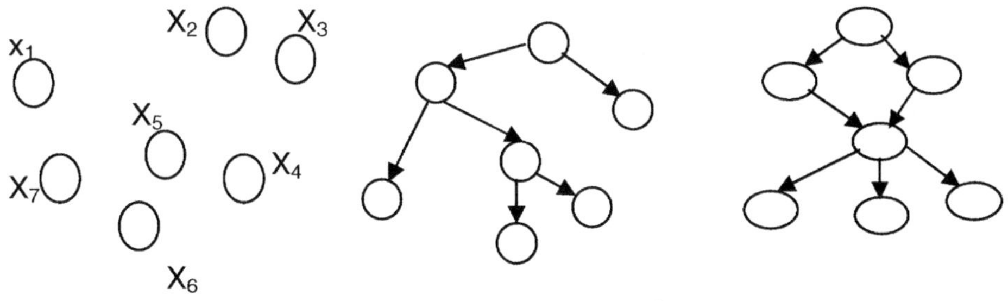

a) Univariate EDA Model (b)Bivariate EDA Model (c)Multivariate EDA Model

Fig. 2. Graphical representation of probability model among variables

The procedure of updating statistical information is as follows:

(1) Initialize fitness table

The size of fitness table is (n + 1) lines and L rows, where n expresses the size of function set, L expresses the total length of all he- ad and body added in the chromosome. For example: Function set = { +, -, *, /, Exp, Log, Sin, Cos, Sqrt }, the chromosome has 4 genes and 1 homeotic gene, the length of head of each gene is 3, the length of body is 7, then the size of total table is (9+1) *(5* (3+ 7)) = 10 *50. The initial value is ε (ε = 0.1) and initial frequency is once each generation.

(2) Initialize statistical information table

Statistical information table has the same size as fitness table, the initial value is also ε (ε = 0.1).The frequency is once every run.

(3) Renew fitness table

Renewing method is as follows: The fitness of each individual is added to the corresponding position of fitness table.

(4) Renew statistical information table

It directly adds the value of fitness table to original value of the corresponding position of statistical information table.

2.2 Fitness Function and Termination Condition

The fitness reflects whether the individual is good or bad. In this paper, the following fitness function is used:

$$f_i = \max(1000 - \sum_{j=1}^{n}\left(\left|\frac{C_{ij} - T_j}{T_j}\right| \cdot factor\right),0)$$

Where n is the number of computation cases, Cij is the value returned by the individual program i for fitness case j and Tj is the target value for fitness case j. Factor is a proportional constant and is equal to 1000 dividing the number of records. Obviously, if the bigger fitness value is, then the better individual is.

The termination condition is the actual Evolutionary generation exceeding reserved evolutionary generation (Maxgenerations) or the best fitness having no change for L continuous generations：

If generation≥Maxgenerations1||unchange_generation≥L

Then Terminates

The best solution or the approximate best solution is what we wanted when terminates.

2.3 Algorithm

```
PROCEDURE  GEGEP algorithm
begin
    Initialize P(0) ;
    Evaluate the fitness of P(0);
    Renew statistical information table;
    t:=0;
    repeat
      Pm(t):=mutation{P ( t) } according to statistical information;
      Pc(t):=crossover{P ( t) };
      Pt(t):=transpose{P ( t) };
      P(t+ 1):=selection{P( t) };
     Renew statistical information table;
```

```
        t:= t+ 1;
    until  termination condition ;
    output solution of Pbest;
end
```

3 Experiment

In order to justify the advantage of GEGEP as compared to traditional GP and basic GEP ,we first experimented on three algorithms with two function modeling . Later a suite of Symbolic Regression were conducted on the testing datasets .Last we experimented on another function modeling.

(1) Function modeling.

We select two cases from relevant references to check the validity of this method. Considering different impact of homeotic gene and conventional genes, the homeotic gene determines which gene is expressed in which cell and how they interact with one another. So the mutation rate is separated into homeotic gene rate and conventional generate.

The parameter setting of two experiments is same and is shown as Table1:

Table 1. Parameter Setting for GEGEP

Evolutionary generation	100000	One-point recombination rate	0.3
Population size	80	Two-point recombination rate	0.3
Mutation rate of homeotic gene	0.044	IS transposition rate	0.1
Mutation rate of conventional genes	0.044	RIS transposition rate	0.1
Gene recombination rate	0.1	Gene transposition rate	0.1
Conventional gene Numbers	4	Homeotic gene's head length	4
Conventional genes' head length	2	Homeotic gene's body length	20
Conventional genes' body length	15		

The calculation result of Y by GP and GEP are respectively obtained from [6] and [7]. There are six main factors: the depth of coalface x 1 (m) ,the height of coalface x 2 (m) , the amount of coalfaces x 3 (m3/t) , the layer interval between working coalface and neighboring coalface x 4 (m) , average day progress of working face x 5 (m /d) and average output per day x 6 (t/d) as the main input variates ,the amount of gas emitted from coalfaces(Y) as the output variate. So the termination set is defined as T = {x 1, x 2, x 3, x 4, x 5, x 6, ? },where ? represents random numerical constant and is chosen from the interval [-10, 10].The function set is defined as F = {+, −, *, ÷, sqrt, exp, sin, cos, ln}.The aim is to obtain a prediction function for Y with x 1, x 2, x 3, x 4, x 5 and x 6 being the parameters: Y =μ (x 1, x 2, x 3, x 4, x 5, x 6). Fifteen cases from sample 1 to 15 in table2 are used as training datum to build the prediction model while three cases from sample 16 to 18 in table3 are used as test datum.

The function evolved by GEGEP is：

y=Ln(Exp((Sin((Sin(((Exp(x3)+x3)-x6))/2.3661))+x2)))+Exp((Sqrt(Sin(Sin(Ln(x4)))) +(Exp((Sin(Sin((x3*x6)))+Sin(x4)))/((x3-Ln (Sin(Sqrt(x5))))+ x4))))

Table 2. Comparison of fitting results by GP ,GEP and GEGEP

sample	x_1 (m)	x_2 (m)	x_3 (m^3/t)	x_4 (m)	x_5 (m/d)	x_6 (t/d)	Calculation result of y (m^3/min)				Relative error of Y		
							Y	GP	GEP	GEGEP	GP	GEP	GEGEP
1	408	2.0	1.92	20	4.42	1825	3.34	3.18	3.27	3.37	0.048	0.020	0.00885
2	411	2.0	2.15	22	4.16	1527	2.97	2.79	3.29	2.93	0.061	0.108	0.01420
3	420	1.8	2.14	19	4.13	1751	3.56	3.89	3.47	3.56	0.092	0.024	1.62E-05
4	432	2.3	2.58	17	4.67	2078	3.62	3.46	3.71	3.71	0.044	0.026	0.02398
5	456	2.2	2.40	20	4.51	2104	4.17	3.98	4.10	4.15	0.046	0.018	0.00579
6	516	2.8	3.22	12	3.45	2242	4.60	4.78	4.68	4.64	0.039	0.017	0.00913
7	527	2.5	2.80	11	3.28	1979	4.92	4.88	5.03	4.98	0.008	0.022	0.01178
8	531	2.9	3.35	13	3.68	2288	4.78	4.38	4.84	4.87	0.084	0.012	0.01833
9	550	2.9	3.61	14	4.02	2325	5.23	5.1	5.19	5.21	0.025	0.008	0.00291
10	563	3.0	3.68	12	3.53	2410	5.56	5.78	5.42	5.57	0.040	0.024	0.00102
11	590	5.9	4.21	18	2.85	3139	7.24	7.09	7.35	7.24	0.021	0.015	0.00056
12	604	6.2	4.03	16	2.64	3354	7.80	8.15	7.49	7.78	0.045	0.040	0.00291
13	607	6.1	4.34	17	2.77	3087	7.68	7.36	7.58	7.68	0.042	0.013	3.1E-05
14	634	6.5	4.80	15	2.92	3620	8.51	8.91	8.30	8.52	0.047	0.024	0.00105
15	640	6.3	4.67	15	2.75	3412	7.95	7.43	7.94	7.98	0.065	0.001	0.00428

Table 3. Comparison of predicted results by GP,GEP and GEGEP

sample	x_1 (m)	x_2 (m)	x_3 (m^3/t)	x_4 (m)	x_5 (m/d)	x_6 (t/d)	Calculation result of y (m^3/min)				Relative error of Y		
							Y	GP	GEP	GEGEP	GP	GEP	GEGEP
16	450	2. 2	2. 43	16	4. 32	1996	4. 06	4. 32	4.11	4.01	0.064	0.012	0.01268
17	544	2. 7	3. 16	13	3. 81	2207	4. 92	5. 15	4.93	4.91	0.046	0.002	0.00105
18	629	6. 4	4. 62	19	2. 80	3456	8. 04	7. 79	8.01	7.80	0.031	0.004	0.03037

From Table2, we can see that the largest relative errors of GP, GEP and GEGEP are 9.2%, 10.8% and 2.398% respectively. Meanwhile, among the 15 training samples, the result of GEP has smaller relative errors than the one of GP on 13 samples, and the result of GEGEP has smaller relative errors than the one of GEP on 13 samples. It's obvious that concerning the goodness of fit, GEP outperforms GP and GEGEP outperforms GEP.

From table3, we can see that GEGEP has the higher prediction results and a smaller relative error of Y than the one of GP and GEP on three and one sample respectively.

(2) Function modeling.

The data comes from paper[8], which describes the GDP of China in some periods . The termination set is defined as T = {K, L, ? }, where ? represents random numerical constant and is chosen from the interval [-10, 10].The function set is defined as F = {+, —, *, ÷, sqrt, exp, sin, cos, ln, tan}. The parameter setting is same as experiment1.

The aim is to obtain a prediction function for GDP with K and L being the parameters: GDP =μ (K, L). Fifteen cases from the year 1980 to 1994 in table4 are used as training datum to build the prediction model, while two cases from 1995 to 1996 in table5 are used as test datum.

The second column(K) in table3 and table4 describes the sum of net value of fixed assets and the average balance of floating assets in the corresponding year. The third column(L) describes the number of employed person, including all kind of workers and peasants. The fourth column(calculation result of GDP) describes the actual statistical data of GDP in the same year.From Table4, we can see that the largest relative errors of GP, GEP and GEGEP are 8.29%, 3.182% and 3.177% respectively. Meanwhile, among the 15 training samples, the result of GEP has smaller relative errors than the one of GP on 13 samples, and the result of GEGEP has smaller relative errors than the one of GEP on 6 samples. It's obvious that concerning the goodness of fit, GEP outperforms GP and GEGEP outperforms GEP.From table5, we can see that GEGEP has the higher prediction results and a smaller relative error of GDP than the one of GP and GEP on one sample.

Table 4. Comparison of fitting results by GP ,GEP and GEGEP

Year	K (109 yuan)	L (104)	Statistical data of GDP (109 yuan)	Calculation result of GDP(109 yuan)			Relative error of GDP		
				GP	GEP	GEGEP	GP	GEP	GEGEP
1980	461.67	394.79	103.52	100.56	103.521	103.800	0.0286	0.00001	0.00271
1981	176.32	413.02	107.69	111.29	107.696	107.504	0.0309	0.00006	0.00145
1982	499.13	420.50	114.10	120.02	114.687	115.421	0.0519	0.00514	0.01158
1983	527.22	435.60	123.40	133.63	123.374	123.460	0.0829	0.00021	0.00049
1984	561.02	447.50	147.47	147.95	142.777	147.238	0.0032	0.03182	0.00157
1985	632.11	455.90	175.71	171.68	173.509	175.821	0.0229	0.01253	0.00063
1986	710.51	466.94	194.67	199.85	193.524	193.802	0.0266	0.00589	0.00446
1987	780.12	470.93	220.00	222.45	217.766	217.706	0.0114	0.01015	0.01043
1988	895.66	465.15	259.64	252.39	260.763	256.438	0.0279	0.00433	0.01233
1989	988.65	469.79	283.34	282.81	287.707	286.217	0.0019	0.01541	0.01015
1990	1075.37	470.07	310.15	308.42	310.990	309.670	0.0056	0.00271	0.00155
1991	1184.58	479.67	342.75	349.05	346.464	347.097	0.0184	0.01084	0.01268
1992	1344.14	485.70	411.24	403.05	410.710	410.960	0.0199	0.00129	0.00068
1993	1688.02	503.10	536.10	529.49	531.169	553.134	0.0123	0.00920	0.03177
1994	2221.42	513.00	725.14	714.81	728.118	729.770	0.0142	0.00411	0.00639

The function evolved by GP is:

GDP=exp (ln (K)-576.032/L)-ln (K+L)

The function evolved by GEP is:

GDP=tan(sin(sin(L)/cos(K)))+sin(tan(L))*9.16+log(abs(tan(L)))+K*log(log(abs (sqrt(L-6.44)+L-K)))/6.32+log(14.46*sqrt(L)+sin(K))/sin(log(K+L))+tan(log(L+5.08) *L3/-9.58)

The function evolved by GPGEP is:

GDP=2*((tan(-2.95022)*K)-sqrt((2*K)))-8.92026-2*(((cos(sqrt(((-0.36897)+K)))* (-8.17927))/(2.44667))-tan(ln(K-5.137215)))-(tan(0.450852-K)-0.99622)+tan(exp(cos (ln(K))))+exp(tan(K))/cos(cos(L))+tan(((cos(sqrt(K-0.36897))*(-8.17927))/(2.44667)) -tan(ln(K-5.137215)))

Table 5. Comparison of fitting results by GP ,GEP and GEGEP

Year	K (109 yuan)	L (104)	Statistical data of GDP (109 yuan)	Calculation result of GDP (109 yuan)			Relative error of GDP		
				GP	GEP	GEGEP	GP	GEP	GEGEP
1995	2843.48	515.30	920.11	921.64	922.012	921.342	0.0017	0.00207	0.00134
1996	3364.34	512.00	1102.10	1083.91	1107.010	1126.604	0.0165	0.00446	0.02223

4 Conclusions

This paper puts forward a new algorithm GEGEP, which takes GEP as foundation, improves gene structure and introduces the learning mechanism. The learning mechanism benefits preserving good structure which leads to a better evolutionary process. So it has the potential of identifying useful structural information emergent in the evolutionary process. Practical examples show that this alg- orithm possesses the advantages of automation of the modeling process, more flexible and various model structures, wider range of applications , faster speed of convergence and higher precision of fitting and predicted datum.

Future research on GEGEP will mainly focus on the following:

(1) Here we only consider Univariate EDA Model and do not consider Bivariate EDA Model and Multivariate EDA Model. So next we will try to use Bivariate EDA Model or Multivariate EDA Model. (2) The speed of algorithm has decreased because of large quantities of computation spending for statistics. We will adopt some methods to improve. (3)We will realize parallelism which can further improve the ability of solving complex problems. (4)Here we only realize one-layer call model , next we will realize multilayer call model.

Acknowledgments

This work is supported by the National Natural Science Foundation of China (Nos. 60473081, 60473037), the National Natural Science Foundation of Education Department of Hebei Province of China under Grant (Nos.2004454, 2005338) and the High Science and Technology Research of Hebei Province of China under Grant (No.05213567, 06213562).

References

1. Ferreira, C. Gene Expression Programming: a New Adaptive Algorithm for Solving Problems. Complex Systems, 13, 2(2001), 87-129.
2. Ferreira C. Gene expression programming[M] . Portugal, An- gra do Heroismo ,2002.
3. Ferreira C. Gene expression programming in problem solving [A]. 6th Online World Conference on Soft Computing in Indu- strial Applications[C] . 2001.
4. Xin Li, Chi Zhou, Weimin Xiao, Peter C. Nelson. Prefix Gene Expression Programming. Genetic and Evolutionary Computa- tion Conference (GECCO)'05, Washington, DC, USA., June 25-29, 2005,
5. P. Larra˜naga and J.A. Lozano. Estimation of distribution alg- orithms. A new tool for evolutionary computation. Kluwer Ac- ademic Publishers, 2001.
6. Zhao CY, Yuan XG , Sun JB. Application of Genetic Progr- amming to Predicting the Amount of Gas Emitted from Coal Face [J]. Journal of Basic Science and Engineering. 1999,7(4): 387-392.
7. Li Q, Cai ZH, Zhu L, Zhao SY. Application of Gene Expr- ession Programming in Predicting the Amount of Gas Emitted from Coal Face[J].Journal of Basic Science and Engineering. 2004.3(12).49-54.
8. CAI Zhihua, JIANG Siwei, Zhu Li , GUO Yuanyuan . A Novel Algorithm of Gene Expression Programming Based on Simul-ated Annealing. International Symposium on Intelligence C-omputation & Applications [C], 2005.605-610

Constrained Optimization Using Organizational Evolutionary Algorithm*

Jing Liu and Weicai Zhong

Institute of Intelligent Information Processing, Xidian University, Xi'an 710071, China
neouma@163.com

Abstract. This paper designs a new kind of structured population and evolutionary operators to form a novel algorithm, Organizational Evolutionary Algorithm (OEA), for solving constrained optimization problems. A simple and non problem-dependent technique is incorporated into OEA to handle the constraints. In OEA, a population consists of organizations, and an organization consists of individuals. All evolutionary operators are designed to simulate the interaction among organizations. In experiments, 4 well-studied engineering design problems are used to test the performance of OEA. The results show that OEA obtains good results both in the solution quality and the computational cost.

Keywords: Evolutionary algorithms, organization, constrained optimization.

1 Introduction

In economics, R. H. Coase explains the sizing and formation of organizations from the framework of transaction costs [1]. The basic idea is that the organization exists because it reduces the overhead transaction costs associated with exchanging goods and services. This concept was introduced to the classifier based on genetic algorithms by Wilcox in 1995 [2], which put emphasis on inventing an autonomous mechanism using transaction costs for forming the appropriately sized organizations within a classifier. Actually, in the real world situation, to achieve their purposes, organizations will compete or cooperate with others so that they can gain more resources. As a result, the resources will be reasonably distributed among all organizations little by little. This process plays an important role in human societies. From the viewpoint of computing, we think such a process can be viewed as a kind of optimization, and is an interesting new source for designing EAs.

Taking inspiration from the aforementioned process, this paper proposes a new frame for evolutionary optimization, namely, first composing organizations by members, which are equivalent to the individuals in traditional EAs, and then composing population by organizations, so that a structured population results. In this frame, all evolutionary operations are performed on organizations. Therefore, three evolutionary operators are developed for organizations. These operators are composed of several

* This work was supported by the National Natural Science Foundation of China under Grant 60502043.

T.-D. Wang et al. (Eds.): SEAL 2006, LNCS 4247, pp. 302–309, 2006.

crossover and mutation operations in common use so that the effect of the new frame can stand out. This new evolutionary frame is named as Organizational Evolutionary Algorithm (OEA).

Being different from [2], OEA does not put emphasis on forming the appropriately sized organizations, but on simulating the interaction among organizations. This is realized by the three evolutionary operators. Therefore, in OEA, all evolutionary operators are not directly performed on individuals, but on organizations. As a result, there is no global selection at all, so the global fitness distribution is not required. An organization interacts with others so that the information can be transferred to them. Obviously, such a model of the population is closer to the real evolutionary mechanism in nature than the traditional model.

In [3], we also propose a new EA, OCEC, for solving classification problems on the basis of organizations. But the organizations in OCEC are different from those of OEA. OCEC is designed with the intrinsic properties of classification in mind. Although both the organizations in OCEC and in OEA are composed of members, the members in OCEC stand for the examples in the training set while the members in OEA stand for the real-valued vectors in the search space. Moreover, the fitness in OCEC is especially designed so that the different importance of the attributes can stand out, while the fitness in OEA is the value of the objective functions. Of course, OCEC and OEA also have some similarities. Both of them simulate the interaction among organizations to solve problems.

2 Organizational Evolutionary Algorithm

2.1 Problem Definition

A constrained optimization problem can be formulated as solving the objective function

$$\text{minimize } f(\boldsymbol{x}), \quad \boldsymbol{x}=(x_1, x_2, \cdots, x_n)\in \mathcal{S}\cap\mathcal{F} \tag{1}$$

where $\mathcal{S}\subseteq\mathcal{R}^n$ defines the search space which is an n-dimensional space bounded by the parametric constraints $\underline{x}_i \le x_i \le \overline{x}_i$, i=1, 2, ..., n. Thus, $\mathcal{S}=[\underline{\boldsymbol{x}}, \overline{\boldsymbol{x}}]$, where $\underline{\boldsymbol{x}}=(\underline{x}_1, \underline{x}_2, ..., \underline{x}_n)$ and $\overline{\boldsymbol{x}}=(\overline{x}_1, \overline{x}_2, ..., \overline{x}_n)$. The feasible region $\mathcal{F}$ is defined as

$$\mathcal{F}=\left\{\boldsymbol{x}\in\mathcal{R}^n \,\middle|\, g_j(\boldsymbol{x})\le 0,\ \ j=1, 2, \cdots, m\right\} \tag{2}$$

where $g_j(\boldsymbol{x})$, j=1, 2, …, m are constraints.

In this paper, the constraints are handled using a comparison mechanism based on the following criteria [4]: 1) Between two feasible solutions, the one with the highest fitness value wins; 2) If one solution is feasible and the other one is infeasible, the feasible solution wins; 3) If both solutions are infeasible, the one with the lowest sum of constraint violation is preferred. This technique is non problem-dependent, and no parameter need to be tuned.

2.2 Organization Definition

In OEA, an organization consists of several members, and a member just is a solution of the objective function. We define **Member** as follows,

Definition 1. A ***Member*** is a real-valued vector that belongs to the search space, namely, $\boldsymbol{Member} \in S$. There are two measures, $\boldsymbol{Member}^F$ and $\boldsymbol{Member}^V$, to evaluate the member's quality. $\boldsymbol{Member}^F$ stands for the fitness, and it is computed as follows,

$$\boldsymbol{Member}^F = -f(\boldsymbol{Member}) \tag{3}$$

$\boldsymbol{Member}^V$ stands for the sum of constraint violation, and it is computed as follows,

$$\boldsymbol{Member}^V = \sum_{j=1}^{m} \max\{0,\ g_j(\boldsymbol{Member})\} \tag{4}$$

Thus, the purpose of OEA is maximizing $\boldsymbol{Member}^F$ of the members. When we compare the qualities of two members, both $\boldsymbol{Member}^F$ and $\boldsymbol{Member}^V$ should be considered. Therefore, the members are compared according to Definition 2.

Definition 2. If two members, $\boldsymbol{Member}_1$ and $\boldsymbol{Member}_2$, satisfy (5) or (6) or (7), then $\boldsymbol{Member}_1$ is better than $\boldsymbol{Member}_2$, and labeled as $\boldsymbol{Member}_1 \rhd \boldsymbol{Member}_2$.

$$\left(\boldsymbol{Member}_1^V = 0\right) \text{ and } \left(\boldsymbol{Member}_2^V = 0\right) \text{ and } \left(\boldsymbol{Member}_1^F > \boldsymbol{Member}_2^F\right) \tag{5}$$

$$\left(\boldsymbol{Member}_1^V = 0\right) \text{ and } \left(\boldsymbol{Member}_2^V > 0\right) \tag{6}$$

$$\left(\boldsymbol{Member}_1^V > 0\right) \text{ and } \left(\boldsymbol{Member}_2^V > 0\right) \text{ and } \left(\boldsymbol{Member}_1^V < \boldsymbol{Member}_2^V\right) \tag{7}$$

Definition 3. An **Organization**, *org*, is a set of members, and the best member is called **Leader**, which is labeled as $\boldsymbol{Leader}_{org}$, that is, an organization and the leader satisfy (8) and (9), where $|org|$ indicates the number of elements in *org*.

$$\left(org = \{\boldsymbol{Member}_1, \boldsymbol{Member}_2, \ldots, \boldsymbol{Member}_{|org|}\}\right) \text{ and } \left(org \neq \varnothing\right) \tag{8}$$

$$\forall \boldsymbol{Member} \in org,\ \boldsymbol{Leader}_{org} \rhd \boldsymbol{Member} \tag{9}$$

When comparing the qualities of two organizations, we just compare the qualities of their leaders. Namely, we say org_1 is better than org_2 if $\boldsymbol{Leader}_{org_1} \rhd \boldsymbol{Leader}_{org_2}$, and labeled as $org_1 \rhd org_2$.

2.3 Evolutionary Operators for Organizations

In the real world situation, there is a fierce competition among organizations, and those with feeble strength always are annexed by others. In OEA, the strength of an organization manifests in the fitness of the leader. So the purpose of each organization is to increase the leader's fitness as much as possible. To achieve this purpose, each organization must interact with others. On the basis of such interaction, three evolutionary operators are designed.

Splitting operator. In human societies, an organization whose size is too large usually is split into several small organizations so that they can be easily managed. In OEA, if most of the members belong to the same organization, the evolutionary operations would be disabled. So when an organization's size exceeds a limit, this organization must be divided. This is the function of the splitting operator. In this operator, Max_{OS} is the parameter controlling the maximum size of an organization, and $Max_{OS}>1$. If a parent organization, org_p, satisfies (10), org_p will be split into two child organizations, org_{c1} and org_{c2}.

$$(|org| > Max_{OS}) \text{ or } \left\{(|org| \le Max_{OS}) \text{ and } \left(U(0,1) < {|org|}/{N_o}\right)\right\} \tag{10}$$

Where $U(\cdot, \cdot)$ is a uniform random number generator, and N_o the number of organizations in initialization. To keep the randomicity and a small difference between the sizes of org_{c1} and org_{c2}, $\frac{|org_p|}{3}$ to $\frac{2|org_p|}{3}$ members are first randomly selected from org_p to form org_{c1}, and the remainder forms org_{c2}.

Annexing operator. This operator realizes the competition between two organizations. The better organization is the winner, and the other one is the loser. The winner will annex the loser to form a larger organization. The members of the winner can directly go into the new organization while those of the loser must die. But the members of the loser perhaps still have useful information, so the leader of the winner interacts with them to generate members for the new organization. The detail of this operator is given as follows.

Two parent organizations, $org_{p1}=\{\boldsymbol{x}_1, \boldsymbol{x}_2, \ldots, \boldsymbol{x}_M\}$ and $org_{p2}=\{\boldsymbol{y}_1, \boldsymbol{y}_2, \ldots, \boldsymbol{y}_N\}$, are randomly selected from the current generation. Without loss of generality, let $org_{p1} \rhd org_{p2}$. Thus, org_{p1} will annex org_{p2} to generate a child organization, $org_c=\{z_1, z_2, \ldots, z_M, z_{M+1}, z_{M+2}, \ldots, z_{M+N}\}$. Where $z_i=\boldsymbol{x}_i$, $i=1, 2, \ldots, M$. Let $AS\in(0, 1)$ be a predefined parameter. Then, if $U_j(0,1)<AS$, z_j, $j=M+1, M+2, \ldots, M+N$ are generated by Annexing Strategy 1 (AnStr1); otherwise, they are generated by Annexing Strategy 2 (AnStr2). The subscript j in $U_j(0,1)$ indicates that the random number is generated anew for each value of j, and AnStr1 and AnStr2 are given by (11) and (12), respectively. Given that the leader of org_{p1} is $(x_1, x_2, \ldots, x_n)$ and new members are $\boldsymbol{r}_j=(r_{j,1}, r_{j,2}, \ldots, r_{j,n})$, $j=1, 2, \ldots, N$.

In AnStr1, $\boldsymbol{r}_j$, $j=1, 2, \ldots, N$ are determined by (11),

$$r_{j,k} = \begin{cases} \underline{x}_k & \beta_k < \underline{x}_k \\ \overline{x}_k & \beta_k > \overline{x}_k \\ \beta_k & \text{otherwise} \end{cases}, \quad \beta_k = x_k + \alpha_k \times (x_k - y_{j,k}), \quad k = 1, 2, \ldots, n \tag{11}$$

where α_k is a uniformly distributed random number over [0, 1], and is generated anew for each value of k.

In AnStr2, $\boldsymbol{r}_j$, j=1, 2, …, N are determined by (12),

$$r_{j,k} = \begin{cases} \underline{x}_k + U(0,1) \times (\overline{x}_k - \underline{x}_k) & U_k(0,1) < \frac{1}{n} \\ x_k & \text{otherwise} \end{cases}, \quad k = 1, 2, ..., n \tag{12}$$

After $\boldsymbol{r}_j$, j=1, 2, …, N are calculated out, z_{j+M}, j=1, 2, …, N are determined by (13),

$$z_{j+M} = \begin{cases} \boldsymbol{r}_j & \boldsymbol{r}_j \triangleright \boldsymbol{y}_j \\ \boldsymbol{r}_j & (\boldsymbol{y}_j \triangleright \boldsymbol{r}_j) \text{ and } \{U_j(0, 1) < \exp(\boldsymbol{r}_j^F - \boldsymbol{y}_j^F)\} \\ \boldsymbol{y}_j & \text{otherwise} \end{cases} \tag{13}$$

As can be seen, when $\boldsymbol{r}_j$ is better than $\boldsymbol{y}_j$, $\boldsymbol{r}_j$ gets into org_c so as to improve the quality of org_c. When $\boldsymbol{r}_j$ is worse than $\boldsymbol{y}_j$, in order to maintain the diversity, $\boldsymbol{r}_j$ gets into org_c with a probability. The more $\boldsymbol{r}_j$ is close to $\boldsymbol{y}_j$, the greater the probability is. For more clarity, this operator is shown in Fig.2. org_c has (M+N) members, where Member$_i$, i=1, 2, …, M, come from org_{p1} and Member$_i$, i=M+1, M+2, …, M+N are generated by the leader of org_{p1} and the members of org_{p2} together. After org_c is generated, its leader is also selected. In this case, the leader is the ith member, that is, the leader of org_{p1}, or the best member among Member$_{M+1}$, Member$_{M+2}$, …, Member$_{M+N}$.

Cooperating operator. This operator realizes the cooperation between two organizations. The leaders of the organizations interact with each other to generate two new members. Then, in each organization, a member which is worse than the new member is replaced. As a result, two new organizations are generated. The detail of this operator is given as follows.

Two parent organizations, org_{p1}={$\boldsymbol{x}_1$, $\boldsymbol{x}_2$, …, $\boldsymbol{x}_M$} and org_{p2}={$\boldsymbol{y}_1$, $\boldsymbol{y}_2$, …, $\boldsymbol{y}_N$}, are randomly selected from the current generation. They will cooperate with each other to generate two child organizations, org_{c1} and org_{c2}. Let $CS \in (0, 1)$ be a predefined parameter. Then, if $U(0, 1)<CS$, then org_{c1} and org_{c2} are generated by Cooperating Strategy 1 (CoStr1); otherwise, they are generated by Cooperating Strategy 2 (CoStr2), where CoStr1 and CoStr2 are given by (14) and (15), respectively. Given that the leader of org_{p1} is (x_1, x_2, …, x_n), the leader of org_{p2} is (y_1, y_2, …, y_n), and two new members are $\boldsymbol{q}$=(q_1, q_2, …, q_n) and $\boldsymbol{r}$=(r_1, r_2, …, r_n).

In CoStr1, $\boldsymbol{q}$ and $\boldsymbol{r}$ are determined by (14),

$$\begin{cases} q_k = \alpha_k \times x_k + (1-\alpha_k) \times y_k \\ r_k = (1-\alpha_k) \times x_k + \alpha_k \times y_k \end{cases}, \quad k = 1, 2, ..., n \tag{14}$$

where α_k is the same with that of (11).

In CoStr2, $\boldsymbol{q}$ and $\boldsymbol{r}$ are determined by (15),

$$\begin{cases} \boldsymbol{q} = (x_1, x_2, \cdots, x_{i_1-1}, y_{i_1}, y_{i_1+1}, \cdots, y_{i_2}, x_{i_2+1}, x_{i_2+2}, \cdots, x_n) \\ \boldsymbol{r} = (y_1, y_2, \cdots, y_{i_1-1}, x_{i_1}, x_{i_1+1}, \cdots, x_{i_2}, y_{i_2+1}, y_{i_2+2}, \cdots, y_n) \end{cases} \tag{15}$$

where $1<i_1<n$, $1<i_2<n$, and $i_1<i_2$.

After $\boldsymbol{q}$ and $\boldsymbol{r}$ are calculated out, org_{c1} and org_{c2} are determined by (16) and (17), respectively,

$$org_{c1} = \begin{cases} \{\boldsymbol{x}_1, \boldsymbol{x}_2, \cdots, \boldsymbol{x}_{i-1}, \boldsymbol{q}, \boldsymbol{x}_{i+1}, \boldsymbol{x}_{i+2}, \cdots, \boldsymbol{x}_M\} & \exists \boldsymbol{x}_i \in org_{p1}, \ \boldsymbol{q} \rhd \boldsymbol{x}_i \\ org_{p1} & \text{otherwise} \end{cases} \quad (16)$$

$$org_{c2} = \begin{cases} \{\boldsymbol{y}_1, \boldsymbol{y}_2, \cdots, \boldsymbol{y}_{j-1}, \boldsymbol{r}, \boldsymbol{y}_{j+1}, \boldsymbol{y}_{j+2}, \cdots, \boldsymbol{y}_N\} & \exists \boldsymbol{y}_j \in org_{p2}, \ \boldsymbol{r} \rhd \boldsymbol{y}_j \\ org_{p2} & \text{otherwise} \end{cases} \quad (17)$$

2.4 Implementation of OEA

In OEA, the population consists of organizations, and the above three operators must be performed on organizations reasonable so that high quality members can be obtained. In this paper, we control the three operators by means of evolution. In each generation, the size of each organization in the population is first checked. If the size of an organization satisfies (10), then the splitting operator is performed on this organization. Next, two organizations are randomly selected from the population, and annexing operator or cooperating operator are performed on them with the same probability. This process is continued until the number of organizations in the population is less than 2. The population evolves generation by generation. At last, the best member among all organizations is output as the result. The details are shown in ALGORITHM 1.

ALGORITHM 1 Organizational Evolutionary Algorithm

```
begin
    Initializing population P0 with No organizations,
    and each organization has one member;
    t←0;
    while(the termination criteria are not reached)do
    begin
        For each organization in Pt, if it satisfies
        (10), performing the splitting operator on
        it, deleting it from Pt, and adding the child
        organizations into Pt+1;
        while(the number of organizations in Pt is
                greater than 1)do
        begin
            Randomly selecting two parent organiza-
            tions, orgp1 and orgp2, from Pt;
            Performing the annexing operator or the
            cooperating operator on orgp1 and orgp2
            with the same probability;
            Deleting orgp1 and orgp2 from Pt, and add-
            ing the child organizations to Pt+1;
        end;
        Moving the organization in Pt to Pt+1;
        t←t+1;
```

```
        end;
        Output the best member among all organizations as
        the result;
    end.
```

As can be seen, in the initialization, each organization has only one member, and the population has total N_o organizations. During the evolutionary process, the number of organizations varies from generation to generation due to the effects of the splitting operator and the annexing operator, but the number of members in the population remains constant.

3 Experimental Studies

In this section, 4 well-studied engineering design problems (Welded Beam Design, Spring Design, Speed Reducer Design, and Three-Bar Truss Design) [5] are used to test the performance of OEA. Moreover, OEA is compared with Society and Civilization Algorithm (SCA) [5]. OEA is terminated after 50 generations for Three-Bar Truss Design and 80 generations for the 3 other problems such that the number of function evaluations of OEA is less than that of SCA. The other parameters of OEA are set as follows: N_o=1500, Max_{OS}=20, AS=0.8, and CS=0.6.

3.1 Experimental Results of OEA

Table I summarizes the experimental results of OEA over 50 trials, which include the best, the median, the mean, the standard deviation, and the worst function value found. OEA provides 100% feasible solutions for all functions. As described in Table I, the standard deviations of the 4 problems is relatively small, so the performance of OEA is very stable in solving these problems.

Table 1. The experimental results of OEA on the 4 engineering design problems over 50 trials

f	Best FV	Median FV	Mean FV	St. dev.	Worst FV	Mean NFE
Welded Beam Design	2.3842157	2.3987177	2.4022077	1.09×10^{-2}	2.4330366	25 549
Spring Design	0.01267551864	0.012719265	0.012715621	9.05×10^{-6}	0.012721341	25 650
Speed Reducer Design	2994.580437	2994.705544	2994.723808	8.75×10^{-2}	2994.910336	25 590
Three-Bar Truss Design	263.8958	263.8972	263.8973	1.26×10^{-3}	263.90055	17 306

3.2 Comparison Between OEA and SCA [5]

The comparison is shown in Table II where the results of SCA are obtained from [5]. As can be seen, except the Best FV of OEA for the Spring Design Problem is little worse than that of SCA, OEA significantly outperforms SCA for the 4 problems in terms of all the four criteria.

Table 2. The comparison between OEA and SCA over 50 trials

f	Method	Best FV	Median FV	Mean FV	St. dev.	Worst FV	NFE of Best
Welded Beam Design	OEA	**2.3842157**	**2.3987177**	**2.4022077**	$\mathbf{1.09{\times}10^{-2}}$	**2.4330366**	25 668
	SCA	2.3854347	3.0025883	3.2551371	$9.59{\times}10^{-1}$	6.3996785	33 095
Spring Design	OEA	0.01267551864	**0.012719265**	**0.012715621**	$\mathbf{9.05{\times}10^{-6}}$	**0.012721341**	25 293
	SCA	**0.01266924934**	0.012922669	0.012922669	$5.92{\times}10^{-4}$	0.016717272	25 167
Speed Reducer Design	OEA	**2994.580437**	**2994.705544**	**2994.723808**	$\mathbf{8.75{\times}10^{-2}}$	**2994.910336**	25 919
	SCA	2994.744241	3001.758264	3001.758264	4.01	3009.964736	54 456
Three-Bar Truss Design	OEA	**263.8958450**	**263.8972**	**263.8973**	$\mathbf{1.26{\times}10^{-3}}$	**263.90055**	17 557
	SCA	263.8958466	263.8989	263.9033	$1.26{\times}10^{-2}$	263.96975	17 610

References

[1] R. H. Coase, *The Firm, the Market, and the Law*, Chicago: University of Chicago Press, 1988.

[2] J. R. Wilcox, *Organizational Learning within a Learning Classifier System*, Master thesis, University of Illinois, 1995. Also Tech. Report No. 95003 IlliGAL.

[3] L. Jiao, J. Jiu, and W. Zhong, "An organizational coevolutionary algorithm for classification," *IEEE Trans. Evol. Comput.*, vol. 10, no. 1, pp.67-80, Feb. 2006.

[4] E. M. Montes, C. A. C. Coello, "A simple multimembered evolution strategy to solve constrained optimization problems," *IEEE Trans. Evol. Comput.*, vol. 9, no. 1, pp.1-17, Feb. 2005.

[5] T. Ray and K. M. Liew, "Society and civilization: An optimization algorithm based on the simulation of social behavior," *IEEE Trans. Evol. Comput.*, vol. 7, no. 4, pp.386-396, Aug. 2003.

Rotationally Invariant Crossover Operators in Evolutionary Multi-objective Optimization

Antony Iorio and Xiaodong Li

School of Computer Science and IT,
RMIT University, Melbourne VIC 3000, Australia
iantony@cs.rmit.edu.au
http://goanna.cs.rmit.edu.au/~iantony
xiaodong@cs.rmit.edu.au
http://goanna.cs.rmit.edu.au/~xiaodong

Abstract. Multi-objective problems with parameter interactions can present difficulties to many optimization algorithms. We have investigated the behaviour of Simplex Crossover (SPX), Unimodal Normally Distributed Crossover (UNDX), Parent-centric Crossover (PCX), and Differential Evolution (DE), as possible alternatives to the Simulated Binary Crossover (SBX) operator within the NSGA-II (Non-dominated Sorting Genetic Algorithm II) on four rotated test problems exhibiting parameter interactions. The rotationally invariant crossover operators demonstrated improved performance in optimizing the problems, over a non-rotationally invariant crossover operator.

1 Introduction

Traditional genetic algorithms that use low mutation rates and fixed step sizes have significant trouble with problems with interdependent relationships between decision variables, but are perfectly suited to many of the test functions currently used in the evaluation of genetic algorithms [1]. These test functions are typically linearly separable and can be decomposed into simpler independent problems. Unfortunately, many real-world problems are not linearly separable, although linear approximations may sometimes be possible between decision variables.

Interdependencies between variables can be introduced into a real-coded functional problem by rotating the coordinate system of a test function. A rotated problem cannot be solved efficiently by the directionless step-sizes and low mutation rates that Genetic Algorithms typically use [1]. Although the NSGA-II is a very robust multi-objective optimization algorithm it suffers from similar limitations as traditional Genetic Algorithms on these problems.

Previous work has reported on the poor performance of a number of Multi-objective Evolutionary Algorithms, including the NSGA-II, on a rotated problem [2]. NSGA-II uses a crossover technique called Simulated-Binary Crossover (SBX) [3,4], combined with a uniform crossover operator in which half the time parameters of an offspring solution are replaced with parameters from a parent solution. This crossover technique searches effectively along the principle

T.-D. Wang et al. (Eds.): SEAL 2006, LNCS 4247, pp. 310–317, 2006.

coordinate axes of the decision space. This makes finding more optimal solutions difficult when the decision space is large, and the problem has parameter interactions. Problems which are rotated, and not aligned with the coordinate axes typically require correlated self-adapting mutation step sizes in order to efficiently search for optimal solutions [1].

Differential Evolution (DE) has previously demonstrated rotationally invariant behaviour in the single objective and multiobjective domain [5,6,7]. Simplex Crossover (SPX), Parent Centric Crossover (PCX), and unimodal Normal Distribution Crossover (UNDX-m) have also demonstrated rotationally invariant behaviour on single objective test problems. This provides the motivation to study the worth of these multi-parent crossover techniques on rotated multi-objective optimization problems, where such characteristics are desirable.

Experiments have been conducted on rotated problem from [7]. These problems were rotated arbitrarily and uniformly in the decision space in order to test the rotationally invariant behaviours of the crossover operators.

In Section 2 we will briefly introduce the crossover operators used in this study, followed by Section 3, where the methodology and parameters associated with the experiments are discussed. Section 4 discusses the results of these experiments, followed by the conclusions drawn from this study in Section 5.

2 Background

The NSGA-II uses a simulated binary crossover operator [4] with uniform crossover to generate offspring parameter values. The SBX operator takes two parents and produces two offspring, but does not have the property of rotational invariance because the correlation between the location of parents, and the location of offspring which are generated, is lost under a rotation of the decision space. The discrete crossover of variables also results in non-rotationally invariant behaviour. For example, if an offspring vector has a parameter replaced by a parent parameter, as it might under some uniform crossover scheme, rotational invariance is destroyed [8]. It has been shown that the SBX has a zero probability of generating some points in the space between two parents [9], although in the new version of SBX implemented in the latest revision of NSGA-II, this problem has been addressed by generating offspring in quadrants adjacent to the location of the parents, as well as surrounding the parents.

The UNDX crossover [10] has demonstrated excellent performance in optimizing highly epistatic functions [11]. It generates offspring around a centroid region specified by a number of parents. It has been applied to some difficult real world problems such as design of optical lens systems [12]. A multi-parent variant of the UNDX was proposed, called UNDX-m [13]. The UNDX-m covers the search space more effectively by having a greater diversity of offspring generated, and it is this variant that we will be considering.

The PCX [14,15] is similar to the UNDX-m, but instead of distributing the offspring around the centroid of a number of parents, the offspring distribute around the parents themselves.

The Simplex crossover (SPX) was originally proposed in [16]. It generates a simplex from a number of parents. This simplex is expanded, and offspring are generated inside the expanded region.

The other reproduction technique studied here is Differential Evolution, which differs from other EAs in the mutation and recombination phase. Differential Evolution has also been applied to multi-objective problems [17,18,19,20,6,7]. Unlike stochastic techniques such as Genetic Algorithms and Evolutionary Strategies, where perturbation occurs in accordance with a random quantity, Differential Evolution uses weighted differences between solution vectors to perturb the population. The variant of differential evolution used in this study is known as *DE/current-to-rand/1* [8]. In order to maintain diversity, a Log-normal *dithering* operator was employed as well [21]. This operator maintains rotational invariance, while helping to randomly perturb individuals within the population.

3 Experiments

In order to test each of the crossover techniques, the crossover operator of NSGA-II was replaced with one of the rotationally invariant crossover operators. For the SPX, UNDX-m, and PCX, a single set of parents was randomly selected each generation from the mating pool. These parents were used to generate 100 new offspring individuals. For the DE and SBX variants, parents were randomly selected from the mating pool, and this was repeatedly done in a single generation for each of the 100 offspring generated. A population size of 100 individuals was used for each of the algorithms on each of the test problems. A number of the crossover techniques investigated here have not previously been studied within the NSGA-II framework, and we expect that some of the choices of our parameter settings to be sub-optimal for the problems explored. It is not our intention to perform a comparative study in order to find the best parameter settings of these crossover techniques, but we do expect non-specifically tuned settings to demonstrate improvements in the performance of the NSGA-II with the rotationally invariant behaviour of the DE, SPX, UNDX-m, and PCX operators. A number of appropriate parameter settings have been reported for these operators and we have utilised these reported settings where possible. It should be noted that these settings were reported with respect to single-objective optimization problems. We leave a more detailed comparative study of these operators on rotated multi-objective problems, as an area of future study.

For the DE variant of NSGA-II, F was set to 0.8 and K was set to 0.6. Suggestions from the literature helped guide our choice of parameter values for the NSDE [5]. The factor which controls the spread of the distribution for F in the dithering operator was set to 0.5 [21].

In the SPX operator, the simplex size, ϵ, determines the size of the expanded simplex, and we have used $\sqrt{n+1}$ which was used in previous studies in the single objective domain, where n is the decision space dimension. A mutation rate of 0.1 was also used with the UNDX-m, PCX, SPX, and SBX variants, using the mutation operator in NSGA-II. For the UNDX-m operator, the parameters

$\sigma_\xi = \frac{1}{\sqrt{m}}$ and $\sigma_\eta = \frac{0.35}{\sqrt{n-m}}$ were recommended in [13] and we have used these values in this study. For the PCX, the σ_ξ and σ_η parameters were set to 0.7 and 0.2 respectively. The PCX variant is sensitive to the σ_ξ parameter. If σ_ξ is too small the offspring generated do not spread across the Pareto-optimal front. In the UNDX-m and PCX version of NSGA-II, m was assigned a value of 3. The NSGA-II used a crossover rate of 0.9, but the other variants each used a crossover rate of 1.0.

Experiments were conducted on the unimodal problem R1, R2, R3, and R4, from [7]. Each of these problems incorporates features which are designed to trap points from progressing along the Pareto-optimal front. Problem R1 is unimodal. Problem R2 is discontinuous in the objective space. Problem R3 has a non-uniform mapping between the decision and objective spaces. Problem R4 is deceptive and has a non-local Pareto-optimal front which can trap points from progressing to the global Pareto-optimal front. These problems are described in more detail in [7]. The problems are also 10-dimensional in the decision space. Rotations were performed in the decision space, on each plane, using a random uniform rotation matrix generated using the technique described in [7]. The rotation introduces non-linear dependencies between all parameters. Each algorithm was run 50 times on each test problem, for a total of 800 generations (80,000 problem evaluations) for each run. A new random uniform rotation matrix was generated for each run of each algorithm. For the purposes of evaluating the algorithms, the generational distance metric was employed, as well as its inverse, in order to measure both the convergence to the Pareto-optimal front, and the diversity of solutions across the front [6]. The GD(Q,P*) metric measure the convergence of the non-dominated set Q, towards the Pareto-optimal set P. Similarly, the GD(P*,Q) metric measure the average distance of P to Q, thereby quantifying the degree that Q covers the set P. As both measures approach zero, one can expect good coverage of the Pareto-optimal set, as well as good convergence.

4 Discussion and Results

Previous work has reported on problem R1 [7] and the tendency of the NSGA-II to migrate non-dominated solutions away from the Pareto-optimal region, as well as the difficulty in expanding across the Pareto-optimal front because of these easily favoured non-dominated solutions.

Each of the rotationally invariant crossover operators apparently yields superior performance over the SBX on this relatively simple problem, with respect to both convergence to the Pareto-optimal front and coverage across the front. The boxplots in Figure 1 also demonstrates that the variation in both the GD(Q,P*) which measures convergence, and the GD(P*,Q) which measure spread, is relatively low for the SPX, DE, UNDX, and PCX variants.

Over successive generations, the SBX operator generates non-dominated solutions which skew away from the Pareto-optimal front [7]. In Problem R2, the Pareto-optimal front is discontinuous. This characteristic exacerbates this behaviour further, because the infeasible regions do not allow better non-dominated

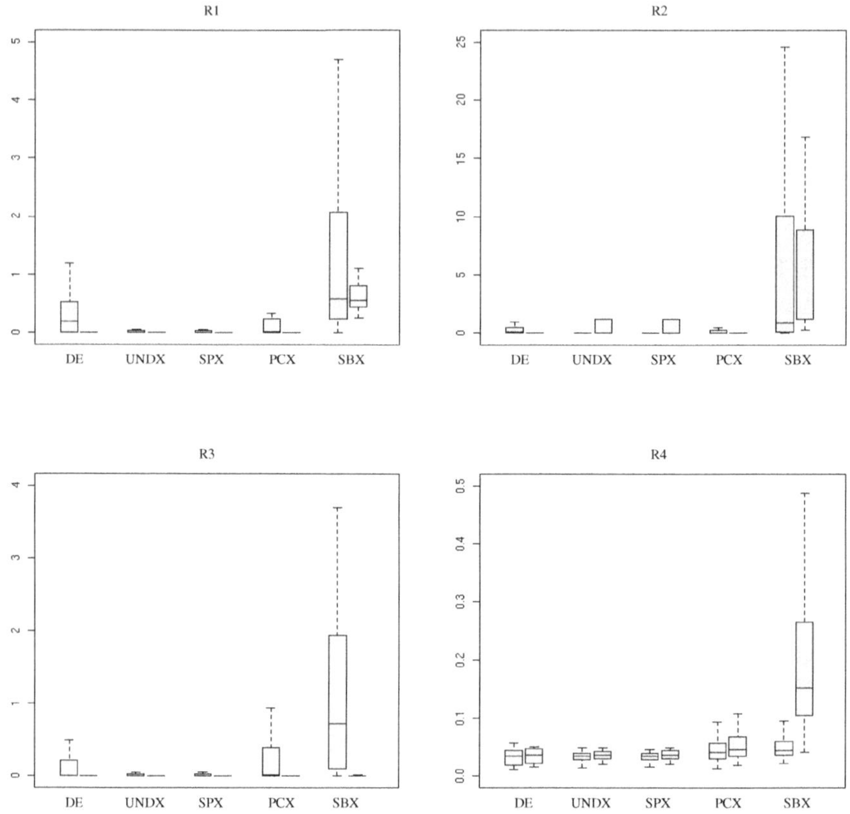

Fig. 1. Boxplots of the $GD(Q, P*)$ and $GD(P*, Q)$ for problem R1, R2, R3, and R4. The $GD(Q, P*)$ boxplot is always to the left of the $GD(P*, Q)$ boxplot.

solutions to be found through independent perturbations of the decision variables when the problem is rotated, as can be seen with the SBX in Figure 2. Contrasting this, from Figure 2, DE does manage to find a variety of solutions close to the Pareto-optimal region. In the boxplots of Figure 1 it is apparent that there is a high variation in the convergence and diversity of the non-dominated solutions found with SBX on R2. The measured convergence and diversity is also worse than the other rotationally invariant operators.

For Problem R3, the NSGA-II wih SBX achieved a rather good spread of non-dominated solutions, but was not able to converge sufficiently to the Pareto-optimal front. Each of the rotationally invariant crossover operators outperformed the SBX on this problem as well.

Problem R4 is highly multimodal, and deceptive. When this problem is rotated, a number of regions can hamper SBX from increasing the spread of solutions across a non-dominated front. This is apparent in the boxplot of Figure 1 as well, where the average of the GD(P*,Q) metric is significantly worse than the rotationally invariant schemes.

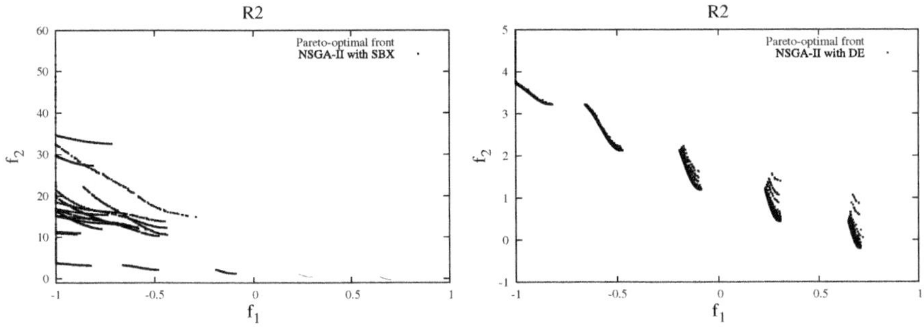

Fig. 2. 50 runs of the NSGA-II with SBX and DE Problem R2

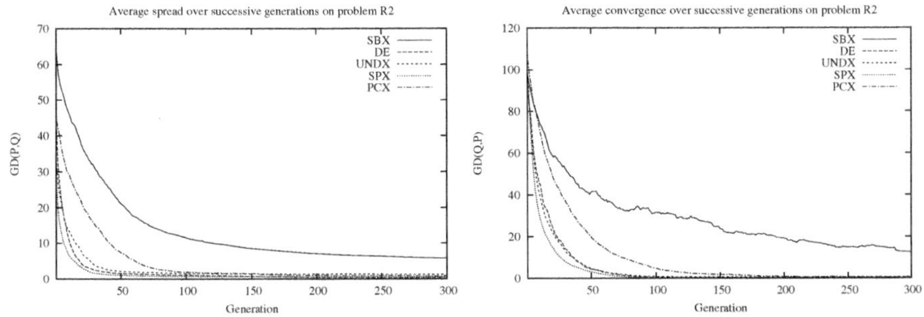

Fig. 3. Average $GD(Q, P*)$ and $GD(P*, Q)$ for problem R2 over 300 generations

From the boxplots in Figure 1 it is apparent that the DE variant demonstrated very competitive performance with respect to the spread of solutions on Problem R1, R2, and R3. It is also of relevance that the number of evaluations required can be significantly reduced through the use of rotationally invariant crossover operator in the presence of parameter interactions. This is demonstrated by the plots in Figure 3. These plots demonstrate the superior performance on Problem R2 of the DE, SPX, UNDX, and PCX variants, in comparison with the baseline NSGA-II with SBX. It is apparent that the rotationally invariant crossover operators have faster convergence to the Pareto-optimal front, while also maintaining a diverse set of solutions across the front. This has important practical relevance to real-world problems which often exhibit parameter interactions and also have computationally expensive evaluations of solutions.

5 Conclusion

This paper has described an empirical study of the effects that rotation of problems has on the NSGA-II. Rotation can trap the search on four problems with the properties of uni-modality, discontinuous Pareto-optimal fronts, a non-uniform mapping between the objective and decision space, and a problem with a deceptive front. We have demonstrated that on these four problems, that the UNDX,

SPX, and DE outperformed the SBX, taking into consideration the performance metrics for convergence to the Pareto-optimal front and distribution of solutions across the front.

There are a number of future avenues of work which may be worthwhile considering, such as the effect of rotation on problems with more than two objectives, and when the dimensionality of the decision space increases. One would expect an increase in difficulty with an increase in the decision space dimension, with degraded non-dominated solutions becoming even more likely with non-rotationally invariany algorithms, because there will be far more non-dominated solutions generated which are not Pareto-optimal.

Secondly, it would be useful to conduct further tests on problems which do not have a linearly distributed Pareto-optimal set. This could be achieved using the Okabe framework for constructing multiobjective test problems [22].

References

1. R. Salomon, "Re-evaluating genetic algorithm performance under coordinate rotation of benchmark functions: A survey of some theoretical and practical aspects of genetic algorithms," *Bio. Systems*, vol. 39, no. 3, pp. 263–278, 1996.
2. K. Deb, A. Pratap, S. Agarwal, and T. Meyarivan, "A fast and elitist multiobjective genetic algorithm: NSGA-II," *IEEE Trans. Evol. Comput.*, vol. 6, no. 2, pp. 182–197, 2002.
3. K. Deb and R. B. Agrawal, "Simulated binary crossover for continuous search space," *Complex Systems*, vol. 9, no. 2, pp. 115–148, 1995.
4. K. Deb and A. Kumar, "Real-coded genetic algorithms with simulated binary crossover: Studies on multi-modal and multi-objective problems," *Complex Systems*, vol. 9, no. 6, pp. 431–454, 1995.
5. D. Corne, M. Dorigo, and F. Glover, Eds., *New Ideas in Optimization.* London UK: McGraw-Hill, 1999.
6. A. Iorio and X. Li, "Incorporating directional information within a differential evolution algorithm for multiobjective optimization," in *Proceedings of the 2006 Genetic and Evolutionary Computation Conference (GECCO-06).* IEEE Press, 2006.
7. ——, "Rotated test problems for assessing the performance of multiobjective optimization algorithms," in *Proceedings of the 2006 Genetic and Evolutionary Computation Conference (GECCO-06).* IEEE Press, 2006.
8. K. Price, *New Ideas in Optimization.* McGraw Hill, 1999, p. 98.
9. P. Ballester and J. N. Carter, "Real-parameter genetic algorithms for finding multiple optimal solutions in multi-modal optimization," in *Proceedings of the 2003 Genetic and Evolutionary Computation Conference (GECCO-03)*, 2003, pp. 707–717.
10. I. Ono, H. Kita, and S. Kobayashi, *A Real-coded Genetic Algorithm using the Unimodal Normal Distribution Crossover*, ser. Advances in Evolutionary Computing : Theory and Applications. Springer, 2003, pp. 213–237.
11. H. Kita, "A comparison study of self-adaptation in evolution strategies and real-coded genetic algorithms," *Evolutionary Computation*, vol. 9, no. 2, pp. 223–241, 2001.
12. I. Ono, S. Kobayashi, and K. Yoshida, "Optimal lens design by real-coded genetic algorithms using UNDX," *Computer Methods in Applied Mechanics and Engineering*, vol. 186, pp. 483–497, 2000.

13. H. Kita, I. Ono, and S. Kobayashi, "Multi-parental extension of the unimodal normal distribution crossover for real-coded genetic algorithms," in *Proc. 1999 Congress on Evolutionary Computation (CEC'99)*, 1999, pp. 1581–1587.
14. K. Deb, D. Joshi, and A. Anand, "Real-coded evolutionary algorithms with parent-centric recombination," Indian Institute of Technology, Kanpur, Tech. Rep. KanGAL Report No.2001003, 2001.
15. ——, "Real-coded evolutionary algorithms with parent-centric recombination," in *Proceedings of the 2002 Congress on Evolutionary Computation CEC2002*. IEEE Press, 2002, pp. 61–66.
16. S. Tsutsui and M. Yamamura, "Multi-parent recombination with simplex crossover in real coded genetic algorithms," in *Proceedings of the 1999 Genetic and Evolutionary Computation Conference (GECCO-99)*, 1999, pp. 657–664.
17. C. S. Chang and D. Y. Xu, "Differential evolution of fuzzy automatic train operation for mass rapid transit system," in *IEEE Proceedings of Electric Power Applications*, vol. 147, no. 3, 2000, pp. 206–212.
18. H. A. Abbass, R. Sarker, and C. Newton, "PDE: A Pareto-frontier differential evolution approach for multi-objective optimization problems," in *Proceedings of the 2001 Congress on Evolutionary Computation (CEC'2001)*, vol. 2, 2001, pp. 971–978.
19. F. Xue, A. C. Sanderson, and R. J. Graves, "Pareto-based multi-objective differential evolution," in *Proceedings of the 2003 Congress on Evolutionary Computation (CEC'2003)*, vol. 2. IEEE Press, 2003, pp. 862–869.
20. N. K. Madavan, "Multiobjective optimization using a Pareto differential evolution approach," in *Proceedings of the 2002 Congress on Evolutionary Computation (CEC'2002)*, vol. 2. IEEE Press, 2002, pp. 1145–1150.
21. K. Price, R. M. Storn, and J. A. Lampinen, *Differential Evolution - A Practical Approach to Global Optimization.* Springer, 2005, p. 88.
22. T. Okabe, Y. Jin, M. Olhofer, and B. Sendhoff, "On test functions for evolutionary multi-objective optimization," in *Parallel Problem Solving from Nature - PPSN VIII : 8th International Conference Proceedings.* Springer, 2004, pp. 792–802.

A Hybrid of Differential Evolution and Genetic Algorithm for Constrained Multiobjective Optimization Problems

Min Zhang, Huantong Geng, Wenjian Luo, Linfeng Huang, and Xufa Wang

Nature Inspired Computation and Applications Laboratory, Department of Computer Science and Technology, University of Science and Technology of China, 230027, Hefei, Anhui, China
{zhangmin, lfhuang}@mail.ustc.edu.cn, htgeng@ustc.edu, {wjluo, xfwang}@ustc.edu.cn

Abstract. Two novel schemes of selecting the current best solutions for multiobjective differential evolution are proposed in this paper. Based on the search biases strategy suggested by Runarsson and Yao, a hybrid of multiobjective differential evolution and genetic algorithm with (N+N) framework for constrained MOPs is given. And then the hybrid algorithm adopting the two schemes respectively is compared with the constrained NSGA-II on 4 benchmark functions constructed by Deb. The experimental results show that the hybrid algorithm has better performance, especially in the distribution of non-dominated set.

1 Introduction

Genetic Algorithm (GA) for Multiobjective Optimization Problems (MOPs) was suggested by Rosenberg in his dissertation as early as 1967 [1]. However, until 1985, the first genetic algorithm for MOPs, namely VEGA, was proposed by Shaffer [2]. Because of the deficiencies of VEGA, Multiobjective Evolutionary Algorithms (MOEAs) have been paid more and more attention. Two generations of Evolutionary Multiobjective Optimization have been classified by Coello Coello [3]. The first generation (1985-1998) emphasizes the simplicity of algorithms, where the most representative are NSGA [4], NPGA [5] and MOGA [6]. The second generation started when elitism became a standard mechanism, which was firstly adopted by Zitzler [7] in SPEA. In this generation, efficiency is stressed and the most representative are SPEA [7], SPEA2 [8], PAES [9] and NSGA-II [10]. Meanwhile, the ε-dominance MOEAs [11], particle swarm optimization [12] and differential evolution [13] for MOPs are also proposed in this generation. These MOEAs are mainly for unconstrained MOPs. However, the real-world MOPs are often with constraints. To solve these problems, the crucial problem is how to handle the constraints in MOPs, i.e., how to balance the search between the feasible and infeasible regions.

Runarsson and Yao [14] proposed the search biases strategy and introduced the differential evolution to their algorithm [15] in order to achieve a good compromise

T.-D. Wang et al. (Eds.): SEAL 2006, LNCS 4247, pp. 318–327, 2006.

between feasible and infeasible regions in constrained single objective optimization. In this paper, the search biases strategy is introduced to solve constrained MOPs. Firstly, two novel schemes of selecting the current best solutions for multiobjective differential evolution (MODE) in constrained MOPs are proposed. And then a hybrid of MODE and GA with the (N+N) framework for constrained MOPs is given. Finally, the hybrid algorithm is implemented on NSGA-II [10] with the two schemes respectively, and is compared with a state-of-the-art MOEA, i.e., constrained NSGA-II (CNSGA-II) [20].

2 Preliminary

2.1 Problem Definition

Definition 1 (***Constrained MOP***). A general constrained MOP includes a set of n decision variables, a set of m objective functions, and a set of p inequality constraints (equality ones may be approximated by inequalities [14]). The goal of optimization is

$$\begin{cases} \min f(\mathbf{x}) = \{f_1(\mathbf{x}), f_2(\mathbf{x}), \cdots, f_m(\mathbf{x})\} \\ s.t. g_i(\mathbf{x}) \le 0, i = 1,2,\cdots, p \end{cases} . \tag{1}$$

Where $\mathbf{x}$ is the ***decision vector*** in decision space $\mathbf{X}$, and $f(\mathbf{x})$ is the ***objective vector*** in objective space $\mathbf{Y}$.

In constrained optimization problems, p constraints are usually transformed to a constraint violation function, which is defined in (2).

$$\varphi(g(\mathbf{x})) = \sum_{i=1}^{p} w_i (\max(g(\mathbf{x}),0))^{\beta} \tag{2}$$

Where the exponent β is usually 1 or 2 and the weights w_i>0 (j=1,…,p) would be tuned during search (β=1 and **w=1** in this study). And the feasible set $\mathbf{X}_f$ is defined as the set of decision vector $\mathbf{x}$ which satisfy the p constraints, i.e., $\varphi(g(\mathbf{x}))=0$.

Definition 2 (***Pareto Dominance*** "$\prec$"). For any two decision vectors **a** and **b,**

$$\mathbf{a} \prec \mathbf{b} \Leftrightarrow \forall 1 \le i \le m \; f_i(\mathbf{a}) \le f_i(\mathbf{b}) \wedge \exists 1 \le j \le m \; f_j(\mathbf{a}) < f_j(\mathbf{b}) \; . \tag{3}$$

The non-dominated set P in $\mathbf{X}_f$ is ***Pareto-optimal set*** (***POS***) and the set $f(P)$ is ***Pareto-optimal front*** (***POF***).

2.2 Constraint Pareto Dominance

Among constraint handling methods for MOPs, the most promising one is the ***Constraint Pareto Dominance*** proposed by Deb [10] [20], which is defined as follows.

Definition 3 (***Constraint Pareto Dominance*** "$\prec_c$") For any two decision vectors **a**, **b,**

$$\mathbf{a} \prec_c \mathbf{b} \Leftrightarrow \begin{cases} (\mathbf{a} \prec \mathbf{b} \wedge \varphi(g(\mathbf{a})) = \varphi(g(\mathbf{b})) = 0) \\ \mathbf{or} (\varphi(g(\mathbf{a})) = 0 \wedge \varphi(g(\mathbf{b})) > 0) \\ \mathbf{or} \; (\varphi(g(\mathbf{b})) > \varphi(g(\mathbf{a})) > 0) \end{cases} . \tag{4}$$

3 Hybrid of MODE and GA for Constrained MOPs

3.1 Differential Evolution

Storn and Price [16] proposed the differential evolution method which is a simple and efficient adaptive scheme for global optimization over continuous spaces, and gave two most promising schemes of differential evolution. Differential Evolution (DE) is a population-based evolutionary algorithm with simple mutation and crossover operators to create next generation. DE has similarities with traditional Evolutionary Algorithms. However, it doesn't employ binary encoding like a simple GA and doesn't utilize a probability density function to self-adapt its parameters like an ES [21].

Runarsson and Yao [14] suggested the search biases in constrained single objective optimization, and proposed the corresponding method as follows. When creating the next generation based on ES (μ, λ), λ individuals are generated according to

$$\mathbf{x}'_k \leftarrow \mathbf{x}_i + \gamma(\mathbf{x}_1 - \mathbf{x}_{i+1}) \ . \tag{5}$$

Where $\mathbf{x}_1$ denotes the top one after stochastic ranking for the whole individuals, i.e., the ***current best solution***, $\mathbf{x}_i$ and $\mathbf{x}_{i+1}$ are random samples from the population, and γ is the parameter of search step length. It is suggested by the authors [19] that (5) can be deduced from the second scheme of the differential evolution method in [16]. And (5) has obtained satisfying results for constrained single objective optimization [14].

3.2 Two Schemes of MODE

According to (5), λ individuals generated newly are close to $\mathbf{x}_1$, which denotes the ***current best solution*** in the population. However, there are usually more than one solutions in the non-dominated set of MOPs. Furthermore, the diversity of population is crucial for MOPs, which determines whether the pure POF could be found or not. Therefore the diversity of population and the pureness of the non-dominated set will be affected seriously if (5) is applied to constrained MOPs directly. So how to choose the "***current best solutions***" in MOPs becomes a key problem.

In order to preserve the population diversity, the "***current best solutions***" should be extended to a solution set, which could denote the main search biases in MOPs. A simple way is to select the boundaries of current non-dominated solutions as the "***current best solutions***". So the number of "***current best solutions***" is no more than the dimension of objective space. Individuals will be generated by sampling the boundaries uniformly, which is described in Algorithm 1.

Algorithm 1. B-Scheme (Boundaries as the ***current best solutions***)

1. **for** $k = 1$ **to** αN **do**
2. $\mathbf{x}_{best}$=*Uniform_Sample*(*Boundary*(**Non_Dominated_Set**))
3. i=*rand*[1,*N*]
4. $\mathbf{x}'_k \leftarrow \mathbf{x}_i + \gamma(\mathbf{x}_{best} - \mathbf{x}_{i+1})$
5. **end for**

In Algorithm 1, *N* denotes the size of population and α is the percentage of the individuals generated by B-Scheme. The αN individuals are close with the boundaries of current non-dominated set. Then the boundary search ability will be improved, and the population diversity could be maintained. So the pure POF will be located with larger probability.

However, it may be still hard to find the pure POF just by enhancing the boundary search ability when the distribution of POF is non-continuous. Because the boundaries of non-dominated set can only represent part of the search biases in such situations. To solve this problem, the representative individuals should be picked out from current population, and offspring are generated by differential evolution around the representatives to improve the search ability. So a better distribution of non-dominated set could be obtained. In this paper, a fitness function based on constraint Pareto dominance and crowding distance [10] is proposed to pick out the representative individuals, which is given as Algorithm 2.

Algorithm 2. Representative Individuals Selection

1. Split the population P according to "$\prec_c$", and get $P = F_1 \cup \cdots \cup F_k$
2. Calculate the crowding distances I of P: the feasible individuals' values are calculated by the definition in [10] while the infeasible ones' values are set to 0
3. Calculate the fitness of the individuals in P: $fitness(\mathbf{x}) = i + 1/(2 + I(\mathbf{x})), \mathbf{x} \in F_i$
4. Sort P by fitness values and the top M are selected as the representatives

The representative individuals selected by Algorithm 2 are feasible solutions or infeasible ones with smaller constraint violations. The diversity of these representative individuals is good because the crowding distance is employed. And then the offspring generated by differential evolution are close to these representatives. So the search ability to feasible region during evolution will be improved and the diversity of the offspring will be better. Treating the M representative individuals (**RI**) as the "***current best solutions***", the offspring will be generated according to Algorithm 3.

Algorithm 3. R-Scheme (Representative Individuals as the ***current best solutions***)

1. **for** k = 1 **to** αN **do**
2. $best$=mod(k-1, M)+1
3. i=$rand$[1,N]
4. $\mathbf{x}'_k \leftarrow \mathbf{x}_i + \gamma(\mathbf{RI}_{best} - \mathbf{x}_{i+1})$
5. **end for**

According to the definition of crowding distance in [10], the I values of boundary individuals in each layer are ∞. Therefore, Algorithm 1 and Algorithm 3 are identical when M equals the number of the non-dominated set boundaries. But when M is less than the number of the boundaries, the representatives are part of the boundary individuals. Here the representatives may not denote the main search biases, and the diversity may be deteriorated in this condition. Thus the value of M should be greater than the number of non-dominated set boundaries to make a distinct difference otherwise the performance of R-Scheme will be similar to B-Scheme or even worse.

3.3 Hybrid Algorithm with the Framework of NSGA-II

The hybrid algorithm (DE-MOEA) proposed in this paper is based on the (N+N) framework of NSGA-II [10]. And N individuals in the next generation are created as: αN individuals are generated by MODE and the left ones are by genetic operators (crossover and mutation) like NSGA-II. DE-MOEA is described in Algorithm 4.

Algorithm 4. DE-MOEA with (N+N) Framework

1. Initialization: create the initial population P_0, t=0, $|P_0|=N$
2. Evaluate the population P_t
3. P_{t+1}=***generate_next_pop***(P_t)
3.1. Calculate the fitness of individuals in P_t according to Algorithm 2
3.2. Sort P_t by the fitness values and select the top N as population Q_t
3.3. Generate αN individuals with MODE on Q_t and put them to P_{t+1}
3.4. Generate $(1-\alpha)N$ individuals from Q_t with binary tournament selection, crossover and mutation, and put the offspring to P_{t+1}
3.5. $P_{t+1}=P_{t+1}+Q_t$
4. $t=t+1$, if termination satisfied then output non-dominated set, else go to step 2

In Algorithm 4, the MODE in step 3.3 can be implemented by Algorithm 1 or Algorithm 3 and the corresponding algorithms are denoted as HBGA and HRGA.

In NSGA-II the termination is satisfied when the number of already split individuals is not less than N. However, it seems that it is needed to split the entire population in DE-MOEA, but actually it doesn't. For any two individuals $\mathbf{a}\in F_i, \mathbf{b}\in F_j (i \neq j)$

$$i \leq fitness(\mathbf{a}) < i+1,\ j \leq fitness(\mathbf{b}) < j+1\ . \tag{6}$$

So (7) follows easily

$$fitness(\mathbf{a}) < fitness(\mathbf{b}) \Leftrightarrow i < j\ . \tag{7}$$

From (7) the terminating condition of splitting population in DE-MOEA is the same as NSGA-II, so the time complexity of algorithm 4 is equal to the NSGA-II.

4 Experimental Results and Discussions

4.1 Test Functions and Performance Measures

In this section, 4 benchmark functions (CTP1, CTP2, CTP6 and CTP7) are chosen from [20] to evaluate the performance of DE-MOEA, which are described in (8), (9).

$$\text{CTP1}\begin{cases}\text{Min. } f_1(x)=x_1, f_2(x)=g(x)\exp(-f_1(x)/g(x)) \\ s.t.\ c_j(x)=f_2(x)-a_j\exp(-b_j f_1(x))\geq 0,\ j=1,2,\cdots,J\end{cases} \tag{8}$$

$$\begin{matrix}\text{CTP2}\\ \text{CTP6}\\ \text{CTP7}\end{matrix}\begin{cases}\text{Min. } f_1(x)=x_1, \text{Min. } f_2(x)=g(x)\left(1-f_1(x)/g(x)\right) \\ s.t.\ c(x)=\cos(\theta)\left(f_2(x)-e\right)-\sin(\theta)f_1(x)\geq \\ a\,|\sin(b\pi(\sin(\theta)(f_2(x)-e)+\cos(\theta)f_1(x))^c)|^d\end{cases} \tag{9}$$

Where the decision space of each function has 5 dimensions, which are defined as: $0 \le x_1 \le 1$, $-5 \le x_{2,3,4,5} \le 5$, and $g(x) = 41 + \sum_{i=2}^{5}(x_i^2 - 10\cos(2\pi x_i))$. For CTP1, J=2, $a_{1,2}$= (0.858, 0.728), $b_{1,2}$= (0.541, 0.295), and the parameters chosen to the different CTP2, CTP6 and CTP7 functions are listed in Table 1.

Table 1. Parameter Settings in CTP2, CTP6 and CTP7

Function	θ	a	b	c	d	e
CTP2	-0.2π	0.2	10	1	6	1
CTP6	0.1π	40	0.5	1	2	-2
CTP7	-0.05π	40	5	1	6	0

Performance measures for MOPs are analyzed and classified by Zitzler [17]. And the unary indicator $D1_R$ [17] and binary indictor V(A, B) [18] are selected here, which are defined as follows.

$$D1_R(\mathbf{X}) = \sum_{\mathbf{a} \in \mathbf{X}^*} \min\{\|\mathbf{a} - \mathbf{b}\|; \mathbf{b} \in \mathbf{X}\} / |\mathbf{X}^*| , \tag{10}$$

where $\mathbf{X}^*$ denotes a reference set (1,000 uniform samples from POF in this paper), and $\mathbf{X}$ is the non-dominated set found by the algorithm.

The definition of V(A, B) is the percentage of objective space dominated exclusively by A in the smallest hypercube which contains the both non-dominated set A and B. Like [18], 50,000 Monte Carlo samples are taken to calculate the values. When calculating the hypercube we desire that the solutions with the first objective less than 10^{-7} should be rejected in order to obtain more distinct differences between the two non-dominated sets by finite samples. The two situations are illustrated in Fig. 1 and Fig. 2, where the two non-dominated sets are obtained from CNSGA-II [20] and HBGA on CTP1 in a run.

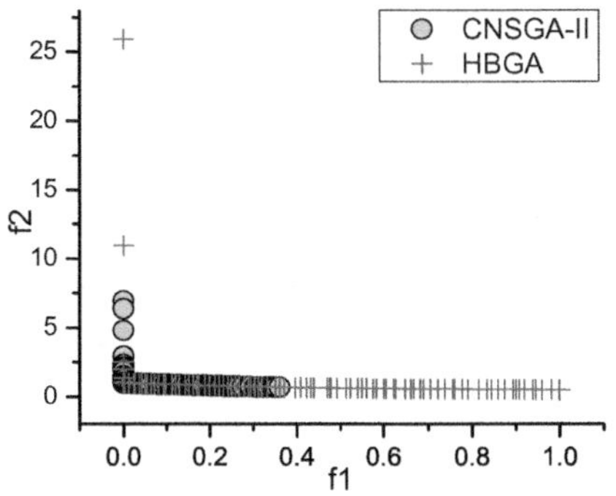

Fig. 1. Two non-dominated set without rejection

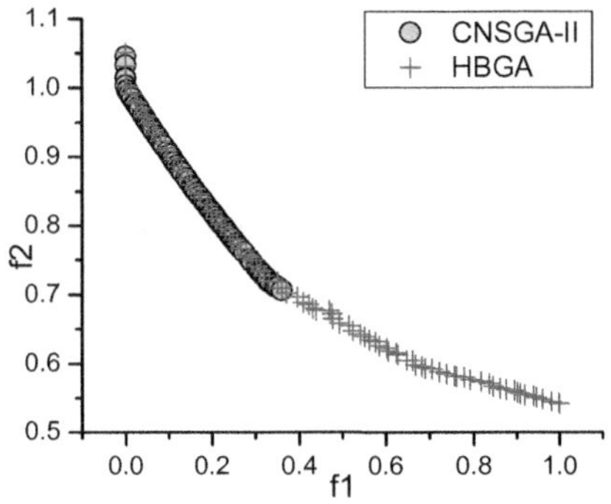

Fig. 2. Two non-dominated set with rejection

From Fig. 1, it can be found that there exist a few solutions with very large f_2 values and quite little f_1 values in each set. This will influence the quality of the V(A,B) seriously because the great majority of samples are located in the area which is dominated by both sets and few samples can find out the slight but significant differences between the both sets. The values of V(A,B) in Fig.1 is (0.0000%, 0.2180%), so we can assert that HBGA outperforms CNSGA-II on CTP1 [17]. However, from the corresponding value (0.1900%, 11.8260%) in Fig.2, this assertion can't come into existence. This is because the differences between the both sets can be captured by this appropriate scale in Fig. 2 with 50,000 samples. Meanwhile, the rejection influences the experimental results little. The f_1 values of all the benchmark functions are ranged from 0 to 1, and then the probability of sample points in the rejected area is 10^{-7}. So the mean and standard deviation values of the times in the rejected area by 50,000 samples are 0.005 and 0.0707 respectively, which can hardly influence the final experimental results. Therefore, this rejection is employed here in order to obtain distinct results without increasing sample times.

4.2 Results and Discussions

To the best of our knowledge, CNSGA-II [20] is the most promising method for constrained MOPs and is selected to compare with DE-MOEA (HBGA and HRGA). All the algorithms are performed in Matlab 7.0, and the source code may be obtained from the author upon request. Real-coded GA (simulated binary crossover "SBX" and polynomial mutation "PM") is adopted in the implementation and the parameter values are the same as [20], which are listed in Table 2.

Table 2. Parameter Settings for CNSGA-II and DE-MOEA

N	α	p_c	p_m	SBX η_c*	PM η_m*	FE*
100	10%	0.9	$1/n$*	20	20	50,000

* n denotes the dimension of decision space, and FE is the total number of function evaluations.

Table 3. Mean and Standard Deviation of the Unary Indicator $D1_R$

Function	CTP1	CTP2	CTP6	CTP7
HBGA	**0.0043** (0.0003)	**0.0023** (0.0003)	0.2234 (0.9297)	**0.0097** (0.0125)
HRGA	**0.0056** (0.0007)	**0.0037** (0.0021)	0.1103 (0.6587)	**0.0241** (0.0932)
CNSGA-II	0.0970 (0.0570)	0.1889 (0.1236)	0.3525 (1.1969)	0.0610 (0.0847)

Results highlighted in bold signify significantly better than CNSGA-II at α=0.05 by a two-tailed test.

Table 4. Mean and Standard Deviation of the Binary Indicator V(A, B)

	CTP1	CTP2	CTP6	CTP7
V(HBGA, CNSGA-II)	**4.4609%** (0.0349)	**12.5761%** (0.0892)	5.4083% (0.1757)	**2.6680%** (0.0467)
V(CNSGA-II, HBGA)	0.2617% (0.0007)	0.0497% (0.0002)	3.2613% (0.1345)	0.1681% (0.0024)
V(HRGA, CNSGA-II)	**4.0455%** (0.0346)	**12.1040%** (0.0881)	5.4424% (0.1766)	**2.9616%** (0.0506)
V(CNSGA-II, HRGA)	0.3684% (0.0014)	0.0961% (0.0006)	1.7138% (0.0965)	0.1814% (0.0021)
V(HBGA, HRGA)	**0.4046%** (0.0015)	**0.2008%** (0.0023)	1.7404% (0.0958)	0.4740% (0.0109)
V(HRGA, HBGA)	0.2000% (0.0009)	0.0442% (0.0003)	3.5931% (0.1382)	0.7050% (0.0168)

Results highlighted in bold signify significantly better than the other at α=0.05 by a two-tailed test.

The values of search step length parameter γ in HBGA and HRGA are set to 1.1 and 0.85 respectively. And the number M in HRGA is set to 5, i.e., 5% of the population size. All the algorithms for each benchmark function perform 100 independent runs, and the initial populations of all the algorithms in each run are the same for fair comparison. Table 3 and Table 4 show the means and standard deviations of the $D1_R$ and V(A, B) indicators obtained by all the algorithms, where the standard deviations are in the parenthesizes.

From Table 3 and Table 4 it can be observed that HBGA and HRGA have better results than CNSGA-II in all the 4 benchmark functions, and especially the results are significant better except CTP6. It can also be seen that HBGA performs better than HRGA in CTP1 and CTP2 while the latter has better results in CTP6.

To illustrate the pureness of the POF found by each algorithm, the statistical values in 100 independent runs are listed in table 5.

Table 5. Pureness Statistic of POF Found by the Three Algorithms

Benchmark Function	Algorithm	Number of runs produce pure POF	Numbers of runs produce partial POF	Number of runs produce local Pareto front
CTP1	HBGA	100	0	0
	HRGA	100	0	0
	CNSGA-II	4	96	0
CTP2	HBGA	100	0	0
	HRGA	100	0	0
	CNSGA-II	10	90	0
CTP6	HBGA	94	2	4
	HRGA	97	1	2
	CNSGA-II	89	4	7
CTP7	HBGA	48	52	0
	HRGA	80	20	0
	CNSGA-II	15	85	0

From table 5, for CTP1 and CTP2, HBGA and HRGA converge to the pure POF with probability 1 in 100 runs, while the probability of CNSGA-II is not greater than 10%. For CTP6, all the three algorithms converge to the partial POF or local Pareto front with a low probability, but it is somewhat lower for HBGA and HRGA. For CTP7 it is hard for CNSGA-II to find the pure POF while the situation becomes easier for HBGA, and especially for HRGA.

The search ability in the boundaries of non-dominated set is improved in both HBGA and HRGA, so it is easier for both HBGA and HRGA to locate the pure POF in CTP1, which has a continuous POF. The CTP2 and CTP7 have non-continuous POF, and the intervals in CTP7's POF are much larger. So it could be still efficient to find the pure POF of CTP2 for HBGA, but it is not the case in CTP7. However, the HRGA could solve the problem by choosing representatives from population for differential evolution. And the experimental results of CTP2 and CTP7 give an evidence for this explanation. CTP6 has several local Pareto fronts for the infeasible holes towards the Pareto-optimal region in objective space. By introducing the different evolution, a better tradeoff between feasible and infeasible regions could be achieved. So the number of convergence to local Pareto front is less than CNSGA-II. Compared with HBGA, the HRGA has better diversity during search for choosing more "current best solutions" for differential evolution, and the probability of its getting into local

Pareto front is much lower than HBGA. However, the approximating and diversity of non-dominated set are two (possible) conflicting objectives [8]. So when the diversity exceeds the actually required, the approximating will be deteriorated. This could be the reason why HBGA performs better than HRGA on CTP1 and CTP2.

With the above results and analysis, the algorithm DE-MOEA proposed in this paper has superior performance compared with CNSGA-II, especially in the distribution of the non-dominated set.

5 Conclusion

This paper proposes two novel schemes of selecting the current best solutions for MODE. And then based on the search biases strategy suggested by Runarsson and Yao, a hybrid algorithm of MODE and GA is put forward here for constrained MOPs. We implement the hybrid algorithm based on NSGA-II with the two schemes of MODE respectively, named HBGA and HRGA. HBGA and HRGA are compared with CNSGA-II on 4 benchmark functions constructed by Deb. Experimental results show that the quality of non-dominated set obtained by the both algorithms is better than that of CNSGA-II on all the benchmark functions. The future work is to apply the hybrid algorithm to more complex problems and applications.

Acknowledgement

This work is partially supported by the National Natural Science Foundation of China (Grant No. 60428202), and the authors would like to thank Prof. X. Yao from University of Birmingham for his helpful comments.

References

1. R.S. Rosenberg: Simulation of genetic populations with biochemical properties. Ph.D. thesis, University of Michigan, Ann Harbor, Michigan, 1967
2. J. David Schaffer: Multiple objective optimization with vector evaluated genetic algorithms. In Genetic Algorithms and their Applications: Proceedings of the First International Conference on Genetic Algorithms, 93–100, Lawrence Erlbaum, 1985
3. C.A. Coello Coello: Evolutionary multi-objective optimization: a historical view of the field. IEEE Computational Intelligence Magazine, 1(1): 28-36, Feb. 2006
4. N. Srinivas, K. Deb: Multiobjective optimization using nondominated sorting in genetic algorithms. Evolutionary Computation, 2(3): 221–248, Fall 1994
5. J. Horn, N. Nafpliotis, and D.E. Goldberg: A niched pareto genetic algorithm for multiobjective optimization: In Proceedings of the 1st CEC, 1: 82–87, June 1994
6. C.M. Fonseca, P.J. Fleming: Genetic algorithms for multiobjective Optimization: Formulation, discussion and generalization. In Proceedings of the Fifth International Conference on Genetic Algorithms, 1993, 416–423
7. E. Zitzler, L. Thiele: Multiobjective evolutionary algorithms: A comparative case study and the strength pareto approach. IEEE Trans. Evol. Comput., 3(4): 257–271, Nov. 1999

8. E. Zitzler, M. Laumanns, L. Thiele: SPEA2: Improving the strength pareto evolutionary algorithm. In EUROGEN 2001. Evolutionary Methods for Design, Optimization and Control with Applications to Industrial Problems, 2002, 95–100
9. J.D. Knowles, D.W. Corne: Approximating the nondominated front using the pareto archived evolution strategy. Evolutionary Computation, 8(2): 149–172, 2000
10. K. Deb, A. Pratap, S. Agarwal, and T. Meyarivan: A fast and elitist multiobjective genetic algorithm: NSGA–II. IEEE Trans. Evol. Comput., 6(2): 182–197, Apr. 2002
11. M. Laumanns, L. Thiele, K. Deb, and E. Zitzler: Combining convergence and diversity in evolutionary multi-objective optimization. Evolutionary Computation, 10(3): 263–282, Fall 2002
12. C.A. Coello Coello, G. Toscano Pulido, and M. Salazar Lechuga: Handling multiple objectives with particle swarm optimization. IEEE Trans. Evol. Comput., 8(3): 256–279, June 2004
13. T. Robič, B. Filipič: DEMO: Differential Evolution for Multiobjective Optimization. EMO 2005, 520-533
14. T. P. Runarsson, X. Yao: Search Biases in Constrained Evolutionary Optimization. IEEE Trans. Syst. Man Cybern. Part C-Appl. Rev., 35(2): 233-243, May 2005
15. T. P. Runarsson, X. Yao: Stochastic Ranking for Constrained Evolutionary Optimization. IEEE Trans. Evol. Comput., 4(3):284-294, Sep. 2000
16. R. Storn, K. Price: Differential evolution-A simple and efficient heuristic for global optimization over continuous spaces. J. Global Optimiz., 11(4): 341–359, Dec. 1997
17. E. Zitzler, L. Thiele, M. Laumanns, C.M. Fonseca, and V. Grunert da Fonseca: Performance Assessment of Multiobjective Optimizers: An Analysis and Review. IEEE Trans. Evol. Comput., 7(2): 117–132, Apr. 2003
18. J.E. Fieldsend, R.M. Everson, and S. Singh: Using unconstrained elite archives for multiobjective optimization. IEEE Trans. Evol. Comput., 7(2): 305-323, June 2003
19. M. Zhang, H.T. Geng, W.J. Luo, L.F. Huang, X.F. Wang: A Novel Search Biases Selection Strategy for Constrained Evolutionary Optimization. CEC 2006, to appear
20. K. Deb, A. Pratap, T. Meyarivan: Constrained Test Problems for Multi-objective Evolutionary Optimization. EMO 2001, 284-298
21. E. Mezura-Montes, J. Velázquez-Reyes, and C.A. Coello Coello: Promising infeasibility and multiple offspring incorporated to differential evolution for constrained optimization. GECCO 2005, 225-232

The Research on Repurchase Announcements of Open-Market Stock

Weimin Tang[1] and Juanjuan Peng[2]

[1] Tang Weimin, Wuhan University of Technology, School of Science,
Zhongnan University of Economics and Law , School of Accountancy
430063 Hubei, China
Tangweimin001@126.com
[2] Peng Juanjuan, Wuhan University of Technology, School of Science,
430063 Hubei, China
Xiaqing19812004@yahoo.com.cn

Abstract. In a repurchase program, different firms can obtain different gains when they announce a stock repurchase, so a firm needs to know whether announcement is an optimal choice. This paper presents a dynamic, two-player game model with imperfect information, analyzes its further equilibrium condition. The model shows that repurchase announcements have various effects on the firms. High-earnings firms choose to make announcements, whereas low-earnings ones are inclined not to announce. Finally, it gives empirical test for the model to validate the conclusion according to the data of China.

1 Introduction

The repurchase program of open-market stock is commonly used for distributing corporate earnings to shareholders. In a repurchase program, a firm buys back its shares in the market over a period time from months to years. During recent years, repurchase programs have become increasingly popular relative to other common forms of payout, approximately accounting for 90% of the dollar value of all announced repurchases[1]. Stock repurchases have become an important financial policy for listing firms these years, they can adjust the market if market undervalues the stock prices of a firm. If companies are unsatisfied with the price of shares, the firms usually choose announcing the repurchase announcements.

Grinblatt and Hwang(1989)[2] provide a foundation for a repurchase-signaling game. They show that withholding part of a new share issue can signal greater earnings, because entrepreneurs of firms with high earnings derive less marginal disutility from the additional risk of holding more firm's stock. Risk-averse entrepreneurs of low-profit firms derive greater marginal disutility from the added risk and so optimal repurchase is less than owners of high-profit firms. So the low-profit firms don't want to simulate high-profit firms because of their aversion to the added risk. Therefore, in contrast to the low-profit firms, high-profit firms are inclined to announce.

William J McNally (1999) [3]constructs a signaling model in open market repurchase. He simulates the effects of a repurchase on the shareholders when repurchase proportion is continuous and concludes that if insiders refrain from tendering, their

T.-D. Wang et al. (Eds.): SEAL 2006, LNCS 4247, pp. 328 – 335, 2006.

choice of the repurchase proportion will reveal much to the market: 1. firms that repurchase more have higher earnings, 2. riskier firms have higher earnings, and 3.firms where insiders have a greater ownership stake have higher earnings, ceteris paribus. McNally demonstrates that signaling and agent problems can explain why firms choose stock repurchase in perfect markets. This paper uses the model to prove that the utility of high earnings type is higher if it signals than if it doesn't signal based on the signaling theory and agent problems, it explains why firms choose a specific payout form(open market stock repurchase) in particular, and extend his conclusion. On the other hand, high earnings firms choose to make a repurchase announcement of open-market and a low earnings firms choose not to announce. But repurchase announcements is a discrete signal here.

2 The Model of Repurchase

2.1 The Variables of Economics

Assumption 1. The open market repurchase is modeled as a dynamic, two-player game with imperfect information.

Assumption 2. The market is assumed to be risk neutral, and the risk free rate is assumed to be zero.

Assumption 3. Figure1 summarizes the sequence of events and decisions over four dates.

2.2 The Analysis of Model (Figure1)

In the first state 0: A risk-averse entrepreneur controls an already existing firm that has cash of C . Most of the action occurs at the beginning: though nature reveals new level of u to firm, entrepreneur announces open market repurchase. Finally, market accepts or rejects, then announces its conditional valuation.

In the second state 1: The firm will generate the intermediate cash flow:

$$x_1 \sim N(0,\sigma^2) \tag{1}$$

In the third state 2: Uncertainty regarding price is resolved, then nature reveals u to market with probability π , and the owner have opportunity to trade.

In the fourth state 3: the firm generates terminal cash flow.

$$x_3 \sim N(u,\sigma^2) \tag{2}$$

The open market repurchase is modeled as a dynamic; the objective of the game is the valuation of these cash flows by the risk-neutral market.

In the first state, nature assigns the firm a new level of period 3 earnings, since all firms start with a common level of expected earning, $u > 0$. Because the intermediate cash flow has a mean of zero, it does not affect the market value of the firm. But the owner has to hold the stock over period 1 and experiences the uncertainty of the

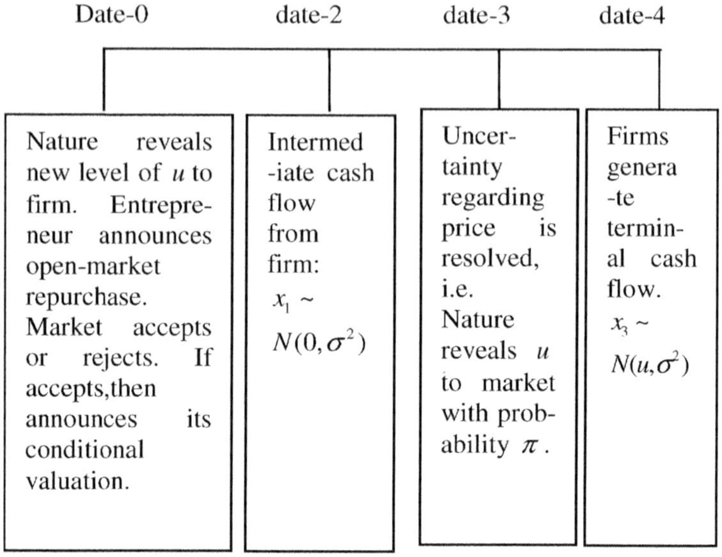

Fig. 1.

cash flow. The random cash flow at time period 1 is critical to the model because it is the source of risk associated with stock ownership. At the same time, the terminal cash flow is very important, because the dominant cash flow from the firm is the terminal payout at time period 3.It is as if the firm is liquidated with an expected payout of $E(x_3) = u$. In the model, we suppose that the owners are interested in maximizing the market value of the firm at the periods2.The entrepreneur essentially behaves as if he or she is going to sell his or her shares at time period 2. Owner 's objective is that they want to maximize expected utility on a date prior to the terminal payout (period 2). From the equation (1) we know that risk is reintroduced by having a period1(x_1) cash flow with a mean of zero and a variance equal to the variance of the terminal cash flow:

$$\mathrm{Var}\,(x_1) = \mathrm{Var}\,(x_3) = \sigma^2 \tag{3}$$

Entrepreneur's prior proportional holdings are α .The market also holds posterior beliefs about the firm's type, which is denoted $P(T \mid NCIB)$ .Entrepreneur announces (or no-announce) the repurchase signaling by the means of $NCIB$ (Normal Course Issuer Bids) . William McNally proved that $NCIB$ was a discrete sign by the conditional event study model, and market calculated its valuation of the firm. The set of expected earnings types are generated from set T , good earnings G , bad earnings N :

$$T = G \cup N \text{ , } N = \{u \mid u = u_N\} \text{ , } G = \{u \mid u_L \le u \le u_H\} \text{ and } u_L < u_H \tag{4}$$

Earnings in subset G can be interpreted as high expected earnings, and earnings in subset N can be interpreted as low expected earnings. At this moment, entrepreneur will choose $NCIB$ to increase their utility. Entrepreneur's utility is dependent on the earnings of firm and the valuation of the firm by the market, and the market calculates its valuation of the firm by the announcements. Therefore, entrepreneur's utility can be denoted as $E[U(u, NCIB)]$, when $NCIB = 1$ for an announcement and $NCIB = 0$ for no announcement. In following equilibrium, market can infer the firm's type. The market observes the repurchase announcements, when the firm in subset G , the price of share can be increased because of signaling, while the firm in subset N , there will be no change about the price. Where u_L is the lower bound and u_H is the upper bound, there is no fixed value.

3 Equilibrium and Market's Strategy

We will aim at the model of open-market stock repurchase signaling, proving that the equilibrium exists by using the mean/variance expected utility function, and obtaining the market's strategy.

Because the market value of the firm is equal to the sum of cash and expected earnings-the market's strategy function, it is denoted by:

$$V_0(\delta) = C + V(NCIB) \tag{5}$$

Making an unbiased valuation of the firm through drawing an unbiased inference about the firm's type is the market's objective. The entrepreneur maximizes his or her wealth expected utility of date 2 by choosing to announce $NCIB$, knowing that investors infer the value of the firm from their decisions. Before expressing the entrepreneur's objective function, some notation and assumptions are explained. $\alpha =$ entrepreneur's prior proportional holdings

$$\alpha \in \{\alpha \mid \alpha_L \le \alpha < 1\} \tag{6}$$

π =probability of truth revelation at date 2. Where π is assumed to be quite small, $\pi < \alpha_L$ (7)

$$\tilde{V}_2 \sim \begin{cases} C + V(NCIB), 1-\pi \\ C + u, \pi \end{cases} \tag{8}$$

Date 2 firm values, $\tilde{V}_2$ is equal to the true value with probability π , and equal to the market valuation with probability $1-\pi$ .The mean and variance of this random variable are:

$$E(\tilde{V}_2) = (C + V(NCIB))(1-\pi) + (C+u)\pi \tag{9}$$

$$\begin{aligned} \operatorname{Var}(\tilde{V}_2) &= E(\tilde{V}_2)^2 - (E\tilde{V}_2)^2 \\ &= \pi(1-\pi)V^2(NCIB) + 2(1-\pi)\pi(2C+u)V(NCIB) \\ &+ C\pi(5C + 4u - 3\pi C - 2u\pi - 1) \end{aligned} \tag{10}$$

On date 2, the entrepreneur's wealth is equal to the value of his or her stock holdings plus the random date 1 intermediate cash flow:

$$W_2 = \alpha(\tilde{V}_2 + x_1) \tag{11}$$

Substituting the (9), (10), and (11) into the mean/variance expected utility function:

$$E[U(W)] = E(W) - (b/2)\,\mathrm{Var}\,(W) \tag{12}$$

We can obtain the entrepreneur's objective function:

$$\begin{aligned} E[U(W_2)] = &-\frac{b\alpha^2}{2}\pi(1-\pi)V^2(VCIB) + \alpha(1-\pi)[1-b\alpha\pi \\ &(2C+u)]V(NCIB) + \{-\frac{b\alpha^2}{2}[C\pi(5C+4u- \\ &3\pi C - 2u\pi - 1) + \sigma^2] + \alpha C(1-\pi) + \alpha\pi(C+u)\} \end{aligned} \tag{13}$$

Theorem. If b satisfies

$$b > [\alpha\pi(3u_L + u_N + 4C)]^{-1} \tag{14}$$

Then beliefs and strategies constitute a Bayesian perfect separating equilibrium.

Firm's strategy

1 Type= G announce ($NCIB = 1$)
2. Type= N do not announce ($NCIB = 0$)

Market's beliefs and strategy:

1. $P(T = G \mid NCIB = 1) = 1 \quad V(NCIB = 1) = u_L$
2. $P(T = N \mid NCIB = 0) = 1 \quad V(NCIB = 0) = u_N$

Proof: 1) If the high-earnings type announces repurchase, the returns is higher than it doesn't announce, high earnings type prefers to signal. Substituting the objective function (13) into the

$$E[U(NCIB = 1, u_G)] > E[U(NCIB = 0, u_G)]$$

$$\begin{aligned} &-\frac{b\alpha^2}{2}\pi(1-\pi)u_L{}^2 + \alpha(1-\pi)[1-b\alpha\pi(2C+u_G)]u_L + \{-\frac{b\alpha^2}{2}[C\pi \\ &(5C+4u_G-3\pi C-2u_G\pi-1)+\sigma^2] + \alpha C(1-\pi) + \alpha\pi(C+u_G)\} \\ &> -\frac{b\alpha^2}{2}\pi(1-\pi)u_N{}^2 + \alpha(1-\pi)[1-b\alpha\pi(2C+u_L)]u_N + \{-\frac{b\alpha^2}{2}[\\ &C\pi(5C+4u_G-3\pi C-2u_G\pi-1)+\sigma^2] + \alpha C(1-\pi) + \alpha\pi(C+u_G)\} \end{aligned}$$

Therefore

$$\begin{aligned} &-\frac{b\alpha^2}{2}\pi(1-\pi)u_L{}^2 + \alpha(1-\pi)[1-b\alpha\pi(2C+u_G)]u_L \\ &> -\frac{b\alpha^2}{2}\pi(1-\pi)u_N{}^2 + \alpha(1-\pi)[1-b\alpha\pi(2C+u_G)]u_N \end{aligned}$$

Since $u_L \le u_G \le u_H$, we have

$$b > [\alpha\pi(3u_L + u_N + 4C)]^{-1}$$

So this inequality holds $\forall u \in G$ if the above condition on b holds and if $u_N < u_L$, then the strategies will be held.

2) If a low earnings type's utility is higher if it refrains from signaling than if it signals to simulate the high earnings type, then for the low earnings type, we have:

$$E[U(NCIB = 0, u_N)] > E[U(NCIB = 1, u_N)] \quad (15)$$

Or

$$u_N > (3b\alpha\pi)^{-1}[2 - b\alpha\pi(4C + u_L)] \quad (16)$$

We have

$$-\frac{b\alpha\pi}{2}(u_L + u_N)(u_N - u_L) + (u_N - u_L) > b\alpha\pi(2C + u_N)(u_N - u_L)$$

Since $u_N < u_L$, so

$$1 - \frac{b\alpha\pi}{2}(u_L + u_N) > b\alpha\pi(2C + u_N)$$

therefore

$$u_N > (3b\alpha\pi)^{-1}[2 - b\alpha\pi(4C + u_L)]$$

So if this condition holds with the equilibrium beliefs and strategies, the beliefs and strategies will be held. That is the same as the high earnings firm.

From the equilibrium, we can know that false signaling increases the insider's uncertainty about future wealth, and correct signaling reduces that uncertainty. Insiders of high earnings firms receive a significant increase in expected wealth and reduce uncertainty about their future wealth if they signal, and insiders of unchanged earnings firms receives a smaller increase in expected wealth and an increase in uncertainty if they signal. If insiders are sufficiently risk averse, then the firms with unchanged earnings types finds out that the uncertainty associated with falsely signaling offsets the potentially greater wealth, and chooses not to simulate a high earnings type. All firms with high earnings types in subset G find out that it is worth to signal because that can increase their expected wealth and reduce the potential variability in their wealth. The high earnings firm's utility is higher if they make the announcements than that if they reject to announce, after they announces the signals, they can obtain returns from buying the cheap shares from the out stockholdings.

From the above analysis, we can know that the equilibrium will be held: that is high earnings firms are inclined to announce, and low earnings firms would better not announce.

4 Empirical

In this section, we make an empirical analysis, using the data from the securities business of China to test the signaling model's implications. The data comes from the

Wind ZIXUN. Because there are not so many companies which take repurchase measures in recently, it causes difficult for people to collect data ,so we take twelve companies which have announced the repurchases for example from 2005 to now ,we will make some simple statistical analysis for them. All the companies below are going on in this way which is open market stock repurchases.

Table 1. The Analysis of Data

Company	Cash-flow (yuan)	Return Rate(%)	Company	Cash-flow (yuan)	Return Rate(%)
000932	**1.5319**	**7.55**	**600165**	**0.0854**	**8.79**
600801	1.5321	-25.55	600521	0.0868	10.64
000090	1.5208	-17.62	600987	0.0867	7.35
600583	1.5198	-2.16	600455	0.0838	0.28
600726	**1.263**	**22.83**	**600810**	**0.0815**	**24.2**
600102	1.2812	12.59	000627	0.0818	9.84
600859	1.2669	9.47	000078	0.0817	10.32
600556	1.2572	19.01	000602	0.0818	19.79
600839	**0.6568**	**18.37**	**600315**	**0.01**	**7.96**
600727	0.6618	12.31	000509	0.0101	-13.46
000950	0.6614	6.91	000883	0.0095	-5.2
600367	0.662	13.45	600845	0.0092	6.42
600637	**0.3806**	**19.69**	**000848**	**-0.042**	**-0.44**
600966	0.3846	8.03	600984	-0.0394	3.4
002006	0.3837	2.07	000801	-0.041	15.67
600638	0.3788	14.18	000906	-0.0428	11.57
600031	**0.308**	**-7.67**	**600121**	**-0.0513**	**-5.29**
000068	0.3086	-12.3	002036	-0.0531	12.68
001696	0.3071	-9.12	600705	-0.0577	-1.51
000917	0.3097	-8.6	000691	-0.0453	-4.44
600220	**0.2059**	**12.44**	**600001**	**-0.1516**	**-31.66**
600095	0.2046	6.81	002020	-0.149	-5.61
600456	0.206	11.78	000967	-0.1493	-10.9
000635	0.2055	12.47	600689	-0.1557	1.06

Firstly we make some sequence for 1354 companies (SHANGHAI or SHENZHEN STOCK EXCHANGE) according to cash-flow per stock, then we define the first 451 companies (one third out of total companies) as high-quality company, at this time CFPS of company is over or equal to 0.2782. And the last 451 companies are low-quality companies, CFPS of company is under or equal to 0.013. In our samples, there are five companies in high-quality company, four companies in low-quality company and three companies between high-quality and low-quality company .so we can draw a conclusion that high-quality company is more apt to announce the repurchases.

In order to explain that the return of announcing the repurchases of high-quality companies is more than that of not announcing the announcements, we stochastically choose some companies which are equal to each repurchase company according to CFPS. We compared stock returns of these companies (total 48) in the same period of announcing the repurchases. Since financial market in chain is still defective maturity, there exist serious asymmetry between inside of the company and outside of the stockholder .So return calculated during the announcing the repurchases is based on the data before one month which are formal announcd by company .we can know that from the table1 below , (the overstriking stands for company which announcd repurchase, the not overstriking stands for three stochastically chosen company which not announcd repurchase), the return of announcing the repurchases of seven high-quality companies before is more than that of not announcing the repurchases ,but the return of announcing the repurchases of there low-quality companies is less than that of not announcing the repurchases, this is in agreement with theory above. There are three companies in the middle, the return of 600810 is the same to high-quality company, as for 600220 and 600165, the return of two companies is more and of one companies is less. The reason for leading to these phenomena is mainly that our sample is relative small and there is not accurate definition for high- and low-quality company. We not consider the coefficient of market risk when we calculate the rate of return else. Therefore, our date conclusion is not so perfect. So we are going to scale-up our sample and make some more particular demonstrative analysis in the future.

5 Summary and Conclusions

The stock repurchases of open market are paid more and more attention at present. Although many companies get large returns when they repurchase, many firms suffer from loss, because they repurchase blindly. So different firms should choose different strategies, if the bad firms simulate good ones to announce a repurchase program, they cannot deal with the firms' problems and they will suffer from larger cost of signaling. Therefore, in order to maximize the firm's utility, we should know the type of the firm and choose the optimal strategy. This paper adopts the William J McNally's (1999)[3] signal model, combines with the mean/variance expected utility function, analyses the same model from different aspects, and extends the conclusion. The model shows that repurchase announcements have various effects on the firms. High-earnings firms choose to make announcements, whereas low-earnings ones are inclined not to announce.

References

1. Bhattacharya.U. and A.Dittma:Costless versus Costly Signaling: Theory and Evidence.working paper, MIT and University of Michigan (2003)
2. M. Grinblatt and C.Y. Hang:Signaling and the Pricing of New Issues.Journal of Finance (June)(1997)393-420
3. William. J. McNally:Open market stock repurchase signaling.Financial Management.vol.28, (1999)55-67
4. Jacob Oded:Why do firms announce open-market repurchase programs?The Review of financial Studies,Vol.18 (2005)
5. Nobuyuki Isagawa:Open-market repurchases Announcements and Stock price behavior in inefficient markets. Financial Management (2002)5-20
6. Leland, H and D. Pyle:Information Asymmetries, Financial Structure and Financial Intermediation. Journal of Finance (May) (1979)371-388
7. William. J. McNally:Information Signaling or Agency Conflicts:What Explains Canadian Open Market Share Repurchase? (2000)

Infeasible Elitists and Stochastic Ranking Selection in Constrained Evolutionary Multi-objective Optimization

Huantong Geng, Min Zhang, Linfeng Huang, and Xufa Wang

Nature Inspired Computation and Applications Laboratory, Department of Computer Science and Technology, University of Science and Technology of China
htgeng@ustc.edu, zhangmin@mail.ustc.edu.cn,
lfhuang@mail.ustc.edu.cn, xfwang@ustc.edu.cn

Abstract. To handle the constrained multi-objective evolutionary optimization problems, the authors firstly analyze Deb's constrained-domination principle (DCDP) and point out that it more likely stick into local optimum on these problems with two or more disconnected feasible regions. Secondly, to handle constraints in multi-objective optimization problems (MOPs), a new constraint handling strategy is proposed, which keeps infeasible elitists to act as bridges connecting disconnected feasible regions besides feasible ones during optimization and adopts stochastic ranking to balance objectives and constraints in each generation. Finally, this strategy is applied to NSGA-II, and then is compared with DCDP on six benchmark constrained MOPs. Our results demonstrate that distribution and stability of the solutions are distinctly improved on the problems with two or more disconnected feasible regions, such as CTP6.

Index Terms. Constraint multi-objective optimization, infeasible elitists, stochastic ranking.

1 Introduction

Since Schaffer firstly applied evolutionary algorithms to solve MOPs in 1985 [1], multi-objective evolutionary algorithms (MOEAs) have amply demonstrated the advantage of using population-based search algorithms for solving multi-objective optimization problems [2]-[8]. In MOPs of varying degrees of complexities, elitist MOEAs have demonstrated their abilities in converging close to the true Pareto-optimal front and maintaining a diverse set of solutions, such as SPEA [2], PESA [3], SPEA2 [4], NSGA-II [5] and RDGA [6]. Despite all these rapid developments, there seem to be not enough studies concentrating on handling constraints. Constraint handling is a crucial part of real-world problem solving and it is time that MOEA researchers focus on solving constrained MOPs [10] [11].

Stochastic ranking strategy [9] [14] has solved constrained single-objective optimization problems successfully, whereas the strategy can't be applied to constrained MOPs directly. This is because the individuals are fully ordered in single-objective optimization but only partially ordered in multi-objective. Despite some MOEAs can transform partial order into full order by the fitness function designed, this kind of full order depends on the algorithm. In addition, the goal of multi-objective optimization is to achieve a set of Pareto optimal solutions and a good distribution of solutions.

T.-D. Wang et al. (Eds.): SEAL 2006, LNCS 4247, pp. 336 – 344, 2006.

At present, one of the most effective constraint handling methods in MOPs is DCDP [5] [10]. In the principle of DCDP, a feasible solution is always superior to an infeasible one, so it probably results in premature convergence. Therefore, by both keeping infeasible elitists during optimization and incorporating stochastic ranking to select individuals in each generation, a novel constraint handling strategy is put forward in this paper to handle constraints to avoid premature convergence.

2 Constrained MOP

A general MOP consists of a number of objectives and is associated with a number of inequality and equality constraints. Mathematically, without loss of generality, a minimization problem can be written as follows:

$$\begin{cases} \min f(\vec{x}) = \{f_1(\vec{x}), f_2(\vec{x}), \cdots, f_m(\vec{x})\} \\ \quad s.t.\ g_i(\vec{x}) \le 0, i = 1,2,\cdots,n \\ \qquad h_j(\vec{x}) = 0, j = 1,2,\cdots,p \end{cases} \tag{1}$$

Where $\vec{x}$ is the l-dimension vector of solutions, $f_k(\vec{x})(1 \le k \le m)$ is the k-th objective function, $g_i(\vec{x})(1 \le i \le n)$ is the i-th inequality constraint, $h_j(\vec{x})(1 \le j \le p)$ is the j-th equality constraint.

Having several objective functions, the notion of "optimum" changes because in MOPs the aim is to find a good tradeoff rather than a single solution as in single-objective optimization. The decision vectors that are non-dominated and satisfy constraints in entire search space are denoted as "Pareto optimal set" or "Pareto optimal solutions". And the corresponding objective vectors are denoted as "Pareto optimal front". The Pareto dominance solution is presented as follows.

Definition (Pareto Dominance Solution). *A solution vector* $\vec{u} = (u_1, \cdots, u_l)$ *is said to dominate a solution* $\vec{v} = (v_1, \cdots, v_l)$ *(denoted by* $\vec{u} \prec \vec{v}$ *) iff*

$$\forall i \in \{1, \cdots, m\},\ f_i(\vec{u}) \le f_i(\vec{v}) \wedge \exists j \in \{1, \cdots, m\},\ f_j(\vec{u}) < f_j(\vec{v}).$$

In constrained MOPs, all equality constraints may be approximated by inequality constraints by $|h_j(\vec{x})| - \varepsilon \le 0$, where$\varepsilon$is a very small value. This allows us to deal only with inequality constraints. Constraint violation function is defined as follows.

$$\begin{cases} \phi(g^+(\vec{x})) = \sum_{i=1}^{n+p} w_i (g_i^+(\vec{x}))^\beta \\ g_i^+(\vec{x}) = \max[\,0, g_i(\vec{x})] \end{cases} \tag{2}$$

Where $\phi(g^+(\vec{x}))$ is the constraint violation function, β is normally 1 or 2, $w_i, i = 1, \cdots, n+p$ are the penalty parameters.

3 Drawbacks of DCDP

In MOPs, DCDP ($\prec'$) [5] [10] is one of the most effective constraint handling methods by far, which is defined as follows:

$$\vec{u} \prec' \vec{v} \Leftrightarrow \begin{cases} \vec{u} \prec \vec{v} \wedge \phi(g^+(\vec{u})) = \phi(g^+(\vec{v})) = 0 \\ \text{or } \ \phi(g^+(\vec{u})) = 0 \wedge \phi(g^+(\vec{v})) > 0 \\ \text{or } \ 0 < \phi(g^+(\vec{u})) < \phi(g^+(\vec{v})) \end{cases}$$

From the above, in DCDP each feasible solution has a better rank than any infeasible one. Obviously the main drawback of this principle is DCDP probably results in premature convergence. The drawback is illustrated in Fig.1 and Fig.2 by 2-objective minimization optimization problems.

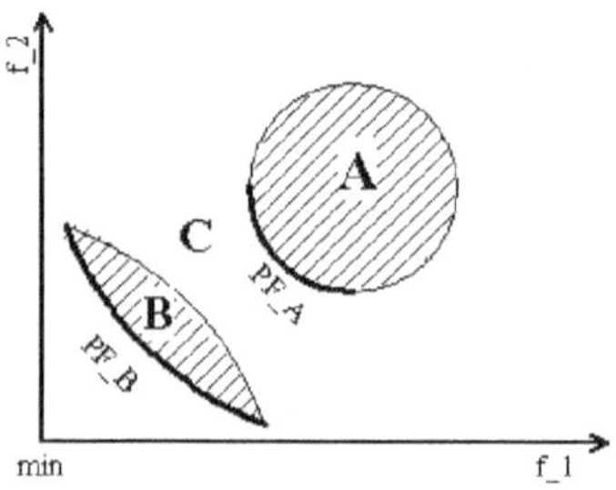

Fig. 1. Outline of local optimum using DCDP

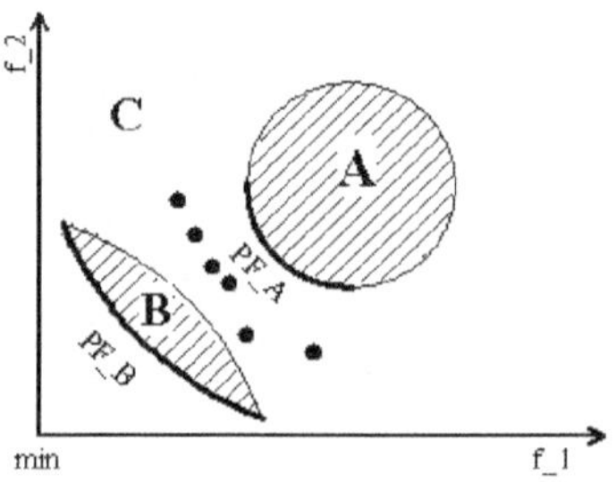

Fig. 2. Outline of global optimum using infeasible elitists

In Fig.1, the regions A and B are feasible ones but C is infeasible, the bold curve PF_A and PF_B are the Pareto optimal fronts in region A and B respectively, but PF_B is the global Pareto optimal front. In DCDP the individuals of region A will dominate those of region C, so the search will be guided to region A. Region C is as a "wall" between region A and region B. As can be seen region B is very small and it is probable that no individual of this region will be generated in the initial population. So finding an individual in region B would be a hard task and evolutionary population will be filled with the individuals in region A, leading to converge to the front PF_A not PF_B.

To avoid the drawback, some infeasible solutions (black nodes in Fig.2) can be kept to act as bridges connecting region A and region B during optimization, and it is more probable that population will be filled with individuals in region B not region A. Finally, optimization will converge to global optimum PF_B.

Considering the analysis above, a novel constraint handling strategy is proposed in next section, which keeps infeasible elitists besides feasible ones during optimization and adopts stochastic ranking to obtain a good balance between non-dominated feasible and infeasible individuals in each generation.

4 Infeasible Elitists and Stochastic Ranking Selection

In DCDP infeasible solutions are always dominated by feasible ones, so it probably steps into local optimum for problems where many feasible regions in the whole search space are disconnected. To avoid premature convergence in DCDP, a novel constraint handling strategy, namely IE-SRS (infeasible elitists and stochastic ranking selection), is given. In IE-SRS, infeasible elitists act as bridges connecting two or more different feasible regions during optimization, and that stochastic ranking selection is applied to balance objectives and constraints in each generation.

Infeasible Elitists Preservation. Ever since Zitzler and Thiele firstly raised elitist population in SPEA [2], elitist strategy has become an effective major strategy to maintain evolutionary population diversity and enhance global convergence [11]. The IE-SRS strategy still adopts elitist preservation scheme, but it includes feasible and infeasible elitist strategies. Elitists are divided into feasible and infeasible elitists. Feasible elitists originate in non-dominated feasible individuals and infeasible elitists are defined as the infeasible individuals that have better objective function values and/or less constraint violation function value.

The process of keeping and updating infeasible elitists is the following:

Step 1) For each individual $\vec{p}$ in evolutionary population.

Step 2) Considering constraint, remove the individuals dominated by $\vec{p}$ from an infeasible elitist set R.

Step 3) If $\vec{p}$ is infeasible ($\phi(g^+(\vec{p}))>0$) then

if $\vec{p}$ is dominated by any of individuals in R, then reject $\vec{p}$ and stop.

if R is not full, then add $\vec{p}$ into R, else replace the individual $\vec{x}$ with the maximum $\phi(g^+(\vec{x})), \vec{x} \in R$ with $\vec{p}$.

Stochastic Ranking Selection. To achieve a better tradeoff in selection between non-dominated feasible and infeasible individuals, the IE-SRS strategy also adopts Runarsson and Yao's [9] stochastic ranking to select individual in each generation. Moreover, to better preserve population diversity, the first step in this selection strategy is to copy all the non-dominated feasible solutions to the next population.

The process of generating the next population based on the stochastic ranking selection is described as follows:

Step 1) Copy all the non-dominated feasible individuals to the next population and sort by ascend *fitness*.

Step 2) If the size of the next population exceeds a given maximum, maintain the population according to the crowded comparison approach [5] and stop.

Step 3) Remainder individuals except the non-dominated feasible individuals are denoted as Pr, let $|Pr|=Nr$, $Pr\{i\}$ is the i-th individual of Pr, set generation counter K *as* 1.

Step 4) Set swapped flag *sflag* as 0, i=1.

Step 5) If both $Pr\{i\}$ and $Pr\{i+1\}$ are feasible and $Pr\{i\}$'s *fitness* is greater than $Pr\{i+1\}$'s, then swap them, *sflag*=1 and go to Step 7.

Step 6) Generate a random number μ between 0 and 1.

If $\mu \leq P_f$ (comparison probability) and $Pr\{i\}$'s *fitness* is greater than $Pr\{i+1\}$'s

or $\mu > P_f$ and $Pr\{i\}$'s $\phi(g^+(\vec{x}))$ is greater than $Pr\{i+1\}$'s

Then swap them, *sflag*=1.

Step 7) If $i \leq Nr$, then i increases 1 and go to Step 5.

Step 8) If $K \geq Nr$ or $sflag$= 0, then stop, else K increases 1 and go to Step 4.

Step 9) Add the front individuals into the next population until population size is equal to a given maximum.

5 Constrained MOEA with IE-SRS (IS-MOEA)

To solve constrained MOPs, a constrained MOEA combining IE-SRS strategy with NSGA-II algorithm (real-coded), named IS-MOEA, is described as follows.

Step 1) ***Parameter setting:*** population size N, infeasible elitist maximum size M, crossover probability P_c, mutation probability P_m, distribution indices for crossover η_c, distribution indices for mutation η_m, comparison probability P_f, penalty parameters $w_i(j=0,1,\ldots,n+p)$, β, maximum generation gen_max.

Step 2) ***Initialization:*** Generate an initial population P, create child population $Q=\Phi$ and infeasible elitists $R=\Phi$, set generation counter gen=1.

Step 3) ***Calculation:*** Calculate m objective function values $(f_1(\vec{x}), f_2(\vec{x}), \cdots, f_m(\vec{x}))$ and constraint violation function value $\phi(g^+(\vec{x}))$ for each individual $\vec{x}$ in P and Q, and then calculate Pareto dominance value for each pair of individuals without considering constraint.

Step 4) ***Fitness assignment:*** According to Pareto dominance value of individuals in P and Q, calculate their ranks and distances using non-dominated sorting and crowding distance assignment algorithms [5], and then calculate fitness values by formula $fitness=rank+distance$.

Step 5) ***Update and selection***: Update infeasible elitists R according to the keeping and updating strategy presented in Section 4, and then generate next population P from P and Q using the stochastic ranking selection strategy referred in Section 4.

Step 6) ***Evolutionary operation:*** According to the individual order in P, select individuals from P to mating pool by binary tournament selection, add R into mating pool, and generate new population Q by performing Simulated Binary Crossover (SBX) and polynomial mutation on the individuals of mating pool.

Step 7) ***Termination:*** If $gen \geq gen\_max$ or another stopping criterion is satisfied then stop, else increment generation counter ($gen=gen+1$) and go to step 3.

6 Experimental Results and Discussions

Test Problems. To evaluate the performance our IE-SRS strategy, six benchmark constrained problems are chosen, which are taken from [5] (CONSTR, SRN, TNK) and [10] (CTP1, CTP6, CTP7), and are summarized in table 1.

Performance Measures. The goals of constrained multi-objective optimization are approximation and diversity, so comparison of Pareto optimal solution sets is difficult [13]. According to [12], $v(A,B)$ measure is also adopted in this study. $v(A,B)$ is the fraction of the volume of the minimum hypercube containing both solution set A and B that is strictly dominated by members of A but is not dominated by members of B, and is calculated by Monte Carlo sampling and counting the fraction of samples that dominated exclusively by A. we take the same 50000 samples as [12].

Results and Discussions. Run both IS-MOEA (denoted as *IE-SRS*) and NSGA-II based on DCDP (denoted as *DCDP*) in MATLAB6.5, both of which adopt real-coded (simulated binary crossover and polynomial mutation). The source code may be obtained from the authors upon request. The parameter settings of both algorithms are as

Table 1. Six constrained minimized benchmark problems

Problem	n	Variable bounds	Objective functions	Constraints
CONSTR	2	$x_1 \in [0.1, 1.0]$ $x_2 \in [0,5]$	$f_1(x)=x_1$ $f_2(x)=(1+x_2)/x_1$	$g_1(x)=x_2+9x_1 \geq 6$ $g_2(x)=-x_2+9x_1 \geq 1$
SRN	2	$x_i \in [-20,20]$ $i=1,2$	$f_1(x)=(x_1-2)^2+(x_2-1)^2+2$ $f_2(x)=9x_1-(x_2-1)^2$	$g_1(x)=x_2^2+x_1^2 \leq 225$ $g_2(x)=x_1-3x_2 \leq -10$
TNK	2	$x_i \in [0,\pi]$ $i=1,2$	$f_1(x)=x_1$ $f_2(x)=x_2$	$g_1(x)=-x_1^2-x_2^2+1+0.1\cos(16\arctan(x_1/x_2)) \leq 0$ $g_2(x)=(x_1-0.5)^2+(x_2-0.5)^2 \leq 0.5$
CTP1	5	$0 \leq x_1 \leq 1, -5 \leq x_i \leq 5$ $i=2,3,4,5$	$f_1(x)=x_1$ $f_2(x)=c(x)\exp(-f_1(x)/c(x))$	$g_1(x)=f_2(x)-0.858\exp(0.541f_1(x)) \geq 0$ $g_2(x)=f_2(x)-0.728\exp(0.295f_1(x)) \geq 0$
CTP6	5	$0 \leq x_1 \leq 1$ $-5 \leq x_i \leq 5$ $i=2,3,4,5$	$f_1(x)=x_1$ $f_2(x)=c(x)(1-f_1(x)/c(x))$	$g_1(x)=\cos(\theta)(f_2(x)-e)-\sin(\theta)f_1(x) \geq a\lvert\sin(b\pi\sin(\theta)$ $(f_2(x)-e)+\cos(\theta)f_1(x))^c)\rvert^d$ $\theta=0.1\pi$, $a=40$, $b=0.5$, $c=1$, $d=2$, $e=-2$
CTP7	5	$0 \leq x_1 \leq 1$ $-5 \leq x_i \leq 5$ $i=2,3,4,5$	$f_1(x)=x_1$ $f_2(x)=c(x)(1-f_1(x)/c(x))$	$g_1(x)=\cos(\theta)(f_2(x)-e)-\sin(\theta)f_1(x) \geq a\lvert\sin(b\pi\sin(\theta)$ $(f_2(x)-e)+\cos(\theta)f_1(x))^c)\rvert^d$ $\theta=0.05\pi$, $a=40$, $b=5$, $c=1$, $d=6$, $e=0$

* $c(x) = 41 + \sum_{i=2}^{5}(x_i^2 - 10\cos(2\pi x_i))$

follows [5][9][10]: population size N=100, infeasible elitist maximum size M=20, crossover probability P_c=0.9, mutation probability $P_m=1/n$ (where n is the number of decision variables), distribution indices for crossover operations η_c=20, distribution indices for mutation operations η_m=20, comparison probability P_f=0.45, penalty parameters $w_j=1$ ($j=1,\ldots,n+p$), $\beta=1$, maximum generation *gen_max*=500. For each benchmark function, 30 independent runs are performed using *IE-SRS* and *DCDP* respectively. Table 2 summarizes the experimental results. Assume U and C are the optimal fronts obtained separately by *IE-SRS* and *DCDP*.

Table 2. Comparison between end-of–run optimal fronts from *IE-SRS* and *DCDP* using *v* measure. Mean are over 30 runs, standard deviation in parentheses, values as a percentage. Results highlighted in bold signify significantly better results under the Wilcoxon nonparametric signed ranks test.

Prob.	$v(U,C)$	$v(C,U)$	Prob.	$v(U,C)$	$v(C,U)$
CONSTR	0.330%(0.00064)	0.370%(0.00059)	CTP1	1.720%(0.00021)	1.653%(0.00020)
SRN	0.380%(0.00018)	0.420%(0.00130)	CTP6	**2.460**%(0.04950)	0.023%(0.00019)
TNK	0.250%(0.00020)	0.270%(0.00038)	CTP7	**0.077**%(0.00120)	0.007%(0.00007)

As is shown in table 2, as for functions CONSTR, SRN, TNK and CTP1, the performance of *IE-SRS* is almost equal to that of *DCDP*. While for CTP6 and CTP7, the performance of *IE-SRS* is significantly better than that of *DCDP*, because problems CTP6 and CTP7 have many disconnected feasible regions [10]. In order to highlight effectively the remarkable results of *IE-SRS*, the Pareto optimal fronts of functions CTP1, CTP6 and CTP7 are illustrated in Fig.3, Fig.4 and Fig.5.

As for problem CTP1, there appear three Pareto optimal fronts in the results of *IE-SRS* and *DCDP*: case 1 is the front of $f_1 \leqslant 0.6$; case 2 is the front of $0.6 < f_1 \leqslant 0.8$; case 3 is the front of $f_1 > 0.8$. In 30 independent runs, for *IE-SRS*, case 1 appears six times, case 2 sixteen times and case 3 eight times; for *DCDP*, case 1 appears twenty-two times, case 2 six times, and case 3 two times. Though the Pareto front of problem CTP1 isn't disconnected, *IE-SRS* increases the population diversity, so *IE-SRS* is better than *DCDP* in solution distribution.

As for problem CTP6, there are four Pareto optimal fronts in the results of *IE-SRS* and *DCDP*: case 1 is the front of $f_1 \leqslant 0.9$ and $5 < f_2$; case 2 is the front of $f_1 \leqslant 0.9$ and $f_2 \leqslant 5$; case 3 is the front of $f_1 > 0.9$ and $5 < f_2$; case 4 is the front of $f_1 > 0.9$ and $f_2 \leqslant 5$. In 30 independent runs, *IE-SRS* wholly gets case 4; for *DCDP*, case 1 appears eight times, case 2 fifteen times, and case 3 four times and case 4 three times. Because the entire search space consists of infeasible patches parallel to the Pareto optimal front, and the situation in Fig. 2 is fitted (where f_2 lies between 5 and 10, there is infeasible region [10]) and considering the results, *IE-SRS* can always converge to Pareto optimum front. However *DCDP* has fallen into local optimum for 12 times, and has found global optimum for 18 times, three of which are complete. That is because *IE-SRS* adopts stochastic ranking and infeasible elitist preservation mechanism, which can maintain the population diversity during evolution, and achieves global optimum. IE-SRS does not get stuck onto the local Pareto fronts, and it is also proved that the effectiveness of this new constraint handling strategy.

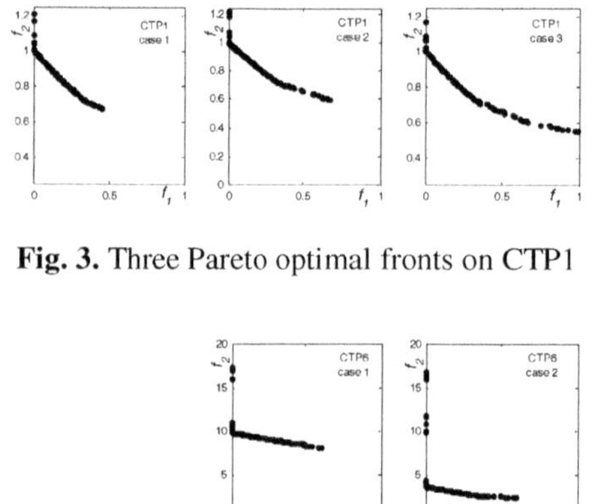

Fig. 3. Three Pareto optimal fronts on CTP1

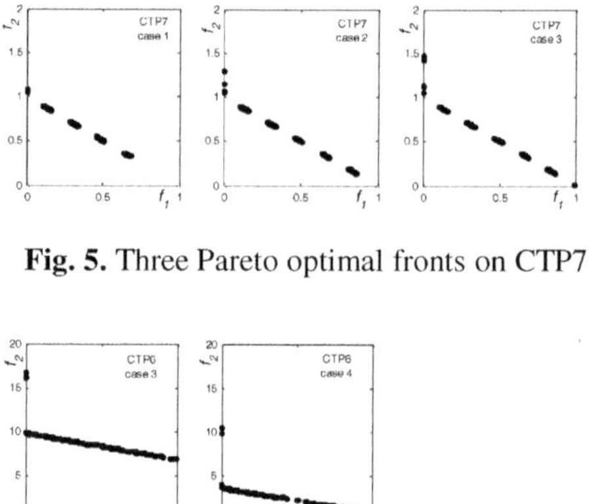

Fig. 5. Three Pareto optimal fronts on CTP7

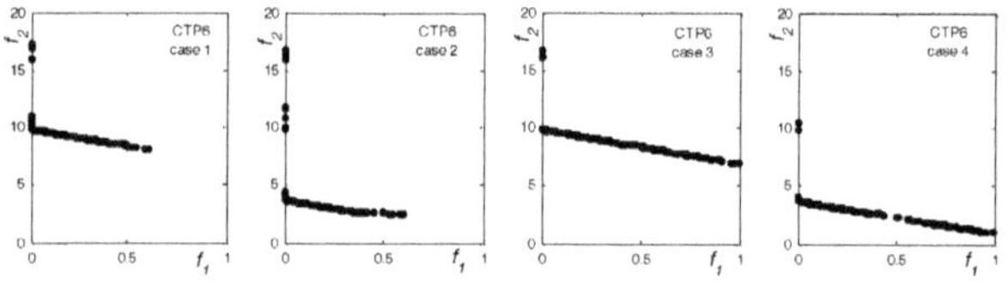

Fig. 4. Four Pareto optimal fronts on CTP6

As for problem CTP7, there appear three Pareto optimal fronts in the results of *IE-SRS* and *DCDP*: case 1 is the front with 5 feasible regions; case 2 the front with 6 feasible regions; and case 3 the front with 7 feasible regions. In 30 independent runs, for *IE-SRS*, case 1 appears zero time, case 2 fifteen times, case 3 fifteen times; for *DCDP*, case 1 appears twelve times, case 2 fifteen times, and case 3 three times. The CPT7 problem makes some portions of the unconstrained Pareto optimal region feasible, thereby making many disconnected feasible regions according to f_1 value [10]. In the results, *IE-SRS* can find at least 6 infeasible regions in the Pareto optimal front every time, in which it finds the entire regions for 15 times, while *DCDP* finds 6-above regions for only 18 times, of which 3 are entire. The above results validate again the effectiveness of using infeasible elitist preservation mechanism and stochastic ranking to maintain population diversity.

To sum up, in light of the above comparisons, the performance of *IE-SRS* is equal to that of *DCDP* for the first four benchmark problems. However for CTP6 and TP7 problems with many disconnected feasible regions, it is obvious that the solution distribution and stability of *IE-SRS* are better than that of *DCDP*.

7 Conclusions

This paper analyses DCDP and points out that it probably converges to local optimum. By keeping infeasible elitists during optimization, a new constraint handling strategy with stochastic ranking approach, namely IE-SRS, has been proposed to handle the constrained MOPs. Compared with DCDP on six benchmark constrained MOPs, IE-SRS has made distinct improvement in distribution and stability of the solutions on the problems with two or more disconnected feasible regions [10], such as CTP6. The future work of this study includes the application of IE-SRS to other MOEAs.

Acknowledgments. This work is supported by the National Natural Science Foundation of China (Grant No. 60428202). The authors would like to thank X. Yao from University of Birmingham and Wenjian Luo from University of Science and Technology of China for their helpful comments.

References

1. Schaffer J. D.: Multiple Objective Optimization with Vector Evaluated Genetic Algorithms. In: Proc. 1st International Conference on Genetic Algorithms, Hillsdale, 1985, 93-100
2. Zitzler E., Thiele L.: Multi-Objective evolutionary algorithms: A comparative case study and the strength pareto approach. IEEE. Trans. Evol. Comput., 1999,3(4):257-271
3. Corne, D.W., Knowles, J.D.: The Pareto-envelope based selection algorithm for multiobjective optimization. In: PPSN VI, Springer, Berlin, 2000:839-848
4. Zitzler E., M.,Laumanns, L.Thiele: SPEA2: Improving the Strength Pareto Evolutionary Algorithm. Tech. Rep. 103, Gloriastrasse 35, CH-8092 Zurich, Switzerland 2001
5. Deb K.,Pratap A.,Agarwal S.,Meyarivan T.: A fast and elitist multiobjective genetic algorithm: NSGA-II. IEEE Trans. Evol. Comput., 2002,6(2):182-197
6. Haiming Lu, Yen G. G.: Rank-density-based multiobjective genetic algorithm and benchmark test function study. IEEE Trans. Evol. Comput., 2003,7(4): 325-343
7. Shinn-Ying Ho, Li-Sun Shu, Jian-Hung Chen: Intelligent evolutionary algorithms for large parameter optimization problems. IEEE Trans. Evol. Comput., 2004,8(6):522-541
8. Lam T. Bui, Branke J., Abbass H. A.: Multiobjective optimization for dynamic environments CEC, 2005, 3:2349- 2356
9. Runarsson T. P., Yao X.: Stochastic Ranking for Constrained Evolutionary Optimization. IEEE Trans. Evol. Comput., 2000,4(3):284-294
10. Deb K.,Pratap A., Meyarivan T.: Constrained Test Problems for Multi-objective Evolutionary Optimization. EMO 2001: 284-298

11. Coello Coello, C.A.: Evolutionary multi-objective optimization: a historical view of the field. Computational Intelligence Magazine, IEEE, 2006, 1(1): 28- 36
12. Jonathan Fieldsend, Richard M. Everson, Sameer Singh: Using unconstrained elite archives for multiobjective optimization. IEEE Trans. Evol. Comput., 2003,7(3): 305-323
13. E. Zitzler, L. Thiele, M. Laumanns, C. M. Fonseca, et al.: Performance assessment of multiobjective optimizers: an analysis and review. IEEE Evol. Comput., 2003,7(2): 117-132
14. Min Zhang, Huantong Geng, Wenjian Luo, et al.: A Novel Search Biases Selection Strategy for Constrained Evolutionary Optimization. CEC, 2006, to appear.

A New Strategy for Parameter Estimation of Dynamic Differential Equations Based on NSGA II*

Yingzi Shi[1], Jiangang Lu[2], and Qiang Zheng[2]

[1] School of Education Science, Hangzhou Teachers College, Hangzhou 310036, China
syz89@163.com

[2] National Laboratory of Industrial Control Technology, Zhejiang University, Hangzhou 310027, China
jglu@iipc.zju.edu.cn

Abstract. A new strategy for parameter estimation of dynamic differential equations based on nondominated sorting genetic algorithm II (NSGA II) and one-step-integral Treanor algorithm is presented. It is adopted to determine the exact model of catalytic cracking of gas oil. Compared with those conventional methods, for example, quadratic programming, the method proposed in this paper is more effective and feasible. With the parameters selected from the NSGA II pareto-optimal solutions, more accurate results can be obtained.

1 Introduction

When attempting to achieve the exact model of industrial process plant, the identification of unknown parameters in dynamic differential equations is always involved. Generally, the way of the identification is to obtain some observed data through sampling randomly, and then resort to the methods of mathematical optimization. The conventional mathematical optimization is to transform the differential equations into nonlinear programming problem denoted by algebraic equations. Ref. [1] solves such problem by using rSQP(reduced Sequential Quadratic Programming) and hybrid automatic differential algorithm, and Ref.[2] by a scheme of hybrid successive quadratic programming. The conventional methods work well in solving the problems, however, they have some drawbacks, such as complicated algorithm, single result, and insufficient precision.

In order to solve the dynamic optimization problem better, we regard such problem as the multi-objective optimization problem and resort to intelligent algorithm such as genetic algorithm. NSGA [3] proposed by Srinivas and Deb, which is a genetic algorithm based on the concept of pareto-optimal [4], has been applied to the problem of multi-objective optimization extensively, while NSGA II [5] developed from NSGA, which accelerates the computation as well as improving the robustness, has become a

* This work was supported by the National Natural Science Foundation of China (No. 20206027), the Key Technologies R&D Program in the 10th Five-year Plan of China (No. 2004BA210A01), and the Technologies R&D Programs of Zhejiang Province (No. 2006C31051 and No. 2006C33059).

T.-D. Wang et al. (Eds.): SEAL 2006, LNCS 4247, pp. 345 – 352, 2006.

more powerful alternative. By adopting NSGA II together with using one-step-integral Treanor algorithm [6], a new approach to solve the problem of dynamic differential equations' parameters identification is presented.

This paper is organized as follows: Section II gives an example of dynamic differential equations and its sampling data. Section III provides a brief introduction to the principle of NSGA II and one-step-integral Treanor algorithm. Section IV presents a new strategy for solving the nonlinear model parameters estimation problem. Conclusion is made in Section V.

2 Example of Dynamic Differential Equations

The model of catalytic cracking of gas oil, proposed by Froment and Bischoff, describes an overall reaction of catalytic cracking of gas oil (A) to gasoline (Q) and other byproducts (S), and is of the form [7]:

$$\left.\begin{array}{l} A \xrightarrow{k_1} Q \\ A \xrightarrow{k3} S \\ Q \xrightarrow{k2} S \end{array}\right\} \tag{1}$$

According to Ref.[7], the model (1) can be denoted by dynamic differential equations as follows:

$$\begin{cases} \dfrac{dy_1}{dt} = -(k_1 + k_3) y_1^{\,2} \\ \dfrac{dy_2}{dt} = k_1 y_1^{\,2} - k_2 y_2 \end{cases} \tag{2}$$

where $y_i (i = 1, 2)$ is the concentrations of A , Q and $k_i (i = 1, 2, 3)$ is the reaction coefficients. Initial condition for dynamic differential equations (2) is $y^0 = (1, 0)^T$, and the constraint condition is $k_i \geq 0, 0 \leq y_i \leq 1$.

The coefficients $k_i (i = 1, 2, 3)$ in the equations are the parameters to be determined. As mentioned above, the acquisition of observed data of model is always a necessity. Through sampling randomly under certain conditions, the 21 observed data of $y_i (i = 1, 2)$, according to Ref. [7], is shown in Table 1.

The problem is to minimize

$$\sum_{j=1}^{21} | y_i(t_j; k) - c_{ij} |^2, i = 1, 2 \tag{3}$$

where c_{ij} is the observed data of y_i at time points t_j , which is given in Table 1, while $y_i(t_j; k)$ is the fitted data of c_{ij} $(i = 1, 2; j = 1, 2, ..., 21)$.

Table 1. Observed Values c_i of y_i at Time t_j

t/s	c_1	c_2
0.000	1.0000	0.0000
0.025	0.8105	0.2000
0.050	0.6208	0.2886
0.075	0.5258	0.3010
0.100	0.4345	0.3215
0.125	0.3903	0.3123
0.150	0.3342	0.2716
0.175	0.3034	0.2551
0.200	0.2735	0.2258
0.225	0.2405	0.1959
0.250	0.2283	0.1789
0.300	0.2071	0.1457
0.350	0.1669	0.1198
0.400	0.1530	0.0909
0.450	0.1339	0.0719
0.500	0.1265	0.0561
0.550	0.1200	0.0460
0.650	0.0990	0.0280
0.750	0.0870	0.0190
0.850	0.0770	0.0140
0.950	0.0690	0.0100

3 NSGA II and One-Step-Integral Treanor Algorithm

In the problem of multi-objectives optimization, what is expected is one solution that fit all the objectives. To attain the goal, the conventional way is always to aggregate the multiple objectives into a singular scalar objective, and this way involves the determining of the objectives' relative importance, which reduces the precision of result. To avoid the disadvantage, several optimization algorithms have been proposed, among which NSGA shows great advantage in several aspects. The multiple objectives are invariably unable to reach the optimal point simultaneously due to the conflicts between them, and there may exist a set of solutions, which are partially superior or inferior to each other, and all the other solutions are dominated by those solutions. These solutions are called Pareto-optimal solutions or nondominated solutions. With multi-objective optimization problem optimized by NSGA, we would obtain a set of feasible solutions, from which we may pick up one solution, which is a compromise after considering the practical requirements.

The only difference between NSGA and SGA is the way that the selection operator works, while the crossover and mutation operators remain the same. Before performing the selection operator, ranking manipulation is conducted where all the population are ranked using the nondominated sorting algorithm based on the individuals' nondomination. Following shows the ranking procedure. Firstly, pick up all the nondominated individuals from the current population to fill in the first nondominated front, and assign a large dummy fitness value to this front. Secondly, ignore the individuals already picked out and choose the nondominated individuals from the rest of the population to constitute the second nondominated front with a relatively smaller dummy fitness value assigned to it. Repeat the procedure until the entire population is divided into several fronts in terms of dummy fitness value.

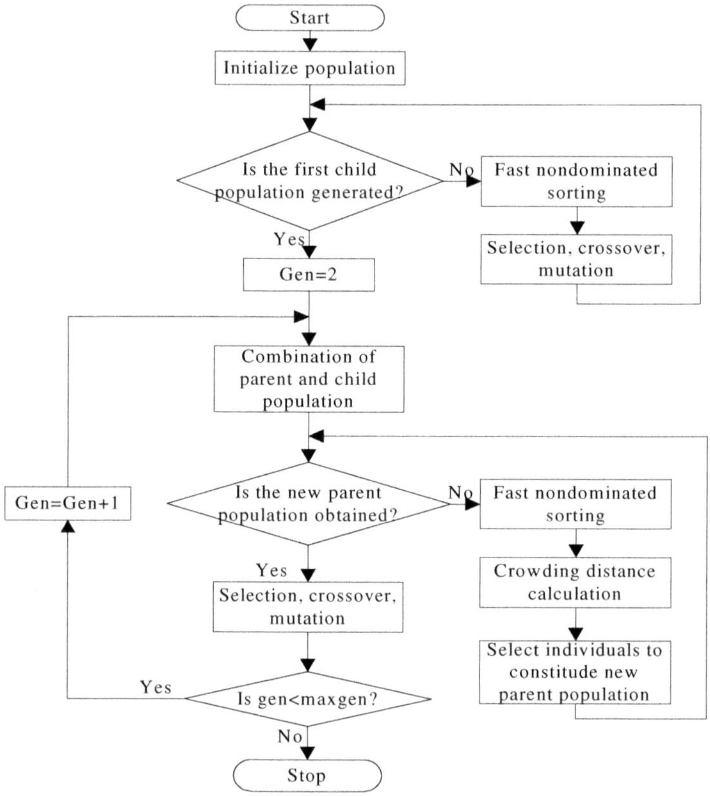

Fig. 1. NSGA II Main Algorithm

NSGA reserves the most excellent offspring population, and is capable of obtaining a set of evenly distributed pareto-optimal solutions, in addition, it results in quick convergence of the population towards the optimal points of the entire region, rather than towards the local region. However, NSGA still has criticisms due to its defects. NSGA II makes improvements on the basis of NSGA in three aspects [5]. Firstly, fast nondominated sorting approach is presented, with the computational complexity reduced from $O(mN^3)$ to $O(mN^2)$ (where m is the number of objectives and N is the population size). Secondly, crowded-comparison approach and its corresponding operator are introduced, which guarantee the diversity of nondominated individuals as well as avoiding the appointment of σ_{share} value, a key parameter of the sharing function in NSGA. Thirdly, elitism is also introduced, that is, to generate a new parent population through selecting individuals from the combined population of the current parent population and its child population.

NSGA II is proved to be an effective alternative in the problem of multi-objective optimization in the Ref. [5] through several experiments and simulation, and the main algorithm of it shows in Fig. 1.

One-step-integral Treanor algorithm is frequently adopted to integrate the dynamic differential equations as follows [6]:

$$\frac{dy_i}{dt} = f_i(t, y_1, ..., y_n), y_i(t_0) = y_{i0}, i = 1, 2, ..., n \tag{4}$$

If the values of $y_{ij}(i = 1,2,...,n)$ in differential equations (4) at the time t_j are given, then at the time $t_{j+1} = t_j + l$,

$$y_{i,j+1} = y_{ij} + \Delta y_i, i = 1, 2, ..., n \tag{5}$$

$$\Delta y_i = \begin{cases} \frac{h}{6}[g_i^{(1)} + 2(g_i^{(2)} + g_i^{(3)}) + g_i^{(4)}], p_i \le 0 \\ h\{g_i^{(1)} r_i^{(2)} + [-3(g_i^{(1)} + p_i Z_i^{(1)}) + 2(g_i^{(2)} + p_i Z_i^{(2)}) \\ +2(g_i^{(3)} + p_i Z_i^{(3)}) - (g_i^{(4)} + p_i Z_i^{(4)})] r_i^{(3)} \\ +4[(g_i^{(1)} + p_i Z_i^{(1)}) - (g_i^{(2)} + p_i Z_i^{(2)}) \\ -(g_i^{(3)} + p_i Z_i^{(3)}) + (g_i^{(4)} + p_i Z_i^{(4)})] r_i^{(4)}\}, p_i > 0 \end{cases} \tag{6}$$

where

$p_i = \frac{g_i^{(3)} - g_i^{(2)}}{Z_i^{(3)} - Z_i^{(2)}}$, $r_i^{(1)} = e^{-p_i l}$, $r_i^{(2)} = \frac{r_i^{(1)} - 1}{-p_i l}$, $r_i^{(3)} = \frac{r_i^{(2)} - 1}{-p_i l}$, $r_i^{(4)} = \frac{r_i^{(3)} - 0.5}{-p_i l}$, $Z_i^{(1)} = y_{ij}$,

$Z_i^{(2)} = Z_i^{(1)} + 0.5 l g_i^{(1)}$, $Z_i^{(3)} = Z_i^{(1)} + 0.5 l g_i^{(2)}$, $Z_i^{(4)} = Z_i^{(1)} + q_i l$,

$g_i^{(1)} = f_i(t_j, Z_1^{(1)}, Z_2^{(1)}, ..., Z_m^{(1)})$, $g_i^{(2)} = f_i(t_j + 0.5l, Z_1^{(2)}, Z_2^{(2)}, ..., Z_m^{(2)})$,

$g_i^{(3)} = f_i(t_j + 0.5l, Z_1^{(3)}, Z_2^{(3)}, ..., Z_m^{(3)})$, $g_i^{(4)} = f_i(t_j + l, Z_1^{(4)}, Z_2^{(4)}, ..., Z_m^{(4)})$,

$$q_i = \begin{cases} g_i^{(3)}, p_i \le 0 \\ 2(g_i^{(3)} - g_i^{(1)}) r_i^{(3)} + (g_i^{(1)} - g_i^{(2)}) r_i^{(2)} + g_i^{(2)}, p_i > 0 \end{cases} \quad (i = 1, 2, ..., n)$$

4 Optimal Estimation of Parameters Based on NSGA II

The key procedure of determining unknown parameters in model (2) using NSGA II is to obtain the value of $y_i(i = 1, 2)$ at $t_j(j = 1, 2, ..., 21)$ by resorting to the one-step-integral Treanor algorithm before calculating the fitness function of this model. Obviously, the objective functions (3) aiming at the minimum values, can be regarded as fitness functions of NSGA II without transformation.

Individuals are represented by (k_1, k_2, k_3), and the real-coded NSGA II is adopted. As mentioned above, the constraint condition of parameters $k_i (i = 1,2,3)$ is $k_i \geq 0$. And it has been proved by the practical experiments that the model could reach a set of pareto-optimal solutions with adequate accuracy when setting the scope of $k_i (i = 1,2,3)$ as [0,100]. The one-step-integral step of Treanor algorithm is $l = 0.001$, and the parameters of NSGA II are in Table 2. Table 3 shows a set of pareto-optimal solutions after optimization. Using solution (11.816,8.353,0.985) from Table 3, the fitted data of $y_i(t_j; k)$ $(i = 1,2; j = 1,2,...,21)$ at time point t_j are listed in Table 4.

Table 2. Algorithm Parameters of NSGA II

Random Seed	0.6
Population size	100
Generation size	200
Real variable number	3
Probability of crossover (real code)	0.9
Probability of mutation (real code)	0.1
Boundary of $k_i (i = 1,2,3)$	[0,100]

Table 3. A Set of Pareto-optimal Solutions

k_1	11.816	11.815	11.815	11.813	11.814 ......
k_2	8.353	8.353	8.353	8.354	8.354 ......
k_3	0.985	0.985	0.986	0.984	0.984 ......

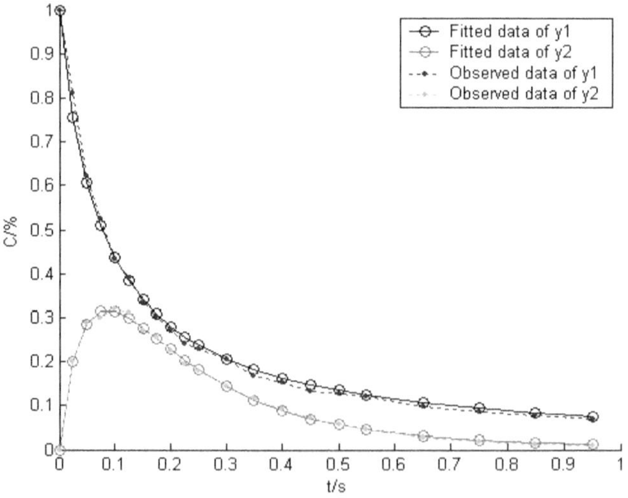

Fig. 2. Comparison of Fitted Data and Observed Data

Table 4. Fitted Data of $y_i(t_j;k)$ at t_j

t/s	y_1	y_2
0.000	1.0000	0.0000
0.025	0.7565	0.2007
0.050	0.6086	0.2852
0.075	0.5091	0.3137
0.100	0.4376	0.3138
0.125	0.3837	0.2992
0.150	0.3416	0.2777
0.175	0.3079	0.2533
0.200	0.2802	0.2285
0.225	0.2571	0.2046
0.250	0.2376	0.1835
0.300	0.2062	0.1444
0.350	0.1821	0.1131
0.400	0.1631	0.0887
0.450	0.1477	0.0700
0.500	0.1349	0.0557
0.550	0.1242	0.0447
0.650	0.1071	0.0298
0.750	0.0942	0.0209
0.850	0.0841	0.0153
0.950	0.0759	0.0117

In order to verify the effectiveness of the result obtained by using NSGA II, two comparisons are conducted here. Firstly, compare the fitted data in Table 4 with the observed data in Table 1, and the comparison result shows in Fig. 2. Secondly, compare the value of the objective functions between the result obtained in this paper and that of Ref. [2], where the objective function values are worked out through one-step-integral Treanor algorithm, and the comparison result shows in Table 5.

Table 5. Comparison of k in Terms of Objective Functions Value

Derivation	Value of $k_i (i = 1,2,3)$	Value of objective functions
Ref. [2]	$(11.948, 7.993, 2.024)$	$(0.0094, 0.0011)$
This paper	$(11.816, 8.353, 0.985)$	$(0.0047, 0.0006)$

As shown in the Fig. 2 and Table 5, the fitted data in Table 4 match the observed data in Table 1 more closely. It can be concluded that the method proposed in this paper is more effective and feasible.

5 Conclusion

A new strategy for parameter estimation of dynamic differential equations based on NSGA II, which is adopted to determine the exact model of catalytic cracking of gas oil, is proposed. This method is proved to be more effective and feasible than the traditional methods. With the parameters selected from the NSGA II pareto-optimal solutions, more accurate results can be obtained.

References

1. Jiang, A.P., Shao, Z.J., Qian, J.X.: Optimization of Reaction Parameters Based on rSQP and Hybrid Automatic Differentiation Algorithm. Journal of Zhejiang University (Engineering Science), 38 (2004) 1606–1610
2. I.-B.Tjoa, L.T.Biegler: Simultaneous Solution and Optimization Strategies for Parameter Estimation of Differential-algebraic Equations Systems. Ind.Eng.Chem.Res.. 30 (1991) 376-385
3. Srinivas, N., Deb, K.: Multiobjective Function Optimization Using Nondominated Sorting Genetic Algorithms[J]. Evolutionary Computation, 2(3) (1995) 221-248
4. Goldberg, D.E.: Genetic Algorithm in Search, Optimization and Machine Learning [M], Addison-Wesley, (1989)
5. Deb, K., Agrawal, S., Pratap, A., et al.: A Fast Elitist Nondominated Sorting Genetic Algorithm For Multi-objective Optimization: NSGA II[A]. Proc of the Parallel Problem Solving from Nature VI Conf[C]. Paris, (2000) 849-858
6. Xu, S.L.: Common Algorithm Set by FORTRAN. Tsinghua University Press, (1995)
7. Froment G.F., Bischoff K.B.: Chemical Reactor Analysis and Design. Wiley, NewYork, (1979)

Vector Prediction Approach to Handle Dynamical Optimization Problems

Bojin Zheng[1,2], Yuanxiang Li[2,⋆], and Ting Hu[3]

[1] College of Computer Science,South-Central University For Nationalities, Wuhan, 430074, China
[2] State Key Lab. of Software Engineering,Wuhan University, Wuhan, 430072, China
[3] Dept. of Computer Science, Memorial University of Newfoundland, NL, Canada
zhengbojin@gmail.com, yxli@whu.edu.cn, tingh@cs.mun.ca

Abstract. The Dynamical Optimization Evolutionary Algorithms (DOEAs) have been applied to solve Dynamical Optimization Problems which are very common in real-world applications. But little work focused on the convergent DOEAs. In this paper new definitions of convergence are proposed and a new algorithm named Vector Prediction Approach is designed. This algorithm firstly analyzes the genes of best individuals from the past, then predicts the next genes of best individual in every tick by Gene Programming, such that the algorithm tracks the optima when time varying. The numerical experiments on two test-bed functions show that this algorithm can track the optima when time varying. The convergence of this algorithm under certain conditions is proved.

1 Introduction

Real-world optimization problems are often time-varying, so they are called Non-Stationary Optimization Problems or Dynamical Optimization Problems(DOPs). At present, applying Evolutionary Computation to solve Static Optimization Problems is under extensive investigation, but as to DOPs, many foundational questions are not answered yet.

Since 1960s, various Dynamical Optimization Evolutionary Algorithms (DOEAs) have been proposed. Kazuko Yamasaki et al. [1] pointed out that all the algorithms base on those typical characteristics of DOPs as follows:

1. Reappearance. The representative DOEAs are memory-based DOEAs[2, 3].
2. Continuity. For example, some DOEAs employ neighborhood search operators.
3. Rarity. The representative DOEAs are Diversity-based DOEAs. The Triggered Hyper-mutation Approach would be the representation of this kind of algorithm[4, 5].
4. Predictability. "Futurist Approach"[6] is the representative algorithm. Currently, prediction-based DOEAs focus on how to rebuild the dynamical environment[7].

T.-D. Wang et al. (Eds.): SEAL 2006, LNCS 4247, pp. 353–360, 2006.

Furthermore, some algorithms tried to combine the characteristics to improve the performance of DOEAs. For example, Multi-population Approach bases on the reappearance and rarity[8].

So far, the convergence of DOEAs has not been well considered. Little work has focused on this topic. Actually, the DOEAs mentioned above do not converge except some memory-based DOEAs under some special conditions. Even if these memory-based DOEAs, they are just capable of dealing with some specific Dynamical Optimization Problems with small periods.

In this paper, we focus on how to design a convergent DOEA. Firstly, we propose the definitions of convergence. Secondly, we introduce a new approach named Vector Prediction Approach which analyzes the track of every gene of the chromosome of the best solutions from the past, and predict the next locations of the genes, such that the algorithm can track the optima of DOPs. Thirdly, we theoretically prove the convergence of the proposed algorithm under some certain conditions. Fourthly,we represent the experimental results which experimentally prove the convergence. At last, we discussed some important issues and future work.

2 Introduction to the Algorithm

Definition 1 (Strong Convergence). *For function* $f(\overrightarrow{x},t)$*,if one algorithm can obtain an infinite sequence of* $\overrightarrow{x}^*{}_t$ $(\overrightarrow{x}^*{}_{t_1}, \overrightarrow{x}^*{}_{t_2}, \cdots)$ *to satisfy*

$$\forall\varepsilon > 0 \wedge i > 0\, |\min f(\overrightarrow{x}, t_i) - f(\overrightarrow{x}^*, t_i)| < \varepsilon \tag{1}$$

then the algorithm is strongly convergent to $f(\overrightarrow{x},t)$.

Here $\overrightarrow{x}$ *is independent variable vector,*$f(\overrightarrow{x},t)$ *is a time-varying function, and the functional value is real number,* t *is time variable and* ε *is an positive arbitrary small constant,*i *is a positive integer.*

Definition 2 (Weak Convergence). *For function* $f(\overrightarrow{x},t)$*,if one algorithm can obtain an infinite sequence of* $\overrightarrow{x}^*{}_t$ $(\overrightarrow{x}^*{}_{t_1}, \overrightarrow{x}^*{}_{t_2}, \cdots)$ *to satisfy*

$$\forall\varepsilon > 0, \exists N > 0 \text{ such that } \forall i > N\, |\min f(\overrightarrow{x}, t_i) - f(\overrightarrow{x}^*, t_i)| < \varepsilon \tag{2}$$

then the algorithm is weakly convergent to $f(\overrightarrow{x},t)$.

The strong convergence is very hard to achieve. So we focus on the weak convergence to the predictable Dynamical Optimization Problems.

It is a good choice to employ meta-learning approach to learn from the past and construct the $\overrightarrow{x}^*_t$'s expressions and predict the next value: $\overrightarrow{x}_{t+1}$.

We suggest that two populations be used, one for evolving, and the other for predicting.

The algorithm can be depicted as follows,

Procedure VPA
Initializing the parameters;
Initializing the population;

Initializing the predicting population;
Evaluate and store the best;
while (not termination-condition) do
begin
if environment changed then
record the number of changes
if prediction-condition then
fitting
end if
predict the current best solution
copy the prediction solution into population
for each solution in population
do crossover and mutation operators to generate the new solution
if the new solution is better than current solution
replace the current solution with the new solution
end if
end for
store the best solution
end if
end

In this algorithm, the population(also called evolutionary population) means the population to be used to solve problem in the current tick and the predicting population means the population to be used to predict.

2.1 Environmental Detection

Environmental detection would contribute to the convergence of DOEAs. Here we assume that the algorithm knows the ticks. That is, the algorithm knows when the environment would change.

2.2 Prediction Procedure

In this algorithm, we use the improved GEP (Gene Expression Programming) [9, 10] to predict the functions, e.g.,use the GEP to perform symbolic regression for obtaining some expressions, and then compute $\overrightarrow{x}_{t+1}$

The procedure would be depicted as follows,

Procedure fitting;
Begin
While(not fitting- termination-condition) do
for each individual in predicting population
evaluate the current individual
do crossover, mutation operator to generate a new individual
evaluating the new individual
if the new individual is better than current individual in predicting population then

replace current individual in predicting population with the new one
end if
end for
end while
store the best individual in the predicting population
end;

In this algorithm, every individual's chromosome is a vector, and the dimension number is the same as the dimension number of $\overrightarrow{x}$. Every element in this vector is a functional expression, and every expression is divided into two parts, the first is symbolic sequence, the second is a array to store the real constants. Therefore, the chromosome of every individual can be depicted as follows,

$$\begin{cases} expression_1 \\ expression_2 \\ \ldots \\ expression_n \end{cases} \tag{3}$$

And the evaluation function would be

$$\begin{array}{l} Fitness = f(x_1, x_2, \ldots, x_n, t) \\ x_i = expression_i(\overrightarrow{x}_{best}, t) \end{array} \tag{4}$$

We store the symbolic sequence with the suffix expression. This make the encoding, decoding and evaluation easy, because it is unnecessary to construct the grammatical trees.

2.3 Prediction Conditions

When the individual which is generated by the predicting population is worse than the best individual which is generated by the evolutionary population, it is obvious that the prediction process should continue.But when the individual which is generated by the predicting population is better than the best individual which is generated by the evolutionary population and continue for several times, we can think the prediction results are good enough, and the prediction process would pause until the prediction is necessary again.

3 Convergence Analysis

Here we denote predicting population as P1 and evolutionary population as P2. The basic definitions and theorems on Markov Chain are not introduced here, please refer to [11].

Definition 3 (**$K - M$ Predictable**). *Given a sequence **S** and a universal machine, if there exists a code **C** whose length is K bit, for any **t**, such that the machine should obtain **S**(t) in **M** step, then the sequence **S** is (K-M) predictable.*

Lemma 1. *A homogeneous Markov chain with finite state space and irreducible transition matrix visits every state infinitely often with probability one regardless of the initial distribution.*

Proposition 1. *Assumed that the variable space is discrete and for any* $i \in [1, n]$*, here* n *is the number of genes, the sequences of genes are* $S_i(K_i, M_i)$ *predictable ,the populations sequence* $(P1, P2)_t (t \geq 0)$ *is a homogeneous finite Markov chain with irreducible transition matrix.*

Theorem 1. *If the variable space is discrete, the environmental changes are known and for any* $i \in [1, n]$*, here* n *is the number of genes, the sequences of genes are* $S_i(K_i, M_i)$ *predictable, then VPA should weakly converge with probability one.*

Proof. 1. According to lemma 1, as the sequences of genes are $S_i(K_i, M_i)$ predictable, the maximum search space of predicting population would be $2^{\sum_{i=1}^{n} K_i}$. Assumed that the variable space is 2^l, the total size of state space would be $2^{\sum_{i=1}^{n} K_i + l}$. Because the transition matrix is irreducible, every state would be infinitely often visited with probability one regardless of the initial distribution. 2. Because the elitism is employed in this algorithm, once the best individual in evolutionary population and the best individual in predicting population emerge simultaneously, they would be stored and would not be replaced. 3. Because the best predicting individual and the best evolutionary individual would emerge simultaneously with probability one and would be stored forever, so this algorithm would weakly converge with probability one.

4 Numerical Experiments and Analysis

4.1 Benchmark Functions

Currently, the moving parabola problem[12] is one of the foundations of the other problems. Here we design two test-bed functions based on it.

Function Z1

$$f(x_1, x_2, t) = 100 * ((x_1 - \sin(t)^2 + (x_2 - \cos(t))^2) \quad (5)$$

Here t= (current generation) - (current generation) mod 2, this means, the environment will change with a 2-generation interval.

Function Z2

$$\begin{aligned} f(x_1, x_2, x_3, t) &= \sqrt{y_1^2 + y_2^2 + y_3^2} \\ y_1 &= x_1 - \sin(\tfrac{t}{2}) \\ y_2 &= x_2 - t \\ y_3 &= x_3 - x_1 \cos(\tfrac{t}{2}) \end{aligned} \quad (6)$$

Here t is the same as in Z1.

Function Z1 has a small period, but Z2 is not periodic.

4.2 Parameters Setting

For comparing with the proposed DOEA, we re-implemented a DOEA with Memory-Enhanced Strategy (MES). The size of memory is set to 60. And the best individual in current tick would replace the eldest individual in the memory just after the environmental change. For both DOEAs, the size of population is set to 300 and the generation number is set to 5000. For every function, the DOEAs are carried out 30 times.

4.3 Results and Analysis

For direct representation, we plot the best fitness sequences of MES and VPA on Z1 and Z2 (The data come from only one run).

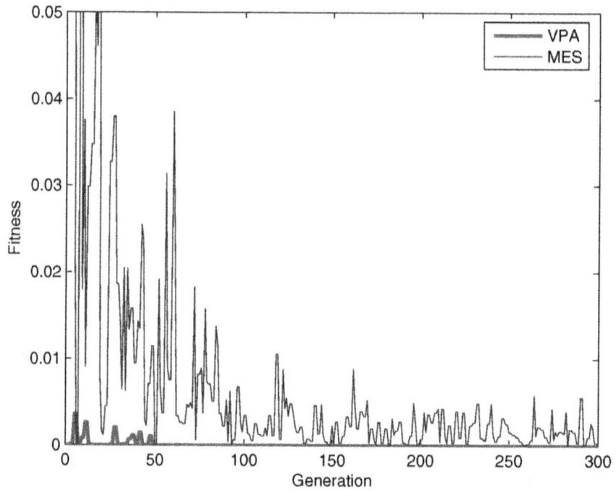

Fig. 1. The comparison on Z1

From the figures we can see that this algorithm can obtain very nice results as the red lines have shown: after some generations(the very generation would be called as the first hitting generation), this algorithm obtained the optima at every tick, but MES did not, though it obtained very good results when dealing with Z1. Furthermore, we did not find the convergence trend of MES. Actually,theoretically, it would not converge when dealing with such problems.

First Hitting Generation (FHG) and Off-Line Performance (OLP) are used to measure the performance of both the algorithms. In Table 1, the average FHG (AVG_FHG) and the standard error of FHG (SE_FHG), average OLP (AVG_OLP) and the standard error of OLP (SE_OLP) are compared.

The data in table 1 imply that the performance of VPA is much better than MES. Actually, VPA converged in all the runs, but MES did not.

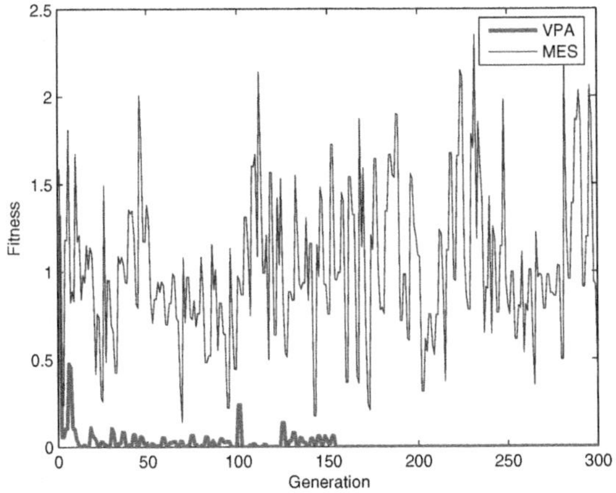

Fig. 2. The comparison on Z2

Table 1. Comparisons of Performance Measures

		AVG_FHG	SE_FHG	AVG_OLP	SE_OLP
Z1	MES	5000	-	0.0355543	0.0062121
	VPA	50.6666667	1.3474066	0.0000334	0.0000008
Z2	MES	5000	-	1.0205465	0.0005568
	VPA	651.4666667	22.5594604	0.0035794	0.0001091

5 Discussions and Future Work

In this paper, we propose a new approach: Vector Prediction Approach. The experimental results and theoretical analysis imply that the algorithm has a good performance. In contrast to the memory-based approach, the proposed approach could deal with non-periodic Dynamical Optimization Problems.

We concentrate on only the dynamical optimization problems without randomness, because the "domino effects" of randomness are very difficult to be categorized. Actually, the randomness probably make a mass of DOPs unsolvable or meaningless. Even if some of them are solvable, the "online" and "off-line" performance measures which are commonly used are somehow doubtable because the measures base on a hypothesis: the comparative algorithms should converge. Anyhow, the randomness would be an interesting topic and we leave it as future work.

Notwithstanding its limitation, this study does imply that Vector Prediction Approach should be feasible to deal with the DOPs which are predictable.

Acknowledgements

The authors gratefully acknowledge the financial support of the National Natural Science Foundation of China under Grant No.60473014, and the Specialized

Research Fund for the Doctoral Program of Higher Education of China under Grant No. 20030486049. The authors gratefully thank to Ms. Lingling Wang, Feng Wang and Shujuan Peng for the modification.

References

1. Yamasaki, K., Kitakaze, K., Sekiguchi, M.: Dynamic optimization by evolutionary algorithms applied to financial time series. In: Proceedings of the 2002 Congress on Evolutionary Computation, CEC '02. Volume 2. (2002) 2017–2022
2. Bendtsen, C.N., Krink, T.: Dynamic memory model for non-stationary optimization. In: Proceedings of the 2002 Congress onEvolutionary Computation, CEC '02. Volume 1. (2002) 145–150
3. Branke, J.: Memory enhanced evolutionary algorithms for changing optimization problems. In: Proceedings of the 1999 Congress on Evolutionary Computation, CEC 99. Volume 3. (1999) 1875–1882
4. Cobb, H.G.: An investigation into the use of hypermutation as an adaptive operator in genetic algorithms having continuous, time-dependent nonstationary environments. Technical Report Memorandum Report 6760, Navy Research Laboratory,Washington, D.C., USA (1990)
5. Morrison, R.: Design evolutionary algorithms for dynamic environments. PhD thesis, George Mason University, Fairfax, Virgina (2002)
6. Hemert, J., Hoyweghen, C., Lukshandl, E., Verbeeck, K.: A futurist approach to dynamic environments. In Branke, J., Bäck, T., eds.: Evolutionary Algorithms for Dynamic Optimization Problems, San Francisco, California, USA (2001) 35–38
7. Dorigo, M., Gambardella, L.M., Middendorf, M., Stutzle, T.: Special section on ant colony optimization. IEEE Transactions on Evolutionary Computation **6**(4) (2002) 317–320
8. Branke, J.: A multi-population approach to dynamic optimization problems. In Parmee, I., ed.: Fourth International Conference on Adaptive Computing in Design and Manufacture (ACDM 2000), Plymouth, Grossbritannien, Springer Verlag (2000) 299–308
9. Ferreira, C.: Gene expression programming in problem solving. In: the 6th Online World Conference on Soft Computing in Industrial Applications(WSC6). Volume 1. (2001) 1–22
10. Ferreira, C.: Gene expression programming: A new adaptive algorithm for solving problems. Complex Systems **13**(2) (2001) 87–129
11. Iosifescu, M.: Finite Markov Processes and Their Applications. Chichester: Wiley (1980)
12. Branke, J.: Evolutionary Optimization in Dynamic Environments. Kluwer Academic Publishers (2002)

Benchmarking Evolutionary and Hybrid Algorithms Using Randomized Self-similar Landscapes

Cara MacNish

University of Western Australia, Perth WA 6009, Australia
cara@csse.uwa.edu.au
http://www.csse.uwa.edu.au/~cara

Abstract. The success (and potential success) of evolutionary algorithms and their hybrids on difficult real-valued optimization problems has led to an explosion in the number of algorithms and variants proposed. This has made it difficult to definitively compare the range of algorithms proposed, and therefore to advance the field.

In this paper we discuss the difficulties of providing widely available benchmarking, and present a solution that addresses these difficulties. Our solution uses automatically generated fractal landscapes, and allows user's algorithms written in any language and run on any platform to be "plugged into" the benchmarking software via the web.

Keywords: real-valued optimisation, evolutionary algorithms, hybrid algorithms, benchmarking.

1 Introduction

The success of population-based optimization algorithms, such as genetic, evolutionary and swarm-based algorithms, along with the steady increase in accessible computing power, has allowed previously impractical search and optimization problems to be addressed across many application domains. This has led to an explosion in the number of algorithms and their variants proposed by researchers. In addition there has been renewed interest in more traditional algorithms, such as hill-climbing and other local searches, as part of hybrid or "memetic" algorithms combining distributed and traditional approaches.

Effective means of comparing the multitude of proposed algorithms are limited. This in turn makes it difficult for the field to progress, since it is difficult to evaluate the relative merits of proposed algorithms and modifications.

In this paper we discuss the difficulties of providing effective and widely available benchmarking, and present a solution that addresses some of these difficulties. Our solution uses a series of automatically generated fractal landscapes, and allows user's algorithms written in any language and run on any platform to be "plugged into" the benchmarking software via the web. The benchmarking system is available on-line at `http://gungurru.csse.uwa.edu.au/cara/huygens/` [1].

T.-D. Wang et al. (Eds.): SEAL 2006, LNCS 4247, pp. 361–368, 2006.

2 Difficulties in Universal Benchmarking

We begin by outlining some of the issues that inhibit the development and adoption of universal benchmarking facilities.

2.1 Programming Languages and Architectures

Researchers develop implementations for testing their algorithms in programming languages and development environments they are familiar with. Problem sets to test the algorithms tend to be developed in the same language, and often as part of the same code. Where other researchers have used different languages or coding structures, it often requires many hours of programming and testing, and in many cases rewriting in a new language, in order to compare algorithms on the same problems.

Our approach to this problem uses Simple Object Access Protocol (SOAP) as an interlingua between the user's code and the benchmarking suite, and is discussed further in Section 4.

2.2 Choice of Problem Domains

Optimization algorithms are developed for a huge range of problem domains. Many of these are not widely known or understood. In order for benchmarking problems to be widely accepted they must be readily available and familiarized.

Some defacto standards have emerged through common use. Well-known examples are those by De Jong [2] and Schaffer [3]. These are typically mathematical functions that are designed to be in some way challenging to the solver. An example is Schaffer's F6 function, shown in Figure 1.

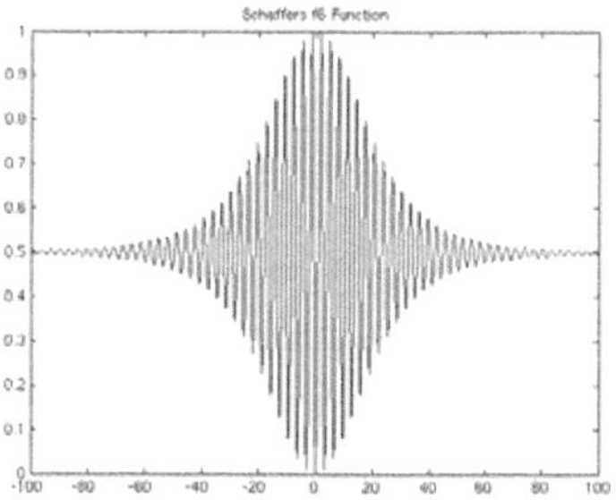

Fig. 1. A cross section of Schaffer's F6 function

While a focus on these problems has raised many important issues, they are arguably poorly representative of naturally occurring optimization problems. F6, for example, is difficult because the closer a candidate solution gets to the global minimum, the bigger the hill that must be climbed or "jumped" to move from one local minimum to the next. However we are not aware of any naturally occurring phenomena which clearly have this property.

In this paper we choose self-similar "landscapes" as our problem domain. This domain is appealingly familiar and intuitive while raising many difficulties for solvers that generalise to other domains, and is discussed futher in Section 3.

2.3 Ready Access to Benchmark Problems and Comparative Statistics

In recent years various more natural benchmarking problems and problem generators have been made available on the web (see for example [4, 5]). However, even if one overcomes the coding issues raised above, it remains difficult to benchmark an algorithm against those of a large number of peers. Aside from the overheads of obtaining or implementing other authors' algorithms, without a centralised system, the number of comparisons needed increases combinatorially with the number of algorithms proposed.

2.4 A Word on the No Free Lunch Theorem

The *No Free Lunch (NFL) Theorem* [6, 7] states, roughly speaking, that no algorithm performs better than any other when averaged over *all* fitness functions. The theorem therefore implies that algorithms cannot be benchmarked to find a best general algorithm. It does not take account, however, of *structure* in naturally occurring classes of problems. The theorem assumes a uniform probability distribution over fitness functions, and closure under permutation. It appears that naturally occurring classes of problems are unlikely to be closed under permutation. Igel and Toussaint [6] show that the fraction of non-empty subsets that are closed under permutation rapidly approaches zero as the cardinality of the search space increases, and that constraints on steepness and number of local minima lead to subsets that are not closed under permutation. (Similar results are provided for non-uniform probability distributions.)

Thus for naturally occurring classes of structured problems, one might expect (as is intuitive) that some algorithms generally perform better than others.

3 Randomised Self-similar Landscapes

In this section we highlight some of the main issues involved in selecting the problem domain and generating instances for evaluation. The reader is referred to [8] for a more detailed treatment.

3.1 Why Landscapes?

As well as being familiar and intuitive, (suitably detailed) landscapes exhibit many of the properties that raise difficulties in search and optimization algorithms. For example, landscapes have a very large number of local minima (sometimes referred to as *highly multimodal*). However these are achieved naturally and at random, unlike mathematical functions such as F6 mentioned earlier.

Secondly, landscapes are more complex than functions such as F6 in that detail does not diminish with scale. Natural landscapes are considered to exhibit (or at least approximate) the fractal property of statistical self-similarity. As one zooms in, rather than reaching a scale in which the landscape is smooth, successively more detail reveals itself. This is particularly important in regard to the following two issues that arise in population-based and hybrid algorithms.

Exploration versus Exploitation. One of the most difficult issues in designing optimisation algorithms is managing the balance between *exploration* (searching for better regions) and *exploitation* (converging on local minima). In many contrived problem domains knowledge of scale of the fitness function allows one to implicitly "cheat" in tuning the parameters of the algorithm. As a simple illustration consider a bowl function (such as F1 in [2]). Any reasonable algorithm can solve this problem quickly using either local search or population convergence. This is because prior information is (implicitly) provided that the minimum lies within the given bounds, and therefore parameters can be chosen to favour convergence.

If we assume, however, that the bowl *could* be a small perturbation in a much larger picture, the implications of the parameter choice change entirely and sufficient exploratory action is required. If the available computing resource is, for example, number of evaluations, we are faced with a very difficult problem of how best to use them.

Population-based Search Versus Local Search. One approach to dealing with the exploration/exploitation problem has been to propose hybrid algorithms combining population-based methods with local search. Again, when the scale at which the detail occurs is known or bounded, these algorithms can perform above their general potential.

To continue the bowl illustration, consider say an egg box (or muffin tray). It would be very easy to parameterize a hybrid algorithm so that the population operators search "across" the egg cups, while local search is used to rapidly find minima within the cups. This may report excellent results for this problem, but perform extremely poorly when the number of minima ("cups") within each cup (and in turn within those cups) is increased — suddenly the scales at which the population-based and local search are operating are inappropriate.

We have chosen automatically generated pseudo-random self-similar surfaces as our problem domain to bring out these challenges. Algorithms that do well on these problems should generalize easily to problems with more limited scale.

3.2 Generating the Landscapes

A Meteor Impact Algorithm. The most widely used algorithms for random fractal landscape generation are *midpoint displacement algorithms*. However they cannot be used for this purpose since the number of midpoints that must be stored or generated increases exponentially with the depth of iteration. This may not be a problem if landscapes are being generated for human viewing, however we wish to generate the landscapes with detail to the limit of the computer's resolution (which we take to be 64 bit IEEE floating point).

We have developed an alternative approach in which the depth of a point can be generated independently of its neighbours. The approach is based on a

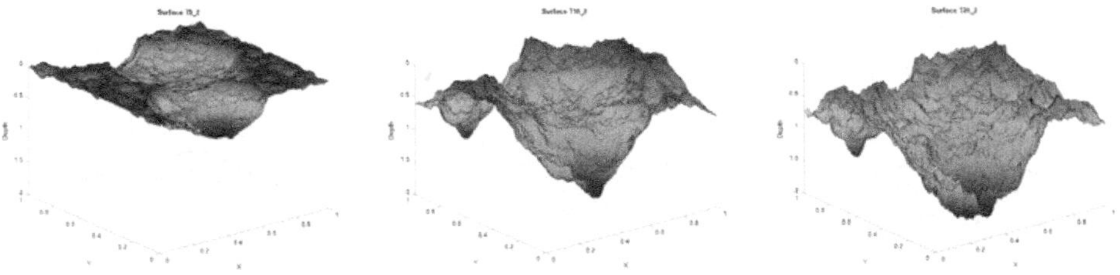

Fig. 2. Training series surfaces 5_2, 10_2 and 20_2 showing successive maturation of the moon surface by meteor impact. Each surface is twice as old as its predecessor.

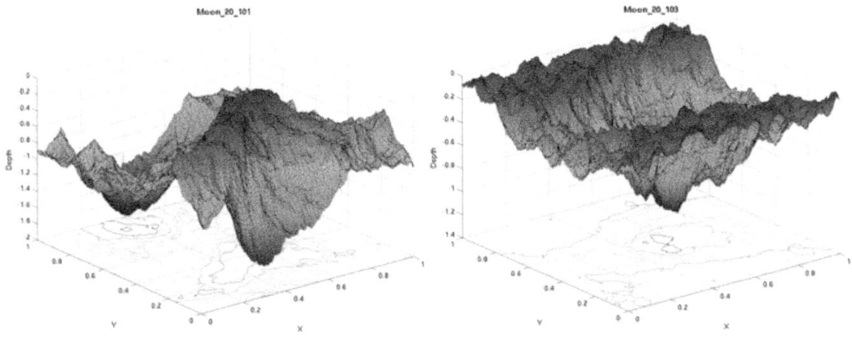

Fig. 3. Two example landscapes from the benchmarking series (20_101 and 20_103). Notice (a) that both wrap in both x and y directions, and (b) that a small difference in seed (101 vs 103) generates entirely unrelated landscapes.

simplified model of the natural process of meteors impacting a moon (or planetary) surface. Intuitively a new moon begins as a sphere, although for practical reasons we use a surface that wraps in both x and y directions. As each meteor or boulder hits the surface it leaves a crater proportional to the size of the boulder. This can be seen in the sequence of surfaces shown in Figure 2.

Seeding a Suite of Landscapes. Using a single landscape for benchmarking would reward optimizers that are overly specific, or overtrained, and may generalize poorly. We therefore use a large suite of landscapes, each of which can be (re)generated from a unique index using a portable (multiplicative congruential) pseudo-random generator [9]. Since sequential seeds produce related random deviates in multiplicative congruential algorithms, we first scatter or hash the index to produce the actual seed used by the generator for each landscape.

Figure 3 shows two example landscapes illustrating the distinctly different outcomes from different indices or seeds.

3.3 Implementation

As with midpoint displacement algorithms, it is not possible to generate and store the surfaces in advance as the storage required is many orders of magnitude

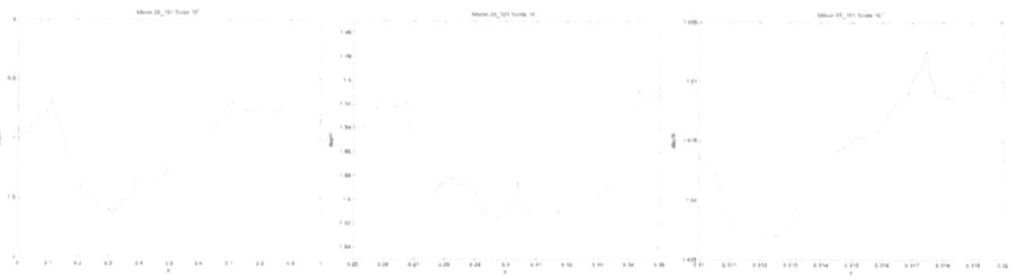

Fig. 4. A one dimensional slice of surface 20_101 shown at magnifications of 10^0 to 10^2. The detail at the minimum of each plot is shown in the subsequent plot.

too great. It is not possible to store all the boulders for the same reason — the number increases exponentially with resolution. Similarly, it is not possible to generate all the boulders when evaluating the surface. What is needed is an algorithm that generates just the boulders that can affect the point (or co-ordinates) of evaluation. This number increases only linearly with resolution (since self-similarity requires that the *average* number of boulders per unit square at any resolution is the same).

We achieve this using a recursive approach. The trick is to associate a recursively generated seed with each unit square at each scale. That is, the seed associated with a square at scale $10^{-(m+1)}$ is uniquely determined by its parent seed at scale 10^{-m}, along with its relative position within that parent square. Thus all seeds are uniquely determined from the surface index. The contribution of boulders within that unit square to the depth of a co-ordinate point can then be determined from that seed, and the contribution from different resolutions summed.

To illustrate that the self-similarity property holds we have included in Figure 4 cross sections of the first landscape from Figure 2 at successively smaller scales. Notice that there is no systematic change in the amount of detail (number of local minima, height of peaks, etc) as one progresses through the sequence. Indeed the images could be shuffled in any order without looking out of place.

4 The Huygens Server

Now that we have addressed the content of the benchmarking suite, we briefly address the question of making it readily accessible to all users. The benchmark server is designed to satisfy the following principles:

1. Access to the benchmarking software and results should be freely available through the web.
2. The user should be able to develop their algorithm in the programming language of their choice, and run it on their platform of choice.
3. The user should be free of any need to understand (or even see) the benchmarking code.

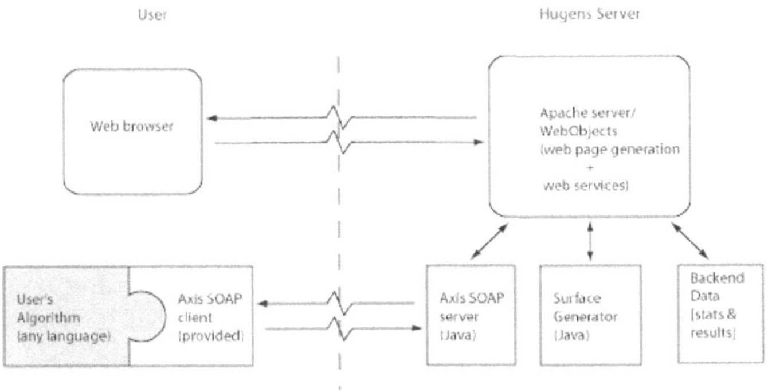

Fig. 5. Architecture of the Huygens Benchmark Server

4. The user should be able to initiate the benchmarking process. (For example, they should not have to submit an algorithm and rely on another human at the "other end" to run it.)

The primary design decisions that allow these principles to be satisfied are the separation of the user and benchmarking (server) environments, and communication via the language independent SOAP protocol [10] for method invocation. The user's algorithm runs in his or her own environment, and "plugs into" the SOAP client. This structure is illustrated in Figure 5.

4.1 Comparing Algorithms

For the purposes of training or fine-tuning algorithms, the server allows unlimited access to a series of training surfaces. For benchmarking, however, some measure of computing resource consumed is required for comparison.

In evolutionary algorithm research a wide variety of measures have been used. Examples include number of epochs or iterations to reach the global minimum (which in our case is unknown) to within a given accuracy, best value achieved in a given number of evaluations, average population fitness, through to environment-dependent measures such as CPU time.

In order to allow comparison of a very broad range of algorithms, including hybrid and traditional search algorithms, we have steered clear of paradigm-specific concepts such as epochs. Similarly to support multiple computing environments we have steered clear of environment-dependent measures. We have taken the view that in most practical applications the expensive operation is fitness evaluation. We therefore allow each algorithm a fixed number of evaluations, or probes, for each landscape to produce its best solution.

We also need to make a decision on how to score algorithms. Because of the fractal nature of the landscapes, the *differences* in raw minimums achieved will decrease exponentially as improvements are made at higher resolutions. Therefore we need to "stretch" those differences as the scale decreases. We achieve this by mapping them onto an exponential function. Further details on scoring and comparison can be found in [8].

5 Conclusion

While arguments such as the No Free Lunch Theorem mean that it will never be possible to identify the "universal best" general purpose optimization algorithm, it is nevertheless vital to develop some definitive and widely accessible means of comparing algorithms. This paper presents one attempt at such a system.

We have argued that the problem domain chosen is intuitively simple, naturally appealing, and challengingly complex, while overcoming technical issues to provide efficient, high resolution fitness functions. We have also presented an architecture for accessing the system that overcomes problems of language and environment incompatibilities. The system allows the user to plug in their algorithm and initiate automated benchmarking and subsequent scoring.

We hope that this system will provide a valuable resource to the evolutionary algorithms research community.

References

1. MacNish, C.: Huygens benchmarking suite. http://gungurru.csse.uwa.edu.au/cara/huygens/ (2006)
2. De Jong, K.A.: Analysis of the behaviour of a class of genetic adaptive systems. PhD thesis, University of Michigan, Ann Arbor, MI (1975)
3. Schaffer, J.D., Caruana, R.A., Eshelman, L.J., Das, R.: A study of control parameters affecting online performance of genetic algorithms for function optimization. In Schaffer, J.D., ed.: Proc. 3rd International Conference on Genetic Algorithms, San Mateo, CA, Morgan Kauffman (1989) 51–60
4. Spears, W.M.: Genetic Algorithms (Evolutionary Algorithms): Repository of test functions (2006) http://www.cs.uwyo.edu/~wspears/functs.html.
5. Spears, W.M., Potter, M.A.: Genetic Algorithsm (Evolutionary Algorithms): Repository of test problem generators (2006) http://www.cs.uwyo.edu/~wspears/generators.html.
6. Igel, C., Toussaint, M.: A no-free-lunch theorem for non-uniform distributions of target functions. J. Mathematical Modelling and Algorithms **3** (2004) 313–322
7. Wolpert, D.H., Macready, W.G.: No Free Lunch theorems for optimization. IEEE Transactions on Evolutionary Computation **1** (1997) 67–82
8. MacNish, C.: Benchmarking evolutionary and hybrid algorithms using randomised self-similar landscapes: Extended version. Technical report, University of Western Australia, School of Computer Science & Software Engineering (2006)
9. Press, W.H., Teukolsky, S.A., Vetterling, W.T., Flannery, B.P.: Numerical Recipes in C. 2nd edn. CUP (1992)
10. W3C XML Protocol Working Group: SOAP Version 1.2. http://www.w3.org/TR/soap12-part0/ (2006)

Interactive Genetic Algorithms Based on Implicit Knowledge Model

Yi-nan Guo, Dun-wei Gong, and Ding-quan Yang

School of Information and Electronic Engineering, China University of Mining and Technology, Xuzhou, 221008 Jiangsu, China
nanfly@126.com

Abstract. Interactive genetic algorithms depend on more knowledge embodied in evolution than other genetic algorithms for explicit fitness functions. But there is a lack of systemic analysis about implicit knowledge of interactive genetic algorithms. Aiming at above problems, an interactive genetic algorithm based on implicit knowledge model is proposed. The knowledge model consisting of users' cognition tendency and the degree of users' preference is put forward, which describes implicit knowledge about users' cognitive and preference. Based on the concept of information entropy, a series of novel operators to realize extracting, updating and utilizing knowledge are illustrated. To analyze the performance of knowledge-based interactive genetic algorithms, two novel measures of dynamic stability and the degree of users' fatigue are presented. Taking fashion design system as a test platform, the rationality of knowledge model and the effective of knowledge induced strategy are proved. Simulation results indicate this algorithm can alleviate users' fatigue and improve the speed of convergence effectively.

1 Introduction

Interactive genetic algorithms(IGAs) are a kind of genetic algorithms in which fitness of individuals is evaluated by human subjectively. They combine canonical genetic algorithms with human intelligent evaluations. Now they have been applied to optimization problems whose objectives can not be expressed by explicit fitness functions, such as music composition, production design and so on[1].However there exits users' fatigue in evaluation, which limits population size and the number of evolutionary generations. Moreover, human subjective fitness mainly reflects users' cognition and preference, which is related to domain knowledge. So IGAs need more knowledge than other genetic algorithms for explicit optimization functions.

Domain knowledge was classified into priori knowledge and dynamic knowledge according to the extracted methods by Giraldez[2].The former is obtained from problems directly before starting a search. So they are relatively constant. The latter is extracted from an evolution and updated throughout the evolution. In this paper, we are interested in the latter one. In IGAs, this kind of knowledge is implicit and reflects users' cognition and preference, called implicit knowledge. The key problems about

T.-D. Wang et al. (Eds.): SEAL 2006, LNCS 4247, pp. 369 – 376, 2006.

implicit knowledge are how to represent, extract and utilize it so as to alleviate users' fatigue and improve the speed of convergence. Up to now, many knowledge-induced evolutionary strategies have been proposed. A knowledge model based on artificial neural networks was adopted to update fitness of bad individuals so as to drive them to good solutions[3][4]. Sebag and Fan presented a rule-based knowledge model generalized by induction learning to improve the generalization of solutions[5][6]. Based on dual structure, cultural algorithms and co-evolutionary GA were proposed, which extract schema to induce the evolution in search space[7][8]. However, the fitness functions of above knowledge-based algorithms are explicit. It is necessary to build a knowledge model aiming at IGAs for implicit optimization functions. So an interactive genetic algorithm based on implicit knowledge model(IK-IGA) is proposed in this paper.

2 Description of Implicit Knowledge

In IGAs, users' cognition and preference embodied in the evolution are extracted from samples and other information about evolution. Samples consist of genotype of individuals and corresponding fitness evaluated by users. In order to simplify the description, an implicit knowledge model based on characteristic-vector is adopted. Define a characteristic-vector C as

$$C=[W \quad P \quad \zeta]^{\mathrm{T}} \tag{1}$$

where $W=[w_1 \; w_2 \ldots w_M]$ and $P=[p_1 \; p_2 \ldots p_M]$ expresses users' cognition tendency and the degree of users' preference to each locus respectively. ζ is the reliability of a characteristic-vector. M is the length of an individual.

In many problems solved by IGAs, individuals consist of some building blocks with special meaning which consist of some neighboring loca. They are called gene-meaning-unit(GM-unit)[9].Suppose individuals take the form of bit strings. Assuming n is the number of GM-unit and m_i is the number of loca included in i^{th} GM-unit which satisfy $M=\sum_{i=1}^{n} m_i$ .Each unit $V_i=\{v_{i1},v_{i2},\ldots,v_{il_i}\}(i<n)$ has $l_i=2^{m_i}$ possible meanings,called attributes. Based on GM-unit, implicit knowledge model is extended as $W=[w_1 \; w_2 \ldots w_n]$ and $P=[p_1 \; p_2 \ldots p_n]$ where w_i denotes the distribution of attributes in i^{th} GM-unit and p_i expresses the importance of ith GM-unit among all units.

There exits users' fatigue during the evaluation in IGAs. More individuals evaluated by users and more similar individuals in each generation will increase users' fatigue. When users feel tired, fitness evaluated by users can not reflect users' true preference any more. This lowers the reliability of extracted implicit knowledge. So the reliability of characteristic-vector is defined as

$$\zeta^t = e^{-t \bullet Sd^t} \tag{2}$$

where t is the number of generation. $Sd^t = \frac{N(N-1)}{2} \Big/ \sum_{i=1}^{N-1} \sum_{j=i+1}^{N} d(x_i,x_j)$ denotes the similarity of population where $d(x_i,x_j)=|x_i-x_j|$ is the Hamming distance between x_i and x_j.

3 Operators for Knowledge

3.1 Extraction

This operator is adopted to extract implicit knowledge described by characteristic-vector from sample-population $\overline{S}$ by statistical learning method.

Extraction of Users' Cognition Tendency. In a GM-unit, attributes which users most like have more probability to exist in offspring. Define the survival probability $\rho_{v_{ij}}$ of jth attribute in i^{th} GM-unit as

$$\rho_{v_{ij}} = \frac{1}{K}\sum_{k=1}^{K} f\left(s_i^k, v_{ij}\right), s_i^k \in \overline{S} \tag{3}$$

where $f\left(x_i, v_{ij}\right) = \begin{cases} 1 & d\left(x_i, v_{ij}\right) = 0 \\ 0 & d\left(x_i, v_{ij}\right) \neq 0 \end{cases}$ is the distance of an attribute of an individual. s_i^k is the value of ith GM-unit of kth sample. $K=INT(\rho_s \cdot N)$ is the number of samples. ρ_s is the probability of sampling. $INT(x)$ rounds the elements of x to the nearest integers. It is obvious that $\rho_{v_{ij}}$ is larger when more samples include v_{ij}.

Define average survival probability of ith GM-unit as $\eta_{V_i} = 1/l_i$ .According to the relationship between $\rho_{v_{ij}}$ and η_{V_i}, users' cognition tendency is extracted as follows.

Rule A: $\left\{w_i = \rho_{v_{ij}} \mid \left(\rho_{v_{ij}} > \eta_{V_i}\right) \wedge \left(\rho_{v_{ij}} > \rho_{v_{ik}}\right), \forall v_{ij}, v_{ik} \in V_i, j \neq k, j,k \leq l_i\right\}$

Rule B: $\left\{w_i = 0 \mid \left(\rho_{v_{ij}} \leq \eta_{V_i}\right), \forall v_{ij} \in V_i,, j \leq l_i\right\}$

Rule C: $\left\{w_i = 0 \mid \left(\rho_{v_{ij}} = \rho_{v_{ik}}\right) \wedge \left(\rho_{v_{ij}} > \eta_{V_i}\right) \wedge \left(\rho_{v_{ik}} > \eta_{V_i}\right), \forall v_{ij}, v_{ik} \in V_i, j \neq k, j,k \leq l_i\right\}$

Rule A shows that users favor the attributes which have the maximum survival probability exceeding average survival probability. In other words, users' cognition tendency about this unit is clear. So these attributes are extracted as good schema which will direct the evolution to good solutions. But if one more attributes with the maximum survival probability as shown in Rule B, it indicates users can not clearly judge which attribute is better. So these attributes can not give specific direction during the evolution. Rule C indicates there is no valuable information embodied in this GM-unit and users' cognition tendency is not clear.

Extraction of The Degree of Users' Preference. When users pay more attention to a GM-unit, users' cognition tendency of this unit is more accurate. That is, this unit is in higher proportion to all units. So the degree of users' preference is formed as follows.

$$p_i = w_i \Big/ \sum_{i=1}^{n} w_i \tag{4}$$

Schema Based on GM-unit. Based on users' cognition tendency, characteristic-individual $Vo=[vo_1\ vo_2 \ldots vo_n]$ is extracted in which $Vo=\begin{cases} v_{ij} & w_i \neq 0 \\ (*)^{m_i} & w_i = 0 \end{cases}$. Combing with the definition of schema proposed by Holland[10], schema based on GM-unit is defined as

$$CH=\{a|a \in SV^i, \forall i, CH_i \neq v_{ij}^* \Rightarrow CH_i=a_j, i=1,2,\ldots n, j=1,2,\ldots l_i\} \tag{5}$$

where $a=(a_1,a_2,\ldots a_n)$, $CH=(CH_1,CH_2,\ldots CH_n)$, $a_i \in SV^i$, $CH_i \in SV_e^i$, $v_{ij}^*=(*)^{m_i}$. $SH^{m_i}=\{0,1\}^{m_i}$ is gene space. $SH_e^{m_i}=\{0,1,*\}^{m_i}$ is extended gene space. $SV^i=\{V_i\} \subset SH^{m_i}$ is attribute space. $SV_e^i=\{V_i, v_i^*\} \subset SH_e^{m_i}$ is extended attribute space.

3.2 Update

Users' cognition varies and becomes clearer during the evolution. So characteristic-vector which reflects users' preference is variable and needs to be updated dynamically so as to exactly track users' cognition in time. Two key problems which influence the speed of track and the efficiency of computation are how to update samples in sample-population and the frequency of extracting knowledge.

Update Strategy for Samples. Individuals in evaluated population are added to sample-population if they improve the diversity of sample-population and provide more information which contributes to the evolution than before. In order to describe the quantity of information, information entropy of sample-population before and after adding x_i in t^{th} generation is define as $H(\bar{s}^{t-1})$ and $H^t(x_i)$.

$$H^t(x_i) = \frac{1}{n}\sum_{i=1}^{n}\left(\frac{1}{l_i}\sum_{j=1}^{l_i} -\overline{\rho_{v_{ij}}}\, lg\, \overline{\rho_{v_{ij}}}\right) \tag{6}$$

$$\overline{\rho_{v_{ij}}} = \frac{1}{K+1}\sum_{k=1}^{K+1} f\left(\bar{s}_i^k, v_{ij}\right), \bar{s}_i^k \in \left\{\bar{S}^{t-i}, x_i\right\}, x_i \in I \tag{7}$$

Letting $F^t(x_i)$ denotes fitness of x_i. A novel update strategy is shown as follows.

Step1: Rank population according to $F^t(x_i)$ and $H^t(x_i)$ respectively.
Step2: Individual with maximum fitness is added to sample-population directly.
Step3: Other individuals which satisfy $H^t(x_i)>H(\bar{s}^{t-1})$ are added to $\bar{s}$.
Step4: Rank new sample-population according to fitness.
Step5: Individuals with minimum fitness are move from new sample-population.

The number of moved individuals is equal to the number of added individuals.

The Frequency of Extracting Knowledge ω_u. ω_u is a key parameter which decides the speed of updating knowledge. It depends on the similarity among samples. While similar samples are more, the diversity of $\bar{s}$ is worse and redundant information embodied in $\bar{s}$ is more. ω_u is defined as

$$\omega_u = e^{-ds(\bar{s})} \tag{8}$$

where $ds(\bar{s})$ denotes the similarity of $\bar{s}$.

3.3 Direction

Each genetic operation can be influenced by implicit knowledge. Here, a novel knowledge-inducing strategy for replacing individuals is proposed. Characteristic-individual *Vo* is adopted as the constraint to limit the number of similar individuals in the population of next generation. Considering users' preference to each GM-unit, the weighted Hamming distance is defined as

$$d_s = \zeta \sum_{i=1}^{n} p_i f\left(x_i, vo_i\right) \tag{9}$$

The detailed steps of the strategy are shown as follows.

Step1: Calculate d_s between *Vo* and each individual in offspring population.
Step2: Move individuals which satisfy d_s=0.Go back to Step1 with a new individual generated randomly.
Step3: Individuals which satisfy $d_s \neq 0$ are placed in new population.

4 Two Measures for IK-IGA

Convergence and the speed of convergence are main measures to analyze the performance of genetic algorithms. But there is no measure for analyzing transition performance of an evolution using IK-IGAs. Moreover, users' fatigue is a key problem in IGAs. But there is no measure to describe the degree of users' fatigue directly. Aiming at above problems, two novel measures including the measure of dynamic stability and the degree of users' fatigue are proposed.

The Degree of Users' Fatigue. The main factors, which reflect users' fatigue, include the number of generation *Ns*, the similarity of population *Sd* and waiting time during the evaluation *tw*. So the degree of users' fatigue *Fa* is defined as follows.

$$Fa=1-e^{-Ns \cdot Sd \cdot tw} \tag{10}$$

The Measure of Dynamic Stability. Combing with the transition analysis in automation control[11],a novel measure of dynamic stability including error and the cost of computation is proposed. The increment of computation time caused by operations about knowledge is called the cost of computation. Letting Ts^t,Ta^t,Tu^t and Ti^t denote the time spent on sampling, extraction, update and direction respectively. The cost of computation is defined as $Ct^t=Ts^t+Ta^t+Tu^t+Ti^t$. It reflects the performance of transition time of IK-IGAs. Assuming x^* is the optimal solution. Error is defined as $e^t=min\{d(x_i^t,x^*)|\forall x_i^t \in X^t\}$. It reflects the spatial performance of IK-IGAs.

These measures are related to the frequency of extracting knowledge ω_u. Under extremely large ω_u, knowledge is updated continually which causes increasing of *Ct*. But error fluctuates in a small range. So this state is called over-induced evolution. Under extremely small ω_u, knowledge can not timely track users' cognition which misguides the evolution and causes large fluctuation of error. So this state is called under-induced evolution. It is obvious that too large or too small ω_u will lead to increasing of evolution generation and slow down the speed of convergence.

5 Simulations and Analysis

In this paper, fashion design system is adopted as a typical background to validate the rationality of IK-IGAs. Visual Basic6.0 as programming tool and Access as database are utilized. In fashion design system, each dress is composed of collar, skirt and sleeve. Each part has two factors including pattern and color. If each factor is described by two bits binary code, each dress will be expressed by 12 bits, which act as 6 GM-units. They are collar's color and pattern, skirt's color and pattern, sleeve's color and pattern orderly[9]. The attributions of each GM-unit are shown in Table.1.

Table 1. The attributions of GM-units

code of attributions		name collar's pattern	sleeve's pattern	skirt's pattern	color
1	name	medium collar	long sleeve	formal skirt	pink
	code	00	00	00	00
2	name	high collar	medium sleeve	long skirt	blue
	code	01	01	01	01
3	name	wide collar	short sleeve	medium skirt	black
	code	10	10	10	10
4	name	gallus	non-sleeve	short skirt	white
	code	11	11	11	11

To validate the rationality of implicit knowledge model, the algorithm proposed in this paper is compared with canonical IGA in two groups of experiments. These experiments have different desired objectives which reflect different psychological requirements of human. Desired objective in Group1 is to find a favorite dress fitting for summer without the limit of color. Desired objective in Group2 is to find a favorite dress fitting for summer and the color is pink. During the experiments, the values of main parameters using in IK-IGA is: crossover rate Pc=0.5, mutation rate Pm=0.01, N=8, stop evolution generation T=30, K=20, ρ_s=0.6.

Table 2. Testing results by canonical IGA and IK-IGA($\overline{Ns}$ and $\overline{T}$ denote average number of individuals evaluated by users and average evolution generation in experiments. x^* and Vo denote optimum individual and characteristic-individual which appear in most of results).

Experiment No.	Group1		Group2	
algorithms	IGA	IK-IGA	IGA	IK-IGA
$\overline{T}$	28	10	40	16
$\overline{Ns}$	224	80	320	128
x^*	111011001111	111011001111	111011001111	110011001111
Vo	-	****11**11**	-	****110011**

20 persons are divided into two testing group. Each group does one experiment. Testing results are shown in Table.2. It is obvious that IK-IGA converges faster than IGA averagely. When IK-IGA is adopted, $\overline{Ns}$ reduces 62.1% averagely. These indicate implicit knowledge model and corresponding inducing strategies are beneficial to

alleviating users' fatigue. *Vo* in different experiments are different, which reflect key characters of desired objectives accurately .This shows *Vo* can track users' preference exactly. Moreover, comparison of results between two groups show while the constraint of desired objectives are more, the speed of convergence is slower. This matches the rules of users' cognition. All above indicate knowledge-inducing strategy is effective and rational.

The degree of users' fatigue *Fa* in above experiments is plotted in Fig1.It is obvious that user feel more tired along with the evolution. But *Fa* is different in different experiments. Based on the same algorithms, *Fa* under desired objective in Group1 is lower than it in Group2 on each generation, as shown in Fig.1(a) and (b).In Group1, desired objective have less constraint than it in Group2.This means users are easy to feel tired when there are more GM-units concerned by them. Under the same desired objectives, users quickly feel tired adopting IGA, as shown in Fig.1(c) and (d). It is obvious that IK-IGA can effectively alleviate users' fatigue.

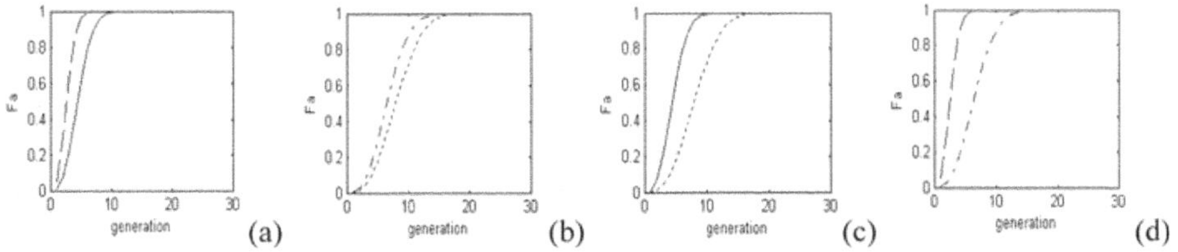

Fig. 1. The degree of users' fatigue(Solid line and dashed denote *Fa* using IGA under different desired objectives. Dotted line and dashdotted line denote *Fa* using IK-IGA under different desired objectives).

To validate the influence of ω_u to IK-IGA,15 persons are divided into 3 groups to do three groups of experiments with different ω_u and same desired objective in Group2.Testing results shown in Table.3 indicate that too large or too small frequency can lead to bad convergence. However, ω_u adopted in IK-IGA is self-adaptive and varies according to the state of sample-population. Results show this updating strategy can improve the speed of convergence effectively.

Table 3. Testing results of different ω_u

the frequency of extracting knowledge ω_u	1	$e^{-ds(\overline{S})}$	0.2
$\overline{T}$	21	17	24

6 Conclusions

Aiming at systemic analysis and effective application about implicit knowledge embodied in IGAs, an interactive genetic algorithm based on implicit knowledge model is proposed. A novel knowledge model based on characteristic-vector is adopted to describe implicit knowledge about users' cognitive tendency the degree of users' preference. Four types of operators are given to realize extracting, updating and utilizing knowledge. Two novel measures about dynamic stability of evolution and the degree of users' fatigue are proposed to analyze the performance of knowledge-based interactive genetic algorithms. Taking fashion design system as a testing platform, the rationality of the knowledge model and the validity of knowledge-inducing strategy are proved. Simulation results indicate this algorithm can alleviate users' fatigue and improve the speed of convergence effectively. The knowledge migration strategy for multi-user based on IK-IGAs is the future research.

Acknowledgements

This work was supported by the China National Natural Science Foundation under Grant 60304016, the National Postdoctoral Science Foundation of China under grant 2005037225, the Postdoctoral Science Foundation of Jiangsu under grant 2004300, the Youth Science Foundation of CUMT under grant OC4465.

References

1. Takagi H.Interactive Evolutionary Computation:Fusions of The Capabilities of EC Optimization and Human Evaluation, Proc. of the IEEE CEC, (2001) 1275-1296
2. R.Giraldez,J.S.Aguilar-ruiz,J.C.Riquelme.Knowledge-based Fast Evaluation for Evolution-ary Learning.IEEE trasaction on SMC-Part C:Application and Review, 35(2005) 254-261
3. Furuya H. Genetic Algorithm and Multilayer Neural Network. Proc. of Calculation and Control, (1998)497-500
4. Gu Hui, Gong Yu-en, Zhao Zhen-xi. A Knowledge Model Based Genetic Algorithm. Computer engineering, 26 (2000)19-21
5. Sebag M, Ravise C, Schoenauer M. Controlling Evolution by Means of Machine Learning. Evolutionary Programming, (1996)57-66
6. Fan Lei, Ruan huai-zhong, Jiao Yu,et al. Conduct Evolution Using Induction Learning. Journal of University of Science and Technology of China, 31 (2001)565-634
7. H.Handa, T.Horiuchi,O. Katai, et al. Co-evolutionary GA with Schema Extraction by Machine Learning Techniques and Its Application to Knapsack Problem. IEEE Conference of Evolutionary Computation, (2001)1213-1219
8. Reynolds,G.R. An Introduction to Cultural Algorithms. Proc. of the 3rd Annual Conference on Evolutionary Programming, (1994)131-139
9. Hao Guo-sheng, Gong Dun-wei, Shi You-qun. Interactive Genetic Algorithm Based on Landscape of Satisfaction and Taboos. Journal of China University of Mining & Technology, (2005)204-208
10. Zhang Wen-xiu,Liang Yi.Mathematical Foundation of Genetic Algorithms. Xi'an Jiaotong University Press, (1999)
11. Morris Driels. Linear Control Systems Engineering. McGraw-Hill. (1996)

An Evolutionary Fuzzy Multi-objective Approach to Cell Formation

Chang-Chun Tsai[1], Chao-Hsien Chu[2,*], and Xiaodan Wu[3]

[1] Department of Industrial and Information Management, National Cheng Kung University, Tainan 701, Taiwan
jimtsai@mail.ncku.edu.tw
[2] College of Information Sciences and Technology, The Pennsylvania State University, 301K IST Building, University Park, PA 16802, USA
chu@ist.psu.edu
[3] School of Management, Hebei University of Technology, Tianjin 300130, China
wxdan2000@yahoo.com.cn

Abstract. Fuzzy mathematical programming (FMP) has been shown not only providing a better and more flexible way of representing the cell formation (CF) problem of cellular manufacturing, but also improving solution quality and computational efficiency. However, FMP cannot meet the demand of real-world applications because it can only be used to solve small-size problems. In this paper, we propose a heuristic genetic algorithm (HGA) as a viable solution for solving large-scale fuzzy multi-objective CF problems. Heuristic crossover and mutation operators are developed to improve computational efficiency. Our results show that the HGA outperforms the FMP and goal programming (GP) models in terms of clustering results, computational time, and user friendliness.

1 Introduction

Mathematical programming (MP) is a common approach to modeling combinatorial optimization problems. However, most elements in MP such as goals, constraints, and parameters, cannot be decided or defined precisely that reduce its practical value. For this reason, fuzzy set theory has been applied widely to solve a variety of real world optimization problems, such as transportation, facility layout, reliability, scheduling, project management, communication network design, etc. [7, 19].

Cellular Manufacturing (CM) is a potential approach that can be used to enhance both flexibility and efficiency in today's small-to-medium lot production environment. CF, which concerns the problem of grouping parts with similar design features or processing requirements and corresponding machines into manufacturing cells, is the key step in implementing CM. Over the past few decades, many analytical methods have been proposed for solving the CF problems. A thorough review of literature can be found, for example, in [4, 5, 16]. Applying fuzzy set theory to CF is a relatively new attempt [14, 15]. A major factor causing the clustering bottleneck in CF is the existence of exceptional elements (EEs), the parts that need to be processed by machines

* Corresponding author. The paper was completed while the author was on leave to Singapore Management University (8/2005 - 7/ 2006).

T.-D. Wang et al. (Eds.): SEAL 2006, LNCS 4247, pp. 377 – 383, 2006.

not in the same cell. Most prior research removes EEs manually or deals with them after forming the initial cells [5, 12]. In [14] and [15] the authors used a FMP approach to dealing with EEs while forming manufacturing cells. They show that FMP performed better than traditional MP in terms of solution quality and executing efficiency. However, when problem size increases, FMP performance deteriorates [14].

Genetic Algorithms (GAs) have been used extensively as an alternative method for solving optimization problems in a wide variety of application domains including engineering, economics, business, telecommunications, and manufacturing [7, 8]. Notwithstanding, GA, is a powerful heuristic search approach that can solve complicated CF problems [6, 7, 18]. Despite of its effectiveness, several issues in GA yet need to be explored in order to get good solutions: (1) The selection of fitness function; (2) The selection and design of heuristic genetic operators: reproduction, crossover, and mutation; and (3) The determination of system parameters: population size, # of generations, crossover probability, and mutation probability. In this paper, an efficient HGA, considering the domain-specific characteristics of CF, was developed to solve the fuzzy multi-objective CF problems. Its performance was compared with FMP and GP approaches.

2 Mathematical Formations

In a fuzzy environment, fuzzy constraints, vague goals, and ambiguous parameters can be taken into consideration in the MP model. Several membership functions can be used to incorporate fuzziness for fuzzy objective functions or parameters [19], among which, non-increasing linear function has shown to perform better than the others for CF. Also, the min-add operator has shown to perform better than other fuzzy operators for CF and the results of asymmetric and symmetric cases of the FMLP models are the same [14, 15]; therefore, in this study, we will use the linear non-linear membership function, min-add operator, and the asymmetric case of FLMP for our comparisons.

2.1 Traditional Models

The proposed bi-objective model is an extension of the model used in Tsai et al. [15], where a second objective (2) is added. Objective function (1) is to minimize the total cost of duplicating machines, inter-cell transfer, and subcontracting. Objective (2) is to maximum total similarity between a pair of parts or machines. The constraint sets and notation can be found in [15]. The single objective model can have either objective (1) or (2). The GP model is the same as that in [14].

$$\text{Min } \sum_k \sum_i A_i R_{ik} + \sum_k \sum_{(i,j)\in sp} I_j Z_{ijk} + \sum_k \sum_{(i,j)\in sp} S_j O_{ijk} \quad (1)$$

$$\text{Max } \sum_{k=1}^{c} \sum_{j=1}^{n} \sum_{j'\neq j}^{n} SC_{jj'} Y_{jk} Y_{j'k} \quad (2)$$

2.2 Fuzzy Multi-objective Linear Programming (FMLP) Model

Applying the non-increasing linear membership function and min-add operator, the objective functions (1) and (2) can be easily modified as:

$$\text{Min } \sum_{k}\sum_{i} A_i R_{ik} + \sum_{k}\sum_{(i,j)\in sp} I_j Z_{ijk} + \sum_{k}\sum_{(i,j)\in sp} S_j O_{ijk} - \lambda P_c \tag{3}$$

$$\sum_{k=1}^{c}\sum_{j=1}^{n}\sum_{j'\neq j}^{n} SC_{jj'} Y_{jk} Y_{j'k} - \lambda P_s \geq Z_s^1 \tag{4}$$

Where (3) is the objective function of minimizing the total cost of dealing with EEs. The λ value for (3) can be obtained from (4)

3 The Proposed Heuristic GA

GAs applications start by encoding the solution of the problem into a chromosome string, followed by population initialization, fitness evaluation, reproduction, crossover, mutation, replacement. The process continued until a predefined termination condition was satisfied [8]. The performance of GAs is highly dependent on the chromosome representation scheme, reproduction operators used, and the systems parameters specified. If these operators and parameters are not properly specified the computational results may not be as good as expected [7, 8]. The proposed HGA uses a reproduction operator similar to that of the traditional GA approach, but replaces the crossover and mutation operators with a new heuristic for each. To explain the major logic of the proposed algorithm, we use an example with 9 machines and 10 parts from [12].

3.1 Chromosome Representation Scheme

The chromosome strings of CF consists of $m + n$ integer genes, in which the first m genes relate to machines and the next n genes correspond to parts; each value of gene stands for the cell number to which a machine or part to be assigned. For example, the chromosome for the illustrated example can be represented as (1 1 2 3 1 2 3 2 1, 2 3 1 3 2 1 2 1 2 3). Here, this chromosome indicates that cell #1 contains machines 1, 2, 5, 9 and parts 3, 6, 8; cell #2 has machines 3, 6, 8 and parts 1, 5, 7, 9 and the cell #3 has machines 4, 7 and parts 2, 4, 10.

3.2 Initial Population

In this study, the initial population was randomly generated to satisfy two problem constraints: (1) the total number of machines in each cell has to be less than or equal to *NM* and (2) the number of machines and parts in each cell have to be greater than one. If the chromosome is infeasible, we simply discard it and regenerate a new one.

3.3 Fitness Function and Reproduction Operator

Similar to the traditional GA approach, we used a roulette wheel selection method to reproduce chromosomes. We used the objective function directly from the FMLP model as the fitness function for evaluation. That is, equation (3) is the main fitness function for FMLP. But since we also need to consider the objective of maximizing total similarity (4), λ value can be calculated from (4) and insert into (3) to obtain the fitness values.

3.4 Heuristic Crossover

Similarity coefficient (SC) has been used in CF to evaluate part similarity. See [14] for details about the SC index that we used. We apply the part similarity idea to improve crossover performance. The process consists of two phases: In the first phase, a traditional single-point crossover is used for part genes. In the second phase, a SC-based heuristic crossover is used to improve the grouping efficiency of part genes.

Single-point crossover. Suppose that the two chromosomes selected for crossover are: Parent 1: 1 1 3 1 1 3 2 3 2, 1 3 1 1 * **2 1 2 3 3 2** and Parent 2: 1 1 2 3 1 2 3 2 1, 2 3 1 3 * 2 1 2 1 2 3. Assuming that the position we randomly selected for crossover is 13 (marked as *), then the two children produced by this phase are: Child 1: 1 1 3 1 1 3 2 3 2, 1 3 1 1 * 2 1 2 1 2 3 and Child 2: 1 1 2 3 1 2 3 2 1, 2 3 1 3 * **2 1 2 3 3 2**.

Heuristic crossover. The logic of our proposed heuristic crossover works as follows:

1. Generating a random number, from the range of [average of *SCs* ± allowance], as the threshold similarity for comparison. The average SC for the example is -0.41. If the allowance for *SCs* is 0.5, then the possible range for generating the threshold similarity is [-0.91, 0.09]. Let us assume that the generated threshold similarity is 0.04.
2. Selecting the children to which the crossover will be performed. This can be done by presetting a probability (normally 0.5), then randomly generate a number for each child. If the random number for a child is larger than the preset probability, that child needs to perform the heuristic crossover; otherwise, that child remains unchanged. Suppose that only the number randomly generated for child 1 is larger than 0.5, then in this case, only child 1 is selected for crossover and child 2 will remain unchanged.
3. Starting from cell #1, randomly select a part j', from a list of existing parts which belong to that cell, as the seed, comparing the similarity value between the seed part and the other parts (i.e., $SC_{jj'}$, $j = 1, 2, ..., n; j \neq j'$) with the threshold similarity. If $SC_{rj'}$ ($r \neq j'$) is larger than the threshold value, merging part r with the seed part j'. This process continues until all cells have been considered. Currently, cell #1 contains four parts 1, 3, 4, and 6. Assuming that part 3 was selected (randomly) as the seed part. Because there is only one part, part 1 ($SC_{13} = 0.14 > 0.04$), qualified, the new cell #1 will contain parts 1 and 3. Assuming that part 7 was selected as the seed part for cell #2, the new cell #2 will contains parts 7 and 10. Consequently, cell #3, will contain the remaining parts, that is, parts 2, 4, 5, 6, 8, and 9. After the heuristic crossover, the new child 1 becomes: 1 1 3 1 1 3 2 3 2, 1 3 1 3 3 3 2 3 3 2.

3.5 Heuristic Mutation

The traditional mutation operator mutates the gene's value randomly according to a small probability of mutation; thus, it is merely a random walk and does not guarantee a positive direction to reach the optimal point. The proposed heuristic mutation remedies this deficiency. The heuristic consists of two modules:

Machine Mutation. The purpose of this module is to mutate gene value for machines, which works as follows:

1. Computing the TP_{ik} values, the total numbers of parts need to be processed by every machine *i* in each cell *k*, and
2. Assigning the cell number having the largest TP_{ik} for each machine gene. If there is a tie, randomly select a cell number having the same number of *TP*.

Part Mutation. The purpose of this module is to mutate the gene value for parts. The process is similar to machine mutation except that we replace TP_{ik} with TM_{jk} - the total number of machines used by each part *j* in each cell *k*.

The heuristic mutation needs to be applied to all children. Let us use child 1 as an example for illustration. In machine mutation, there are three parts, 1, 3 and 10, need to be processed by machine 1; since parts 1 and 3 belong to cell #1 and part 10 belongs to cell #2; the total number of parts to be processed by machine 1 at cell #1 (i.e., TP_{11}) is 2 and the value for cell #2 (i.e., TP_{12}) is 1. Therefore, machine 1 will be assigned to cell #1. The process continued until all machines for this child are considered. For part mutation, there are five machines, 1, 2, 3, 4, and 6, need to be used to process part 1. Since machines 1, 2, and 4 belong to cell #1 and machines 3 and 6 belong to cell #3, part 1 should be assigned to cell #1 (instead of cell #3). The process continued until all parts for this child are considered. The process continued until all children were considered. The new children so generated are: Child 1': 1 3 1 3 3 1 2 3 2, 1 1 1 1 2 1 2 3 3 2 and Child 2': 2 1 3 3 1 2 2 3 2, 2 2 1 1 3 1 3 3 2 1.

4 Computational Experience

Nine data sets from open literature were used to evaluate the relative performance of the proposed HGA. Since most of these data sets did not include relevant information, such as part processing times, costs involved, and part demand, we generated them randomly using a computer program based on the mean value and the variance given in data set 2 [12]. We used the LINDO (linear interactive and discrete optimizer) package running on a PC with Pentium 4 processor to solve all models. The formulations required for running LINDO were generated by a generator coded in BASIC language. The GA was coded in ANSI C. Performances were measured in terms of total cost, CPU time (in seconds), and total similarity. Table 1 summarizes our computational results.

Several observations can be made from the computational results. First, in terms of clustering results, HGA and GP allows us to consider a trade-off between two conflicting objectives while FLP can only consider either objective, thus, HGA and GP is more flexible than FLP. HGA can always produce the same results as those of FMLP and GP, but it is more computationally efficient and can be used to solve larger-size problem within an acceptable timeframe.

Table 1. Summary of computational results

Data Set	Size (m x n)	NM (Pr)	Ref	FMLP		GP		HGA		
				Cost	CPU	Cost	CPU	Cost	CPU	SCs
1	9 x 9	4 (2)	[9]	168,200	822	168,200	2,244	168,200	2.857	3.21
2	9 x 10	4 (2)	[12]	325,892	5,138	325,892	265,982	325,784	7.198	-2.66
3	12 x 10	4 (3)	[11]	220,908	1,359	220,908	43,231	220,947	5.824	5.263
4	12 x 19	4 (3)	[17]	–	–	–	–	788,385	24.945	7.391
5	14 x 24	4 (2)	[10]	67,758	67,534	–	–	67,774	8.791	2.38
6	16 x 30	5 (3)	[13]	–	–	–	–	939,332	10.67	-4.293
7	16 x 43	5 (4)	[1]	–	–	–	–	1,102,827	155.28	-21.85
8	24 x 40	5 (3)	[3]	–	–	–	–	520,178	249.94	49.81
9	40 x 100	6 (3)	[2]	–	–	–	–	2,108,586	229.55	310.31

–: Computer running time exceeds one day (86,400 seconds).

Second, in terms of computational efficiency, the HGA outperformed FMLP and GP models in term of CPU time. Although FMP approach is more computationally efficient than goal programming, when the problem size increases, it becomes impracticable to use FMP or GP to solve the problem. In contrast, the HGA took less than four minutes to solve the problem with 40 machines and 100 parts. Thus, only the HGA is efficient enough to tackle real-world applications.

Third, the HGA is more user-friendly than traditional GAs. The HGA does not use crossover and mutation probabilities in reproducing offspring, thus it can reduce the troublesome of determining these parameters. Also, our experience shown that, for most data sets, a population of 10 or 40 chromosomes and two generations are sufficient to converge to a good solution. The only exception is data set #9, which requires 100 chromosomes as its population.

Finally, the HGA is a flexible design that it can be easily adapted to solve other MP or FMP models for CF, by changing the corresponding fitness function (s). For instance, using objective (1) or (2) as the fitness function, we can solve large-scale single objective CF problems modeling with LP or FLP. Similarly, by changing fitness function to the objective function of goal programming model, we can solve large-scale goal programming CF problems in only just few minutes.

5 Conclusions

In this paper, an efficient heuristic GA, integrating the domain-specific characteristics of CF, was developed to solve the fuzzy multi-objective CF problems. Our computational results indicated that the proposed HGA is fairly efficient and its solution quality is the same as the optimal solution for small-size data sets. But the most distinguished feature is that HGA is more user-friendly. It reduces the troublesome of determining the appropriate crossover and mutation rates, needed by most traditional GA approaches. Moreover, the HGA is flexible that it can be easily adapted to solve CF problems with other MP models, by changing the corresponding fitness function. The significant improvement in computational efficiency can be attributed to the heuristic crossover and mutation operators used.

This study shows that a hybrid genetic algorithms and fuzzy modeling is a practical approach for solving large-scale CF problems. It may have potential for solving other combinatorial optimization problems.

References

1. Burbidge, J.L.: The Introduction of Group Technology. John Wiley and Sons, New York (1975)
2. Chandrasekharan, M.P., Rajagopalan, R.: ZODIAC-An Algorithm for Concurrent Formation of Part-Families and Machine-Cells. Int. J. of Prod. Res. 25(6) (1987) 835-850
3. Chandrasekharan, M.P., Rajagopalan, R.: Groupability: An Analysis of the Properties of Binary Data Matrices for Group Technology. Int. J. of Prod. Res. 27(6) (1989) 1035-1052
4. Chu, C.H.: Clustering Analysis in Manufacturing Cellular Formation. OMEGA. 17 (1989) 289-295
5. Chu, C.H.: Recent Advances in Mathematical Programming for Cell Formation. In: Kamran, A.K. and D.H. Liles (eds.): Planning, Design and Analysis of Cellular manufacturing System. Elsevier science B. V., The Netherlands (1995) 3-46.
6. Chu, C.H., Tsai, C.C.: A Heuristic Algorithm for Grouping Manufacturing Cells. Proceedings of the 2001 IEEE Congress on Evolutionary Computation. Seoul, Korea (2001) 310-317
7. Gen, M., Cheng, R.: Genetic Algorithms and Engineering Design. Wiley Interscience Publication, MA. (1997)
8. Goldberg, D.E.: Genetic Algorithms: in Search, Optimization & Machine Learning. Addison-Wesley, Inc., MA. (1989)
9. Gongaware, T.A., Ham, I.: Cluster Analysis Applications for Group Technology Manufacturing Systems. Proceedings of the 9th North American Manufacturing Research Conference (NAMRC). (1981) 503-508
10. King, J.R.: Machine-Component Group Formation in Group Technology. OMEGA. 8(2) (1980) 193-199
11. McAuley, J.: Machine Grouping for Efficient Production. The Production Engineering. 51 (1972) 53-57
12. Shafer, S.M., Kern, G.M., Wei, J.C.: A Mathematical Programming Approach for Dealing with Exceptional Elements in Cellular Manufacturing Int. J. of Prod. Res. 30 (1992) 1029-1036
13. Srinivasan, G.S., Narendarn, T.T., Mahadevan, B.: An Assignment Model for the Part-Families Problem in Group Technology. International Journal of Production Research. 28(1) (1990) 145-152
14. Tsai, C.C., Chu, C.H.: Fuzzy Multiobjective Linear Programming Model for Manufacturing Cell Formation. Proceedings. of the National Decision Science Conference, November (1996) 1270-1272
15. Tsai, C.C., Chu, C.H., Barta, T.: Analysis and Modeling of a Manufacturing Cell Formation Problem with Fuzzy Integer Programming. IIE Transactions. 29(7) (1997) 533-547
16. Wemmerlöv, U., Hyer, N. L.: Procedures for the Part Family, Machine Group Identification Problem in Cellular Manufacturing. J. of Operations Management. 6 (1986) 125-147
17. Witte, J.D.: The Use of Similarity Coefficients in Production Flow Analysis. Int. J. of Prod. Res. 18(4) (1992) 503-514
18. Wu, X.D., Chu, C.H., Wang, Y.F., Yan, W.L.: Concurrent Design of Cellular Manufacturing Systems: A Genetic Algorithm Approach. Int. J. of Prod. Res. 44(6) (2006) 1217-1241
19. Zimmermann, H.J.: Fuzzy Set Theory and Its Applications. Kluwer Academic Publishers, Boston (1991)

Dual Guidance in Evolutionary Multi-objective Optimization by Localization

Lam T. Bui[1], Kalyanmoy Deb[2], Hussein A. Abbass[1], and Daryl Essam[1]

[1] The Artificial Life and Adaptive Robotics Laboratory, School of ITEE, UNSW@ADFA, Canberra, Australia
{l.bui, h.abbass, d.essam}@adfa.edu.au
[2] Mechanical Engineering Department, Indian Institute of Technology, Kanpur, PIN 208 016, India
deb@iitk.ac.in

Abstract. In this paper, we propose a framework using local models for multi-objective optimization to guide the search heuristic in both the decision and objective spaces. The localization is built using a limited number of adaptive spheres in the decision space. These spheres are usually guided, using some direction information, in the decision space towards the areas with non-dominated solutions. We use a second mechanism to adjust the spheres to specialize on different parts of the Pareto front using the guided dominance technique in the objective space. With this dual guidance, we can easily guide spheres towards different parts of the Pareto front while also exploring the decision space efficiently.

1 Introduction

Evolutionary multi-objective optimization has been applied in numerous domains [4,5]. Researchers have been investigating theoretically as well as empirically the performance of evolutionary multi-objective optimization algorithms (EMOs) on a wide range of artificial optimization problems from combinatorial, real-valued to dynamic and noisy problems. To date, there exists a number of algorithms, such as VEGA [7], SPEA2 [9], PDE [1] and NSGA-II [5]. These algorithms are still being continuously analyzed, compared, and tested under various problems and criteria.

EMOs (or Evolutionary Algorithms in general) are usually blind search techniques in the sense that they do not usually use any auxiliary functions like derivatives (as in traditional deterministic optimization techniques). To reduce the effect of the "blindness", there are increasing number of attempts to incorporate some guidance techniques into EMOs. Basically, some guidance is employed to direct the search towards promising areas satisfying specific criteria, such as avoiding infeasible areas or of approaching particular parts of the Pareto front. Guidance can be done in either decision or objective spaces. In this paper, we hypothesize that interleaving guidance in both the decision and objective spaces can help to accelerate the search process.

T.-D. Wang et al. (Eds.): SEAL 2006, LNCS 4247, pp. 384–391, 2006.

Our proposed idea is to localize the search in the decision space by using the framework of local models [3] that divide the decision search space into a number of local search areas, where each area is seen as a hyper-sphere. In other words, we transform searching in EMOs on the original search space into a sphere-oriented space, where each sphere is running its own version of EMOs. These spheres move following some direction information to improve their local Pareto front. When we apply the guided dominance mechanism [6], they also tend to specialize and move towards different parts of the Pareto Optimal Front (POF).

The remainder of the paper is organized as follows: background information are the methodology and presented in Section 2. An experimental study is carried out in section 3. The last section is devoted to the conclusion.

2 Background

2.1 Guided Dominance Approach

The motivation for guided dominance [2] is that in practical problems we usually need a limited number of sample points of the POF, rather than the whole POF. With guided dominance, the dominance relation is determined from the transformed functions Ω of the original objective functions F, in which the points in the POF area of interest dominate all others in the remaining areas of POF. For the details of this approach, readers are referred to [2].

Based on this approach, Deb et al [6] proposed a technique to divide the POF into a number of parts where each part is tracked by a subpopulation. In order to do this, an equivalent number of weighted matrices is defined to transform the original objective functions. In other words, the dominance relation in each population is defined by a separate weighted matrix. With the partition of the POF in the objective space, the search process is easily guided. Note that this approach only works on problems with a convex POF; and also that is the class of problems we are addressing in this paper.

2.2 Local Models

In local models, the decision search space S is divided up into a number of non-overlapping spheres, where each sphere s_i is defined by a pair of a centroid and radius: $S = [s_0, s_1, ..s_n]$ and $s_i = (c_i, r_i)$. Initially, all r_i are set to be the same value r. Inside each sphere, points are generated uniformly, except for the restriction that they are kept distant from each other by a predefined distance threshold β.

More formally, let D_A^B be the Euclidean distance between two centroids of arbitrary spheres A and B and d_i^j be the Euclidean distance between two arbitrary points i and j inside any arbitrary sphere A where c_A is the centroid of that sphere A. The following condition must then hold:

$$D_A^B \geq 2r, d_i^{c_A} \leq r, d_j^{c_A} \leq r, d_i^j \geq \beta \tag{1}$$

To initialize a sphere s_i, we use a spherical coordinate system. Assume $x = (x_1, x_2, ...x_n)$ is a point in the Cartesian coordinate system. We can calculate x from the parameters of an equivalent spherical coordinate system, including a radial coordinate r, and *n-1* angular coordinates $\alpha_1, \alpha_2...\alpha_{n-1}$ as follows:

$$\begin{aligned}
x_1 &= c_1^i + r\cos(\alpha_1) \\
x_2 &= c_2^i + r\sin(\alpha_1)\cos(\alpha_2) \\
&\cdots \\
x_{n-1} &= c_{n-1}^i + r\sin(\alpha_1)...\sin(\alpha_{n-2})\cos(\alpha_{n-1}) \\
x_n &= c_n^i + r\sin(\alpha_1)...\sin(\alpha_{n-2})\sin(\alpha_{n-1})
\end{aligned} \tag{2}$$

Therefore, for a point x in a sphere, we generate randomly a set of *n-1* angular values (α) and then apply Eq. 2 to calculate Cartesian coordinate values for x. Each sphere is run by its own EMO algorithm. Over time, these spheres move and are being guided towards the global Pareto front. The general steps for the framework of local models are as follows:

- **Step 1:** *Define spheres: Number of spheres, Radius for spheres, and Minimal distance between two points*
- **Step 2:** *Calculate the initial positions of the centroids of the spheres, complying with the rules in Eq. 1*
- **Step 3:** *Initialize spheres: using an uniform distribution, while following Eq. 2 and complying with the rules in Eq. 1.*
- **Step 4:** *Run one evolutionary cycle with the EMO on each sphere.*
- **Step 5:** *Start the moving operator to move spheres.*
- **Step 6:** *If Stop condition is not met Goto step 4, otherwise Stop the process.*

An issue associated with the above framework is how to balance exploration and exploitation. This model might not have an advantage over the global model in the case of single-modal problems since it heavily focuses on exploration. However, in the case of multi-modal problems, the exploration ability in combination with a suitable adaptation strategy for spheres can help the system to approach the global optima quickly. Obviously, the adaptation strategy for spheres is the central point of the proposed local models, as it defines how to suitably move the spheres (including their speed and direction) and how to adjust the radius of the spheres to be suitable with the current state. We refer the readers to [3] for more details on localization.

2.3 Guidance of Spheres

All spheres are initialized following the conditions in Equation 1. Centroids are then recalculated for all spheres after every round of evolution - a round is completed when all spheres finish one cycle of their own evolutionary process. The new and old centroids are used to determine the direction of improvement. We use PSO-V2 (a version of the local models to guide spheres in the decision space), which in simple terms, applies a weak stochastic pressure to move the spheres towards the global optima. Details of how the direction of improvement is implemented as well as PSO-V2 are given in [3] instead for space limitations.

Note that the direction of improvement in the local models exploits both local and global information. However, in problems such as the ZDTs, the use of global information might cause the spheres to quickly move closely to each other as they are approaching the POF. Thus, searching time might be wasted since the spheres search the same areas of the POF. It seems better to instead guide each sphere to occupy a different part of the POF, or at least reduce as much as possible the overlapping of the POF's parts that are discovered by the spheres.

To implement this idea, we need to divide the POF into a number of parts. Each part is then used to guide a sphere. In this way, we use *a number of global centroids* instead of only one as in PSO-V2. The number of parts, spheres, and global centroids are kept the same. However, we want to use a "*soft*" division, in the sense that the parts are allowed some overlapping, but the overlapping is kept as small as possible. For this, we select the guided dominance approach in [6].

In this approach, each sphere is associated with a POF part the one that it contributes the most non-dominated solutions to, in comparison with other parts. Each POF part is assigned only one sphere. When a sphere needs to be guided by global information, the sphere's centroid is determined from both the new local centroid, which is calculated based on all individuals of the sphere that belong to its associated POF part and the global centroid, which is considered to be the centroid of that POF part.

3 Experiments

In order to validate the proposed method, which we call GUIDED, we carried out a comparative study in which we tested the method on three ZDT problems: ZDT1, ZDT3 and ZDT4 which all have similar convex shapes of their POFs, but different types of difficulties, namely continuous, discontinuous, and multi-modal (See [11] for more details). We selected NSGA-II as the algorithm to run in each sphere of our models. Also, all the results will be analyzed and compared with two other systems, one is built on an equivalent number of sub-populations, which are defined on the global search space as well as NSGA-II itself.

We use the hypervolume ratio [4,10] to measure the comparative performance of the different techniques. Note that hypervolume is used to indicate both the *closeness and diversity* of the obtained POFs. The reference point for each problem is seen as the worst point in the objective space obtained by all comparing methods.

We initialize the parameters as follows, and apply non-dominated sorting to update the global archive. All cases were tested on 30 separate runs with different random seeds. We used 5 spheres, total population size is 200, maximum global archive size 100, update frequency for each centroid is every 5 generations, crossover rate 0.9 and mutation rate 0.1.

We will compare the guided version with NSGA-II with the same population of 200 individuals and also with NSGA-II with five - possibly initially overlapping - populations, called 5NSGA-II. All models were run with the same number of evaluations in order to make a fair comparison.

3.1 Results and Discussion

Convergence is one of the most important characteristics of an optimization technique since its main use is to assess the performance of the algorithm. However, the way of looking at convergence of single objective and multiobjective optimizations are quite different [8]. If some measurements of the objective function, with regard to the number of generations, are experimentally considered as an indication for convergence in single objective optimization, it is not a suitable method for multi-objective optimizations since they do not involve the mutual assessment on all objective functions.

Further, the consideration of convergence is not only on how close the obtained POF, at the last generation, is in comparison to the true Pareto optimal front, but also the rate of convergence which is convergence over time. We consider both issues in this section.

For the closeness of the obtained POF, we use the hyper–volume ratio, denoted H. However, it should be noted that this measurement is not always possible in practice, since the true POF is not always known. With test problems, we calculate H of the last generation for all models, and report the best, median, worst, mean and standard derivation among 30 runs. All the results are reported in Table 1.

Table 1. Values of the hyper-volume ratio for each method on ZDT1, ZDT3, and ZDT4 (bold indicates the best results obtained on a problem)

Prob	Models	Best	Median	Worst	Mean(STD)
	GUIDED	**0.9998**	**0.9996**	**0.9988**	**0.9995 (0.0002)**
ZDT1	NSGA-II	0.9989	0.9987	0.9984	0.9987 (0.0001)
	5NSGA-II	0.9937	0.9925	0.9900	0.9923 (0.0009)
	GUIDED	**1.0000**	**0.9998**	**0.9993**	**0.9998 (0.0002)**
ZDT3	NSGA-II	0.9989	0.9980	0.9974	0.9981 (0.0003)
	5NSGA-II	0.9924	0.9890	0.9872	0.9895 (0.0013)
	GUIDED	**1.0000**	0.9984	0.9948	0.9982 (0.0015)
ZDT4	NSGA-II	**1.0000**	**0.9997**	**0.9974**	**0.9994 (0.0008)**
	5NSGA-II	0.9994	0.9951	0.9923	0.9954 (0.0016)

It is obvious that allowing populations to run concurrently without guidance (even on the global scale) does not help to improve the performance of the optimization process. That is the reason for the inferior performance of 5NSGA-II on all problems.

For ZDT1, an easy problem, there is no contradiction between the local and global information; hence GUIDED was able to get closer to the POF and achieve the best overall performance. Moreover, the localization in the objective space resulted in some sort of a division of labor thus allowing the system to smoothly converge to the POF. ZDT3 shows similar behaviour.

However, in the case of ZDT4, the GUIDED approach was not as good as we thought it should be. Despite that one of the runs obtained the best overall hyper-volume ration, the average and overall performance is inferior to NSGA-II. We conjecture that the small population size in each sphere, is the reason for these inferior results, since ZDT4 is highly multi-modal. This point will be analyzed later in this section.

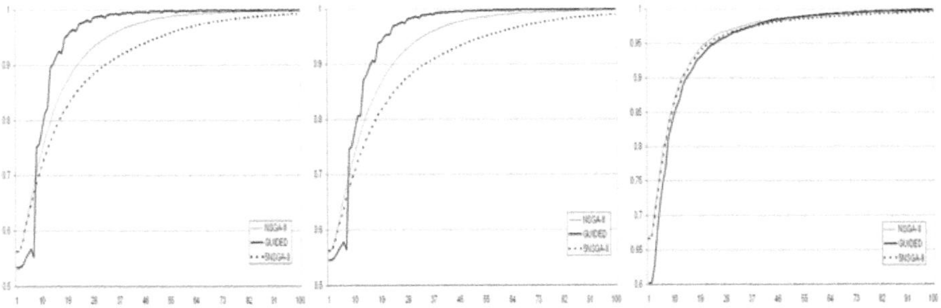

Fig. 1. The hypervolume ratio of differing techniques (up to 20000 evaluations) over time in ZDT1, ZDT3, and ZDT4

To track the convergence of an EMO, there are several techniques in the literature. However, in this analysis, we have used another simple mechanism for tracking the convergence, by measuring the hypervolume ratio (H) over time, since it is consistent with the performance measure used above. We can compare H of all models; as an example in Figure 1, we visualize the averaged H of 30 runs for all models. It is very clear that in the first few generations, GUIDED is quite slow. This is because localization starts with an exploration phase which consumes time. However, the adaptive strategy of GUIDED helps the method to adjust to move faster if the search space seems smooth enough. It then becomes clearer to see that GUIDED converges very quick to the optimal. However, for ZDT4, it seems that GUIDED gets trapped in the local optima a bit longer than NSGA-II does.

As we hypothesized above, that the small sub-population sizes might have caused the poor performance of the local models, since they were unable to capture the local fitness landscape well enough when the problem is highly multi-modal such as for ZDT4. To test this hypothesis, we increased the population size for each sphere to 100 individuals (500 individuals overall). The hyper–volume ratio that each method achieved over time (up to 50000 evaluations) is visualized in Figure 2.

It is clear from Figure 2 that GUIDED achieves faster convergence than the other two methods. Again, the pattern is repeated in which GUIDED starts slowly in the exploration phase and increases the speed over time. It is also

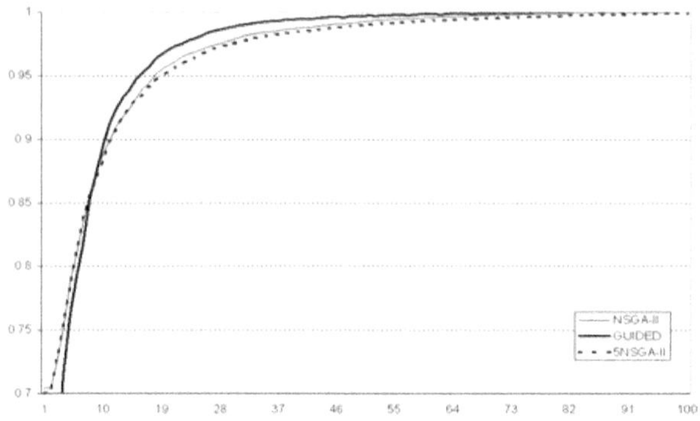

Fig. 2. The hyper–volume ratio of the three techniques over time for ZDT4 with a total population size of 500 individuals and up to 50000 objective evaluations

possible to see a regular series of drops in the curve for GUIDED (also in Figure 1). These drops reflect some sort of loss of diversity in the POF. This sometimes happens when the sphere's motion changes (with the update frequency), and then the sub-population has to adjust to that change. Therefore, this is reflected in the drop-recovery cycles shown in the figure, which seems to be part of the process of first moving to a new better area in the search space, followed by an exploration phase of that area.

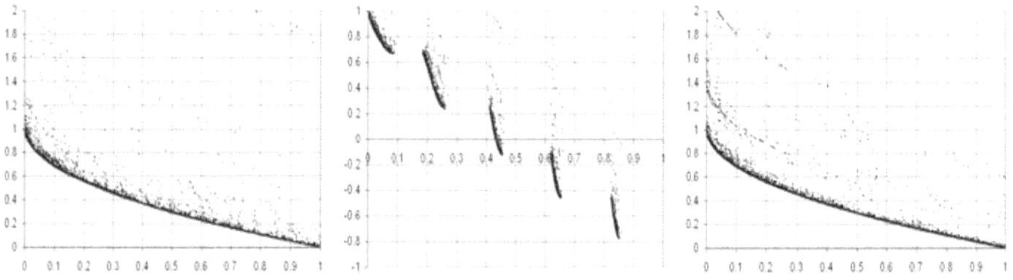

Fig. 3. POFs that GUIDED obtained over time for ZDT1, ZDT3, and ZDT4

All in all, the dual guidance technique shows a good ability to quickly approach the true POF. With the above test problems, it was able to obtain converged and diverse POFs (see Figure 3).

4 Conclusion

This paper proposed a novel technique (GUIDED) to guide a localized - using hyper–spheres - version of NSGA-II. Each sphere is simultaneously focusing on separate areas of the decision and objective space. The technique was tested against NSGA-II with one population, as well as with multi-populations searching on the global space. The experimental results showed the superior performance of GUIDED in comparison with NSGA-II and 5NSGA-II. For future work, we intend to validate the approach with different schemes of dividing the POF and also with the problems that have more than two objectives.

Acknowledgements

This work is supported by the Australian Research Council (ARC) Centre for Complex Systems grant number CEO0348249 and the Postgraduate Research Student Overseas Grant from UNSW@ADFA.

References

1. H. A. Abbass, R. Sarker, and C. Newton. PDE: A Pareto frontier differential evolution approach for multiobjective optimization problems. In *Proceedings of CEC-2001*, volume 2, pages 971–978. IEEE Press, 2001.
2. J. Branke, T. Kaufler, and H. Schmeck. Guiding multi-objective evolutionary algorithms towards interesting regions. technical report no. 399. Technical report, Institute AIFB, University of Karlsruhe, Germany, 2000.
3. L.T. Bui, H.A. Abbass, and D. Essam. Local models: An approach to disibuted multi-objective optimization. technical report no. 200601002. Technical report, ALAR, ITEE,UNSW@ADFA, Australia, 2006.
4. C. A. C. Coello, D. A. V. Veldhuizen, and G. B. Lamont. *Evolutionary Algorithms for Solving Multi-Objective Problems.* Kluwer Publisher, New York, USA, 2002.
5. K. Deb. *Multiobjective Optimization using Evolutionary Algorithms.* John Wiley and Son Ltd, New York, 2001.
6. K. Deb, P. Zope, and A. Jain. Distributed computing of pareto optimal solutions using multi-objective evolutionary algorithms. Technical report, No. 2002008, KANGAL, IITK, India, 2002.
7. J.D. Schaffer. Multiple objective optimization with vector evaluated genetic algorithms. In *Proceedings of the First International Conference on Genetic Algorithms*, pages 93–100, Hillsdale, New Jersey, 1985.
8. K.C. Tan, T.H. Lee, and E.F. Khor. Evolutionary algorithms with dynamic population size and local exploration for multiobjective optimization. *IEEE Transactions on Evolutionary Computation*, 5(6):565–588, 2001.
9. E. Zitzler, M. Laumanns, and L. Thiele. SPEA2: Improving the strength pareto evolutionary algorithm. Technical report, ETH in Zurich, Swiss, 2001.
10. E. Zitzler and L. Thiele. Multi-objective optimization using evolutionary algorithms - a comparative case study. In *PPSN*. Springer, 1998.
11. E. Zitzler, L. Thiele, and K. Deb. Comparision of multiobjective evolutionary algorithms: Emprical results. *Evolutionary Computation*, 8(1):173–195, 2000.

Hybridisation of Particle Swarm Optimization and Fast Evolutionary Programming*

Jingsong He[1,2], Zhengyu Yang[1,3], and Xin Yao[1,4]

[1] Nature Inspired Computation and Applications Laboratory
[2] Department of Electronic Science and Technology
[3] Department of Computer Science and Technology, University of Science and Technology of China
[4] School of Computer Science, University of Birmingham
hjss@ustc.edu.cn[1,2], zhyuyang@mail.ustc.edu.cn[1,3], x.yao@cs.bham.ac.uk[1,4]

Abstract. Particle swarm optimization (PSO) and fast evolutionary programming (FEP) are two widely used population-based optimisation algorithms. The ideas behind these two algorithms are quite different. While PSO is very efficient in local converging to an optimum due to its use of directional information, FEP is better at global exploration and finding a near optimum globally. This paper proposes a novel hybridisation of PSO and FEP, i.e., fast PSO (FPSO), where the strength of PSO and FEP is combined. In particular, the ideas behind Gaussian and Cauchy mutations are incorporated into PSO. The new FPSO has been tested on a number of benchmark functions. The preliminary results have shown that FPSO outperformed both PSO and FEP significantly.

1 Introduction

Evolutionary computation for numerical/parameter optimization can be divided into two ideas, one is mainly with probability-based technique, such as the classical evolutionary programming (CEP) and its improved version[1]; the other is mainly with notion-based vector operation, such as particle swarm optimization[2]. One of the way of recent studies on evolutionary programming (EP) is clearly along with the direction of analyzing the characteristics of different mutations with different probability density functions, e.g. from the Gaussian mutations to the Cauchy mutations[1], and the Lévy mutations[3], and then to multiple representations[4]. Historically, heuristic methods were also introduced into studies of evolutionary programming algorithms, such as 1/5 rule[5].

However, notion-based algorithms, such as the particle swarm optimization (PSO) at which this paper aims specially, has few of introducing the advantageous mechanism of probability-based studies included in evolutionary programming. Obviously, it is an interesting and valuable issue for improving the performance of notion-based optimization algorithms with probability factors, if

* This work is partially supported by the National Natural Science Foundation of China through Grant No. 60573170 and Grant No. 60428202.

T.-D. Wang et al. (Eds.): SEAL 2006, LNCS 4247, pp. 392–399, 2006.

there is a certain conjunction between them could be found. For the problem of introducing probability factors into notion-based algorithms to make improvement of optimization performance, the educed concept in studies of evolutionary programming deserves more attentions.

In this paper, the similarity and difference between PSO and EP were investigated, and the common feature between them thereby had been generalized. Through analyzing the expected length of PSO jumps with probability method, and possible shortages of PSO under some special conditions, we present a fast PSO (FPSO) algorithm with hybrid using of Gaussian mutations and Cauchy mutations in the update equation simply. Experimental results show that FPSO is efficient in integration of different optimization ideas: the proposed FPSO has better performance than both the standard PSO and FEP.

2 Preliminaries

2.1 Particle Swarm Optimization

PSO[2] consists of a swarm of particles flying in an n-dimensional, real-valued search space of possible problem solutions. Every particle has a pair of real-valued vectors $(\boldsymbol{x}_i, \boldsymbol{v}_i)$, $\forall i \in \{1, \cdots, \mu\}$, where $\boldsymbol{x}_i$ denotes the ith position vector, and $\boldsymbol{v}_i$ denotes the ith velocity vector, μ is the number of particles (similar to the population size of EPs). At each time (similar to the generation of EPs) the velocity is updated and the particles is moved to a new position. The new position of a particle is calculated by the sum of previous position and the new velocity as follows.

$$\boldsymbol{x}_i' = \boldsymbol{x}_i + \boldsymbol{v}_i \tag{1}$$

The update of the velocity from the previous velocity to the new velocity is determined by the following equation

$$\boldsymbol{v}_i' = \varpi \boldsymbol{v}_i + U(0, \phi_1)(\boldsymbol{p}_i - \boldsymbol{x}_i) + U(0, \phi_2)(\boldsymbol{g} - \boldsymbol{x}_i) \tag{2}$$

where $U(a, b)$ is a uniformly distribute number between a and b. The parameter ϖ is called the inertia weight and controls the magnitude of the old velocity $\boldsymbol{v}_i$ in the calculation of new velocity $\boldsymbol{v}_i'$, where ϕ_1 and ϕ_2 determine the significance of the memory stored own best position of the ith particle $\boldsymbol{p}_i$ and the global best position of swarm $\boldsymbol{g}$ respectively.

2.2 Fast Evolutionary Programming

Optimization by Evolutionary Programming (EP) can be summarized into two major steps: 1) mutate the solutions in the current population; 2) select the nest generation from the mutated and the current solutions. Each individual is taken as a pair of real-valued vectors $(\boldsymbol{x}_i, \boldsymbol{\eta}_i)$, $\forall i \in \{1, \cdots, \mu\}$, where $\boldsymbol{x}_i$'s are objective vectors variables and $\boldsymbol{\eta}_i$'s are standard deviations for mutations (also known as strategy parameters in self-adaptive evolutionary algorithms). For each current individual $(\boldsymbol{x_i}, \boldsymbol{\eta}_i)$, the update equation are

$$\boldsymbol{x}'_i(j) = \boldsymbol{x}_i(j) + \boldsymbol{\eta}_i(j)N_j(0,1) \tag{3}$$

$$\boldsymbol{\eta}'_i(j) = \boldsymbol{\eta}_i(j)exp(\tau' N(0,1) + \tau N_j(0,1)) \tag{4}$$

where $\boldsymbol{x}_i(j)$, $\boldsymbol{x}'_i(j)$, $\boldsymbol{\eta}_i(j)$, and $\boldsymbol{\eta}'_i(j)$ denote the j-th component of the vectors $\boldsymbol{x}_i$, $\boldsymbol{x}'_i$, $\boldsymbol{\eta}_i$ and $\boldsymbol{\eta}'_i$, respectively. $N(0,1)$ denotes a normally distributed one-dimensional random number with mean zero and standard deviation one. $N_j(0,1)$ indicates that the random number is generated anew for each value of j. The factors τ and τ' are commonly set to $(\sqrt{2\sqrt{n}})^{-1}$ and $(\sqrt{2n})^{-1}$[1,5]. Study by Gehlhaar and Fogel[6] showed that swapping the order of (3) and (4) may improve EP's performance.

Usually, evolutionary programming using (3) and (4) is called the classical evolutionary programming (CEP). In [1], Cauchy mutation is introduced into (3) to substitute Gaussian mutation. The one-dimensional Cauchy density function centered at origin is defined by

$$f_t(x) = \frac{1}{\pi}\frac{t}{t^2 + x^2}, \quad -\infty < x < \infty \tag{5}$$

where $t > 0$ is a scale parameter. The update equation of (3) is replaced by

$$\boldsymbol{x}'_i(j) = \boldsymbol{x}_i(j) + \boldsymbol{\eta}_i(j)\delta_j \tag{6}$$

where δ_j is a Cauchy random variable with the scale parameter $t = 1$ and is generated anew for each of j.

EP using Cauchy mutations is called the fast evolutionary programming (FEP). In general, Gaussian mutation is more suitable for fine search, and Cauchy mutation is more suitable for coarse search when individuals are relative far away from the global optimum and for escaping from local minima.

3 Improving PSO by Introducing Reduced Factors

As related in Section 2, it can be found out that the computational circle of PSO and EP are similar. To make it clear, we use an uniform update equation for PSO and EPs as follows.

$$\boldsymbol{x}'_i(j) = \boldsymbol{x}_i(j) + \boldsymbol{\xi}_i(j)\ \psi \tag{7}$$

where $\boldsymbol{\xi}_i(j)$ means $\boldsymbol{v}_i(j)$ in PSO, and means $\boldsymbol{\eta}_i(j)$ in EP. ψ means a constant number (i.e. $\psi = 1$) in PSO, and means a Gaussian random number $N_j(0,1)$ in CEP, and means a Cauchy random number δ_j in FEP.

It is apparent, the PSO can also subjoin a reduced factor with velocity in the update equation for changing position, i.e. take ψ as a random number other than a constant one. Since the velocity of a moving object usually may be damped in the air, there is no reason to take ψ as one in a movement equation in intuition.

From the viewpoint of evolutionary programming, the search step size of PSO with update equation (1) may have relative fewer chance of fine search when a particle is just at the neighborhood of the global minimum.

Generalize the analysis method for the mean search step size in [1], the expected length of ψ jumps in the universal equation (7) can be calculated as follows:

$$E_{\psi} = \int_{-\infty}^{+\infty} x \cdot \Psi(x)\, dx \tag{8}$$

where $\Psi(x)$ is the distribution density function for generating the number ψ. Obviously, if $\Psi(x)$ takes Gaussian function, the mean step size of CEP is

$$E_{Gaussian} = 2\int_{0}^{+\infty} x \frac{1}{\sqrt{2\pi}} e^{-\frac{x^2}{2}} dx = 0.8,$$

and if $\Psi(x)$ takes Cauchy function, the mean step size of FEP is

$$E_{Cauchy} = 2\int_{0}^{+\infty} x \frac{1}{\pi(1+x^2)} dx = +\infty,$$

and if $\Psi(x)$ takes $\Psi = \delta(x-1)$, the mean step size of PSO is

$$E_{PSO} = \int_{-\infty}^{+\infty} x\ \delta(x-1)dx = 1.$$

That is, if the update equation (2) of PSO can be taken as a kind of self-adaptive mechanism like evolutionary programming, the expected length of PSO jumps is within that of Gaussian and Cauchy.

To make the PSO jumps more localized, Eq.(1) can be improved directly by follows:

$$\boldsymbol{x}_i' = \boldsymbol{x}_i + \boldsymbol{v}_i N(0,1) \tag{9}$$

where $N(0,1)$ denotes a normally distributed one-dimensional random number with mean zero and standard deviation one.

Now another problem appears immediately. Compare to Cauchy mutation, Eq.(1) and Eq.(9) are both with relative smaller search step size, therefor there are relative more less chance for PSO to generate an offspring (or moving a particle to new position) at the neighborhood near the global optimum. Look into the velocity adjust equation (2), which can be thought as an another self-adaptive mechanism corresponding to EP, the change of the new velocity of a particle $\boldsymbol{x}_i$ can be known easily. Usable hints are included in two case studies:

1) $\boldsymbol{x}_i$ is just the global position of the swarm thus also its own best position, i.e. $\boldsymbol{x}_i = \boldsymbol{p}_i$ and $\boldsymbol{x}_i = \boldsymbol{g}$. In this case, the new velocity becomes

$$\boldsymbol{v}_i' = \varpi \boldsymbol{v}_i.$$

To simplify the analysis, assume the optimization problem is unimodal, and the moving direction of the particle $\boldsymbol{x}_i$ is correct. Thereby,

$$\boldsymbol{v}(t+k) = \varpi^k \boldsymbol{v}(t), \quad \varpi < 1$$

where t denotes the time step, and k denotes the time delay. It is obviously that even in this kind of ideal circumstance, the performance of PSO may be worse, since the moving speed of $\boldsymbol{g}$ would tend to zero.

2) $\boldsymbol{x}_i$ is just its own best position, but not the global position of the swarm, i.e. $\boldsymbol{x}_i = \boldsymbol{p}_i$ and $\boldsymbol{x}_i \neq \boldsymbol{g}$. In this case, the new velocity becomes

$$\boldsymbol{v}_i' = \varpi \boldsymbol{v}_i + U(0, \phi_2)(\boldsymbol{g} - \boldsymbol{p}_i).$$

Succeed the assumption of the first case, the global best position $\boldsymbol{g}$ can be a constant vector after some generations (time step), thus at the next time step

$$\begin{aligned} \boldsymbol{x}_i' &= \boldsymbol{x}_i + \boldsymbol{v}_i' = \boldsymbol{x}_i + \left[\varpi \boldsymbol{v}_i + U(0, \phi_2)(\boldsymbol{g} - \boldsymbol{p}_i)\right] \\ &= \left[\boldsymbol{x}_i + U(0, \phi_2)(\boldsymbol{g} - \boldsymbol{x}_i)\right] + \varpi \boldsymbol{v}_i \end{aligned}$$

The first item in above equation means the new position is generated more nearly at neighborhood of the global best position, and the second item means the velocity is still the old direction but with reduced numerical value. It is obviously detrimental to escaping the trap of PSO itself that the best own position is converged to the global best position too fast, since the case is similar to the case of the first study.

Although the above two case studies have some special condition, the possible problem is trouble indeed. To make the global best position $\boldsymbol{g}$ and those best own positions $\boldsymbol{p}$ have more diversity, the following adjustment equation can be used for remedy:

$$\boldsymbol{x}_i'(j) = \boldsymbol{x}_i(j) + \boldsymbol{v}_i(j)\boldsymbol{\delta}_i(j), \quad s.t. \;\; \boldsymbol{x}_i = \boldsymbol{p}_i, \;\; or \; \boldsymbol{x}_i = \boldsymbol{g}. \tag{10}$$

where δ_j is a long jumps Cauchy random variable with the scale parameter $t = 1$ and is generated anew for each of j.

Using Eq.(9) and (10) together to replace Eq.(1) can improve the standard PSO much faster. In addition, recombination technique in evolutionary programming can still be introduced into PSO to replace Eq.(1) randomly as follows.

$$\boldsymbol{x}_i' = \frac{1}{2}(\boldsymbol{x}_i + \boldsymbol{g})U(0, \phi_1) + \frac{1}{2}(\boldsymbol{x}_i + \boldsymbol{p_i})U(0, \phi_2) + \boldsymbol{v}_i, \tag{11}$$

where the first and the second items denote recombination between the current particle and the global best, so as to the own best respectively. The above equation denotes a twice recombination between two items. Here the improved PSO with hybrid adjustment of velocity is noted as fast PSO, or FPSO.

4 Experiments

The algorithms used for comparison were FEP, the standard PSO, and the FPSO. For all algorithms, the parameter settings are fixed as follows without any manually tuning in the later.

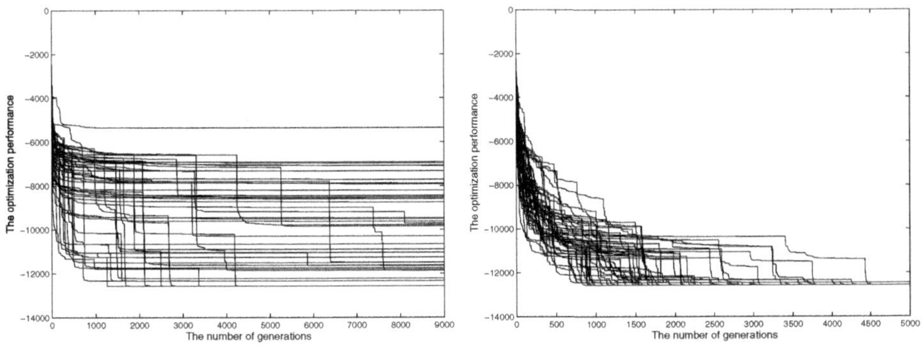

Fig. 1. The optimization performance on f_8 by PSO (left) and FPSO (right), where each line denotes one trial of optimization. It is apparent, although PSO is a non-selection optimization method, there are many traps out of the mechanism of PSO itself, while FPSO can escape traps of standard PSO efficiently.

Table 1. Comparison between FPSO and the standard PSO. All results have been averaged over 50 runs, where the value of t with 49 degrees of freedom is significant at $\alpha = 0.05$ by a two-tailed test.

Test	FPSO		Std. PSO		FPSO-PSO
Func.	Generation	Mean Best(*Std. Dev*)	Generation	Mean Best (*Std. Dev*)	t-test
f_1	1,500	2.23e-6 (*2.23e-6*)	1,500	8.63e-3 (*1.05e-2*)	−5.8102†
f_3	1,500	7.58e-4 (*8.15e-4*)	5,000	6.35 (*5.55*)	−8.0894†
f_5	1,500	1.25e-4 (*1.33e-4*)	1,500	0.48 (*0.62*)	−5.4729†
f_8	5,000	-12567.12(*16.59*)	9,000	-10052 (*2044.6*)	−8.6980†
f_9	1,000	8.06e-4 (*7.02e-4*)	5,000	19.11 (*24.26*)	−5.5698†
f_{10}	1,500	1.10e-3 (*4.66e-4*)	1,500	1.99e-2 (*1.13e-2*)	−11.754†
f_{11}	300	8.00e-6 (*7.02e-6*)	2,000	0.22 (*0.27*)	−5.7614†
f_{12}	300	6.86e-6 (*6.34e-6*)	1,500	5.82e-5 (*6.14e-5*)	−5.8812†
f_{13}	300	2.33e-5 (*2.44e-5*)	1,500	4.92e-4 (*5.11e-4*)	−6.4784†
f_{21}	100	-10.135 (*0.0015*)	100	-10.10 (*0.0503*)	−4.9180†
f_{22}	100	-10.386 (*0.0134*)	100	-10.35 (*0.0637*)	−3.9110†
f_{23}	100	-10.52 (*0.0216*)	100	-10.49 (*0.03985*)	−4.6800†

For FEP, the parameter settings are the same as in [1]. That is, the population size $\mu = 100$, the initial $\eta = 3$, the tournament size $q = 10$ for selection. For the standard PSO and FPSO, the same number of particles $\mu = 10$, the same initial $v = 3$ as the η in FEP. The inertia weight ϖ for the standard PSO and FPSO were just the random number bounded in $[0, 1]$ (doing such was to eliminate man-made factors as much as possible). For the standard PSO, its update equations were Eq.(1) and Eq.(2), where $\phi_1 = \phi_2 = 2$. This kind of settings can make better results for the standard PSO than the tested results in [7]. For FPSO, its updated equations are Eq.(9) and Eq.(10), where ϕ_1 and ϕ_2 just take one.

Table 2. Comparison between FPSO and FEP. All results have been averaged over 50 runs, where the value of t with 49 degrees of freedom is significant at $\alpha = 0.05$ by a two-tailed test

Test Func.	FPSO Generation	FPSO Mean Best(*Std. Dev*)	FEP Generation	FEP Mean Best (*Std. Dev*)	FPSO-FEP t-test
f_1	1,500	2.23e-6 (*2.23e-6*)	1,500	5.7e-4 (*1.3e-4*)	−30.878†
f_3	1,500	7.58e-4 (*8.15e-4*)	5,000	1.6e-2 (*1.4e-2*)	−7.6854†
f_5	1,500	1.25e-4 (*1.33e-4*)	20,000	5.06 (*5.87*)	−6.0952†
f_8	5,000	-12567.12(*16.59*)	9,000	-12554.5 (*52.6*)	−1.618
f_9	1,000	8.06e-4 (*7.02e-4*)	5,000	4.6e-2 (*1.2e-2*)	−26.585†
f_{10}	1,500	1.10e-3 (*4.66e-4*)	1,500	1.8e-2 (*2.1e-3*)	−55.554†
f_{11}	300	8.00e-6 (*7.02e-6*)	2,000	1.6e-2 (*2.2e-2*)	−5.14†
f_{12}	300	6.86e-6 (*6.34e-6*)	1,500	9.2e-6 (*3.6e-6*)	−2.2695†
f_{13}	300	2.33e-5 (*2.44e-5*)	1,500	1.6e-4 (*7.3e-5*)	−12.558†
f_{21}	100	-10.135 (*0.0015*)	100	-5.52 (*1.59*)	−20.524†
f_{22}	100	-10.386 (*0.0134*)	100	-5.52 (*2.12*)	−16.23†
f_{23}	100	-10.52 (*0.0216*)	100	-6.57 (*3.14*)	−8.8949†

Benchmark functions for testing are taken from [1], here we still use their original number. f_1, f_3, and f_5 are unimodal functions, f_8–f_{13} are multimodal function with many local minima, and f_{21}–f_{23} are multimodal functions with a few of local minima. All of them are widely used in optimization algorithms researches. The results and comparisons for benchmark problems of PSO, FEP, and FPSO are summarized in Table 1 and Table 2. It is clear, FPSO, the hybrid of the standard PSO and FEP, has shown better performance both than PSO and FEP.

The explanation why FPSO is better than FEP (as shown in Table 2) is relative simple: since pure-probability-based FEP has no referenced direction for generating offsprings, there are many non-directed offsprings have been generated. In this sense, that introducing heuristic bias into probability-based searches to improve its performance may be an option.

The expectant comparative result between PSO and FPSO confirms our analytical results in Section 3, that is, the expected length of the standard PSO jumps is within Gaussian mutations and Cauchy mutations, thereby that introducing more smaller and more larger jumps into PSO is necessary. To the standard PSO using Eq.(2), the search method for each particle $\boldsymbol{x}_i$ can be explained as a random search in half of a random selected hyper-ellipsoid region with pole $\boldsymbol{g}$ and pole $\boldsymbol{p}_i$. To FPSO using Eq.(9) and Eq.(10), the search method for each particle $\boldsymbol{x}_i$ can be explained as a probability-based search in a random selected hyper-ellipsoid region, or in a random selected hyperboloidal region, or in a random selected hyper-sphere region since both Gaussian and Cauchy random number have negative value, where Gaussian factor makes searches in relative small size with different density, while Cauchy factor enlarges this region much more. In addition, recorders of the optimization performance of PSO and FPSO on the typical multimodal function f_8 are shown in Fig.1, which can illustrate the difference between them clearly.

5 Conclusions

This paper proposes FPSO, and evaluates its optimization performance on a number of benchmark problems. FPSO is the result of hybridisation between the standard PSO and the fast evolutionary programming (FEP), which inherits two kinds of different ideas on optimization problems: heuristic bias and probabilistic search.

On the issue of hybrid evolutionary algorithm, the analytic methodology on the problem of expected length of probability-based jumps and the concept of neighborhood for generating offsprings are introduced into the analysis of the notion-based PSO with heuristic search. The experiment results show that the hybrid approach provides an effective and efficient way to connect two kinds of optimization ideas as well as methods.

References

1. X. Yao, Y. Liu, G. Lin: Evolutionary Programming Made Faster. IEEE Trans. on Evolutionary Computation, **2**(2), (1999) pp. 82–102.
2. J.Kennedy and R.C.Eberhart: Particle Swarm Optimization. In proceedings of the 1995 IEEE International Conference on Neural Networks, IEEE press, **4**, (1995) pp. 1942–1948.
3. C.Y. Lee and X. Yao: Evolutionary programming using the mutations based on Lévy probability distribution. IEEE Transactions on Evolutionary Computation, **8**(5), (October 2004) pp. 456–470.
4. T. Schnier and X. Yao: Using Multiple Representations in Evolutionary Algorithms. Proceedings of the 2000 Congress on Evolutionary Computation, IEEE Press, Piscataway, NJ, USA, July 2000. pp.479-486.
5. T. Bäck and H. -P. Schwefel: An overview of evolutionary algorithms for parameter optimization. Evol. Comput., **1**(1), (1993) pp. 1–23.
6. D. K. Gehlharr and D. B. Fogel: Tuning evolutionary programming for conformationally flexible molecular docking. Evolutionary Programming V: Proc. of the Fifth Annual Conference on Evolutionary Programming, L. J. Fogel, P. J. Angeline, and T. Bäck, EDs. Cambridge, MA: MIT Press, (1996) pp. 419–429.
7. Zhenguo Tu and Yong Lu: A Robust Stochastic Genetic Algorithm (StGA) for Global Numerical Optimization. IEEE Trans. on Evolutionary Computation. **8**(5), (Oct. 2004) pp. 456–470.

An Improved Particle Swarm Pareto Optimizer with Local Search and Clustering

Ching-Shih Tsou[1], Hsiao-Hua Fang[2], Hsu-Hwa Chang[1], and Chia-Hung Kao[2]

[1] Department of Business Administration, National Taipei College of Business
10051 Taipei, Taiwan
{cstsou, hhchang}@mail.ntcb.edu.tw
[2] Department of Information Management, Shih Hsin University
11604 Taipei, Taiwan
fsh@cc.shu.edu.tw, gary.kaoo@msa.hinet.net

Abstract. In this paper, the local search and clustering mechanism are incorporated into the Multi-Objective Particle Swarm Optimization (MOPSO). The local search mechanism prevents premature convergence, hence enhances the convergence of optimizer to true Pareto-optimal front. The clustering mechanism reduces the nondominated solutions to a handful number such that we can speed up the search and maintain the diversity of the nondominated solutions. The performance of this approach is evaluated on metrics from literature. The results against a three objectives optimization problem show that the proposed Pareto optimizer is competitive with the strength Pareto evolutionary algorithm (SPEA) in converging towards the front and generates a well-distributed nondominated set.

1 Introduction

Many optimization models in real-world involve single objective function only. However, the assumption that decision makers pursue the single objective of cost minimization (or wealth maximization) rather than multiple objectives has been questioned in many literatures. For example, in product design a firm may wish to minimize its manufacturing cost while also trying to maximize the performance of the prototype. These objectives are usually incommensurate and in conflict with one another. A multi-objective optimization problem (MOP), hence, does not have a single solution that could optimize all objectives simultaneously.

Instead of aggregating the objectives into a scalar function and solving the resulting single objective optimization problem, we are concerned with finding a set of optimal trade-offs, the so called Pareto-optimal set. The curve (for two objectives) or surface (more than two objectives) that exhibits the optimal tradeoff possibilities between objectives is known as the Pareto-optimal front. Solutions lying on the Pareto front can not improve any objective without degrading at least one of the others. Therefore, the goal of a multi-objective optimizer (or Pareto optimizer) is to find the Pareto front of these nondominated solutions.

Because of their population-based approach, evolutionary algorithms, or genetic algorithms, are able to find several nondominated solutions in a single run. Over

T.-D. Wang et al. (Eds.): SEAL 2006, LNCS 4247, pp. 400–407, 2006.

many years of research has produced a number of efficient multi-objective evolutionary algorithms (MOEAs), which are ready to be applied to real-world problems. Several MOEAs, such as VEGA (vector evaluated genetic algorithm), NSGA (nondominating sorting genetic algorithm), NSGA II, and NPGA (niched Pareto genetic algorithm), have been suggested since a decade ago [1]. In addition to MOEAs, scientists have been studying ants, bees, and wasps because of the amazing efficiency of social insects in the real world. Particle swarm optimization (PSO) inspired by the flocking and schooling behavior of birds and fishes, differs from evolutionary computation methods in that the population members, called particles, are flown through the problem hyperspace [2]. Since PSO and evolutionary algorithms have structural similarities, it is a natural extension to apply PSO to MOPs. One of the successful applications of PSO to MOP, named multi-objective PSO (MOPSO), is the seminal work of Coello and Lechunga [3]. Other famous algorithms which apply PSO to find nondominated solutions of an MOP are Nondominated Sorting Particle Swarm Optimization (NSPSO) [4], Vector Evaluated PSO (VEPSO) [5], and the algorithm proposed by Fieldsend and Singh [6].

No matter what specific algorithm you prefer, two conflicting ends in applying evolutionary-like algorithm to MOPs are that ensuring convergence closer to the true Pareto-optimal front and maintaining a diverse and handful nondominated set. So, constructing a Pareto optimizer also falls in a dilemma – how to converge to the true Pareto-optimal front while achieving a well-distributed front in a reasonable time. This paper proposes an approach called MOPSO-LC, which incorporates the local search and clustering mechanism into the MOPSO, trying to escape from the dilemma described above. The rest of this paper is organized as follows. Section 2 reviews basic concepts of Pareto optimality. MOPSO-LC is described in Sections 3. Section 4 reports the computational results of comparing the proposed Pareto optimizer to a famous MOEA named strength Pareto evolutionary algorithm (SPEA). Finally, conclusions and future research directions are drawn out in Section 5.

2 Pareto Optimality

An MOP (also Pareto optimization problem) with K objectives and bounded decision variables can be stated as follows:

$$\text{Minimize } \vec{\mathbf{f}}(\vec{\mathbf{x}}) = [f_1(\vec{\mathbf{x}}), f_2(\vec{\mathbf{x}}), \ldots, f_K(\vec{\mathbf{x}})]^T \tag{1}$$

$$\text{Subject to } \vec{\mathbf{x}} \in \Omega \tag{2}$$

$$\Omega = \left\{ \vec{\mathbf{x}} \middle| l_i \le x_i \le u_i, i = 1, 2, \ldots, D \right\}, \tag{3}$$

where $\vec{\mathbf{x}} = [x_1, x_2, \ldots, x_D]^T$ is a D dimensional vector.

A decision vector $\vec{\mathbf{u}} = (u_1, u_2, \ldots, u_D)$ is said to strictly dominate $\vec{\mathbf{v}} = (v_1, v_2, \ldots, v_D)$ (denoted by $\vec{\mathbf{u}} \prec \vec{\mathbf{v}}$) if and only if $\vec{\mathbf{f}}(\vec{\mathbf{u}})$ is partially less than $\vec{\mathbf{f}}(\vec{\mathbf{v}})$, i.e., $f_i(\vec{\mathbf{u}}) \le f_i(\vec{\mathbf{v}})\ \forall i \in \{1, 2, \ldots, K\}$ and $f_i(\vec{\mathbf{u}}) < f_i(\vec{\mathbf{v}})\ \exists i \in \{1, 2, \ldots, K\}$. Less stringently, $\vec{\mathbf{u}}$ weakly dominates $\vec{\mathbf{v}}$ (denoted by $\vec{\mathbf{u}} \preceq \vec{\mathbf{v}}$) if and only if $f_i(\vec{\mathbf{u}}) \le f_i(\vec{\mathbf{v}})\ \forall i \in \{1, 2, \ldots, K\}$.

A set of decision vectors is said to be a nondominated set if no member of the set is dominated by any other member. The true Pareto-optimal front, $\tilde{P}$, is the nondominated set of solutions which are not dominated by any feasible solution. One way to solving an MOP is to approximate the Pareto-optimal front by the nondominated solutions generating from the solution algorithm.

3 Proposed Approach

The proposed algorithm called MOPSO-LC incorporates the mechanism of local search (L) and clustering (C) into the MOPSO specifically on enhancing convergence to true Pareto-optimal front and reducing the nondominated solutions to a handful and diverse set. The scheme of the proposed approach is presented in Table 1.

To run the proposed algorithm in Table 1 we need to dictate the number of particles (N) and the number of iterations (T). After initializing each particle's position $\vec{\mathbf{x}}_n^D$, velocity $\vec{\mathbf{v}}_n^D$, the individual best solution $\vec{\mathbf{p}}_n^D$, and the global best solution $\vec{\mathbf{g}}_n^D$, the nondominated set $\tilde{A}$ will collect current nondominated solutions.

Table 1. The pseudo-code of MOPSO-LC

01 $\tilde{A} = \varnothing$

02 $\left\{\vec{\mathbf{x}}_n^D, \vec{\mathbf{v}}_n^D, \vec{\mathbf{p}}_n^D, \vec{\mathbf{g}}_n^D\right\}_{n=1}^{N} = \mathit{Initialize}()$

03 for $t = 1$ to T

04 $\delta = ((\max\_\delta - \min\_\delta)\times(T - t) / T) + \min\_\delta$

05 *LocalSearch*(***LSITER***, δ)

06 for $\boldsymbol{n} = 1$ to N

07 $\vec{\mathbf{g}}_n^D = \mathit{Random}(\tilde{A})$

08 for $\boldsymbol{d} = 1$ to D

09 $v_{nd}^{new} = w \cdot v_{nd}^{old} + c_1 \cdot r_1 \cdot (p_{nd} - x_{nd}) + c_2 \cdot r_2 \cdot (g_{nd} - x_{nd})$

10 $x_{nd}^{new} = x_{nd}^{old} + v_{nd}^{new}$

11 end for

12 $\vec{\mathbf{f}}(\vec{\mathbf{x}}_n^D) = \left[f_1(\vec{\mathbf{x}}_n^D), f_2(\vec{\mathbf{x}}_n^D), \ldots, f_K(\vec{\mathbf{x}}_n^D)\right]$

13 *Update Nondominated Set* $\tilde{A}$

14 if $(\vec{\mathbf{x}}_n^D \preceq \vec{\mathbf{p}}_n^D)$ or $(\vec{\mathbf{x}}_n^D \not\prec\succ \vec{\mathbf{p}}_n^D)$

15 $\vec{\mathbf{p}}_n^D = \vec{\mathbf{x}}_n^D$

16 end if

17 end for

18 *Cluster Nondominated Solutions in* $\tilde{A}$

19 end for

At each iteration, the local search subroutine (line 05 in Table 1) described in the next subsection will be called in to gather local information for each nondominated solution and update $\tilde{A}$ if needed. To search in the decision space, we randomly choose a solution from the nondominated set, $\tilde{A}$, as a global best solution for each particle (line 07 in Table 1) [7]. Then the velocities $\vec{\mathbf{v}}_n^D$ and positions $\vec{\mathbf{x}}_n^D$ of each particle are updated according to lines 09 and 10. Notice that each particle might have a different global guide for its next flight. This is why the global best solutions carry a subscript n. After evaluating the objective vector for each particle in line 12, we update the nondominated set ($\tilde{A}$) in line 13 and its individual best solution ($\vec{\mathbf{p}}_n^D$) in lines 14-16. If $\vec{\mathbf{x}}_n^D$ is not weakly dominated by all the nondominated solutions in $\tilde{A}$, then $\vec{\mathbf{x}}_n^D$ is added into $\tilde{A}$ and the solutions in $\tilde{A}$ which are strongly dominated by $\vec{\mathbf{x}}_n^D$ are deleted. If $\vec{\mathbf{x}}_n^D$ weakly dominates $\vec{\mathbf{p}}_n^D$ or they are not strongly dominated by each other, then $\vec{\mathbf{p}}_n^D$ is set to the current position ($\vec{\mathbf{x}}_n^D$) (lines 14-16 in Table 1). Finally, the nondominated solutions in $\tilde{A}$ is clustered in order to reduce the size of nondominated set while maintaining its diversity (line 18 in Table 1).

Table 2. The *LocalSearch*() procedure

LocalSearch(*LSITER*, δ)
$L = \delta \cdot \left(\max_{d} \{ u^d - l^d \} \right)$
for i = 1 to $|\tilde{A}|$ do
  counter = 1
  while *counter* < *LSITER* do
    $\vec{\mathbf{z}}^D = \vec{\mathbf{x}}_i^D$
    for d = 1 to D do
      $\lambda_1 = U(0,1)$; $\lambda_2 = U(0,1)$
      if $\lambda_1 > 0.5$ then
        $z^d = z^d + \lambda_2 L$
      else
        $z^d = z^d - \lambda_2 L$
      end if
    end for
    if $\vec{\mathbf{z}}^D \prec \vec{\mathbf{x}}_i^D$ then
      Update Nondominated Set $\tilde{A}$
      $\vec{\mathbf{x}}_i^D = \vec{\mathbf{z}}^D$
    end if
    counter++
  end while
end for

3.1 Local Search

The parameters, *LSITER* and δ, represent the number of iterations and the range for the local search, respectively. The local search procedure iterates as follows. First, the maximum feasible step length (L) is calculated according to the parameter of local search range δ (line 02 in Table 2). Similar to the simulated annealing, the local search range δ decreases from $\max\_\delta$ to $\min\_\delta$ as the number of iterations, T, increases (line 04 in Table 1). Second, for each nondominated solution $\vec{\mathbf{x}}_i^D$ in $\tilde{A}$, improvement is sought dimension by dimension (lines 7-14 in Table 2). The temporary D-dimensional vector, $\vec{\mathbf{z}}^D$, first holds the initial information of each particle. Next, two random numbers are generated to set moving direction and step length for each dimension, respectively. If the vector $\vec{\mathbf{z}}^D$ observes a better nondominated solution within *LSITER* iterations, the nondominated set $\tilde{A}$ is updated and the local search for particle i ends (lines 15-18 in Table 2).

3.2 Clustering

In general, the Pareto-optimal front is comprised of extremely large number of solutions. Presenting all nondominated solutions found is not necessary because of the bounded rationality of decision makers. Moreover, the size of nondominated set impose heavy computational burden on the algorithm each time when a newly found solution is put into a big set for verification. On the other hand, preserving the diversity of a nondominated set by clustering, for each particle, could lift the probability of flying through the less crowded area in the current set. Consequently, pruning the nondominated set and maintaining its diversity might be necessary or even imperative.

Morse [9] might be the first one to apply the clustering technique to this problem. Generally speaking, clustering is concerned with the division of data into homogenous subgroups. The objective of this division is twofold: data items within one cluster are required to be similar to each other, while those within different clusters should be dissimilar. Because the average linkage method has been proved to perform well on pruning the nondominated set in SPEA [8], it is also adopted in our approach.

4 Computational Results

The following test problem was used to validate the MOEM.

$$\text{Minimize} \quad f_1(x_1, x_2) = \frac{2000}{x_2} + 25\left(\frac{x_1}{2} + 2x_2\right) \tag{4}$$

$$\text{Minimize} \quad f_2(x_1, x_2) = \frac{100}{x_1}\left(1 - g(x_2)\right) \tag{5}$$

$$\text{Minimize} \quad f_3(x_1, x_2) = \frac{200}{x_1}\left(h(x_2) - x_2\left(1 - g(x_2)\right)\right) \tag{6}$$

$$\text{Suject to } 0 \le x_1 \le 100\,,\ 3 \le x_2 \le 50\,, \tag{7}$$

where

$$g(x_2) = 1 - (0.5(1 + 0.196854x_2 + 0.115194x_2^2 + 0.000344x_2^3 + 0.019527x_2^4)^{-4}) \tag{8}$$

$$h(x_2) = 1/\sqrt{2e} \cdot \exp(-x_2^2/2) \tag{9}$$

The benchmark is the well-known SPEA which outperformed four population-based approaches including VEGA [10], Aggregation by Variable Objective Weighting (AVOW) [11], NPGA [12], and NSGA [13] on a 0/1 knapsack problem.

The population size for SPEA and MOPSO-LC are set to 40. The number of iterations and nondominated solutions for both algorithms are set to 60 and 50, respectively. For SPEA, the crossover and mutation probabilities are set to 0.9 and 0.2, respectively. For MOPSO-LC, the number of local search iterations, maximum and minimum range of local search is set to 1, 0.3 and 0.1, respectively. To compare our results with SPEA in a quantitative way we use the following three performance measures: set coverage metric $C(U,V)$, spacing (S), and maximum spread (D) [14].

$$C(U,V) = \frac{\left|\{b \in V \mid \exists\, a \in U : a \preceq b\}\right|}{|V|} \tag{10}$$

where $|\cdot|$ means the number of components in the set.

$$S = \sqrt{\frac{1}{|\bar{A}|}\sum_{i=1}^{|\bar{A}|}\left(d_i - \bar{d}\right)^2} \tag{11}$$

where $d_i = \min_{j \in \bar{A} \wedge j \neq i} \sum_{k=1}^{K} \left| f_k^i - f_k^j \right|$ and $\bar{d}$ is the mean value of the absolute distance measure $\bar{d} = \sum_{i=1}^{|\bar{A}|} \frac{d_i}{|\bar{A}|}$.

$$D = \sqrt{\sum_{k=1}^{K}\left(\max_{i=1}^{|\bar{A}|} f_k^i - \min_{i=1}^{|\bar{A}|} f_k^i\right)^2} \tag{12}$$

Tables 3-6 are the results of 10 independent runs of both algorithms. From the results in these tables, some findings are explained in the following.

Table 3. Results of the set coverage (*C*) metric for MOPSO-LC and SPEA

Set Coverage	C(MOPSO-LC, SPEA)	C(SPEA, MOPSO-LC)
Average	0.74	0.228
Coefficient of variation	0.14216	0.144576
Standard Deviation	0.377041	0.380232
Median	0.94	0.03

Table 4. Results of the spacing (*S*) metric for MOPSO-LC and SPEA

Spacing	MOPSO-LC	SPEA
Average	0.010599152	0.009006427
Coefficient of variation	9.74735E-06	4.78817E-06
Standard Deviation	0.003122074	0.002188189
Median	0.010197755	0.0085064

Table 5. Results of the maximum spread (D) metric for MOPSO-LC and SPEA

Maximum Spread	MOPSO-LC	SPEA
Average	0.517781	0.492949
Coefficient of variation	0.000615	0.000512
Standard Deviation	0.024794	0.022618
Median	0.514683	0.494712

Table 6. Results of the computational time for MOPSO-LC and SPEA

Time	MOPSO-LC	SPEA
Average	17.9	20.1
Coefficient of variation	8.89	6.69
Standard Deviation	2.98161	2.586503
Median	18	20

1. In view of the set coverage metric in Table 3, MOPSO-LC exhibit much better results than SPEA. The nondominated solutions generated by MOPSO-LC are closer to the Pareto front than those by SPEA in our test problem.
2. For the spacing and maximum spread in Tables 4 and 5, both algorithms are nearly equal. SPEA slightly outperforms MOPSO-LC in the spacing metric. However, MOPSO-LC gets better score than SPEA in the maximum spread metric. This implies MOPSO-LC perform as well as SPEA do in the distribution and the spread of nondominated solutions in our test problem.
3. Although the coefficient of variation and the standard deviation of execution time of MOPSO-LC are somewhat bigger than those of SPEA in Table 6, the average and median execution time of MOPSO-LC are smaller than those of SPEA. So, in terms of computational time, MOPSO-LC is also competitive with SPEA in our test problem.

5 Concluding Remarks

A Pareto optimizer has incorporated the local search and clustering mechanism into the MOPSO. The local search mechanism prevents premature convergence, hence enhances the convergence of optimizer to true Pareto-optimal front. The clustering mechanism reduces the nondominated solutions to a handful number while maintaining the diversity of nondominated set in which the global guide for each particle is generated. Computational results show that the MOPSO-LC performs well in our test problem, although more experiments should be conducted.

This study intends to point out a direction for combining different algorithms into a Pareto optimization scheme. The local search employed here is a linear random search and the clustering method is also in its basic form, but they indeed strengthen the possibilities of particles in MOPSO flying towards the Pareto front and generating a well-distributed nondominated set. Future research to this work include thorough comparisons with SPEA or other MOEAs using more difficult test problems, and accommodating linear or nonlinear constraints in the Pareto optimizer.

References

1. Deb, K.: Multi-Objective Optimization using Evolutionary Algorithms. John Wiley & Sons, New York (2001).
2. Kennedy, J. and Eberhart, R.C.: Particle Swarm Optimization. Proceedings of IEEE International Conference on Neural Networks, Vol. IV. Piscataway. NJ (1995) 1942-1948.
3. Coello, C., Lechunga, M.: MOPSO: A Proposal for Multiple Objective Particle Swarm Optimization. In: Proceedings of the 2002 Congress on Evolutionary Computation, IEEE Press (2002) 1051-1056.
4. Li, X.: A Nondominated Sorting Particle Swarm Optimizer for Multiobjective Optimization. In Lecture Notes in Computer Science, vol. 2723, Proc. Genetic and Evolutionary Computation, GECCO 2003, Part I, E. Cantú-Paz *et al.*, Eds. Berlin, Germany, pp. 37–48, July 2003.
5. Parsopoulos, K. and Vrahatis, M.: Particle Swarm Optimization Method in Multiobjective Problems. In Proc. 2002 ACM Symp. Applied Computing (SAC'2002), Madrid, pages 603–607, Spain, 2002.
6. Fieldsend, J. and Singh, S.: A Multi-Objective Algorithm Based upon Particle Swarm Optimization, an Efficient Data Structure and Turbulence. In Proc. 2002 U.K. Workshop on Computational Intelligence, Bir
7. Alvarez-Benitez, J.E., Everson, R.M., Fieldsend, J.E.: A MOPSO Algorithm Based Exclusively on Pareto Dominance Concepts. Lecture Notes in Computer Science 3410 (2005) 459-473.
8. Zitzler, E. and Thiele, L.: Multiobjective Evolutionary Algorithms: a Comparative Case Study and the Strength Pareto Approach. IEEE Transactions on Evolutionary Computation 3(4) (1999) 257-271.
9. Morse, J.N.: Reducing the Size of the Nondominated Set: Pruning by Clustering. Computers and Operations Research 7(2) (1980) 55-66.
10. Schaffer, J.: Multiple Objective Optimization with Vector Evaluated Genetic Algorithms. Genetic Algorithms and their Applications: Proceedings of the First International Conference on Genetic Algorithms (1985) 93–100.
11. Hajela, P., Lin, C.: Genetic Search Strategies in Multicriterion Optimal Design. Structural Optimization 4 (1992) 99–107.
12. Horn, J., Nafpliotis, N., Goldberg, D.: A Niched Pareto Genetic Algorithm for Multiobjective Optimization. Proceedings of the First IEEE Conference on Evolutionary Computation 1 (1994) 82–87.
13. Srinivas, N., Deb, K.: Multiobjective Optimization Using Nondominated Sorting in Genetic Algorithms. Evolutionary Computation 2 (1994) 221– 248.
14. Okabe, T., Jin, Y., Sendhoff, B.: A Critical Survey of Performance Indices for Multi-Objective Optimization, Proc. of 2003 Congress on Evolutionary Computation (2003) 878-885.

A Hybrid Genetic Algorithm for Solving a Class of Nonlinear Bilevel Programming Problems

Hecheng Li[1] and Yuping Wang[2]

[1] Department of Mathematics Science,
Xidian University, Xi'an 710071, China
lihecheng@qhnu.edu.cn
[2] School of Computer Science and Technology,
Xidian University, Xi'an 710071, China
ywang@xidian.edu.cn

Abstract. In this paper, a special nonlinear bilevel programming problem (BLPP), in which the follower's problem is a convex quadratic programming in y, is transformed into an equivalent single-level programming problem by using Karush-Kuhn-Tucker(K-K-T) condition. To solve the equivalent problem effectively, firstly, a genetic algorithm is incorporated with *Lemke* algorithm. For x fixed, the optimal solution y of the follower's problem can be obtained by *Lemke* algorithm, then (x, y) is a feasible or approximately feasible solution of the transformed problem and considered as a point in the population; secondly, based on the best individuals in the population, a special crossover operator is designed to generate high quality individuals; finally, a new hybrid genetic algorithm is proposed for solving this class of bilevel programming problems. The simulation on 20 benchmark problems demonstrates the effectiveness of the proposed algorithm.

1 Introduction

The bilevel programming problem(BLPP) can be defined as

$$\begin{cases} \min\limits_{x\in X} F(x,y) \\ s.t.\ G(x,y) \leq 0 \\ \min\limits_{y\in Y} f(x,y) \\ s.t.\ g(x,y) \leq 0 \end{cases} \tag{1}$$

and $x \in X \subseteq R^n$, $y \in Y \subseteq R^m$. Here $x(y)$ is called the leader's (follower's) variable. In the same way, F(f) is called the leader's (follower's) objective function. The bilevel programming problem is a mathematical model of the leader-follower game, it can be viewed as a static version of the noncooperative, two-person game introduced by Von Stackelberg [1] in the context of unbalanced economic markets. As an optimization problem with a hierarchical structure, BLPP has a wide variety of applications, such as network design, transport system planning, and

T.-D. Wang et al. (Eds.): SEAL 2006, LNCS 4247, pp. 408–415, 2006.

management and economics [2,3,4]. However, owing to the complex structure, the vast majority of researches on BLPP are concentrated on the linear version of the problem [5,6,7,8], and a few works on the nonlinear BLPP. Moreover, most of existing algorithms for nonlinear BLPP are usually based on the assumption that all of the functions are convex and twice differentiable [9,10,11]. In this paper, we extend the linear case to a special class of nonlinear BLPP, where the follower's problem is a convex quadratic programming in y and the leader's functions may be nonconvex and nondifferentiable. In order to solve the transformed problem effectively by using Genetic Algorithm(GA), firstly, for x fixed, we apply *Lemke* algorithm to obtain an optimal solution y of the follower's problem, and (x, y) is regarded as an individual in the population, which guarantees that we can get an approximately feasible point that at least satisfies the follower's problem; secondly, for the leader's constraint $G(x, y) \leq 0$, we design a fitness function in a magnified feasible region. In the new feasible region, the fitness function makes any feasible solutions better than all infeasible solutions; finally, a new hybrid genetic algorithm with specifically designed crossover operator is proposed to solve the class of bilevel programming problems.

2 Bilevel Programming Problem

We consider the following nonlinear bilevel programming problem:

$$\begin{cases} \min\limits_{x \in X} F(x, y) \\ s.t.\ G(x, y) \leq 0 \\ \min\limits_{y \in Y} f(x, y) = 1/2 \times y^T Q(x) y + c(x)^T y \\ s.t.\ A(x)y + b(x) \leq 0,\ y \geq 0 \end{cases} \tag{2}$$

where $F : R^n \times R^m \to R$, $G : R^n \times R^m \to R^p$, for each $x \in X$, $Q(x) \in R^{m \times m}$ is symmetric and positive definite, $A(x) \in R^{q \times m}$, $b(x) \in R^q$ and $c(x) \in R^m$. For the purpose of convenience, denote $Q(x)$, $c(x)$, $A(x)$ and $b(x)$ by Q, c, A, and b, respectively. The sets of X and Y may represent upper and lower bounds on elements of the vectors x and y. Now we introduce some related definitions [14].

1) Search space: $\Omega = \{(x, y) | x \in X, y \in Y\}$.
2) Constraint region: $S = \{(x, y) \in \Omega | G(x, y) \leq 0, A(x)y + b(x) \leq 0,\ y \geq 0\}$.
3) For x fixed, the feasible region of follower's problem: $S(x) = \{y \in Y | A(x)y + b(x) \leq 0,\ y \geq 0\}$.
4) Projection of S onto the leader's decision space: $S(X) = \{x \in X | \exists y, (x, y) \in S(x)\}$.
5) The follower's rational reaction set for each $x \in S(X)$: $M(x) = \{y \in Y | y \in argmin\{f(x, v), v \in S(x)\}\}$.
6) Inducible region: $IR = \{(x, y) \in S | y \in M(x)\}$.

In the remainder, we always assume that for $x \in X$ fixed, $S(x)$ is nonempty as well as S. Since, for x fixed, all functions in the follower's problem are differentiable and convex in y, K-K-T stationary-point problem of follower's programming can be written as following

$$\begin{cases} Qy + A^T\lambda_1 + \lambda_2 + c = 0 \\ Ay + b + u = 0 \\ \lambda_1^T u = 0 \\ \lambda_2^T y = 0 \\ \lambda_i, y, u \geq 0, i = 1, 2. \end{cases} \tag{3}$$

where $\lambda_1 = (\lambda_{11}, \lambda_{12}, \ldots, \lambda_{1q})^T$, $\lambda_2 = (\lambda_{21}, \lambda_{22}, \ldots, \lambda_{2m})^T$ are Lagrangian multipliers, and $u \in R^q$ is an slack vector. Replace the follower's programming in (2) by (3), then (2) is transformed into (4) as follows

$$\begin{cases} \min\limits_{x \in X} F(x, y) \\ s.t.\ G(x, y) \leq 0 \\ E(x, y, \lambda_1, \lambda_2, u) = 0 \\ y, \lambda_1, \lambda_2, u \geq 0 \end{cases} \tag{4}$$

where $E(x, y, \lambda_1, \lambda_2, u) = 0$ stands for all equations in (3). In order to solve the transformed programming (4), at first, we rewrite (3) as the following linear complementary problem

$$\begin{cases} w + Mz = \hat{b} & (c1) \\ w, z \geq 0 & (c2) \\ w^T z = 0 & (c3) \end{cases} \tag{5}$$

where

$$w = \begin{pmatrix} \lambda_2 \\ u \end{pmatrix}, \quad z = \begin{pmatrix} y \\ \lambda_1 \end{pmatrix}, \quad M = \begin{pmatrix} Q & A^T \\ A & 0 \end{pmatrix}, \quad \hat{b} = \begin{pmatrix} -c \\ -b \end{pmatrix}$$

If (w, z) satisfying (c3) is a basic feasible solution of (c1-c2) in (5), (w, z) is called a complementary basic feasible solution (CBFS) of (5)[13]. For x fixed, if one wants to get the optimal solution y of the follower's problem, what he needs to do is to solve problem (5) via *Lemke* algorithm for a CBFS (w, z) [13]. Take y from z, then (x, y) satisfies all the follower's constraints and for x fixed, $f(x, y)$ gets its optimal value at y. Such points (x, y) form the initial population of GA. In the process of evolving, genetic offspring are generated in two steps. At first, the variable x is acted by crossover or mutation operator to get $\bar{x}$; For $\bar{x}$ fixed, to solve problem (5) for $\bar{y}$. Such $(\bar{x}, \bar{y})$ is regarded as the offspring and satisfies the follower's optimization problem. When the populations are generated with the method, all that we need to do is to solve the following single programming problem

$$\begin{cases} \min\limits_{x \in X} F(x, y) \\ s.t.\ G(x, y) \leq 0 \end{cases} \tag{6}$$

3 New GA

3.1 Fitness Function

Let $\epsilon = (\epsilon_1, \ldots, \epsilon_p)^T$,where ϵ_i are small positive numbers and tend to zero with the increasing of the generations, $\bar{S} = \{(x, y) \in \Omega | G(x, y) \leq \epsilon\}$, $v(x, y) = \max(\max(G(x, y) - \epsilon), 0)$, and K be an upper-bound of $F(x, y)$ on the set $\{(x, y) | v(x, y) = 0\}$. The fitness function $\bar{F}(x, y)$ is defined as following

$$\bar{F}(x, y) = \begin{cases} F(x, y), & v(x, y) = 0; \\ K + v(x, y), & v(x, y) \neq 0. \end{cases}$$

The fitness function implies that any point in $\bar{S}$ always is better than all points out of $\bar{S}$, and between two points out of $\bar{S}$, the point with the smaller v is better than the other.

3.2 Crossover Operator

In the process of evolving, the best point found always is recorded as (x_{best}, y_{best}). Let (x, y) be a crossover parent. We, at first, get an $\bar{x}$ by the formula $\bar{x} = x + diag(\tau) \times r \times (x_{best} - x)$, where $\tau \in R^n$ is a positive vector determined by the upper and lower bounds of variable x, and $r \in [0, 1]$ is a random number. For $\bar{x}$ fixed, we solve the problem (5) for $\bar{y}$. The resulting $(\bar{x}, \bar{y})$ is the crossover offspring of (x, y).

3.3 Mutation Operator

Let (x, y) be a mutation parent. We, at first, get an $\hat{x}$ by Gaussian mutation $\hat{x} = x + \triangle(0, \sigma^2)$, where each component of $\triangle(0, \sigma^2)$ obeys Gaussian distribution $N(0, \sigma_i^2)$, $i = 1, 2, \ldots, n$. For $\hat{x}$ fixed, we solve the problem (5) for $\hat{y}$. The resulting $(\hat{x}, \hat{y})$ is the mutation offspring of (x, y).

3.4 Proposed Algorithm

Algorithm 1

Step 1 (Initialization). Randomly generate a set *pop* of N_p points in X. For each $x \in$ *pop* fixed, we solve the problem (5) for y. If $y \notin Y$, randomly generate another x to replace the original one, the process doesn't stop until $y \in Y$ is satisfied. All such points (x, y) form the initial population *pop*(0) with N_p individuals. Let N stand for the set of all the best points found by *Algorithm* 1 so far and $k = 0$.

Step 2. Evaluate the fitness value $\bar{F}(x, y)$ of each point in *pop*(k), and let $N = \{p_{i_1}, \cdots, p_{i_k}\}$.

Step 3 (*Crossover*). Randomly select a parent p from *pop*(k) according to the crossover probability p_c. We can obtain a crossover offspring $\bar{p}$ via the crossover operator. Let $O1$ stand for the set of all these offspring.

Table 1. Comparison of the best results found by *Algorithm*1 and by the related algorithms in references for Problems 1-20

No.	CPU(s)	$F(x^*, y^*)$		$f(x^*, y^*)$	
		*Algorithm*1	Ref.	*Algorithm*1	Ref.
$F1$([15]E4.1)	10.0115	0.5015	0.5	−15.7475	−14.4879
$F2$ ([15]E4.2)	8.1894	0.5015	0.5	−4.5000	−4.5
$F3$ ([15]E4.3)	12.0063	1.8605	1.859	−10.9321	−10.9310
$F4$ ([15]E4.4)	10.6969	0.8485	0.919	−22.9505	−19.4686
$F5$ ([15]E4.5)	9.3870	0.8975	0.897	−14.9249	−14.9311
$F6$ ([15]E4.6)	10.1625	1.5629	1.562	−11.6826	−11.6808
$F7$ ([15]E3.1)	10.1292	−8.9172	−8.92	−6.1179	−6.054
$F8$ ([15]E3.2)	9.2068	−7.5784	−7.56	−0.5723	−0.5799
$F9$ ([15]E3.3)	11.4052	−11.9985	−12	−438.4165	−112.71
$F10$ ([15]E3.4)	10.3932	−3.6	−3.6	−2	−2
$F11$ ([15]E3.5)	9.5588	−3.92	−3.15	−2	−16.29
$F12$ ([16]E5.2)	10.625	225	225	100	100
$F13$ ([17]E3.2)	11.2531	0	5	200	0
$F14$ ([9]E3.2)	10.2531	−12.6787	−12.6787	−1.016	−1.016
$F15$ ([18]E4.2)	23.2859	−6600	−6599.99	$f_1 = 23.6358$	$f_1 = 23.47$
				$f_2 = 30.5833$	$f_2 = 30.83$
$F16$ ([12]E6.2)	6.0922	81.3279	82.44	−0.3359	0.271
$F17$ ([12]E6.3)	6.4875	100	100.58	0	0.001
$F18$ ([12]E6.4)	13.6151	−1.2099	3.57	7.6172	2.4
$F19$	23.0260	0	NA	13.3283	NA
$F20$	11.7354	0	NA	200	NA

Sept 4 (Mutation). Randomly select parents from *pop*(k) according to the mutation probability p_m. For each selected parent p, we can get its offspring $\hat{p}$ via the mutation operator. Let $O2$ stand for the set of all these offspring.
Step 5 (*Selection*). Let $O = O1 \cup O2$ and evaluate the fitness values of all points in O. Select the best n_1 points from the set *pop*(k) $\cup\, O$ and randomly select $N_p - n_1$ points from the remaining points of the set. All these selected points form the next population *pop*($k+1$).
Step 6. If the termination condition is satisfied, then stop; otherwise, renew N, let $k = k + 1$ and $\epsilon = \theta\epsilon$, $\theta \in [0, 1]$, go to *Step* 3.

4 Simulation Results

In this section, 20 benchmark problems are used for simulation, in which the first 18 problems are selected from the related references. In order to illustrate the efficiency of the proposed algorithm for solving complex nonlinear BLPPs with nondifferentiable , even nonconvex, leader's objective function, we construct 2 new benchmark problems, $F19$ and $F20$ as follows

Table 2. Comparison of the best solutions found by *Algorithm*1 in 30 runs and by the related algorithms in references for problems 1-20

	(x^*, y^*)	
No.	*Algorithm*1	Ref.
$F1$([15]E4.1)	$(2.25.5, 2.9985, 2.9985)$	$(2.07, 3, 3)$
$F2$ ([15]E4.2)	$(0, 2.9985, 2.9985)$	$(0, 3, 3)$
$F3$ ([15]E4.3)	$(3.4564, 1.7071, 2.5685)$	$(3.456, 1.707, 2.569)$
$F4$ ([15]E4.4)	$(2.8563, 3.8807, 3.0402)$	$(2.498, 3.632, 2.8)$
$F5$ ([15]E4.5)	$(3.9977, 1.6651, 3.8865)$	$(3.999, 1.665, 3.887)$
$F6$ ([15]E4.6)	$(1.9095, 2.9786, 2.2321)$	$(1.90, 2.979, 2.232)$
$F7$ ([15]E3.1)	$(1.2046, 3.0972, 2.5968, 1.7922)$	$(0.97, 3.14, 2.6, 1.8)$
$F8$ ([15]E3.2)	$(0.2785, 0.4759, 2.3439, 1.0327)$	$(0.28, 0.48, 2.34, 1.03)$
$F9$ ([15]E3.3)	$(46.7128, 124.9863, 2.9985, 2.9985)$	$(20.26, 42.81, 3, 3,)$
$F10$ ([15]E3.4)	$(2, 0.0296, 2, 0)$	$(2, 0.06, 2, 0)$
$F11$ ([15]E3.5)	$(-0.4050, 0.7975, 2, 0)$	$(2.42, -3.65, 0, 1.58)$
$F12$ ([16]E5.2)	$(20, 5, 10, 5)$	$(20, 5, 10, 5)$
$F13$ ([17]E3.2)	$(0, 0, -10, -10)$	$(25, 30, 5, 10)$
$F14$ ([9]E3.2)	$(0, 2, 1.8750, 0.9063)$	$(0, 2, 1.8750, 0.9063)$
$F15$ ([18]E4.2)	$(7.034, 3.122, 11.938, 17.906, 0.25,$ $9.906, 29.844, 0)$	$(7.05, 4.13, 11.93, 17.89, 0.26,$ $9.92, 29.82, 0)$
$F16$ ([12]E6.2)	$(10.0164, 0.8197)$	$(10.04, 0.1429)$
$F17$ ([12]E6.3)	$(10.0000, 10.0000)$	$(10.03, 9.969)$
$F18$ ([12]E6.4)	$(1.8889, 0.8889, 0)$	NA
$F19$	$(13.6508, 8.5111, 10.0000, 8.5111)$	NA
$F20$	$(0, 0, -10, -10)$	NA

F19).The problem is the same as $F12$ except for

$$F(x, y) = |\sin((x_1 - 30)^2 + (x_2 - 20)^2 - 20y_1 + 20y_2 - 225)|$$

F20). The problem is the same as $F13$ except for

$$F(x, y) = |\sin(2x_1 + 2x_2 - 3y_1 - 3y_2 - 60)|$$

In all problems, the follower's problems are quadratic programming problems in y for x fixed, while the leader's functions may be nonlinear and nondifferentiable.

The parameters are chosen as follows: the population size $N_p = 30$, $p_c = 0.8$, $p_m = 0.3$, $n_1 = 10$, the initial $\epsilon = (1, \cdots, 1) \in R^p$, $\theta = 0.7$ for $k \leq k_{max}/2$, while $\theta = 0$ for $k > k_{max}/2$, where k represents the generation number, while k_{max} the maximum generation number. For $F1 - F20$, The algorithm stops after 50 generations. We execute *Algorithm* 1 in 30 independent runs on each problem on a computer(Intel Pentium IV-2.66GHz), and record the following data: (1) best solution (x^*, y^*), and the leader's(follower's) objective values $F(x^*, y^*)(f(x^*, y^*))$ at the best solution; (2) worst solution $(\bar{x}, \bar{y})$ and the leader's (follower's) objective function value $F(\bar{x}, \bar{y})$ $(f(\bar{x}, \bar{y}))$ at the point $(\bar{x}, \bar{y})$; (3) mean value of CPU time(denoted by CPU in short).

Table 3. The worst results found by *Algorithm*1

No.	$F(\bar{x},\bar{y})$	$f(\bar{x},\bar{y})$	$(\bar{x},\bar{y})$
$F6$ ([15]E4.6)	1.5633	−11.6859	(1.9101, 2.9785, 2.2319)
$F8$ ([15]E3.2)	−7.5782	−0.5726	(0.2827, 0.4668, 2.3432, 1.0307)
$F15$ ([18]E4.2)	−6599.6	$f_1 = 25.1041$	(6.9191, 3.0261, 12.0210, 18.0337,
		$f_2 = 28.4778$	0.1417, 9.8035, 30.0525, 0)

All results are presented in Tables 1-3, in which Table 1 provides the comparison of the best results found by *Algorithm*1 in 30 runs and by the related algorithms in references for problems 1-20 , as well as mean CPU time needed by *Algorithm*1 in a run. Table 2 shows the best solutions found by *Algorithm*1 in 30 runs and by the related algorithms in references for problems 1-20. For those functions for which *Algorithm*1 finds different results in 30 runs, Table 3 gives the worst solutions $(\bar{x},\bar{y})$ and the related objective function values. The first volume of Tables 1-3 consists of two parts, Fi and $([l]Ej.k)$. Fi stand for test function numbers $(i = 1, 2, \cdots, 20)$, and $([l]Ej.k)$ the kth example in Section j of reference $[l]$. NA means that the result is not available for the algorithm and Ref. stands for the related algorithms in references in Tables 1-3.

It can be seen from Table 1 that for problems $F4, F11, F13, F16, F17$ and $F18$, the best results found by *Algorithm* 1 are better than those by the compared algorithms in the references, which indicates these algorithms can't find the optimal solutions of these problems; For the constructed $F19$ and $F20$, *Algorithm*1 can find the best results; For other problems, the best results found by *Algorithm*1 are almost as good as those by the compared algorithms.

In all 30 runs, *Algorithm*1 finds the best results of all problems except the problems 6, 8, and 15. For the three problems, the proposed algorithm also found the best solutions in 25, 24 and 19 runs, respectively. The worst results are shown in Table 3. From Table 3, we can find the worst results are close to the best results. This means that the proposed *Algorithm* 1 is stable and robust.

5 Conclusion

For the nonlinear bilevel programming problem, whose follower level problem is a convex quadratic programming in y, we transform it into a single-level programming problem by using K-K-T condition. To solve the equivalent problem effectively, we propose a new hybrid genetic algorithm incorporated with *Lemke* algorithm, in which an efficient crossover operator is designed to generate high quality individuals. The simulation on 20 benchmark problems demonstrates the robustness and the effectiveness of the proposed algorithm.

Acknowledgments

This work is supported by the National Natural Science Foundation of China under Grant No. 60374063.

References

1. Stackelberg, H. Von: The Theory of the Market Economy. William Hodge, Lond, UK (1952)
2. Marcotte, P.: Network optimization with continuous parameters. Trans. Sci. 17 (1983) 181–197
3. Suh, S. , Kim, T.: Solving nonlinear bilevel programming models of equilibrium network design problem: A comparative review. Ann. Oper. Res. 34 (1992) 203–218
4. Milier, T. , Friesz, T., Robin, R.: Heuristic algorithms for delivered price spatially competitive net work facility location problems. Ann. Oper. Res. 34 (1992) 177–202
5. Hansen, P., Jaumard, B., Savard, G.: New branch-and-bound rules for linear bilevel programming. SIAM J. Sci. Stat. Comput. 13(5) (1992) 1194–1217
6. Amouzegar, M. A., Moshirvaziri, K.: A penalty method for linear bilevel programming problems. In Multilevel optimization: Algorithms, complexity and Applications. A. Migdalas, M. Pardalos, and P. Varbrand, Eds. Norwell,Kluwer MA (1997) ch.11
7. Bard, J. ,Moore, J.: A branch and bound algorithm for the bilevel programming problem. SIAM J. Sci. Stat. Comput. 11(5) (1990) 281–292
8. Candler, W. , Townsley, R.: A linear bi-level programming problem. Comput. Oper. Res. 9(1) (1982) 59–76
9. Amouzegar, M. A.: A global optimization method for nonlinear bilevel programming problems. IEEE Trans. Syst., Man, Cybern. B, Cybern. 29(6) (1999) 771–777
10. Vicente, L., Savard, G., Judice, J.:Descent approach for quadratic bilevel programming. J. Optim. Theory Appl. 81(2) (1994) 379–399
11. Al-Khayyal, F., Horst, R., Pardalos, P.: Global optimization of concave function subject to quadratic constraints: An application in nonlinear bilevel programming. Ann. Oper. Res. 34 (1992) 125–147
12. Oduguwa, V., Roy, R.: Bi-level optimization using genetic algorithm . In Proc. IEEE Int. Conf. Artificial Intelligence Systems (2002) 123–128
13. Bazaraa, M. S., Shetty, C. M.: Nonlinear Programming: Theory and Algorithms. Wiley, Chichester (1979)
14. Bard, J. F.: Practical Bilevel Optimization. Norwell, Kluwer, MA(1998)
15. Outrata, J. V.: On the numerical solution of a class of Stackelberg problem. Zeitschrift Fur Operational Reseach 34 (1990) 255–278
16. Shimizu, K., Aiyoshi, E.: A new computational method for Syackelberg and minmax problems by use of a penalty method. IEEE Trans. Autom. Control AC-26(2) (1981) 460–466
17. Aiyoshi, E., Shimuzu, K.: A solution method for the static constrained Stackelberg problem via penalty method. IEEE Trans. Autom. Control AC-29(12) (1984) 1111–1114
18. Bard, J. F.: Covex two-level optimization. Math. Programming 40 (1988) 15–27

A Synergistic Selection Strategy in the Genetic Algorithms

Ting Kuo

Takming College
No. 56, Sec. 1, Huanshan Rd., Neihu District, Taipei, Taiwan, R.O.C.
tkuo@takming.edu.tw

Abstract. According to the Neo-Darwinist, natural selection can be classified into three categories: directional selection, disruptive selection, and stabilizing selection. Traditional genetic algorithms can be viewed as a process of evolution based on directional selection that gives more chances of reproduction to superior individuals. However, this strategy sometimes is myopic and is apt to trap the search into a local optimal. Should we restrict genetic algorithms to direction selection? No! First, we show that stabilizing selection and disruptive selection are complementary and that hybridize them may supersede directional selection. Then, we adopt an island model of parallel genetic algorithms on which two types of selection strategies are applied to two sub-populations that both evolve independently and migration is allowed between them periodically. Experimental results show that the cooperation of disruptive selection and stabilizing selection is an effective and robust way in the genetic algorithms.

1 Introduction

The motivation of this study is try to design an effective and robust genetic algorithm (GA) to solve problems. The three primary operators in the GA are selection, crossover, and mutation. What we emphasize on is the selection operator. That is to say, our goal is to design an effective and robust selection strategy in the GA. According to the Neo-Darwinist, three types of natural selection can be distinguished: directional selection, disruptive selection, and stabilizing selection [11]. Directional selection has the effect of increasing (or decreasing) the mean value of the entire population. Disruptive selection tends to eliminate individuals of mediocre values and prefers to those individuals of extreme values. In contrast, stabilizing selection tends to favor individuals of mediocre values. Throughout this paper, we will use the TGAs stands for traditional genetic algorithms that adopt directional selection, the DGAs stands for genetic algorithms that adopt disruptive selection and the SGAs stands for genetic algorithms that adopt stabilizing selection. The only difference between these three types of genetic algorithms relies on the fitness function that maps the object function value to the fitness measure. It must be noted that we still adopt proportionate selection

T.-D. Wang et al. (Eds.): SEAL 2006, LNCS 4247, pp. 416 – 423, 2006.

with weighted roulette wheel sampling method to implement a generational production process.[1]

Following the "survival-of-the-fittest" principle, the TGAs offer more chances of reproduction to superior individuals.[2] But this strategy sometimes is myopic and is likely to guide the search to a misleading direction. The key of success is what kind of solution should be viewed as a "fitter"solution. Kuo and Hwang introduced the DGAs that favor to both superior and inferior individuals simultaneously, and confirmed that it can solve some problems that the TGAs perform inefficiently or even do not solve [7], [9].[3]

We believe that it should be a good idea to view a solution as"fitter"as it is closer to the optimal solution no matter what function value it has. This belief is based on the fact that for any solution x that is away from the optimal solution with d bits, there are 2^{L-d} schemata out of its 2^L schemata contain the optimal solution [8].

With this realization, we have confidence in arguing that the function values of solutions in the neighborhood (in the Hamming distance, hence thereafter) of the optimal solution act as an important role in the genetic algorithms. That is to say, in some problem instances, even a contemporary mediocre solution also may has a greater potential of "evolving" to a superior solution, in the future. Why we discard these mediocre solutions rather than exploit them? In fact, just like the NFL tells us, we can foresee that different types of selection strategies are good for different types of optimization problems [15]. Unfortunately, to deal with an optimization problem in practice, we have no prior knowledge of the solution space to which we are faced. Why we confine us to a certain selection strategy? Maybe we can syncretize different types of selection strategies to enhance the searching power of genetic algorithms. In brief, we do not expect to looking for a universal strategy that can ensure to guide the search towards the area where the optimal solution locates in, what we try to do is not miss the chance of exploiting and exploring those solutions that are mediocre or even inferior in a contemporary population but actually seat around the optimal solution. This is the philosophy of our study.

To achieve our goal, we will accomplish three tasks. First, we will demonstrate that the SGAs, just like the DGAs do, promise to be helpful in solving some problem instances that are hard for the TGAs to optimize. Secondly, we will demonstrate that the SGAs and the DGAs are complementary to each other and hybridize them may

---

1 Although there are other selection schemes, for example, such as tournament selection and rank selection, proportionate selection is the mostly used method that gives an individual the probability of reproduction according to the ratio of the individual's fitness to the entire population's fitness [3].There are two general types of probabilistic reproduction process: generational reproduction and steady-state reproduction [14]. Generational reproduction replaces the entire population with a new one in each generation. In contrast, steady-state reproduction only replaces part of individuals in each generation.

2 Traditionally, we use the function value of an individual as its fitness measure or use a monotonic mapping between function value and fitness measure, i.e., $g(\boldsymbol{x}_1)<= g(\boldsymbol{x}_2)$ iff $f(\boldsymbol{x}_1)<= f(\boldsymbol{x}_2)$, where $g(\boldsymbol{x})$ is fitness measure of an individual $\boldsymbol{x}$, $f(\boldsymbol{x})$ is its function value of a maximization problem [4].

3 The mapping between function value and fitness measure is non-monotonic: $g(\boldsymbol{x}) = abs\ (f(\boldsymbol{x}) - \bar{f}(t))$, where $\bar{f}(t)$ is average function value of the population in generation t, and abs is an operation of taking absolute value of its argument.

supersede the TGAs. Thirdly, we will demonstrate that the cooperation of the SGAs and the DGAs is an effective and robust way in the genetic algorithms.

2 Stabilizing Selection

We adopt a stabilizing selection that, unlike the TGAs, takes a non-monotonic mapping between function value and fitness measure as follows

$$g(x) = 1/\,(abs\,(f(x) - f(t))). \tag{1}$$

Here, all notations are same as those in footnotes 2 and 3. In short, a stabilizing selection eliminates those extreme solutions and prefers to give more chances to those mediocre solutions. We construct two multimodal problem instances that are functions of unitation. A function of unitation is a function that the function value of a binary string depends only on the total number of 1's in that string and not on the position of those 1's [2]. Figs. 1 and 2 show the landscapes of these two problem instances in the unitation space, where $u(\boldsymbol{x})$ is the total number of 1's in the solution $\boldsymbol{x}$, $f(\boldsymbol{x})$ is function value of the solution $\boldsymbol{x}$. The optimal solution of these two problem instances is $\boldsymbol{x}$ = 111111111, that is $u(\boldsymbol{x}) = 9$, with a function value of 1024. From Fig. 2, it is clear that the optimal solution of problem instance F_2 is surrounded by mediocre solutions; whereas the superior and inferior solutions are far away from the optimal solution. As we know that the TGAs prefer to those superior individuals, problem instance F_2 is hard to optimize by the TGAs.

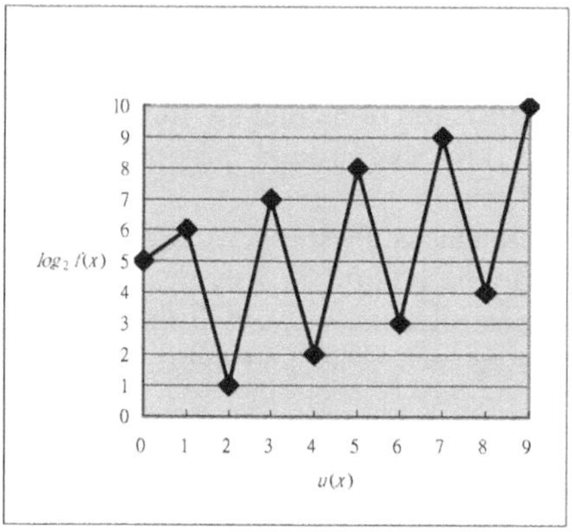

Fig. 1. Landscape of problem instance F_1

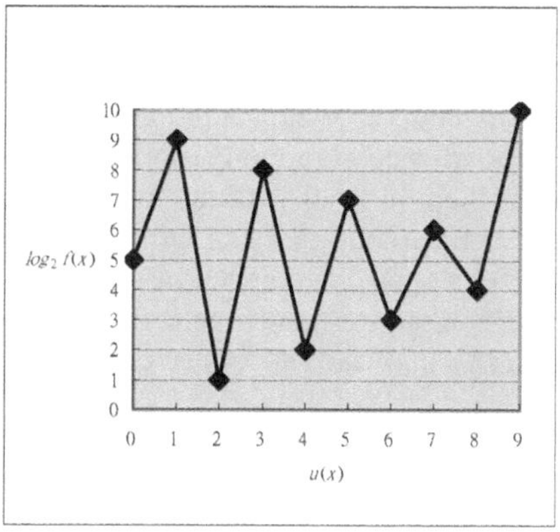

Fig. 2. Landscape of problem instance F_2

To measure the hardness of these two problem instances we need an estimator that can classify them into the type of GA-hard or GA-easy. Jones and Forrest proposed to use a statistics, the fitness distance correlation (FDC), as a measure of problem difficulty [6].[4] The FDC indicator $\boldsymbol{r}$ for problem instances F_1 is -0.42. It is straightforward (i.e., easy) for the TGAs to optimize. In contrast, the FDC indicator $\boldsymbol{r}$ for problem

[4] Here, actually, the fitness measure stands for the function value.

instances F_2 is 0.326.It is misleading (i.e., hard) for the TGAs to optimize, and the degree of difficulty is corresponding to several fully deceptive functions [2], [6].

In this study, a population size of twenty and the following parameter settings were used: the crossover rates (P_c) 0.9, 0.7, 0.5, 0.3, and 0.1; and the mutation rates (P_m) 0.01, 0.05, 0.1, 0.2, and 0.3. That is, there are twenty-five settings of parameters, and for each setting of parameters, we tested the SGAs and the TGAs, respectively. We replicated twenty re-initialized runs for each setting of parameters, and each run lasts for one hundred generations. The performance of a single run was taken to be the best solution ever found during the search.

Experimental results show that the TGAs perform very well on problem instance F_1. On the other hand, the SGAs do not solve this problem instance successfully. Conversely, the SGAs perform very well on problem instances F_2, but the TGAs do

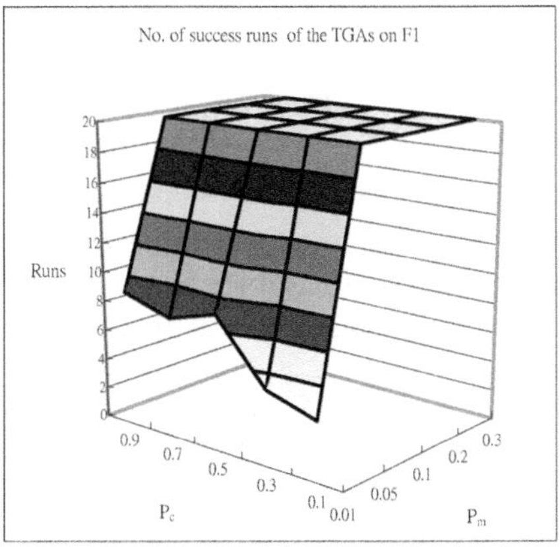

Fig. 3. Performance of the TGAs on F_1

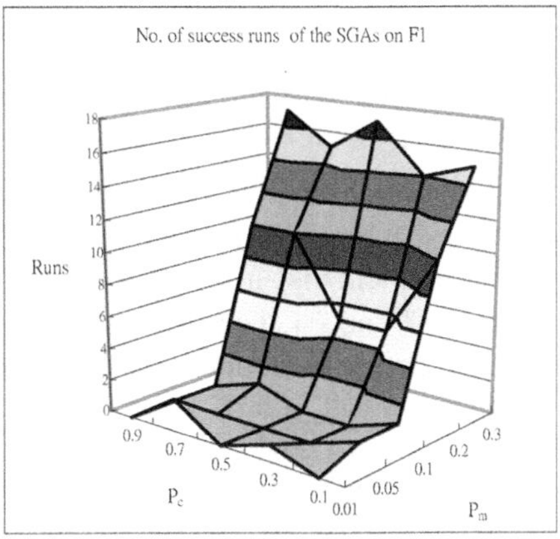

Fig. 4. Performance of the SGAs on F_1

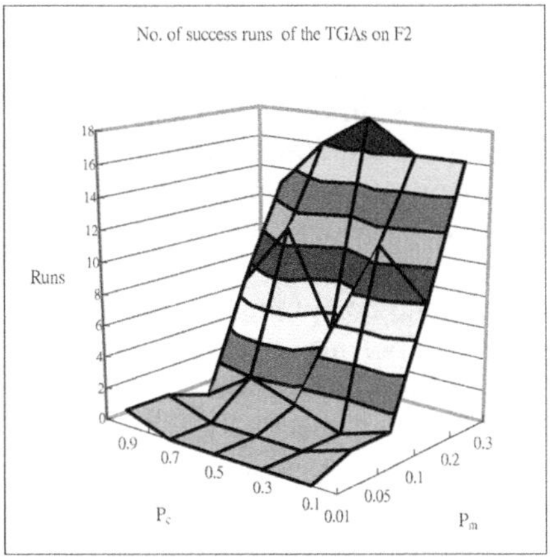

Fig. 5. Performance of the TGAs on F_2

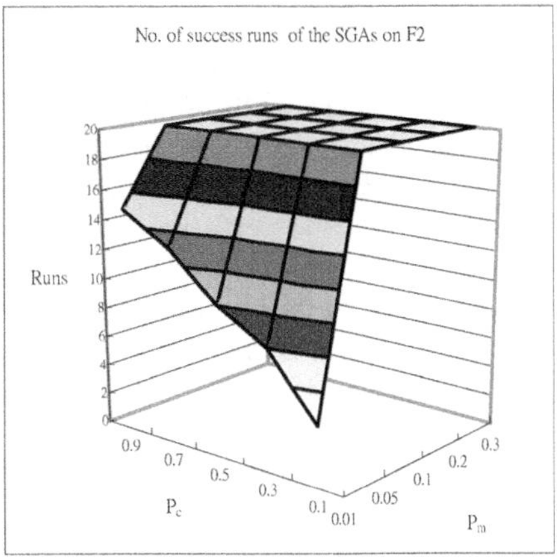

Fig. 6. Performance of the SGAs on F_2

not solve this problem instance successfully. These results are presented in Figs. 3 to 6. In short, these experimental results verify that the selection strategy is of paramount importance in the genetic algorithms and that the function values of solutions that located in the neighborhood of the optimal solution actually play an important role in the genetic search.

3 Should We Restrict Genetic Algorithms to the Direction Selection?

Now, the question is should we restrict genetic algorithms to the direction selection? To answer this question, we will show that the SGAs and the DGAs are complementary to each other and that the SGAs combine with the DGAs may supersede the TGAs. The test problems used here are also functions of unitation. In the case of a representation which use a binary string of length l, the function values of all 2^l feasible solutions can be classified into l+1 different groups, each group maps to the strings that contain 0,1,2,..., or l total number of 1's in that string.

To be fair, we should consider all possible ways of the mapping, in other words, we should test all possible landscapes of the search spaces and this number is $(l+1)!$ Thus, we use a binary string of length five to construct the test problem instances. For these test problem instances, we map the function values of feasible solutions to 2 、4 、8 、16 、32 and 64. For the sake of simplicity and without loss of generality, we assume the optimal solution of these problem instances is x = 11111, that is $u(x)$ = 5, with a function value of 64. Thus, we test 5! = 120 possible problem instances.

In all cases, a population size of six and the following parameter settings were used: the crossover rates 0.9 、0.7 、0.5 、0.3 、0.1and 0; and the mutation rates 0.001 、0.01 、0.05 、0.1 and 0.3. That is, there are thirty settings of parameters. For each setting, we replicated ten re-initialized runs that each run lasts for twelve generations to test the SGAs, the DGAs and the TGAs, respectively. The performance of a single run was taken to be the best solution ever found during that run.

The results indicate that, on the average, the SGAs and the DGAs perform slightly better than the TGAs. With carefully checkup, we find that different types of selection strategies are good for different types of search spaces. In other words, each type of selection strategy has its own upsides and downsides. Furthermore, the SGAs and the DGAs are complementary to each other. Fig. 7 manifests this finding explicitly. It is worth noting that the performance of the TGAs varies widely over all 120 problem instances.

In order to solidify this finding, we went further into computing the coefficients of correlation of the data presented in Fig 7. The coefficients of correlation between every two types of GAs are -0.85 for the SGAs and the DGAs, -0.245 for the SGAs and the TGAs, and 0.636 for the DGAs and the TGAs, respectively. The first figure shows a strong complementary relationship between the SGAs and the DGAs. The last figure indicates that the DGAs may substitute for the TGAs. In brief, as we have seen, these findings encourage us to study how to hybridize different types of selection strategies for enhancing the searching power of genetic algorithms. We will address this issue in next section.

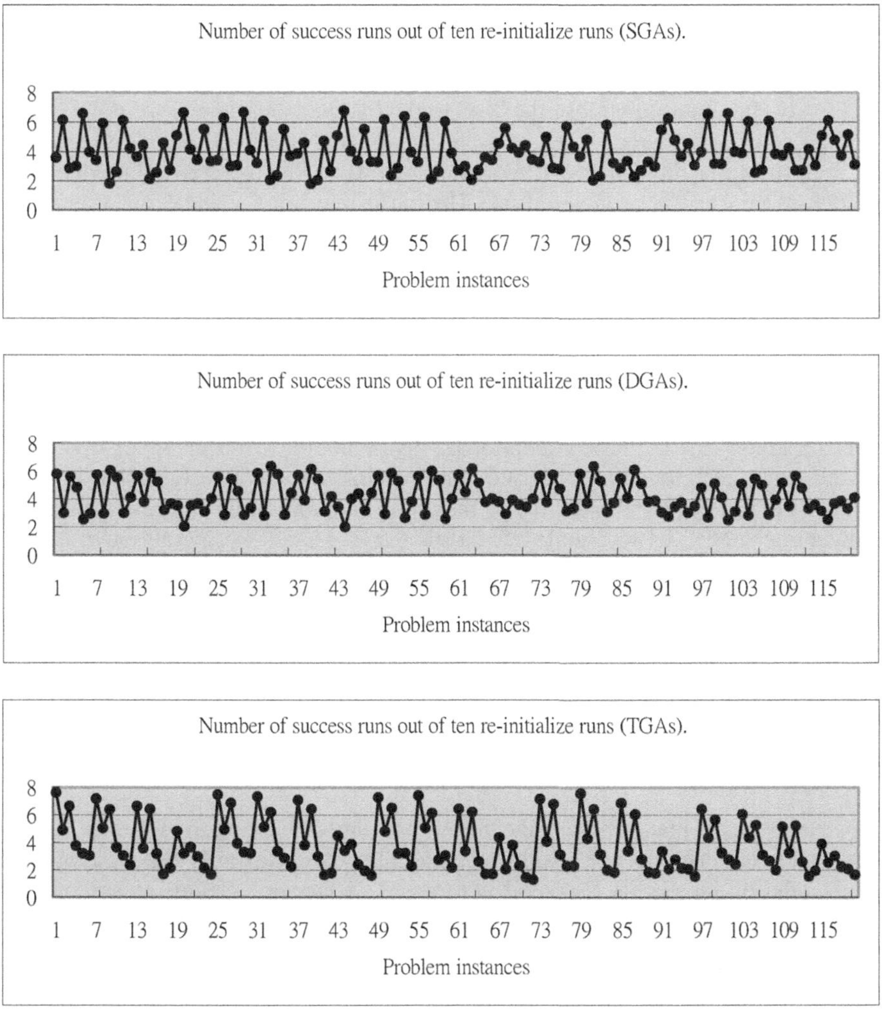

Fig. 7. Performance of the SGAs, the DGAs, and the TGAs from the perspective of problem instances (i.e., average over thirty settings of parameters)

4 A Synergistic Selection Strategy

It is advisable to implement a parallel GAs on which different types of selection strategies are applied to different GAs. The two models of parallel GAs that most being cited in literatures are island model and grid model (or neighborhood models) [1], [10], [12], [13]. In this study, we adopt an island model of parallel GAs on which two types of selection strategies are applied to two subpopulations that both evolve independently and migration of the best individual is allowed between them periodically to substitute the worst individual in a subpopulation with the best individual from the other one.

The test problems and parameter settings used here are same as those mentioned in the preceding section. For each setting of parameters we replicated ten re-initialized runs by three island models and a TGA model (with a population size of twelve), respectively. In notation, we use the T+S to denote the island model of the TGAs and the SGAs, the T+D of the TGAs and the DGAs, and the S+D of the SGAs and the DGAs. Each run lasting for twelve generations and for the three island models migration took place every four generations. The performance of a single run was taken to be the best ever found individual of that run.

Experimental results give evidence to claim that the cooperation of disruptive selection and stabilizing selection is an effective and reliable combination of selection strategies in the genetic algorithms. Overall, the model of S+D performs better than other models. Especially, we can find that the S+D model outperforms the T model, in the criteria of *avg.*, by 10.45% on the average. In details, for each one of all thirty settings of parameters, the *avg.* of the S+D model is largest and its *std.* is smallest except for some cases. These exceptional situations almost occur under lower mutation rates, such as 0.01 and 0.001. We interpret this to mean that in such a low mutation rate, the number of flipped bit of the entire population during one generation is less than one (6*5*0.01=0.3 and 6*5*0.001=0.03). That is to say, although a mediocre or even an inferior individual has been intently exploited by stabilizing or disruptive selection to produce more offspring, these new born individuals still can not evolve to a superior individual without the exploration of mutation. Of cause, lower crossover rates, such as 0.3, 0.1 and 0, may be is one of the causes of these exceptional situations.

5 Conclusion and Suggestions

Let us stress again that, whereas the SGAs and the DGAs just like the TGAs or all other search schemes do not ensure to guide the search towards the area where the optimal solution locates in, they will not miss the chance of exploiting and exploring the solutions that actually locate surrounding the optimal solution but are relative worse than other superior solutions in a contemporary population.

Despite the encouraging results of this study, future research is obviously required. We intend to continue pursing this line of work in a series of experimental studies. To address some real problems will be the first step. Besides, since the selection mechanism is the kernel of evolutionary computation techniques, the proposed model deserves an in-depth study to extend its scope of applications.

References

1. Cohoon, J. P., Hedge, S. U., Martin, W. N., Richards, D.: Punctuated equilibria: A parallel Genetic Algorithms. In: *Proceedings of the 2nd International Conference on Genetic algorithms* (1987) 148-154
2. Deb, K., Goldberg, D. E.: Sufficient conditions for deceptive and easy binary functions. *Annals of Mathematics and Artificial Intelligence*, 10 (1994) 385-408
3. Goldberg, D. E., Deb, K.: A Comparative Analysis of Selection Schemes Used in Genetic Algorithms. In: *Proceedings of the 1st Foundations of Genetic Algorithms* (1991) 69-93

4. Grefenstette, J. J., Baker, J. E.: How Genetic Algorithms Work: a Critical Look at Implicit Parallelism. In: *Proceedings of the 3rd International Conference on Genetic algorithms* (1989) 20-27
5. Holland, J. H.: *Adaptation in Natural and Artificial Systems*. The University of Michigan Press, Ann Arbor, MI (1975)
6. Jones, T. C., Forrest, S.: Fitness Distance Correlation as a Measure of Problem Difficulty for Genetic Algorithms. In: *Proceedings of the 6th International Conference on Genetic algorithms* (1995) 184-192
7. Kuo, T., Hwang, S. Y.: A Genetic Algorithm with Disruptive Selection. *IEEE Trans. on System, Man, and Cybernetics*, 26(2) (1996) 299-307
8. Kuo, T., Hwang, S. Y.: Why DGAs Work Well on GA-hard Functions?. New Generation Computing, 14 (1996) 459-479
9. Kuo, T., Hwang, S. Y.: Using Disruptive Selection to Maintain Diversity in Genetic Algorithms. *Applied Intelligence*, 7(3) (1997) 257-267
10. Manderick, B., Spiessens, P.: Fine-grained Parallel Genetic Algorithms. In: *Proceedings of the 3rd International Conference on Genetic algorithms* (1989) 428-433
11. Manly, B. F. J.: *The Statistics of Natural Selection on Animal Populations*. Chapman and Hall, London, UK (1984)
12. Petty, Chrisila B., Leuze, Michael R., Grefenstette, J. J.: A Parallel Genetic Algorithm. In: *Proceedings of the 2nd International Conference on Genetic algorithms* (1987) 155-161
13. Tanese, R.: Distributed Genetic Algorithms. In: *Proceedings of the 3rd International Conference on Genetic algorithms* (1987) 434-439
14. Whitley, L. D., Kauth, J.: GENITOR: a Different Genetic Algorithm. In: *Proceedings of the Rocky Mountain Conference on Artificial Intelligence*, Denver, CO (1988) 118-130
15. Wolpert, David H., William G. Macready: No Free Lunch Theorems for Optimization. *IEEE Transactions on Evolutionary Computation*, 1(1) (1997) 67-82

Priority-Based Genetic Local Search and Its Application to the Traveling Salesman Problem

Jyh-Da Wei and D.T. Lee

Institute of Information Science, Academia Sinica, Taipei, Taiwan
{jdwei, dtlee}@iis.sinica.edu.tw

Abstract. Genetic algorithms and genetic local search are population based general-purpose search algorithms. Nevertheless, most of combinatorial optimization problems have critical requirements in their definition and are usually not easy to solve due to the difficulty in gene encoding. The traveling salesman problem is an example that requires each node to be visited exactly once. In this paper, we propose a genetic local search method with priority-based encoding. This method retains generality in applications, supports schema analysis during searching process, and is verified to gain remarkable search results for the traveling salesman problem.

1 Introduction

Genetic algorithms (GAs) are population-based and problem independent search algorithms. Figure 1(a) shows the main processes of genetic algorithms, i.e., initialization, evaluation, selection, and reproduction. Hybrid with local search heuristics, Genetic Local Search (GLS) is an upgraded version that replaces each individual with its local optimal neighbor. As shown in Fig. 1(b), a local search process is launched in evaluation. GLS is thereby regarded as a method to mimic the cultural evolution instead of biological evolution, and also referred to as Memetic Algorithm (MA) or Lamarckian Evolutionary Algorithm [1].

Genetic local search is aimed to improve search quality of traditional genetic algorithms and has been examined to search efficiently for the near-optimal solutions to certain combinatorial optimization problems, such as the constraint satisfaction problem [2], flowshop scheduling problem [3], constraint minimum spanning tree problem (dMST) [4], and traveling salesman problem (TSP) [5]. Notably, these optimization problems usually have critical requirements that have forced researchers to develop new genetic operators. For example, for the dMST we have an upper bound on the node degrees and for the TSP we require that each city be visited exactly once.

These critical requirements put a great constraint for us to encode the genes in a GA. In the case of the TSP instance, for example, directly encoding cities into the chromosomes (the order presentation as shown in Fig. 2) does not work altogether with traditional crossover methods — the offspring may become an invalid tour if two chromosomes crossover using traditional 1-point or 2-point

T.-D. Wang et al. (Eds.): SEAL 2006, LNCS 4247, pp. 424–432, 2006.

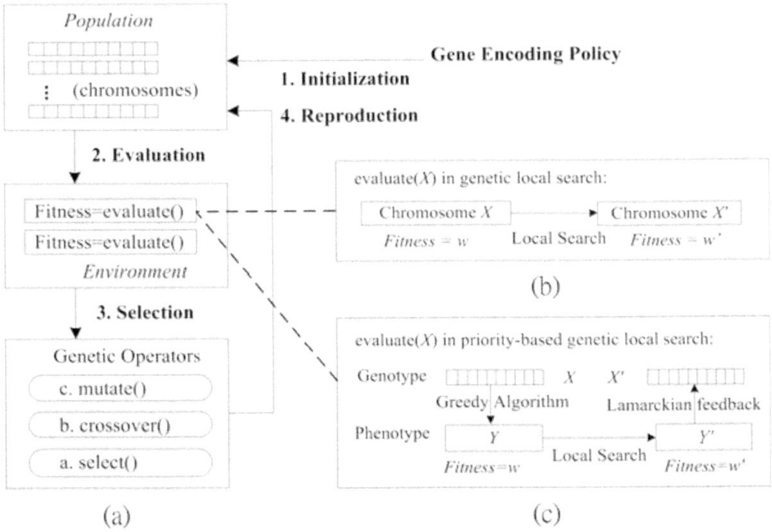

Fig. 1. Genetic Algorithm (GA) and Genetic Local Search (GLS). (a) GA flowchart, (b) GLS is a combination of GA and local search heuristics; (c) Priority-Based GLS (PB-GLS) proposed in this paper uses a greedy algorithm and a Lamarckian feedback process to alternate between genotype and phenotype.

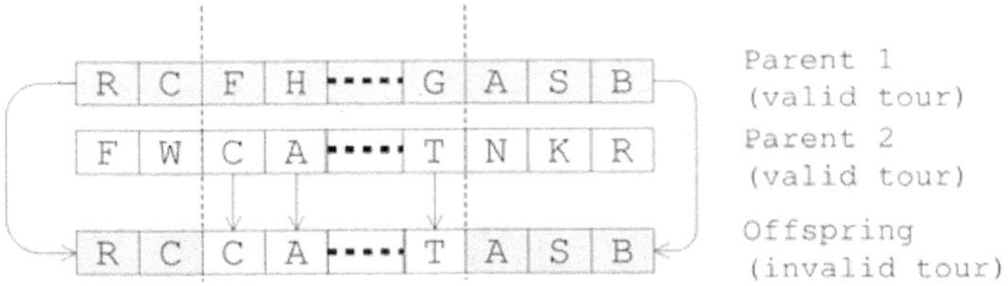

Fig. 2. Directly encoding cities into the chromosomes does not work altogether with traditional crossover methods. The offspring may become an invalid TSP tour.

operators. To overcome this difficulty, some renowned crossover operators were developed, many of which depend on the tour representation, such as the order presentation (partially matched crossover, PMX), the adjacency presentation (matrix crossover, MX), and the locus representation (natural crossover). There are also genetic operators independent of tour representations, such as greedy crossover (GX) and distance preserving crossover (DPX). For more information, the reader is referred to [6] for a comprehensive overview.

Although the-above mentioned approaches make genetic algorithms and genetic local search algorithms applicable to the TSP, they have some drawbacks as follows: The order representation does not ensure a unique representation for a TSP tour. This situation usually retards the evolutionary process. The adjacency matrix representation is time-consuming and does not have a significant performance. The locus representation is much limited in that it can only be applicable for the Euclidean TSP. Most importantly, these crossover operators are specialized mainly for the TSP and involve repair procedures to generate a valid tour. In contrast with the original intention of genetic algorithms, these operators are short of general practical values. In this paper, we propose a Genetic Local Search method with Priority-Based encoding, dubbed the "PB-GLS" model. This model retains generality in applications, supports schema analysis during searching process, and has been empirically proven to gain remarkable search results for the traveling salesman problem.

2 Priority-Based Genetic Local Search

For a combinatorial optimization problem for which a near-optimal solution can be obtained by using a greedy algorithm, certain entities, such as the nodes of the dMST and TSP problems and the jobs in the flowshop scheduling are selected step by step. Herein, the links between two consecutively selected entities are called "consecutive selections". We propose a priority-based encoding policy that assigns priorities to all the links between entities and the policy is expected to set high priority to the correct consecutive selections. The greedy algorithm employed remains the same, except that we select the next entity in consideration of the link priority prior to the original ranking key. By doing so, the greedy algorithm leads to a valid solution and the priority encoding makes it possible to follow traditional genetic evolutions. This approach does not lose generality in applications because we only need to provide a chromosome conformation that is simply a priority assignment.

2.1 Priority-Based Encoding with Local Search Method

As Fig. 1(c) shows, our idea is based on Mendelian inheritance that distinguishes genotype and phenotype in inheritance process. A greedy algorithm plays the role as the biochemical process that transfers the genotype encoding to the phenotype of each individual. The PB-GLS model further conducts a local search method to improve this phenotype. After that, we need a Lamarckian feedback process for encoding the local optimal solution and converting it back to its genotype. This process can be done if we enable all consecutive selections in the given solution by assigning them with higher priorities and disable potentially incorrect links by setting lower priorities. The range of priorities can be determined experimentally, although two priority levels are sufficient in our case.

2.2 Characteristics of the Priority-Based GLS

The priority-based genetic local search has three main features, i.e., broad applicability, problem transformation, and simulation of Mendelian inheritance theory.

1. Broad applicability: The priority-based encoding policy suits to any problem whose optimal solution can be approximated by a greedy algorithm, because

the greedy algorithm is characterized by two features, i.e., (a) the candidate entities are selected one after another sequentially, and (b) the selected entities are not discarded thereafter.
2. Problem transformation: The PB-GLS transforms combination and permutation problems into priority assignment problems. This problem transformation suggests a new direction to tackle the given problems. Imagine that the perfect optimal solution contains some crucial consecutive selections of problem entities (e.g. crucial edges in the TSP). Assigning higher priorities to these links leads to a near-optimal solution. Naturally, priority-based encoding allows us to analyze searching schema during the search process.
3. Simulation of Mendelian inheritance theory: We use greedy algorithms to simulate chemical processes, and use the priority-based encoding policy to simulate the gene codes in inheritance procedure. These priorities control the biochemical processes to "enable" and "disable" some biological functions, and finally develop a phenotype that fits the definition of the genotype.

3 Application to the TSP

We demonstrate using the PB-GLS model to solve the TSP. A greedy algorithm known as double-ended nearest neighbor (DENN) is used in this experiment.

3.1 Experiment Design

The DENN algorithm is described as follows. Let $E(A, B)$ denote the edge between city A and city B, and assume $E(A, B)$ is identical to $E(B, A)$ for any two distinct cities A and B:

1. Sort the edges by their costs into a sequence S.
2. Initialize a partial tour $T = \{S[1]\}$. Let $S[1] = E(A, B)$ be the current subtour from A to B.
3. Suppose the current subtour is from X to Y. We trace the sequence S to find the first edge $E(P, Q)$ that could extend the subtour at either end city X or city Y without creating a cycle, i.e., a complete tour that does not visit all the cities.
4. If the above edge $E(P, Q)$ is found, add it into T to extend the current subtour and repeat step 3; otherwise, add $E(Y, X)$ into T and return T as the searching result.

Note that the current state is extended by adding new nodes (cities) repeatedly. We now add priorities to the edges and change the sorting step by considering priorities of these edges first and then their costs in the first step. This change never affects the validity of tours, because the other steps of this greedy algorithm remain unchanged. The most concerned question is whether any tour can be represented by this encoding method. Considering that a greedy algorithm never discards an object once this object is selected, we can construct any given tour "T" by a greedy algorithm as the following formula describes:

$$\begin{aligned}&(\ \forall\ (r,s,t,k,|\{r,s,t,k\}|=4\ )\\&\qquad (\{E(r,s),E(s,t)\}\subseteq T\ \wedge\ C(k,s)\leq min\{C(r,s),C(s,t)\}\\&\quad \Rightarrow P(k,s)>max\{P(r,s),P(s,t)\})\ )\\&\Rightarrow \textit{The greedy algorithm constructs } T,\end{aligned}\tag{1}$$

where $C(a,b)$ and $P(a,b)$ are the cost and priority of edge $E(a,b)$ respectively, and a lower priority value $P(a,b)$ reflects a higher priority of edge $E(a,b)$ to be included in the tour.

The above description implies that the priority-based encoding can be used to search the global optimal solution. Two levels of priorities are sufficient to guarantee such an optimal solution. Formula (1) is also used in the Lamarckian feedback process of our GLS model. We apply the LK heuristic [7] as the local search method.

3.2 Complexity Analysis

It is well-known that an exhausted search for a TSP has an exponential time complexity. Suppose that n is the number of vertices. An exact solution takes $O(n!)$ time, which is prohibitively long. Therefore, polynomial-time heuristic search approaches are proposed. Heuristic or local search algorithms have complexities ranging from $O(n^2)$ (e.g., nearest neighbor, double ended nearest neighbor, and nearest insertion), $O(n^2 \log(n))$ (e.g., shortest edge first) to $O(n^{2.2})$ (e.g., LK) or higher order (e.g., k-opt).

The genetic algorithms with specialized crossover operators have time complexity $O(kmn^2)$, where k is the generation number and m is the population size. The n^2 factor is due to the fact that all the repair procedures need to scan all the possible pairs of the vertices which is $O(n^2)$. If we combine genetic algorithms with a local search algorithm, the latter affects the time complexity. For example, the genetic local search algorithm incorporating the DPX operator with the LK heuristic [5] has time complexity $O(kmn^{2.2})$. This model, referred to as DPX-LK model, is currently known as the most efficient model and will be compared with our model in the next section.

The time complexity of our model depends on the selected greedy algorithm, the Lamarckian feedback process, and the local search heuristic. When the DENN algorithm and LK heuristic are used, the PB-GLS needs $O(n^2)$ time to crossover the parent chromosomes and $O(n^2 + n^{2.2} + n^2)$ time to construct the tour, search the local optimum, and feedback the upgraded gene information. Therefore, the total time complexity is also $O(kmn^{2.2})$.

4 Experimental Results

In this section, we conduct three parts of experiments applying PB-GLS to solve the TSP. The first part used our own data to demonstrate how priority encoding is used and how it supports schema analysis. The second part used benchmark instances released by the *TSPLIB* [8] and proved that the proposed model can find near-optimal solutions identical to the best known results, in cases where the

Table 1. Cities in the first part of experiments

city	coordinates		city	coordinates		city	coordinates		city	coordinates	
1	0.047	0.692	6	0.259	0.085	11	0.436	0.367	16	0.684	0.292
2	0.054	0.577	7	0.326	0.748	12	0.525	0.222	17	0.694	0.585
3	0.095	0.004	8	0.336	0.530	13	0.534	0.031	18	0.732	0.811
4	0.146	0.855	9	0.399	0.143	14	0.542	0.602	19	0.891	0.082
5	0.157	0.319	10	0.399	0.275	15	0.586	0.982	20	0.946	0.728

number of cities was no more than 400. In the last part, we generated sparsely connected maps from the TSPLIB data instance and compared the experimental results between our model and the currently most efficient DPX-LK hybrid searching model. The PB-GLS model in these experiments uses Holland's simple genetic algorithm model with the uniform crossover operator and conducts the LK heuristic for local searching every five generation. To reduce the searching space and simplify the result discussion, priority encoding in these experiments used only two-level priority, i.e., the high priority was 1 and the low priority was 2. The population size and mutation rate were set as 100 and 0.2 respectively.

Table 1 lists the locations of the cities in the first part of experiments. These cities were randomly generated in the $[0, 1] \times [0, 1]$ square. Figure 3(a) and (b) show the city map and the near-optimal solution, found by both the DPX-LK hybrid model and our model within 100 generations. According to repeated experimental results, we believe that this tour with cost=4.35 is very close to the optimal tour. Table 2 lists the searching results of gene values partially, and the edges by their costs in ascending order. Columns from $p1$ to $p4$ are converged near-optimal chromosomes. All these chromosomes can develop the near-optimal TSP tour that is shown in Fig. 3(b). The final column denotes whether the edge under consideration is selected to be part of the tour.

Edges $E(3, 5)$, $E(7, 15)$ and $E(13, 19)$ in the 45th, 53rd, and 57th rows are the longest three edges contained in the tour. We can observe that they all receive a high priority. They are likely the crucial edges in the optimal tour. Interestingly, the three edges excluded from the tour, $E(9, 10)$, $E(9, 12)$ and $E(11, 12)$, are also remarkable because they are quite short and all receive a low priority. This result demonstrates that priority encoding allows schema explanation in searching optimal TSP tours. This schema is also likely to be useful when the number of cities increases or decreases.

In the second part of experiments, we used the instances released on the *TSPLIB* website to test our method. Experimental results reveal that the PB-GLS can find near-optimal solutions in maps with no more than 400 cities, such as the $st70$, $ch150$, $a280$ and $rd400$ data instances. The solutions obtained are identical to the presently best known results. For example, Fig. 3(c) is the experimental result for the $rd400$ data instance with the tour cost equal to 15281.

The time complexity of the proposed model is described in the previous section as $O(k\,m\,n^{2.2})$, where k, m and n are respectively the generation number, population size and city size. The running time increases quickly as more cities are added. For the first part of experiments, the searching result converged within 100 generations and took less than 3 seconds running on Sun's Ultra SPARC

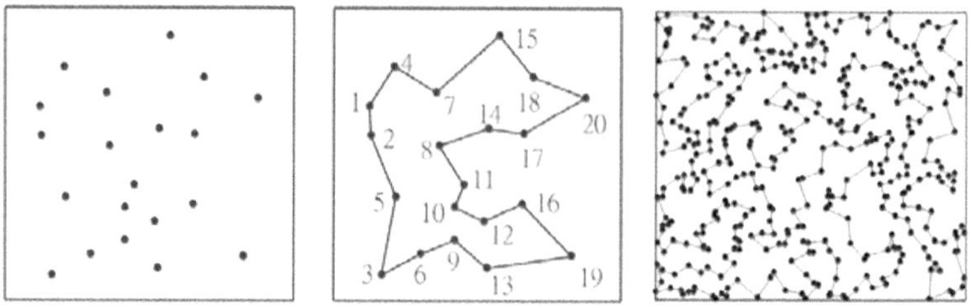

(a) (b) (c)

Fig. 3. Experimental results of the TSP. (a) map of the cities of Table 1; (b) near-optimal solution of the above listed cities; (c) near-optimal solution of the rd400 data instance.

Table 2. Searching results of the first part experiment (shown only partially). The edges are sorted by their costs. Columns from $p1$ to $p4$ are the converged priorities on four distinct chromosomes.

	cost	adjacency		p1	p2	p3	p4	T		cost	adjacency		p1	p2	p3	p4	T
1	0.100	10	11	1	1	1	1	Yes	41	0.295	16	19	1	2	1	1	Yes
2	0.115	1	2	1	1	1	1	Yes	42	0.299	5	9	2	2	2	1	-
3	0.132	9	10	2	2	2	2	-	43	0.299	6	12	2	2	1	2	-
4	0.137	10	12	1	1	1	1	Yes	44	0.301	13	16	1	2	1	1	-
5	0.149	9	12	2	2	2	2	-	45	0.320	3	5	1	1	1	1	Yes
6	0.151	6	9	1	1	1	1	Yes	46	0.321	2	7	2	2	2	1	-
7	0.152	14	17	1	1	1	1	Yes	47	0.322	9	16	1	2	2	1	-
8	0.170	11	12	2	2	2	2	-	48	0.331	1	8	2	2	2	2	-
9	0.174	12	16	1	1	1	1	Yes	49	0.333	6	11	2	2	1	2	-
10	0.176	9	13	1	1	1	1	Yes	50	0.334	3	9	1	1	1	2	-
11	0.183	3	6	1	1	1	2	Yes	51	0.337	11	17	1	2	1	1	-
12	0.190	8	11	1	1	1	1	Yes	52	0.340	14	16	1	2	2	1	-
13	0.191	1	4	1	1	1	1	Yes	53	0.350	7	15	1	1	1	1	Yes
14	0.191	12	13	2	1	1	2	-	54	0.351	11	13	1	1	1	2	-
15	0.209	4	7	1	1	1	1	Yes	55	0.357	10	14	2	1	1	2	-
16	0.218	8	14	1	1	1	1	Yes	56	0.361	8	12	2	1	1	1	-
17	0.218	7	8	2	1	1	2	-	57	0.361	13	19	1	1	1	1	Yes
18	0.225	15	18	1	1	1	1	Yes	58	0.361	8	17	2	1	2	2	-
19	0.228	9	11	1	1	2	2	-	59	0.376	4	8	2	1	2	1	-
20	0.230	17	18	2	2	2	2	-	60	0.380	12	14	2	2	2	1	-

III Workstation with 750-MHz clock rate. In case of 400 cities, it takes an average of 6216 generations and almost 3000-minute CPU time before the evolution converges to the best known solution.

If we do not want to enlarge the population size and the generation number, it could be necessary that we prune the longest edges from the chromosomes to improve the performance for a large scale TSP. In fact, fully connected maps are not usual in the real word. In the third part of experiments, we generated sparsely connected maps from *rd*400 data instance. In addition to the 400 edges in the best-known optimal tour, another 1600, 2600, 3600, 4600, and 5600 edges were randomly selected and added into the testing bed. We then conducted five experiments using these 2000, 3000, 4000, 5000 and 6000 edges as the test data respectively; these 400 nodes each had 10, 15, 20, 25 and 30 adjacent nodes in average. Our model was compared with the DPX-LK hybrid model.

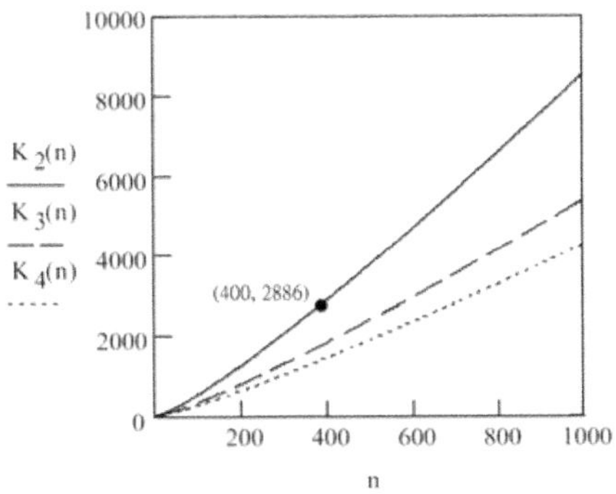

Fig. 4. Assuming p priority levels are used for searching in a sparsely connected map with k edges and n cities, $K_p(n) = log_p(n!)$ represents the highest tolerable values of k to ensure the searching space of PB-GLS less than that of the DPX-LK model ($p^k < n!$)

Table 3. Applying DPX-LK and PB-GLS models to find TSP tours in sparsely connected $rd400$ maps

	DPX-LK		Priority-based GLS	
#(Edges)	Generations	CPU time (sec)	Generations	CPU time (sec)
6000	268	11768	695	20949
5000	205	8933	389	11663
4000	144	6357	207	6195
3000	93	4102	74	2203
2000	35	1512	27	804

Assuming the p priority levels are used for testing a data instance with k edges and n nodes, the searching spaces of our model and the DPX-LK hybrid model are of sizes p^k and $n!$ respectively. The condition to let $p^k < n!$ can be derived as

$$p^k < n! \Longleftrightarrow k < log_p(n!) = \sum_{v=1}^{n}(log_p v) \tag{2}$$

Figure 4 draws the right part of formula (2), denoted as function $K_p(n)$, with $n \in \{1, 2, .., 1000\}$ and $p \in \{2, 3, 4\}$. Given $p = 2$ and $n = 400$, k must be less than 2886 to ensure search space $p^k < n!$. However, the experimental results listed in Table 3 reveal that our model converged efficiently than the DPX-LK model even with $k = 4000$. This result implies that using permutation-based algorithms in search sparsely connected maps may suffer an overhead that does not occur when we use priority-based algorithms.

5 Conclusion

Genetic algorithms and Genetic Local Search methods encounter great difficulties in solving certain combinatorial optimization problems, when they have critical requirements which cannot be easily encoded in a genetic algorithm or genetic local search method. Previous results made use of specialized genetic operators to enhance the GA and GLS. We have proposed a priority-based encoding method in conjunction with greedy algorithms. We encode link priorities as chromosomes, and then use the underlying greedy algorithms to construct the corresponding solution as the phenotype. By doing so, traditional genetic algorithms can be exploited as usual.

Priority-based encoding supports not only broad applications but also schema analysis. The priority-based genetic local search is empirically tested to achieve remarkable searching results for the TSP by iteratively converging to crucial edges. According to experimental results, this model found near-optimal solutions to TSPLIB instances — in cases where the number of cities is no more than 400, the results are identical to the best previously known results. Experimental results also reveal that the permutation-based algorithms using specialized GA operators have an overhead in searching sparsely connected maps. This overhead does not occur when we use priority-based algorithms, because it is not necessary to encode the disconnected links into the chromosome.

Acknowledgments

This work was supported in part by the National Science Council under the Grants NSC94-2213-E-001-004, NSC95-2221-E-001-006-MY3 and NSC-94-2752-E-002-005-PAE.

References

1. Digalakis, J., Margaritis, K.: Performance comparison of memetic algorithms. Applied Mathematics and Computation **158** (2004) 237–252
2. Marchiori, E., Steenbeek, A.: A genetic local search algorithm for random binary constraint satisfaction problems. In: SAC (1). (2000) 458–462
3. Arroyo, J., Armentano, V.: Genetic local search for multi-objective flowshop scheduling problems. European Journal of Operational Research **167** (2005) 717–738
4. Zeng, Y., Wang, Y.P.: A new genetic algorithm with local search method for degree-constrained minimum spanning tree problem. In: Proc. Fifth International Conference on Computational Intelligence and Multimedia Applications. (2003) 218–222
5. Freisleben, B., Merz, P.: A genetic local search algorithm for solving symmetric and asymmetric traveling salesman problems. In: Proceedings of the 1996 IEEE International Conference on Evolutionary Computation. (1996) 616–621
6. Larranaga, P., Kuijpers, C.M.H., Murga, R.H., Inza, I., Dizdarevic, S.: Genetic algorithms for the travelling salesman problems: A review of representations and operators. In: Artificial Intelligence Review. Volume 13. Kluwer Academic Publishers, Netherlands (1999) 129–170
7. Lin, S., Kernighan, B.: An effective heuristic algorithm for the traveling salesman tours. Operations Research **21** (1973) 498–516
8. Reinelt, G.: TSPLIB website. (http://www.iwr.uni-heidelberg.de/groups/comopt/software/TSPLIB95/tsp/)

Integrated the Simplified Interpolation and Clonal Selection into the Particle Swarm Optimization for Optimization Problems

Jing Wang, Xiaohua Zhang, and Licheng Jiao

Institute of Intelligent Information Processing and National Key Laboratory for Radar Signal Processing,
Xidian University, Xi'an, Shaanxi 710071, China
w_jingcn@163.com, xh_zhang@mail.xidian.edu.cn,
lchjiao@mail.xidian.edu.cn

Abstract. Particle Swarm Optimization (PSO) is gaining momentum as a simple and effective optimization technique. However, its performance on complex problem with multiple minima falls short of that of the Ant Clony Optimization (ACO) algorithm. The new algorithm, which we call Hybrid Particle Swarm Optimization, combines the ideas of particle swarm optimization with clonal selection strategy and simplified quadratic interpolation (SQI), which is used to improve its local search ability, and to escape from the local optima. Simulation results on 14 benchmark test functions show that the hybrid algorithm is able to avoid the premature convergence and find much better solutions with high speed.

1 Introduction

The Particle Swarm Optimization Algorithms (PSO), originally introduced in terms of social and cognitive behavior, such as flocks of birds and school of fish ,by Kennedy and Eberhart in 1995 [1]. It has a number of desirable properties, including simplicity of implementation, scalability in dimension, and good empirical performance, and has widely applied to many optimization problems [2]. Even so, it is not without problems. PSO suffers from premature convergence, tending to get stuck in local minima, we also found that it suffers from an ineffective strategy, especially around local minima. Moreover, adjusting the tunable parameters of PSO to obtain good performance can be a difficult task [2].

Here we integrated the simplified quadratic interpolation (SQI) method and clonal selection into the real-coded PSO, and presented a Hybrid PSO (HPSO). The SQI method is used to improve its local searching ability of the algorithm, and to escape from the local optima. We compared the performance of the HPSO to that of Standard Genetic Algorithms (SGA)[7], Clonal Selection Algorithms (CSA), Cloning Selection PSO (CSPSO), Chaos Mutation Evolutionary Algorithms (CMEA) and Self-adaptive Chaos Clonal Algorithms (SCCA) [3] in solving many benchmark test functions, the simulation results show that the HPSO is efficient.

T.-D. Wang et al. (Eds.): SEAL 2006, LNCS 4247, pp. 433 – 440, 2006.

In the remainder of this paper, the algorithm we proposed is described in Section 2, and Section 3 is the experiments of the HPSO on benchmark test functions comparing with the available algorithms, we finally conclude our paper in Section 4.

2 The Hybrid Particle Swarm Optimization

2.1 The Particle Swarm Optimization

The basic PSO algorithm begins by scattering a number of "Particles" in the function domain space. Each particle represents a candidate solution, $\vec{x_i} = (x_{i1}, x_{i2}, \ldots, x_{iD})$, D is the dimension of the search space. the i-th particle of the swarm population knows: **a)** its personal best position $\vec{P_i} = (P_{i1}, P_{i2,} \ldots P_{gD})$, i.e., the best position this particle has visited so far; and **b)** the global best position, $\vec{P_g} = (P_{g1}, P_{g2}, \ldots, P_{gD})$, i.e., the position of the best particle that gives the best fitness value in the entire population; and **c)** its current velocity, $\vec{v_i} = (v_{i1}, v_{i2}, \ldots v_{iD})$, which presents its position change. The following equation (2-1) uses the above information to calculate the new update velocity for each particle in the next iteration step. Equation (2-2) updates each particle's position in the search space.

$$v_{id}^{k+1} = wv_{id}^{k} + c_1 r_1 (P_{id}^{k} - X_{id}^{k}) + c_2 r_2 (P_{gd}^{k} - X_{id}^{k}) \tag{2-1}$$

$$x_{id}^{k+1} = x_{id}^{k} + v_{id}^{k+1} \tag{2-2}$$

where $i = 1, 2, \ldots, N, d = 1, 2, \ldots, D$; N is the size of the swarm population; w is the inertia weight, which is often used as a parameter to control exploration/exploitation in the search space, and it can be set to a constant, a linear or nonlinear variance which is constructed on linearly decreasing weight (LDW) as follows:

$$w(t) = (w_{\text{int}} - w_{end})(T_{\max} - t) / T_{\max} + w_{end} \tag{2-3}$$

where t and $T_{\max}$ is the current and maximal generation. c_1 and c_2 are two coefficients (positive constants); r_1 and r_2 are two random number within the range $[0, 1]$. There is also a $v_{\max}$, which sets the upper and lower bound for velocity values.

2.2 Clonal Selection Algorithm [3]

The famous antibody clonal selection theory was put forward by Burnet in 1958. Clonal operator intrinsically produces a new sub population in the neighborhood of every individual based on its fitness, and thus expands the search space to improve its learning capability. An inconsistent mutation operator is embeds into the clonal selection instead of crossover in normalized real-coded individual. we will illustrate the process in following steps:

Step 1: Rank all of the individuals according to their fitness values

Step 2: Decide the clonal size of every individual, produce a new sub population *sub* of this individual U by mutation with the operator Δ. Supposed that $u_t = (u_1, u_2, \ldots, u_k, \ldots)$ is an individual of the t-th generation, if we mutate u_k, the result will be $u_t' = (u_1, u_2, \ldots, u_k', \ldots)$,where,

$$u_k' = \begin{cases} u_k + \Delta(t, UB - u_k) & if\ r > 0 \\ u_k - \Delta(t, u_k - LB) & else \end{cases} \tag{2-4}$$

and r is a random number within the range $[0,1]$, $\Delta(t, y) = y \times (1 - \delta^{(1-t/T)^b})$, δ is a random number within $[0,1]$, b is a constant, UB, LB is the upper and lower bound of u_k, t and T are the current and total generation.

Step 3: Update the population: calculate the fitness values of the sub population's individuals, select the best one of the *sub* and renew U.

Step 4: Repeat step 2 and step 3 with $U(i)$ $(i = 1, 2, \ldots N)$, produce the new population.

2.3 The Simplified Quadratic Interpolation

SQI is one method of polynomial approximation, which constructs a quadratic interpolation polynomial closing to the object function by using the information and function values which the object function lies in many points, then calculate the optima of this polynomial as approximate solutions of the object function.

Here, we introduced the SQI method, which will be integrated into PSO. Denote three individuals by $x^a = (x_1^a, \ldots, x_n^a)^T$, $x^b = (x_1^b, \ldots, x_n^b)^T$, $x^c = (x_1^c, \ldots, x_n^c)^T$ and calculate their fitness values $f_a = fit(x^a)$, $f_b = fit(x^b)$, $f_c = fit(x^c)$, Suppose that $fit(x^a) > fit(x^b)$ and $fit(x^c) > fit(x^b)$. Then, derived from the threepoint quadratic interpolation, the approximate minimal point $\bar{x} = (\bar{x}_1, \ldots, \bar{x}_n)^T$ is calculated according to:

$$\bar{x}_i = \frac{1}{2}\left\{ \frac{[(x_i^b)^2 - (x_i^c)^2]f_a + [(x_i^c)^2 - (x_i^a)^2]f_b + [(x_i^a)^2 - (x_i^b)^2]f_c}{(x_i^b - x_i^c)f_a + (x_i^c - x_i^a)f_b + (x_i^a - x_i^b)f_c} \right\} \tag{2-5}$$
$$i = 1, \ldots, n$$

2.4 The Hybrid PSO Algorithm

Now, we present the HPSO algorithm for optimization problems as follows.

Data. The parameters we employed are: population size N, the dimension of the high dimensional function Nd, the maximal time of running cs, maximal number of generation ge, two positive constants coefficients $c_1 = c_2 = 2$.

Step 1. (Initialization): set $cs=1$, generate randomly the initial population set $P(k) = \{x^1(k), \ldots x^N(k)\}$, and compute the fitness values $fit(x^i(k))$, $i = 1, \ldots, N$, $k = 1, \ldots Nd$ and iterate cs times.

Step 2. (Cloning): if $ga < ge$, execute clonal mutation by the strategy we introduced in section 2.2, and generate a new population.

Step 3. (PSO Operation [7]):

Step 3.1 In above step we generate a new population p_{best}, the optima of all the individuals of the population is g_{best}.

Step 3.2 $i = 1, 2, ..., Nd$, update population by the two equations (2-1) and (2-2), they stand for the renewal of velocity and individual respectively.

Step 3.3 If the individual surpass the upper and lower bound, substitute it with the boundary value closing to it.

Step 3.4 Compare the new population with the old one, preserve the better one, and renew the g_{best} as well.

Step 3.5 If $i < Nd$, then go to step 3.2, otherwise go to step 4.

Step 4. (SQI): Rank the new population according to their fitness values, order ascending and relabel them. The ordered fitness value set is denoted by $FIT = \{F(1), F(2), ..., F(N)\}$, the corresponding individual set is denoted by $X = \{x(1), x(2), ... x(N)\}$, let $x_b = x(1), f_b = F(1)$.

Step 4.1 Generate at random two integers $r_1 \in [2, 2+L]$, $r_2 \in [3, 3+L]$, such that $r_1 \neq r_2$, here, we chose $L = 3$, let $x^a = x(r_1),\ f_a = fit(r_1),\ x^c = x(r_2),\ f_c = fit(r_2)$.

Step 4.2 If $(x_i^b - x_i^c)f_a + (x_i^c - x_i^a)f_b + (x_i^a - x_i^b)f_c < \varepsilon \ (\varepsilon = 10^{-6})$, then go to step 4.1, otherwise, go to step 4.3.

Step 4.3 Calculate $\bar{x}$ by using (2-5), and then the fitness value $fit(\bar{x})$.

Step 4.4 If $fit(\bar{x}) < f_b$, then replace the worst solution in current population with $\bar{x}$.

Step 5. (Stopping Criterion): if the stopping criterion is met, then stop, and record the best individual as the approximate global optimal solution of problem. Otherwise, set $ga = ga + 1$, and go to step 2.

3 Experiments

To evaluate the performance of the proposed HPSO, we test 14 benchmark test functions described in [3], which include 6 2-D functions and 8 difficult test functions (functions with high-dimensional decision variables), which often used to test the performance of proposed algorithm. Table 3-1 and table 3-2 give the main characteristics of these functions.

From the two tables above, we can see evidently that almost all of them have a host of local optima, and some even have the amount beyond our calculation, especially for high dimensional function. So these functions can certify the validity, reliability and stability of the algorithm.

We compare HPSO against CSA, CSPSO, and other better algorithms such as CMEA, SCCA, and SGA. For all functions and algorithms, we set $Nd = 10$ (the number of the dimension), the iteration $cs = 30$, and chose the total evolutionary generation $ge = 100$, $ge = 300$ for 2-D and high-dimensional functions respectively. The final results are shown in table 3-3 and table 3-4.

Table 3-1. The main characteristics of 6 2-D test functions."TF" stands for test function, "CGO"stands for the coordinate of Global optimum, "NGO" stands for the number of global optimum and "NLO" stands for the number of local optimum. *NA* stands for the number is innumerable.

TF	optimum	CGO	NGO	NLO
f_1	−2.1188	$(\pm 0.64, \pm 0.64)$	4	32
f_2	−3600	(0,0)	1	4
f_3	0	(0,0)	1	*NA*
f_4	0	(0,0)	1	*NA*
f_5	−1	(0,0)	1	4
f_6	−1	(0,0)	1	4

Table 3-2. The main characteristics of 8 high-dimensional functions

TF	optimum	CGO	NLO
f_7	0	$x_i = 0 \quad i = 1, 2, ..., N$	1
f_8	0	$x_i = 0 \quad i = 1, 2, ..., N$	1
f_9	$10N$	$x_i = 0 \quad i = 1, 2, ..., N$	$N!$
f_{10}	0	$x_i = 0 \quad i = 1, 2, ..., N$	*NA*
f_{11}	0	$x_i = 0 \quad i = 1, 2, ..., N$	*NA*
f_{12}	78.3324	$x_i = -2.9035 \quad i = 1, 2, ..., N$	2^N
f_{13}	$418.9828N$	$x_i = 421 \quad i = 1, 2, ..., N$	*NA*
f_{14}	0	$x_i = 0 \quad i = 1, 2, ..., N$	*NA*

In order to evaluate the ability of convergence of the HPSO, we list the statistic results of SGA,CEA (Chaos Evolutionary Algorithms) [3] and HPSO, in which we set $\varepsilon = 0.001$, the maximum generation $ge = 1500$,and running 30 times. The one which failing to find the solution in the given maximal generation is regarded as an individual not receiving the optima, and excluded from the domain of the statistic results.

Analyzing the results shown in table 3-3 and table 3-4, we conclude that HPSO surpasses other algorithms. They indicate that the qualities of the solutions found by other algorithms are significantly worse than those by HPSO for all the benchmark test functions. Additionally, the high accuracy of HPSO in finding the better solution should be put forward especially. We can also see from table 3-5 that HPSO mostly converging to a good solution with just a few steps comparing to SGA and CEA. To recapitulate, the performance of HPSO is better than that of other algorithms, and it is a powerful algorithm in resolving single objective optimization problems.

Table 3-3. Comparison of the mean solutions found by SGA, CMEA and SCCA

TF	SGA		CMEA		SCCA	
	mean	std	mean	std	mean	std
f_1	-2.0889	2.7876 e-004	-2.0650	0.0033	-2.1187	1.1877 e-009
f_2	-2.7488 e+003	6.7262 e+004	3.2972 e+003	1.5298 e+005	-2.6774 e+003	46.3113
f_3	0.3282	0.7521	0.2682	0.0097	0.1033	4.0255 e-004
f_4	0.7702	2.3861	0.5420	0.3149	0.0111	1.9720 e-005
f_5	-0.9607	0.0414	-0.9980	6.9722 e-006	-0.9825	4.5686 e-004
f_6	-1.0000	3.5108 e-016	-0.9997	9.1115 e-008	-1.0000	6.0065 e-012
f_7	21.2244	60.6353	18.1085	86.6769	0.0413	8.7442 e-005
f_8	8.2545	57.7729	9.6750	31.9932	0.0040	1.8892 e-006
f_9	-85.9322	25.2125	-89.4161	6.7282	-99.4818	0.2236
f_{10}	11.5421	17.7708	2.4163	0.2549	0.1027	8.2656 e-005
f_{11}	-44.4523	702.4695	-24.0003	137.6679	-0.0557	0.0055
f_{12}	-73.4419	12.5048	-77.4795	0.2181	-78.2917	2.4116 e-004
f_{13}	-3.2529 e+003	1.9807 e+004	-3.7919 e+003	3.0715 e+004	-4.1684 e+003	26.9978
f_{14}	9.1435	58.0884	14.3562	35.2720	0.6607	0.0034

Table 3-4. Comparison of the mean solutions found by HPSO against CSPSO and HPSO

TF	CSA		CSPSO		HPSO	
	mean	std	mean	std	mean	std
f_1	-2.1187	4.3903 e-010	-2.1188	4.0373 e-005	-2.1188	1.6729 e-005
f_2	-2.7488 e+003	0.0301	-3600	0	-3600	0
f_3	0.1098	0.1512	0.0047	0.0181	1.7049 e-013	4.7738 e-013
f_4	0.0067	0.0270	0	0	0	0
f_5	-0.9869	0.0633	-0.9913	0.0552	-1	0
f_6	-1.0000	4.4133 e-006	-1	0	-1	0
f_7	2.4761 e-005	2.1744 e-010	1.1700 e-014	5.5171 e-01	5.7309 e-038	1.7662 e-037

Table 3-4. (*continued*)

f_8	8.4397 e-010	3.3055 e-018	1.0794 e-009	5.1656 e-009	2.2959 e-016	6.8150 e-016
f_9	-99.9005	0.0891	-100.0000	4.7580 e-013	-100	0
f_{10}	1.2745 e-009	2.7026 e-019	1.6206 e-009	2.6588 e-009	4.1585 e-025	9.3232 e-025
f_{11}	-13.7967	9.4155	-0.0155	0.0856	-0.0068	0.0277
f_{12}	-73.4419	12.5048	-78.3323	1.1080 e-007	-78.3323	1.2073 e-007
f_{13}	-3.2529 e+003	1.9807 e+003	-4.1898 e+003	3.0583 e-005	-4.1898 e+003	3.5544 e-005
f_{14}	0.0561	0.0190	0.0161	0.1214	0.0060	0.0569

Table 3-5. Comparison of the optimal performance of SCA , CEA and HPSO. N1, N2/N3/N4 stand for the number of achieving the optimum for every algorithm, the maximal/minimal/mean number of the generation achieving the optimum. "-" means the experiment is not done , "/ "means no feasible solutions were found.

TF	HPSO		SGA		CEA	
	N1	N2/N3/N4 (std)	N1	N2/N3/N4 (std)	N1	N2/N3/N4 (std)
f_1	30	15/2/8 (1.57e-02)	10	202/28//112 (1.97e-02)	22	478/3/118 (1.86e-02)
f_2	30	4/2/2.3 (1.3941e-04)	2	13/13/12.5 (1.86e-02)	0	/
f_3	30	749/10/195 (1.1853e-04)	—	—	—	—
f_4	30	7/4/5 (2.8747e-04)	30	56/9/21 (3.19e-04)	30	35/5/18 (2.20e-04)
f_5	26	9/2/4 (1.5e-02)	9	95/1/29 (1.89e-02)	15	250/1/59 (2.58e-03)
f_6	30	3/2/2 (2.7622e-04)	—	—	—	—
f_7	30	69/58/59 (1.5974e-04)	30	135/86/101 (2.42e-04)	30	58/34/44 (1.49e-04)
f_8	30	76/40/59 (1.4968e-04)	30	337/92/228 (1.61e-04)	30	201/146/174 (1.14e-04)
f_9	30	32/24/28 (2.0059e-04)	24	223/104/179 (4.17e-01)	30	138/36/61 (3.53e-03)
f_{10}	30	59/43/51 (1.0407e-04)	30	104/78/91 (1.76e-1)	30	36/26/31 (1.29e-03)
f_{11}	30	499/101/269 (1.9540e-04)	30	1247/456/923 (3.34e-03)	30	126/61/96 (1.45e-02)

Table 3-5. *(continued)*

f_{12}	30	49/36/43 (9.21e-02)	—	—	—	—
f_{13}	30	839/359/611 (5.8e-03)	—	—	—	—
f_{14}	29	134/2/6 (8.9e-03)	16	134/71/115 (3.93e-01)	30	179/32/70 (1.57e-04)

4 Conclusion

In this paper, HPSO algorithm is proposed by integrated simplified quadratic interpolation and clonal selection into the improved PSO. The SQI method is a powerful heuristic way for improving the local search ability of the algorithm, and making the algorithm escape from the local optima. Clonal selection is an effective strategy in maintaining the diversity of the population. The simulation results indicated that HPSO performs remarkably well against others algorithms. It increases local search ability and exploitation of PSO while retaining its key properties of sociality, momentum, exploration and stability. It represents a substantial improvement over the other algorithms not only in the resulting solution, but also in the speed and accuracy with which they are found. However, it still exists some problems, and we will concentrate on it in the further work.

References

1. Kennedy, J. and Eberhart, R., "Particle Swarm Optimization", IEEE International Conference on Neural Networks, 1995, Perth, Australia.
2. Christopher K. Monson and Kevin D. Seppi: The Kalman Swarm -A New Approach to Particle Motion in Swarm Optimization. In: K. Deb et al. (Eds.): GECCO 2004, LNCS 3102, pp. 140–150, 2004. Springer-Verlag Berlin Heidelberg 2004
3. DU Haifeng:The Study and Application with Immune Clonal Compution and the Artificial Immune Net .The Research Report of Postdoctoral Study of Xidian University, 2003
4. Hongli, Yongchang Jiao, and Yuping Wang: Integrating the Simplified Interpolation into the Genetic Algorithm for Constrained Optimization Problems. In: Y.Hao et al. (Eds.): CIS 2005, Part I, LNAI 3801, pp.247-254, 2005. Springer-Verlag Berlin Heidelberg 2005
5. Licheng Jiao, Jing Liu, Weicai Zhong: An organizational coevolutionary algorithm for classification. IEEE Trans. Evolutionary Computation, vol. 10, no. 1, pp. 67 – 80,2006
6. S. L. Ho, Shiyou Yang, Guangzheng Ni, Edward W. C. Lo, and H. C. Wong: A Particle Swarm Optimization-Based Method for Multiobjective Design Optimizations. IEEE Trans. on Magnetics, Vol. 41, no. 5, pp. 1756-1759, 2005
7. Minqiang Li et al. Basic Theory and Application of Genetic Algorithm.Beijing: Scientific Publishing House, 2002.3

Hybrid Model of Genetic Algorithm and Cultural Algorithms for Optimization Problem

Fang Gao[1], Hongwei Liu[1], Qiang Zhao[2], and Gang Cui[1]

[1] School of Computer Science and Technology, Harbin Institute of Technology, 150001 Harbin, China
gaofang@hit.edu.cn, {cg,liuhongwei}@ftcl.hit.edu.cn
[2] School of Traffic Transportation Engineering, Northeast Forestry University, Harbin 150040, P.R. China
qyangzhao@163.com

Abstract. To solve constrained optimization problems, we propose to integrate genetic algorithm (GA) and cultural algorithms (CA) to develop a hybrid model (HMGCA). In this model, GA's selection and crossover operations are used in CA's population space. A direct comparison-proportional method is employed in GA's selections to keep a certain proportion of infeasible but better (with higher fitness) individuals, which is beneficial to the optimization. Elitist preservation strategy is also used to enhance the global convergence. GA's mutation is replaced by CA based mutation operation which can attract individuals to move to the semi-feasible and feasible region of the optimization problem to improve search direction in GA. Thus it is possible to enhance search ability and to reduce computational cost. A simulation example shows the effectiveness of the proposed approach.

1 Introduction

Cultural algorithms were introduced by Reynold as new evolutionary computational approaches [1]. It utilizes culture as a vehicle and took a dual inheritance mechanism. The cultural interaction between its two spaces in cultural algorithms can ensure that the information acquired by an individual can be shared with the entire population. Thus culture can be taken as evolving knowledge to influence each individual's behavior in the population space, and it also developed further as the process of evolution. It can be seen that cultural algorithms support self-adaptation at both the individual and the population level. Now Cultural Algorithms have been developed for many application fields due to their noticeable advantages [2, 3, 4, 5, 6].

Genetic algorithm is a highly paralleled and efficient method based on natural evolution. It has enjoyed increasing popularity in solving optimization problems. Particularly, the real coded version of GA (called RGA), has the advantages of high precision and easy to search in large space, meanwhile it avoids the troublesome encoding and decoding process of computing the objective function. Therefore RGA has been widely studied and applied in different fields, such as controller design [7, 8], engineering optimization [9, 10] and neural network training [11].

T.-D. Wang et al. (Eds.): SEAL 2006, LNCS 4247, pp. 441 – 448, 2006.

However, how to solve nonlinear constrained optimization problem remains a difficult and open question for GA and other evolutionary algorithms. The existing constraint-handling methods includes: (1) discarding infeasible solution, (2) intelligent encoding based method, (3) repair method [12, 13], and penalty function methods and so on. Penalty function is now the most used one in all methods. It imposes penalties on infeasible individuals that violate the problem constraints. The major difficulty in using penalty functions is still how to design an appropriate penalty function for a specific problem.

In recent years, CA shows its powerful ability to solve constrained optimization problems because its unique characteristics. In CA, constraint of optimization problem can be extracted and stored in belief space in forms of knowledge. This constraint knowledge is further used to guide the evolution in the population, and enhance the search ability and efficiency. Here, we propose to integrate real coded genetic algorithm and cultural algorithm to develop a more efficient algorithm, called HMGCA, in which GA's selection and crossover are still used, and its mutation operation is replaced by CA based mutation, which are guided by the constraint knowledge of cultural algorithms and able to 'attract' the individuals to move to the feasible region of the optimization problem. It is possible to enhance the global search ability and to reduce the computational cost.

2 Cultural Algorithms

2.1 Overview

Cultural algorithms provide for interaction and mutual cooperation of two distinct levels of evolution: a population space and a belief space. It models cultural evolution by allowing a collection of information at a macro evolutionary level. Such information can be shared among single individuals whose evolutionary dynamics constitute the micro level. The two levels influence each other through the communication protocol. The presence of the belief space provides for a global knowledge repository and can guide the search towards better solutions, by using the knowledge to prune large portions of the state space. A framework for a CA can be depicted in Fig. 1.

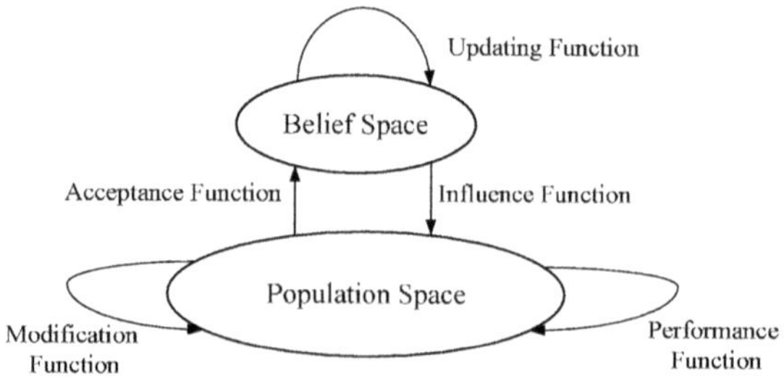

Fig. 1. Framework of cultural algorithms

As shown in Fig.1, the two spaces are connected together by an explicit communication protocol composed of two functions: an acceptance function and an influence function. The contents of the belief space can be altered via an updating function. In the population space, like traditional evolutionary population models, individuals are first evaluated by a performance function. Then, new individuals are created by a modification function.

2.2 Cultural Algorithms Using Regional Schemata

Jin and Reynolds explored the use of cultural algorithms for global optimization with very encouraging results [14]. They proposed a new framework based on cultural algorithms and regional schemata to solve constrained optimization problems. The regional schemata are the extension of classic symbolic schemata, and it represents the regional knowledge that can be used to provide the functional landscape information and to guide the algorithm search in the population. In detail, a n-dimensional regional-based schema, called belief-cell, is as an explicit mechanism that supports the acquisition, storage and integration of knowledge about non-linear constraints in a cultural algorithm. This belief-cell can be used to guide the search of an evolutionary computation technique by pruning the infeasible individuals in the population space and promoting the exploration of promising regions of the search space [4, 14].

3 Genetic Cultural Algorithm for Constrained Optimization

When CA is used for a constrained optimization problem, the domain space of the optimization problem can be partitioned into many feasible, infeasible and semi-infeasible regions, which is stored and continuously updated in belief space as constraint knowledge [14]. The knowledge is further used to conduct individual evolution in population space by adjust it into feasible regions of problem. By contrast with CA, GA is a random search algorithm lacking of similar intelligence technique of CA. In our genetic cultural algorithm, GA is supervised by CA, and intelligently adjusts its individuals into feasible and semi-feasible regions to avoid blindness searching and to reduce computation cost.

3.1 Constrained Optimization Problem Description

The general forms of constrained optimization problem can be described as follows:

$$\min f(\boldsymbol{x})$$

$$s.t.\begin{cases} g_j(\boldsymbol{x}) \le 0, & j = 1,2,\dots,q \\ h_k(\boldsymbol{x}) = 0, & k = q+1,\cdots,m \\ a_i \le x_i \le b_i, & i = 1,2,\cdots,n \end{cases} \tag{1}$$

where $\boldsymbol{x} = (x_1, x_2, \cdots, x_n) \in R^n$, $f(\boldsymbol{x})$ is the objective function. $g_k(\boldsymbol{x})$ and $h_k(\boldsymbol{x})$ are inequality and equality constraints respectively.

3.2 Genetic Operators

Our HMGCA algorithm also uses three operators, namely selection (reproduction), crossover and mutation, in the population space.

(1) Selection. The selection operation is a random selection process. However, the individuals with best fitness values would have more chance to be chosen for reproduction. To realize a better combination with DCPM, competition selection method is employed here: two parents individuals are selected from the population, they are compared according to their fitness values, and the one with higher fitness value is selected into the next generation. The above operation will be repeated for pop_size times to form a new population, where pop_size is the population size.

For the constraint handling, we employ the strategy called direct comparison-proportional method (DCPM) as proposed in literature [15]. For general constrained optimization problem, some infeasible individuals' maybe exists near the global optimum and holds high fitness values. Although they are infeasible in current iteration, further GA operations (for example crossover and mutation) maybe make them to create new feasible offspring with higher fitness value. Thus it's helpful for the optimization to keep one small part of infeasible but good individuals. Based on the above idea, DCPM presents the comparison rules among individuals, as well as the adaptation strategy to keep certain proportion of infeasible individual in the population.

To describe the magnitude that all inequality constraints are violated, a measuring function is defined as follows:

$$viol(x) = \sum_{j=1}^{m} f_j(x), \tag{2}$$

where $f_j(x)$, j =1,2,...,m, is a series of penalty functions to measure what extent each constraint is violated to, it is defined as:

$$f_j(x) = \begin{cases} \max\{0, g_j(x)\} & if\ 1 \le j \le q \\ |h_j(x)| & if\ q+1 \le j \le m \end{cases}, \tag{3}$$

For a predefined constant ε ($\varepsilon > 0$), DCPM compares two individuals according to the follows rules:

(a) When two individuals are both feasible, select the one with higher fitness value.
(b) When two individuals are both infeasible, select neither.
(c) When an individual x is feasible and another individual x' is infeasible, and if $viol(x) \le \varepsilon$, compare their fitness, and select the one with higher fitness value.

To maintain a rational proportion of infeasible individuals in the whole population, an adaptive updating strategy for ε can be done when every K generations evolution is completed, the update is as:

$$\varepsilon_{new} = \begin{cases} 1.2 \cdot \varepsilon_{old} & if\ a/pop\_size \le p \\ 0.8 \cdot \varepsilon_{old} & if\ a/pop\_size > p\,, \\ \varepsilon_{old} & else \end{cases} \quad (4)$$

where a is the number of infeasible individuals, p is a predefined constant to determine the proportion of infeasible individuals.

In order to increase the probability of global convergence, an elitist strategy is added into the end of selection. The best individual in current population will be directly duplicated into next generation population by replacing its worst individual.

(2) Crossover. After selection operations, two parent individuals exchange information to produce a new offspring. This process is called crossover. Crossover is governed by a crossover probability P_c, which determines the GA program whether to execute crossover. In this operation, two individuals X_i and X_j are randomly selected as parents from the pool of individuals formed by the selection procedure and crossover operations between them are executed according to the following equations:

$$\begin{aligned} X_i' &= \alpha \cdot X_i + (1-\alpha) \cdot X_j\,, \\ X_j' &= (1-\alpha) \cdot X_i + \alpha \cdot X_j\,, \end{aligned} \quad (5)$$

where α is a random number extracted from region (0, 1). X_i' and X_j' are new offsprings created by crossover.

A crossover operation can thus yield better solutions by combining the good features of existing solutions. If no crossover is performed due to the crossover probability, the offspring is the exact copy of the parent solution.

(3) Mutation. Each offspring created from crossover is altered in a process called mutation. The effect of this operation is to create diversity in the solution population, thus the operation can avoid trapping in local minimum. In the real coded GA, mutation alters each offspring as the follow:

$$X_i'' = X_i' + \beta \cdot d\,, \quad (6)$$

where X_i' is the parent individual selected for mutation. β and d are the step size and random direction of mutation, respectively. X_i'' is the new offspring by mutation.

Using the above mutation operation, the offspring generated could violate the constraints of the problem. In this paper, we introduce the intelligent mutation scheme used in the regional-based sliding window cultural algorithm [4, 14]. Instead of Eq. (6), the new offspring can be created as

$$X_i''' = \begin{cases} X_i'' + \gamma \cdot (u_j - l_j) \cdot N_{i,j}(0,1) & if\ X_i'' \in \{non-infeasible\ cells\} \\ moveTo(choose(Cell[l]) & if\ X_i'' \in \{infeasible\ cells\} \end{cases}, \quad (7)$$

where l_j and u_j represent the current lower bound and upper bound, respectively, for the j th dimension of X_i. γ is a certain positive number. $Cell[\bullet]$ is a r-dimensional template, called regional schemata, which is maintained in the belief space. This template is used to record the constraint characteristics of a certain region in the search space. For a given cell i, namely a small region in domain space of optimization problem, it may be:

(1) a feasible cell including only valid individuals;
(2) an infeasible cell including only invalid individuals;
(3) a semi-feasible cell including both valid individuals and invalid ones;
(4) a unknown cell with no individuals in it, so no knowledge about it now.

Based on which types all individuals in the cell belongs to, the i th cell can be assigned a different weight as follows:

$$W_i = \begin{cases} w_1 & \text{if } Cell[i] \in \{unknown\ cells\} \\ w_2 & \text{if } Cell[i] \in \{feasible\ cells\} \\ w_3 & \text{if } Cell[i] \in \{semi-feasible\ cells\} \\ w_4 & \text{if } Cell[i] \in \{infeasible\ cells\} \end{cases} . \tag{8}$$

The *choose* (cell[l]) function is a function in Eq. (7) used to select a target cell from all cells for the *moveTo*() function. This process can be implemented by roulette wheel selection based on the weights values in Eq. (8), just similar to the previous selection operation.

Assuming that the kth cell C_k is selected by roulette wheel selection, *moveTo* (C_k) creates a new offspring as follows:

$$X_i''' = Left_k + \text{uniform}(0,1) \cdot Csize_k \tag{9}$$

where $Left_k$ is a $1 \times r$ array which denotes the left-most position of cell C_k, $Csize_k$ is a $1 \times r$ array which represents the sizes of C_k in all dimensions, and $\text{uniform}(0,1)$ generates a $1 \times r$ array of numbers within $[0,1]$ by uniform distribution.

It can be seen that the above mutation operation is guided by constraint knowledge preserved in belief space. The individuals continuously adjust its direction to semi-feasible and feasible area of the problem space. These cultural based intelligences will enhance the search ability of HMGCA.

4 Numerical Example

The general optimization problems, minimization or maximization, can occur in many economic, technical and scientific projects. Solving it remains a difficult and open problem for evolutionary algorithms. The common opinion about evolutionary algorithms is that they are good optimization methods but can't handle constraints well [4, 14]. Cultural algorithms can learn constrain knowledge during the search instead of having to acquire it beforehand and benefit from this acquired knowledge. The example of nonlinear constrained optimization problem in literature [4] is used to illustrate the approach. It is given as follows:

Objective function: $\min f(x, y) = -12x - 7y + y^2$,

Constraints: $0 \le x \le 2$;

$$0 \le y \le 3;$$
$$y \le -2x^4 + 2.$$

The GA population size is $pop\_size = 50$, and other main parameters setting are: $p = 0.1$, initial value of ε equals 10, P_c =0.3, γ =0.3, $w_1 = w_2 = 2$, w_3 =3, $w_4 = 1$. Simulations are performed with Matlab on 2.6G Pentium PC. The illustrated test run is shown in Fig. 2-(c). The proposed algorithm is tested statistically using the above example as listed in Table 1. To examine the performance, a standard real coded GA (called RGA) and CA by X. D. Jin are used as the comparison as shown in Fig. 2-(a),(b) and Table 1.

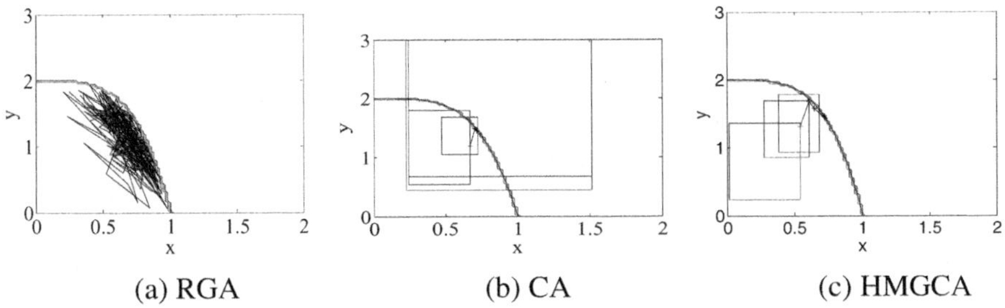

Fig. 2. Iteration process of algorithms

Table 1. Statistical contrast results of algorithms

Algorithms	Mean of number of generations	Standard deviation of number of generations	Mean of running times
RGA	728	223	5.2
CA	5.4	1.58	4.8
HMGCA	4.9	1. 61	4.5

From Table 1, it can be seen that RGA needs a large number of iteration generations because of its less intelligence in search process, but its running takes no long time since its simplicity algorithm results in lower computation cost of each iteration. CA has much fewer iteration generations and less time because its population space evolves under guidance of the belief space and has better adjusting direction to the optimal solution. Further, the HMGCA shows better performance to some extent because it synthesizes both the advantages of GA and CA, which is a GA-based search scheme meanwhile under intelligent guidance.

5 Conclusions

We present a hybrid computation model by integrating GA and CA. GA's selection and crossover operators are used in CA's population space, and GA's mutation is replaced by CA based mutation under guidance of the constraint knowledge of cultural algorithm. In GA's selection operation, introduction of DCPM can keep a certain proportion of infeasible but better individuals into next generation to speed up the convergence. Besides elitist preservation strategy can strengthen the global convergence. This hybrid model can synthesize both the advantages of genetic algorithm and cultural algorithms and shows better performance for constrained optimization.

References

1. Reynolds, R. G.: An Introduction to Cultural Algorithms. Proceedings of the 3rd Annual Conference on Evolutionary Programming, World Scientific Publishing, (1994)108–121
2. Reynolds, R. G., Chung, C. J.: A Self-adaptive Approach to Representation Shifts in Cultural Algorithms. IEEE. 3(96)94–99
3. Becerra, R. L., Carlos A. Coello Coello.: Culturizing Differential Evolution for Constrained Optimization. Proceedings of the Fifth Mexican International Conference in Computer Science, IEEE,(2004)304–311
4. Jin, X.D., Reynolds, R. G.: Mining Knowledge in Large Scale Databases Using Cultural Algorithms with Constraint Handling Mechanisms, Proceeding of the 2000 congress on evolutionary computation., IEEE ,(2000)1498–1506
5. Ho, N. B., Tay. J. C.:GENACE: An Efficient Cultural Algorithm for Solving the Flexible Job-Shop Problem, Proceeding of 2004 Congress on Evolutionary Computation, 2(2004) 1759–1766
6. Yuan, X. H., Yuan, Y. B.; Application of Cultural Algorithm to Generation Scheduling of Hydrothermal Systems. Energy Conversion and Management. 47(2006) 2192–2201
7. Lin, F., Shieh, H., Shyu, K., Huang, P.: On-line Gain-tuning IP Controller Using Real-Coded Genetic Algorithm. Electric Power Systems Research, 72(2004)157–169
8. Arfiadi, Y., Hadi, M. N. S.: Optimal Direct (static) Output Feedback Controller Using Real Coded Genetic Algorithms. Computers and Structures, 79(2001)1625–1634
9. Ha, J., Fung R., Han, C.: Optimization of an Impact Drive Mechanism Based on Real-coded Genetic Algorithm, Sensors and Actuators, 121(2005)488–493
10. Yan, S.Z., Zheng, K., Zhao, Q., Zhang, L.: Optimal Placement of Active Members for Truss Structure Using Genetic Algorithm. Lecture Notes in Computer Science. 3645(2005) 386–395.
11. Blanco, A., Delgado, M., Pegalajar, M. C.: A Real Coded Genetic Algorithm for Training Recurrent Neural Networks. Neural Networks, 14(2001)93–105
12. Myung, H., Kim, J. H.: Hybrid Evolutionary Programming for Heavily Constrained Problems. BioSystems. 38(1996) 29–43
13. Michalewicz, Z.: Genetic Algorithms + Data Structures = Evolution Programs, 3rd edition. New York: Springer-Verlag,1996
14. Li, M. Q., Kou, J. Z., Lin, D., Li, S. Q.: Based theory and Application of Genetic Algorithm, 3rd edition. Science Press, 2004
15. Jin, X.D., Reynolds, R. G.: Using Knowledge-Based Evolutionary Computation to Solve Nonlinear Constraint Optimization Problems: a Cultural Algorithm Approach, IEEE, (1999)1672–1678

Selection Enthusiasm

A. Agrawal and I. Mitchell

Middlesex University, UK
{a.agrawal, i.mitchell}@mdx.ac.uk

Abstract. Selection Enthusiasm is a technique that allows weaker individuals in a population to compete with stronger individuals. In essence, each time a individual is selected its enthusiasm for being selected again is diminished relatively; the converse happens to the unselected individuals i.e. their raw fitness is adjusted. Therefore the fitness of an individual is based on two parameters; objectiveness and Selected Enthusiasm. The effects of such a technique are measured and results show that using selection enthusiasism yields fitter individuals and a more diverse population.

1 Introduction

Premature convergence is defined as prevention of exploration in the population; thus a Genetic Algorithm, GA, with premature convergence has a population with low diversity. This is caused by fitter individuals taking over the population too quickly. Selection pressure set too high causes fitter individuals to be selected more often. If this high selection pressure is maintained over several generations a population will emerge with poor sub-optimal solutions. The balance can be maintained by changing the five basic properties of the standard Genetic Algorithm: Population Size, Selection Methods, Recombination Techniques, Replacement Methods and Mutation. This paper investigates Selection Methods and introduces a new selection operator, Selection Enthusiasm, to maintain diversity.

To investigate if the Selection Enthusiasm has an effect, the diversity of each population was measured. The diversity of the phenotype–not to be confused with the fitness or objective of an individual–was considered and a new diversity metric is introduced. All experiments in this paper use the standard GA, as stated in [4,7].

The structure of this paper is as follows: Section 2 looks at the problem domain studied and the new diversity metric. Section 3 discusses the newly propose Selection Enthusiasm method. Section 4 discusses the experimental design and evaluates the Selection Enthusiasm method. Section 5 discusses the contribution of this paper and some of the conclusions reached.

2 Diversity

The Symmetric Traveling Salesman Problem (STSP) [9] is a suitable domain to test Selection Enthusiasm for two reasons; firstly, because its NP-completeness provides a non-trivial problem solving area and standard GA's find it difficult to solve. None of the

T.-D. Wang et al. (Eds.): SEAL 2006, LNCS 4247, pp. 449–456, 2006.

results are optimum, and finding optimality is not the reason for stagnation. Secondly, STSP provides a simple population diversity metric. Given a symmetric graph $\mathbf{G}(\mathbf{V}, \mathbf{E})$ the total number of possible edges (see Equation 1).

$$\sum_{\forall e \in \mathbf{E}} e = \frac{n(n-1)}{2} \tag{1}$$

where E is the set of all edges, and n is the number of vertices. Initially it was thought that a reasonable measure of phenotypic diversity metric would be the edge density, δ, for any given population i.e.count the number of unique edges in the population divided by the total number of edges (see Equation2).

$$\delta = \frac{\sum_{\forall e \in \mathbf{P}}}{\sum_{\forall e \in \mathbf{E}}} \tag{2}$$

where $\mathbf{P}$ is the set of all edges in the population. The edge density, δ, is a value between $[0, 1]$.

There are many diversity metrics for Genetic Programs, see [1], but, fewer diversity metrics for Genetic Algorithms. This is due to the many different representations available to GA ([6] for many different GA representations for TSP), whereas the tree formation is predominant in GP's. Many researchers take various measurements of the fitness of the individuals. This does not inform us of the diversity of the population, but, the diversity of the performance. Therefore, this leaves two options to measure: i)genotype or ii) phenotype.

Koza [5] suggests 4 different types of fitness: Raw Fitness; the evaluation and ultimately the distance of the route taken. Standardize Fitness; minimization occurs whereby the shorter routes are better. Adjusted Fitness and Normalized Fitness combine to give fitness proportionate selection. It is important here to consider what is the phenotype? For this we can refer to biological terms for some guidance; the phenotype is the physical representation of the genotype. Therefore this does not refer to an individual's performance, but rather, to an individual's physical representation. In terms of the STSP: the genotype is a list of numbers; the phenotype is the route taken; the performance is the distance. There is very little difference between the genotype and phenotype and this is due to the order-based representation that makes the mapping simplistic. Just to labour on this point, the difference between the two stages is that the genotype states the order of the cities; the phenotype includes the edges between each of the cities; the evaluation is the addition of all this edge values.

Since the number of cities in each Hamiltonian cycle is always going to be the same for every route then the study of the genotype is of little interest. The evaluation of the phenotype, raw fitness, does not indicate how different each route is e.g. a route with 2 cities swapped could produce a much higher result. This does not suggest that the population has a higher diversity since the individual concerned uses the same edges apart from 2. It is the order of these cities that is of interest and this is represented in the phenotype and the number of edges. In an n-city STSP problem the total number of edges are displayed in Eq.1. When investigating initial populations it was discovered that they contained an instance of every edge; Therefore in a 130 STSP problem there are $(130 * 129)/2 = 8395$ edges. If the population is set large enough, say 2000, then

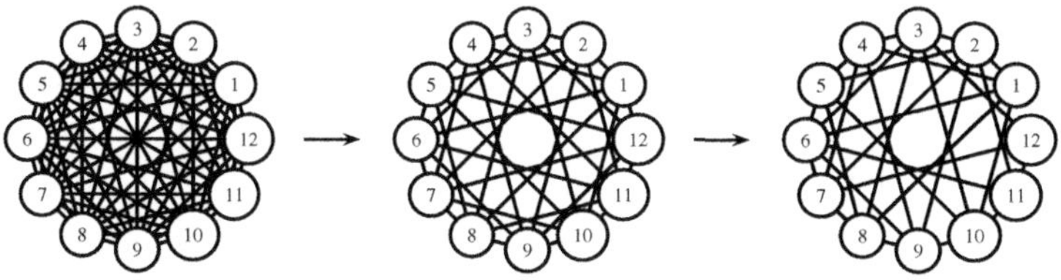

Fig. 1. Investigation of edge density for a 12 city STSP

there are $2000 * 130 = 260,000$ possible edges in a population. The diversity metric is similar to Daida's [2] tree visualisation for genetic programming.

Let us consider a simple 12 city problem in Fig.1. If we were to plot each of the edges in the population initial generations would reveal a dense graph–the more edges the better the diversity–as on the left of Fig.1. It is expected that later generations will reveal a less dense graph indicating that there are less edges and hence less diversity. This, however, is not enough information since the graphs do not tell us the distribution of the edges in the population and therefore the frequency of each edge is recorded and the mean, μ, and the standard deviation, σ, is also measured.

Figure 1 shows approximately what was expected for edge density for a 12 city problem. It is expected that as the number of generations tends to M, then the edge density decreases, where M is the maximum number of generations; however, it was discovered that with large populations the density did not change significantly. This is due to the high number of edges in a population exceeding the number of edges in a given STSP instance and hence it is highly probable that at least one instance of one edge is likely to occur.

As seen edge density, δ does not take into consideration the edge frequency. Therefore the stdev., σ, was taken of the edge frequency. Thus, the higher the stdev. the lower the diversity and the lower the stdev. the higher the diversity e.g. if the stdev. is 0 then every edge has been used in the population a mean number of times and hence high diversity, and when the stdev. is high, say 300, we can infer that there is a greater variance in the edge frequency and hence a low diversity.

3 Selection Enthusiasm

Selection Enthusiasm is combined with Tournament Selection. Firstly the Tourmanent, $\mathbf{T}$, is selected from the population and ranked according to their adjusted fitness, $f(t_i)$, as in Eqn.3.

$$Rank(\mathbf{T}) = \{t_1, t_2, \ldots, t_{\theta-1}, t_\theta\} \quad (3)$$

where θ is the size of the tournament and $Rank(\mathbf{T})$ orders individuals such that $f(t_1) < f(t_2) < \ldots < f(t_{\theta-1}) < f(t_\theta)$.

The principle behind Selection Enthusiasm is to marginally improve the fitness of individuals that have not been selected; thus if and when the individual competes in another tournament it will have marginally improved its likelihood of being selected. The rate of enthusiasm, λ, is the percentage by which the fitness is increased.

Table 1. Parameters of the GA

Parameter	Value
Parent Selection	tournament;generational
Crossover	two-point;PMX
Tournament Size	10
Initial Generation	random(Knuth)
Population Size	2000
Replacment Percentage	0%
Mutation Rate	0.05%
Uniqueness Test	none
Stopping criteria	2000 generations
Number of Runs	20

The two "winners" in the tournament are selected as parents. To reduce high selection pressures, these individuals are not altered. The $\theta - 2$ individuals in the tournament that are not selected have their fitness increased. This process is explained in Eqn.4.

$$f(t_i) = \begin{cases} f(t_i) & \forall i = 1, 2 \\ \lambda f(t_i) & \forall i = 3, 4, \ldots, \theta \end{cases} \quad (4)$$

where λ is the rate of enthusiasm and $f(t_i)$ is the adjusted fitness of the i-th individual in the tournament.

The value of the rate of enthusiasm, λ, is set to a constant, 0.95; thus adjusting the fitness of non-selected individuals in the tournament by 5%. The emergent properties of Selection Enthusiasm allow an individual to improve its fitness until it is selected as a parent from the tournament. After the individual has been selected as a parent Eqn.4 resets the fitness to raw fitness.

4 Experimental Design

Unless stated Table 1 shows the parameter settings for each experiment. The experiment is simple, investigate diversity and fitness of populations over time. All experiments are run on STSP's atta48, berlin52, eil76 and ch130 for 2000 generations, 20 times. The difference between Experiment 1 & 2 is the additional of selection enthusiasm technique. Each experiment is run 20 times on 4 STSP problems, taken from [9], giving an overall total of 200 tests. A random block was implemented as suggested in [8], to make the statistical analysis valid.

4.1 Hypotheses

The two hypotheses tested in this paper are to see if Selection Enthusiasm has an effect on the GA and are as follows:

Table 2. Investigating Evaluations

STSP	atta48		berlin52		eil76		ch130	
Generation	without	with	without	with	without	with	without	with
50	39399	**38967**	10968	**10576**	**1202**	1363	31149	**28171**
100	36380	**35363**	9945	**8157**	**796**	838	23636	**20241**
200	36107	**35363**	9945	**8157**	726	**681**	15272	**13345**
500	36107	**35363**	9945	**8157**	726	**681**	10804	**9925**
1000	36107	**35363**	9945	**8157**	726	**681**	10729	**9807**
1500	36107	**35363**	9945	**8157**	715	**670**	10652	**9807**
2000	35942	**35363**	9945	**8157**	715	**667**	10444	**9763**

Table 3. Investigating Diversity

STSP	atta48	berlin52	eil76	ch130
hyp.mean	0	0	0	0
mean, μ	0.219150666	17.03154779	-26.70991049	12.69621759
median	0.24038497	4.865679491	-29.46434485	14.12411478
stdev.,σ	18.09401087	22.81404362	15.65289069	9.597465705
t-test	0.148338395	9.143184417	-20.89890393	18.70823367
p-value	0.882276278	4.18885E-16	3.84068E-46	9.93322E-46

H1: Selection Enthusiasm yields fitter individuals.
H2: Selection Enthusiasm increases diversity.

4.2 Measuring Perfomance

Table 2 compares results between using standard tournament selection and standard tournament selection with Selection Enthusiasm. Table 2 shows the best individual of a Generation (mean over 20 runs, to 0 d.p.). Results in bold show the fitter solutions; all cases show that the GA with Selection Enthusiasm perfroms better. From this it can be inferred that the Selection Enthusiasm improves the performance of the standard GA.

From these results it can be concluded that Hypothesis 1, H1, is true. The next subsection investigates the diversity using the diversity metric, σ, and tests the second hypothesis.

4.3 Measuring Diversity

From Table 2 results show there is very little change after 150 generations. There is one exception, ch130 which show little change after 200 generations. Therefore, the investigation will only look at the first 150 and 200 generations with concerns to diversity.

Table 3 shows four two-way t-tests between Experiments 1 & 2 for each STSP. Null Hypothesis, $N_0 : \mu = 0$, indicating that Selection Enthusiasm does not change the diversity of a population and the Alternate Hypothesis, $N_a : \mu \neq 0$. Degrees of

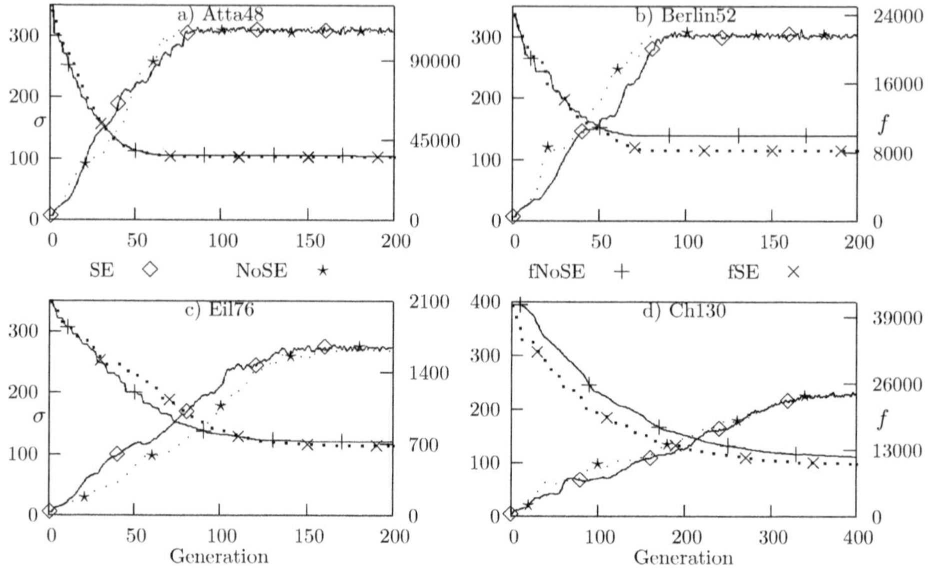

Fig. 2. a-d) compares diversity and fitness

Freedom is 150 for atta48, berlin52 and eil76, and 200 for ch130. From Table 3 with the exception of the first result, all t-tests yield small p-values: thus rejecting N_0, and therefore accepting the Alternate Hypothesis. Two of these results, berlin52 and ch130, have positive means indicating that Selection Enthusiasm increases diversity. These results show that diversity is not the only contributing factor to fitness, since Table 2 clearly indicates that selection enthusiasm performs better.

4.4 Measuring Diversity and Fitness

Figures 2a-d) show that as the solution converges the diversity remains constant. The keys are as follows:

- NoSE – Stdev. of diversity for Tournament Selection with No Selection Enthusiasm;
- SE–Stdev of diversity for Tournament Selection with Selection Enthusiasm;
- fNoSe – Raw Fitness of best individual (mean over 20 runs) in Tournament Selection with No Selection Enthusiasm;
- fSE – Raw Fitness of best individual (mean over 20 runs) in Tournament Selection with Selection Enthusiasm.

Stdev.,σ, of edge frequency is plotted on the left y-axis. Raw fitness, f, of the best individual (mean over 20 runs) is plotted on the right y-axis. It can be inferred from this graph that there is a correlation between the diversity metric and the fitness converging.

5 Conclusion

Selection Enthusiasm does have an effect on the overall performance of GA. As expected over long-term, 1000-2000 Generations, the solutions have converged, but, over short-term, between 0-500 generations, selection enthusiasm improves fitness of solutions.

The diversity metric, σ, for STSP gives a precise measure of the diversity. As expected figures 2a-d) indicate clearly that there is a correlation between the diversity metric and fitness of the population. This diversity metric can be used in future runs to determine when a GA is converging.

5.1 Contribution

There are two major contributions of this paper:

- **Selection Enthusiasm;** a proposed selection method that adjusts the raw fitness value of the individual.
- **Diversity Metric;** a diversity metric that is based on the phenotype properties.

Two hypotheses were proposed and tested. H1: Selection Enthusiasm improved fitness; results in Table 2 give evidence that this hypothesis is true. This is a significant result considering that $\lambda = 0.95$ and therefore made small changes to the individual's fitness. This combined with the tournament size set to 10 limited the effect of Selection Enthusiasm on the population; however, the emerging properties are such that Selection Enthusiasm allows weaker individuals to be selected from a tournament. H2: Selection Enthusiasm improved diversity; results in Table 3 were inconclusive; however, the diversity metric is an improvement on existing measures and from Figs 2a-d) show that it clearly correlates with fitness.

5.2 Future Work

There are several improvements that could be considered with Selection Enthusiasm and different values assigned to λ. Future work will include finding an optimum setting of λ. Currently, the value is set to a constant, 0.95. This value could be changed for future runs to see if improved fitness or diversity is reached. Ekárt in [3] employs various techniques to reduce bloat in GP's, similarly these techniques could be used in GA's to change the diversity in a population e.g making λ variable and dependend on the diversity of the population.

References

1. E.K. Burke, Steven Gustafson, and G. Kendall. Diversity in genetic programming: An analysis of measures and correlation with fitness. *IEEE Transactions on Evolutionary Computation*, 8(1):47–62, 2004.
2. Jason M. Daida, Adam M. Hilss, David J. Ward, and Stephen L. Long. Visualizing tree structures in genetic programming. In E. Cantú-Paz, J. A. Foster, K. Deb, D. Davis, R. Roy, U.-M. O'Reilly, H.-G. Beyer, R. Standish, G. Kendall, S. Wilson, M. Harman, J. Wegener, D. Dasgupta, M. A. Potter, A. C. Schultz, K. Dowsland, N. Jonoska, and J. Miller, editors, *Genetic and Evolutionary Computation – GECCO-2003*, volume 2724 of *LNCS*, pages 1652–1664, Chicago, 12-16 July 2003. Springer-Verlag.

3. Aniko Ekárt. *Genetic programming: new performance improving methods and applications.* PhD thesis, Eötvös Lorand University, 2001.
4. J. H. Holland. *Adaptation in Natural and Artificial Systems.* MIT Press, 1992.
5. John R. Koza. *Genetic Programming: On the Programming of Computers by Means of Natural Selection.* MIT Press, Cambridge, MA, USA, 1992.
6. P. Larrañaga and *et al.* Genetic algorithms for the travelling salesman problem: A review of representations and operators. *Artificial Intelligence Review*, 13:129–170, 1999.
7. Mitchell M. *An introduction to genetic algorithms.* MIT Press, 1996.
8. Czarn A. MacNish C. Kaipillil V. Turlach B. & Gupta R. Statistical exploratory analysis of genetic algorithms. *IEEE Trans. on Evolutionary Computation*, 8:405–421, 2004.
9. G. Reinelt. Tsplib - a traveling salesman library. *ORSA Journal on Computing*, 3:376–384.

Spatially-Structured Evolutionary Algorithms and Sharing: Do They Mix?

Grant Dick and Peter A. Whigham

Department of Information Science,
University of Otago,
Dunedin, New Zealand
gdick@infoscience.otago.ac.nz

Abstract. Spatially-structured populations are one approach to increasing genetic diversity in an evolutionary algorithm (EA). However, they are susceptible to convergence to a single peak in a multimodal fitness landscape. Niching methods, such as fitness sharing, allow an EA to maintain multiple solutions in a single population, however they have rarely been used in conjunction with spatially-structured populations. This paper introduces *local sharing*, a method that applies sharing to the overlapping demes of a spatially-structured population. The combination of these two methods succeeds in maintaining multiple solutions in problems that have previously proved difficult for sharing alone (and vice-versa).

1 Introduction

Traditional evolutionary algorithms (EAs) have difficulty in discovering and maintaining multiple solutions to multimodal fitness landscapes. Through a combination of selection and genetic drift, an EA tends to converge its population onto a single point in a fitness landscape. Previous work has identified this problem and has looked to nature for to provide possible solutions [1]. One method, *fitness sharing*, has received particular attention [2]. This method uses the concept of finite resources to encourage elements of the population to explore different regions of the fitness landscape. Sharing has been shown to generally perform well, although it is known to have difficulty on at least one class of multimodal problems.

Another method which claims to encourage population diversity is to impose a structure on the population and restrict selection and mating to geographically close individuals. Previous researchers have noted that, while *spatially-structured evolutionary algorithms* (SSEAs) do not prevent population convergence to a single phenotype, they tend to increase the time required for diversity loss.

Previous work has explored the concept of merging fitness sharing with SSEAs [3]. The author noted that a sharing-like operator, combined with a spatially-structured population, increased an EA's ability to maintain multiple peaks within a single population. However, the work was peculiar in that sharing was applied globally, while selection and mating was confined to demes.

T.-D. Wang et al. (Eds.): SEAL 2006, LNCS 4247, pp. 457–464, 2006.

A method in which sharing acted locally was proposed, however it has never materialised in subsequent work.

This paper introduces the concept of *local sharing*. The proposed method applies sharing to each deme prior to selection and offspring replacement. This method reduces the overall complexity of sharing, and is applicable to problems that sharing has difficulty handling.

The remainder of this paper is structured as follows: §2.1 describes the concepts of sharing and SSEAs as used in this paper; the proposed local sharing method is described in §3 and a comparison between this method and existing SSEAs is given in §4. Finally, a brief discussion of the findings of this paper and a suggested path for future work is presented in §5.

2 Sharing Methods and Spatially-Structured EAs

The concept of sharing finite resources in an evolutionary algorithm was first proposed by Holland [4], but *fitness sharing* [2] was the first successful attempt at modelling resource contention within a simple EA. With sharing, each optima in the fitness landscape is allocated a finite number of resources in relation to its size. The fitness of a given optimum in space must then be shared by the number of individuals representing this solution. This introduces two benefits over a simple EA:

1. Rather than attempt to crowd around a single peak, individuals within the population will actively seek out less populated optima. This ensures the constant presence of a selection pressure and counteracts the effects of genetic drift.
2. The effective value of a peak is reduced as the number of individuals exploiting it increases. This in turn makes the lesser-valued optima in the fitness landscape more attractive and drives selection towards them. This reduces the likelihood of losing the lesser peaks from the population.

In order for sharing to work, some assumptions need to be made; it is assumed that the optima in the problem occupy similar-sized areas and that they are evenly distributed throughout the fitness landscape. This is due to the mechanism sharing uses to determine which resources are being consumed by each individual. Sharing uses a radius measure, called the *sharing radius*, to determine the amount of similarity between individuals and hence the *niche count* of an individual, given by:

$$sh(d) = \begin{cases} 1 - \left(\frac{d}{\sigma_{share}}\right)^{\alpha}, & \text{if } d < \sigma_{share} \\ 0, & \text{otherwise.} \end{cases}$$

The value of σ_{share} is the sharing radius and α is a value that alters the shape of the function (typically, this is set to 1). In the above function, d represents the distance between two individuals in either a genotypic or phenotypic space.

The distance measure is preferably in the phenotype space, in other words, the distance metric should be applied as closely to the problem space as possible.

Implementing fitness sharing is a simple extension to standard EAs. Its sole purpose is to dynamically alter the fitness of individuals after evaluation; in terms of implementation, it plays no direct part in the fitness function. The other operators in an EA, such as selection and recombination, are equally unaffected.

2.1 Spatially-Structured Evolutionary Algorithms

Fitness sharing is one method in which an EA can incorporate speciation. Another method is to impose a spatial structure on the population so that there is a concept of geographic distance between individuals. Geographically close individuals form localised subpopulations, or *demes*, within the global population. Mating is confined to demes and offspring are placed within the vicinity of their parents. Evolution potentially acts on demes in different ways so that they explore different regions of the fitness landscape. Given enough time, individuals within a deme will possess sufficiently different genotypes from those of other geographically distant demes that they could be considered a different species.

A spatially-structured evolutionary algorithm (SSEA) [5] is essentially an implementation of the parapatric speciation concept; there are no complete boundaries within the population to restrict gene flow. Instead, SSEAs use the vast distances between some individuals to slow the rate of exchange of genetic material to a sufficiently low level so as to promote local divergence of genotypes [6, 7]. Though each SSEA implements a slightly different algorithm, the overall method for using a spatially-structured population within an EA is shown in Algorithm 1. Essentially, there are three basic steps to an SSEA; first, for a given location, construct a deme. After this, select the parents from this deme. Finally, the newly created offspring must be inserted into the deme.

Algorithm 1. The general sequence for a spatially-structured evolutionary algorithm

```
input : A given problem
output: A spatially-structured population of evolved candidate solutions to the problem.
population ← {};
foreach location in space do
  | population[location] ← initialIndividual();
end
while not done do
  | generation ← {};
  | foreach location in space do
  |   | deme ← constructDeme(location);
  |   | parents ← select(deme);
  |   | offspring ← breedOffspring(parents);
  |   | evaluate(offspring);
  |   | generation[location] ← pickSurvivor(offspring, deme);
  | end
  | population ← generation;
end
return population;
```

3 Local Sharing Methods for SSEAs

Spears, as part of his simple subpopulation schemes concept, used a spatially-structured population in conjunction with a sharing method operating of tag bits [3]. The sharing component operated once per generation and considered the population as a whole. After this step, selection and reproduction continued as for a typical SSEA. Spears noted that this scheme was more effective at maintaining multiple peaks in a fitness landscape than a simple subpopulation scheme that did not impose a population structure. It is not entirely clear that the increase in performance was down to the combination of an SSEA and the sharing method, or if the improvement was purely through the adoption of spatial constraints in the population. The issue is further complicated by the fact that the sharing component of the algorithm and the local mating did not interact; the population topology played no part in determining the shared fitness of the population members. Spears himself noted this peculiarity and suggested a third subpopulation scheme which used the population topology within the sharing process [3]. However, this new method has not appeared in any subsequent work.

This paper extends Spears' idea to create a new niching method, *local sharing.* The concept is actually rather simple; at the start of a generation, each location calculates the shared fitness of its occupant by determining the individual's niche count with respect to the other members of the deme. After this, selection and reproduction are performed as per a normal SSEA except that individuals are selected via their shared fitness. Upon completion of mating, the shared fitness of offspring are calculated. Offspring are then inserted into the population if their shared fitness exceeds that of the current occupant.

One benefit of the local sharing method is a reduced complexity compared to that of panmictic fitness sharing. Typically, fitness sharing is of $O(N^2)$ complexity, as all pairs of individuals must be considered. With local sharing, only individuals that share a common deme are compared, therefore the complexity of local sharing given a deme size of d is $O(2Nd)$. For small deme sizes, this represents a considerable saving in the number of distance comparisons required. Another benefit of local sharing is the number of sharing instances that are performed per generation. Local sharing performs multiple, smaller instances of sharing in parallel per generation. The isolation of demes allows each sharing instance to concentrate of different parts of the fitness landscape, increasing the chance of discovering multiple optima. Finally, local sharing permits the use of elitism, which helps to preserve optima once they are discovered.

4 Empirical Analysis

The local sharing SSEA was compared with a traditional SSEA and a panmictic EA incorporating fitness sharing. Each algorithm was tested against two problems. The first problem, referred to as *M6*, is the Shekel's foxhole problem as used in De Jong's thesis on genetic algorithms [8]. This problem is frequently used in comparisons of niching methods [9] as it has 25 local optima in the fitness landscape. The peaks vary in value, so a typical SSEA will have difficulty maintaining all but the global optimum. The second problem, *M7*, is a massively multimodal, deceptive problem [10]. This problem has over five million optima in the fitness landscape, of which only 32 are global and of interest. Fitness sharing struggles with this problem as the deceptive optima lie within the sharing radius of the global peaks and hence compete for the same resources. Together, *M6* and *M7* form a good test suite for comparing the hybrid local sharing SSEA. The fitness landscapes of *M6* and *M7* are shown in Figure 1.

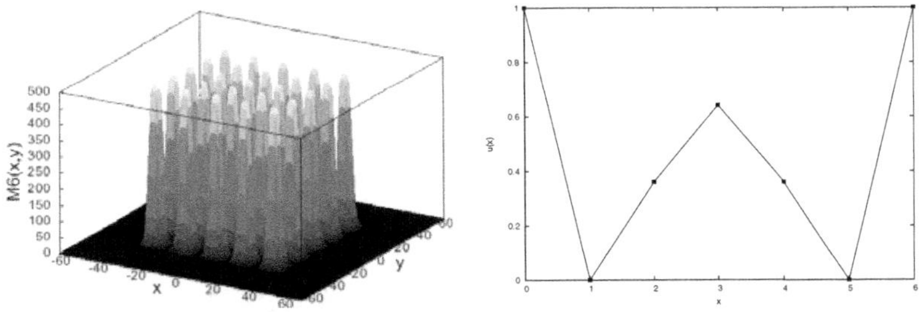

(a) *M6* (b) *M7*

Fig. 1. The fitness landscapes of the test problems used in this paper

Population size plays an important role in the performance of EAs. The population sizes for each of the test problems was taken from previous work. The population size for *M6* was determined via a population sizing model developed for fitness sharing [11] and was 289 (a 17×17 torus for the SSEAs). The population size for *M7* was taken from empirical studies [9] and was 676 (a 26×26 torus).

Selection in all EAs was via fitness proportional selection. Selection in the sharing EA was via stochastic universal sampling [12] to remain consistent with previous work [9]. Roulette-wheel selection was used for the SSEAs, again for consistency with previous work [13]. Two parents were selected (without replacement) to create offspring. A single offspring was created via one-point crossover for the SSEAs, while the fitness sharing EA created two offspring for each reproduction instance.

The parameters for each test were as follows: One-point crossover with probability 1.0; mutation was via bit-flipping applied with probability 0.002 per locus; the sharing radius σ_{share} was 8 for *M6* and 6 for *M7*; each SSEA used a torus population structure and a Von-Neumann neighbourhood was used to determine demes; elitist replacement of offspring was used in both SSEAs (via shared fitness for local sharing); each run of an EA lasted for 500 generations.

4.1 Results

The performance of each SSEA on the test problems was measured through two statistics used in previous niching studies [14]. The first measurement records the number of peaks discovered and maintained by the population over 500 generations. Higher values for the number of maintained niches indicates better algorithm performance. The second measure is the Chi-Square-Like performance which measures the deviation of the population from that of an "ideal" population in which all individuals reside on a peak in distributions relative to a peak's value. Lower values for the Chi-Square-Like measure indicate better performance (a value of zero being ideal). These measurements were averaged over 100 runs and are reported below.

The number of peaks maintained over time is shown in Figure 2. On *M6*, local sharing maintains approximately half of the 25 optima present in the fitness landscape after 500 generations. However, local sharing on *M7* is far superior to that of either panmictic sharing of the traditional SSEA, with nearly all 32 optima present in the population after 500 generations. The Chi-Square-Like performance of the three EAs, as shown in Figure 3, tells a similar story to the number of peaks performance; local sharing distributes individuals among peaks more efficiently than the traditional SSEA on both *M6* and *M7* and is superior to panmictic sharing on the *M7* problem.

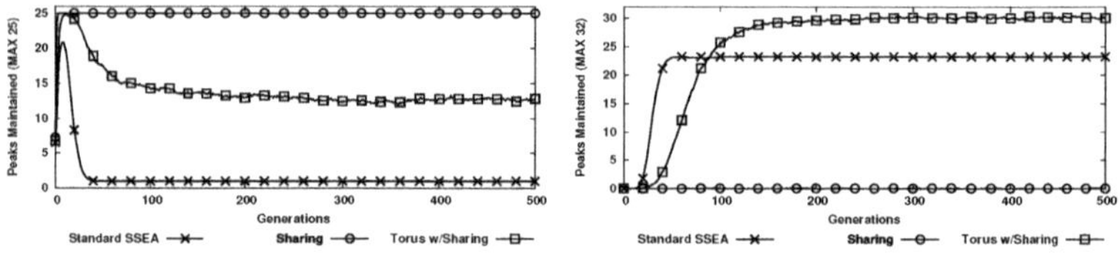

(a) *M6* (b) *M7*

Fig. 2. Number of peaks maintained by each algorithm on *M6* and *M7*

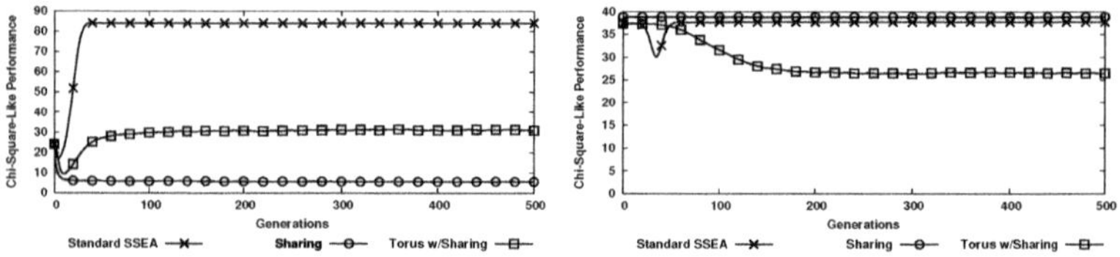

(a) *M6* (b) *M7*

Fig. 3. Chi-Square-Like performance of the tested EAs on *M6* and *M7*

4.2 Discussion

The results presented here indicate that local sharing is able to discover and maintain more optima than a traditional SSEA. Local sharing is also able to support multiple optima on *M7*, a problem that has traditionally been difficult to handle using sharing methods. However, the performance of local sharing on *M6* does not match that of panmictic sharing. This may be in part due to the elitist replacement of local sharing; too much emphasis is placed on the global optimum by elitism at the expense of other niches. Elitism plays an important role in local sharing's performance on *M7*, so elitist replacement cannot be simply removed from the local sharing method. Instead, we propose a change in elitism from a strict, "always better than" approach to a probabilistic method. In the proposed strategy, offspring enter a tournament with the current occupant to determine who is passed into the next generation. The probability of an offspring o winning the tournament over the current occupant i is:

$$p(o) = \frac{f_{sh}(o)}{f_{sh}(o) + f_{sh}(i)}$$

where f_{sh} is the shared fitness of an individual. Experiments on *M6* were repeated using this alternative elitism method and the results are shown in Figure 4. The probabilistic elitism method allows local sharing to support more optima within the population, and the distribution of individuals among those peaks is much closer to that of panmictic sharing (as indicated by the Chi-Square-Like performance).

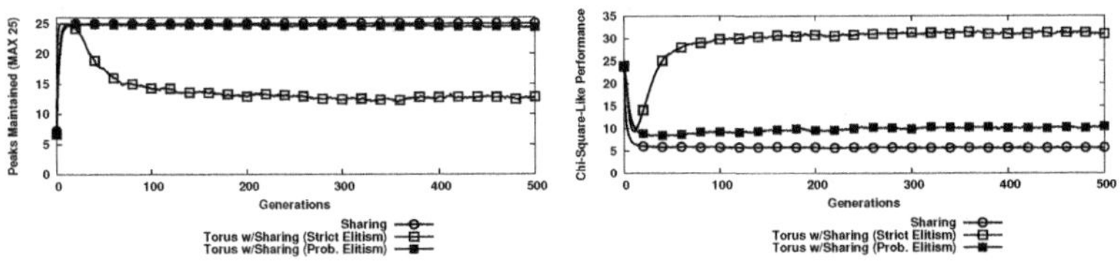

(a) Peaks Maintained (b) Chi-Square-Like Performance

Fig. 4. Comparison of elitism strategies for the local sharing method on *M6*

5 Conclusion and Future Work

Space often plays an important role in speciation of populations. Traditionally, EAs have used individual-based comparisons to perform niching in the absence of spatial population structure. This paper presents a new method for niching, local sharing, which takes the traditionally panmictic niching method of fitness sharing and applies it within the demes of a spatially-structured population. Initial results suggest that local sharing is applicable to problems that have proved difficult to solve by either sharing or SSEAs alone.

Local sharing is clearly superior to panmictic sharing on the *M7* problem. This is possibly in part due to local sharing's elitist replacement policies. However, the same elitist strategy appears to prevent local sharing from maintaining all optima in *M6*. An alternative elitism strategy greatly improved the performance of local sharing on *M6*. Future work should investigate the elitism strategies of local sharing in greater detail. One possible direction might be to implement elitism as a Boltzmann tournament with annealing [15] so that the higher valued optima do not dominate elitism in earlier generations of a run.

References

1. Mahfoud, S.W.: Niching methods for genetic algorithms. PhD thesis, University of Illinois at Urbana-Champaign, Urbana, IL, USA (1995) IlliGAL Report 95001.
2. Goldberg, D.E., Richardson, J.: Genetic algorithms with sharing for multi-modal function optimisation. In: Proc of the 2nd Int. Conf. on Genetic Algorithms and Their Applications. (1987) 41–49
3. Spears, W.M.: Simple subpopulation schemes. In Sebald, A.V., Fogel, L.J., eds.: Evolutionary Programming: Proc. of the Third Annual Conf., Singapore, World Scientific Press (1994) 296–307
4. Holland, J.H.: Adaptation in Natural and Artificial Systems. 2 edn. The MIT Press, Cambridge, Massachusetts (1992)
5. Tomassini, M.: Spatially structured evolutionary algorithms. Springer (2005)
6. Wright, S.: Isolation by distance. Genetics **28**(2) (1943) 114–138
7. Mayr, E.: Populations, species and evolution; an abridgment of Animal species and evolution. Harvard University Press (1970)
8. De Jong, K.A.: An Analysis of the Behavior of a Class of Genetic Adaptive Systems. PhD thesis, University of Michigan, Ann Arbor, MI (1975) Dissertation Abstracts International 36(10), 5140B, University Microfilms Number 76-9381.
9. Mahfoud, S.W.: A comparison of parallel and sequential niching methods. In Eshelman, L., ed.: Proceedings of the Sixth International Conference on Genetic Algorithms, San Francisco, CA, Morgan Kaufmann (1995) 136–143
10. Goldberg, D.E., Deb, K., Horn, J.: Massive multimodality, deception, and genetic algorithms. In R. Männer, Manderick, B., eds.: Parallel Problem Solving from Nature, 2, Amsterdam, Elsevier Science Publishers, B. V. (1992) 37–46
11. Mahfoud, S.W.: Population size and genetic drift in fitness sharing. In Whitley, L.D., Vose, M.D., eds.: Foundations of genetic algorithms 3, San Francisco, Morgan Kaufmann (1995) 185–224
12. Baker, J.E.: Reducing bias and inefficiency in the selection algorithm. In Grefenstette, J.J., ed.: Genetic Algorithms and their Applications (ICGA'87), Hillsdale, New Jersey, Lawrence Erlbaum Associates (1987) 14–21
13. Sarma, J.: An Analysis of Decentralized and Spatially Distributed Genetic Algorithms. PhD thesis, George Mason University, Fairfax VA, USA (1998) double sided.
14. Deb, K., Goldberg, D.E.: An investigation of niche and species formation in genetic function optimization. In Schaffer, J.D., ed.: Proc. of the Third Int. Conf. on Genetic Algorithms, San Mateo, CA, Morgan Kaufmann (1989) 42–50
15. Goldberg, D.E.: A note on Boltzmann tournament selection for genetic algorithms and population–oriented simulated annealing. Complex Systems **4**(4) (1990) 445–460

Suggested Algorithms for Blood Cell Images

Weixing Wang[1], Qi Zhao[2], and Wang Luo[1]

[1] School of Electronic Engineering, University of Electronic Science and Technology of China, Post code: 610054, China
[2] Chongqing University of Posts & Telecommunications, Post code: 400065, China
wxwang@ee.uestc.edu.cn, znn525d@yahoo.com

Abstract. Morphological mathematics is a powerful tool for image segmentation. The watershed is popularly used for multiple object images. In this paper, a new watershed algorithm without considering accurate boundary information is presented, for grey scale image segmentation, and Morphological mathematics is also used for cluster splitting. The result of running this algorithm shows that blood cell image can be well segmented using the proposed algorithm. A genetic algorithm is suggested for the further study.

Keywords: Blood Image segmentation, Watershed, Splitting, genetic algorithm, Region merging.

1 Introduction

In blood analysis, doctors look for three different kinds of cells, red, white and blood platelets, hence the cells are needed to delineate [1]. One of example image is shown in Fig. 1, the images are original ones; the light color is for background and dark color for blood cell. In the image, a part of cells include one or two white sports in their interior, and blood cells touch each other. It is difficult to use one gray or color value to separate the touching objects, because the touching parts have gray values closed to the gray or color values of the cells, but the cell color or grey values are different from background. The white sports have gay values equal to or higher than image background. These characteristics make image segmentation difficult, and hard to identify blood cells in the image. According to the blood cell image investigation and segmentation testing, the need for a set of new segmentation algorithms for touching problem was obvious. There is a number of image segmentation algorithms have been studied by other researchers [1-3], but it is difficult to directly to use one of them for the blood cell images.

Numerous segmentation methods for cell image from peripheral blood smears have been proposed. Histogram threshold methods and clustering methods are often used. Boundary detection method using local edge information is another approach; however it is difficult to integrate all the edge information to make satisfactory object boundaries. Region growing methods have been widely used in connection with local and global object properties; but it is hard to determine a stopping rule. Here we propose a segmentation method using watershed of mathematical morphology. It focuses on the image's geometrical characters and has the virtue of speediness,

T.-D. Wang et al. (Eds.): SEAL 2006, LNCS 4247, pp. 465–472, 2006.

exactness and robustness. In this paper we focus on the problem of segmentation of blood cells by microscope images. The proposed system firstly preprocesses the image to reduce the noise and other influencing factors. Secondly it explains how to segment the cells by watershed algorithm and finally it using the region merging to overcome the flaw of over-segmentation.

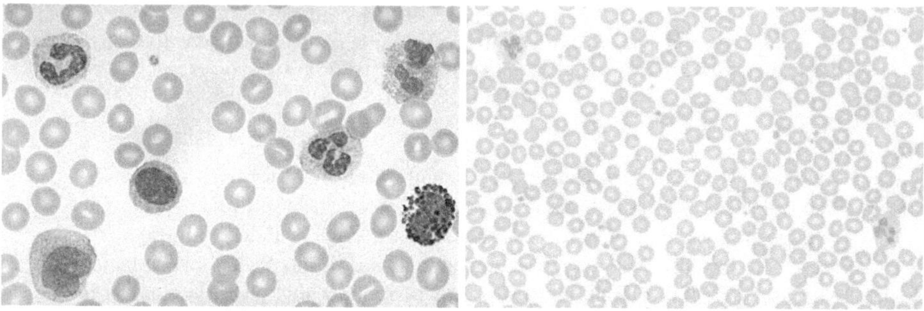

(a) (b)

Fig. 1. Two blood cell images: (a) parsley located cells, (b) densely located cells

2 Image Segmentation

For the grey level image segmentation, among the existing algorithms, watershed transformation has proved to be a very powerful tool for morphological image segmentation because of its moderate computational complexity and ability to identify the important closed contours of a given image. Watershed algorithms are widely used in fields like biomedical signal processing and medical image processing.

A gradient image is considered as a topographical surface where the numerical value of each pixel stands for the elevation at that point [4-7]. Smooth surfaces could be decomposed into hills and dales by studying the critical points and slope lines of a surface. By viewing grayscale in an image as elevation and simulating rainfall, it is possible to decompose an image into watershed regions. Approaches to computation of watersheds vary ranging from iterative methods to flooding methods.

Some watershed algorithms were developed to process digital elevation models and were based on local neighborhood operations on square grids. Other approaches use "immersion simulations" to identify watershed regions by flooding the image with water starting at intensity minima. Here, we use a method integrating watersheds segmentation and region merge which overcome the flaw of over-segmentation due to the noise and other irregularities of the gradient.

Extracting watersheds from digital image is far from an easy task. Our investigation of watersheds has three phases. First, we describe how watersheds and their boundaries can be computed in the grayscale images. Then, we apply a region-merge procedure according to the properties of the catchments including. This algorithm proves to be effective in blood cell image segment by the experiment.

Watersheds are traditionally defined in terms of the drainage patterns of rainfall [5]. Regions of terrain that drain to the same point are defined to be part of the same watershed. The same analysis can be applied to images by viewing intensity as height. In this case, the image gradient is used to predict the direction of drainage in an image. By following the image gradient downhill from each point in the image, the set of points, which drain to each local intensity minimum, can be identified [5-6]. These disjoint regions are called the watershed regions (catchment basins) of the image.

Before making segmentation, let us assume the gray scale image F as digital grids which have M rows and N columns. Each pixel has two spatial coordinates and gray levels. We denote (x, y) the coordinate of each pixel, denote f(x, y) the gray value of this pixel. So each pixel stands for a grid, and its gray value stand for elevation. The detail segment process can be express as follows. First, we compute the watersheds for an image is identifying the local intensity minima. These are the points which define the bottoms of watersheds. To distinguish these critical points, each pixel is compared with its eight nearest neighbors. If all neighbors are greater than the central pixel, it is identified as an intensity minimum. After all the intensity minima are identified, the number of watershed regions is identified at the same time, since each watershed region corresponds to one intensity minimum. Next, we calculate the image gradient. The goal here is to identify the drainage directions for each pixel in the image. Rather than calculate the gradient based on the partial derivatives of the image or morphological gradient theory, each point is searched to determine most steeply downhill directions according to the approach following.

Let Dist (x, y) denote the distance between any pixel and any intensity minimum. Similarly, $Fall(x, y)$ denote the elevation fall between any pixel and any intensity minimum (the intensity difference between them). Then we compute:

$$Grad = \frac{Dist}{Fall} . \tag{1}$$

For each pixel, we obtain all the values of Gradient associated with all of intensity minima. Later, we select the maximum among all the Gradient values which corresponds to a given intensity minimum. The drainage direction for each pixel in the image is identified because the direction from each pixel to the selected intensity minimum is the mostly steeply downhill direction. These directions may or may not be in opposite directions due to discreteness. There are nine possible direction for each pixel (the central pixel could be a minimum), which are encoded and stored in a temporary image for use in the gradient following step of our algorithm [6]. Take a 5×5 region in the image for example.

Partitioning the input image into watersheds begins by marking the locations of intensity minima with unique region identifiers in an output image. For each of the remaining points in the image, the gradient information is used to follow the image downhill to some intensity minimum. The identifier of this minimum is then recorded in the output pixel corresponding to this starting point. Once all pixels in the image have been associated with their respective minima, the output image will contain the watershed regions of the image [7-9]. We can locate the watershed boundaries by accessing each pixel and its neighbors (its 4 neighbors in 4-connectivity). If some of its neighbor pixels aren't in the same watershed region as given pixel, we mark this pixel as well as the neighbor pixels which are in different regions. The watershed boundaries can be located after all the pixels and its neighbors be detected.

30	23	21	29	32
24	19	16	18	21
18	17	15	17	19
24	20	16	18	12
29	21	18	13	11

↘	↘	↓	↙	↓
↘	↘	↓	↙	↓
→	→		↘	↓
↗	↗	↘	↘	↓
↗	→	→	→	

Fig. 2. An example of the watershed regions and gradient vectors indicate the direction toward the lowest value 8-neighbor at each point. All points drain to intensity minima 11 or 15, defining two watershed regions.

Since watershed boundaries are defined in terms of the global drainage patterns of the image rather than local differential geometry, not all boundaries detected in the image should be marked as watershed boundaries. Only those boundaries that separate drainage basins are identified. In more realistic images there are thousands of intensity minima and associated watershed regions. In this case, the image is over-segmented. So the problem is identifying which watershed boundaries mark significant image structures and which catchments should be merged. Some methods on controlling over-segmentation reduction use only fractional regional minima of the input image in the flooding step rather than all of them. These selected regional minima are referred to as markers. Internal markers are associated with objects of interest, and external markers are associated with the background [7-10]. Prior to the application of the watershed transform, the image can be modified so that its regional minima are identical to a predetermined set of markers. Although markers have been successfully used in segmenting many types of images, their selection requires either careful user intervention or explicit prior knowledge on the image structure.

In addition to the above over-segmentation reduction method, we apply a new region-merging procedure according to the property of the catchment basins. Let us express the process of merge more formally. First, we calculate the parameters of the catchment basins in the image: CB being the catchment basin in the over-segmentation image, Surface (CB) denotes the surface area of every catchment basin. Similarly, Depth (CB) denotes its depth; Volume (CB) denotes its volume. Meanwhile we define the $f(x,y)$ $(0 \leqslant f(x,y) \leqslant 255)$ as the gray value in position (x, y).

$$\text{Surface (CB)} = \sum_{\forall (X,Y) \in CB} 1 \,. \tag{2}$$

$$\text{Depth (CB)} = \max_{\forall (x,y) \in CB} f(x,y) \,. \tag{3}$$

$$\text{Volume (CB)} = \sum_{\forall (x,y) \in CB} f(x,y) \,. \tag{4}$$

We merge the catchment basins according to these three parameters. Before this processing, we set the threshold of surface, depth and volume, so that we can deal with these catchment basins dissatisfy the threshold as the background and merge the adjacent catchment basins satisfy the threshold. Set mark (WS) be the Bool style, and ST, DT, VT are token of surface, depth and volume. Similarly, ΔST , ΔDT , ΔVT are token of the increment of the three threshold.

Condition 1 $(Suiface(CB) \geq ST) \cup (Depth(CB) \geq DT) \cup (Volume(CB) \geq VT)$ $\forall \ mark(CB) = True$;

Condition 2 $Depth(CB) \geq DT$;

Condition 3 $Surface(CB) \geq ST$;

Condition 4 $Volume(CB) \geq VT$.

The emerging can be followed as four steps:

1) If CB can't satisfy condition 1, set mark (CB) = False, and emerge the adjacent CB according to $mark(CB) = True$. (PRI is Depth>Surface>Volume).
2) $DT = DT + \Delta DT$, if CB can't satisfy condition 2, set mark (CB) =False, and emerge the adjacent CB according to the mark (CB) =True.
3) $ST = ST + \Delta ST$, if CB can't satisfy condition 3, set mark (CB) =False, and emerge the adjacent CB according to the mark (CB) =True.
4) $VT = VT + \Delta VT$, if CB can't satisfy condition 2, set mark (CB) =False, and emerge the adjacent CB according to the mark (CB) =True. We get the final result through the iterative steps above (See Fig.4 (b)).

For under-segmentation problem, the studied algorithm can be summarized as: (1) using morphological mathematics to detect skeletons of blood cells or clusters; (2) using the distance information of the skeletons to detect valley points which can be the candidates of starting points of splitting curves; (3) finding out concave points on the boundaries of the blood cells; and (5) tracing the splitting curves. Fig. 4 shows the general sequence of splitting algorithm. Fig. 5 shows two examples of cell delineation results.

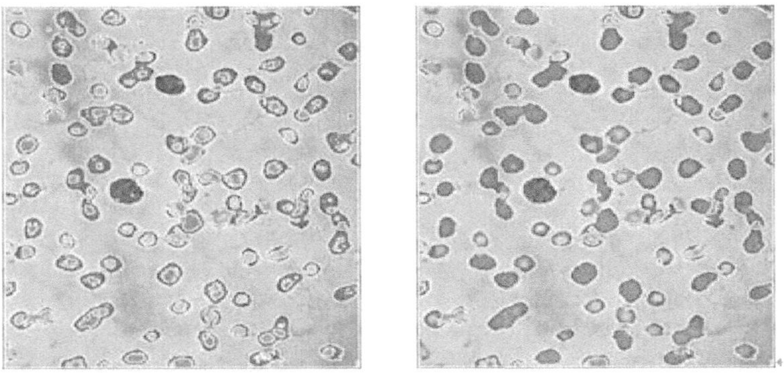

(a) (b)

Fig. 3. (a) The image after watershed segmentation, (b) the image after merging

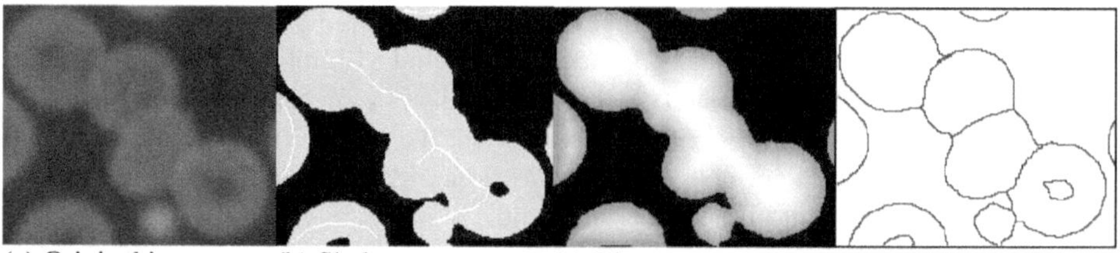

(a) Original image (b) Skeletons (c) Distance mapping (d) Decomposing result

Fig. 4. Decomposing sequence for touching cells, based on cell shape information

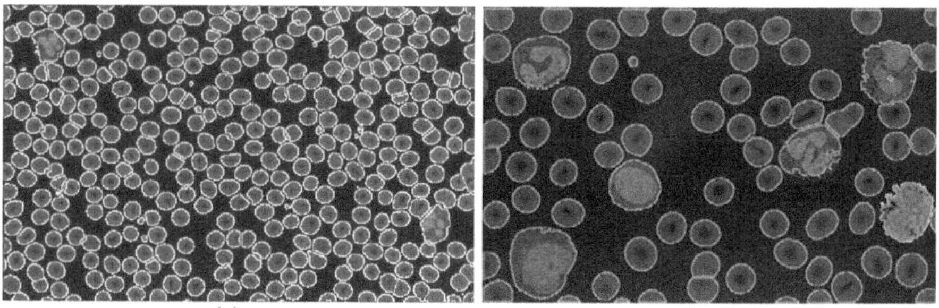

(a) (b)

Fig. 5. The image after splitting based on shape information: (a) splitting result on the image in Fig. 1(b); (b) cell delineation result on the image in Fig. 1(a), where image background splitting; cluster decomposing based on local light grey variation; and decomposing clusters on cell shape information are used subsequently

3 Further Study

The genetic algorithm will be useful to enhance the developed blood cell image segmentation results for the further study. The basic idea is described as the follows.

Give a genetic input image, X, of size n × m, X (i, j) will indicate the gray-level value of the pixel located in (i, j).

The genetic chromosome $\alpha_{i,j}$ is coded by a 32 bit binary string that codes the pixel-label, λ in the 8 less significant bits and the pixel position (i, j) in the 24 most significant bits. Here, λ identifies the clusters to which the pixel belongs. The functions used to codify label and position are:

$$L(\lambda)=\left(\frac{2^8-1}{K}\right)\times\lambda\,,\; S(k)=\left(\frac{2^{24}-1}{N\times M}\right)\times k \tag{5}$$

where, k = i × m + j and K is the maximum number of clusters. It follows that $\alpha_{i,j}=\alpha_k\equiv b_{31}b_{30}\ldots b_0$ with $b_{31}b_{30}\ldots b_{23}=L(\lambda)$ and $b_{22}b_{21}\ldots b_0=S(k)$. Each chromosome can be denoted with the ordered pair $\delta(L,S)=(L(\lambda),S(k))$.

The inverse function of L and S, $L^{-1}(\delta)$ and $S^{-1}(\delta)$, return the label, $\lambda = L^{-1}(\delta)$, of a pixel in position $(i,j) \equiv S^{-1}(\delta)$. Each segment P_j is characterized by the mean value, mv_j, of the gray levels:

$$mv_j = \frac{\sum_{\delta \in P_j} X\left(S^{-1}(\delta)\right)}{|P_j|} \tag{6}$$

The fitness function, *f*, has been defined on the basis of the similarity function, ρ, computed between a given chromosome $\delta(L,S) = (L(\lambda), S(k))$ and the corresponding segment P_λ:

$$f(\delta) = \rho\left(\delta, mv_{L^{-1}(\delta)}\right) \tag{7}$$

To evolve the system the classical single point crossover with bit mutation has been used and will be denoted by Γ. Random labels are assigned to the starting population of chromosomes. The evolved population at the iteration *t* is:

$$P(t) = \{\alpha_1(t), \alpha_2(t), \ldots, \alpha_{n\times m}(t)\} \tag{8}$$

Applying the genetic operators we obtain the population

$$\Gamma(P(t)) = \{\beta_1(t), \beta_2(t), \ldots, \beta_{n\times m}(t)\} \tag{9}$$

where, $\beta_r = \Gamma(\alpha_r)$. The new population is determined by the selection process as follows:

$$P(t+1) = \{\gamma_1, \gamma_2, \ldots, \gamma_{n\times m}\} \tag{10}$$

such that

$$\gamma_r = \begin{cases} \beta_r \text{ if } f(\beta_r) < f(\alpha_r) \\ \alpha_r \text{ otherwise} \end{cases} \tag{11}$$

The genetic operator and the selection process are applied until a halting conditions, based on the convergence of the total variance ($V_{ar_t} = \sum_k^K \sigma_t(k)$), is satisfied: Halt computation if $\left|V_{ar_{t-1}} - V_{ar_t}\right| \leq \phi$, where, $\sigma_t(k)$ is the internal variance of the cluster *k* at the iteration *t* and $\phi \geq 0$. The condition $\phi = 0$ is not usually reached and value of ϕ is determined by the heuristics $\phi \approx \sqrt{\min\left(V_{ar_{t-1}}, V_{ar_t}\right)}$.

The above idea can be used for supplementary of the developed watershed algorithm, we are in the continuous study for this topic by using this genetic algorithm idea.

4 Conclusions

Morphological mathematics is used for image processing and image segmentation for about 30 years. A number of special algorithms were developed in the last three decades, but there are very few researchers studied the algorithms for blood cell images. We proposed a method of watershed-based region merging for gray level image segmentation. Catchment basins were used as initial segments and conflicting regions, which were found among the watersheds by using gradient changes, guided the region merging with dissimilarity function and terminated the iterative process according to the conditions we given. Our algorithm is focused on the catchment basins according to the region merging last step, not the watersheds, so we can get the satisfied output image quickly.

In order to resolve a common problem – object touching, we studied a splitting algorithm. The algorithm development is much different to normal splitting algorithms, we decompose clusters on cell shape information. For the further study, a genetic algorithm idea is suggested.

Acknowledgements

This contribution is part of the SCST (Science and Technology Bureau of Sichuan Province) project, Project No.05JY029-070.

References

1. Cecilia Di Ruberto et al., Analysis of infected blood cell images using morphological operators, Image and Vision Computing 20 (2002) 133-146.
2. Fu, K.S., Mu, J.K., 1981, A survey on image segmentation. Pattern Recognition, Vol. 13, 3-16.
3. Pal, N.R., Pal, S.K., 1993, A review of image segmentation techniques. Pattern Recognition, Vol. 26, No. 9, 1277-1294.
4. Vicent L, Solille P, Watershed in digital spaces: An efficient algorithm based immersion simulations [J]. IEEE Trans PAMI, 1991, 13(6): 538~598.
5. Rolf Adams, Leanne Bischof, Seeded Region Growing, IEEE Trans PAMI, 1994, 16(6):641~647.
6. John M. Gauch, Image Segmentation and Analysis via Multiscale Gradient Watershed Hierrchies, IEEE Trans IP, 1999, 8(1):69~79.
7. Kostas Haris, Serafim N. Efstratiadis, Hybrid Image Segmentation Using Watersheds and Fast Region Merging, IEEE Trans IP, 1998, 7(12):1684~1699.
8. Rafael C. Gonzalez, Richard E. Woods, Digital Image Processing, Second Edition Publishing House of Electronics Industry, 2002.7.
9. Shao-Yi Chien, Yu-Wen Huang, Shyh-yih Ma, Predictive watershed for image sequences segmentation, IEEE, 2000,9/02:3196~319.
10. Fernando Soares, Watershed lines suppression by waterfall marker improvement and linneighbourhood analysis, IEEE, 2002, ICPR'04.

An Adaptive Multi-objective Particle Swarm Optimization for Color Image Fusion

Yifeng Niu and Lincheng Shen

College of Mechatronic Engineering and Automation
National University of Defense Technology
Changsha, 410073, China
{niuyifeng, lcshen}@nudt.edu.cn

Abstract. A novel algorithm of adaptive multi-objective particle swarm optimization (AMOPSO-II) is proposed and used to search the optimal color image fusion parameters, which can achieve the optimal fusion indices. First the algorithm of AMOPSO-II is designed; then the model of color image fusion in YUV color space is established, and the proper evaluation indices are given; and finally AMOPSO-II is used to search the optimal fusion parameters. AMOPSO-II uses a new crowding operator to improve the distribution of nondominated solutions along the Pareto front, and uses the uniform design to obtain the optimal combination of the parameters of AMOPSO-II. Experimental results indicate that AMOPSO-II has better exploratory capabilities than MOPSO and AMOPSO-I, and that the approach to color image fusion based on AMOPSO-II realizes the Pareto optimal color image fusion.

1 Introduction

At present, multi-objective evolutionary algorithms include Pareto Archive Evolutionary Strategy (PASE) [1], Strength Pareto Evolutionary Algorithm (SPEA2) [2], Nondominated Sorting Genetic Algorithm II (NSGA-II) [3], Multiple Objective Particle Swarm Optimization (MOPSO) [4], etc, where MOPSO has a higher convergence speed and better optimization capacities [5]. However, MOPSO uses an adaptive grid [1] to record the searched particles, once the number of the objectives is greater than 3, MOPSO will need too much calculation time. So we presented an adaptive multi-objective particle swarm optimization (AMOPSO-I) in [6], in which the adaptive grid is discarded, and a crowding distance [3], an adaptive inertia weight and an adaptive mutation are introduced to improve the search capacity. But the crowding distance needs too much time and the optimal combination of the parameters is difficult to obtain in AMOPSO-I, so we propose AMOPSO-II. AMOPSO-II adopts a new distance and introduces the uniform design to obtain the optimal combination of parameters. In contrast to AMOPSO-I and MOPSO, AMOPSO-II has a higher convergence speed and better exploratory capabilities.

T.-D. Wang et al. (Eds.): SEAL 2006, LNCS 4247, pp. 473–480, 2006.

Color image fusion can be defined as the process of combining two or more color images into a single composite image with extended information content [7]. If one image is regarded as one information dimension, image fusion can be regarded as an optimization problem in several information dimensions. A better result, even the optimal result, can be acquired through searching the optimal parameters. In fact, there are various kinds of evaluation indices. The traditional solution is to change the multi-objective problem into a single-objective problem using a weighted linear method. However, the relation of the indices is often non-linear, and this method needs to know the weights of different indices in advance. In order to realize the optimal image fusion, it is highly necessary to introduce multi-objective optimization algorithms to search the optimal parameters. In [6], an approach to image fusion in the gray-scale space based on multi-objective optimization was explored. However, for the color image fusion, a nave approach might include performing image fusion separately and independently on each color channel, then providing the resulting three color channels as a single color image. In practice, this does not work for two reasons: interpretation of color scale space for feature selection and dependencies between the color components. Therefore, the approach to color image fusion in YUV color space is presented, and AMOPSO-II is used to optimize the fusion parameters.

The remainder of this paper is organized as follows. The AMOPSO-II algorithm is designed in Sect. 2. The methodology of color image fusion is introduced in Sect. 3. In addition, the evaluation indices also are given in section Sect. 3. The experimental results and analysis are given in Sect. 4. Finally, a review of the results and the future research areas are discussed in Sect. 5.

2 AMOPSO-II Algorithm

Kennedy J. and Eberhart R.C. brought forward particle swarm optimization (PSO) inspired by the choreography of a bird flock in 1995 [8]. PSO has shown a high convergence speed in multi-objective optimization [4], [5], [6], [9]. In order to improve the performance of the algorithm, we make an improvement and propose "AMOPSO-II" (adaptive multi-objective particle swarm optimization), in which not only the adaptive mutation operator and the adaptive inertia weight is used to raise the searching capacity, but also a new crowding operator based on Manhattan distance is used to improve the distribution of nondominated solutions along the Pareto front and maintain the population diversity, and the uniform design is used to obtain the optimal combination of the algorithm parameters.

2.1 AMOPSO-II Flow

The flow of AMOPSO-II is shown as Fig. 1. First the position and velocity of each particle in the population are initialized, and the nondominated particles is stored in the repository; second the velocity and position of each particle are updated [6], the partly particles adaptively mutate [4] and the particles is maintained within the decision space; third each particle is evaluated and their records and the repository

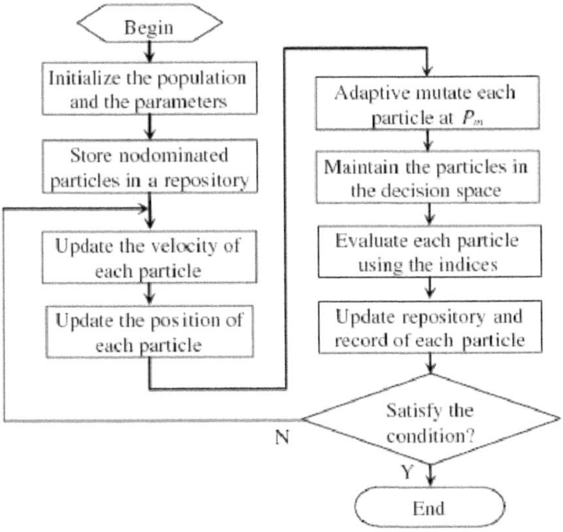

Fig. 1. Illustration of AMOPSO-II algorithm

are updated; then the cycle begins. When the maximum cycle number is reached, the Pareto solutions in the repository are output.

2.2 Crowding Operator

In order to improve the distribution of nondominated solutions along the Pareto front, we introduce a concept of crowding distance from NSGA-II [3] that indicates the population density. When comparing the Pareto optimality between two individuals, we find that the one with a higher crowding distance (locating the sparse region) is superior. In [3], the crowding distance has $O(mn\log n)$ computational complexity, and may need too much time because of sorting order. Here we propose a new crowding distance that can be calculated using the Manhattan distance between the points and the barycentre of their objectives. It is defined as

$$Dis[i] = \sum |f_{ij} - G_j| \tag{1}$$

where $Dis[i]$ is the distance of particle i in the population, f_{ij} is objective j of particle i, G_j is the barycentre of all the objectives j. The new crowding distance doesn't need to sort order and has less complexity, and it is superior to the grid [1], [4] because the latter may fail to allocate memory when there exist too many objectives.

2.3 Uniform Design of Parameters

We introduce the uniform design [10] to optimize the parameters of AMOPSO-II. The main objective of uniform design is to sample a small set of points from a given set of points, such that the sampled points are uniformly scattered. Let there be n factors and q levels per factor. When n and q are given, the uniform design selects q combinations out of q^n possible combinations, such that these q combinations are scattered uniformly over the space of all possible combinations. The selected combinations are expressed in terms of a uniform array $U(n,q) = [U_{i,j}]_{q\times n}$, where $U_{i,j}$ is the level of the jth factor in the ith combination. When q is prime and $q > n$, $U_{i,j}$ is given by

$$U_{i,j} = (i\sigma^{j-1} \text{mod} q) + 1 \tag{2}$$

where σ is determined by the number of factors and the number of levels [10].

3 Multi-objective Color Image Fusion in YUV Space

3.1 Color Image Fusion Model

The result of the color image fusion should preserve color blending consistency and color spatial consistency [7]. In the case of fusing source images, it is desired that the structural details be emphasized while the color and its saturation are preserved. For our methods, we choose the YUV color space, which has components representing luminance, saturation, and hue. As shown in Fig. 2, the approach to multi-objective color image fusion in YUV space is as follows.

Step 1: Input the source images A and B and convert the two images from RGB color space into YUV color space respectively. Since the source images can be assumed to have similar saturation and hue, the average of the U and V components from source images can be substituted for the U and V components in the fused image respectively, which can also reduce the computation complexity.

Step 2: Component Y represents the luminance, hence both the selection functions are based on the discrete wavelet transform (DWT) of the luminance components. Find the DWT of component Y of A and B to a specified number

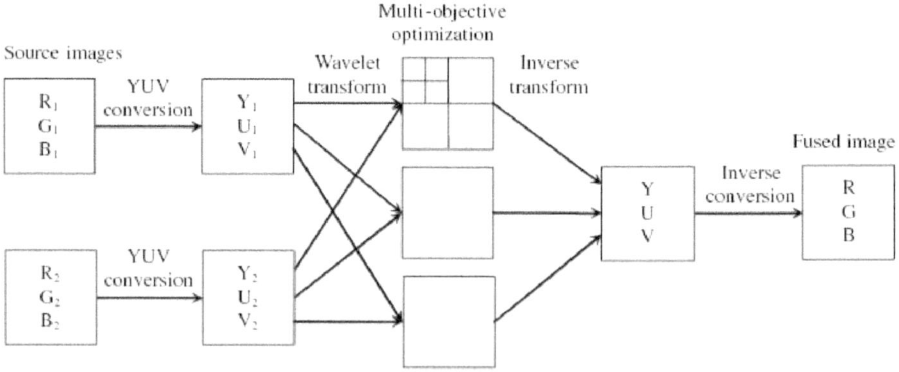

Fig. 2. Illustration of multi-objective color image fusion in YUV space

of the decomposition level, we will get one approximation and $3 \times J$ details at each level, where J is the decomposition level.

Step 3: For the details of component Y in DWT domain, salient features in each source image are identified. The coefficient with the largest salience is substituted for the fused coefficient [11].

Step 4: For approximations of component Y in DWT domain, let C_F, C_A, and C_B be the approximations of F, A, and B respectively, two different fusion rules will be adopted. One rule called"uniform weight method (UWM)" is given by

$$C_F(x, y) = w_1 \cdot C_A(x, y) + w_2 \cdot C_B(x, y) \tag{3}$$

where w_1 and w_2 are the decision variables with the values in the range [0, 1].

The other called "adaptive weight method (AWM)" is given by

$$C_F(x, y) = w_1(x, y) \cdot C_A(x, y) + w_2(x, y) \cdot C_B(x, y) \tag{4}$$

where $w_1(x, y)$ and $w_2(x, y)$ are decision variables.

Step 5. Using AMOPSO-II, we can find the optimal decision variables of the Y component in DWT domain, and realize the optimal color image fusion.

Step 6: The new sets of coefficients are used to find the inverse transform to get the Y component of the fused image F.

Step 7: The fused image in RGB color space can be attained using the inverse YUV conversion.

3.2 Evaluation Indices of Color Image Fusion

The establishment of an evaluation index system is the basis of the optimization that determines the performance of the final fused image [12]. The evaluation indices of color image fusion can be divided into two categories with respect to the subjects reflected. One category reflects the image features, such as gradient and entropy. The second reflects the relation of the fused image of F to the reference image of R, such as peak signal to noise ratio (PSNR) and structural similarity.

Gradient. Gradient reflects the change rate in the details that can be used to represent the clarity degree of an image. The higher the gradient of the fused image is, the clearer it is. In the YUV representation, Y represents the luminance of the image; hence gradient in the luminance component can also reflect the characteristics of human vision system.

Entropy. Entropy is an index to evaluate the information quantity contained in an image. If the value of entropy becomes higher after fusing, it indicates that the information quantity increases and the fusion performances are improved. The entropy in a color image is given by

$$E = (E_R + E_G + E_B)/3 \tag{5}$$

where R, G, and B denote the three color channel respectively, E_R denotes the entropy in channel R.

PSNR. The higher the value of PSNR (Peak Signal to Noise Ratio) is, the better the fused image is. PSNR in a color image is given by

$$PSNR = (PSNR_R + PSNR_G + PSNR_B)/3 \tag{6}$$

Structural Similarity. Structural similarity (SSIM) is designed by modeling any image distortion [13]. The higher the value of SSIM is, the more similar the F is to R. If two images are identical, the similarity is maximal and equals one. SSIM for color image fusion is given by

$$SSIM = (SSIM_R + SSIM_G + SSIM_B)/3 \tag{7}$$

4 Experiments

The performances of the proposed color image fusion approach using AMOPSO-II is tested and compared with those of different fusion schemes. The image "fishman" from Lab. for Image and Video Engineering of Texas University is selected as the reference image of R with 256 × 256 pixels in size, each pixel being represented by three bytes (one for each of the R, G, and B channels). The two source images of A and B is shown in Fig. 3(a) and Fig. 3(b) We use AMOPSO-II to search the Pareto optimal weights of the color image fusion model and compare the results with those of SWM (Simple Wavelet Method), MOPSO, and AMOPSO-I. In order to get common evaluations, the sum of the weights at each position in source images are limited to 1.

4.1 Uniform Design of AMOPSO-II Parameters

The more important and representative parameters of AMOPSO-II include the number of particles, the number of cycles, the size of the repository, and the mutation probability. We construct a uniform array with four factors and five levels as follows, where σ is equal to 2. We compute $U(4,5)$ based on (2) and get

$$U(4,5) = \begin{bmatrix} 2\,3\,5\,4 \\ 3\,5\,4\,2 \\ 4\,2\,3\,5 \\ 5\,4\,2\,3 \\ 1\,1\,1\,1 \end{bmatrix}$$

The value range of the number of particles is [40, 120]; the range of the number of cycles is [50, 250]; the range of the size of the repository is [100, 300]; the range of the mutation probability is [0.02, 0.06]. All combinations are run for a maximum value of 100 evaluations. Results show that the third combination is the optimal in the problem. Thus, the parameters of AMOPSO-II are as follow: the particle number is 100; the maximum cycle number is 100; the allowed maximum capacity is 200; the mutation probability is 0.06. The parameters of MOPSO and AMOPSO-I are the same.

4.2 Comparison Among Different Fusion Schemes

Since the solutions to color image fusion are nondominated by one another, we give preference to the four indices so as to select the Pareto optimal solutions, e.g. one order is SSIM, Gradient, PSNR, Entropy. The Pareto optimal fused images are shown in Fig. 3(c) and Fig. 3(d). Table 1 shows the evaluation indices of the fused images from different schemes, where MOPSO denotes the AWM based on MOPSO, AMOPSO-I denotes AWM based on AMOPSO-I, AWM and UWM denote methods based on AMOPSO-II, SWM takes a fixed weight of 0.5 for the approximations.

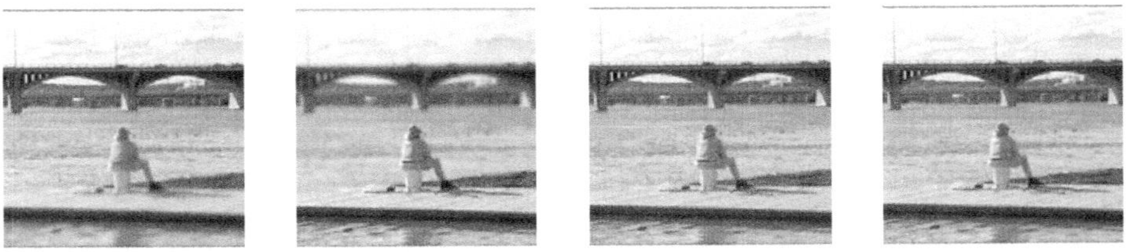

(a) Source image A (b) Source image B (c) UMW image (d) AWM image

Fig. 3. Source and fused images

Table 1. Evaluation indices of the fused images from different schemes

Schemes	Level	Gradient	Entropy	PSNR	SSIM
UMW	0	8.8186	7.6890	29.2640	0.9930
AWM	0	9.3027	7.7108	28.3477	0.9912
SWM	3	12.0254	7.7316	33.6438	0.9968
UMW	3	12.0236	7.7286	33.6572	0.9973
MOPSO	3	12.0293	7.7324	33.6974	0.9976
AMOPSO-I	3	12.0304	7.7329	33.6795	0.9976
AWM	3	12.0312	7.7394	33.8427	0.9977

From Table 1, we can see that the indices of AWM are not superior to those of UWM when the decomposition level is equal to 1. The reason is that the decision variables of AWM are too many and AWM can't reach the Pareto optimal front in limited time, e.g. the number of iteration is 100. The run time of AWM should increase with the number of decision variables. When the level is greater than one in DWT domain, the indices of AWM are superior to those of UWM because the weights of AWM are adaptive in different regions. The higher the decomposition level is, the better the fused image is, for a higher level decreases the decision variables and improves the adaptability (the maximum value of the level is limited to 3). The indices of AMOPSO-I and MOPSO are inferior to those of AMOPSO-II, which indicates that the new crowding distance can achieve better results than the distance in [3], that the uniform design can help to improve the performances of AMOPSO-II, and that MOPSO needs too much memory because the grid [4] is worse for too many objectives, e.g. 4.

Therefore, the approach to color image fusion that uses AMOPSO-II to search the adaptive fusion weights at level 3 in DWT domain is the optimal. This approach can overcome the limitations of given fusion parameters, obtain the optimal fusion results, and effectively enhance the features of the color image.

5 Conclusions

AMOPSO-II proposed is an effective algorithm to solve the parameter optimization of color image fusion, which can effectively enhance the features of the color image. One aspect that we would like to explore in the future is to analyze the evaluation indices system to acquire a meaningful measurement. We are also considering improving the optimization performances of AMOPSO-II.

References

1. Knowles, J.D., Corne, D.W.: Approximating the Nondominated Front Using the Pareto Archived Evolution Strategy. Evol. Comput. 2 (2000) 149-172
2. Zitzler, E., Laumanns, M., Thiele L.: SPEA2: Improving the Strength Pareto Evolutionary Algorithm, TIK-Report 103, ETH, Zurich, Switzerland (2001)
3. Deb, K., Pratap, A., Agarwal, S., Meyarivan, T.: A Fast and Elitist Multiobjective Genetic Algorithm: NSGA-II. IEEE Trans. Evol. Comput. 2 (2002) 182-197
4. Coello, C.A., Pulido, G.T., Lechuga, M.S.: Handling Multiple Objectives with Particle Swarm Optimization. IEEE Trans. Evol. Comput. 3 (2004) 256-279
5. Huang, V.L., Suganthan, P.N., Liang, J.J.: Comprehensive Learning Particle Swarm Optimizer for Solving Multiobjective Optimization Problems. Int. J. Intell. Syst. 2 (2006) 209-226
6. Niu, Y.F., Shen, L.C.: A Novel Approach to Image Fusion Based on Multi-Objective Optimization. In: Proc. WCICA06, Dalian, (2006) 9911-9915
7. Bogoni, L., Hansen, M.: Pattern-Selective Color Image Fusion. Pattern Recogn. 8 (2001) 1515-1526
8. Kennedy, J., Eberhart, R.C.: Particle swarm optimization. In: Proc. IEEE Int. Conf. on Neural Networks, Perth (1995) 1942-1948.
9. Kennedy, J., Eberhart, R.C.: Swarm Intelligence. Morgan Kaufmann, San Mateo (2001)
10. Leung, Y.W., Wang, Y.P.: Multiobjective Programming Using Uniform Design and Genetic Algorithm. IEEE Trans. Syst. Man Cybern. Pt. C: Appl. Rev. 3 (2000) 293-304
11. Huang, X.S., Chen, Z.: A Wavelet-Based Image Fusion Algorithm. In: Proc. IEEE TENCON, Beijing (2002) 602-605
12. Toet, A., Lucassen, M. P.: A Universal Color Image Quality Metric. In: Proc. SPIE, Vol. 5108 (2003) 13-23
13. Wang, Z., Bovik, A.C., Sheikh, H.R., Simoncelli, E.P.: Image Quality Assessment: From Error Visibility to Structural Similarity. IEEE Trans. Image Process. 4 (2004) 600-612

A New Dynamic Particle Swarm Optimizer

Binbin Zheng[1], Yuanxiang Li[1], Xianjun Shen[1], and Bojin Zheng[2]

[1] State Key Lab. of Software Engineering, Wuhan University, Wuhan, 430072, China
[2] College of Computer Science, South-Central University For Nationalities, Wuhan, 430074, China
luckzbb@163.com, yxli@whu.edu.cn, xjshen@mail.ccnu.edu.cn, zhengbojin@gmail.com

Abstract. This paper presents a new optimization model— Dynamic Particle Swarm Optimizer (DPSO). A new acceptance rule that based on the principle of minimal free energy from the statistical mechanics is introduced to the standard particle swarm optimizer. A definition of the entropy of the particle system is given. Then the law of entropy increment is applied to control the algorithm. Simulations have been done to illustrate the significant and effective impact of this new rule on the particle swarm optimizer.

1 Introduction

The particle swarm optimization (PSO) is an optimization algorithm developed by Kennedy and Eberhart in 1995[1] (We will refer to it as standard PSO) .It is developed through simulating social behavior. In a particle swarm optimization, the individuals are "evolved" by cooperation and competition among themselves through generations. Each particle adjusts its flying according to its own flying experience and its companions' flying experience. Each individual is named as a "particle" which, in fact, represents a potential solution to a problem. Each particle is treated as a point in a D dimensional space. The $i-$th particle is represented as X_i $(x_{i1}, x_{i2} \ldots x_{iD})$. The best previous position (the best position giving the best fitness value) of particle i is represented as $P_i(p_{i1}, p_{i2} \cdots \quad p_{iD})$. The index of the best particle among all the particles in the population is represented by $P_g(p_{g1}, p_{g2} \cdots \quad p_{gD})$. The rate of the position change (velocity) for particle i is represented as $V_i(v_{il}, v_{i2} \ldots Vi_D)$.

From iteration k to the following, each particle X_i moves according to a rule that depends on three factors, as follows.

$$V_i^{k+1} = w * V_i^k + c1 * r1 * (P_i^k - X_i^k) + c2 * r2 * (P_g^k - X_i^k) \tag{1}$$

$$X_i^{k+1} = X_i^k + V_i^{k+1} \tag{2}$$

where w is called inertia weights which fixed in the beginning of the process, c1 and c2 are two positive constants, r1 and r2 are two random numbers sampled from a uniform distribution in $[0, 1]$. The second part of the equation (1) is the

T.-D. Wang et al. (Eds.): SEAL 2006, LNCS 4247, pp. 481–488, 2006.

"cognition" part, which represents the private thinking of the particle itself. The third part is the "social" part, which represents the collaboration among the particles. The equation (1) is used to calculate the particle's new velocity according to its previous velocity and the distances of its current position from its own best experience (position) and the group's best experience. Then the particle flies toward a new position according to equation (2). The performance of each particle is measured according to a predefined fitness function, which is related to the problem to be solved.

Now PSO has been around for over ten years. Already, it is being researched and utilized in over a dozen countries. And some researchers found that the stochastic nature of the particle swarm optimizer makes it more difficult to prove (or disprove) like global convergence. Solis and R.Wets [2] have studied the convergence of stochastic search algorithms, most notably that of pure random search algorithms, providing criteria under which algorithms can be considered to be global search algorithms, or merely local search algorithms. Frans Van Den Bergh [3] used their definitions extensively in the study of the convergence characteristics of the PSO and the guaranteed convergence PSO (GCPSO), he proved the PSO is not even guaranteed to be local extrema, and GCPSO can converge on a local extremum.

In this paper, we introduce a new dynamic acceptance rule in order to get rid of this shortcoming. The new acceptance rule is based on the principle of minimal free energy from statistical mechanics. It utilizes the temperature and the Shannon Information Entropy, which is defined according to the particle system, to get a hint on controlling the diversity of the system. So the particles fly randomly and the optimizer gets out of the local optimal easily.

2 Theoretical Foundation of the DPSO

2.1 The Principle of Minimal Free Energy

Entropy is a very important concept in thermodynamics. The second law of thermodynamics is that the isolated system is always evolving towards the direction of entropy increasing. It means that in the process of an isolated system evolving towards the stable status, the entropy of the system increases continually. It is described that,

$$dS \geq 0 \tag{3}$$

where S is defined as the entropy, and when the entropy reaches the maximum, Eq.(3) gains the equal mark.

Joish Willard Gibbs proposed the concept of free energy in 1879. The formula is,

$$F = E - TS \tag{4}$$

where E is the interned energy, T is the temperature of the system, and S is the entropy of the system. The principle of minimal free energy is described as follows:

For a closed system that keep the temperature unchanged through exchanging heat with the ambience, the spontaneous diversifications of the states are always going along the reducing direction of the free energy. When the free energy gets to the minimum value, the systems come to equilibrium.

Consider that there are two states of constant temperature system, when the system changes from the first state to the second one, the change of free energy is that,

$$\triangle F = \triangle E - T\triangle S \tag{5}$$

It shows that in the process of the system evolving from unstable status to stable one, there is the competition between energy and entropy, and temperature decides the relative weight.

From the viewpoint of the DPSO, by regarding -E as the mean fitness value of the particle system, minimization of the free energy can be interpreted as taking a balance of the minimization of the energy function (the first term in the RHS of Eq.(4)), or equivalently maximization of the fitness function in the DPSO and maintenance of the diversity (the second term in the RHS of Eq.(4)).

2.2 The Shannon Information Entropy

Entropy defined by Shannon in information theory is closely related to thermodynamic entropy as defined by physicists and chemists. Boltzmann and Gibbs did considerable work on statistical thermodynamics. This work was the inspiration for adopting the term entropy in information theory. There are deep relationships between entropy in the thermodynamic and informational senses.

In general, for a source = (S, P) with source alphabet $S=$ $\{a_1 ... \ a_n\}$ and discrete probability distribution $P = \{p_1 ... \ p_n\}$, the entropy is,

$$H(S) = -\sum p_i log \ p_i \tag{6}$$

where p_i is the probability of a_i (say $p_i = p(a_i)$).

2.3 The Simulated Annealing Algorithm

In the simulated annealing (SA), the state of the system X is perturbed and a candidate of the new state X' is generated. If the energy value of the candidate $E(X')$ is smaller than that of the current state $E(X)$, the candidate is accepted as a new state. If $E(X')$ is larger than $E(X)$, the transition probability is controlled by a parameter T called the temperature. By repeating the above probabilistic procedure, the system is expected to attain the global minimum of the energy function [4].

In the DPSO, we also set a temperature, which is used to decide the weight of the diversity during the searching process. The temperature would decrease while the algorithm runs. So we can use some temperature schedule like in the SA to decrease the temperature T in Eq. (4) as the generation increase.

3 A New Dynamic Particle Swarm Optimizer

The idea behind DPSO is to try to keep the particles from converging at a certain local extremum at the early stage of the optimization. Refer to equation (1), the right side of which consists of three parts. The third part is velocity memory. Without this part, every particle will be "flying" towards its weighted centroid of its own best position and the global best position of the population, and it is more likely to exhibit a local search ability. However, even with the first part, classical PSO is not even guaranteed to be local extrema [3].The most striking point of classical PSO is the fact that for the essential of the movement rule, the performance of classical PSO heavily depends on the initial seeds. It is intuitive that if the particle swarm is not properly initialized, the algorithm risks to be trapped at some local minimum. Therefore some more tuning is needed .To avoid this kind of problem, some authors have suggested procedures of "re-seeding" the research by generating new particles at distinct places of the search space.

The effect of our new acceptance rule is to solve this problem. Since it keeps the population well distributed in the solution space, it is better than random re-seeding. The new acceptance is as follows.

In general, we consider the Global Optimization Problems such as finding $x^* \in S$ such that

$$f(x) \leq f(x^*), x \in S \tag{7}$$

where $S \subset R^d$ is compact, and f: $S \to R$ is a continuous function defined on S. Under these assumptions, the GOP is solvable, because f attains its minimum on S.

Then we give the definition of how to calculate the information entropy in DPSO.

The variable space VSP of the GOP is divided into M grids [5].We choose different M for different problems depending on the variable space of each function. They can be marked as $G_1, G_2, \ldots, G_M$, which satisfy,

$$G_i \cap G_j = \emptyset, i, j \in \{1, 2, \ldots, M\}, i \neq j \tag{8}$$

$$\cup G_i = VSP \ , i = 1, 2, \ldots, M \tag{9}$$

Then the number N_i should be calculated as the particles that locate in the grid G_i, $i \in \{1,2,\ldots,M\}$, and apparently they satisfy,

$$\sum N_i = N, i = 1, 2, \ldots, M \tag{10}$$

The possibility of each grid is defined as

$$q_i = N_i/N, i \in \{1, 2, \ldots, M\} \tag{11}$$

and they will satisfy

$$\sum q_i = 1, i \in \{1, 2, \ldots, M\} \tag{12}$$

Therefore, the information entropy S of the particles system can be calculated as

$$S = \sum -q_i log\ q_i, i \in \{1, 2, \ldots, M\} \tag{13}$$

Then the free energy of the particle system can be calculated as

$$F =< E > - \ TS \tag{14}$$

where $< E >$ is the mean energy of the system. And in DPSO, the fitness of a particle is considered as its energy.

At the t-th iteration, after a particle flying according to the movement rule, the new position will not replace the particle's previous one, but be used to take place of another particle's position in the current population $P(t)$. Compute the free energy for a trial population $P(t+1,\ k)$, where the new position replace the k-th particle of $P\ (t)$ (k is from 1 to N). Select the $P(t+1,k)$ for which the free energy is minimal. Then use the new position to replace the k-th particle's position of $P(t)$.

At a given iteration, after particle x_i flies, we get a new x_i', the general scheme of DPSO is as the following:

(I) Set k =1;

(II) compute the free energy for a trial population $P(t+1,k)$, where the new x_i' replace the k-th particle of $P(t)$;

(III) i=i+1, return to II;

(IV) Select the $P(t+1,k)$ for which the free energy is minimal, and use the new x_i' to replace the k-th particle.

Since there is a lot of computation work to do, we introduce a new parameter called dynamical selector [6], to the algorithm to reduce it. In the following we will give two definitions to make the new parameter easier to understand.

The momentum $m\ (t,\ x_i)$ of a particle x_i at iteration t is defined as

$$m(t, x_i) = f(t, x_i) - f(t-1, x_i) \tag{15}$$

where $f(t,\ x_i)$ denotes the function value of the particle x_i at iteration t.

The activity of particle x_i at iteration t is defined as

$$a(t, x_i) = a(t-1, x_i) + 1 \tag{16}$$

if particle x_i is selected to fly at iteration t. Otherwise x_i keeps unchanging.

Incorporating these two quantities the dynamical selector is defined as

$$slct(t, x_i) = \sum |m(t, x_i)| + a(t, x_i) \tag{17}$$

In DPSO, $slct\ (t,\ x_i)$ (i =1, 2... N) is sorted in the order from small to large expressed as $slct\ (t)$ for short.

At a given iterationwe only choose a certain number of particles on the forefront of $slct(t-1)$.Then at every iteration, the number of flying particles is smaller, so there will be less computation work to do, and also all the particles have a chance to be chosen according to reference [6].

4 Testing DPSO

The significance of a new optimization algorithm depends on the effectiveness of solving practical problems. In this section, the new DPSO is used to solve two typical hard global minimization problems, which are often used to test the performance and reliability of optimization algorithms. The results can be easily compared with other methods.

Test 1: Six-hump camel back function[7]

$f_1(X) = (4- 2.1x_1^2 + x_1^4/3)\, x_1^2 + x_1\, x_2 + (-4 + 4\, x_2^2)\, x_1^2$,

$(-3 \le x_1 \le 3,\ -2 \le x_2 \le 2)$.

It is known that this function has six local minima, just as its name suggests, and has one global minimum value *-1.031628* at two points (*-0.089842*, *0.712656*) and (0.089842, 0.712656). Because of this, many algorithms only can find the local minima, but lose the global optimization solutions.

Test 2: Shubert function III[8]

$f_2(X)=\{\sum i\cos\ [(i+1)\ x_1 + i]\}\{\sum i\cos[(i+1)\ x_2 + i]\} + (x_1+1.42513)^2 + (x_2+0.80032)^2$ $(-10 \le x_1,\ x_2 \le 10)$ $(i=1,2\ldots5)$.

This function has 760 local minima and only one global minimum*-186.7309* at the point (−1.42513,−0.80032). Because of the large number of local minima and the steep slope around the global minimum, this function has been widely recognized as an important test function.

In our program, the parameters are as follows:

the number of the particles (the size of the population) $N = 100$;

the maximal iterative number *MAXGENS=10000;*

the initial temperature *INITEMPER=100;*

the number of particles that are chosen to fly at each iterative step $m = 20$;

c1=c2=1.494;

$w = 0.729$;

the number of the grids that the variable space are divided into $M = 6$ in test 1, and $M = 8$ in test 2;

the temperature schedule is defined as

$$T(t+1) = 0.95T(t) \tag{18}$$

which decreases the temperature slowly while the iterative number increases.

The program was executed 30 times for each problem. Table 1 shows the numerical results of test 1, and Table 2 demonstrates the results of test 2. Though the tables, we can see that DPSO is quite an efficient optimization algorithm for GOP, and the successful rate of finding the optimal solutions is very high.(Average of best fitness means the global minimum value of the function.)

Table 1. Experiment results for test 1

	Average of best fitness	Standard deviation for the samples (STDEV)
PSO	-1.0316284	1.82574E-07
DPSO	-1.0316285	0

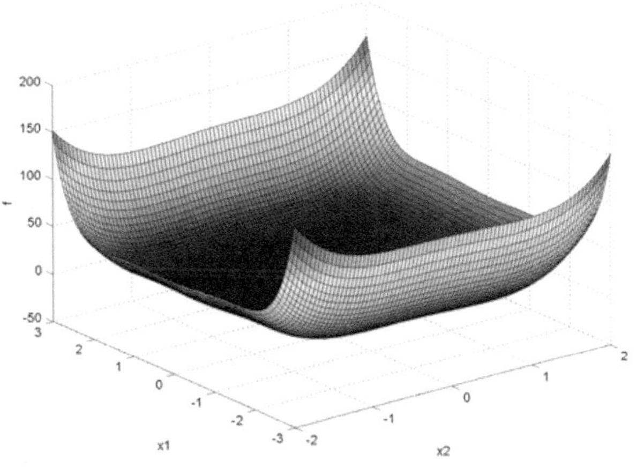

Fig. 1. The figure of function 1

Table 2. Experiment results for test 2

	Average of best fitness	Standard deviation for the samples (STDEV)
PSO	-186.4187377	0.388865957
DPSO	-186.5956853	0.301233649

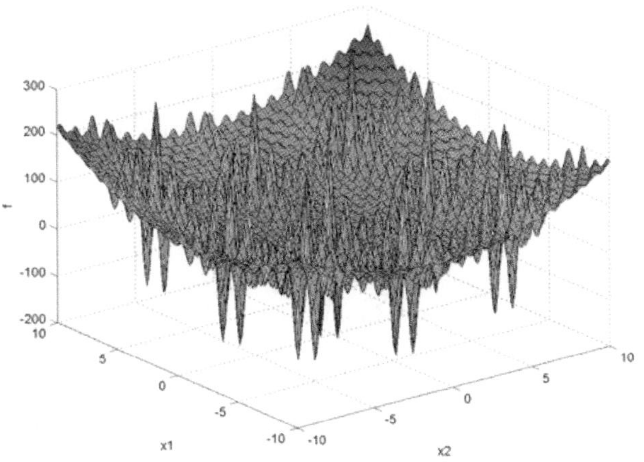

Fig. 2. The figure of function 2

5 Conclusion

In this paper, attempting to keep the particles from converging at a certain local extremum at the early stage of the optimization, a new DPSO is proposed based on the theory of statistical mechanics and Information Theory. And a new selector and acceptance strategy is presented. Some temperature schedule from the Simulated Annealing algorithms is also used in this DPSO.

By solving some typical testing problems, the efficiency and effectiveness of the DPSO are tested.

Although we get some achievement though this work, we still have some problems to research. An exactitude analyse in theory to explain the effectiveness of the new selector and acceptance strategy is needed. How to set some adaptive values for the parameters in the DPSO, and how to apply it to real world problem domains, where there are many difficult problems existing: these are all the problems left for further studies.

After all, putting use of some theories in natural science into computation has an extensive prospect. For the long run, we should pay much more attention to this field, and may get more and more useful and interesting conclusions.

Acknowledgements. We would like particularly to thank State Key Laboratory of Software Engineering, Wuhan University, for supporting our research work in this field. Also, we are grateful to the National Natural Science Foundation under Grant No. 60473014 and the Specialized Research Fund for the Doctoral Program of Higher Education of China under Grant No. 20030486049.

References

1. Kennedy, J., Eberhart, R.: Particle swarm optimization. In: IEEE International Conference on Neural Networks. (1995) 1942–1948
2. Solis, F., Wets, R.: Minimization by random search techniques. Mathematics of Operations Research **6**(11) (1981) 19–30
3. Van den Bergh, F.: An Analysis of Particle Swam Optimizers. PhD thesis, University of Pretoria (2000)
4. Davis, L.: Genetic Algorithms and Simulated Annealing. San Francisco:Morgan Kaufmann (1987)
5. Hu, T., Li, Y., Ding, W.: A new dynamical evolutionary algorithm based on the principle of minimal free energy. In Kang, L., ed.: International Symposium on Intelligent Computation and its Application, ISICA 2005, Wuhan, China (2005) 749–754
6. Li, Y., Zou, X., Kang, L., Michalewicz, Z.: A new dynamical evolutionary algorithm based on statistical mechanics. Journal of Computer Science and Technology **18**(3) (2003) 361–368
7. Zbigniew, M.: Genetic algorithms + Data structures= Evolution programs. Berlin:Springer-Verlag (1996)
8. Ge, R.: The globally convexized filled functions for global optimization. Applied Mathematics and Computation **1**(35) (1990) 131–158

Mean-Contribution Ant System: An Improved Version of Ant Colony Optimization for Traveling Salesman Problem*

Anzuo Liu[1], Guishi Deng[2], and Shimin Shan[3]

[1,3] Dalian University of Technology, Institute of Systems Engineering, Dalian, Liaoning Province, China, 116023
{anzuoliu, ssm}@student.dlut.edu.cn
[2] Dalian University of Technology, Institute of Systems Engineering, Dalian, Liaoning Province, China, 116023
denggs@dlut.edu.cn

Abstract. To enhance the diversity of search space, an improved version of Ant Colony Optimization (ACO), Mean-Contribution Ant System (MCAS) which is derived from Max-Min Ant System (MMAS), is presented in this paper. A new contribution function introduced in MCAS is used to improve the selection strategy of ants and the mechanism "pheromone trails smooth" mentioned by MMAS. Influenced by the improvements, the diversity of search space can be enhanced, which leads to better results. A series of benchmark Traveling Salesman Problems (TSPs) were utilized to test the performances of MCAS and MMAS respectively. The experiment results indicate that MCAS can outperform MMAS in most cases.

1 Introduction

Ant Colony Optimization, a novel approach to solve combinatorial optimization problems, was originally introduced in [1, 3], and had been applied widely [2]. It is inspired by the behavior of real ants [1], which can establish the shortest path from food sources to their nests. In the search process, a moving ant leaves the substances, which are called pheromone trails, to communicate among ants and help others decide which arc should be chosen to construct the whole tour. Meanwhile, the intensity of pheromone trails was declining due to evaporation over time unless more pheromone was laid down by other ants. An ant's tendency to choose specific arc is positively correlated to the trails intensity, i.e., the more times an arc is chosen, the more pheromone trails would be left, which would attract more ants consequently.

ACO uses simple artificial ants, called ants for short, to imitate real characteristics of ant colonies. Ants can communicate with each other via a mechanism influenced by the intensity of pheromone trails and the length of arcs [4, 5]. This mechanism allows the collaboration among ants in order to search the solution space of combinatorial optimization problems. At present, there are many effective and efficient

* This work is partially supported by the National Natural Science Foundation of China, No.70272050.

T.-D. Wang et al. (Eds.): SEAL 2006, LNCS 4247, pp. 489–496, 2006.

improved versions including MMAS [11, 12] which is regarded as one of the most effective ACO algorithms for TSPs. However, it is inevitable that MMAS cannot be satisfying in the process of maintaining the diversity of search space, which would deeply influence the qualities of results.

In this paper, we present Mean-Contribution Ant System, an improved version of ACO based on MMAS, and prove that the validity of MCAS is better than that of MMAS through experiments on symmetric and asymmetric TSPs in TSPLIB (http://www.iwr.uni-heidelberg.de/groups/comopt/software/TSPLIB95/index.html).

The rest of this paper is organized as follow. To keep the integrality of this paper, the TSP and MMAS will be firstly introduced in Section 2. In Section 3, we present the Mean-Contribution Ant System algorithm and its improvements compared with MMAS. The results of experiments will be shown in Section 4 and conclusion with further work in Section 5.

2 ACO and MMAS for TSPs

The TSP can be formulated as the problem of finding a cyclic route of minimal length visiting every city exactly once in a complete weighted graph A. TSP can be represented by a complete weighted graph G = (V, A, d) where V is a set of cities, $A = \{(i,j) | (i,j) \in V \times V\}$ is a set of arcs, and $d(i,j) \in R$ is a weight function, i.e. Euclidian distance for each pair of cities. Ψ is the solution space of a TSP instance. The aim is to find a tour $t \in \Psi$ with the minimal cost function $f(t)=\sum d(i,j)$, $(i,j) \in t$.

ACO makes use of ants that are set randomly on selected cities to construct candidate solutions for TSPs [8]. Each ant from the city i selected a city j which has not been visited according to probability P(i,j) which is influenced by the inverse of the distance $\eta(i,j)$ and the trails intensity $\tau(i,j)$ [1]. The equation is given by (1) [7].

$$Pij=\tau(i,j)^{\alpha}\times\eta(i,j)^{\beta} \;/\; \sum\tau(i,j)^{\alpha}\times\eta(i,j)^{\beta} \quad (1)$$

Where α and β are parameters, which determine the relative importance of pheromone and distance respectively.

The pheromone trails are updated according to evaporation coefficient and increments of trails after all ants having constructed a complete tour. The equation is given by (2).

$$\tau_{ij}^{after}=\rho\times\tau_{ij}^{before}+\sum\triangle\tau_{ij}^{k} \quad (2)$$

Where$\rho$is the persistence of trails, and $\triangle\tau_{ij}^{k}=1 \;/\; L^{k}$ only if arc(i,j) is chosen by ant k, and $1-\rho$ represents the evaporation.

Because ACO is a positive feedback search algorithm [6], the arcs used frequently and contained in shorter tours will accumulate a larger amount of pheromone and in turn will be selected more often in future. If ants only select cities with the highest intensity of pheromone trails, the stagnation of search will happen and final results will be obtained. The process of tour construction associating with trails update will be repeated for a given number of iterations.

MMAS algorithm improves the ACO by introducing restriction mechanism, which confines the trail strengths between maximum and minimum to efficiently prevent search from premature stagnation. The maximum trail limit $\tau_{max}=1 / L_{min}$, where L_{min} is the minimal tour length found during the current iteration of search process. The minimum trail limit changing with maximum is represented as $\tau_{min}=\tau_{max} / 2n$ roughly (more details see [12]), where n is the number of cities. In MMAS, ants simply choose cities according to the lengths of arcs and the intensity of trails. In fact, arcs with much shorter lengths will be chosen with higher probability and left with more pheromone trails by ants, which results in that other longer arcs, which should be selected more often, are chosen with such a low probability during the process of search that it is difficult to warrant the diversity of search space. That is the primary problem which MCAS wants to solve.

To characterize the diversity of search space, the mean λ-branching factor can be used, for more details see [9]. If the mean λ-branching factor is very low (be equal to 1 approximately), ants will select no more arcs except ones with the highest trail intensity.

3 Mean-Contribution Ant System

In our experiments with MMAS, we observed that although some arcs with longer lengths and lower trail intensities were chosen seldom by ants, they can be contained in some evidently better tours occasionally. At the same time, however, the differences of trail intensity between these arcs and those with the highest trail intensity will be greater and greater due to the increasing iterations. The foregoing situation will result in that these longer arcs, which may be more beneficial to satisfying results than those ones with much shorter lengths, cannot obtain enough reinforcement of pheromone trails to attract ants to explore new potential tours. In this way, algorithm will be stagnant with suboptimal results.

We use 3 sample instances including eil76.tsp, kroA100.tsp and kro124p.atsp in TSPLIB to demonstrate this tendency (see Fig. 1). The x-coordinate represents the iteration of search process, and the y-coordinate represents the average difference of trails intensity between arcs contained in better tours and arcs with the highest trails intensity. It is obvious that these arcs which are beneficial to better results were seldom selected due to the lower intensity of pheromone trails and the difference of trail intensity between them and those with the highest trails intensity is increasing sharply. If ants select arcs only according to length of arcs and intensity of pheromone trails, the chance these longer arcs being selected will be smaller and smaller accompanying with the process of search, because selection strategy neglecting other information from which the final result will benefit is unilateral. The increasing differences also cannot warrant the diversity of search space, and suboptimal results will be found. Therefore arcs with longer lengths and lower trails intensity should not be completely neglected by ants in searching process. Thus, it is reasonable to introduce other factors which can represent arcs' characteristics better than length and trails intensity do.

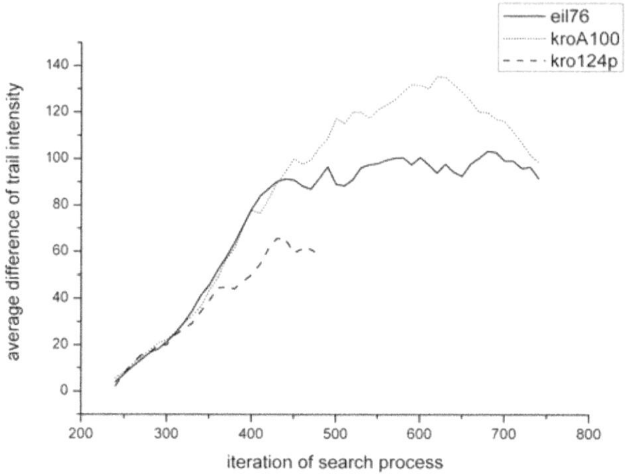

Fig. 1. The average difference of trail intensity between arc with the highest trail intensity and others increases obviously accompanying with the process of search on the examples eil76, kroA100 and kro124p, respectively. The x-coordinate represents the process of search by using the number of iterations, and the y-coordinate represents the average difference of trail intensity.

In the MCAS, we call arcs which are contained in the best tours of iteration, "contributive arcs" in spite of their lengths and trail intensities. The reason is that ants only see the lengths of arcs when they select the next city and cannot objectively estimate the values of arcs without a global vision. So ants should selected cities not only according to arcs' lengths and trails intensities but also according to their contribution to better solutions. Although not all of arcs contained in best tours of iteration are actually helpful to obtain better solutions, ants will not select them only by their contribution in spite of their lengths and trails intensities. Thus, if an arc belongs to "contributive arcs", it may be beneficial to global optimization or at least not be detrimental to better solutions, so it should be selected later with a higher probability. We can use this idea to improve the selection strategy of ants. There are two obvious differences between MMAS and MCAS, as follow:

1. In the MCAS, ants will choose arcs according to a selection strategy which not only considers the lengths and trails intensities of arcs, but also bases on the contribution of arcs to good results.
2. MMAS introduces a mechanism called "pheromone trails smooth," [12] to avoid suboptimal results when algorithm closes premature stagnancy. This mechanism increases arcs' trails intensity proportionally without discrimination. To the contrary, in MCAS, this mechanism will be changed and rely on the contribution of arcs to obtain good results.

Therefore, these contributive arcs will obtain more opportunities to be explored in order to attain some competitive tours. The reason for making this improvement is dual. On the one hand, arcs which are contributive for good results could be explored with a great probability. On the other hand, the useless arcs could be selected as little as possible to make useful arcs be selected with relative more chances in finite iterations. Arcs with greater contributions for obtaining better results will be explored more often. The improved equation is given by (3).

$$P_{ij}=\tau(i,j)^{\alpha}\times\eta(i,j)^{\beta}\times\omega(i,j)^{\gamma} \,/\, \sum\tau(i,j)^{\alpha}\times\eta(i,j)^{\beta}\times\omega(i,j)^{\gamma} \tag{3}$$

Where γ is the parameter of relative importance, contribution function $\omega_{ij}=L_{max} / L_{ij}$. L_{ij} and L_{max} represent the quality of tours consisting of arc(i,j) and the quality of tours consisting of the arc with the highest trail intensity among the arcs which begin with city i respectively. L_{max} equals to mean length of tours which contain the arc with the highest trail intensity. The values of L_{ij} and L_{max} will increase with an inverse proportion according to qualities of tours they construct.

L_{ij} could be either the mean length or the shortest length of the best tours consisting of arc(i,j) in each iteration. If the quality of tours consisting of arc(i,j) is better than that of tours consisting of the arc with the highest trail intensity among the arcs which begin with city i, $\omega_{ij}>1$, i.e., arc(i,j) which contributes more to good results should be selected with greater probability, vice versa. If arc(i,j) is not contained in any best tour after each iteration, its contribution will have no influence, then $\omega_{ij}=1$. Therefore ants will continue to choose contributive arcs with relatively higher probabilities and pay less attentions to those arcs which are longer and hence useless to good results; meanwhile the diversity of search space can be retained in order to find better results, because ants will construct tours with a global vision.

3.1 Mean Length Versus Shortest Length for L_{ij}

As mentioned above, L_{ij} can be the mean length of the best tours consisting of arc(i,j). But some bad tours make the average value of lengths bigger than it should be, because the length of the best tour after each iteration is much longer when the mean λ-branching factor is not low enough. To diminish the counterproductive effects brought by long tours, in our experiment we calculate the L_{ij} when the value of mean λ-branching factor is less than 2.0.

Also, L_{ij} can be the shortest length of tour consisting of arc(i,j) to eliminate bad results. Although some L_{ij} are shorter than L_{max} ($\omega_{ij}>1$), they are not helpful to find better tour at all. In practice, these arcs with little contribution may be selected more frequently, which is not expected and leads to a suboptimal result.

These two approaches maybe lead to different results. The experiment results of these two approaches are shown in Table 1.

According to Table 1, it is obvious that most results with L_{ij} as mean length are better than the ones with L_{ij} as the shortest length at all aspects (the best result, average results and standard deviation). So mean length will be chosen to represent the quality of tours consisting of arc(i,j), and then ω_{ij} represent the mean contribution of arc(i,j). This is the reason why the improved algorithm was named Mean-Contribution Ant System.

Table 1. Average experiment results with 10 independent runs and 10000 iterations, obtained by using mean length and the shortest length to respresent L_{ij} respectively, Dev. is deviation

Problem	Mean Length			Shortest Length		
	best	average	Dev.	best	average	Dev.
eil51.tsp	428	428.7	1.16	427	429.4	1.78
eil76.tsp	539	541.5	2.07	539	542.2	2.86
kroA100.tsp	21296	21395.1	47.16	21379	21411	33.73
bier127.tsp	119026	119383.4	186.31	119074	119680.3	369.39
ch150.tsp	6555	6579.5	17.48	6559	6576.3	16.57
ftv64.atsp	1854	1861.7	13.87	1854	1862.1	14.45
ry48p.atsp	14459	14641.9	127.48	14549	14660.3	120.3
kro124p.atsp	36695	37411.3	370.54	36882	37640.9	463.12

3.2 Smoothing of the Pheromone Trails with Contribution

In MMAS, an important improvement is the smoothing of the pheromone trails (PTS). When algorithm has converged or is very close to convergence which is indicated by the mean λ-branching factor, this mechanism increases the pheromone trails proportionally based on their differences from the maximum pheromone trail limit (for more details see [12]). The main idea of PTS is to prevent ants from completely stopping searching the solution components with low pheromone trail.

Nevertheless, the trail increasing only according to the difference of pheromone trails between arcs and the one with the highest trail intensity is not reasonable. Without considering whether and how many arcs contribute to good results, the trail intensity of arcs which are not beneficial to shorter tours will be enhanced mistakenly as much as or even more than the one of useful arcs. Thus, new algorithm has to be proposed to correct this mistake. In MCAS, the arcs with great contribution should obtain correspondingly more enhancements, and others with small contribution should obtain less. So we improve the PTS by increasing the pheromone trails of arcs according to their contribution function ω and make the trail intensity of arcs with more contribution higher. In this way, the contributive arcs will obtain more enhancements than others. The equation is given by (4).

$$\tau_{ij}^{*}=\tau_{ij}+\varepsilon\times(\tau_{max}-\tau_{ij})\times\omega_{ij} \tag{4}$$

Where τ_{ij}^{*} and τ_{ij} are the pheromone trails before and after the smoothing. The value of ε is between 0 and 1.

4 Experiments with MCAS

We choose equation (3) as the selection strategy and use the iteration-best ants for pheromone trails update (for more details see [11, 12]). The number of ants m is chosen equal to the number of cities n, and ants are set averagely on each city. Other parameters are set as follow: α=1, β=2, γ=3, ρ=0.98 and iteration = 10000.

In order to avoid the premature stagnation of search, MCAS continue to use the restriction mechanism for the intensity of pheromone trails, and only the best tour in each iteration is allowed to update pheromone trails. Furthermore, a 2-opt local search is added for higher quality solutions [10], and the average performance of these two algorithms (MCAS and MMAS) achieved over 10 independent runs. The experiment results are showed in Table 2.

Table 2. Average experiment results with 10 independent runs and 10000 iterations, obtained by using MMAS and MCAS respectively. Obviously, the latter can achieve better results than the former, Dev. is deviation.

Problem	MMAS			MCAS		
	best	average	Dev.	best	average	Dev.
eil51.tsp	426	427.3	0.67	426	426.8	0.79
eil76.tsp	538	539.9	2.64	538	538.1	0.32
kroA100.tsp	21282	21349.7	41.36	21282	21347.6	38.36
bier127.tsp	118803	118909	54.96	118759	118882.3	48.61
ch150.tsp	6550	6559	6.32	6554	6560.9	7.08
ftv64.atsp	1854	1854.2	0.63	1850	1853.6	1.26
ry48p.atsp	14459	14494	38.02	14422	14489.1	31.22
kro124p.atsp	36619	36914.2	137.82	36561	36828.7	273.16

In Table 2, both the best results and average results obtained by MCAS are better than those obtained by MMAS for all problems except ch150, because of the maintenance of diversity of search space. The experiment results indicate that the performance of MCAS is better than that of MMAS.

5 Conclusion and Further Work

In this paper, we presented the increasing difference of trail intensities between contributive arcs and ones with the highest trail intensity, which will leads to a relative shallow diversity of search space. To enhance diversity, we introduced a new contribution function, according to which ants' selection strategy is improved. New strategy depends on not only the lengths and trail intensity of arcs but also their contributions to good results. There are two representations of contribution of arcs, one of which is the mean length of tours consisting of arc(i,j) and the other is the shortest length of tours consisting of arc(i,j). Throughout our experiments, we find that the former can achieve better results than the latter. Another improvement is that algorithm dynamically diminishes the difference of trail intensity among arcs on the basis of their contribution. The experiment results indicate that MCAS avoids the sparse diversity of search space effectively and increases possibility of acquirement of better solutions. Its performance is better than or at least the same as MMAS.

The results obtained by ACO and its improved versions are closed to the best ones which are known at present. In future, a competitive tour improvement algorithm will be considered to enhance the performance of Ant Colony algorithm by making use of the better solution obtained by MCAS. How to improve the efficiency of algorithm, especially when the number of cities is extremely large, is another problem which should be solved.

References

1. Colorni A, Dorigo M, Maniezzo V. Distributed Optimization by Ant Colonies. In: Proceedings of the First European Conference on Artificial Life, Paris, France, pp. 134-142, 1991, Elsevier.
2. Christian Blum. Ant Colony Optimization: Introduction and recent trends. Physics of Life Reviews, 2(2005):353-373.
3. Dorigo M. Optimization, Learning, and Natural Algorithms. PhD thesis, Dipartimento di Elettronica, Politecnico di Milano, 1992.
4. Dorigo M, Maniezzo V, Colorni A. The Ant System: Optimization by a colony of cooperating agents. IEEE Transactions on Systems, Man Cybernetics Part B:1996;26(1):29-41.
5. Dorigo M, Gambardella LM. Ant Colony System: A Cooperative Learning Approach to the Traveling Salesman Problem. IEEE Transactions on Evolutionary Computation 1997;1(1):53-66.
6. Dorigo M, Maniezzo V, Colorni A. Positive Feedback as a Search Strategy. Technical Report 91-016, Dipartimento di Elettronica, Politecnico di Milano, Italy, 1991.
7. Dorigo M, Gambardella LM. Ant Colonies for the Traveling Salesman Problem. Biosystems,43:73–81, 1997.
8. Gambardella LM, Dorigo M. Solving Symmetric and Asymmetric TSPs by Ant Colonies. In: Baeck T, Fukuda T, Michalewicz Z, editors. Proceedings of the 1996 IEEE international Conference on Evolutionary Computation(ICEC'96). Piscataway, NJ:IEEE Press; 1996. p.622-627.
9. Gambardella LM, Dorigo M. Ant-Q: A Reinforcement Learning Approach to the Traveling Salesman Problem. In Proceedings of the 11th International Conference on Machine Learning, Morgan Kaufmann, San Francisco, CA, 1995, pp. 252-260.
10. M.G.A. Verhoeven, E.H.L. Aarts, P.C.J. Swinkels. A parallel 2-opt algorithm for the Traveling Salesman Problem. Future Generation Computer System, 11(1995): 175-182.
11. Stützle T, Hoos HH. The MAX-MIN Ant System and Local Search for the Traveling Salesman Problem. In: T. Bäck, Z. Michalewicz, X. Yao (Eds.), Proceedings of the IEEE International Conference on Evolutionary Computation(ICEC'97), IEEE Press, Piscataway, USA, 1997, pp. 309-314.
12. Stützle T, Hoos HH. MAX-MIN Ant System. Future Generation Computer System 2000;16(8):889-914.

A Diversity-Guided Quantum-Behaved Particle Swarm Optimization Algorithm

Jun Sun, Wenbo Xu, and Wei Fang

School of Information Technology, Southern Yangtze University,
No. 1800, Lihudadao Road, Wuxi Jiangsu 214122, China
{sunjun_wx, xwb_sytu, wxfangwei}@hotmail.com

Abstract. One of the primary complaints toward Particle Swarm Optimization (PSO) is the occurrence of premature convergence. Quantum-behaved Particle Swarm Optimization (QPSO), a novel variant of PSO, is a global convergent algorithm whose search strategy makes it own stronger global search ability than PSO. But like PSO and other evolutionary optimization technique, premature convergence in the QPSO is also inevitable and may deteriorate with the problem to be solved becoming more complex. In this paper, we propose a new Diversity-Guided QPSO (DGQPSO), in which a mutation operation is exerted on global best particle to prevent the swarm from clustering, enabling the particle to escape the sub-optima more easily. The DGQPSO, along with the PSO and QPSO, is tested on several benchmark functions for performance comparison. The experiment results show that the DGQPSO outperforms the PSO and QPSO in alleviating the premature convergence.

1 Introduction

The Particle Swarm Optimization (PSO), first introduced by Kennedy and Eberhart [4], is a population-based optimization technique that can be likened to the behavior of a flock of birds or the sociology behavior of an organism. In PSO with population size M, each particle i ($1 \le i \le M$) represents a possible solution to the optimization problem at hand, flies in D-dimensional search space with the following attributes: A current position in the search space $X_i = (X_{i,1}, X_{i,2}, \cdots, X_{i,D})$; a current velocity $V_i = (V_{i,1}, V_{i,2}, \cdots, V_{i,D})$, and a personal best (***pbest***) position $P_i = (P_{i,1}, P_{i,2}, \cdots, P_{i,D})$. During each iteration, each particle is updated according to

$$V_{i,j}(t+1) = w \cdot V_{i,j}(t) + c_1 \cdot r_{1,i}(t) \cdot [P_{i,j}(t) - V_{i,j}(t)] + c_2 \cdot r_{2,i}(t) \cdot [P_{g,j}(t) - X_{i,j}(t)] \quad (1)$$

$$X_i(t+1) = X_i(t) + V_i(t+1) \quad (2)$$

where $r_1 \sim U(0,1)$, $r_2 \sim U(0,1)$ are elements from two uniform random sequences in the range (0,1), and c_1 and c_2 denote the acceleration coefficients which typically are both set to a value of 2.0. The personal best position of each particle (the position with the best fitness value experienced by the particle so far) P_i is updated at each

T.-D. Wang et al. (Eds.): SEAL 2006, LNCS 4247, pp. 497 – 504, 2006.

iteration and P_g is the global best position found by any particle during all previous steps. The value of each component in every V_i vector can be restricted to the range $[-V_{\max}, V_{\max}]$ to reduce the likelihood of particles leaving the search space. The variable w in (1) is called the inertia weight; this value is typically setup to vary linearly from 0.9 to near 0.4 during the course of training run ([14], [15]). Since its origin, there have been many improvement of PSO. For more detailed information of these improved versions, one may see literatures such as [1], [2], [5], [6] and [7] etc.

In the previous work [8], [9], [10], we proposed a novel PSO called Quantum-behaved Particle Swarm Optimization (QPSO), which was inspired by quantum mechanics. It has been shown that QPSO outperforms the original PSO on several widely known benchmark function problems. However, like other evolutionary algorithm, the PSO as well as QPSO, confront the problem of premature convergence, which result in great performance loss and sub-optimal solutions. In the PSO or QPSO, the fast information flow between particles seems to be the reason for clustering of particles. In this paper, we propose a Diversity-Guided Quantum-behaved Particle Swarm Optimization (DGQPSO). In the DGQPSO, a threshold value was set for population's diversity measure to prevent premature convergence and therefore enhance the overall performance of the QPSO. The rest part of the paper is organized as follows. In the next section, the QSO is introduced. The DGQPSO is proposed in Section 3. Section 4 is the experiment results and discussion. Some conclusion remarks are given in Section 5.

2 Quantum-Behaved Particle Swarm Optimization

Trajectory analyses in [3] demonstrated that, to guarantee convergence of the PSO algorithm, each particle must converge to its local attractor $p_i = (p_{i,1}, p_{i,2}, \cdots p_{i,D})$, of which the coordinates are:

$$p_{id} = (c_1 P_{i,i} + c_2 P_{g,j})/(c_1 + c_2), \text{ or } p_{i,j} = \varphi \cdot P_{i,j} + (1-\varphi) \cdot P_{g,j}, \tag{3}$$

for j=1,2,...D where $\varphi \sim U(0,1)$ .Assume that there is one-dimensional δ potential well on each dimension at point p_i and each particle has quantum behavior. The probability distribution function of particle's position on each dimension is

$$F(X_{i,j}) = e^{-2|p_{i,j} - X_{i,j}|/L_{i,j}} \tag{4}$$

where $L_{i,j}$ determines search scope of each particle. Employing Monte Carlo method, we can obtain the position of the particle by

$$X_{i,j} = p_{i,j} \pm \frac{L_{i,j}}{2} \ln(1/u) \qquad u = rand(0,1) \tag{5}$$

where u is a random number uniformly distributed in (0, 1). The value of $L_{i,j}$ is evaluated by $L_{i,j} = 2\beta \cdot |C_j - X_{i,j}(t)|$, where C called Mean Best Position, is defined as the mean of the *pbest* positions of all particles. That is

$$C_j = \frac{1}{M}\sum_{i=1}^{M} P_{i,j}, \quad (j = 1,2,\cdots,D) \tag{6}$$

Hence, the position of the particle updates according to the following equaiton

$$X_{i,j}(t+1) = p_{i,j} \pm \beta \cdot \left|C_j - X_{i,j}(t)\right| \ln(1/u) \qquad u = rand(0,1) \tag{7}$$

where parameter β is called Contraction-Expansion Coefficient, which can be tuned to control the convergence speed of the algorithms. The PSO with equation (7) is called Quantum-behaved Particle Swarm Optimization (QPSO).

3 Diversity-Guided QPSO

In a PSO system, with the fast information flow between particles due to its collectiveness, diversity of the particle swarm declines rapidly, leaving the PSO algorithm with great difficulties of escaping local optima. In QPSO, although the search scope of an individual particle at each iteration is the whole feasible solution space of the problem, diversity loss of the whole population is also inevitable. Inspired by works undertaken by Ursem and Riget *et al* ([12], [13]), we propose a Diversity-Guided Quantum-behaved PSO (DGQPSO) in this paper.

As Uresem and Riget did, the diversity in DGQPSO is measured by average Euclidean distance from the particle to their centroid. However, in PSO or QPSO system, there are two populations or swarms, one of which is composed of the current positions of the particles, and the other of which is the set of the personal best positions of the particles,

$$S_X = (X_1, X_2, \cdots, X_M),\ S_P = (P_1, P_2, \cdots, P_M) \tag{8}$$

Correspondingly, there are two forms of diversity measure for a QPSO or PSO system described as follows,

$$diversity\ (S_X) = \frac{1}{M \cdot |A|} \cdot \sum_{i=1}^{M} \sqrt{\sum_{j=1}^{D} (X_{i,j} - \overline{X_j})^2},\quad \overline{X_j} = \frac{1}{M}\sum_{i=1}^{M} X_{i,j} \tag{9}$$

$$diversity\ (S_P) = \frac{1}{M \cdot |A|} \cdot \sum_{i=1}^{M} \sqrt{\sum_{j=1}^{D} (P_{i,j} - \overline{P_j})^2},\quad \overline{P_j} = \frac{1}{M}\sum_{i=1}^{M} P_{i,j} = C_j \tag{10}$$

where $|A|$ is the length of longest the diagonal in the search space, D is the dimensionality of the problem. Hence, we may guide the search of the particles with one of the above diversity measures when the algorithm is running.

At the beginning of the search, the diversity of the particle swarm in QPSO is high after initialization. With the development of evolution, the convergence of the particle makes the diversity be declining, which, in turn, is enhancing the local search ability (exploitation) but weakening the global search ability (exploration) of the algorithm. At early or middle stage of the evolution, the declination of the diversity is necessary for the particle swarm to search effectively. However, after middle or at later stage, the particles may converge into such a small region that the diversity of the swarm is very low and further search is impossible. At that time, if the particle with global best position is at local optima or sub-optima, premature convergence occurs.

To avoid the premature convergence and improve the performance of QPSO, we proposed a Diversity-Guided QPSO (DGQPSO), in which a low bound for *diversity* (S_X) or *diversity* (S_P) is set to prevent the diversity from constantly declining. The procedure of the algorithm is as follows. After initialization, the algorithm is running in convergence mode that is realized by varying β from 1.0 to 0.5 on the course of running. This control method of the parameter is also adopted in the original QPSO and can result in good performance of QPSO generally. On the course of evolution, if the diversity measure of the swarm declines to below the low bound d_{low}, the particles will explode to increase the diversity until it is larger than d_{low}.

In [11], we proposed two methods of making the particles explode. One method is to control the value β, that is, we can set $\beta=\beta_0$ ($\beta_0>1.79$) once *diversity*(S_X) is lower than the threshold value d_{low}. The other method of increasing the diversity is re-initializing the Mean Best Position C of the population across the search space once *diversity*(S_X) is smaller than d_{low}. The reason for the re-initialization of C is that when the diversity is low, the distance between the particle and C is too small for the particle to escape the local optima as can be seen from equation (7). Thus re-initializing C could enlarge the gaps between particles and C, consequently making particles explode temporarily. These two methods are efficient when we adopt *diversity* (S_X) as the diversity measure, but fail to result in good performance if diversity measure is *diversity*(S_P).

In this paper, we propose a new method of exerting the following mutation operation on the particle with global best position if the diversity measure is smaller than d_{low},

$$P_{g,j} = P_{g,j} + \gamma \cdot |A| \cdot \varepsilon, \quad \varepsilon \sim N(0,1), \ (j=1,2,\cdots,D) \tag{11}$$

where ε is a random number with standard normal distribution N(0,1), γ is a parameter.

When the mutation operation is exerted, the displacement of the global best particle will make increase the average distance of the particles' personal best positions from their mean best position C and thus enhance the *diversity*(S_P). In the meanwhile, the position C will be pulled away from its original position by the displaced global best particle, which, in turn, enlarges the gaps between particles' current position and the position C, consequently making particles' search scope extended and resulting in the gain of *diversity* (S_X). Therefore, we may see that this mutation method can work on both *diversity* (S_X) and *diversity*(S_P). In this paper, we present two versions of Diversity-Guided QPSO, one use *diversity* (S_X) as diversity measure and called $DGQPSO_X$, the other employ *diversity*(S_P) and named $DGQPSO_P$. The DGQPSO is outlined as follows

```
DGQPSO
Initialize particles with random position Xi=X[i][:]and
set the personal position of particles to Pi=Xi;
for t=1 to MAXITER
```

```
    Compute the mean best position C[:] by equation (6);
    β=(1.0-0.5)*(MAXITER-t)/MAXITER+0.5;
    Measure the diversity by (9) (DGQPSO_X)or (10)(DGQPSO_P)
    if (diversity<dlow)(in explosion mode)
        for j=1 to D
            P[g][:]=P[g][:]+γ*|A|*ε; X[g][:]=P[g][:];
            f(Pg)=f(P[g][:]); f(Xi)=f(Pg);
        endfor
    endif
    for i = 1 to swarm size M
        If f(Xi)<f(Pi) then Pi=Xi; Endif
        Find the Pg=arg min f(P[g][:]);
        for j=1 to D
            φ=rand(0,1); u=rand(0,1);
            p=φ*P[i][j]+(1-φ)*P[g][j];
            if (rand(0,1)>0.5)
                X[i][j]=p-β*abs(mbestd-xid)*log(1/u);
            Else
                X[i][j]=p+β*abs(mbestd-xid)*log(1/u);
            Endif
        Endfor
    Endfor
Endfor
```

4 Experiment Results and Discussion

We have tested the QPSO, DGQPSO as well as original PSO with inertia weight (called Standard PSO or SPSO) on four widely known benchmark functions that are used for testing the performance of different evolutionary optimization strategies. These functions are all minimization problems with minimum value zero. The four test functions are listed in Table 1. In all experiments, the initial range of the population in all cases also listed in Table 1 is asymmetry.

The population size is set to be 20 for all cases and we had 50 trial runs for every instance and recorded mean best fitness over 50 runs and the standard deviations. The maximum numbers of iterations are set as 1000, 1500 and 2000 corresponding to the dimensions 10, 20 and 30 for first there functions, respectively. The dimension of the last functions is 2 and the maximum number of iterations is 2000 for this function. In performance test of the SPSO, the inertia weight w is decreases linearly from 0.9 to 0.4 and V_{max} is set to be X_{max} as in [14] and [15]. In performance tests for QPSO and DGQPSO, the Contraction-Expansion Coefficient β varies from 1.0 to 0.5 linearly when the algorithms are running. For both DGQPSO, the threshold value (low bound) of the diversity measure d_{low} is to be 0.0001 and the value of coefficient γ is set to 0.00001. This configuration for DGQPSO seems to generate better results than other parameter settings as shown by our preliminary experiments. Moreover, the length of longest the diagonal in the search space can be computed by $|A| = (X_{\max} - X_{\min}) \cdot \sqrt{D}$.

Table 1. The four benchmark functions tested for performance comparison

	Function Expression	(X_{max}, X_{min})	Initial Range
Rosenbrok	$f_1(X)=\sum_{i=1}^{n-1}(100\cdot(x_{i+1}-x_i^2)^2+(x_i-1)^2)$	(-100, 100)	(15, 30)
Rastrigrin	$f_2(X)=\sum_{i=1}^{n}(x_i^2-10\cdot\cos(2\pi x_i)-10)$	(-10, 10)	(2.56, 5.12)
Greiwank	$f_3(X)=\frac{1}{4000}\sum_{i=1}^{n}(x_i-100)^2-\prod_{i=1}^{n}\cos\left(\frac{x_i-100}{\sqrt{i}}\right)+1$	(-600, 600)	(300, 600)
Shaffer's	$f_4(X)=0.5+\frac{(\sin(\sqrt{x_1^2+x_2^2})^2}{(1.0+0.001(x_1^2+x_2^2))^2}$	(-100, 100)	(30, 100)

The mean values and standard deviations of best fitness values for 50 runs for each case are recorded in Table 2 to Table 5. It can be seen that both $DGQPSO_X$ and $DGQPSO_P$ outperform SPSO and QSPO considerably on the first three functions. On Rosenbrock function, the mean best fitness values resulted from DGQPSO are less than those from QPSO by at least two times. When dimensionality is 30, among 50 runs, $DGQPSO_X$ and $DGQPSO_P$ can search out the best fitness values 2.41×10^{-5} and 2.57×10^{-5}, respectively, which are the optimal value of the function and never found out by SPSO and QPSO with the same parameter settings. On Rastrigrin function, the advantages of DGQPSO over QPSO and SPSO are also remarkable. When the dimensionality is 30, $DGQPSO_X$ and $DGQPSO_P$ can hit the best values 2.69×10^{-7} and 1.54×10^{-8} respectively among 50 such that QPSO and SPSO can never found in the experiments. On Greiwank function, both DGQPSO have more successful searches than the other two algorithms. On Shaffer's function, both DGQPSPO have the similar performance with original QPSO, which means that diversity controlling does not work on this function.

Table 2. Average best fitness and standard deviation of all algorithms on Rosenbrock function

	Dimension	Generation	Mean Best Fitness	St. Deviation
SPSO	10	1000	94.1276	194.3648
	20	1500	204.337	293.4544
	30	2000	313.734	547.2635
QPSO	10	1000	59.4764	153.0842
	20	1500	110.664	149.5483
	30	2000	147.609	210.3262
$DGQPSO_X$	10	1000	10.6058	24.0337
	20	1500	50.9962	48.1224
	30	2000	55.3429	55.3429
$DGQPSO_P$	10	1000	11.4846	18.9461
	20	1500	40.1360	34.5854
	30	2000	68.5157	91.4298

Table 3. Average best fitness and standard deviation of all algorithms on Rastrigrin function

	Dimension	Generation	Mean Best Fitness	St. Deviation
SPSO	10	1000	5.5382	3.0477
	20	1500	23.1544	10.4739
	30	2000	47.4168	17.1595
QPSO	10	1000	5.2543	2.8952
	20	1500	16.2673	5.9771
	30	2000	31.4576	7.6882
DGQPSO$_X$	10	1000	3.2979	3.5853
	20	1500	7.8673	8.3829
	30	2000	14.9209	12.9739
DGQPSO$_P$	10	1000	3.2503	4.1251
	20	1500	6.5896	6.9187
	30	2000	14.9766	13.1449

Table 4. Average best fitness and standard deviation of all algorithms on Greiwank function

	Dimension	Generation	Mean Best Fitness	St. Deviation
SPSO	10	1000	0.09217	0.0833
	20	1500	0.03002	0.03255
	30	2000	0.01811	0.02477
QPSO	10	1000	0.08331	0.06805
	20	1500	0.02033	0.02257
	30	2000	0.01119	0.01462
DGQPSO$_X$	10	1000	0.0703	0.0599
	20	1500	0.0141	0.0215
	30	2000	0.0049	0.0088
DGQPSO$_P$	10	1000	0.0693	0.0648
	20	1500	0.0140	0.0162
	30	2000	0.0056	0.0085

Table 5. Average best fitness and standard deviation of all algorithms on Shaffer's function

	Dimension	Generation	Mean Best Fitness	St. Deviation
SPSO	2	2000	2.78E-04	0.001284
QPSO	2	2000	0.001361	0.003405
DGQPSO1	2	2000	0.001360	0.003405
DGQPSO2	2	2000	0.001361	0.003405

5 Conclusion

In this paper, we proposed a Diversity-Guided QPSO (DGQPSO), in which low bound value is set for the diversity to prevent the particles from clustering. The diversity controlling in DGQPSO is realized by exerting mutation operation on global best particle once the diversity declines to below the low bound. This method can be used to control both forms of diversity. The results of experiments on benchmark functions show that DGQPSO outperforms SPSO and QPSO considerably. DGQPSO is also superior to the diversity control method for QPSO in [11], because the mutation operation on global best particle may guide the swarm into promising regions with better solutions.

References

1. Angeline, P.J.: Using Selection to Improve Particle Swarm Optimization. Proc. 1998 IEEE International Conference on Evolutionary Computation. Piscataway, NJ (1998) 84-89
2. Clerc, M.: The Swarm and Queen: Towards a Deterministic and Adaptive Particle Swarm Optimization. Proc. 1999 Congress on Evolutionary Computation. Piscataway, NJ (1999) 1951-1957
3. Clerc, M., Kennedy, J.: The Particle Swarm: Explosion, Stability, and Convergence in a Multi-dimensional Complex Space. IEEE Transactions on Evolutionary Computation, Vol. 6, No. 1. Piscataway, NJ (2002) 58-73
4. Kennedy, J., Eberhart, R.C.: Particle Swarm Optimization. Proc. IEEE 1995 International Conference on Neural Networks, IV. Piscataway, NJ (1995) 1942-1948
5. Kennedy, J.: Small worlds and Mega-minds: Effects of Neighborhood Topology on Particle Swarm Performance. Proc. 1999 Congress on Evolutionary Computation. Piscataway, NJ (1999) 1931-1938
6. Kennedy, J.: Bare Bones Particle Swarm. Proc. IEEE 2003 Swarm Intelligence Symposium, Indianapolis, IN (2003) 80-87
7. Suganthan, P.N.: Particle Swarm Optimizer with Neighborhood Operator. Proc. 1999 Congress on Evolutionary Computation, Piscataway, NJ (1999) 1958-1962
8. Sun, J., Feng, B., Xu, W.-B.: Particle Swarm Optimization with Particles Having Quantum Behavior. Proc. 2004 Congress on Evolutionary Computation, Piscataway, NJ (2004) 325-331
9. Sun, J., Xu, W.-B., Feng, B.: A Global Search Strategy of Quantum-behaved Particle Swarm Optimization. Proc. 2004 IEEE Conference on Cybernetics and Intelligent Systems, Singapore (2004) 111-115
10. Sun, J., Xu, W.-B., Feng, B.: Adaptive Parameter Control for Quantum-behaved Particle Swarm Optimization on Individual Level. Proc. 2005 IEEE International Conference on Systems, Man and Cybernetics. Piscataway, NJ (2005) 3049-3054
11. Sun, J., Xu, W.-B., Fang, W: Quantum-Behaved Particle Swarm Optimization Algorithm with Controlled Diversity. Proc. 2006 International Conference on Computational Science (3), (2006): 847-854
12. Ursem, R. K.: Diversity-Guided Evolutionary Algorithms, Proc. 2002 The Parallel Problem Solving from Nature Conference, (2001) 462-471
13. Riget, J, Vesterstrøm, J.S.: A Diversity-Guided Particle Swarm Optimizer-the ARPSO. Technical Report, University of Aarhus, Denmark (2002)
14. Shi, Y., Eberhart, R.: Empirical Study of Particle Swarm Optimization. Proc. 1999 Congress on Evolutionary Computation. Piscataway, NJ (1999) 1945-1950
15. Shi, Y., Eberhart, R.C.: A Modified Particle Swarm. Proc. 1998 IEEE International Conference on Evolutionary Computation. Piscataway, NJ (1998) 69-73

Multimodal Optimisation with Structured Populations and Local Environments

Grant Dick and Peter A. Whigham

Department of Information Science,
University of Otago,
Dunedin, New Zealand
gdick@infoscience.otago.ac.nz

Abstract. Spatially-structured evolutionary algorithms are frequently implemented using a homogeneous environment throughout space. Such a configuration does not promote local adaptation of individuals in space. This paper introduces an evolutionary algorithm using space and localised environments to promote speciation. Surprisingly, a randomly generated "rugged" landscape appears to best support speciation by encouraging crossover between niches, while maintaining locally distinct species.

1 Introduction

Multimodal fitness landscapes present a difficult challenge for traditional evolutionary algorithms (EAs) as they are typically unable to discover and maintain multiple optima during the course of a single run. This is due to two factors: genetic drift (in the case of equal-valued optima) will converge the population onto a single point in the fitness landscape and selection (in the case of unequal peaks) will adapt the population towards the fittest peak.

One analogy with discovering multiple optima is to view the evolutionary process as a form of speciation. Speciation models often rely on some form of breeding separation between subpopulations either spatially or through assortative mate preference. Spatial structure separates individuals into geographically isolated groups (*demes*); this increases inbreeding and the resultant changes in genetic drift promote local genetic divergence in the population [1]. In addition, a spatially structured population often presents resources (eg. food, shelter) that vary with location. Natural selection in space tends to encourage adaptation to these local *environments*, again encouraging local divergence of gene frequencies.

Spatially-structured evolutionary algorithms (SSEAs) are known to delay convergence of a population and increase overall genetic diversity [2]. This is primarily through the change in genetic drift that is brought about through local mating, but is also due in part to the fact that overall selection pressure in an SSEA is qualitatively similar but quantitatively less than an unstructured, panmictic EA [3]. The environment presented by a typical SSEA is typically homogeneous; the environment does not change with location and hence the second

T.-D. Wang et al. (Eds.): SEAL 2006, LNCS 4247, pp. 505–512, 2006.

property that spatial structures permit (local adaptation), is not present in an SSEA.

Previous work has explored the use of SSEAs with localised environments in the context of multiobjective optimisation [4,5]. This paper extends this work to the field of multimodal function optimisation. Each location in space is initialised with a vector of "ideal" phenotypic traits. The mismatch between this vector and the phenotype of an individual occupying the location is used to determine the occupant's *localised fitness.* Selection for parents is based around this localised fitness, which helps to promote local adaptation. Empirical testing of this method on several benchmark problems indicates that it is able to sustain multiple optima more effectively than existing spatially-structured techniques. In addition, the proposed model is robust and requires no additional problem-specific information than a typical evolutionary algorithm.

The remainder of this paper is structured as follows: §2 describes the SSEA used in this paper and a brief review on the success of SSEAs at searching multimodal fitness landscapes; the proposed environmental gradient for SSEAs is given in §3 and a comparison between this method and existing SSEAs is given in §4. Finally, a brief discussion of the findings of this paper and a suggested path for future work is presented in §5.

2 Niching with Spatially-Structured Populations

Previous research has developed many methods that extend traditional EAs and allow them to maintain multiple optima for sustained periods [6,7,8]. Collectively, these solutions are often referred to as *niching* EAs. This paper discusses one specific group of niching EAs called *spatially-structured evolutionary algorithms* (SSEAs). These methods impose a structure on the population of an EA and restrict selection and mating to geographically "close" individuals [2].

SSEAs are said to naturally promote speciation. Many studies have investigated the performance of SSEAs for multimodal optimisation, although usually in the context of searching for one globally optimal solution. Davidor [9] created the ECO GA and claimed that areas of homogeneity presented themselves in space through the local interactions promoted by the population structure. Davidor tested this framework on a multimodal problem with five optima of varying value. The ECO GA produces localised "niches" that remained for a period of time until the were eventually overwritten by the global optimum. The author claims that the niches remain for thousands of generations, however a steady state population was used in which only two individuals were created per generation. The niching ability of this SSEA therefore seems somewhat overstated. Davidor *et al.* [10] modified the ECO GA in later work and demonstrated better-than-benchmark performance on a job scheduling problem. However, the focus of this later work was not on niching and the ability of the modified ECO GA to maintain multiple solutions was not reported.

Others have suggested that spatially-structured populations are able to support multiple optima [11,3]. Collins and Jefferson [11] noted that their SSEA maintained homogeneous groups of optimal solutions on a bimodal graph partitioning problem and that regions of "lethal" (low fitness) individuals existed between these groups. The formation of lethals is not surprising as the two optima in this problem are bitwise compliments and hence crossover between peaks should produce low-fitness offspring with high probability. The ability of this SSEA to support more than two optimal species was not reported. Sarma [3] examined an SSEA's performance on two multimodal fitness landscapes each with five optimal phenotypes. In one landscape, the optima were of equal value, in the other they varied. Sarma noted that the SSEA always converged to the single global optima in the unequal-peak fitness landscape. However, she was able to support three peaks in the fitness landscape with equal value optima.

2.1 SSEA Implementation

This paper uses a two-dimensional population structure (a torus) as the basis of its SSEA. Selection is limited to demes, which are constructed using the Von-Neumann neighbourhood. The demes are overlapping in space, so any individual belongs to five demes. This allows propagation of good solutions throughout the population without the need for migration. Fitness proportionate selection is used to remain consistent with previous work. Two parents are selected (without replacement) from the deme and a single offspring is produced via one-point crossover. Elitist replacement is also used; offspring replace the current occupant of a location under the strict requirement that they are of greater localised fitness.

3 Localised Environments for SSEAs

An SSEA usually presents the same environment throughout space. However, evidence from theoretical population biology suggests that a series of varying local environments is more likely to promote speciation [12]. To this end, Murata *et al.* [4] implemented a multiobjective SSEA in which each location possessed a vector of weights corresponding to the emphasis to be placed on each objective. Certain locations placed a higher importance on one objective, while others treated each objective with equal importance. This encouraged certain areas to optimise specific objectives within the overall problem. A similar environmental gradient concept was used by Kirley [5] in his metapopulation evolutionary algorithm.

This paper introduces localised environmental conditions into an SSEA using three methods. The first method applies a linear gradient for each phenotypic trait. An example of such a gradient is shown in Figure 1(a). In the case of multiple phenotypic traits, each gradient is placed onto the population with a random offset. This is to increase the probability that a set of locations will exist that are reasonably close to the ideal conditions for optimal phenotypes in the fitness landscape. The second approach creates a gradient that covers the space radially (Figure 1(b)). As with the linear gradient approach, the gradient for each phenotypic trait is offset by a random amount so as to increase the likelihood of creating suitable habitats for optimal phenotypes. The final approach is to create a "rugged" landscape. Each location in space contains a randomly generated environment (Figure 1(c)). This method differs from the other two in that the environments presented at a given location are not highly correlated with those of its neighbours.

(a) Linear (b) Radial (c) Rugged

Fig. 1. The three varying environments implemented in this paper

Simply introducing a varying environment into the space of an SSEA is not enough to encourage local adaptation. A function relating the phenotype of an individual to the ideal phenotype for a local environment is needed. This paper uses a function similar to the Gaussian function proposed by Doebeli and Dieckmann [12]. The mismatch between the phenotype p and the environment p_0 is defined as:

$$K\left(p, p_0\right) = e^{-\frac{1}{2}\left(\frac{p-p_0}{\sigma}\right)^2} \tag{1}$$

where σ determines the severity of the environmental response. High values of σ negate the importance of the environment, while lower values encourage local adaptation.While this parameter can obviously be tuned on a per-problem basis, this paper uses a value of $\frac{1}{4}$ of the range of permissible phenotypes. This is in order to demonstrate the robustness of the proposed method. In the case of multiple phenotypic traits, the response of a phenotype to an environment is the mean value of Equation 1 for each trait.

The combination of environment, response function and phenotype can be used to determine an individual's local fitness. An individual with a phenotype p occupying a location with environment e has the localised fitness of:

$$f_{local}\left(p, e\right) = f\left(p\right) \cdot K(p, e) \tag{2}$$

where $f\left(p\right)$ is the fitness of an individual in the absence of the environment.

4 Empirical Analysis

The proposed environmental SSEA was compared with two SSEAs from previous work: the ECO GA [9,10] and an SSEA similar to that used by Collins and

Jefferson [11] and Sarma [3]. In the case of the ECO GA, both the original implementation and the later method were tested.

Each SSEA was tested on four benchmark problems taken from a well-known test suite [13]. The first two problems, *M1* and *M4*, are sinusoidal problems. *M1* has five equally spaced peaks, all of equal value. *M4* also has five peaks, however they are of decreasing value and the distance between neighbouring optima is not equal. These two problems are very simple, although most SSEAs have trouble maintaining all five peaks. The third problem, *M6* is the Shekel's foxholes problem. This has 25 unequal-value peaks in the two-dimensional fitness landscape. The final test problem, *M7*, is designed to be deliberately difficult for a niching EA. *M7* has 32 global, desirable optima and over five million undesired, deceptive optima. The reader is referred to the original work describing the test suite for more detailed information on these functions [13].

Population sizes for each problem were taken from previous work. In the case of *M1* and *M4* the population size was 225, taken from Sarma's study [3]. The population size for *M6* was 289 and was determined from Mahfoud's study on population sizing for niching methods [14]. Finally, the population size of *M7* was 676 taken from empirical studies in niching by Mahfoud [13].

Each SSEA used a similar set of parameters for crossover and mutation. Since each algorithm implemented elitism, full crossover (probability 1.0) was used. Bit-flipping mutation was applied at a per-locus level with probability 0.002. Reproduction of two parents created a single offspring, except in the case of the ECO GA, which creates two offspring per crossover. Also, the ECO GA used a steady state population. A generation for the ECO GA was considered to be the number of iterations required to create a number of offspring equal to the population size. This allowed for direct comparison to the other SSEAs.

4.1 Results

The performance of each SSEA on the test problems was measured through two statistics used in previous niching studies [15]. The first measurement records the number of peaks discovered and maintained by the population over 500 generations. The second measure is the Chi-Square-Like performance which measures the deviation of the population from that of an "ideal" population[1]. Higher values for the number of maintained niches indicates better algorithm performance, while lower values for the Chi-Square-Like measure indicate better performance (a value of zero being ideal). These measurements were averaged over 100 runs.

The number of peaks maintained over time is shown in Figure 2. All environmental SSEAs are able to support nearly all optima on the *M1* and *M4* problems while the "traditional" SSEAs quickly lose all but one optimum from the population. A similar trend is observed on the *M6* problem. In the case of the *M7* problem, all SSEAs (except the original ECO GA) support multiple optima, however the most peaks are supported by an SSEA using a rugged environment.

[1] A population in which all individuals reside on a peak in distributions relative to a peaks value.

The Chi-Square-Like performance of the various algorithms is shown in Figure 3. All three environmental SSEAs exhibit better distribution of individuals on peaks than the SSEAs from previous work.

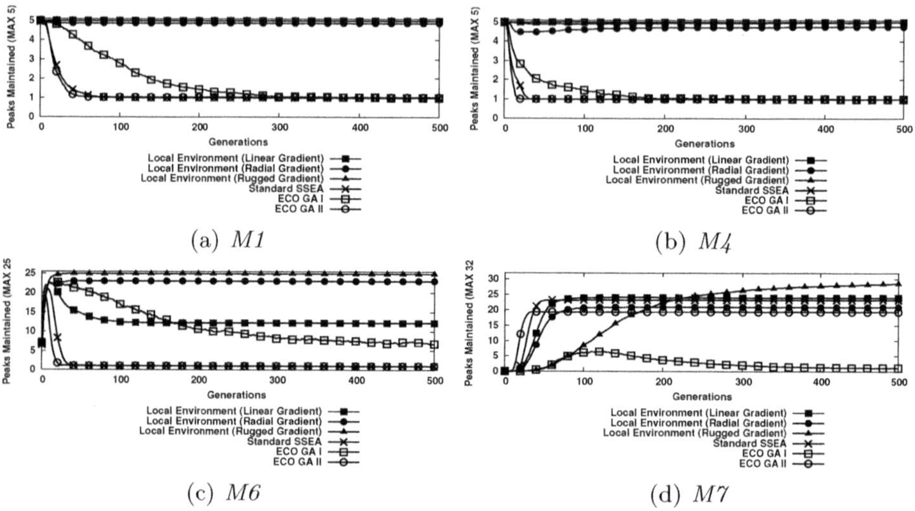

Fig. 2. Number of peaks maintained for each problem using the different SSEAs

Two statements can be made about the results presented here: the niching abilities of the traditional SSEAs presented here are supprisingly poor; conversely, SSEAs incorporating a localised environments are effective at maintaining multiple peaks of a multimodal fitness landscape. The behaviour of the SSEA using the rugged landscape is of particular interest, as most theories on speciation on a cline assume gradual environmental changes are required for local adaptation [12]. A possible explaination for this could be that a rugged environment increases the likelihood of mating between two distinct optima. This scenario would be particularly beneficial to the *M6* and *M7* problems, as crossover of two distinct optima in these problems will produce a third optima with high probability. The linear and radial environments promote breeding between genetically close individuals. This restricts the flow of genetic information between subpopulations and reduces the crossover interactions between niches.

If the above theory is true, then one would expect an SSEA using the rugged environment to exhibit lower levels of inbreeding than the other SSEAs. This is easily testable by calculating the inbreeding coefficient, F, of the population [11]. The inbreeding coefficient of SSEAs using the three environments is shown in Figure 4. In this figure, a higher value for F indicates a greater level of inbreeding (and hence intra-niche mating) and a value close to zero implies more frequent crossover between individuals from different niches. As can be seen, in both cases the rugged environment demonstrates less inbreeding than the other two environmental configurations.

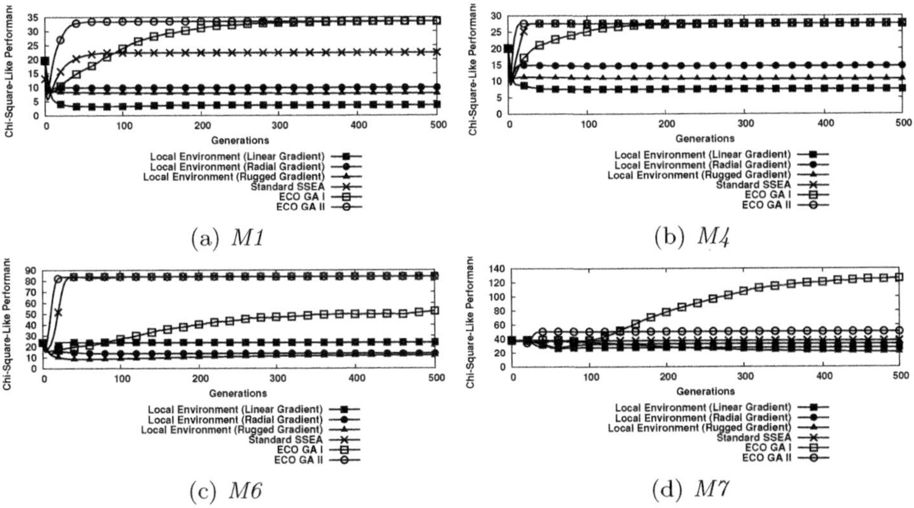

Fig. 3. Number of peaks maintained for each problem using the different SSEAs

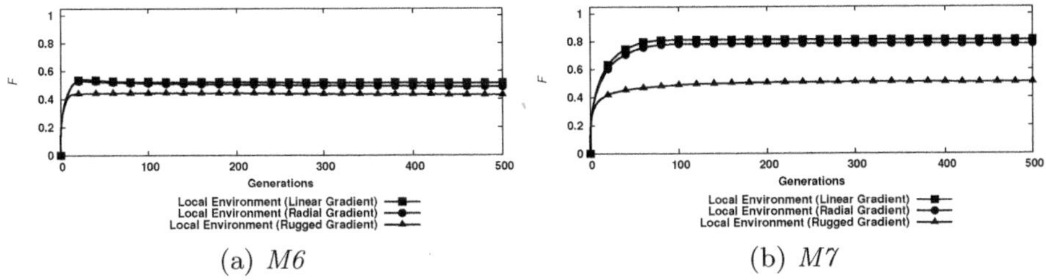

Fig. 4. Inbreeding coefficient over time for the three gradients on the *M6* and *M7* problems

5 Conclusion and Future Work

This paper presents an SSEA customised for use in searching multimodal fitness landscapes. It does this through the use of local adaptation via environments that vary with location. A suppising outcome of this work is that a randomly generated, rugged environment promotes and maintains the most speciation. This could be in part to the increase in crossover between niches that such a landscape promotes.

The proposed algorithm is an extension of existing SSEAs that is simple to implement, requires little computational overhead and requires no additional problem-specific information to operate. Future work in this field should explore the progression of environments from highly ordered to completely random. Doing so may discover a point of criticality that maximises the benefit of incorporating localised environments into SSEAs. In addition, the effect alternative response functions would have on the behaviour of environmental SSEAs needs to be considered in more detail.

References

1. Dick, G., Whigham, P.A.: The behaviour of genetic drift in a spatially-structured evolutionary algorithm. In Corne, D., Michalewicz, Z., McKay, B., Eiben, G., Fogel, D., Fonseca, C., Greenwood, G., Raidl, G., Tan, K.C., Zalzala, A., eds.: Proceedings of the 2005 IEEE Congress on Evolutionary Computation. Volume 2., Edinburgh, Scotland, UK, IEEE Press (2005) 1855–1860
2. Tomassini, M.: Spatially structured evolutionary algorithms. Springer (2005)
3. Sarma, J.: An Analysis of Decentralized and Spatially Distributed Genetic Algorithms. PhD thesis, George Mason University, Fairfax VA, USA (1998)
4. Murata, T., Ishibuchi, H., Gen, M.: Cellular genetic local search for multi-objective optimization. In Whitley, D., Goldberg, D., Cantu-Paz, E., Spector, L., Parmee, I., Beyer, H.G., eds.: Proceedings of the Genetic and Evolutionary Computation Conference (GECCO-2000), Las Vegas, Nevada, USA, Morgan Kaufmann (2000) 307–314
5. Kirley, M.: MEA: A metapopulation evolutionary algorithm for multi-objective optimisation problems. In: Proceedings of the 2001 IEEE Conference on Evolutionary Computation, Seoul, Korea, IEEE Press (2001)
6. Goldberg, D.E., Richardson, J.: Genetic algorithms with sharing for multi-modal function optimisation. In: Proc of the 2nd Int. Conf. on Genetic Algorithms and Their Applications. (1987) 41–49
7. Mahfoud, S.W.: Crowding and preselection revisited. In Männer, R., Manderick, B., eds.: Parallel problem solving from nature 2, Amsterdam, North-Holland (1992) 27–36
8. Pétrowski, A.: A clearing procedure as a niching method for genetic algorithms. In: Proceedings of the 1996 IEEE International Conference on Evolutionary Computation. (1996) 798–803
9. Davidor, Y.: A naturally occuring niche & species phenomenon: The model and first results. In Belew, R.K., Booker, L.B., eds.: Proceedings of the Fourth International Conference on Genetic Algorithms (ICGA'91), San Mateo, California, Morgan Kaufmann Publishers (1991) 257–263
10. Davidor, Y., Yamada, T., Nakano, R.: The ECOlogical framework II: Improving GA performance at virtually zero cost. In Forrest, S., ed.: Proc. of the Fifth Int. Conf. on Genetic Algorithms, San Mateo, CA, Morgan Kaufmann (1993) 171–176
11. Collins, R.J., Jefferson, D.R.: Selection in massively parallel genetic algorithms. In Belew, R.K., Booker, L.B., eds.: Proceedings of the Fourth International Conference on Genetic Algorithms, San Mateo, CA, Morgan Kaufmann Publishers (1991)
12. Doebeli, M., Dieckmann, U.: Speciation along environmental gradients. Nature **421**(6920) (2003) 259–264
13. Mahfoud, S.W.: A comparison of parallel and sequential niching methods. In Eshelman, L., ed.: Proceedings of the Sixth International Conference on Genetic Algorithms, San Francisco, CA, Morgan Kaufmann (1995) 136–143
14. Mahfoud, S.W.: Population size and genetic drift in fitness sharing. In Whitley, L.D., Vose, M.D., eds.: Foundations of genetic algorithms 3, San Francisco, Morgan Kaufmann (1995) 185–224
15. Deb, K., Goldberg, D.E.: An investigation of niche and species formation in genetic function optimization. In Schaffer, J.D., ed.: Proc. of the Third Int. Conf. on Genetic Algorithms, San Mateo, CA, Morgan Kaufmann (1989) 42–50

A Time Complexity Analysis of ACO for Linear Functions

Zhifeng Hao[1,2], Han Huang[3], Xili Zhang[1], and Kun Tu[1]

[1] College of Mathematical Science, South China University of Technology, Guangzhou 510640, P.R. China
[2] Nation Mobile Communications Research Laboratory Southeast University, Nanjing, 210096, China
[3] College of Computer Science and Engineering, South China University of Technology, Guangzhou 510640, P.R. China
hanhuang_scut@hotmail.com

Abstract. The time complexity analysis of ant colony optimization (ACO) is one of the open problems in ACO research. There has been little proposed work on this topic recently. In the present paper, two ACO algorithms (ACO I and ACO II) for linear functions with Boolean input are indicated, and their time complexity is estimated based on drift analysis which is a mathematical tool for analyzing evolutionary algorithms. It is proved that the algorithm ACO II can find the optimal solution with a polynomial time complexity. It is a preliminary work about estimating the time complexity of ACO, which should be improved in the future study.

1 Introduction

ACO was first proposed by M. Dorigo and his colleagues as a multi-agent approach to deal with difficult combinatorial optimization problems such as TSP [1]. Since then, a number of applications to the NP-hard problems have shown the effectiveness of ACO [1]. Up till now, Ant Colony System (ACS) [2] and MAX-MIN Ant System (MMAS) [3] are so successful and classical that their strategies such as pheromone global-local update and Maximum-Minimum of pheromone are widely used in recent research [1].

At present, the study on the speed of convergence is a hot topic [4]. As for the ACO's convergence, W. J. Gutjahr [4-6], T. Stützle [7], M. Dorigo[7-9], A. Fahmy [10], and S. Fidanova [11, 12] have done a lot of work, which only paid attention to the convergence of ACO. However, there have been few progresses in the study of ACO's time complexity. In the latest survey of ACO [9], the NO.1 open problem is proposed as follows:

Open problem 1. The proofs that were presented in this section do not say anything about the time required to find an optimal solution, which can be astronomically large. It would be interesting to obtain results on the convergence speed for ACO algorithms, in spirit similar to what has been done in evolutionary computation for relatively simple problems.

T.-D. Wang et al. (Eds.): SEAL 2006, LNCS 4247, pp. 513 – 520, 2006.

For this interesting topic, this paper introduces two ACO algorithms for the linear function [13] solved by (1+1)EA [14]. With drift analysis by J. He and X. Yao[15-19], we prove that the algorithms can find the optimal solution with the time complexity $O(n^k)$ and $O(n^2)$ respectively. According to the analysis of the algorithms, it is concluded that heuristic information is crucial to the performance of ACO in solving problems of linear functions.

2 Drift Analysis for Mean First Hitting Times

Drift analysis was applied for estimating the computation time of evolutionary algorithms by J. He and X. Yao [15, 20]. Let $\{\xi_t; t = 0,1,...\}$ be a Markov chain associated with an EA. Its first hitting time to the optimal set E_{opt}, or the number of generations for the EA to find an optimal solution first time, is defined by $\tau := \min\{t \geq 0; \xi_t \in E_{opt}\}$ [15]. The upper bound of the first hitting time can be estimated through the lower bound of the mean drift [21].

Lemma 1. Given a distance function $V(x)$, if $\{V(\xi_t); t = 0,1,2,...\}$ satisfies: for any time $t \geq 0$ and any population ξ_t with $V(\xi_t) > 0$,

$$E[V(\xi_t) - V(\xi_{t+1}) | \xi_t] \geq c_{low} \tag{1}$$

where $c_{low} > 0$, then the mean first hitting time satisfies

$$E[\tau | \xi_0] \leq \frac{V(\xi_0)}{c_{low}} \tag{2}$$

Based on this lemma, we analyze the time complexity of the ACO algorithms which are introduced in the following sections.

3 Time Complexity of ACO for Boolean Linear Function

Consider the following linear function [13]:

$$f(x) = w_0 + \sum_{i=1}^{n} w_i s_i \tag{3}$$

where $x = (s_1 ... s_n)$ is a binary string, weights $w_1 \geq w_2 \geq ... \geq w_n > 0$ and $w_0 \geq 0$. This function is maximal at (1,...,1). There are two ACO algorithms indicating in this section.

The framework of the ACO algorithms for this problem is:

```
Algorithm: ACO for linear function
Input: (w_1, w_2, ..., w_n)
begin
      Initialization;
      t = 0
   while termination conditions not met do
      Solution construction;
      Solution updating;
      Pheromone updating;
      t = t + 1;
   until meeting the condition to stop
end.
Output: The best-so-far solution
```

3.1 ACO I for Boolean Linear Function and Its Time Complexity

ACO I is a single-ant algorithm, which is designed in the following.

Solution construction:

A solution $x(t) = (s_1, s_2, ..., s_n)$ is constructed by the artificial ant based on the transition rule: $P^{(t)}\{s_1 = a\} = \frac{\tau_1^{(t)}(a)}{\sum_{j=0}^{1}\tau_1^{(t)}(j)}$ and $P^{(t)}\{s_i = a | s_{i-1} = b\} = \frac{\tau_i^{(t)}(b,a)}{\sum_{j=0}^{1}\tau_i^{(t)}(b,j)}$ ($i=2,...,n$), where $a=0,1$ and $b=0,1$.

Pheromone vector is defined as $T^{(t)} = (T_1^{(t)}, ..., T_n^{(t)})$ where $T_1^{(t)} = (\tau_1^{(t)}(0), \tau_1^{(t)}(1))$ and $T_i^{(t)} = (\tau_i^{(t)}(0,0), \tau_i^{(t)}(0,1), \tau_i^{(t)}(1,0), \tau_i^{(t)}(1,1))$.

Initialization:

$T_1^{(0)} = (\tau_1, \tau_1)$ and $T_i^{(0)} = (\tau_i, \tau_i, \tau_i, \tau_i)$ ($i=2,...,n$) where $\tau_i = \frac{w_i}{n-2}$ ($i = 1, ..., n$).

Solution updating:

If $f(r(t)) < f(x(t))$ then $r(t+1) = x(t)$, otherwise, $r(t+1) = r(t)$, where $r(t)$ is the current best solution from iteration 0 to t.

Pheromone Updating:

$$T_1^{(t+1)} = \left(\tau_1^{(t)}(0),\ \tau_1^{(t)}(1)\right) + \left(\Delta\tau_1^{(t)}(0),\ \Delta\tau_1^{(t)}(1)\right) \tag{4}$$

If $s_1 \in x(t)$ and $f(r'(t)) > f(r(t))$ then $\Delta\tau_1^{(t+1)}(s_1) = f(r'(t)) - f(r(t))$ where $r(t) = (r_1, r_2, ..., r_n)$ and $r'(t) = (s_1, r_2, ..., r_n)$, otherwise, $\Delta\tau_i^{(t+1)}(s_1) = 0$.

$$T_i^{(t+1)} = \begin{pmatrix} \tau_i^{(t)}(0,0) & \tau_i^{(t)}(0,1) \\ \tau_i^{(t)}(1,0) & \tau_i^{(t)}(1,1) \end{pmatrix} + \begin{pmatrix} \Delta\tau_i^{(t)}(0,0) & \Delta\tau_i^{(t)}(0,1) \\ \Delta\tau_i^{(t)}(1,0) & \Delta\tau_i^{(t)}(1,1) \end{pmatrix} \quad (5)$$

If $s_i \in x(t)$ and $f(r'(t)) > f(r(t))$ then $\Delta\tau_i^{(t)}(s_{i-1}, s_i) = f(r'(t)) - f(r(t))$ where $r(t) = (r_1, ..., r_{i-1}, r_i ..., r_n)$ and $r'(t) = (r_1, ..., r_i, s_i ..., r_n)$, otherwise, $\Delta\tau_i^{(t)}(s_{i-1}, s_i) = 0$. Let $\{\xi_t; t = 0,1,...\}$ be a Markov chain associated with an ACO, where $\xi_t = (x(t), T^{(t)})$. We denote $V(\xi_t) = V(x(t))$ because $V(\xi_t) = 0$ when $x(t) \in E_{opt}$.

Theorem 1. The worst-case time complexity of ACO I is $O(n^{l+2})$ when the distance of the initial solution found is l.

Proof. A distance function is denoted as $V(x(t)) = \sum_{i=1}^{n} |s_i - 1|$. So we have:

$$\begin{aligned} & E[V(r(t)) - V(r(t+1)) | r(t) = x] \\ &= E^+[V(r(t)) - V(r(t+1)) | r(t) = x] + E^-[V(r(t)) - V(r(t+1)) | r(t) = x] \\ &= (V(x) - \sum_{\{y:V(y)<V(x)\}} P(x,y;t)V(y)) + (V(x) - \sum_{\{y:V(y)>V(x)\}} P(x,y;t)V(y)) \end{aligned}$$

Similar to J. He and X. Yao's study [15], we consider the positive drift $E^+[V(r(t)) - V(r(t+1)) | r(t) = x]$. Given $V(r(t)) = l$, the events that k bits ($k = 1,...,l$) among zero-valued bits of $r(t)$ change into 1 and other bits keep unchanged after solution construction, will lead to a positive drift. There are $\binom{l}{k}$ kinds of the events denoted as Ω_i ($i = 1,...,\binom{l}{k}$) for $k = 1,...,l$.

$$E^+[V(r(t)) - V(r(t+1)) | r(t)] = \sum_{k=1}^{l} k [\sum_{i=1}^{\binom{l}{k}} q_{0,1}(k, \Omega_i) \cdot q_{0,0}(l-k, \Omega_i) \cdot q_{1,1}(n-l, \Omega_i)]$$

where $q_{a,b}(k, \Omega_i)$ is the probability that k bits among a-valued bits of $r(t)$ change into b value in the event Ω_i ($a, b = 0,1$).

Then we consider the negative drift. Only the following event can produce the negative drift $E^-[V(r(t)) - V(r(t+1)) | r(t) = x]$: k bits ($k = 1,..., \min(l, n-l-1)$) in zero-valued bits of $r(t)$ change into 1, $k+m$ bits ($m = 1,...,n-l-k$) of one-valued bits of $r(t)$ change into 0, and all other bits keep unchanged after solution construction of artificial ant. There are $\binom{n-l}{k+m}$ kinds of events that $k+m$ bits of one-valued bits of $r(t)$ change into 0, which are denoted as Θ_i ($i=1,\ldots,\binom{n-l}{k+m}$).

$$E^{-}[V(r(t))-V(r(t+1))\,|\,r(t)]$$

$$=-\sum_{k=1}^{l}\{\sum_{i=1}^{\binom{l}{k}}[q_{0,1}(k,\Omega_i)\cdot q_{0,0}(l-k,\Omega_i)\sum_{m=1}^{n-l-k}m\sum_{j=1}^{\binom{n-l}{k+m}}q_{1,0}(k+m,\Theta_j)\cdot q_{1,1}(n-l-k-m,\Theta_j)]\}$$

where $q_{a,b}(k+m,\Theta_j)$ is the probability that $k+m$ bits among a-valued bits of $r(t)$ change into b value in the event Θ_j ($a,b=0,1$).

$$E[V(r(t))-V(r(t+1))\,|\,r(t)]=D_1\cdot D_2 \tag{6}$$

where

$$D_1=\sum_{k=1}^{l}\sum_{i=1}^{\binom{l}{k}}q_{0,1}(k,\Omega_i)\cdot q_{0,0}(l-k,\Omega_i)\cdot q_{1,1}(n-l,\Omega_i) \tag{7}$$

and

$$D_2=k-\sum_{m=1}^{n-l-k}m\sum_{j=1}^{\binom{n-l}{k+m}}q_{1,0}(k+m,\Theta_j)\cdot\frac{1}{q_{1,1}(k+m,\Theta_j)} \tag{8}$$

According to the solution construction and updating rules, we have:

$q_{1,0}(k+m,\Theta_j)=\prod_{i=N_i}^{N_{k+m}}\frac{\tau_i}{\tau_i+\tau_i+w_i}$ and $q_{1,1}(k+m,\Theta_j)=\prod_{i=N_i}^{N_{k+m}}\frac{\tau_i+w_i}{\tau_i+\tau_i+w_i}$.

where $1\le N_i\le n$ ($i=1,\ldots,k+m$).

Therefore, $D_2\ge k-\sum_{m=1}^{n-l-k}m\binom{n-l}{k+m}\left(\frac{1}{n-1}\right)^{k+m}$ because $\tau_i=\frac{w_i}{n-2}$ ($i=1,\ldots,n$).

By the same mean,

$$D_1\ge\sum_{k=1}^{l}k\binom{l}{k}\left(\frac{1}{n}\right)^{l}\left(1-\frac{1}{n}\right)^{n-l}\ge l\left(\frac{1}{n}\right)^{l}\left(1-\frac{1}{n}\right)^{n} \tag{9}$$

$$\therefore E[V(r(t))-V(r(t+1))\,|\,r(t)]=D_1\cdot D_2$$

$$\geq l\left(\frac{1}{n}\right)^{l}\left(1-\frac{1}{n}\right)^{n}(1+\frac{1}{2(n-1)}-\sum_{m=2}^{+\infty}\frac{m}{(m+1)!})\geq\frac{l}{2(n-1)}\left(\frac{1}{n}\right)^{l}=c_{low}>0$$

$$\therefore E[\tau|\xi_0]\leq\frac{V(\xi_0)}{c_{low}}=\frac{2(n-1)}{l}n^{l+1}$$

Thus, the worst-case time complexity of ACO I is $O(n^{l+2})$, which depends on the value of the distance of the initial solution l. We design ACO II as an improvement in the following section.

3.2 ACO II for Boolean Linear Function and Its Time Complexity

ACO II is a single-ant algorithm with heuristic information, which is designed in the following.

Solution construction:

A solution $x(t)=(s_1,s_2,...,s_n)$ is constructed by the ant based on the transition rule:

$$P^{(t)}\{s_1=a\}=\frac{\tau_1^{(t)}(a)+\eta_1(a)}{\sum_{j=0}^{1}(\tau_1^{(t)}(j)+\eta_1(j))} \tag{10}$$

$$P^{(t)}\{s_i=a|s_{i-1}=b\}=\frac{\tau_i^{(t)}(b,a)+\eta_i(a)}{\sum_{j=0}^{1}(\tau_i^{(t)}(b,j)+\eta_i(j))} \tag{11}$$

where ($i=2,...,n$), and $a=0,1$ and $b=0,1$.

Pheromone vector is defined as the one of ACO I. Heuristic vector is defined as $G^{(t)}=(G_1^{(t)},...,G_n^{(t)})$ where $G_i=(\eta_i(0),\eta_i(1))=(0\times w_i,1\times w_i)$ ($i=1,...,n$).

The initialization, solution updating and pheromone updating of ACO II are the same as ACO I.

Theorem 2. The worst-case time complexity of ACO II is $O(n^2)$.

Proof. According to Exp. (4)-(7), we have:

$$D_1\geq\sum_{k=1}^{l}k\binom{l}{k}\left(\frac{n-1}{n}\right)^{k}\left(\frac{1}{n}\right)^{l-k}\left(1-\frac{1}{2n}\right)^{n-l}\geq l\left(\frac{n-1}{n}\right)^{l}\left(1-\frac{1}{n}\right)^{n-l}=l\left(1-\frac{1}{n}\right)^{n}$$

$$D_2\geq k-\sum_{m=1}^{n-l-k}m\binom{n-l}{k+m}\left(\frac{1}{2n-1}\right)^{k+m}\geq k-\sum_{m=1}^{n-l-k}m\binom{n-l}{k+m}\left(\frac{1}{n-1}\right)^{k+m}>0$$

$$\therefore E[V(r(t))-V(r(t+1))|r(t)]$$

$$= D_1 \cdot D_2 \geq l\left(1-\frac{1}{n}\right)^n (1+\frac{1}{2(n-1)}-\sum_{m=2}^{+\infty}\frac{m}{(m+l)!}) \geq \frac{l}{2(n-1)}$$

$$\therefore E[\tau|\xi_0] \leq \frac{V(\xi_0)}{c_{low}} = \frac{2n(n-1)}{l}$$

Therefore, heuristic information is crucial to the performance of ACO in solving problems of linear functions, which accords with our intuitive judgment.

4 Discussions and Conclusions

This paper proposes a preliminary work about estimating the time complexity of ACO which is one of the most popular evolutionary computation methods. Two single-ant ACO algorithms are designed specifically for linear functions with Boolean input. ACO II is the algorithm of ACO I with heuristic information, their time complexity is analyzed with drift analysis, which is the mathematical tool applied to the computational time analysis of evolutionary algorithm. Through the drift analysis, some conclusions can be drawn as follows:

First, the ACO I and ACO II can find the optimal solution with the time complexity $O(n^k)$ and $O(n^2)$ respectively. Second, heuristic information is crucial to the performance of ACO in solving problems of linear functions, which prompts ACO II to perform better than ACO I. Third, besides evolutionary algorithms, drift analysis can be used for studying ACO algorithms for linear functions.

Future study is suggested to analyze ACO and EA for other problems, such as ONE-MAX problem, TSP problem and so forth with drift analysis. Furthermore, new mathematical tools for estimating the computational time of ACO and EA should be studied because the encoding of the optimal solution is a necessity to drift analysis.

Acknowledgements

This work has been supported by the National Natural Science Foundation of China (60433020, 10471045), Program for New Century Excellent Talents in University (NCET-05-0734), Social Science Research Foundation of MOE (2005-241), Natural Science Foundation of Guangdong Province (031360, 04020079), Excellent Young Teachers Program of Ministry of Education of China, Fok Ying Tong Education Foundation (91005), Guangdong Key Laboratory of Computer Network Foundation (CN200403), Key Technology Research and Development Program of Guangdong Province (2005B10101010), Key Technology Research and Development Program of Tianhe District (051G041) , State Key Lab. for Novel Software Technology, Nanjing University (200603), open research fund of National Mobile Communications Research Laboratory, Southeast University (A200605) and Natural Science Foundation of South China University of Technology (E512199, D76010).

We would like to extend our thanks to Mr. Ruichu Cai and other colleagues in the seminar of intelligent algorithm, who helped offered a lot of constructive suggestions.

References

1. Dorigo, M., Caro, G.D., Gambardella, L.M.: Ant algorithms for Discrete Optimization. Massachusetts Institute of Technology, Artificial Life 5: (1999) 137-172
2. Dorigo, M., Gambardella, L.M.: Ant Colony System: A Cooperative Learning Approach to the Traveling Salesman Problem. IEEE Transactions on Evolutionary Computation, 1 (1), (1997) 53-66
3. Stützle, T., Hoos, H.H.: MAX-MIN ant system, Future Gener. Comput. Syst., Vol.16, No. 8, (2000) 889–914
4. Gutjahr, W.J.: Ageneralized convergence result for the graph-based ant system metaheuristic. Tech. Report 99-09, Department of Statistics and Decision Support Systems, University of Vienna, Austria (1999)
5. Gutjahr, W.J.: Agraph-based ant system and its convergence. Future Gen. Comput. Systems 16 (9), (2000) 873–888
6. Gutjahr, W. J.: ACO algorithms with guaranteed convergence to the optimal solution. Information Processing Letters 82, (2002) 145–153
7. Stützle, T., Dorigo, M.: A Short Convergence Proof for a Class of Ant Colony Optimization Algorithms. IEEE Transactions on Evolutionary Computation, 6 (4), (2002) 358-365
8. Dorigo, M., Stützle, T.: Ant Colony Optimization. MIT Press, Cambridge, MA (2004)
9. Dorigo, M., Blum, C.: Ant colony optimization theory: A survey. Theoretical Computer Science, 344, 243–278, (2005)
10. Badr, A., Fahmy, A.: A proof of convergence for Ant algorithms. Information Sciences 160, (2004) 267 –279
11. Fidanova, S.: ACO Algorithm with Additional Reinforcement. M. Dorigo et al. (Eds.): ANTS 2002, LNCS 2463, (2002) 292-293
12. Fidanova, S.: Convergence Proof for a Monte Carlo Method for Combinatorial Optimization Problems. M. Bubak et al. (Eds.): ICCS 2004, LNCS 3039, (2004) 523-530
13. Droste, S., Jansen T., I. Wegener: A rigorous complexity analysis of the (1+1)-evolutionary algorithms. Evolutionary Computation 6(2), (1998) 185-196
14. Droste, S., Jansen T., Wegener I.: On the analysis of the (1+1)-evolutionary algorithms. Theoretical Computer Science 276(1-2), (2002) 51-81
15. He, J., Yao, X.: A study of drift analysis for estimating computation time of evolutionary algorithms. Natural Computing, 3, (2004) 21–35
16. He, J., Yao, X.: From an individual to a population: An analysis of the first hitting time of population-based evolutionary algorithms. IEEE Transactions on Evolutionary Computation. 6 (5) (2002) 495–511
17. He, J., Yao, X.: Erratum to: Drift analysis and average time complexity of evolutionary algorithms. Artificial Intelligence 127 (2001) 57–85, Artificial Intelligence 140 (1) (2002) 245–248
18. He, J., Huang, H.: The computational time analysis of genetic algorithms, in: Proc. Fifth Chinese Joint Conference on Artificial Intelligence. Xi'an Jiaotong University Press, Xi'an, (1998) 440–443
19. He, J., Huang, H., Kang, L.: The computational time of genetic algorithms for fully deceptive problem. Chinese J. Comput. 21 (9) (1999) 999–1003
20. He, J., Yao, X., Li, J.: A Comparative Study of Three Evolutionary Algorithms Incorporating Different Amounts of Domain Knowledge for Node Covering Problem. IEEE Transactions on systems, man, and cybernetics—Part C: Applications and Reviews, Vol. 35, NO. 2, MAY (2005)
21. He, J., Yao, X.: Towards an analytic framework for analyzing the computation time of evolutionary algorithms. Artificial Intelligence, 145(1-2), (2003) 59-97

Particle Swarm Optimization Based on Information Diffusion and Clonal Selection*

Yanping Lv[1], Shaozi Li[1], Shuili Chen[2], Qingshan Jiang [1], and Wenzhong Guo[3]

[1] Intelligent Multimedia Technology Lab, Department of Computer Science, Xiamen University, Xiamen, Fujian 3610005, China
catlet.lyp@gmail.com, szlig@xmu.edu.cn, QJiang@xmu.edu.cn
[2] Department of Mathematics, Jimei University, Xiamen, Fujian 361021, China
sgzx@jmu.edu.cn
[3] Department of Computer Science, Fuzhou University, Fuzhou, Fujian 350001, China

Abstract. A novel PSO algorithm called InformPSO is introduced in this paper. The premature convergence problem is a deficiency of PSOs. First, we analyze the causes of premature convergence for conventional PSO. Second, the principles of information diffusion and clonal selection are incorporated into the proposed PSO algorithm to achieve a better diversity and break away from local optima. Finally, when compared with several other PSO variants, it yields better performance on optimization of unimodal and multimodal benchmark functions.

1 Introduction

Particle swarm optimization is one of the evolutionary computation techniques based on swarm intelligence. In PSO, each solution is a point in the search space and may be regarded as a particle. The particle could find a global optimum through its own efforts and social cooperation with the other particles around it. Each particle has a fitness value and a velocity. The particles fly through the problem space by learning from the best experiences of all the particles. Therefore, the particles have a tendency to fly towards better search area over the course of search process.

The velocity and position updates of the ith particle are as follows:

$$V_i(t+1) = w \cdot V_i(t) + c1 \cdot r1 \cdot (pBest_i - X_i) + c2 \cdot r2 \cdot (gBest - X_i) \quad (1)$$

$$X_i(t+1) = X_i(t) + V_i(t+1) \quad (2)$$

where X_i is the position of the ith particle, V_i presents its velocity, $pBest_i$ is the best previous position yielding the best fitness value for it. *gBest* is the best position discovered by the whole population and *w* is the inertia weight used to balance between the global and local search abilities. *c1* and *c2* are the acceleration constants, which

* This paper is supported by the National Natural Science Foundation (10471083) and the Natural Science Fund, Science & Technology Project of Fujian Province (A0310009, A0510023, 2001J005, Z0511008), the 985 Innovation Project on Information Technique of Xiamen University (2004-2007), Academician Fund of Xiamen University, China.

T.-D. Wang et al. (Eds.): SEAL 2006, LNCS 4247, pp. 521 – 528, 2006.

represent the weighting of stochastic acceleration terms that pull each particle towards *pbest* and *gbest* positions. *r*1 and *r*2 are two random functions in the range [0, 1].

The PSO algorithm is simple in concept, easy to implement and computationally efficient. Since its introduction in 1995 [1], PSO has attracted a lot of attention. Many researchers have worked on improving its performance in various ways. Some researchers investigated the influence of parameters in the PSO to improve its performance. A new inertia weight parameter is incorporated into the original PSO algorithms by Shi [2]. Another parameter called constriction coefficient is introduced to ensure the convergence of the PSO [3]. In [4], Ratnaweera et al. introduced to PSO the time varying acceleration coefficients in addition to the time varying inertia weight.

Many researchers have worked on improving PSO's performance by designing different types of topologies. Kennedy [5] claimed that PSO with a small neighborhood might perform better on complex problems, vice versa. Kennedy [6] tested PSOs with neighborhoods of regular shape. In [7], a dynamic neighborhood concept for their multi-objective PSO is proposed. FDR-PSO [8] with near neighbor interactions selects one particle with a higher fitness value to be used in the velocity updating equation.

Some researchers investigated hybridization by combining PSO with other evolutionary to improve the standard PSO's performance. Evolutionary operators like selection, crossover and mutation have been applied to PSO in [9]. In [10], each of the PSO, genetic algorithm and hill-climbing search algorithm was applied to a different subgroup and an individual can dynamically be assigned to a subgroup considering its recent search progress. In [11], CPSO employed cooperative behavior to significantly improve the performance of the original PSO algorithm by splitting the solution variables into a smaller number of variables in the search space.

The paper is organized as follows. Section2 analyses causes of premature convergence of traditional PSO algorithm and proposes an improved PSO algorithm. Results of experimental evaluation are given in Section 3, which contains the description of benchmark continuous optimization problems used for comparison of algorithms, the experimental setting for each algorithm, and discussions about the results. Section 4 gives conclusions and future work.

2 PSO Based on Information Diffusion and Clonal Selection

2.1 Premature Convergence

Though there are numerous versions of PSO, premature convergence when solving multimodal problems is still the main deficiency of the PSO. In the original PSO, each particle learns from its *pBest* and *gBest* simultaneously. Restrict the same social cognition aspect to all particles in the swarm appears to be somewhat an arbitrary decision. Furthermore, each particle obtains the same information from the *gBest* with others even if a particle is far from the *gBest*. In such situations, particles may be fleetly attracted and easily trapped into a local optimum if the *gBest* is in a complex search environment with numerous local solutions. Another cause of premature convergence of PSO is that the *pBest* and *gBest* have no contributions to the *gBest* from the velocity update equation. The current best particle in the original PSO always flies in its direction of previous velocity, which makes it easy to trap into a local optimum and unable to break away from it.

2.2 PSO Based on Information Diffusion and Clonal Selection (InformPSO)

In fact, information diffusion among biological particles is a time process. Particles, close to the current best particle (*gBest*), change the direction and rate of velocities fleetly towards it, while particles, far from it, move more slowly towards it. On the assumption that information is diffused among particles in a short time, information received by particles close to the *gBest* is more than that received by those far from it. Therefore, an information function , related to membership degrees with respect to its "surrounding", is incorporated into the PSO to adjust the variable "social cognition" aspect of particles.

In this improved version of PSO, the velocity update is expressed as follows:

$$V_i(t+1) = w \cdot V_i(t) + c1 \cdot r1 \cdot (pBest_i - X_i) + F\{\mu(d_i)\} \cdot c2 \cdot r2 \cdot (gBest - X_i) \quad (3)$$

where, $F\{\mu(d_i)\}$ is an information diffusion function, $\mu(d_i)$ represents the membership degree of ith particle with respect to the "surrounding" of the *gBest*, d_i is the distance between particle i and the *gBest.* Here, the distance is measured by their position difference. By inspecting the expression in (3), we understand that particles perform variable-wise velocity update to the *gBest.* This operation improves the local search ability of PSO, increases the particles' diversity and enables the swarm to overcome premature convergence problem.

In order to pull the *gBest* to another direction if it is trapped in local optima, we use clonal selection operation on it. In multimodal problems, the *gBest* is often a local optimum, which may give other particles wrong information and lead to bad results. In our algorithm, we use a new method for the *gBest* to move out of local optima. First, the *gBest* is clonald [13] into a sub-swarm, then this sub-swarm mutates into a new one with different fitness values according to Cauchy distribution, finally the one with the highest fitness value is chosen as the *gBest* for velocity update of next generation. As a result, the *gBest* is improved in a local region by clonal selection operation, which enables the PSO algorithm effectively to break away from local optima.

3 Experimental Results

3.1 Benchmark Functions

In the experiments, nine different D dimensional benchmark functions [8] [14] [15] with different properties are chosen to test InformPSO's performance. The equations are listed below:

Table 1. Nine benchmark functions

Name	Function	Search Range	x^*	$f(x^*)$
Sphere	$f1 = \sum^{D} x_i^2$	[-100, 100]	0	[0,0,…0]
Hyper-ellipsoid	$f2 = \sum_{i=1}^{D} i \cdot x_i^2$	[-5.12, 5.12]	0	[0,0,…0]

Table 1. (*continued*)

Sum of different powers	$f_3 = \sum_{i=1}^{D} \lvert x_i \rvert^{i+1}$	[-1, 1]	0	[1,1,…1]
Rotated hyper-ellipsoid	$f_4 = \sum_{i=1}^{D} \left[\sum_{j=1}^{i} x_j \right]^2$	[-65, 65]	0	[0,0,…0]
Rosenbrock	$f_5 = \sum_{i=1}^{D} [100 \cdot (x_{i+1} - x_i^2)^2 + (x_i - 1)^2)]$	[-2.048, 2.048]	0	[1,1,…1]
Griewank	$f_6 = \frac{1}{4000} \sum_{i=1}^{D} x_i^2 - \prod_{i=1}^{n} \cos(\frac{x_i}{\sqrt{i}}) + 1$	[-600, 600]	0	[0,0,…0]
Ackley	$f_7 = 20 + e - 20\exp(-0.2\sqrt{\frac{1}{D}\sum_{i=1}^{D} x_i^2}) - \exp[\frac{1}{D}\sum_{i=1}^{D} \cos(2\pi x_i)]$	[-32.768, 32.768]	0	[0,0,…0]
Rastrigin	$f_8 = \sum_{i=1}^{D} [x_i^2 - 10 \cdot \cos(2\pi x_i) + 10]$	[-5.12, 5.12]	0	[0,0,…0]
Weierstrass	$f_9 = \sum_{i=1}^{D} (\sum_{k=0}^{k_{max}} [a^k \cdot \cos(2\pi b^k (x_i + 0.5))]) - D \sum_{k=0}^{k_{max}} [a^k \cdot \cos(2\pi b^k \cdot 0.5)]$ $a = 0.5, b = 3, k_{max} = 20$	[-0.5, 0.5]	0	[0,0,…0]

3.2 Parameters Setting

To optimize these test functions, the error goal is set at 1e-40 and w is reducing with increasing generations from 0.9 to 0.4. $c1$ and $c2$ are both set at 2. Particles' initial positions are restricted by the search range and their velocities are restricted by V_{max}, which is equal to the search range. The number of generations, for which each algorithm is run, is set at *Max_Gen*. Except these common parameters used in PSOs, there

Table 2. The same parameters setting used for InformPSO, PSO_w with that used for traditional PSO, FDR_PSO [8]. *Initial Range*, variable range in biased initial particles; V_{max}, the max velocity; *Size*, the number of particles; *Max_Gen*, the max generation; *Dim*, the number of dimensions of functions.

Function	*Initial Range*	V_{max}	*Max_Gen*	*Size*	*Dim*
f1	[-5.12 ,5.12]	10	1000	10	20
f2	[-5.12, 5.12]	10	1000	10	20
f3	[-1, 1]	2	1000	10	10
f4	[-65.536, 65.536]	110	1000	10	10
f5	[-2.048, 2.048]	4	1500	10	2
f6	[-600, 600]	600	1000	10	10

Table 3. The same parameters setting used for InformPSO and PSO_w on four multimodal functions with that used for PSO_cf_local, UPSO, PSO_H, DMS_PSO [15] except for the parameter, *Max_Gen*

Function	*Initial Range*	V_{max}	*Max_Gen*	*Size*	*Dim*
f6	[-600, 600]	600	2000	30	30
f7	[-32.768, 32.768]	64	2000	30	30
f8	[-5.12, 5.12]	10	2000	30	30
f9	[-0.5, 0.5]	1	2000	30	10

is an additional parameter in InformPSO that needs to be specified. It is sub-swarm's population size, *m*. Suppose not to know if the function to be optimized is unimodal or multimodal, *m* is set at 10 for all functions.

3.3 Comparison with Other PSOs

In this part, we tested the same nine benchmark test functions using PSO_w. Parameters are the same setting with parameters setting for InformPSO in Table 2 and Table 3. Other interesting variations of the PSO algorithm (described below) have recently been proposed by researchers. Although we have not implemented all these algorithms, we conducted comparison with them using the results reported in the publications cited below:

- The original PSO [1].
- The modified PSO with inertia weight (PSO_w) [2].
- Local Version of PSO with constriction factors (PSO_cf_local) [7].
- Unified Particle Swarm Optimization (UPSO) [16].
- Fitness-distance-ratio based particle swarm optimization (FDR_PSO) [8].
- Cooperative PSO (CPSO_H) [11].
- Dynamic multi-swarm PSO (DMS_PSO) [15].
- Our improved PSO (InformPSO).

Figure1 presents the results of InformPSO and PSO_w on the first six optimization functions introduced in the previous section. The two algorithms were continuously performed for 30 trials on each function. The best fitness values of each generation have been plotted in graphs.

From Figure1 and graphs displayed in the literature [8], we see that InformPSO surpasses the three PSOs (PSO_w, traditional PSO, FDR_PSO) on each of the six functions. InformPSO has a significantly better global convergence ability. It achieves a sub-optima solution within 200 iterations for the first four unimodal functions. For Rosenbrock's and Griewank's functions, it gets global minima on most of trials. In each case, we find that the other three algorithms perform well in initial iterations, but particles easily reach the same fitness, consequently move into a local optimum. Furthermore, they fail to make further progress in later iterations, especially in multimodal problems. This does not happen in InformPSO as shown in all the graphs above, where the best fitness continues to differ for many iterations. InformPSO is able to move out of local optima in later iterations even being trapped in it. This effect was most marked for the last graph in Figure 1.

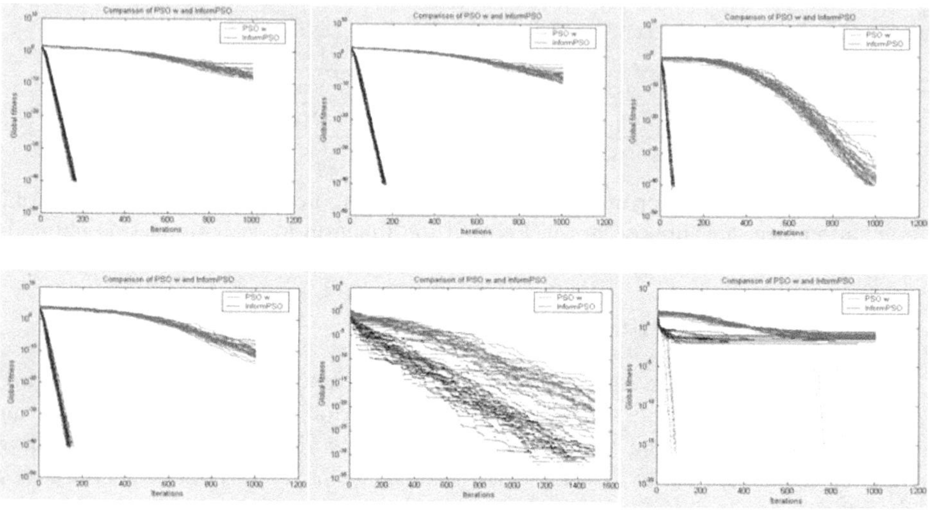

Fig. 1. InformPSO vs. PSO_w on f1-f6 functions

Table 4. The best and average results achieved on the first six test functions using traditional PSO and FDR_PSO [8], PSO_w and InformPSO

Algorithm Function	Best (Average) Fitness Values Achieved					
	Trad_PSO	FDR_PSO	PSO_w		InformPSO	
f1	0.0156	2.2415e-006	1.2102e-10	5.9724e-006	3.9843e-041	7.2610e-041
f2	2.1546e-007	1.8740e-015	1.7493e-009	8.0364e-005	2.3428e-041	6.9662e-041
f3	2.3626e-011	9.3621e-034	5.1684e-041	2.8907e-022	1.2685e-042	4.0353e-041
f4	0.0038	3.7622e-007	1.401e-014	1.9369e-008	2.8450e-041	6.6157e-041
f5	5.3273e-008	4.0697e-012	1.9722e-029	2.5544e-017	0	1.7027e-026
f6	1.3217e-008	1.1842e-016	0.0418	0.1152	0	8.2867e-003

Table 4 also shows that InformPSO yields better results for the six test functions than the other three PSOs. On average, InformPSO achieves the best and average fitness values in each case. It is able to converge to global optima on unimodal problems. It can find global optima or sub-optima for Rosenbrock's function and multi-modal Griewank's function. As an example in point, it attains to the global minimum of Rosenbrock's function for 20 out of 30 trials and that of Griewank' function for 25 out of 30 trials.

Figure2 gives the results of InformPSO and PSO_w on four multimodal benchmark functions. The two algorithms were performed for 20 trials on each multimodal function. The best fitness values of each generation have been plotted in graphs.

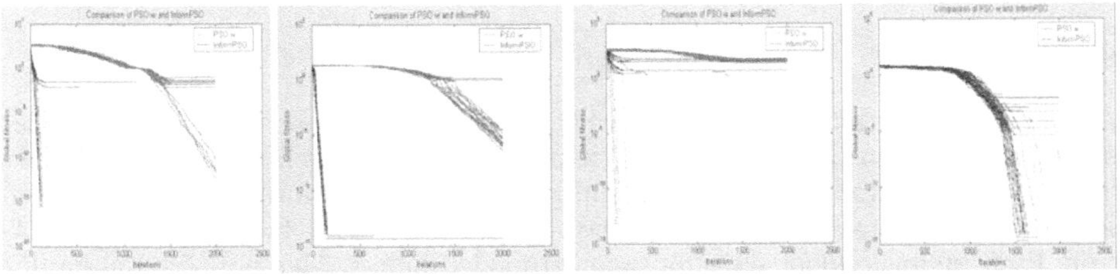

Fig. 2. InformPSO vs. PSO_w on the four multimodal functions

From graphs above, we observe that InformPSO yields better results for the four multimodal functions than PSO_w. Our proposed PSO algorithm rapidly finds global minima for three out of four multimodal functions except for Ackley's function. On multimodal functions, PSO_w rapidly loses the population diversity, and easily converge to a local optimum. And it is unable to improve the best fitness in later iterations. The population diversity of InformPSO results from its use of an information diffusion function in velocity update. The clonal selection operation contributes the best fitness to differing for many iterations. Thus, InformPSO is much less likely than the three PSO_w to get stuck in a local optimum and more effectively breaks away from a local optimum if being trapped in it. But our algorithm only attains a local solution to Ackley's function. PSO_w needs much more iterations to reach the same local solution. It seems that the two PSOs are unable to find the global optimum region for Ackley's function.

Table 5. The best and average results achieved by different PSOs [16]

<table>
<tr><td>Function</td><td colspan="8">Best (Average) Fitness Values Achieved</td></tr>
<tr><td>Algorithm</td><td colspan="2">f6</td><td colspan="2">f7</td><td colspan="2">f8</td><td colspan="2">f9</td></tr>
<tr><td>PSO_w</td><td colspan="2">4.308e-013
2.262e-002</td><td colspan="2">7.268e-007
3.700e-001</td><td colspan="2">26.824
43.828</td><td colspan="2">0
1.291e-002</td></tr>
<tr><td>PSO_cf_local</td><td colspan="2">4.971e-002</td><td colspan="2">9.763e-008</td><td colspan="2">9.769</td><td colspan="2">1.812e-001</td></tr>
<tr><td>UPSO</td><td colspan="2">3.453e-002</td><td colspan="2">3.379e-009</td><td colspan="2">14.231</td><td colspan="2">2.226</td></tr>
<tr><td>CPSO_H</td><td colspan="2">1.527e-001</td><td colspan="2">1.715e+003</td><td colspan="2">21.392</td><td colspan="2">4.208</td></tr>
<tr><td>DMSPSO</td><td colspan="2">2.338e-002</td><td colspan="2">4.062e-009</td><td colspan="2">3.277</td><td colspan="2">0</td></tr>
<tr><td>InformPSO</td><td>0</td><td>0</td><td>4.441e-015</td><td>4.441e-015</td><td>0</td><td>7.661</td><td>0</td><td>1.806e-004</td></tr>
</table>

From the results, we can observe that among the six PSO algorithms, InformPSO performs the best results for Griewank's (f6) and Ackley's (f7) multimodal functions. Though when compared with DMSPSO, InformPSO achieves worse results for Rastrigin's (f8) and Weierstrass's (f9) function on average, it reaches the global minimum for 15 out of 20 trials on Rastrigin's function and for 19 out of 20 trials on Weierstrass's function. However, on Ackley's function, InformPSO and PSO_w fail to arrive in the global minimum region. These comparisons suggest that InformPSO surpasses many of the recent improvements of the PSO algorithm.

4 Conclusions

A novel PSO algorithm called InformPSO and one of its applications are introduced and discussed in this paper. In order to improve the local search ability and achieve a better diversity, information diffusion function is given to InformPSO. Particles perform variable-wise velocity update to the current best particle. In order to break away from local optima, clonal selection is incorporated into it. This new PSO algorithm gives better performance on unimodal and on complex multimodal problems when compared with other PSO variants.

References

[1] RC Eberhart, J Kennedy, "A new optimizer using particle swarm theory", *P. 6th on Micromachine and Human Science,* Japan, pp. 39-43, 1995.

[2] Y Shi, RC Eberhart, "A modified particle swarm optimizer", *CEC*, pp. 69-73, 1998.

[3] M Clerc, J Kennedy, "The particle swarm-explosion, stability, and convergence in a multidimensional complex space", *TEC*, Vol. 6, pp. 58-73, 2002.

[4] A Ratnaweera, S Halgamuge, "Self-organizing hierarchical particle swarm optimizer with time varying accelerating coefficients", *TEC*, Vol. 8, pp. 240 - 255, 2004.

[5] J Kennedy, "Small worlds and mega-minds: effects of neighborhood topology on particle swarm performance", *CEC*, pp. 1931-1938, 1999.

[6] J Kennedy, R Mendes, "Population structure and particle swarm performance", CEC'02, USA.

[7] X Hu, RC Eberhart, "Multiobjective optimization using dynamic neighborhood particle swarm optimization", CEC'02, pp. 1677-1681.

[8] T Peram, K Veeramachaneni, "Fitness-distance-ratio based particle swarm optimization'', *P. IEEE Swarm Intelligence Symp.,* USA, pp. 174-181, 2003.

[9] M Lovbjerg, TK Rasmussen, "Hybrid particle swarm optimizer with breeding and subpopulations", *P. Genetic Evol. Comput. Conj (GECCO)*, 2001.

[10] T Krink, M Lovbjerg, "The life cycle model: combining particle swarm optimization, genetic algorithms and hill Climbers", *P. Parallel Problem Solving from Nature VII*, pp. 621-630, 2002.

[11] F Bergh, AP Engelbrecht, "A cooperative approach to particle swarm optimization", *TEC*, Vol. 8, pp. 225 - 239, 2004.

[12] Huang Chongfu, "Principle of information diffusion", *Fuzzy Sets and Systems*, Vol. 91, pp. 69-90, 1997.

[13] FM Burnet, *The Clonal Selection Theory of Acquired Immunity*, Cambridge University Press, 1959.

[14] JJ Liang, AK Qin, "Particle Swarm Optimization Algorithms with Novel Learning Strategies", *SMC2004*, Netherlands, 2004.

[15] JJ Liang, PN Suganthan, "Dynamic Multi-Swarm Particle Swarm Optimizer", *Proc. of IEEE Swarm Intelligence Symp.*, pp. 124-129, 2005.

[16] KE Parsopoulos, MN Vrahatis, "UPSO- A Unified Particle Swarm Optimization Scheme", *LNCS*, pp. 868-873, 2004.

Evolutionary Bayesian Classifier-Based Optimization in Continuous Domains

Teresa Miquélez, Endika Bengoetxea, and Pedro Larrañaga

Intelligent Systems Group
University of the Basque Country, Computer Engineering Faculty
P.O. Box 649, E-20018 San Sebastian, Spain
http://www.sc.ehu.es/isg
{teresa, endika, ccplamup}@si.ehu.es

Abstract. In this work, we present a generalisation to continuous domains of an optimization method based on evolutionary computation that applies Bayesian classifiers in the learning process. The main difference between other estimation of distribution algorithms (EDAs) and this new method –known as Evolutionary Bayesian Classifier-based Optimization Algorithms (EBCOAs)– is the way the fitness function is taken into account, as a new variable, to generate the probabilistic graphical model that will be applied for sampling the next population.

We also present experimental results to compare performance of this new method with other methods of the evolutionary computation field like evolution strategies, and EDAs. Results obtained show that this new approach can at least obtain similar performance as these other paradigms[1].

1 Introduction

Evolutionary computation techniques have undergo a great development with the extensive use of paradigms such as Genetic Algorithms (GAs) [5,8], Evolution Strategies (ES) [6], and Estimation of Distribution Algorithms (EDAs) [10,13,17,18,20]. Many other evolutionary computation paradigms have also been recently proposed such as Learnable Evolution Model (LEM) [14], and Evolutionary Bayesian Classifier-based Optimization Algorithms (EBCOAs) [16].

The main difference between them is the way of improving the population of individuals, in order to obtain fitter solutions to a concrete optimization problem. In GAs and ES the evolution is based on using crossover and mutation operators, without expressing explicitly the characteristics of the selected individuals within a population. EDAs

[1] This paper has been partially supported by the Spanish Ministry of Science and Technology and the Basque Government with grants TIN2004-21428-E and SAIOTEK-S-PE04UN25 respectively. The authors would like to thank Alexander Mendiburu, and the rest of the members of the Intelligent Systems Group of the University of the Basque Country for their useful advise and contribution to this work. We would also like to thank Francisco Herrera and the rest of the Soft Computing and Intelligent Information Systems research group of the University of Granada for their support with the use of CMA-ES.

T.-D. Wang et al. (Eds.): SEAL 2006, LNCS 4247, pp. 529–536, 2006.

take into account these explicit characteristics by considering the interdependencies between the different variables that form an individual, learning a probabilistic graphical model to represent them. The approach of LEM, EBCOAs, and other similar proposals in this direction [12] is different in the sense of applying classification techniques to build models that represent the main characteristics for which an individual in the population appears in the group of the best (or worst) individuals within the population of a generation. In that sense, these paradigms also take into consideration less fit individuals of the population in order to enhance and estimate the differences between the best and worst cases. This knowledge is later used for instantiating new individuals for a new population.

This paper introduces EBCOAs (Evolutionary Bayesian Classifier-based Optimization Algorithms) for continuous domains, which are motivated as an improvement of EDAs for the need to avoid them to fall into local optima in very complex optimization problems. EBCOAs evolve to a fitter generation by constructing models that take into account more differences than simply a subset of the fittest individuals. Continuous EBCOAs generalize the ideas presented in [16] by supervised classification paradigms in the form of conditional Gaussian networks to improve the generation of individuals every generation.

2 Bayesian Classifiers for Continuous Domains

In EBCOAs for continuous domains, the classifiers are based on the learning of probabilistic graphical models, more concretely Bayesian classifiers based on conditional Gaussian networks [11]. The literature contains several examples of classifiers combined with evolutionary computation techniques. One of the first examples is the LEM algorithm [14] which makes use of rules to build a classifiers that records the main differences between the groups of best and worst individuals of each population.

The *supervised classification* problem with n continuous predictor variables consists in assigning any vector $\boldsymbol{x} = (x_1, \ldots, x_n) \in \mathcal{R}^n$ to one of the $|C|$ classes of a class variable C that is known. The class value is denoted by c and therefore we have that $c \in \{1, 2, \ldots, |C|\}$. As a result, a classifier in supervised classification is defined as a function $\gamma : (x_1, \ldots, x_n) \rightarrow \{1, 2, \ldots, |C|\}$ that assigns class labels to observations.

Next, we provide some examples of the classifiers from the ones considering less interdependencies to the ones considering most of them.

2.1 Naive Bayes

The Bayesian classifier that considers all the variables $X_1, \ldots, X_n$ to be conditionally independent given the class value C is known as naive Bayes [15]. In this case, the probabilistic graphical model can be considered to be a fixed structure as illustrated in Figure 1(a). In continuous domains it is usual to assume that the joint density function follows a n–dimensional normal distribution, and since independence between the variables –given the class variable C– is assumed, this is factorized by a product of unidimensional and conditionally independent normal densities. Therefore, when classifying a new individual using the naive Bayes classifier we have that:

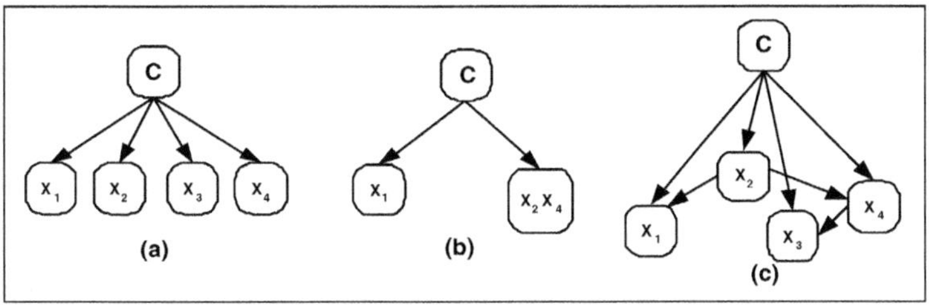

Fig. 1. Example of structures of Bayesian classifiers that can be obtained as a result of the different classification model building algorithms in a problem with four variables $X_1, \ldots, X_4$ and the class-variable C: (a) naive Bayes (b) seminaive Bayes (c) tree augmented naive Bayes.

$$p(C = c|X_1 = x_1, \ldots X_n = x_n) \propto p(c) \cdot f(x_1|c) \cdot f(x_2|c) \cdot \ldots \cdot f(x_n|c)$$

where

$$f(x_i|c) = \frac{1}{\sqrt{2\pi}\sigma_{ic}} e^{-\frac{1}{2}(\frac{x_{ic}-\mu_{ic}}{\sigma_{ic}})^2}$$

for all $i = 1, \ldots, n$ and $c = 1, \ldots, |C|$ with μ_{ic} and σ_{ic} representing the mean and the standard deviation of $X_i|C = c$ respectively.

In order to apply a naive Bayes classifier in EBCOAs, the estimation of the a priori probability of the class, $p(c)$, as well as the parameters μ_{ic} and σ_{ic} of the conditional density functions, $f(x_i|c)$, are carried out from the database of selected individuals at each generation.

2.2 Seminaive Bayes

The seminaive Bayes classifier [9] provides more complexity than the former since it is able to take into account dependencies between groups of variables. This paradigm represents the variables found to be related as a *fused* node in the conditional Gaussian network, that is the seminaive Bayesian classifier proposed to group some variables in a single node of the structure. Figure 1(b) illustrates the structure of a seminaive Bayesian classifier for a problem with four variables, treating each of these grouped variables as a single super-variable regarding the factorization of the probability distribution. When grouping variables all the inter-dependencies between them are taken into account implicitly in the Bayesian classifier. In a seminaive Bayesian classifier it is also possible to ignore some variables and therefore not to include them in the final probabilistic graphical model, which has the effect of considering these variables not to be relevant for labeling vectors to a particular c class. [19] introduced a greedy algorithm to detect irrelevant as well as dependent variables (susceptible to be grouped) and to propose variables that are likely to be ignored in a Bayesian classifier, although this is described only for discrete domains. We propose to adapt it to the case of continuous domains by grouping dependent continuous variables as a single multidimensional variable in the form of one node in the conditional Gaussian network.

Considering the example in Figure 1(b) to be the seminaive Bayes model structure learned from one supervised classification problem, an individual $\boldsymbol{x} = (x_1, x_2, x_3, x_4)$ will be assigned to the following class:

$$c^* = \arg\max_c p(c) f(x_1|c) f(x_2, x_4|c) \tag{1}$$

Following this approach, in cases in which a variable X_i is estimated to be conditionally independent of the rest given the class variable (such as variable X_1 in the latter example), $f(x_i|c)$ will be computed as

$$f(x_i|c) = \frac{1}{\sqrt{2\pi}\sigma_{ic}} e^{-\frac{1}{2}\left(\frac{x_{ic}-\mu_{ic}}{\sigma_{ic}}\right)^2}$$

Analogously, for the case of the dependencies with a number p of grouped variables over a single node in the structure, similarly as variables X_2 and X_4 in our example, the corresponding factor will be assumed to follow a p-dimensional normal distribution for each value of variable C. Considering that $\mathbf{z}_j$ represents the jth group of p variables, we would have that

$$f(\mathbf{z}_j|c) = \frac{1}{\sqrt{(2\pi)^p |\Sigma_{jc}|}} e^{-\frac{1}{2}(\mathbf{z}_j-\boldsymbol{\mu}_{jc})^T \Sigma_{jc}^{-1} (\mathbf{z}_j-\boldsymbol{\mu}_{jc})}.$$

where Σ_{jc} is a $p \times p$ matrix representing the variance-covariance matrix of the jth group of p variables when $C = c$ and $\boldsymbol{\mu}_{jc}$ denotes its corresponding expectation.

2.3 Tree Augmented Naive Bayes

Another example of a Bayesian classifier that is able to take into account different dependencies between variables than the previous seminaive approach is the tree augmented naive Bayes classifier [4]. Its name comes from the fact that the structures obtained as a result of its learning approach have the form of a tree. This algorithm constitutes an adaptation of the Chow-Liu algorithm [3] for predictor continuous variables by estimating the mutual information between two univariate normal distributions.

Figure 1(c) shows the type of structures that could be obtained when applying the tree augmented naive Bayes algorithm for a problem similarly as for the two previous Bayesian classifiers. Following this particular example, an individual $\boldsymbol{x} = (x_1, x_2, x_3, x_4)$ will be assigned to the class

$$c^* = \arg\max_c p(c) f(x_1|c, x_2) f(x_2|c) f(x_3|c, x_4) f(x_4|c, x_2) \tag{2}$$

Note that in this case the calculation of $f(x_i|c, x_{k(i)})$ –where $X_{k(i)}$ represents the predictor parent variable of variable X_i in case that this parent exists– can be computed as

$$f(x_i|c, x_{k(i)}) = \frac{p(c) \cdot f(x_i, x_{k(i)}|c)}{p(c) \cdot f(x_{k(i)}|c)} = \frac{f(x_i, x_{k(i)}|c)}{f(x_{k(i)}|c)}$$

3 The Evolutionary Bayesian Classifier-Based Optimization Algorithm Approach

This approach combines Bayesian classifiers such as the ones presented in the previous section and evolutionary computation to solve optimization problems. The main idea is that having a population of solutions for the optimization problem, we will evolve to a next population of fitter individuals by constructing a Bayesian classifier that will represent the main characteristics between the fittest and the least fit individuals. The EBCOA approach contains the following steps:

1. Firstly, the initial population D_0 of R individuals is generated. This initial population is generated similarly as EDAs, usually by assuming an uniform distribution on each variable. Each of the created individuals is evaluated.
2. Secondly, each of the individuals in D_l are given a label $|K| < R$ to classify them following their respective fitness value. This is the supervised classification step, as each of the R individuals is assigned a k label, and as a result the class variable K is created in the database, forming D_l^K.
3. Thirdly, D_l^C is created by selecting from D_l^K only the $|C| \leq |K|$ classes that will be used for building the Bayesian classifier, usually taking into account at least the best and worst classes of individuals uniquely. A similar scheme of selecting individuals with extreme fitness know as Stabilizing Selection is presented in [2]. The rest of the classes in D_l^K could be discarded to facilitate the learning by enhancing the differences between the most distant classes. The individuals which are in $D_l^K \backslash D_l^C$ are simply ignored.
4. A Bayesian classifier is build based on D_l^C by applying techniques such as the ones described in the previous section. This classifier estimates the probability distribution $p_l(c|\boldsymbol{x}) \propto p_l(c) f_l(\boldsymbol{x}|c)$ which represents the probability of any individual $\boldsymbol{x}$ to be classified in the any of the different possible $|C|$ classes.
5. Finally, the new population D_{l+1} constituted by the R new individuals is obtained by carrying out the simulation of the probability distribution $p_l(c) f_l(\boldsymbol{x}|c)$. This step can be performed very similarly as in EDAs[2].

Steps 2, 3, 4 and 5 are repeated until a stopping criterion is satisfied. Examples of stopping conditions are: achieving a fixed number of populations or a fixed number of different evaluated individuals, uniformity in the generated population, and the fact of not obtaining an individual with a better fitness value after a certain number of generations.

The step of learning the Bayesian classifier is the most critical one regarding the performance of EBCOAs in terms of convergence speed and computation time.

4 Experiments on Standard Continuous Optimization Problems

Experiments were carried out in order to test the performance of continuous EBCOAs compared to such of some continuous EDAs and ES. For this comparison, we have chosen continuous EDAs that take into account different number of dependencies between

[2] The reader can find a detailed description on this topic as well as a review of some of the possible techniques that can be applied in this step in [10].

Table 1. Mean results after 10 runs with each algorithm and objective function. The V and E columns represent the best fitness value obtained and the evaluations number respectively.

	Ackley		Griewangk		Rosenbrock		Sphere		SumCan	
$n = 10$	V	E	V	E	V	E	V	E	V	E
$EBCOA_{NB}$	7.7E-6	18116	3.0E-2	33597	4.4E+3	18235	4.4E-6	19632	1.0E+05	34116
$EBCOA_{S_{NB}}$	6.1E-6	11891	4.7E-6	8061	9.0E+0	7821	3.9E-6	7422	1.0E+05	16599
$EBCOA_{TAN}$	6.7E-6	11333	5.3E-6	10495	3.8E+3	6704	5.1E-6	9657	1.0E+05	31163
$UMDA_c$	8.8E-6	23063	5.4E-2	58814	8.7E+0	52589	7.3E-6	15003	1.6E+04	60250
$MIMIC_c$	7.8E-6	23382	8.2E-2	59093	8.7E+0	44968	6.7E-6	15163	1.7E+04	60250
$EGNA_{ee}$	7.9E-6	22983	5.4E-2	58654	8.6E+0	28889	6.7E-6	14884	1.0E+05	37108
$EGNA_{BGe}$	8.5E-6	22904	9.2E-2	54784	8.6E+0	26375	7.0E-6	14884	1.0E+05	37826
CMA-ES	1.9E-7	23962	2.7E-8	14562	3.9E-8	44082	3.7E-8	13802	1.0E+05	45682
$n = 50$										
$EBCOA_{NB}$	5.5E-6	29328	5.8E-6	24619	5.1E+5	2914	3.5E-6	16839	1.0E+05	34036
$EBCOA_{S_{NB}}$	6.9E-6	11851	5.7E-6	9976	4.9E+1	9976	5.7E-6	9976	1.0E+05	27293
$EBCOA_{TAN}$	7.2E-6	10136	5.2E-6	8500	4.9E+1	7343	5.9E-6	10016	1.0E+05	30086
$UMDA_c$	1.6E-5	56819	8.3E-6	35751	4.9E+1	59412	8.6E-6	42175	6.9E-1	60250
$MIMIC_c$	1.7E-5	56819	8.4E-6	35392	4.9E+1	59133	9.2E-6	42135	6.6E-1	60250
$EGNA_{ee}$	5.6E-2	38345	1.4E-4	33517	6.0E+1	50913	7.8E-3	39462	8.9E+0	27572
$EGNA_{BGe}$	2.5E-2	58694	5.0E-5	34116	5.7E+1	60250	1.6E-3	48519	1.8E+1	56340
CMA-ES	2.4E-3	60002	1.5E-6	60002	5.5E+1	60002	1.2E-5	60002	7.0E-1	60002

variables: $UMDA_c$, $MIMIC_c$, $EGNA_{ee}$ and $EGNA_{BGe}$[3]. In addition, the overall performance was compared to such of ES [21]. As an example of the latter we chose CMA-ES (Evolution Strategy with Covariance Matrix Adaptation)[7] which is considered as one of those with better performance[4].

The optimization problems selected are the ones proposed in [1] to compare evolutionary computation algorithms in continuous domains, namely Ackley, Griewangk, Rosenbrock generalized, Sphere Model, and Summation Cancelation.

The size of the population was decided to be $R = 400$ since we consider that this is the smallest reasonable size required to allow EBCOAs discern characteristics of the fittest and less fit individuals. The experiments have been carried out with individuals of 10 and 50 variables ($n = 10$ and $n = 50$ respectively). For EBCOAs, we set the following parameters: $|K| = 3$ and $|C| = 2$ (hence, we consider only the best and worst classes of individuals). The size of the best and worst classes was chosen to be $R/4$, where R is the size of the population as previously mentioned. In the case of continuous EDAs, the learning of the model is done after selecting the $R/2$ best ones of the population, keeping the same size of the population of $R = 400$ and an elitist approach. Finally, for CMA-ES we apply again the same population size and 200 as maximum number of parents, while the rest of parameters have been left as the default values suggested by the original authors.

The stopping criterion for all the algorithms and fitness functions is satisfied when the optimum solution is found (assuming this case when the result obtained was closer

[3] See [10] for a deep review on these algorithms.

[4] For the simulation step we have applied the MATLAB program cmaes.m version 2.34 available at `http://www.icos.ethz.ch/software/evolutionary_computation/cmaes.m`

than 10^{-6} from the global optimum fitness), when a maximum of 150 generations was reached, or simply when no improvement has been obtained in the generation of the last population. Each algorithm is run 10 times for each of the optimization problems. Table 1 shows mean values of these 10 runs, illustrating the fitness value of the best individual obtained (V) and the number of evaluations (E) required for each of the experiments. The results of Table 1 evidence that EBCOAs perform quite well in most of the optimization problems proposed, and also that the results obtained at least are comparable (in some cases even improved) to the ones of EDAs and CMA-ES. If we focus individually in each of the proposed optimization problems, we appreciate that in those containing no local optima (i.e. Sphere problem) or in the ones having many small local optima (such as Ackley and Griewangk) EBCOAs show a performance comparable to such of EDAs and CMA-ES. On the other hand, in optimization problems containing local optima with much higher size (i.e. Summation Cancellation) EBCOAs manage to reach the optimum for $n = 10$ and $n = 50$, while both EDAs and CMA-ES are also capable of solving them when the size of the problem is of 10 variables but show a poorer performance when the size of the problem increases to 50. The Rosenbrock function is especially difficult to optimize regarding the fact that the global optimum is located in the middle of a quite flat region. This is the reason for EDAs not to manage to find this global optimum and to fall in a local optimum. EBCOAs do not either manage to improve these results and perform similarly as EDAs. CMA-ES is able to obtain the global optimum uniquely in small problem sizes (n=10), although when increasing the size of individuals to n=50 it also shows the same behavior as EDAs and EBCOAs.

At the light of the results we can conclude that continuous EBCOAs is a new paradigm that is able to obtain comparable results as continuous EDAs and ES, although its bigger computation cost makes it more suitable for complex problems since mainly in those the results outperform EDAs and ES.

5 Conclusions and Further Work

The original contribution of this paper is to generalize EBCOAs to continuous domains by applying Bayesian classifiers in these type of domains. EBCOAs combine evolutionary computation techniques and Bayesian classifiers in order to solve optimization problems. Experimental results to compare the performance of this new approach on typical optimization problems in continuous domains have been shown, and they have been compared with such of continuous EDAs and ES.

Future research trends also include the study and experimentation of other Bayesian classifiers, even more complex ones capable to take into account more interdependencies between variables that could be useful for more complex problems.

References

1. E. Bengoetxea, T. Miquélez, P. Larrañaga, and J. A. Lozano. Experimental results in function optimization with EDAs in continuous domain. In P. Larrañaga and J. A. Lozano, editors, *Estimation of Distribution Algorithms. A New Tool for Evolutionary Computation*, pages 181–194. Kluwer Academic Publishers, 2001.

2. G. Cervone. LEM2 Theory and Implementation of the Learnable Evolution. Technical report, Machine Learning and Inference Laboratory, George Mason University, 1999.
3. C. Chow and C. Liu. Approximating discrete probability distributions with dependence trees. *IEEE Transactions on Information Theory*, 14(3):462–467, 1968.
4. N. Friedman, D. Geiger, and M. Goldsmidt. Bayesian network classifiers. *Machine Learning*, 29(2):131–163, 1997.
5. D. E. Goldberg. *Genetic Algorithms in Search, Optimization, and Machine Learning*. Addison-Wesley, Reading, Massachusetts, USA, 1989.
6. N. Hansen. The CMA evolution strategy: A comparing review. In I. Inza J.A. Lozano, P. Larrañaga and E. Bengoetxea, editors, *Towards a New Evolutionary Computation. Advances in Estimation of Distribution Algorithms*, pages 75–102. Springer, 2006.
7. N. Hansen and S. Kern. Evaluating the CMA evolution etrategy on multimodal test functions. In *Eighth International Conference on Parallel Problem Solving from Nature – PPSN VIII*, pages 282–291, 2004.
8. J. H. Holland. *Adaptation in Natural and Artificial Systems*. The University of Michigan Press, Michigan, 1975.
9. I. Kononenko. Semi-naïve Bayesian classifiers. In *Proceedings of the 6th European Working Session on Learning*, pages 206–219, Porto, Portugal, 1991.
10. P. Larrañaga and J. A. Lozano. *Estimation of Distribution Algorithms. A New Tool for Evolutionary Computation*. Kluwer Academic Publishers, 2001.
11. S. L. Lauritzen and N. Wermuth. Graphical models for associations between variables, some of which are qualitative and some quantitative. *Annals of Statistics*, 17:31–57, 1989.
12. X. Llorà and D.E. Goldberg. Wise breeding GA via machine learning techniques for function optimization. In Cantú-Paz et al., editor, *Proceedings of the Genetic and Evolutionary Computation Conference – GECCO-03, Part I*, Lecture Notes in Computer Science 2723, pages 1172–1183, Chicago, Illinois, 2003. Springer.
13. J.A. Lozano, P. Larrañaga, I. Inza, and E. Bengoetxea. *Towards a New Evolutionary Computation. Advances in Estimation of Distribution Algorithms*. Springer, 2006.
14. R.S. Michalski. Learnable evolution model: Evolutionary processes guided by machine learning. *Machine Learning*, 38:9–40, 2000.
15. M. Minsky. Steps toward artificial intelligence. *Transactions on Institute of Radio Engineers*, 49:8–30, 1961.
16. T. Miquélez, E. Bengoetxea, and P. Larrañaga. Evolutionary computation based on Bayesian classifiers. *International Journal of Applied Mathematics and Computer Science*, 14(3): 335–349, 2004.
17. H. Mühlenbein, T. Mahning, and A. Ochoa. Schemata, distributions and graphical models in evolutionary optimization. *Journal of Heuristics*, 5(2):215–247, 1999.
18. H. Mühlenbein and G. Paaß. From recombination of genes to the estimation of distributions: I. Binary parameters. In M. Voigt et al., editor, *Parallel Problem Solving from Nature - PPSN IV. Lecture Notes in Computer Science 1411*, pages 178–187, 1996.
19. M. Pazzani. Searching for dependencies in Bayesian classifiers. In D. Fisher and H.-J. Lenz, editors, *Learning from Data: Artificial Intelligence and Statistics V*, pages 239–248, New York, NY, 1997. Springer–Verlag.
20. M. Pelikan. *Hierarchical Bayesian Optimization Algorithm: Toward a New Generation of Evolutionary Algorithms*. Springer, 2005.
21. H.-P. Schwefel. *Evolution and Optimum Seeking*. Wiley InterScience, 1995.

A Simple Ranking and Selection for Constrained Evolutionary Optimization

Ehab Z. Elfeky, Ruhul A. Sarker, and Daryl L. Essam

School of ITEE, University of New South Wales, ADFA, Canberra 2600, Australia
{e.elfeky, r.sarker, d.essam}@adfa.edu.au

Abstract. Many optimization problems that involve practical applications have functional constraints, and some of these constraints are active, meaning that they prevent any solution from improving the objective function value beyond the constraint limits. Therefore, the optimal solution usually lies on the boundary of the feasible region. In order to converge faster when solving such problems, a new ranking and selection scheme is introduced which exploits this feature of constrained problems. In conjunction with selection, a new crossover method is also presented based on three parents. When comparing the results of this new algorithm with four other evolutionary based methods, using nine benchmark problems from the relevant literature, it shows very encouraging performance.

1 Introduction

Many optimization problems are nonlinear and constrained. These problems can be represented as follows:

$$\begin{aligned} &\min \quad f(X) \\ &\textit{subject to} \quad && g_i(X) \le 0, && i = 1,2,\dots, m, \\ & && h_j(X) = 0, && j = 1,2,\dots, p, \\ & && L_i \le x_i \le U_i, && i = 1,2,\dots, n. \end{aligned} \tag{1}$$

where $X \in R^n$ is the vector of solutions $X=[x_1,x_2,\dots,x_n]^T$ The objective function is $f(X)$, m is the number of inequality constraints, $g_i(X)$ is the ith inequality constraint, p is the number of equality constraints, and $h_j(X)$ is the jth equality constraint. Each decision variable x_i has a lower bound L_i and an upper bound U_i.

Over the last two decades, Evolutionary Algorithms (EAs) have proved themselves as global optimization techniques. Among the evolutionary algorithms, Genetic Algorithms (GAs) are the most widely used technique for solving optimization problems. Constrained optimization problems have been considered as difficult problems. Many researchers and practitioners (including Mezura and Coello [5], Barbosa and Lemonge [1], Deb [2], Koziel and Michalewicz [4] and others) attempted to solve constrained problems using traditional GAs. Later, Runarsson and Yao [8] developed an Evolutionary Strategy (ES) based approach for solving constrained optimization problems. They have shown that their algorithm outperforms all GA based approaches for

T.-D. Wang et al. (Eds.): SEAL 2006, LNCS 4247, pp. 537 – 544, 2006.

13 well-known test problems. From these studies, a question may arise of why GA-based approaches cannot perform better or at least equal to the ES-based algorithm - which is the motivation of this research. In this paper, we have investigated whether a GA based approach can perform as good as an ES-based approach, for solving certain classes of constrained optimization problems. To carry out this investigation, we have designed a new ranking and selection method and a new crossover as part of our proposed GA.

It is a common situation for many constrained optimization problems that some constraints are active at the global optimum point, thus the optimum point lies on the boundary of the feasible space [4]. This is usually true for business applications with limited resources. In such cases, it seems natural to restrict the search of the solution to near the boundary of the feasible space [7]. However, this may not be the situation for some other problems such as engineering design and specially generated problems. As the first step of our algorithm development, we restrict ourselves only to inequality constraints.

As we know, crossover is the heart of the storm for GAs. Many studies had been carried out to show how this operator affects the evolutionary process. The most widely used crossovers are k-point, uniform, intermediate, global discrete, order-based, and matrix-based [9]. Most of these crossovers are based on two parents. However, Eiben et al. [3] introduced a multi-parent reproduction process to GAs. They have indicated that using more parents makes it harder for good individuals to transfer their fittest genes to the resulting offspring. However, this property is good for population diversity, which in turn would be expected to slow down the convergence. In practice, the result of crossover with more parents is relatively better than what may be expected in terms of the quality of offspring. This means that multi-parent crossover provides better convergence than generally expected [3]. Keeping this in mind, we have designed a three parents crossover in this paper.

Without mutation, GAs could be trapped in local optimum while solving complex optimization problems. Among others, uniform and non-uniform mutations are well known in GA applications. Uniform mutation uses uniform random changes to individuals - which favors diversity but slows down convergence. Michalewicz [6] proposed a dynamic non-uniform mutation to reduce the disadvantage of uniform mutation in real-coded GA. Zhao et. al. [11] reported that a non-uniform mutation operator has the feature of searching the space uniformly at the early stage and very locally at the later stage. In other words, the non-uniform mutation has the common merits of a higher probability of making long jumps at early stages and much better local fine-tuning ability at later stages [11]. We have introduced a mutation that uses both uniform and non-uniform mutation, to exploit the advantages of both of them.

In GA applications, there are many different ways to select good individuals to survive and reproduce new offspring. There are three most commonly used ranking and selection methods. First, proportionate reproduction (also known as roulette wheel selection [9]), in this scheme individuals are chosen for selection in proportion to their fitness value. In the second method, the population is sorted from best to worst; then the higher ranked individuals are given higher probabilities to survive [9]. The third method is tournament selection where a random number of individuals are chosen from the population (with or without replacement) and the best individual from this group is chosen as a parent for the next generation [9]. This process is repeated until the mating pool is filled. There are a variety of other selection methods including stochastic methods [8]. In this paper, we introduce an algorithm which uses a tournament selection in some stages of the evolution process, and uses a new ranking method in other stages. This will be discussed later in more details.

The penalty function is the most widely used method to deal with constraints in constrained optimization. The penalty techniques used are: static, dynamic, annealing, adaptive, death penalties, superiority of feasible points, and faster adaptive methods. A good comparison and analysis of these methods can be found in Sarker and Newton [10]. In this paper, we have calculated the constraint violation, but have not penalized the individuals similarly to most GA algorithms. Instead, we have used this information to rank and select the individuals as parents as detailed later.

The developed algorithm was tested using nine benchmark problems from the specialized literature and was compared with three other GA-based and one ES-based algorithms. From the comparisons, we can claim that the proposed algorithm performing very well for the nine inequality constrained optimization problems tested.

This paper is organized as follows, Section 2 presents the proposed algorithm, illustrative examples and comparison are stated in Section 3 and finally, Section 4 contains the conclusion.

2 The Proposed Algorithm

In this section, we present our proposed algorithm. This algorithm uses a floating point representation, and the main steps are follows:

1. Create random initial population.
2. Check feasibility of all individuals.
3. Evaluate the population.
4. If the stopping criterion has been met, stop; otherwise continue.
5. Rank the individuals and make the selection.
6. Apply the triangular crossover.
7. Apply the mutation.
8. Apply elitism by replacing the worst current individual with the overall generations' best individual.
9. Go to step 4.

The details of the components of our algorithm are discussed below:

2.1 Ranking

To exploit the feature that optimal solutions exist on the boundary of the feasible region, the selection and search process should concentrate on the individuals in that region. Therefore, the ranking scheme is designed as follows:

- The feasible individuals are ordered from the best to the worst based on their objective function value.
- Then, those solutions are divided into two groups. Group (a) has a fixed proportion of those with higher quality (smaller objective function value), and group (b) has lower quality feasible solutions. Hence the group (b) individuals are on the worse

side of the feasible region. Consequently, they should not be considered in the selection process, as they will slow down the convergence of the algorithm.

- The infeasible individuals are arranged in ascending order of the constraint violations. A given proportion of the individuals with the highest violations (high summation of the violation in all constraints) are discarded from the selection process (see group (e) in figure 1), because they are further away from the feasible region and will slow down the convergence of the algorithm. We are fully aware that these discarded individuals may diversify the search process, but consider that this diversity requires more computational time.
- The rest of the infeasible individuals are then arranged in ascending order of their objective function values. All infeasible individuals who have worse objective function values than the best feasible individual should be discarded from the selection (see group (d) in figure 1), because they will guide the search process in the wrong direction away from the optimal solution.
- The remaining individuals are the target of the selection, because they are in the right space near to both the optimal and feasible space (see group (c) in figure 1).

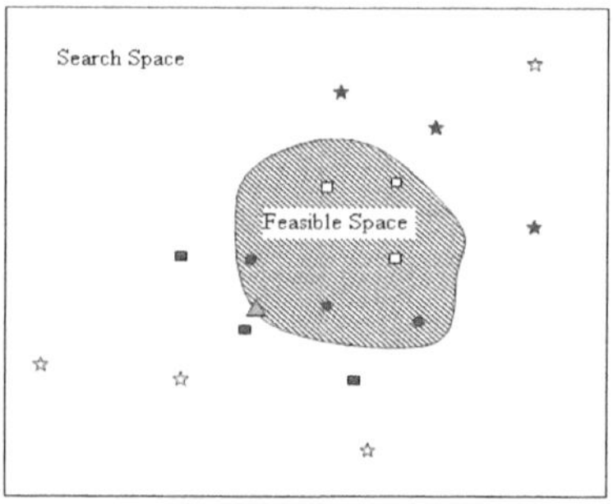

Type	Criteria		Group	Features	Action
Feasible		Objective Value ↓	(a) •	O(a) < O(b) V(a) = 0	Consider
			(b) ▫	O(b) > O(a) V(b) = 0	Discard
Infeasible	Constraints Violation ↓	Objective Value ↓	(c) ▪	O(c) < O(a) V(c) >≈ 0	Consider
			(d) ★	O(d) > O(a) V(d) >≈ 0	Discard
			(e) ☆	V(e) >>> 0	Discard

Fig. 1. The ranking scheme in the proposed algorithm. O(x) is the objective value of individuals in group (x). V(x) is the constraints violation of individuals in group (x). is the optimal solution.

2.2 Selection

Up to now, there are two groups of individuals to still be considered, group (a) which includes the individuals on the feasible side of the boundary of the feasible region, and the other group (c) that includes the individuals on the infeasible side. If there are individuals in group (c), then in the selection process, two feasible individuals will be selected from the first group, and one infeasible individual will be selected from the second group, these three individuals then undergo the triangular crossover process, otherwise all three individuals are chosen from group (a).

This ranking and selection scheme needs a reasonable amount of both feasible and infeasible individuals to work; therefore this mechanism is applied only if the ratio between the feasible individuals and the population size (feasibility ratio) is between 0.3 and 0.8. Otherwise, the regular tournament selection is used to select two individuals from the tournament set and the third one randomly. In cases where the optimal solution is not on the boundary, group (b) can be used instead of group (c).

2.3 Triangular Crossover

Consider the three individuals selected as parents p_1, p_2, p_3, and choose any three random numbers r_1, r_2, r_3, each of them in the interval $[0,1]$ where $r_1 + r_2 + r_3 = 1$, the resulted offspring will be constructed as a linear combination of the three parents as follows:

$$o_1=(r_1 * p_1)+ (r_2 * p_3)+ (r_3 * p_2),\ o_2=(r_1 * p_2)+ (r_2 * p_1)+ (r_3 * p_3),\ o_3=(r_1 * p_3)+ (r_2 * p_2)+ (r_3 * p_1).$$

Selecting two feasible parents and one infeasible parent gives a higher probability for the offspring to be feasible. The resulting offspring from this crossover of such selected individuals should ideally be nearer to the boundary of the feasible region than their parents were. This method of crossover increases the diversity of the population in good locations in the search space, more than as two parents would, as stated earlier in section 1. Therefore, any loss in diversity which may come from discarding many individuals during the selection could be compensated by using this crossover.

2.4 Mutation

We have introduced a mixed uniform and non-uniform mutation to exploit the advantages of both mutation methods. The probability of doing mutation is fixed during the whole evolution, but the step size is nonlinearly decreased over time as stated in Michalewicz [6]. There is also a low probability of doing a uniform mutation that moves an individual to any part of the search space. In this way, the algorithm is converging during most of the evolution using the non-uniform mutation, at the same time there is some chance to explore the rest of the feasible space in later generations using that part of the uniform mutation. This method helps with sophisticated multi-modal problems or problems with complicated feasible fitness landscapes.

2.5 Constraints Handling

Deb [2] introduced a method to handle constraints in an efficient way. In that method all infeasible individuals were penalized, and the best infeasible individual was assigned worse fitness than the worst feasible individual. We used this method when the feasibility ratio was too low. After that, the constraint handling was done implicitly during the ranking and selection scheme, where we exclude high violation individuals, therefore it converges to the feasible space over time.

3 Results and Discussion

In order to measure the performance of our developed algorithm, we have compared it denoted as TC (Triangular Crossover) with four existing algorithms by solving nine benchmark problems out of a set of thirteen commonly used by other researchers. The four test problems that were excluded involved equality constraints, which are not included in the focus of this paper. The characteristics of the test problems can be found in Runarrson and Yao [8]. The algorithms compared are Barbosa and Lemonge [1] (denoted as ACM), Deb's [2] GA (DEB), and Koziel and Michalewicz's [4] GA (KM). Although these three methods were GA based, Runarsson and Yao [8] introduced an evolution strategy method (RY) which depends on stochastic ranking, so far, the results of this method are considered the best among the existing algorithms.

We have solved each test problem 30 times with different random seeds. We have presented the best fitness values over 30 runs in Table 1. In all cases, only 350,000 objective function evaluations had been made before stopping the algorithm, as a similar evaluation budget was used in the other four algorithms. The population size was set to 30, except in g02, g07, and g10, where because they have a higher number of variables, instead they used 150 individuals, but still with the same number of fitness evaluations. The probability of using cross-over is 0.8 for using the whole arithmetic triangular crossover discussed earlier. The overall mutation probability is 0.1 and among the mutated individuals we have chosen a probability of 0.9 to use the non-uniform step size and 0.1 for the uniform step size.

Considering the best results out of 30 independent runs, we can claim that our proposed algorithm achieved superior results than the three GA-based algorithms (KM, DEB and ACM) presented. Note that ACM and DEB results are available only for 4 and 5 test problems respectively [1, 2]. If we compare the solutions of our algorithm with the RY algorithm, both of them achieved the exact optimal solution in 5 test problems g01, g04, g06, g08, and g12. Our algorithm achieved significantly better solutions for test problems g02 and g10, while RY obtained the optimal solution for g09 and our algorithm failed to do so, although the difference between the results is insignificant. Moreover, RY is significantly better for g07. Therefore, we can claim that based on the best solutions obtained, our algorithm is comparable to RY.

Since GA and ES are stochastic algorithms, it is logical to make a stochastic comparison instead of a static comparison using the best fitness value. Such a static comparison could be misleading as the best fitness value could simply be an outlier. For this purpose, we have analyzed the mean and standard deviations of the 30 independent runs for the two comparable algorithms.

Table 1. The best results out of 30 run to each algorithm. RY = Runarrson & Yao, KM = Koziel & Michalewicz, DEB = Deb, ACM = Adaptive Penalty, TC = Proposed Algorithm.

Fcn	Optimal	Best				
		RY	KM	DEB	ACM	TC
g01	-15.000	**-15.000**	-14.786	**-15.000**	**-15.000**	**-15.000**
g02	-0.803619	-0.803515	-0.799530	-	-	**-0.803616**
g04	-30665.539	**-30665.539**	-30664.500	-30665.537	-30665.403	**-30665.539**
g06	-6961.814	**-6961.814**	-6952.100	-	-	**-6961.814**
g07	24.306	**24.307**	24.620	24.373	-	24.505
g08	-0.095825	**-0.095825**	**-0.095825**	-	**-0.095825**	**-0.095825**
g09	680.630	**680.630**	680.910	680.635	680.667	680.633
g10	7049.331	7054.316	7147.900	7060.221	-	**7049.474**
g12	-1.000000	**-1.000000**	-0.999900	-	-	**-1.000000**

As per our analysis, both the algorithms have the same performance in g01, g08, and g12. The proposed algorithm has slightly better performance with respect to both measures of performance in g02 and is significantly better in g06 and g10. RY has a better mean, but not better standard deviation in g09, RY has slightly better performance in g04 and is significantly better in g07. By calculating in how many problems the proposed algorithm is better, we found for the mean they are 6 for each algorithm, but for the standard deviation 6 for TC and only 5 for RY.

Table 2. Percentage of variation, Mean and Std. Deviation of 30 runs

Fcn	Optimal	Percentage of variation		Mean		STD Deviation	
		RY	TC	RY	TC	RY	TC
g01	**-15.000**	**Optimal**	**Optimal**	**-15.000**	**-15.000**	**0.00E+00**	**0.00E+00**
g02	**-0.803619**	0.01294%	**0.00037%**	-0.781975	**-0.791345**	2.00E-02	**9.42E-03**
g04	**-30665.539**	**Optimal**	**Optimal**	**-30665.539**	-30665.531	**2.00E-05**	9.16E-03
g06	**-6961.814**	**Optimal**	**Optimal**	-6875.940	**-6961.814**	1.60E+02	**3.70E-12**
g07	**24.306**	**0.00411%**	0.81873%	**24.374**	25.057	**6.60E-02**	2.38E-01
g08	**-0.095825**	**Optimal**	**Optimal**	**-0.095825**	**-0.095825**	**2.60E-17**	4.23E-17
g09	**680.630**	**Optimal**	0.00044%	**680.656**	680.659	3.40E-02	**1.98E-02**
g10	**7049.331**	0.07072%	**0.00203%**	7559.192	**7493.719**	5.30E+02	**3.87E+02**
g12	**-1.000000**	**Optimal**	**Optimal**	**-1.000000**	**-1.000000**	**0.00E+00**	**0.00E+00**

In this paper, we have empirically shown that our approach is able to deal with a variety of constrained optimization problems (i.e., with both linear and nonlinear constraints and objective function, and with inequality constraints). The benchmark adopted includes test functions with both small and large feasible spaces. We also argue that our proposed approach is very simple to implement and can solve a variety of problems. Finally, we have shown that GA-based approaches can perform as well as ES-based approaches for inequality constrained optimization problems.

4 Conclusions

In this paper, new ranking, selection, and crossover methods were introduced to solve constrained optimization problems. The idea behind these new methods is the exploitation of some of the features of constrained problems. The performance of the proposed algorithm has been compared with four existing evolutionary algorithms using nine benchmark test problems. The results of the proposed algorithm are clearly better than the three GA-based approaches and are competitive with the best known ES-based approach. In two test problems, we have better results than other existing EA-based solutions. The superiority of our algorithm is the combined effect of our proposed ranking, selection, crossover and mutation methods. Finally, we have found that genetic algorithms can achieve similar results to those of evolution strategy in such types of problems. As immediate future work, we will test our algorithm for problems with equality constraints, and a mix of equality and inequality constraints. In addition, we will investigate the individual contributions of our proposed ranking, selection, crossover and mutation operators to the problem solving process.

References

1. Barbosa, H.J.C., Lemonge, A.C.C.: A new adaptive penalty scheme for genetic algorithms. Inf. Sci. 156 (2003) 215-251
2. Deb, K.: An efficient constraint handling method for genetic algorithms. Computer Methods in Applied Mechanics and Engineering 186 (2000) 311-338
3. Eiben, A.E., M., C.H., Kemenade, V., Kok, J.N.: Orgy in the Computer: Multi-Parent Reproduction in Genetic Algorithms. In: F. Moran, A.M., J. J. Merelo, and P. Chacon (ed.): 3rd International Conference on Artificial Life, Vol. 929. Lecture Notes in Artificial Intelligence (1995) 934-945
4. Koziel, S., Michalewicz, Z.: Evolutionary Algorithms, Homomorphous Mappings, and Constrained Parameter Optimization. Evolutionary Computation 7 (1999) 19-44
5. Mezura-Montes, E., Coello, C.A.C.: A Simple Multimembered Evolution Strategy to Solve Constrained Optimization Problems. Evolutionary Computation, IEEE Transactions 9 (2005) 1-17
6. Michalewicz, Z.: Genetic algorithms + data structures = evolution programs. Springer-Verlag, Berlin; New York (1996)
7. Michalewicz, Z.: Genetic Algorithms, Numerical Optimization, and Constraints. In: Eshelman, L. (ed.): 6th International Conference on Genetic Algorithms. Morgan Kaufmann, San Francisco, CA (1995) 151-158
8. Runarsson, T.P., Xin, Y.: Stochastic ranking for constrained evolutionary optimization. Evolutionary Computation, IEEE Transactions on 4 (2000) 284
9. Sarker, R., Kamruzzaman, J., Newton, C.: Evolutionary optimization (EvOpt): a brief review and analysis. International Journal of Computational Intelligence and Applications 3 (2003) 311-330
10. Sarker, R., Newton, C.: A Comparative Study of Different Penalty Function-Based GAs for Constrained Optimization. Proceedings of the 4th Australia-Japan Joint Workshop on Intelligent and Evolutionary Systems, Japan (2000) 64-70
11. Zhao, X., Gao, X.S., Hu, Z.: Evolutionary Programming Based on Non-Uniform Mutation. In: Sciences, A.o.M.a.S. (ed.): Mathematics-Mechanization Research Preprints. Chinese Academy of Sciences, Beijing, China (2004) 352–374

Optimizing Continuous Problems Using Estimation of Distribution Algorithm Based on Histogram Model*

Nan Ding[1], Shude Zhou[2], and Zengqi Sun[2]

[1] Department of Electronic Engineering, Tsinghua University, Beijing, China
ding-n04@mails.tsinghua.edu.cn
[2] Department of Computer Science and Technology, Tsinghua University, Beijing, China
zsd03@mails.tsinghua.edu.cn, szq-dcs@tsinghua.edu.cn

Abstract. In the field of estimation of distribution algorithms, choosing probabilistic model for optimizing continuous problems is still a challenging task. This paper proposes an improved estimation of distribution algorithm (HEDA) based on histogram probabilistic model. By utilizing both historical and current population information, a novel learning method – accumulation strategy – is introduced to update the histogram model. In the sampling phase, mutation strategy is used to increase the diversity of population. In solving some well-known hard continuous problems, experimental results support that HEDA behaves much better than the conventional histogram-based implementation both in convergence speed and scalability. Compared with UMDA-Gaussian, SGA and CMA-ES, the proposed algorithms exhibit excellent performance in the test functions.

1 Introduction

Recently, estimation of distribution algorithms have become the hot topic in the field of evolutionary computation [1]. The contribution of EDAs not only lies in its ability to explicitly learn the linkage relationship among variables, but also provides a novel macroscopical evolutionary paradigm, in which without any conventional operators, the population evolves by iteratively learning and sampling the probabilistic distribution model that describes the movements of population. Theoretical and empirical researches have shown that EDAs are a class of efficient black-box optimization algorithms.

The core of EDAs is the probabilistic model.

For 0-1 domain problem, the basic probabilistic model is very simple: $P(0) = p$ and $P(1) = 1 - p$, where $0 \le p \le 1$. All the existent 0-1 EDAs are based on the basic model. During the last decade, large amounts of different versions of 0-1 EDAs were developed, which can be classified into three categories [1]: without interaction, pairwise interaction and multi-interaction. PBIL, UMDA and cGA are estimation of distribu-

* The work is funded by National Key Project for Basic Research of China (G2002cb312205).

T.-D. Wang et al. (Eds.): SEAL 2006, LNCS 4247, pp. 545 – 552, 2006.

tion algorithms which do not take the interaction among variables into account; BMDA and MIMIC use probabilistic model that can represent the relationship between two variables; BOA and FDA can describe the distribution of solutions by Bayesian Network, which can model complex interaction among variables [1-3]. The amazing success in discrete domain attracts people to design efficient EDAs for continuous problems.

However, choosing probabilistic model for continuous domain is still a challenging problem, even though several attempts have been made to extend the research results from discrete to continuous problems. The complexity of continuous fitness landscape makes it impossible to choose an almighty probabilistic model that fits any problem. In general, continuous EDAs can be classified into two approaches – indirect and direct. The former employs transform methods such as discretization [4] and the latter estimates the parameters of the predefined distribution [3,6-9]. It should be noted that the "direct" approach occupies a predominant position because the "indirect" approach fails to scale with problem size and solution precision [12]. According to the probabilistic model employed in "direct" approach, continuous EDAs can be further classified into two categories: Gaussian-based EDAs[1,3,6-9] and Histogram-based EDAs [5]. Most of the work concentrates on Gaussian probabilistic model. These include $PBIL_C$, $UMDA_C$, EMNA, EGNA, IDEA and so on [1,3]. Histogram-based EDAs mainly refer to the work by Tsutsui, S. et al.[5], in which marginal histogram model was used to model the population in continuous domain for the first time.

The purpose of the paper is to further study the continuous EDAs based on histogram model. A novel estimation of distribution algorithm (HEDA) based on histogram model is developed. Accumulation strategy is used to update the probabilistic model. In sampling phase, mutation strategy is designed to enhance the population diversity. Experimental analyses will show the performance of HEDA compared with FWH, SGA, UMDA-Gaussian[7] and CMA-ES[11].

The present paper is organized as follows. Next section will give the detailed description of HEDA. In Section 3, numerical experiments of HEDA, FWH, CMA-ES, UMDA-Gaussian and SGA are described. The paper is concluded in Section 4.

2 HEDA – Histogram-Based Estimation of Distribution Algorithm

2.1 The General Description of HEDA

(1) Set the generation counter $t := 1$
(2) Divide the searching space of each variable into a certain number of bins. These bins should be of the same width and do not overlap with each other.
(3) Initialize the histogram model in which the height of each bin is same. The histogram generated should be normalized.
(4) Generate population $P(t)$ using sampling method described in subsection 2.3.
(5) Evaluate and rank the population $P(t)$. Save the elitist.
(6) Update the histogram model using accumulation learning strategy.

(7) Update the generation counter $t := t+1$.
(8) If the termination conditions have not been satisfied, go to step 4.
(9) HEDA is finished and solutions are obtained in $P(t)$.

2.2 Learning Method in HEDA

Accumulation learning strategy is proposed to update the histogram model. The histogram model is updated according to two kinds of information: historical and current information. In each generation, for each variable i, the selected individuals will be used to construct a histogram model H_C^i. The old histogram for variable i is denoted as H_H^i. The height of a certain bin j of the renewed histogram for variable i is:

$$H^i(j) = \alpha H_H^i(j) + (1-\alpha) \bullet H_C^i(j), \tag{1}$$

where $H_H^i(j)$ is the height of bin j in the old model, α $(0 \le \alpha \le 1)$ is the accumulation factor which determines the effect of the old model on the new model, $H_C^i(j)$ is the height of the bin j in model $H_C^i(j)$, and $\sum_j H_C^i(j) = 1$. It is clear that new histogram H^i has been normalized. Accumulation strategy used in HEDA is a method to reserve the information of historical model. In comparison with FWH [5] which reserves some good individuals in each generation, HEDA emphasizes the importance of the model building. The reservation of the historical model also reserves the information of the good individuals in the past generations. If α =0, HEDA is reduced to the FWH [5].

In addition, the contribution of different individuals with different fitness values is considered in the learning process. The histogram model is learned according to relative ranking among different individuals. The height of each bin is modified according to the ranking and the position of each individual. Different rankings of the individuals lead to different increments of the bins. In the paper, the relationship is linear. If N best individuals of the population are selected, the k-th best individual ($k \le N$) will make an increment of the corresponding bin which it belongs to by:

$$\Delta h_k^i = (N-k+1) / \sum_{l=1}^{N} l = \frac{2(N-k+1)}{N(N+1)} \tag{2}$$

So,

$$H_C^i(j) = \sum_{k=1}^{N} \Delta h_k^i \bullet \delta_{jk}^i \tag{3}$$

where $\delta_{jk}^i = 1$ for $\{\delta_{jk}^i \mid k \in \{1,2,...,N\} \wedge \min_j^i \le v_k^i < \max_j^i\}$, and $\delta_{jk}^i = 0$ otherwise. v_k^i denotes the value of variable i of the k-th best individual, $\min_j^i$ and $\max_j^i$ denote the lower and upper bound of bin j of variable i. The better individuals will have more effect on the new model. Updating H_C^i based on the ranking information helps improve the convergence property of HEDA.

2.3 Sampling Method in HEDA

In HEDA, the population is sampled according to the model as follows: First, the bin j is selected according to the probability of $H^i(j)$; then an individual is generated in the searching space of the bin j with uniform distribution.

In order to enhance the diversity of population, mutation strategy is used. In HEDA, the mutation strategy means that each variable of an individual has a probability to be generated randomly. Here "randomly" means that the variable of an individual is generated with uniform distribution in the whole searching space. If the mutation rate is set p_m, that means there is a possibility of p_m for each variable of each individual to be generated randomly and a possibility of $1-p_m$ for it to be generated according to the histogram model.

3 Numerical Experiments

3.1 Experimental Settings

Several well-known continuous test functions, which include the 20-variable two-peak function, the 20-variable Rastrigin function, the 10-variable Griewank function and the 5-variable Schfewel function, are used to verify the performance of HEDA. Table 1 lists these experimental functions and their respective optimal solutions.

Table 1. Test functions

Function name	Formulation	Domain	Optimal solution
Schwefel	$\sum_{i=2}^{n}((x_1-x_i^2)^2-(x_i-1)^2)$	$[-2,2]^5$	[1,1,1,...,1]
Rastrigin	$10n+\sum_{i=1}^{n}(x_i^2-10\cos(2\pi x_i))$	$[-5,5]^{20}$	[0,0,0,...,0]
Griewank	$\sum_{i=1}^{n}\frac{x_i^2}{4000}-\prod_1^n\cos(\frac{x_i}{\sqrt{i}})+1$	$[-5,5]^{10}$	[0,0,0,...,0]
Two-peak	$5n-\sum_{i=1}^{n}f_i$, twopeak f_i [5]	$[0,12]^{20}$	[1,1,1,...,1]

In the experiments, the proposed HEDA has been compared with several well-known continuous evolutionary algorithms including FWH, UMDA-Gaussian, SGA and CMA-ES. FWH is the original proposed continuous EDA based on histogram probability model [5]. The objective of comparison with FWH is to show that HEDA is a superior optimization method based on histogram model. UMDA-Gaussian is an EDA based on Gaussian probability model [7]; SGA refers to simple genetic algorithm; CMA-ES is an advanced evolution strategy with covariance matrix adaptation [11]. The objective of the comparison with UMDA-Gaussian, SGA and CMA-ES is to show that, among many kinds of continuous evolutionary algorithms, HEDA hold an outstanding position in solving the above test functions.

To evaluate the performance of HEDA, two criteria are used: convergence property and scalability. The convergence property is to measure the ability of the algorithm to reach the global optimum. In our experiments, we evaluate the convergence property by measuring the number of runs in which algorithm succeeds in finding the global optimum and the mean number of function evaluations (MNE) to find the global optimum in those successful runs. We define the successful detection of the solution as being within $\pm\varepsilon$ of the actual optimum point ($\varepsilon = 0.1$). The scalability of HEDA in solving two-peak problem is used to see how the behavior of algorithms changes when the dimension of the problem increases.

All the algorithms run 20 independent times on each problem. The termination condition for the 5 algorithms is detection of optimal solution or maximum 100,000 function evaluations. In all the algorithms, the initial population is generated uniformly. The parameter settings HEDA, FWH, SGA, UMDA-Gaussian and CMA-ES are as follows.

HEDA: the width of bin is set 0.1, mutation rate is set 0.05, and accumulation rate 0.2. 50% of population is selected for model updating. Elitist strategy is used.

FWH: we directly use the results obtained by FWH published in [5].

SGA: crossover probability 0.8 and mutation probability 0.05. Selection method is tournament selection. Elitist strategy is used.

UMDA-Gaussian: the Gaussian probability model is updated using method introduced in [9]. 50% of parent population is selected for model updating. Elitist strategy is used.

CMA-ES: $\mu = \lambda/4$, where is λ is the size of parent population and μ is the number of descendants. The Matlab code of CMA-ES in [13] is used in the experiments.

3.2 Convergence Property

Table 2 illustrates the convergence property of HEDA, UMDA-Gaussian, FWH, SGA, and CMA-ES in solving the above test functions.

Firstly, we compare the performance of HEDA and FWH, both of which have the common feature: estimation of distribution algorithms based on marginal histogram probability model. The results of FWH have been published in [1]. From table 2, it is obvious that HEDA performs much better than FWH. For example, in solving Griewank function, with population size 100, HEDA can always obtain the optimal solution in the 20 runs, while FWH fails in all the 20 runs. Excellent behavior in solving the 4 problems shows the effectiveness of accumulation learning strategy and mutation strategy in HEDA.

And then, let's see the comparison with SGA, UMDA-Gaussian and CMA-ES. As shown in Table 2, for Schwefel functions, CMA-ES exhibits very good performance. For example, with population size 100, CMA-ES can obtain optimal solution in all the 20 runs with less MNE 1205.6. UMDA-Gaussian performs very well in solving Griewank function and it can always succeed in converging to optimum in all the 20 runs. However, UMDA-Gaussian and CMA-ES performs worse than HEDA in solving other test functions. It can be observed that the convergence property of HEDA

Table 2. Convergence property of HEDA, FWH, SGA, UMDA-Gaussion and CMA-ES in solving a clsss of continuous optimization problems

		Population Size											
		100		200		300		400		600		800	
		#OPT	MNE	#OPT	MNE	#OPT	MNE	#OPT	MNE	#OPT	MNE	#OPT	MNE
HEDA	Schwefel	20	2607.5	20	3221.4	20	2475.3	20	4124.5	20	3985.6	20	5426.7
	Rastrigin	20	23756.0	20	19601.2	20	17638.2	20	16968.2	20	17888.4	20	16397.5
	Griewank	20	6164.3	20	5709.4	20	5687.7	20	6972.0	20	9218.7	20	10081.2
	Two-peak	20	15422.7	20	12442.2	20	11874.4	20	12060.0	20	10456.3	20	12112.3
FWH	Schwefel [5]	16	1342.6	19	1962.3	19	3302.9	20	4012.7	20	5924.8	20	7597.9
	Rastrigin[5]	1	3478.0	13	6770.6	20	9720.2	20	12616.4	20	18518.5	20	24113.5
	Griewank[5]	2	2779.0	7	5783.3	15	9220.9	19	12439.8	19	17517.2	19	22502.3
	Two-peak[5]	1	2562.0	14	4926.5	20	7178.3	19	9609.8	19	14016.6	20	18537.9
UMDA-Gaus	Schwefel	18	1460.0	6	3666.7	9	4560.0	7	4666.8	4	7240.8	5	8056.0
	Rastrigin	0	----	6	18600.0	12	28450.4	11	36320.4	13	57632.0	14	68572.4
	Griewank	20	1840.5	20	3690.4	20	5490.6	20	7250.4	20	10635.0	20	13349.8
	Two-peak	0	----	0	----	0	----	0	----	1	9948.0	4	9680.0
SGA	Schwefel	4	2375.6	14	2957.2	17	4923.6	15	6826.8	20	9090.0	20	10142.5
	Rastrigin	0	----	0	----	0	----	5	27834.6	6	39212.8	6	50467.3
	Griewank	0	-----	1	7564.0	2	14581.2	3	15067.8	5	30075.0	6	44612.5
	Two-peak	0	----	0	----	0	----	0	----	4	54326.0	7	93672.4
CMA-ES	Schwefel	20	1205.6	20	2035.4	20	2728.3	20	3181.5	20	4038.5	20	5352.2
	Rastrigin	0	----	1	19200.0	1	28600	0	----	6	55811.5	7	75220.3
	Griewank	12	3860.0	20	7429.5	20	10740.6	20	13525.0	20	18465.2	20	22450.6
	Two-peak	0	----	0	----	0	----	0	----	0	----	0	----

Note: "----" indicates that the algorithm failed to obtain the optimum in all the 20 runs.

is consistently good in solving all the 4 functions. CMA-ES often fails in solving Two-peak function and Rastrigin function; UMDA-Gaussian often fails in solving Two-peak function, Schwefel function and Rastrigin function. Compared with SGA, HEDA performs much better both in MNE and #OPT. Experimental results show that HEDA is a stable, robust and efficient algorithm in solving the test functions.

3.3 Scalability of HEDA

The scalability of HEDA is tested on Two-peak function. The problem dimension starts at 10, increased to 100 with step 10. For each dimension, 20 independent runs are executed. Fig.1 shows the mean number of fitness evaluations until HEDA finds the optimal solution. The number of fitness evaluations can be approximated by $O(n^{1.256})$. Therefore, the results indicate that HEDA can solve Two-peak function in sub-quadratic number of evaluations.

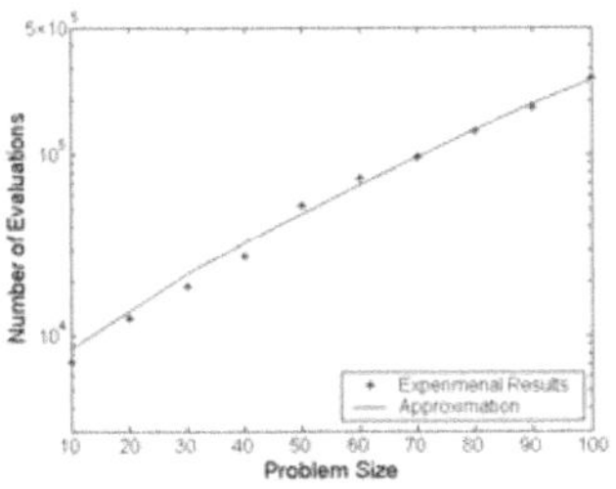

Fig. 1. Scalability of HEDA for Two-peak problem

3.4 Drawback of HEDA

In the previous subsections, experiments demonstrate that HEDA can efficiently solve the test functions and performs much better than some of well-known continuous evolutionary algorithms. However, in the procedure of HEDA, the relationship among variables is not taken into consideration. This will limit the ability of HEDA in solving complicated problems with strong linkage information. Here, 20-variable Rosenbrock function with optimum [1,1,...,1] is used to test the limited performance of HEDA:

$$f(x)=\sum_{i=2}^{n}100(x_i-x_{i-1}^2)^2+(1-x_{i-1})^2 .$$

There are strongly correlated variables in Rosenbrock function. In the experiment, 10 runs are executed for each population setting. The experimental results in Table 3 demonstrate it is hard to find optimal solution using HEDA. The drawback of HEDA is due to the fact that HEDA does not encode the linkage information.

Table 3. Convergence property of HEDA in solving Rosenbrock function

100		200		400		800	
OPT#	MNE	OPT#	MNE	OPT#	MNE	OPT#	MNE
1	29232.6	2	18625.4	0	----	1	62816.0

4 Conclusion

In this paper, we have proposed an improved estimation of distribution algorithm (HEDA) based on histogram probabilistic model. In the algorithm, a novel learning strategy named accumulation strategy is proposed, which considers the historical and current information of population at the same time. Mutation strategy is brought into the sampling phase to enhance the population diversity. HEDA is tested to solve a class of well-known continuous problems. Experimental results demonstrate that HEDA significantly outperforms FWH, and exhibits excellent capability compared to other evolutionary algorithms, e.g. CMA-ES, SGA and UMDA-Gaussian.

It is also noted that the capability of HEDA is limited by the fact that HEDA does not take the relationship among variables into account. Future work will focus on extensions of HEDA that can model the relationship among variables.

References

1. Larrañaga, P., Lozano, J. A.: Estimation of Distribution Algorithms: A new Tool for Evolutionary Computation. Kluwer Academic Publishers (2002)
2. Pelikan, M.: Hierarchical Bayesian optimization algorithm: Toward a new generation of evolutionary algorithms. Springer-Verlag (2005)
3. Očenášek, J.: Parallel Estimation of Distribution Algorithms. Ph.D. Dissertation in Brno University of Technology (2002)
4. Pelikan, M., David E.G., Tsutsui, S.: Combining the Strengths of the Bayesian Optimization Algorithm and Adaptive Evolution Strategies. IlliGAL Report No. 2001023 (2001)
5. Tsutsui, S., Pelikan, M., Goldberg, D.E.: Evolutionary Algorithm Using Marginal Histogram Models in Continuous Domain. IlliGAL Report No. 2001019, UIUC (2001)
6. Larrañaga, P., Etxeberria, R., Lozano, J. A., Peña, J. M.: Optimization by Learning and Simulation of Bayesian and Gaussian Networks. Technical Report EHU-KZAA-IK-4/99, University of the Basque Country (1999)
7. Larrañaga, P., Etxeberria, R., Lozano, J. A., Peňa, J. M. Optimization in Continuous Domains by Learning and Simulation of Gaussian networks. In: Proceedings of the Genetic and Evolutionary Computation Conference. Las Vegas, Nevada (2000)
8. Sebag, M., Ducolombier, A.: Extending population-based incremental learning to continuous search spaces. Parallel Problem Solving from Nature – PPSN V, Springer-Verlag (1998) 418–427
9. Larrañaga, P., Lozano, J. A., Bengoetxea, E.: Estimation of Distribution Algorithms based on multivariate normal and Gaussian networks. Technical report KZZA-IK-1-01, University of the Basque Country (2001)
10. Bosman, P., Thierens, D.: Expanding from Discrete to Continuous Estimation of Distribution Algorithms: IDEA. Parallel Problem Solving From Nature- PPSN VI (2000) 767–776
11. Hansen, N., Mueller S.D., Koumoutsakos, P.: Reducing the Time Complexity of the Derandomized Evolution Strategy with Covariance Matrix Adaptation (CMA-ES). Evolutionary Computation, Vol. 11. (2003) 1–18
12. Ahn, C.W.: Real-coded Bayesian Optimization Algorithm. Advances in Evolutionary Algorithms: Theory, Design and Practice. Springer Publication, Vol.18. (2006) 85–124
13. URL: http://lautaro.bionik.tu-berlin.de/user/niko/

Population Climbing Evolutionary Algorithm for Multimodal Function Global Optimization

Chen Ziyi and Kang Lishan

School of Computer, China University of Geoscience, Wuhan 430074, Hubei, China
chenziyi@263.net, kang_whu@yahoo.com

Abstract. This paper presents a population climbing evolutionary algorithm (PCEA) for solving function optimization containing multiple global optima. The algorithm combines a multi-parent crossover operator with the complete local search. The multi-parent crossover operator can enables individual to draw closer to each optimal solution,thus the population will be divided into subpopulations automatically , meanwhile, the local search is adopted to enable individual to converge to the nearest optimal solution which belongs to the same attractor. By this way, each individuals can converge to a global optima, then the population can maintain all global optima. Comparing with other algorithms, it has the following advantages.(1) The algorithm is very simple with little computation complexity .(2) Proposed algorithm needs no additional control parameter which depends on a special problem. The experiment results show that PCEA is very efficient for the optimization of multimodal functions, usually it can obtain all the global optimal solutions by running once of the algorithm.

1 Introduction

Many real world problems could be transformed to multimodal function optimal problems.However, when attempting to optimise a multi-modal function, the simple Genetic Algorithm (SGA) only converges to a single solution of the multiple optima due to genetic drift. So many researchers have proposed some effective methods to solve multimodal function optimization.DeJong(1975) in his doctoral dissertation used a crowding model to introduce diversity among solutions in a GA population [1].The main idea of DeJong's study was the suggestion of replacing one solution by a similar solution in maintaining optimum solutions in an evolving population.Mahfoud improved standard crowding of De Jong by introducing competition between children and parents of identical niches [2].After crossover and eventually mutation,each child replaces the nearest parent if it has an higher fitness.Other researchers have developed some different crowding schemes,for example,Restricted Tournament Selection (RTS)[3],adapts tournament selection for multimodal optimization,Mengshoel et al.[4] proposed probabilistic crowding as a probabilistic extension of the original Deterministic crowding (DC) method.Goldberg and Richardson proposed another revolutionary concept, sharing function model [5],where instead of replacing a solution by a similar solution, the focus was more on degrading the fitness of similar solutions.Most subsequent GA studies have used this model in solving multimodal optimization problems [6].However, there are two difficulties with the sharing function

T.-D. Wang et al. (Eds.): SEAL 2006, LNCS 4247, pp. 553 – 559, 2006.

approach. On the one hand, fitness sharing is computationally expensive because it requires the comparison of each couple of individuals to measure their distance at each generation. On the other hand, it is difficult to choose a suitable parameter Q_{share} .

Many other people's researches fall into parallel subpopulation method [7].One important class of parallel subpopulation method is island model parallel GAs (IMGAs)[8].Species conservation proposed by Jian-ping Li et al[7] also is a technique of parallel subpopulation method.The main drawback lies in these algorithms is that some additional control parameters have been introduced that need careful selection to ensure good algorithm performance.If we lack of the background of the multimodal function optimization problems,choosing suitable parameters is a difficult task .

This paper presents a population climbing evolutionary algorithm (PCEA) for solving function optimization containing multiple global optima. The algorithm combines a multi-parent crossover operator with the complete local search. The multi-parent crossover operator can enables individual to draw closer to each optimal solution,thus the population will be divided into subpopulations automatically , meanwhile, the local search is adopted to enable individual to converge to the nearest optimal solution which belongs to the same attractor. By this way, each individuals can converge to a global optima, then the population can maintain all global optima. The experiment results show that PCEA is very efficient for the optimization of multimodal functions, usually it can obtain all the global optimal solutions by running once of the algorithm.

The rest of this paper is organized as follows:Section 2 gives the structure of PCEA and describes the implementation of the algorithm.The experiments and comparisons with other algorithms are illustrated by solving several benchmark problems in section 3.Finally,section 4 concludes with some brief remarks.

2 Description of Population Climbing Evolutionary Algorithm(PCEA)

The algorithmic description of PCEA is as follows:

PROCEDURE PCEA ALGORITHM
BEGIN
  Randomly initialize population P={ $X_1, X_2, ..., X_N$ };
  generation=0;
  t=0 ;
  while(the termination condition is not satisfied) do
  BEGIN
    select M points $X'_1, X'_2, ..., X'_M$ from population ;
    X=GCMX($X'_1, X'_2, ..., X'_M$) ;
    IF $better(X, X_{worst})$ THEN BEGIN
      $X_{worst} = X$;

```
            X= LocalSearch(X);
            IF better(X, X_worst) THEN  BEGIN
               X_worst = X;
            END;
      END;
      generation = generation + 1;
   ENDWHILE;
   Output Result;
END;
END.
```

2.1 Multi-parent Crossover Operator (GCMX)

Multi-parent crossover operators,which mean that more than two parents are involved when generating offspring,are a more flexible version,generalizing the traditional two-parent crossover of nature.A novel Multi-parent crossover operator is introduced to accelerate the convergence of the evolution[12]. Fig.1 shows how to generate a new individual using Multi-Parent crossover operator.

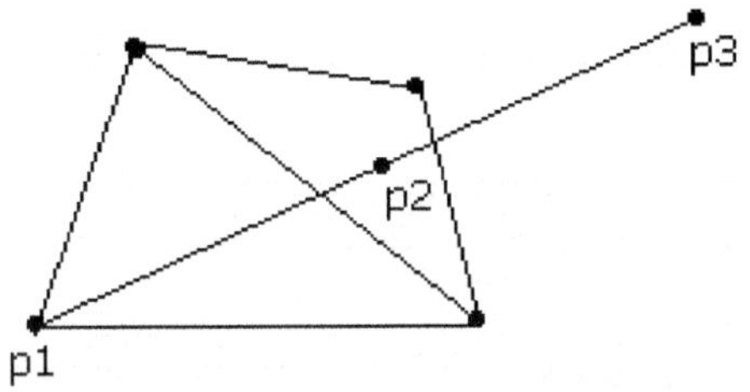

Fig. 1. Multi-Parent Crossover Operator

The flow of generating a new child is as follows:

1. M individuals $x_j, j = 1,2,...M$ are randomly selected in the population.
2. find the worst individual p1 in these M individuals (by comparing with the value of their fitness)
3. Denote the center of the remanent M-1 individuals as p2:

$$p2 = \frac{\sum_{j=1}^{M} x_j - p1}{M - 1} \tag{1}$$

4. find the reflecting point p3 of p1 corresponding to p2:

$$p3 = 2 \times p2 - p1 \tag{2}$$

This Multi-Parent crossover operator has two main advantages :

1. The new offspring generated by GCMX will near the region where individuals may have higher fitness value,so there is rather larger probability that the new individual has higher fitness value than that of the worst individual in the population.Thus,this crossover operator can accelerate the constringency because of its strong direction.In fact,it can instruct its search direction using statistical information.
2. This operator can enables each individual to climb to one optimal solution in evolution, thus the population will be divided into subpopulations automatically .

2.2 Local Search

The local search is adopted when a new child is generated, as a result, the new individual will be converge to a optimal solution. To avoid the disadvantages of the traditional optimization algorithms, we adopt SIMPLEX algorithm[11] to implement the complete local search algorithm, because SIMPLEX algorithm belongs to the class of direct search methods, a class of optimization algorithms which neither compute nor approximate any derivatives of the objective function.

3 Numerical Experiments and Analysis

When testing the algorithm on well understood problems,there are two measures of performance:

1. Success rate: the ratio of found modals and actual modals.
2. The average number of objective function evaluations required to find these optima.

The algorithm SCGA proposed in [7] and the algorithm TSEA proposed in [10] are also evolutionary algorithms for solving multimodal function global optimization problems which proved to be very effective. Some test problems are selected to compare the performance of the proposed algorithm with the algorithm SCGA and TSEA.

Example 1[9]. Humpback function (the function has six local optimal solutions,two of which are global optimal solutions)

$$\min f(x_1,x_2) = (4-2.1x_1^2 + x_1^4/3)x_1^2 + x_1x_2 + (-4+4x_2^2)x_2^2 \tag{3}$$

where $x_1 \in [-3,3], x_2 \in [-2,2]$

Example 2[9]. N-dimension Shubert function(when n=2, the function has 720 local optimal solutions,18 of which are global optimal solutions. When n=3,the function has 81 global optimal solutions) .

$$\min f(x_1,x_2,\cdots,x_n) = \prod_{i=1}^{n}\sum_{j=1}^{5} j\cos((j+1)x_i + j) \tag{4}$$

where $x_i \in [-10,10], i = 1,2,\cdots,n$.

Example 3 two-peak trap [7]
The fitness function of the two-peak trap is defined by:

$$F(c)=\begin{cases}\frac{160}{10}c \quad for\ 0\le c<10\\ \frac{160}{5}(15-c) \quad for\ 10\le c<15\\ \frac{200}{5}(c-15) \quad for\ 15\le c\le 20\end{cases} \tag{5}$$

This function has a global maximum of 200 at c = 20,but it has a "central" false maximum of 160 for c= 10.

To compare the performance of the proposed algorithm with the algorithm SCGA and the algorithm TSEA,we use the same population size adopted in SCGA and TSEA for each test problem.Table 1 summarizes the experimental results we obtained by using PCEA and statistics for the 30 independent runs. Table 1 also gives the comparison between our results and the lasted results [7] and [10] that we can find in the literature.

From the Table 1,we can see that the experimental results using the new algorithm are surprisingly good. For the benchmark problems,PCEA can find all optimal solutions by running once of the algorithm.Note that the average number of function evaluations of PCEA is larger than that of the algorithm SCGA and TSEA for every test problems, the reason lies that PCEA has gained all exact global optima by running once of the algorithm, while SCGA and TSEA can only gain the approximate

Table 1. The comparison of this algorithm PCEA and the algorithms SCGA in the [7] and the algorithm TSEA in the [10]

Example No	Algorithm	Population size	Actual modals	Success rate	Average number of function evaluations
Example 1	TSEA	50	2	100%	1824
	SCGA	50	2	100%	1836
	PCEA	50	2	100%	12752
Example 2 (n=2)	TSEA	1000	18	100%	35016
	SCGA	1000	18	100%	64178
	PCEA	1000	18	100%	460918
Example 2(n=3)	TSEA	4000	81	100%	352486
	SCGA	4000	81	100%	850338
	PCEA	4000	81	100%	2552564
Example 3	TSEA	50	1	100%	522
	SCGA	50	1	100%	625
	PCEA	50	1	100%	2496

global optima . To compare with the performance of PCEA and SCGA and TSEA fairly, the additional local search should be added to gain the exact global optima for the algorithm SCGA and TSEA. Generally, we can conclude that the average number of function evaluations of PCEA is less than that of the algorithm SCGA and TSEA for those difficult problems containing many global optima.

Moreover,there is no additional control parameter in PCEA which is sensitive to a special problem,but the choice of the species distance has a significant effect on the performance of the algorithm SCGA.Meanwhile, it is no need to adopt a special mechanism to maintain the diversity of population in our algorithm , so the algorithm PCEA is very simple with little computational cost .The results show that our algorithm is efficient for the global optimization of multimodal functions.

4 Conclusions

This paper presents a population climbing evolutionary algorithm (PCEA) for solving function optimization containing multiple global optima. Comparing with other algorithms, it has the following advantages.

(1) Our algorithm need no special mechanism to maintain the diversity of population, so the algorithm is very simple with little computation complexity.
(2) The algorithm can gain all exact global optima by running once of the algorithm.
(3) Proposed algorithm need no additional control parameter which depends on a special problem. This is a most important feature which shows the generality of our algorithm.

Acknowledgment

This research is supported by the National Natural Science Foundation of China (No. 60473081,60133010).

References

1. De Jong, K. A.:An Analysis of Behavior of a Class of Genetic Adaptive Systems. Ph.D. thesis,University of Michigan, Ann Arbor, Michigan(1975)
2. Mahfoud, S.W. :Niching methods for genetic algorithms. IlliGAL Technical Report 95001, Illinois Genetic Algorithms Laboratory, University of Illinois, Urbana, Illinois(1995)
3. G. Harik: Finding multimodal solutions using restricted tournament selection. Proc. 6th Int. Conf. Genetic Algorithms, L. J. Eshelman (Ed.), Morgan Kaufmann, San Mateo, CA (1995) 24-31
4. O.J. Mengshoel, D.E. Goldberg.:Probabilistic crowding: Deterministic crowding with probabilistic replacement. In W. Banzhaf et al. (Eds.) Proc. Of the Genetic and Evolutionary Computation Conference GECCO-99, Morgan Kaufmann Publishers, San Francisco, CA (1999) 409-416

5. Goldberg D E, Richardson J.:Genetic algorithms with sharing for multi-modal function optimization. In: Grefenstette eds. Proceedings of the second International Conference on Genetic Algorithms. NJ: Lawrence Erlbaum Associates(1987)41-49
6. Bellomo D, Naso D, B.Turchiano.:Improving genetic algorithms: an approach Based on multi-elitism and lamarckian mutation. In: A. E. Kamel, K.Mellouli, P. Bome, eds. Proceedings of 2002 IEEE International Conference on Systems, Man and Cybernetics. NJ:IEEE Press(2002)4:6
7. Jianping Li, Marton E. Balizs et al.:A Species Conserving Genetic Algorithm for multimodal function optimization. Evolutionary Computation(2002)10(3): 207-234
8. Gordon V S, Whitley D et al. :Dataflow parallelism in genetic algorithms. In: Männer, R., Manderick, B. eds. Parallel Problem Solving from Nature 2. Amsterdam: Elsevier Science(1992)533-542
9. Michalewicz, Z. :Genetic Algorithms + Data Structures = Evolution Programs. Springer-Verlag, New York(1996)
10. Chen ZY,Kang LS:Steady-state Evolutionary Algorithm for Multimodal Function Global Optimization .Computation Intelligence and Security, PT 1,Proceedings,3801:200-207, Lecture Notes in Artificial Intelligence,Springer-Verlag, Berlin,2005
11. J. A. Nelder and R. Mead, A Simplex Method for Function Minimization, Computer Journal,1965, Vol. 7, pp.308-313
12. Chen ZY,Kang LS,Zeng SY:An Algorithm for Solving Constrained Function Optimization Problems Based on Multi-Parent Crossover Operator .International conference of progress in Intelligence Computation &Applications,wuhan(ISICA 2005),China University of Geosciences of Press, 2005,4

Nucleus Classification and Recognition of Uterine Cervical Pap-Smears Using Fuzzy ART Algorithm

Kwang-Baek Kim[1], Sungshin Kim[2], and Kwee-Bo Sim[3]

[1] Dept. of Computer Engineering, Silla University,
Gwaebop-dong, Sasang-gu, Busan, Korea
gbkim@silla.ac.kr
[2] School of Electrical Engineering, Pusan National University,
Jangjeon-dong, Geumjeong-gu, Busan, Korea
sskim@pusan.ac.kr
[3] School of Electrical and Electronic Engineering, Chung-Ang University,
Heukseok-dong, Dongjak-gu, Seoul, Korea
kbsim@cau.ac.kr

Abstract. Segmentation for the region of nucleus in the image of uterine cervical cytodiagnosis is known as the most difficult and important part in the automatic cervical cancer recognition system. In this paper, the region of nucleus is extracted from an image of uterine cervical cytodiagnosis using the HSI model. The characteristics of the nucleus are extracted from the analysis of morphemetric features, densitometric features, colorimetric features, and textural features based on the detected region of nucleus area. The classification criterion of a nucleus is defined according to the standard categories of the Bethesda system. The fuzzy ART algorithm is used to the extracted nucleus and the results show that the proposed method is efficient in nucleus recognition and uterine cervical Pap-Smears extraction.

1 Introduction

Cervical cancer is a malignant tumor developed in the epithelial tissue of the cervix. Conquering the disease is a very important matter. A cell is gathered from cervix uteri of a patient for a uterine cervical cancer check first. Pathologist inspects the dyed sample by a naked eye through a microscope for the uterine cervical cancer consists of observing a cell and an action of a nucleus [1]. The best method for a completely curing cervix cancer is to prevent the cell from developing into cervical cancer. For this purpose, there have been many efforts to completely or at least partially automate the process of cytodiagnosis during the last 40 years [2][3].

Diagnosis of the region of interest in a medical image is largely consisted of area segmentation, feature extraction and characteristic analysis. A medical doctor diagnoses a disease by using character analysis which is deciphering the extracted features to analyze and compare clinical information. The approaches can be largely divided into the pixel-center method and the area-center method [4][5]. Pixel-center method assigns an independent meaning to each pixel according to a predefined criterion. Pixel-center method can use the overall characteristic [4]. Although area-center method relatively needs more calculation time than the pixel-center method, it provides the usage of regional characteristics [5].

T.-D. Wang et al. (Eds.): SEAL 2006, LNCS 4247, pp. 560–567, 2006.

In this paper, the following simplification process allows the nucleus to be more easily detected in the image: (i) converting the extracted image of cervix uteri cytodiagnosis to a grey scaled image, (ii) removing noise using brightness information, and (iii) applying a 5×5 fuzzy grey morphology operation. Extracted information is categorized into 4 degrees based on the extent of abnormality in each nucleus by the fuzzy ART (adaptive resonance theory) algorithm.

2 Nucleus Area Segmentation of Cervix Uteri Cytodiagnosis

The proposed algorithm to extract the nucleus of cervix uteri cytodiagnosis is shown in Fig. 1.

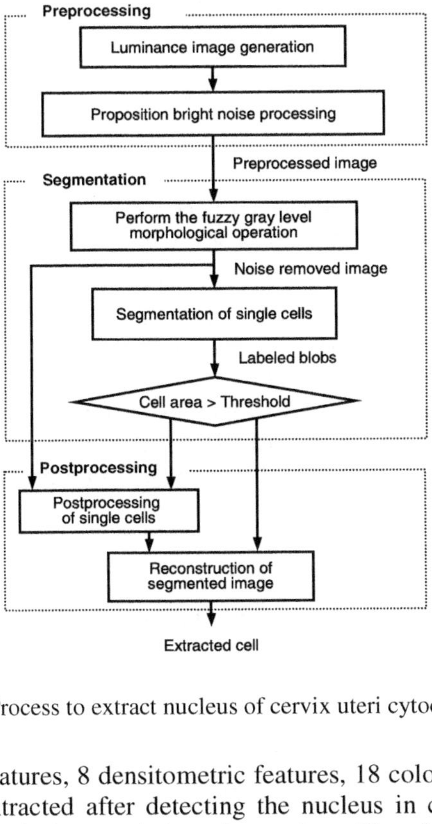

Fig. 1. Process to extract nucleus of cervix uteri cytodiagnosis

13 morphometric features, 8 densitometric features, 18 colorimetric features, and a textural feature are extracted after detecting the nucleus in cervix uteri in order to analysis the changes and characteristics of a nucleus. The classification criterion of extracted nucleus is defined according to the standard categories of the Bethesda system. The fuzzy ART algorithm is applied to classify the malignancy degree of the extracted nucleus according to the standard criterion.

Color images are changed into grey images as a pre-treatment process. Noise is removed by Eq. (1) to improve the quality of the image.

$$z' = \frac{b' - a'}{b - a} \times (z - a) + a' \tag{1}$$

In Eq. (1), a is the lowest brightness value + (30% of the highest brightness value), b is 80% of the highest brightness value, and z is $a \le z \le b$. a' and b' are 0 and 255 respectively. The image of cervix uteri cytodiagnosis is in Fig. 2(a). The grey image in which noise is removed by the proposed pre-treatment process is in Fig. 2(b).

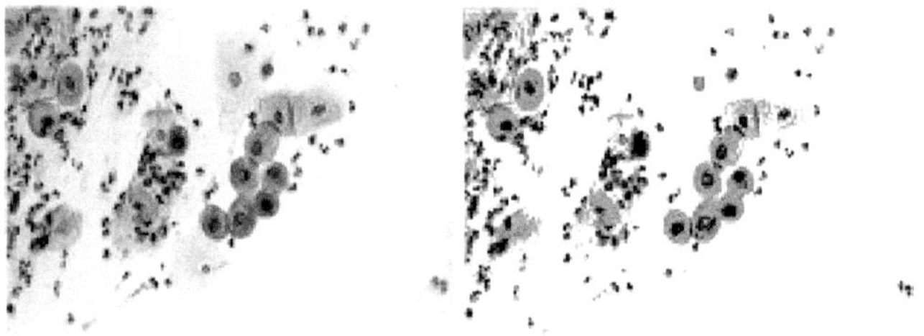

(a) (b)

Fig. 2. Image of cervix uteri cytodiagnosis with noise reduction: (a) Image of cervix uteri cytodiagnosis, (b) Image of cervix uteri cytodiagnosis which noise is removed

If noise is removed by the proposed pre-treatment process, partial information of the normal cell nucleus and cancer cell nucleus is lost. The extracted nucleus of the normal cell and abnormal cell can be precisely extracted by applying a 5×5 fuzzy grey morphology operation. Fuzzy morphology operation is in Eq. (2) and (3). a is the original image and b is the 5×5 mask. The values of mask related to the brightness are described in Fig. 3. The image that is applied to a 5×5 fuzzy grey morphology operation is shown in Fig. 4.

$$\begin{aligned} & A \odot B = \{(x, \mu_{A-B}(x)) \mid x \in E^N\} \\ & \mu_{A-B}(x) = \inf_{z \in E^N} \min\,[1, MRF(\mu_A(z), \mu_{(B:x)}(z))] \\ & MRF(a,b) = \begin{cases} 0 & \text{if } b = 0 \\ a/b & \text{otherwise} \end{cases} \end{aligned} \tag{2}$$

$$\begin{aligned} & A \oplus B = \{(x, \mu_{A \oplus B}(x)) \mid x \in E^N\} \\ & \mu_{A \oplus B}(x) = \sup_{z \in E^N} [\mu_A(z) \times \mu_{(B:x)}(z)] \end{aligned} \tag{3}$$

Threshold is chosen by using a repeat threshold selection method in 45% to 100% section of the histogram that is based on the simplified image using the proposed pre-treatment process.

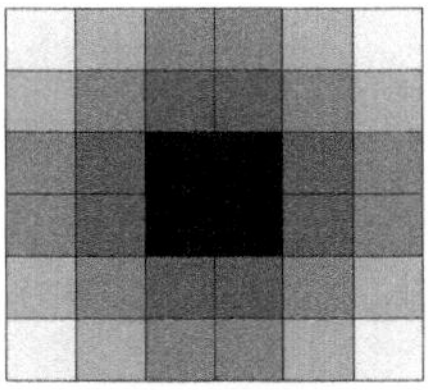

0.137	0.341	0.769	0.769	0.341	0.137
0.341	0.769	0.983	0.983	0.769	0.341
0.769	0.983	1	1	0.983	0.769
0.769	0.983	1	1	0.983	0.769
0.341	0.769	0.983	0.983	0.769	0.341
0.137	0.341	0.769	0.769	0.341	0.137

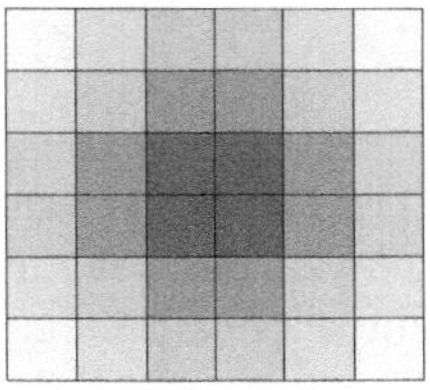

0.098	0.102	0.214	0.214	0.102	0.098
0.102	0.214	0.543	0.543	0.214	0.102
0.214	0.543	0.769	0.769	0.543	0.214
0.214	0.543	0.769	0.769	0.543	0.214
0.102	0.214	0.543	0.543	0.214	0.102
0.098	0.102	0.214	0.214	0.102	0.098

Fig. 3. The values of mask related to the brightness of 200 and 150, respectively

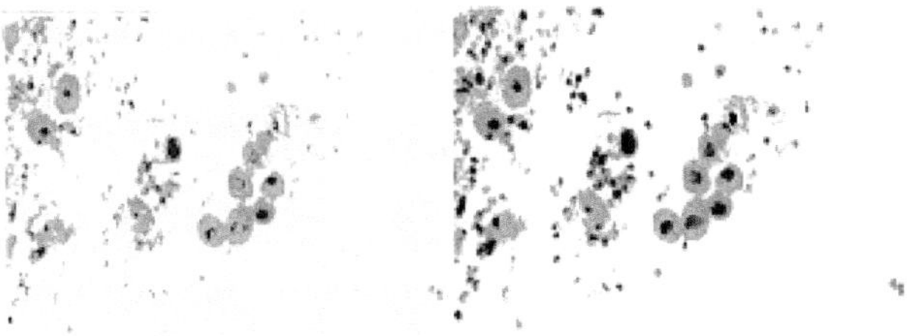

(a) Erosion (b) Dilation

Fig. 4. Images of the closing results

3 Nucleus Characteristic Extraction for Cancer Cell Recognition

A normal nucleus in cervix uteri cytodiagnosis appears small and pale and the nucleus, cytoplasm ratio is also small. On the other hand, an abnormal cell nucleus has a large size and longish or irregular shape compared to a normal cell [6][7].

In this paper, the characteristics of the nucleus and cell image were extracted to classify these characteristics. First, the following features are extracted for nucleus characteristic: area of nucleus, circumference of nucleus, ratio between circumference of nucleus and circumference of quadrilateral, degree of roundness in nucleus' shape, reciprocal of degree of roundness in nucleus' shape, log10 (height/width) in the smallest area of quadrilateral, the longest interior line in horizontal and vertical directions, ratio between area of nucleus and area of quadrilateral. Second, we calculate the area that includes out area of nucleus and area of convex hull wrapped the nucleus in the convex range. HVS divides texture information into channels in which the energy vector and energy deviation are calculated. Texture characteristic vector that uses energy is calculated by the following.

$$q_{mn} = C_{mn} \sum_{\omega} \sum_{\theta} [p_{\theta}(\omega)]^2 \tag{4}$$

$$e_{mn} = \log(1 + p_{mn}) \tag{5}$$

Here $p_\theta(\omega)$ represents the value in the frequency space of each channel. C_{mn} is a constant for the normalization value. The calculation of the texture feature using energy deflection is as follows [8]:

$$q_{mn} = \sqrt{D_{mn} \sum_w \sum_\theta [(p_\theta(w))^2 - p_{mn}]^2} \tag{6}$$

$$d_{mn} = \log(1 + q_{mn}). \tag{7}$$

Here, D_{mn} is a constant for the normalization value. The calculated values from equation (4) to (7) and the texture representation that displays texture feature of a nucleus using the average value and standard deviation of an image is expressed in Eq. (8).

$$Descriptor_{texture} = \begin{bmatrix} dc & std & e_{00} & e_{01} & \dots \\ e_{45} & d_{00} & d_{01} & \dots & d_{45} \end{bmatrix} \tag{8}$$

4 Nucleus Classification and Recognition

The ART2 architecture is evolved to perform learning for binary input patterns and also accommodate continuous valued components in input patterns. The averaged mean value of the difference between input vector and connection weight is used for comparison with the vigilance factor [9].

Fuzzy ART algorithm is an autonomous learning algorithm combined fuzzy logic and ART learning model. Similarity measure in fuzzy ART uses Min operator ($\wedge$) of the fuzzy logic intersection operator as shown in Eq. (9).

$$\|X \wedge W\| / \|X\| \tag{9}$$

Output of the fuzzy ART algorithm is calculated with Eq. (10) and a node that has the biggest value is selected as a winner node. O_j is the output value, O_{j*} is the output of the j-th winner node. α is a designer's choice parameter.

$$O_j = \{\| X \wedge W \|\} / \{\alpha + \| W \|\}, \quad O_{j*} = \vee(O_j) \tag{10}$$

The improved fuzzy ART algorithm that modifies the learning parameter according to the winning occurrence can classify and recognize features of the extracted nucleus. Yager's intersection operator is defined in Eq. (11) and the parameter p is a monotonic decline function.

$$\mu(X_i) = 1 - \min\left[1, \{(1 - X_1)^p + \dots + (1 - X_n)^p\}^{1/p}\right] \tag{11}$$

The proposed fuzzy ART algorithm in order to classify and recognize features of the extracted nucleus is shown in Fig. 5.

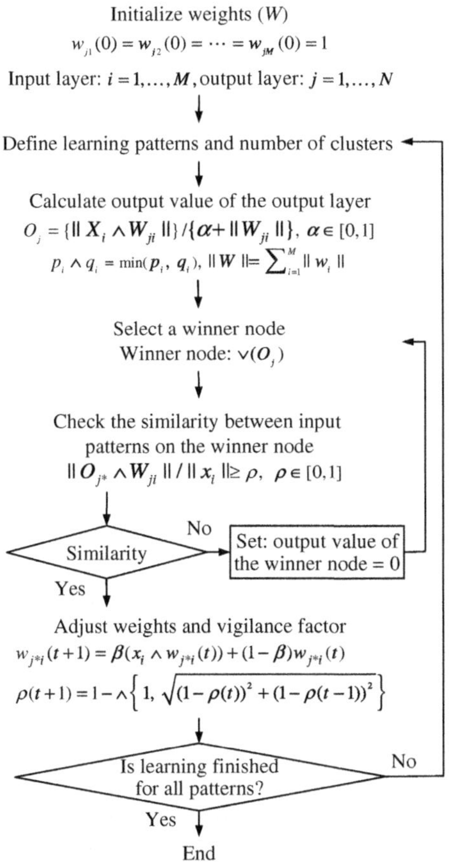

Fig. 5. Algorithm for the selection of the critical value

5 Experiment and Result Analysis

The environment of the experiment is embodied by Visual C++ 6.0 and C++ Builder 6.0. Specimen is 20 samples of 640*480 cervix uteri cytodiagnosis image size and it is acquired in Pusan university hospital. The results of the proposed method were compared with the diagnostic results of a medical specialist The nucleus number of cervix uteri cytodiagnosis extracted from a medical specialist in 20 samples is 316, and the number extracted nucleus is 284. The accuracy of extraction rate is 89.8% in this research. Table 1 shows a part of the characteristic information of the extracted nucleus based on the clinical trials and cell diagnostics of a medical specialist. The classification based on the Bethesda System is in Fig. 6. By using standards of nucleus classification information to identify normal cells and cancer cells such as in Fig. 6, the accuracy of cancer diagnosis can be much more improved compared to classifying the whole nucleus that appears in the image.

Table 1. A part of the characteristic information of the extracted nucleus

Characteristic information of nucleus	Characteristic value of nucleus
Girth ratio between nucleus and rectangular region	0.5791
Degree of roundness of the nucleus shape	0.8571
Area ratio between nucleus and rectangular region	0.6457
Area of convex hull around nucleus	0.5424
Standard deviation of brightness	0.8721
Variance of brightness	0.7452
Mean of wavelet LL2	0.7611
Variance of wavelet LL2	0.7562

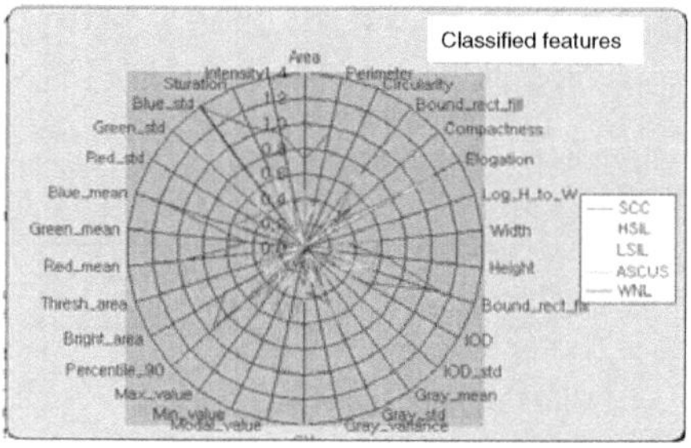

Fig. 6. Information of the standard category in the Bethesda system

The proposed fuzzy ART algorithm is applied to the characteristic information of the 284 extracted nucleuses in order to classify and distinguish normal cell, abnormal cell and cancerous cell. After learning process based on the characteristic information of nuclei, the number of cluster is set to 152. The results from classifying and distinguishing the state of cell using fuzzy ART algorithm based on the Bethesda System is in Table 2.

As can be concluded from Table 2, much more frequently in the proposed method were normal cells classified as abnormal cells than in the diagnosis of a medical specialist. This is because the extracted nucleus is dyed, making it open to being classified as an abnormal cell. But, it can be confirmed that accuracy between abnormal cell and cancer cell that a medical specialist diagnosis has a little performance but classification grade of abnormal cell is different. But it can be concluded through Table. 3 that the proposed method is comparatively efficient in classifying abnormal cells and cancer cells, and can help a medical specialist in diagnosis.

Table 2. Classification and recognition results of cell by fuzzy ART algorithm

Medical specialist			Proposed method		
Normal cell(WNL)	92		Normal cell(WNL)	72	
Abnormal cell	ASCUS	82	Abnormal cell	ASCUS	94
	LSIL	34		LSIL	39
	HSIL	56		HSIL	58
Cancer cell(SCC)	20		Cancer cell(SCC)	21	

6 Conclusion

The nucleus is easily detected by using the following simplification process of the image: converting the extracted image of cervix uteri cytodiagnosis to a grey-scaled image, removing noise using compression and density transformation, and applying a 5×5 fuzzy grey morphology operation. Extracted information is categorized and recognized into normal cells, abnormal cells with 4 degrees, and cancer cells by the fuzzy ART algorithm. The results show that the proposed method is efficient in recognizing abnormal cells and cancer cells. In the future, research should be conducted to correctly extract characteristic information of nucleus by a fuzzy neural network and to clearly define a criterion of classification to reduce presumption error in nucleus classification.

References

1. Seo C. W., Choi S. J., Hong M. K., Lee H. Y., Jeong W. G.: Epidemiologic Observation of Diagnosis of Cervical Cancer and Comparative Study of Abnormal Cytologic Smear and Biopsy Result. Journal of The Korean Academy of Family Medicine 17(1) (1996) 76-82
2. Mark R. Rutenberg: Neural Network Based Automated Cytological Specimen Classification Systems and Method. United States Patent, Patent No.4965725 (1990)
3. Heinz K. Grohs, O. A. Nassem Husain.: Automated Cervical Cancer Screening. Igaku-shoin. (1994)
4. Hugo B. G., Ian R., Alistair C., Kudair H. James H. Tucker and Nasseem H.: Automation in cervical cytology: an overview. Analytical Cellular Pathology 4 (1992) 25-48
5. Lee J. D.: Color Atlas Diagnostic Cytology: Press of Korea Medical Publishing Company. (1989)
6. Kim H. Y., Kim S. A., Choi Y. C., Kim B. S., Kim H. S., Nam S. E.: A Study on Nucleus Segmentation of Uterine Cervical Pap-Smears using Multi Stage Segmentation Technique. Journal of Korea Society of Medical Informatics 5(1) (1999) 89-95
7. Kim K. B., Yun H. W.: A Study on Recognition of Bronchogenic Cancer Cell Image Using a New Physiological Fuzzy Neural Networks. Japanese Journal of Medical Electronics and Biological Engineering 13(5) (1999) 39-43
8. Manjunath B. S., Ma W. Y.: Texture Features for Browsing and Retrieval of Image Data. IEEE Trans. On Pattern Analysis and Machine Intelligence 18(8) (1996)
9. K. B. Kim, Y. J. Kim: Recognition of English calling cards by using enhanced fuzzy radial basis function neural networks, IEICE Trans. Fundamentals of Electronics, Communications and Computer Sciences, E87-A (6) (2004) 1355-1362

Learning Bayesian Networks Structures Based on Memory Binary Particle Swarm Optimization

Xiao-Lin Li[1], Shuang-Cheng Wang[2,3], and Xiang-Dong He[4]

[1] National Laboratory for Novel Software Technology
Nanjing University, Nanjing 210093, China
lixl_126@126.com
[2] Department of Information Science
Shanghai Lixin University of Commerce, Shanghai, China
[3] China Lixin Risk Management Research Institute
Shanghai Lixin University of Commerce, Shanghai, China
wangsc@lixin.edu.cn
[4] Nanjing Research & Development Center
ZTE Corporation, Nanjing 210012, China
hexd_163@163.com

Abstract. This paper describes a new data mining algorithm to learn Bayesian networks structures based on memory binary particle swarm optimization method and the Minimum Description Length (MDL) principle. An memory binary particle swarm optimization (MBPSO) is proposed. A memory influence is added to a binary particle swarm optimization. The purpose of the added memory feature is to prevent and overcome premature convergence by providing particle specific alternate target points to be used at times instead of the best current position of the particle. In addition, our algorithm, like some previous work, does not need to have a complete variable ordering as input. The experimental results illustrate that our algorithm not only improves the quality of the solutions, but also reduces the time cost.

1 Introduction

The Bayesian belief network is a powerful knowledge representation and reasoning tool under conditions of uncertainty. Recently, learning the Bayesian network from a database has drawn noticeable attention of researchers in the field of artificial intelligence. To this end, researchers have developed many algorithms to induct a Bayesian network from a given database [1], [2], [3], [4], [5], [6].

Particle swarm optimization (PSO), rooting from simulation of swarm of bird, is a new branch of Evolution Algorithms based on swarm intelligence. The concept of PSO, which can be described with only several lines of codes, is more easily understood and realized than some other optimization algorithms. PSO has been successfully applied in many engineering projects.

In this paper, we have developed a new data mining algorithm to learn Bayesian networks structures based on an improved swarm intelligence method and the Minimum Description Length (MDL) principle. An important characteristic of our algorithm is that, in order to prevent and overcome premature convergence, memory

T.-D. Wang et al. (Eds.): SEAL 2006, LNCS 4247, pp. 568 – 574, 2006.

feature are introduced into binary particle swarm optimization. The memory influence conceptually derives from the pheromone trail of Ant Colony Optimization (ACO). Furthermore, our algorithm, like some previous work, does not need to impose restriction of having a complete variable ordering as input.

We'll begin with a brief introduction to Bayesian network and MDL principle. In section 2, the memory particle swarm optimization algorithm will be discussed. The in section 4 and 5, the performance of our algorithm will be demonstrated by conducting a series of experiments as well as a summary of the whole paper be made.

2 Bayesian Networks and MDL Metric

2.1 Bayesian Networks

A Bayesian network is a directed acyclic graph (DAG), nodes of which are labeled with variables and conditional probability tables of the node variable which is given its parents in the graph. The joint probability distribution (JPD) is then expressed in the following formula:

$$P(x_1,\cdots\cdots,x_n)=\prod_{k=1\cdots n}P(x_k\mid\pi(x_k)) \tag{1}$$

where $\pi(x_k)$ is the configuration of X_k's parent node set $\Pi(X_k)$.

2.2 The MDL Metric

The MDL metric [7] is derived from information theory and incorporates the MDL principle. With the composition of the description length for network structure and the description length for data, the MDL metric tries to balance between model accuracy and complexity. Using the metric, a better network would have a smaller score. Similar to other metrics, the MDL score for a Bayesian network, S, is *decomposable* and could be written as in equation 2. The MDL score of the network is simply the summation of the MDL score of $\Pi(X_k)$ of every node X_k in the network.

$$MDL(S)=\sum_k MDL(X_k,\Pi(X_k)) \tag{2}$$

According to the resolvability of the MDL metric, equation 2 can be written when we learn Bayesian networks form complete data as follows:

$$\begin{aligned}MDL(S)=N\sum_{k=1}^{N}\sum_{X_k,\Pi(X_k)}P(X_k,\Pi(X_k))\log P(X_k,\Pi(X_k))\\ -\sum_{k=1}^{N}\frac{\log N}{2}\|\Pi(X_k)\|(\|X_k\|-1)\end{aligned} \tag{3}$$

Where N is database size, $\| X_k \|$ is the number of different values of X_k, and $\| \Pi(X_k) \|$ is the number of different parent value combinations of $\Pi(X_k)$.

3 Memory Binary Particle Swarm Optimization Algorithm

Particle swarm optimization (PSO), originally developed by Kennedy and Elberhart [8], is a method for optimizing hard numerical functions on metaphor of social behavior of flocks of birds and schools of fish. It is an evolutionary computation technique based on swarm intelligence. A swarm consists of individuals, called particles, which change their positions over time. Each particle represents a potential solution to the problem. In a PSO system, particles fly around in a multi-dimensional search space. During its flight each particle adjusts its position according to its own experience and the experience of its neighbors, making use of the best position encountered by itself and its neighbors. The effect is that particles move towards the better solution areas, while still having the ability to search a wide area around the better solution areas. The performance of each particle is measured according to a predefined fitness function, which is related to the problem being solved and indicates how good a candidate solution is. The PSO has been found to be robust and fast in solving non-linear, non-differentiable, multi-modal problems. The mathematical abstract and executive steps of PSO are as follows.

Let the i th particle in a D -dimensional space be represented as $X_i = (x_{i1}, \ldots, x_{id}, \ldots, x_{iD})$. The best previous position (which possesses the best fitness value) of the i th particle is recorded and represented as $P_i = (p_{i1}, \ldots, p_{id}, \ldots, p_{iD})$, which is also called *pbest*. The index of the best *pbest* among all the particles is represented by the symbol g. The location P_g is also called *gbest*. The velocity for the i th particle is represented as $V_i = (v_{i1}, \ldots, v_{id}, \ldots, v_{iD})$. The concept of the particle swarm optimization consists of, at each time step, changing the velocity and location of each particle towards its *pbest* and *gbest* locations according to Equations (4) and (5), respectively:

$$V_i(k+1) = \omega V_i(k) + c_1 r_1 (P_i - X_i(k))/\Delta t + c_2 r_2 (P_g - X_i(k))/\Delta t \tag{4}$$

$$X_i(k+1) = X_i(k) + V_i(k+1)\Delta t \tag{5}$$

where ω is the inertia coefficient which is a constant in interval [0, 1] and can be adjusted in the direction of linear decrease [9]; c_1 and c_2 are learning rates which are nonnegative constants; r_1 and r_2 are generated randomly in the interval [0, 1]; Δt is the time interval, and commonly be set as a unit; $v_{id} \in [-v_{\max}, v_{\max}]$, and $v_{\max}$ is a designated maximum velocity. The termination criterion for iterations is determined according to whether the maximum generation or a designated value of the fitness is reached.

The method described above can be considered as the conventional particle swarm optimization, in which as time goes on, some particles become inactive quickly because they are similar to the *gbest* and lost their velocities. In the following generations, they will have less contribution for their very low global and local search capability and this problem will induce premature convergence. Kennedy and Eberhart also developed the discrete binary version of the PSO. Then the particle changes its value by [10]

$$V_i(k+1) = \omega V_i(k) + c_1 r_1 (P_i - X_i(k))/\Delta t + c_2 r_2 (P_g - X_i(k))/\Delta t \tag{6}$$

$$\text{if } \rho_i(k+1) < sig\left(v_i(k+1)\right) \text{ then } x_i(k+1) = 1 \text{ ; else } x_i(k+1) = 0 \tag{7}$$

In this paper, a memory binary particle swarm optimization is proposed. The purpose of the added memory feature is to prevent and overcome premature convergence by providing particle specific alternate target points to be used at times instead of the best current position of the particle [11]. To optimize this effect each particle in the swarm maintains its own memory. The maximum size of the memory and the probability that one of the points it contains will be used instead of the current local optimal point. The local optimum point will be added to the memory if the fitness of this point is better than the least fit stored point. It may also be required to differ by at least a specified amount from any point already in the memory. The new memory point replaces the least fit point if the memory is full. There is a certain probability that a point from the memory will be used instead of the *pbest* in equation 6 above. When a point from a particle memory is to be used the point may be chosen randomly or the probability of selection may be fitness based (with better fitness producing a higher probability of selection).

The memory binary particle swarm optimization algorithm we propose is shown below.

1. Create an initial population, *Pop(t)*. The initial population size is *N*.
2. If the current population contains the optimal individual, then the evolutionary stop; otherwise, continues.
3. Create *N* offspring according to formula (6) and (7), then randomly select a point from a particle memory and use it to instead of *pbest*.
4. Go to 2.
5. Display the result.

4 Experimental Results and Analyses

We have conducted a number of experiments to evaluate the performance of the immune binary particle swarm optimization algorithm. The learning algorithms take the data set only as input. The data set is derived from ALARM network (http://www.norsys.com/netlib/alarm.htm).

Firstly, we generate 5,000 cases from this structure and learn a Bayesian network from the data set ten times. Then we select the best network structure as the final structure. The population size *N* is 30 and the maximum number of generations is

5,000. We employ our learning algorithm to solve the ALARM problem. Some parameters in the experiment are taken as: c_1 =1.9, c_2 =0.8, ω =0.5, ρ =0.5, $v_{\max}$ =8. In order to achieve the algorithm conveniently, we select a point from a particle memory randomly. Table 1 is the performance variation with different memory depth.

Table 1. Performance variation with different memory depth

Memory depth	Generation	MDL metric
No memory	4083.7	81268.3
100	4052.8	81252.5
200	4033.1	81240.6
300	4022.4	81231.7

We also compare our algorithm with binary particle swarm optimization and classical GA algorithm [6]. The algorithms run without missing data and memory depth is 300. The MDL metric of the original network structures for the ALARM data sets of 5,000 cases is 81,219.74.

We also implemented a classical GA to learning the ALARM network. The one-point crossover and mutation operations of classical GA are used. The crossover probability p_c is 0.9 and the mutation probability p_m is 0.01. The MDL metric for memory binary particle swarm optimization algorithm, binary particle swarm optimization algorithm and the classical GA are delineated in Figure 1.

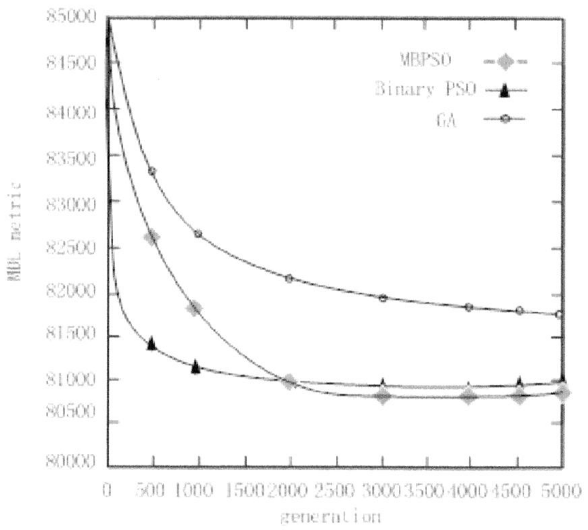

Fig. 1. The MDL metric for the ALARM network

From Figure 1, we see that the value of the average of the MDL metric for memory binary particle swarm optimization algorithm is 81231.7, the value of the average of the MDL metric for binary particle swarm optimization algorithm is 81268.3 and the value of the average of the MDL metric for the GA is 8,1789.4. We find immune binary particle swarm optimization algorithm evolves good Bayesian network structures at an average generation of 4022.4. Binary particle swarm optimization algorithm and GA obtain the solutions at average generation of 4083.7 and 4495.4. From Figure 1, we can also find that the proposed algorithm performs more poorly than Binary particle swarm optimization algorithm does at the early generations. But the performance of the proposed algorithm is better at the end of the generations. The reason of the phenomenon is that the proposed algorithm adds memory feature in order to prevent premature convergence. Thus, we can conclude that memory binary particle swarm optimization algorithm finds better network structures at earlier generations than binary particle swarm optimization algorithm and the GA does.

5 Conclusions

In this paper we propose a memory binary particle swarm optimization algorithm. A memory influence is added to binary particle swarm optimization. The purpose of the added memory feature is to prevent and overcome premature convergence. Further, the proposed algorithm is used to learn Bayesian networks from data. The experimental results show that the proposed algorithm is effective for learning Bayesian networks from data.

Acknowledgements

This research was supported by Shanghai Academic Discipline Project under Grant No. P1601 and Key Research Project from Shanghai Municipal Education Commission under Grant No. 05zz66.

References

1. J Suzuki.: A construction of Bayesian networks from databases based on a MDL scheme, In: *Proceedings of the 9th Conference of Uncertainty in Artificial Intelligence*. San Mateo, CA: Morgan Kaufmann, (1993) 266-273,
2. Y Xiang, S K M Wong.: Learning conditional independence relations from a probabilistic model, Department of Computer Science, University of Regina, CA, (1994) Tech Rep: CS-94-03.
3. D Heckerman.: Learning Bayesian network: The combination of knowledge and statistic data, *Machine Learning*, (1995) 20: 197-243,.
4. J Cheng, R Greiner, J Kelly.: Learning Bayesian networks from data: An efficient algorithm based on information theory, *Artificial Intelligence*, (2002) 137: 43-90.

5. Lam, W. and Bacchus, F.: Learning Bayesian belief networks: An algorithm based on the MDL principle, *Computational Intelligence*, (1994) 10.
6. P. Larrañaga, M. Poza, Y. Yurramendi, R. Murga, and C. Kuijpers.: Structure Learning of Bayesian Network by Genetic Algorithms: A Performance Analysis of Control Parameters, *IEEE Trans. Pattern Analysis and Machine Intelligence,* (1996) 18: 912-926.
7. W. Lam and F. Bacchus.: Learning Bayesian belief networks: an algorithm based on the MDL principle, *Computational Intelligence*, (1994) 10: 269–293.
8. J Kennedy and R.C. Eberhart.: Particle swarm optimization. In: *Proceedings of IEEE International Conference on Neural Networks*, (1995) 1942-1948,.
9. Y. Shi and R.C. Eberhart.: A modified particle swarm optimizer. In: *Proceedings of IEEE International Conference of Evolutionary Computation*, Anchorage, Alaska, May, (1998)69-73.
10. J. Kennedy and R. Eberhart.: A discrete binary version of the particle swarm optimization algorithm. In: *Proceedings of the Conference on Systems, Man, and Cybernetics*. (1997) 4104-4109.
11. T. Hendtlass.: Preserving Diversity in Particle Swarm Optimization. IEA/AIE, LNAI 2718, (2003) 31-40.

Automated Feature Selection Based on an Adaptive Genetic Algorithm for Brain-Computer Interfaces

Guo-zheng Yan, Ting Wu, and Bang-hua Yang

School of Electronic, Information and Electrical Engineering,
Shanghai Jiao Tong University, 200240 Shanghai, China
{gzhyan, wu_ting, ybh}@sjtu.edu.cn

Abstract. In brain-computer interfaces (BCIs), a feature selection approach using an adaptive genetic algorithm (AGA) is described in this paper. In the AGA, each individual among the population has its own crossover probability and mutation probability. The probabilities of crossover and mutation are varied depending on the fitness values of the individuals. The adaptive probabilities of crossover and mutation are propitious to maintain diversity in the population and sustain the convergence capacity of the genetic algorithms (GAs). The performance of the AGA is compared with those of the Standard GA (SGA) and the Filter method in selecting feature subset for BCIs. The results show that the classification accuracy obtained by the AGA is significantly higher than those obtained by other methods. Furthermore, the AGA has a higher convergence rate than the SGA.

1 Introduction

Brain-computer interfaces (BCIs) are devices intended to help disabled people communicate with a computer using the brains' electrical activity. The electrical activity can be measured by electroencephalogram (EEG) [1]. Most BCIs make use of spontaneous mental activities (e.g., thinking on moving a finger, the hand, or the whole arm, etc.) to produce distinguishable electroencephalogram (EEG) signals [2], [3]. The distinguishable EEG signals are then transformed into external actions. Over the past years a variety of evidences have evaluated the possibility to recognize a few mental tasks from EEG signals [4], [5]. However, how to improve the recognition performance of EEG signals in signal processing is still a key problem. This paper will focus on a feature selection approach.

Feature selection is the problem of selecting a subset of d features from a set of D ($D>d$) features based on some optimization criterion. An automated feature selection is crucial for classification because irrelevant features or redundant information are known to cause the classifier to have poor generalization, increase the computational complexity and require more training samples. Various kinds of possible features (autoregressive parameters, power spectral density, averages, wavelet packet energy, etc.) are used for classifying the EEG signals, but the most effective features remain unclear. So, the algorithms which can find a good approximation to the best subset need to be developed.

The most common algorithms for feature selection include Filter algorithms and Wrapper algorithms. The main disadvantage of the Filter algorithms is that it selects

T.-D. Wang et al. (Eds.): SEAL 2006, LNCS 4247, pp. 575 – 582, 2006.

feature subsets that are independent of classification algorithms and ignores the effects of the selected feature subset on the performance of the classification algorithm. As a kind of wrapper method, the genetic algorithm (GA) is often used to perform feature selection.

In the feature selection algorithms for BCIs, there are some reported applications [6], [7] which are based on SGA, but there are very few reported application based on AGA. However, in most cases, the AGA outperforms the SGA significantly [8]. We will explore an adaptive GA (AGA) method. It is compared with those of the Standard GA (SGA) and the Filter method in selecting feature subset for BCIs.

2 Dataset and Feature Extraction

2.1 Dataset

All data were acquired from six healthy subjects (three male and three female, 22-35 years old). The subjects were asked to move a cursor up and down (two mental activities) on a computer screen, while his slow cortical potentials (SCPs) were taken. Each trial lasted 6s and consisted of three phases: a 1s rest phase, a 1.5-s cue presentation phase, and a 3.5-s feedback phase. The cue presentation is a visual target appearing either at the top or bottom. Data were recorded during the 3.5-s feedback at a sampling rate 256Hz. The feedback is provided by a cursor whose vertical position indicated the current level of SCPs (Cz-Mastoids). The following six channels of EEG data were recorded (denotation follows the 10/20 system):

Ch1: A1-Cz (A1 = left mastoid) Ch2: A2-Cz (A2 = right mastoid)
Ch3: (2 cm frontal of C3)-Cz Ch4: (2 cm parietal of C3)-Cz
Ch5: (2 cm frontal of C4)-Cz Ch6: (2 cm parietal of) C4-Cz

2.2 Feature Extraction

Three common types of feature extraction methods were used in this paper.

(1) Autoregressive model coefficients (AR): The autoregressive coefficients of 3 orders, obtained using the Yule-Walker method [9]. We can write the AR features of Ch1 as $\{f_1, f_2, f_3\}$, Ch2 as $\{f_4, f_5, f_6\}$, … , Ch6 as $\{f_{16}, f_{17}, f_{18}\}$.

(2) Average coefficients (AC): We select db4 wavelet functions to decompose the EEG signals up to sixth level giving 64 (2^6) sub-bands. The first 25 sub-bands whose frequencies lower than 50Hz are adopted and the other sub-bands are discarded as useless information. Average coefficients values of the 25 sub-bands are calculated. So, we can obtain 25-Dimensional AC features for each single channel. The AC features from Ch1 to Ch6 can be written as $\{f_{19}, \ldots, f_{43}; f_{44}, \ldots, f_{68}; f_{144}, \ldots, f_{168}\}$.

(3) Average energies (AE): The process of obtaining AE is similar with AC. We also can obtain 25-Dimensional AE features for each single channel. The AE features from Ch1 to Ch6 can be written as $\{f_{169}, \ldots, f_{193}; f_{194}, \ldots, f_{218}; f_{294}, \ldots, f_{318}\}$.

All the features can be written as $U=\{f_1, f_2, \ldots, f_{318}\}$.

3 AGA Method

3.1 SGA

The schematic of the EEG recognition procedure is shown in Fig.1. We can see from Fig.1 that the recognition procedure consists of three steps and the feature selection is one of the three steps. The GA algorithm starts to search from a set of initial solutions in a population. An individual (also called chromosome) represents a possible solution to the problem to be solved and consists of many genes. The procedures in designing GAs mainly include the following five steps.

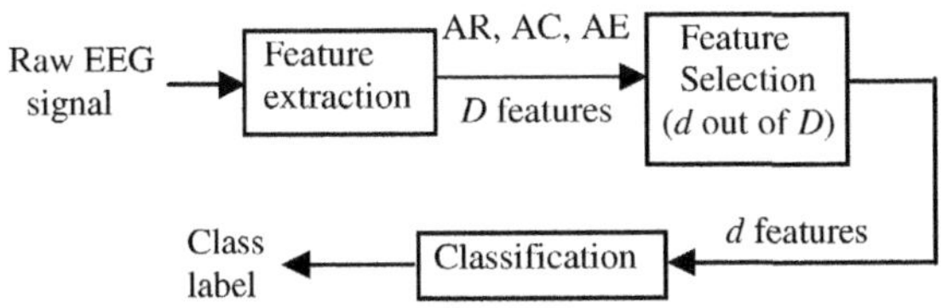

Fig. 1. The schematic of the EEG recognition

(1) Chromosome Encoding

Firstly, let us denote the D (to our problem, D=318) features uniquely by distinct variables from f_1 to f_D, as $U=\{f_1, f_2, \ldots, f_D\}$. For the feature selection problem, a string with D binary digits is used. A binary digit represents a feature, values 1 and 0 meaning selected and removed respectively.

(2) Design of the Fitness Function

A chromosome C represents a selected feature subset X and the evaluation function is $f(X_C)$. In this paper, $f(X_C)$ evaluates the classification performance and controls the required subset size d. The fitness can be defined as:

$$f(X_C) = f_1(X_C) - f_2(X_C) \tag{1}$$

Where $f_2(X_C) = \omega * \| |X_C| - d \|$ with a penalty coefficient ω, $|X_C|$ is subset size of X_C and $f_1(X_C)$ is the classification accuracy of testing samples using 5-fold Cross Validation (5-CV) of all trials. We use the probabilistic neural network (PNN) as our classifier. Because of easy training and a solid statistical foundation in Bayesian estimation theory, PNN has become an effective tool for solving many classification problems [10]. In the PNN, we take the value of the spread of the radial basis functions 1.

(3) Selection Operator

Our design adopts the rank-based roulette-wheel selection scheme. In addition, we use elitist selection.

(4) Crossover Operator

Crossover operation is one of the main methods of producing new generation. Crossover occurs only with some probability P_c (crossover probability). When the solutions are not subjected to crossover, they remain unmodified. Notable crossover techniques include heuristic crossover, arithmetic crossover and so on. Each crossover operator has its own advantages and disadvantageous. Therefore, heuristic crossover and arithmetic crossover are adopted synchronously in order to avoid defects of single operator.

(5) Mutation Operator

Mutation operation is an assistant method of producing new generation. Mutation involves the modification of the value of each gene of a solution with some probability P_m(mutation probability). There are many mutation operators used in the GA, such as multi-boundary, multi-uniform, multi-Gaussian. Similar reasons with selecting crossover operator, the above three mutation operators are adopted synchronously.

3.2 Adaptive Selections of Probabilities of the Crossover and the Mutation

Fixed crossover probability P_c and mutation probability P_m may result in premature and local convergence. On the one hand, a larger P_c and P_m makes GAs have a better search capability. However, it also enlarges the fluctuation of fitness values among individuals, which is detrimental to the convergence. On the other hand, a smaller P_c and P_m makes the GA have better development capability and decrease the fluctuation of fitness values among individuals. However, it maybe causes premature [11]. The choice of probabilities of crossover and mutation, P_c and P_m, are known to critically affect the behavior and performance of GAs. So, an adaptive selection strategy of P_c and P_m is very important to improve the performance of GAs. We will discuss an adaptive selection method of P_c and P_m.

We define four variables f_{max}, f^* , f', f'', where f_{max} and f^* are the maximum fitness and the average fitness of a population respectively; f' is the larger one of fitness values of the two individuals used to cross; f'' is the fitness of the individual used to mutate. It is obvious that $(f_{max} - f^*)$ represents the stability of the whole population. A small $(f_{max} - f^*)$ means that the difference among individuals of population is small. So, P_c and P_m can be determined by$(f_{max} - f^*)$. In order to avoid premature, P_c and P_m should be increased when $(f_{max} - f^*)$ is small, while P_c and P_m should be decreased when $(f_{max} - f^*)$ is large. Therefore, we adopt the parameters selection strategy that P_c and P_m are inversely proportional to $(f_{max} - f^*)$.

However, when the population is close to global optimum, P_c and P_m will increase according to the strategy mentioned above because of the smaller$(f_{max} - f^*)$. As a result, the strategy enlarges the probability of being destroyed of the best individual though it can overcome premature. So, we need not only overcome premature but also preserve excellent individuals. An effective way is that P_c and P_m varies among individuals within the same generation. Individuals with higher fitness values should be preserved and their P_c and P_m should be decreased, while P_c and P_m of individuals with lower fitness values should be increased. Finally, we can conclude that P_c and P_m are not only related to $(f_{max} - f^*)$ but also related to $(f_{max} - f')$ and $(f_{max} - f'')$. So, we can define the following formulas:

$$P_c = k_1(f_{max} - f')/(f_{max} - f^*) \tag{2}$$

$$P_m = k_2(f_{max} - f'')/(f_{max} - f^*) \tag{3}$$

Where k_1, k_2 are constants $k_1, k_2 \leq 1.0$, the two parameters should be adjusted according to a given problem.

In addition, the AGA also provides an effective method to the problem of deciding the optimal values of P_c and P_m.

We adopt the following parameters: 1) population size P=100; 2) the maximum number of generation T=150; 3) penalty coefficient w=0.2; It should be noted that the first two parameters are selected according to common suggestion [10].

4 Results and Analysis

4.1 Results

In order to choose a small number of key features for practical application, we take feature subset size d=5, 10, 15, 20, …, 45, 50 in our experiment. We discuss the following three methods:

(1) The AGA method

For the AGA, P_c and P_m are determined according to expressions (2) and (3) given in section 3. The AGA has introduced new parameters k_1 and k_2 for controlling the adaptive nature of P_c and P_m. We evaluate the performance of the AGA by varying values of k_1 (0.1,0.2,0.3,…,0.8) and k_2 (0.1,0.2,0.3,…,0.8). We notice that no significant difference of the AGA in terms of classification accuracy. In other words, the AGA is not sensitive to parameters k_1 and k_2. So, we finally choose k_1=0.5 and k_2 =0.3 for the AGA.

(2) The SGA method

For SGA, Moderately large values of crossover probability P_c (0.5-1.0) and small values of P_m (0.001-0.05) are commonly employed in GA practice [10]. We use P_c =0.8 and P_m=0.03.

It should be noted that we run the program 10 times to each subset size d and to each subject. Fig.2 shows the average classification accuracy (firstly average the classification accuracy of the 10 times and then the six subjects) versus the number of generation at d = 20.

(3) The Filter method based on Fisher criterion (FFC)

FFC is independent of the classification algorithm. The Fisher criterion function, $F(w)$, can be written as

$$F(\omega) = (\omega^T S_B \omega)/(\omega^T S_W \omega) \tag{4}$$

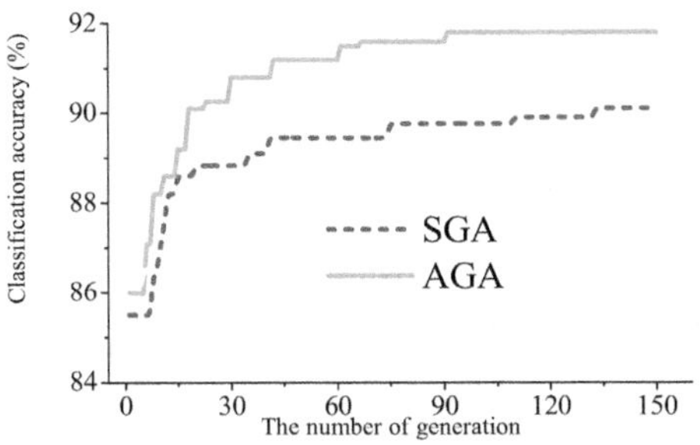

Fig. 2. The classification accuracy versus the number of generation (d = 20)

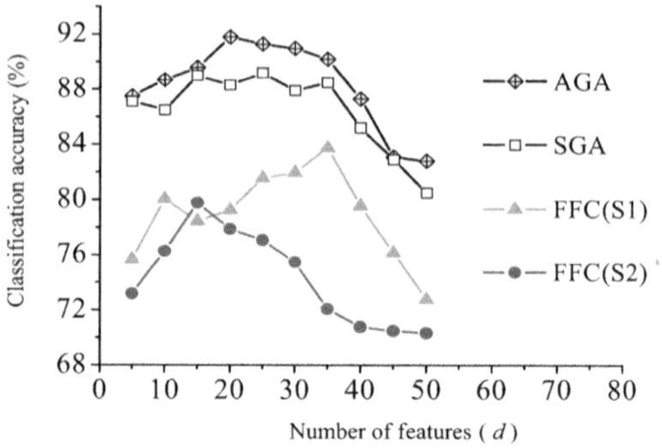

Fig. 3. Classification accuracy versus the number of features used for classification. This figure is the average classification of testing samples accuracy for the six subjects with different methods (AGA, SGA, FFC).

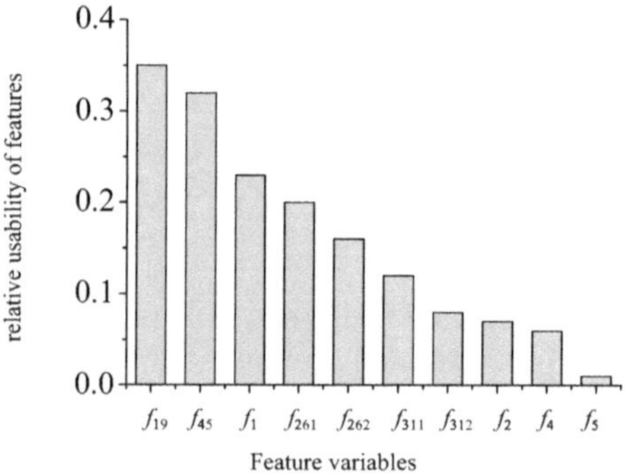

Fig. 4. Relative usability of the ten best features

where S_B is the between-class scatter matrix, S_w is the within-class scatter matrix, and ω is Fisher weight vector which can be obtained by maximizing the value of F. F can be used as a means of assessing the separability of two classes of data. The higher the value of F, the more separable the data are. We adopt two strategies based on the criterion. 1) Strategy1 (S1): it evaluates each feature individually based on the value of F and then selects the d features with the highest values. 2) Strategy2 (S2): it is called the forward sequential method that starts by choosing the best individual feature. Then the feature subset is built from the ground up, by repeatedly adding the next feature that works best with the previously chosen features.

Fig.3 shows the classification accuracy versus the number of features (d) used for classification with different methods. The average standard deviations of classification accuracy with AGA, SGA, FFC (S1), and FFC (S2) are 0.53, 0.59, 0, and 0 respectively. We examine the ten best features that are most frequently used for classification by the AGA method. The relative usability (the ratio between times being selected of each feature and times being selected of all features) of the ten features (averaged over all the running times, subset size, and subjects) is shown in Fig.4. At the same time, we analyse the corresponding channels of the features and the relative usage of the channels is shown in Fig.5.

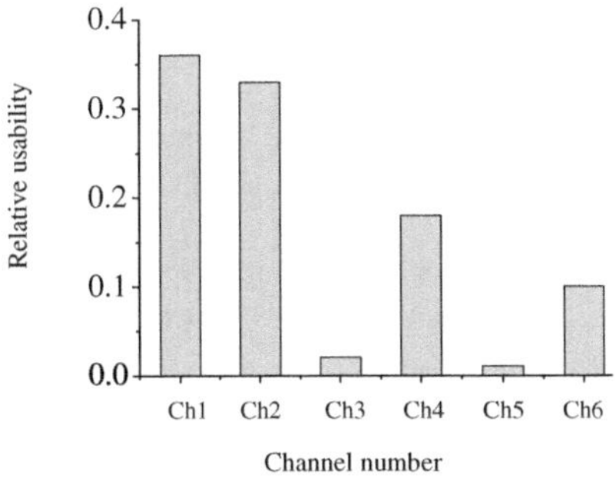

Fig. 5. Relative usability of the six channels

4.2 Analysis

We can see from Fig.3 that the classification accuracies of the features obtained by the AGA and SGA are significantly higher than those obtained by FFC methods. The FFC(S1) is a simple ranking method and it obtains the lowest classification accuracy since it ignores the correlation between features. The FFC(S2) obtains higher classification accuracy than that obtained by the FFC(S1) since the FFC(S2) find *d* features that work well together. Although the average standard deviations of GAs is higher than that of FFC, the GAs are still excellent feature selection methods owing to their classification performances.

We can see from Fig.2 that the AGA obtains higher classification accuracy and convergence rate than the SGA. The AGA and SGA has similar standard deviations. Although the evidence is not conclusive, it appears that the AGA outperforms the SGA for feature selection of BCIs. By using adaptively varying P_c and P_m, the AGA not only improves the performance of GAs but also provides a solution to select the optimal values of P_c and P_m.

In conclusion, the AGA is an efficient feature selection algorithm and obtains the highest classification accuracy among all the methods.

5 Conclusion

The AGA method can pick the most promising features from vast number of available features automatically.

The proposed method provides a good solution for feature selection in BCIs - especially for the development of new paradigms and the use of more EEG recording positions.

The most frequently selected features and channels will provide some useful information for the design of BCIs.

The experiment results show that the AGA is a promising method in feature selection for BCIs. However, the AGA method needs to be handled properly for good performance. More researches and experiments are still needed to test its robustness.

References

1. Guger, C., Neuper, C., Walterspacher, D., Strein, T., and Pfurtscheller, G.: Rapid prototyping of an EEG-based brain–computer interface (BCI). IEEE Trans. Neural Syst. Rehab. Eng., vol. 9, no. 1 (2001) 49–58
2. Lal, T.N., Schröder, M., Hinterberger, T., Weston, J., and Bogdan, M.: Support vector channel selection in BCI. IEEE Trans. Neural Syst.Rehab. Eng, vol. 5, no. 6 (2004) 1003–1010
3. Pfurtscheller, G., Neuper, C., Schlogl, A., and Lugger, K.: Separability of EEG signals recorded during right and left motor imagery using adaptive autoregressive parameters. IEEE Trans. Rehab. Eng., vol. 6, no. 3 (1998) 316–325
4. Millán, J.: Adaptive brain interfaces. Commun. ACM., vol. 46 (2003) 74–80
5. Millán, J., and Mouriño, J.: Asynchronous BCI and local neural classifiers: An overview of the adaptive brain interface project. IEEE Trans. Neural Syst. Rehab. Eng., vol. 11 (2003) 159–161
6. Schroeder, M., Bogdan, M., Rosenstiel, W., Hinterberger, T., and Birhaumer, N.: Automated EEG feature selection for brain computer interfaces. In Proc. 1st int. IEEE EMBS Conf. Neural Eng (2003) 626–629
7. Garrett, D., Peterson, D.A., Anderson, C.W., and Thaut, M.H.: Comparison of linear, nonlinear, and feature election methods for EEG signal classification. IEEE Trans. Neural Syst. Rehab. Eng., vol. 11, no. 2 (2003) 141–144
8. Srinivas, M., Patnaik, L. M.: Adaptive Probabilities of Crossover and Mutation in Genetic Algorithms. IEEE Transactions on Systems, Man, and Cybernetics, vol. 24, no.4 (1994) 656–667
9. WANG, H.Y.: Nonstationary random signal analysis and processing. Defence Industry Press, China (1999)
10. Mao, K.Z., Tan, K.C., and Ser, W.: Probabilistic neural-network structure determination for pattern classification. IEEE Tran. Neural Networks, vol. 11, no. 4 (2000) 1010–1017
11. Zhang, S.H., and Ju, G.: A real-coded adaptive genetic algorithm and its application research in thermal process identification. Proceedings of CSEE. China, vol. 24, no.l.2 (2004) 210–214

Pricing Strategy in an Arborescent Supply Chain System

P.C. Yang[1], H.M. Wee[2], and S.L. Chung[3]

[1] Industrial Engineering and Management Department, St. John's University, Tamsui, Taipei 25135
pcyang@mail.sju.edu.tw
[2] Industrial Engineering Department, Chung Yuan Christian University, Chungli, Taiwan
[3] Information Management Department, St. John's University, Tamsui, Taipei

Abstract. This study develops an optimal pricing and replenishment policy in an arborescent (tree-like) supply chain system. Since it benefits the up-streams more than the down-streams when a producer, a distributor and multiple retailers are integrated, a pricing strategy with price reduction is derived to entice the down-streams to accept the incentive system with minimum total cost. Negotiation factors are incorporated to balance the cost saving between the players. A numerical example solved by GA (genetic algorithm) is provided to illustrate the theory and the problem-solving tool. The result shows that the percentage of total cost reduction incurred from the integration is quite impressive.

1 Introduction

The term leagile has been used by Naylor et al. [8] to include two important concepts: leanness and agility. Leanness emphasizes cost reduction with total waste removal to maximize profits and providing service through a level schedule. Agility requires design of total flexibility to maximize profits through providing exactly what the customer requires at minimum cost. Finch [2] has studied the relationship between the JIT concept and quality improvement.

Monahan [7] was one of the early authors who analyzed a vendor-oriented optimal quantity discount policy that maximized the vendor's gain, but did so at no additional cost to the buyer. Lee and Rosenblatt [5] generalized Monahan's model and developed an algorithm to solve the vendor's ordering and price discount policy. Weng and Wong [10] developed a general all-unit quantity discount model to determine the optimal pricing and replenishment policy. Wee [9] developed a lot-for-lot discount pricing policy for deteriorating items with constant demand rate. Chen et al. [1] addressed a coordination mechanism for a distribution system with one supplier and multiple retailers. None of them considered the general replenishment and pricing policies for an integrated arborescent supply chain system.

GA (genetic algorithm) is a powerful tool to solve the complex-structure problem of many variables. John Holland and his team applied their understanding of the adaptive processes of natural systems to design software for creating artificial systems that retained the robustness of natural systems (Holland [3]). During the last decade, GA, which is a search technique based on the mechanics of natural selection and natural genetics, has

T.-D. Wang et al. (Eds.): SEAL 2006, LNCS 4247, pp. 583 – 591, 2006.

been commonly used to solve global optimization problems. Jinxing and Jiefang [4] studied the application of GA for solving lot-sizing problems. Li et al. [8] demonstrated that GA is effective for dealing with production planning and scheduling problems.

In this study, a general replenishment and joint pricing policy is developed in an integrated arborescent system considering price reduction for reciprocity. GA is used to solve the problem. A numerical example using GA as a solution tool is provided to illustrate the theory.

2 Mathematical Modeling and Analysis

Three scenarios are discussed. The first scenario neglects integration and price reduction. The second scenario considers the integration of all the players without considering price reduction. The last scenario considers integration and price reduction of all players simultaneously.

The mathematical model in this study is based on the following assumptions:

(a) The replenishment rates of the distributor and the retailers are instantaneous, while the producer's replenishment rate is finite.
(b) Each retailer has constant demand rate.
(c) All-unit price reduction is considered.
(d) All players have complete knowledge of each other's information.
(e) A producer, a distributor and multiple retailers are considered.
(f) Unit purchase price is assumed to be the same for all retailers in scenario 1 and 2

The producer's parameters are defined as follows:

P Production rate
T_{p1j} Replenishment period for retailer j in scenario 1
T_{pi} Replenishment period in scenario i, i= 2, 3
C_p Setup cost, $ per cycle
C_{pd} Fixed cost to process order of any size
P_p Unit cost
F_p Inventory carrying cost percentage per time per unit dollar
I_{p1j} Stairs-shaped average inventory level in scenario 1 and retailer j
I_{pi} Stairs-shaped average inventory level in scenario i, i= 2, 3
n_{p1j} Number of deliveries from producer to distributor for retailer j per T_{p1j} in scenario 1
n_{pi} Number of deliveries from producer to distributor per T_{pi} in scenario i, i= 2, 3
TC_{pi} Total cost in scenario i, i= 1, 2, 3
S_p Cost saving of TC_{p3} with respect to TC_{p1}

The distributor's parameters are defined as follows:

T_{d1j} Replenishment period for retailer j in scenario 1
T_{di} Replenishment period in scenario i, i= 2, 3
C_d Setup cost, $ per cycle
C_{dr} Fixed cost to process order of any size
P_{di} Unit cost in scenario i, i= 1, 2, 3
F_d Inventory carrying cost percentage per time and per unit dollar
I_{di} Stairs-shaped average inventory level in scenario i, i= 1, 2, 3
n_{ij} Integral number of deliveries to retailer j per T_{di} in scenario i
TC_{di} Total cost in scenario i, i= 1, 2, 3
S_d Cost saving of TC_{d3} with respect to TC_{d1}

The retailers' parameters are defined as follows:

T_{rij} Replenishment period for retailer *j* in scenario *i*, *i*= 1, 2, 3 and retailer *j*
C_{rj} Ordering cost for retailer *j*
P_{rij} Unit purchased price for retailer *j* to distributor in scenario *i*, *i*= 1, 2, 3
F_{rj} Inventory carrying cost percentage for retailer *j* per year per dollar
I_{rij} Average inventory level for retailer *j* in scenario *i*, *i*= 1, 2, 3
TC_{ri} Total cost for all retailers in scenario *i*, *i*= 1, 2, 3
TC_{rij} Total cost for retailer *j* in scenario *i*, *i*= 1, 2, 3
S_r Cost saving of TC_{r3} with respect to TC_{r1}
S_{rj} Cost saving of TC_{r3j} with respect to TC_{r1j}
N Number of retailers

The common parameters for the producer, the distributor, and the retailers are:

d_j Demand rate for retailer *j*
D Total demand rate of all retailers
TC_i Integrated total cost including TC_{pi}, TC_{di} and TC_{ri}

An example of single producer, single-distributor and multiple retailers is given. This example can be extended to other problem.

The distributor's inventory level and the producer's time-weighted inventory are depicted in Figure 1. The retailer *j*'s, the distributor's and the producer's replenishment intervals are $T_{pi}/(n_{pi}n_{ij})$, T_{pi}/n_{pi} and T_{pi} respectively for *i*= 2, 3. In scenario *1*, the retailer *j*'s, the distributor's and the producer's replenishment intervals are $T_{p1j}/(n_{p1j}n_{1j})$, T_{p1j}/n_{p1j} and T_{p1j} respectively. The retailer *j*'s average inventory level is:

$$I_{rij} = \frac{d_j T_{rij}}{2}, \quad i= 1, 2, 3 \tag{1}$$

Since both the distributor's and the retailer's replenishment rates are instantaneous, the actual distributor's average inventory level, I_{di} in the integrated system is the difference between the distributor's single-echelon average inventory level and all retailers' average inventory level. It is demonstrated as follows:

$$I_{di} = \frac{DT_{di}}{2} - \sum_{j=1}^{N} \frac{d_j T_{rij}}{2}, \; i= 2, 3; \; I_{dij} = \frac{d_j T_{dij}}{2} - \frac{d_j T_{rij}}{2}, \; i= 1 \tag{2}$$

From Yang and Wee [11], the producer's average inventory level is demonstrated in Figure 1 and derived as follows:

$$\begin{aligned} I_{pi} &= (time - weighted \quad inventory \;)/(\; producer \quad cycle \quad time\;) \\ &= \frac{1}{T_{pi}}\{[triangle \quad area\;(a_1a_2b_1,\ldots a_4a_5b_5)] + [recta\;ngular \quad area\;(b_2b_3b_4a_3,\ldots b_5b_6b_7a_5)]\} \\ &= \frac{1}{T_{pi}}[\frac{D^2T_{di}^{\;2}n_{pi}}{2P} + DT_{di}^{\;2}(1-\frac{D}{P})(1+2+3+\ldots+(n_{pi}-1))] \\ &= \frac{DT_{di}}{2}[(n_{pi}-1)(1-\frac{D}{P})+\frac{D}{P}], \qquad i = 2, 3 \end{aligned} \tag{3}$$

And

$$I_{pij} = \frac{d_j T_{dij}}{2}[(n_{pij}-1)(1-\frac{d_j}{P})+\frac{d_j}{P}], \; i= 1 \tag{4}$$

The annual retailer j's and all retailers' total costs are

$$TC_{rij} = P_{rij} d_j + \frac{C_{rj}}{T_{rij}} + \frac{d_j T_{rij} P_{rij} F_{rj}}{2} \tag{5}$$

and

$$TC_{ri} = \sum_{j=1}^{N} TC_{rij} = \sum_{j=1}^{N} P_{rij} d_j + \sum_{j=1}^{N} \frac{C_{rj}}{T_{rij}} + \sum_{j=1}^{N} \frac{d_j T_{rij} P_{rij} F_{rj}}{2} \tag{6}$$

Inventory level and time-weighted inventory

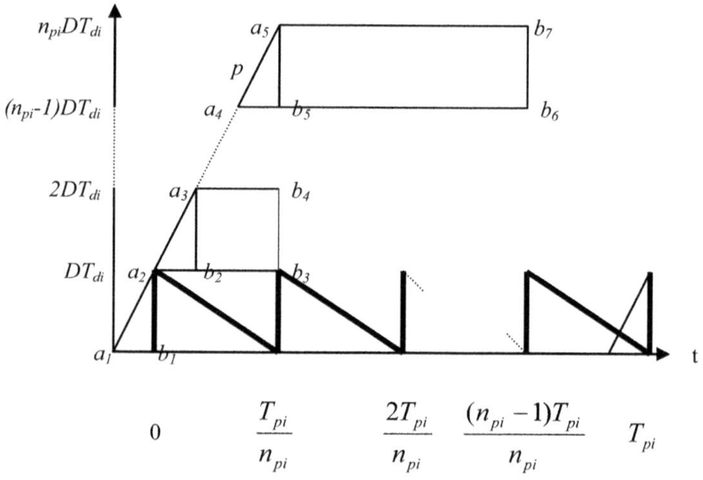

Fig. 1. DC's inventory level and producer's time-weighted inventory in scenario i= 2, 3

2.1 Scenario 1: Integration and Price Reduction Are Not Considered

The distributor's annual cost is

$$TC_{d1} = P_{d1} D + \sum_{j=1}^{N} \frac{C_d + n_{1j} C_{dr}}{T_{d1j}} + \sum_{j=1}^{N} \left(\frac{T_{d1j} - T_{r1j}}{2}\right) d_j P_{d1} F_{dj} \tag{7}$$

The relation between T_{d1j} and T_{r1j} is

$$T_{d1j} = n_{1j} T_{r1j} \tag{8}$$

Using (4), the producer's annual cost is

$$TC_{p1} = P_p D + \sum_{j=1}^{N} \frac{C_p + n_{p1j} C_{pd}}{n_{p1j} T_{d1j}} + \sum_{j=1}^{N} I_{p1j} P_p F_p \tag{9}$$

The relation between T_{p1j} and T_{d1j} is

$$T_{p1j} = n_{p1j} T_{d1j} \tag{10}$$

In the buyer market, the retailers make the first-step decision (11), the distributor makes the second-step decision (12) and the producer makes the last-step decision (13). The retailers, the distributor and the producer make their own independent decision instead of joint decision. Their individual and total costs are:

$$TC_{r1}^* = \sum_{j=1}^{N} \underset{T_{r1j}}{Min} TC_{r1j} \tag{11}$$

$$TC_{d1}^* = \sum_{j=1}^{N} \underset{n_{1j}}{Min} TC_{d1} \tag{12}$$

$$TC_{p1}^* = \underset{n_{p1j}}{Min} TC_{p1} \tag{13}$$

$$TC_1^* = TC_{r1}^* + TC_{d1}^* + TC_{p1}^* \tag{14}$$

2.2 Scenario 2: The Integration of the Producer, Distributor and Retailers Without Price Reduction

The distributor's cost, the producer's cost and the integrated cost of all players' cost are

$$TC_{d2} = P_{d2}D + \frac{1}{T_{d2}}(C_d + \sum_{j=1}^{N} n_{2j}C_{dr}) + (\frac{DT_{d2}}{2} - \sum_{j=1}^{N}\frac{d_jT_{r2j}}{2})P_{d2}F_d \tag{15}$$

$$TC_{p2} = P_pD + \frac{(C_p + n_{p2}C_{pd})}{n_{p2}T_{d2}} + \frac{1}{2}DT_{d2}[(n_{p2}-1)(1-\frac{D}{P}) + \frac{D}{P}]P_pF_p \tag{16}$$

And

$$TC_2 = TC_{r2} + TC_{d2} + TC_{p2} \tag{17}$$

The optimal value of the integrated total cost in scenario 2 is

$$TC_2^* = \underset{all\ n_{p2}, n_{2j}, T_{r2j}}{Min} (TC_{r2} + TC_{d2} + TC_{p2}) \tag{18}$$

For scenario 2, integration is considered. The variables, T_{r2j}, n_{2j} and n_{p2} are optimized jointly.

2.3 Scenario 3: The Integration of the Producer, Distributor and Retailers with Price Reduction

The retailer *j*'s discount price, P_{r3j}, is smaller than P_{r1j}. The annual distributor's cost is:

$$TC_{d3} = P_{d3}D + \frac{1}{T_{d3}}(C_d + \sum_{j=1}^{N} n_{3j}C_{dr}) + I_{d3}P_{d3}F_d + \sum_{j=1}^{N}(P_{r1j}d_j - P_{r3j}d_j) \tag{19}$$

The first, second, third and last terms in (19) are the purchased cost, the ordering cost and the order processing cost, the stairs-shaped carrying cost, and the increased cost when the distributor offers price reduction respectively.
The annual producer's cost is:

$$TC_{p3} = P_p D + \frac{1}{T_{p3}}(C_p + n_{p3}C_{pd}) + I_{p3}P_p F_p + (P_{d1} - P_{d3})D \tag{20}$$

Let the retailer j's cost saving be defined as the difference between TC_{r3j} and TC_{r1j}, one has

$$S_{rj} = TC_{r1j} - TC_{r3j} \tag{21}$$

All retailers' total cost saving is:

$$S_r = \sum_{j=1}^{N} S_{rj} \tag{22}$$

The distributor's cost saving is defined as the difference between TC_{d3} and TC_{d1}. One has

$$S_d = TC_{d1} - TC_{d3} \tag{23}$$

The producer's cost saving is defined as the difference between TC_{p3} and TC_{p1}. One has

$$S_p = TC_{p1} - TC_{p3} \tag{24}$$

The values of S_{rj}, S_r, S_d and S_p are greater than zero. Their relationship is defined as:

$$S_p = \beta\, S_r,\ S_d = \alpha S_r \text{ and } S_{rj} = \gamma_j S_r \tag{25}$$

where

β, α and γ_j are positive negotiation factors and $\gamma_1 + \gamma_2 + \ldots + \gamma_N = 1$

When each negotiation factor is zero, it means all saving are accrued to the retailers; when $\beta = 1$ and $\alpha = 1$, it implies that the total cost saving is equally distributed between the producer, the distributor, and the retailers. A large β means that benefit is accrued mainly to the producer. The nonlinear constrained function optimizing the integrated system is:

$$Min\ TC_3 = TC_{r3} + TC_{d3} + TC_{p3} \tag{26}$$

Subject to the constraint (25)

Using (25), by means of the Variable-Reduction Method, each P_{r3j} and P_{d3} can be derived as a function of variables, n_{p3}, n_{3j} and T_{r3j}. Therefore, the integrated total cost, TC_3 is a function of variables, n_{p3}, n_{3j} and T_{r3j}.

3 GA Solution Procedure

Using a direct analogy to this natural evolution, GA presumes a potential solution in the form of an individual that can be represented by strings of genes. Throughout the genetic evolution, some fitter chromosomes tend to yield good quality offspring inherit from their parents via reproduction.

This study derives the number of deliveries per period to minimize the total cost. The objective function is $TC_i(T_{rij}, n_{pi}, n_{ij})$ with decision variables T_{rij}, n_{pi} and n_{ij}. GA deals with a chromosome of the problem instead of decision variables. The values of T_{rij}, n_{pi} and n_{ij} can be determined by the following GA procedure:

(a) Representation: Chromosome encoding is the first problem that must be considered in applying GA to solve an optimization problem. Phenotype could represent a real numbers and an integer numbers here. For each chromosome, real numbers and integer numbers representation are used as follows:

$$x = (T_{rij}, n_{pi}, n_{ij}) = (T_{ri1}, T_{ri2} \ldots T_{riN}, n_{pi}, n_{i1}, n_{i2} \ldots n_{iN})$$

(b) Initialization: Generate a random population of n chromosomes (which are suitable solutions for the problem)

(c) Evaluation: Assess the fitness $f(x)$ of each chromosome x in the population. The fitness value $f_k = f(x_k) = TC_i(x_k)$ where $k = 1, 2 \ldots n$

(d) Selection schemes: Select two parent chromosomes from a population based on their fitness using a roulette wheel selection technique, thus ensuring high quality have a higher chance of becoming parents than low quality individuals.

(e) Crossover: Approximately 50%-75% crossover probability exists, indicating the probability that the parents will cross over to form new offspring. If no crossover occurs, the offspring are an exact copy of the parents.

(f) Mutation: About 0.5%-1.0% of population mutation rate mutate new offspring at each locus (position in the chromosome). Accordingly, the offspring might have genetic material information not inherited from either parent, thus avoiding falling into the local optimum.

(g) Replacement: An elitist strategy and a steady-state evolution are used to generate a new population, which can be used for an additional algorithm run.

(h) Termination: If the number of generations exceeds 200, then stop; otherwise go to (b).

4 A Numerical Example

The following numerical example is used to illustrate the GA solution procedure given in section 3.

One producer, one distributor and eight retailers; Annual demand rate (units per year): d_1=6,000, d_2=8,000, d_3=15,000, d_4=12,000, d_5=14,000, d_6=16,000, d_7= 18,000, d_8= 20,000; Production rate, P=300,000 units per year; Retailer's ordering cost: C_{r1}= \$100, C_{r2}=\$110, C_{r3}=\$50, C_{r4}=\$80, C_{r5}=\$90, C_{r6}=\$40, C_{r7}=\$110, C_{r8}=\$45; Retailer's percentage carrying cost per year per dollar: F_{r1}=0.2, F_{r2}=0.21, F_{r3}=0.26, F_{r4}=0.22, F_{r5}=0.25, F_{r6}=0.24, F_{r7}=0.26, F_{r8}=0.3; Each retailer's purchased unit price before price discount, P_{r1j}=P_{r2j} =\$25; Distributor's setup cost, C_d=\$5,000; Distributor's fixed cost to process retailer's order of any size, C_{dr}= \$100; Distributor's percentage carrying cost per year per dollar, F_d=0.2; Distributor's unit cost, P_{d1}=P_{d2}=\$20; Producer's setup cost, C_p=\$10,000; Producer's fixed cost to process distributor's order of any size, C_{pd}=\$100; Producer's percentage carrying cost per year per dollar, F_p=0.2; Producer's unit cost,

P_p=\$15; Negotiation factors: β=0.25,α=0.25, $\gamma_1 = \gamma_2 = \gamma_3 = \gamma_4 = 0.15$ and $\gamma_5 = \gamma_6 = \gamma_7 = \gamma_8 = 0.1$.

Using the GA solution procedure given in section 4 and Evolver 4 (a genetic algorithm-based optimizer), the results are given in Table 1. Table 1 illustrates the solutions in various scenarios. In scenario 1, the unit purchased price from the distributor and the retailers are P_{dI}=\$20 and P_{rIj}= \$25 respectively. The retailers' optimal replenishment intervals in scenario 1 are T_{rI1}^*=0.0816, T_{rI2}^* =0.0724... and T_{rI8}^*=0.0245 years. The retailers' regular price minimum total annual cost is TC_{rI}=\$2,752,349. With eight, eight... and fourteen times of deliveries from the distributor to the eight retailers per cycle (n_{I1}=8, n_{I2}=8... and n_{I8}=14), the distributor's annual total cost is TC_{dI}= \$2,374,755. With two times of deliveries from the producer to the distributor per cycle (n_{pIj}=2, j=1, 2...8), the producer's annual total cost is TC_{pI}= \$1,799,695, and the total annual cost is TC_I= \$6,924,799.

In scenario 2, the producer, the distributor and the retailers are integrated without price reduction. The minimum integrated cost intervals are T_{r21}^* =0.1673 and T_{r22}^*=0.1673...and T_{r28}^*=0.0558 years, the retailers', the distributor's, the producer's and the integrated total cost are TC_{r2}^*=\$2,763,383, TC_{d2}^*=\$2,233,253, TC_{p2}^*=\$1,692,837 and TC_2^*=\$6,689,473 respectively. The integrated total cost saving in scenario 2 with respect to scenario 1 is \$235,326. Since the producer benefits \$104,855, the distributor benefits \$141,502 and the retailers lose \$11,034, the retailers will resist integrating in such condition.

Table 1. The optimal solution in various scenarios

Scenario i	i= 1	i= 2	i= 3
$P_{ri1}, P_{ri2}, P_{ri3}, P_{ri4}$ $P_{ri5}, P_{ri6}, P_{ri7}, P_{ri8}$	25, 25, 25, 25 25, 25, 25, 25	25, 25, 25, 25 25, 25, 25, 25	20.99, 21.93, 23.33, 22.82 23.61, 23.91, 24.06, 24.15
$n_{i1}, n_{i2}, n_{i3}, n_{i4}$ $n_{i5}, n_{i6}, n_{i7}, n_{i8}$	8, 8, 13, 9 9, 14, 9, 14	1, 1, 2, 1 2, 2, 2, 3	1, 1, 2, 1 2, 2, 2, 3
$T_{ri1}, T_{ri2}, T_{ri3}$, $T_{ri4}, T_{ri5}, T_{ri6}$, T_{ri7}, T_{ri8}	0.0816, 0.0724, 0.0320 0.0492, 0.0454, 0.0289 0.0434, 0.0245	0.1673, 0.1673, 0.0837 0.1673, 0.0837, 0.0837 0.0837, 0.0558	0.1705, 0.1705, 0.0852 0.1705, 0.0852, 0.0852 0.0852, 0.0568
P_{di}	20	20	19.402
n_{pij} or n_{pi}	n_{plj} = 2, j= 1...8	n_{p2} = 2	n_{p3} = 2
TC_{ri};(TC_{rI}-TC_{ri})	2,752,349	2,763,383; (-11,034)	2,593,671; (158,677)
TC_{di};(TC_{dI}-TC_{di})	2,374,755	2,233,253; (141,502)	2,335,086; (39,669)
TC_{pi};(TC_{pI}-TC_{pi})	1,799,695	1,692,837; (104,858)	1,758,026; (39,6669)
TC_i;(TC_I-TC_i)	6,924,799	6,689,473; (235,326)	6,686,783; (238,016)
$PTCR_i$	-------------	3.40%	3.44%

To entice the retailers to cooperate in implementing the integrated system, the upstream offers some discount to the downstream in the selling price or permissible delay in payment. The optimal retailers' replenishment intervals are T_{r31}^*=0.1705, T_{r32}^*=0.1705...and T_{r38}^*=0.0568 years. The reduction of the total annual cost from scenario 1 to scenario 3 is $238,016. The reduced transaction prices for retailer 1, retailer 2…and the distributor are $20.99 (16.04% price discount), $21.93 (12.04% price discount) and $19.40 (3.00% price discount) respectively.

The percentage total cost reduction ($PTCR_i$) of TC_i with respect to TC_1 be defined as $PTCR_i = (TC_1 - TC_i)/TC_1$, i= 2, 3 are 3.4% and 3.44% respectively.

5 Concluding Remark

Using GA (genetic algorithm), an optimal pricing and replenishment strategy is derived in an arborescent supply chain system. This paper shows that the integration effect results in an impressive percentage total cost reduction, and the price reduction provides an incentive system to entice the retailers to order more quantity for mutual benefits. Negotiation factors are incorporated to share the cost saving benefits.

References

1. Chen, F., Federgrun, A., Zheng, Y.S.: Coordination mechanisms for a distribution system with one supplier and multiple retailers. Management Science **47** (5) (2001) 693-708.
2. Finch, B.: Japanese management techniques in small manufacturing companies: a strategy for implementation. Production and inventory management **27** (3) (1986) 30-38.
3. Holland, J. H.: Adaptation in Natural and Artificial Systems**,** Ann Arbor: University of Michigan Press (1975).
4. Jinxing, X., Jiefang, D.: Heuristic genetic algorithm for general capacitated lot-sizing problems. Computers and Mathematics with Applications **44** (2002), 263-276.
5. Lee, L.E., and Rosenblatt, M.J.: A generalized quantity discount pricing model to increase supplier's profits. Management Science **32** (1986) 1177-85.
6. Li, Y., Man, K. F., and Tang, K. S.: Genetic algorithm to production planning and scheduling problems for manufacturing systems. Production Planning & Control **11** (5) (2000) 443-458.
7. Monahan, J.P.: A quantity discount pricing model to increase vendor profits, Management Science **30** (1984) 720-726.
8. Naylor, J.B., Naim, M.M., and Berry, D.: Leagility: integrating the lean and agile manufacturing paradigm in the total supply chain. International Journal of Production Economics **62** (1998) 107-118.
9. Wee, H.M.: Optimal buyer-seller discount pricing and ordering policy for deteriorating items. The Engineering Economist Winter **43** (2) (1998) 151-168.
10. Weng, Z.K., Wong, R.T.: General models for the supplier's all-unit quantity discount policy. Navel Research Logistics **40** (1993) 971-991.
11. Yang, P.C., Wee, H.M.: The economic lot size of the integrated vendor-buyer inventory system derived without derivatives. Optimal Control Applications and Methods **23** (3) (2002) 163-169.

Online Program Simplification in Genetic Programming

Mengjie Zhang[1,2], Phillip Wong[1], and Dongping Qian[2]

[1] School of Mathematics, Statistics and Computer Science
Victoria University of Wellington, P.O. Box 600, Wellington, New Zealand
[2] Artificial Intelligence Research Centre, Agricultural University of Hebei, China
{mengjie, phillip}@mcs.vuw.ac.nz, {zmj, qdp}@hebau.edu.cn

Abstract. This paper describes an approach to online simplification of evolved programs in genetic programming (GP). Rather than manually simplifying genetic programs after evolution for interpretation purposes only, this approach automatically simplifies programs during evolution. In this approach, algebraic simplification rules, algebraic equivalence and prime techniques are used to simplify genetic programs. The simplification based GP system is examined and compared to a standard GP system on a regression problem and a classification problem. The results suggest that, at certain frequencies or proportions, this system can not only achieve superior performance to the standard system on these problems, but also significantly reduce the sizes of evolved programs.

1 Introduction

Since the late 1990s, genetic programming (GP) has already been applied to many fields, including image analysis [1], object detection [2], regression problems [3] and even control programs for walking robots [4], and achieved quite a reasonable level of success.

While showing promise, current GP techniques are limited, often require a very long evolution time, and frequently do not give satisfactory results for difficult tasks. One problem is the redundancy of programs. Typically, the programs are not simplified until the end of the evolutionary process to enable analysis. However, the redundancies also affect the search process. They force the search into exploring unnecessarily complex parts of the search space [5,2]. The redundancies and complexities have the undesirable consequences that the search process is very inefficient in execution, and the programs are very difficult to understand and interpret. However, the redundant components of the evolving programs can provide a wider variety of possible program fragments for constructing new programs.

The goal of this paper is to invent a method in GP that does online program simplification during the evolutionary process. We will investigate the effect of performing online simplification of the programs during the evolutionary process, to discover whether the reduction in complexity outweighs the possible benefits of redundancy. This approach will be examined and compared with the standard GP without simplification on a regression problem and a classification problem.

T.-D. Wang et al. (Eds.): SEAL 2006, LNCS 4247, pp. 592–600, 2006.

Table 1. Typical simplification rules

No. Precondition	Effective Result	No. Precondition	Effective Result
(1) if<0(A, b, c)	→ b if A < 0, else c	(2) if<0(a, b, b)	→ b
(3) A + B	→ C, C = A + B	(4) A - B	→ C, C = A - B
(5) A × A	→ C, C = A × B	(6) A ÷ B	→ C, C = A ÷ B
(7) A + (B + c)	→ C + c, C = A + B	(8) A + (B - c)	→ C - c, C = A + B
(9) A - (B + c)	→ C - c, C = A - B	(10) A - (B - c)	→ C + c, C = A - B
(11) A × (B × c)	→ C × c, C = A × B	(12) A × (B ÷ c)	→ C ÷ c, C = A × B
(13) A ÷ (B ÷ c)	→ C × c, C = A ÷ B	(14) A + (b + C)	→ B + b, B = A + C
(15) A + (b - C)	→ B + b, B = A - C	(16) A - (b + C)	→ B - b, B = A - C
(17) A - (b - C)	→ B - b, B = A + C	(18) A × (b × C)	→ B × b, B = A × C
(19) A × (b ÷ C)	→ C × b, B = A ÷ C	(20) A ÷ (b ÷ C)	→ B ÷ b, B = A × C
(21) a ÷ 1	→ a	(22) a ÷ a	→ 1
(23) 0 ÷ a	→ 0	(24) 0 × a = a × 0	→ 0
(25) a × 1 = 1 × a	→ a	(26) a + 0 = 0 + a	→ a
(27) a - 0	→ a	(28) a - a	→ 0
(29) a × $\frac{1}{b}$ = $\frac{1}{b}$ × a	→ $\frac{a}{b}$	(30) a × $\frac{b}{a}$ = $\frac{b}{a}$ × a	→ b

2 The Approach

In the tree-based GP, a genetic program looks like an algebraic expression. The function set consists of the commonly used four arithmetic operators and a conditional operator `+`, `-`, `×`, `÷`, `if`. The terminal set consists of a number of feature terminals from the task and several constant terminals. The task of the simplification method is to obtain a smaller program, by removing the redundancy of a program, that yields the same output as the original program. In this approach, we use this idea to construct simplification rules, apply these rules using a postfix search to the genetic programs, and use hashing to estimate the algebraic equivalence to simplify the genetic programs during evolution.

2.1 The Simplification Rules

As in algebraic expression simplification, we use multiple rules (*ruleset*) to simplify a given genetic program, as shown in table 1. A specific rule might only be suitable for removing/reducing a particular part of the genetic program. In this table, constants are represented by upper-case letters (e.g. `A`, `B`), and variables are represented by lower-case letters (e.g `a`, `b`).

2.2 Algebraic Equivalence of Two Subtrees

In a simplification system, it is important to determine whether two subtrees have the same role (or are *equivalent*). For two single nodes, this is fairly trivial; for multi-node subtrees, this will be more difficult. Our goal is to allow for not only noticeably similar expressions (e.g. `(x + y + z)` and `(z + x + y)`) to be identified as equivalent, but also seemingly dissimilar expressions, for example, `(/ (+ (- (* w x) (* x y)) (* (- w y) y)) (- (* x x) (* y y)))` and `(/ (- w y) (- x y))` as well, which is a hard problem.

We use hashing techniques to address the equivalence of two subtrees [6,7] to cope with all common terminals and functions in the evolved programs. In this work, p is used to denote the *hashing order* for the hash function. It is important that the collection of hash values qualify as a finite field [8] and so p should be a prime number as any finite field with p elements is isomorphic to $\mathbb{Z}_p$ [8].

Feature Terminals. In GP, feature terminals represent inputs from the task environment and always keep the same value for a particular fitness case for all genetic programs during evolution. Accordingly, in this approach, we assign the feature terminals certain random hash values at the beginning of a GP run, which remain unchanged for the entire evolution.

Constant Terminals. In GP, constants are usually represented by floating point numbers. We handle this by approximating the floating point with a rational number, thus converting it to a simple division of two integers.

Calculating accurate and irreducible rationals can be very time consuming, so a quick approximation is used. The numerator is formed by multiplying the floating point by a predefined precision constant (δ) and truncating the leftover fractional part. Using the same precision constant as a denominator, a rational representation can be very quickly found.

$$Hash(c) = \frac{c \times \delta}{\delta} \bmod p = (c \times \delta) \times \frac{1}{\delta} \bmod p \tag{1}$$

This approach, of course, requires *modular division*. Now, the division of two numbers $\frac{x}{y}$ is equivalent to the multiplication of the first number with the multiplicative inverse of the second number $x \times \frac{1}{y}$. So to perform division, one needs only to calculate the multiplicative inverse of y and multiply by x. The key point here is to find the integer equivalence of the inverse of δ mod p. In this approach, this is done using the *Extended Euclidean Algorithm* [9,10].

The Arithmetic Operators. Because the hashing method takes place in a finite field, hashing these arithmetic operators are handled using *modulo arithmetic* within the field as follows, where the division hashing follows the rule of the extended Euclidean algorithm mentioned earlier.

$$Hash(A + B) = (A + B) \bmod p \tag{2}$$

$$Hash(A - B) = (A - B) \bmod p \tag{3}$$

$$Hash(A \times B) = (A \times B) \bmod p \tag{4}$$

$$Hash(A \div B) = (A \div B) \bmod p \tag{5}$$

The `if` operator. The `if` conditional operator is a more difficult case, as it is *not* an arithmetic function and so cannot simply be converted to a *modulo arithmetic* equivalent. We use the following approach to handle this operator, which uses division and addition to take into account the position of the three parameters.

$$Hash(\texttt{if}(A, B, C)) = (\frac{A}{B} + C) \bmod p \tag{6}$$

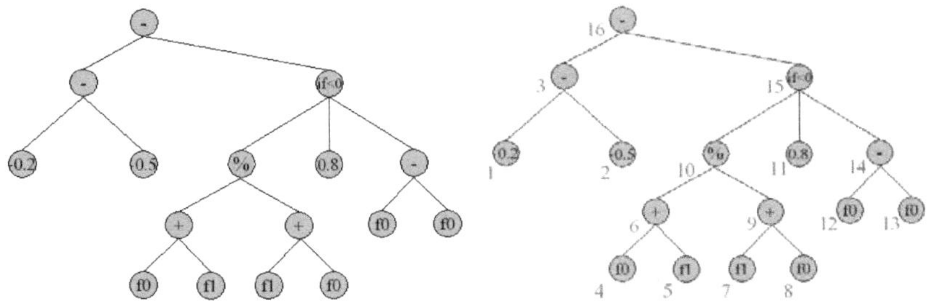

(a) (b)

Fig. 1. An example program. (a) The original program tree; (b)traversal order.

Operator Closure. All of the functions supported are closed, meaning that for any of the functions $\diamond \in \{+, -, \times, \div, \texttt{if} < 0\}$, $(Hash(A) \diamond Hash(B)) \bmod p = Hash(A \diamond B)$ in $\mathbb{Z}_p$. Using this property, one does not need to recalculate the hash values of subtrees each time a tree is to be hashed, as stored hash values of subtrees can be combined to give correct hash values of the whole tree.

2.3 Simplification Process

To apply the ruleset to a genetic program for simplification, we use a kind of "greedy" engine, which is a recursive algorithm. It recursively travels through the program tree in a bottom-up fashion by the postfix order traversal mode. For each node it processes, the algorithm checks each simplification rule in the ruleset. If a rule matches, it is applied to the partial tree associated with the node to make simplification. If *none* of the rules can be applied at a node, the algorithm moves to the next (either neighbouring or parent) node. In this way, the algorithm guarantees that each node in the program tree is visited once only.

Here, we use an example to show the simplification process for a given genetic program. The example program `(- (- -0.2 -0.5) (if<0 (% (+ f0 f1) (+ f1 f0)) 0.8 (- f0 f0)))` can be represented in the tree shown in figure 1 (a). Assume that the hashing order is `17`, `f0` and `f1` are "randomly" assigned the values `3` and `5` respectively. The algorithm traverses the program tree in a "bottom-up" fashion using a post-fix traversal. This means that the algorithm processes the program nodes in the order depicted by integers in figure 1 (b).

The first node inspected by the algorithm is "`-0.2`", followed by "`-0.5`". As no simplification rule exists in the ruleset that governs single nodes, these nodes (and indeed the entire bottom layer of nodes) are left unchanged. Next, the algorithm moves to the parent node of "`-0.2`" and "`-0.5`", which is "`-`". The subtree formed by this node and its children `(- -0.2 -0.5)` matches the precondition for rule (4) `A`$-$`B`. The system applies this rule, replacing the subtree with the rule's effective result: "`0.3`".

Now, the subtrees `(+ f0 f1)` and `(+ f1 f0)` do not match the preconditions for any of the rules, so are left unchanged. Note however, that they both have the same algebraic equivalence hash value (shown in figure 2 (a)). Therefore, when node *10* ("%") is inspected, the subtree `(% (+ f0 f1) (+ f1 f0))` does indeed match the precondition for rule (22) `a÷a`. The entire subtree is replaced using the rule to a single node `1`. Similarly, the subtree `(- f0 f0)` matches rule (28) `a−a` and is replaced by the single node `0` when the algorithm processes "`-`". Figure 2 (b) shows the tree after processing nodes *1* through *14*.

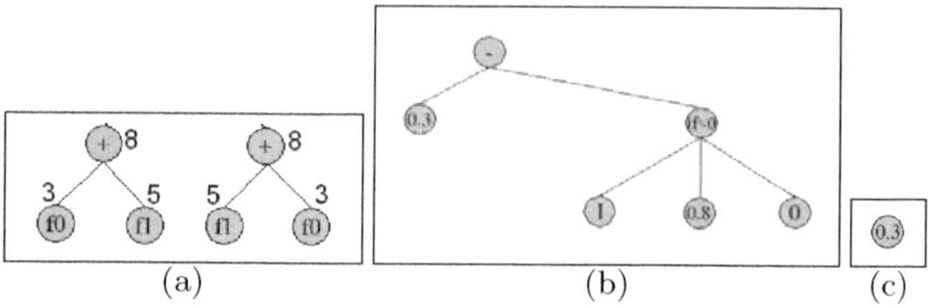

Fig. 2. Program simplification. (a)Hashing of two subtrees with same value; (b) Partial simplification; (c) final program.

At this stage, the program is already reduced to 6 nodes in size, and there are still two nodes left to be processed. Inspecting the `if<0` node, the algorithm matches it with rule (1) `if<0(A b c)`, as the first parameter of the `if<0` operator is a constant. In this case, the constant is 1, which will obviously never be less than 0. The system then, following the rule, replaces this subtree with its third parameter, which is 0.

Lastly the root node is processed, which again matches rule (4) `A−A`. Applying it yields the final result, a single numerical constant node "`0.3`" (figure 2 (c)).

3 Experimentation Setup

3.1 Data Sets

We used two data sets, a symbolic regression task and a object classification tasks, to examine the simplification method. The regression task is a quite complicated piecewise function. We used 200 data points in [-10, 10] as the fitness cases and the task is to evolve a genetic program that conform the curve, as shown in figure 3 (a).

The classification task uses a subset of the Yale Database B Face Dataset [11]. It consists of face images of 5 subjects taken from a single position under 65 different lighting conditions. This creates a set of 325 instances. The backgrounds of the faces are very complicated and different, making the classification problem more difficult. Example images for the task are shown in figure 3 (b). Due to a small number of data examples, 10-fold cross validation method is applied.

3.2 Terminal Set, Function Set and Fitness Function

The terminal set consists of a number of *feature terminals* as well as several randomly generated *constant terminals*. For the symbolic regression task, the feature

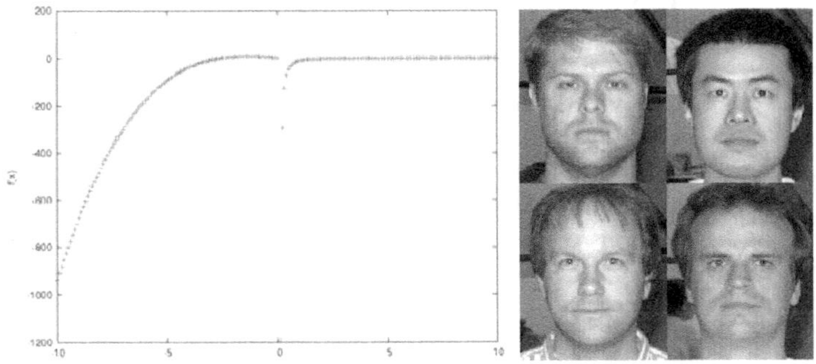

(a) (b)

Fig. 3. Two data sets. (a) regression; (b) classification.

terminal corresponds to the single independent variable. In the face data set, we used 18 feature terminals representing the extracted pixel statistic features from the various facial regions. We use the four basic arithmetic operators and a conditional operator to form the function set $\{+, -, \times, \div, \texttt{if<0}\}$. For the symbolic regression task, the fitness of a program is governed by the mean squared error of the desired output and the actual output of the program on all the fitness cases. For the classification task, the fitness of a program is governed by the classification accuracy in the training set.

3.3 Parameters

The population size is 500, the rates used for crossover, mutation and reproduction are 60%, 30% and 10%, respectively, and the maximum program depth is 8. The evolution will run 50 generations unless an ideal solution program is found, in which case the evolution was terminated early. The *hash order* p was 1000077157, the *constant precision* δ is 1000000. In addition, we used different *proportions*, the percentage of programs in a population to be applied to simplification, and different *frequencies*, how often (in generations) the simplification process is applied, to examine property of the simplification algorithm. The proportions tested here are 0%, 5%, 10%, 20%, 50% and 100%. The frequency used here are every 0, 1, 2, 4, and 6 generations. All single experiments were repeated 50 runs to get the means and standard deviations as results.

4 Results and Discussion

4.1 Overall Results

Table 2 shows the typical best results of the two GP approaches on the two data sets in terms of the effectiveness (best fitness—mean squared error for regression and classification accuracy for classification), training efficiency (training time), and average size of all the programs in the systems in number of nodes.

Table 2. Average best results for the two tasks

Task	Frequency	Proportion	Best Fitness	Time(s)	Avg. Prog Size
Regression	Without	Without	83.774 ± 75.283	5.141 ± 1.019	104.436 ± 22.171
	Every 1	20%	60.354 ± 44.365	4.743 ± 1.383	77.392 ± 24.393
	Every 2	100%	67.346 ± 59.315	4.270 ± 0.759	74.841 ± 13.886
Classification	Without	Without	85.5% ± 11.7%	2.646 ± 0.578	37.861 ± 8.755
	Every 1	20%	87.3% ± 7.2%	2.445 ± 0.484	29.436 ± 5.855
	Every 2	100%	86.7% ± 11.7%	2.367 ± 0.460	29.364 ± 5.966

Efficiency. As expected, it is always possible to find certain proportions and/or frequencies at which the GP approach with the simplification spent shorter time to evolve good programs than the standard GP without simplification. This is mainly because the simplification process removes the redundancy, makes the genetic programs shorter, and accordingly reduces the search space.

Effectiveness. According to table 2, it is always possible to find certain proportions or frequencies at which the GP approach with the proposed simplification achieved superior fitness, either in mean square error or accuracy, on these data sets than the basic GP approach without simplification.

We hypothesised that the simplification process during evolution might destroy the existing good building blocks of the genetic programs, which might result in worse performance. However, these results are clearly different from the original hypothesis. After checking the evolutionary process, we identify the following reasons. At the beginning of evolution, although the simplification algorithm might destroy some potentially good building blocks, this effect was very much offset by the powerful crossover operator, which can preserve good, even form larger, building blocks. At the later stage, when the programs are getting larger, the crossover operator starts to destroy good existing building blocks. The simplification algorithm, however, can generate new genetic materials which might contain good building blocks by *reorganising* the entire genetic programs.

Program Size. According to our experiments, the average size of the programs is significantly reduced for the GP system with simplification at all frequencies and all proportions over the basic GP without simplification. The small size programs have a big advantage in that the actual computation time of the solution program will be short. This is particularly useful in the situations that has a strict time requirement such as in some industrial control and security systems.

4.2 Simplification Frequency/Proportion Analysis

According to our experiments, applying simplification to *all* programs at *every* generation (the last line of table 2) led to a slight loss in fitness and/or a slightly higher computational cost in most cases. This suggests that, if we apply the simplification too often, the programs will not have sufficient chances for evolution and the simplification overhead will be increased. If the evolution chances are reduced to some extent or the overhead outweighs the time saved from processing smaller simplified programs, the performances will deteriorate.

5 Conclusions

This paper aimed to develop an online program simplification approach in GP during the evolutionary process. This goal was successfully achieved by defining a set of algebraic simplification rules, traversing the program tree in a bottom-up fashion by a postfix order, and applying the simplification rules along with an algebraic equivalence component to non-terminal nodes in the evolved programs.

The GP system with the simplification algorithm was examined and compared with the basic GP approach without simplification on a regression problem and a classification problem. The results suggest that, at certain proportions or certain frequencies, the new simplification approach always outperformed the basic GP approach in terms of system effectiveness, efficiency and program size on these data sets.

The results also suggest that performing simplification of all programs in the population at every generation is not recommended. While the approach seems to be able to reduce the search space, it is not clear whether and/or how it destroys good building blocks in the early stage of evolution, which needs to be further investigated.

Acknowledgement. This work was supported in part by the Marsden Fund at Royal Society of New Zealand under grant No. 05-VUW-017 and University Research Fund 6/9 at Victoria University of Wellington.

References

1. Poli, R.: Genetic programming for image analysis. In Koza, J.R., Goldberg, D.E., Fogel, D.B., RioloOB, R.L., eds.: Genetic Programming 1996: Proceedings of the First Annual Conference, Stanford University, CA, USA, MIT Press (1996) 363–368
2. Zhang, M., Ciesielski, V.: Genetic programming for multiple class object detection. In Foo, N., ed.: 12th Australian Joint Conference on Artificial Intelligence. Volume 1747 of LNAI., Sydney, Australia, Springer-Verlag (1999) 180–192
3. Koza, J.R.: Genetic Programming: On the Programming of Computers by Means of Natural Selection. MIT Press, Cambridge, MA, USA (1992)
4. Busch, J., Ziegler, J., Aue, C., Ross, A., Sawitzki, D., Banzhaf, W.: Automatic generation of control programs for walking robots using genetic programming. In: EuroGP '02: Proceedings of the 5th European Conference on Genetic Programming, London, UK (2002) 258–267
5. Tackett, W.A.: Genetic programming for feature discovery and image discrimination. In Forrest, S., ed.: Proceedings of the 5th International Conference on Genetic Algorithms, ICGA-93, University of Illinois at Urbana-Champaign, Morgan Kaufmann (1993) 303–309

6. Martin, W.A.: Determining the equivalence of algebraic expressions by hash coding. j-J-ACM **18**(4) (1971) 549–558
7. Gonnet, G.H.: Determining equivalence of expressions in random polynomial time. In: STOC '84: Proceedings of the sixteenth annual ACM symposium on Theory of computing, New York, NY, USA, ACM Press (1984) 334–341
8. Lidl, R., Niederreiter, H.: Introduction to finite fields and their applications. Cambridge University Press, New York, NY, USA (1986)
9. Trappe, W., Washington, L.C.: Introduction to Cryptograpy with Coding theory. 2ed edn. Prentice-Hall (2006)
10. Cherowitzo, B.: (2006) Lecture Notes. http://www-math.cudenver.edu/~wcherowi/courses/m5410/exeucalg.html. Visited on 7 January 2006.
11. Georghiades, A., Belhumeur, P., Kriegman, D.: From few to many: Illumination cone models for face recognition under variable lighting and pose. IEEE Trans. Pattern Anal. Mach. Intelligence **23**(6) (2001) 643–660

Refining Fitness Functions and Optimising Training Data in GP for Object Detection

Mengjie Zhang[1,2], Malcolm Lett[1], and Yuejin Ma[2]

[1] School of Mathematics, Statistics and Computer Science
Victoria University of Wellington, P.O. Box 600, Wellington, New Zealand
[2] College of Mech. and Elec. Eng., Agricultural University of Hebei, China
{mengjie, malcolm}@mcs.vuw.ac.nz, {zmj, myj}@hebau.edu.cn

Abstract. This paper describes an approach to the refinement of a fitness function and the optimisation of training data in genetic programming for object detection particularly object localisation problems. The approach is examined and compared with an existing fitness function on three object detection problems of increasing difficulty. The results suggest that the new fitness function outperforms the old one by producing far fewer false alarms and spending much less training time and that some particular types of training examples contain most of the useful information for object detection.

1 Introduction

Genetic programming (GP) is a relatively recent and fast developing approach to automatic programming [1,2,3]. Since the 1990s, GP has been applied to a range of object recognition tasks with some success [1,4,5,6,7].

Finding a good fitness function for a particular object detection problem is an important but difficult task in developing a GP system. Various fitness functions have been devised for object detection, with varying success [1,5,7,8,9]. They tend to combine many parameters using scaling factors which specify the relative importance of each parameter, with no obvious indication of what scaling factors are good for a given problem. Many of these fitness functions for localisation require clustering to be performed to group multiple localisations of a single object into a single point before the fitness is determined [10,9,8]. Other measures are then incorporated in order to include information about the pre-clustered results. While some of these systems achieved good detection rates, many of them resulted in a large number of false alarms.

Organising training data is critical to any learning approaches. The previous approaches in object detection tend to use all possible positions of the large image in training an object detector [10,8], which often require a very long training time due to the use of a large number of positions on the background.

This paper aims to investigate a new fitness function and a new way to optimise the training data in GP for object localisation, with the goal of improving the detection performance. The approach will be examined on a sequence of object detection problems of increasing difficulty.

T.-D. Wang et al. (Eds.): SEAL 2006, LNCS 4247, pp. 601–608, 2006.

2 Overview of the Approach

In object detection, a raw image is taken and a trained localiser applied to it, producing a set of points found to be the positions of these objects. Single objects could have multiple positions ("localisations"), however ideally there would be exactly one localisation per object. Regions of the image are "cut out" at these specified positions and then classified using a trained classifier.

The object localisation stage is performed by means of a window which sweeps over the whole image, and for each position, the features are extracted and passed to the trained localiser. The localiser then determines whether each position is an object or not (i.e. background).

GP for Object Localisation. This work will focus on object localisation using genetic programming, which has a learning process and a testing procedure. In the learning/evolutionary process, the evolved genetic programs use a square input field which is large enough to contain each of the objects of interest. The programs are applied at many sampled positions within the images in the training set to detect the objects of interest. If the program localiser returns a value greater than or equal to zero, then this position is considered the centre of an object of interest; otherwise it is considered background. In the test procedure, the best evolved genetic program obtained in the learning process is then applied, in a moving window fashion, to the whole images in the test set to measure object detection performance.

Data Sets. We used three image data sets of New Zealand 5 and 10 cent coins in the experiments. Examples are shown in Figure 1. The data sets are intended to provide object detection problems of increasing difficulty. The first data set (*easy*) contains images of tails and heads of 5 and 10 cent coins against an almost uniform background. The second set (*medium difficulty*) consists of heads and tails of 10 cent coins against a noisy background, making the task harder. The third data set (*hard*) contains tails and heads of both 5 and 10 cent coins against a noisy background.

We used 24 images for each data set in our experiments and equally split them into three sets: a training set, a validation set, and a test set.

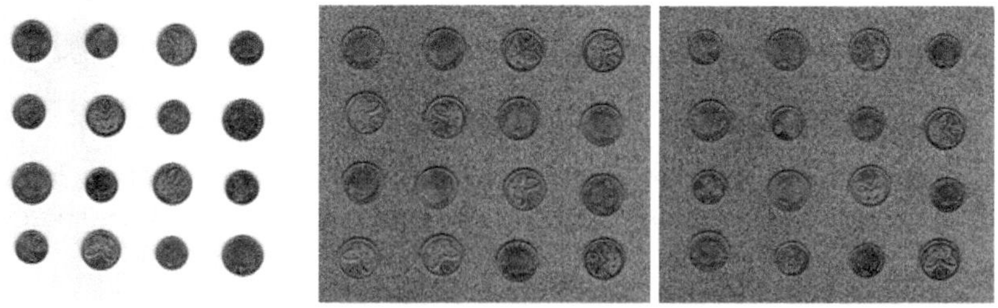

(a) (b) (c)

Fig. 1. Sample images in the three data sets. (a) Easy; (b) Medium difficulty; (c) Hard.

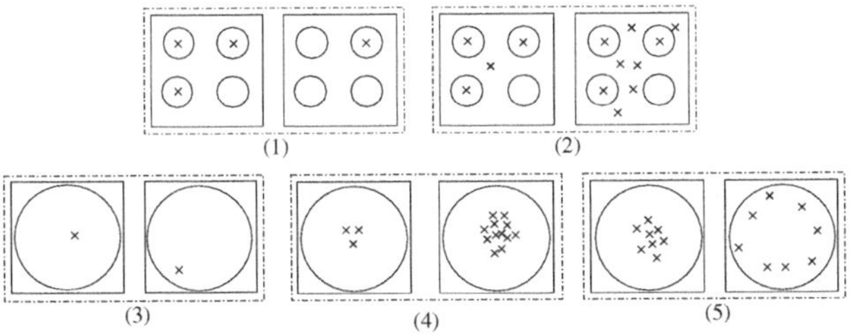

Fig. 2. Examples of the design considerations of the fitness function

GP System Configurations. In this system, we used tree structures to represent genetic programs [2]. The ramped half-and-half method [1] was used for generating programs in the initial population and for the mutation operator. The proportional selection mechanism and the reproduction, crossover and mutation operators were used in evolution.

The terminals use image features extracted by calculating the mean and standard deviation of pixel values within several circular regions. This set of features has the advantages of being rotationally invariance. In addition, we also used a constant terminal. The function set contains the four standard arithmetic and a conditional operation {`+, -, *, /, if`}.

We used a population of 500 genetic programs in each experiment run. The reproduction rate, crossover rate and mutation rate were 5%, 70% and 25%, respectively. The maximum program size was 8. The system run 50 generations unless it found a solution, in which case the evolution was terminated early. A total number of 100 runs were performed on each data set and the average results are calculated and presented.

3 Fitness Function

Different evolved programs typically result in different numbers of false alarms and such differences should be reflected by the fitness function. For example, some requirements shown in figure 2 should be considered. In this figure, the circles are target objects and squares are large images or regions. A cross (x) represents a detected object. In each of the five cases, the program associated with the left figure should be considered better than that with the right.

As the goal is to detect the target objects with no or a small number of false alarms, many GP systems uses a combination of detection rate and false alarm rate or recall and precision as the fitness function. For example, a previous GP system uses the following fitness function [6]:

$$fitness_{CBF} = A \cdot (1 - DR) + B \cdot FAR + C \cdot FAA \tag{1}$$

where DR, FAR, and FAA are detection rate, false alarm rate (also called *false alarms per object*), and false alarm area, respectively, and A, B, C are constant weights which reflect the relative importance of detection rate versus false alarm rate versus false alarm area. Since this method used clustering before calculating the fitness, we refer to it as *clustering based fitness*, or CBF for short. While this fitness function has considered case 1, and partially considered cases 2 and 4, it does not take into accounts of cases 3 and 5.

3.1 A New Fitness Function — RLWF

To deal with all the situations in the five design requirements, we developed a new fitness function based on a "Relative Localisation Weighted F-measure" (RLWF), which attempts to acknowledge the worth/goodness of individual localisations made by a genetic program. Instead of using either correct or incorrect to represent a localisation, each localisation is allocated a weight (referred to as the *localisation fitness, LF*) which represents its individual worth.

Each weight is calculated based on its relative location, or the distance of the localisation from the centre of the closest object, as shown in Equation 2.

$$\mathrm{LF}(x,y) = \begin{cases} 1 - \frac{\sqrt{x^2+y^2}}{r} & \text{, if } \sqrt{x^2+y^2} \le r \\ 0 & \text{, otherwise} \end{cases} \tag{2}$$

where $\sqrt{x^2+y^2}$ is the distance of the localisation position (x, y) from target object centre, and r is called the "localisation fitness radius", defined by the user. In this system, r is set to a half of the square size of the input window. The localisation fitness is then used to construct the new fitness function, as shown in Equations 3 to 4. The precision and recall are calculated by taking the localisation fitness for all the localisations of each object and dividing this by the total number of localisations or total number of target objects respectively.

$$\mathrm{WP} = \frac{\sum_{i=1}^{N}\sum_{j=1}^{L_i} \mathrm{LF}(x_{ij}, y_{ij})}{\sum_{i=1}^{N} L_i}, \qquad \mathrm{WR} = \frac{\sum_{i=1}^{N} \frac{\sum_{j=1}^{L_i} \mathrm{LF}(x_{ij}, y_{ij})}{L_i}}{N} \tag{3}$$

$$\mathrm{fitness}_{RLWF} = \frac{2 \times \mathrm{WP} \times \mathrm{WR}}{\mathrm{WP} + \mathrm{WR}} \tag{4}$$

where N is the total number of target objects, (x_{ij}, y_{ij}) is the position of the j-th localisation of object i, L_i is number of localisations made to object i, WP and WR are the weighted precision and recall.

3.2 Results

To give a fair comparison, the "localisation recall (LR) and precision (LP)" were used to measure the final object detection accuracy on the test set. LR is the number of objects with one or more correct localisations within the localisation fitness radius at the target object centres as a percentage of the total number of target objects, and LP is the number of correct localisations which fall within the localisation radius at the target object centres as a percentage of the total number of localisations made. In addition, we also check the "Extra Localisations" (ExtraLocs) for each system to measure how many extra localisations were made for each object.

Table 1. Results of the GP systems with the two fitness functions

Dataset	Fitness function	Test Accuracy			Training Efficiency	
		LR (%)	LP (%)	ExtraLocs	Generations	time(sec)
Easy	CBF	99.99	98.26	324.09	13.69	178.99
	RLWF	99.99	99.36	98.35	36.44	111.33
Medium	CBF	99.60	83.19	804.88	36.90	431.94
	RLWF	99.90	94.42	95.69	34.35	105.56
Hard	CBF	98.22	75.54	1484.51	31.02	493.65
	RLWF	99.53	87.65	114.86	33.27	107.18

Table 1 shows the results of the GP systems with the two fitness functions. The results on the easy data set show that both the fitness functions achieved good test accuracy. Almost all the objects of interest in this data set were successfully localised with very few false alarms (both LR and LP are very close to 100%), reflecting the fact that the detection task in this data set is relatively easy. However, the new fitness function (RLWF) produced a far fewer number of extra localisations per object than the clustering based fitness function (CBF). Although CBF used a considerably smaller number of generations than the new RLWF, it actually spent about 50% longer training time. This confirms our early hypothesis that the clustering process in CBF is time consuming and the approach with RLWF is more efficient than that with CBF.

The results on the other two data sets show a similar pattern in terms of the number of extra localisations and training time. The systems with RLWF always produced a significantly fewer number of extra localisations and a much short training time than CBF. In addition, although almost all the objects of interest in the large images were successfully detected (LRs are almost 100%), the localisation precisions achieved by RLWF were significantly better than CBF, suggesting that the new fitness function outperforms the existing one in terms of reducing false alarms.

4 Optimising Training Data

This approach focuses on investigating whether some examples are better than others and how to pick them up.

4.1 Four Training Data Types

The traditional approaches usually use *positive* and *negative* examples. The former refers to the exact object examples and the latter refers to those for the background [5,6,9]. However, this did not consider those with some pieces of objects and some pieces of background. In this approach, we identified four basic types of training examples, as shown in figure 3. The *exact centre* type (figure 3a) refers to the positive object examples which sit exactly the centre of the sweeping window. The *background* type (figure 3d) refers to the positions (x)

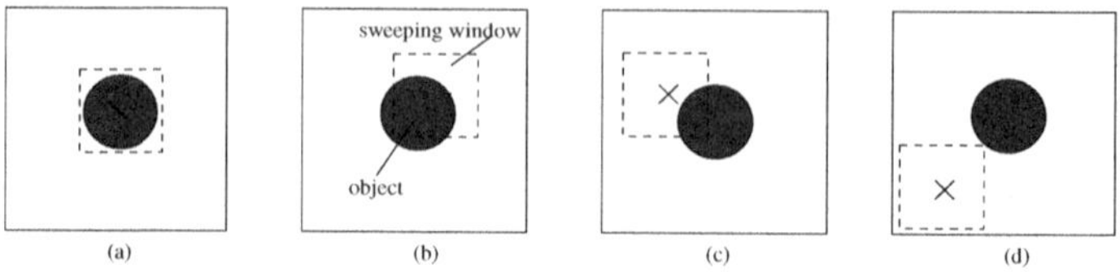

Fig. 3. Examples of training data types caused by different input window positions

which do not contain any piece of objects. The *close to center* type refers to the examples that have the centre of the sweeping window falling down within the bounds of an object (figure 3b). The *include objects* type refers to those that contain some pixels of an object but are not considered as *close to centre.*

4.2 Optimisation of Training Data

We assume that there is some proportion of these four types which is optimal (or close to optimal) for an object detection problem. From previous research, we found that the exact centre type was always important for object detection. As the number of examples of this type is very small, we will always use this type of examples in the experiments and vary the proportions among the rest three types to find the optimal combinations.

If C, I and B represent the percentages of the examples for the three types *close to centre, include objects* and *background*, then we have $C+I+B = 100\%$, which has the nice feature that it represents only a plane effectively reducing the parameter search space from 3D to 2D, as shown in figure 4 (left). We experimented with 28 separate proportions sampled from the plane, as shown in figure 4 (right), where each entry represents value for I for a given C and B. For example, the first two entries in the first row show that, using no background ($B = 0$), we will examine 100% C with 0% I, and 83% C with 17% I type objects.

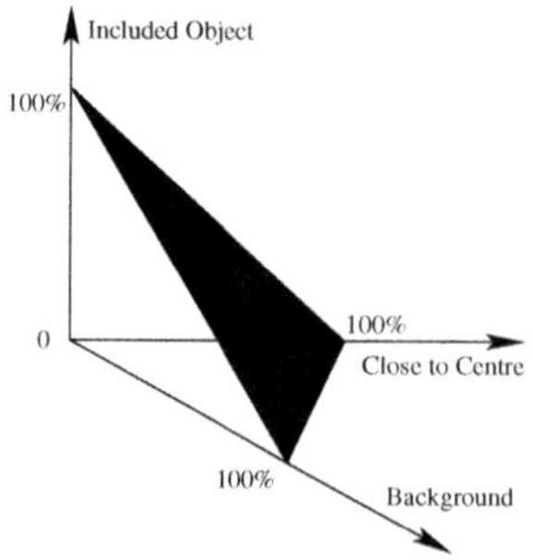

	I for each C and B (%)						
B\C	100	83	67	50	33	17	0
0	0	17	33	50	67	83	100
17		0	17	33	50	67	83
33			0	17	33	50	67
50				0	17	33	50
67					0	17	33
83						0	17
100							0

Fig. 4. Training data proportions set

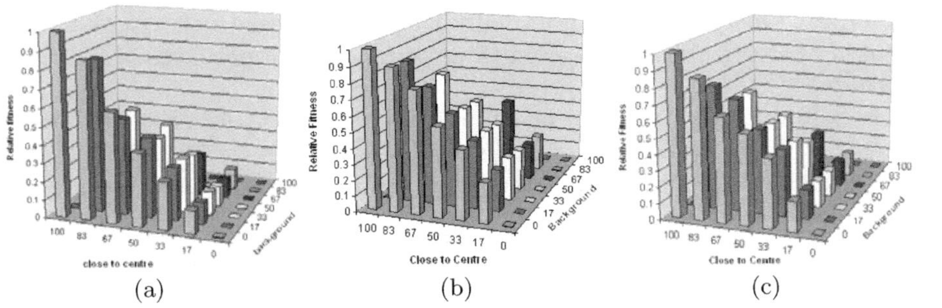

Fig. 5. Results of optimisation. (a) Easy; (b) Medium difficulty; (c) Hard.

4.3 Results

The average results of 100 independent runs on the *test set* are shown in figure 5. In the figure, the x and y axises are the C and B, and the z is the *relative* fitness for the these problems (1.0 or 100% means the ideal case). For all the three data sets, the percentage of the objects for the *Close to Centre* type (C) played an important role using our new fitness function. The best detection results were achieved at 100% C examples and the worst results were produced when we do not use any example of this type. The more object examples used in this type, the best results achieved. However, the *Background* type objects were not critical for these data sets. These results suggest that, when using the new RLWF fitness function for object detection, good results can be achieved with only the two types, *Exact Centre* and *Close to Centre*, and most (if not all) object examples for the other two types *Include Object* and *Background* can be taken out from the training set.

This suggests that the new RLWF fitness function is capable of learning well from these two types of examples and can cope well with the goal of finding object centres from large images. This is mainly due to the fact that RLWF has considered the relative effect of the detected "objects" in different locations.

A further inspection of the use of the old fitness function reveals that the old fitness function must use object examples from all the four types. This is because the old fitness function cannot capture the relative effect information from the objects of the first two types only. This also suggests that the new fitness function is more effective than the old one for object detection.

5 Conclusions

The goal of this paper was to develop a new fitness function for object detection and investigate its influence on optimising the training data. Rather than using a clustering process to determine the number of objects detected by the GP systems, the new fitness function used the localisation fitness and weighted F-measures to reflect the goodness of the detected objects. To invetigate the training data with this fitness function, we categorised the training data into four types. This approach was examined and compared to that with the old clustering based fitness function on three coin detection problems of increasing difficulty.

The results suggest that the new fitness function outperforms the old one by producing far fewer false alarms and spending much less training time. Further investigation on the four types of the training object examples suggests that the new fitness function is effective in optimising training data for object detection.

Acknowledgement

This work was supported in part by the Marsden Fund at Royal Society of New Zealand under grant No. 05-VUW-017 and University Research Fund 6/9 at Victoria University of Wellington.

References

1. Banzhaf, W., Nordin, P., Keller, R.E., Francone, F.D.: Genetic Programming: An Introduction on the Automatic Evolution of computer programs and its Applications. Morgan Kaufmann Publishers (1998)
2. Koza, J.R.: Genetic programming : on the programming of computers by means of natural selection. Cambridge, Mass. : MIT Press, London, England (1992)
3. Koza, J.R.: Genetic Programming II: Automatic Discovery of Reusable Programs. Cambridge, Mass. : MIT Press, London, England (1994)
4. Song, A., Ciesielski, V., Williams, H.: Texture classifiers generated by genetic programming. In Fogel, D.B., El-Sharkawi, M.A., Yao, X., Greenwood, G., Iba, H., Marrow, P., Shackleton, M., eds.: Proceedings of the 2002 Congress on Evolutionary Computation CEC2002, IEEE Press (2002) 243–248
5. Tackett, W.A.: Genetic programming for feature discovery and image discrimination. In Forrest, S., ed.: Proceedings of the 5th International Conference on Genetic Algorithms, ICGA-93, University of Illinois at Urbana-Champaign, Morgan Kaufmann (1993) 303–309
6. Zhang, M., Andreae, P., Pritchard, M.: Pixel statistics and false alarm area in genetic programming for object detection. In Cagnoni, S., ed.: Applications of Evolutionary Computing, Lecture Notes in Computer Science, LNCS Vol. 2611, Springer-Verlag (2003) 455–466
7. Zhang, M., Ciesielski, V., Andreae, P.: A domain independent window-approach to multiclass object detection using genetic programming. EURASIP Journal on Signal Processing, Special Issue on Genetic and Evolutionary Computation for Signal Processing and Image Analysis **2003**(8) (2003) 841–859
8. Smart, W., Zhang, M.: Classification strategies for image classification in genetic programming. In Bailey, D., ed.: Proceeding of Image and Vision Computing Conference, Palmerston North, New Zealand (2003) 402–407
9. Howard, D., Roberts, S.C., Brankin, R.: Target detection in SAR imagery by genetic programming. Advances in Engineering Software **30** (1999) 303–311
10. Bhowan, U.: A domain independent approach to multi-class object detection using genetic programming. Master's thesis, BSc Honours research project/thesis, School of Mathematical and Computing Sciences, Victoria University of Wellington (2003)

A Novel Hybrid System for Dynamic Control*

Byung Joo Kim[1] and Il Kon Kim[2]

[1] Youngsan University Dept. of Information and Communication Engineering, Korea
bjkim@ysu.ac.kr
[2] Kyungpook National University Department of Computer Science, Korea
ikkim@knu.ac.kr

Abstract. In this paper we propose a hybrid model which includes both first principles differential equations and a least squares support vector machine (LS-SVM). It is used to forecast and control an environmental process. This inclusion of the first principles knowledge in this hybrid model is shown to improve substantially the stability of the model predictions in spite of the unmeasurability of some of the key parameters. Proposed hybrid model is compared with both a hybrid neural network(HNN) as well as hybrid neural network with extended kalman filter(HNN-EKF). From experimental results, proposed hybrid model shown to be far superior when used for extrapolation compared to HNN and HNN-EKF.

Keywords: Hybrid Least Squares Support Vector Machine, First Principles, Kalman Filter, Neural Network.

1 Introduction

The treatment of polluted water is one of the most important environmental problems worldwide. Unfortunately, the control of such bioprocess is difficult because the inability to measure key parameters of the process. In these cases the solution is of the type of inference control [1,2,3] which uses an machine learning method to model the unmeasurable parameters, for use in conjunction with the known part of the model. This hybrid model has the advantage of the flexibility of nonparametric curve estimation for capturing the dynamics of the unmeasurable part of the process along with a constraint imposed by the first principles model which acts to keep the predictions both physically plausible and stable[4]. Hybrid model can be trained with fewer data and extrapolate better than pure machine learning method. In recent years, there has been a spate of research[5,6] showing the utility of these hybrid model in a variety of applications with particular attention to fed-batch bioreactors[7]. The aim of this paper is to propose a hybrid model being able to perform on-line estimation of directly unmeasurable process variables. Main attention is paid to hybrid models that is combination of first principles(FP) models with least squares

* This study was supported by a grant of the Korea Health 21 R&D Project, Ministry of Health & Welfare, Republic of Korea (A05-0909-A80405-05N1-00000A).

T.-D. Wang et al. (Eds.): SEAL 2006, LNCS 4247, pp. 609–616, 2006.

support vector machine(LS-SVM) as well as with neural network(NN) based models, with NN combined extended Kalman filter(HNN-EKF) is discussed. We review the dynamics of microbial growth in the next section. In Section 3, we introduce a hybrid LS-SVM model which uses a LS-SVM to estimate the growth rate within the governing equations for the kinetics of process. In Section 4, we compare the performance of this model with various other strategies. Finally we investigate experimental results.

2 Dynamics of Microbial Growth

Biological reactors exhibit a wide range of dynamic hehaviors and offer many challenges to modeling a complex kinetic expressions[8]. A bioreactor typically consists of a large vessel containing an aqueous solution of biomass and one or more substances. Bioreactors operating in a fed-batch manner that is, the vessel is given an initial charge of biomass and substrate, and periodic additions of substrate is made. Bioreactors operated in a fed-batch manner and are quite difficult to model, since their operation involves microbial growth under constantly changing conditions. Nevertheless, knowledge of process parameters(such as growth rate kinetics) under a wide range of operating conditions is very important in efficiently designing optimal reactor operation policies. A fed-batch bioreactor can be described by the following equations

$$\frac{dB}{dt} = \mu(t)B(t) - \frac{Q(t)}{V(t)}B(t) \tag{1}$$

$$\frac{dS}{dt} = -\mu(t)B(t) + \frac{Q(t)}{V(t)}[S_i(t) - S(t)] \tag{2}$$

$$\frac{dV}{dt} = Q(t) \tag{3}$$

where $B(t)$ is the biomass concentration at time t, $S(t)$ is the substrate concentration, $Q(t)$ is the volumetric flow into the system, $V(t)$ is the volume, and $\mu(t)$ is the specific growth rate. The differential equations(1)-(3) can be viewed as a partial model of the system consisting of simple mass balances on the biomass, substrate and volume in the reactor. The dynamics of the process are contained in the kinetics parameter $\mu(t)$, known as the specific growth rate, which governs the conversion of substrate to biomass. It is the parameter that complicates the process, as it is a time-varying function of the biochemical variables of the system. In the literature, many models have been developed for this unmeasurable growth rate of which the most widely used are the Monod and Haldane function of the amount of substrate in the system

$$\mu(t) = \frac{\mu^* S(t)}{K_m + S(t) + \frac{S(t)^2}{K_i}} \tag{4}$$

Above expression will only be used to simulate the *true* process model; for all modeling techniques described in the remainder of the paper, the above expression describing the cell growth rate will be completely unknown.

The differential equations (1)-(3), can be discretized in unit time as follows

$$B_{t+1} = B_t + (\mu_t - \frac{Q_t}{V_t})B_t \tag{5}$$

$$S_{t+1} = -\mu_t B_t + S_t(1 - \frac{Q_t}{V_t}) + \frac{Q_t}{V_t} S_i(t) \tag{6}$$

$$V_{t+1} = V_t + Q_t \tag{7}$$

If error terms ε, v are added to reflect measurement error and model uncertainty, then the system takes the form of a nonlinear dynamical system in discrete time:

$$x_{t+1} = f_t(x_t, \Psi_t, \mu_t(x_t)) + \varepsilon_t \tag{8}$$

$$z_{t+1} = H(x_t) + v_t \tag{9}$$

Note that the form of μ_t is unknown in equation (8) and it is unmeasurable. Because not all the state variables may be observable, the function H in equation (9) is used to show the transformation from the full state x to the observable state z. This step may also involve error, as shown by v. Were the μ_t observable, an extended Kalman filter[9] could be applied for prediction of the next state, using the current state estimates. Even when μ_t is not known, one can, through a two step estimation scheme, iteratively estimate the next state and the μ_t. The method is cumbersome, and requires a number of choices of turning parameters[10]. Moreover, while this method captures the dynamics of the system in the short run, it is unable to forecast changes in the system in the long term unless the growth rate stays constant. Thus, it is useless for process scheduling or for extrapolation under different conditions. The hybrid LS-SVM described below accomplishes the same estimation, but with the added feature that is can provide reasonable extrapolation by directly estimating the growth rate.

3 Hybrid Least Square Support Vector Machine

3.1 Least Square Support Vector Machine

In the last decade, neural network have been successfully used as *black-box* models of dynamic systems and, more specifically, as process variable estimators in bioreactor modeling applications[11]. In these efforts the process was operating in a continuous mode; however, identification of batch processes is much more difficult, since a wide range of operating regimes is involved and less data may be available. Major breakthroughs are obtained at this point with new class of neural networks called support vector machines(SVM)[12]. Solving QP in SVM needs complex computation and difficulty in implementation. Moreover, training time is long and memory requirement is square. To solve this problem Suykens developed least squares support vector machine(LS-SVM)[13]. In this paper, we focus on function estimation using LS-SVM.

3.2 Hybrid Least Square Support Vector Machine

As discussed in the previous section, it is quite straightforward to derive an approximate model of the bioreacotr (Eqs. 1-3) from simple first principles considerations such as mass balances on the process variables. However, the critical factor in determining the dynamic behavior of the process is the unknown kinetics(growth rate), of the conversion of substrate to biomass. The central idea of this paper is to integrate the available approximate model with a LS-SVM which approximates the unknown kinetics, in order to form a combined model structure which can be characterized as a hybrid LS-SVM process model. The first principles partial model specifies process variable interactions from physical considerations; the LS-SVM complements this model by estimating unmeasured process parameters in such a way as to satisfy the first principles constraints; nonparametric estimation is needed since no knowledge is available about these parameters. Such structured models are expected to perform better than black-box models in process identification tasks, since generalization and extrapolation are confined only to the uncertain parts of the process while the basic model is always consistent with first principles and does not allow a physical variable interactions. A schematic representation of the hybrid LS-SVM model is shown in Figure 1. LS-SVM component receives as inputs the process variables and provides an estimate of the current parameter values, in this case growth rate. LS-SVM's output serves as an input to the first principles component, which produces as output the values of the process variables at the end of each sampling time. The combination of these two building blocks yields a complete hybrid LS-SVM model of the bioreaction system. An important issue for LS-SVM is model selection. For real world problem one often employs an RBF kernel. In this case, the kernel width and regularization parameter need to be selected. In [14] it is shown that the use of 10-fold crossvalidation for hyperparameter selection of LS-SVM consistently leads to very good results on a large number of UCI benchmark data sets[15] in comparision with many other methods reported in the literature.

4 Experiment

We evaluate this hybrid LS-SVM(HLS-SVM) modeling scheme by comparing its prediction accuracy with hybrid neural network(HNN) proposed by Ungar[9] and hybrid neural network with extended Kalman filter(HNN-EKF). In order to test the behavior of the HLS-SVM, we simulated data via the known first principles differential equations (1)-(3) under a realistic condition. The specific growth rate was generated using the Haldane model (equation (4)) with the parameter

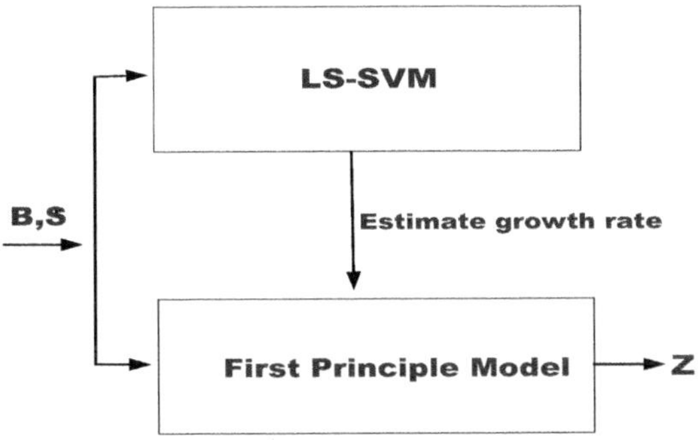

Fig. 1. A schematic representation of the hybrid LS-SVM model. The LS-SVM estimates the unobservable growth rate. This information plus the estimated growth rate is then fed into the known model to obtain the next state.

values listed in[10], namely $\mu^* = 5, K_m = 10, K_i = 0.1$. The inlet concentration $S_i = 3.5$ and the flowrate, Q, was held constant at 0.1. Future state variables were then generated using the differential equations with random noise. We take the same initial values B_0, S_0, V_0 to generate the training, validation, test data set used in [10]. Table 1 shows initial parameter value and number of data in train, validation and test data respectively.

Table 1. Initial parameter value of train, validation and test data set

	B_0	S_0	V_0	Noise	Number of data
Train	0.1	0.5	10	$N(0,0.01^2)$	20
Validation	0.3	0.7	4	$N(0,0.01^2)$	80
Test	0.001	1.5	4	$N(0,0.01^2)$	80

For the HNN, number of input, hidden and output node is 2,3 and 1 and weight update is done by backpropagation method to estimate the growth rate. The weights that produced the smallest SSE(Sum of Square Error) for the validation data were selected for making the predictions for the test set. Because learning and generalization ability of backpropagation neural network model is sensitive to initial weight space, we try 10 times experiment and use the average SSE value. In case of HNN-EKF, initial value of weight is set to 1. Covariance and error matrix in process state set to identity matrix. Variance of error in measurement state is 0.01. Forward pass to compute the output in HNN-EKF is the same procedure as backpropagation neural network model does. Weight update is done by EKF method. In HLS-SVM, we take the RBF kernel and kernel parameter $\sigma = 1.96$ and $\lambda = 0.479$ are obtained by cross-validation method. Figure 2 shows estimated and true $\mu(t)$ for training data by three methods. Asterisk(*) in each graph is estimated value of $\mu(t)$. As we can see HNN model

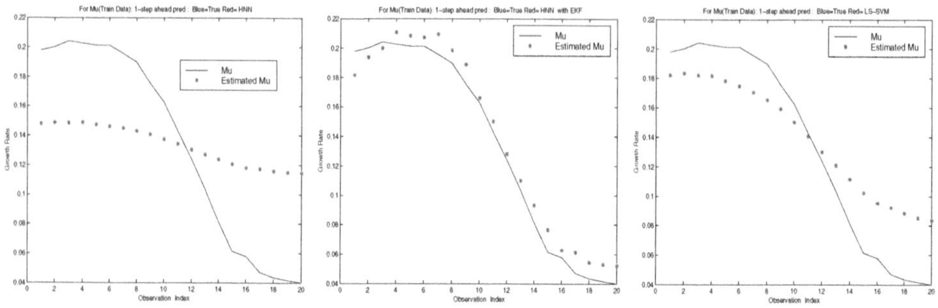

Fig. 2. From left to right estimated growth rate by HNN, HNN-EKF, HLS-SVM respectively

Table 2. Comparision of sum of square error(SSE) on B and S by HNN, HNN-EKF and HLS-SVM method

Method	Train		Validation		Test	
	B	S	B	S	B	S
HNN	0.22	0.22	19.57	19.52	23.5280	23.30
HNN-EKF	0.00310	0.02375	0.01846	0.01584	0.67861	0.67558
HLS-SVM	0.1206	0.03356	0.03759	0.03433	0.18469	0.19361

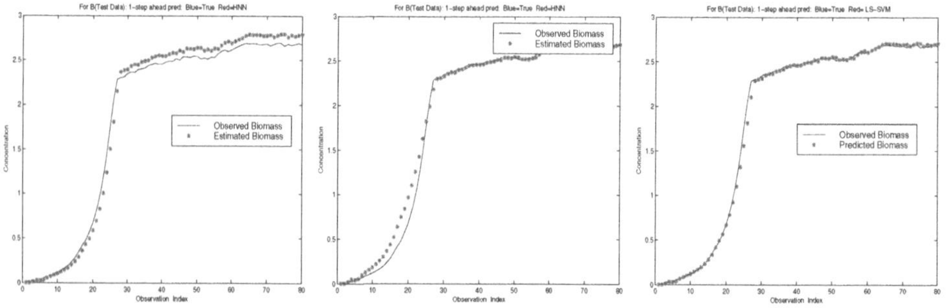

Fig. 3. From left to right estimated Biomass by HNN, HNN-EKF, HLS-SVM respectively

can't estimate $\mu(t)$ well, whereas HNN-EKF and HLS-SVM model estimate well. On the contrary Figure 3 and 4 show the generalization performance on B and S respectively. From table 2 experimental results shows that in estimating B and S, HLS-SVM is about 100 times outperform than HNN because LS-SVM overcomes the overfitting problem and guarantees the global minima. Comparing HLS-SVM to HNN-EKF the former is about 3 times outperform than the later. Generalization ability of HNN-EKF is decreased because overfitting occurs more or less during learning process.

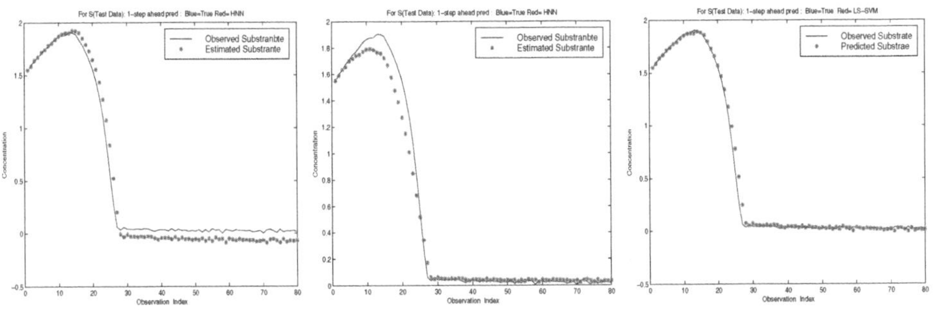

Fig. 4. From left to right estimated Substrate by HNN, HNN-EKF, HLS-SVM respectively

5 Summary and Conclusion

In this paper we propose a new hybrid model comprised of a LS-SVM, together with first principles differential equations for the forecasting and simulation an environmental process. When sufficiently large training sets of data are available, traditional black-box model gives an accurate model of the process. However limited data are available, the hybrid model gives significantly better accuracy, particularly on extrapolation. The hybrid model, once learned, can be used for process control and optimization. The concept of combining LS-SVM with first principle knowledge is a powerful one, and goes well beyond the example presented in this paper.

References

1. B. Joseph, and C. Brosilou, "Inferential Control of Processes, Part I Steady State Analysis and Design," AICHE J., Vol. 24, No. 3, 1978.
2. P. Lindskog, and L. Ljung, "Ensuring Certain Physical Properties in Black Box Model by Applying Fuzzy Techniques," Technical Report, 1996.
3. C. Brosilow, and M. Tong, "Inference Control of Processes, Part II The Structure and Dynamics of Inferential Control Systems," AIChE J.,Vol. 24, No. 3, 1978.
4. Acceessible at http://www.ici.ro/ici/revista/sic99-1/art04.html
5. E.J. Molga, and K.R. Westerterp, "Neural network based model of the kinetics of catalytic hydrogeneration reactions," Stud. Surf. Sci. Catal. 109,379-388, 1997.
6. A.Y.D. Tesen et al "Predictive control of quality in batch polymerization using hybrid ANN models," AICHE Journal 42(2),455-465, 1996.
7. B. Saxen and H. Saxen, "A neural network based model of bioreaction kinetics," Canadian Journal of Chemical Engineering 74(1), 124-131, 1996.
8. G. Weich, and G. Bishop, "An Introduction to the Kalman
9. C. Dimitris, Psichogios and H. Ungar, "A Hybrid Neural Network-First Principles Approach to Process Modeling," AICHE Journal 38(10),14991511, 1992.
10. P.A. Lant, M.J. Williams, G.A. Montague, M.T. Tham, and A.J. Morris, "A Comparision of Adaptive Estimation With Neural Based Techniques for Bioprocess Application," Proc. of the American Control Conf., 2713(1990).

11. S. Gunn, Support Vector Machines for Classification and Regression, ISIS Technical Report, U. of Southampton, 1998.
12. J.A.K. Suykens, "Nonlinear Modeling and Support Vector Machines," Accessible at http://www.kdiss.or.kr/kdiss/
13. X. Shao, Model Selection Using Statistical Learning Theory, Ph. D. Thesis, U. of Minnesota, 1999.
14. Accessible at http://www.ics.uci.edu/ mlearn/MLRepository.html

SVM Based Speaker Selection Using GMM Supervector for Rapid Speaker Adaptation

Jian Wang, Jianjun Lei, Jun Guo, and Zhen Yang

School of Information Engineering
Beijing University of Posts and Telecommunications
100876 Beijing, China
Wangjian200810@sohu.com

Abstract. In this paper, we propose a novel method for rapid speaker adaptation called speaker support vector selection (SSVS). By taking gaussian mixture model (GMM) as speaker model, the speakers acoustically close to the test speaker are selected .Different from other selection method, just computing the likelihood between models, we utilizing support vector machines (SVM) to obtain a 'more optimal speaker subset'. Such selection is dynamically determined according to the distribution of reference speakers close the test. Furthermore, a single-pass re-estimation procedure conditioned on the selected speakers is shown. This adaptation strategy was evaluated in a large vocabulary speech recognition task. The presented method improves the relative accuracy rates by 13% compared to the baseline system.

1 Introduction

The main challenge of speaker adaptation is to improve speech recognition performance by adjusting the speaker independent (SI) recognition system toward speaker dependent (SD) system. Dominant speaker adaptation technologies such as MLLR [1] and MAP [2] may become inefficient because of the lack of enough enrollment data. Recently, the adaptation based on the technique of speaker clustering and selection has been shown to be a successful solution for the sparseness of enrollment data [3]. The motivation of such system is based on considering that the training data contains a number of training speakers, some of whom are closer to the test speaker than the others. If the model is re-estimated from such selected speakers, it should be reasonably close to the SD parameters, but the successes greatly depend on not only the number of selected speakers but also whether these statistics are sufficient for describing the distribution of the reference speakers. How to make a trade off between good coverage and small variance among the cohorts selected is still a very trick problems relied on the experiments. Dynamic instead of fixed number of close speaker selection seems to be a good alternative.

In this paper, agreeing with the assumption above, we try to find subset of training speakers who are acoustically close to the test speaker using support vector machine (SVM) and then re-estimate the speaker's model. Experimental results shown that the adaptation algorithm can obtain relatively accurate model based on two schemes as follow: 1) SVM outperforms general speaker selection method since it uses a smart way to choose an optimal set of reference models as well as save computation

T.-D. Wang et al. (Eds.): SEAL 2006, LNCS 4247, pp. 617–624, 2006.

time; 2) The adapted model can be calculated by using the previously stored hidden markov model (HMM) statistics, by which, a quick adaptation can be done.

This paper is organized as follows. In the next section, the basic idea for speaker selection is described. Our proposed method, speaker support vector selection (SSVS) is explained in section 3. In section 4, we show our experiments in detail. Conclusions and discussion are provided in section 5.

2 Speaker Adaptation Based on Speaker Selection

The basic idea for speaker selection is that in speaker recognition field, one utterance with three seconds has achieved good accuracy on speaker identification. Then we can make full use of the statistics from the selected speaker subsets. There are various implementations of selection in practice which need to consider the following issues:

- Efficient speaker representations.
- Reliable similarity measurement between speakers.
- Number of close speakers (cohorts) to be picked out.
- Number of utterances from test speaker when considering the performance.

Experiments in [4] showed that gaussian mixture model (GMM) based speaker representation and likelihood score based similarity measurement is the most efficient strategy (GMM-LR). The main procedure for selection is as follows.

- Train one GMM for each training speaker ready for selection.
- Calculate the likelihood of adaptation data of test speaker in each GMM.
- Select training speakers with the K largest likelihood as cohorts.
- Retraining the SI model using data of selected cohorts.

However, retraining the SI model by data of selected reference speakers is very time-consuming. Model combination, that is, interpolating the Gaussian mean vectors of cohort models to obtain a new model, which can save more computation time, though only pre-calculated statistics are used [5].

3 The Algorithm

3.1 SVM Based Speaker Selection

For creating a speaker model, a typical strategy is to use MAP adaptation on a background model which is created by training a GMM on a large population of speakers which is called universal background model (UBM) [6]. Then we can select the speakers who are close to the test speaker as described in Section 2. But it must build one model for each reference speaker and compute the likelihood of all speaker models during selection. Furthermore, the fact that the optimal number of cohorts for each test speaker is different motivates us to conduct dynamic speaker selection.

SVM is a promising machine learning technique developed from the theory of Structural Risk Minimization [7, 8]. It is typically constructed as a two class classifier. Fig.1 shows a typical two-class problem in which the examples are perfectly separable using a linear decision region. H1 and H2 define two hyperplanes. The closest in-class and out-of-class examples lying on these two hyperplanes are called the support vectors obtained from the training set by an optimization process. The optimization condition relies upon a maximum margin concept. For a separable data set, the system places a hyperplane in a high-dimensional space so that the hyperplane has maximum margin. The focus, then, of the SVM training process is to model the boundary between classes. The ideal outputs are either 1 or -1, depending upon whether the corresponding support vector is in class 0 or class 1, respectively. Optimization on the input data in this case involves the use of a kernel-based transformation:

$$K(x_i, x_j) = \Phi(x_i)^T \Phi(x_j) \tag{1}$$

where $\Phi(x)$ is a mapping from the input space (where x lives) to a possibly infinite dimensional space.

Kernels allow a dot product to be computed in a higher dimensional space without explicitly mapping the data into these spaces. A kernel-based decision function has the form:

$$f(x) = \sum_{i=1}^{N} \alpha_i y_i K(x, x_i) + b \tag{2}$$

Different kernel functions form different SVM algorithms, here are three kernel functions that are often used [9, 10]:

1) linear:

$$k(x_i, x_j) = x_i^T x_j . \tag{3}$$

2) polynomial:

$$k(x_i, x_j) = (\gamma x_i^T x_j + r)^d . \tag{4}$$

3) radial basis function (RBF):

$$k(x_i, x_j) = \exp(-\gamma \| x_i - x_j \|^2), \gamma > 0 \tag{5}$$

here, γ, r, and d are kernel parameters.

As support vectors carry all relevant information about the classification problem, the speakers who are acoustically close to the target speaker can be selected as support vectors. Particularly, such training method can give a dynamic decision of defines how complex the classifier needs to be by the reference speaker set itself. This is in stark contrast to systems such as GMM-LR, where the complexity of the system is typically predetermined.

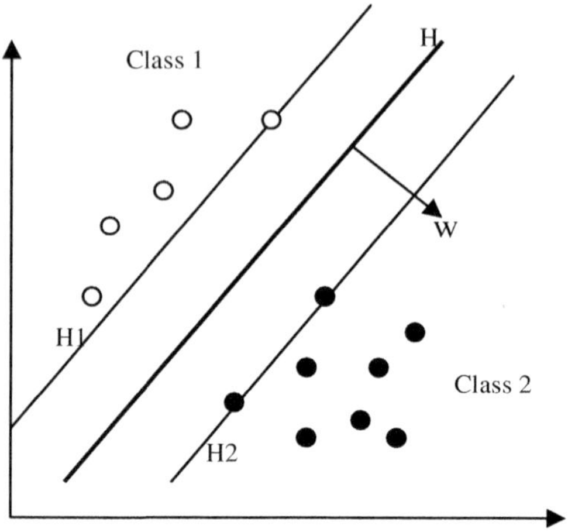

Fig. 1. H is the optimal hyperplane because it maximizes the margin the distance between the hyperplanes H1 and H2. Maximizing the margin indirectly results in better generalization.

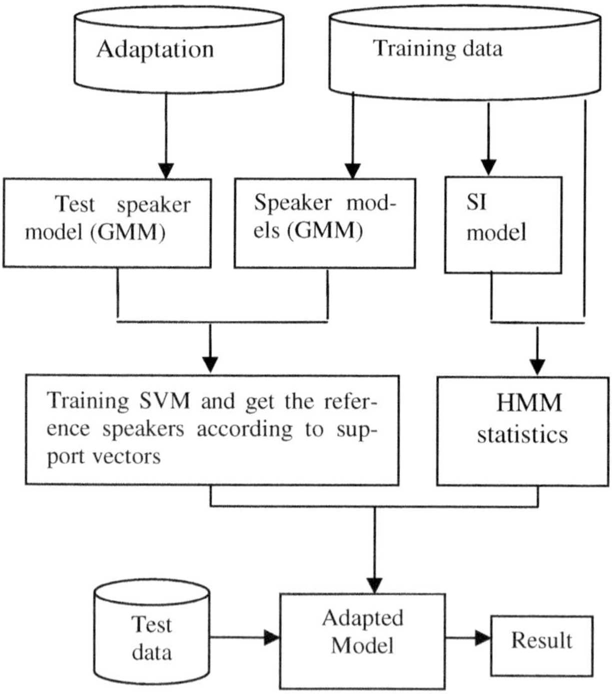

Fig. 2. Main adaptation procedure of selection can be considered as the training of SVM

3.2 Main Procedure of Proposed Flow

The detailed procedure is depicted in Figure 2. Before performing selection, we need to train the HMM model for each reference speakers, the statistics of training will be stored for further utilizing. As described in section 2.1, the GMM models are adapted by using MAP technique [2]. These adapted models can be considered as the approximation of the reference speakers.

For implementing reason, we extract the mean vectors of the GMM of M reference speaker to form M supervector $S_m (m = 1,2,\cdots M)$. We also apply MAP adaptation to the test utterance based UBM and extract the mean vectors of the Gaussians to form the supervector of test speaker $\hat{S}_t$, every dimension of the $\hat{S}_t$ has the same length and order as S_m.

Using these M+1 supervectors to represent speakers of two classes(one class is reference speakers and the other is test speaker), a SVM can be trained with the training set $\{(\hat{S}_t, y), (S_m, y_m)\}$, where $m = 1,2,\cdots,M$, $y = -1$ and $y_m = 1$ .Then we are able to select a subset of speakers corresponding to the support vectors in SVM. Finally, a speaker adapted acoustic model is calculated from the HMM statistics of the selected speakers using a statistical calculation method.

3.3 HMM Model Estimation

In this part, a single-pass re-estimation procedure, conditioned on the speaker-independent model, is adopted. In the adaptation procedure, there has no inherent structure's limitation of transformation-based adaptation schemes such as MLLR .A speaker adapted acoustic model is calculated from the HMM statistics of the selected speakers using a statistical calculation method.

The process of re-estimation would update the value of each parameter. the posteriori probability of occupying the *m*'th mixture component, $L_m^{i,r}(t)$, conditioned on the SI model, at time *t* for the *r*'th observation of the *i*'th cohort can be stored in advance .

The one-pass re-estimation formula may be expressed follows:

$$\begin{aligned} \tilde{\mu}_m &= \sum_{i=1}^{N}\sum_{r=1}^{R_i}\sum_{t=1}^{T_r}(L_m^{i,r}(t)\cdot O^{i,r}(t)) \Big/ \sum_{i=1}^{N}\sum_{r=1}^{R_i}\sum_{t=1}^{T_r} L_m^{i,r}(t) \\ &= \sum_{i=1}^{N} Q_m^i \Big/ \sum_{i=1}^{N} L_m^i \end{aligned} \tag{6}$$

where:

$$L_m^i = \sum_{r=1}^{R_i}\sum_{t=1}^{T_r} L_m^{i,r}(t) \tag{7}$$

$$Q_m^i = \sum_{r=1}^{R_i}\sum_{t=1}^{T_r} L_m^{i,r}(t) Q^{i,r}(t) \tag{8}$$

Among them, $O^{i,r}(t)$ is the observation vector of the *r*'th observation of the *i*'th speaker at time *t*, $\tilde{\mu}_m$ is the estimated mean vector of the *m*'th mixture component of the target speaker. The variance matrix and the mixture weight of the m'th mixture component can also be estimated in a similar way.

4 Experiment Results

4.1 Experiment Setup

The database we used in these experiments is selected from the mandarin Chinese corpus provided by the 863 plan (China High-Tech Development Plan).About 60 hours of speech data from Chinese male speakers are used to train a gender-dependent SI model. We use 39 dimensional features consisting of 12 cepstral coefficients and log energy feature with the corresponding delta and acceleration coefficients. A five-state structure of a left-to-right HMM model with eight continuous density mixtures is trained. Then triphone-based HMM models are used in this continuous speech recognition.

The speakers ready for selection consist of 100 male speakers, with 250 utterances each. Typically one utterance, both in training and test set, lasts 3~5 seconds. Test set consists of 20 male speakers from the same accent with training set, 20 utterances each. 10 of them are used for selecting and adaptation. The other 10 are used for testing. It should be noted that we focus on very rapid adaptation of large-vocabulary system in this paper. All the adaptation methods in experiments are performed with only one adaptation sentence.

4.2 Experimental Results

When considering different kernels of SVM in SSVS, we have experimented with three kinds of kernels for speaker selection: linear, polynomial and RBF. The GMM-LR system is chosen as our baseline. Table 1 shows average recognition rates of SSVS (with 60 reference speakers) with different kernels of SVM. SV represents the number of selected speakers. We take the conventional methods, the MAP and MLLR as an comparison.

From Table 1 we can see that GMM-LR based selection can get a better perform then traditional methods and linear kernel SVM based SSVS obtains the best recognition accuracy. As we known, although the polynomial kernel and RBF kernel can handle the case when the relation between class labels and attributes is nonlinear, but they use more parameters(see equation 4, 5) during training which influences the complexity of model selection. Then the support vectors obtained by linear kernel may bring a better performance in speaker selection.

Table 1. Word accuracy percentage of different kernel of SVM

Accuracy %	Baseline N=	MLLR	MAP	SSVS		
				Linear SV=33	Poly SV=13	RBF SV=35
AWP Rel imp	46.81 NA	44.95 -3.97	44.33 -5.3	52.92 13.05	52.5 12.16	50.83 8.59

From Figure.3 we can conclude that as the number of reference speakers grows, the proposed method can select more accurate support vectors which are acoustically close to the test. So the performance will be improved. It is make out dynamically choosing the optimal cohort speakers for each target speaker is one of the key concerns in order to keep the balance between good coverage of phone context and acoustic similarity to the target speaker.

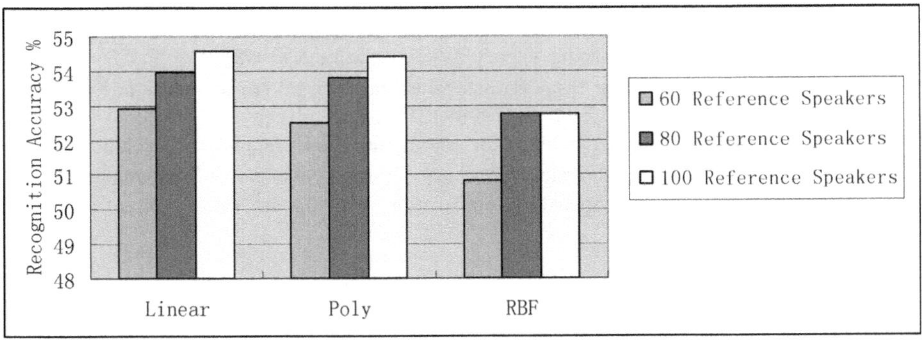

Fig. 3. Comparison with different number of reference speakers and different kernels

5 Conclusion

Generally, performance of selection based speaker adaptation is very sensitive to the choice of initial models, the number of selected speakers is always fixed. A novel selection method based on SVM is proposed in this paper. It realizes dynamic speaker selection by finding the support vector and its corresponding speaker in the subset of reference speakers. Our experiments have shown the proposed scheme can significantly improve the performance and robustness of adaptation, even few adaptation sentence is available. Further work will be focus on how to measure the relative contribution of each phone model of each specific speaker selected.

Acknowledgements

This research is partially supported by NSFC (National Natural Science Foundation of China) under Grant No.60475007, Key Project of Chinese Ministry of Education under Grant No.02029, and the Foundation of Chinese Ministry of Education for Century Spanning Talent.

References

1. C. J. Legetter and P. C. Woodland : Maximum Likelihood Linear Regression for Speaker Adaptation of Continuous Density HMM's in Compute. Speech Lang., vol. 9, (1996) 171–186
2. J.L. Gauvain and C.H. Lee: Maximum a posterior estimation for multivariate Gaussian observations of Markov chains. IEEE Trans. Speech Audio Processing, vol. 2, (1994)291–298

3. A. Sankar, F. Beaufays, and V. Digalakis: Training data clustering for improved speech recognition. in Proc. Eurospeech, (1995)502–505
4. C. Huang, T. Chen and E. Chang: Speaker Selection Training for Large Vocabulary Continuous Speech Recognition. Proc. ICASSP (2002)
5. D. A. Reynolds and R. C. Rose: Robust text dependent speaker identification using Gaussian mixture speaker models. IEEE Transactions on Speech and Audio Processing, vol. 3 (1995) 72-83
6. D. A. Reynolds, T. Quatieri and R. B. Dunn : Speaker Verification Using Adapted Gaussian Mixture Models. Digital Signal Processing, vol. 10, (2000) 19-41
7. V. N. Vapnik: Statistical Learning Theory. Wiley New York (1998)
8. C. J. C. Burges: A tutorial on support vector machines for pattern recognition. Knowledge Discovery Data Mining, vol. 2, no. 2, (1998) 121–167
9. S. R. Gunn: Support vector machines for classification and regression. Technical Report Image Speech and Intelligent Systems Research Group, University of Southampton(1997)
10. M.Schmidt and H. Gish: Speaker identification via support vector classifiers. Proc. ICASSP(1996) 105–108

Using Simulation to Improve the Flexibility of Adaptive Workflow Models Based on Temporal Logic

Jianchuan Xing[1,2], Zhishu Li[1], and Liangyin Chen[1]

[1] School of Computer, Sichuan University, Chengdu, 610065, China
xingjianchuan@tom.com, lzshu5@yahoo.com.cn, chenliangyin@gmail.com

[2] Development Center (Chengdu), Forlink Technologies, Chengdu 610016, China

Abstract. Many researchers focus on enhancing the flexibility of the workflow management system. This paper shows how simulation plays an important role in improving the flexibility of temporal logic based workflow specification (TLWS) model. Detecting the TLWS model's demand for changing and deciding what changes to carry out are very difficult. Simulation analysis can help to do these. After a task is finished, its time, cost and quality will be computed. And each will be compared with dual threshold values. If the value is below the bottom threshold, an exception will be thrown. If it is above this threshold, but below the top threshold, a warning will be sent out. If the value is above the top threshold, the execution is excellent. Workflow specification documents also need to be translated into simulation specification documents by using XLST. The utility of using simulation in improving the flexibility of TLWS model is outstanding.

Keywords: simulation, adaptive workflow, temporal logic, flexibility.

1 Introduction

Many organizations make use of workflow management systems for automating their business processes. People widely recognize that a workflow management system should provide flexible ways of managing workflows [1].

At present, flexible workflow management systems are based on various workflow models. These models can provide convenience to users. But there are many disadvantages in these models. The flexible workflow management system based on TLWS model [2] can overcome the disadvantages of other models.

Because of the complexity of modifying workflows, simulation analysis can be used to check the desirability of the changes.

2 Workflow Model

2.1 TLWS Model

The kernel of workflow model is the specification of temporal synchronization among activities. In general, there are some kinds of workflow models as follows: Graph model, Petri-Net model, Event-Condition-Action model, CTR model and Hypermedia model [3].

T.-D. Wang et al. (Eds.): SEAL 2006, LNCS 4247, pp. 625–631, 2006.

Based on the linear temporal logic theory [4], Z.S. Tang [5] designed a temporal logic based CASE environment, XYZ System, which can support various ways of programming. Based on temporal logic and the XYZ system, Ma proposes a workflow model based on temporal logic, called TLWS model [2].

TLWS model is defined as follow:

Definition 1. Temporal Logic based Workflow Specification (TLWS) model is defined as a six tuple, TLWS = (Σ, F, R, T, O, S0), where

Σ is a set of scene states;

R is a set of roles;

T is a set of tasks;

F is called as a set of scene state transition rules. The formula of a rule is formed as: $\text{lb}= S_i \wedge P_i \Rightarrow @_i(Q_i \wedge \text{lb})= S_j$;

O: F→{activity expression}∪{ω};

S0 is the start scene state.

Compared with the previous models, TLWS model has some excellent features:

(1) Process of workflow service;

(2) Step-wise refinement design of a workflow service, thus the model supports the specification of workflow process and its evolution;

(3) Abstraction of activities for a workflow service;

(4) Synchronizations among activities.

Those features make TLWS model a powerful model. TLWS model is compared with the previous models in Table 1 [6].

Table 1. Comparing workflow models

Characteristic	Graph	Petri-Net	ECA	CTR	TLWS
Semantic	No	Yes	No	Yes	Yes
Activity abstraction	No	No	No	No	Yes
Refinement design	No	No	No	No	Yes

2.2 Structure of the System

The system based on TLWS model includes several parts as follows: XPDL editor, workflow model library, workflow modifier, workflow instance library, workflow engine. The structure of the system is shown in Figure 1.

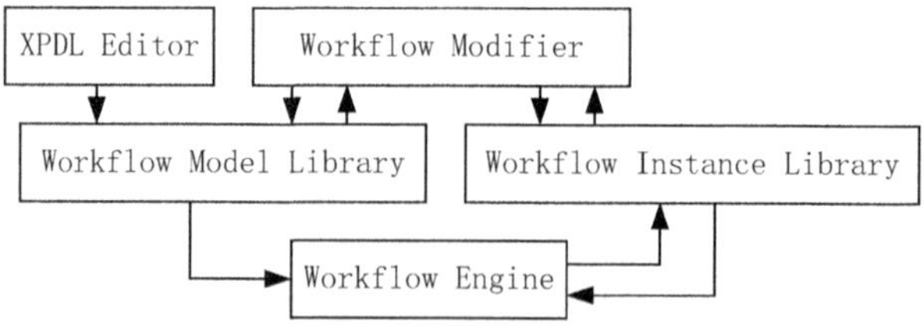

Fig. 1. Structure of the system based on TLWS model

This system can provide both the flexibility by selection and the flexibility by modification. In build-time stage, users can design the flexible workflow process by XPDL editor. They can also use the abstract activity and exception handling to provide the flexibility by selection.

In run-time stage, the workflow engine will provide the optional processes in the abstract activity for users. Moreover, when an exception is thrown, the engine will turn to the part of the exception handling.

Users can modify the workflow model or workflow instance dynamically in the library. The system uses the workflow modifier to provide the flexibility by modification. Users can also modify the workflow process in XPDL editor.

3 Improving the Flexibility

The flexible workflow management system based on TLWS model improves flexibility in two ways [6]: by selection and by modification.

3.1 By Selection

When workflow runs normally, abstract activity provides the optional sub-flow; when exception happens, the system will provide the optional exception handling.

Normal Running. In one business process, there often are many ways to choose. For example, if one machine part is needed in an enterprise business process, there are two possible ways to gain it: purchase one or produce one. The different ways to gain the part means different workflows to be chosen. If users want to purchase one, the workflow may be like: Apply → Approve → Purchase. If users want to produce it, the workflow may be like: Apply → Approve → Design → Produce. Figure 2 shows this process.

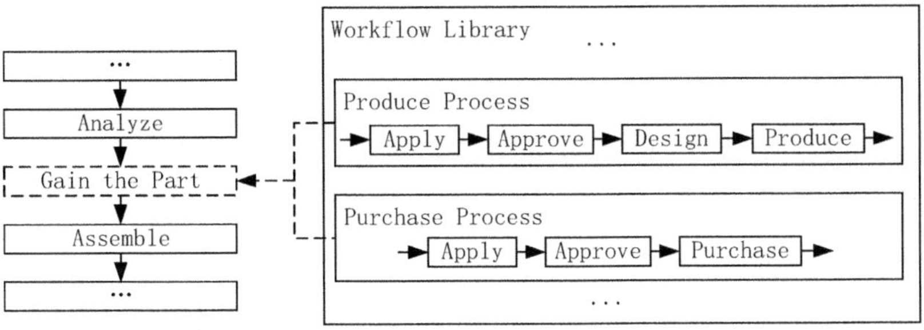

Fig. 2. Process selection in abstract activity

We can use abstract activity to describe such selection in the process. Abstract activity includes different sub-flow. In the run-time phase, the system will list all the optional process in a workflow instance. According to the requirement, workflow participant choose one of these optional process to complete the abstract activity. Figure 3 shows the structure of abstract activity.

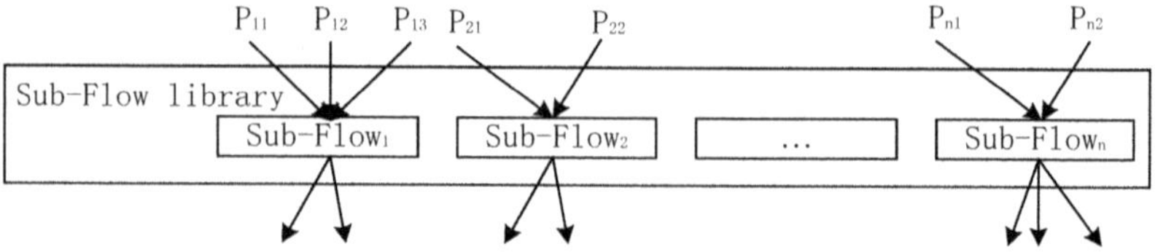

Fig. 3. Structure of abstract activity

Exception Handling. In the flexible workflow management system based on TLWS model, the flexibility by selection is also supported on the exception handling. There are many errors caused by external or internal events in run-time stage. When these errors occur, the system must provide the exception handling mechanism in order not to crash. With such good mechanism, the system can be robust.

When the exception is thrown, the workflow engine will call the corresponding activity to handle this exception. For example, when a process is overtime, we can inform the manager, wait or go on the process ignoring the exception.

3.2 By Modification

We provide static modification and dynamic modification to improve flexibility by modification in the flexible workflow management system.

Static Modification. In the workflow management system, the modification can be done by a set of basic primitive functions. They are as follows:

(1) addEdge(A, B, Condition): Add a directional edge between activity A and activity B, and Condition is the condition of transfer.
(2) AddNode(A): Add the activity A into workflow.
(3) DeleteEdge(A, B): Delete the directional edge between activity A and activity B.
(4) DeleteNode(A): Delete the activity A from workflow.

We can compose very complicated modification by assembling these primitive functions. We can modify the workflow statically or dynamically with these primitive functions.

Dynamic Modification. An important problem in dynamic modification is how the modification to the workflow process affects the running workflow instance. There are two important conceptions: execution point and modification point. Execution point indicates the position of a task node where a workflow instance is running currently. Modification point indicates the position of a task node we want to change.

According to the relation between execution point and modification point, we will discuss how to change the running workflow instance by the following three situations:

(1) Modification point is prior to execution point. It means the change to the workflow instance will not take effect until a new workflow process restarts.
(2) Modification point is the same as execution point. This situation can be subdivided into three cases. If the modification adds a node before the

execution point, the change will not take effect like the case (1). If the modification adds a node after the execution point, the change will take effect. If the modification modifies the current execution point, it depends on the state of the current execution point. If current execution point is in "wait" state, the change will take effect. If current execution point is in "active" state, it does not permit to be modified because the node is running.

(3) Modification point is later than execution point. It means the change to the workflow instance will take effect after the modification.

4 Workflow Simulation

Simulation analysis plays an important role in detecting the need for changing and deciding what changes to carry out. Before a change is actually made, its possible effects can be explored with simulation.

4.1 Workflow Modeling

For modeling purpose, a workflow can be abstractly represented by using directed graphs (e.g., one for control flow and one for data flow, or one for both). The directed graph consists of a nonempty set of nodes and edges with the following properties:

(1) There are two types of nodes (AND node and XOR node).
(2) One of the nodes in the directed graph must be designated to be a START node and one must be designated to be a STOP node.
(3) The directed graph must be weakly connected.
(4) Edge labels indicate the probability of edges being selected as the outgoing edge.

4.2 Change Evaluation

After a task is finished, its time, cost and quality metrics will be computed. These values will be sent to the monitor and saved in the monitor's log. Each of these three metrics will be compared with dual threshold values [7]. There are three results: (i) the value is below the bottom threshold, (ii) it is above this threshold but below the top threshold, (iii) the value is above the top threshold. For the first case, an appropriate exception will be thrown. For the second case, a warning will be sent to the monitor. For the third case, the execution is considered to be satisfactory.

If an exception is thrown, it will be handled by exception handling process. There are several options such as ignore the exception, use an alternative task or adapt the workflow. As warnings are accumulated in the monitor, a pattern may indicate a helpful change. Such a pattern could be detected by a human monitoring the workflow or by an agent examining the monitor's log. Identifying and classifying a pattern of substandard quality are consisted of in the detecting process. Then possible corrective actions are determined.

Simulation is useful in setting the thresholds and in creating workflow modifications. Owing to simulation, the question what happens if one makes this change to the workflow is considered safely.

5 System Integration

In order to facilitate rapid simulation of workflow, it is necessary to translate workflow design specifications into simulation specifications [7]. The workflow designer saves workflow designs as XML documents in an XML Model Repository. Such a design can be retrieved and translated into simulation specifications. The translation is simplified since both systems represent their models with AND node and XOR node. Before the simulation is actually run, this default mapping from a workflow node to a simulation node can be adjusted by using the workflow model design tool.

Workflow designs and simulation designs follow their different Document Type Definition (DTD). Therefore, a workflow design document must be mapped to a simulation design document. This is accomplished using an XLST specification. The integration of workflow model with simulation is depicted in Figure 4.

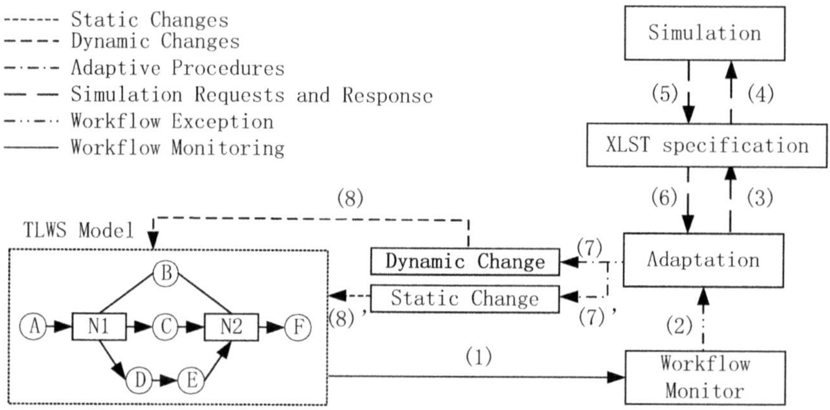

Fig. 4. Overall System Architecture

6 Conclusions

This paper has shown how simulation can be useful for supporting and improving the flexibility of TLWS models. We can answer "what-if" questions via a simulation instead of actually trying the change.

Because both workflow and simulation have the analogical conceptual frameworks, interoperability is facilitated. Since design tools for both systems use XML for saving designs in a repository, design specifications can be translated using XSLT. Currently, translation is in the direction from workflow specification to simulation specification. This is because that is what is needed and that the simulation specification is more abstract.

Quality metrics are very important to decide whether a workflow should be adapted and what change should actually be made. These metrics are important for both the workflow and the simulation.

Acknowledgments. We wish to thank Forlink Technologies for providing an excellent scientific research environment all along this work.

References

1. J. Herbst and D. Karagiannis, "Integrating Machine Learning and Workflow Management to Support Acquisition and Adaptation of Workflow Models", in Proceedings of the Ninth IEEE Conference on Database and Expert Systems Applications, 1998.
2. Huadong Ma, A Workflow Model Based on Temporal Logic, CSCWID2004, pp.327-332 Xiamen, China, May 2004.
3. Hongchen Li, Meilin Shi. Workflow models and their formal descriptions, Chinese Journal of Computers, 2003, 26(11): 1456-1463.
4. Z. Manna, A. Pnueli. Temporal Logic for Reactive and Concurrent System, Springer Verlag, 1992.
5. Z.S. Tang, C. Zhao. A Temporal Logic Language Oriented Toward Software Engineering, Journal of Software, 1994,5(12):1-12.
6. Yi Duan, Huadong Ma. "Modeling Flexible Workflow Based on Temporal Logic". The 9th Conference on Computer Supported Cooperative Work in Design Proceedings, pp.508-513 Convertry, UK, 2005.
7. John A. Miller, Jorge Cardoso, and Gregory Silver. "Using Simulation to Facilitate Effective Workflow Adaptation". Proceedings of the 35th Annual Simulation Symposium(SS'02),2002.

Towards Intrinsic Evolvable Hardware for Predictive Lossless Image Compression*

Jingsong He[1,2], Xin Yao[1,3], and Jian Tang[2]

[1] Nature Inspired Computation and Applications Laboratory (NICAL)
[2] Department of Electronic Science and Technology, University of Science and Technology of China
[3] School of Computer Science, University of Birmingham
hjss@ustc.edu.cn, x.yao@cs.bham.ac.uk, jtang@ustc.edu.cn

Abstract. This paper presents a novel method for predictive lossless image compression via evolving a set of switches, which can be implemented easily by intrinsic evolvable hardware mode. A set of compounded mutations for binary chromosome through combining the local asexually reproducing with multiple mean step size search was proposed, and a gradually approach method for evolving larger scale images was fabricated. Experimental results show that the proposed method can reduce the computing time much more, and can scale up the image size increasing up to 70 times with relative slower increase speed of computing time.

1 Introduction

As the new emerged field, evolvable hardware (EHW) may provides new types of mechanism for automatic circuit design and optimization, and adapting the environment by circuit itself. Recent studies on applications of evolvable hardware conduct an important issue, that is to seek valuable applications of evolvable hardware and to discover new problems and their corresponding solutions. Indeed, the way to study problems through combining intelligent computation with real-world applications is one of the most important methodologies in the field of evolvable hardware.

Among various applications, the adaptive lossless image compression is one of the typical applications of evolvable hardware. T.Higuchi et.al.[1] firstly presented an evolvable chip implemented by a special functional FPGA (F^2PGA) with their special variable length genetic algorithm (VGA) for compressing images, which has been regarded as the merely non-toy problem in the early researches in the field of evolvable hardware[2]. And afterwards, they present a further research on the problem of evolving large scale images (up to 315MB, the computing time has not been mentioned.) in [3]. As a different study, A. Fukunaga et.al.[2][4] presented a new prototype system based on genetic programming (GP) to solute the problem of such implementation on conventional

* This work is partially supported by the National Natural Science Foundation of China through Grant No. 60573170 and Grant No. 60428202.

T.-D. Wang et al. (Eds.): SEAL 2006, LNCS 4247, pp. 632–639, 2006.

FPGA. [5] shows that the GP-based model can enlarge the size of processable image from 64K-bytes up to 2M-bytes, and can reduce the computing time from 2 hours[4] down to 5 minutes (in simulation mode) at the same time.

However, both models above have limitations in their functionality: VGA for optimizing templates is executed on a host computer[3]; The compiling time of converting a chromosomes to an assessable circuit is usually about 0.5 hour[2]. Obviously, problems behind them are seriously. In this sense, intrinsic EHW may be more suitable for real-time applications[6], specially for the task of predictive lossless image compression.

This paper proposes a novel and simple evolutionary technique for predictive lossless image compression, where a genetic algorithm with small size of populations was used for reducing the circuit resources, and the parameters of predictive function was discrete, therefore the chromosome can be corresponded to a set of circuit switches thus can be evolved on chip directly. Experimental results show that the proposed method can process more larger images, while reduce the computing time efficiently.

2 Lossless Image Compression Through Evolving a Set of Switches

The character of extrinsic evolvable hardware mode is to use symbolic expression in chromosome, where each symbol corresponds to a real circuit completed by conventional design. The advantage of using extrinsic mode is that many commercial compilers and circuit resources out of conventional design can be employed directly. The disadvantage of extrinsic mode in real-time control and adaption is that this kind of using may be detrimental to online tasks, since extrinsic mode needs an extra device to accommodate a compiler software, and has to have many compiling time and download time.

On the problem of implementing the self-adaption task of real-time lossless image compression on a chip, if the control parameters could be described as a set of switches, the evolving of states of switches will be easily achieved inside a chip through intrinsic mode.

2.1 The Binary Cording Mechanism

To obtain a fixed length chromosome with binary cording, here the exponential function is used to approach the predictive function. Thus the 4-neighbor predictive mode can be written as

$$x'_k = \sum_{i=1}^{4} x_{k,i} e^{-\alpha_i d(x_k, x_{k,i})}, \tag{1}$$

where x'_k denotes the kth predicted value, $x_{k,i}$ denotes the pixel value of the ith neighbor of x_k, α denotes the parameter of the interpolating function, and $d(x_k, x_{k,i})$ denotes the distance between the current point and its ith neighbor.

Therefore, the task of prediction becomes to minimize the following objective function

$$f(x) = \sum_{k=1}^{N} ||x_k - x'_k||^2 = \sum_{k=1}^{N} \big[x_k - \sum_{i=1}^{4} x_{k,i} e^{-\alpha_i d(x, x_i)}\big]^2 \qquad (2)$$

where N is the number of pixels for prediction. Conveniently, assume the effect of interpolating function only affects its neighboring pixels, i.e. $d(x_k, x_{k,i})$ equals to one uniformly. Hence (2) becomes

$$f(x) = \sum_{k=1}^{N} \big[x_k - \sum_{i=1}^{4} x_{k,i} e^{-\alpha_i}\big]^2. \qquad (3)$$

Hence the problem of lossless image compression becomes a parameter optimization problem with four variables α_1, α_2, α_3, α_4. Obviously, there are many numerical/function optimization algorithms, e.g., IFEP[7], and StGA[8], can solute this kind of problem well. However, the task of optimization problem in the field of evolvable hardware (specially in intrinsic EHW mode) is quite different from tasks implemented by software.

To make the problem of Eq.(3) be easily implemented by intrinsic EHW, we can make parameters in Eq.(3) coded with a set of binary string. Thus the value of exponential function can be calculated on chip by querying an embedded LOOK-UP Table (LUT). e.g., we can use "101101,011100,101100,010101" represent "0.8125, -1.7500, 0.7500, -1.3125" for the values of α_1, α_2, α_3, α_4 respectively, thus the candidate solution of predictive function with formulation of Eq.(1) can be expressed as linear function as $x' = 2.25353478721321x_1 + 0.173773943450445x_2 + 2.11700001661267x_3 + 0.269146348729184x_4$.

It is apparent, 1) The size of LUT is equal to $2^{L/4}$, where L is the length of binary string; 2) This kind of cording method can make the task of predictive lossless image compression being implemented on chip easily.

2.2 The Binary Evolutionary Programming Algorithm

Two important notions can make help in achieving a binary evolutionary programming algorithm. The one is the technique of asexually reproducing for local selection presented in StGA, and the other is the idea discussed in IFEP that long jumps can help to generating an offspring at the neighborhood of the global minimum. For the sake of minimizing the population size to reduce circuit resources as much as possible, here the notion of asexually reproducing and the idea of using long jumps are fabricated to archive the binary evolutionary programming algorithm.

Technically, the asexually reproducing can repeat one by one with a series of compounding mutations as shown in Fig.1, where h_0 is the parent, h_1–h_5 are offsprings, p_{m1} and p_{m2} are probability for mutations. Empirically, p_{m1} and p_{m2} can usually take $\frac{1}{3}$ and $\frac{1}{9}$ respectively. Thereby, the averaged differences (search step size with hamming distance) between the parent and each offspring can

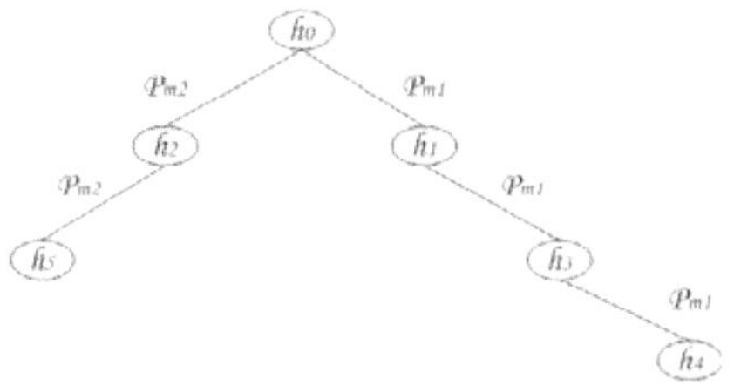

Fig. 1. The illustration of the asexually dendriform reproducing, where h_0 is the parent, h_1, h_2, h_3, h_4, and h_5 are offsprings, p_{m1} and p_{m2} are the probability of mutations

be calculated to be $\frac{1}{3}$, $\frac{1}{9}$, $\frac{4}{9}$, $\frac{1}{27}$, and $\frac{16}{81}$ of the length of bit string respectively. It is apparent, the characteristic of the local selection mechanism in StGA is absorbed but the random generation method is substituted to a mixed search with five different mean step size of mutations, the characteristic of using two kinds different mean step size (Gaussian mutation and Cauchy mutation) for mutation presented in IFEP is also absorbed but the jumping has been made more times and more deeply.

Since the local dendriform reproducing as shown in Fig.1 is mixed with relative large mutations and relative small mutations, it can be thought as a hybrid mechanism between local and global search. For scalability reason, here we just take it as an independent evolution procedure, and denote it as binEP. The basic procedure of binEP is summarized as follows.

Step 1. Initial a L-bits binary string randomly, and denote it as h_0. Set $C = 1$. Each bit of h_0 is taken as a circuit switch for controlling the real number of parameter in the predictive function. Evaluate the fitness of h_0.

Step 2. Generate five offsprings asexually as follows.
Mutate h_0 with the probability $p_{m1} \rightarrow h_1$;
Mutate h_0 with the probability $p_{m2} \rightarrow h_2$;
Mutate h_1 with the probability $p_{m1} \rightarrow h_3$;
Mutate h_3 with the probability $p_{m1} \rightarrow h_4$;
Mutate h_2 with the probability $p_{m2} \rightarrow h_5$;
where $1 > p_{m1} > p_{m2}$, p_{m1} and p_{m2} are the probability of mutation happened on each binary bit with the value changes from 0 to 1 or from 1 to 0.

Step 3. Evaluate the fitness of each offspring by Eq.(3).

Step 4. Select the best one out of $\{h_1, h_2, h_3, h_4, h_5\}$, and denoted as h_0'. If h_0' is better than or equal to h_0, using h_0' substitute h_0 to be the parent of next generation.

Step 5. Stop if the halting criterion is satisfied; otherwise, $C = C + 1$, go to Step 2.

The sampling and interpolating mechanism in the field of signal processing can make the problem of scalability solved simply and directly. To reduce the computational time on large amount data, here we use a pyramidal fitness evaluation as follows. For convenient reason, we denote it as BinEP.

Step 1. Segment the predictive area of a $m \times n$ size image to a set of $l \times k$ templet, where $l < m$, $k < n$.

Step 2. Take the set of mean value of each $l \times k$ templet as the input data, execute *binEP*.

Step 3. If the fitness has not been improved after some generations, reduce l and k randomly, go to Step 2. Stop if the fitness has not been improved after some generations, when $l = 1$ and $k = 1$.

Obviously in BinEP, the changing of input data can change the environment of binEP from coarse granularity to fine granularity. Since the coarse expression of search space can make the objective problem be approached more easily, therefor BinEP can supply a kind of greedily and gradually approach through disposing the scalability problem at the same time.

3 Simulation Experiments

The proposed BinEP is evaluated by comparing with the method of Huffman coding, and the lossless JPEG. Images used here are Lena (as shown in Fig.2) and a number of science images from NASA[9] (we use these images because they are open, thus the work in this paper can be validated by others). In experiments, the initial parameters of BinEP are that $l = 10$, $k = 10$, $p_{m1} = \frac{1}{3}$, $p_{m2} = \frac{1}{9}$, and the length of chromosome is 24 bits. The size of compressed image is calculated as: *Sizeof(compressed Error)+Sizeof(compressed Bord)+Sizeof(Binary string).*

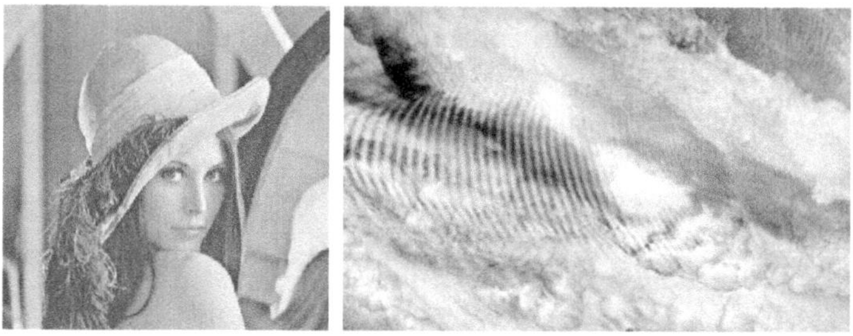

Fig. 2. The 256×256 image of Lena, and the 1374×889 science image PIA04349

The experiments are divided into two parts based on different points of view. The first is for comparing with the GP-based method on Lena image, since this kind of comparison can exhibit the efficiency of methods basically. The second is for evaluation on large scale images to see the scalability of BinEP. Although the proposed evolutionary technique can be implemented by on our DSP development platform (with TMS320VC5402), to put down the evolving time for computational efforts comparison, all results have been averaged over 50 runs of software simulation on AMD 1.2G CPU.

3.1 Evaluation on Small Size Image

Experiment results on Lena are summarized in Table 1, where two kinds of parameter setting for BinEP with 3×3 and 1×1 templet (which means BinEP is run with no technique of gradually approach, i.e., binEP) are also listed for analysis. It is clear, the optimization performance of BinEP are better both than the GP-based method and the lossless JPEG2000.

It is also clear that BinEP and binEP have the same performance of optimization, while the computing time of BinEP is much less than that of binEP. That is, the proposed technique for large amount of input data with gradually approach method is validated to be feasible on small size image.

Table 1. Comparison between BinEP and GP-based method on Lena image (256×256, grey, 66536 bytes), where BinEP is with 3×3 templet, and binEP means BinEP with 1×1 templet (i.e., BinEP with no technique of gradually approach). Here the method for coding error matrix are both the Huffman code technique.

Methods	Compressed Size	Computing Time
GP	$43,154$ *bytes*	169.0 ± 0.1 *sec.*
binEP	$40,247$ *bytes*	11.23 ± 0.1 *sec.*
BinEP	$40,247$ *bytes*	3.080 ± 0.1 *sec.*
JPEG2000	$44,090$ *bytes*	< 1 *sec.*

3.2 Evaluation on Large Scale Images

In experiment, the templet used in BinEP is 10×10. Two lossless coding techniques were used to code the evolved error matrix, the one is Huffman code which was used in the past approaches in [2] and [4], the other is the lossless JPEG2000 which is investigated newly in this paper. The experimental results are listed in Table 2, where the results of using Huffman code and the lossless JPEG2000 compress the original image directly are also summarized. It is apparent, the compressive results and the computing time of BinEP on large scale images is better and acceptable. The difficulty of BinEP on vary large images, such as PIA07335, is at the later period of evolution when the templet is shrunk to one pixel. At this time, the computing time can not be reduced since the evolved data is total of an image.

Take the result of binEP (with no technique of gradually approach) on 256×256 size image Lena as the reference, the efficiency of BinEP on larger scale images can be summarized approximately as shown in Fig.3. It seems that the satisfying of BinEP is up to 70 times increase of image size, where the increase speed of images is relative slower than the increase speed of computing time.

3.3 General Comparison Between BinEP and DRC

The *Dispersed Reference Compression* (DRC)[3] is the newly developed method comes of [1]. Essentially, the idea of proposed method and DRC is quite different.

Table 2. Comparison between BinEP, Huffman code, and the lossless JPEG2000 on large scale images. BinEPv1 means that Huffman code is used for coding the error matrix, and BinEPv2 means the lossless JPEG2000 is used for coding the error matrix.

Image Name	Huffman (bytes)	JPEG2000 (bytes)	BinEPv1 (bytes)	BinEPv2 (bytes)	Evolving Time (sec.)
$PIA07335$	5,578,020	2,515,657	2,311,920	2,389,501	266.57 ± 3
$PIA07217$	6,286,500	2,031,297	1,918,890	1,983,626	214.89 ± 3
$PIA05578$	5,522,256	1,540,111	1,698,480	1,444,500	161.81 ± 4
$PIA07225$	4,919,459	2,108,906	2,151,157	2,016,007	121.58 ± 2
$PIA07096$	4,247,303	1,938,940	1,637,002	1,886,143	132.92 ± 3
$PIA07343$	2,270,780	1,935,780	1,762,368	1,814,022	75.50 ± 2
$PIA07227$	1,955,670	1,622,853	1,552,944	1,550,086	52.23 ± 3
$PIA04349$	1,115,492	789,619	720,341	735,189	32.32 ± 3
$PIA05202$	816,127	576,951	534,536	541,220	25.14 ± 2
$PIA06322$	731,328	529,634	481,381	492,653	37.03 ± 2

[1]The size of images from up to down are: 3000×2400, 3000×2400, 2400×2400, 2841×1846, 3000×1688, 2104×1726, 1320×1840, 1374×889, 1065×771, and 1239×805 respectively.

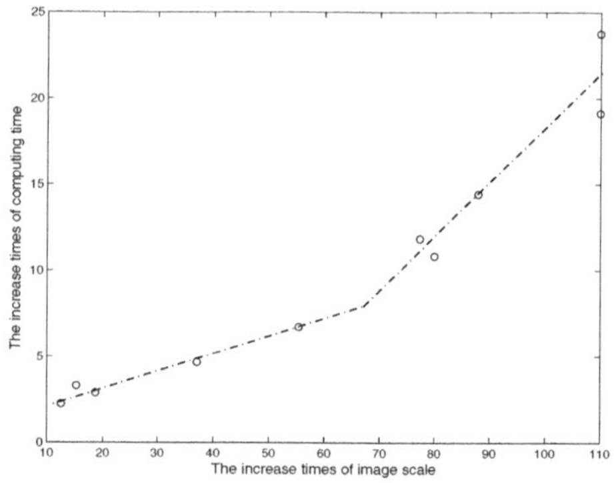

Fig. 3. The approximately estimation of the efficiency of BinEP on larger scale images, where the size of image for contrast is 256×256. The vertical axe is the increase times of computing time, and the horizontal axe is the increase times of image scale.

On the issue of evolving large scale images, [3] presents a number of results of compression ratio, however had not shown details about computing time. Thus it is hard to give an exact experimental comparison between BinEP and DRC, since the computing time is the most important issue for evolvable hardware, special for real-time applications.

For the issue of predictive lossless image compression itself, intrinsic EHW is more attractive than extrinsic EHW. The reason is vary simple and clear: using extrinsic EHW, images have been evolved by software (simulator) already, that repeat the last work done by software on a hardware has few realistic meaning but for study. On this issue, the proposed method has more advantageous, since its problem expression is suit for intrinsic EHW mode.

4 Conclusions

This paper proposes an intrinsic EHW model for predictive lossless image compression, which can be implemented on a chip directly, other than extrinsic EHW model used by [1-4] and [5]. For the problem of solving large scale of images, this paper presents a binary evolutionary programming (BinEP) by combining the technique of local asexually reproducing with the idea of using mixed mutations.

The proposed evolutionary technique is suit for implementation with intrinsic evolvable hardware mode. Experimental results show that the proposed method can reduce the computing time much more, and can scale up the processed image size up to 70 times larger with relative slower increase speed of computing time.

References

1. T. Higuchi, M. Murakawa, M. Iwata, I. Kajitani, W. Liu, and M. Salami: Evolvable hardware at function level. In IEEE Intemational Conference on Evolutionary Computation, (1997) 187–192.
2. A.Fukunaga, K.Hayworth, A.Stoica: Evolvable Hardware for Spacecraft Autonomy. Aerospace Conference, IEEE, **3** (1998) 135–143.
3. H. Sakanashi, M. Iwata, and T. Higuchi: A Lossless Compression Method for Halftone Images using Evolvable Hardware. Evolvable Systems: From Biology to Hardware, Lecture Notes in Computer Science 2210, pp. 314-326, Springer Verlag, 2001.
4. A.Fukunaga, A.Stechert, "Evolving Nonlinear Predictive Models for lossless Image Compression with Genetic Programming", Proc.of the Third Annual Genetic Programming Conference,Winsconsin,1998.
5. J.He, X.Wang, M.Zhang, J.Wang, Q.Fang: New Research on Scalability of Lossless Image Compression by GP Engine. The 2005 NASA/DoD Conference on Evolvable Hardware, June 29 - July 1, 2005, The Westin Grand, Washington DC, USA. 160–164.
6. X. Yao, T.higuchi: Promises and challenges of evolvable hardward. IEEE Trans. On Systems, Man, and Cybernetics - Part C: Applications and Reviews. **29**(1), (Feb. 1999).
7. X. Yao, Y. Liu, G. Lin. Evolutionary Programming Made Faster. IEEE Trans. on Evolutionary Computation, **2**(2), (1999) 82–102.
8. Zhenguo Tu and Yong Lu: A Robust Stochastic Genetic Algorithm (StGA) for Global Numerical Optimization. IEEE Trans. on Evolutionary Computation. **8**(5), (Oct. 2004) 456–470.
9. http://photojournal.jpl.nasa.gov/gallery/universe

Using Qualitative Description for the Dynamic Analysis of Virus Inflection*

Hailin Feng[1], Chenxi Shao[1,2], and Zicai Wang[2]

[1] Computer Science Dept. of University of Science and Technology of China, Hefei 230027,China
hlfeng@mail.ustc.edu.cn, cxshao@ustc.edu.cn
[2] Simulation Center, Haerbin Institute of Technology, Haerbin, 150001, China

Abstract. A method based on qualitative description is introduced to analyze the dynamic system of viral infection, and to predict the qualitative behaviors of the system by describing qualitative states and the transition between the states. The algorithm of building qualitative states is presented and then the qualitative states graph can be created from the dynamic model of viral infection. The reasoning algorithm is given to accomplish the transitions between the qualitative states. The qualitative trends of the state variables and their relative positions of the viral infection model are obtained, and then the corresponding transition graph of these qualitative states is gained. The qualitative simulation of the system is also given without precise formulation of the model.

1 Introduction

The behavior analysis and prediction of dynamic system is a difficult task due to the absence of sufficient quantitative information and the lack of efficient methods. Especially in those fields that the models being researched are not fully known, the problem is more complicated. For a nonlinear dynamic model, the system may exhibit complex character even in low dimensions[5].

The dynamical models in biological fields suffer this problem frequently. For the durative viral infection, the pathogenic bodies of virus propagate in the environments of the host organicity, and these bodies may be restricted by the physical and chemical barrier of the host, such as the mucous membrane of skin, the phagocyte, T cell, and so on[4,11,16]. For these complicated relationships among the terms in the environment, the dynamical model of the viral infection has not been known completely, which increase the difficulty to describe and analyze the viral infection model.

Researchers have shown much interest in such nonlinear dynamical systems with incompletely known knowledge. If the dynamical system is given as ordinary differential equation, B.Kuipers writes its qualitative model in the formula of qualitative differential equation (QDE) and reasons the process qualitatively[10,14]. If the dynamical system is linear or piecewise linear, the corresponding qualitatively simulation can be used to analyze the characters of the system[7]. The method of phase plane analysis can draw some useful conclusions to the behavior prediction in the case of

* Supported by Key Project of Natural Science Foundation of China (60434010).

T.-D. Wang et al. (Eds.): SEAL 2006, LNCS 4247, pp. 640 – 647, 2006.

low dimensions system[6]. J.L.Gouze analyzes the dynamical system of biology by considering the positive and negative circuits in the system[9]. And V.Volterra does some researches on the variations and fluctuations of the dy-namical model, and implements it in coexisting animal species[8]. If the system is certain nonlinear dynamical model with dimension differential equations, a qualitative method can analyze the be-haviors of system by studying the existing and sequence of every variable[3,15]. The subsequent works give some definitions and formal descriptions[13]. In reference [2] the author uses a qualitative method to describe the interaction among the factors existing in viral infection system and build qualitative viral infection model.

Here we introduce the method based on qualitative description[3] to analyze viral infection model. The main idea is to create the qualitative behaviors description and build the transition graph between these qualitative states from the nonlinear dynamical model with incompletely known information.

We present the algorithm of building qualitative states to create the set of almost all possible qualitative characters. The qualitative description is expanded by relating the qualitative behaviors of extrema and those crossing through the equilibrium values. The characters of states are defined by the sign of variables' qualitative trending and all the states are restricted to a specific set.

Based on the set of qualitative states, we also present a reasoning algorithm of state transition, which predict the possible qualitative behaviors of the system according to the qualitative states of velocity and the variable. The transition is represented as a graph with crossing point of extrema and variables to describe the behaviors of dynamical model and the possible transition paths are all given in the graph. At the end of the paper, the application of qualitative analysis method to the viral infection colony dynamical system is given and the conclusion is drawn.

2 The Preliminary Knowledge

Let Ω be open convex domain in $\mathbb{R}^n$, and f is a map from Ω to $\mathbb{R}^n$. Here because the model in our studying is biological viral infection model, so domain Ω should have biologic meaning, which means that the values of variables of the models are restricted in domain and without meaning in outside of the domain.

For $x \in \mathbb{R}^n$, we call $x > 0$ if for all i, $x_i > 0$. For $x \in \mathbb{R}^n$ and $y \in \mathbb{R}$ we consider the function sign: $sign(x) = \begin{pmatrix} sign(x_1) \\ \cdots \\ sign(x_n) \end{pmatrix}$, $sign(y) = \begin{cases} -1 & if\ y < 0, \\ 0 & if\ y = 0, \\ 1 & if\ y > 0, \end{cases}$

We will give some definitions that about the qualitative description[13].

Definition 1. If for all $i \neq j$, the partial derivative $\partial y_i / \partial x_j(x)$ won't cancel in Ω, we call the system Σ has a monotonous interactions in Ω. Obviously, for such monotonous system the off-diagonal terms of the Jacobian matrix all have a fixed sign in Ω.

Definition 2. Consider the set S_n with 2^n elements: $S_n = \{\sigma = (\sigma_1, ..., \sigma_n)^T;\ \sigma_i \in \{-1, 1\}\}$, S_n implies all the sign vector in $\mathbb{R}^n$ with values are ± 1. We can order all the elements and use σ^q to denote the q th element. We will select $\sigma^1 = (1, ..., 1)^T$ for convenience.

Definition 3. For all j in $\{1,...,n\}$ and q in $\{1,...,2^n\}$, if the partial derivative $\partial y_i / \partial x_j(x)$ has fixed sign in the domain: $W_{\sigma^q}(x^*) = \{x \in \Omega;\ diag(\sigma^q)(x - x^*) > 0\}$, then the system (Σ) can be called diagonally x^* monotonous, and $W_{\sigma^q}(x^*)$ is called the orthants of x^* deviation(Here the x^* is the equilibrium point of the system). The velocity domain Z_{σ^q} is defined as: $Z_{\sigma^q} = \{x \in \Omega;\ diag(\sigma^q) f(x) > 0\}$ and Z_{σ^q} is delimited by U_i : $U_i = \{x \in \Omega;\ f_i(x) = 0\}$.

Definition 4. Consider first the consistent set between velocity domain Z_{σ^q} and deviation domain $W_{\sigma^q}(x^*)$, which means the sign of $f(x)$ for a given x in $W_{\sigma^q}(x^*)$. That is: $\Gamma^q = \{sign(f(x)), x \in W_{\sigma^q}(x^*) \setminus U\}$ .The formal definition is to provide the restriction of possible qualitative states. In fact, the elements in Γ^q imply the qualitative events of the velocity $\dot{x}$.

Next we consider the set: $\Omega_{\sigma^q\sigma^p}(x^*) = W_{\sigma^q}(x^*) \cap Z_{\sigma^p}$, and according to the definition above we get: $\Omega_{\sigma^q\sigma^p}(x^*) = \{x \in W_{\sigma^q}(x^*); diag(\sigma^p) f(x) > 0\}$. The sets may be empty, and we call the non-empty sets as possible regions.

3 The Transition Between Qualitative States

After the sets of qualitative states of the system are defined, in this section we will consider how to analyze the behaviors of the dynamic system if the initial state is given, which means the transition between these qualitative states.

The transitions between qualitative states proceed among the neighbors domain, so it touches the notion of neighbor. Two domains are called as neighbors if there is only one sign that is different in the equilibrium deviation vector or the velocity vector.

3.1 Transition Rules

Now consider the transition rules that describe the possible transferring events between qualitative states in the strict domain Ω partitioned by $\Omega_{\sigma^q\sigma^p}(x^*)$. The qualitative behaviors obey the rule:

Considering the system (Σ) with monotonous interaction and equilibrium x^*, and the domains $\Omega_{\sigma^{q1}\sigma^{p1}}(x^*)$ and $\Omega_{\sigma^{q2}\sigma^{p2}}(x^*)$ are strict neighbors, and $t_{k,k+1}$ is the sign of the element $(k, k+1)$ of Jacobian matrix, we have:

(1) Suppose they are strict U-neighbors, and if $t_{k,k+1}\sigma_{k+1}^{p1} = \sigma_k^{p1}$ (or $-\sigma_k^{p1}$) ,then it is only possible from the $\Omega_{\sigma^{q2}\sigma^{p2}}(x^*)$ to $\Omega_{\sigma^{q1}\sigma^{p1}}(x^*)$ (or from $\Omega_{\sigma^{q1}\sigma^{p1}}(x^*)$ to $\Omega_{\sigma^{q2}\sigma^{p2}}(x^*)$), and it corresponds to a minimum (or maximum) of variable x_k.

(2) Suppose they are strict V-neighbors, and if $t_{k,k+1}\sigma_{k+1}^{q1} = \sigma_k^{q1}$ (or $-\sigma_k^{q1}$), then it is only possible from the $\Omega_{\sigma^{q2}\sigma^{p2}}(x^*)$ to $\Omega_{\sigma^{q1}\sigma^{p1}}(x^*)$ (or from $\Omega_{\sigma^{q1}\sigma^{p1}}(x^*)$ to $\Omega_{\sigma^{q2}\sigma^{p2}}(x^*)$), and it corresponds for x_k to cross from bottom to top (or from top to bottom) of variable x_k.

The rule above provides the possible transitions between the qualitative regions, and a global qualitative behaviors rule will be shown to describe the qualitative behavior paths of a dynamical system with monotonous interactions, at the moment the domain Ω is partitioned by $\Omega_{\sigma^q \sigma^p}(x^*)$.

3.2 Algorithm of Creating Qualitative States

In this section the algorithm steps will be presented to create qualitative states graph for a nonlinear dynamical model of biologic system (here the system aims at our studying model and it can be extended to a general case) according to the definition and the rules above. The main idea is to build the qualitative behaviors of the velocity and the qualitative information of the variables' deviation of the equilibrium from the given dynamical system. The algorithm is described as follows:

Input: A nonlinear dynamical model (Σ) $\dot{x} = f(x)$;

Output: The set M including the qualitative behaviors of $(x - x^*)$ and $\dot{x}$;

Steps:

(1) The $n \times n$ matrix J will be created according to the dynamical model;

(2) From the equations $f(x) = 0$,we get the solution $(x_1^*, ..., x_n^*)$ (here assumed that the approximately linearized part A is nonsingular, and which is also satisfy the models we are studying);

(3) Produce all the 2^n elements of S_n , all of which are $n \times 1$ matrix and the values in the matrix are either $+1$ or -1 ,and inserts them in the set S ;

(4) Substitute $(x_1^*, ..., x_n^*)$ for the x in $J_{n\times n}$ to create matrix R ;

(5) Compute $R_{n\times n} \times diag(a)$ for all the elements $a_{n\times 1}$, and insert them in the set L ;

(6) If the signs of elements in the set L conform the judgment condition, then compute the qualitative states of velocity $sign(f(x))$ for all a in S , and insert them in set F ;

(7) Find all elements in S and F that have $a \in S$ and $diag(a) f(x) > 0$, and insert them in M ;

Following the steps above, the qualitative states graph can be created from a given dynamical model.

3.3 Reasoning Algorithm of States Transition

In this section the reasoning algorithm of states transition is presented based on the sets of qualitative states after they are created by the algorithm in Section 3.4. The main idea is to find the next possibly accessible state from the qualitative regions of velocity and those relative to equilibrium, and so all the possible transition paths in the States Transition Graph (STG) are given from reasoning.

The reasoning algorithm is described as follows:

(1) Initialize the system and set the visiting flag of all qualitative states as FALSE;

(2) The initial state is obtained from the actual model, and the analyzing program will represent the state as a qualitative set that can be identified by our system, which is also taken as the input $(x_1, ..., x_n; x_1^*, ..., x_n^*)$;

(3) Choose a state m from the set M according to the initial state; if the visiting flag of m is FALSE, then set it as TRUE and goto (4); if the flag of m is TRUE, then follow the directed arrow to go to next state that has not been visited (the flag is FALSE) , let the state as m and set the flag as TRUE and goto (4); if the visiting flag of all the qualitative states are TRUE, then goto (8);
(4) Find all the strict domain neighbors of m according to the qualitatively restricted rules and insert them in the set N ;
(5) If the neighbors set N is not empty, then select certain element n and delete it from set N , goto(6); If the neighbors set is empty, then follow the directed arrow to go to next state and let this state as the initial one, and goto (3);
(6) Scan all the elements in M : if n is in the set M , then goto (7); else then goto (5);
(7) Using the directed arrow to couple this qualitative state and n according to the Rule 1 and endue the transition symbol for them;
(8) End of the reasoning with creating the States Transition Graph (STG).

4 Qualitative Analysis of the Viral Infection Dynamical Model

To illustrate better the method of dynamical analysis based on qualitative description, we will take the viral infection dynamical model[1] as an application example. The process of virus infection for the viral entry can be shown in the equation as follows. Let T denotes the target cells, V the virus and I the infected cells. The parameters k, s, d, p, c, λ are all positive values. For the biological interests, we select the initial values as $T(0) > 0$, $I(0) = 0$, $V(0) > 0$, and the model can be called as model (Σ_{VIT}) .

The viral infection dynamical model in host body can be described as differential equations:

$$\begin{cases} dV/dt = pI - cV & (1) \\ dI/dt = kVT - sI & (2) \\ dT/dt = \lambda - dT - kVT & (3) \end{cases}$$

The comparison of this class of models with experimental data is given in [11]. The dynamic parameters can be estimated with mathematical tools[12]. Now we will give basis properties of the model that are important.

Property 1. The open domain $\Omega = \{(V,I,T) \in R^3; V > 0, I \geq 0, T > 0\}$ is positively invariant.

Because the variables (including the initial values) will be meaningless in the outside of domain Ω , so we study the model only in this domain.

Property 2. The virus infection model has monotonous interaction in the domain $\Omega = \{(V,I,T) \in R^3; V > 0, I \geq 0, T > 0\}$.

The parameters of the system (Σ_{VIT}) don't rely on the time explicitly, so it is an autonomous system. We can consider the equilibrium (V^*, I^*, T^*) in particularly initial states: $(V^* = (kp\lambda - csd)/csk, I^* = (kp\lambda - csd)/psk, T^* = cs\lambda/kp)$.

From the intersection of $W_{\sigma^q}(x^*)$ and Z_{σ^q} we get the set of qualitative spaces $\Omega_{\sigma^q\sigma^p}(x^*)$. It will be ignored if the set is empty, while the nonempty sets $\Omega_{\sigma^q\sigma^p}(x^*)$ indicate the qualitative states that are admitted by the model, which is shown in Fig.1. The domain $\Omega_{\sigma^q\sigma^p}(x^*)$ can be represented by two column vectors and the values are signs. The one implies σ^q and the other σ^p .

Then we get the graphic representation of $\Omega_{\sigma^q\sigma^p}(x^*)$ and the transition graph between states. The qualitative behaviors of the system are given as a graphic form in Fig.2. With different initial situations the reasoning paths will be different too.

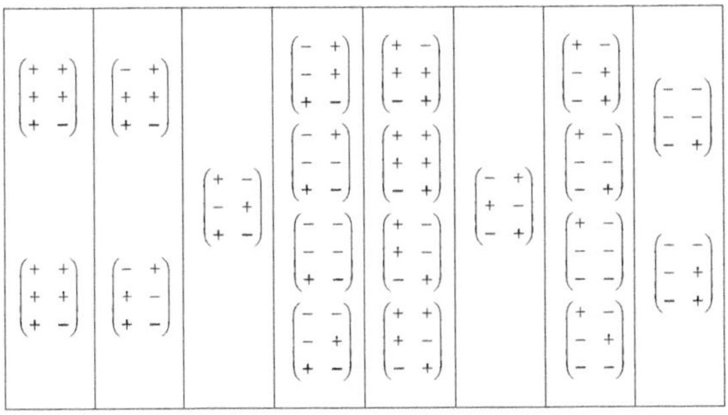

Fig. 1. The graphic representation of qualitative sets

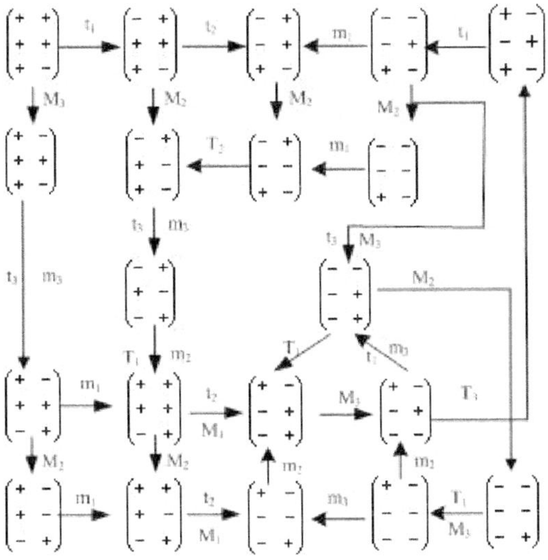

Fig. 2. Basic transition graph of qualitative states. M_i (or m_i)denotes a maximum(or a minimum) for the variable x_i . T_i (respectively t_i)denotes a crossing of its equilibrium point from bottom to top(respectively from top to bottom) for the variable x_i .

The nodes in the figure indicate the possible qualitative states that are compose of two parts: one is the relative position of the variable to the equilibrium point, and the other is the behavior signs of velocity. For example, when the qualitative state of velocity transfer from $(-,+,-)$ to $(+,+,-)$ and the corresponding sign of the coefficient is positive, the variable is admitted to pass a minimum and the letter of the changing state is marked as m_1, and the path is from the first state to the next with passing the minimum. If the qualitative state relative to the equilibrium point is transferring from $(-,+,+)$ to $(-,-,+)$, then the variable will follow the path to the next state by crossing the equilibrium from top to bottom and the letter is marked as t_k.

If the initially qualitative state is known, we obtain a qualitative simulation by following the edges between the nodes, which are determined by the transition rules and reasoning algorithm. It is shown that, at most, one maximum, one minimum, one equilibrium point crossing from top to bottom or reverse for each state variable are possible. Note that the path from the initial state to certain domain or final domain is not unique in qualitative conditions and there may be several paths can be followed. The transition path may be determined more exactly if we have more quantitative knowledge of the model.

5 Conclusion

The quantitative analysis of viral infection dynamical model can't be processed easily due to the lack of complex quantitative knowledge in such biological system; therefore, the methods based on qualitative analysis become our alternative solution to researches in the complicated dynamical system.

In this paper we introduce the method based on qualitative description to analyze the viral infection model, which predict the qualitative events of the system by describing the qualitative states and the possible transitions between those states, and this prediction and analysis don't rely on the precise quantitative knowledge of the model. The building algorithm is presented to produce the qualitative states graph from the dynamical model of viral infection. And the reasoning algorithm of states transition is also given to accomplish automatically the transition between qualitative states graph.

The method of qualitative analysis can be used both to predict the qualitative behavior of biological viral infection model, and to validate the structure of theoretical model. We can compare the qualitative behaviors gained from our simulation with the experimental observations of the temporal scenarios of qualitative transitions. Indeed, if the observed sequences of qualitative events do not correspond to a sequence contained in the graph, it implies a conflict between the model and the data.

However, the dynamical research of biological model based on qualitative is still on the way, there are lots of work should be done further. For instance, the validation of the theoretical model, the combination of quantitative with qualitative knowledge, the formal representation of states graph and filtration, and so on, are our further work in this direction.

References

1. Alan S.Pereson: Modelling Viral and Immune System Dynamics. Nature Reviews. 2(2002)28-36
2. Shao ChenXi, Feng Hailin, Wu Yue, Fan Jinfeng: Qualitative prediction in Virus entry based on feedback. Proceeding of Asia Simulation Conference. 2 (2005) 1343-1347
3. O.Bernard, J.-l.Gouze: Transient behavior of biological loop models, with application to the Droop model. Math. Biosci. 127 (1995) 19-43
4. Michael J Rust, Melike Lakadamyali: Assembly of endocytic machinery around individual influenza viruses during viral entry. Nature structural & molecular biology.11(2004) 567-573
5. L.Perko: Differential Equations and Dynamical Systems. Springer, Berlin. (1991)
6. E.Sacks: Automatic systems perspective on qualitative simulation. Artificial Intelligence. 42 (1990) 349-362
7. Hidde de Jong, Jean-luc Gouze, Cellne Hernandez: Qualitative Simulation of Genetic Regulatory Networks Using Piecewise-Linear Models. Bulletin of Mathematical Biology. 66 (2004) 301-340
8. V.Volterra: Variations and fluctuations in the numbers of coexisting animal species. Springer, Berlin. 22 (1978) 1923-1940
9. J.L.Gouze: Positive and Negative Circuits in Dynamical Systems. J.Biol.Syst.6(1998): 11-15
10. B.Kuipers: Qualitative Reasoning. MIT Press, Cambridge, MA. (1994)
11. Perelson, A.s., Neumann, et al.: HIV-1 dynamics in vivo: virion clearance rate, infected cell life-span, and viral generation time. Science. 271 (1996) 1582-1586
12. Ho, D.D., Deumann, A.U., Perelson, et al.: Rapid turnover of plasma virions and CD4 lymphocytes in HIV-1 infection. Nature. 373 (1996) 123-126
13. Olivier Bernard, Jean-Luc Gouze: Global qualitative description of a class of nonlinear dynamical systems. Artificial Intelligence. 136 (2002) 29-59
14. B.Kuipers: Qualitative Simulation. Encyclopedia of Physical Science and technology. Third Edition. Academic Press. (2001) 287-300
15. O.Bernard, J.L.Gouze: Transient behavior of biological models as a tool of qualitative validation-Aplication to the Droop model and to a N-P-Z model. J.Biol.Syst. 4 (1996) 303-314
16. Sanjay K.Phogat, Dimiter S.Dimitrov: Cell biology of virus entry: a review of selected topics from the 3rd International Frederick meeting. Biochimica et Biophysica Acta. 1614 (2003) 85-88

Emotion Interaction of Virtual Character in 3D Artificial Society

Zhen Liu

Faculty of Information Science and Technology,
Ningbo University, Ningbo 315211
liuzhen@nbu.edu.cn

Abstract. In a 3D intelligent virtual environment, multi 3D virtual characters interact with emotion and construct a 3D artificial society. Modeling emotion interaction is a challenging topic for virtual artificial society. Nonverbal emotion interaction is a direct communication manner for virtual characters. A cognitive model of virtual character is presented. A 3D virtual character has a cognitive architecture with built-in knowledge that control emotion and the response to outer stimuli. Some new concepts on nonverbal social interaction are set up.

1 Introduction

Modern computer technology brings people more simulation methods for life. Artificial life is the research field that tries to describe and simulate life by setting up virtual artificial systems with the properties of life. Emotion interaction is very important in real society. Modeling emotion interaction in a 3D virtual society is an interesting topic. A believable virtual character has the ability of emotion interaction to other virtual characters with verbal and nonverbal manner. In fact, nonverbal interaction is even more important than verbal interaction, particularly in respect of social interaction. Nonverbal social interaction includes a variety of signals, body language, gestures, tough, physical distance, facial expression, and nonverbal vocalization [1][2][3].

There are a lot of researches on emotion modeling, but little on modeling emotion interaction with social status in a 3D virtual environment. Ortony et al. set up an emotion cognitive model that is called OCC model [4]. In the model, emotions are generated in reaction to objects, actions of agents and events. Behavior animation of virtual character is a new animation technique, and a virtual character has perception and behavior [5]. Funge presented a hierarchy of computer graphics modeling [6]: the bottom three layers are geometric, kinematics and physics, while the top layer is behavioral. Behavioral modeling involves characters that perceive environment stimuli and react appropriately. Badler et al.developed the Jack software for simulation human behavior [7]. Thalmann et al. presented a frame of virtual characters[8], a virtual character should not only look visual, but also, they must have behavior, perception, memory and some reasoning intelligence. In recent years, Cassell et al. developed a behavior expression toolkit for virtual characters with XML-based language [9], Pelachaud

T.-D. Wang et al. (Eds.): SEAL 2006, LNCS 4247, pp. 648–655, 2006.

created a subtle method of facial expression for virtual characters [10], Chi et al. built a system called EMOTE to express natural posture movements of a virtual character [11].

On the basis of these researches, the goal of the paper is to build a model of non-verbal emotion interaction process between two virtual characters. A cognitive model of 3D virtual characters is presented in this paper, a 3D virtual character is provided with a cognitive architecture to control emotion interaction.

2 Cognitive Model of 3D Virtual Characters

A 3D virtual character is regarded as agent with a built-in knowledge. This section improves the result in previous paper [12], a cognitive architecture of 3D virtual character is presented in Fig. 1. We can explain all components of cognitive architecture as follows:

(1) Sensors module collects environments information from memory module. In this paper, we only consider visual sensors, which can read from memory module to get current information of a 3D environment.

(2) Perception module is different from sensor module, and a virtual character can perceive the meaning of objects in environment through perception module. The perception module reads and filtrates information from sensor module and collect information of outer stimuli, a simplified attention mechanism can be integrated in perception module. In general, attention mechanism of human vision is derived from the fact that the image in center of an eye is high-acuity region, beyond which the image resolution drops [5]. In a dynamic environment, virtual characters need not focus on all objects in environment. In this paper, an attention object list can be set up beforehand for different virtual characters. If an object is in the scope of perception, and is not in attention object list, character will not perceive the object. Moreover, the perception module reads

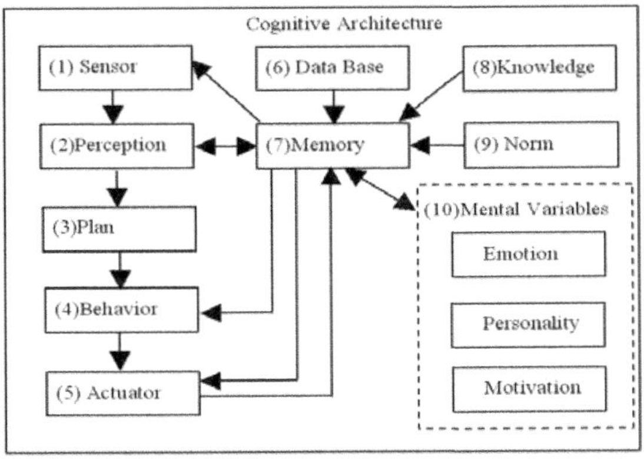

Fig. 1. Architecture of 3D virtual character

the memory module to get mental variables, knowledge, and social norm. Meanwhile, perception module can communicate with mental variables by memory module.

(3) Plan module execute behavior plans by stimuli, knowledge, norm, and mental variables. Behavior plans include some of productive rulers.

(4) Behavior module creates behavior codes by behavior plans. Inhibitory gain and fatigue are time sequence characteristic of behavior. The higher Inhibitory gain, the longer the duration of the behavior is and new behavior is excited only under new stimuli. Fatigue means that behavior with low degree of priority can obtain the chance to carry out, once a certain behavior is carried out, the behavior will stop at some time. We introduce the inhibitory gain coefficient (a real number greater than one) and fatigue coefficient (a real number smaller than one) to measure inhibitory gain and fatigue correspondingly.

(5) Actuator module executes the behavior in behavior code, it includes inverse kinematic arithmetic to drive locomotion, and read motion capture data from memory module (memory module will read motion capture data in database). When actuator module successful executes a behavior code, it will write to memory module with an action sign that indicate whether the character moves to or executes a behavior code.

(6) Database module includes 3D geometry of virtual environment, original information, such as, the original location and parameters of virtual character, motion capture data, 3D model and location of objects, default motion plan scripts that record some goal location.

(7) Memory module serves as a center of information share among all other modules.

(8) Knowledge module includes guiding knowledge in environment for virtual characters. For example, the meanings of objects in environment are part of knowledge.

(9) Norm module includes status, interaction information and interaction rules, it controls the process of a nonverbal social interaction, it provides the social knowledge for virtual character.

(10) Mental variables module includes emotion, personality and motivation. This module read external stimuli from memory (the perception module write stimuli information to memory module). Activation of an emotion is relative to external stimuli and inner mental variables. If an emotion is active, this module will create emotion expression, emotion expression code will be sent to behavior module. Emotion is core of the module, personality is some stable psychological traits of a virtual character, and motivation variables include some physiology parameters of a virtual character.

3 A Model of Social Norm

A virtual character lives in 3D artificial society, and the behavior and emotion of the character will be influenced by its social attribute.In this section, some new definitions about social interaction process of virtual characters are described.

Definition 1. *For a virtual character, a status is a social degree or position. In general, a virtual character may own many status, let* $\boldsymbol{ST}(CA)$ *is a status set for virtual character CA,* $\boldsymbol{ST}(CA)=\{st_1,..., st_N\}$, st_i *is a status (such as mother or son),* $i \in [1, N]$, *N is the number of* $\boldsymbol{ST}(CA)$.

Status plays an important role in a social interaction. For example, in a virtual office, there are two kinds of social status altogether, namely the manager and staff member. The manager's status is higher than the status of the staff member. In general, a person will control emotion expression by one's status.

Definition 2. *Social relationships are connections with other characters in an artificial society, and a social relationships set of a virtual character can describe who is friend or who is an enemy.*

Definition 3. *For two virtual characters* CA_1 *and* CA_2, *let* $FD(CA_1/CA_2)$ *is friendliness degree from* CA_1 *to* CA_2. *If* $FD(CA_1/CA_2)=1$, CA_2 *is a friend of* CA_1; *If* $FD(CA_1/CA_2)=-1$, CA_2 *is a enemy of* CA_1; *If* $FD(CA_1/CA_2)=0$, CA_2 *is a stranger of* CA_1; *If* $FD(CA_1/CA_2)=2$, CA_2 *is a lover of* CA_1; *If* $FD(CA_1/CA_2)=3$, CA_2 *is a mother or father of* CA_1; *If* $FD(CA_1/CA_2)=4$, CA_1 *is a mother or father of* CA_2.

A virtual character judges others with friendliness degree. In general, a virtual character will not interact with a stranger unless in some exceptive conditions (calling help in danger etc.).

Definition 4. *For two virtual characters* CA_1 *and* CA_2, *let* $ET_1(CA_1/CA_2)$ *is default-ending time of interaction from* CA_1 *to* CA_2, *let* $ET_2(CA_2/CA_1)$ *is default-ending time of interaction from* CA_2 *to* CA_1, *and ET is the time from beginning to ending in interaction.*

In general, if $ET>\min(ET_1(CA_1/CA_2), ET_2(CA_2/CA_1))$, *the interaction will end.*

Definition 5. *For two virtual characters* CA_1 *and* CA_2, *let* $IR(CA_1/CA_2)$ *is interaction radius from* CA_1 *to* CA_2, *let* $DS(CA_1/CA_2)$ *is distance from* CA_1 *to* CA_2. *In general, if* $DS(CA_1/CA_2)>IR(CA_1/CA_2)$, CA_1 *will not make interaction to* CA_2; *if* $DS(CA_1/CA_2)\leq IR(CA_1/CA_2)$, CA_1 *may make interaction to* CA_2.

In default condition, when two agents encounter together, interaction radius is critical distance of interaction triggering.

Definition 6. *For two virtual characters* CA_1 *and* CA_2, *let* $PN(CA_1, CA_2)$ *is priority degree of social interaction between* CA_1 *and* CA_2. *If* $PN(CA_1, CA_2)=0$, CA_1 *first interact with* CA_2, CA_1 *is initiator; If* $PN(CA_1, CA_2)=1$, CA_2 *first interact with* CA_1, CA_2 *is initiator; If* $PN(CA_1, CA_2)=2$, CA_1 *and* CA_2 *interact each other at the same time.*

In general, a virtual character acts different status with interaction to others. For instance, there are three virtual characters CA_1, CA_2 *and* CA_3, CA_2 *is mother of* CA_1, CA_3 *is a student of* CA_1, *when* CA_1 *meets* CA_2 *or* CA_3, CA_1 *usually first interacts with* CA_2, CA_3 *usually first interacts with* CA_1, *and* $PN(CA_1, CA_2)=0$, $PN(CA_1, CA_3)=1$.

Fig. 2. Tom is happy when he meets John

Fig. 3. John is also happy to Tom

Definition 7. *For two virtual characters* CA_1 *and* CA_2*, let* ***INS***$(CA_1 \leftarrow CA_2)$ *is an interaction signal set from* CA_2 *to* CA_1*,* ***INS***$(CA_1 \leftarrow CA_2)=\{ins_1,\ldots,ins_M\}$*,* ins_j *is a nonverbal interaction signal (such as "happy face expression"),* $j \in [1, M]$*,* M *is the number of* ***INS*** $(CA_1 \leftarrow CA_2)$*.*

In a virtual environment, when two virtual characters begin to interact each other, we can suppose each of them is able to know interaction signal, in a practical demo system, interaction signals are sent to memory module by social norm module.

Definition 8. *For a virtual characters* CA*, let* $IRU(CA)$ *is an interaction rule for virtual character* CA*,* IRU *control the manner of interaction,* IRU *include some production rulers.*

We can give a demo example to illustrate social norm and emotion interaction. Tom, John and Billy are three virtual characters in the demo system. Billy is an enemy of Tom, John is a friend of Tom. When Tom meets John, Tom will smile to John. When Tom meets Billy, Tom will be angry with Billy.Four snapshots of the demo system are shown in Fig.2-Fig.5, and the social norm for Tom is recorded in a script file as follows:

Status (Tom):=(worker);
Social relationships:=(enemy of Billy, friend of John)
Friendliness degree (to Billy)=-1;
Friendliness degree (to John)=1;
Friendliness degree (to others)=0;
Default-ending time of interaction (to Billy)=2 minutes;
Default-ending time of interaction (to John)=1 minutes;
Default-ending time of interaction (to others)=0.1 minutes;
Interaction radius (to Billy)= 10 meter;
Interaction radius (to John)= 3 meter;
Interaction radius (to others)= 5 meter;
Priority degree of social interaction(to Billy)=0;
Priority degree of social interaction(to John)=0;
Priority degree of social interaction(to others)=1;
Interaction signal set=(angry, happy, ...);
Emotion Interaction rules of sending information to others

Fig. 4. Tom meets Billy and stops walking

Fig. 5. Tom is angry when he meets Billy

If Friendliness degree=-1 then Emotion to other = angry
Else If Friendliness degree=1 then Emotion to other = happy
Else
Emotion to other =Null; //no any emotion to others
End
Emotion Interaction rules of receiving information from others
If Emotion from enemy = sad then Emotion to enemy =happy
Else Emotion to enemy =angry
End
If Emotion from friend = sad then Emotion to friend =sad
Else
Emotion to friend =happy
End
End

4 Conclusion

3D virtual characters are graphics entities that totally produced by computer system. A believable 3D virtual character should be provided with architecture that includes the mechanism of emotion interaction. In a certain virtual environment, multi virtual characters interact with emotions and construct a 3D artificial society. Some new concepts of social norm for virtual characters are presented. A social norm includes status information, interaction signals and interaction rules. All these new concepts are illustrated.

Simulation of emotion interaction for virtual characters is a very difficult subject. This paper only gives a primary frame for emotion interaction. Social interaction is related to many factors, such as different culture, emotion, personality, motivation etc. In fact, the architecture of a 3D virtual character can integrate all these new components.

Acknowledgements

The work described in this paper was co-supported by science and technology project of Zhejiang Province Science Department (grant no: 2006C33046), forepart professional research of ministry of science and technology of the Peoples Republic of China (grant no: 2005cca04400), University Research Project of Zhejiang Province Education Department (grant no: 20051731), and K.C.Wong Magna Fund in Ningbo University.

References

1. Bernstein, D.A., Stewart, A.C., Roy, E.J., Wickens, C.D.: Psychology (forth edition), New York: Houghton Miffin Company (1997) 360-361
2. SATRONGMAN, K.T.:The Psychology of Emotion(fifth edition), chichester, England:John Wiley and sons(2003) 67-68

3. Zastrow, C, Kirst-Ashman,K.K.: Understanding Human Behavior and the Social Envi-ronment, Chicago: Nelson-Hall Publisher (1987) 322-323
4. Ortony, A., Clore, G.L., Collins, A.: The cognitive structure of emotions. New York: Cambridge University Press (1988)
5. Tu, X., Terzopoulos, D.: Artificial fishes: Physics, locomotion, perception, behavior, In Proceedings of SIGGRAPH94 (1994) 43-50
6. Funge, J., Tu, X., Terzopoulos, D.: Cognitive Modeling: Knowledge, Reasoning and Planning for Intelligent Agents, In Proceedings of SIGGRAPH99 (1999) 29-38
7. Badler, N., Phillips, C., Webber, B.: Simulating Humans: Computer Graphics Animation and Control, New York: Oxford University Press (1993) 154-159.
8. Thalmann,N.M., Thalmann,D.: Artifical Life and Virtual Reality, chichester, England: John Wiley and sons Press(1994)1-10.
9. Cassell, J., Vilhjalmsson, H.H., Bickmore, T.: BEAT: the behavior expression Animation toolkit. In: ACM SIGGRAPH. (2001) 477C486
10. Pelachaud, C., Poggi, I.: Subtleties of facial expressions in embodied agents. Journal of Visualization and Computer Animation. 13 (2002) 287C300
11. Chi, D., Costa, M., Zhao, L., Badler, N.: The EMOTE model for effort and shape. In Proceedings of SIGGRAPH00 (2000) 173C182
12. Liu, Z., Pan, Z.G.: An Emotion Model of 3D Virtual Characters In Intelligent Virtual Environment, In First International conference on affective computing and intelligent interaction, ACII2005, LNCS 3784, Beijing ,China, (2005),629-636

Genetic Evolution of the Ant Species in Function Representation Framework

Lukáš Pichl and Yuji Shimizu

Department of Information Science, International Christian University,
Osawa 3-10-2, Mitaka, 181-8585, Japan
lukas@icu.ac.jp
http://cpu.icu.ac.jp/~lukas/

Abstract. This paper explores the feasibility of computer simulation of evolving populations of social animals in nature, both from the anatomical and socio-environmental viewpoints, addressing the gap between the algorithms for evolution of digital objects, and the evolution of species in nature. The main components of ant body are mathematically described within the function representation framework; the parameters directly determine both the visual characteristics of the ant as well as the body characteristics encoded by the genome. The environmental diversification of ant subspecies is studied for fungus-growing ants, in which single-queen mating reproduction couples with large size of accessory male glands, while multiple-queen mating correlates to large size of accessory testes. Our results show that within an environment of restricted resources, both competing modes of sexual reproduction survive. The frequency with which either mode becomes dominant in the population is driven by the value of the mutation probability. The function representation model should be useful also in the simulation of other simple animal species, because of the ease in relating the genome parameters to computer visualization tools.

1 Introduction

Ants are social animals, which have a very specific reproduction cycle. They live in colonies, which consist of a queen, sterile female workers and male drones. The queen lays fertilized eggs, which then metamorphose via larvae and pupae stages to grown up adult female ants. At the same time, a small amount of male drones develops from the unfertilized eggs. The queen mates in the air, once per life, which may last up to 30 years. Few days after mating, the male drones die, while the queen rips off her wings, and starts laying eggs for a new ant colony. She stores all the sperm from mating in spermatheca in her body[1]. In order to complete the reproduction cycle, a long phase of community growth and maturation is required, before the fertile females are born, ready to become the queens of the next ant colonies.

The anatomical features among ant species are very similar; nevertheless, variations has been found in male organs related to the sexual reproduction: the size of

[1] Monoclonal sperm, which carries identical genes of the male in all cells, can in some cases replace the standard sperm competition mechanism.

T.-D. Wang et al. (Eds.): SEAL 2006, LNCS 4247, pp. 656–663, 2006.

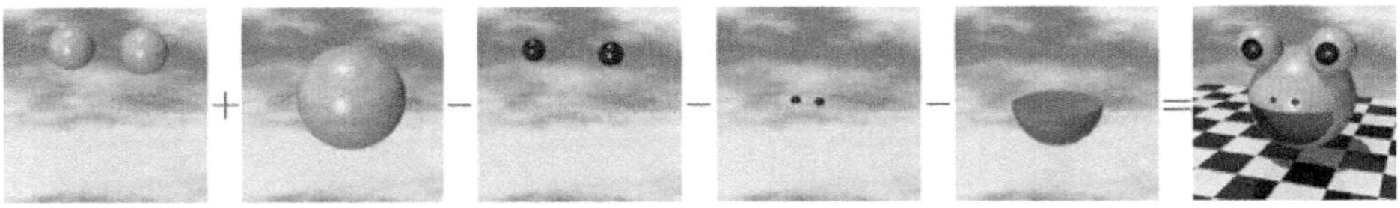

Fig. 1. FRep: creating a spatial model (frog cartoon) from HF graphics primitives [3]

accessory testicles that contain previously formed sperm; and the size of accessory glands, which contain accessory mating fluids secreted earlier: these fluids have a positive impact on queens life-length, fertility, and may also act as anti-aphrodisiacs against queen mating with other drones. It has been shown that multiple queen-mating correlates with large testes and small accessory glands; single queen-mating correlates with small testes and large accessory glans. Ant colonies resulting from the former mechanism are genetically more diverse and resistant to illnesses; the colonies resulting from latter mechanism are more cooperative and stable. Within the environment with limited resources, both mating mechanism can represent a competitive advantage, especially in case the mating mechanism is minor [1].

A reproductive success of queen genes consists in the ergonomic growth of her colony, which allows for multiple male and queen offsprings; and in the mating success of her offsprings. While there is no environmental pressure known for the queens, the drones compete by the size of accessory glands and testes, which correlates to the mating mechanism. The reproductive success of male genes can thus always be reduced to that of the mother queen.

Here we develop a computer simulation model of ant species, that includes (1) genetic encoding of the main body characteristics, (2) function-based visualization of ant creatures, and (3) model of the environment with restricted resources, and the two possible mating mechanisms. Within this evolutional framework, we determine the predominance of mating mechanisms in relation to the genetic algorithm (GA) operators. The rest of the paper is organized as follows. Section 2 explains the mathematical parametrization of ant body within the function representation (FRep) framework, and introduces its visualization tool, the HyperFun software. An educational illustration of GA-run ant evolution from an initial population of random creatures is also included. Section 3 deals with the evolution of ant colonies competing for limited resources and discusses the principal results on mating. We conclude with final remarks in Section 4.

2 Function Representation of Ants

This section show how the ant species can be represented by using the functional representation, parameters of which are subjected to the genetic algorithm. As an illustration, we also evolve a population of random ants into a physiologically consistent species[2].

[2] Although the single optimum GA is not the main subject of the paper, visualization of such evolution may be of educational interest in biology and/or computer science.

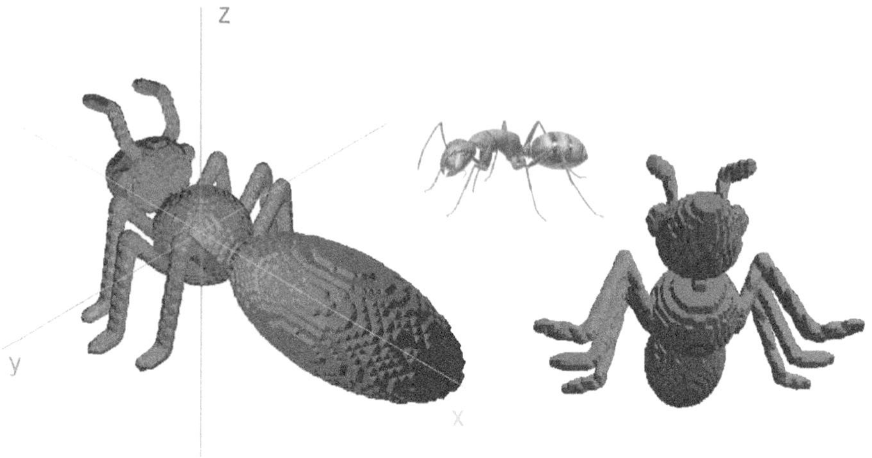

Fig. 2. Parameterization of Ant Model: FRep with HyperFun

FRep was introduced in [2] as a uniform representation of multidimensional geometric objects (shapes). The objects are described by a real-valued function F of the space variables $\boldsymbol{r}$: the points where $F(\boldsymbol{r}) \leq 0$ belong to the object; $F(\boldsymbol{r}) = 0$ defines the surface. One of the advantages of shape modeling with FRep is the availability of an associated computer graphics tool, the HyperFun (HF) [3]. FRep-defined objects are automatically visualized and animated without requiring explicit polygonization. Elementary geometrical objects as well as interpolated voxel clusters naturally express themselves as a function in the format of a parse tree with object primitives in the leaves, and FRep operators in the tree nodes. Operations such as the set-theoretic, blending, offsetting, projections or non-linear deformations have been formulated for FRep within the space of continuous real-valued functions. Figure 1 shows an example how to create a complex spatial model of an animal (a frog) from elementary FRep components, by using summation and subtraction (the sky and chessboard components were added ex post into the picture).

In order to create a digital model of ant species, we must consistently parameterize the ant anatomy. All body components (head, eyes, feelers, trunk, legs and metasoma) must be adjusted for connectivity for all gene values, unlike from the original model [4]. The ideal ant creature, which is going to be the attractor of single-optimum evolutional race is shown in Fig. 2. The body is defined as a sphere (hfSphere centered at [0,0,0] with radius ρ. While the x coordinate for the starting point of any legs can be selected freely between $(-\rho, \rho)$, the next coordinate, y, is restricted to $(-\sqrt{\rho^2 - x^2}, \sqrt{\rho^2 - x^2})$, and the last one, z, is determined as $z = \pm\sqrt{\rho^2 - x^2 - z^2}$. The particular values are binary-encoded as GA parameters in $< -1, 1 >$ with proper normalization factors. The length of leg segments and the orientation of joints are also encoded as parameters; the minimal anatomical restrictions are taken care of as outlined in Fig. 3, so that

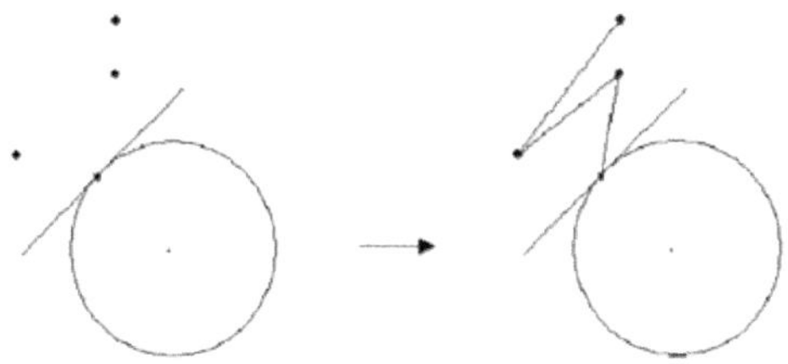

Fig. 3. Ant model: parameterization of legs and joints

the legs are symmetric and do not interfere with the body. The shape of metasoma is modeled as an ellipsoid with the principal half-axis sized a, and shifted by $\rho + a$ e from the center of the trunk. Similarly, the head attaches to the body on the opposite side; its shaped is obtained by spatially-dependent transform of a sphere along the body-symmetry axis (HF function hfTaperX). In a similar manner, spherical eyes are attached; the mathematical model of attaching feelers to the head closely follows the procedure for attaching legs to the trunk (cf. Fig. 3).

Since the HyperFun language is a visualization tool, and does not easily allow for numerical calculations, we have opted to generate the input files for HF graphics from within a C-language program, which runs the GA. The program outputs the full parameterization for each HF ant, including all commands and directives to the HF interpreter; the HF parameters are simply obtained by decoding gene values.

The parameters of the single-attractor GA [5], in which the ant creature in Fig. 2 is evolved, are: 38 float genes normalized to the range in $< -1, 1 >$ and encoded as 20 bits; 200 GA steps; a population of 300 creatures; multiple point crossover operator; and single-bit mutation operator (probability 1%). The fitness function is expressed as the inverse of the least-square error from the parameters of the ideal ant. Figure 4 shows the FRep-consistent population of random digital ants, which was used to initialize the GA. Three evolved creatures are randomly selected at steps 2, 10, and 30 in Fig. 5, when the GA population effectively became homogenous. Most of the initial ant shapes in Fig. 4 would be life-inconsistent. It is however worth noting that all are GA and FRep consistent; we are not aware of any other CG software, in which the genetic evolution and creature visualization is tractable [6].

3 Evolution of Ant-Mating Mechanism

After the GA developed in the previous section illustrated the advantages of FRep in simulated animal evolution, here we develop a simulation of realistic ant species, namely the fungus-growing ants [1]. Variations of the body parts sizes are negligible except for the accessory male organs related to reproduction. The size of body parts is practically frozen; on the other hand, two more objects, accessory glands and accessory testes are added to the GA.

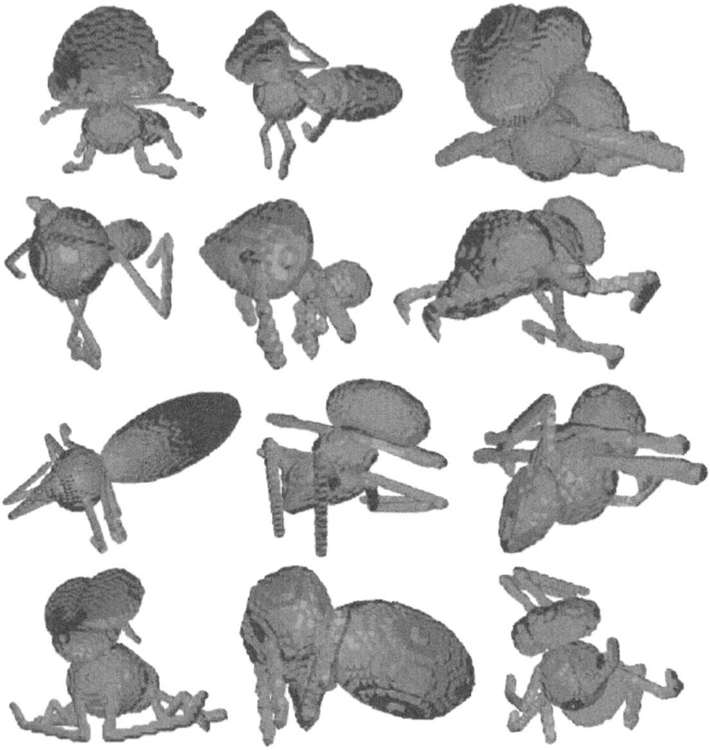

Fig. 4. Random ant creatures in GA evolution

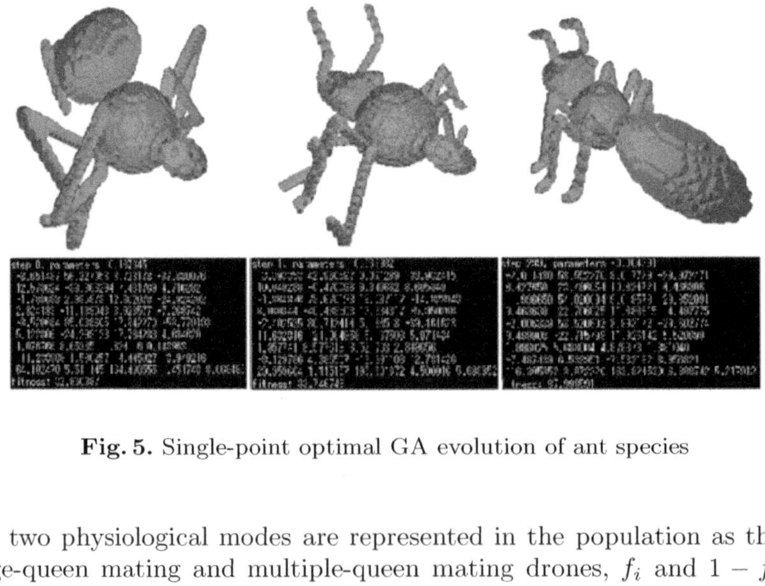

Fig. 5. Single-point optimal GA evolution of ant species

The two physiological modes are represented in the population as the ratio of singe-queen mating and multiple-queen mating drones, f_i and $1 - f_i$ for a

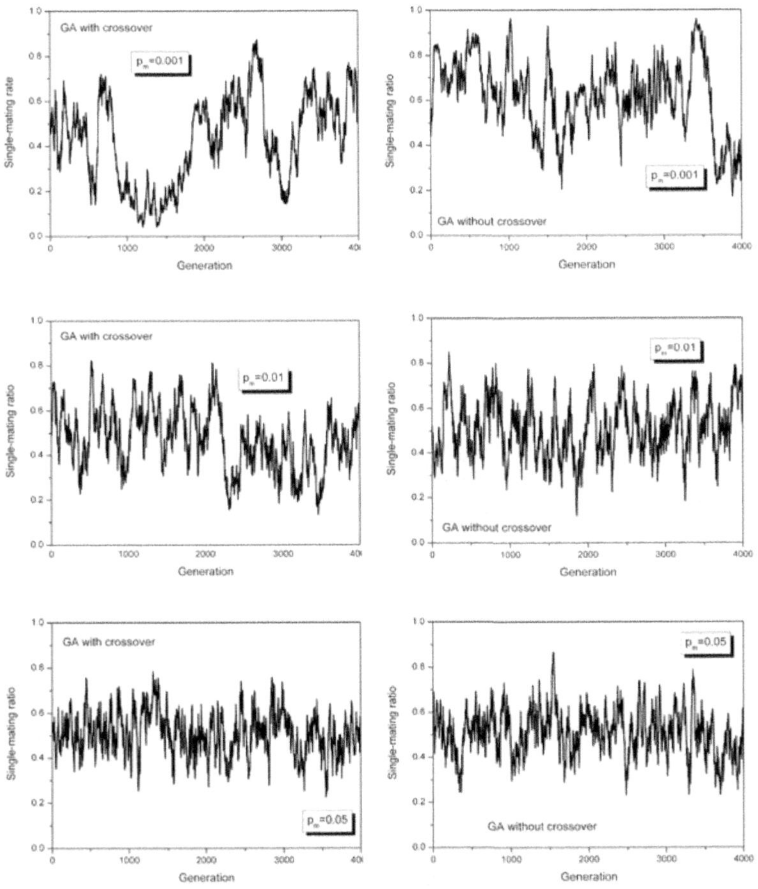

Fig. 6. Single-mating ratio for ants under the various GA scenarios

colony, and $f = \sum_i f_i$ and $1 - f$ for the entire population. The GA models a two-stage reproductive cycle, which requires the queens to undertake investment into mating or feeding of the first ant generation, and selection of single-mating or multiple-mating female offsprings. Both decisions are gene-encoded and evaluated by the environment. Single-mating mode is preferred in the environment when the multiple-mating mode is inflating and vice versa. This is expressed in the model fitness function of the ant colony i,

$$\Phi_i(t) = (a - s_i(t))(b + s_i(t)[f_i(t)\frac{\sum_i f_i(t)s_i(t)}{\sum_i f_i(t_-)s_i(t_-)} + (1 - f_i(t))\frac{\sum_i(1 - f_i(t))s_i(t)}{\sum_i f_i(t_-)s_i(t_-)}]),$$

where t_- denotes the previous half-life generation, and a and b are certain parameters. Through the minority-enhanced mating, the ant colonies are coupled in effect; this represents a resource-limited environment. Except for the additional genes and the above fitness function, the GA employed is the same as in

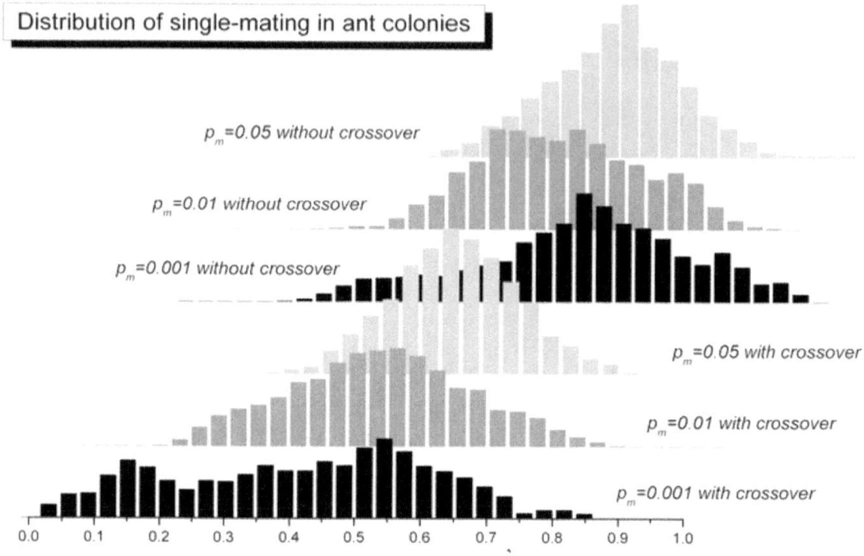

Fig. 7. Histogram the for data distributions from Fig. 6

the previous section. In order to study the dynamics of queen mating modes, we selected three values of the mutation probability, $p_m = 0.001$, $p_m = 0.01$, and $p_m = 0.05$. Unlike from the natural evolution, both the GA simulation with and without crossover operator were performed in 4,000 evolutionary steps. The behavior of a typical single-queen mating ratio in all cases is shown in Fig. 6, and the histograms of the six distributions are given in Fig. 7. Whether the crossover is included in the set of GA operators or not, it is clear from Fig. 6 that an increased value of mutation probability increases the frequency of mating cycle oscillations; their amplitude decreases to non-specific 50% at the same time. In addition, the inclusion of the crossover in accordance with the natural evolution stabilizes fluctuations of the mating cycle in case of the (biologically plausible) low values of mutation probability. The histograms in Fig. 7 clearly shows that the entire space of mating mechanisms is best screened in an crossover-including reproduction cycle with a low incidence of mutations.

4 Conclusion

In this paper, we have developed a functional representation approach to the simulated evolution and visualization of ant species. Within a restricted-resource environment, it was found that low values of mutation probability in conjunction with the genetic crossover operator support interleaving periods of dominance among the single-queen and multiple-queen mating mechanisms. As the value of the mutation probability increases, the oscillations in mating behavior become less profound. At present, this finding is particular to the computer simulation, since the ant species cannot be easily followed on sufficiently long time-scale in the experiment, similar to other computer science studies of ant behavior [7]. The functional representation framework to evolution developed here is also considered useful for various applications in educational software.

Acknowledgements

The authors acknowledge a partial support by a JSPS Grant-in-Aid and the Academic Frontier Program of MEXT.

References

1. B. Baer and J. J. Boomsma, Male reproductive investment and queen mating-frequency in fungus-frowing ants, Behavioral Ecology **15** (3) 426-432 (2004).
2. A. Pasko, V. Adzhiev, A. Sourin, V. Savchenko, "Function representation in geometric modeling: concepts, implementation and applications," The Visual Computer **11** (8) 429–446 (1995).
3. Pasko, Alexander A., "HyperFun Project", http://cis.k.hosei.ac.jp/~F-rep/HF_proj.html, Hosei University, Japan, 2006.
4. K. Masato, An instance of HyperFun ant, http://cis.k.hosei.ac.jp/~F-rep/HF_ant.html
5. Goldberg, David E., "Genetic Algorithms in Search, Optimization, and Machine Learning", Addison-Wesley Publishing Company, Third Edition, 1989.
6. Y. Shimizu and L. Pichl, HF-parameterized ant model, http://cpu.icu.ac.jp/~lukas/hfant/ant1.c.
7. J. A. R. Marshall, T. Kovacs, A. R. Dornhaus, N. R. Franks, "Simulating the evolution of ant behaviour in evaluating nest sites." In: W. Banzhaf et al. (eds.) Advances in Artificial Life (ECAL 2003), Lecture Notes in Artificial Intelligence **2801** Springer Verlag, Heidelberg (2003).

The Dynamics of Network Minority Game

Bing-Hong Wang

Department of Modern Physics, University of Science and Technology of China, Hefei, 230026 China
bhwang@ustc.edu.cn
Shanghai Academy of System Science, Shanghai 200093 China

Abstract. The evolutionary dynamics of minority games based on three generic networks have been investigated : Kauffman's NK networks (Kauffman nets), growing directed networks (**GDNets**), and growing directed networks with a small fraction of link reversals (**GDRNets**). We show that the dynamics and the associated phase structure of the game depend crucially on the structure of the underlying network. The dynamics on **GDNets** is very stable for all values of the connection number K, in contrast to the dynamics on Kauffman's NK networks, which becomes chaotic when $K > K_c = 2$. The dynamics of **GDRNets**, on the other hand, is near critical. Under a simple evolutionary scheme, the network system with a "near" critical dynamics evolves to a high level of global coordination among its agents; this suggests that criticality leads to the best performance. For Kauffman nets with $K > 3$, the evolutionary scheme has no effect on the dynamics (it remains chaotic) and the performance of the MG resembles that of a random choice game (RCG).

1 Introduction

Complex networks have attracted immense interest in recent years, due to their great capability and flexibility in describing a wide range of natural and social systems. The study of the organization of complex networks has attracted intensive research interest [1,2,3] ever since the seminal works of Strogatz on small-world networks [4] and Barabási and Albert [5] on scale-free networks. The power law or scale-free degree distributions have been established in many real-world complex networks [4,5].

The dynamics of a complex network can be studied in the context of a system of interactive elements (agents) on the network; it depends on how the network is organized and how the elements interact. Here we study a network version of the minority game (MG) model proposed by Challet and Zhang [6], which is a simplification of Arthur's El Farol bar attendance model [7]. The MG model serves as an interesting paradigm for a system of adaptive agents competing for limited resources. The phase structures of the original MG [8] and the evolutionary version of the game [9,10,11,12] have been well understood. Note that in the MG models, the agents are not directly linked to one another, but they are influenced by the global environment created by the collective action of all the agents.

T.-D. Wang et al. (Eds.): SEAL 2006, LNCS 4247, pp. 664–671, 2006.

The study of network dynamics is pioneered by Kauffman [13,14] who introduced NK random networks and studied its Boolean dynamics. Recently there are quite a number of studies on different aspects of network dynamics. Aldana and Cluzel demonstrated that the scale-free network favors robust dynamics[15]. Paczuski *et al* [16] considered the MG model on a random network to study the self-organized process which leads to a stationary but intermittent state. Galstyan [17] studied a network MG, focusing on how the change of the mean connectivity K of a random network affects the global coordination of the system of different capacities. Anghel *et al* [18] used the MG model to investigate how interagent's communications across a network lead to a formation of an influence network. Here we address the question of how different network organizations affect the dynamics of the system in the context of the network MG. In particular we would like to know 1) how the dynamics and phase structure of the network minority game depend on the network organization, and 2) how evolution affects the dynamics and phase structure of the game. We will consider three types of rather generic networks: Kauffman's NK random networks (Kauffman nets), growing directed networks (**GDNets**), and growing directed networks with a fraction of link reversals (**GDRNets**) as described in Ref. [20].

2 Network Minority Game Model

We consider two types of dynamics of minority game based on the networks.

(A) MG of network agents: The network minority game model is defined in a similar way as the original MG model [6], except for the input for the strategy of an agent (node). In the original MG, the input for each agent's strategies at time t is a vector of the winning decisions of the game in the previous M time steps. In the network (local) MG, however, the input for each agent's strategies at time t is a vector consisting of the decisions of the K agents she connects to at the previous time step $t-1$. Specifically, the network based MG model consists of N (odd number) agents described by the state variables $s_i = \{0,1\}, i = 1,2,...,N$, each connected to another K agents, $i_1, i_2, ..., i_K$. Each agent has S strategies which are the mapping functions specifying a binary output state (0 or 1) for each possible input vector consisting of the states of her K connected agents. The state or decision of the ith agent at the current time step t is determined by the states/decisions of the K agents it connects to at the previous time step $t-1$, i.e.

$$s_i(t) = F_i^j(s_{i_1}(t-1), s_{i_2}(t-1), ..., s_{i_K}(t-1)) \quad (1)$$

where $s_{i_k}(k = 1,2,...,K)$ is the state of the kth agent that is connected to agent i, and $F_i^j, j = 1,..,S$ are S Boolean functions (strategies) taken from the strategy space consisting of 2^{2^K} strategies. As in the standard minority game, each agent keeps a record of the cumulative wealth $W_i(t)$ as well as the cumulative (pseudo) scores, $Q_i^s(t), i = 1,2,...,N, s = 1,2,..,S$, for each of her S strategies. Before the game starts each agent selects at random one of her S strategies, and the cumulative wealth $W_i(t)$ and strategy scores $Q_i^s(t)$ are initialized to zero. At each time step, each agent decides which of the two groups (0 or 1) to join based on the best-scoring strategy (the strategy that would have made the most winning predictions in the past) among her S strategies. The agent gains (loses) one point in her cumulative wealth for her winning (losing) decision and each strategy gains (loses) one pseudo-point for its winning (losing) prediction. The agents who are among the minority win; those among the majority lose. Let $A(t)$ be the number of agents choosing 1 at time step t. Then a standard measure of utilization of the limited resources (system performance) can be defined as the variance of $A(t)$ over a time period T:

$$\sigma^2 = \frac{1}{T} \sum_{t=t_0}^{t_0+T} (A(t) - \bar{A})^2 \tag{2}$$

where $\bar{A} = \frac{1}{T} \sum_{t=t_0}^{t_0+T} A(t) \sim N/2$ is the mean number of agents choosing 1. Clearly σ^2 measures the global coordination among agents. The optimal (smallest) value of variance σ^2 is 0.25, where the number of winning agents reaches its optimal value, $(N-1)/2$, in every time step. For a random choice game (RCG), where each agent makes decision by coin-tossing, the value of the variance σ^2 is $0.25N$. The game is adaptive as each agent has S strategies to choose from, attempting to increase her chance of winning.

The key difference between the network MG and the original MG is that the agents in the original MG use global information while the agents in the network MG use local information. The evolution of the original MG is based on the M time-step history of global information while the network MG employs a one-step forward dynamics.

(B) Evolutionary Minority Game: We first investigate the adaptive MG models and used the results as a basis for comparison with the evolutionary MG model, which is our main interest. In an evolutionary dynamics, the quenched strategies can be changed and more generic behaviors emerge. We use the standard evolutionary dynamics described in Refs. [9,12]. In this evolutionary scheme, each agent is required to change her S strategies (by choosing S new strategies randomly) whenever her cumulative wealth $W_i(t)$ is below a pre-specified bankruptcy threshold, $-W_c (W_c > 0)$. The bankrupted agents re-set their wealth and strategy pseudo-scores to zero, and the game continues. The network connection, however, does not evolve. We have found that this evolutionary scheme is much more effective than other scheme used for studying MGs [16], in which the evolution happens at the end of every epoch of specified duration (say 10,000 time steps), and only the worst performer is required to change her strategies after each epoch.

The dynamics of the game depends crucially on the network organization. Here we consider three types of rather generic networks for studying network dynamics. To study the dynamics, we need to specify input-output relationship among the nodes; this naturally leads to a directed network, in which the direction of a link indicates an input from the node at the other end of the link. The first directed network we consider is the well-known Kauffman net [13], in which the inputs to a node are randomly selected. The Kauffman net is thus a random directed network. The second one is a growing directed network, in which the new nodes are controlled by the old ones, but not vice versa. This is an extreme case. In many real networks, there is a strong degree of hierarchical dependence; in most cases the older nodes (the nodes at the higher hierarchy) influence the newer nodes (the ones at the lower hierarchy). But there bound to be exceptions. This lead to the third class of networks: a growing directed network with a fraction of link reversals. Below we describe in details these networks. Note that in our model, the network connection, once generated, remain fixed.

The Kauffman NK random network (Kauffman net). The Kauffman net is generated by specifying N agents first, and then connecting each agent randomly to K other agents, whose decisions serve as the input to its strategies.

Growing directed network (GDNet). A growing directed network is generated according to the description given in Ref. [20]. We start with an initial cluster of $K+1$ agents, which are mutually connected (two directed links between each pair of agents). At each stage, we add a new agent and connect it to K other agents already present in the network. The link is directed from the new agent to the existing ones, meaning that the strategies of the new (younger) agent are based on the states of the existing (older) ones. We assume that the probability of connecting a new agent to an existing one with degree k_{in} is proportional to $k_{in}^{\alpha}+1$, where k_{in} is the number of incoming links from the existing agent. The constant 1 is added to give a nonzero starting weight to the agents that have not been connected to. For $\alpha = 0$, we have a *growing directed random network* which we refer to as **GDNet I**. For $\alpha > 0$, we have preferential attachment. The special case of $\alpha = 1$ corresponds to a *scale-free directed network* which we refer to as **GDNet II**. In this network, the out-degree is K for all the agents, but the in-degree follows a power law distribution. The undirected version of this model corresponds to the well-known Barabási-Albert scale-free network [1]. In the growing networks (random or scale-free), the younger agents are influenced by the older ones, except for the initial $K+1$ agents who are mutually influenced.

Growing directed network with a fraction of link reversals (GDRNet). This network is based on the above growing network, but we introduce a small fraction of link reversals. Let p be the probability that each agent has a link reversal: when each new agent is connected to other K agents already present in the network, each link has a probability of $q = p/K$ to have its direction reversed. We consider two **GDRNets**: **GDRNet I** with $\alpha = 0$ and **GDRNet II** with $\alpha = 1$. **GDRNet** is the general case of a generic class of directed networks. Kauffman's net and **GDNet** are two extreme cases, corresponding to $q = 0.5$ and $q = 1$ with respectively.

There are two new features for **GDRNets**:

1) Some agents may have more than K strategy inputs, while others may have fewer than K inputs; but the mean number of inputs for an agent remains as K.
2) Some younger agents can influence the older agents.

3 Numerical Results and Analysis

Let's first examine the performance of MG on the Kauffman net. The simulation results show that when $K = 2$, the variance σ^2 has very large fluctuations (four orders of magnitude for $N = 401$ and five orders of magnitude for $N = 901$); different initial conditions give rise to very different σ^2. This reflects the fact that the system dynamics is critical for $K = 2$. For $K \geq 3$ the system performs like a random choice game. The observation we obtained is consistent with the well-known result for the Boolean dynamics on Kauffman nets: when $K = 2$ the system is at the "edge of chaos" and for $K \geq 3$ the system is chaotic [13,21].

The dynamics in the original MG depends on two variables: N, the system size and M, the memory size of the agents. There are three different phases for different memory value M, described by a Savit curve [8]. The critical value for an optimal global coordination is $M_c \sim \ln(N)$, which depends on N. For the network MG on the Kauffman net, however, the three phases of the dynamics are: stable for $K = 1$, critical for $K = 2$, and chaotic for $K \geq 3$. The critical value is fixed at $K_c = 2$, which does not depend on N. So the dynamics of network MG depends on only one variable, K, and the chaotic regime dominates.

Now let's check how the game performs when the simple evolution scheme described above is applied. The simulation results show that evolution helps dramatically improve the system performance when the connection number K is small ($K \leq 3$), but has virtually no effect for larger $K(> 4)$. This means that for $K \leq 3$ the system is at stable or "critical" state, but for $K \geq 4$ it is chaotic. Note that evolution has shifted the critical point from $K = 2$ to $K = 3$, suggesting that it is more powerful than adaptation (modeled by strategy switching) in bring out order in complex systems.

Let's now examine the performance of MG on growing directed networks. Two limiting cases of the growing directed networks are checked: 1) **GDNet I**, the growing random directed network ($\alpha = 0$); 2) **GDNet II**, the growing directed network with a linear preferential attachment ($\alpha = 1$).

We have the following observations for the MG dynamics from simulation results: 1) there are large fluctuations in the values of the variance σ^2; 2) there seems to be no significant difference in the MG dynamics for the two growing network models. So in terms of a simple (non-evolutionary) dynamical process, all growing networks (irrespective of its value of preferential attachment exponent $\alpha \in [0, 1]$) have similar dynamics and the scale-free network is not special. The stability of the dynamics is due to the construction process of growing networks, which leads to a maximum state cycle length of 2^{K+1} as was pointed out in Ref. [20], irrespective of the value of α.

Comparing the dynamics of growing networks with that of the Kauffman net, we see a lot of differences. In the Kauffman net, the dynamics is stable or critical for $K \leq 2$. In growing networks, however, the dynamics is stable for all the values of K, on both **GDNet I** and **GDNet II**. Thus non-growing and growing networks have very different network dynamics.

The results of system performance for the EMG on **GDNets** show that evolution helps reduce the variance σ^2 dramatically (by more than two orders of magnitude). Although the fluctuations in the variance is still large, the values are all below the value of the variance corresponding to RCG. We can also see that the EMG on **GDNet I** performs better than the EMG on **GDNet II**, but the difference is small. This is not surprising as the essential property of all these growing directed networks are the same: the younger agents are always influenced by the older ones; this gives rise to stable dynamics.

By comparing the results for growing networks with the results for the Kauffman net, we see that the dynamics of these two types of networks are very different. This is due to the differences in network construction process. For the Kauffman net, each agent chooses, at random, other K agents for inputs to her strategies. Any given agent has a potential to influence many other agents. It is not surprising that, for large enough K ($K \geq 3$), the system is virtually chaotic. However, for **GDNets**, the dynamics is driven by the initial cluster of $K + 1$ agents; this results in a stable dynamics in which the maximum cycle of length is 2^{K+1}.

We now examine the performance of the MG on growing networks with link reversals, **GDRNet I** and **GDRNet II**. The results show that, without evolution, the MG on **GDRNets** produces similar results as the MG on **GDNets**. However, the EMG on **GDRNets** produces significantly better results than the EMG on **GDNets**. **GDRNet II**, in particular, gives the best performance among all the network models. These observations suggest that MG dynamics is not very sensitive to different attachment algorithm in a growing directed network. But the attachment algorithm makes some difference in the performance of the EMG. If we examine the results more carefully, we see that for **GDRNet II**, the EMG performance is so good that the variance σ^2 is below 1 most of the time. This means that the difference between the attendance numbers in the majority group and the minority group is less than 2 on average, which is very close to the theoretical bound where the difference is 1 at every time step. This result is generic for small p (we have checked a number of values for $p < 0.1$).

An earlier paper [20] has shown that, the general Boolean dynamics on **GDRNets** is close to critical, in the sense that the distribution of the state cycle lengths is close to a power law. Our result suggests that, under an evolutionary dynamics, criticality makes the system more efficient.

4 Discussion and Conclusion

We have presented an extensive numerical investigation on the dynamics of the MG on three general classes of networks, and we have found a few generic dynamics features of the network MG. The dynamics of the network MG is significantly different from that of the original MG. In the original MG, the critical value for an optimal global coordination is $M_c \sim \ln(N)$, so the dynamics depends on two variables: N, the system size and M, the memory size of the agents. In the network MG, however, the dynamics depends on K only. The MG on the Kauffman net exhibits three phases: stable for $K = 1$, critical for $K = 2$, and chaotic when $K \geq 3$; this is consistent with Kauffman's Boolean dynamics. We have studied the dynamics of the EMG on the Kauffman net, and we have found that, the critical value of K shifts to $K_c = 3$, different from the critical value $K_c = 2$ for the non-evolutionary MG. Evolution makes a significant difference.

The dynamics of network MGs depend crucially on the organization of the underlying network structures. Besides Kauffman's net, we have also investigated the dynamics of the MG on a generic class of growing directed networks. For the extreme case of no link reversal (**GDNet**), we show that, the dynamics is stable on these growing networks, and it is very different from the dynamics on the Kauffman NK random network. This is due to the way the network is constructed. There is no critical K value, beyond which the dynamics is chaotic. In the Kauffman net all the agents are treated equally; every agent has an equal probability to influence others. This results in a very large "influence network" of a given agent, particularly for large K. However, in growing directed networks, the initial cluster of agents dictate the dynamics of the system; the "junior" agents have no influence on the "senior" ones.

We have also studied the MG dynamics on a modified growing directed network model which allows a small fraction of link reversals. Our numerical results show that the best system coordination and performance emerges for the EMG on the scale-free network with link reversals (**GDRNet II**); the variance σ^2 in the EMG on **GDRNet II** reaches to such a low level that it's close to the theoretical bound most of the time. As the dynamics on **GDRNet II** for small p is nearly critical and the network is scale-free, our results suggest that evolution makes the agents best coordinated on critical scale-free networks.

Acknowledgments

This work was supported by the National Natural Science Foundation of China (Grant Nos. 10472116, 10532060 and 10547004), by the Special Research Funds for Theoretical Physics Frontier Problems (NSFC Grant No.A0524701), and by the President Funding of Chinese Academy of Science.

References

1. R. Albert and A.-L. Barabási, Rev. Mod. Pays. **74**, 47 (2002)
2. M. E. J. Newman, SIAM Review **45**, 167 (2003)
3. S. N. Dorogovtsev and J.F.F. Mendes, Advances in Physics **51**, 1079 (2002)
4. S.H. Strogatz, Nature **410**, 268(2001).
5. A.-L. Barabási and R. Albert, Science, **286**, 509-512 (1999)
6. D. Challet and Y.-C. Zhang, Physica A, **246**, 407 (1997)
7. W. Arthur, Am. Econ. Assoc. Paper, Proc. **84**, 406 (1994)
8. R. Savit, R. Manuca, and R. Riolo, Phys. Rev. Lett. **82**, 2203 (1999).
9. N.F. Johnson, P. M. Hui, R. Jonson, and T. S. Lo, Phys. Rev. Lett. **82**, 3360(1999)
10. S. Hod. and E. Nakar, Phys. Rev. Lett. **88**, 238702 (2002)
11. K. Chen, B.-H. Wang, and B. Yuan, Phys. Rev. E. **69**, 025102(R) (2004)
12. B. Yuan and K. Chen, Phys. Rev. E. **69**, 067106 (2004)

13. S. A. Kauffman, "The Origins of Order: Self-Organization and Selection in Evolution", Oxford University Press, New York, 1993.
14. S. A. Kauffman, J. Theor. Biol. **22**, 437 (1969).
15. M. Aldana and P. Cluzel, PNAS, **100**, 8710 (2003)
16. M. Paczuski, K. E. Bassler, and A. Corral, Phys. Rev. Lett. **84**, 3185 (2000)
17. A. Galstyan and K. Lerman, Phys. Rev. E. **66**, 015103(R) (2002)
18. M. Anghel, Z. Toroczkai, K. E. Bassler and G. Korniss, Phys. Rev. Lett. **92**, 058701 (2004).
19. J. D. Watts and S. H. Strogatz, Nature **393**, 440(1998).
20. B. Yuan, K. Chen and B.-H. Wang, arXiv: cond-mat/0408391.
21. B. Derrida and Y. Pomeau, Europhys. Lett. **1**, 45 (1986).

A Parallel Solution to the HIP Game Based on Genetic Algorithms

Tatiana Tambouratzis

Department of Industrial Management & Technology, University of Piraeus,
107 Deligiorgi St, Piraeus 185 34, Greece
tatianatambouratzis@gmail.com
http://www.tex.unipi.gr/dep/tambouratzis/main.htm

Abstract. In this piece of research, genetic algorithms are put forward for solving the HIP game. The proposed parallel approach manipulates candidate solutions via selection and mutation; no crossover has been employed. The population is limited to one candidate solution per generation, thus keeping the computational complexity of the approach to a minimum. It is shown that the proposed approach is superior to the approaches reported in the literature: solutions are more speedily provided while the frequency of finding a solution is significantly higher.

1 Introduction

Game theory [10] constitutes a branch of applied mathematics that studies the complex pattern of interactions among agents. Its capability of modeling problems whose outcome depends not only on the problem constraints (e.g. the market conditions) but also on the strategies of all the agents involved (e.g. the various overlapping and perhaps even conflicting business plans) has rendered game theory a convenient tool for solving problems in a variety of fields (e.g. economics; operations research; psychology, sociology and political science; international relations and military strategy; evolutionary biology; logic and computer science). Game-theoretical solutions aim at increasing/decreasing the gain/loss of each agent at each game such that maximization/minimization of the overall gain/loss (subject to a balance of the individual gains/losses) is also attained.

The family of HIP games [3-4] is played on a checkerboard of dimensions nxn by a number of players sharing n^2 counters. Counter placement is performed at vacant locations of the checkerboard in such a manner that no four counters of the same player form a square; once a square is created the corresponding player loses the game. In a HIP game, a tie is sought, i.e. placement of all the counters of all the players at distinct locations of the checkerboard such that no squares are formed.

In this piece of research, a genetic algorithm [5-6,9] is put forward for providing ties to the family of HIP games. The proposed parallel approach manipulates candidate solutions via selection and mutation; no crossover has been employed. The population is limited to one candidate solution per generation, thus keeping the

T.-D. Wang et al. (Eds.): SEAL 2006, LNCS 4247, pp. 672 – 679, 2006.

computational complexity of the solution to a minimum. It is shown that, even for this limited population size and the simplicity of the evolution scheme, the proposed approach is superior to the existing approaches: ties are more speedily provided while the frequency of finding a tie is significantly higher.

2 The Family of HIP Games

The HIP family comprises:

- The original version HIP1. This involves a checkerboard of dimensions *n*x*n* (*n* even), two players and $n^2/2$ counters per player, where the counters of the two players are distinguishable (e.g. by colour). The game is played by the placement of the counters of the two players at vacant locations of the checkerboard in such a manner that no four counters of the same player form a square on the checkerboard; once a square is created the corresponding player loses the game and the opponent wins. In order for a tie to be reached, placement of all the counters of all the players at distinct locations of the checkerboard is sought such that no squares are formed. A tie is possible for n=4 and 6, while no ties exist for larger values of n.
- The variant HIP2. This was introduced in order to investigate the potential and efficiency (i.e. the scalability) of the various tie-seeking approaches to increasingly complex instances of the HIP family. HIP2 involves a checkerboard of dimensions *n*x*n* (n any even number greater than 4), $n/2$ players and $2n$ distinguishable counters per player; the manner and aim of counter placement are the same as for HIP1, while a tie is possible for all values of n.

Table 1. Efficiency and accuracy results of the GA-based and the ANN approaches concerning HIP1 for n=4 (a) and n=6 (b)

Accuracy/efficiency n=4	Average number of iterations	% success
GA	1.33	100
ANN	25.6	85

(a)

Accuracy/efficiency n=6	Average number of iterations	% success
GA	39.41	100
ANN	85.2	27

(b)

Serial [3-4,8] as well as parallel [2] approaches to the HIP games appear in the literature. A tie may be hard to find serially since counter placement alternates between players and the first player is at a strong advantage of winning the game; by contrast, and owing to the synchronized counter placement by all players, parallel approaches are more adept at finding ties. The only systematic results (concerning efficiency and accuracy) have been reported for the parallel approach of Funabiki & Takefuji [2]. This employs an energy-driven artificial neural network (ANN) of the Hopfield & Tank architecture [7] for providing ties to both HIP1 (n=4 and 6) and HIP2 (n=6, 8, 10 and 12) games. The average number of iterations required until a tie is reached as well as the frequency of finding a tie are shown in the last rows of Tables 1(a-b) and 2(a-d) for the aforementioned instances of HIP1 and HIP2, respectively. The lower than 100% success rate is due to the purely monotonic decrease in energy, according to which it is possible for a local rather than a global[1] energy minimum to be settled upon.

3 Genetic Algorithms

Genetic algorithms (GAs) [5-6,9] are heuristics that employ biologically-derived evolution techniques (natural selection, crossover and mutation) in order to rapidly find optimal or near-optimal solutions to hard-to-solve problems; such problems are characterized by complex quality-of-solution landscapes, i.e. problem spaces where similar/distinct candidate solutions may have a significantly different/very similar quality of solution.

GAs operate in terms of:

- Chromosomes, i.e. candidate solutions encoded as sets of genes; binary encodings are the norm but non-binary encodings are also possible.
- A fitness function, which must be appropriately devised so that the fitness of each chromosome tallies with the quality of the corresponding candidate solution. A chromosome of maximum fitness corresponds to a globally optimal solution of the problem.
- A population, i.e. an ensemble of chromosomes that are simultaneously considered in order to concurrently investigate the problem space and guide the GA towards a chromosome of maximum fitness (i.e. a globally optimal solution).
- Generations of evolution. At the beginning of GA operation, an initial population of chromosomes is created via random gene assignment; owing to their means of construction, the initial chromosomes tend to be of low fitness. At each generation, a new population is derived by collectively evolving the current population; the new population constitutes the current population of the next generation. Evolution is performed via:
  (a) Selection, i.e. the creation of a new population by choosing either exclusively among the new chromosomes or among both the fittest chromosomes of the current population and the new chromosomes.
  (b) Crossover, i.e. the creation of two new chromosomes by globally modifying (i.e. exchanging sets of homologous genes between) a pair of chromosomes of the current population.
  (c) Mutation, i.e. the local modification of the new chromosomes by randomly changing the values of their genes with a small probability.

[1] A small number of squares created by either/both players rather than a tie.

Selection, crossover and mutation create populations of progressively higher average fitness at each subsequent generation[2]. Generations succeed one another until either a chromosome of maximum fitness is found or some other termination criterion (e.g. maximum allowable number of generations) is satisfied.

4 The Proposed GA-Based Approach

4.1 GA-Based Approach for HIP1

The GA-based approach for providing ties to HIP1 involves the following construction and operation characteristics:

- Each chromosome is encoded as a binary nxn matrix such that (i) the chromosome structure directly corresponds to the nxn checkerboard, and (ii) the binary value assigned to each gene encodes the player whose counter is placed at the corresponding location of the checkerboard.
- The fitness function counts the difference between the maximum number of squares $\frac{n^4-n^2}{12}$ that can be formed on the nxn checkerboard and the total number of squares created by the genes of the nxn binary chromosome. The fitness value ranges in the interval of integers $[0,\frac{n^4-n^2}{12}]$, where the maximum fitness value of $\frac{n^4-n^2}{12}$ denotes that a tie has been found.
- Small populations (of 1, 2,..., n chromosomes per generation) have been examined in order to keep the computational complexity of the GA low.
- Two ways of constructing the initial chromosomes have been investigated:
  (a) Non-directed (random) construction; this ensures that a valid number of counters are placed on the checkerboard per player and also implements Templeton's strategy[3], which obliterates the creation of squares that are symmetrical to the centre of the checkerboard.
  (b) Directed construction; this ensures that a valid number of counters are placed on the checkerboard per player, enforces Templeton's strategy, and also limits – if possible - the formation of squares[4] that are not symmetrical to the centre of the checkerboard.

---

[2] The fitness of a given chromosome of the population is not necessarily higher than the average fitness of the chromosomes of the previous population; in other words, the change in fitness between the individual chromosomes of different populations is not always monotonic.

[3] Templeton's strategy accomplishes a four-fold reduction of the problem space by necessitating (a) that the two players place their counters on the nxn checkerboard at locations rotated by 90° relative to each other (e.g. locations (i,j) against $(j,n+1-i)$, $i,j \leq n$) and, subsequently, (b) that each player places pairs of counters at symmetrical locations (rotated by 180° relative to the centre of the checkerboard (e.g. locations (i,j) and $(n+1-i,n+1-j)$, $i,j \leq n$).

[4] This is implemented by checking for such squares during construction and attempting to locally rearrange the counters (binary values) involved.

Table 2. Efficiency and accuracy results of the GA-based and the ANN approaches concerning HIP2 for n=6 (a); n=8 (b); n=10 (c); n=12 (d)

Accuracy/efficiency n=6	Average number of iterations	% success
GA	10.69	100
ANN	65.8	87

(a)

Accuracy/efficiency n=8	Average number of iterations	% success
GA	18.82	99
ANN	128.7	75

(b)

Accuracy/efficiency n=10	Average number of iterations	% success
GA	34.67	98
ANN	147.5	30

(c)

Accuracy/efficiency n=12	Average number of iterations	% success
GA	35.37	97
ANN	234.1	10

(d)

During each generation of evolution, no crossover is performed[5] and four-fold mutation is employed: a gene (i,j) , $i,j \le n$, of the nxn chromosome is selected (either randomly or via roulette-wheel, i.e. such that genes forming more/less squares have larger/smaller probabilities of being selected) and its value, together with the values of genes $(n+1-i,n+1-j)$, $(j,n+1-i)$ and $(n+1-j,i)$ are swapped[6]. Selection of the new chromosomes to be included in the next generation involves either choosing the fittest chromosomes or applying roulette-wheel. GA operation is terminated once a chromosome of maximum fitness is found in the population; in case 300 generations of evolution have been completed and no chromosome of maximum fitness has been found, it is assumed that the GA has failed to find a solution to HIP1[7].

---

[5] The lack of crossover does not constitute an exceptional GA practice: the earliest GAs relied on mutation alone, while a non-negligible number of problems have been tackled by crossover-free GAs (e.g. ANN structural and weight training by mutation only [1]).

[6] While respecting the placement of exactly $n^2/2$ counters per player as well as the enforcement of Templeton's strategy, four-fold mutation aims at reducing the number of squares formed by the new chromosome.

[7] Although stringent, this limit has been set in order to significantly restrict the computational complexity of the GA.

4.2 GA-Based Approach for HIP2

The GA-based approach for solving (providing ties to) HIP2 is similar to that of HIP1. The following modifications have been performed in order to accommodate the larger number of players:

- As for HIP1, each chromosome constitutes an *nxn* matrix, directly expressing the *nxn* checkerboard. Here, however, a transparent non-binary representation has been opted for; each gene is assigned values 1, 2,..., or $n/2$, denoting the player ($player_1$, $player_2$,... or $player_{n/2}$, respectively) whose counter is placed at the corresponding location of the checkerboard.
- Initial chromosome construction (either non-directed or directed) is similar to that of HIP1, but has been formulated for $n/2$ players.

Both the generations of evolution and the termination criterion are implemented as for HIP1. The only difference concerns four-fold mutation: the values of genes (i,j) and $(n+1-i,n+1-j)$ can be exchanged with those of genes (i',j') and $(n+1-i',n+1-j')$ in the *nxn* chromosome only if the two non-symmetrical genes (i,j) and (i',j'), $i,j,i',j' \leq n$, correspond to counters of different players.

5 Results - Comparisons

5.1 Tests for HIP1

The tests performed for HIP1 aim at investigating the potential of the GA approach in providing ties to the basic HIP game. Although the HIP1 instances (n=4, 6) are of relatively low computational complexity, they impart significant insights on the efficiency and accuracy of the GA approach in solving the HIP game.

The following combinations of GA construction and operation factors have been investigated:

(i) Population sizes ranging from 1 to n (n possibilities).
(ii) Random and directed construction of the initial chromosomes (2 possibilities).
(iii) Random and roulette-wheel selection of the gene to be subjected to four-fold mutation (2 possibilities).
(iv) Single application of mutation per chromosome of the population, whereby the new chromosome(s) constitute the population of the next generation; double application of mutation per chromosome, whereby either the fittest new chromosome(s) constitute the population of the next generation or roulette-wheel selection is applied to the new chromosomes (3 possibilities).

Five hundred tests have been performed for each of the $12n$ combinations of the aforementioned factors for the two HIP1 instances (48 and 72 combinations for n=4 and 6, respectively). The most successful (computationally efficient as well as accurate) GA-based approach for HIP1 involves:

(a) a population size of 1,
(b) directed construction of the initial chromosome,
(c) selection of the fittest new chromosome to constitute the new population,

(d) two independent applications of four-fold mutation to the chromosome of the current population, resulting in two new chromosomes; for the creation of each new chromosome, the gene to be subjected to mutation has been selected via roulette-wheel.

The first rows of Tables 1(a-b) illustrate the results of the GA-based approach for the two instances of HIP1 for n=4 and 6, respectively. It can be seen that the GA-based approach is superior to the ANN approach (last rows of the same Tables), both in terms of the average number of iterations required until a tie is found and in terms of accuracy[8]; a 100% success rate is demonstrated by the GA-based approach, i.e. a tie is always found in less than 300 generations of evolution. It should be mentioned that a variety of ties are produced, even when the generations of evolution start from the same initial chromosome. The reduced success rate of the ANN approach is attributed to the purely monotonic descent in terms of energy.

5.2 Tests for HIP2

The tests performed for HIP2 aim at investigating the scalability of the GA-based approach in providing ties to the HIP game. The same GA-based approach as for HIP1 (points (a)-(d) of section 5.1) has been implemented. The first line of Tables 2(a-d) illustrates the cumulative results of 100 tests performed by the GA-based approach for increasing values of n up to 12 (values for which results of the ANN approach are available [2]). An especially gradual form of graceful degradation is observed by the GA-based approach, resulting in a clear superiority over the ANN approach both in terms of the average number of iterations required until a tie is found and in terms of accuracy.

6 Conclusions

The HIP game constitutes an interesting example of game theory, where a tie is sought. HIP is played on a checkerboard of dimensions nxn by a number of players sharing n^2 counters. Counter placement is performed at vacant locations of the checkerboard in such a manner that no four counters of the same player form a square; once a square is created the corresponding player loses the game. In the family of HIP games, a tie is equivalent to placement of all the counters of all the players at distinct locations of the checkerboard such that no squares are formed. Serial as well as parallel approaches to the HIP games have been reported in the literature. A tie may be hard to find serially since counter placement alternates between players and the first player is at a strong advantage of winning the game; by contrast, and owing to the synchronized counter placement by all players, parallel approaches are more adept at finding ties.

In this piece of research, genetic algorithms have been put forward for providing ties to the family of HIP games. Following transparent construction of the chromosomes (such that they directly represent the candidate solutions) and the

[8] For the ANN approach, the proportion of unsuccessful games amounts to tests where a globally optimal solution is never found (i.e. not after 300 iterations, as for the GA-based approach).

design of an appropriate fitness function, the proposed parallel approach manipulates candidate solutions via selection and mutation only. The population has been limited to one chromosome per generation of evolution, thus keeping the computational complexity of the solution to a minimum. It has been found essential to restrict the formation of squares (via directed construction) of the initial chromosome in order to further reduce the computational complexity of the approach.

Despite the limited population size and the simplicity of the evolution scheme, the proposed approach has been found superior to the existing approaches: ties are more speedily provided while the frequency of finding a tie is significantly higher. Future research will focus upon determining the limits of the proposed approach, namely the maximum complexity (in terms of n values) of HIP2 that can be successfully solved.

References

1. Angeline, P.J., Saunders, G.M., Pollack, J.B.: An evolutionary approach that construct recurrent neural networks, IEEE Transactions on Neural Networks 5 (1994) 54-64
2. Funabiki, N., Takefuji, Y,: A parallel algorithm for solving the 'Hip' games, Neurocomputing 3 (1991) 97-106
3. Gardner, M.: New Mathematical Diversions from Scientific American, University of Chicago Press, U.S.A. (1983)
4. Gardner, M.: My Best Mathematical and Logical Puzzles, Dover Pubs., U.K. (1994)
5. Goldberg, D.E.: Genetic Algorithms in Search, Optimization and Machine Learning, Kluwer Academic Publishers, Boston, MA (1989)
6. Goldberg, D.E.: The Design of Innovation: Lessons from and for Competent Genetic Algorithms, Addison-Wesley, Reading, MA (2002)
7. Hopfield, J.J., Tank, D.W.: Neural computation of decisions in optimization problems, Biological Cybernetics 52 (1985) 141-152
8. Langman, H.: Play Mathematics, Hafner, New York (1962)
9. Michalewicz, Z.: Genetic Algorithms + Data Structures = Evolution Programs, Springer-Verlag (1999)
10. von Neumann, J., Morgenstern, O.: Theory of Games and Economic Behavior, Princeton University Press, U.S.A. (1944)

Evolution of Cooperation Using Random Pairing on Social Networks

Sihai Zhang, Shuangping Chen, and Xufa Wang

Department of Computer Science and Technology, University of Science and Technology of China, 230027, Hefei, Anhui, China
{shzhang, shpchen, xfwang}@ustc.edu.cn

Abstract. We studied the evolution of cooperation on social networks based on personal reputation using random pairing rule. Small-world networks and scale-free networks are used as practical network model. The iterated prisoner's dilemma game are adopted as theorotical tool in which players are paired according to the network structure to play the ONE-SHOT prisoner's dilemma game. Computer simulation shows that TIT-FOR-TAT-like strategy pattern will emerge from initial enviroments and cooperation can be maintained even in social networks when players have little chance to play continuous repeated games.

1 Introduction

Many researchers in different fields, such as economics, physics and biology, have been inspired by the evolution of cooperation with great interest. Recent research shows that the most prominant mechanisms of cooperation might be direct reciprocity[1][2], indirect reciprocity[3][4], voluntary participation[5][6][7] and spatial structures[8][9][10].

Current work on spatial structures mainly concertrate on modeling spatially structured populations by confining players to lattice sites. The performance of a player is determined by the payoffs accumulated in his interactions with his neighbors. As we know, social networks exhibit many new statistical properties, such as large cluster coefficient, short average shortest path and power-law degree distribution[11]. Many real networks, especially social networks, have small-world and scale-free properties. People live in societies so that people live on social networks which makes it necessary to study the cooperation in network structured populations.

Another motivation of this research is that randomicity of interactions in real world has not got much attention yet. Communication and transportation technologies have been changing people's life in every respect and the likelihood of interaction between people seperated here and there has been increased greatly. This has changed the conditions of iterated games[12].

The work of this paper has two distinctness with previous research in that:

1. We consider the cooperation mechanisms on social networks in which interactions between players are constrained by network structures.

T.-D. Wang et al. (Eds.): SEAL 2006, LNCS 4247, pp. 680–687, 2006.

2. We also introduce randomicity into the model proposed in this paper which means: (I) If two players meet they just play ONE-SHOT game and (II) The probability that they meet at different generations is independent.

The analysis to our model shows that cooperation is possible and the computer simulation comfirms the analysis with certain TIT-FOR-TAT-like strategy emerging from initial neutral states under determinate conditions.

This paper is organized as follows. In Section. 2, the game model on social networks considering randomicity based on reputation is proposed. In Section 3, computer simulation and experimental results are presented. Discussion and conclusion is presented in Section 4.

2 Game Model

There are N players in the game population and the relations between these N players form certain network structure in it. We aims to study the effect of static social networks on cooperation emergence so that two network models, small-world network and scale-free network, are utilized as the networks in the game model. Here we assume that the interaction relationship between all players in the population form certain network structure in which players are considered as nodes and relationship as edges.

When two players meet they play the ONE-SHOT prisoner's dilemma game. In the prisoner's dilemma game, for mutual cooperation both players will obtain the reward R, but only P for mutual defection. If one player defects and the other cooperates, the defector receives the highest payoff T while the sucker gets the lowest payoff S[2]. Note that the practical values of payoff must satisfy $T > R > P > S$. When extended to iterated cases the additional requirement of $R > (S + T)/2$ is necessary.

In the proposed model the game will be played infinite generation and at each generation N players will propose a game one by one as so-called proposer. When a player becomes the proposer, he randomly selects one of his neighbors in current network to be the responder. The payoffs one player gets are accumulated to evaluate the performance of this player.

2.1 Definition of Reputation

In reciprocal altruism personal reputation contributes great to the cooperation. Modern technology makes it possible to record and access personal information more and more easily thus people's actions or records in modern society could be seen as public information. It is also known that this public information could not be obtained completely due to economical cost. So we use one player's latest three actions in games to present his or her reputation.

Definition 1. *Let A be game population which keeps N players. ith player is denoted by A_i, for all $i \in [1, N]$.*

Definition 2. *Let $H_i(t)$, $i \in [1, N]$ and $t \in (0, \infty)$, be A_i's all historical actions. So $H_i(t)$, $H_i(t-1)$ and $H_i(t-2)$ are supposed to be accessed freely at time t.*

2.2 Strategy Operators

Because each action in prisoner's dilemma has two operations, Cooperation(C) or Defection(D), there are 8 accessable states in historical state space, each representing one kind of reputation. To each reputation every player has two operators, named Cooperator and Defector. Each operator has a property of weight and the ratio of them denotes which operator is predominant when meeting other player with corresponding reputation. When two players interact at some generation, each one decides his action based on his opponent's reputation and his own operators matching that reputation. The Roullte Wheel Selection method is adopted based on their weights to decide which operator will be used.

In this paper each player's reputation is made up of his latest three actions and his operators denote corresponding strategy to deal with certain reputation. As explained above, each action has two possibilities so that reputation has 8 states and each player possess 16 operators. For example, operator $CCCC$ will take Cooperation when facing players with reputation CCC.

The operators's weights are to be adjusted according to the payoffs they get in interactions. The weight will be increased when the operator gets more payoff while decreased when less. The regulation obeys the equation below:

$$W_{i,j,t+1} = \beta * W_{i,j,t} + \frac{P_{i,t}}{\gamma * M}, 1 \leq i \leq N, 1 \leq j \leq 16, 0 < \beta < 1, \gamma > 0; \quad (1)$$

Here, $W_{i,j,t}$ denotes the weight of player i's operator j at time t. $P_{i,t}$ denotes the payoff player i gets at time t and M denotes the largest possible payoff in the prisoner's dilemma game in this paper. γ is used to adjust the increasing step of weight and β denotes the attenuation coefficient.

2.3 Analysis of Dynamics

In this section analysis of dynamics will be given out. According to equation (1) the weight regulation of player i has four cases:

$$W_{i,j,t+1} = \beta * W_{i,j,t} + \frac{S}{\gamma * M}; \quad (2)$$

$$W_{i,j,t+1} = \beta * W_{i,j,t} + \frac{R}{\gamma * M}; \quad (3)$$

$$W_{i,j,t+1} = \beta * W_{i,j,t} + \frac{P}{\gamma * M}; \quad (4)$$

$$W_{i,j,t+1} = \beta * W_{i,j,t} + \frac{T}{\gamma * M}; \quad (5)$$

Here, equation (3) means mutual cooperations and equation (4) mutual defections. Equation (2) means player i cooperates while the opponent defects and equation (5) player i defects while the opponent cooperates. Fig. 1 shows the sample of this regulation. There are four parallel lines denoting equation (2) to (5). The dot

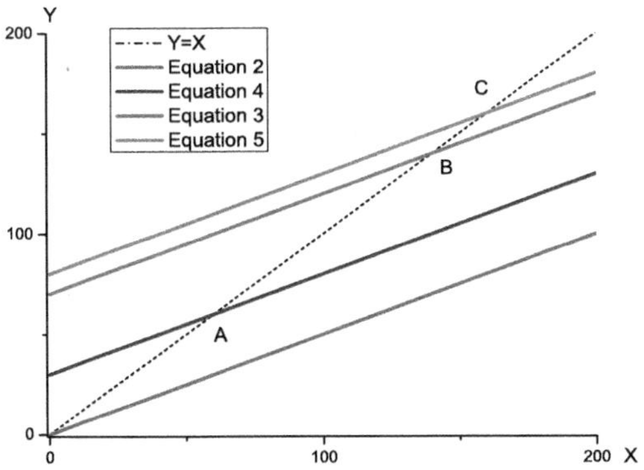

Fig. 1. Sketch Map of Model Analysis

line denotes the equation $Y = X$ which intersects equation (2) at point O, equation (3) at point B, equation (4) at point A and equation (5) at point C.

The weight regulation of equation (2) will make $W_{i,j,t+1}$ move to point O while equation (3) will make $W_{i,j,t+1}$ move to point B. So in equilibrium state the weight of cooperators will be in (X_O,X_B). The same induction holds true for equation (4) and (5) that the equilibria weight of defectors will be in (X_A,X_C). Thus in (X_A, X_B) the cooperators might have the advantage over defectors so that the cooperation is possible in the model proposed in this paper. This analysis also holds true to other operators. In addition, as long as the payoff matrix satisfies the condition of prisoner's dilemma game, cooperation becomes possible.

3 Computer Simulation

For all experiments presented in this paper, the population size is 65535, $\beta = 0.999$ and $\gamma = 2.5$. All results are the average of 10 repeated trials and each trial has 100 generations. The values of T=8, R=7, P=3, S=0 if no extra statement.

Two network models are used as the static network structure for the iterated prisone's games which are small-world networks and scale-free networks. We use the algorithm in [13] to generate the small-world network and the algorithm in [14] for scale-free network.

The initial weights of players' operators are set according to the parameter $ICD(initialcooperationdegree) \in (0, 2)$, which determines the initial probability for each player to take which operator(C or D). With ICD increasing the player will have more initial possibility to take operator C.

3.1 Emergence of Cooperation

Whether cooperation can emerge or not is the main topic in evolutionary game theory. Fig. 1 shows the numerical results of cooperation evolution on social networks. It is obvious that certain cooperative strategy DOES evolve from initial conditions. We notice that different strategy emerges with different conditions

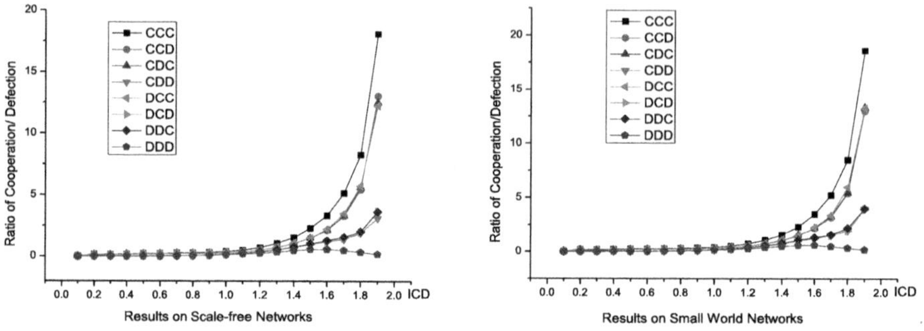

Fig. 2. Ratio of Cooperation/Defection on Social networks. Subgraph on left side is on Scale-free networks and right side is Small-world networks. ICD increases from 0.1 to 1.9 with step=0.1.

and the evolution result is closely related with different initial coopration degree. For example, when ICD=1.9, initial average ratio of all C/D=19 but the final average values of ratio are: CCC=17.6362, CCD=14.0591, CDC=14.8232, CDD=4.2007, DCC=14.8777, DCD=3.5331, DDC=4.1192, DDD=0.3096. These eight ratio under different ICD stand for the strategy of the whole population. We point out that the strategy evolved under ICD=1.9 is similar with TIT-FOR-TAT:

1. ratio of CCC=17.6362 means that one will cooperate when his opponent's reputation is excellent, in Axelrod's words, do not break the boat first.
2. ratio of DDD=0.3096 means that one player will punish those whose reputation is bad by taking defection.
3. ratio of CDC=14.8232, DCC=14.8777 and DDC=4.1192 means that players hold room for forgiving others with one or two defection records.

In addition, the strategy evolved possesses the ability which TIT-FOR-TAT does not have. The strategy can correct and avoid unmeant mistakes or inattentions that lead to defection action. The simulation results in Fig. 1 show that networks model affect little on the final evolution of strategy.

3.2 Influence of Game Generations

The generations of games has notable effect on the evolutionary results. Extreme example is that in ONE-SHOT prisoner's dilemma game the mutual defections of both players becomes the Nash Equilibria. Fig. 3 shows the influence of generations on the ratio of weights with ICD=1.8(left side) and 1.0(right side).

First of all, a remarkable conclusion is that the ratios of each kind of reputation evolve with generations and finally converge to certain equilirium values. But the equilibrium values differ with the ICD. In the case of ICD=1.8, all the ratio

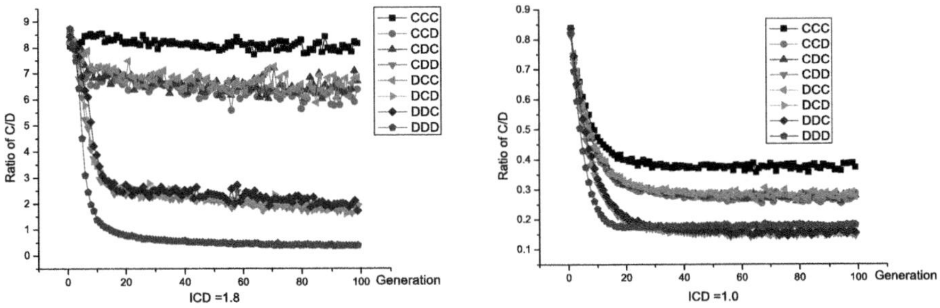

Fig. 3. Ratio of Weights (Cooperation / Defection) with Generation Increasing on Scale-free networks

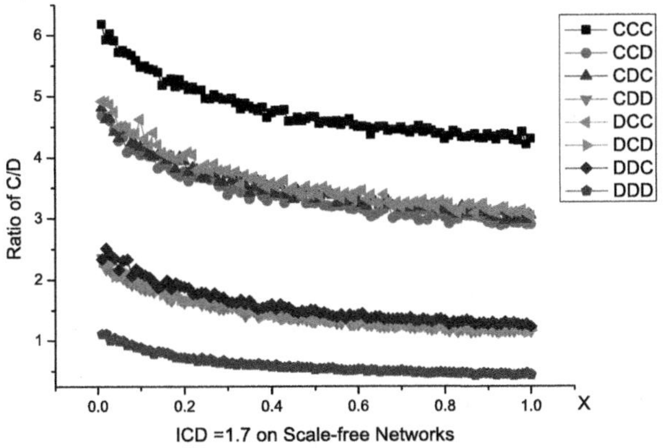

Fig. 4. Equilibrium Ratio of Cooperation / Defection on Social Networks

equal 9 at initial generation but evolve along different curves. After about 80 generations the ratios become steady. Similar evolution also happens to the case of ICD=1.0 except that the steady values of ratios are quite different.

3.3 Influence of Cooperation Cost

Players are all rational in the proposed game model that they choose their actions only according to the calculation of payoff they will get. We import a variable factor $X \in (0, 1)$, called cooperation cost, into the payoff matrix so that the values of T=1+X, R=1, P=0 and S= -X. X is called as cooperation cost just because when one takes C he might suffer the potential lost X if his opponent takes D. It is obvious that this payoff still satisfies the condition of prisoner's dilemma game. Fig.4 shows the typical influence of cost on cooperation emergence on scale-free networks when ICD = 1.7 and two trends can be summarize:

1. All weight ratios(C/D) of these eight reputations decreases with the increase of X. It is natural that when cooperation cost is quite low, players are more prone to cooperate with others otherwise the motivity of cooperating will fall down. With X very close to 0 the strategy evolved will even cooperate with players with DDD reputation just because there are little difference between cooperation and defection.
2. More importantly, the diversification of X do not change the cooperation essence of equilibrium strategies. When facing players with reputation CCC the cooperators are always dominant over defectors, although the cooperative degree are weakened as the cooperation cost grows.

Such trends are also true when using small-world networks and the simulation result are ommitted for saving sapce.

4 Conclusions

The influence of population structure turns out to be important for the evolution of cooperation. We introduce the social network structures and randomicity of interactions into iterated prisoner's dilemma game to meet the new development of real world. The model takes players' partial, if not all, information of previous actions as personal reputation which decide the actual action along with the weights of the corresponding operators. Theoretical analysis prove that the evolution of cooperation are possible in the game model proposed and computer simulation results show that the strategy evolved are cooperative and forgiving under certain conditions.

The presented model and computer simulation put forward a new viewpoint on the evolution of cooperation among players located in social networks. Defection are not the only equilibrium on social networks even when playing ONE-SHOT prisoner's dilemma game if initial cooperation degree are quite high. Although not proved, we believe that there exists one critical value of ICD that cooperation will emerge and maintain if ICD is larger than it.

Acknowledgement

This paper is supported by National Natural Science Foundation of China Joint Research Fund for Overseas Chinese Young Scholars(No.60428202). Author Zhang Sihai would like to thank Santa Fe Institute for providing related materials.

References

1. Trivers R. L.,: The evolution of reciprocal altruism. Q. Rev. Biol. **46**, 35–37 (1971)
2. Axelrod R.: The evolution of Cooperation. Basic Book New York (1984)
3. Nowak M. A. and Sigmund, K.: Evolution of indirect reciprocity by image scoring. Nature(London) **393**, 573–577 (1998)

4. Wedekind C. and Milinski M.: Cooperation through image scoring in humans. Science **288**, 850–852 (2000)
5. Hauert C., Monte S. De, Hofbauer J. and Sigmund, K.: Volunteering as red queen mechanism for cooperation in public good games. Science **296**, 1129–1132 (2002)
6. Szabó C. and Hauert C.: Phase transition and volunteering in spatial public good games. Phys. Rev. Lett. **89**, 118101 1–4 (2002)
7. Semmann D., Krembeck H. J. and Milinski M.: Volunteering leads to rock-paper-scissors dynamics in a public good game. Nature(London) **425**, 390–393 (2003)
8. Nowak, M. A. and May, R. M.: Evolutionary games and spatial chaos. Nature(London). **359**, 826–829 (1992)
9. Nakamaru, M., Matsuda, H., Iwasa, Y.: The evolution of cooperation in a lattice-structured population. J. Theor. Biol. **184** 65–81 (1997)
10. Brauchli, K., Killingback, T. and Doebeli, M.: Evolution of cooperation in spatially structured population. J. Theor. Biol. **200** 405–417 (1999)
11. Newman M. E. J.: Evolution of Networks. Arch. Rat. Mech. Anal. **78**, 315–333 (1982)
12. Axelrod R.: The Complexity of Cooperation. Princeton University Press. (1997)
13. Klemm, K. and Eguiluz, V. M.: Growing scale-free networks with small-world behavior. Phys. Rev. E **65** 057102 (2003)
14. Barabási and Albert A. L.: Emergence of scaling in random networks. Science **286** 509–512 (1999)

Selecting Valuable Stock Using Genetic Algorithm

Chengxiong Zhou[1], Lean Yu[1,2], Tao Huang[3], Shouyang Wang[1], and Kin Keung Lai[2]

[1] Institute of Systems Science, Academy of Mathematics and Systems Science, Chinese Academy of Sciences, Beijing 100080, China
zcx1975@sina.com, {yulean, sywang}@amss.ac.cn
[2] Department of Management Sciences, City University of Hong Kong, Tat Chee Avenue, Kowloon, Hong Kong
{msyulean, mskklai}@cityu.edu.hk
[3] School of Public Policy & Management, Tsinghua University, Beijing, 100084, China
ht01@mails.tsinghua.edu.cn

Abstract. In this study, we utilize the genetic algorithm (GA) to select high quality stocks with investment value. Given the fundamental financial and price information of stocks trading, we attempt to use GA to identify stocks that are likely to outperform the market by having excess returns. To evaluate the efficiency of the GA for stock selection, the return of equally weighted portfolio formed by the stocks selected by GA is used as evaluation criterion. Experiment results reveal that the proposed GA for stock selection provides a very flexible and useful tool to assist the investors in selecting valuable stocks.

1 Introduction

In the stock market, investors are often faced with a large number of stocks. A crucial work of their investment decision process is the selection of stocks. From a data-mining perspective, the problem of stock selection is to identify good quality stocks that are potential to outperform the market by having excess return in the future. Given the fundamental accounting and price information of stock trading, it is a prediction problem that involves discovering useful patterns or relationship in the data, and applying that information to identify whether a stock is good quality.

Obviously, it is not an easy task for many investors when they faced with enormous amount of stocks in the market. With focus on the business computing, applying artificial intelligence to portfolio selection and optimization is one way to meet the challenge. Some research has presented to solve asset selection problem. Levin [1] applied artificial neural network to select valuable stocks. Chu [2] used fuzzy multiple attribute decision analysis to select stocks for portfolio. Similarly, Zargham [3] used a fuzzy rule-based system to evaluate the listed stocks and realize stock selection. Recently, Fan [4] utilized support vector machine to train universal feedforward neural networks to perform stock selection.

However, these approaches have some drawbacks in solving the stock selection problem. For example, fuzzy approach [2-3] usually lacks learning ability, while neural network approach [1, 4] has overfitting problem and is often easy to trap into local minima. In order to overcome these shortcomings, GA is used to perform this task. Some related typical literature can be referred to [5-7] for more details.

T.-D. Wang et al. (Eds.): SEAL 2006, LNCS 4247, pp. 688 – 694, 2006.

The main aim of this study is to select some valuable stocks using GA and to test the efficiency of the GA for stock selection. The rest of the study is organized as follows. Section 2 describes the selection process based on the genetic algorithm in detail. Section 3 presents a simulation experiment. And Section 4 concludes.

2 GA-Based Stock Selection Process

Generally, GA imitates the natural selection process in biological evolution with selection, crossover and mutation, and the sequence of the different operations of a genetic algorithm is shown in the left part of Fig. 1. That is, GA is procedures modeled after genetics and evolution. Genetics provide the chromosomal representation to encode the solution space of the problem while evolutionary procedures are designed to efficiently search for attractive solutions to large and complex problem. Usually, GA is based on the survival-of-the-fittest fashion by gradually manipulating the potential problem solutions to obtain the more superior solutions in population. Optimization is performed in the representation rather than in the problem space directly. To date, GA has become a popular optimization method as they often succeed in finding the best optimum by global search in contrast to most common optimization algorithms. Interested readers can be referred to [8-9] for more details.

The aim of this study is to identify the quality of each stock using GA so that investors can choose some good ones for investment. Here we use stock ranking to determine the quality of stock. The stocks with a high rank are regarded as good quality stock. In this study, some financial indicators of the listed companies are employed to determine and identify the quality of each stock. That is, the financial indicators of the companies are used as input variables while a score is given to rate the stocks. The output variable is stock ranking. Throughout the study, four important financial indicators, return on capital employed (ROCE), price/earnings ratio (P/E Ratio), earning per share (EPS) and liquidity ratio are utilized in this study.

ROCE is an indicator of a company's profitability related to the total financing, which is calculated as

$$\text{ROCE} = (\text{Profit})/(\text{Shareholder's equity}) \times 100\% \tag{1}$$

The higher the indicator (ROCE), the better is the company's performance in terms of how efficient the company utilizes shareholder's capital to produce revenue.

P/E Ratio measures the multiple of earnings per share at which the stock is traded on the stock exchange. The higher the ratio, the stronger is the company's earning power. The calculation of this ratio is computed by

$$\text{P/E ratio} = (\text{stock price})/(\text{earnings per share}) \times 100\% \tag{2}$$

EPS is a performance indicator that expresses a company's net income in relation to the number of ordinary shares issued. Generally, the higher the indicator, the better is the company's investment value. The calculation of the indicator can be represented as

$$\text{Earnings per share} = (\text{Net income})/(\text{The number of ordinary shares}) \tag{3}$$

Liquidity ratio measures the extent to which a company can quickly liquidate assets to cover short-term liabilities. It is calculated as follows:

$$\text{Liquidity Ratio} = (\text{Current Assets})/(\text{Current Liabilities}) \times 100\% \quad (4)$$

If the liquidity ratio is too high, company performance is not good due to too much cash or stock on hand. When the ratio is too low, the company does not have sufficient cash to settle short-term debt.

When the input variables are determined, we can use GA to distinguish and identify the quality of each stock, as illustrated in Fig. 1. The detailed procedure is illustrated as follows.

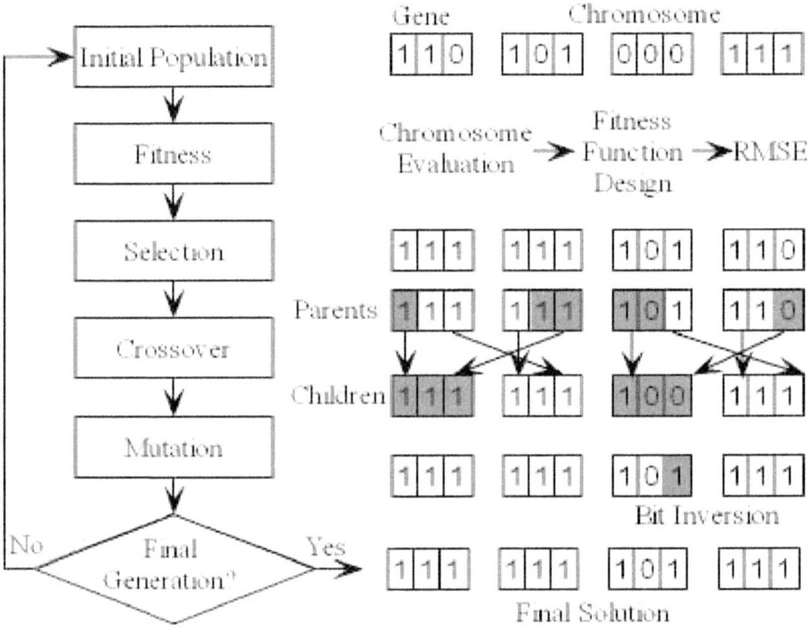

Fig. 1. Stock selection with genetic algorithm

First of all, a population, which consists of a given number of chromosomes, is initially created by randomly assigning "1" and "0" to all genes. In the case of stock ranking, a gene contains only a single bit string for the status of input variable. The top right part of Figure 1 shows a population with four chromosomes, each chromosome includes different genes. In this study, the initial population of the GA is generated by encoding four input variables. For the testing case of ROCE, we design 8 statuses representing different qualities in terms of different interval, varying from 0 (Extremely poor) to 7 (very good). An example of encoding ROCE is shown in Table 1. Other input variables are encoded by the same principle. That is, the binary string of a gene consists of three single bits, as illustrated by Fig. 1.

Table 1. An example of encoding ROCE

ROCE value	Status	Encoding
(-∞, -30%]	0	000
(-30%, -20%]	1	001
(-20%,-10%]	2	010
(-10%,0%]	3	011
(0%, 10%]	4	100
(10%, 20%]	5	101
(20%, 30%]	6	110
(30%,+∞)	7	111

Note that 3-digit encoding is used for simplicity in this study. Of course, 4-digit encoding is also adopted, but the computations will be rather complexity.

The subsequent work is to evaluate the chromosomes generated by previous operation by a so-called fitness function, while the design of the fitness function is a crucial point in using GA, which determines what a GA should optimize. Since the output is some estimated stock ranking of designated testing companies, some actual stock ranking should be defined in advance for designing fitness function. Here we use annual price return (APR) to rank the listed stock and the APR is represented as

$$APR_n = \frac{ASP_n - ASP_{n-1}}{ASP_{n-1}} \tag{5}$$

where APR_n is the annual price return for year n, ASP_n is the annual stock price for year n. Usually, the stocks with a high annual price return are regarded as good stocks. With the value of APR evaluated for each of the N trading stocks, they will be assigned for a ranking r ranged from 1 and N, where 1 is the highest value of the APR while N is the lowest. For convenience of comparison, the stock's rank r should be mapped linearly into stock ranking ranged from 0 to 7 with the following equation:

$$R_{actual} = 7 \times \frac{N - r}{N - 1} \tag{6}$$

Thus, the fitness function can be designed to minimize the root mean square error (RMSE) of the difference between the financial indicator derived ranking and the next year's actual ranking of all the listed companies for a particular chromosome, representing by

$$RMSE = \sqrt{\frac{1}{m} \sum_{t=1}^{m} (R_{derived} - R_{actual})^2} \tag{7}$$

After evolving the fitness of the population, the best chromosomes with the highest fitness value are selected by means of the roulette wheel. Thereby, the chromosomes are allocated space on a roulette wheel proportional to their fitness and thus the fittest chromosomes are more likely selected. In the following crossover step, offspring chromosomes are created by some crossover techniques. A so-called one-point crossover technique is employed, which randomly selects a crossover point within the chromosome. Then two parent chromosomes are interchanged at this point to produce two new offspring. After that, the chromosomes are mutated with a probability of 0.005 per gene by randomly changing genes from "0" to "1" and vice versa. The mutation prevents the GA from converging too quickly in a small area of the search space. Finally, the final generation will be judged. If yes, then the optimized results are obtained. If no, then the evaluation and reproduction steps are repeated until a certain number of generations, until a defined fitness or until a convergence criterion of the population are reached. In the ideal case, all chromosomes of the last generation have the same genes representing the optimal solution.

Through the process of GA optimization, the stocks are ranked according to the fundamental financial information and price return. Investors can select the top n stocks to construct a portfolio.

3 Experiment Analysis

The daily data used in this study is stock closing price obtained from Shanghai Stock Exchange (SSE) (http://www.sse.com.cn). The sample data span the period from January 2, 2002 to December 31, 2004. Monthly and yearly data in this study are obtained by daily data computation. For simulation, 100 stocks are randomly selected. In this study, we select 100 stocks from Shanghai A share, and their stock codes vary from 600000 to 600100.

First of all, the company financial information as the input variables is fed into the GA to obtain the derived company ranking. This output is compared with the actual stock ranking in terms of APR, as indicated by Equations (5) and (6). In the process of GA optimization, the RMSE between the derived and the actual ranking of each stock is calculated and served as the evaluation function of the GA process. The best chromosome obtained is used to rank the stocks and the top n stocks are chosen for the portfolio. For experiment purpose, the top 10 and 20 stocks are chosen for testing according to the ranking of stock quality using GA. The top 10 and 20 stocks selected by GA can construct a portfolio. For convenience, equally weighted portfolios are built for comparison purpose.

In order to evaluate the usefulness of the GA optimization, we compared the net accumulated return generated by the selected stock from GA with a benchmark. The benchmark return is determined by an equally weighted portfolio of all the stocks available in the experiment. Fig. 2 reveals the results for different portfolios.

From Fig. 2, we can find that the net accumulated return of the equally weighted portfolio formed by the stocks selected by GA is significantly outperformed the benchmark. In addition, the performance of the portfolio of the 10 stocks is better that of the 20 stocks. As we know, portfolio does not only focus on the expected return but also on risk minimization. The larger the number of stocks in the portfolio is, the more flexible for the portfolio to make the best composition to avoid risk. However, selecting good quality stocks is the prerequisite of obtaining a good portfolio. That is, although the portfolio with the large number of stocks can lower the risk to some extent, some bad quality stocks may include into the portfolio, which influences the portfolio performance. Meantime, this result also demonstrates that the portfolio with

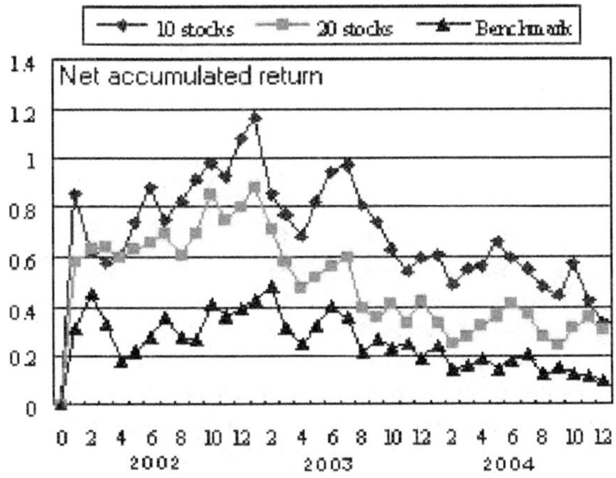

Fig. 2. Accumulated return for different portfolios

the large number of stocks does not necessary outperform the portfolio with the small number of stocks if the investors select good quality stocks. Therefore it is wise for investors to select a limit number of good quality stocks for constructing a portfolio.

4 Conclusions

This study uses genetic optimization algorithm to perform stocks selection for portfolio. Experiment results reveal that the GA optimization approach has shown to be useful to the problem of stock selection, which can help investors select the most valuable stocks for portfolio.

Acknowledgements

This work is partially supported by National Natural Science Foundation of China (NSFC No. 70221001); Key Laboratory of Management, Decision and Information Systems of Chinese Academy of Sciences and Strategic Research Grant of City University of Hong Kong (SRG No. 7001677).

References

1. Levin, A.U.: Stock Selection via Nonlinear Multi-factor Models. Advances in Neural Information Processing Systems (1995) 966-972
2. Chu, T.C. Tsao, C.T. Shiue, Y.R.: Application of Fuzzy Multiple Attribute Decision Making on Company Analysis for Stock Selection. Proceedings of Soft Computing in Intelligent Systems and Information Processing (1996) 509-514
3. Zargham, M.R., Sayeh, M.R.: A Web-Based Information System for Stock Selection and Evaluation. Proceedings of the First International Workshop on Advance Issues of E-Commerce and Web-Based Information Systems (1999) 81-83

4. Fan, A., Palaniswami, M.: Stock Selection Using Support Vector Machines. Proceedings of International Joint Conference on Neural Networks 3 (2001) 1793-1798
5. Lin, L., Cao, L., Wang, J., Zhang, C.: The Applications of Genetic Algorithms in Stock Market Data Mining Optimization. In: Zanasi, A., Ebecken, N.F.F., Brebbia, C.A. (Eds.): Data Mining V, WIT Press (2004)
6. Chen, S.H. Genetic Algorithms and Genetic Programming in Computational Finance. Kluwer Academic Publishers, Dordrecht (2002)
7. Thomas, J., Sycara, K.: The Importance of Simplicity and Validation in Genetic Programming for Data Mining in Financial Data. Proceedings of the Joint AAAI-1999 and GECCO-1999 Workshop on Data Mining with Evolutionary Algorithms (1999)
8. Holland, J. H.: Genetic Algorithms. Scientific American 267 (1992) 66-72
9. Goldberg, D.E.: Genetic Algorithm in Search, Optimization, and Machine Learning. Addison-Wesley, Reading, MA (1989)

Adaptive Learning in Complex Trade Networks

Tomas Klos[1,⋆] and Bart Nooteboom[2]

[1] Center for Mathematics and Computer Science, Amsterdam, The Netherlands
tomas.klos@cwi.nl
http://homepages.cwi.nl/~tomas/
[2] Center for Economic Research (CentER), University of Tilburg, The Netherlands

Abstract. The reinforcement learning paradigm is ideally suited for dealing with the requirements posed by a recent approach to economic modeling called Agent-based Computational Economics (ACE): the application of Holland's Complex Adaptive Systems (CAS) paradigm to economics. In this approach, economic phenomena emerge from the decentralized interactions among autonomous, heterogenous, boundedly rational, adaptive economic agents, rather than from idealized interactions among 'representative agents' or equilibrium analysis over the heads of the agents involved.

In this paper, we study an industrial goods market, where buyers need to decide between making and buying components. Traditionally, Transaction Cost Economics (TCE) has been used to analyze these types of situations. However, a number of criticisms of TCE have been raised, which the ACE approach allows us to resolve. Our resulting Agent-based Computational Transaction Cost Economics (ACTCE) approach allows us to study systems of interacting agents both at the level with which TCE deals (allowing comparison and verification), as well as at the level of individual agents, allowing extension of the theory's predictive power.

1 Introduction

A Complex Adaptive System (CAS) [1] "is a complex system containing adaptive agents, networked so that the environment of each adaptive agent includes other agents in the system" [2, p. 365]. The application of this paradigm to economics is called Agent-based Comptutational Economics (ACE) [3,4],[1] which is the computational study of economies modeled as evolving systems of autonomous interacting agents. As compared to earlier approaches to economies as self-organizing and evolving systems, ACE uses powerful new computational tools to permit "the constructive grounding of economic theories in the thinking and interactions of automous agents" [3, p. 283].

In the current paper, we employ an ACE perspective to the study of inter-firm relations on intermediate goods markets. These are typically studied using Transaction Cost Economics (TCE) [5,6], which takes the 'transaction' as its basic unit of analysis, and analyzes which structural forms should be used for

⋆ Corresponding author.

[1] For a wide variety of materials related to ACE, see Leigh Tesfatsion's ACE website at http://www.econ.iastate.edu/tesfatsi/ace.htm

T.-D. Wang et al. (Eds.): SEAL 2006, LNCS 4247, pp. 695–702, 2006.

organizing transactions. [6, p. 1]: "A transaction occurs when a good or service is transferred across a technologically separable interface. One stage of activity terminates and another begins." If activities (or 'stages of production') are thought of as nodes, and transactions as directed edges between nodes (specifying how the outputs of certain activities are inputs to others) then TCE is concerned with the mapping of transactions to organizational forms: simplistically, the question of which nodes (and consequently, the transactions between them) should be organized *within firms* (the 'make' alternative), or *on the market*, i.e. across firm boundaries (the 'buy' alternative). TCE then proceeds to analyze which organizational structure is most suited (i.e. 'economic') for organizing a transaction given its characteristics (frequency, uncertainty, and asset specificity—see Sect. 2.1), and to hypothesize that in reality, only (transaction cost) economic structural forms are actually used to organize transactions.

Nothwithstanding the value and validity of Transaction Cost Economic reasoning and its results, the theory seems limited in certain ways. Speficically, a number of criticisms of TCE have been raised [7,8], similar to the more general criticisms of economics that spawned the emergence of the ACE approach. In general, TCE has been acknowledged to disregard the role of learning, adaptation and innnovation, including trust (see [8] for a more complete discussion). For example, Williamson admitted that "the study of economic organization in a regime of rapid technological innovation poses much more difficult issues than those addressed here" [6, p. 143]. Furthermore, as Ronald Coase, the founding father of TCE admits [9], "[t]he analysis cannot be confined to what happens within a single firm. The costs of co-ordination within a firm and the level of transaction costs that it faces are affected by its ability to purchase inputs from other firms, and their ability to supply these inputs depends in part on their costs of co-ordination and the level of transaction costs that they face which are similarly affected by what these are in still other firms. What we are dealing with is a complex interrelated structure."

The CAS paradigm of course, is ideally suited for dealing with such a complex interrelated structure. Applying the CAS paradigm to the TCE domain in what we call Agent-based Computational Transaction Cost Economics (ACTCE), we let the distribution of economic activity across different organizational forms emerge from processes of interaction between autonomous boundedly rational agents, as they adapt future decisions to past experiences (cf. [4]). The system may or may not settle down and if it does, the resulting equilibrium may or may not be transaction cost economic (rational). In any case, "[i]t is the process of becoming rather than the never-reached end points that we must study if we are to gain insight" [1, p. 19].

With respect to modeling agents and adaptation, the most popular technique employed in ACE is the Genetic Algorithm (GA) [10], a search heuristic based on the theories of natural selection and genetics. Agent strategies are typically encoded as chromosomes and evolved by means of genetic operators such as selection and reproduction. However, as elaborated in [11], being a population-based search heuristic, a GA may not be the most appropriate mechanism for modeling individual agents' adaptive behavior,[2] to the extent that (1) it uses

[2] A GA may, on the other hand, be used for searching the 'never-reached end points' in the CAS of interest, to be used for benchmarking the adaptive agents' performance.

population-level information about agents' relative fitness which will typically not be available to individual (boundedly rational) agents, and (2) agents' strategies are evolved by recombining other agents' strategies which may result in unrealistic discontinuities in agents' behavior, while on the other hand, agents are more realistically assumed to be unable to perceive (and imitate) each other's strategies directly, but rather only their resulting behavior. Furthermore, an agent in a CAS is defined as being 'adaptive' if "the actions of the agent in its environment can be assigned a value (performance, utility, payoff, fitness, or the like); and the agent behaves in such a way as to improve this value over time" [2, p. 365]. Noting the similarity between this definition and the statement in [12, p. 7], that "[a] reinforcement learning agent's sole objective is to maximize the total reward it receives in the long run," we propose to model adaptive agents in our ACTCE approach as reinforcement learning agents.

The next section (2) describes our model of adaptive agents in complex interrelated systems of inter-firm relations. We performed computational 'in silico' experiments with our artificial economy, results from which are described in Section 3. Section 4 concludes the paper.

2 The Model: Matching Adaptive Buyers and Suppliers

We model interactions between buyers and suppliers on an industrial market, i.e. a market for an intermediate good.[3] The buyers may buy this intermediate good from a supplier ('buy') or produce the good themselves ('make'). In any case, the buyers use the intermediate good to produce a final good which they sell to consumers on a final goods market.

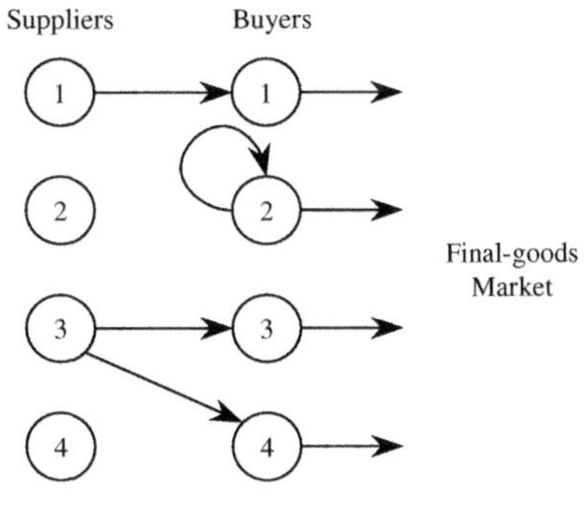

Fig. 1. Buyers are assigned to suppliers or to themselves

Because we choose to model the market as a complex interrelated system of interacting agents, we can not rely on economic theory's standard anonymized randomized matching device. A way of performing a matching based on heterogenous agents' idiosyncratic preferences is provided by the Deferred Choice and Refusal (DCR) algorithm [13], a modification of the original Deferred Acceptance Algorithm [14]. The DCR algorithm matches agents on two sides of a market to each other, based on the preferences each agent has over all agents on the other side of the market. In our model, the agents set preferences by calculating each other's 'scores:' expected payoffs obtained from transacting with each other (see Sect. 2.1).

[3] We will use the terms 'buyer' and 'supplier' to refer to the agents on the industrial market, and the terms 'seller' and 'consumer' to refer to the agents on the final goods market. A buyer on the industrial market is a seller on the final goods market.

We will in turn make a slight modification to the DCR algorithm: since buyers have the option of making rather than buying, a buyer does not *need* to be matched, while a supplier will always rather be matched to a buyer than not be matched at all. This consideration is implemented by letting each buyer also calculate his own score for himself, and by letting all suppliers not scoring higher than his own score be 'unacceptable' for the buyer: he will rather be matched to (and make for) himself than buy from an unacceptable supplier. This effectively endogenizes the buyers' make-or-buy decision in the matching algorithm. One final remark about the matching algorithm concerns quota: the buyers as well as the suppliers have a maximum number of matches each can be involved in at any one time. The buyers have an offer quotum $q_o \geq 1$, and the suppliers have an acceptance quotum $q_a \geq 1$. (See [15] for full details of the matching algorithm used.)

2.1 Scores: Profitability, Trust and Loyalty

On the final goods market, products may be heterogenous or *differentiated*, meaning that the products of individual sellers are treated as being to some extent unique: they are imperfect substitutes for each other, giving sellers a degree of market power, i.e. the ability to set their price independent of their competitors and make a profit. We model this using an exogenous differentiation parameter, $0 \leq d \leq 1$. The most important of a transaction's characteristics (see Sect. 1) is the *specificity of the assets* invested in it. To the extent that assets are specific to a transaction, they can not be used for another transaction. Since a heterogenous product implies that it is different from competitors' products, we assume a 1-to-1 correspondence between the differentiation of a product and the assets required to produce it. TCE predicts that as a transaction requires more specifically invested assets, choosing the market to organize it carries less advantages, as a supplier will more and more be producing exclusively for the buyer: increasing differentiation will thus be expected to lead to more making relative to buying. Furthermore, if un-specific assets are invested in by a supplier, she may accumulate them in the production for multiple buyers, and attain economies in their increased scale. On the other hand, continous uninterrupted use of assets specific for a particular buyer will, over time, generate economies due to learning-by-doing.

These factors (differentiation, and economies of scale and of learning) determine the profit that can *potentially* be made in a transaction between a buyer and a supplier, or between a buyer and himself. The 'scores' mentioned above, on which the agents base their preference ranking of the agents on the other side of the market, are essentially calculated as the *expected* profit in a transaction, which is equal to potential profit multiplied with the probability of attaining it. This probability is each agent's subjective estimation of the partner's reliability or trustworthiness, i.e., the agent's *trust* in the partner. Following [16], we assume trust between partners to increase with longer sequences of transactions and an agent's trust in his partner to break down after the partner has broken off the sequence. Trustworthiness is then the absence of opportunism, another central concept in TCE. Because we want to also allow for the opposite of opportunism, we include loyalty in an agent's score calculation, and let the agents adaptively learn to what extent they should be loyal. Finally, we want to allow the agents to learn how they should weigh profitability vs. trust, so we change the score calculation from a simple expected value calculation to a Cobb-Douglas functional form:

$$\text{score}_{ij} = \text{potential profit}_{ij}^{\alpha_i} \cdot \text{trust}_{ij}^{1-\alpha_i} + p_{ij} \cdot \tau_i, \tag{1}$$

where $\alpha_{\min} \leq \alpha_i \leq \alpha_{\max}$ is the weight agent i assigns to profit versus trust, and $\tau_{\min} \leq \tau_i \leq \tau_{\max}$ is agent i's loyalty: if agent j is agent i's current partner, then $p_{ij} = 1$ (otherwise $p_{ij} = 0$) and agent i adds τ_i to agent j's score to express that other agents scores have to be at least τ_i higher than agent j's score in order for agent i to prefer them to agent j.

2.2 Adaptively Learning α and τ

With these specifications in place, we can simulate the system. However, the specification of the model at the level of individual agents allows us to also, and in particular, specify agents' adaptive learning, for which we employ reinforcement learning (RL) methods [12], as explained in the Sect. 1. In the RL paradigm, an agent's *policy* determines its actions given the state of the environment. In our model, the policy tells the agent, in each round of the simulation, which α- and τ-values to use—these values are the actions the agent can take. A *reward function* maps each state of the environment (or each combination of state and action, if actions are tailored to states) to a reward, indicating the desirability of that state. In our model, the reward is the profit the agent makes in a round of the simulation, depending on the agent's chosen action (α- and τ-values), and on the other agents' actions. Finally, a *Value function* tells the agent what the long run accumulated reward of α- and τ-values are.[4] Whereas rewards are immediate, they can be used to estimate the long-run Value of each action.

In our model, the agents' adaptive learning pertains to the values they use for α and τ. For both α and τ, a number of possible values is entertained by the agent. In each round of the simulation, the agents start by selecting a value to use for both α and τ, giving preference to values with high estimated Value. They use these α- and τ-values to calculate scores (see Eq. 1) and establish their preference ranking. Then, the matching algorithm assigns buyers to suppliers or to themselves. Next, all suppliers who are matched to buyers invest in assets and produce for those buyers, possibly generating economies of scale and/or learning-by-doing in the process. The price at which suppliers deliver to buyers is set in such a way, that profits are shared equally between the agents involved: the suppliers have now made their profit for this round. Buyers who are not matched to a supplier produce for themselves, after which all buyers sell their (differentiated) product on the final goods market and make their profit. Finally, both buyers and suppliers use the profit made in the current round as the reward with which to update their estimate of the Value of the particular values they used for α and τ in the current round.

[4] To avoid terminological confusion between the different values that α and τ can take on the one hand, and the long run *Value* of actions (values for α or τ) on the other hand, we call them α- and τ-values and (capitalized) Values, respectively.

3 Numerical Experiments

3.1 Experimental Setup

We simulated a number of different market settings, varying both the degree of product differentiation on the buyers' final goods market and the ratio between the number of buyers and suppliers (and their quota) on the industrial market. First we describe some of the settings which were not changed during the experiments. (Details of how economies of scale and learning-by-doing are calculated and of how trust is updated, are omitted because of space restrictions, but can be found in an early predecessor of the current paper [15].)

Each experimental run lasted for 500 timesteps, and was replicated 50 times. Results are presented as averages across runs, buyers, suppliers, etc. as indicated. Each of the agents (both buyers and suppliers) was given a total of 11 evenly distributed α- and τ-values between 0 and 1 inclusive (.0, .1, .2, ..., 1.0). The Value estimates of each of these were initialized 'realistically,' i.e. at 0 (see [12, p. 39–41]). Subsequently, Values were estimated as the 'sample average' of previous rewards, and action selection was done using an ϵ-greedy method with $\epsilon = .1$, meaning that the greedy action (the one with the highest current estimated Value) was selected, but with a 10% probability a random action was selected.

By choosing these settings, we have selected a fairly standard implementation of the RL paradigm. We would expect minor sensitivity of our results to changes in this specification. The major challenge, of course, would be to implement a psychologically more plausible 'theory of mind' to fill the agents' black box decision making modules, but this is left for future work.

3.2 Results

We ran all experiments for 2 different values of product differentiation, 0.2 and 0.7. Given the specification of the model, these represent 2 extreme sets of circumstances. In the first set of experiments, the suppliers are collectively unable to meet all buyers' demand for intermediate goods (24 buyers with $q_o = 1$ and 6 suppliers with $q_a = 2$), and in the second set, the situation is reversed (6 buyers with $q_o = 1$ and 12 suppliers with $q_a = 2$). The resulting proportion of intermediate goods that the buyers make rather than buy is shown in Fig. 2, where the top 2 graphs are for experiment 1 and the bottom 2 are for experiment 2. In both cases, the lower graph is for $d = 0.2$ and the upper graph is for $d = 0.7$: if differentiation is low then the largest proportion is bought, since the low level of asset specificity means that suppliers can generate economies of scale immediately. In addition, in the first

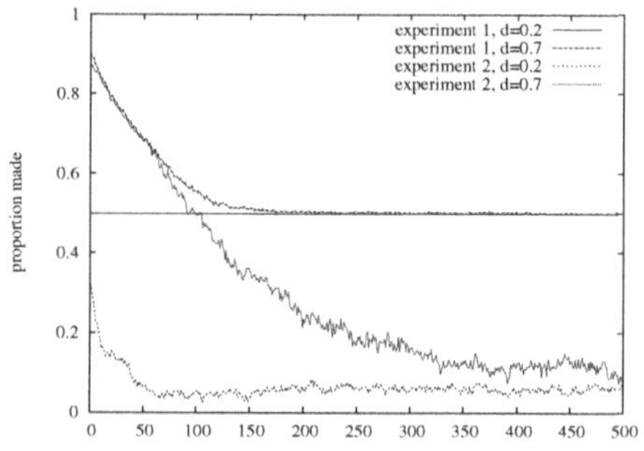

Fig. 2. Emergence of distribution of activities across firms and markets

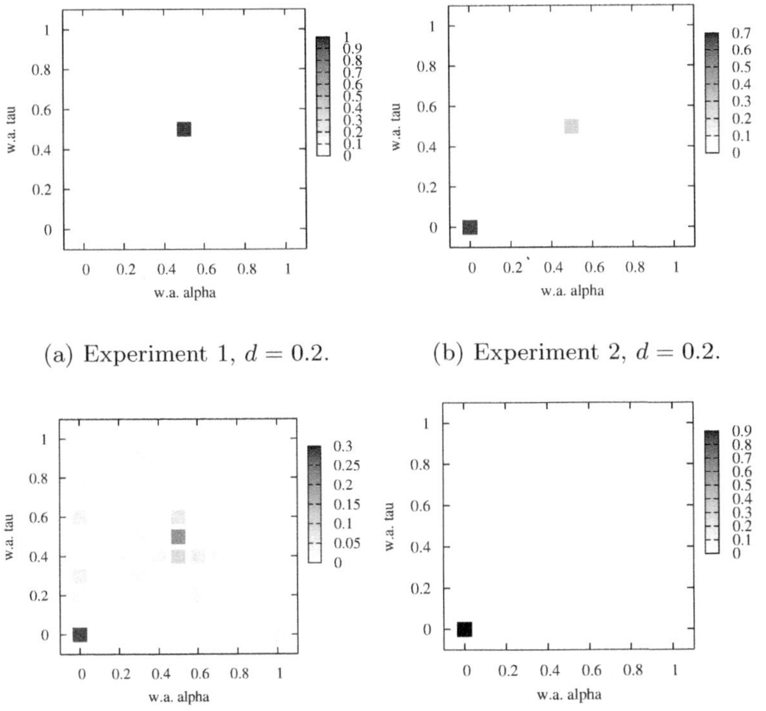

(a) Experiment 1, $d = 0.2$.

(b) Experiment 2, $d = 0.2$.

(c) Experiment 1, $d = 0.7$.

(d) Experiment 2, $d = 0.7$.

Fig. 3. Weighted average α vs. weighted average τ for 1 supplier

experiment the results are bounded by the restricted capacity of the suppliers, which is removed in the second experiment. In both cases, it takes longer for the buyers to learn that it is still profitable to buy when differentiation is high, since in that case, increased profits are due to economies-of-learning, which take a while to build up, and trust also needs to grow first.

Fig. 3 shows combinations of weighted averages for α and τ for one of the suppliers (calculated over all possible values and their Values); the first experiment is on the left and the second on the right. These plots indicate which α- and τ-values have the highest Value for the agent: which combinations this supplier adaptively learns to use. In parcticular, the plots show the proportion of all 50 runs in which this particular supplier's combination of weighted averages for α and τ ended up in each of the 11×11 possible categories (note the different scales in the graphs). Overall, there is a tendency for the supplier to use lower values for α as well as τ: they learn to focus on their trust in a partner, rather than their partner's profit generating potential, but also to be quick to switch to higher trusted partners. The effect is especially strong in the second experiment, where competition among suppliers is high and a focus on trust makes the supplier stick with his current buyer once she has one.

4 Conclusion

We have argued for the application of reinforcement learning techniques when building Agent-based Computational Economics models. As an example, we presented a model of interacting adaptive agents on an industrial market, and illustrated its usefulness through computational experiments. The distribution of economic activity across different organizational forms emerges from the decentralized interactions among autnomous, boundedly rational adaptive agents, in a way that confirms the predictions of Transaction Cost Economics. Furthermore, we have been able, in a very natural way, to incorporate more realistic behavior at the agent level, including adaptive learning and trust. With our model securely in place then, many more experiments are possible, extending well beyond the reach of standard (transaction cost) economic theory.

References

1. Holland, J.H.: Complex adaptive systems. Daedalus **121** (1992) 17–30
2. Holland, J.H., Miller, J.H.: Artificial adaptive agents in economic theory. Am. Ec. Rev. **81** (1991) 365–370
3. Tesfatsion, L.S.: Introduction to the special issue on Agent-based Computational Economics. J. Ec. Dyn. & Control **25** (2001) 281–293
4. Epstein, J.M., Axtell, R.L.: Growing Artificial Societies: Social Science from the Bottom Up. Brookings Institution Press/MIT Press (1996)
5. Coase, R.H.: The nature of the firm. Economica NS **4** (1937) 386–405
6. Williamson, O.E.: The Economic Institutions of Capitalism. Free Press (1985)
7. Nooteboom, B.: Towards a dynamic theory of transactions. J. Evol. Ec. **2** (1992) 281–299
8. Nooteboom, B., Klos, T.B., Jorna, R.J.: Adaptive trust and co-operation: An agent-based simulation approach. In: Trust in Cyber-Societies. Volume 2246 of Lecture Notes in AI. Springer (2001) 83–109
9. Coase, R.H.: The new institutional economics. Am. Ec. Rev. **88** (1998) 72–74
10. Holland, J.H.: Adaptation in Natural and Artificial Systems. MIT Press (1992)
11. Klos, T.B.: Decentralized interaction and co-adaptation in the repeated prisoner's dilemma. Computational & Mathematical Organization Theory **5** (1999) 147–165
12. Sutton, R.S., Barto, A.G.: Reinforcement Learning. MIT Press (1998)
13. Tesfatsion, L.S.: Structure, behavior, and market power in an evolutionary labor market with adaptive search. J. Ec. Dyn. & Control **25** (2001) 419–457
14. Gale, D., Shapley, L.S.: College admissions and the stability of marriage. Am. Math. Monthly **69** (1962) 9–15
15. Klos, T.B., Nooteboom, B.: Agent-based computational transaction cost economics. J. Ec. Dyn. & Control **25** (2001) 503–526
16. Gulati, R.: Does familiarity breed trust? The implications of repeated ties for contractual choice in alliances. Acad. Management J. **38** (1995) 85–112

Genetically Optimized Artificial Neural Network for Financial Time Series Data Mining

Serge Hayward

Department of Finance, Ecole Supérieure de Commerce de Dijon, France
shayward@escdijon.com

Abstract. This paper examines stock prices forecasting and trading strategies' development with means of computational intelligence (CI), addressing the issue of an artificial neural network (ANN) topology dependency.

Simulations reveal optimal network settings. Optimality of discovered ANN topologies' is explained through their links with the ARMA processes, thus presenting identified structures as nonlinear generalizations of such processes. Optimal settings examination demonstrates the weak relationships between statistical and economic criteria.

The research demonstrates that fine-tuning ANN settings is an important stage in the computational model set-up for results' improvement and mechanism understanding. Genetic algorithm (GA) is proposed to be used for model discovery, making technical decisions less arbitrary and adding additional explanatory power to the analysis of economic systems with CI.

The paper is a step towards the econometric foundation of CI in finance. The choice of evaluation criteria combining statistical and economic qualities is viewed as essential for an adequate analysis of economic systems.

Keywords: Artificial Neural Network; Genetic Algorithm; Summary Statistics; Economic Profitability; Stock Trading Strategies.

1 Introduction

Computational or information based complexity is considered as intrinsic difficulty of obtaining approximate solutions to problems due to information being noisy, costly or partial. Given the characteristics of the existing data, we also aim to examine whether there exists optimal complexity for a model necessary to learn the underlying behavior.

A significant part of financial research deals with identifying relationships between observed variables. Conventional financial modeling goes through deciding upon a mechanism and searching for parameters that give the best fit between observations and the model. Econometrics is supposed to direct the choice for the model's functional form. At the same time, the density assumption rests as a controversial and problematic question. CI provides a general data mining structure, particularly suitable for complex non-liner relationships in financial data, without the need to make assumptions about the data generating mechanism and beliefs formation. However, CI tools are often viewed as the 'black-box' structures. Unlike the well-established statistical foundation of econometrics, a search for the foundation of CI

T.-D. Wang et al. (Eds.): SEAL 2006, LNCS 4247, pp. 703 – 717, 2006.

tools in finance is in early stages. This paper is motivated by a search for the evolutionary artificial neural network (EANN) settings, founded statistically and in economic terms, for financial applications.

2 Methodology

For our experiment we build ANN forecasts and generate a posterior optimal rule. The rule, using future information to determine the best current trading action, returns a buy/sell signal (B/S) today if prices tomorrow have increased/decreased. A posterior optimal rule signal (PORS) is then modeled with ANN forecasts, generating a trading B/S signal. Combining a trading signal with a strategy warrants a position to be taken. We consider a number of market timing strategies, appropriate for different strengths of the B/S signal. If we have a buy (sell) signal on the basis of prices expected to increase (decrease) then we enter a long (short) position. Note that our approach is different from standard B/S signal generation by a technical trading rule. In the latter it is only a signal from a technical trading rule that establishes that prices are expected to increase/decrease. In our model we collaborate the signal's expectations of price change (given by PORS) with a time-series forecast.

To apply our methodology we develop the dual network structure, presented in Figure 1. The forecasting network feeds into the action network, from which information set includes the output of the first network and PORS, as well as the inputs used for forecasting, in order to relate the forecast to the data upon which it was based.

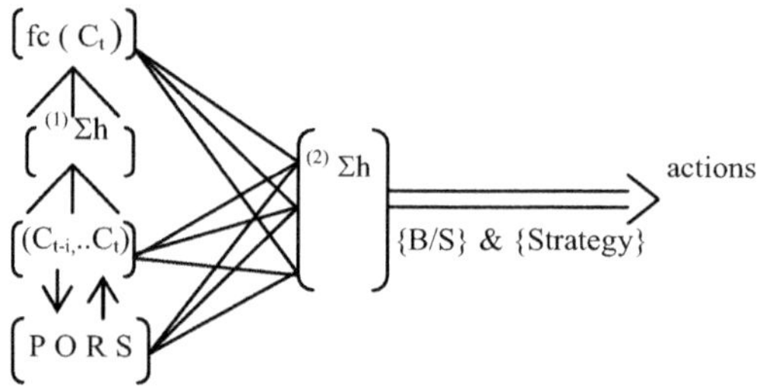

Fig. 1. Dual ANN: (1) forecasting network; (2) acting network

This structure is an effort to relate actions' profitability to forecasting quality, examining this relationship in computational settings. The model is evolutionary in the sense it considers a population of networks (individual agents facing identical problems/instances) that generate different solutions, which are assessed and selected on the basis of their fitness. Backpropagation is used in the forecasting net to learn to approximate the unknown conditional expectation function (without the need to make assumptions about data generating mechanism and beliefs formation). It is also employed in the action net to learn the relationship between forecasts' statistical and actions' economic characteristics. Lastly, agents discover their optimal models with GA; applying it for ANN model discovery makes technical decisions less arbitrary. The structure seems to be intuitive and simple to generate results independent from a chosen architecture. The results produced are sufficiently general, being stable for multiple independent runs with different random seeds for a dual forecasting/action net and a single forecasting net.

2.1 Generating Posterior Optimal Rule Signal

PORS is a function of a trading strategy adopted and based on the amount of minimum profit and the number of samples into the future. Stepping forward one sample at a time, the potential profit is examined. If the profit expected is enough to clear the minimum profit after transaction costs (TC), a PORS is generated. The direction of PORS is governed by the direction of the price movement. Normally, the strength of the signal reflects the size of underlying price changes, although, we also examine signals without this correlation to identify when profit generating conditions begin. Lastly, we consider PORS generated only at the points of highest profit to establish the maximum profit available. Since this type of signal is the most vulnerable to everyday noise in financial data, we add some random noise (up to 10%) to help ANN prediction to overcome just dealing with a constant value of zero.

3 Description of the Environment

Let Y be a random variable defined on a probability space $(\Omega, \mathcal{F}, P)$. Ω is a space of outcomes, $\mathcal{F}$ is a σ-field and P is a probability measure. For a space $(\Omega, \mathcal{F}, P)$ a conditional probability $P[A|\mathcal{F}]$ for a set A, defined with respect to a σ-field $\mathcal{F}$, is the conditional probability of the set A, being evaluated in light of the information available in the σ-field $\mathcal{F}$. Suppose economic agents' utility functions given by a general form:

$$U(\mathbf{W}_{t+s}) = g(Y_{t+s}, \delta(fc_{t+s})) \tag{1}$$

According to (1), agents' utility depends on: a target variable Y_{t+s}; a decision/strategy variable, $\delta(fc_{t+s})$, which is a function of the forecast, fc_{t+s}, where $s \geq 1$ is a forecasting horizon. Setting the horizon equal to 1, we examine the next period forecast (when this simplification does not undermine the results for $s \geq 1$). A reward variable W_{t+s} is sufficiently general to consider different types of economic agents and includes wealth, reputation, etc. $w_{t+1}(y_{t+1}, fc_{t+1})$ is the response function, stating that at time $t+1$ an agent's reward w_{t+1} depends on the realization of the target variable y_{t+1} and on the accuracy of the target's forecast, fc_{t+1}. Forecasting is regarded as a major factor of a decision rule, being close to the reality in financial markets. Also, it has a developed statistical foundation in econometrics allowing its application in evolutionary computation.

Let $fct_{+1} = \theta' X_t$ to be a forecast of Y_{t+1} conditional on the information set $\mathcal{F}_t$, where unknown m-vector of parameters, $\theta \chi \Theta$, with Θ to be compact in $\mathbb{R}^k$ and observable at time t n-vector of variables, X_t. X_t are $\mathcal{F}_t$-measurable and might include some exogenous variables, indicators, lags of Y_t, etc. An optimal forecast does not exclude model misspecification, which can be due to the form of fc_{t+1} or failure to include all relevant information in X_t. With imperfect foresight, the response function and, therefore, the utility function are negatively correlated with forecast error, $e_{t+1} \equiv y_{t+1} - fc_{t+1}$; $|e_{t+1}| > 0$. A mapping of the forecast into a strategy rule, $\delta(fc_{t+1})$ (combined with elements of X_t) determines a predictive density g_y, which establishes agents' actions.

In this setting, maximizing expected utility requires us to find an optimal forecast, fc_{t+1} and to establish an optimal decision rule, $\delta(fc_{t+1})$. Note that optimality is with respect to a particular utility function, implemented through a loss function, in the sense that there is no loss for a correct decision and a positive loss for incorrect one. Given a utility function, expected utility maximization requires minimization of the expected value of a loss function, representing the relationship between the size of the forecast error and the economic loss incurred because of that error. A strategy development (mapping of the forecast into a decision rule) is another way to minimize the expected value of a loss function.

A loss function, L: $\mathbb{R}\delta\ \mathbb{R}^+$, related to some economic criteria or a statistical measure of accuracy, takes a general form:

$$L(p,\ a,\ e) \equiv [a + (1 - 2a)1(e < 0)]e^p\ , \tag{2}$$

where p is a coefficient of risk aversion; e is the forecast error; $\alpha \chi [0,1]$ is the degree of asymmetry in the forecaster's loss function. $L(p, \alpha, e)$ is $\mathcal{F}_t$-measurable. It could also be presented as:

$$L(p,\ a,\ \theta) \equiv [a + (1 - 2a)1(Y_{t+1} - fc_{t+1}(\theta) < 0)]|Y_{t+1} - fc_{t+1}(\theta)|^p\ , \tag{3}$$

where α and p are shape parameters and a vector of unknown parameters, $\theta \chi \Theta$. For given values of p and α an agent's optimal one-period forecast is

$$\min_{\theta \in \Theta} E[L(\rho,\ \alpha,\ \theta)]\ =\ E[L(Y_{t+1} - fc_{t+1})] = \ E[L(e_{t+1})]\,. \tag{4}$$

Training EANN with different settings allows us to examine how models' statistical and economic performances relate to their topology choices.

4 Experimental Designs

We use ANN with GA optimization for the building/evolution of price forecast and trading strategy development/evolution upon relevant forecast. The mechanism appears to be an intuitive way to deal with agents' cognitive limits in forecasting and optimization, modeling the traders' learning process to approximate the unknown conditional expectation function. It also provides a natural procedure to consider decisions' heterogeneity by agents viewing similar information. GA facilitates an optimal choice of network settings and adds additional explanatory power to the analysis.

4.1 Learning Paradigm

To learn a mapping $\mathbb{R}^d \delta \mathbb{R}$ an input/output training set $\mathcal{D}_I = \{x_i, y_i\}_{i=1}^{I}$ is presented to the network. x_i χ $\mathbb{R}^d$ is assumed to be drawn from continuous probability measure with compact support. Learning entails selecting a learning system $\mathcal{L}$ = {$\mathcal{H}$, $\mathcal{A}$}, where the set $\mathcal{H}$ is the learning model and *A* is a learning algorithm. From a collection of candidate functions, $\mathcal{H}$ (assumed to be continuous) a hypothesis function *h* is chosen by a learning algorithm $\mathcal{A}$: $\mathcal{D}_I$ δ $\mathcal{H}$ on the basis of a performance criterion.

Learning law is a systematic way of changing the network parameters (weights) in an automated fashion, such that the loss function is minimized. One of the most common algorithms used in supervised learning is backpropagation. Being simple and computationally efficient, the search here, nevertheless, can get caught in local minima. Backpropagation is also often criticized for being noisy and slow to converge. To improve the original gradient learning, particularly its slowness of convergence, we examine a number of alternatives.

Consider the vector, Ψ as the weight space, we are searching for. The gradient descent is given by $\nabla L = \frac{\partial L}{\partial \psi}$. Expanding the loss function *L* about the current point ψ_o gives:

$$L(\psi) = L_0 + (\psi - \psi_0) \cdot \nabla L(\psi_0) + \tfrac{1}{2}(\psi - \psi_0) \cdot H \cdot (\psi - \psi_0) + \ldots, \quad (5)$$

where *H* is the second derivative Hessian matrix evaluated at ψ_o, $H_{ij} = \frac{\partial^2 L}{\partial \psi_i \partial \psi_j}$. The gradient is obtained by differentiating (5):

$$\nabla L(\psi) = \nabla L(\psi_0) + H \cdot (\psi - \psi_0) + \ldots \quad (6)$$

For the optimization task the minimum $L(\psi)$, where $=L(\psi)=0$ need to be located. A common approach here is to set (6) to zero, disregarding the higher-order terms:

$$\nabla L(\psi) = \nabla L(\psi_0) + H \cdot (\psi - \psi_0) = 0. \quad (7)$$

Solving (7) for ψ gives:

$$\psi = \psi_0 - H^{-1} \nabla L(\psi_0). \quad (8)$$

A popular minimization technique is to use the first derivative information (only) with line searches along selected directions. If *D* is a direction, starting from ψ_o, staying on the line $\psi = \psi_o + \alpha D$, α is chosen to minimize $L(\psi)$.

In the Steepest Descent Method one chose $D = - =L(\psi_o)$, repeating minimization along a line in the gradient direction and re-evaluating the gradient. Since all successive steps are perpendicular, the new gradient descent $=L^{new}$ is also perpendicular to the old direction D^{old}, giving zigzagging path after the line minimization,

$$0 = \frac{\partial}{\partial \alpha} L(\psi_0 + \alpha D^{old}) = D^{old} \cdot \nabla L^{new} . \tag{9}$$

The step size, η determines how far the movement should go before obtaining another directional estimate. For one step ($\sum_{n=1}^{N}$) the weight update with a step size, η is given:

$$\Delta\psi_i(n+1) = \eta_i \nabla\psi_i . \tag{10}$$

With small steps it takes longer to reach the minimum, increasing the probability of getting caught in local minima. On the other hand, large steps may result in overshooting, causing the system to rattle/diverge. Starting with a large step size and decreasing it until the network becomes stable, one finds a value that solves the problem in fewer iterations. We utilize small step to fine-tune the convergence in the later stages of training.

The momentum provides the gradient descent with some inertia, so that it tends to move along the average estimate direction. The amount of inertia (the amount of the past to average over) is given by the parameter, μ. For a given momentum μ and the step size η, the weight update is defined as:

$$\Delta\psi_i(n+1) = \eta_i \nabla\psi_i + \mu\Delta\psi_i(n) \tag{11}$$

The higher the momentum, the more it smoothes the gradient estimate and the less effect a single change in the gradient has on the weight change. It also helps to escape local minima, although oscillations may occur at the extreme.

A second order method, the Conjugate Gradient uses the second derivatives of the performance surface to determine the weight update, unlike the steepest descent algorithm where only the local approximation of the slope of the performance surface is used to find the best direction for the weights' movement. At each step a new conjugate direction is determined and movement goes along this direction to the minimum error. The new search direction includes the gradient direction and the previous search direction:

$$D^{new} = -\nabla L^{new} + \beta D^{old} , \tag{12}$$

where β is the choice parameter, determining the amount of past direction to mix with the gradient to form a new direction.

The new search direction should not change (to first order) the component of the gradient along the old direction. If α is a line search parameter, then:

$$D^{old} \cdot \nabla L(\psi_0 + \alpha D^{new}) = 0 . \tag{13}$$

Therefore, the vectors D^{new} and D^{old} are conjugate in the following expression:

$$D^{old} \cdot H \cdot D^{new} = 0 . \tag{14}$$

β in (12) is chosen such that the new search direction maintains as best as possible the minimization that was achieved in the previous step, for example with the Polak-Ribiere rule:

$$\beta = \frac{(\nabla L^{new} - \nabla L^{old}) \cdot \nabla L^{new}}{(\nabla L^{old})^2} \tag{15}$$

For the quadratic performance surface with information from the Hessian one can determine the exact position of the minimum along each direction, but for non-quadratic surfaces, a line search is often used. In theory, there are only N conjugate directions in a space of N dimensions, thus the algorithm is reset each N iterations. The advantage of conjugate gradient method is that there is no need to store, compute and invert the Hessian matrix. Updating the weights in a direction that is conjugate to all past movements in the gradient, the zigzagging of first order gradient descent methods could be avoided.

The Scaled Conjugate Gradient method without real parameters is based on computing the Hessian times a vector, $H*\Psi$. An offset is added to the Hessian, $H+\delta I$ to ensure that the Hessian is a positive definite, so that the denominator in the expression below is always positive. For the step size α it could be expressed in the following way:

$$\alpha = -\frac{C^T G}{C^T (H + \delta I) C + \delta |C|^2}, \tag{16}$$

where C is the direction vector and G the gradient vector. The parameter δ is set such that for low values the learning rate is large and it is small for high values. δ adjusted in a way that if the performance surface is far from quadratic, δ is increased, resulting in smaller step size. To determine the closeness to quadratic performance surface, Λ is used and is given by:

$$\Lambda = \frac{2(L(\psi) - L(\psi + \alpha C))}{\alpha C^T G}. \tag{17}$$

For example for $\Lambda>0.75$ (very quadratic) δ is multiplied by 5; for $\Lambda<0.25$, δ is multiplied by 4; for $\Lambda<0$, there is no change in weights. By a first order approximation:

$$(H + \delta I) C \approx \frac{L'(\psi + \sigma C) - L'(\psi)}{\sigma} + \delta C. \tag{18}$$

(18) implies that the Hessian calculations could be replaced with additional estimation of the gradients.

Delta-Bar-Delta is an adaptive step-size procedure for searching a performance surface. The step size and momentum are adapted according to the previous values of the error. If the current and past weight updates are both of the same sign, the learning rate increases linearly. Different signs for the updates indicate that the weight has been moved too far and the learning rate decreases geometrically to avoid divergence. Therefore, the step size update is given:

$$\Delta\eta_i(n) = \begin{cases} k & S_i(n-1)\nabla\psi_i(n) > O \\ -\beta\eta_i(n) & S_i(n-1)\nabla\psi_i(n) < O \\ 0 & otherwise \end{cases}, \tag{19}$$

with

$$S_i(n) = (1-\delta)\nabla\psi_i(n-1) + \delta S_i(n-1) , \qquad (20)$$

where k is additive constant; β is multiplicative constant and δ is smoothing factor.

Considering how the data is fired through the network, synchronization in Static, Trajectory and Fixed Point modes are examined. Static learning assumes that the output of a network is strictly a function of its present input (the network topology is static). The gradients and sensitivities are only dependent on the error and activations from the current time step. Training a network in Trajectory mode assumes that each exemplar has a temporal dimension and that there exists some desired response for the network's output over the period. The network is first run forward in time over the entire period, during which an error is determined between the network's output and the desired response. Following that the network is run backwards for a prescribed number of samples to compute the gradients and sensitivities, completing a single exemplar. Fixed Point mode assumes that each exemplar represents a static pattern; that is to be embedded as a fixed point of a recurrent network. Here the terms forward samples and backward samples can be thought of as the forward relaxation period and backward relaxation period, respectively. All inputs are held constant while the network is repeatedly fired during its forward relaxation period. There are no guarantees that the forward activity of the network will relax to a fixed point, or even relax at all. After the network has relaxed, an error is determined and held as constant input to the backpropagation layer. Similarly, the error is backpropagated through the backpropagation plane for its backward relaxation period, completing a single exemplar.

A feedforward network, where the response is obtained in one time step (an instantaneous mapper), can only be trained by fixed point learning. On the other hand, recurrent networks, can be trained either by fixed point learning or by trajectory learning. A static ANN makes decisions based on the present input only; it can not perform functions that involve knowledge about the history of the input signal. On the other hand, dynamic networks are able to process time varying signals. They posses an extended memory mechanism, which is capable to store past values of the input signal. In the time delay neural network the memory is a tap delay line, i.e. a set of memory locations that store the past of the input.

It is possible to use self-recurrent connections as memory, like in Jordan/Elman Network context units. Considering, for example, the gamma memory as a structure with local feedback, it cascades self-recurrent connections and extends the context unit with more versatile storage. It accepts the tap delay line as a special case. A form of temporal learning must be used to adapt the gamma parameter (e.g. real time recurrent learning). The advantage of this structure in dynamic networks is that it provides a controllable memory with a predefined number of taps. Furthermore, as the network adapts the gamma parameter to minimize the output error, the best combination of depth and resolution can be achieved.

4.2 ANN Topology

Proposition: Given the characteristics of the data there exists an optimal network complexity, required to learn the underlying behavior. In this experiment popular

ANN models are considered and their performances for modeling and forecasting the financial time series dynamics are examined.

Multilayer Perceptron (MLP) is the most basic of the ANN topologies for non-linearly separable problems. The data in a MLP follows a single path with no recursion or memory elements. It is viewed that for static pattern classification, the MLP with two hidden layers is a universal pattern classifier. The discriminant functions can take any shape, as required by the input data clusters. In terms of mapping abilities, the MLP with a (non-polynomial) Tauber-Wiener transfer function is believed to be a universal approximator.

We use three layers MLP for prediction and strategy development. For instance, a one-period price forecast takes the following form:

$$Fc(C_{t+1}) = h_2(\psi_0 + \sum_{j=1}^{J} \psi_j h_1(\psi_{0j} + \sum_{i=0}^{I} \psi_{i,j} C_{t-i})) + e_t \; . \tag{21}$$

In (21) the input layer has I inputs, $\{c_{t-0},\ldots, c_{t-I}\}$; the hidden layer has J hidden nodes and the output layer has one output, $Fc(C_{t+1})$. Layers are fully connected by weights, $\psi_{i,j}$; ψ_o and ψ_{oj} are biases. Transfer functions are represented by h_1 and h_2. We run experiments under two transfer functions, the hyperbolic tangent, $h_s(x) = \dfrac{2}{1+e^{-2x}} - 1$, with $-1<h_s(x)<+1$ and the sigmoid, $h_t(x) = \dfrac{1}{1+e^{-x}}$, with $0<h_t(x)<+1$.

Jordan and Elman Networks (J/E) networks based on the concept of context in their processing. A set of context units is a layer (or a part) that receives feedback signals. Unlike the forward propagation the feedback signal occurs with reference to time. A context for processing at time t comes from the network state at time $t-1$ through the context units. Therefore, the state of the network at any time depends on an aggregate of previous states and the current input. It has been claimed that this type of ANN capable not only to recognize sequences, but also generate them in some cases.

Jordan architecture [1, 2] differs from Elman architecture [3], primarily, by having the context units fed from the output layer and from themselves instead of the hidden layer. For the source of the feedback to the context units we consider four options: the input, the 1st hidden layer, the 2nd hidden layer and the output. In linear systems the use of the past of the input signal generates the moving average (MA) representation. They are particularly suited for signals that have a spectrum with sharp valleys and broad peaks. The use of the past of the output results in the autoregressive (AR) representations. They are supposed to model well signals with broad valleys and sharp spectral peaks. In the case of non-linear systems, these two topologies are considered as non-linear MA and AR (NMA and NAR). The Jordan net is viewed as a restricted case of NAR, while the configuration with context units fed by the input layer considered as a restricted case of NMA. Elman's net does not have a counterpart in linear systems.

Different values of the context unit's time-constant are considered in the experiment. It is expected to find a trade-off between extending the memory further back into the past and loosing sensitivity to details. As a rule of thumb, the value of the time-constant should produce an exponential decay rate that matches the characteristic time scale of the input sequence. Since we only control the time-constant (i.e. the exponential decay) the weighting over time is inflexible. Furthermore, a small change in the context unit's time-constant is reflected in a large change in the weighting (due to the exponential relationship between time-constant and amplitude).

Time-Lag Recurrent Network (TLRN) is viewed as MLP's extension with short term memory structures that have local recurrent connections. It has smaller network size required to learn temporal problems when compared to MLP that use extra inputs to represent the past samples. On the other hand the backpropagation through time adopted with TLRN requires a lot of memory. TLRN is characterized by low sensitivity to noise. The recurrence of the TLRN provides the advantage of an adaptive memory depth (it finds the best duration to represent the input signal's past). In our experiment the following memory structures are considered: Time Delay Neural Network Memory (TDNN); Gamma memory (GM) and Laguarre memory (LM). With Focused topology only the past of the input is remembered.

It is noted that using a TLRN with Focused TDNN memory has a similar effect to using multiple samples for the inputs to a basic MLP. The primary difference between the two methods is that, focused TDNN memory only allows for one memory depth to be used for all of the inputs, whereas the lag input settings allow us to specify various memory depths.

Recurrent Network (RN) delays one or more of the processing values in the network so that they will be used in the calculation of the next output, rather than the current output. These are often combined with the memory elements found in TLRN. Fully RN does not include a non-recurrent feedforward processing path. All data flows through the recurrent processing. On the other hand partially RN includes a non-recurrent feedforward processing path. RN contains multiple processing paths. Each processing path has potentials of specializing on a different aspect of the incoming data, allowing to consider multiple conditions.

The support vector machine (SVM) is considered as a classifier capable to transform complex decision surfaces into simpler ones to apply linear discriminant functions. It uses only inputs that are near the decision surface as they provide the most information about the classification[1].

We consider that ANN architecture is an application-dependent. Maximum information available about the problem needs to be built into the network to achieve good learning abilities (accuracy on training and predictability on testing data). At the same time for good generalization abilities parsimonious structure is required (possibly with a complexity penalty). We construct and modify architectures incrementally identifying optimal settings for modeling financial data. Furthermore, GA is used to search for optimal settings.

4.3 Performance Surface

The performance of ANN learning is monitored by observing how the cost changes over training iterations. The learning curve presents the internal error over each epoch of training, comparing the output of the ANN to the desired output. In price

[1] The support vector machine loss function is used to implement the large margin classifier segment of the SVM model and takes the following form: $L_{SVM} = \Gamma^{1}\, \Sigma(Y_i - tanh\Phi_I)^2$.

forecasting, the target is the next day closing price, where in signal modeling, the target is the current strategy. Achieving an accurate representation of the mapping between the input and the target might not necessarily lead to a forecast to be exploitable or a strategy using that forecast to be profitable.

We consider that evaluation criteria should measure not so much absolute effectiveness of the model with respect to the environment but rather its relative effectiveness with respect to other models. Although we train ANN with the goal to minimize internal error function, we test and optimize its generalization ability by comparing its performance with the results of a benchmark, an efficient prediction (EP). In forecasting prices, EP is the last known value. For predicting strategies, it is the buy/hold (B/H) strategy. The degree of improvement over efficient prediction (IEP) is calculated as an error from a de-normalized value of the ANN and a desired output, then normalizing the result with the difference between the target and EP value.

4.4 Profitability as Performance Measure

To make the final goal meaningful in economic terms we use profitability as a measure of overall success. We examine the following forms of cumulative and individual trades return measures: non-realized simple aggregate return (r); profit/loss factor; average, maximum gain/loss. In addition we estimate exit efficiency, measuring whether trades may have been held too long, relative to the maximum amount of profit to be made, as well as the frequency and the length of trades, including out of market position. To assess risk exposure we adopt common 'primitive' statistics, the Sharpe ratio[2] (SR) and the maximum drawdown (ξ). The latter, calculating the percentage loss relative to the initial investment for the date range, measures the size of losses occurred while achieving given gains. It demonstrates how prone a strategy is to losses. To overcome the Fisher effect we consider trading positions with a one-day delay.

TC is assumed to be paid both when entering and exiting the market, as a percentage of the trade value. TC accounts for broker's fees, taxes, liquidity cost (bid-ask spread), as well as costs of collecting/analysis of information and opportunity costs. According to [4] large institutional investors achieve one-way TC about 0.1-0.2%. Often TC in this range is used in computational models. Since TC (defined above) would differ for heterogeneous agents, we report the break-even TC that offsets trading revenue with costs leading to zero profits.

Thus, in this paper profitability is a function of return, risk and transaction costs. The classification of the ANN output as different types of B/S signals determines the capability of the model to detect the key turning points of price movement. Evaluating the mapping of a forecast into a strategy, $\delta(fc_{t+1})$, assesses the success in establishing a predictive density, g_y that determines agents' actions.

4.5 Time Horizons and Trading Strategies Styles

Heterogeneous traders in the experiment use different lengths of past and forward time horizons to build their forecasts/strategies. We have run the experiment on stock indexes from a number of markets and found that 'optimal' length of

[2] Given by the average return divided by the standard deviation of that return.

training/validation period is a function of specific market conditions. In this paper we adopt three memory time horizons, [6; 5; 2½] years. We run the experiment with a one year testing horizon, as it seems to be reasonable from the actual trading strategies perspective and supported by similar experiments.

Both long and short trades are allowed in the simulation. Investing total funds for the first trade, subsequent trades (during a year) are made by re-investing all of the money returned from the previous trades. If the account no longer has enough capital to cover TC, trading stops.

4.6 Genetic Training Optimization

In this research evolutionary computation is used for ANN model discovery, considering GA optimization for: network's topology; performance surface; learning rules; number of neurons and memory taps; weight update; step size and momentum rate. GA tests various settings from different initial conditions (in the absence of a priori knowledge and to avoid symmetry that can trap the search algorithm). Since the overall objective of financial forecasting is to make a trading decision, based on that forecast profitable, economic criteria rather than statistical qualities need to be employed for the final goal. We use GA optimization with the aim to minimize IEP value and profitability as a measure of overall success.

5 Empirical Application

5.1 Data

We consider daily closing prices for the MTMS (Moscow Times) share index obtained from Yahoo Finance. The time period under investigation is 01/01/97 to 23/01/04. There were altogether 1575 observations in the row data set. Examining the data graphically reveals that the stock prices exhibit a prominent upward, but non-linear trend, with pronounced and persistent fluctuations about it, which increase in variability as the level of the series increases. Asset prices look persistent and close to unit root or non-stationarity. Descriptive statistics confirm that the unit-root hypothesis cannot be rejected at any confidence level. The data also exhibits large and persistent price volatility with significant autocovarience even at high order lags.

Changes in prices increase in amplitude and exhibit clustering volatility. The daily return displays excess kurtosis and the null of no skewness is rejected at 5% critical level. The tests statistics lead to rejection of the Gaussian hypothesis for the distribution of the series. It confirms that high-frequency stock returns follow a leptokurtic and skewed distribution incompatible with normality assumed often in the analytical literature.

5.2 Experimental Results

ANN with GA optimization was programmed with various topologies[3]. Altogether we have generated and considered 93 forecasting and 143 trading strategies' settings.

[3] Programs in Visual C++, v. 6.0 are available upon request. We have run tests on TradingSolutons, v. 2.1, NeuroSolutions v. 4.22 and Matlab v. 6.

Effectiveness of search algorithm was examined with multiple trials for each setting. 92% of 10 individual runs produce identical results, confirming the replicability of our models. Efficiency of the search was assessed by the time it takes to find good results. The search with ANN unoptimized genetically took a few minutes, where the search with GA optimization lasted on average 120 minutes on a Pentium 4 processor.

Over a one year testing period 19 trading strategies were able to outperform in economic terms the B/H strategy, with an investment of $10,000 and a TC of 2% of trade value. The average return improvement over B/H strategy was 20%, with the first five outperforming the benchmark by 50% and the last three by 2%. The primary strategy superiority over B/H strategy was 72%.

For the five best performing strategies, the break-even TC was estimated to be 2.75%, increasing to 3.5% for the first three and nearly 5% for the primary strategy. Thus, the break-even TC for at least the primary strategy appears to be high enough to exceed actual TC. Profitability produced by our simple architecture supports computational model development based on economic and statistical foundations.

The examination of the performances of networks with different topologies has identified consistently the best results in economic terms for a one year testing period by a single hidden layer basic MLP and TLRN with Focus Laguarre memory (FLM); all with the hyperbolic tangent transfer function. The seven most profitable strategies are represented by those ANN. They also have good performances in statistical terms. Although, there was no such a clear dominance as in economic performance. Among the ten most accurate predictions nine are basic MLP and TLRN with FLM. At the same time, the best accuracy was achieved by Jordan ANN with the output feedback to the context units.

In price forecasting, among the ten most accurate networks, eight are basic MLP and TLRN with FLM, also sharing the first three positions. Among the five most accurate forecasting ANN are also Generalized Feedforward MLP, producing the accuracy that follow immediately the first three networks.

GA model discovery reveals that MLP and TLRN with FLM, with neurons number in the hidden layer in the range [5, 12] and Conjugate Gradient learning rule generate the best performance in statistical and economic terms for forecasting and acting nets. It is noticed that Conjugate Gradient weight update took twice as long comparing with the Steepest Descent method. GA optimization also establishes the batch training as optimal in most cases for static networks.

We relate satisfactory performances of MLP and TLRN in financial data mining to their established links with ARMA processes. MLP and TLRN could be considered as nonlinear generalizations of those models. Having identified that MLP and TLRN are particularly appropriate for financial time series modeling, we investigate performances of those topologies.

Table 1 presents statistical and economic characteristics of the primary (basic MLP) and secondary (TLRN-FLM) strategies models as well as the statistically best performer (JN). All three models are trained on 6 years of data. JN seems to be a very good directional model, where the primary and secondary strategies exhibit only week DA. Correlation of desired and ANN output show a right sign, but insignificant correlation for JN and a perverse sign correlation for the primary and secondary strategies. IEP shows a good improvement on a random chance for the primary and a satisfactory improvement for the secondary strategies, where the result for the most accurate strategy was less adequate (although, IEP<1 was not expected for JN). These results confirm that statistical criteria, such as correlation and DA, have only weak relationships with economic criteria. Robust links of IEP with annualized return support its fitness for a computational model's performance surface setup, as well as makes it an appropriate evaluation criterion for an economic prediction.

The complexity of the three networks, given by the number of a hidden layer neurons, shows that the profitable strategies have more parsimonious structures than JN. Inferior generalization of JN manifests itself in the model's overspecialization on the training data and poor economic performance during the testing period. Parsimonious optimality is furthermore identified by GA optimization of TLRN, where a moderate optimal length of 16 bars was found for FLM

Table 1. Primary and Secondary Strategies

Measures /Topologies	MLP	TLRN	JN
Accuracy (%)	32.38	37.62	58.10
Correlation	-0.125	-0.049	0.0317
IEP	0.852	0.929	0.956
Hidden Layer Neurons	14	10	22
Return (%)	128.1	113.1	1.15
Sharpe Ratio	0.17	0.16	0.01
Profitable Trades (%)	85.7	61.5	12.4

MLP traded seven times during the test year with overall 85.7% of profitable trades. Four long trades generated 100% wins, where short trades produced 66.7% wins. Annualized return over testing period was 128.1%, significantly exceeding the comparable return of B/H strategy (74.69%). In terms of risk measures, the primary strategy seems to be less risky than B/H benchmark. In comparison TLRN traded more frequently: 13 trades for the test period with overall 61.5% of profitable trades. 7 long trades generated 71.4% wins and 6 short trades produced 50% wins. Although trading frequency of primary and secondary strategies differ by 86%, their annualized return and riskness are close to each other, supporting the idea that an optimal trading frequency is in the range [6, 12]. This conjecture is also confirmed by PORS trading frequency; 12 annual trades are required to generate the maximum profit available. On the other hand JN model produced merely 3 trades. Spending much of the time out of the market the strategy demonstrates consistently low return and high risk exposure. Poor economic performance of JN seems to be related to its notably high complexity and the training data over-fitting. Although, producing superior statistical performance the model has failed in economic terms.

The above results demonstrate that the optimal network structure and complexity are determined by data characteristics. Networks complexity seems to be positively correlated with statistical performance. On the other hand superior economic performance is achieved by parsimonious structures with good generalization abilities. Profitable models might display inadequate characteristics, measured by conventional statistics. A good model for financial time series is considered to be the one with valid statistical foundation, capable of learning the complex dynamics of a socio-economic system and fulfilling the final objective to be viable in economic terms.

6 Conclusion

The system considered in the paper is self-organized, given economic agents' abilities to learn and adapt to changes. The models examined are robust due to agents' ability to determine their future actions (form their expectations) using memory of their previous experiences. The primary strategy generated reveals good economic performance on out of sample data. The bootstrap method, used to test the significance of the profitability and predictive ability, produced p-values, demonstrating that the performance of the models in the experiment is statistically different from a random walk with drift.

Optimal settings' examination demonstrates weak relationships between statistical and economic criteria. A good model for financial time series forecasting is considered to be the one with valid statistical foundation, capable of learning complex dynamics of a socio-economic system and fulfilling the final objective to be viable in economic terms.

This research demonstrates that fine-tuning of ANN settings is an important stage in computational model set-up. GA is particularly appropriate for model discovery, making technical decisions less arbitrary and adding additional explanatory power to the analysis of economic systems with CI.

References

[1] M. I. Jordan, "Attractor Dynamics and Parallelism in a Connectionist Sequential Machine," presented at Proceedings of the 8th annual Conference of the Cognitive Science Society, Hillsdale, 1986.

[2] M. I. Jordan, "Serial Order: A Parallel, Distributed Processing Approach," in *Advances in Connectionist Theory: Speech*, J. L. Elman and D. E. Rumelhart, Eds. Hillsdale: Erlbaum, 1989.

[3] J. L. Elman, "Finding Structure in Time," *Cognitive Science*, vol. 14, pp. 179-211, 1990.

[4] R. J. Sweeney, "Some Filter Rule Tests: Methods and Results," *Journal of Financial and Quantitative Analysis*, vol. 23, pp. 285-301, 1988.

Simulation of Cooperation for Price Competition in Oligopolies

Tzai-Der Wang[1] and Colin Fyfe[2]

[1] Department of Industrial Engineering and Management, Cheng Shiu University, Kaohsiung County, Taiwan, ROC
[2] Applied Computational Intelligence Research Unit, The University of Paisley, Scotland, UK

Abstract. In this research, an agent-based simulation model for price competition in oligopolies is built and Genetic Algorithm is used to evolve the oligopolies' decisions of price while facing the competitors in markets. The experimental results show two factors influencing the price competition situations and 'given' factor that competitor can not control leads strong influence on their decision of price. Total cooperation (Collusion to high prices) seems not to be achieved under the different parameter settings while many competitors involving in the market and a limitation of cooperation forms in which no effect forces the achievement of total cooperation.

1 Introduction

In economics, researchers use game theory to explain competition situations. Unlike other forms of market structure such as perfect competition and monopolies, oligopolies are notoriously unpredictable. Possible outcomes of oligopolistic competition include implicit collusion through tacit recognition of standard industry practices, explicit collusion through the formation of a cartel, and cyclical periods of intense price competition followed by price stability. One way of investigating competition among the few that is used by some researchers is in terms of evolutionary methods, e.g., GAs (Genetic Algorithm) [5], based on the iterated prisoner's dilemma (IPD) [2]. In this approach, the behavior of the firms that participate in the oligopoly is determined by the evolutionarily fittest strategy - i.e. the strategy adapted to the competitive environment.

The IPD or NIPD (N-persons' IPD) [4], [6] are ideal games to describe the deci-sions made when humans meet a dilemma. However, economists have developed models of economic events based on the IPD, but these have not proved to be successful. Therefore, some researchers suggested building other agent-based models to fit the complex features of economic competition [1]. Chen and Ni [3] built a 3-player model using economic concepts, i.e., profits, costs, market shares, etc., to simulate a real competitive market. This model, which will be described as the oligopoly game model, can be used as a base to build N-player models in order to describe not only oligopoly competitions but also others. In the oligopoly game model, cooperators will act together to increase their prices

T.-D. Wang et al. (Eds.): SEAL 2006, LNCS 4247, pp. 718–725, 2006.

while defectors will try to gain customers by decreasing their prices. However, realistically both will be trying to maximize their profits. It is assumed that the agents, in this case the firms involved in competing, are rational, albeit with limited information processing ability. There are a number of objective functions which they could maximize, such as sales or market share, but we will make the standard assumption which is made in economics, that the participant firms are attempting to maximize profit.

In this experiment, we introduce Chen and Ni's oligopoly game model and use to analyze the importance for competitive cooperation of environmentally 'given' factors (such as the number of competitors or market growth), in contrast with those factors which are under the firm's control (i.e. price, cost and profit margin).

2 Competition Game Model

The competition game model is based on the 3-players' oligopoly game model which was established by Chen and Ni [3]. We accept the hypothesis that Chen assumed for the competition environments and extend the player numbers in the game to investigate whether the oligopoly competition might shade into perfect competition.

Different to Chen and Ni's model, we model the change in customer loyalty directly at the current round and this affects company profits immediately since customers are actually very liable not to buy at the high price initially. However, we still model that customers enter and leave the market at the next round as they feel price changes prevail in the market.

Each company has the choice to set a high price (cooperating) or a low price (defecting). If it defects, it will win market share from the other (high price setting) companies though with a reduced profit margin. The profit of one player is,

$$Profit = (Price - Cost) * Customer\ number \tag{1}$$

A proportion of customers will move to other companies with a lower price when one company chooses a high price. Potential customers will move into this market at the next round attracted by low prices in the market and, on the other hand, the current customers also may leave the market at next time because the current prices are high. The factors we use for this game model are,

1. High price (P_h): The price when the company cooperates with others.
2. Low price (P_l): The price when the company defects against others.
3. Cost (C): The cost of one goods or service.
4. Market share loss (α): The percentage of customers that move away when a company chooses a high price.
5. External customer influence (β): A function to modify the customers attracted by this market or leaving this market by the price changes of these companies.

Therefore, the profit of one unit sale for one cooperating company owning one customer is $P_h - C$ and for a defecting one is $P_l - C$. This means the subjective factors are the prices that the company can choose and the profit of single sale is controlled by the company internally. However, the market share loss and external customer influence will be determined by the environment that the company cannot manage directly. When the environment changes, i.e., the market share loss (α) or external customer influence changes (β); the company should adopt strategies to meet this environment situation to maintain its total profit in the market. We call these two parameters, α and β, objective factors (or 'givan' factors).

The matrix to show the customers' movement for the next phase of the competition for the N-companies market is;

$$M_t = \begin{bmatrix} m_{11}^t m_{21}^t m_{31}^t \ldots\ldots m_{N1}^t \\ m_{12}^t m_{22}^t \ldots\ldots\ldots\ldots m_{N2}^t \\ m_{13}^t m_{23}^t \ldots\ldots\ldots\ldots m_{N3}^t \\ \vdots \\ m_{1N}^t \ldots\ldots\ldots\ldots\ldots m_{NN}^t \end{bmatrix} \tag{2}$$

Where m_{ij}^t denotes the proportion of the customers of firm i which will move to firm j in the period t. Let ${n^t}_i$ be the number of customers of firm i at time t, and N_t be the vector $[n_1^t, n_2^t, \ldots . n_N^t]^T$. The customer numbers at the period $t+1$ are as,

$$N_{t+1} = M_{t+1} N_t \tag{3}$$

When there are some defectors among the cooperators, the movement of customers becomes complicated. Let there be m companies cooperating in this N-companies' market, i.e., there are m cooperators and $(N - m)$ defectors and $N > m > 0$. According to the definition of market share loss, when a company chooses cooperation, it will lose a fixed proportion α of customers to other defecting competitors no matter how many defectors unless there none and the defecting companies will share the total customers flowing away from the cooperating companies. The cooperator will maintain $(1 - \alpha)$ customers in the next round and a defector will gain $m/(N - m) * \alpha$ in the next round.

If the total customer numbers in this market is fixed (the external customer influence factor is 0 in a totally isolated competition environment), the customer movement can be explained easily by the above. However, customers outwith the market will be attracted by the decrease in prices and customers in this market will give up buying anything at the next period due to an increase in prices. The external customer influence parameter is used to describe the customer movement into the market and out from it. The external customer influence is concerned with the actions of all the competitors at the current round. Because we assume the effects of the external customer influence for each company is the same, we set β to a constant value for all companies. Therefore, the customer number vector at time t in this market, N_{t+1} can be described as,

$$N_{t+1} = M_{t+1} N_t (1 + \beta_t) \tag{4}$$

The profits that each company in this game can get are:

$$\pi_i^t = (P_i^t - C) n_i^t \tag{5}$$

Where $i = 1, 2....,N$, i.e., N players, and t is the period.

3 Implementation of Competition Game Model

This game is investigated with a simple GA with separate pool for each company. The values of each bit on the strategy chromosome for representation are 1 for cooperation (high price) and 0 for defection (low price). The first bit of the chromosomes decides the actions at the first round of game and the following bits represent the strategy with respect to the competition outcome at the previous round, i.e., $C_{N-1}, C_{N-2}, ..., C_1, C_0, D_0, D_1, ..., D_{N-2}$, and D_{N-1}, here C/D denotes player actions, cooperation/defection and number (0....N) denote the number of cooperators of competitors in that round . Therefore, the length of the strategy chromosome is $1 + 2N$. Even though the number of players increases, the chromosome is still short and the possibility of disrupting the schema is small. For the game part, the parameters we set are,

1. High price ($P_h > 1$) for cooperator and low price ($P_l > 1$) for defector;
2. Cost of each goods or service ($C = 1$);
3. Total customer number at the first round, $N = 1000$;
4. The number of rounds in every game, r;
5. Internal market share loss (α),$0 \leq \alpha \leq 1$;
6. And external customer influence (β). More cooperators in the competition lead lower β value and $-1 \leq \beta \leq 0$ to describe the loss/gain of customers for the whole market.

We are interested in the emergence of total cooperation, i.e., every competitor choose high price in competitions, under the parameter settings influence. Therefore, we run each game by each parameter settings and record the number of competitions with each competition situation, e.g., all the players cooperate with one another, at each generation to find if the total cooperation will be form before 1000 generations of the evolution.

4 Comparative Experiments

According to the previous work [7] of oligopoly game model, we find that the influence on the strategies of competitors dose not rely on the value of high price and it of low price which the company adopt respectively but on relations of high price and low price. Therefore, we simplify the three subjective factors, high price, low price, and cost, as **Profit Ratio**$= (P_h - C)/(P_l - C)$. Firstly, we try to simulate the influence of Market share loss (α) on cooperation so that the β values are set as 0. The environment is set simple firstly in that the customer movements are zero, i.e., $/alpha = 0$. Therefore, that the payoffs to cooperators are more than those to defectors in every case. We can thus predict that the result should tend to cooperation but also use this set to experiment the influence of number of players in the game. The number of game with all cooperators is shown in Table 1 and we can find that high profit ratio ($P_h = 2.0, P_l = 1.2$) leads more cooperation outcomes but the difference between the influence of high and low Profit ratios is not obvious when the cooperation is easy to achieve. More competitors in the competition make the cooperation hard to be achieved. This is very similar to common sense since the collusion become difficult in perfect competition because of difficulty for checking all other competitors.

Table 1. The number of total cooperation after 1000 generations with total loyalty

Company Number	Subjective Factors	
	$P_h = 1.4, P_l = 1.2$	$P_h = 2.0, P_l = 1.2$
5	943.99	943.99
6	931.00	943.99
7	917.69	943.99
8	907.80	943.99
9	890.93	943.99
10	884.54	943.99

Secondary, we find that the emergence of cooperation is similar under the same profit ration even though the actual subjective factors (P_h and P_l) are different. We check profit ration =2, 2.5, 3 5 while the market share loss as 0.2, 0.3, 0.5 and 0.7 and all the simulations here can conclude that the profit ratio could replace P_h, P_l, and C to be a subjective factor.

We investigate the effect of changing the parameter determining the market share loss. We run the 3-players game under the same profit ratio with different market share loss values, 0.2, 0.3, 0.4, 0.5, and 0.7. Table 2 shows some exemplar results. We find that the higher a value of the market share loss, the harder for cooperation. When profit ratio=2, there is no α value which can ensure the emergence of cooperation because the disadvantage of cooperation is too high. However, at lower α values, the number of total defectors decreases. The similar influence of market share loss on cooperation also appears under other values of profit ratio. We can say that the higher the market share loss leads more difficult to emerge cooperation. We also check which profit ratio affects cooperation at which α in the 3- players game and find:

1. With a low α value(0.2), and profit ratio=3, 2-cooperators dominate.
2. When profit ratio=4, 2-cooperators and total cooperation have similar likelihoods.
3. When profit ratio increases to 5, total cooperation dominates.

Table 2. Profit ratio=2, the simulations of the different market share loss

Market share loss (α)	All Defectors	1 Cooperator	2 Cooperators	3 Cooperators
0.2	499.13	290.55	182.16	28.14
0.3	903.95	67.16	22.90	5.98
0.4	994.03	37.84	14.15	3.98
0.5	966.54	20.99	11.50	0.96
0.7	981.47	13.73	3.97	0.83

However, there are still many simulations which cannot achieve total cooperation and there are still many defectors in the games.

We may also investigate different values of α. With $\alpha = 0.3$, profit ratio must increase to 3.5 if we are to get cooperators surviving, at which time 2-cooperators dominate. However, as the profit ratio increase to 5, we still cannot make total coop-eration dominant but the number of 2-cooperators increases to around 500. With $\alpha = 0.4$, total defection dominates when profit ratio ≤ 4. While profit ratio=5, there are many 1-cooperator and 2-cooperators situations appearing in the games with $\alpha = 0.4$. With $\alpha = 0.5$, even if the profit ratio increases to 5, there are still 80% of competition outcomes which show total defection. At $\alpha = 0.7$, the number of total defection outcomes is still more than 900 when profit ratio is increased to 6. There is no chance for cooperators to survive in the competition when the market share loss is so high.

When $\alpha = 0.2$, even when we set the P_h very high (profit ratio=5), there are almost 45% of competition outcomes not achieving total cooperation. We experiment in this competition environment with $\alpha = 0.2$ with even higher profit ratios to check if it is ever possible for total cooperation to break out (See Table 3). We find that the rate of change of increasing total cooperation becomes slow when the profit ratio is set too high. This may suggest that there is a limitation on the maximum possible value of total cooperation with these parameter values. The best result is 70% of total cooperation and 24% of 2-cooperators co-existing in the game.

We now investigate increasing the number of the players involved in the competi-tion to 10. Here, to make comparisons simple, we set all the simulations with =0.2 and change the profit ratio (1.5 to 5). We use 0.2 for the market share loss since, with our experience of simulating the 3-players game, we know that cooperation is hard to evolve when the market share is high. We investigate the 2-players game initially. In the 2-players game, cooperation is easy to achieve. We find that with profit ratio=2.5, total cooperation emerges and dominates. When profit ratio=3, the number of total cooperators is around 800 and it is 870 when profit ratio=5.

In our investigation with the 4-players game, total defection is dominant when the profit ratio< 2.5. Above 2.5, the dominant outcome is 2 cooperators even when the profit ratio increases to 5. However, we also find that 3-cooperators emerge when the profit ratio is set higher. Similar things happen in the 5-players game but the dominant outcome is still 3 players. In the 6-players game, the

Table 3. α=2, a comparison of the different profit ratio effects

Profit ratio	All Defectors	1 Cooperator	2 Cooperators	3 Cooperators
6	35.73	58.78	282.68	622.80
8	26.65	51.12	258.57	633.66
10	24.82	48.18	255.63	671.37
12	22.96	45.71	237.66	693.66
14	22.73	44.81	237.90	694.67
15	24.51	44.75	231.52	699.23

same dominant outcome, 3 cooperators, emerges when the profit ratio=2.5. The number of 4 cooperators increases and the number of 2 cooperators decreases at the same time when the profit ratio increases. The dominant outcomes in the 7-players and 8-players games are 4 cooperators and both appear when profit ratio=2.5. In the 9-players game, 5 cooperators is dominant but the number of 6-cooperators is similar to the number of 5 cooperators when profit ratio=5.

The investigation of the 10-players game gives similar results to the 9-players game. The dominant outcome is 5 cooperators but the number of 6 cooperators increases when the Profit ratio increases. We also investigate if this stable domination can be changed. When profit ratio=15, number of 7 players is the highest even though it is only around 250.

We also investigate the effect of competition by external customer influence but there are not obvious finding to prove these factors affecting the emergence of cooperation strongly.

There are two interesting finding from these investigations:

1. Profit ratio=2.5 seems to be a critical point when $\alpha = 0.2$. Smaller than that, total defection is dominant no matter how many players are involved in the competition. When the profit ratio 2.5, cooperation emerges and the simulation will achieve the dominant outcomes. They may not achieve the total cooperation but the cooperation does exist in the game.
2. When the number of players exceeds 3, we cannot achieve total cooperation with the profit ratio$\leq$ 5. The dominant outcome other than total defection is near to $N/2$ cooperators which seems to be the result of a random search. However, when the profit ratio increases, the more cooperation seems to appear. We can conclude that, when N is too big, most of the simulations which cannot achieve total cooperation fall into a random search but the advantage of cooperation with a high profit ratio still exists.

5 Conclusion

The market competition game model is design to predict the emergence of cooperation for price competition under the market environment and the price set by companies involving. From the experiment results, our first major finding is that the number of firms is not the sole determinant of the formation of a cartel (as is usually taught in industrial economics). We have identified the profit ratio and customers' sensitivity to price ratio as being important factors in determining whether a cartel will form: all other things being equal, we are more likely to observe collusive behavior in commodities with low demand elasticity; secondly, we are more liable to observe collusive behavior in an industry with high profit differential than those with a low profit ratio. The objective factors affect the results more than the subjective factor. This means that the environment of the competition is the most important thing for competition no matter what the company wants subjectively.

There seems to be a critical point in the profit ratio at profit ratio=2.5 when $\alpha = 0.2$. Under profit ratio=2.5, the best strategy is total defection no matter how many competitors there are. Other values of α give other values of the critical point but such values seem always to exist. Otherwise, with a fixed α, it is rarely possible to force all simulations to evolve towards cooperation. We may suggest that there is always the space for defection or cartel/collusion

External influence seems not to affect the results as strongly as internal factors. It is worth investigating more external influences on the market to realize its influence on the decisions in the future research.

References

1. Alchia, A.: Uncertainty Evolution and Economic Theory, Journal of Political Economics. **58** (1950) 211–221
2. Axelroad, R.: The Evolution of Cooperation, Penguin Books. (1990)
3. Chen, S-H., Ni, C-C.: Simulating the Ecology of Oligopolistc Competition with Genetic Algorithms, Knowledge and Information Systems. **2** (2000) 285–309
4. Colman, A. M.: Game Theory and Experimental Games, Pergamon Press, Oxford, England. (1982)
5. Holland, J.: Adaptation in Natural and Artificial Systems, University of Michigan, 2nd Ed. MIT Press. (1992)
6. Yao X., Darwen, P. J.: An Experimental Study of N-person Iterated Prisoners' Dilemma Game, Informatica **18** (1994) 435-450
7. Wang, T. D., Fyfe, C., Marney, J. P.: A Comparison of an Oligopoly Game and the N-iterated Prisoner's Dilemma, The Fifth International Conference of the Society for Computational Economics. (1999)

Exploiting Quotients of Markov Chains to Derive Properties of the Stationary Distribution of the Markov Chain Associated to an Evolutionary Algorithm

Boris Mitavskiy, Jonathan E. Rowe, Alden Wright, and Lothar M. Schmitt

Academic Unit of Mathematical Modelling and Genetic Epidemiology, Division of Genomic Medicine, School of Medicine, University of Sheffield, Sheffield S10 2JF.
B.Mitavskiy@sheffield.ac.uk
School of Computer Science, University of Birmingham, Birmingham B15 2TT, Great Britain
J.E.Rowe@cs.bham.ac.uk
Computer Science, University of Montana, Missoula, MT 59812 USA
wright@cs.umt.edu
School of Computer Science and Engineering, The University of Aizu, Aizu-Wakamatsu City, Fukushima Prefecture, 965-8580, Japan
lothar@u-aizu.ac.jp

Abstract. In this work, a method is presented for analysis of Markov chains modeling evolutionary algorithms through use of a suitable quotient construction. Such a notion of quotient of a Markov chain is frequently referred to as "coarse graining" in the evolutionary computation literature. We shall discuss the construction of a quotient of an irreducible Markov chain with respect to an arbitrary equivalence relation on the state space. The stationary distribution of the quotient chain is "coherent" with the stationary distribution of the original chain. Although the transition probabilities of the quotient chain depend on the stationary distribution of the original chain, we can still exploit the quotient construction to deduce some relevant properties of the stationary distribution of the original chain. As one application, we shall establish inequalities that describe how fast the stationary distribution of Markov chains modelling evolutionary algorithms concentrates on the uniform populations as the mutation rate converges to 0. Further applications are discussed.

1 Introduction

One of the aspects of the theoretical analysis of the evolutionary algorithms is studying the properties of the Markov chains associated with these algorithms. Many research articles in the field of evolutionary computing have been devoted to this subject (see, for instance, [7], [8], [11] and [2] for a survey of known results and open questions). One difficulty that arises with this approach is the fact that the number of states of this Markov chain grows very fast with respect to the size of the search space and the number of elements in a population. Indeed, if Ω denotes the search space, the number of sates of this Markov chain for a population of size m is $|\Omega|^m$. In the current paper we introduce a construction which can be viewed as a "quotient" (or, according to a commonly accepted evolutionary algorithm terminology, a "coarse graining") of a Markov chain with respect to an equivalence relation. This construction is applicable

T.-D. Wang et al. (Eds.): SEAL 2006, LNCS 4247, pp. 726–733, 2006.

to all irreducible Markov chains (which is true of Markov chains modelling evolutionary algorithms with positive mutation rate: see, for instance, [11] or [2]). The "quotient chain" will be shown to be irreducible as well and its unique stationary distribution is coherent with that of the original chain. Although the transition probabilities of the quotient chain depend on the stationary distribution of the original chain (which is the subject of our investigation), we can still use the quotient construction to deduce some interesting properties of the stationary distribution of *both* the quotient chain and the original chain. To illustrate this technique, we establish inequalities that show how fast the stationary distribution of a Markov chain modelling an evolutionary algorithm concentrates on the uniform populations (populations consisting of the repeated copies of the same individual only) with respect to a decreasing mutation rate. This extends the results of [3] and [10] in the sense that it provides an estimate on the rates of concentration of the stationary distribution apart from establishing the limits. Such information may be helpful in deciding on the amount of fitness pressure and/or mutation rate which is most appropriate for a given algorithm. It should be noted that an inequality analogous to the one in corollary 10 has been obtained in [8] using entirely different methods. It is worth mentioning that the method introduced in the current paper is very simple and elementary unlike those in [7], [8], [3] and [10]. To the best of the authors' knowledge, the inequality in corollary 11 is new. This inequality establishes a connection between the rate of concentration of the stationary distribution of the Markov chain modelling an evolutionary algorithm on the uniform populations and the maximum expected waiting time to reach a uniform population starting from any other population. Estimating such expected waiting times for various recombination operators remains an open problem. The authors of this work see its main contributions not so much being the results on the rate of concentration of the stationary distribution on the uniform populations but the innovative quotient of the Markov chain construction which allows for deduction of known and new results in a rather elementary setting. The authors hope that this technique will have further significant applications in future research.

2 Notation and Basic Framework

Let Ω be a finite set which shall be called the *search space*. Let $f : \Omega \to (0, \infty)$ be a function which shall be called the *fitness function*. The goal for an optimization algorithm such as an evolutionary algorithm is to find a maximum of the function f.

Definition 1. A population P of size m is an element of Ω^m. A population is called uniform, if it is of the form $(x, x, \dots, x)$ for some $x \in \Omega$.

Remark 2. We note that there are two primary methods for representing populations: the familiar multi-set representation as used, e.g., in [11] and ordered tuples as in definition 1 above and, e.g., in [7,8]. Each has advantages, depending upon the particular analytical goals. This work uses the ordered tuples method, which is perhaps the more natural method, but may be less familiar to the reader.

In the current paper we shall be dealing with an evolutionary algorithm having a fixed population size m and a fixed fitness landscape. The algorithm cycles through fitness-proportional selection, recombination and mutation *stages*. A *stage* is a probabilistic rule which takes one population as an input and produces another population of the same size as an output. For example, fitness-proportional selection is introduced below:

Definition 3. The stage of fitness-proportional selection takes a given population $P = (x_1, x_2, \ldots, x_m)$ with $x_i \in \Omega$ as an input. A new population $P' = (y_1, y_2, \ldots, y_m)^T$ is obtained where y_i's are chosen independently m times form the individuals of P and $y_i = x_j$ with probability $\frac{f(x_j)}{\Sigma_{l=1}^{m} f(x_l)}$.

In our framework, recombination can be any stage which does not alter uniform populations in any way with probability 1.[1]

The main purpose of mutation is to ensure that the algorithm is *ergodic* in the sense that every population is reached from every other population with some positive (although rather small) probability. We shall therefore define mutation stage as follows:

Definition 4. Mutation is a stage which produces a population $P' = (y_1, y_2, \ldots, y_m)$ from a population $P = (x_1, x_2, \ldots, x_m)$ in such a way that $x_i \neq y_i$ with probability at most δ for some $\delta > 0$ independent of P, P' and i. The greatest lower bound μ over all such δ shall be called the *mutation rate* in the framework of the current paper. In addition, we require that for every $y \in \Omega$ the probability that $y_i = y$ is bounded below by $\Omega(\mu^l)$ for some fixed l. We shell refer to the smallest such l as the *generalized string length*. Finally, we require that the event $y_i = x_i$ is independent of the event $y_j = x_j$ for $i \neq j$.

The following fact is easy to verify:

Proposition 5. *Consider a mutation stage M with mutation rate μ. If a population $P' = (y_1, y_2, \ldots, y_m)$ is an output of the population $P = (x_1, x_2, \ldots, x_m)$ upon the completion of M, then the probability that $y_i \neq x_i$ for fixed i is at most μ.*

In [11], it has been pointed out that heuristic search algorithms give rise to the following Markov process[2] (see also [2]): The state space of this Markov process is the set of all populations of a fixed size m. In our notation, this set, is simply Ω^m. The transition probability $p_{\mathbf{xy}}$ is simply the probability that the population $\mathbf{y} \in \Omega^m$ is obtained from the population $\mathbf{x}$ upon the completion of fitness-proportional selection followed by recombination and then followed by mutation. The aim of the current paper is to illustrate a new technique introduced in the next section for studying, e.g, the rate of concentration on uniform populations of the stationary distribution of a Markov chain modelling an evolutionary algorithm.

3 Quotients of Irreducible Markov Chains

Throughout the current section we shall be dealing with an irreducible Markov chain $\mathcal{M}$ over a finite state space $\mathcal{X}$. $\{p_{\mathbf{x},\mathbf{y}}\}$ denotes the Markov transition matrix with the convention that $p_{\mathbf{x},\mathbf{y}}$ is the probability of getting y in the next stage given x. Suppose

---

[1] Most types of recombination used in practice are pure in the sense of [9] so that they preserve the uniform populations.

[2] Recall that in the current paper, we use the m-tuple representation.

we are given an equivalence relation $\sim$ partitioning the state space $\mathcal{X}$. The main idea of the current section is to construct an irreducible Markov chain over the equivalence classes under $\sim$ (i.e. over the set $\mathcal{X}/\sim$) whose stationary distribution is compatible with that of $\mathcal{M}$. This construction is a slight generalization of the construction in [1]:

Definition 6. Given an irreducible Markov chain $\mathcal{M}$ over a finite state space $\mathcal{X}$ determined by the transition matrix $\{p_{\mathbf{x},\mathbf{y}}\}$ and an equivalence relation $\sim$ on $\mathcal{X}$, let π denote the unique stationary distribution of the Markov chain $\mathcal{M}$. Define the *quotient* Markov chain $\mathcal{M}/\sim$ over the state space $\mathcal{X}/\sim$ of equivalence classes via $\sim$ to be determined by the transition matrix $\{\tilde{p}_{\mathcal{U},\mathcal{V}}\}_{\mathcal{U},\mathcal{V}\in\mathcal{X}/\sim}$ given as

$$\tilde{p}_{\mathcal{U},\mathcal{V}} = \frac{1}{\pi(\mathcal{U})}\sum_{\mathbf{x}\in\mathcal{U}}\pi(\mathbf{x})\cdot p_{\mathbf{x},\mathcal{V}} = \frac{1}{\pi(\mathcal{U})}\sum_{\mathbf{x}\in\mathcal{U}}\sum_{\mathbf{y}\in\mathcal{V}}\pi(\mathbf{x})\cdot p_{\mathbf{x},\mathbf{y}}.$$

Here $p_{\mathbf{x},\mathcal{V}}$ denotes the transition probability of getting somewhere inside of $\mathcal{V}$ given $\mathbf{x}$. Since $\mathcal{V} = \bigcup_{y\in\mathcal{V}}\{y\}$ it follows that $p_{\mathbf{x},\mathcal{V}} = \sum_{y\in\mathcal{V}} p_{\mathbf{x},\mathbf{y}}$ and hence the equation above holds.

Intuitively, the quotient Markov chain $\mathcal{M}/\sim$ is obtained by running the original chain $\mathcal{M}$ starting with the stationary distribution and computing the transition probabilities conditioned with respect to the stationary input. If one starts with an arbitrary distribution and runs the process for a long period of time then the transition probabilities in definition 6 serve as a good approximation to the transition probabilities induced by the corresponding stochastic process. Thus, the following fact should not be a surprise:

Theorem 7. *Let π denote the stationary distribution of an irreducible Markov chain $\mathcal{M}$ determined by the transition matrix $\{p_{\mathbf{x},\mathbf{y}}\}_{\mathbf{x},\mathbf{y}\in\mathcal{X}}$. Suppose we are given an equivalence relation $\sim$ partitioning the state space $\mathcal{X}$. Then the quotient Markov chain $\mathcal{M}/\sim$ is irreducible and its unique stationary distribution $\tilde{\pi}$ is compatible with π in the sense that for every $\mathcal{O}\in\mathcal{X}/\sim$, we have $\tilde{\pi}(\{\mathcal{O}\}) = \pi(\mathcal{O})$.*

Proof: Since the original chain $\mathcal{M}$ is assumed to be irreducible, it follows that there exists an $n\in\mathbb{N}$ such that for all $\mathbf{x}$, $\mathbf{y}\in\mathcal{X}$ we have $p^n_{\mathbf{x},\mathbf{y}} > 0$ where $p^n_{\mathbf{x},\mathbf{y}}$ denotes the probability that $\mathbf{y}$ is reached from $\mathbf{x}$ after exactly n time steps. This, in turn, is equivalent to saying that there exists a sequence of states $\mathbf{x}_1 = \mathbf{x}$, $\mathbf{x}_2,\ldots,\mathbf{x}_n = y$ such that $p_{\mathbf{x_i},\mathbf{x_{i+1}}} > 0$. Let $\mathcal{O}_i$ denote the equivalence class of x_i under $\sim$. Now we see that

$$\tilde{p}_{\mathcal{O}_i,\mathcal{O}_{i+1}} = \frac{1}{\pi(\mathcal{O}_i)}\sum_{\mathbf{x}\in\mathcal{O}_i}\sum_{\mathbf{z}\in\mathcal{O}_{i+1}}\pi(\mathbf{x})\cdot p_{\mathbf{x},\mathbf{z}} \geq \frac{1}{\pi(\mathcal{O}_i)}\cdot\pi(\mathbf{x}_i)\cdot p_{\mathbf{x}_i,\mathbf{x}_{i+1}} > 0.$$

This shows that $\tilde{p}^n_{\mathcal{O}_1,\mathcal{O}_n} > 0$. Since the equivalence classes are nonempty and the choices of $\mathbf{x}$ and $\mathbf{y}$ are arbitrary, it follows that $\tilde{p}^n_{\mathcal{U},\mathcal{V}} > 0\ \forall\mathcal{U},\ \mathcal{V}\in\mathcal{X}/\sim$. This shows that the Markov chain $\mathcal{M}/\sim$ is irreducible and, hence, has a unique stationary distribution $\tilde{\pi}$. The fact that $\tilde{\pi}(\{\mathcal{O}\}) = \pi(\mathcal{O})$ is the stationary distribution of $\mathcal{M}/\sim$ can now be verified by direct computation. Indeed, we obtain

$$\sum_{\mathcal{O}\in\mathcal{X}/\sim}\tilde{\pi}(\{\mathcal{O}\})\cdot\tilde{p}_{\mathcal{O},\mathcal{U}} = \sum_{\mathcal{O}\in\mathcal{X}/\sim}\pi(\mathcal{O})\cdot\frac{1}{\pi(\mathcal{O})}\sum_{\mathbf{x}\in\mathcal{O}}\sum_{\mathbf{z}\in\mathcal{U}}\pi(\mathbf{x})\cdot p_{\mathbf{x},\mathbf{z}} =$$

$$= \sum_{\mathbf{x}\in\mathcal{X}}\sum_{\mathbf{z}\in\mathcal{U}} \pi(\mathbf{x})\cdot p_{\mathbf{x},\mathbf{z}} = \sum_{\mathbf{z}\in\mathcal{U}}\sum_{\mathbf{x}\in\mathcal{X}} \pi(\mathbf{x})\cdot p_{\mathbf{x},\mathbf{z}} \overset{\text{by stationarity of } \pi}{=} \sum_{\mathbf{z}\in\mathcal{U}} \pi(\mathbf{z}) = \pi(\mathcal{U}) = \tilde{\pi}(\{\mathcal{U}\}).$$

This establishes the stationarity of $\tilde{\pi}$ and theorem 7 now follows. □

Although theorem 7 is rather elementary it allows us to make useful observations of the following type:

Corollary 8. *Suppose we are given an irreducible Markov chain $\mathcal{M}$ over the state space $\mathcal{X}$, and let $\mathcal{X} = A \cup B$ with $A \cap B = \emptyset$. Suppose that for every $a \in A$ we have $p_{a,\, B} < \mu$ while for every $b \in B$ $p_{b,\, A} > \kappa$. If π denotes the unique stationary distribution of $\mathcal{M}$, we have $\pi(B) < \mu/\kappa$ and $\pi(A) > 1 - \mu/\kappa$. We also have $\pi(A) \geq (1 + \mu/\kappa)^{-1}$.*

Proof: Let $\sim$ denote the equivalence relation corresponding to the partition $\{A,\, B\}$ of $\mathcal{X}$. Now consider the Markov chain $\mathcal{M}/\!\sim$. This is a Markov chain determined by the 2×2 transition matrix

$$\begin{pmatrix} p_{A,A} & p_{B,\,A} \\ p_{A,B} & p_{B,\,B} \end{pmatrix}$$

where $p_{A,B} = \frac{1}{\pi(A)} \sum_{a\in A} \pi(a) \cdot p_{a,\,B} < \frac{1}{\pi(A)} \sum_{a\in A} \pi(a) \cdot \mu = \mu$ and, likewise, $p_{B,A} = \frac{1}{\pi(B)} \sum_{b\in B} \pi(b) \cdot p_{b,\,A} > \frac{1}{\pi(B)} \sum_{b\in B} \pi(b) \cdot \kappa = \kappa$. According to theorem 7, the Markov chain $\mathcal{M}/\!\sim$ is irreducible and its unique stationary distribution $\tilde{\pi}$ satisfies $\tilde{\pi}(\{A\}) = \pi(A)$ and $\tilde{\pi}(\{B\}) = \pi(B)$. Moreover, by direct computation it is easy to see that the stationary distribution of the Markov chain determined by the 2×2 transition matrix above (i.e. of the Markov chain $\mathcal{M}/\!\sim$) is $\tilde{\pi}(\{A\}) = \frac{p_{B,A}}{p_{A,B}+p_{B,A}}$ and $\tilde{\pi}(\{B\}) = \frac{p_{A,B}}{p_{A,B}+p_{B,A}}$. We finally obtain the desired inequalities $\pi(B) = \tilde{\pi}(\{B\}) = \frac{p_{A,B}}{p_{A,B}+p_{B,A}} < \frac{\mu}{\kappa}$ and $\pi(A) = 1 - \pi(B) > 1 - \frac{\mu}{\kappa}$. Likewise, writing $\pi(A) = \frac{p_{B,A}}{p_{B,A}+p_{A,B}} = \frac{1}{1+\frac{p_{A,B}}{p_{B,A}}}$ and using our previous bounds on $p_{A,B}$ and $p_{B,A}$ we obtain the last assertion. □

Corollary 8 can be somewhat strengthened by observing that any power of a Markov transition matrix determining an irreducible Markov chain also determines an irreducible Markov chain having the same stationary distribution as the original one. Applying corollary 8 to every power of a Markov transition matrix then gives us the following fact:

Corollary 9. *Suppose we are given an irreducible Markov chain $\mathcal{M}$ over the state space $\mathcal{X}$, and let $\mathcal{X} = A \cup B$ with $A \cap B = \emptyset$. Suppose for every $a \in A$ we have $p^n_{a,\, B} < \mu_n$ while for every $b \in B$ $p^n_{b,\, A} > \kappa_n$. Then, if π denotes the unique stationary distribution of $\mathcal{M}$, we have $\pi(B) < \inf\{\mu_n/\kappa_n \,|\, n \in \mathbb{N}\}$ and $\pi(A) > 1 - \inf\{\mu_n/\kappa_n \,|\, n \in \mathbb{N}\}$.*

Corollary 9 readily implies some basic observations about the rate of concentration of the stationary distribution of Markov chains modelling EAs. Our bounds apply to a rather wide class of evolutionary algorithms as described in section 2.

4 Applications of the Quotient Construction

First, let us recall a well-known fact that a Markov chain modelling an evolutionary algorithm described in section 2 is irreducible thanks to the ergodicity of mutation (see for instance [2] and [11]). The main idea of what follows is to apply corollaries 8 and 9 to the subset $H \subseteq \Omega^m$ of the uniform populations and its complement in Ω^m.

Corollary 10. *Suppose we are given an evolutionary algorithm $\mathcal{A}$ with mutation rate μ. Let π denote the unique stationary distribution of the Markov chain associated to the algorithm $\mathcal{A}$. Then we have $\pi(H) \geq 1 - m^{m+1}(1-\mu)^{-m}\mu$.*

Proof: This is an immediate application of corollary 8. Indeed, the event of destroying a given uniform population is equivalent to the event of applying the nonidentity transformation to either one of the elements of that population and so is a union of events happening with probability μ each. Hence the probability of destroying a given uniform population is bounded above by $m\mu$. The probability of passing from a non-uniform population to a uniform one is at least as large as the probability of consecutive m independent drawings of the most fit individual. The probability of picking the most fit individual in a population is bounded below by the probability of picking a given individual form a population where all the individuals have the same fitness, which is $1/m$. Doing so consecutively and independently m times is m^{-m}. Afterwards, with probability $(1-\mu)^m$ everyone stays the same. The desired equation now follows immediately from corollary 8 □

The bound in corollary 10 is a rather weak one. This is not too surprising the more so that it applies to a wide class of algorithms. One should be able to improve the bound in corollary 10 for specific types of algorithms using corollary 9 instead of corollary 8.

Corollary 11. *Suppose we are given an evolutionary algorithm $\mathcal{A}$ with mutation rate μ. Let $T(\mathbf{x})$ denote the random variable measuring the number of steps it takes for an EA to reach a uniform population starting with the population $\mathbf{x}$. Let π denote the unique stationary distribution of the Markov chain associated to the algorithm $\mathcal{A}$. Then we have $\pi(H) \geq 1 - \alpha(1-\mu)^{-\alpha}\mu$, where $\alpha = 2m \cdot \max\{E(T(\mathbf{x})) \mid \mathbf{x} \in \Omega^m - H\}$.*

Proof: First note that

$$\forall \mathbf{x} \in \Omega^m - H, \text{ we have } p^L_{\mathbf{x},H} \geq P(T(\mathbf{x}) < L) \cdot (1-\mu)^{Lm}.$$

By Markov's inequality we have

$$P(T(\mathbf{x}) < L) \geq 1 - \frac{E(T(\mathbf{x}))}{L} \geq \frac{1}{2} \quad \text{for } L \geq 2E(T(\mathbf{x})).$$

We then deduce that

$$p^{\beta}_{\mathbf{y},H} \geq \frac{1}{2}(1-\mu)^{\alpha} \quad \forall\, \mathbf{y} \in \Omega^m - H$$

where $\beta = 2\max\{E(T(\mathbf{x})) \mid \mathbf{x} \in \Omega^m - H\} = \alpha/m$. Just like in the proof of corollary 10, we have

$$p^{\beta}_{\mathbf{y},\Omega^m - H} \leq m \cdot \max\{E(T(\mathbf{x})) \mid \mathbf{x} \in \Omega^m - H\}\mu$$

which finally gives $\pi(H) \geq 1 - \alpha(1-\mu)^{-\alpha}\mu$.

□

We end this paper with one more application of corollary 8 to a question which has been extensively studied in [3], [7] and in [8] and in [10]. Hopefully the reader will appreciate the simplicity and explicitness of the argument presented below comparing to the technique introduced in [3]. Also unlike the methods in [3] and in [10], the argument below provides an estimate on the convergence rates. See also [7] and [8] in this regard. The authors hope that the method of the current paper can be useful in the future for deriving similar interesting results. Throughout the following we shall assume for the simplicity of presentation that there is only a unique fittest individual in our search space, Ω.[3] We shall need the following notion of the selection pressure first:

Definition 12. We define *selection pressure* t to be any positive parameter which changes fitness in the order preserving manner so that if $f(t)$ is the minimal probability of going from a population containing a copy of the fittest individual x to the uniform population consisting of a single copy of x only after fitness-proportional selection then $f(t) \to 1$ as $t \to \infty$.

Corollary 13. *Consider a parameterized family of algorithms with common generalized string length l, $\{\mathcal{A}(\mu, t)\}_{\mu \in (0, 1), t > 0}$ where μ is the mutation rate and t is the selection pressure of the algorithm $\mathcal{A}(\mu, t)$. Denote by $\pi_{\mu, t}$ the stationary distribution of the Markov chain associated to the algorithm $\mathcal{A}(\mu, t)$. Suppose $m > l$. Then we have $\lim_{\mu \to 0} \lim_{p \to \infty} \pi_{\mu, t}(\mathbf{u}) = 1$ where $\mathbf{u}$ denotes the uniform population consisting of the unique fittest individual only.*

Proof: Let A denote the set of all populations of size m containing at least one copy of the unique fittest individual x, and let B denote the set consisting of the rest of the populations of size m. (i. e. is the set of these populations which do not contain x). Now, for every small mutation rate $\mu > 0$ select the selection pressure p large enough so that $1 - f(p) < \mu^m$. Now observe that for every $\mathbf{a} \in A$ after selection we end up with $\mathbf{u}$ with probability $f(p)$ (close to 1). Notice that we used the uniqueness of the fittest individual here. Provided this is the case, recombination does not alter anything and we end up with $\mathbf{u}$ again. The only way to get out from A after mutation is only to mutate at least one bit in every individual of $\mathbf{u}$ which happens with probability $O(\mu^m)$ (since there are m individuals, the probability of mutating an individual is bounded above by $O(\mu)$ and individuals are mutated independently). The only other way to get out from A is to avoid getting $\mathbf{u}$ upon completion of the selection stage which happens with probability $1 - f(p) < \mu^m$ by the choice of p. Therefore we have $\forall\, \mathbf{a} \in A$ $p_{\mathbf{a} \to B} \leq f(p)O(\mu^m) + (1 - f(p)) < f(p)\mu^m + \mu^m = O(\mu^m)$. Likewise, observe that for any given population $\mathbf{b} \in B$ it suffices to mutate some individual in $\mathbf{b}$ into the fittest individual x to obtain a population in A after mutation. According to definition 4 this happens with probability $\Omega(\mu^l)$. Now, applying corollary 8, we conclude that $\pi_{\mu, t}(A) \geq (1 + O(\epsilon^N)/\Omega(\epsilon^l))^{-1} = 1/(1 + O(\mu^{m-l})) \to 1$ as $\mu \to 0$ as long as p is chosen so that $1 - f(p) < \mu^m$ (notice that choosing p so that $1 - f(p) < \mu^\kappa$ for $\kappa > l$ would suffice as well, but would give slower convergence rate) and so it follows that $\lim_{\mu \to 0} \lim_{p \to \infty} \pi_{\mu, t}(A) = 1$. The desired conclusion now follows by combining this fact with corollary 10, carefully noting that the bound in corollary 10 is independent of the selection pressure. □

[3] This assumption can be relaxed by requiring that the set of fittest individuals is closed under recombination. The proof presented below goes through with only minor modifications.

5 Conclusion

In the present paper, we constructed a quotient (or, in the language used by the evolutionary computation community, a "coarse graining") of an irreducible Markov chain with respect to an arbitrary equivalence relation on the search space. As an illustration of how this simple construction can be applied, we established some inequalities that show how fast the stationary distribution of a Markov chain modelling an evolutionary algorithm concentrates on the uniform populations (populations consisting of the repeated copies of a single individual only). It was shown (see corollary 8 and 9 that the stationary distribution value of the set of uniform populations is bounded below by $1 - k\mu$ where μ is the mutation rate and k is a multiplicative constant depending on the population size.[4] In addition, we presented a new estimate in Corollary 9 for the expected time it takes for an evolutionary algorithm to reach a uniform population starting with the population $\mathbf{x}$. These results based upon our new method show that the two bounds and estimates mentioned above are closely related.

References

1. Aldous, D. and Fill, J. (2001) Reversible Markov Chains and Random Walks on Graphs (unpublished book) http://www.stat.berkeley.edu/ aldous/RWG/book.html
2. Coffey, S. (1999) An Applied Probabilist's Guide to Genetic Algorithms. *A Thesis Submitted to The University of Dublin for the degree of Master in Science.*
3. Davis, T. and Principe, J. (1991) A Simulated Annealing Like Convergence Theory for the Simple Genetic Algorithm. In R. Belew, and L. Bookers, editors, *Proceedings of the Fourth International Confernence on Genetic Algorithms*, pages 174-181, Morgan Kaufmann.
4. Mitavskiy, B., Rowe, J. (to appear). An Extension of Geiringer Theorem for a wide class of evolutionary search algorithms. *Evolutionary Computation.*
5. Mitavskiy, B. (2005) A schema-based version of Geiringer's Theorem for nonlinear Genetic Programming with homologous crossover. *Foundations of Genetic Algorithms 8*, pages 156-175, Springer-Verlag.
6. Mitavskiy, B., Rowe, J. (accepted) Some Results about the Markov Chains Associated to GPs and General EAs. *Theoretical Computer Science*
7. Schmitt, L. (2001) "Theory of Genetic Algorithms." *Theor. Computer Science*, 259: 1-61.
8. Schmitt, L. (2004) "Theory of Genetic Algorithms II: Models for Genetic Operators over the String-Tensor representation of Populations and Convergence to Global Optima for Arbitrary Fitness Function under Scaling." *Theoretical Computer Science*, 310: 181-231.
9. Radcliffe, N. (1994). The algebra of genetic algorithms. *Annals of Mathematics and Artificial Intelligence*, 10:339-384. http://users.breathemail.net/njr/papers/amai94.pdf
10. Suzuki, J. (1998). A Further Result on the Markov Chain Model of Genetic Algorithm and Its application to Simulated Annealing-Like strategy. *IEEE Transactions on Systems, Man and Cybernatics*, 28(1).
11. Vose, M. (1999). The simple genetic algorithm: foundations and theory. *MIT Press* .

[4] A similar result was obtained in [8] by completely different methods.

Accessibility and Runtime Between Convex Neutral Networks

Per Kristian Lehre and Pauline C. Haddow

Department of Computer and Information Science
Norwegian University of Science and Technology
{lehre, pauline}@idi.ntnu.no

Abstract. Many important fitness functions in Evolutionary Computation (EC) have high degree of neutrality i.e. large regions of the search space with identical fitness. However, the impact of neutrality on the runtime of Evolutionary Algorithms (EAs) is not fully understood. This work analyses the impact of the accessibility between neutral networks on the runtime of a simple randomised search heuristic. The runtime analysis uses a connection between random walks on graphs and electrical resistive networks.

1 Introduction

Evolutionary algorithms are successful on a wide range of problems, but often notoriously hard to analyse. The classical metaphor of populations climbing towards higher points in a fitness landscape is often insufficient for explaining the types of evolutionary dynamics experienced in practice. In the late sixties, Kimura proposed an alternative view termed *neutral evolution* [1]. He argued that most of the mutations in biological organisms are neutral. Hence, populations are randomly drifting on so-called *neutral networks* (or plateaus) until a port to a more fit neutral network is found.

Fitness functions may often exhibit some form of neutrality. Naturally, one may ask if neutrality is beneficial in EC. However, there is no clear answer in the literature. Some experimental evidence suggests that neutrality may be beneficial [2], but the results are not consistent [3]. It seems that very large neutral networks i.e. NEEDLE [4] cause an exponential increase in runtime. However neutral networks with other shapes i.e. polynomially long paths, are easy to overcome [5]. A further complicating factor is that neutral networks of different shapes and sizes may occur within the same fitness function or genotype-phenotype mapping [6].

Theoretical research on neutrality raise the question on which mathematical formalism to use. For example, in a fitness function with neutrality, it has been shown that the use of a distance metric to measure the *progress rate* of an EA has important limitations [5].

An alternative approach is a new formalism called *accessibility* [7]. Informally, the accessibility from a neutral network X into another neutral network Y is the

T.-D. Wang et al. (Eds.): SEAL 2006, LNCS 4247, pp. 734–741, 2006.

fraction of the border of network X which falls within network Y. Accessibility between neutral networks is not necessarily symmetric. Ideas for calculating accessibilities in indirect genotype-phenotype mappings were discussed in [8]. This paper looks at the use of accessibility for characterising which neutral networks are hard to overcome for a simple randomised search heuristic.

The analysis uses standard notation e.g. O and Θ for asymptotic growth of functions. The symbol $H_n = \sum_{i=1}^{n} 1/i$ denotes the harmonic series; symbol Σ denotes a finite alphabet, and Σ^n the set of strings of length n over Σ. The Hamming distance between strings x and y is denoted $d(x, y)$ and the symbol Γ denotes the neighbourhood relation over graphs and strings.

Definition 1 (Accessibility). *Given two subsets $X, Y \subset S$, the* accessibility *from X to Y is defined as $A(Y \leftarrow X) := |\partial X \cap Y|/|\partial X|$, where the* border ∂X *is defined as $\partial X := \{b \in S \setminus X \mid \exists x \in X \text{ st. } b \in \Gamma(x)\}$.*

Definition 2 (Convexity). *A subset X of Σ^n is called* convex *if $[x, y] \subseteq X$ for all $x, y \in X$, where $[x, y] := \{z \in \Sigma^n \mid d(x, z) + d(z, y) = d(x, y)\}$. If X is convex, then there exist sets $X_i \subseteq \Sigma$ such that $X = X_1 \times \cdots \times X_n$ [8].*

The iterations of an EA running on a highly neutral problem can be divided into time epochs [9]. An epoch starts when the population has entered a neutral network and ends when the population has found another, higher fit neutral network. During an epoch, the population drifts randomly on the neutral network and the fitness of the population remains constant. The duration of an epoch depends partly on characteristics of the current neutral network, and that of its neighbouring networks. The goal of this paper is to better understand how one such characteristic, accessibility, influences the duration of a single epoch.

In order to focus on a single epoch a simple fitness function called TwoNets is defined, consisting of two neutral networks X and Y. The optimum is any point on network Y and the points on network X are local optima, and it is assumed that the EA has found network X and needs to find network Y.

Definition 3 (TwoNets). *Given a search space S and two non-empty subsets X and Y of this space, the fitness function is defined as* $\text{TwoNets}_{X,Y}(s) := 2$ *if $s \in Y$,* $\text{TwoNets}_{X,Y}(s) := 1$ *if $s \in X$, and* $\text{TwoNets}_{X,Y}(s) := 0$ *otherwise.*

This work considers two types of search spaces for TwoNets, the two-dimensional integer lattice $\mathbb{Z} \times \mathbb{Z}$, and the set of strings Σ^n.

The goal is to understand how a standard genetic algorithm behaves on TwoNets. However, investigations herein, are conducted using a simple algorithm called Random Local Search (RLS). RLS works similar to a hill climber. A randomly chosen search point is selected. In each iteration, a randomly chosen neighbour is selected and replaces the current search point if its fitness is equal or better than the current search point. Here, it is assumed that the initial search point belongs to network X. RLS on TwoNets will, therefore, behave as a random walk on network X, only accepting search points on network X, until a point on network Y is found.

2 Random Walks and Electrical Resistive Networks

A *simple random walk* on an undirected graph $G = (V, E)$ is a discrete time stochastic process $W_1, W_2, ...$ with the node set V of the graph as state space. Given the state $W_t = u$ of the process at time t, the next state W_{t+1} is sampled uniformly at random from the set of neighbour nodes $\Gamma(u)$ of node u. Given a cost function $c : E \rightarrow \mathrm{I\!R}$ on the edge set, a simple random walk can be generalised into a *general random walk* with state transition probabilities $Prob(W_{t+1} = v \mid W_t = u) = c(u,v)/\sum_{w\in\Gamma(u)} c(u,w)$. The *hitting time* h_{xy} between nodes x and y is the expected number of time steps to reach node y starting from node x. Hitting times are not necessarily symmetrical i.e. $h_{xy} \neq h_{yx}$.

Problems in random walks are mathematically connected to the theory of electrical resistive networks [10,11]. A graph $G = (V, E)$ is transformed into a resistive network by replacing each edge with a resistor with unit resistance. Given two nodes, the *effective resistance* R_{xy} is defined as the resistance between node x and node y when one ampere current is inserted in node x and removed from node y. Tetali showed [12] that the hitting time h_{xy} can be calculated as

$$h_{xy} = \frac{1}{2} \sum_{z\in V} d(z)(R_{xy} + R_{yz} - R_{xz}), \tag{1}$$

where $d(z)$ is the degree of node z. The connection is also valid for for general random walks. Given a cost function c, each edge (u, v) is replaced with a resistor with resistance $r_{uv} = 1/c(u, v)$. The hitting time h_{xy} can now be calculated for general random walks by replacing the degree $d(z)$ in Eq. (1) with the generalised degree $c(z) = \sum_{v\in\Gamma(z)} 1/r_{zv}$.

Finding these effective resistances is not always trivial, so one is interested in techniques that can be used to give good estimates. A simple technique, called Rayleigh's principle [10] is to either remove edges from the network, thus increasing the effective resistance in the network, or short-circuiting nodes thus decreasing the effective resistance in the network. An alternative technique called Thompson's principle can be applied by defining a flow on the network [11]. A flow is a function $f : E \rightarrow \mathrm{I\!R}$ defined on the edges which satisfies $f(u, v) = -f(u, v)$. The *net flow* out of a node is defined as $f(u) := \sum_{v\in\Gamma(u)} f(u, v)$. The *power* of a flow f is defined as $P(f) := \sum_{\{u,v\}\in E} r_{u,v} \cdot f(u, v)^2$. If the flow is defined such that the net flows are $f(x) = 1, f(y) = -1$, and $f(z) = 0$ for all $z \neq x, y$, then the effective resistance can be upper bounded by $R_{xy} \leq P(f)$. In such flows, node x is called the source and node y is called the sink.

3 Runtime on Integer Lattice

In the two-dimensional integer lattice $\mathbb{Z} \times \mathbb{Z}$, convex neutral networks are rectangles, and neighbouring neutral networks can therefore only touch each other on one of the four sides. To simplify further, it is assumed that neutral network X is a square of size $n \times n$, spanning the area between point $(1, 1)$ and point (n, n). Furthermore, assume that network Y neighbours network X along a part of the right side of X. With these assumptions, network X and Y have minimal accessibility of $A(Y \leftarrow X) = 1/4n$ when they touch each other only in one corner and maximal accessibility of $A(Y \leftarrow X) = 1/4$ when an entire side of network X neighbours network Y as in Fig. 1.

The objective is to find the expected runtime RLS needs to find a node in network Y. To find the expected runtime on the worst case scenario, it is assumed that the initial search point is the point $x = (1, 1)$, which is the point in network X furthest away from network Y. All the nodes in network Y are combined into a single node labelled y as the particular node found in network Y is not relevant. Nodes at the edges of network X that neighbours nodes outside network Y have self-loops because search points outside networks X and Y have inferior fitness and will never be accepted as the new search point. RLS will behave similar to a random walk on the graphs shown in Fig. 1 (left and centre), and the expected runtime of RLS on TwoNets equals the hitting time h_{xy}.

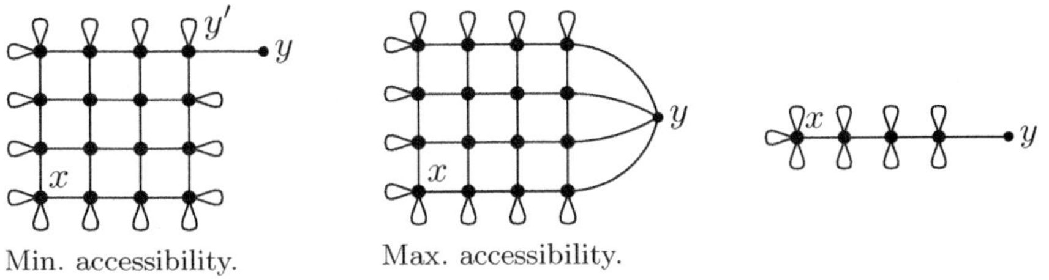

Fig. 1. Graphs for integer lattice

Proposition 1. *Let X be a convex neutral network in $\mathbb{Z} \times \mathbb{Z}$ of size $n \times n$, and let Y be a convex neutral network which neighbours network X such that they have maximal accessibility $A(Y \leftarrow X) = 1/4$. Then the expected runtime of RLS on* TwoNets$_{X,Y}$ *is $2n^2 + O(n)$.*

Proof. In this case, all nodes (n, j), for $1 \leq j \leq n$ neighbour node y. The vertical position in the lattice is therefore not essential, and the graph can be simplified as shown in Fig. 1 (right).

$$h_{xy} = \frac{1}{2}\sum_z d(z)(R_{xy} + R_{yz} - R_{xz}) = \frac{1}{2}R_{xy}\sum_z d(z) + \frac{1}{2}\sum_z d(z)(R_{yz} - R_{xz}) \quad (2)$$

$$= \frac{1}{2}R_{xy}\sum_z d(z) = \frac{1}{2}R_{xy}(4n + 1). \quad (3)$$

The effective resistance R_{xy} in the simplified graph is n, hence the asymptotic hitting time is $h_{xy} = 2n^2 + O(n)$.

Proposition 2. *Let X be a convex neutral network in $\mathbb{Z} \times \mathbb{Z}$ of size $n \times n$, and let Y be a convex neutral network which neighbours network X such that they have minimal accessibility $A(Y \leftarrow X) = 1/4n$. Then the expected runtime of RLS on* TwoNets$_{X,Y}$ *is $\Theta(n^2 \ln n)$.*

Proof. From Fig. 1 (left), it is easy to see that $R_{yz} = R_{yy'} + R_{y'z}$ for all nodes z in X.

$$h_{xy} = \frac{1}{2}\sum_{z} d(z)(R_{xy} + R_{yz} - R_{xz}) = \frac{1}{2}d(y)(R_{xy} + R_{yy} - R_{xy}) \tag{4}$$

$$+\frac{1}{2}\sum_{z\neq y} d(z)(R_{xy'} + R_{y'y} + R_{yy'} + R_{y'z} - R_{xz}) \tag{5}$$

$$= \frac{1}{2}(R_{xy'} + 2R_{yy'}) \cdot \sum_{z\neq y} d(z). \tag{6}$$

The effective resistance $R_{yy'}$ is 1, so it is only necessary to calculate bounds for $R_{xy'}$. The lower bound is calculated using Rayleigh's principle, and the upper bound is calculated using Thompson's principle. To find a lower bound of resistance $R_{xy'}$, diagonal nodes are short-circuited. For example, nodes (1,2) and (2,1) are short-circuited, nodes (1,3), (2,2) and (3,1) are short-circuited etc. Short-circuiting nodes can only decrease the resistance in the network. There are $2i$ resistors in parallel between diagonal $i-1$ and i for $1 < i \leq n$. Using the rule for resistors in parallel, a lower bound for the effective resistance is therefore $R_{xy'} \geq 2\sum_{k=1}^{n-1} 1/2k = H_{n-1} \geq \ln(n-1)$.

To find an upper bound of resistance $R_{xy'}$, Thompson's principle is applied by defining a flow f on the network, similarly to in [11]. For all nodes (i,j) with $2 \leq i+j \leq n$, define a function f as $f((i,j),(i+1,j)) = -f((i+1,j),(i,j)) := \frac{i}{(i+j)(i+j-1)}$ and $f((i,j),(i,j+1)) = -f((i,j+1),(i,j)) := \frac{j}{(i+j)(i+j-1)}$. For all nodes (i,j) with $n+1 \leq i+j \leq 2n-1$, the flows are defined symmetrically along the diagonal line from point $(1,n)$ to point $(n,1)$, for example, $f((n-1,n),(n,n)) = f((1,1),(1,2))$. It can easily be verified that function f is a flow with source $(1,1)$ and sink (n,n). Furthermore, the flow out of any node (i,j) is $f((i,j),(i+1,j)) + f((i,j),(i,j+1)) = 1/(i+j-1)$. Therefore, the flow along any single edge incident on a node (i,j) at distance $d = i+j-2$ from node $(1,1)$ is never more than $1/(i+j-1) = 1/(d+1)$. Furthermore, there are $2(d+1)$ edges out of nodes at distance $d \leq n-2$ from node $(1,1)$. Resistance $R_{xy'}$ can now be bounded by the power of flow f which is calculated along the diagonals.

$$R_{xy'} \leq P(f) = 2\sum_{d=0}^{n-2} \cdot \sum_{i+j=d+2} f((i,j),(i+1,j))^2 + f((i,j),(i,j+1))^2 \tag{7}$$

$$\leq 2\sum_{d=0}^{n-2} 2(d+1) \cdot \frac{1}{(d+1)^2} = 4H_{n-1} = O(\ln(n-1)). \tag{8}$$

Asymptotically, the effective resistance $R_{xy'}$ is $\Theta(\ln n)$. Hence, the asymptotic hitting time is $h_{xy} = \frac{1}{2}(R_{xy'} + 2R_{yy'}) \cdot \sum_{z\neq y} d(z) = \frac{1}{2}(\Theta(\ln n) + 2) \cdot \Theta(n^2) = \Theta(n^2 \ln n)$.

4 Runtime on Hypercube

The graph G_{XY} of the general random walk for the hypercube is defined as follows: If $X = X_1 \times \cdots \times X_n$ and $Y = Y_1 \times \cdots \times Y_n$ are neighbouring, convex neutral networks, then there exists an index i^* such $X_{i^*} \cap Y_{i^*} = \emptyset$, and $X_j \cap Y_j \neq \emptyset$ for all other indices $j \neq i^*$ [8]. For notational convenience, assume that the sets are configured such that $X_j \not\subset Y_j$ for all j, $1 \leq j \leq \ell$, $X_j \subset Y_j$ for all j, $\ell + 1 \leq j \leq n-1$, and $X_n \cap Y_n = \emptyset$. For any bit string $u \in \{0,1\}^\ell$, define $X^u := X_1^u \times \cdots \times X_\ell^u \times X_{\ell+1} \times \cdots \times X_n$, where for all $i, 1 \leq i \leq \ell$, set X^u is defined as $X_i^u := X_i \setminus Y_i$ if $u_i = 0$, and $X_i^u := X_i \cap Y_i$ if $u_i = 1$. Clearly, $X^u \cap X^v = \emptyset$ for all $u \neq v$, and $X = \bigcup_{u \in \{0,1\}^\ell} X^u$, so $\{X^u\}_{u \in \{0,1\}^\ell}$ is a set partition of X. The graph $G_{XY} = (V, E)$ is now defined with nodes $V := \{X^u \mid u \in \{0,1\}^\ell\} \cup \{\partial X \cap Y\}$, and edges $E := \{\{X^u, X^v\} \mid u, v \in \{0,1\}^\ell, d(u,v) \leq 1\} \cup \{\{X^{\mathbf{1}}, \partial X \cap Y\}\}$, where $\mathbf{1} = 1^\ell$. The graph is essentially a binary hypercube of dimension ℓ, with the node $\partial X \cap Y$ attached to it. Furthermore, define a cost function c over the edges as $c(X^u, X^v) := |X^u| \cdot |X_i^v|$ for all $u, v \in \{0,1\}^\ell$, with $d(u,v) = 1$ and $u_i \neq v_i$. On edges $\{X^u, X^u\}$, the cost function is defined as $c(X^u, X^u) := |X^u| \cdot n \cdot (m-1) - \sum_{Z \in \Gamma(X^u) \setminus X^u} c(X^u, Z)$, and finally for the last edge, $c(X^{\mathbf{1}}, \partial X \cap Y) := |X^{\mathbf{1}}| \cdot |Y_{i^*}|$. The generalised degree $c(X^u)$ of a node in the graph is now $c(X^u) = \sum_{Z \in \Gamma(X^u)} c(X^u, Z) = |X^u| \cdot n \cdot (m-1)$. Hence, the transition probabilities in a general random walk over G_{XY} becomes $Prob(W_{t+1} = X^v \mid W_t = X^u) = |X_i^v|/(n \cdot (m-1))$. This is exactly the probability that RLS moves to a search point in set X^v, given that the current search point belongs to set X^u. The expected runtime of RLS on TwoNets$_{X,Y}$ is therefore h_{xy}, where $x = X^{\mathbf{0}}$ and $y = \partial X \cap Y$.

For the integer lattice, the behaviour of RLS corresponded to a simple random walk with identical transition probabilities. Calculation of the hitting time h_{xy} in Eq. (6) exploited the fact that the term $\sum_{z \neq y} d(z)(R_{y'z} - R_{xz})$ vanishes, which does not hold for the general random walk used in the hypercube. The expression for the hitting time in the hypercube therefore becomes more complicated. In the following, let $x = X^{\mathbf{0}}, y' = X^{\mathbf{1}}$ and $y = \partial X \cap Y$ and note that $R_{zy} = R_{zy'} + R_{y'y}$ for any $z \neq y$.

$$h_{xy} = \frac{1}{2} \sum_z c(z)(R_{xy} + R_{yz} - R_{xz}) = \frac{1}{2} c(y)(R_{xy} + R_{yy} - R_{xy}) \tag{9}$$

$$+ \frac{1}{2} \sum_{z \neq y} c(z)(R_{xy'} + R_{y'y} + R_{yy'} + R_{y'z} - R_{xz}) \tag{10}$$

$$= R_{yy'} \sum_{z \neq y} c(z) + \frac{1}{2} \sum_{z \neq y} c(z)(R_{xy'} + R_{y'z} - R_{xz}) \tag{11}$$

$$= R_{yy'} \cdot n \cdot (m-1) \cdot |X| + h_{xy'}. \tag{12}$$

The value of $R_{yy'}$ is known, so the main difficulty in the expression above is the term $h_{xy'}$. This term vanishes in the case of maximal accessibility, allowing the hitting time to be calculated easily for this case.

Proposition 3. *Let X and Y be two convex neutral networks in Σ^n, $|\Sigma| = m$ having maximal accessibility $A(Y \leftarrow X) = 1$. Then the expected runtime T_n of RLS on* TwoNets$_{X,Y}$ *is bounded by $n \leq T_n \leq n \cdot (m-1)$.*

Proof. The accessibility $A(Y \leftarrow X)$ has the maximal value of 1 for configurations of X and Y satisfying $X_i = Y_i = \Sigma$ for all i, $1 \leq i \leq n-1$, $X_n \cup Y_n = \Sigma$ and $X_n \cap Y_n = \emptyset$. In such configurations, $X = X^1$, hence the graph corresponding to the network consists of only two nodes, X and $\partial X \cap Y$. Therefore $x = y'$ and the term $h_{xy'}$ in Eq. (12) vanishes. The resistance $R_{yy'} = c(X^1, \partial X \cap Y)^{-1}$ is $1/(|X| \cdot |Y_n|)$, so the hitting time h_{xy} is therefore at least n and at most $n \cdot (m-1)$.

Omitting the second term $h_{xy'}$ in Eq. (12) gives a lower bound on the hitting time h_{xy}. This lower bound can in turn be shown to be dependant on the accessibility, giving the main result of this work.

Theorem 1. *Let X and Y be two convex neutral networks in Σ^n, $|\Sigma| = m$. Then the expected runtime T_n of RLS on* TwoNets$_{X,Y}$ *is bounded from below by $T_n \geq \frac{1}{(m-1)} \cdot A(Y \leftarrow X)^{-1}$.*

Proof. The expected runtime of RLS on the problem is given by the hitting time h_{xy} of the general random walk in the graph G_{XY} defined above. First note that the cardinality of ∂X is given by $|\partial X| = \sum_{i=1}^{n} |X_1| \cdots |\overline{X}_i| \cdots |X_n| = |X| \sum_{i=1}^{n} |\overline{X}_i|/|X_i|$. The fraction $|\overline{X}_i|/|X_i|$ is at most $(m-1)$ for all i, $1 \leq i \leq n$, therefore the cardinality of the border is bounded from above by $|\partial X| \leq n \cdot (m-1) \cdot |X|$. Combining this with Eq. (12) now yields the desired result

$$h_{xy} \geq R_{yy'} \cdot n \cdot (m-1) \cdot |X| = \frac{n \cdot (m-1) \cdot |X|}{c(X^1, \partial X \cap Y)} = \frac{n \cdot (m-1) \cdot |X|}{c(\partial X \cap Y, X^1)} \tag{13}$$

$$\geq \frac{|\partial X|}{|\partial X \cap Y| \cdot |X_n|} \geq \frac{1}{m-1} \cdot A(Y \leftarrow X)^{-1}. \tag{14}$$

The following corollary to Theorem 1 complements the result in Proposition 3 with the case of minimal accessibility.

Corollary 1. *Let X and Y be two convex neutral networks in Σ^n, $|\Sigma| = m$ having minimal accessibility $A(Y \leftarrow X)$. Then the expected runtime T_n of RLS on* TwoNets$_{X,Y}$ *is bounded from below by $T_n \geq m^{n-1}$.*

Proof. The accessibility $A(Y \leftarrow X)$ has the minimal value for configurations of networks X and Y satisfying $|Y_j| = 1$ for all $1 \leq j \leq n$, $|X_n| = 1$ and $|X_j| = m$ for $1 \leq j \leq n-1$. For such networks, the accessibility is $A(Y \leftarrow X) = 1/(m^{n-1} \cdot (m-1))$, which by Theorem 1 gives a lower bound of m^{n-1} on the expected runtime of RLS.

5 Conclusion

This paper analysed the runtime of a simple search heuristic called RLS on TwoNets, a simple problem with two neutral networks. The objective was to determine the importance of the accessibility between the networks. The analysis used a connection between random walks on graphs and electrical resistive networks. Two different search spaces were considered, the two-dimensional integer lattice and the set of strings over a general alphabet, i.e. the hypercube.

For square convex networks in the two-dimensional integer lattice, changing the accessibility from maximal to minimal value only increases the runtime by an $\ln n$-factor. For convex neutral networks in the hypercube, changing the accessibility from maximal to minimal value increases the runtime from linear to exponential time in the length of the string. Furthermore, in the hypercube, the accessibility value can be used to give a lower bound on the runtime of RLS on the problem. These results show that accessibility can be useful to understand evolutionary dynamics in neutral search spaces.

References

1. Kimura, M.: Evolutionary rate at the molecular level. Nature **217** (1968) 624–626
2. Ebner, M., Shackleton, M., Shipman, R.: How neutral networks influence evolvability. Complexity **7** (2001) 19–33
3. Knowles, J.D., Watson, R.A.: On the Utility of Redundant Encodings in Mutation-Based Evolutionary Search. In: Proceedings of the 7th International Conference on Parallel Problem Solving from Nature. Volume 2439 of LNCS. (2002) 88–98
4. Droste, S., Jansen, T., Wegener, I.: On the analysis of the (1+1) evolutionary algorithm. Theoretical Computer Science **276** (2002) 51–81
5. Jansen, T., Wegener, I.: Evolutionary algorithms - how to cope with plateaus of constant fitness and when to reject strings of the same fitness. IEEE Transactions on Evolutionary Computation **5** (2001) 589–599
6. Lehre, P.K., Haddow, P.C.: Phenotypic Complexity and Local Variations in Neutral Degree. In: Proceedings of the Sixth International Workshop on Information Processing in Cells and Tissues. (2005) To appear in BioSystems Journal.
7. Stadler, B.M.R., Stadler, P.F., Wagner, G.P., Fontana, W.: The topology of the possible: Formal spaces underlying patterns of evolutionary change. Journal of Theoretical Biology **213** (2001) 241–274
8. Lehre, P.K., Haddow, P.C.: Accessibility between Neutral Networks in Indirect Genotype-Phenotype Mappings. In: Proceedings of the Congress on Evolutionary Computation. Volume 1., IEEE Press (2005) 419–426
9. Barnett, L.: Evolutionary Search on Fitness Landscapes with Neutral Networks. PhD thesis, CCNR & COGS, University of Sussex (2002)
10. Doyle, P.G., Snell, J.L.: Random Walks and Electric Networks. Mathematical Association of America (1984)
11. Chandra, A.K., Raghavan, P., Ruzzo, W.L., Smolensky, R., Tiwari, P.: The electrical resistance of a graph captures its commute and cover times. Computational Complexity **6** (1996) 312–340
12. Tetali, P.: Random walks and the effective resistance of networks. Journal of Theoretical Probability **4** (1991) 101–109

The Emergence of Cooperation in Asynchronous Iterated Prisoner's Dilemma

David Cornforth[1] and David Newth[2]

[1] School of ITEE, University of NSW @ ADFA
Northcott Drive, Canberra ACT 2600, Australia
david.cornforth@adfa.edu.au

[2] CSIRO Centre for Complex Systems Science
CSIRO Marine and Atmospheric Research
Canberra ACT, Australia
david.newth@csiro.au

Abstract. The Iterated Prisoners Dilemma (IPD) has received much attention because of its ability to demonstrate altruistic behavior. However, most studies focus on the synchronous case, where players make their decisions simultaneously. As this is implausible in most biological contexts, a more generalized approach is required to study the emergence of altruistic behavior in an evolutionary context. Here, we take previous results and present a generalized Markov model for asynchronous IPD, where both, one, or neither player can make a decision at a given time step. We show that the type of asynchronous timing introduced into the model influences the strategy that dominates. The framework presented here is a more biologically plausible scenario through which to investigate altruistic behavior.

1 Introduction

Throughout the animal kingdom, individuals frequently show altruistic cooperative behavior in situations where it would seem, on cursory examination, to be more beneficial to act in an egocentric way. Many of these scenarios can be modeled as a game in which two players interact, and each elects to either cooperate (i.e. play C) or defect (i.e. play D). If both players choose to co-operate, both earn a payoff or "reward" R, which is larger than the "punishment" payoff P, that is received if both players defect. Should the players choose different strategies, the player who defects receives the maximum payoff T "temptation", while the player who cooperates receives the "suckers" payoff S. The payoff must also satisfy the following conditions: $T > R > P > S$ and $2R > (T + S)$. From this, the dilemma becomes apparent: regardless of the player's choice, an individual is better off defecting in a game consisting of a single round. Formally, this game is known as the Prisoners Dilemma, and has been used as the canonical example studying the cooperation behavior in political, social, ethical and biological contexts [1,2].

Should the game be repeated an unknown number of times, however, mutual cooperation becomes a realistic option. This is known as the iterated Prisoner's

T.-D. Wang et al. (Eds.): SEAL 2006, LNCS 4247, pp. 742–749, 2006.

Dilemma (IPD). The most successful strategy is no longer Always Defect (AD), but belongs to a family of less egocentric strategies, of which Tit-For-Tat (TFT) is the most renowned [1,2]. Others such as Win-Stay-Lose-Shift (WSLS) and Firm-But-Fair (FBF) belong to this family and can even outperform TFT [7].

One of the underlying assumptions of IPD is that players move in synchrony. That is, both players must decide on a strategy for the next round at exactly the same time: neither player is allowed to evaluate the other's strategy for the next round. However, as we have shown in [3], asynchrony can lead to a diverse range of behaviors even in very simple environments. The results of that work suggest that the relative timing of decision making can have a huge impact on the resulting behavior. While several accounts have investigated alternative timing schemes in IPD [4,7,8], we will investigate the effect of more generalized asynchronous timing schemes on the emergence of cooperation in this model.

In this account, we develop a generalized Markov model for describing the moves in the asynchronous IPD. As an initial study of the effect of asynchrony on the evolution of cooperation, we study three simple cases. These cases show that while cooperation seems inevitable, how players respond to defections differs depending upon the timing involved. Biological systems are unlikely to employ strict relative timing of decisions, and so our experiments provide a more realistic scenario under which to investigate the evolution of cooperation.

2 Prisoner's Dilemma

In this study we use the approach outlined in [7]. A player's strategy p is defined by a set of four parameters p_1, p_2, p_3, and p_4, these being the player's probabilities of cooperating based on the joint behaviors on the previous move CC, CD, DC and DD respectively (the first letter refers to the players own behavior, while the second letter refers to their opponents behavior). The value $p_i \in [0.001, 0.999]$ denotes the probability of playing C on the next round given the outcome of the previous moves. Limiting the range of values that p_i can take, means the system contains no perfect strategies. This approach takes into account errors in implementing a strategy, which is almost unavoidable in any biological context [5]. If a player p meets a player with strategy $p'_i = (p'_1, p'_2, p'_3, p'_4)$, then the transition from one round to the text is given by the Markov chain:

$$\begin{pmatrix} p_1p'_1 & p_1(1-p'_1) & (1-p_1)p'_1 & (1-p_1)(1-p'_1) \\ p_2p'_3 & p_2(1-p'_3) & (1-p_2)p'_3 & (1-p_2)(1-p'_3) \\ p_3p'_2 & p_3(1-p'_2) & (1-p_3)p'_2 & (1-p_3)(1-p'_2) \\ p_4p'_4 & p_4(1-p'_4) & (1-p_4)p'_4 & (1-p_4)(1-p'_4) \end{pmatrix}. \tag{1}$$

Given the values p_i and p'_i are positive, then matrix 1 has a unique left eigenvector $\mathbf{S} = (s_1, s_2, s_3, s_4)$ associated with the eigenvalue 1. The asymptotic frequency of outcome i is then given by s_i so that the payoff for the p-player in the PD, i.e. the limit of the mean payoff per round, is simply given by

$$s_1R + s_2S + s_3T + s_4P. \tag{2}$$

Throughout the remainder of this paper the Alexrod payoff values $R = 3$, $S = 0$, $T = 5$ and $P = 1$ are used.

2.1 Asynchronous Moves

The Markov matrix in eqn. 1 depicts the transition matrix for the synchronous case. We need to define a a generalized Markov model that describes the chances of taking a given transition. To do this we need to treat four cases: (1) When both players simultaneously move; (2) When p moves, and p' does not; (3) When p' moves, and p does not; and (4) When neither player moves.

Let q_1, q_2, q_3, and q_4 be the probability of the four cases occurring (for case 1, 2, 3 and 4, respectively), such that $\sum(q_i) = 1$. For each case there is also a corresponding Markov matrix, $\mathbf{Q_1}$, $\mathbf{Q_2}$, $\mathbf{Q_3}$, and $\mathbf{Q_4}$, which describes the system state transition. Markov matrix $\mathbf{Q_1}$ is the synchronous case and is equivalent to eqn. 1.

Cases 2 and 3, are the situations where one player moves, while the other does not. The Markov matrix $\mathbf{Q_2}$ is therefore:

$$\begin{pmatrix} p_1 & 0 & (1-p_1) & 0 \\ 0 & p_2 & 0 & (1-p_2) \\ p_3 & 0 & (1-p_3) & 0 \\ 0 & p_4 & 0 & (1-p_4) \end{pmatrix}, \tag{3}$$

and $\mathbf{Q_3}$ is:

$$\begin{pmatrix} p'_1 & (1-p'_1) & 0 & 0 \\ p'_3 & (1-p'_3) & 0 & 0 \\ 0 & 0 & p'_2 & (1-p'_2) \\ 0 & 0 & p'_4 & (1-p'_4) \end{pmatrix}. \tag{4}$$

Finally case 4, where neither player moves, $\mathbf{Q_4} = I$ (identity matrix). The general case is given by:

$$q_1.\mathbf{Q_1} + q_2.\mathbf{Q_2} + q_3.\mathbf{Q_3} + q_4.\mathbf{Q_4}, \tag{5}$$

With this result, we now briefly mention the special cases covered in [4,8]:

- The synchronous case: Here $q_i = (1, 0, 0, 0)$. The general Markov model reduces to $\mathbf{Q_1}$. This is represented in Fig.1A.
- The strict alternating case [7]: Here the Markov model alternates between $q_i = (0, 1, 0, 0)$ and $(0, 0, 1, 0)$ so the model becomes either $\mathbf{Q_2}.\mathbf{Q_3}$ or $\mathbf{Q_3}.\mathbf{Q_2}$, depending on who moves first.
- The asynchronous random alternating case [4]: Here, $q_i = (0, 0.5, 0.5, 0)$, for one round, then becomes $(0, 1, 0, 0)$ or $(0, 0, 1, 0)$ depending on the previous outcome, so the model becomes $\frac{1}{2}(\mathbf{Q_2}.\mathbf{Q_3} + \mathbf{Q_3}.\mathbf{Q_2})$. This is represented in Fig.1B.
- The asynchronous random with replacement case: Here, $q_i = (0, 0.5, 0.5, 0)$, and the model converges to the same result as the asynchronous random alternating case, and the model is effectively $\mathbf{Q_2}.\mathbf{Q_3}$. This is represented in Fig.1C.

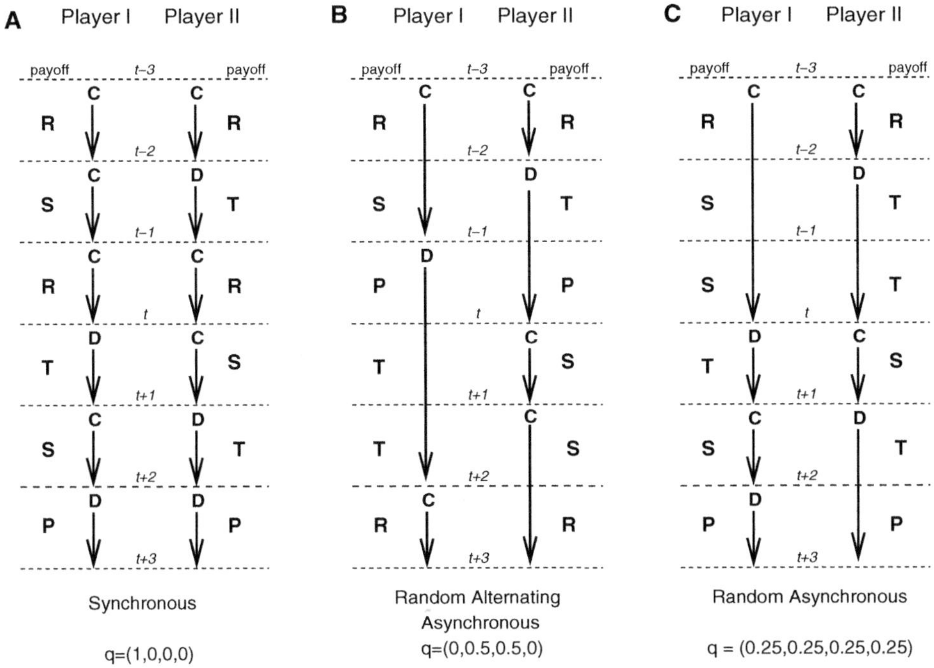

Fig. 1. Different methods for updating game state. A. In the Synchronous case, both players take a turn (make their decision regarding their next state) at exactly the same time. B. In the Asynchronous Random Alternating case, each player take exactly one turn in every 2 time steps. Although the order may change, each player gets exactly the same number of turns. C. In the Asynchronous Random with Replacement case, players take turns according to a uniform distribution of four equally likely possibilities: at $t-3$, t, and $t+1$ both players take turns together; at $t+2$ Player I takes a turn; at $t-2$ Player II takes a turn; and at $t-1$ neither player takes a turn. Although there is no guarantee that either player will receive any given number of turns, the expectation is that each player will take a turn every two time steps.

Our model is capable of reproducing the systems already studied in the literature, however our approach provides a much more flexible approach to the analysis of asynchrony.

2.2 Evolutionary Dynamics

To place the game in an evolutionary context, payoff is equated with fitness. Simulations are initially seeded with a single strategy of $(0.5, 0.5, 0.5, 0.5)$, and we keep track of relative proportion of strategies rather than the number of individuals. At each time step, all strategies play one another. The proportion of the population that a strategy occupies is adjusted in accordance with dynamics outlined in [7]. At each generation there is a 1% chance that a mutant strategy occupying 0.11% of the population will attempt to invade the pool. Those strategies occupying less than 0.1% of the population are deemed to be non-viable, and are removed from the pool. The space occupied by strategy i is calculated as per [6], with fitness determined by eqn. 2 and 5. Mutant strategies are drawn from a U-shaped distribution $[\pi x(1-x)]^{\frac{1}{2}}$, to get bias as strategies such as AD, TFT, FBF, WSLS are located toward the boundary of the four dimensional space [8].

3 Experiments and Results

We tested three examples: q_i =(1,0,0,0); $q_i = (0, 0.5, 0.5, 0)$; and $q_i = (0.25, 0.25, 0.25, 0.25)$. Each experiment was run 100 times for 10^6 generations. We measured the number of strategies occurring in the population, the average payoff, and the average values of the four p_i parameters. Results are presented in Figures 2–4. All graphs represent the average from 100 runs.

Figure 2 shows the behavior of the model for $q_i = (1, 0, 0, 0)$, which corresponds to the Synchronous case. The average values of p_i settled at $(1, 0.1, 0, 0.8)$, indicating dominance of the Win Stay Lose Shift strategy in the population. Our results for the Synchronous case validate our model by obtaining very similar results to those of [8]. This position was reached after a large transition at approximately $t = 300,000$, accompanied by a temporary rise in the number of strategies appearing in the simulation, and a large increase in the average payoff.

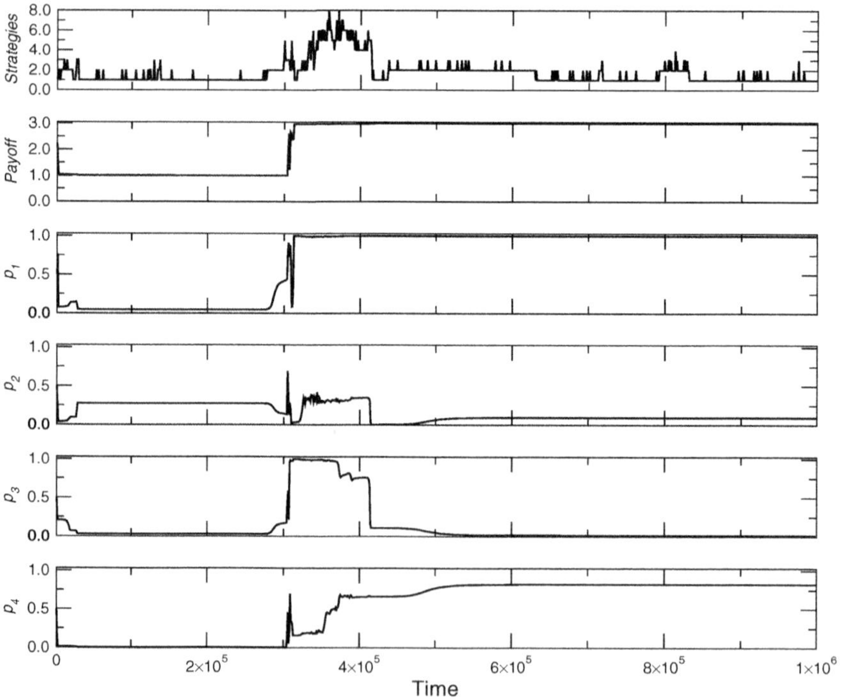

Fig. 2. Evolution of the Synchronous case, $q_i = (1, 0, 0, 0)$

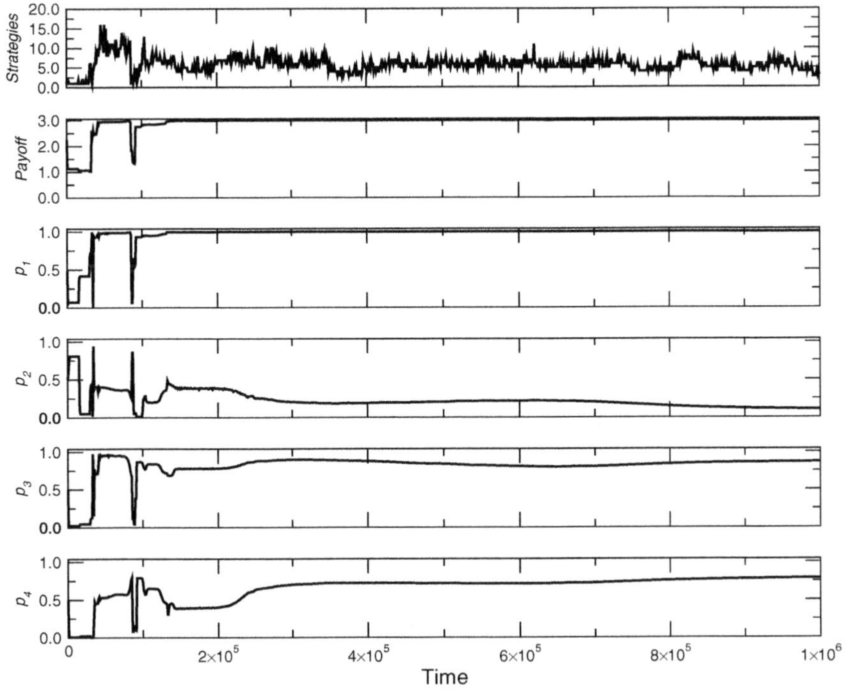

Fig. 3. Evolution of the Asynchronous case, $q_i = (0, 0.5, 0.5, 0)$

Figure 3 shows the behavior of the model for $q_i = (0, 0.5, 0.5, 0)$, the Asynchronous Random with Replacement case. The model settles down relatively quickly to final values of $p_i = (1, 0.1, 0.9, 0.8)$, indicating dominance of the FBF strategy. Note that this cannot be compared with the result in [8] for their asynchronous case, as they took a very different approach.

Figure 4 shows the behavior of the model for $q_i = (0.25, 0.25, 0.25, 0.25)$, which is asynchronous, but with the possibility of both players choosing their strategy together every fourth time step, on average. The model appears to settle at the values $p_i = (1, 0, 0.7, 1)$, which again suggests a dominance by the FBF strategy. However, transient behavior lasts longer, and alternate strategies were being examined later in time (around $t = 800,000$), when compared to the two cases examined above. It is interesting to note that the FBF strategy is probably dominating at around $t = 600,000$, but alternatives were examined after this time. These long transients suggest chaotic behavior, which may confer a degree of adaptability upon the model: individuals may be able to quickly adapt to new strategies of their opponents.

4 Discussion and Conclusion

In this work, we have presented, for the first time, a generalized Markovian model of the asynchronous iterated prisoners' dilemma. Using this model, we

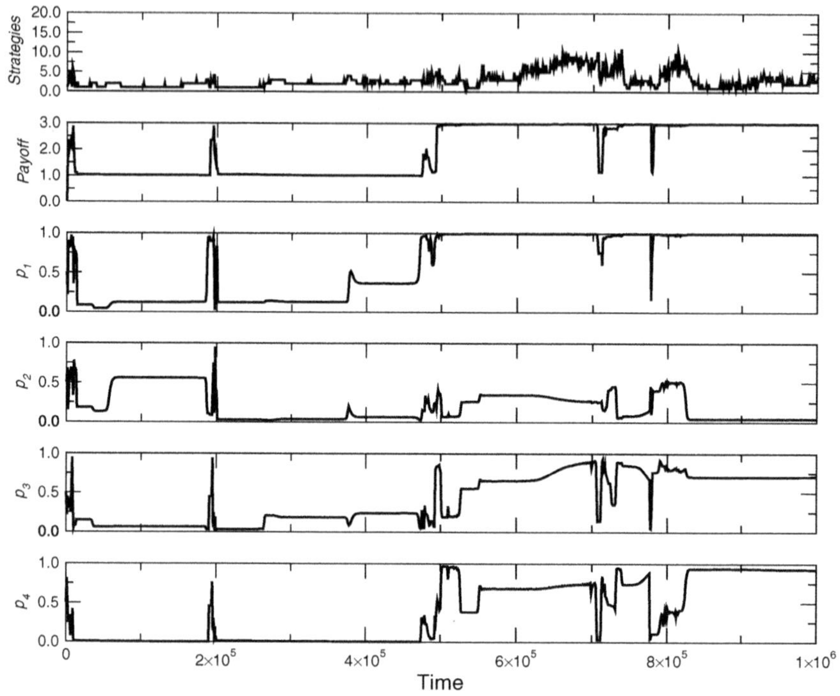

Fig. 4. Evolution of the model using $q_i = (0.25, 0.25, 0.25, 0.25)$

have shown that the synchronous IPD is a special case, and our results agree with previously published analysis. We have then shown how our model may be used under a spectrum of assumptions about the relative timing of decisions in the asynchronous IPD, where this spectrum is given by the parameter space q_i. This enables researchers to experiment with a wider range of scenarios, which could include more biologically realistic models of emerging altruism.

Comparing the three cases we examined, we have shown that the timing strategy has a big impact upon the behavior of the model. The implication is that the pursuit of a biologically unrealistic model (such as assuming that organisms in an evolutionary context make decisions in synchrony) could produce lower confidence in results obtained, depending on the context and objectives of the model.

We also conclude that cooperation seems to be inevitable in an iterated game, regardless of the timing used. In all three cases examined, the average payoff approached 3, indicating cooperation, in contrast to AD, which would produce a payoff of 1. In all three cases, the average payoff began at 1, and increased to 3 during evolution of the model. This suggests a clear evolutionary advantage of altruistic over egocentric behavior.

However, there are marked differences in the three cases we examined. The second case, (3), showed the fastest dominance of one strategy over the others. This is surprising, since one would expect that a synchronous game (proceeding in a more controlled, orderly fashion) would resolve the problem of strategy choice more quickly. The second case also seems to have maintained the most diversity in the population, examining a peak of 15 strategies at $t = 50,000$, compared to 8 for the Synchronous case, and 10 for the third case examined. In addition, the number of strategies remained relatively high right up to the end of the simulation.

In future work, we plan to consider more cases, and also to expand our work to examine asynchrony in other games such as hawk-dove, battle of the sexes, and the stag hunt game. Each of these games have different Nash and dominating equilibria, suggesting that asynchrony may lead to novel cooperative dynamics.

The main contribution of this work is a generalized model that provides a wider range of possible scenarios, in order to prepare models of cooperation that can be more biologically realistic. In providing this generalized model, we have shown that asynchrony is associated with greater cooperation between players.

References

1. Alexrod, R.: The evolution of cooperation. New York: Basic Books. (1984)
2. Axelrod, R. and Hamilton, W.D.:The Evolution of Cooperation. Science. **211** (1981) 1390–1396
3. Cornforth, D., Green, D. G., and Newth, D.: Ordered asynchronous processes in multi-agent systems. Physica D. **204** (2005) 70–82
4. Frean, M. R.: The prisoner's dilemma without synchrony. Proceedings of the Royal Society, Series B. **257** (1994) 75–79
5. May, R. M.: More evolution of cooperation. Nature. **327** (1987) 15–17
6. Maynard-Smith, J.: 1982 Evolution and the theory of games. Cambridge University Press. (1982)
7. Nowak, M. and Sigmund, K.: A strategy of win-stay, lose-shift that outperforms tit-for-tat in the Prisoner's Dilemma game. Nature. **364** (1993) 56–58
8. Nowak, M. and Sigmund, K.: The alternating Prisoner's Dilemma. Journal of Theoretical Biology. **168** (1994) 219–226

Comparison of Two ICA Algorithms in BSS Applied to Non-destructive Vibratory Tests

Juan-José González de-la-Rosa[1], Carlos G. Puntonet[2], Rosa Piotrkowski, Antonio Moreno, and Juan-Manuel Górriz

[1] University of Cádiz, Research Group TIC168 - Computational Electronics Instrumentation and Physics Engineering, EPSA, Av. Ramón Puyol S/N. 11202, Algeciras-Cádiz, Spain
juanjose.delarosa@uca.es

[2] University of Granada, Department of Architecture and Computers Technology, ESII, C/Periodista Daniel Saucedo. 18071, Granada, Spain
carlos@atc.ugr.es

Abstract. Two independent component analysis (ICA) algorithms are applied for blind source separation (BSS) in a synthetic, multi-sensor situation, within a non-destructive pipeline test. CumICA is based in the computation of the cross-cumulants of the mixtures and needs the aid of a digital high-pass filter to achieve the same SNR (up to -40 dB) as Fast-ICA. Acoustic Emission (AE) sequences were acquired by a wide frequency range transducer (100-800 kHz) and digitalized by a 2.5 MHz, 8-bit ADC. Four common sources in AE testing are linearly mixed, involving real AE sequences, impulses and parasitic signals modelling human activity.

1 Introduction

AE and vibratory signal processing usually deals with separation of multiple events which sequentially occur in several measurement points during a non-destructive test. In most situations, the test involves the study of the behavior of secondary events, or reflections, resulting from an excitation (the main event). These echoes carry information related with the medium through which they propagate, as well as surfaces where they reflect [1].

But, in almost every measurement scenario, an acquired sequence contains information regarding not only the AE under study, but also additive noise processes (mainly from the measurement equipment) and other parasitic signals, e.g. originated by human activity or machinery vibrations. As a consequence, in non-favorable SNR cases, BSS should be accomplished before characterization [2], in order to obtain the most reliable *fingerprint* of the AE event.

The purpose of this paper is twofold. First we show how two ICA algorithms separate the true signal from the parasitic ones taking a multi-sensor array of inputs (SNR=-40). Secondly, we compare performances of Cum-ICA and Fast-ICA, resulting that Cum-ICA needs the aid of a post high-pass filter to achieve the same SNR as Fast-ICA. This comparison could be interesting for a future implementation of the code in an automatic analysis system.

T.-D. Wang et al. (Eds.): SEAL 2006, LNCS 4247, pp. 750–755, 2006.

The paper is structured as follows: in Section 2 we make a brief progress report on the characterization of vibratory emissions. Section 3 summarizes the ICA model and outlines its properties. Results are displayed in section 4. Finally, conclusions and achievements are drawn in section 5.

2 Acoustic Emission Signal Processing

Elastic energy travels through the material as a stress wave and is typically detected using a piezoelectric transducer, which converts the surface displacement (vibrations) to an electrical signal. AE signal processing is used for the detection and characterization of failures in non-destructive testing and identification of low-level biological signals [2]. Most AE signals are non-stationary and they consist of overlapping bursts with unknown amplitude and arrival time. These characteristics can be described by modelling the signal by means of neural networks, and using wavelet transforms [1]. These second-order techniques have been applied in an automatic analysis context of the estimation of the time and amplitude of the bursts. Multiresolution has proven good performance in de-noising (up to SNR=-30 dB, with modelled signals) and estimation of time instances, due to the selectivity of the wavelets filters banks [3].

Higher order statistics (HOS) have enhanced characterization in analyzing biological signals due to the capability for rejecting noise [4]. This is the reason whereby HOS could be used as part of an ICA algorithm.

3 The ICA Model and Algorithms

3.1 Outline of ICA

BSS by ICA is receiving attention because of its applications in many fields such as speech recognition, medicine and telecommunications [5]. Statistical methods in BSS are based in the probability distributions and the cumulants of the mixtures. The recovered signals (the source estimators) have to satisfy a condition which is modelled by a contrast function. The underlying assumptions are the mutual independence among sources and the non-singularity of the mixing matrix [2],[6].

Let $\mathbf{s}(t) = [s_1(t), s_2(t), \ldots, s_m(t)]^T$ be the transposed vector of sources (statistically independent). The mixture of the sources is modelled via

$$\mathbf{x}(t) = \mathbf{A} \cdot \mathbf{s}(t) \tag{1}$$

where $\mathbf{x}(t) = [x_1(t), x_2(t), \ldots, x_m(t)]^T$ is the available vector of observations and $\mathbf{A} = [a_{ij}] \in \Re^{m \times n}$ is the unknown mixing matrix, modelling the environment in which signals are mixed, transmitted and measured [7]. We assume that $\mathbf{A}$ is a non-singular n×n square matrix. The goal of ICA is to find a non-singular n×m separating matrix $\mathbf{B}$ such that extracts sources via

$$\hat{\mathbf{s}}(t) = \mathbf{y}(t) = \mathbf{B} \cdot \mathbf{x}(t) = \mathbf{B} \cdot \mathbf{A} \cdot \mathbf{s}(t) \tag{2}$$

where $\mathbf{y}(t) = [y_1(t), y_2(t), \ldots, y_m(t)]^T$ is an estimator of the sources. The separating matrix has a scaling freedom on each row because the relative amplitudes of sources in $\mathbf{s}(t)$ and columns of $\mathbf{A}$ are unknown [6]. The transfer matrix $\mathbf{G} \equiv \mathbf{BA}$ relates the vector of independent (original) signals to its estimators.

3.2 CumICA

High order statistics, known as cumulants, are used to infer new properties about the data of non-Gaussian processes. Before, such processes had to be treated as if they were Gaussian, but second order statistics are phase-blind. The relationship among the cumulant of r stochastic signals and their moments of order $p, p \leq r$, can be calculated by using the *Leonov-Shiryayev* formula [8]:

$$\begin{aligned} Cum(x_1, ..., x_r) = \sum (-1)^k \cdot (k-1)! \cdot E\{\prod_{i \in v_1} x_i\} \\ \cdot E\{\prod_{j \in v_2} x_j\} \cdots E\{\prod_{k \in v_p} x_k\} \end{aligned} \tag{3}$$

where the addition operator is extended over all the set of v_i $(1 \leq i \leq p \leq r)$ and v_i compose a partition of *1,...,r*.

A set of random variables are statistically independent if their cross-cumulants are zero. This is used to define a contrast function, by minimizing the distance between the cumulants of the sources s(t) and the outputs y(t). As sources are unknown, it is necessary to involve the observed signals. Separation is developed using the following contrast function based on the entropy of the outputs [2]:

$$H(\mathbf{z}) = H(\mathbf{s}) + log[det(\mathbf{G})] - \sum \frac{\mathbf{C}_{1+\beta,y_i}}{1+\beta} \tag{4}$$

where $\mathbf{C}_{1+\beta,y_i}$ is the $1+\beta$th-order cumulant of the ith output, $\mathbf{z}$ is a non-linear function of the outputs y_i, $\mathbf{s}$ is the source vector, $\mathbf{G}$ is the global transfer matrix of the ICA model and $\beta > 1$ is an integer verifying that $\beta + 1$-order cumulants are non-zero.

Using equation 4, the separating matrix can be obtained by means of the following recurrent equation [7]

$$\mathbf{B}^{(h+1)} = [\mathbf{I} + \mu^{(h)}(\mathbf{C}^{1,\beta}_{y,y}\mathbf{S}^{\beta}_{y} - I)]\mathbf{B}^{(h)} \tag{5}$$

where $\mathbf{S}^{\beta}_{y}$ is the matrix of the signs of the output cumulants. Equation 5 is interpreted as a quasi-Newton algorithm of the cumulant matrix $\mathbf{C}^{1,\beta}_{y,y}$. The learning rate parameters $\mu^{(h)}$ and η are related by:

$$\mu^{(h)} = \min\left(\frac{2\eta}{1+\eta\beta}, \frac{\eta}{1+\eta\|\mathbf{C}^{1,\beta}_{y,y}\|_p}\right) \tag{6}$$

with $\eta < 1$ to avoid $\mathbf{B}^{(h+1)}$ being singular; $\|.\|_p$ denotes de p-norm of a matrix. The adaptative equation 5 converges, if the matrix $\mathbf{C}^{1,\beta}_{y,y}\mathbf{S}^{\beta}_{y}$ tends to the identity.

3.3 FastICA

One of the independent components is estimated by $y = \mathbf{b}^T\mathbf{x}$. The goal of FastICA is to take the vector $\mathbf{b}$ that maximizes the non-Gaussianity (independence)of y, by finding the maxima of its negentropy [6]. The algorithm scheme is an approximative Newton iteration, resulting from the application of the *Kuhn-Tucker* conditions. This leads to the equation 7

$$E\{\mathbf{x}g(\mathbf{b}^T\mathbf{x}) - \beta\mathbf{b} = 0\} \tag{7}$$

where g is a non-quadratic function and β is an iteration parameter.

Provided with the mathematical foundations the experimental results are outlined.

4 Experimental Results

The sensor is attached to the outer surface of the pipeline, under mechanical excitation. Each sequence comprises 2502 points (sampling frequency of 2.5 MHz and 8 bits of resolution), and assembles the main AE event and the subsequent reflections (echoes). Four sources have been considered and linearly mixed. A real AE event, an uniform white noise ($SNR = -40\ dB$), a damped sine wave and an impulse-like event. The damping sine wave models a mechanical vibration which may occur, i.e. as a consequence of a maintenance action. It has a damping factor of 2000 and a frequency of 8000 Hz. Finally, the impulse is included as a very common signal registered in vibration monitoring.

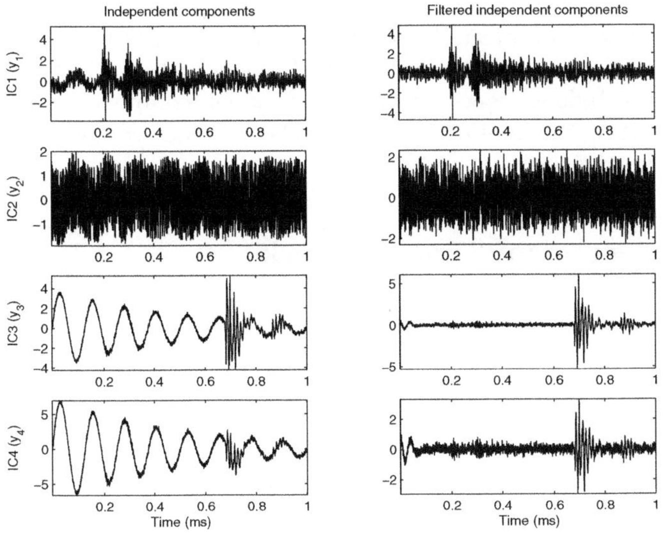

Fig. 1. Estimated and filtered sources via CumICA (ICs; Independent Components). Left column: AE event, noise, damping sine wave plus impulse, idem. Right column: filtered signals.

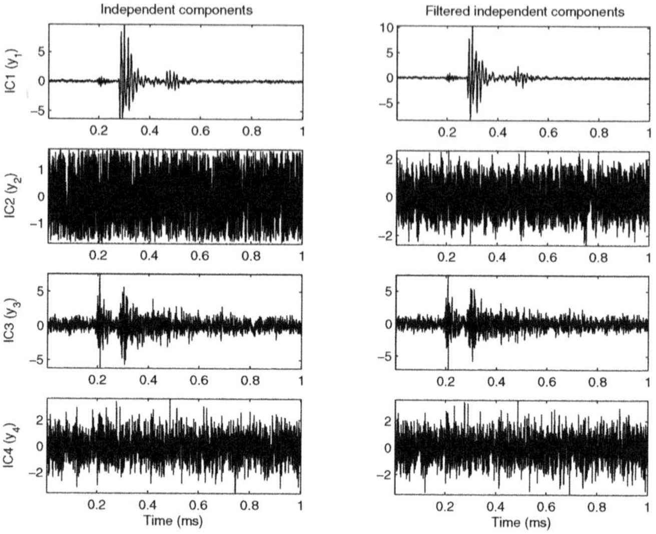

Fig. 2. Estimated and filtered sources via FastICA. Right column (very similar to the left) top to bottom: impulse, noise, AE event, noise. Post-filtering is not necessary to recover the AE event and the impulse.

The results of CumICA are depicted in Fig. 1. The damping sinusoid is considered as a frequency component of the impulse-like event because IC3 and IC4 are almost the same. The final independent components are obtained filtering the independent components by a 5th-order *Butterworth* high-pass digital filter (20000 kHz). A typical result from FastICA been included in Fig. 2.

Finally, the independence of the independent components have been performed by getting the joint distributions. These results lead us to conclude about the use of the algorithms.

5 Conclusions and Future Work

ICA is far different from traditional methods, as power spectrum, which obtain an energy diagram of the different frequency components, with the risk that low-level sounds could be masked. This experience shows that the algorithm is able to separate the sources with small energy levels in comparison to the background noise. This is explained away by statistical independence basis of ICA, regardless of the energy associated to each frequency component. The post filtering action let us work with very low SNR signals. FastICA maximizes the non-Gaussianity, so it is not necessary a filter stage. The next step is oriented in a double direction. First, a stage involving four real mixtures will be developed. Second, and simultaneously, the computational complexity of the algorithms have to be reduced to perform an implementation.

Acknowledgement

The authors would like to thank the *Spanish Ministry of Education and Science* for funding the projects DPI2003-00878, TEC2004-06096 and PTR1995-0824-OP.

References

1. de la Rosa, J.J.G., Lloret, I., Ruzzante, J., Piotrkowski, R., Armeite, M., Pumarega, M.L.: Higher-order characterization of acoustic emission signals. In: CIMSA 2005, Proceedings of the 2005 IEEE International Conference on Computational Intelligence for Measurement Systems and Applicacions, Giardini Naxos, Italy, 20-22 July 2005, ISBN 0-7803-9026-1; IEEE Catalog Number 05EX1037 (2005) 296–300 Paper CM5027. Oral Presentation in the Session 16 Advanced Signal Processing 2.
2. de la Rosa, J.J.G., Puntonet, C.G., Lloret, I.: An application of the independent component analysis to monitor acoustic emission signals generated by termite activity in wood. Measurement (Ed. Elsevier) **37** (2005) 63–76 Available online 12 October 2004.
3. de la Rosa, J.J.G., Puntonet, C.G., Lloret, I., Górriz, J.M.: Wavelets and wavelet packets applied to termite detection. Lecture Notes in Computer Science (LNCS) **3514** (2005) 900–907 Computational Science - ICCS 2005: 5th International Conference, GA Atlanta, USA, May 22-25, 2005, Proceedings, Part I.
4. Puntonet, C.G., de la Rosa, J.J.G., Lloret, I., Górriz, J.M.: Recognition of insect emissions applying the discrete wavelet transform. Lecture Notes in Computer Science (LNCS) **3686** (2005) 505–513 Third International Conference on Advances in Pattern Recognition, ICAPR 2005 Bath, UK, August 22-25, 2005, Proceedings, Part I.
5. Mansour, A., Barros, A.K., Onishi, N.: Comparison among three estimators for higher-order statistics. In: The Fifth International Conference on Neural Information Processing, Kitakyushu, Japan (1998)
6. Hyvärinen, A., Oja, E.: Independent Components Analysis: A Tutorial. Helsinki University of Technology, Laboratory of Computer and Information Science (1999)
7. de la Rosa, J.J.G., Puntonet, C.G., Górriz, J.M., Lloret, I.: An application of ICA to identify vibratory low-level signals generated by termites. Lecture Notes in Computer Science (LNCS) **3195** (2004) 1126–1133 Proceedings of the Fifth International Conference, ICA 2004, Granada, Spain.
8. Swami, A., Mendel, J.M., Nikias, C.L.: Higher-Order Spectral Analysis Toolbox User's Guide. (2001)

An Application of Intelligent PSO Algorithm to Adaptive Compensation for Polarization Mode Dispersion in Optical Fiber Communication Systems

Xiaoguang Zhang[1,2], Lixia Xi[1,2], Gaoyan Duan[1,2], Li Yu[1,2], Zhongyuan Yu[1,2], and Bojun Yang[1,2]

[1] Department of Physics, Beijing University of Posts and Telecommunications, Beijing 100876, China
{Zhang}zhang.x.g@263.net
[2] Key Laboratory of Optical Communication and Lightwave Technologies, Ministry of Education, Beijing 100876, China

Abstract. In high bit rate optical fiber communication systems, Polarization mode dispersion (PMD) is one of the main factors to signal distortion and needs to be compensated. Because PMD possesses the time-varying and statistical properties, to establish an effective control algorithm for adaptive or automatic PMD compensation is a challenging task. Widely used control algorithms are the gradient-based peak search methods, whose main drawbacks are easy being locked into local sub-optima for compensation and no ability to resist noise. In this paper, we introduce a new evolutionary approach, particle swarm optimization (PSO), into automatic PMD compensation as feedback control algorithm. The experiment results showed that PSO-based control algorithm had unique features of rapid convergence to the global optimum without being trapped in local sub-optima and good robustness to noise in the transmission line that had never been achieved in PMD compensation before.

1 Introduction

In high bit rate optical fiber communication systems, when bit rate is beyond 10Gb/s, polarization mode dispersion (PMD) has become one of the main limiting factors preventing capacity increase, because of PMD induced signal distortion. So PMD compensation has become one of hot topics in recent years [1~2]. An ordinary feedback type automatic PMD compensator can be divided into three subparts: the PMD monitoring unit, the compensation unit, and the logic control unit. The details of the three subparts will be discussed in Section 2.1. Roughly speaking, the procedure of automatic feedback controlled PMD compensation is that: the PMD monitoring unit detects the PMD correlated information as feedback signal. The logic control unit automatically controls the compensation unit by an intelligent and rapid control algorithm through analyzing the feedback signal, and the compensation completes as result. In [1] and [2] the algorithm used as the control part of a PMD compensator employed gradient based peak search methods. However, we found that as the

T.-D. Wang et al. (Eds.): SEAL 2006, LNCS 4247, pp. 756 – 765, 2006.

numbers of control parameters increased, the gradient based algorithm often became locked into local sub-optima, rather than the global-optimum. Besides, it would be less effective for a system with a relatively high noise level in the PMD monitor, because the gradient information between neighboring signals would be submerged in noise. We introduced the particle swarm optimization (PSO) into logic control unit for the adaptive PMD compensator for the first time, and realized a series of compensation experiments. With PSO algorithm we give some of the feasible and effective solutions for some critical problems that have been headaches in the field of PMD compensation for a long time.

2 A Brief Introduction to Polarization Mode Dispersion and PMD Compensation

2.1 Polarization Mode Dispersion

Polarization mode dispersion has its origins in optical birefringence [3]. In a single mode fiber, an optical wave traveling in the fiber can be represented as the linear superposition of two orthogonal polarized HE_{11} modes. In an ideal fiber, with a perfect circular symmetric cross-section, the two modes HE_{11}^x and HE_{11}^y are indistinguishable (degenerate) with the same time group delay. However, real fibers have some amount of asymmetry due to imperfections in manufacturing process or mechanical stress on the fiber after manufacture as shown in Fig. 1. The asymmetry breaks the degeneracy of the HE_{11}^x and HE_{11}^y modes, resulting in birefringence with a difference in the phase and group velocities of two modes.

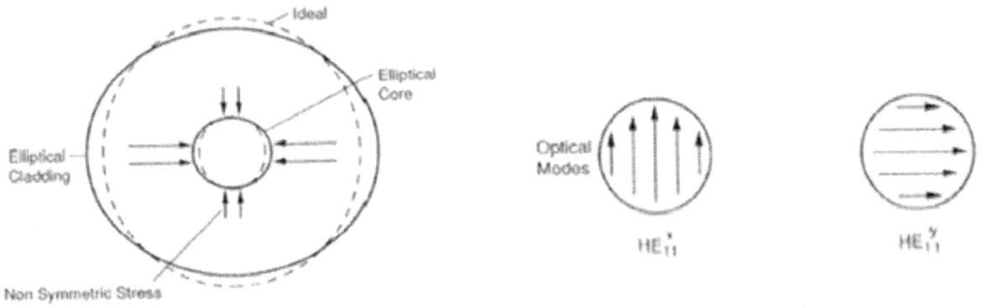

Fig. 1. Asymmetry of a real fiber and degeneracy of two orthogonal HE_{11} modes

If a pulsed optical wave that is linearly polarized at 45° to the birefringence axis is launched into a birefringent fiber, the pulse will be splitted and separated at output end of the fiber due to the different group velocities of two HE_{11} modes, as shown in Fig. 2, resulting in a signal distortion in optical transmission system. The time separation between two modes is defined as differential group delay (DGD) $\Delta\tau$. Roughly speaking, the fast and slow axis is called principal states of polarization. This phenomenon is called polarization mode dispersion.

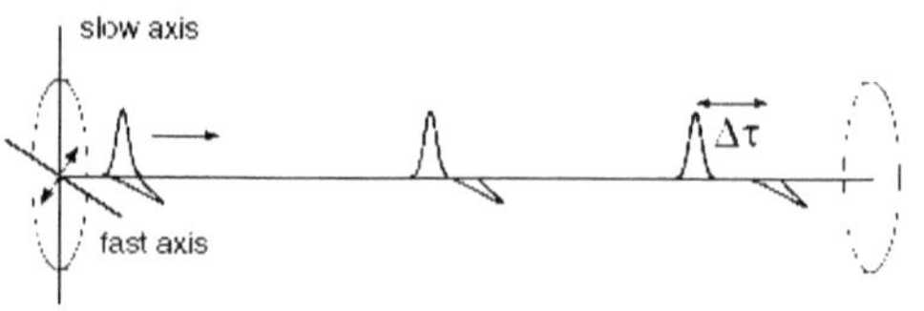

Fig. 2. Pulse splitting due to birefringence

2.2 The Configuration of Automatic PMD Compensator

Polarization mode dispersion can be divided into first-order and high-order PMD according to its Taylor-series expansion with frequency deviation $\Delta\omega$ from the carrier frequency ω_0. The first-order and second-order PMD are the two dominant impairment factors to the optical fiber transmission systems.

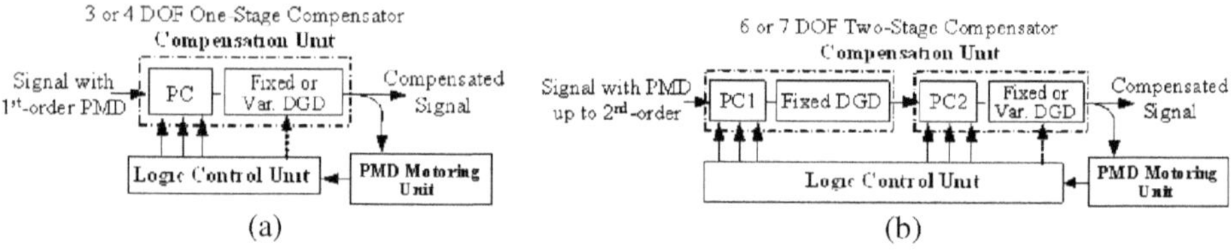

Fig. 3. The configuration of one-stage and two-stage compensators

There are two compensation schemes, pre-compensation and post-compensation. As mentioned in the section of Introduction, for the scheme of optical feedback post-compensation, the compensator has three subparts: the PMD motoring unit, the compensation unit, and the logic control unit. It is widely believed that the one-stage compensators are able to compensate PMD to the first-order. For the one-stage compensator, the compensation unit is composed of a polarization controller (PC) whose function is to transform the state of polarization (SOP) of input optical wave into output state, and a differential group delay (DGD) line with the purpose of eliminating the DGD of the input optical signals (Fig. 3 (a)). One-stage compensator have 3 or 4 control parameters (or degrees of freedom (DOF)), three for PC and one for DGD line, to be controlled depending on whether the DGD line is fixed or varied. The two-stage compensators, composed of two segments of PC+DGD, can compensate the PMD up to the second-order [4]. They have two compensation units and 6 or 7 control parameters (or DOF) to be controlled depending on whether the second delay line is fixed or varied (Fig. 3 (b)).

The automatic PMD compensation is a process for a control algorithm to find an optimal combination of control parameters, in order for the feedback signal (PMD motoring signal) to reach a global optimum, in an intelligent, fast, and reliable manner. In our experiment, the degree of polarization (DOP), obtained by an in-line polarimeter in the PMD monitoring unit, was used as feedback signal. The DOP of light wave is defined as follows using Stokes parameters S_0, S_1, S_2, S_3.

$$\mathrm{DOP} = \frac{\sqrt{S_1^2 + S_2^2 + S_3^2}}{S_0} \tag{1}$$

The DOP of any light wave varies in the range of 0 to 1. The optical pulses at the receiving end have a DOP of 1 when there is no PMD in the fiber link, and the DOP value decreases as PMD increases [2]. The polarization controller used in the compensation unit is the electrically controlled whose three cells were adjusted by controlling voltages in the experiment. In our experiment, fixed delay line was adopted. Therefore the control parameters for the one-stage compensator were 3 voltages (V_1, V_2, V_3) of PC, and the control parameters for the two-stage compensator were 6 voltages (V_1, V_2, V_3, V_4, V_5, V_6) of PC1 and PC2.

The procedure of the PMD compensation is: the control algorithm in logic control unit automatically adjusts 6 voltages (V_1, V_2, V_3, V_4, V_5, V_6) of PC1 and PC2 until the feedback signal DOP reaches its maximum.

3 Automatic PMD Compensation Using PSO Algorithm

3.1 The Theory of PSO-Based Control Algorithm

As mentioned in Section 2.2, the DOP value that is taken as the feedback signal decreases as PMD in the fiber link increases. In the feedback post-compensator, the task of the control algorithm in logic control unit is automatically searching for global maximum DOP through adjusting the multi-voltages of the polarization controllers (PCs) in the compensation unit, in an intelligent, fast, and reliable manner, which can be described mathematically as:

$$\underset{V_1, V_2, V_3 \cdots}{\mathrm{MAX}}(\mathrm{DOP}) \tag{2}$$

There is no simple method to predict DOP function in (2) in an automatic compensation system. A good algorithm is, therefore, required to solve problem (2), which is the problem of searching for the global maximum of DOP in a multi-dimensional hyperspace. The number of parameters (or degree of freedom) is the number of dimensions of the hyperspace, and is 3 for our one-stage compensator and 6 for our two-stage compensator.

Generally, more degrees of freedom result in more sub-maxima existing, which will increase the hard task of the searching algorithm. Unfortunately there exist several DOP sub-maxima in the compensation process. Fig. 4 is a typical DOP surface map for our PMD compensation system.We can see in Fig. 4 that, there are several sub-maxima beside a global maximum in the searching space. We can also find that, the DOP surface is not smooth because of the noise in the fiber link.

In most of the related literature, the adopted control algorithms have not been explicitly characterized. In [1] and [2] the algorithm used employed gradient based peak search methods. However, with the numbers of control parameters increasing, the gradient based algorithm often became locked into local sub-maxima, rather than the global-maximum. Besides, it would be less effective for a system with a relatively high noise level as shown in Fig.4, because the gradient information between neighboring signals would be submerged in noise. Therefore finding a practical feedback control algorithm with the desirable features is still a challenging task. A competitive searching algorithm in PMD compensation should at least satisfy following features: (1) rapid convergence to the global optimum rather than being trapped in local sub-optima; (2) good robustness to noise.

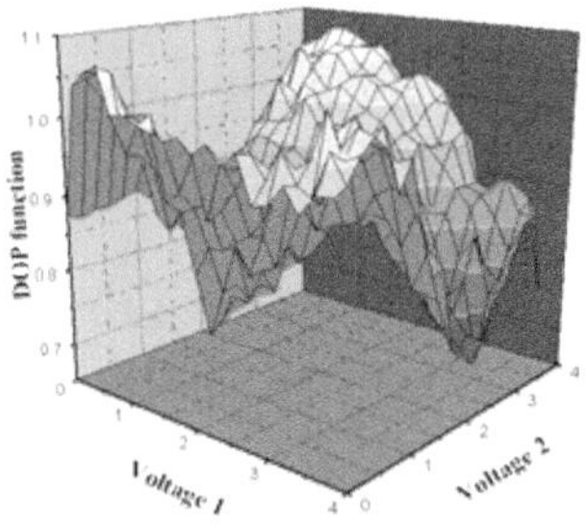

Fig. 4. The DOP surface map in the PMD compensation system

The PSO algorithm, proposed by Kennedy and Eberhart [5], has proved to be very effective in solving global optimization for multi-dimensional problems in static, noisy, and continuously changing environments [6]. We introduced for the first time the PSO technique into automatic PMD compensation in a series of experiments [7].

At the beginning, the PSO algorithm randomly initializes a population (called swarm) of individuals (called particles). Each particle represents a single intersection of multi-dimensional hyperspace. The position of the *i*-th particle is represented by the position vector $X_i = (x_{i1}, x_{i2}, \cdots, x_{iD})$. In the D-dimensional-DOF PMD compensation scheme depicted in Fig.3, the components of the *i*-th particle are represented by the combination of D voltages (V_1, V_2, …, V_D). The particles evaluate their position relative to a goal at every iteration. In each iteration every particle adjusts its trajectory (by its velocity $V_i = (v_{i1}, v_{i2}, \cdots, v_{iD})$) toward its own previous best position, and toward the previous best position attained by any member of its topological neighborhood. If any particle's position is close enough to the goal function, it is considered as having found the global optimum and the recurrence is ended.

Generally, there are two kinds of topological neighborhood structures: global neighborhood structure, corresponding to the global version of PSO (GPSO), and local neighborhood structure, corresponding to the local version of PSO (LPSO). For the global neighborhood structure the whole swarm is considered as the neighborhood (Fig.5 (a)), while for the local neighborhood structure some smaller number of adjacent members in sub-swarm is taken as the neighborhood [8] (Fig.5 (b)). The detail of process for implementing the global version of PSO can be found in [9]. In the global neighborhood structure, each particle's search is influenced by the best position found by any member of the entire population. In contrast, each particle in the local neighborhood structure is influenced only by parts of the adjacent members. Therefore, the local version of PSO (LPSO) has fewer opportunities to be trapped in sub-optima than the global version of PSO (GPSO).

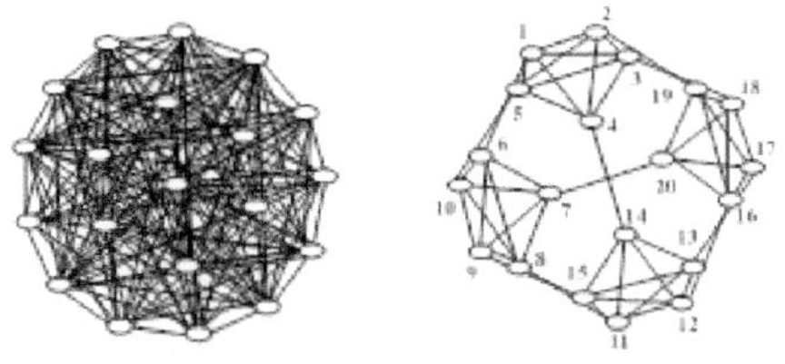

(a) (b)

Fig. 5. One of two topologic structures for (a) global neighborhood and (b) local neighborhood

In our experiment 20 particles are used either in GPSO or LPSO, which is a balance between the accuracy required in searching for the global optimum and time consumed. For LPSO neighborhood, it is found that having 5 neighbors for every particle gives the highest success rate in finding the global optimum [8] (Fig.5 (b)). The relationship for structure of LPSO we adopted is shown in Fig. 5 (b).

3.2 The Results of the Automatic PMD Compensation Using PSO

We will describe here the results of one of experiments we have done, the automatic second-order PMD compensation using two-stage compensator in 40Gb/s time-division-multiplexing (OTDM) transmission system by using PSO algorithm. We employed both GPSO and LPSO as the control algorithm respectively, in order to make a comparison of effectiveness of them.

We conducted 18 times of compensation by controlling 6 voltages of PC1 and PC2 shown in Fig.3 (b) through the GPSO and LPSO algorithms, respectively. We randomly selected the 18 different initial PMD states of the PMD emulator (corresponds to 18 different initial DOP values) for 18 different experiments. The function of PMD emulator is to emulate PMD as same as in real fiber. In every process of global DOP maximum searching, we recorded the variation of best DOP values in each iteration and, with the maximum iteration number set to 50, the results are shown in Fig. 6.

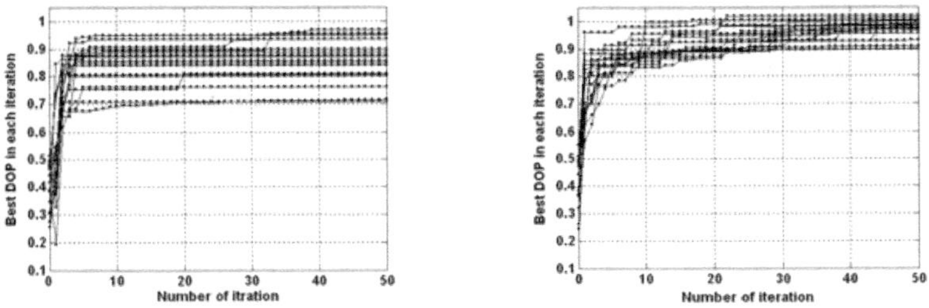

(a) (b)

Fig. 6. The best DOP vs. iteration recorded in 6-DOF second-order PMD compensation using LPSO (a) and GPSO algorithm (b)

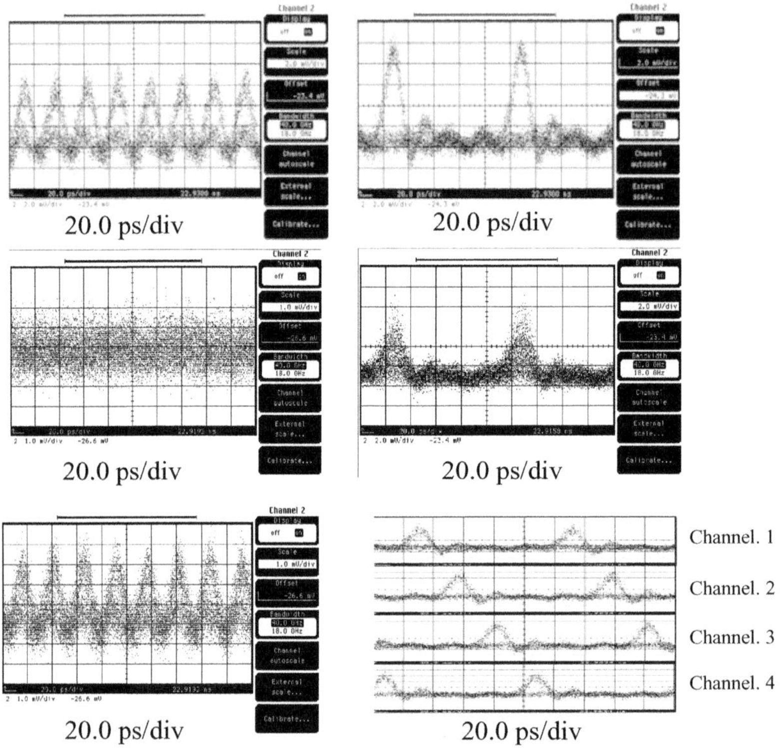

Fig. 7. Eye diagrams to show the procedure of automatic PMD compensation in 40Gb/s OTDM optical transmission system. (a) Back-to-back 40Gb/s OTDM signal. (b) Back-to-back demultiplexed 10Gb/s signal. (c) 40Gb/s signal without PMD compensation. (d) Demultiplexed 10Gb/s signal without PMD compensation. (e) 40Gb/s signal with PMD compensation. (f) Four demultiplexed 10Gb/s signals with PMD compensation.

Because of more local sub-maxima and relative high level noise in 6-DOF system, for the GPSO case there are some initial PMD states for which DOP only achieves the value of 0.7 (Fig. 6(a)), corresponding to being trapped in local sub-maxima and failure of compensation. In contrast, for the LPSO case all final searched DOP values exceed 0.9, no matter what the initial PMD state is (Fig. 6(b)). Furthermore, if we set DOP value of 0.9 as the criterion which is considered to achieve the compensation, all the DOP values reach that criterion within about 25 iterations. We can draw the conclusion that LPSO can better undertake the task of solving multi-dimensional problems, and that it is a better searching algorithm for adaptive PMD compensation up to high-order.

Fig. 7 shows the eye diagrams displayed on the screen of the oscilloscope at receiver end, in the whole procedure of automatic PMD compensation in 40Gb/s OTDM optical transmission system. The eye diagrams in left column of Fig. 7 are the 40Gb/s OTDM signals in situations of back-to-back, before and after PMD compensation. The eye diagrams in right column are the 10Gb/s demultiplexed signals with the same meaning. When we adjusted the PMD emulator with the result that the eyes were closed, implying severe PMD induced signal distortion with DOP = 0.23. After switching on the compensator, the eyes opened and DOP reached close to 1 within about 500 milliseconds through optimum searching by LPSO algorithm.

3.3 The PSO Technique Used in Tracking Process of the Control Algorithm

The algorithm for real-time adaptive PMD compensation should include two stages. First, the searching algorithm finds the global optimum from any initial PMD condition. Then the tracking algorithm starts to track the changed optimum, because the PMD in the real fiber link always randomly changes, due to changes in the environment such as temperature fluctuations etc.

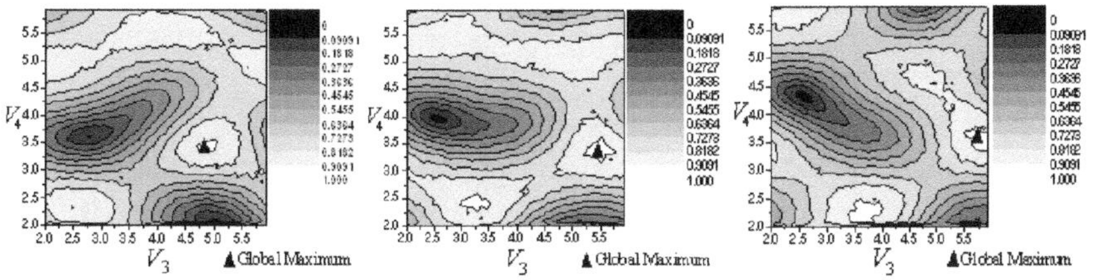

Fig. 8. Location drifting of global DOP maximum from (a) to (c) indicating the PMD changes with time

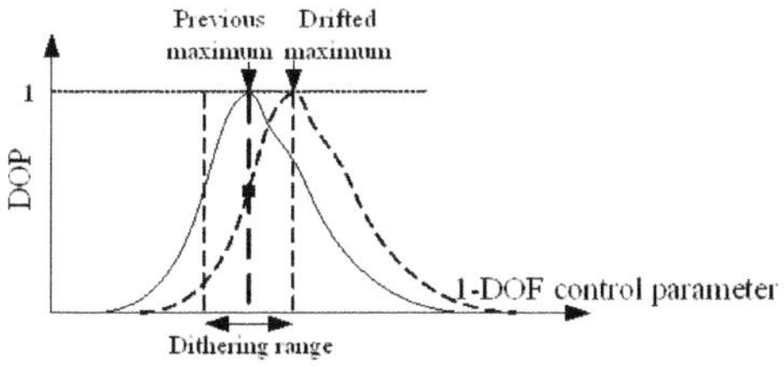

Fig. 9. The dithering solution for tracking the varied DOP maximum

When the PMD in the fiber link changes, the global DOP maximum just drifts away from the previous location as shown in Fig. 8. A natural thought of solution is a tracking method of slight disturbances or dithering around the previous DOP maximum as shown Fig. 9, which was adopted in [2]. This was also gradient-based control algorithm which would not adequate for the systems with a relatively high noise level in the PMD monitoring unit. Furthermore, for a one-DOF control system, there are two directions (positive and negative) for dithering. For a two-DOF system, there will be 8 directions (east, west, south, north, southeast, southwest, northeast, northwest), and for D-DOF, 3^D-1 directions. In conclusion, for multi-DOF systems the amount of calculation will become comparatively large, making it unsuitable for real-time tracking.

Because of its good performance in the presence of noise, and its multi-dimensional searching capability, we used the PSO searching technique in the smaller 6-dimensional local space around the previous optimum location to achieve the goal of tracking changing optimum. After the global optimum search process is completed, the tracking algorithm starts to work according to the DOP values. When the DOP in the fiber link is higher than 98% of that obtained for the previous optimum, the algorithm does nothing. Otherwise, as long as the DOP is lower than this criterion, local space searching is initiated. The size of the local searching space is adjusted with time according to the deviation from the criterion DOP, which is set to 0.9×98%=0.88 for the experiment. For the tracking algorithm, 5 particles and GPSO were adopted because of the faster speed needed for tracking and the smaller space in which to search. The flow chart of the control program is shown in Fig. 10.

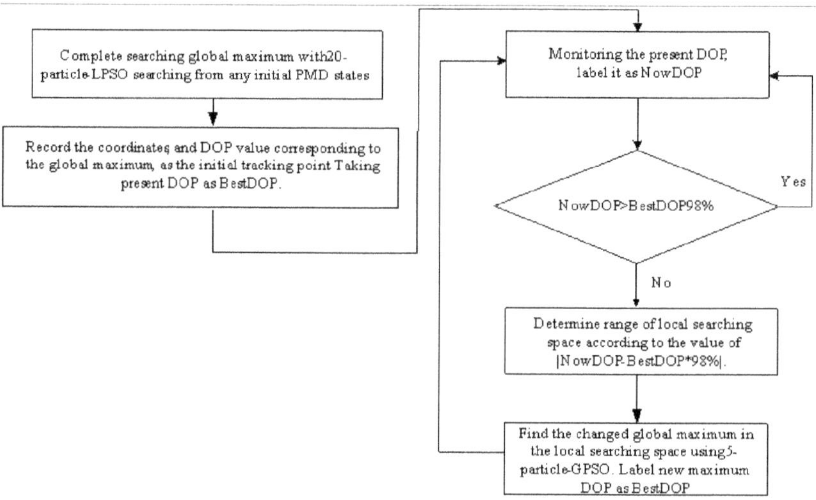

Fig. 10. The flow chart of the control program based on PSO

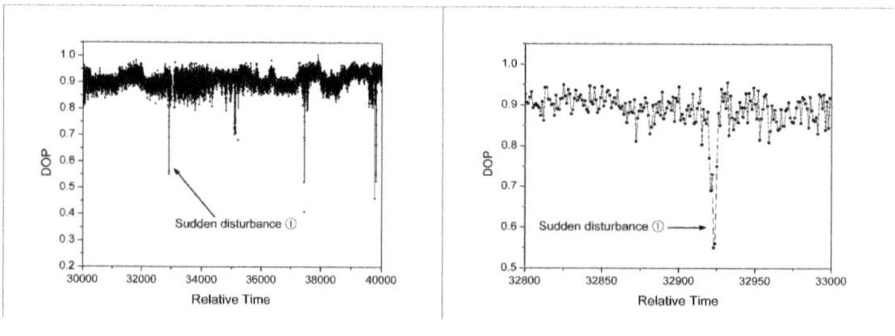

Fig. 11. The performance of the tracking algorithm for tracking the changed optimum DOP. (a)In relative long time, there are some sudden disturbances by sudden rotating the PC of emulator. (b)Details of sudden disturbance ①.

In the experiment, the tracking algorithm worked well when the PMD in the fiber link varied slowly and smoothly with the environment. The eye diagrams are nearly unchanged. Fig.11 shows the tracking results with small vibration of DOP values around the criterion (0.88). But if there is a sharp disturbance in the fiber link, the tracking algorithm will force the system rapidly to recover to the condition beyond criterion.

4 Conclusions

For the fist time, we have introduced the particle swarm optimization into automatic polarization mode dispersion compensation. The experiment showed that PSO exhibited the desirable features for automatic PMD compensation of rapid convergence to the global compensation optimum searching without being trapped in local sub-optima that corresponded to the failure of compensation, and good robustness to noise in the transmission line. However, all these problems that PSO can solve have been headaches in the field of PMD compensation for a long time. By comparison of global version of PSO (GPSO) and local version of PSO (LPSO), it was shown that LPSO is better solution for automatic PMD compensation.

Acknowledgements

This work was supported by the National Natural Science Foundation of China under Grant 60577064, and the National "863" High Technology Project of China under Grant 2001AA122041.

References

1. Noé, R., Sandel, D., Yoshida-Dierolf, M., Hinz, S., Mirvoda, V., Schöpflin, A., Glingener, C., Gottwald, E., Scheerer, C., Fischer, G. Weyrauch, T., Haase, W.: Polarization Mode Dispersion Compensation at 10, 20, and 40Gb/s with Various Optical Equaliziers. J. Lightwave Technol. 17 (1999) 1602-1616
2. Rasmussen, J. C.: Automatic PMD and Chromatic Dispersion Compensation in High Capacity Transmission. In: 2003 Digest of the LEOS Summer Topical Meetings, (2003) 47-48.
3. Kogelnik, H., Jopson, R. M., Nelson, L.: Polarization-Mode Dispersion, In: Kaminow, I. P., Li, T. (eds): Optical Fiber Telecommunications, IV B. Academic Press, San Diego San Francisco New York Boston London Sydney Tokyo, (2002) 725-861
4. Kim, S.: Schemes for Complete Compensation for Polarization Mode Dispersion up to Second Order. Opt. Lett. 27 (2002) 577-579
5. Kennedy, J., Eberhart, R. C.: Paticle Swarm Optimization. In: Proc. of IEEE International Conference on Neural Networks. Piscataway, NJ, USA, (1995) 1942-1948
6. Laskari, E. C., Parsopoulos, K. E., Vrahatis, M. N.: Particle Swarm Optimization for Minimax Problems," In: Proc. of the 2002 Congress on Evolutionary Computation. Vol.2. (2002) 1576-1581
7. Zhang, X. G., Yu, L., Zheng, Y., Shen, Y., Zhou, G. T., Chen, L., Xi, L. X., Yuan, T. C., Zhang, J. Z., Yang, B. J.: Two-Stage Adaptive PMD Compensation in 40Gb/s OTDM Optical Communication System Using PSO Algorithm. Opt. Quantum Electron. 36 (2004) 1089-1104
8. Kennedy, J., Mendes, R.: Population Structure and Particle Swarm Performance. In: Proc. of the 2002 Congress on Evolutionary Computation. Vol.2. (2002) 1671-1676
9. Eberhart, R. C., Kennedy, J.: A New Optimizer Using Particle Swarm Theory. In: Proc. of the Sixth International Symposium on Micro Machine and Human Science. (1995) 39-43

Coordinated Resource Allocation by Mutual Outsourcing in Decentralized Supply Chain

Kung-Jeng Wang[1], H.M. Wee[2], Jonas Yu[3], and K.Y. Kung[4]

[1] Department of Industrial Management, National Taiwan University of Science and Technology, Taipei, Taiwan, R.O.C.
kjwang@mail.ndhu.edu.tw
[2] Department of Industrial Engineering Chung Yuan Christian University, Chungli, Taiwan
[3] Logisitcs Management Department, Takming College, Taipei, Taiwan
[4] Department of Mechanical Engineering, Nanya Institute of Technology, Chungli, Taiwan

Abstract. Lumpy demand forces capacity planners to maximize the profit of individual factories as well as simultaneously take advantage of outsourcing from its supply chain and even competitors. This study examines a capacity planning business model in which consists of many profit-centered factories (autonomous agents). We propose an ant algorithm to solve a set of non-linear mixed integer programming models with different economic objectives and constraints. The proposed method allows a mutually acceptable capacity plan for a set of customer tasks to be allocated by the negotiating parties, each with information on company objectives, cost and price. Experiment results reveal that near optimal solutions for both isolated (a single factory) and negotiation-based (between factories) environments are obtained.

1 Introduction

Lumpy demand forces capacity planners to maximize the profit of an individual factory as well as taking advantage of outsourcing to its supply chain's partners and even competitors. Although a competitive situation exists among those companies, a collaborative integration for resource and demand sharing is highly attractive to those industries.

Unfortunately, conventional capacity models only deal with the capacity planning problem for a single factory. It fails to match the capacity and order-sharing decision-making requirement among companies. Asymmetric information (i.e., resource capacity and order status) further results in inefficiency on capacity utilization and poor profitability for individual factory. As a result, the excess or insufficient resource capacity will lead to extra cost or customer service reduction.

In order to trade off between low production cost and high level of service, it is critical for managers to plan resource capacity from a broader perspective through mutual outsourcing. Such a capacity and order sharing strategy has been found in many industries such as transportation industry with shared carriers, semiconductor manufacturing industry with shared machines, and food production industry with shared outlets.

T.-D. Wang et al. (Eds.): SEAL 2006, LNCS 4247, pp. 766 – 773, 2006.

This study proposed an inter-factories capacity planning model and the corresponding negotiation in resource capacity coping procedure. The following issues are the focuses of this study. (i) What is the best cost-effectiveness portfolio and allocation of resources to fulfill orders? (ii) How to develop a mutually acceptable resource allocation plan for individual factory under the information asymmetry? (iii) How is the performance of the proposed algorithm for solving this inter-factories capacity problem?

Focusing on capacity requirement planning instead of scheduling, this study will propose an inter-factories capacity negotiation framework and solve the capacity planning problem by an ant algorithm.

2 Literature Survey

Modeling technology-economy trade-off to decide the best capacity planning of a single factory is a basis in dealing with this type of problems. Only limited studies proposed strategic concepts (Hsu 1998, Mayer 1993). Rajagopalan (1994) presented a mix integer linear programming (MILP) model to handle capacity planning of a single product. Wang and Lin (2002) developed a capacity planning model for two simultaneous resources.

The research of cooperative based resource planning usually focuses on equilibrium of the system instead of optimality of individuals. There are several studies in the literature directly dealing with capacity trading through autonomous coordination and negotiation among factories. Cachon and Zipkin (1999) addressed competitive and cooperative relationships of factories and their effects in a two-stage supply chain from the perspective of inventory policies. Jiang (2000) proposed a methodology of capacity trading for solving short-term capacity shortage incurred in wafer foundries. Chang (2001) developed a simple Internet-based auction scheme to sell foundry capacity. Their system acts as a capacity manager of a foundry that automatically negotiated with customers. Huang (2002) proposed a capacity adjustment method to build an agent-based order exchange system.

Due to trade-off between efficiency and solution quality, soft-computing based methods emerge rapidly to solve the resources allocation and expansion problem. One of recently developed, population-based, heuristic algorithms is the Ant Algorithm. Ant algorithms are popular because of their capability to solve discrete NP-hard combinational optimization problems in quickly. Many ant algorithms are inspired by ant colony optimization (ACO) meta-heuristics such as the Ant-Solver, the ant colony system (ACS) (Dorigo and Di Caro, 1999), and the MAX-MIN ant system (MMAS) (Dorigo and Stützle 2000). In many industrial situations, ACO algorithm has been shown to offer successful solution strategies for large and complex problems of production systems. For instance, Ying and Liao (2003) presented an ant colony system approach for scheduling problems.

3 Modeling the Coordinated Resource Allocation Problem

The proposed model considers a single factory capacity planning problem and an inter-factories capacity issue. Several economic models with individual factories and an inter-factories capacity negotiation model are considered to improve resources utilization among factories by sharing excess capacity.

In order to illustrate the framework, three economic models are presented, i.e., a capacity planning model with excess capacity, a capacity planning model with excess orders and an inter-factories capacity planning model.

The notations of the capacity planning models are defined as following: a_{tj} : Configuration relationship between orders and working cells. $a_{tj} \in \{0,1\}$; 1 if working cell t, $t=\{1, 2, \ldots, v\}$, can manufacture order type j, $j=\{1, 2, \ldots, n\}$, 0 otherwise. d_{kj} : Demand quantity (in pieces) of order j in planning period k, $k = \{1, 2, \ldots, p\}$. μ_{tj} : Throughput of working cell t when used to produce order type j (in pieces per period). W: Working hours of each period of time. u_{kt} : Target utilization of working cell t in period k. c_{kt} : Costs of purchasing working cell t in period k. I: Capital interest rate. $C_t^{'}$: Unit price of selling capacity of remaining working cells t of a capacity seller. C_{tj}: Capacity seller's unit price of working cell t to produce capacity buyer's order j (or, equivalently, the unit cost of resource usage of the capacity buyer). CC: Upper bound of initial budget. R : Adjustable parameter of price of resource capacity. P_j : Unit profit of order j.

Decision variables in the models are as follows: N_{kt} : Number of working cell t in period k. x_{ktj} : Quantity produced by working cell t to meet order type j in period k. δ_{kt} : Increment (or decrement) number of working cell t from period k-1 to period k. $N_{kt}^{'}$: Number of remaining working cell t in period k. d'_{kj} : Number of remaining quantities of order j in period k after task allocation of an individual factory is done. d''_{kj} : Number of remaining quantities of order j in period k after task allocation is done by an inter-factories capacity negotiation model.

A set of MILP models, each with individual objectives to formulate the resource portfolio problem is presented below. Two typical models are developed—one is for a factory with excess resources capacity and the other is for a factory with extra order capacity. Finally, an inter-factories negotiation model is developed.

3.1 Individual Factory Capacity Planning Model (Capacity over Demand)

This model assumes all demands must be satisfied with additional reserve capacity. Investment in new resources is allowed using a finite budget. The goal is to maximize profit of the working cells. The model is denoted as follows:

$$\text{Maximize } z = \sum_{k=1}^{p} \sum_{t=1}^{v} \frac{N_{kt}^{'} \cdot C_t^{'}}{(1+I)^k} \tag{1}$$

Subject to

$$\sum_{t=1}^{v} a_{tj} x_{ktj} = d_{kj}, \ \forall k, j \tag{2}$$

$$0 \le \sum_{k=1}^{p}\sum_{t=1}^{v}\frac{c_{kt}\delta_{kt}}{(1+I)^k} \le CC \tag{3}$$

$$N_{kt} - \sum_{j=1}^{n}\frac{a_{tj}x_{ktj}}{Wu_{kt}\mu_{tj}} \ge 0,\ \forall k,t \tag{4}$$

$$N'_{kt} = N_{kt} - \sum_{j=1}^{n}\frac{a_{tj}x_{ktj}}{Wu_{kt}\mu_{tj}} \tag{5}$$

$$\delta_{kt} = N_{kt} - N_{(k-1)t} \forall k,t \tag{6}$$

$$N_{kt}, x_{ktj} \ge 0,\ N_{kt},\ x_{ktj} \text{ and } \delta_{kt} \text{ are of integers}$$

Constraint (2) is a capacity balancing equation specifying the allocated quantity to each type of cells should be equal to the capacity required by the orders. Constraint (3) confines the upper bound of budget. Constraint (4) specifies the capacity limit of working cells used in each period. Constraint (5) computes the remanding capacity of working cells. Constraint (6) states the change in the number of machines of type t from period k-1 to period k, due to resource replacement.

Note that although new purchased resources may have higher efficiency than existing resources, they are also more expensive. A capacity planner thus must determine how to trade-off between investing in new resources and deploying existing ones to meet delivery dates of orders.

3.2 Individual Factory Capacity Planning Model (Capacity of Resources Provided by a Factory with Demand over Capacity)

This model assumes that orders will not be fulfilled except they are profitable. Working cells may not have enough capacity and require outsourcing for extra capacity of resources. Besides, a factory can invest in resources using a finite budget, and phase out old ones. The goal is to maximize the profit by fulfilling orders. The model is denoted as Model II.

$$\text{Maximize } z = \sum_{k=1}^{p}\sum_{t=1}^{v}\frac{\sum_{j=1}^{n} P_j \cdot x_{ktj} - c_{kt}\delta_{kt}}{(1+I)^k} \tag{7}$$

Subject to

$$\sum_{t=1}^{v} a_{tj}x_{ktj} \le d_{kj},\ \forall k,j \tag{8}$$

$$0 \le \sum_{k=1}^{p}\sum_{t=1}^{v}\frac{c_{kt}\delta_{kt}}{(1+I)^k} \le CC \tag{9}$$

$$N_{kt} - \sum_{j=1}^{n}\frac{a_{tj}x_{ktj}}{Wu_{kt}\mu_{tj}} \ge 0,\ \forall k,t \tag{10}$$

$$\delta_{kt} = N_{kt} - N_{(k-1)t}, \forall k,t \tag{11}$$

$$N_{kt}, x_{ktj} \in Z^+,\ \delta_{kt} \in Z$$

Constraints (8) are the capacity balancing equation ensuring the capacity allocated to each type of working cell not larger than that required by the orders. The remaining constraints are the same as (7).

3.3 Inter-factories Capacity Planning Model

The third model is formulated to solve an inter-factories, supply-demand negotiation problem. A mediator considers both the excess capacity and excess orders of two factories. The goal of the mediator is to maximize the net profit by completing excess orders using excess capacity of the factories. The model is denoted as follows:

$$\text{Maximize} \quad z = \sum_{k=1}^{p}\sum_{t=1}^{v}\sum_{j=1}^{n}(P_j - C_{tj})\cdot x_{ktj}/(1+I)^k \tag{12}$$

Subject to

$$\sum_{t=1}^{v} a_{tj} x_{ktj} \le d'_{kj}, \ \forall k, j \tag{13}$$

$$N'_{kt} - \sum_{j=1}^{n} \frac{a_{tj} x_{ktj}}{Wu_{kt}\mu_{tj}} \ge 0, \ \forall k, t \tag{14}$$

$$N'_{kt}, x_{ktj} \in Z^+$$

The meaning of the other constraints is the same as the ones in (1)~(11). In this model, the two individual factories may bargain (i.e., the price of resources, C_{tj}) to reach a deal.

4 Solving the Problem by Ant Algorithm

Due to the complexity of problems, this study proposes an ant heuristic algorithm to find an efficient resource portfolio plan in which the resource investment decision, capital usage plan, resource configuration, and task allocation are determined simultaneously. The algorithm not only reduces the total cost of producing all orders but also improves total profit in a factory level as well as the system level.

The proposed modified ant algorithm (called MAA) follows the classical ACO algorithmic scheme and improves its efficiency by incorporating a constraint propagation procedure for solving the problem as follows:

MAA ranks variables x_js by the constrained variable rule. The ant searching strategy begins with the repair mechanism that allows an artificial ant to construct a complete non-violated assignment of values from $D(x_j)$ (the domain of variable x_j) to variables x_js.

```
Procedure: Modified Ant Algorithm (MAA)
Begin
Set parameters and initialize pheromone trails
Sort variables by the most constrained variable rule.
Repeat
     For c  from 1 to MaxCycle
        For n from 1 to N_ants
           A ← φ
          While |A|<|X| Do
Select a variable x_j ∈ X that is not assigned in A
Choose a value v ∈ D(x_j) with probability P_A(<x_j,v>) using
the repair mechanism to guarantee all solutions are
feasible;
A← A∪{<x_j , >}
     End While
   End For
Update pheromone trails using the best ant of cycles
(the cycle best) {A_k}
If (several cycles pass by) then reinforce pheromone
trails using the best ant trail (the global best) {A_l}
End For
Until max trials reached
End
```

The proposed constraint propagation procedure together with the ant algorithm fixes the value domains of the variables that have not yet been searched. Hence, each artificial ant walks in the search space of feasible solution regions. The procedure uses formulas (2), (8) and (13) to confine values assigned to x_{ktj}. After the end of a cycle, each x_{ktj} is assigned a value and N_{kt} is computed using the ceiling integer of the right hand side of formula (15) as:

$$N_{kt} \geq \sum_{j=1}^{n} \frac{a_{tj} x_{ktj}}{Wu_{kt} \mu_{tj}} \geq 0, \forall k, t \quad (15)$$

5 Experiments

A case with ten types of machines, ten types of orders and ten periods of production horizon problems are considered to verify the performance of proposed MAA algorithm. The parameters of the proposed MAA algorithm were carefully investigated and tuned in a sensitivity analysis. Major parameters include the evaporation rate (ρ,γ)=(0.01,0.01), the number of artificial ant (N_{ants} =75), the number of reinforcement cycles (*RF*=25) and the number of cycles in a trial (*C*=1,000). JAVA language is used to develop the code of the proposed ant algorithm run on a P4 CPU with 256MB RAM.

The first case is presented for illustrating model I. The available operational time of each working cell, *W*, is 1,800 hours for each period of time. The target utilization of each kind of working cells is 100%. The interest rate is 6% in all periods. The upper bound of budget, *CC*, is 7 million USD. The initial number of working cells is N_0=(4,4,4,4,4,4,3,3,3,3) in which each column represents the number of machine in a cell. The unit price of selling capacity of remaining working cells *t* of a capacity seller, $C_t^{'}$, is set to 1250, 1975, 775, 1350, 2700, 1325, 2350, 2450, 2375, 11,200 respectively in this case.

The problem is highly complex. ILOG OPL optimization software (2005) was applied but failed to solve the case in 32 hours. The best solutions obtained by MAA (in 5000 CPU seconds) are 3.920 x10^5.

The other case is presented for illustrating model II. The initial portfolio of working cells is (1,1,1,1,1,1,1,1,1,1). The unit profits of orders are (20,30,40,20,30,40,20,30,40,40) respectively. The available operational time, target utilization, interest rate, budget are the same as those in case I. The best solutions obtained by MAA (in 2000 seconds) are 7.379 x10^7.

The remaining capacity of the working cells obtained in the first case and the remaining orders obtained in the second case are used. It is assumed that the capacity unit price (C_{tj}) at the capacity seller is proportional to μ_{tj} by multiplying a ratio *R*. Note that *R*=0 imply that resources are free. The other parameters are the same as the ones in case I.

The algorithms can solve the problem in a few CPU seconds. Figure 1 depicts the cost-profit structure for different resource outsourcing cost (in a ratio of *R*). The gross profit of the capacity buyer decreases as the price of the required capacity price increases. When *R*=1.4 the gross profit converges to zero, which means no deal is reached. Note that when *R*= 0.6, the system will reach balance when two factories have equal profit.

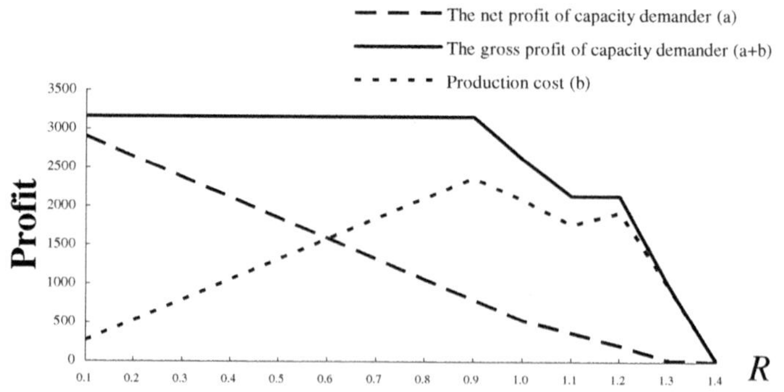

Fig. 1. Cost-profit structure for different resource outsourcing cost (in a ratio of R)

6 Conclusions

This study develops a negotiation-based supply-demand framework where the issues of coordinated resource allocation between factories tasks can be resolved economically. Experiments reveal that the proposed ant algorithm derive good solutions in both of isolated (a single factory level) and negotiation-based (at a mediator level) environments.

Further researches can be done for semiconductor manufacturing and testing, alternative means of acquiring resources, purchasing new facilities, renting from competitors, transferring from other plants and selling equipment. Beside the proposed ant algorithm, one can incorporate local search methods to improve further the solution efficiency.

References

1. Cachon, G. P. and Zipkin, P. H. (1999) Competitive and cooperative inventory policies in a two-stage supply chain, Management Science, 45(7), 936-953.
2. Chang, T. S. (2001) An auction system for resource capacity, Taiwan Patent No. 503361, 29 (27).
3. Dorigo, M. and Di Caro, G. (1999) The ant colony optimization meta-heuristic, In D. Corne, M. Dorigo and F. Glover, editors, New Ideas in Optimization, McGraw-Hill, 11-32.
4. Dorigo, M. and Stützle, T. (2000) The ant colony optimization metaheuristic: algorithms, applications and advances, Technical Report IRIDIA/2000-32, IRIDIA, Universite Libre de Bruxelles, Belgium.
5. Hsu, J. S. (1998) Equipment replacement policy- A survey, Journal of Production and Inventory Management, 29 (4), 23-27.
6. Huang, S. Z. (2002) An agent-based order exchange model of wafer fab, National Chiao Tung University, Master Thesis.
7. ILOG (2005) Optimization suite white paper.
8. Jiang, I. S. (2000) A framework of capacity trading in semiconductor manufacturing industry, Department of Industrial engineering, National Taiwan University, Master Thesis.
9. Mayer, B. C. (1993) Market obsolescence and strategic replacement models, The Engineering Economist, 38, 209-222.
10. Rajagopalan, S. (1994) Capacity expansion with alternative technology choices, European Journal of Operational Research, 392-402.
11. Wang K.-J. and Lin S. H. (2002) Capacity expansion and allocation for a semiconductor testing facility with a constrained budget, Production Planning and Control, 13(5), 429-437.
12. Ying K.-C. and Liao C.-J. (2003) An ant colony system approach for scheduling problems, Production Planning and Control, 14(1), 68–75.

Modeling and Optimization of the Specificity in Cell Signaling Pathways Based on a High Performance Multi-objective Evolutionary Algorithm

Xiufen Zou[1], Yu Chen[1], and Zishu Pan[2,*]

[1] College of Mathematics and Statistics, Wuhan University, Wuhan 430072, China
xfzou@whu.edu.cn, chy_math@163.com
[2] College of Life Science, Wuhan University, Wuhan 430072, China
zspan@whu.edu.cn

Abstract. A central question in cell and developmental biology is how signaling pathways maintain specificity and avoid erroneous cross-talk so that distinct signals produce the appropriate changes. In this paper, a model system of the yeast mating, invasive growth and stress-responsive mitogen activated protein kinase (MAPK) cascades for scaffolding-mediated is developed. Optimization with respect to the mutual specificity of this model system is performed by a high performance multi-objective evolutionary algorithm (HPMOEA) based on the principles of the minimal free energy in thermodynamics. The results are good agreement with published experimental data. (1) Scaffold proteins can enhance specificity in cell signaling when different pathways share common components; (2) The mutual specificity could be accomplished by a selectively-activated scaffold that had a relatively high value of dissociation constant and reasonably small values of leakage rates; (3) When Pareto-optimal mutual specificity is achieved, the coefficients, deactivation rates reach fastest, association and leakage rates reach slowest.

1 Introduction

Cells respond to a plethora of signals using a limited set of intracellular signal transduction components. Surprisingly, pathways that transduce distinct signals can share protein components, yet avoid erroneous cross-talks. An important unsolved problem in cell biology is to understand how specificity from signal to cellular response is maintained between different signal transduction pathways that share similar (or identical) components, particularly when this occurs in the same cell[1-3].

A quantitative analysis of intracellular signal processing will substantially increase our understanding of biological systems and may provide insight into how diseases are initiated and treated. A major obstacle to this goal, however, is the challenge of obtaining a broad and integrated appreciation of the mechanisms that promote signaling specificity. Numerous publications attempted to address this issue. However, at least in part due to lack of strict or unified definition of pathway specificity, no firm conclusion has been drawn and the mechanism of specificity

* Corresponding author.

T.-D. Wang et al. (Eds.): SEAL 2006, LNCS 4247, pp. 774 – 781, 2006.

maintenance remains elusive. Very recently, a precise mathematical definition of specificity in interconnected biochemical pathways was developed [2], and the mechanisms enhancing or diminishing specificity by considering in detail the roles of crosstalk, compartmentalization, scaffolds in steady state were analyzed in a quantitative and rather general way.

This paper is an extension to these works. We choose MAPK cascade signaling in the yeast as a model system and investigate how reaction rates for the scaffold and its complexes influence the specificity. We were particularly interested in the optimal design principles that we can identify from the point of view of the specificity. Hence, in this paper, we focus on searching the optimal specificity of this nonlinear system by using a high performance multi-objective evolutionary algorithm (HPMOEA) based on the principle of the minimal free energy in thermodynamics. In section 2, we will use the recent experimental data on yeast MAPK cascades to develop a more complicated scaffolding-mediated model containing conservation law. In section 3, we will define the specificity and mutual specificity for the two cascades, which closely follow those discussed in ref [2], and describe the HPMOEA for optimizing specificity of model system. In section 4, numerical experiments are conducted. Finally, some conclusions and future work are addressed in section 5.

2 Modeling of Yeast MAPK Cascade Signaling

MAPK cascades exist in all animals, plants and fungi, where they participate in the regulation of normal and pathological aspects of cell growth, division, differentiation, and death [4]. In yeast, elements of the same MAPK cascade regulate three distinct processes in the same cell: mating, filamentous invasive growth, and the response to osmotic stress. Mating and haploid invasive growth use the same MAPKs, Fus3 and Kss1, the same MAPK kinase (MKK, or MEK) Ste7, and the same MEK kinase (MEKK), Ste11. Ste11 also activates the Pbs2 MEK and the Hog1 MAPK in response to osmotic stress (Fig. 1A). These three pathways react to different stimuli (pheromone, nutrient status, and osmotic stress, respectively) and regulate distinct endpoints (mating, invasive growth, and glycerol production, respectively). Both Fus3 and Kss1 are activated during mating, while Kss1 is preferentially activated during invasive growth, and Hog1 only by stress; these patterns make sense given the distinct functions of these MAPKs. How is selective MAPK activation achieved when distinct signals get funneled through a common set of components? In yeast, Ste5, the prototypical MAPK scaffold protein, is thought to enhance specificity via multiple mechanisms. First, by concentrating the components of a single pathway, Ste5 may selectively channel within-pathway signaling (Fig. 1B). For example, Ste5 is thought to promote Ste11 activation of Ste7, but not of Pbs2, because Ste5 binds Ste7 but not Pbs2. Second, by holding on tightly to bound, activated components, Ste5 and other scaffolds may sequester these components, preventing them from straying into other pathways (Fig. 1B). Third, again by sequestering bound components, scaffolds may isolate and protect them from misdirected signals when other pathways are active (Fig. 1C). These models are thought to be relevant to many other scaffold proteins in many organisms. To our best knowledge, none of these popular models of scaffold function have been mathematically modeled before.

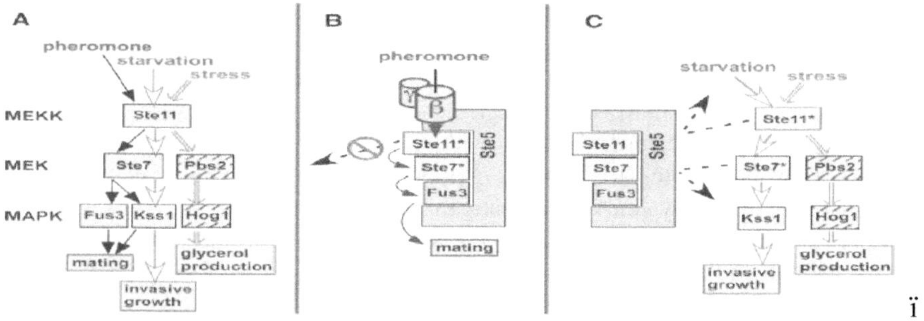

Fig. 1. (A) Shared MAPK cascade components signal to three distinct endpoints. (B) and (C) Models for how the Ste5 scaffold protein may promote signaling specificity (Copyright by [4]).

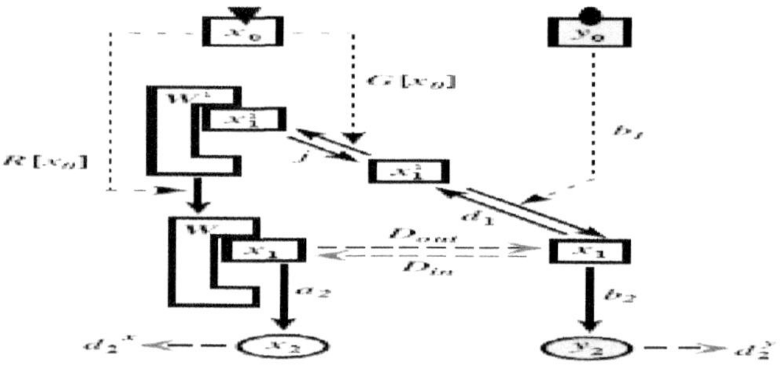

Fig. 2. Scaffold Model

To capture some key features of the experimental data, we takes account of several processes essential for scaffolding and model 2 of 3 yeast pathways(specific mating pheromone signaling pathway with input $x_0(t)$ and the invasive growth pathway with input $y_0(t)$). Hence we modify the previous scaffold model [2] to include two forms of the scaffold: inactive scaffold and active scaffold, shown in Fig. 2. Let us suppose that species x_1^i (the inactive form of x_1) can form a complex with another species (the scaffold), which we term W^i (here i stands for ``inactive"). The rate of this process may depend on the signal x_0, so we call it $G[x_0]$. The complex W^i can release free inactive x_1^i with the rate j. The complex W^i can be activated with a rate defined by the presence of signal x_0, $R[x_0]$. The active form of W gives rise to the final product of the X cascade, x_2. It can also release free active x_1 with the rate D_{out}; the reverse of this reaction, that is, the binding of activated x_1 with the scaffold, happens at the rate D_{in}. In other words, D_{out} is the dissociation constant for the movement of x1 out/off of the scaffold, and D_{in} is a first-order association constant for the binding of x1 into the scaffold complex. On the other hand, for the Y cascade, x_1 can be activated in its free form by the presence of input signal y_0 with the rate b_1, and x_1 in turn activates y_2. By using the Mass Action or Michelis-Menton formulations, the system of equations about these dynamics is described in equations(1) .

$$\begin{cases} \dot{W}^{i} = G[x_0]x_1^i - R[x_0]W^{i} - jW^{i} \\ \dot{W} = R[x_0]W^{i} - D_{out}W + D_{in}x_1 \\ \dot{x}_1 = b_1 y_0(t)x_1^i + D_{out}W - d_1 x_1 - D_{in}x_1 \\ \dot{x}_1^i = -G[x_0]x_1^i - b_1 y_0(t)x_1^i + jW^{i} + d_1 x_1 \\ \dot{x}_2 = a_2 W - d_2^x x_2 \\ \dot{y}_2 = b_2 x_1 - d_2^y y_2 \end{cases} \tag{1}$$

Suppose that $G[x_0] = gx_0(t) + h_1$, $R[x_0] = rx_0(t) + h_2$, where h_1, h_2 are leakage rates. The solution of equations (1) is determined by 13 coefficients, which affect the specificity of pathway or mutual specificity of this network. From real biological phenomena, higher specificity means a better and more stable biochemical network, so the problem is converted into the following multi-objective optimization problem:ï

(P) For each biochemical network, find the optimal combination of the on/off rates and the length of the cascade that maximizes the specificity S_x, S_y for input x_0, y_0.

3 Description of the HPMOEA

3.1 The Two Objective Functions

First, let us denote by $\overline{x}|X = \int_0^\infty x_f(t)\,dt\Big|_{x_0>0, y_0=0}$ the total amount of the final product x_f when the cell is exposed to signal x_0 but not to signal y_0. Similarly, $\overline{y}\,|\,X$ denotes the total amount of y_f under the action of signal x_0. Then we can define the signal specificity of cascade X as the ratio

$$S_x = \frac{\overline{x}|X}{\overline{y}|X} \tag{2}$$

Thus, if pathway X is activated by a given signal and this does not affect the output from pathway Y, the specificity of X with respect to Y in response to that signal is infinite, or complete. However, if there are some cross-talks between the pathways, then activation of Y will result in some output from X, and the specificity will be finite. Similarly, we define the specificity of cascade Y as

$$S_y = \frac{\overline{y}|Y}{\overline{x}|Y} \tag{3}$$

A pathway is called "**has specificity**" if S > 1, and "**has specificity of degree k**" if S > k for some real number k > 1. A network is called "**has Mutual specificity**" if both of the two pathways with specificities greater than 1 [2]. Therefore, S_x, S_y are the two objective functions in our algorithm.

3.2 The Fitness Assignment Strategy

Based on the principles of the minimal free energy in thermodynamics, HPMOEA was proposed to solve multi-objective optimization problems [5]. We combined the rank value $R(i)$ calculated by Pareto-dominance relation with Gibbs entropy $S(i)$, and the crowding distance $d(i)$ to assign a new fitness $F(i)$ for each individual i in the population, that is

$$F(i) = R(i) - TS(i) - d(i) \tag{4}$$

Where $R(i)$ is the rank value of individual i, which is equal to the number of solution n_i that dominates solution i (see [6] for details). The rank values can be computed as follows.

$$R(i) = |\Omega_i|, \text{ where } \Omega_i = \{\vec{x}_j \mid \vec{x}_j \prec \vec{x}_i, 1 \le j \le N, j \ne i\} \tag{5}$$

In this way, R(i) = 0 corresponds to a nondominated individual, whereas a high R(i) value means that i is dominated by many individuals. $S(i) = -p_T(i)\log p_T(i)$, where $p_T(i) = (1/Z)\exp(-R(i)/T)$ is the analog of the Gibbs distribution, $Z = \sum_{i=1}^{N} \exp(-R(i)/T)$,T is a parameter called the temperature, N is the population size, and d(i) is calculated by using a density estimation technique (which proposed by Deb et al[7]) .

In HPMOEA, the fitness values are sorted in increasing order. The individual in the population which the fitness value is smallest is called "the best individual", and the individual in population which the fitness value is largest is called "the worst individual". The structures of HPMOEA and RANK are described as follows:

```
Procedure1 HPMOEA
Step1: t=0, generate randomly an initial population P(t);
Step2: Calculate the rank values of all individuals in
P(t)by using Procedure RANK;
Step3: Save the individuals whose rank values are zero;
Step4: Calculate the fitness of all individuals according
to equation (4),and sort them in increasing order;
Step5: Repeatedly execute step6 to step11 until the
termination conditions are satisfied;
Step6: t=t+1;
```

Step7: Randomly select m_1 individuals to do multi-parent crossover and m_2 individuals to mutate, and to generate n new individuals;
Step8: Compare the new individuals with the worst individuals, and accept new individuals;
Step9: Calculate the rank values of all individuals in new population P(t) by using **Procedure RANK**;
Step10: Save the individuals whose rank values are zero;
Step11: Calculate the fitness of all individuals according to equation (4), sort them in increasing order, and record the worst individuals;
Step12: Output the all results.

Procedure2 RANK
Step1: Use four-order Runge-Kutta method to solve equatioins(1) for given parameters and inputs;
Step2: Use Composite trapezoid rule in numerical integration to obtain two objective values S_x, S_y;
Step3: Calculate the Rank values of each individual according to equation(5)

4 The Numerical Results

In our model system, there are 13 coefficients: $a_2, d_2^x, b_2, d_2^y, b_1, D_{out}, d_1, D_{in}, g, j, r, h_1, h_2$. According to the biological experimental data in yeast, the coefficients are arranged as following: $a_2, b_2 \in [2,10]$ $d_2^x, d_2^y \in [0.001, 2]$, other coefficients belong to [0.001,10]. The algorithm has been coded in C language and implemented on a Pentium PC 700MHz in double precision arithmetic. The main parameter setting is: population size N=30, temperature T=10000. We run the program 10 different

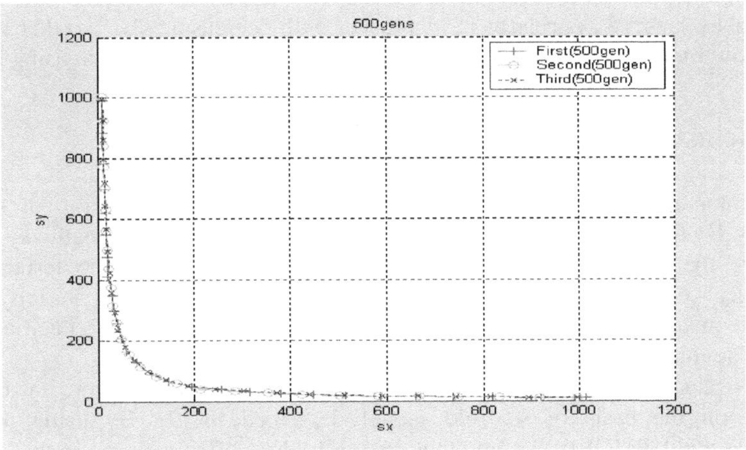

Fig. 3. Pareto-optimal mutual specificity for running 500 generations

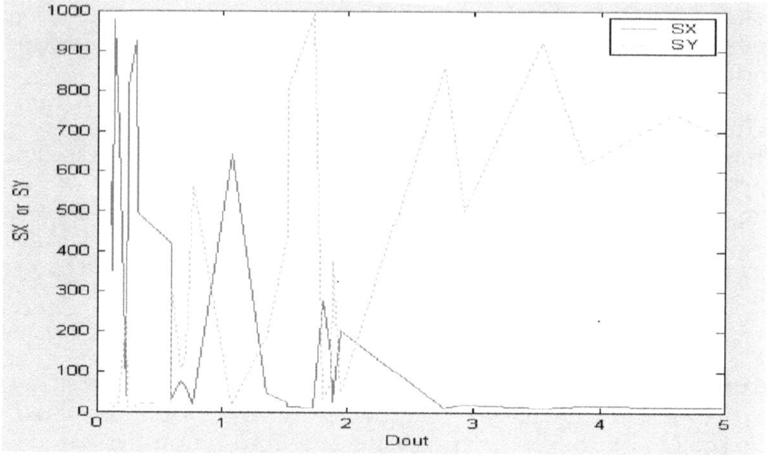

Fig. 4. The relationship between D_{out} and S_x,S_y in last generation

times for 500 generations, the results of three of 10 different times are showed in Fig.3, where X-axis is the specificity of X cascade S_X, Y-axis is the specificity of Y cascade S_y. The above results indicate that HPMOEA can be used to obtain stable Pareto-optimal fronts in all runs.

As can be seen from the results: (1) In the initial stages of searching, mutual specificity could be achieved by a selectively-activated scaffold that had relatively high values of D_{out} and D_{in}, and a reasonably small value of h. In fact, this is situation is quite close to that suggested by recent experiments [4]. (2) It is obvious that increasing the deactivation rate d_l and decreasing the leakage rates h_l, h_2, together with decreasing D_{in} will lead to high degrees of mutual specificity. (3) In last generation for running HPMOEA, that is, when Pareto-optimal mutual specificity is achieved, the coefficients D_{in}, h_l, h_2 are very close to the lower bound 0.001, and d_l are close to the upper bound 10 in the setting of coefficients. The relationship between dissociation constant D_{out} and mutual specificity S_x, S_y are depicted in Fig4. This result makes sense: S_X is relatively increasing with decreasing D_{out} (so that very little x_l leaks out by bound scaffold). Conversely, S_Y is increasing with increasing D_{out}.

5 Conclusions and Future Work

The maintenance of specificity is a critical factor in the evolution of signaling networks. By duplication and divergence of preexisting parts, a new pathway emerged Therefore, the requirement for specificity has undoubtedly shaped the design logic of biochemical networks. A quantitative description of signaling specificity is also important in understanding tumorigenesis because a breakdown in specificity often leads to the initiation or progression of cancer .

In this paper, we presented a more accurate model of scaffold-mediated MAKP cascades on the basis of scaffold model described in [2]. By using numerical techniques and HPMOEA, we find that for scaffolds to effectively promote specificity by sequestration, deactivation rates must be fast. Furthermore, scaffold binding rate must be slow, which is likely to constrain signal speed and amplification. Our results provide insight into the regulatory roles of the signaling components and may help to explain the optimal design of pathways from the viewpoint of specificity.

Our future work will include examining specificity and fidelity simultaneously in complex biochemical networks and trade-offs between specificity and signal amplitude, rate and duration.

Acknowledgement

We would like to thank professors Q. Nie, N. Komarova and L. Bardwell at UCI for helpful discussions. This work was supported by NNSF of China (No.60573168).

References

1. Elaine A. Elion, Maosong Qi, Weidong Chen. SIGNAL TRANSDUCTION: Signaling Specificity in Yeast, Science, Vol 307,(2005) 687-688.
2. Natalia L Komarova, Xiufen Zou, Qing Nie and Lee Bardwell. A Theoretical Framework For Specificity In Cell Signaling, Molecular Systems Biology (2005), report.
3. Monica A. Schwartz and Hiten D. Madhani. Principles of MAP Kinase Signaling Specificity In Saccharomyces Cerevisiae. Annu. Rev. Genet. 2004. 38:725–748.
4. Flatauer LJ, Zadeh SF, Bardwell L Mitogen-activated protein kinases with distinct requirements for Ste5 scaffolding influence signaling specificity in Saccharomyces cerevisiae. Mol Cell Biol 25: (2005)1793–1803
5. Xiufen Zou, Minzhong Liu, Lishan Kang, Jun He. A High Performance Multi-objective Evolutionary Algorithm Based on the Principle of Thermodynamics, Lecture Notes in Computer Science, Vol 3242. Springer-Verlag, 2004, 922-931,
6. Zitzler E, Thiele L, Laumanns M et al.: Performance Assessment of Multiobjective Optimizers: An Analysis and Review. IEEE Transactions on Evolutionary Computation 7(2003) 117-132
7. Deb K, Pratap A, Agarwal S and Meyarivan T. A Fast and Elitist Multi-objective Genetic Algorithm :NSGA II, IEEE Transaction on Evolutionary Computation 6, 2002,182-197.

Elastic Image Registration Using Attractive and Repulsive Particle Swarm Optimization*

Yang Xuan[1,2] and Pei Jihong[1]

[1] Intelligent Information Processing Laboratory, Shenzhen University, 518060, China
jhpei@szu.edu.cn
[2] College of Information and Engineering, Shenzhen University, 518060, China
pdwxyang@263.net

Abstract. Elastic image registration plays an important role in medical image registration. For elastic image registration based on landmarks of sub-images, optimization algorithm is applied to extract landmarks. But local maxima of similarity measure make optimization difficult to convergence to global maximum. The registration error will lead to location error of landmarks and lead to unexpected elastic transformation results. In this paper, an elastic image registration method using attractive and repulsive particle swarm optimization (ARPSO) is proposed. For each subimage, rigid registration is done using ARPSO. In attractive phase, particles converge to promise regions in the search space. In repulsive phase, particles are repelled each other along opposition directions and new particles are created, which might avoid premature greatly. Next, thin plate spline transformation is used for the elastic interpolation between landmarks. Experiments show that our method does well in the elastic image registration experiments.

1 Introduction

Image registration is the process of overlaying two or more images of the same scene to achieve biological, anatomical or functional correspondence. Elastic registration is a process for aligning spatial correspondence of two images within the constraints of an image deformation model. One principal approach to elastic image registration is based on point landmarks, whose advantage is the transformation can be stated in analytic form and lead to efficient computational schemes [1-4]. Landmark-based schemes first extract landmarks from images and then compute a transformation based on these features [1]. Because the precision of landmarks affects the effectiveness of the registration greatly, extraction of corresponding points in two images plays an important role in image registration. Likar [2] combined prior and floating information on the joint probability to improve the subdivided local registration. Maintz [3] proposed to use a global joint histogram based on optimized mutual information combined with a local registration measure to enable local elastic registration.

* Sponsored by: National Natural Science Funds(No. 60572101) and Guangdong Natura Science Funds (No. 31789).

T.-D. Wang et al. (Eds.): SEAL 2006, LNCS 4247, pp. 782 – 789, 2006.

No matter what similarity measure used, the optimization plays an important role in searching for the registration parameters of subimages and extraction of landmarks. When optimization algorithm converges to the local maxima, mis-registration parameters of subimages are obtained. Correspondingly, landmarks extracted are incorrect and result to registration error in elastic transformation. Powell's method is a popular optimization method in multimodality image registration. However, Powell's method could run into local maxima and reach local optimal results easily. Particle Swarm Optimization (PSO) is another form of Evolutionary Computation introduced by Eberhart and Kennedy[5], which has been applied to registration [6]. A major problem with PSO is premature convergence (PC), which results in great performance loss and sub-optimal solutions. In this paper, we propose an elastic image registration method using attractive and repulsive particle swarm optimization. In our method, images are divided into subimages and rigid registration is done using feature efficiency coefficient as the similarity measure. Attractive and repulsive particle swarm optimization (ARPSO) is applied to obtain rigid registration parameters of each subimage. Landmarks are extracted as the center of each rigid registered subimage pairs. Finally, thin plate spline transformation is applied to deform the float image. ARPSO performs optimization using attractive phase and repulsive phase. In attractive phase, particles converge to promise regions in the search space, which is like basic PSO. In repulsive phase, particles are repelled each other along opposition directions and new particles are created. Premature might be avoided by new particles. ARPSO could get away from local maxima easily than basic PSO and converge to global maxima.

2 Similarity Measure of Sub-images

In our method, image is partitioned into sub-images in same size, such as 32×32. In order to extract point landmarks, each subimage is registered rigidly by translation and rotation transformation. The most important process in our method is choosing a similarity measure of sub-images, which is the objective function of optimization.

Mutual information has been widely used to measure image similarity in recent years [10, 11]. However, its application on sub-images with small samples is questionable, as many local maximums might happen, or the global maximum might be away from the actual max. The feature efficiency coefficient defined by Butz[7] is a general concept to qualify image features based on information theoretical framework. Mutual information, Normalized entropy and overlap-invariant entropy can be seen as a particular case of the feature efficiency coefficient with order n [7].

$$e(X,Y)=\frac{I(X,Y)^{n}}{H(X,Y)^{1-n}} \qquad n\in[0,1] \tag{1}$$

where $I(X,Y)$ is the mutual information of random variables X and Y. $H(X,Y)$ is the joint entropy. n is the order. The feature efficiency coefficient is derived from the thought that features that best capture the relationship between the two random variables it is necessary to chose those with the highest mutual information. But simply maximizing the mutual information is dangerous to add superfluous information, which increases the joint entropy $H(X,Y)$[7]. Maximize the feature efficiency coefficient is to select features which are highly related with less information. But pointed out that feature pairs which carry information that is present in both signal (large mutual information), but only information that is present in both signals (lov joint entropy), are the most adapted features for multimodal image registration [9].

When the order n chose appropriately, the feature efficiency coefficient will d well in medical image registration [7,8,11]. For sub-images with small samples, th feature efficiency coefficient with order n can be used to register images. In ou method, we choose the feature efficiency coefficient with $n = 2/3$ as the similarit measure of sub-images.

3 Attractive and Repulsive PSO

Particle Swarm Optimization (PSO) has been applied to image registration [6]. I differs from other evolution motivated evolutionary computation techniques in that i is motivated from the simulation of social behavior [5]. PSO is another form of evolu- tionary computation and is stochastic in nature much like Genetic Algorithms. In a PSO system, particles fly around in a multidimensional search space. During flight each particle adjusts its position according to its own experience, and according to the experience of a neighboring particle, making use of the best position encountered by itself and its neighbor. A minimization (or maximization) of the problem topology is found both by a particle remembering its own past best position (*pbest*) and the com- panions' best overall position (*gbest*).

In PSO the particle simulates a bird's behavior where social sharing of informatior takes place and individuals can profit from the discoveries and previous experience of all other companions during the search for food. Each particle in the population flies over the search space in order to find promising regions in the search space. The par- ticle swarm optimization concept consists of, at each time step, changing the velocity of each particle toward its *pbest*. Acceleration is weighted by a random term, with separate random numbers being generated for acceleration toward *pbest*.

In PSO, the ith particle x_i, $i = 1, \cdots, N$ moves by addition of a velocity vector v_i, which is a function of the best position found by the particle and of the best positior found so far among all particles.

$$v_i(k+1) = \phi(k) v_i(k) + \alpha_1 \left[\gamma_{1i} \left(p_i - x_i(k) \right) \right] + \alpha_2 \left[\gamma_{2i} \left(G - x_i(k) \right) \right] \tag{2}$$

$$x_i(k+1) = x_i(k) + v_i(k+1) \tag{3}$$

where, i is particle index, k is iteration index, $\phi(k)$ is the inertial weight. v_i is the velocity of i th particle, x is the position of i th particle, p_i is the best position found by ith particle, G is the best position found by swarm, $\gamma_{1,2}$ is the random numbers on the interval [0,1] applied to i th particle, $\alpha_{1,2}$ is acceleration constants. In order to perform PSO, we set inertial weights $\phi(k)$ be monotonically decreasing function of the iterations index. The iteration number is up to 250. The population size is 20 and acceleration constants are $\alpha_1 = \alpha_2 = 2$. Our search space is three dimensions, one rotation and two translations. $x_i = (t_x, t_y, \theta)$, where t_x and t_y are translation parameter, θ is rotation parameter.

A major problem with PSO is premature convergence (PC), which results in great performance loss and sub-optimal solutions. The main reason for premature convergence is a too high selection pressure or a too high gene flow between population individuals [13], such as decreasing of diversity in search space that leads to a total implosion and ultimately fitness stagnation of the swarm.

Riget et.al [12] defined the attractive and repulsive PSO (ARPSO) to overcome the premature convergence by modifying the diversity of particles in the search space. When the particles diversity is high, the particles will attract each other just as the basic PSO algorithm. The information of good solutions will flow between particles. When the particles diversity is low, that means all particles are similar to each other, the particles are no longer attracted to but instead repelled by the best known particle position and its own previous best position. The velocity-update formula is modified by multiplying the direction coefficient *dir*, which decides directly whether the particles attract or repel each other [12].

$$v_i(k+1) = \phi(k) v_i(k) + dir\{\alpha_1[\gamma_{1i}(p_i - x_i(k))] + \alpha_2[\gamma_{2i}(G - x_i(k))]\} \tag{4}$$

The direction coefficient *dir* is set as:

$$dir = -dir \quad if\ diversity < dLow\ or\ diversity > dHigh \tag{5}$$

where *dLow* and *dHigh* are the lower bound and the up bound of the diversity alternatively. The initial value of *dir* is 1. Equation (5) means that when *dir* is set to be 1, the swarm is contracting and consequently the diversity decreases. Particles are attracted each other to find promising regions in the search space. When the diversity drops below a lower bound, it switches to the repulsion phase, *dir* is set to be -1 where particles are repelled to increase the diversity of population, in which the swarm expands gradually. When the diversity of *dhigh* is reached, it switches back to the attraction phase. In ARPSO, the repulsive phase plays an important role. When all particles are similar to each other, the diversity of population could be increased using repulsive phase, which means potential particles are created and premature convergence might be avoided greatly using these new particles.

The diversity measure of the swarm is set as fellow [12],

$$diversity(S) = \frac{1}{|S| \cdot |L|} \sum_{i=1}^{|S|} \sqrt{\sum_{j=1}^{N} (p_{ij} - \overline{p}_j)^2} \tag{6}$$

where S is the particle set, $|S|$ is the swarm size, $|L|$ is the length of longest the diagonal in the search space, N is the dimensionality of the problem, p_{ij} is the *j*th value of the *i*th particle and $\overline{p}_j$ is the *j*th average value of p_j.

For elastic medical image registration, image is partitioned into 32×32 sub-images. Each subimage is registered rigidly by translation and rotation transformation using ARPSO, where the feature efficiency coefficient is the objective function. For each sub-image pair, the center points are extracted as the point landmarks pair. Next, Thin plate spline transformation is applied to deform the float image to obtain the elastic registration result.

4 Thin Plate Spline Transformation (TPS)

TPS models the deformations by interpolating displacements between source and target points [4]. Given two images: the source image (S) and the target image (T), one set of n landmark pairs $P_0:\{(p_i,q_i)\mid p_i\in S, q_i\in T, i=1,\cdots,n\}$, TPS transformation $f(x,y)$ has the form

$$f(x,y)=\Phi_s(x,y)+R_s(x,y)=a_1+a_x x+a_y y+\sum_{i=1}^{n}\omega_i U(r_i) \tag{7}$$

where $U(r_i)=r_i^2\log r_i^2$, $r_i=|p_i-(x,y)|=\sqrt{(x_i-x)^2+(y_i-y)^2}$ .It consists of two parts: the affine part $\Phi_s(x,y)$, a sum of polynomials with coefficients $a=(a_1,a_x,a_y)$, and the elastic parts $R_s(x,y)$, a sum of radial basis functions (RBFs) with coefficients $\omega=(\omega_1,\omega_2,\ldots,\omega_n)$. It fulfills the interpolation conditions $f(x_i,y_i)=q_i\ (i=1,\cdots,n)$ and minimizes the bending energy $E_{TPS}(f)$:

$$E_{TPS}(f)=\iint\left|\left|\frac{\partial^2 f}{\partial x^2}\right|^2+\left|\frac{\partial^2 f}{\partial x\partial y}\right|^2+\left|\frac{\partial^2 f}{\partial y^2}\right|^2\right|dxdy \tag{8}$$

Only need to solve the linear system for TPS coefficients:

$$\begin{cases} K\omega+Pa=v \\ P^T\omega\ =O \end{cases} \tag{9}$$

where $K_{ij}=U(r_{ij})$, the ith row of P is $P_i=(1,x_i,y_i)$, O is a 3×3 matrix of zeros, $A=(a_1,a_x,a_y)$, $v=(q_1,\ldots,q_n)$.

5 Experiments

We will show several examples for medical image registration which show the method validity for deformation images. In our method, the diversity parameters *dlow* and *dhigh* were set at 0.02 and 0.25 respectively.

In order to illustrate the registration precision of our method, artifical images are used to do registration. At first, an artifical image (circle) is created as the reference image (figure 1 (a)). The reference image is deformed in manual to be the float image (figure 1 (b)). Point landmarks are extracted and marked (figure 1 (c)) by our method. TPS is applied to transform the float image based on extracted landmarks. The elastic registered image is display in figure 1 (d). It can be seen that the registration precision of our method is satisfied because of accurateness of landmarks using ARPSO. Figure 2 is another example of artifical image (rectangle) elastic registration. The elastic registration results of our method are satisfied also.

We used synthetic MR-scans from the BrainWeb database to represent the quality of elastic registration results of our method. The results for T2-PD registration are shown in Fig.3. Figure 3 (a) is a T2 image, which is the reference image. The float image PD is deformed in manual. The corresponding point landmarks are labeled in figure 3 (c) also. The elastic registered image is display in figure 3 (d). In order to show the elastic registration results of our method, edge differences of the reference image and elastic registered image are displayed in figure 3(e). It can be seen that the registration precision of our method is satisfied. Figure 4 is the elastic registration results of PD-T1, where elastic deformation appeared on the float image T1. Figure 5

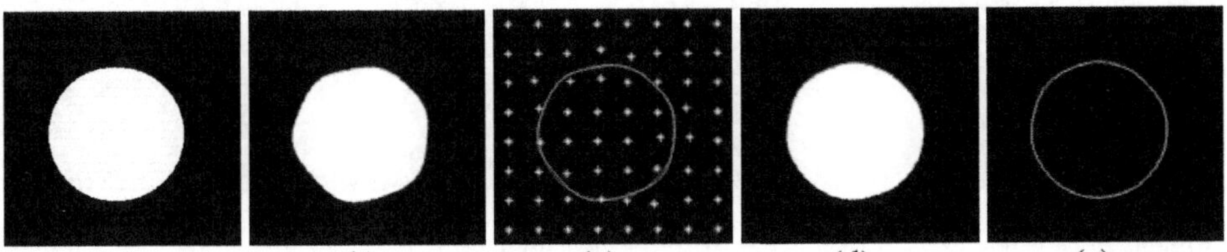

(a) (b) (c) (d) (e)

Fig. 1. Elastic registration of an artifical image. (a) reference image, (b) float image, (c) extracted landmark pairs, (d) deformed image of our method, (e) edge error of reference image and deformed image.

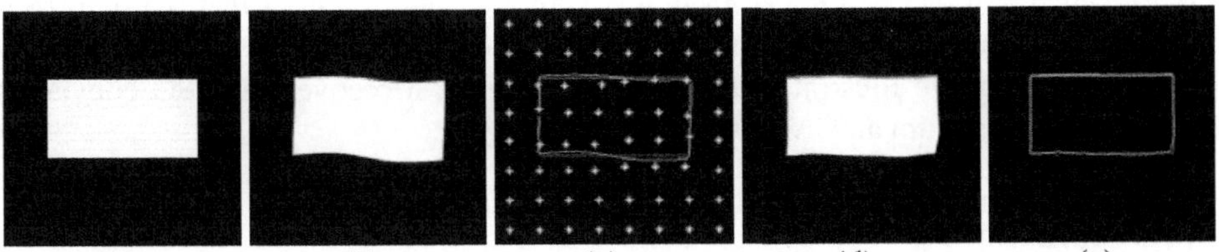

(a) (b) (c) (d) (e)

Fig. 2. Elastic registration of an artifical image. (a) reference image, (b) float image, (c) extracted landmark pairs, (d) deformed image of our method, (e) edge error of reference image and deformed image.

(a) (b) (c) (d) (e)

Fig. 3. Elastic registration of T2-PD. (a) reference image T2, (b) float image PD, (c) extracted landmark pairs, (d) deformed image of our method, (e) edge error of reference image and deformed image.

(a) (b) (c) (d) (e)

Fig. 4. Elastic registration of PD-T1. (a) reference image PD, (b) float image T1, (c) extracted landmark pairs, (d) deformed image of our method, (e) edge error of reference image and deformed image.

Fig. 5. Elastic registration of T1-T2. (a) reference image T1, (b) float image T2, (c) extracted landmark pairs, (d) deformed image of our method, (e) edge error of reference image and deformed image.

is the elastic registration results of T1-T2, where elastic deformation appeared on the float image T2. It can be seen from figure 3, figure 4 and figure 5 that the deformed images of our method matched the reference image well, which means the similarity measure of subimage illustrated the alignment degree. Moreover, ARPSO converges to the global maximum and avoids premature easily.

6 Conclusion

In multimodal medical image elastic registration, optimization algorithm plays an important role in extracting point landmarks of subimage pairs. In this paper, an elastic image registration method using attractive and repulsive particle swarm optimization is presented. ARPSO performs optimization using attractive phase and repulsive phase. When the particles diversity is high, the particles will attract each other just as the basic PSO algorithm. The information of good solutions will flow between particles. When the particles diversity is low, it turns to repulsive phase, particles are no longer attracted to but instead repelled each other along opposition directions and new potential particles are created. Premature convergence might be avoided by new potential particles. Moreover, feature efficiency coefficient is applied to be the objective function of optimization. Experiments of artifical images and multimodal medical images show that our method is feasible and robust.

References

1. ROHR,K.: Elastic Registration of Multimodal medical images: A Survey. Auszug aus: Kunstliche Intelligenz, Heft3, (2000)11-17
2. Bostjan Likar and Franjo Pernus: A hierarchical approach to elastic registration based on mutual information. Image Vision and Computing. 19 (2001) 33-44
3. J B Antoine Maintz, Erik H W Meijering, M A viegever: General multimodal elastic registration based on mutual information. In: K. M. Hanson (ed.): Proceedings of SPIE Medical Imaging 1998. vol. 3338. (1998) 144-154
4. Fred L Bookstein: Principal warps: thin-plate splines and the decomposition of deformations. IEEE Trans. Pattern Analysis and Machine Intelligence. Vol. 11, No.6, (1989) 567-585
5. Eberhart, R.C., and Kennedy, J.: A new optimizer using particle swarm theory. In: Proc. Sixth International Symposium on Micro Machine and Human Science (Nagoya, Japan), IEEE Service Center, Piscataway, NJ, (1995) 39-43
6. Mark P. Wachowiak, Renata Smolíková, Yufeng Zheng, Jacek M. Zurada, and Adel S. Elmaghraby: An Approach to Multimodal Biomedical Image Registration Utilizing Particle Swarm Optimization. IEEE Trans. Evolutionary Computation. Vol.8, No.3 (2004) 289-301
7. T. Butz, J. P. Thiran: Feature-space mutual information for multi-modal signal processing with application to medical image registration. In: European Signal Processing Conference 2002, Toulouse, France. 1(2002) 3-10
8. T. Butz, J.P. Thiran: Affine registration with feature space mutual information. In: Medcial Image Computing and Computer Assisted Intervention. Lecture Notes in Computer Science, Vol. 2208 (2001) 549-556
9. T. Butz, J.P. Thiran: From error probability to information theoretic (multi-modal) signal processing. Signal Processing. 85 (2005) 875–902
10. P. Viola and W.M. Wells III: Alignment by maximization of mutual information. In: Proceedings of the Fifth International Conference on Computer Vision. IEEE Computer Society Press, Los Alamitos, CA. (1995) 16-23.
11. C.Studholme, D.J.Hawkes, and D.L.G. Hill: An overlap invariant entropy measure of 3D medical image alignment. Pattern Recognition. 32 (1999) 71-86
12. Riget, J., Vesterstroem, J.S.: A diversity-guided particle swarm optimizer - the ARPSO. Technical Report 2002-02, Department of Computer Science, University of Aarhus (2002)
13. Parsopoulos, K. E. and Vrahatis, M. N.: Modification of the particle swarm optimizer for locating all the global minima. In: Proceedings of the International Conference on Artificial Neural Networks and Genetic Algorithms (ICANNGA 2001), Prague, Czech Republic. (2001) 324—327

An Immune Clonal Selection Scheduling Algorithm for Input-Queued Switches

Liu Fang and Zhao Jing

School of Computer Science and Engineering, Xidian University, Xi' an 710071, China
f63liu@163.com

Abstract. Immune Clonal Selection Algorithm (ICSA) is a new intelligent algorithm that can effectively overcome the prematurity and has fast convergence speed. An Immune Clonal Selection Scheduling Algorithm (ICSSA) is proposed by applying ICSA to the input-queued packet switch scheduling in this paper. ICSSA is compared with other previous algorithms about two performance measures: the average delay and the maximum throughput of the switch. Closed-form expressions for these measures are derived under uniform i.i.d. Bernoulli, diagonal and bursty traffic model. The experimental results show that better performances can be obtained by ICSSA, and 100% throughput can be guaranteed for these traffic models.

1 Introduction

The commercial machines popular today utilize *iterative round-robin matching with slip* (iSLIP) algorithm widely [1]. The main characteristic of iSLIP is its simplicity: it is readily implemented in hardware and can operate at high speed. For uniform i.i.d. Bernoulli arrivals, iSLIP has the appealing property that it is stable for any admissible load. iSLIP, however, can become unstable for admissible nonuniform traffic and bursty traffic[2].

Researchers have proposed several other good switch scheduling algorithms, such as Dual Round-Robin Matching (DRRM)[3], Exhaustive Service Dual Round-Robin Matching (EDRRM) algorithm [4], iterative longest queue first (iLQF)[5], reservation with preemption and acknowledgment (RPA)[6], and matrix unit cell scheduler (MUCS)[7]. With centralized implementations, the runtime of these algorithms is $O(N^2)$ or more. But by adopting parallelism and pipelining (which means adding spatial complexity in hardware) these algorithms can considerably reduce their time complexity. However, under non-uniform input traffic the performance of these algorithms is poor compared to MWM: They induce long delays, and their throughput can be less than 100 percent. Furthermore, solutions that intrinsically possess a $O(N^2)$ runtime complexity are unlikely to scale for implementation in high-speed and large-sized switches.

The *maximum weight matching* (MWM) algorithm delivers a throughput of up to 100 percent [8][9] and provides low delays by keeping queue sizes small. However, it is too complex to implement because it requires $O(N^3)$ iterations in the worst case.

T.-D. Wang et al. (Eds.): SEAL 2006, LNCS 4247, pp. 790 – 797, 2006.

Noticing the existing disadvantages of iterative scheduling algorithms and MWM, we studied a novel genetic algorithm based on immune [10], a multi-agent genetic algorithm for global numerical optimization [11], a multiagent evolutionary algorithm for constraint satisfaction problems [12] and an artificial immune system [13][14]. In the end, we introduced, describe and evaluate a novel approach, the Immune Clonal Selection Algorithm (ICSA) to approximate maximum matchings for bipartite graphs. In this paper we quantitatively evaluate ICSSA approximating *maximum size matching* and *maximum weight matching* for scheduling cells in an input-queued switch. In particular, we compare the performance of ICSSA to RRM and iSLIP. We also compare ICSSA against iLQF and MWM. The experiment results have shown that better performance can be obtained by ICSSA for different traffic models.

2 Basic Idea of Immune Clonal Selection Algorithm

The basic process of immune clonal selection algorithm is denoted as follows [13]:

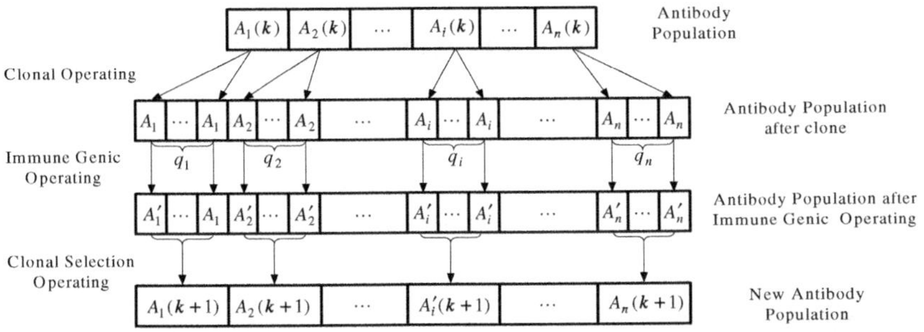

Fig. 1. Basic process of the immune clonal selection algorithm

Clonal Operating T_c^C **:**

$$T_c^C(\bar{A}(k)) = [T_c^C(A_1(k)), T_c^C(A_2(k)), \cdots, T_c^C(A_n(k))]^T \quad (1)$$

where $T_c^C(A_i(k)) = I_i * A_i(k)$, $i = 1,2,...,n$; I_i is a q_i dimension row vector whose elements are all 1

$$q_i(k) = g(n_c, f(A_i(k))) \quad (2)$$

where $n_c \succ n$ is a given integer relating to clone size, $f(A_i(k))$ is the affinity function.

Immune Genic Operating T_g^C **:** We apply the Gauss mutation here. Mutate the population clone operated with the probability p_m

$$\overline{A}(k) = T_g^C(\overline{A'}(k)) \tag{3}$$

Clonal Selection Operating T_s^C **:** $\forall i = 1,2,\ldots,n$, exit the mutated antibody $B = \{A'_{ij}(k) \mid \max f(A'_{ij}), j = 1,2,\ldots,q_i - 1\}$, then B replaces the antibody $A_i(k) \in \overline{A}(k)$ in the original population by the possibility p_s .

$$p_s = \begin{cases} 1 & f(A_i(k)) < f(B) \\ exp(-\dfrac{f(A_i(k)) - f(B)}{\alpha}) & f(A_i(k)) \geq f(B) and A(k) \text{ is the best antibody in the population} \\ 0 & f(A_i(k)) \geq f(B) and A(k) \text{ is not the best antibody in the population} \end{cases} \tag{4}$$

where $\alpha > 0$ is a value related to the population diversity. So the antibody population is updated, and the information exchanging among the antibody population is realized.

3 The Immune Clonal Selection Scheduling Algorithm

Consider an input-queued switch with m inputs and n outputs, ICSSA described in this thesis attempt to match the set of inputs *I*, of an input-queued switch, to the set of outputs *J*. An efficient algorithm is one that serves as many input-queues as possible. This bipartite graph matching problem can be regarded as an optimization problem, thus try to find a matching *M* with maximum total size or total weight.

3.1 Encoding

For applying immune clonal selection algorithm to *maximum size* matching problem, a particular encoding form is required to represent a feasible solution of this problem.

An antibody code is a string with finite length $A = \alpha_1\alpha_2 \cdots \alpha_l$, where α_i s gene code and l is the length of genes. We will assume that the number of input ports equals the number of output ports, i.e., $|I| = |J| = N$, where *N* is the number of ports. For our application, *l*=*N*, and α_i is the input port which is selected by output *i*. In order to obtain a conflict-free match *M* between the set of inputs and outputs, an antibody is a permutation of input port index from 1 to *N*. This measure assures that every antibody generated in this way indicate a valid solution of matching problem.

3.2 Decoding and Affinity Function

As discussed before, the *maximum size* matching problem can be translated to an optimization problem: $max\{f(e^{-1}(A)) : A \in I\}$, where $e^{-1}(A)$ is the decoding of antibody *A*; set *I* is antibody space, and *f* is a positive real function which is called antibody-antigen affinity function. Antibody-antigen affinity function indicates the objective function value of candidate solution. Gene code α_i of antibody *A* means, of this solution, input port α_i have been chosen by output port i. For maximum size matching problem, if there is a request of input port α_i for forwarding packet to output port i, size of this edge $s_{\alpha_i,i}(t)=1$, else $s_{\alpha_i,i}(t)=0$. Size of matching: $S=\sum_{i=1}^{N} s_{\alpha_i,i}(t)$, i.e., affinity function: $f\left(e^{-1}(A)\right)=\sum_{i=1}^{N} s_{\alpha_i,i}(t)$. For maximum weight match, we adopted the LQF strategy: preferential service is given to input queues that are more heavily occupied. This is achieved by defining weight of edge $w_{\alpha_i,i}(t)$ to be equal to the queue occupancy $L_{\alpha_i,i}(t)$, therefore total weight of the match $W=\sum_{i=1}^{N} w_{\alpha_i,i}(t)=\sum_{i=1}^{N} L_{\alpha_i,i}(t)$, i.e., affinity function $f\left(e^{-1}(A)\right)=\sum_{i=1}^{N} L_{\alpha_i,i}(t)$.

3.3 Mutation Operating

For every antibody representing a valid solution, mutation operation is expected to avoid destroying the validity of antibody. Therefore we designed a proper mutation approach, to exchange the value of two genes that are randomly selected from the antibody.

3.4 Description of Algorithm

Step 1. Set population scale, clone population scale parameter n_c and mutation probability p_m;

Step 2. Randomly generate initial antibody population $\vec{A}(0)$ and initialize k = 0;

Step 3. Calculate affinity function value of antibody population;

Step 4. Each antibody in the antibody population is reproduced according to formula (1), clonal scale of each antibody is determined by formula (6):

$$q_i(k)=Int\left(n_c * \frac{f(A_i(k))}{\sum_{j=1}^{popscale} f(A_i(k))} * \theta_i\right) \quad ,i=1,2,\cdots,popscale \tag{5}$$

where $Int(x)$ indicates the minimum integer which is bigger than x, $n_c > n$ is a given integer relating to clonal scale. θ_i is a value which indicates the affinity between antibody i and other antibody. It is defined as formula (6):

$$\theta_i = \min\left\{\exp(\|A_i - A_j\|)\right\}, i \neq j; i,j = 1,2,\ldots,n \tag{6}$$

where $\|*\|$ is a arbitrary norm, commonly adopt Euclidean distance between the two antibodies, at the same time $\|*\|$ should be normalized, i.e. $0 \le \|*\| \le 1$.

Step 6. Clonal mutation is used to the clone population operated with the probability p_m;

Step 7. The clonal selection operator is performed to generate the new antibody population according to the formula (4);

Step 8. k=k+1;

Step 9. If the termination condition is met, the calculation is terminated; otherwise, return to step 3.

The computational complexity to identify the scheduling algorithm is $O(g \times n)$ for a $N \times N$ switch, where g is the number of generations and n is the scale of antibody population. Since the ICSSA algorithm described above requires the computation of the affinity function value of each antibody, it can be done in parallel by n modules. Once the algorithm is identified, its on-line computational complexity is $O(g)$.

4 Experiment Results and Analyses

To validate the performance of ICSSA, an exhaustive simulation is done under various traffic modes including uniform i.i.d. Bernoulli, nonuniform diagonal and bursty traffic.

4.1 Bernoulli Traffic

We consider first the case where cells arrive as a Bernoulli process with cell destinations distributed uniformly over all outputs. In Fig. 2, the results were obtained using the simulation for a 8×8 switch. Fig. 2(a) shows the average cell latency (measured in cells) versus the average offered load for ICSSA with FIFO, RRM, iSLIP, iLQF and MWM.

This is the case: Because of HOL blocking, FIFO queuing achieves maximum offered load slightly more than 58%. For an offered load of just 63% the round-robin algorithm becomes unstable. The iSLIP algorithm is a variation on RRM designed to reduce the synchronization of the output arbiters, then achieve 100% throughput. Under low load, the performance of maximum-size ICSSA (the curve labeled msICSSA) is almost identical to iSLIP: arriving cells usually find empty input queues, and on average there are only a small number of inputs requesting a given output. As the load increases, the number of input queues that keep cells increases, leading to a large sized match. In fact, under uniform 100% offered load the maximum-size - ICSSA providing a better match and 100% throughput. Meanwhile, the curve labeled mwICSSA indicates maximum-weight ICSSA is the only algorithm able to approximate MWM wonderfully.

4.2 Diagonal Traffic

Diagonal traffic is a kind of nonuniform traffic mode. All inputs are equally loaded on a normalized scale, and $\rho \in (0,1)$ denote the normalized load. The arrival process is Bernoulli i.i.d.. Let $|k| = (k \bmod N)$. $\lambda_{ii} = 2\rho / 3N$, $\lambda_{i|i+1|} = \rho / 3N$. $\forall i$, and $\lambda_{ij} = 0$ for all other i and j. This is a very skewed loading, in the sense that input i has packets only for outputs i and $|i + 1|$. It is more difficult to schedule than uniform loading.

Fig. 2(b) shows the average cell latency (measured in cells) versus the average offered load for ICSSA with other algorithms under diagonal traffic mode. It is observed that under load lower than 75%, the performance of ICSSA is almost identical to that of iterative scheduling algorithms. When the load keeps on increasing, ICSSA shows better performance. Especially, the maximum-weight ICSSA is the only algorithm that can achieve throughout up to 100% under non-uniform traffic mode, for the curve approximate MWM wonderfully.

4.3 "Bursty" Traffic

Real network traffic is highly correlated from cell to cell [15] and so in practice, cells tend to arrive in bursts, corresponding perhaps to a packet that has been segmented or a packetized video frame.

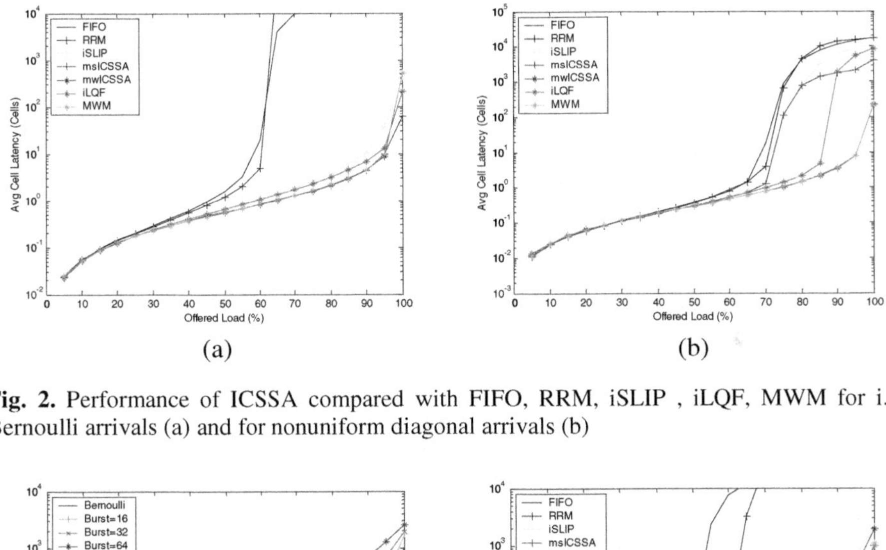

(a) (b)

Fig. 2. Performance of ICSSA compared with FIFO, RRM, iSLIP , iLQF, MWM for i.i.d. Bernoulli arrivals (a) and for nonuniform diagonal arrivals (b)

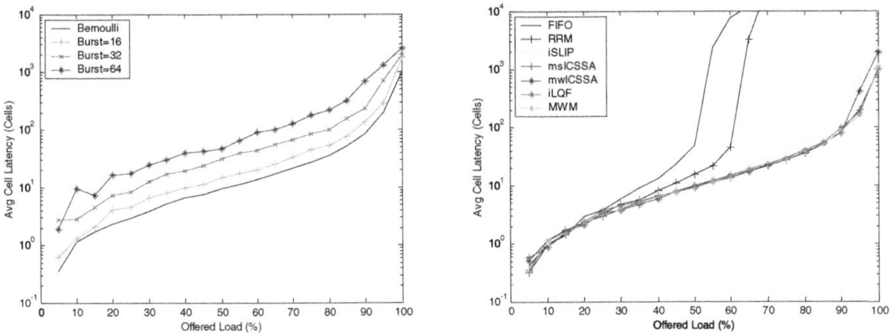

(a) (b)

Fig. 3. The performance of ICSSA (a) and the performance of ICSSA compared with FIFO, RRM, iSLIP, iLQF and MWM (b) under 2-state Markov-modulated Bernoulli arrivals

We illustrate the effect of burstiness on ICSSA using an on-off arrival process modulated by a 2-state Markov-chain. The source alternately produces a burst of full cells (all with the same destination) followed by an idle period of empty cells. The bursts and idle periods contain a geometrically distributed number of cells.

Fig. 3(a) shows the performance of ICSSA under this arrival process for a 8×8 switch, comparing it with the performance under uniform i.i.d. Bernoulli arrivals. As we would expect, the increased burst size leads to a higher queuing delay. In fact, the average latency is proportional to the expected burst length.

As Fig. 3(b) shows, we compared the performance of ICSSA with other algorithms under this traffic model. The default of bust length is 10. Burstiness tends to concentrate the conflicts on outputs rather than inputs: each burst contains cells destined for the same output and each input dominated by a single burst at a time. As a result, the performance is limited by output contention. The experimental results show that ICSSA performs better than other algorithms under bursty traffic.

5 Conclusion

ICSA is a new intelligent algorithm intended to integrate the local searching with the global and the probability evolution searching with the stochastic searching [14].

Compared to iterative algorithms using "round-robin" arbitration, the novel ICSSA has the following properties: (1) No request is needed to send between inputs and outputs, as a result the communication load is reduced. (2) More importantly, it needs only iteration per time slot, regardless of the size of the switch. (3) ICSSA converges approximately to maximum match leading to a better throughput, whereas algorithms using "round-robin" arbitration like iSLIP usually find, at best, a maximal match: the largest size match without removing connections made in earlier iterations. (4) Once implemented in hardware, antibodies of one generation can compute in parallel, which leads to good compromise between low on-line computational complexity and goodness of throughput and delay performance.

Acknowledgement. This work is supported by the National Natural Science Foundation of China under Grant No. 60372045 and No. 60133010, the Defence Pre-Research Project of China under Grant No.51406020104DZ0124, the Key Science-Technology Project of Higher Education of China under Grant No. 0202A022 and the National Research Foundation for the Doctoral Program of Higher Education of China No. 20030701013.

References

1. Mitko Gospodinov, Evgeniya Gospodinova, "Analysis of iSLIP scheduling algorithm for input-queuing switches", International Conference on Computer Systems and Technologies-CompSysTech'2004.
2. Nick McKoewn, "iSLIP: A Scheduling Algorithm for Input-Queued Switches", *IEEE Transactions on Networking, Vol 7, No.2, April 1999.*
3. H. J. Chao, "Saturn: a terabit packet switch using Dual Round-Robin",IEEE Communication Magazine, vol. 38 12, pp. 78-84, Dec. 2000.

4. Li Y, Panwar S, Chao HJ., "The dual Round-Robin matching switch with exhaustive service". In: Gunner C, ed. Proc. of the IEEE Workshop on High Performance Switching and Routing. Kobe: IEEE Communications Society, 2002. 58–63.
5. N. McKeown, *Scheduling Algorithms forInput-Queued Cell Switches,* doctoral dissertation, Dept. of EECS, Univ. of California, Berkeley, 1995.
6. M.M. Ajmone et al., "RPA: A Flexible Scheduling Algorithm for Input Buffered Switches," *IEEE Trans. Communications*, vol. 47, no. 12, Dec. 1999, pp. 1921-1933.
7. H. Duan et al., "A High Performance OC12/ OC48 Queue Design Prototype for Input Buffered ATM Switches," *INFOCOM 97: 16th Ann. Joint Conf. of the IEEE Computer and Comm. Societies* (Infocom 97), IEEE CS Press, Los Alamitos, Calif., 1997, pp. 20-28.
8. N. McKeown, V. Anantharan, and J. Walrand, "Achieving 100% Throughput in an Input-Queued Switch," *Proc. 15th Ann. Joint Conf. of the IEEE Computer and Comm. Societies* (Infocom 96), IEEE CS Press, Los Alamitos, Calif., 1996, pp. 296-302.
9. J. Dai and B. Prabhakar, "The Throughput of Data Switches with and without Speedup," *IEEE Infocom 2000*, IEEE Press, Piscataway, N.J., 2000, pp. 556-564.
10. Jiao Licheng, Wang Lei: A novel genetic algorithm based on immune, IEEE Trans. on System, Man, and Cybernetics—Part A, 30 (2000) 552–561
11. Zhong Wei-Cai, Liu Jing, Xue Ming-Zhi, Jiao Li-Cheng. A Multi-Agent Genetic Algorithm for Global Numerical Optimization. IEEE Trans.System,Man and Cybernetics—Part B. 34(2), (2004)1128-1141
12. Liu Jing, Zhong Wei-Cai, Jiao Li-Cheng. A multiagent evolutionary algorithm for constraint satisfaction problems. IEEE Trans. Syst., Man, and Cybern. B, 36(1), (2006) 54-73
13. Haifeng DU, Licheng JIAO, Sun'an Wang.: Clonal Operator and Antibody Clone Algorithms. Proceedings of the First International Conference on Machine Learning and Cybernetics, Beijing, (2002) 506–510
14. Jiao Licheng, Du Haifeng, "An Artificial Immune System: Pogress and Prospect", ACTA ELECTRONICA SINICA, 31(10), (2003) 1540-1549
15. Leland, W.E.; Willinger, W.; Taqqu, M.; Wilson, D. "On the self-similar nature of Ethernet traffic", Poc. of Sigcomm, San Francisco, 1993, pp.183-193

Prototyping Direction Optimization of Points Data Oriented Rapid Prototyping Based on Genetic Algorithm*

Cheng Xiaomin, Cheng Feng, and Yang wei

School of Mechanical Engineering, Ningbo University of Technology, 315016, Ningbo, China
chxm369@hotmail.com

Abstract. A new approach was proposed as a prototyping direction optimization of points data. Based on perspective theory, a curve surface was built up on the peak point of the produced point lattice of an entity. A lattice grid was used to represent the represent volume of the entity, which was employed in rapidly calculating the represent volume in real time. After analyzing the optimal object functions and strategy, authors adopted the genetic algorithm on selection of operators, cross breeding operators, mutation operators, iteration termination condition and colony scales, etc. The optimization program was set up using Matlab and the optimization was obtained for prototyping direction. At the end of the paper, the aforementioned approach was verified using an actual example, which was further validated on the rapid prototyping machine. The simulation results show that a three-dimensional reconstruction was not necessary based on this proposed points data prototyping direction optimization. The results also indicate that a perfect optimization has been achieved for real time optimization in a space with 360 degree. On the basis of the aforementioned optimization approach, the best position of an entity can be located for rapid prototyping, which can increase prototyping efficiency and reduce the time spending on prototyping. It then can lower the cost.

Keywords: points data, prototyping direction, genetic algorithm, optimization.

1 Introduction

Large amount of manufactures need representing volume in order to utilize rapid prototyping technologies, such as FDM (Fused Deposition Modeling), SLA (Stereolithography Apparatus), SDM (Shape Deposition Manufacturing) and RFP (Rapid Freezing Prototyping), etc. In general, representing volume can be produced during the process of original shape manufacturing. Such representing volume not only impacts the manufacturing accuracy of the original shape and lengthening in duration, but also increase manufacturing cost. Therefore, during the rapid prototyping, the necessary representing volume is expected to be reduced as much as possible in addition to ensure prototyping quality. At present, during rapid prototyping

* The Project Supported by Zhejiang Provincial Natural Science Foundation of China: 502140.
The Project Supported by Ningbo City Youth Foundation of China: 2003A62021.

T.-D. Wang et al. (Eds.): SEAL 2006, LNCS 4247, pp. 798 – 805, 2006.

manufacturing, most of the approaches of prototyping-oriented optimization are based on the CAD models of entity[1]. The actually steps include set-up of an CAD model of the entity based on the points data, selection of a specific angle, then calculation of the representing volume. After comparison of multiple scenarios, one of the best scenarios was picked up as the best prototyping angle. Based on the approach proposed in this study, no three-dimensional model of re-construction is necessary. The points data was directly processed and the optimization of entity prototyping is achieved in a direction of 360 degree in space[2]. The actual steps can be described as follows:

Acquiring points data of entity → setting up points data three-dimensional lattice → setting up mathematical model of the representing volume and doing calculation → prototyping direction optimization using genetic algorithm.

2 Relationship Between Transmitting Illumination Model and Lattice of Entity

2.1 Transmitting Illumination Model

At first, points data is supposed being surrounded by a surrounding box. Then this surrounding box is gradually divided into smaller three-dimensional grid. Such three-dimensional grid is then traced using light ray. Surrounding box is a kind of closing surface of points data, such as cuboid surface. Within the scenario, a piece of light ray will not intersect with the surface of an entity if this piece of ray does not intersect with the surrounding box. If this piece of ray does intersect with the surrounding box, then this intersection can be calculated between the ray and the surface. Because of the simplicity of describing a surrounding box, the testing of the above mentioned intersection is such a comparatively simple method that reduce large amount of complicated intersection calculation[3].

The piece of light ray can be expressed by:

$$O= Q + E, t\in[0, \infty] \tag{1}$$

where,

$Q=[x_Q,y_Q,z_Q]^T$ starting point of the ray;
$E=[x_E,y_E,z_E]^T$ unit vector, representing the direction of the ray;
$O=[x,y,z]^T$ random point along the ray.

Assuming the length of a three-dimensional grid as Δ, then its six surface can be expressed by $x=i\Delta$, $x=(i+1)\Delta$, $y=j\Delta$, $y=(j+1)\Delta$, $z=k\Delta$, $z=(k+1)\Delta$. If the ray intersects with this three-dimensional grid, in general, two intersection points can be obtained, which are inputting point of Pin, and outgoing point of Pout. These two points is possible overlapped as one point under specific situation. Based on the ray direction, the six surface of the three-dimensional grid can be divided into two groups. One group is a starting surface, the other one is an end surface. When $x_E>0$, $x=i\Delta$ is the starting surface and $x=(i+1)\Delta$ is the end surface. The other surfaces can be treated similarly. The intersection of the light ray and the starting surface is Pin, while the intersection with the end surface is Pout. The location of Pout determines which next three-dimensional grid the light ray will locate after outgoing from the previous grid. Now the end surface is set up as $x=(i+1)\Delta$, $y=j\Delta$, $z=(k+1)\Delta$, then:

If Pout is located on the x=(i+1)Δ surface, the next three-dimensional grid will be (i+1, j, k); If Pout is located on the y=(j+1)Δ surface, the next three-dimensional grid will be(i, j-1, k); If Pout is located on the z=(k+1)Δ surface, the next three-dimensional grid will be (I, j, k+1).

In order to determine which surface Pout is going to locate, inputting equation (1) into the formulas of the above three end surfaces and the following three parameters was obtained, including:

$$t_1 = \frac{(i+1)\Delta - x_Q}{x_E}, \quad t_2 = \frac{j\Delta - y_Q}{y_E}, \quad t_3 = \frac{(k+1)\Delta - z_Q}{z_E} \tag{2}$$

Then, P_{out} should be located on the end surface having the smallest parameter.

The light ray from the source will firstly intersect with the front surface of a specific three-dimensional grid. If there is no data within this grid, the value of zero will represent this grid. If there is data in the grid, the value should be 1. The light ray then goes into the next grid. The above procedures will repeat. At the end, all the obtained data can be represented as a lattice.

2.2 Expression of Three-Dimensional Entity Using Lattice

During prototyping optimization, the calculation of representing volume is difficult because of the complication of the entity shape and the variation of representing volume at various positions. It is therefore necessary searching for an approach with which the representing volume can be computed rapidly in an increasing in efficiency. The purpose of expressing points data in a format of lattice is for calculating the representing volume.

Setting up of a three-dimensional mesh. The scanned points data of an entity firstly is represented in a format of matrix. Based on the order of points data, the matrix is also presented in the same order of x、y、z.

The maximum and minimum values of points data can be found in the matrix and can be expressed as x_{max}、x_{min}、y_{max}、y_{min}、z_{max}、z_{min}. A surface normal to x axis can be obtained through points of x_{max}、x_{min}. A surface normal to y axis can be obtained through points of y_{max}、y_{min}. A surface normal to z axis can be obtained through points of z_{max}、z_{min}. A surround box is set up by these six normal surfaces. The points data is included within this box.

Based on the scanning step length of △x、△y、△z and the size of the surrounding box, the structure of the three dimensional lattice of the points data can be determined. The scanning step length can be input during set-up of scanning. It can also be obtained within the points data. X_{num}、Y_{num}、Z_{num} can be computed by equation (3) as follows:

$$\begin{aligned} x_{num} &= \frac{x_{max} - x_{min}}{\Delta x} + 1 \\ y_{num} &= \frac{y_{max} - y_{min}}{\Delta y} + 1 \\ z_{num} &= \frac{z_{max} - z_{min}}{\Delta z} + 1 \end{aligned} \tag{3}$$

Within a surrounding box, a surface normal to x axis was set up on the layer of (X_{num}-1), a surface normal to y axis was set up on the layer of (Y_{num}-1), and a surface normal to z axis was set up on the layer of (Z_{num}-1). The above three layers constitute a three dimensional grid of $X_{num} \times Y_{num} \times Z_{num}$, and the area of each grid can be expressed as $\triangle x \times \triangle y \times \triangle z$.

Setting up of 01 point lattice. The coordinates of (i, j, k) within the three dimensional grid can be determined by the point coordinates of (x, y, z) within the matrix by including the data of matrix into the created three-dimensional grid. Such created three-dimensional grid is then processed using transmitting illumination model. Based on the above method, 01 lattice can be created, and the entity can be represented by 01 point lattice rather than coordinates.

Large amount of unnecessary data are available within the points data and some of which have similar coordinate values. During the process of creating point lattice, it is possible that these kinds of data are transferred into one small grid and can be processed as one data. The size of the data is then naturally compressed. If the space between grids is big enough, the compressed data will not impact the calculation accuracy.

3 Calculation of Represent Volume

Within laser scanning system, its original coordinate does not coincide with the center of working platform of the machine. The X-Y-Z coordinate system of the scanning machine is Cartesian Coordinate System. The model of scanning system for this study is 3DFAMILY. The value range of unit vector (a, b, c) can be expressed by a unit spherical of $a^2+b^2+c^2=1$. Representing volume of V (a, b, c) has no linear relationship because of the various shape of processed entity. Such volume can also be expressed by a convex function. Therefore, prototyping direction optimization of an entity can focus on solving the non-linear problem of the following single function of:

$$\begin{aligned} &\min z=f(x_1,x_2,\ldots, x_n) \\ &\quad \text{s.t. } h_i(x_1,x_2,\ldots, x_n)\leq 0 \qquad i=1,2,\ldots,m \\ &\qquad g_i(x_1,x_2,\ldots, x_n)=0 \qquad j=1,2,\ldots, p \\ &\qquad (x_1,x_2,\ldots,x_n)\in R^n \end{aligned} \tag{4}$$

During prototyping process, the representing volume will be treated as an objective function. The prototyping direction with the smallest representing volume is the best prototyping direction. As described, an entity can be represented as a 01 lattice. Therefore, the representing volume can be obtained based on computing the amount of grids of points data. A program was set up using Matlab language for computing the representing volumes corresponding to various locations of the entity.

4 Optimization Based on Genetic Algorithm

Prototyping direction optimization is a comparatively complicated optimization problem. Its objective function can be expressed as non-convex, non-linear, multiple waves and noisy format. Analysis method is improper for this kind of problem. Traditional optimization methods will also have many difficulties for solving this problem. The best optimization maybe obtained within specific range but not the whole range[4]. Therefore, it is necessary to find a new and more stable optimization method for solving such prototyping direction problems of entity.

Genetic algorithm is established on the basis of natural choice and group random genetic theory. Such algorithm has features of random, iteration and evolution, and also flexible and extensive. So it is suitable for solving non-convex objective functions of prototyping direction optimization problems[5].

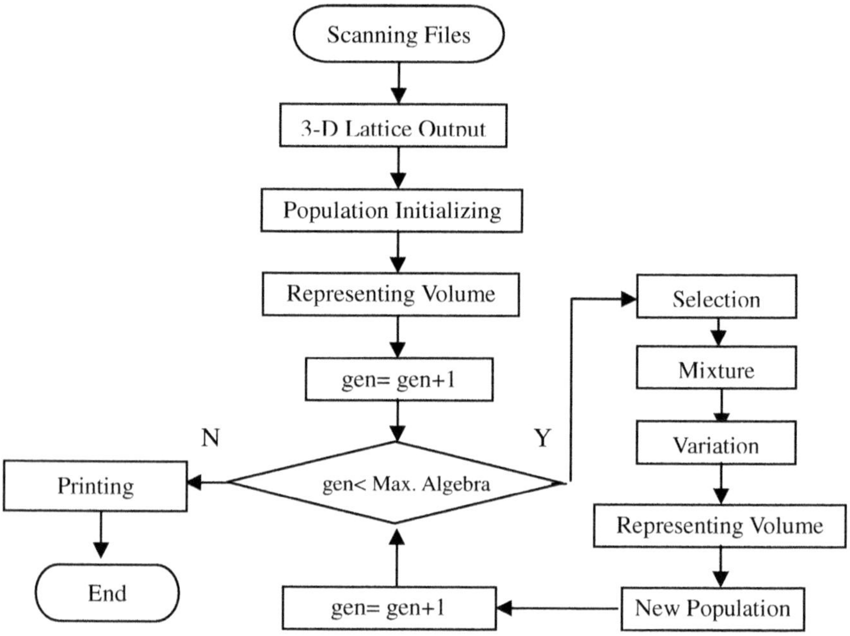

Fig. 1. Main program flow

Coding of real number has features of high accuracy and easy searching. So such coding method has high efficiency for solving function optimization problem. It can keep better population diversity. At the same time, the topology structure of real number coding remains the same as that of representing space, which can make it easy to find effective hereditary operators based on traditional optimization methods. Therefore, real number coding was used limiting optimization problem in this study for. In order to randomly produce initial population, the required equation can be expressed as $a^2+b^2+c^2=1$. So, every individual in initial population was located within feasible region formed as a unit sphere. Standard geometry distribution was used in selecting operators and also corrected arithmetic hybrid. Variation operator adopts uneven variation. The non-changed variation parameters was randomly determined between a, b and c. Then, changed parameters were also determined. Variation direction (upward or downward) was randomly determined. Variation quantity was produced randomly. Then the c value was adjusted according to unit vector, $|c| = \sqrt{1-a^2-b^2}$, c symbol does not change.

The main program can repeat multi-processing genetic algorithm. The users should input different parameters before running the main program each time. After running, the statistical results relating the evolution process were input into the output data files. The first step is to initialize, including acquiring algorithm parameters, distribution of data space, initialization of random number producing machine as well as producing initial population, and exporting statistical information of each generation, etc. Then the evolution of each generation is calculated. The main program flow chart is shows on Fig.1.

5 Prototyping Direction Optimization Actual Example

In this section, the proposed prototyping direction optimization was proved using the points data of a vase as an example. A 360^0 revolving scanning method was used in the vase sampling. Because the height of the entity exceeded that of the scanner, the whole scanning process was divided into several phases. The data of the final scanning was synthesis, which resulted in the points data of the revolving vase, as shown on Fig.2. The sampling span is 0.5 mm along X direction, Y direction and Z direction. Based on 360^0 revolving of the vase, the sampling was carried out every 2^0. After sampling, a 01 lattice was set up based on the points data. The required representing volume of the entity was calculated along various prototyping direction. Finally, optimization program of genetic algorithm was used in prototyping direction optimization, which located the best position of the entity. Such obtained best position was verified by the calculating result of prototyping machine.

During transferring points data into 01 lattice, the division of 3-D grid kept consistent with the sampling span. Based on the maximum size of the entity, a 01 lattice was obtained with 97 mm long, 97 mm wide and 83 mm high. The height was corresponding to the Z direction of scanner. The length was corresponding to the Y direction of scanner. The width was corresponding to the X direction of scanner. The prototyping direction was represented by the angels between unit vectors of a, b and c and X axle positive direction, Y axle positive direction, and Z axle positive direction. The final results were expressed in a format of angle.

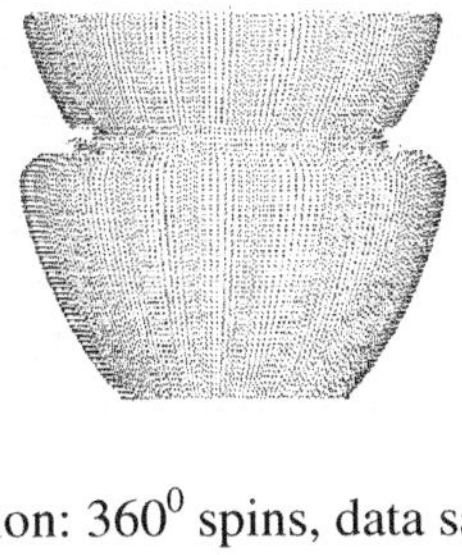

Sampling condition: 360^0 spins, data sampling every 2^0,
Sampling span along X Y, Z direction is 0.5 mm

Fig. 2. Points data of vase scanning

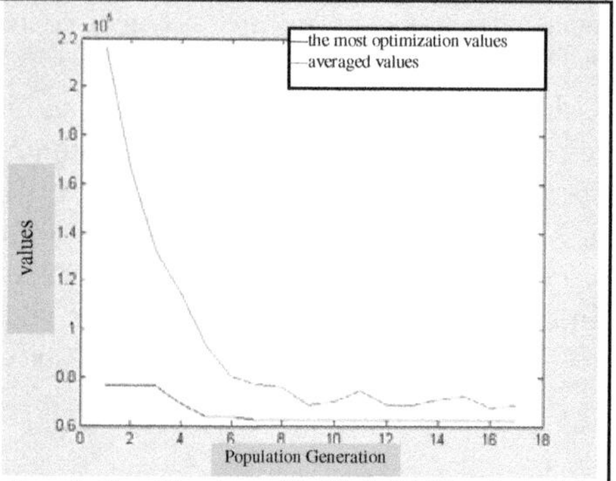

Population Number=80, Hybrid Number=18, Population Generation=16, b=1/3, q=0.05 Variation Number=18

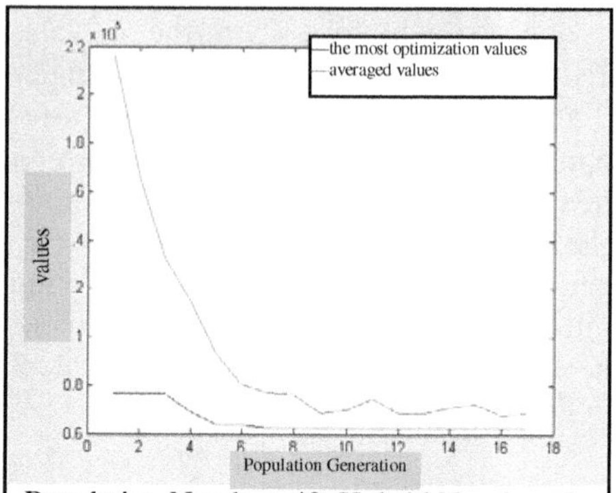

Population Number=40, Hybrid Number=8, Population Generation=16, b=5.0e-001, q=0.05, Variation Number=8

Fig. 3. The change of the most optimization values and its averaged values

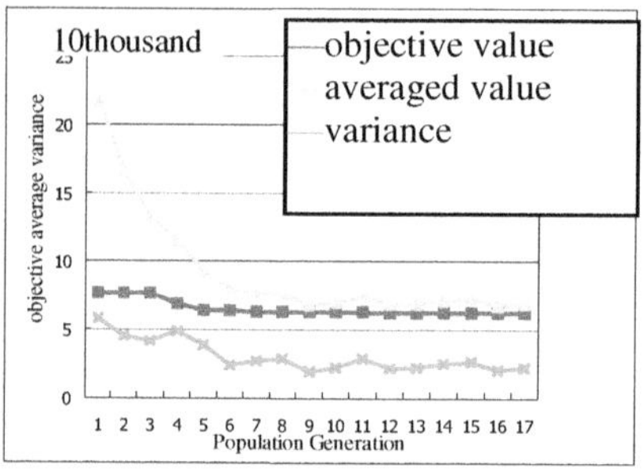

Population Number=80, Hybrid Number=18, Population Generation=16, b=1/3, Variation Number=18 q=0.05

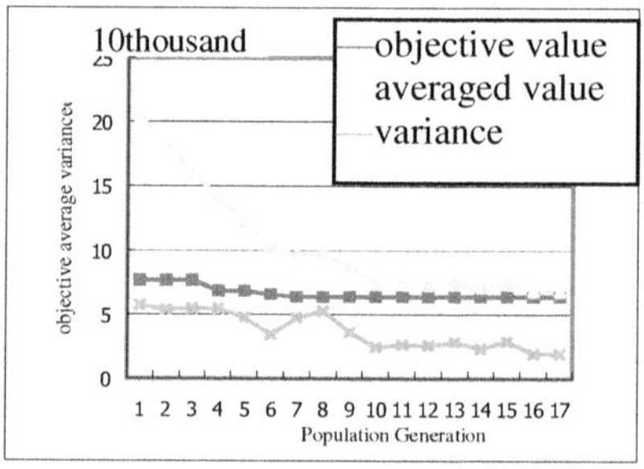

Population Number=40, Hybrid Number=8, Population Generation=16, b=5.0e-001, Variation Number=8, q=0.05

Fig. 4. Vase prototyping direction optimization record of each generation

The output results should include most updated optimization records, optimization record of each generation, the best position of the entity during prototyping, the number of objective function of V (a, b, c), as well as running time.

Variance analysis was carried out. Fig.3 presents the optimization values of all revolving scanning generations and the average values development. Based on the presented curves, increasing operating algebra resulted in a gradually closing of these two curves, which indicates an approaching of the best solution. The output files certified the above results. Fig.4 shows the maximum optimization change of each generation during prototyping direction optimization. According to the curves, increasing the population generations resulted in the averaged values closing to the objective value. On the basis of the analysis to the multiple running results and also consideration of the possible errors, the prototyping direction was determined based on the averaged values of the multiple running results. Therefore, the prototyping direction optimization of the vase is:

The angles between the normal direction of the entity and the axis of X, Y and Z are 90^0, 90^0, 180^0, respectively. This conclusion was verified by the rapidly prototyping process result. So it is practical to do points data prototyping direction optimization using genetic algorithm. Such optimization result has important meaning for fast prototyping technology.

References

1. Feng Lin, Wei Sun and Yongnian Yan. Optimization with minimum process error for layered manufacturing fabrication, Rapid Prototyping Journal, 2001, 7(2): 73-81
2. Hu Ying, Xu Xin He. The fast light reflection algorithm based on light correlation. Graphics and Imagine Journal, 2004, 9(2): 234-239
3. Jorg Schwerdt, Michiel Smid, Man Chung Hon and Ravi Janardan.Computing an optimal hatching direction in Layered Manufacturing.2001, 1
4. Dong Tao, Zhang Wei Yi and Wang Chun Hui. Entity Layers Direction Optimization Update Technology During Fast Prototyping Manufacturing. Computer Engineering and Application, 2003(1): 45-48
5. Kamal C.Sarma.Bilevel Parallel Genetic Algorithms for Optimization of Large Steel Structures.Computer-Aided Civil and Infrastructure Engineering, 2001(16): 295-304

Generation of a Large Variety of 3-Dimensional Maze Problems with a Genetic Algorithm

Ryujiro Akashi[1] and Yoshiji Fujiimoto[2]

[1] Department of Applied Mathematics and Informatics,
Faculty of Science and Technology, Ryukoku University
[2] Department of Media Informatics,
Faculty of Science and Technology, Ryukoku University
Seta Ohtsu Shiga, Japan
fujimoto@rins.ryukoku.ac.jp

Abstract. This study generates a large variety of 3-dimensional maze (3-D maze) problems with a range of through-path lengths by a genetic algorithm. The 3-D mazes consist of 27 cubes containing a T-shaped cavity stacked into a 3×3×3 cube. When the cubes are stacked with the appropriate orientations, a 3-D maze is formed by the cavities. About 2,000 3-D maze problems with through-path lengths from 18 to 54 segments (two segments per cube) were generated by the genetic algorithm using two evaluation functions generating long and short path lengths respectively.

1 Introduction

The purpose of this study is to generate a large variety of 3-dimensional maze (3-D maze) problems using a genetic algorithm (GA), and in particular, to generate 3-D maze problems with different path lengths from the entrance to the exit.

The 3-D maze that is the object of this study is an educational toy recently developed by the educational–industrial complex program of REC (Ryukoku Extension Center) at Ryukoku University. The 3-D maze consists of 27 small transparent plastic cubes with a T-shaped cavity (T-hole) through each cube, called a "T-cube". The cubes stack to form a larger 3×3×3 cube called an "M-cube", and are packed in a transparent plastic case as shown in Fig. 1. The entrance is set at a corner cube and the exit is located at the corner cube diagonally opposite the entrance cube. When the 27 T-cubes are stacked with appropriate orientations, a path (T-path) from the entrance to the exit is formed in the M-cube

Fig. 1. 3-D maze (M-cube) and T-cubes

T.-D. Wang et al. (Eds.): SEAL 2006, LNCS 4247, pp. 806 – 813, 2006.

connecting the T-holes of the T-cubes. A wide variety of 3-D maze problems can be generated, depending on the directions of the T-cube placements.

The 3-D maze problem is played by putting a small ball bearing into the entrance hole and moving the ball towards the exit through the paths of the maze by slanting and turning the M-cube by hand. When the ball reaches the exit of the maze, the problem is solved.

To generate a variety of 3-D maze problems, sequences of the directions of the 27 T-cubes can be evolved as chromosomes by a genetic algorithm under two evaluation functions. One evaluation function generates maze problems with long through-paths and the other generates maze problems with short paths.

Bentley has classified evolutionary design by computers into four types[1,2]: evolutionary design optimization, creative evolutionary design, evolutionary art and evolutionary artificial life-forms. However, this 3-D maze application does not precisely belong to any of the four types; it is considered to lie between design optimization[3] and creative evolutionary design[4,5], because this application is to generate a wide variety of 3-D maze problems with the diversity and emergency of GAs.

2 Graphical Expression of 3-D Maze (M-cube)

In order to evaluate the direction sequences of the T-cubes in an M-cube using a GA, the 3-D maze (M-cube) is expressed by a graph. A T-cube has seven nodes, one in the center of the T-cube and one at each the centers of the six surfaces of the cube. It also has three edges that correspond to the three holes from the center of the T-cube to three of its surfaces, as shown in Fig. 2(a). When the T-cubes are stacked as an M-cube, the two nodes on the face-to-face surfaces of adjacent T-cubes are unified as one node, as shown in Fig. 2(b). Therefore, an M-cube can be expressed by a graph (M-graph) with 135 nodes and $27 \times 3 = 81$ edges.

There are 12 possible directions for T-cube placement, as shown in Fig. 3. By numbering these directions, the connections between the nodes by the edges in the M-graph can be determined for a given sequence of directions of the 27 T-cubes in an M-cube. The connections in the M-graph can then be expressed as an adjacent matrix $\boldsymbol{A}$ with 135×135 elements, With each element of $\boldsymbol{A}$ showing whether or not a connection between a pair of nodes exists by "1" or "0" respectively.

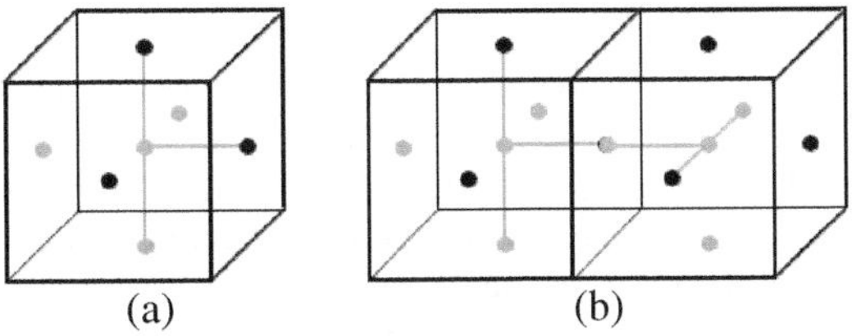

(a) (b)

Fig. 2. Graphical expression of T-cube

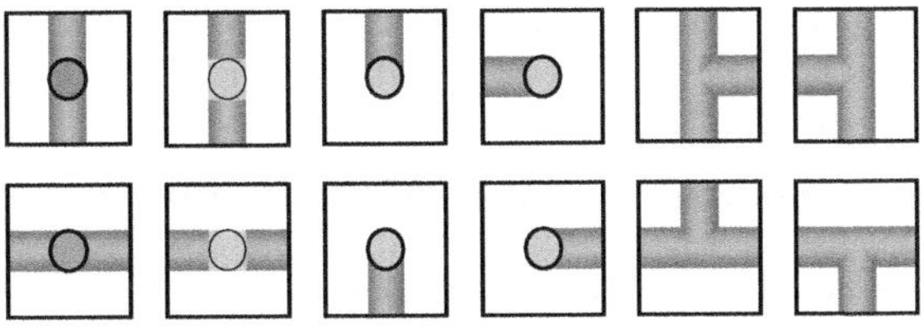

Fig. 3. 12 directions of T-cube placement

2.1 Detecting a Through-Path and Its Length

The method for detecting a T-path and its length from $\boldsymbol{A}$ is as follows.

The k-th power of $\boldsymbol{A}$ is expressed as $\boldsymbol{A}^k$. The normal matrix product is used for the product operation of adjacent matrices. In $\boldsymbol{A}^k$ that is shown in Fig.4, the start row vector is defined as $\boldsymbol{ST}_k$, corresponding to the starting node that is the entrance of the 3-D maze and indicates the number of paths with length k that can reach each node from the starting node. The goal row vector is defined as $\boldsymbol{GL}_k$, corresponding to the goal node that is the exit of the 3-D maze and indicates the number of paths that can reach each node from the goal node with length k. Therefore, a T-path from the starting node to the goal node exists when nonzero elements appear at the same position in both $\boldsymbol{ST}_k$ and $\boldsymbol{GL}_k$. Such a path with the smallest value of k is considered to be the T-path. The length of the T-path is $2k$ when the length of an edge is assumed to be 1; k shows the number of T-cubes that the T-path passes through.

$$A^k = \begin{pmatrix} 0 & 0 & 1 & \cdots & 0 & 2 & 0 \\ 0 & 4 & 0 & \cdots & 1 & 0 & 1 \\ 1 & & \ddots & & & & \\ \vdots & & & \ddots & & \vdots & \vdots \\ 0 & 1 & & & \ddots & & \\ 2 & 0 & 1 & \cdots & 1 & 0 & 0 \\ 0 & 1 & & \cdots & & & \end{pmatrix} \begin{matrix} \\ \leftarrow ST_k \\ \\ \\ \\ \leftarrow GL_k \\ \\ \end{matrix}$$

Fig. 4. Adjacent matrix and start and goal vectors

2.2 Reach Length and Connectivity

When a T-path does not exist, the S-reach is defined as the length of the longest path without duplicate nodes that reaches from the start node to each node in the sub-graph (S-graph) connected to the start node. The G-reach is defined similarly to the S-reach starting from the goal node. The sum of $\boldsymbol{ST}_k$ ($k = 1$ to m) is expressed as $\boldsymbol{SST}_m$ and the sum of $\boldsymbol{GL}_k$ ($k = 1$ to m) is expressed as $\boldsymbol{SGL}_m$:

$$SST_m = \sum_{k=1}^{m} ST_k , \qquad SGL_m = \sum_{k=1}^{m} GL_k .$$

The numbers of non-zero elements in $\boldsymbol{SST}_m$ and $\boldsymbol{SGL}_m$ are defined as $\boldsymbol{NZS}(m)$ and $\boldsymbol{NZG}(m)$. The smallest M_1 and M_2 that satisfy $\boldsymbol{NZS}(M_1) = \boldsymbol{NZS}(M_1+1)$ and $\boldsymbol{NZG}(M_2) = \boldsymbol{NZG}(M_2+1)$ give the S-reach and G-reach, respectively, i.e. the S-reach is M_1 and the G-reach is M_2. The sum of the S-reach and G-reach is the SG-reach. Furthermore, the number of non-zero elements in the vector sum of $\boldsymbol{SST}_m$ and $\boldsymbol{SGL}_m$ at the position corresponding to the center nodes of the T-cubes is defined as $\boldsymbol{NZSG}(m)$. $\boldsymbol{NZSG}(M)$ for the minimum M that satisfies $\boldsymbol{NZSG}(M) = \boldsymbol{NZSG}(M+1)$ gives the number of center nodes that are reached from the start node or goal node and is defined as the connect number of the 3-D maze. When $\boldsymbol{NZSG}(M)$ is 27, all the T-cubes are reached from the start node or the goal node and the variable ***con*** is set to 1; otherwise ***con*** is set to 0.

3 Generation of 3-D Maze Problems by GA

To generate 3-D maze problems with GA, chromosomes (individuals) and evaluation functions are first prepared.

3.1 Chromosome

A chromosome is comprised of a sequence of the direction numbers of the 27 T-cubes. The 12 directions of T-cube placement, as shown in Fig. 3, must be coded as four-bit binary numbers. Therefore, a chromosome consists of a binary sequence of 108 (27×4) bits.

In coding the placement directions, the 12 directions are assigned codes from 0 to 11 and four directions are picked at random without duplications and assigned the codes from 12 to 15 in each experiment. This coding method emphasizes the diversity of genetic algorithms.

3.2 Evaluation Functions

Two evaluation functions for generating a large variety of 3-D mazes with different T-path lengths are prepared as follows

Evaluation Function A:
This evaluation function generates 3-D mazes with long T-path lengths and is defined as follows.

If (T-path does not exist),
  Fitness = SG-reach + NZSG(M),
else if (T-path exists)
  Fitness = TP-length + NZSG(M) + 27 *con.

When a T-path does not exist, this evaluation function accelerates the formation of a T-path by expanding the S-reach and G-reach and increasing the number of connected T-cubes.

When a T-path exists, it facilitates the extension of the T-path and increases the number of connected T-cubes, preserving complete maze problems. A complete maze problem is one for which the M-cube graph has a T-path and in which the center nodes of all the T-cubes are connected to the start node.

Evaluation Function B:
This evaluation function generates 3-D mazes with short T-path lengths and is defined as follows.

If (T-path does not exist and SG-reach < 14),
  Fitness = 40 + SG-reach + NZSG(M),
else if (T-path does not exist and 14 <= SG-reach < 54)
  Fitness = 68 - SG-reach + NZSG(M),
else if (T-path exists)
  Fitness = 68 - TP-length + NZSG(M) + 27 *con.

When a T-path does not exist, this function has a peak at SG-search = 14. The function facilitates the appearance of the shortest T-path.

When a T-path exists, a higher fitness is assigned for a shorter T-path in the M-graph. This makes it easy for a T-path to shorten towards the shortest T-path. It also works to increase the number of T-cubes connected to the start node increase and preserves complete maze problems.

4 Empirical Results

Experiments generating 3-D maze problems with a GA were conducted using the two evaluation functions above under the following parameters.

Number of experiments: 10
Maximum number of trials: 100,000
Seeds of random numbers: variable for each experiment
Population: 400
Length of chromosome: $4 \times 27 = 108$
Crossover: 2-point crossover
Crossover rate: 1.0
Mutation: bit flip
Mutation rate: 0.01
Selection method: ranking method for mixed populations of parents and children [6,7]

The experimental results are shown in Table 1. In this table, TPL is the T-path length in the M-cube graph and the number of T-cubes that the T-path passes through is denoted by TPL/2. TP is the number of generated 3-D maze problems with different T-paths, and CMP is the number of different complete maze problems generated by the GA.

Table 1. Number of generated 3-D maze problems

Evaluation Function A			Evaluation Function B		
TPL	TP	CMP	TPL	TP	CMP
14	0	0	14	0	0
18	1	1	18	141	114396
22	7	20	22	126	14660
26	63	150	26	97	9031
30	159	1657	30	47	276
34	244	3509	34	5	13
38	327	4876	38	1	1
42	375	9365	42	0	0
46	283	17216	46	0	0
50	136	76935	50	0	0
54	20	129169	54	0	0
Total	1615	239389	Total	417	138377

T-paths are differentiated by the number sequence of the T-cubes that the T-path passes through. In the results, however, the directions of the T-cube placements are disregarded. Complete maze problems are also discriminated by the direction sequence of the 27 T-cubes.

The data of Table 1 is aggregated through 10 experiments by the following procedure.

(1) Whenever a complete maze problem is newly generated in the evolutionary process, its chromosome information is converted into a direction code sequence

(DC-sequence) of the 27 T-cubes and the number sequence of the T-cubes (TP-sequence) that the T-path passes through.

(2) The direction code sequence of a new complete maze problem is compared with each direction code sequence already stored in the complete maze problem pool (CMP-pool). If a complete maze problem with the same direction code sequence is found, the new complete maze problem is discarded. Otherwise, the DC-sequence and TP-sequence of the new complete maze problem are stored in the CMP-pool and the CMP count is increased for the TPL of the new complete maze problem.

(3) After 10 experiments, the CMP-pool is sorted by TP-sequence. The number of different TP-sequences for each TPL is counted in the CMP-pool.

For evaluation function A, the value of CMP increases as TPL increases. However, the value of TP decreases after a peak at TPL = 42. When TPL = 54, that is, when the T-path passes through all the T-cubes, CMP is very large compared with TP. This is because when a T-cube is used as an I-type aisle in the T-path, there are four T-cube placement directions and when a T-cube is used as an L-type aisle, there are two directions. Namely, there are many combinatorial variations of directions for the placement of the T-cubes. Evaluation function A gives the same value for complete maze problems with the same TPL, because ***NZSG***(*M*) is constant. Therefore, the power of evolution is consumed in changing the placement directions for the same T-path and hence it is hard to change the T-path length. This prevents the evolution of chromosomes towards longer T-paths.

For evaluation function B, both TP and CMP increase for decreasing TPL. CMP increases more than TP for the same reason as for evaluation function A. In the experiments, no complete maze problem with a shortest TLP of 14 was generated. It is considered that the shortest T-path is formed at an early stage of evolution but the value of the evaluation function is small because ***NZSG***(M) is small at that stage and hence it is selected out.

In order to generate only complete maze problems with a shortest TLP, other type of evaluation functions defined as follows are tried.

If (T-path does not exist),

$$\textbf{\textit{Fitness}} = \textbf{\textit{LN}} - \textbf{\textit{a}}\,(|\,14 - \textbf{\textit{SG-reach}}\,|\,)^{p} - \textbf{\textit{b}}\;(27 - \textbf{\textit{NZSG(M)}}\,)^{p},$$

else if (T-path exists)

$$\textbf{\textit{Fitness}} = \textbf{\textit{LN}} - \textbf{\textit{a}}\,(|\,14 - \textbf{\textit{SG-reach}}\,|\,)^{p} - \textbf{\textit{b}}\,(27 - \textbf{\textit{NZSG(M)}}\,)^{p} + \textbf{\textit{c}}.$$

Where ***LN*** is a large value enough to makes the ***Fitness*** positive, ***p*** is 0.5 or 2.0, and ***a***, ***b***, ***c*** are coefficients for weight and reward.

However, the shortest complete maze problem has not been generated for the experiments with various combinations of parameters, because the fitness value is trapped in local optimums such as (***SG-reach, NZSG(M)***) = (14, 23) or (18, 27).

Three examples of 3-D maze problems that have T-path with shorter length, middle length or longer length are shown in Fig. 5, respectively. In Fig. 5, a ▲ mark shows the up-ward connection and a ▼ mark shows the down-ward connection.

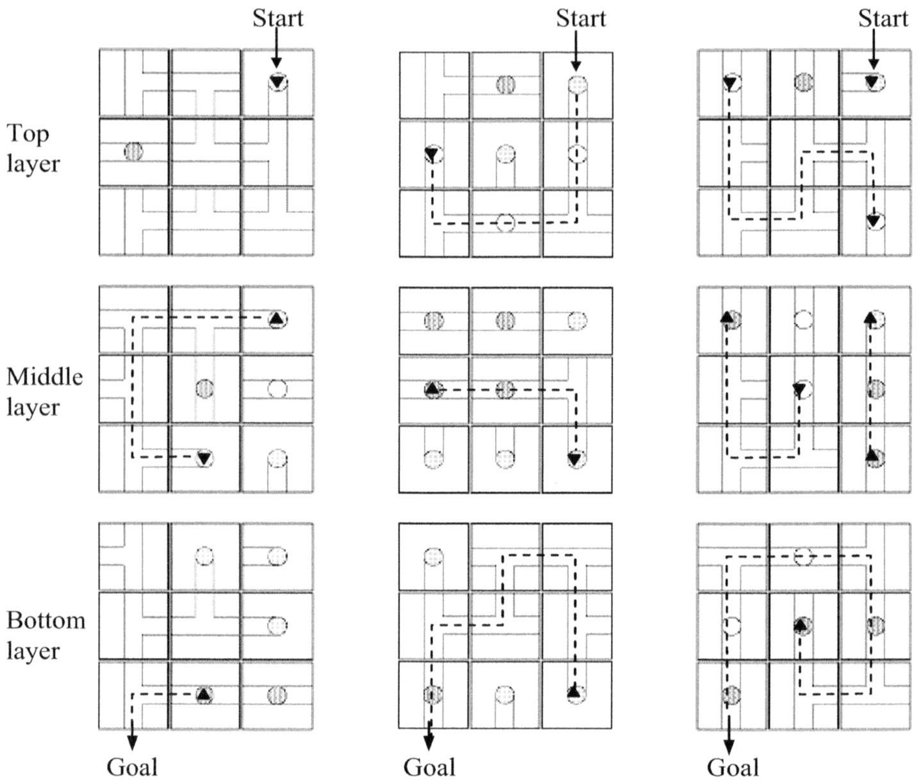

(a) Problem with a shorter T-path

(b) Problem with a middle T-path

(c) Problem with a longer T-path

Fig. 5. examples of 3-D maze problems

5 Conclusion

In this study, about 2,000 3-D maze problems with different T-path lengths were generated by a GA using two evaluation functions, one for long T-path lengths and one for short T-path lengths. However, no 3-D maze problems with the shortest T-path length were generated.

Future work will attempt to elucidate why no 3-D maze problems with the shortest T-path length were generated and improve the evaluation functions to make it possible to generate such problems. A further challenge is to generate 3-D maze problems that consist of 4×4×4 M-cubes.

References

1. Bentley, P. ed.: Evolutionary Design by Computer. Morgan Kaufmann Publishers (1999).
2. Bentley, P. J. and Corne D. W. eds.: Creative Evolutionary System. Morgan Kaufmann Publishers (2002).

3. Eby, D., Averill, R., Gelfand, B., Punch, W., Mathews, 0. and Goodman, E.: An Injection Island GA for Flywheel Design Optimization. 5thEuropian Congress on Intelligent Techniques and Soft Computing FUFIT '97, vol. 1, Verlag Mainz, Aachen, pp. 687-691(1997)
4. Funes, P. and Pollack, J.: Computer Evolution of Buildable Objects. Fourth European Conference on Artificial Life, Cambridge, MA: MIT Press. pp. 358-367(1997)
5. Mazza, R. H. and Congdon, C. B. : Towards a Genetic Algorithms Approach to Design 3D Polygonal Tree Models. The 2002 IEEE Congress on Evolutionary Computation Proceedings, vol. 2, pp. 1842-1847(2002)
6. Tsutsui, S. and Fujimoto, Y.: Forking Genetic Algorithm with Blocking and Shrinking Modes. Proc. 5th ICGA, pp. 206–213 (1993)
7. Tsutsui, S., Fujimoto, Y. and Ghosh, A.: Forking Genetic Algorithms: GA with Search Space Division Schemes. Evolutionary Computation, Vol. 5, No. 1, pp. 61–80 (1997)

MPC-Based Control Methodology in Networked Control Systems

Ke Zhang, Hai Huang, and Jianming Zhang*

National Laboratory of Industrial Control Technology,
Institute of Advanced Process Control, Zhejiang University, Hangzhou 310027, P.R. China
kzhang@iipc.zju.edu.cn, jmzhang@iipc.zju.edu.cn

Abstract. Network random delays directly cause the degradations of networked control systems. A methodology based on model predictive control (MPC) is proposed to overcome the non-deterministic delays in data communication of the network. The general algorithm of dynamic matrix control (DMC) has been improved to make it suitable for network condition. When data packets can not arrive at target nodes in sequence, the predictive value of the system output can be used to take the place of the actual measure value by the controller, and the predictive value of control input will be acted as the required control value which coordinate the whole control system. The experiment results based on Motor Ethernet Control Open Platform (MECOP) show the effectiveness of the real-time networked control strategy.

1 Introduction

Networked control system (NCS) which takes advantages of simple structure, distributed control and low cost, will be applied more in remote control systems [1]. NCS also shows disadvantages in some applications, especially in motor speed control application. In NCS, sensors, controllers and other nodes share the limited network bandwidth, which causes queuing, frame collision and network-induced delay. Network random delay degrades system performance, even destabilizes the system.

In order to solve the problems above, various methodologies have been formulated to treat the problems. Luck and Ray developed a queuing methodology, which can be used to reshape random network delay to deterministic delay [2]. Nilsson proposed the optimal stochastic control methodology [3]. The sampling time scheduling methodology was developed by Hong, in which a sampling period is appropriately selected to keep the system stable [4].

Besides above methodologies, many researchers have shown interest in applying model predictive control (MPC) for NCS. Model predictive control has already shown reliability and availability on petrochemical industry, electrical power and aerospace [5]. This paper proposes an improved model predictive control methodology for

* Corresponding author.

T.-D. Wang et al. (Eds.): SEAL 2006, LNCS 4247, pp. 814 – 820, 2006.

network condition, in which moving optimization and feedback compensation are used to deal with the negative effect caused by network random delay.

2 Analysis of Network Induced Delay

In a typical networked control system as shown in Fig. 1, there are sensor- to-controller delay τ^{sc}, controller-to-actuator delays τ^{ca} and controller calculation delay τ^{c}. All these delays are non-deterministic. Usually the calculation delay τ^{c} is negligibly small compared with τ^{sc} and τ^{ca}.

In order to deal with network random delay, some assumption may be required: (a) The sampling time skew between nodes is assumed to be a negligibly small value, so that all the nodes can be assumed synchronized. (b) The data packet carrying measurement includes a time-stamp of when the sensor takes sample of the plant output. (c) The data packet carrying control signal includes a time-stamp of when the controller sends the packet. Timing diagram of NCS is shown in Fig. 2.

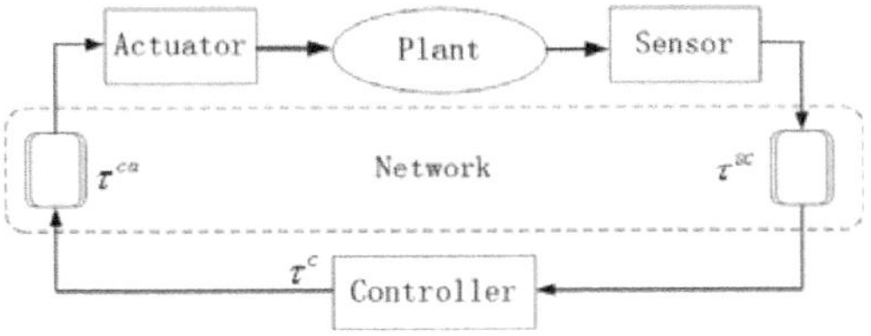

Fig. 1. The Structure of Networked Control System

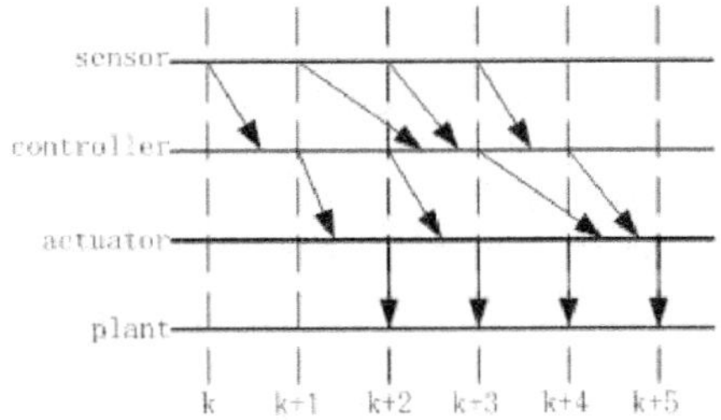

Fig. 2. Timing Diagram of Networked Control System

If τ^{sc} and τ^{ca} are both less than sampling period T, the control process from sensor's sending plant output to control signal's showing effect on plant costs $3T$, as the process starting at kT in Fig. 2. If τ^{sc} is greater than T, during the sampling interval starting from $(k+1)T$ in Fig. 2, controller can not receive any measurement, and during one interval later there may be more than one measurement getting to controller. It is similar when τ^{ca} is greater than T.

3 The Improved DMC Algorithm

The general Dynamic Matrix Control (DMC) algorithm has been improved to make it suitable for the network condition. Similar to the general DMC algorithm, controller must store the future trend vector $\hat{y}_{N1}(k)=[\hat{y}_1(k\mid k) \quad \dots \quad \hat{y}_1(k+N-1\mid k)]^T$. Here, the vector $\hat{y}_{N1}(k)$ starts with $\hat{y}_1(k\mid k)$, which is the predictive value of system output for sampling point kT, and estimated at sampling point kT.

As shown in Fig. 2, at $(k+1)T$, controller obtain the adjusting error

$$e(k)=y(k)-\hat{y}_1(k\mid k). \tag{1}$$

Then $\hat{y}_{N1}(k)$ can be corrected with the adjusting error and shifted to be the future trend vector without control input as

$$\begin{aligned}\hat{y}_{N0}(k+1)&=S(\hat{y}_{N1}(k)+he(k))\\&=[\hat{y}_0(k+1\mid k+1) \quad \dots \quad \hat{y}_0(k+N\mid k+1)]'\end{aligned} \tag{2}$$

where h is the correcting vector, and $S=\begin{bmatrix}0 & 1 & \dots & 0\\ 0 & 0 & \ddots & \vdots\\ \vdots & \vdots & \dots & 1\\ 0 & 0 & \dots & 1\end{bmatrix}$ is the shift matrix.

The objective function for sampling point $(k+1)T$ is following:

$$\begin{aligned}\min J(k+1)=&[w_P{}^{k+3}(k+1)-\hat{y}_{PM}{}^{k+3}(k+1)]^T Q_{P\times P}[w_p{}^{k+3}(k+1)-\hat{y}_{PM}{}^{k+3}(k+1)]\\&+[\Delta u_M(k+1)]^T R_{M\times M}[\Delta u_M(k+1)]\end{aligned} \tag{3}$$

where P is prediction horizon, and M is control horizon. $Q_{P\times P}$ and $R_{M\times M}$ are weight matrix. $w_P{}^{k+3}(k+1)=[w(k+3) \quad \dots \quad w(k+P+2)]^T$ is the setpoint. And $\hat{y}_{PM}{}^{k+3}(k+1)=\hat{y}_{P0}{}^{k+3}(k+1)+A\cdot\Delta u_M(k+1)$ is the future trend vector if all the control inputs in $\Delta u_M(k+1)$ are moved in sequentially, in which $\hat{y}_{P0}{}^{k+3}(k+1)=[\hat{y}_0(k+3\mid k+1) \quad \dots \quad \hat{y}_0(k+P+2\mid k+1)]^T$, and $A_{P\times M}$ is dynamic matrix. So the optimal control input vector can be calculated as

$$\begin{aligned}\Delta u_M(k+1) &= [\Delta u(k+1) \quad \ldots \quad \Delta u(k+M))]^T \\ &= (A^T QA+R)^{-1} A^T Q[w_P{}^{k+3}(k+1) - \hat{y}_{P0}{}^{k+3}(k+1)] \\ &= F[w_P{}^{k+3}(k+1) - \hat{y}_{P0}{}^{k+3}(k+1)]\end{aligned} \tag{4}$$

The controller gain matrix F can be calculated off-line. Controller sends the whole control vector to actuator instead of $\Delta u(k+1)$. Then the future trend vector should be updated for the calculation of the snext sampling point,

$$\hat{y}_{N1}(k+1) = \hat{y}_{N0}(k+1) + [0 \quad 0 \quad a_1 \quad \ldots \quad a_{N-2}]^T \Delta u(k+1), \tag{5}$$

where $\{a_i\}$ are the step response coefficients.

At sampling point $(k+2)T$ in Fig. 2, actuator stores $\Delta u_M(k+1)$ into internal registers, and selects $\Delta u(k+1)$ in the first register unit as the input of plant.

In the case of either τ^{sc} or τ^{ca} being greater than T, the algorithm should be improved further. When τ^{sc} is greater than T, for example in Fig. 2, in the sampling interval from $(k+1)T$ to $(k+2)T$, controller does not receive the measurement, instead, it selects $\hat{y}_1(k+1 \mid k+1)$ from $\hat{y}_{N1}(k+1)$ to replace the actual measurement $y(k+1)$, and then performs the calculation normally. When controller receives more than one measurement in one sampling interval, τ^{sc} s can be obtained by subtracting time-stamp from current instant, then the measurement whose τ^{sc} is less than T is used for calculation.

The algorithm of actuator is similar to that of controller. For example in Fig. 2, during the sampling interval between $(k+3)T$ and $(k+4)T$, actuator does not receive $\Delta u_M(k+3)$, then it loads $\Delta u_M(k+2)$ which is stored at $(k+3)T$, and shifts it to be $\Delta u_M(k+3) = S \cdot \Delta u_M(k+2)$. Then value in the first register is selected as the input to plant at sampling time $(k+4)T$. If more than one control vector arrives at actuator, the latest vector is stored into registers and selected to plant.

4 Experiment Study

In order to verify the effectiveness of the improved DMC algorithm, a Motor Ethernet Control Open Platform (MECOP) has been implemented. In MECOP, a 3-phase PMSM, is driven by a module which is equipped with TI's TMS320LF2407A-16-bit fixed point DSP, RTL8019AS, and Mitsubishi's PM20CSJ060 IPM. And a PC is used

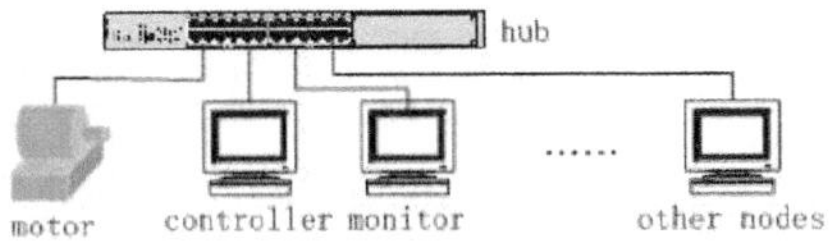

Fig. 3. The structure of MECOP

as the master controller. A monitor based on Labview is constituted on another PC, to display the real-time speed of the motor and the network delay. In addition, lots of disturbing nodes are set up to access the shared transmission medium, to causes random delay. These components are all connected to a hub, and the communication between these nodes is based on UDP/IP. The structure of MECOP is shown in Fig. 3.

In the application of the improved DMC algorithm, the sampling period is set to be 10ms. With parameters chosen suitably, the controller gain matrix is computed off-line and stored. In the experiments, the setpoint is a square wave, and its period is 5s, its amplitude is from 500rpm to 1000rpm. The network delay becomes greater, when the number of disturbing nodes increases.

In order to compare the control performance, three different algorithms: PID, DMC and the improved DMC have been applied under three network conditions: (a) there is no disturbing node, (b) there are a few disturbing nodes to access the shared bus, (c) there are lots of disturbing nodes. The control performances of the three algorithms are shown in Fig. 4, Fig. 5 and Fig. 6. In every figure, the three curves above indicate the speed of motor, and the below indicate the loop delays.

As shown in all figures, when there is no disturbing node, the control performances of the three algorithms are all satisfying. If a few disturbing nodes are connected to the

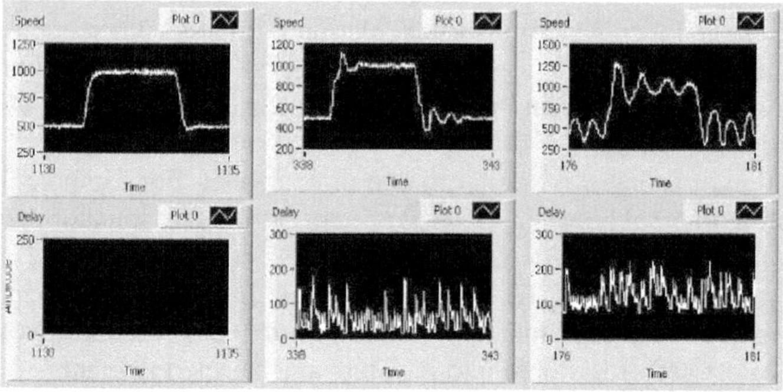

Fig. 4. The Control Performance of the PID Algorithm

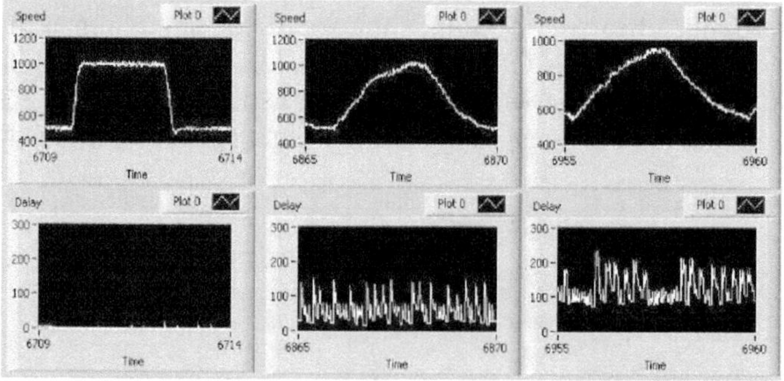

Fig. 5. The Control Performance of the General DMC Algorithm

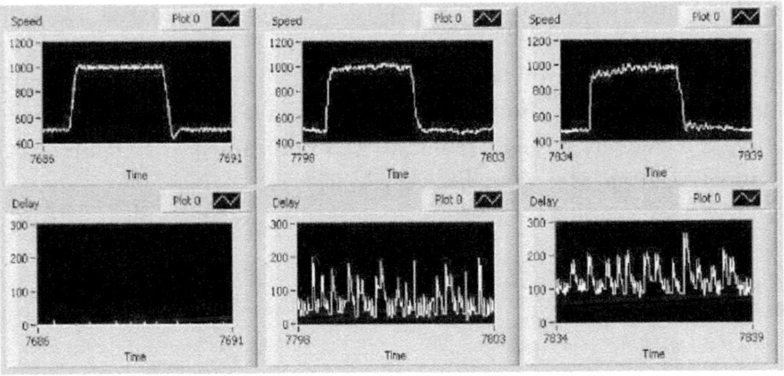

Fig. 6. The Control Performance of the Improved DMC Algorithm

network, the delays become to 20~160ms. The increased delays degrade the performances of PID and DMC, while the improved DMC remains satisfying. When the number of disturbing nodes increases, the delays are ranging between 80 and 200ms. The maximum overshoot and settling time of PID become larger and longer, and the output curve of DMC fails to follow the track of the setpoint. While the performance of the improved DMC still shows good tracking capacity.

5 Conclusions

An improved DMC algorithm has been proposed to deal with the control performance degradation caused by network-induced delay. In the improved DMC algorithm, the moving optimization and feedback compensation have been adjusted to be suitable for network condition, and the predictive value of system output is used to take place of the unreached measurement, and the control vector is shifted to take place of the unreached control vector. The result of the experiments based on MECOP has shown the effectiveness of the improved DMC algorithm. And compared with the PID and the general DMC algorithms, the improved DMC algorithm maintains good tracking capacity and satisfying performance, when the delay increases.

Acknowledgements

This work was supported by the National Natural Science Foundation of China No. 60421002.

References

1. Yodyium Tipsuwan, Mo-Yuen Chow. Control methodologies in networked control systems. Control engineering practice. 2003, 11(10): 1099-1111
2. Luck R & Ray A. An observer-based compensator for distributed delays. Automatica. 1990, 26(5): 903-908

3. Nilsson, J. Real-time control systems with delays. Ph.D.dissertation, Lund Institute of Technology. 1998
4. Hong. Scheduling algorithm of data sampling times in the integrated communication and control systems. IEEE Transactions on Control Systems Technology. 1995, 3(2): 225-230
5. Cutler, C. R., & Ramaker, B. L. Dynamic matrix control——a computer control algorithm. AICHE national meeting, Houston, TX, April 1979

A Dynamic Clonal Selection Algorithm for Project Optimization Scheduling*

Xiaoying Pan, Fang Liu, and Licheng Jiao

Institute of Intelligent Information Processing, Xidian University, Xi'an, China, 710071
xiaoying_pan@163.com

Abstract. A Dynamic Clonal Selection Algorithm for the Resource-Constrained Project Scheduling Problem (RCPSP-DCSA) is proposed in this paper. Based on the mechanism of nature immune system and characteristic of project scheduling, the encoding of solution, some operators (including crossover, mutation, deriving and death) and the function of affinity are given. Through a thorough computational study for a standard set of project instances in PSPLIB, the impact of the parameters on the performance of algorithm are analyzed. Experimental results show RCPSP-DCSA has a good performance and it can reach near-optimal solutions in reasonable time.

1 Introduction

The Resource-Constrained Project Scheduling Problem (RCPSP) is a very important problem in manufacture project and project manage. RCPSP is to schedule the activities such that precedence and resource-constrained are satisfied while optimizing some managerial objective. Being an NP-hard problem, RCPSP is an interesting research area and many mathematical models have been used to tackle this problem, such as Integer Programming, Implicit Enumeration and Branch-and-cut[1], etc. Certainly, some heuristic procedures were proposed. They involve simulated annealing[2], ant colony[3],genetic algorithm and its improved algorithm[4][5][6]. Kolisch[8] summarized past work and gave some new directions. From his work, we can see some GA-based heuristic approaches have a good performance. Considering a good characteristic of clonal selection algorithm for the optimized problem, this paper proposes a dynamic clone selection algorithm for the precedence and resource-constrained single-mode project optimization scheduling whose objective is project makespan.

2 Problem Formulation

Assume the non-preemptable activities in a project are numbered from 1 to J, where the dummy activities 1 and J respectively mark the beginning and the ending of the project. The duration of an activity j is denoted by d_j where duration is zero for dummy

* Supported by the National Natural Science Foundation of China under Grant Nos. 60372045 and the National Grand Fundamental Research 973 Program of China under Grant No. 2001CB309403.

T.-D. Wang et al. (Eds.): SEAL 2006, LNCS 4247, pp. 821–828, 2006.

activities ($d_1 = 0 = d_J$) and non-zero for others ($d_j > 0, j = 2, \ldots, J-1$). The start time of each activity is denoted by $S_j (1 \leq j \leq J)$. There are K renewable resource types, with $r_{jk}(1 \leq j \leq J, 1 \leq k \leq K)$ being the resource requirements of activity j with respect to resource k. $R_k(1 \leq k \leq K)$ is the constant availability of resource k throughout the project duration. We denote the set of immediate predecessors of an activity j by P_j.

Based on the above definitions of the variables, the mathematical model of RCPSP can be formulated as equation (1):

$$\begin{cases} min\ S_J & (a) \\ s.t.\ \ S_j - S_i \geq d_i,\ i \in P_j & (b) \\ \quad \sum\limits_{j \in A_t} r_{jk} \leq R_k & (c) \\ \quad t = 1, 2, \ldots, S_J;\ k = 1, 2, \ldots, K & (d) \end{cases} \tag{1}$$

The objective function (a) minimizes the project duration determined by the completion time. Constraints (b) and (c) ensure the precedence and resource constraints.

3 A Dynamic Clonal Selection Algorithm for Project Optimization Scheduling

Based on the mechanism of nature immune system, a dynamic clonal selection algorithm for project optimization scheduling is designed. It designs the affinity to testing the antibody's performance, crossover and mutation operators to generate new antibody, deriving operator to simulate the mature B-cell in marrow, and death operator to remove the aging units.

3.1 Antibody Encoding

In general, encoding has two types. In direct representation, a chromosome represents a solution of the original problem. But it is too complicated to represent and manipulate. In indirect representation, binary system to represent a solution is adopted. The immune genic operation of which is very simple, but it is difficult to find an appropriate encoding. This paper adopts the scheduling sequence and the start time of each activity as the antibody. An additional gene F to represent the direction of the parallel scheduling.

$$I = \begin{pmatrix} J \\ S \end{pmatrix} \mid F \Bigg) = \begin{pmatrix} j_1, j_2, \ldots, j_J, \\ s_{j_1}, s_{j_2}, \ldots, s_{j_J}, \end{pmatrix} \mid F \Bigg). \tag{2}$$

The J in the first part is a valid sequence of activities. S is the corresponding start time of each activity by forward($F = 1$) or backward($F = 0$) parallel scheduling[5]. So, every antibody here expresses an only scheduling I.

3.2 Affinity

In resource-constrained project scheduling problem, antibody-antigen affinity describes the performance of scheduling (i.e. the duration of the project). We will find the project makespan, so the shorter the duration, the higher the Ab-Ag affinity. The affinity $f(i)$ to the antibody i is defined as (3).

$$f(i) = F_{max} - F(i) + 1. \tag{3}$$

where F_{max} is the longest duration in current population. $F(i)$ is the duration of antibody i, i.e. the start time of work J, $F(i) = S_J(i)$.

The similarity of two antibodies is considered in the affinity of them. To keep the diversity in population and make the antibodies distributing uniformly, the same antibody must be deleted. The affinity s_{ij} between antibody i and antibody j is defined as (4).

$$s_{ij} = \begin{cases} 1, & \text{if } I_i = I_j \\ 0, & else \end{cases} \tag{4}$$

If and only if two antibodies are identical, the affinity is equal to 1, else is 0.

3.3 Initial Population

To ensure the antibodies contain good antibody and distributing in the solution area as equably as possible, priority rule based generation and random selecting are adopted.

A priority rule based algorithm generally consists of a schedule generation scheme and a priority rule. The scheduling generation scheme determines what tasks are eligible to be added to the partial schedule. A priority rule is used to determine the best task to be added to the partial schedule.In general, the parallel method saves computational effort relative to the serial method. In fact, no priority rule is absolutely good. So in this paper, parallel generation scheme and shortest process time rule are adopted to generate scheduling sequences.

3.4 Crossover and Mutation

In the process of clone crossover, roulette was adopted to select parents. Several crossover operators that preserve precedence relationships have been developed, such namely one-point, two-point and multi-point operator.

•*One-Point Operator:*
Randomly generate an integer r, $1 < r < n$.
For $i \in [1, r]$, $c[i] = p_1[i]$.
For $j \in [1, n]$, if $p_2[j] \notin c$ then $r = r + 1$ and $c[r] = p_2[j]$.
Interchanging p_1 and p_2 produces another child.

•*Two-Point Operator:*
Randomly generate two integers r_1 and r_2, $1 < r_1 < r_2 < n$.
For $i \in [1, r_1]$ and $i \in [r_2 + 1, n]$, $c[i] = p_1[i]$.
For $j \in [1, n]$, if $p_2[j] \notin c$ then $r_1 = r_1 + 1$ and $c[r_1] = p_2[j]$.
Again, interchanging p_1 and p_2 produces another child.

•*Multi-Point Operator:*
This operator is effected by randomly dividing each parent into R sections $R \geq 2$. Starting with the first, each section of the child inherits all its tasks and their order from one of the parents alternately, ignoring the tasks already chosen.

Since the crossover operators described preceding valid children from valid parents, it follows that they all preserve any subsequences common to both parents. For this reason, relying on these operators alone is likely to lead to convergence too quickly. There is need, therefore, to develop an effective mutation operator that can reintroduce diversity.

Choose a task v at random by a defined probability and move it to a possible position randomly without violating any precedence constraint. In the same time, change the direction of parallel scheduling.

3.5 Deriving Operator and Clone Death

In immune system, to keep the diversity of whole system, the mature B-cells in marrow enter into the antibody colony. To simulate this function, deriving operator is adopted. After clone selection, some new antibodies generated randomly enter the existing antibody colony as the next generation's antibodies.

Similarity, the aging B-cells will died step by step in immune system. To keep the antibodies distributing uniformly, the similar antibodies will died by clone death. Antibody i removed when $s_{ij} = 1$.

3.6 End Criteria

To achieve a good result, several different criteria are proposed.

1. Stop iterating after a pre-defined number of steps.
2. Stop the iterative process when the duration is equal to the optimal.
3. When the number of search sequences reach predefined number.

3.7 Algorithm RCPSP-ADPCSA

RCPSP-ADPCSA works as follows:

1. $k = 0$, select forward or backward direction by $p = 0.5$ and generate initial scheduling sequences $A(0)$. In the same time, antibody $b(k)$ is generated by priority rule based approach.
2. Determine its affinity $f(i)$ to all the individuals. $f(i) = F_{max} - F(i) + 1$. And Select 30% highest affinity individuals and $b(k)$ to $M(k)$, others to $A_b(k)$.
3. Mutation probability p_m^M, $p_m^{A_b}$ and clone size nc^M, nc^{A_b} are initialized by affinity.
4. Clonal operation. Population $Y(k)$ is obtained.
5. Crossover operation and generate $B(k)$. Mutation and produce population $C(k)$.
6. Apply clone selection to $B(k) \cup C(k) \cup A(k)$ and deriving operator is introduced. Generate new sequences randomly. Put these sequences to population and get $Z(k)$.
7. Determine s_{ij} of all the antibodies in $Z(k)$. If $s_{ij} = 1$, clone death is applied. Delete the antibody i from $Z(k)$. A new population $A(k + 1)$ is gained. and $b(k + 1)$ is the best individual in the iteration k.
8. Judge whether it satisfied stop criteria or not, if satisfied then stop, else go to 2.

4 Experimental Results

4.1 Benchmark Problem

To testing the performance of RCPSP-DCSA, the single-mode resource-constrained project scheduling problems in PSPLIB[9] were selected. Testing problems can be divided to four groups by the task's number in project. Using a full factorial design of the parameters *NC*, *RF* and *RS* in Table 1, with ten replications for each combination, a total of $3 \times 4 \times 4 \times 10 = 480$ test problems can be generated. If three levels for the number of tasks are selected at 30, 60 and 90 for the *J*30, *J*60 and *J*90 problem sets respectively. The *J*120 problem sets is generated with the parameters shown in table 1 and comprises $3 \times 4 \times 5 \times 10 = 600$ instances.

Table 1. Parameter settings for factorial design

Benchmark Set	*NC*	*RF*	*RS*
*J*30, *J*60, *J*90	{1.5, 1.8, 2.1}	{0.25, 0.5, 0.75, 1.0}	{0.2, 0.5, 0.7, 1.0}
*J*120	{1.5, 1.8, 2.1}	{0.25, 0.5, 0.75, 1.0}	{0.1, 0.2, 0.3, 0.4, 0.5}

All the *J*30 instances have been solved optimally and the optimal solutions are available for comparison. Some of the *J*60, *J*90 and *J*120 instances have been solved optimally, but for most, only upper and lower bounds are available. It is important to note that these upper bounds have been complied from the results achieved by different researchers using a variety of different algorithms, i.e., there is no single method that can achieve all the best solutions.

4.2 Comparison Criteria

In view of the randomized nature of the algorithm, all experiments involve a certain number of replications (for example 20). In order to evaluation to be comprehensive, the following criteria have been used:

1. Average Error Rate $AveErr\%$:

$$AveErr\% = \frac{all(\frac{solution\ makespan - best\_known\ makespan}{best\_known\ makespan} \times 100\%)}{the\ number\ of\ problems} \tag{5}$$

2. Maximum Error Rate $MaxErr\%$: the largest error rate of all the problems.

3. Average Deviation $AveDeviation$:

$$AveDeviation = \frac{all(solution\ makespan - best\_known\ makespan)}{the\ number\ of\ problems} \tag{6}$$

4. Optimal Proportion $OptPro\%$:

$$OptPro\% = \frac{the\ number\ of\ optimal\ solution}{the\ number\ of\ testing\ instances \times replication's\ time} \tag{7}$$

4.3 Preliminary Experiments

Experimental result can be influenced by the type of crossover. The testing data involve 6 groups, $J30\_3$, $J30\_13$, $J30\_42$, $J120\_7$, $J120\_33$ and $J120\_45$. Every group consists of 10 instances. The 10 instances of $J30\_13$ will average spend 1469.08 seconds to find optimal solutions. $J30\_13$ is the group which spend most time in $J30$ and the average spend time in $J30$ is only 28.97 seconds. $J120\_7$ and $J120\_33$ are groups which can not find optimal solution at present. Others are selected randomly.

In algorithm RCPSP-DCSA, deriving operator and clone death are adopted. So the initial population size has little effect on the results. The number of activities in a project as the initial population size here. In addition, the clone size is relation to the affinity. A general experience value $pi = 3$ is used. The experimental results for different crossover types are shown in table 2 ($P_m^M = 1/J$, $P_m^{A_b} = 0.6$). When the number of search sequences reached 1000, the iteration stop.

Table 2. Results for different crossover

measure	crossover	$J30\_3$	$J30\_13$	$J30\_42$	$J120\_7$	$J120\_33$	$J120\_55$
	one-point	0.11	0.79	0.14	5.01	3.73	1.24
AveErr%	two-point	**0.00**	**0.45**	**0.03**	**4.76**	**3.36**	**1.01**
	multi-point	0.08	0.62	0.07	4.84	3.41	1.53
	one-point	4.17	3.77	3.19	18.33	**13.99**	9.89
MaxErr%	two point	**0.00**	**3.45**	**1.22**	**17.56**	**13.99**	**7.77**
	multi-point	3.22	3.95	2.38	**17.56**	14.26	9.89
	one-point	0.06	0.58	0.13	8.92	5.61	1.30
Avedeviation	two-point	**0.00**	**0.33**	**0.02**	**5.17**	**4.97**	**1.06**
	multi-point	0.05	0.46	0.06	5.21	5.36	1.61
	one-point	95.00	61.00	90.50	42.50	47.00	68.50
OptPro%	two-point	**100.0**	**76.00**	**97.50**	**47.50**	**51.50**	**72.50**
	multi-point	96.00	68.00	94.50	46.00	49.00	65.00

From table 2, we know the result can reach optimal when adopts two-point crossover. This type will reduce the average error rate and increase the optimal proportion as much as possible. In the latter experiments, two-point crossover is adopted.

In clone selection algorithm, mutation is the main immune genic operation. So the mutation probability has many impact on the results. As to the memory unit which save the near-optimal individuals, it requires a little mutation probability to complete local

Table 3. Results of different mutation probability

measure	mutation probability							
	0.3	0.4	0.5	0.6	0.7	0.8	0.9	1.0
AveErr%	1.16	1.07	0.93	**0.84**	1.10	1.10	0.99	1.00
MaxErr%	6.00	6.67	6.00	**4.67**	6.67	5.33	5.33	5.33
AveDeviation	1.23	1.13	0.98	**0.89**	1.17	1.17	1.04	1.05
OptPro%	81.00	85.50	84.50	**87.50**	80.50	86.50	**87.50**	82.00

search and $p_m^M = 1/J$. To individuals in the general unit, a much more mutation probability was required to complete randomly search. To determine the value of mutation probability, $J60\_23$ was selected at random in $J60$. Its parameters are $NC = 1.80$, $RF = 0.50$ and $RS = 0.70$ respectively. Table 3 shows the results. (Initial population size is the number of tasks, two-point crossover, $p_m^{A_b} = 0.3, 0.4, 0.5, 0.6, 0.7, 0.8, 0.9, 1.0$, the number of search sequences is 1000)

As table 3 is shown, the performance of this algorithm is good when $p_m^{A_b} = 0.6$. All the rest experiments adopt this value.

4.4 Main Experiment Results

Kolisch[8] proposed that when testing the benchmark instances in PSPLIB, the iteration can stop with the search sequences reached $1,000$ or $5,000$. The testing in here consist of 4 groups, $J30$, $J60$, $J90$ and $J120$ respectively. The experimental results are shown in table 4. In this table, the results of $J30$ were compared with optimal solutions and the others were compared with near-optimal solutions which solved by heuristic methods.

From table 4, we can make a conclusion that the RCPSP-DCSA has a good performance on solving the single mode resource-constrained project scheduling problem. To the $J30$ which involves 32 works, the percentage of find optimal solutions achieves 98.73%; And the average deviation is only 0.09 to that instances which fail to find optimal solutions. It is important that the best-known solutions for these instances have been contributed by several authors and several methods, which suggests that no

Table 4. Testing results of benchmark instances

measure	$J30$		$J60$		$J90$		$J120$	
	1,000	5,000	1,000	5,000	1,000	5,000	1,000	5,000
$AveErr\%$	0.12	0.10	0.86	0.78	1.89	1.63	4.27	3.75
$MaxErr\%$	2.73	2.73	4.28	3.25	5.24	4.63	8.63	7.29
$AveDeviation$	0.11	0.09	1.23	1.07	2.36	2.18	3.96	3.65
$OptPro\%$	98.67	98.73	88.63	88.92	82.67	83.02	69.65	70.30

Table 5. Compare results with other algorithms

algorithm	reference	set	$AveErr\%$	$AveDeviation$	$OptPro\%$
SGA	[4](2002)	$J30$	0.239	0.208	92.29
		$J60$	0.987	1.079	80.21
		$J120$	4.173	4.648	54.83
BPGA	[5](2005)	$J30$	0.187	0.170	93.96
		$J60$	1.025	1.145	81.88
		$J120$	3.968	3.429	62.33
FBSP_GA	[6](2005)	$J30$	0.146	0.165	93.18
		$J60$	1.037	1.167	81.54
		$J120$	3.853	3.989	63.25
RCPSP-DCSA	/	$J30$	0.102	0.092	98.73
		$J60$	0.781	1.073	88.92
		$J120$	3.753	3.654	70.30

algorithm has been able to generate an overwhelming proportion of these solutions. The algorithm RCPSP-DCSA can suit most instances and can find near-optimal solutions. In addition, this algorithm also has a good performance for *J*60, *J*90 and *J*120. It can find good solutions in reasonable time.

To verify the performance of RCPSP-DCSA further, some approaches are compared. The results are shown in table 5. And Compared with these good heuristic approaches, RCPSP-DCSA also has some advantages.

5 Conclusions

An immune clone selection algorithm for solving the single-project, single-mode, resource-constrained project scheduling problem has been presented. This problem is characterized by the presence of precedence relationships among the works, as well as resource capacity constraints. By design the encoding, operators and affinity function according to the characteristic of problems, it can solve problem successfully. The experimental results show RCPSP-DCSA has a good performance and it can solves problems include different number works.

References

1. Tam*á*s Kis. A branch-and-cut algorithm for scheduling of projects with variable-intensity activities. Mathematical Programming, 2005(103): 515-539
2. Paul R. Thomas, Said Salhi. A Tabu Search Approach for the Resource Constrained Project Scheduling Problem. Journal of Heuristics, 2001(4): 123-139
3. Joaqułn Bautista, Jordi Pereira. Ant colonies for the RCPSP Problem. Topics in Artificial Intelligence: 5th Catalonian Conference on AI, CCIA 2002, Castell'on, Spain, October 24-25, 2002: 257-268
4. Khalil S. Hindi, Hongbo Yang, Krzysztof Fleszar. An Evolutionary Algorithm for Resource-Constrained Project Scheduling. IEEE Transactions on Evolutionary Computation, 2002(6): 512-518
5. Dieter Debels, Mario Vanhoucke. A Bi-population Based Genetic Algorithm for the Resource-Constrained Project Scheduling Problem. Computational Science and Its Applications-ICCSA 2005: International Conference, Singapore, May 9-12, 2005, Proceedings, Part IV. Pages: 378-387
6. Hong Wang, Dan Lin. A Genetic Algorithm for Solving Resource-Constrained Project Scheduling Problem. Advances in Natural Computation: First International Conference, ICNC 2005, Changsha, China, August 27-29, 2005, Proceedings, Part III, Pages: 185-193
7. Vicente Valls, Francisco Ballestłn. A Population-Based Approach to the Resource- Constrained Project Scheduling Problem. Annals of Operations Research, 2004(131): 305-324
8. Kolisch, R., Hartmann, S. Experimental Investigation of Heuristics for Resource-Constrained Project Scheduling: An Update. European Journal of Operational Research, to appear (2005)
9. http://129.187.106.231/psplib/

Pareto Meta-heuristics for Generating Safe Flight Trajectories Under Weather Hazards

Sameer Alam, Lam T. Bui, Hussein A. Abbass, and Michael Barlow

Defence and Security Applications Research Centre,
and The ARC Center for Complex Systems,
University of New South Wales at The Australian Defence Force Academy,
Canberra, ACT 2600, Australia

Abstract. This paper compares ant colony optimization (ACO) and evolutionary multi-objective optimization (EMO) for the weather avoidance in a free flight environment. The problem involves a number of potentially conflicting objectives such as minimizing deviations, weather avoidance, minimizing distance traveled and hard constraints like aircraft performance. Therefore, we modeled the problem as a multi-objective problem with the aim of finding a set of non dominated solutions. This approach is expected to provide pilots the additional degree of freedom necessary for self optimized route planning in Free Flight. Experiments were conducted on a high fidelity air traffic simulator and results indicate that the ACO approach is better suited for this problem, due to its ability to generate solutions in early iterations as well as building better quality non dominated solutions over time.

1 Introduction

Weather is identified as the single largest contributor to delay in the air traffic control (ATC) system and is a major factor in aircraft safety incidents and accidents [1]. The future concept of Free Flight [2] air traffic management system, the task of separation assurance and weather avoidance will move from air traffic controllers to the pilots and airborne systems. Given the autonomous control of aircraft to the pilots, safety is a major issue in Free Flight specially in hazardous weather conditions. Efficient and robust weather avoidance algorithms are needed to ensure safety and better fuel economics resulting from optimal weather avoidance trajectories. We investigate the problem of finding weather avoidance routes in the Free Flight as a multi objective optimization problem and defined it in terms of Pareto optimality. Two classes of meta heuristic search techniques were investigated for generating a set of non-dominated trajectories.

The first is evolutionary multi-objective optimization techniques (EMOs) [3]. EMOs, give decision makers more options to choose the best solution according to post analysis preference information [4]. EMOs, offer a set of solutions, where user preferences can be elicited afterwards to decide which solution to execute.

The second class is ant colony optimization techniques (ACOs) that has been used for multi objective optimization problems as well [5,6]. We undertook recent

T.-D. Wang et al. (Eds.): SEAL 2006, LNCS 4247, pp. 829–836, 2006.

studies on the use of ACO for weather avoidance [7,8], where it was shown that the ACO approach using A* heuristics as visibility parameter and pheromone update based on solution quality gives high quality non dominated solutions. However, there has not been any empirical evidence to prefer ACO over EMO in this context; hence, this paper.

This paper is organized as follows, we briefly discuss weather hazardous in aviation, and some avoidance algorithms in the literature, then we present our design of the weather avoidance algorithms using the two techniques, followed by experiments, results, some conclusions and future directions.

2 Aviation Weather Hazard

Weather disturbances can severely damage the airframe of an aircraft [9], and can potentially damage the navigational and electronic equipment leading to pilot's loss of control and the endangerment of the life of passengers and crew members. The NTSB (U.S. National Transport Safety Board) aviation accident database [1] for data collected between 1991 to 2001 reveals that weather was the contributing factor or cause in 21% of accidents. Clearly the availability of onboard weather avoidance system and an auto generated optimized route capability can greatly enhance the safety and efficiency of a flight and its crew [10].

The problem of hazardous weather avoidance can be seen as optimal path planning on a 3D grid, and several heuristic based approaches have been applied like A*, weighted regions, potential field etc. [11,12]. These approaches have several drawbacks such as becoming trapped in local minima, high computational cost and memory requirements as search space increases. In previous studies of weather avoidance systems [7,11,13], the heuristic recommends a single optimized route, leaving pilots with no choice but to execute the proposed solution trajectory. In Free Flight, pilots will face complex situations and besides weather avoidance they will have to optimize their routes on other objectives as well. This makes previous approaches less attractive and thus the need arises for an algorithm which can generate a set of optimal solutions to choose from, and be able to search multiple routes in the presence of multiple airspace constraints including bad weather. A Pareto approach may provide the pilots with flexibility and speedy response, while also providing the industry with a generic approach for weather avoidance system design.

3 Weather Avoidance Algorithm Design

3.1 Problem Definition

The problem of weather avoidance in a Free Flight environment is considered in 3D which is approximated with grid cells in hyper-rectangular discrete space, where hazardous weather cells are present in a distributed manner. We have modeled bad weather as thunderstorm cells and used the broad classification of the US National Weather System (NWS) for assigning radar reflectivity and corresponding weather information in different colors as follows:

Color	Radar Reflectivity(dBZ)	Weather Type
None	I < 5	None
Blue	5 < I <20	Light
Green	20 < I <30	Moderate
Yellow	30 < I <40	Heavy
Red	40 < I	Intense

The algorithm receives the following inputs:

1. x × y × z grid: Rectangular discrete space of 100nm × 100nm × 3000ft,
2. Start node in the grid: The start node is that mesh point of the search grid which is closest from the current position of the aircraft,
3. End node in the grid: The end node is the point in the mesh which is closest to the next trajectory change point,
4. Weather Cells: 6-12 distributed weather cells each of dimensions 10nm X 10nm X 3000ft(typical of a single thunderstorm cell dimension) with severity in terms of radar reflectivity range 5dBZ - 50dBZ.

Given these grid points, the state space is enumerated. This state space is then pre-processed by removing those states which violate hard constraints such as aircraft performance parameters. This ensures that all the trajectories generated are within aircraft safety parameters. The details of this pre-processing can be found in [8].

The output of the algorithm is a non dominated set of routes which minimize weather impact, heading changes, and distance traveled.

3.2 Algorithm Design

For ACO, we used our ACO algorithm for weather avoidance [8] with a cost function based on the A* algorithm and a dynamic weight allocation for obtaining a non–dominated set of solutions.

For EMO we used NSGA2 which is an elitism EMO algorithm [3]. We modified the approach suggested by [14] for 3D path planning to encoding the path in a bit string representation, by adding a penalty (distance remaining to destination) term to an individual which failed to reach the destination node. For generating a complete path from XY monotone path and XZ monotone path we did interweaving based on the lower array bounds of the two paths.

The three objectives defined for EMO and ACO were:

1. Minimize changes in aircraft heading: This is the sum total of all heading changes required to follow a particular route. Large value of heading change indicates many turns leading to passengers discomfort and higher fuel burn.
2. Minimize distance traveled: The distance traveled by the aircraft in the search grid to maneuver a particular route.
3. Minimize Weather cells severity measure: The sum total of all weather cells severity which an aircraft encounters for a particular path.

In EMO, if an individual fails to reach the exit point of the grid due to violation of the search envelop or the maximum number of moves allowed then the remaining distance is treated as a penalty. Any individual that violates the aircraft performance parameter constraints, is eliminated from selection in future generations. The penalty term is added to the distance and heading objectives.

Penalty value is scaled up by a factor of 5.0, and weather by a factor of 3.0 to balance the magnitude of the three objectives. If an individual reaches the destination node, then the penalty distance will be zero, thus making the objective function similar to the ACO objective function.

In ACO, an ant which fails to reach the exit point within a certain number of moves is eliminated. Only those ants that successfully make it to the exit point of the mesh are allowed to update the pheromone levels (based on solution quality) on their respective paths. Ants work on a preprocessed state space which is free of any transition move that may violate the aircraft's performance parameters, this eliminates the need for checking whether the path generated by an ant violates aircraft performance constraints.

4 Experiments

For optimal parameter configuration for EMO we performed initial experiments for a population size of 30 individuals and 300 generations with the following combinations of crossover rate and mutation rate: Crossover Rate = {0.1,0.2,0.5, 0.8,1.0} and Mutation rate = {0.01,0.02,0.03,0.04,0.05}. The best combination of crossover and mutation rates is Crossover Rate = 0.5 and Mutation Rate = 0.02. For ACO we used the parameter configurations suggested in our previous work [8].

Results, averaged over 10 runs are presented here. Based on experimental data we tried to answer the following questions to compare the two approaches.

1. *How many solutions were generated by the two approaches over generations/iterations time?*
   Due to approach controller's stringent time schedule for last trajectory change point, sometimes pilots prefer quick solutions even if they are sub optimal rather than wait a long time for an optimized solution to appear.

   Figure 1, shows the number of solutions proposed by the two approaches over 300 iterations. Solutions that are only complete (i.e starts from entry point, ends at exit point and do not violate aircraft performance parameters) were recorded. As can be seen from Figure 1, EMO can take as much as 30 generation before any complete solution is found, whereas ACO produces solutions in the very first iteration. This can be attributed to the fact that ACO works on a pre processed state space, which has only the valid space to explore, so there are less pitfalls for ants whereas in EMO, any individual that violates the constraints, doesn't get selected in the next generation.
2. *How many unique solutions(trajectories) were generated by both approaches?*
   Large number of solutions in terms of unique trajectories are desirable, since they provide the system or pilots with more options based on the situation.

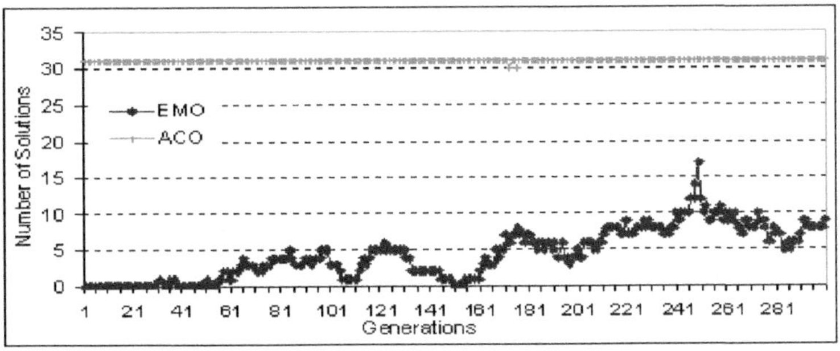

Fig. 1. Set of solutions generated over iterations/generation, by the two approaches

As shown in Figure 2 (right), ACO generated a very high number of unique solutions as compared to EMO. A high value for the exploration parameter (0.9) makes the ants explore more space, resulting in diverse routes to the destination. Whereas EMO follows an evolutionary approach to guide the solutions towards the destination. The penalty function of EMO does not let individuals explore more of the search space and is guided towards routes that are complete and valid. However this advantage that ACO has, may be eliminated by designing a better representation for the evolutionary algorithm, which we leave for future work.

3. *How many non dominated solutions(NDS) were generated by both approaches?* The non dominated set of solutions generated by each approach is investigated to see if there is any advantage for ACO for being able to generate a large number of unique solutions. As shown in Figure 2 (left), ACO despite a finding high number of unique solutions, finds a few non dominated solutions as compared to EMO. Whereas the EMO approach finds few unique solutions but most of them are non dominated solutions.

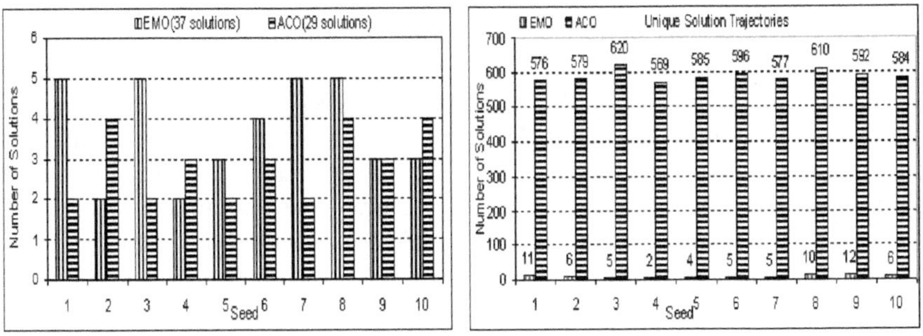

Fig. 2. On left: Set of non dominated solutions generated by ACO and EMO; On right: Set of unique solutions(trajectories) generated by ACO and EMO

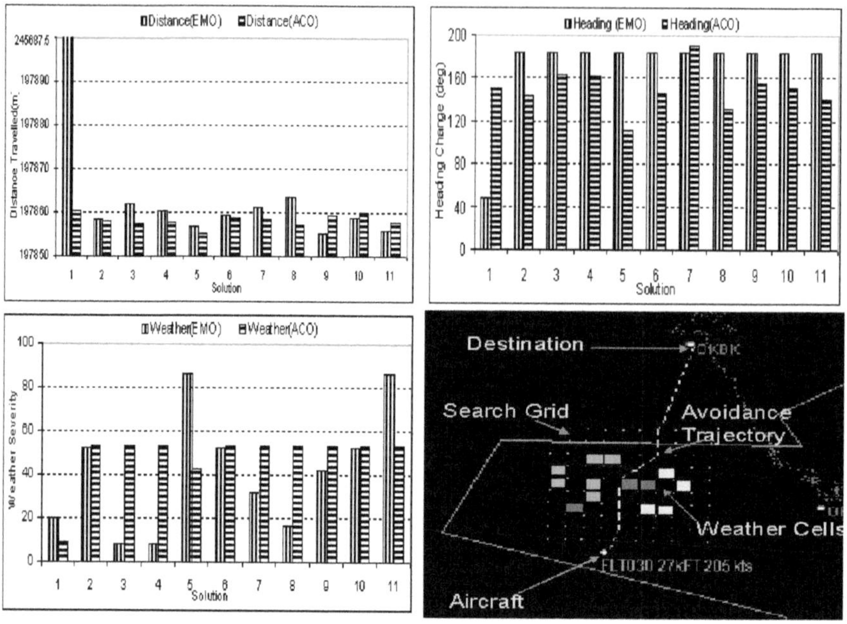

Fig. 3. Non dominated solutions on three objectives generated by ACO and EMO. top left: distance traveled, Top right: heading change, bottom left: weather severity encountered and bottom right:Air traffic simulator snapshot of a weather avoidance trajectory.

4. *How many non dominated solutions of EMO dominated ACO non dominated solutions and vice versa?*
   When the non dominated solutions for the 10 runs were consolidated for both the approaches and duplicate trajectories were eliminated, we found that both approaches generated a total of 11 unique non dominated solutions. However EMO dominated only one NDS solution generated by ACO, whereas ACO dominated 7 NDS solutions of EMO. Figure 3 shows the performance of both algorithms in terms of NDS on the three objectives separately. It can be seen that ACO has advantage in terms of heading minimization and distance travelled, however in terms of weather cells encounter EMO proposes four solutions which have low weather impact whereas ACO proposed only two. However this advantage of EMO comes at a heavy price in terms of large deviation in trajectory resulting from frequent climbs and descents maneuvers.

We further visualize the NDS trajectories generated by the two approaches. It can be seen from Figure 4 top, EMO solutions involves a lot of climb and descend maneuvers resulting in large heading changes and distance travelled, which makes them less desirable for problem investigated as compared to ACO.

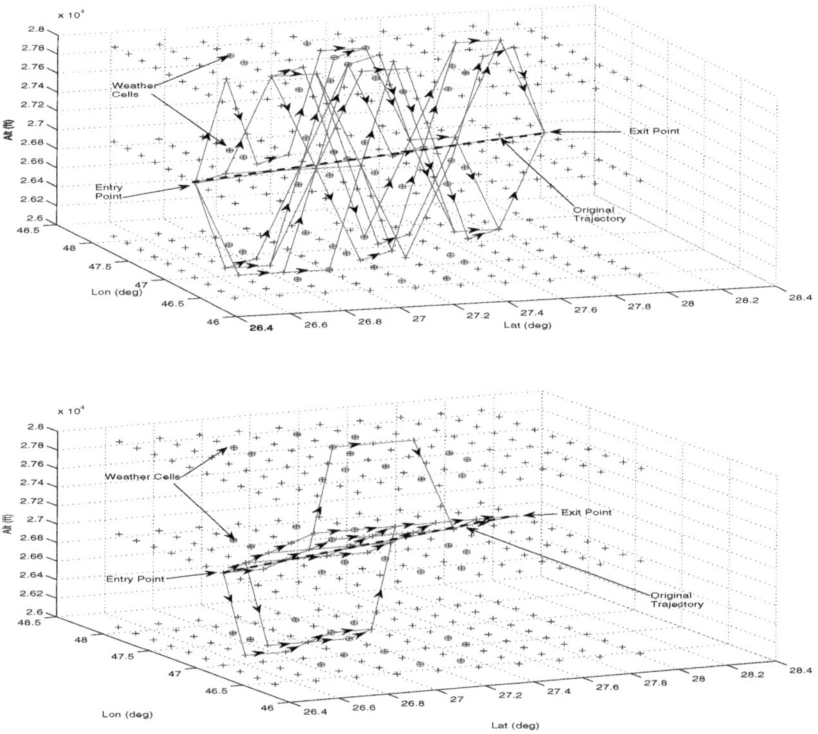

Fig. 4. 3D view of non dominated set of solution generated by ACO(Bottom) and EMO(Top) algorithm.The arrow markers shows the direction of travel. Bad Weather cells are shown as circles, however they do not indicate the weather cell severity, which can be seen in the air traffic simulator snapshot.

5 Conclusions and Future Directions

We investigated two meta heuristic search techniques viz. ACO and EMO for generating a non dominated set of routes in a weather constrained airspace. The results indicate that the ACO approach can generate quick solutions (sub optimal initially) whereas EMO took several generations before any valid routes was generated. This feature gives the ACO algorithm an advantage in the sense that some times pilots prefer quick but suboptimal solutions due to time constraints. The ACO algorithm given proper tuning of exploration and exploitation parameters, searches a large state space as compared to EMO this gives ACO an increased likelihood of finding a better set of NDS. EMO generated a diverse set of solutions, however many of them involved uneconomic maneuvers for the aircraft. In the final set of NDS, ACO was able to dominate a larger number of EMO based NDS, and generated better quality solution set. We are currently conducting experiments on our Free Flight air traffic simulator for dynamic and noisy weather environment and with multi aircraft scenario to study these two approaches further.

Acknowledgement

This work is supported by the Australian Research Council (ARC) Centre for Complex Systems grant number CEO0348249.

References

1. NTSB: (Aviation accidents statistical report:1991-2000) Washington D.C.(2001).
2. RTCA: Report of the RTCA board of director's select committee on Free Flight. Technical report, RTCA Inc., Washington, DC (1995)
3. Deb, K.: Multi-Objective Optimization Using Evolutionary Algorithms. John Wiley & Sons (2001)
4. Zitzler, E., Thiele, L., Deb, K.: Comparision of multiobjective evolutionary algorithms: Emprical results. Evolutionary Computation **8** (2000) 173–195
5. Dorigo, M., Stutlez, T.: Ant Colony Optimization. MIT Press, Cambridge (2004)
6. Lopez-Ibanez, M., Paquete, L., Stutzle, T.: On the Design of ACO for the Biobjective Quadratic Assignment Problem. In: ANTS 2004. Volume 3172 of LNCS. Springer-Verlag, Berlin Heidelberg (2004) 214–225
7. Alam, S., Abbass, H., Barlow, M., Lindsay, P.: Mapping lessons from ants to free flight: An ant-based weather avoidance algorithm in free flight airspace. In Bender, A., ed.: Proc. of of SPIE (Complex Systems). Number 6039, Brisbane, Qld (2005)
8. Alam, S., Abbass, H., Barlow, M.: Multi-objective ant colony optimization for weather avoidance in a free flight environment. Technical Report TR-ALAR-200603004, School of ITEE, UNSW@ADFA, Canberra, Australia (2006)
9. Peter, F.: Aviation Weather. Jepesson Sanderson, Inc. (1995)
10. Krozel, J.: Free Flight Research Issues and Literature Search. Technical Report NAS2-98005, NASA Ames Research Center, Moffett Field, CA 94035 (2000)
11. Krozel, J., Penny, S., Prete, J., Mitchell, J.: Comparison of algorithms for synthesizing weather avoidance routes in transition airspace. In: Proc. of the AIAA Guidance, Navigation, and Control Conference, 2004, Providence, RI, (2004)
12. Mata, C., Mitchell, J.: A new algorithm for computing shortest paths in weighted planar subdivisions. In: Proc. of the 13th Annual ACM Symposium on Computational Geometry, Nice, France, ACM (1997) 264–273
13. Bokadia, S., Valasek, J.: Severe weather avoidance using informed heuristic search. In: AIAA Guidance, Navigation, and Control Conference and Exhibit. Number AIAA-2001-4232, Montreal, Canada (2001)
14. Sugihara, K., Smith, J.: A genetic algorithm for 3-d path planning of a mobile robot. Technical Report 96-09-01, Software Engineering Research Laboratory, Department of Information and Computer Sciences, Univ. of Hawaii, Manoa (1996)

Land Combat Scenario Planning: A Multiobjective Approach

Ang Yang, Hussein A. Abbass, and Ruhul Sarker

Defence and Security Applications Research Centre (DSA)
University of New South Wales, Australian Defence Force Academy
Canberra, ACT 2600, Australia
{a.yang, h.abbass, r.sarker}@adfa.edu.au

Abstract. The simulation of land combat operations is a complex task. The space of possibilities is exponential and the performance criteria are usually in conflict; thus finding a sweet spot in this complex search space is a hard task. This paper focuses on the effect of population size and mutation rate on the performance of NSGA–II, as the evolutionary multiobjective optimization technique, to decide on the composition of forces using a complex land combat multi-agent scenario planning tool.

1 Introduction

Land combat is a complex task. Identifying a suitable composition of forces - in terms of size of each group in a mission, type of weapons and communication used by each group, and ammunition load - is normally planned by a team of experts, human–based simulation, or computer simulation. Military has used computer simulation for a long time as a cheap way for testing complex military concepts. Recently, the use of agent–based simulation (ABS) is replacing traditional simulations. In an ABS, agents can be autonomous, grouped in teams, have different capabilities and personalities, and use different strategies.

ABS has opened many opportunities for testing military concepts. However, the search space of these black–box simulations is very complex. Moreover, the success of a military operation is almost always determined through a set of conflicting objectives. Thus, it is natural to look at the use of evolutionary multiobjective optimization (EMO) techniques to search these complex landscapes. However, the performance of EMO, like other evolutionary computation techniques, can be sensitive to parameters and the optimal set of parameters is normally problem dependent [1]. In this paper, we investigate the effect of two parameters, in particular, for NSGA–II [2].

In the following section, we present the experimental setup followed by the results and analysis. Finally conclusion and future work are discussed.

2 Experimental Setup

2.1 The Land Combat Simulation System

We use the warfare intelligent system for dynamic optimization of missions (WISDOM–II), which is based on the Network centric multi-agent architecture

T.-D. Wang et al. (Eds.): SEAL 2006, LNCS 4247, pp. 837–844, 2006.

(NCMAA) [3], as the multi–agent combat simulation engine. WISDOM has been used in many experiments for capability planning. It has five distinct components: the command, control, and communication (C3) component, the sensor component, the engagement component, the visualization component and the reasoning component.

Four agent-types are supported in WISDOM–II: combatant, group leader, team leader and swarm commander. Agents are defined by their characteristics and personalities. Each agent has nine types of characteristics: health, skill, probability to respect - thus follow - the command, visibility, vision, communication, movement and engagement. The swarm commander can build plans and give orders to combat groups. The personality in WISDOM–II is a defined by two values: a magnitude and a direction vector representing the attraction–repulsion direction and weight for each agent.

The movement of each agent is determined by its situation awareness and personality vector. In each time step, the agent can only move to its neighbor cells based on the overall influence of all perceived agents. A strategic decision is made by the swarm leader of each force based on the common operating picture (COP), which is the global view of the battle field for that force. For more details of WISDOM-II, please refer to Yang *et al.*[3].

2.2 The Scenario Setup

The scenario used in this paper includes two forces - blue and red force - playing against each others (Figure 1). The simulation environment is 30x30 cells and the destination flag (each force has a goal to occupy an area) is located at the middle of the environment. Both blue and red forces are composed of four groups, each of which consists of a set of homogenous agents. The capability of the red force is fixed during the simulation while that of the blue force is determined by the evolutionary process as shown in the table 1. The fourth group in both forces (R4 and B4) is for surveillance.

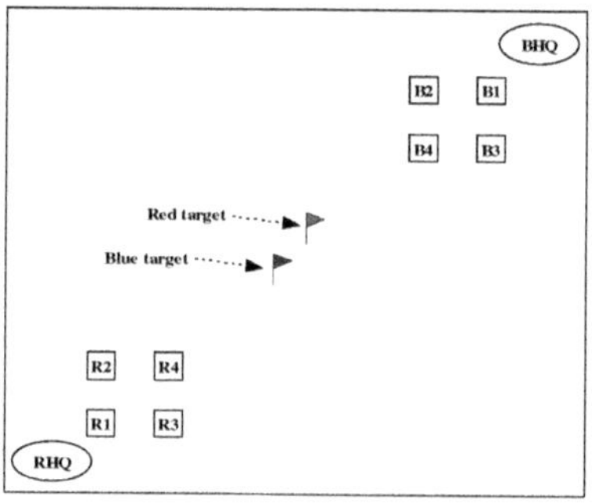

Fig. 1. The combat scenario. BHQ and RHQ is the headquarter for the blue and red force respectively.

2.3 The Evolutionary Setup

A chromosome is composed of 4 parts, each part corresponds to the capability of one group in the blue force. Each of the first three parts consists of eight variables corresponding to vision, communication and weapon parameters. The last part has four variables only because the surveillance agent does not have weapons. The total number of variables in the chromosome is 28.

Table 1. The capability of forces

RED FORCE		Group R1	Group R2	Group R3	Group R4
# of Agents		5	20	20	1
Vision Range		4	5	5	10
Communication	Range	5	5	5	10
	Lost Probability	0	0	0	0
	Latency	0	0	0	0
Weapon	Minimal Distance	3	0	0	-
	Range	4	4	4	-
	Strength	3	3	3	-
	Radius	1	0	0	-
BLUE FORCE		**Group B1**	**Group B2**	**Group B3**	**Group B4**
# of Agents		10	10	10	1
Vision Range		1 - 10			
Communication	Range	1 - 10			
	Lost Probability	0 - 0.5			
	Latency	0 - 2			
Weapon	Minimal Distance	1 - 8			
	Range	1 - 8			
	Strength	2 - 8			
	Radius	0 - 2			

Yang *et al.* [4, 3] argued that taking a simple linear combination of the objectives may hide some information which is crucial in understanding the dynamics within a warfare simulation. Therefore, we explicitly represent the two objectives in the problem and use the real variable mode of the non-dominated Sorting Genetic Algorithm – II (NSGA–II) [2], a multi-objective optimization algorithm. Two objectives are defined: minimizing the cost of the blue force and minimizing the casualty of the blue force.

We run each experiment for 1000 generations, four different population sizes - 20, 40, 80 and 100, and three different mutation probabilities - 0.01, 0.036 and 0.1, respectively. The reason for choosing 0.036 is to follow some literatures which are suggesting the mutation rate to be the reciprocal of the number of variables ($\frac{1}{28} \approx 0.036$) [6, 7, 8, 2, 9, 10, 11].

Therefore we have 12 different configurations. We fix the probability of crossover to 0.9, distribution index for crossover to 15, and the distribution index for mutation to 20 [2, 9, 10, 11]. Each individual is simulated 30 times with different seeds, each for 150 time steps. The fitness of each individual will be the average of 30 simulations. Each configuration is repeated 10 times with different seeds.

3 Results and Analysis

To compare the performances of different configurations, we adopt a statistical comparison method proposed by Knowles and Corne [14] in 2000, which extended the method introduced by Fonesca and Fleming in 1996 [13].

First, we try to understand the convergence during the search process by comparing the performance of the pareto-optimal sets of generation n with that of generation $n - 50$ to see when the improvement stops (see Figure 2).

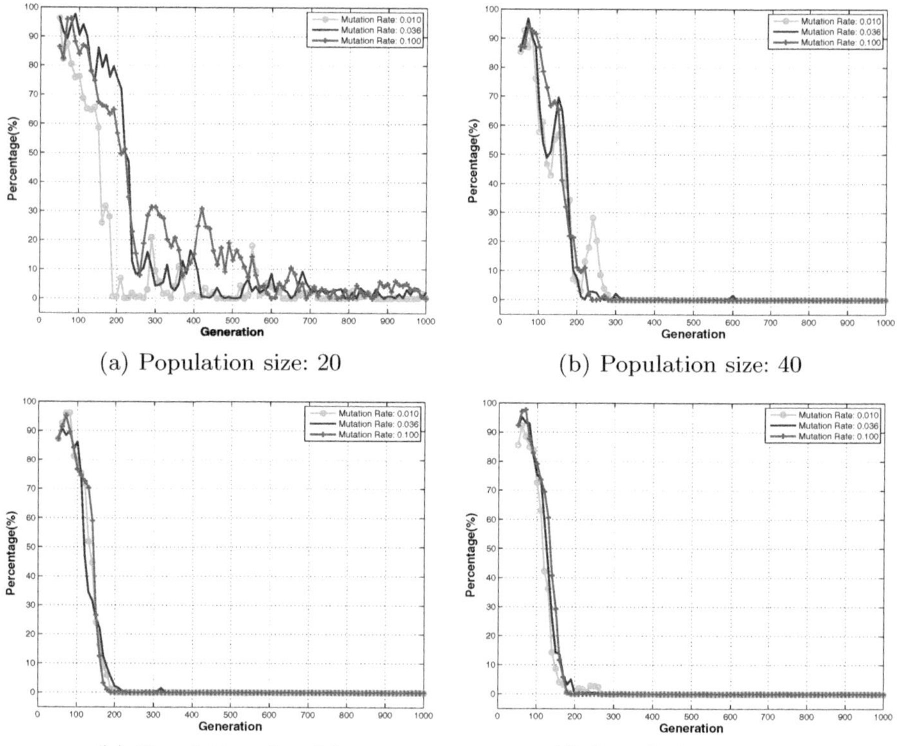

(a) Population size: 20

(b) Population size: 40

(c) Population size: 80

(d) Population size: 100

Fig. 2. The percentage of the objective space where the pareto-optimal set of the generation x outperforms that of the generation $x - 50$ within a single configuration

The figure shows that for different mutation probabilities, stagnation does not occur over the 1000 generations for population size 20 while it occurs at around generation 320, 220 and 280 for population sizes 40, 80 and 100 respectively. One may also notice that the level of mutation probability does not influence

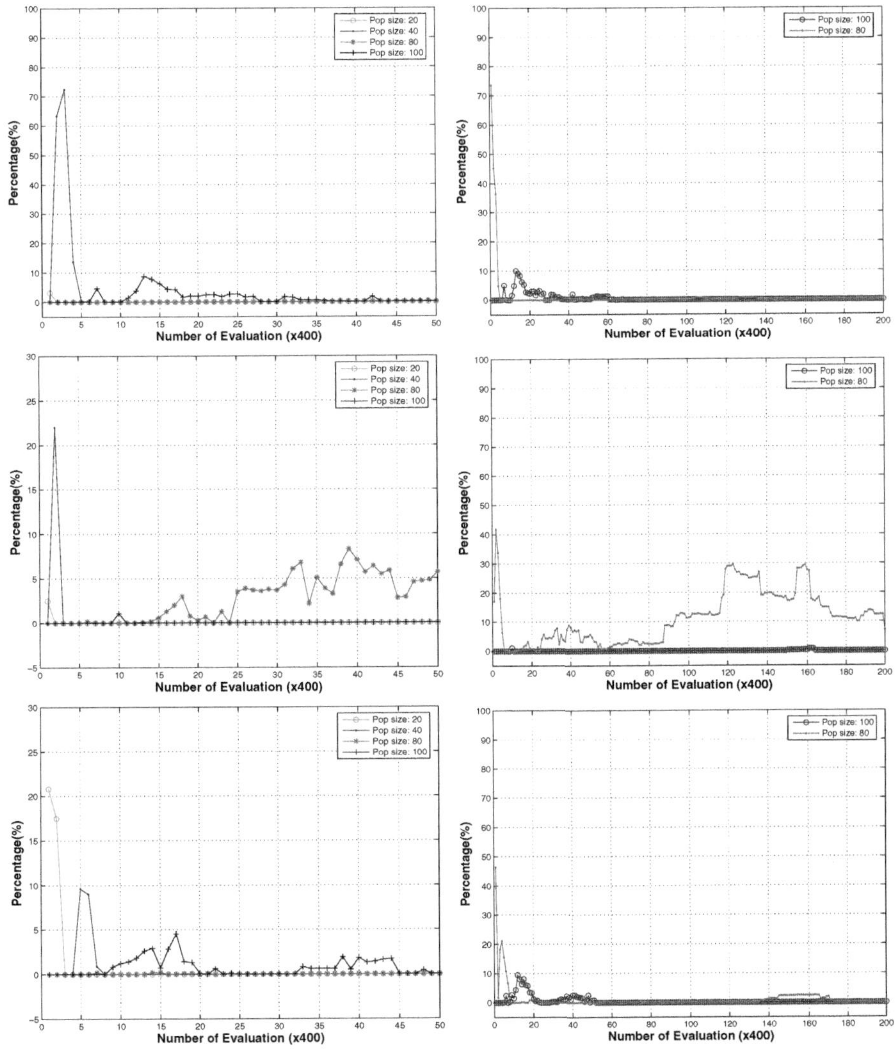

Fig. 3. The percentage of the objective space where the non–dominated set of one population size outperforms that of other population sizes at the same number of objective evaluations. Form top to bottom, the mutation rates are: 0.010, 0.036, 0.10.

the convergence for population sizes except for population size 20, where there are some insignificant differences.

Stagnation can be seen as an indication for the convergence of the algorithm. However, it is not enough to know when it occurs. It is also important to know the quality of each non–dominated set obtained by each setup as presented in Figure 3. The sub–figures on left hand side represent a comparison among all four setups with different population sizes. Each sub-figure corresponds to a

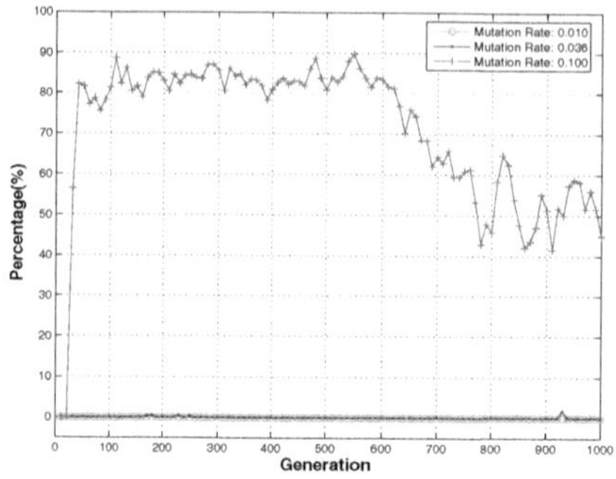

(a) Population size: 20

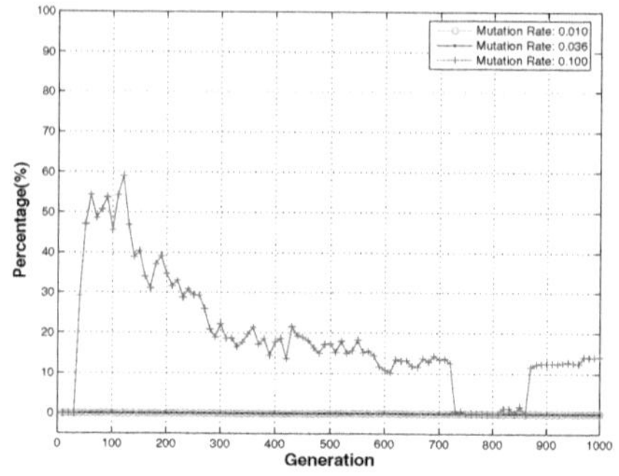

(b) Population size: 40

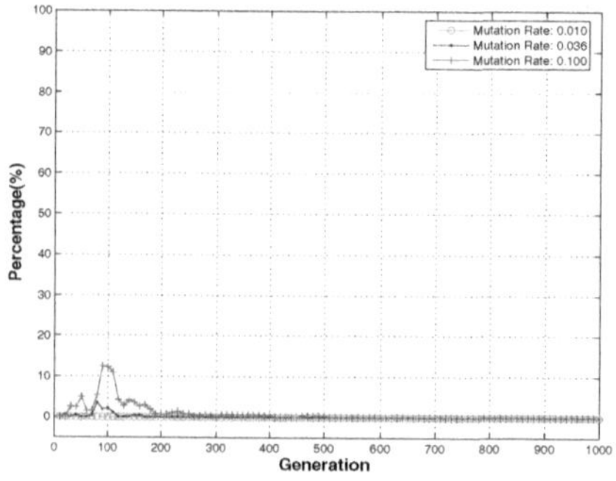

(c) Population size: 80

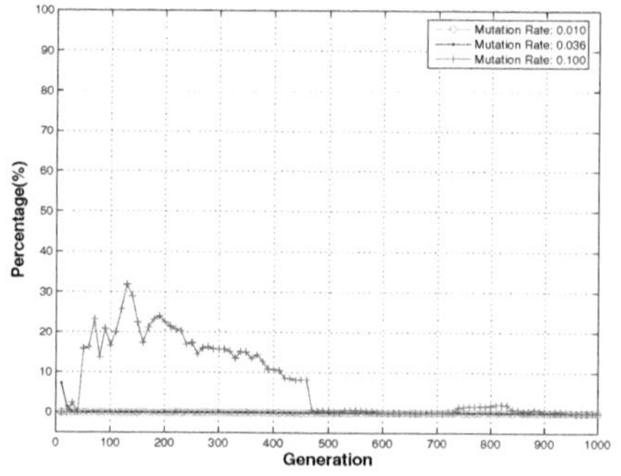

(d) Population size: 100

Fig. 4. The percentage of the objective space where the pareto-optimal set from one mutation probability outperforms that of any others

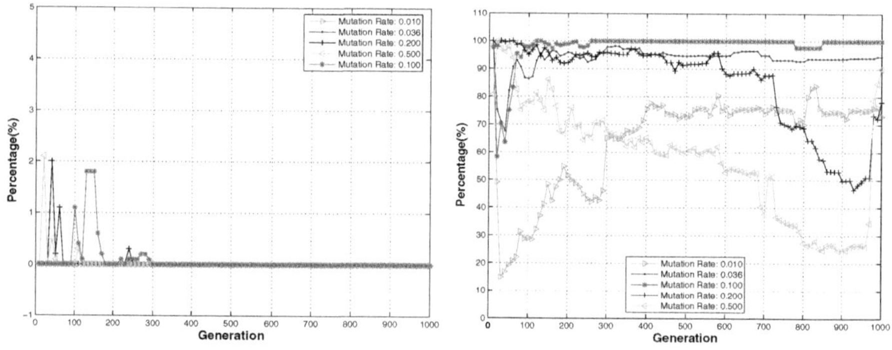

Fig. 5. Left: The percentage of the objective space where the pareto-optimal set from one mutation probability outperforms that from all other mutation probability with the population size of 80. **Right**: The percentage of the objective space where the pareto-optimal set from one mutation probability is not outperformed by that from any other mutation probabilities with the population size of 80.

different mutation level. Each comparison is done after 400 objective evaluations (this is a common factor for the four population size). The non–dominated sets achieved by each setup after each 400 objective evaluations are collected and compared.

All figures show consistently that population size 80 has the best overall performance. The sub–figures on the right hand side make a more precise comparison between the two best performing population sizes: 80 and 100. One can see that with mutation rate of 0.036, population size 80 is consistently better than 100.

We now turn our attention to comparing the different mutation rates. It seems from Figure 4 that the performance of mutation probability 0.1 is better than any of the other. This may suggest that a higher mutation probability is beneficial for this problem. We thus test a mutation probability of 0.2 and 0.5 for population size 80.

Figure 5 presents the performance with population size 80 at five different mutation probabilities. We can now have more confident to suggest that a population size of 80 with mutation probability of 0.1 gives the best performance for this scenario, and the convergence occurs at generation 220.

4 Conclusions and Future Work

In this paper, we tested the effect of population size and mutation probability on the evolution of force compositions in a multi–objective setting. It was found that population size 80 with mutation probability 0.1 gave us the best performance.

A key question in this work is how to generalize this result to other land combat problems. Fortunately, the military usually has a small number of generic scenarios to test different concepts (normally less than a dozen). Thus repeating this process for each scenario is not a difficult task. However, once we can identify the parameter setting for each scenario, we can undertake more experiments with different scenario setups. These setups may change the fitness landscape in terms of signals, but from our previous experiments, they do not change the characteristics of the landscape. In other words, the characteristics of the landscape vary from one scenario to another, but do not vary much when a scenario is run with different settings. Thus, we expect that the parameter setting for one scenario will continue to be useful when running this scenario with different scenario parameters.

Our current work is focusing on studying approximation methods to reduce the computationally very expensive task of evaluating on the actual simulator which can be very time consuming.

Acknowledgement

This work is supported by the University of New South Wales grant PS04411 and the Australian Research Council (ARC) Centre on Complex Systems grant number CEO0348249. The authors also wish to thank Mr. Lam Bui for useful discussions.

References

1. Ursem, R.K.: Models for Evolutionary Algorithms and Their Applications in System Identification and Control Optimization. Ph.d. thesis, University of Aarhus, Denmark (2003)
2. Deb, K., Agrawal, S., Pratap, A., Meyarivan, T.: A fast and elitist multi-objective genetic algorithm: NSGA-II. IEEE Transactions on Evolutionary Computation **6** (2002) 182–197
3. Yang, A., Abbass, H.A., Sarker, R.: WISDOM-II: A network centric model for warfare. In: Ninth International Conference on Knowledge-Based Intelligent Information & Engineering Systems (KES 2005), LNCS 3683, Melbourne, Australia (2005)
4. Yang, A., Abbass, H.A., Sarker, R.: Landscape dynamics in multi-agent simulation combat systems. In: Proceedings of 17th Joint Australian Conference on Artificial Intelligence, LNAI 3339, Cairns, Australia, Springer-Verlag (2004)
5. Yang, A., Abbass, H.A., Sarker, R., Curtis, N.J.: Evolving capability requirements in WISDOM-II. In Abbass, H.A., Bossamier, T., Wiles, J., eds.: Advances in Artificial Life, Proceeding of The Second Australian Conference on Artificial Life (ACAL05), Sydney, Australia, World Scientific Publisher (2005) 335–348
6. Deb, K., Agrawal, R.: Simulated binary crossover for continuous search space. Complex Systems **9** (1995) 115–148
7. Deb, K., Goyal, M.: A combined genetic adaptive search (GeneAS) for engineering design. Computer Science and Informatics **26** (1996) 30–45
8. Deb, K., Agrawal, S.: Understanding interactions among genetic algorithm parameters. In Banzhaf, W., Reeves, C., eds.: Foundations of Genetic Algorithms 5. Morgan Kaufmann, San Francisco, CA (1999) 265–286
9. Ballester, P.J., Carter, J.N.: Real-parameter genetic algorithms for finding multiple optimal solutions in multi-modal optimization. In: Proceedings of the Genetic and Evolutionary Computation Conference, Lecture Notes in Computer Science 2723, Chicago, USA, Springer-Verlag (2003) 706–717
10. Khare, V., Yao, X., Deb, K.: Performance scaling of multi-objective evolutionary algorithms. In Fonseca, C.M., Fleming, P.J., Zitzler, E., Deb, K., Thiele, L., eds.: Proceedings of the 2nd International Conference on Evolutionary Multi-Criterion Optimization (EMO'03), LNCS 2632, Faro, Portugal, Springer-Verlag (2003) 376–390
11. Iorio, A., Li, X.: A cooperative coevolutionary multiobjective algorithm using non-dominated sorting. In: Proceeding of Genetic and Evolutionary Computation Conference 2004 (GECCO'04), Lecture Notes in Computer Science (LNCS 3102), Seattle, USA, Springer-Verlag (2004) 537–548
12. Zitzler, E., Thiele, L., Deb, K.: Comparision of multiobjective evolutionary algorithms: Emprical results. Evolutionary Computation **8** (2000) 173195
13. Fonseca, C.M., Fleming, P.J.: On the performance assessment and comparison of stochastic multiobjective optimizers. In Ebeling, V.W., Rechenberg, I., Schwefel, H.P., eds.: Proceedings of the 4th International Conference on Parallel Problem Solving from Nature, LNCS 1141, Springer-Verlag (1996) 584593
14. Knowles, J.D., Corne, D.W.: Approximating the nondominated front using the pareto archived evolution strategy. Evolutionary Computation **8** (2000) 149–172

Automatic Seizure Detection Based on Support Vector Machines with Genetic Algorithms*

Jinfeng Fan[1], Chenxi Shao[1], Yang Ouyang[1],
Jian Wang[1], Shaobin Li[1], and Zicai Wang[2]

[1] Department of Computer Science and Technology, University of Science and Technology of China, Hefei, Anhui, 230027, P.R. China
jffan@mail.ustc.edu.cn, cxshao@ustc.edu.cn,
{ouyangy, jianw3, cplee}@mail.ustc.edu.cn
[2] Control & Simulation Center, Harbin Institute of Technology, Harbin, Heilongjiang, 150001, P.R. China

Abstract. The electroencephalogram (EEG) machine is the most influential tool in the diagnosis of epilepsy, which is one of the most common neurological disorders. In this paper, a new seizure detection approach, which combined the genetic algorithm (GA) and the support vector machine (SVM), is proposed to improve visual inspection of EEG recordings. Genetic operations are utilized to optimize the performance of SVM classifier, which includes three aspects: feature subset selection, channel subset selection and parameter optimization of SVM. These optimization operations are performed simultaneously during the training process. The epileptic EEG data acquired from hospital are divided into two parts of training set and testing set. The results from the test on EEG data show that the method may more effectively recognize the spike and sharp transients from the EEG recording of epileptic patients than those without using optimal operations.

1 Introduction

The electroencephalogram (EEG) signal, which is the recordings of the bioelectrical and biomechanical activities of brain, is widely used in clinical applications. It provides a great deal of valuable information for diagnosing, monitoring and managing of neurological disorders related to epilepsy. Traditionally, visual epilepsy evaluation of the EEG recordings is carried out by an experienced EEGer. This operation is time consuming and tedious, especially for long recordings. In addition, disagreement among readers of the same record is possible due to the subjective nature of the analysis [1]. Therefore, an automation or semiautomatic method for seizure detection must be developed to reduce doctor's labor and improve the results of traditional visually analysis.

Several Methods have been proposed for automatic detection of spikes or seizure occurrences. Weng and Khorasani [2] introduced an approach for adaptively adjusting the structure of a multi-layer back-propagation network for automatic seizure

* This work was supported by Key Project of Natural Science Foundation of China (60434010).

T.-D. Wang et al. (Eds.): SEAL 2006, LNCS 4247, pp. 845 – 852, 2006.

detection. Subasi [3] proposed automatic seizure detection using discrete wavelet transform and ANN. Acir and Guzelis [1] proposed a two-stage procedure based on support vector machines (SVMs) for the automatic detection of epileptic spikes. These methods used all the available channels to evaluate features which serve as the input information of the classification. However, it is not necessary because some channels are redundancy and irrelevant. It is important to select optimal channels for improving the efficiency of seizure detection.

Parameter optimization on SVM model is also plays a crucial role in real-world applications with high accuracy and stability. The parameters such as C and σ^2 presented in the radial basis function (RBF) kernel can be determined during training process by using genetic algorithms (GAs).

In this paper, a method based on the combination of genetic algorithm and support vector machine (termed GA-SVM) is proposed. The three aspects of optimization operation: optimal feature subsets, optimal channel subsets and optimal parameters are performed by using this hybrid method simultaneously. The application on the epileptic seizure detection has demonstrated its high potential as a powerful tool for the resolution of multi-criterion problem.

2 Methods

2.1 Data Preprocessing

EEG data obtained from sampling equipment regularly contain certain artifacts that make it difficult to extract and interpret the interesting information efficiently. In an applicable system, the stage of data preprocessing is often required to improve the reliability of the classification. Our objective is to maximally distinguish the seizure signals from background noise and other potentials. In order to achieve this purpose, three preprocessing operations are applied. First, a band-pass filter is used for the reason that a meaningful phase is only given on narrowband signals. Second, EEG recordings are divided into sequential, non-overlapping, fixed length segments of 1024 samples. The data quality assures it feasible to evaluate various features and preserves relatively stationary of waveform in one segment. Last, a part of segments must be selected and marked with labels for training classifier by specialist with rich clinical experience.

2.2 Feature Extraction

For general seizure detection systems, features are extracted to serve as the input information. So far many approaches for the extraction of quantitative features from EEG signal were introduced. The methods based on probabilistic and statistical theories are widely applied. Spectrum analysis is also a popular method which is used to extract frequentation-domain from time or space domain signals using transform methods such as fast Fourier transform (FFT). The application of nonlinear dynamics methods to the problem of the description of an EEG has got relatively successful results. In particular, some researches have shown strong indications of nonlinear deterministic and finite-dimensional structures in epileptic EEG signals [4].

In this study, all these methods are used to extract distinct features. 20 statistical parameters are selected. Spectrum analysis parameter set consist 5 power values of frequency band. The permutation entropy (PE) [5], Hurst exponent (HE) [6], correlation dimension (CD) [7] and the largest Lyapunov exponent (LLE) are selected as the nonlinear features because of their well discriminative ability. These 29 features compose a feature pool [8], from which the dominate features are selected for the classifier.

2.3 Feature Selection

There is a possibility that some features in the pool might be redundant or even irrelevant for seizure detection. Redundant features do not add significant information for class distinction. Similarly, irrelevant features do not participate in class distinction and even can mix up boundaries between classes. Therefore, in order to reduce computation time and enhance the detection accuracy, it is necessary to devise a method for selecting the most discriminatory features with few redundancy or irrelevancy. The procedure of feature selection is wrapped in GA-SVM seizure detection system in this study. In training phase, once the learning algorithm got the optimal resolution, the best feature subset is determined.

2.4 Channel Selection

The channels used to record EEG signals may contain redundant or irrelevant information. So it is also attractive to reduce the number of channels necessary for seizure detection without increasing detection error. Channel selection may not only reduce computation time of seizure detection, but also help surgeons localize the epileptogenic foci. In this study, the selection operation of optimal channels carrying significant information for detection purposes is processed during training phase.

2.5 Support Vector Machines

Support vector machines (SVMs) based on statistical learning theory are first introduced by Vapnik [9]. In the design of an SVM model, a proper kernel function must be selected. After a number of empirical experiments on the EEG data, the results indicated that Gaussian RBF kernel seemed to offer the better performance than others. So, the Gaussian RBF kernel is selected in our implementation.

In order get the optimal performance of SVM, the process of parameter optimization plays an important roles. Here are two primary parameters, penalty parameter C and the kernel parameter σ^2, to be carefully determined. GA is utilized to search for optimal parameters of the SVM model with the minimum output errors.

2.6 Evolutionary Algorithms

Evolutionary algorithms (EAs) are a class of problem solving techniques based on the principles of natural selection in the biological world. Genetic algorithms (GAs) are popular evolutionary search methods that focus on optimizing general combinatorial problems. In the study, GAs are applied to searching for the global optimal solution containing the best feature subsets, the optimal channels and the optimal parameter values simultaneously in our study. Two important issues in the design of GA are the genetic coding used to define the problem and the evaluation function, called fitness.

2.6.1 Encoding

In our implementation, the popularly used binary encoding method is selected. The chromosome code is composed of four parts, i.e. features and channels and the two parameters C and σ^2. The code length is the sum of the lengths of the four parts (see Fig. 1). For the part of feature encoding, each binary gene code represents whether the corresponding feature is selected or not. For the part of channel encoding, 1 denotes the corresponding channel is active, whereas 0 means it is inactive. For the two parameters of SVM, the values are converted to the binary string, which is used as the corresponding codes.

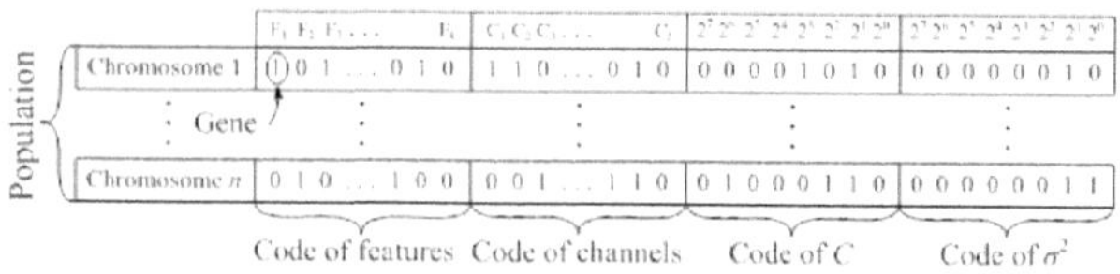

Fig. 1. Encoding structure of chromosome composed of channels, features and two parameters

2.6.2 Fitness Function

Fitness function, also called adaptation function, is another vital issue of the GA. It assesses the fitness ability of each individual and maps it to a fitness value. All individuals are ordered according their fitness ability, and the individuals with high fitness will have more opportunities to survive to the next generation during the operation of selection.

In the study, we encounter a problem of multi-criterion optimization, which contains three sub problems for searching optimal solution, i.e. the optimizations of feature subset, channel subset and the parameters of SVM. The purpose of feature selection is to choose fewer features capturing the essential information of seizure activity. The target of channel selection is to select principle channels which associate to the epileptogenic foci. The parameter optimization aims to obtain the minimization of recognition error. Then, by a comprehensive consideration of these criteria, a simple fitness function is defined as a linear combination of them. It is given by

$$\text{Fitness} = A \cdot \frac{\text{Feature}_{\text{unused}}}{\text{Featue}_{\text{total}}} + B \cdot \frac{\text{Channel}_{\text{incative}}}{\text{Channel}_{\text{total}}} + C \cdot \frac{\text{Seizure}_{\text{correct}}}{\text{Seizrue}_{\text{total}}}, \tag{1}$$

where A, B and C are constant coefficients. Here, we set $A = B = 1$, $C = 3$ by some empirical experiments.

2.6.3 Genetic Operation

The GA uses three genetic operations (selection, crossover and mutation) to generate the next population from current one. The values of parameters settings for genetic operations use the results of previous researches for reference and are tuned appropriately by a number of experiments on the training data set.

2.6.3.1 Selection. The operation of selection chooses individuals of population for survival from existing population according to their fitness values. First, in order to keep the best chromosomes that may produce the fittest individuals during crossover and mutation operations, the selection rate is set to allow the simple duplication them to the next generation population. The standard roulette wheel random procedure, then, is employed as the selection mechanism. The individuals that survive to next generation are placed in a mating pool for following crossover and mutation operations.

2.6.3.2 Crossover. The crossover operation combines two chromosomes with high fitness values to produce children with the intention to keep the good characteristics of their parents and to get rid of the bad characteristics. The single point crossover operation is applied to randomly paired individuals selected from the mating pool. Since the chromosome consists of four parts, each part is respectively performed and all crossover points are determined randomly. Fig. 2 illustrates the crossover operation.

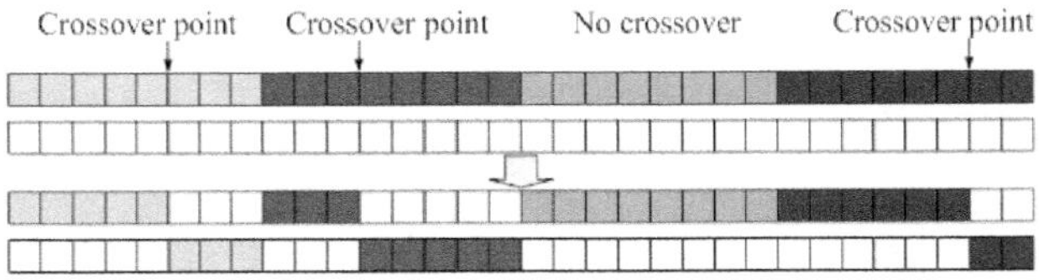

Fig. 2. Single point crossover for each part of the chromosome with the random crossover point and the crossover rate of 0.8

2.6.3.3 Mutation. The mutation operation is utilized to search for further solution space and to avoid local convergence of GA. In this study, a multi-point bit-flip mutation, based on a specified mutation probability of 0.01, is applied to the four parts of the chromosome. The locations of mutation are randomly determined every time mutation operator is performed. A simple example of mutation operation is shown in Fig. 3.

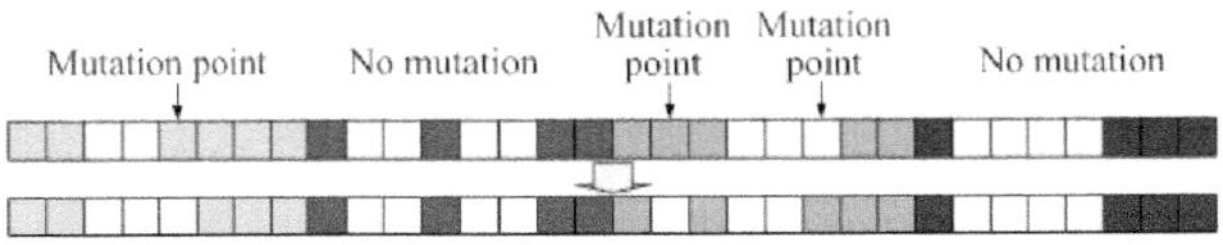

Fig. 3. Multi-point bit-flip mutation for each part of the chromosome with the random mutation point and the mutation probability of 0.01

2.7 Integration of GA and SVM

As mentioned earlier, the integration of GA-SVM is proposed to optimize features selection, channel selection and parameters of SVM. The method searches for the best solution from a large and possibly multi-modal search space. Traditional methods widely use two major classes of optimization techniques, i.e. calculus-based techniques and enumerative techniques. The former methods use gradient-based search mechanisms. The latter ones may be implemented by dynamic programming [10]. In our study, a multi-criteria problem makes the common methods very difficult to achieve the expected targets or reach the solution with very high computational complexity. The GA provides a non-conventional nonlinear search algorithm to obtain fast results. Thus, it is very appropriate and required that the GA-based methods are applied to the multi-criteria problem.

For the SVM model, parameter optimization is the crucial factor to determine its performance. The GA is utilized to achieve theses optimal values of features, channels and parameters simultaneously for epileptic seizure detection. The combined GA-SVM method use supervised learning mechanism.

The architecture of the GA-SVM model is illustrated in Fig. 4. The EEG data are divided into two parts. One part is used to train GA-SVM, and the other part is used to test. The figure shows that the testing process only evaluates special features on optimal channels which are determined during training process.

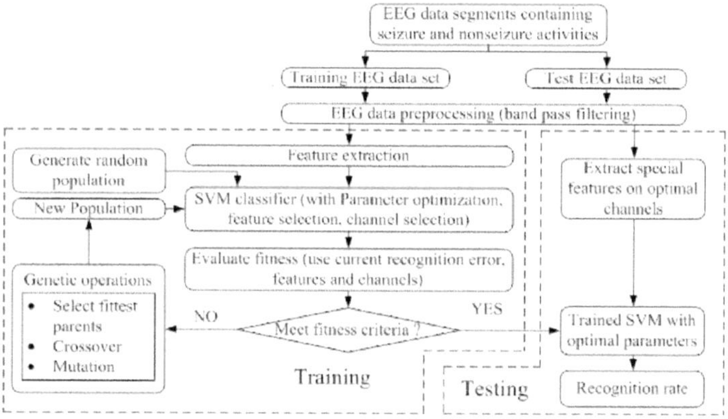

Fig. 4. Architecture of the GA-SVM model and its flow chart

3 Materials and Results

3.1 Data Acquisition

The EEG data used in our study were collected from 12 epileptic patients who had been diagnosed epilepsy in hospital. The EEG signals were recorded using digital EEG equipment with the sample rate of 256Hz. 19 Ag/AgCl disk electrodes were placed using 10-20 international electrode placement system. Band-pass filter with cut-off frequencies of 0.5Hz and 45Hz was selected to preprocess these raw data. The length of each recording is beyond two hours. For each patient, data are separated for segments with length of 4 seconds. One-third segments ware randomly selected for training, and the remaining ones were used for testing.

The EEG recordings were inspected by the neurologists with experience in the clinical analysis of EEG signals, and seizure events were masked as labels in a new file. Then, these labels were revised by other experts to solve disagreements.

3.2 Results

The GA-SVM is trained and tested on each patient. Fig. 5 illustrates the curves of recognition accuracy versus patient for different optimization settings. It can be seen that the curve are higher by using optimal features and channels than others, and the recognition accuracy is lowest when no optimization operations are performed.

Table 1 gives the results of the numeric values. When all channels and features are used, the recognition accuracy is 83.9%. It rises to 86.9% with all channels and optimal feature subset, and it is 87.7% with optimal channels and all features. When the two optimization operations are all performed, the recognition accuracy reaches the highest value, i.e. 91.3%. The table also gives the number of channels and the number of features which are selected to test at different optimization conditions.

Table 2 gives the results of optimized SVM kernel parameters for every patient EEG data after training with GA-based method.

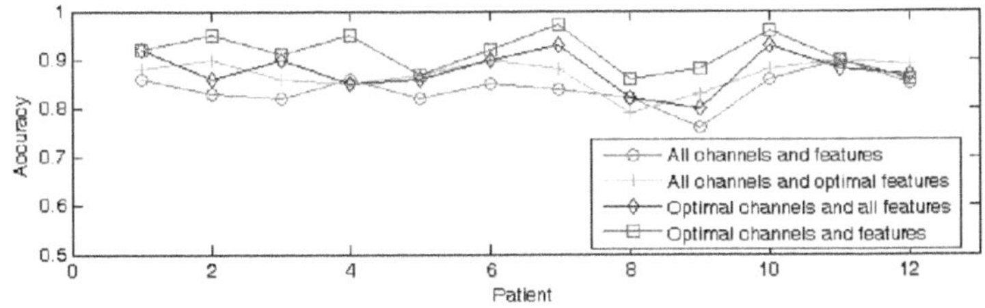

Fig. 5. Curves of recognition accuracy versus patient for different optimization settings

Table 1. Results of channel selection and feature selection and recognition accuracy

	All channels and all features	Optimal channels and all features	All channels and dominant features	Optimal channels and dominant features
Average number of channels	19	3.8	19	4.7
Average number of features	29	29	8.3	7.4
Recognition accuracy	0.839 ± 0.034	0.869 ± 0.033	0.877 ± 0.042	0.913 ± 0.039

Table 2. Values of SVM kernel parameter optimized by using GA-based method

Patient	1	2	3	4	5	6	7	8	9	10	11	12
C	863	630	1013	921	1105	586	739	419	1425	894	927	1179
σ^2	5.46	4.3	3.20	4.17	3.81	3.52	3.21	3.86	4.22	3.29	5.50	4.82

3.3 Discussion

According to the above results, it can be found that the recognition accuracy and the generalization ability of the SVM model can be significantly enhanced by GA-based method. The performance of the SVM is influenced greatly by the optimization of kernel parameters. The GA model provides an error-feedback mechanism to adaptively adjust the kernel parameters. The recognition accuracy is also sensitive to the structure of the input information of SVM. Increasing the number of features can increase the flexibility of the model, but it also increase irrelevancy and redundancy of the input information. Reducing the useless information, the optimizations of features selection and channel selection increase both the recognition accuracy and computational performance for seizure detection with the trained model.

4 Conclusion

In this paper, we have proposed a method combined GA-SVM to detect the epileptic seizure. The optimal operations of feature selection, channel selection and kernel parameters of SVM are performed simultaneously to maximize the performance of the automatic seizure detection. The results show that the proposed method may more effectively recognize the spike and sharp transients from the EEG recording of epileptic patients. This method may be also employed to improve the visual inspection of EEG recordings and help the diagnosis of some other neural diseases.

References

1. Acir N., Guzelis C.: Automatic Spike Detection in EEG by a Two-Stage Procedure based on Support vector machines. Computers in Biology and Medicine, Vol. 34.7. (2004) 561-575
2. Weng W., Khorasani K.: An Adaptive Structure Neural Networks with Application to EEG Automatic Seizure Detection. Neural Network, Vol. 9.7. (1996) 1223-1240
3. Subasi A.: Epileptic seizure detection using dynamic wavelet network. Expert System with Applications, Vol. 29.2. (2005) 343-355
4. Gautama T., Mandic D.P., Van Hulle M.M.: Indications of Nonlinear Structures in Brain Electrical Activity. Phys. Rev. E, 67(2003): 046204
5. Bandt C., Pompe B.: Permutation Entropy — a Natural Complexity Measure for time series. Physical Review Letters, Vol. 88.17. (2002) 174102
6. Natarajan K., Acharya U.R., Alias F., et al: Nonlinear Analysis of EEG Signals at Different mental States. Biomedical Engineering Online, Vol. 3.7. (2004)
7. Shao C.X., Fan J.F., Feng H.L., et al: Fuzzy Analysis of Epileptic EEG Based on Qualitative Simulation. Asia Simulation Conference/6th International Conference on System Simulation and Scientific Computing, Beijing, China, Vol. 2. (2005) 1348-1352
8. Cho S.Y., Kim B., Park E., et al: Automatic Recognition of Alzheimer's Disease Using Genetic Algorithms and Neural Network. Lecture Notes in Computer Science, 2658 (2003) 695-702
9. Vapnik V.N.: The Nature of Statistical Learning Theory. New York: Springer (1995)
10. Saxena A., Saad A.: Evolving an Artificial Neural Network Classifier for Condition Monitoring of Rotating Mechanical systems. Applied Soft Computing, (2006)

Articulated Body Tracking by Immune Particle Filter

Zhaohui Gan[1] and Min Jiang[2,3]

[1] School of Information Science and Technology,
Wuhan University of Science and Technology, 430072, Wuhan China
ganzhl1@yahoo.com.cn
[2] College of Computer Science and Technology,
Wuhan University of Science and Technology, 430081, Wuhan, China
[3] School of Computer, Wuhan University, 430072, Wuhan China

Abstract. The tracking of articulated body in images sequences is a challenging problem due to complexity and high dimensionality of the configuration space. In this paper, we propose a new algorithm to combine Artificial Immune and particle filter for articulated body motion tracking, fusing the strengths of both approaches. Compared with previous optimization based particle filter, our method overcomes the disadvantages of inefficiency by incorporating artificial immune algorithm into particle filter. Evaluations on MOCAP dataset show that immune particle filter algorithm performs better than anneal particle filter.

1 Introduction

Articulated body tracking is currently one of the most active research topics in computer vision. It has various applications in human animation, human-computer interaction, video surveillance and sports video analysis.

The previous optimization based particle filter algorithms [5][6] for articulated object motion tracking are not population based optimization algorithm, each particle perform its optimization procedure individually, and the computational resources is evenly allocated to each particle.

In order to reach a good accuracy with sparse sampling and overcome the disadvantages of the previous optimization based particle filter algorithms, we propose a novel approach for body motion capture, called immune particle filter (IPF). In IPF, the optimization procedure is carried out by artificial immune algorithm. Artificial immune algorithm is a population-based algorithm. Compared to the formal non-population based algorithms, IPF can efficiently eliminate the previous with low affinity by clone selection, and allocated more computational resources to more probable hypothesizes. Moreover, compare the other population based optimization algorithm, such as GA and SA etc., Artificial immune algorithm increases efficiency by mutating inversely proportional to affinity.

2 Related Works

Particle filter applies a recursive Bayesian filter based on propagation of sample set over time, maintains multiple hypotheses at the same time and uses a stochastic

T.-D. Wang et al. (Eds.): SEAL 2006, LNCS 4247, pp. 853 – 857, 2006.

motion model to predict the position of the object. The state of the modeled object at time t is denoted by S_t, where $S_t = \{s_1, s_2, \ldots, s_t\}$, the measurement of the object at time t is Z_t, denotes all the observations $\{z_1, z_2, \ldots, z_t\}$ up to time t. The process density $p(s_t \mid s_{t-1})$ and observation density $p(s_t \mid Z_t)$ are not required to be Gaussian. Particle filtering is to approximate the probability distribution by a weighted sample set $\{(s_t^{(n)}, \pi_t^{(n)})\}_{n=1}^{N}$. Each sample $s_t^{(n)}$ represents one hypothetical state of the object, with a corresponding discrete sampling probability $\pi_t^{(n)}$.

In human body tracking problems, the dimensionality of the parameter space is far too high to represent accurately the true posterior distribution everywhere. Recent approaches to handling this problem can roughly be classified into two types: lower the dimensionality [1][2] or introduce particles with a more sophisticated behavior, which was named as 'smart particles' by Bray [4].

3 The Priori Knowledge of Human Body Motion

The priori knowledge of human body motion includes human body model, model constraints and motion model etc. Human body model is a kind of structure model of human body based on the knowledge of anatomy human body, which it represents Articulated model and pose express of human body, and also determines matching features in tracking process. The motion model (dynamic model) denotes the law of human body motion as mathematical formula. The motion constraints reflect constraints which human body motion model should follow, for example, limb rotation range, the constraints which one part of human body couldn't penetrate through other parts etc.

3.1 Human Body Model and Constraints

We select a 3D articulated model in [11] as body model. The model use simple 3D ellipsoids and generalized cylinders to represent 10 rigid segment of human body, and those segment was organized as a kinematical tree, which consists of 15 joints (articulations). The model has 31 degrees of freedom that include local joint angles, the position and orientation of the root of the kinematical tree, thus, Each pose can be represented by a 31-dimensional vector.

3.2 Motion Model

Sophisticated motion prior models learnt from training sets of 3D human motions have been proposed in [3][9][10], but in our case, for the generality of our approach, we choose first order derivative motion model

$$s_t = s_{t-1} + w \tag{1}$$

Where s_t is a 31-dimensional state vector of 3D articulated model at time t, w is the diffusion variable $w \sim N(0, \Sigma)$, Σ is a diagonal matrix which learnt from training sets of 3D human motions.

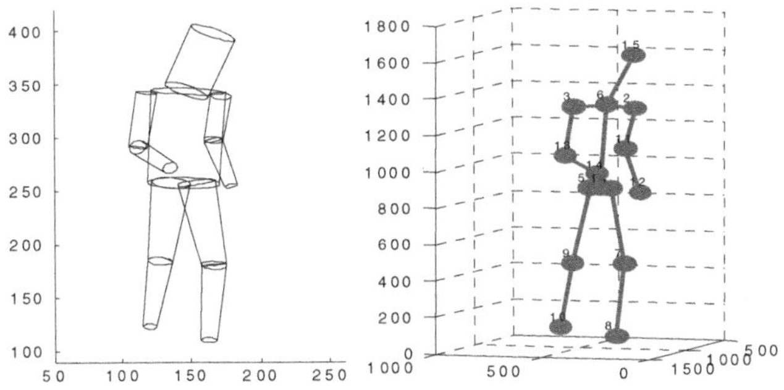

(a) 10 components of body (b) 15 articulations

Fig. 1. Articulated model of human body, it use simple 3D ellipsoids and generalized cylinders to represent rigid segment of human body

4 Immune Particle Filter

Fitting and tracking the articulated model to the detected humans in the video streams requires the definition of a observation model allowing to project this model onto the available image properties.

Given an image observation z_t , the likelihood measure for model hypothesis x is

$$p(z_t \mid x_t = x) = \exp(-E^t(x)) \tag{2}$$

where the total model fit error $E^t(x)$ consist of silhouettes fit error $E^t_{sihouette}(x)$ and edges fit error $E^t_{edge}(x)$, $E^t(x) = \sum_{i=1}^{C} (\alpha E^{t,i}_{edge}(x) + (1-\alpha) E^{t,i}_{sihouette}(x))$, C is the number of cameras, and , α is a weight factor, $0 \le \alpha \le 1$.

The framework of our immune particle filter algorithm is based on particle filter, and incorporates the immune optimization into the four steps of traditional particle filter(**Resampling, Predict, Measure, Estimate**),and it includes six steps: **Resample, Predict, Feature extraction, Immune optimization, Measure, Estimate**. The flow of immune particle filter is as follows.

Algorithm: Immune Particle Filter
Input: old sample-set $\{s^{(n)}_{t-1}, w^{(n)}_{t-1}\}^N_n$ at time t-1.
Output: new sample-set $\{s^{(n)}_t, w^{(n)}_t\}^N_n$ and pose estimate $\hat{s}_t$ for time t ,.
begin

Resample particles $s^{(n)}_{t-1}$ with probability $w^{(n)}_{t-1}$ to obtain N random particles $s'^{(n)}_{t-1}$ according to $\{w_{t-1}\}^N_n$.

Predict sample $\{s_t^{(n)}\}_n^N$ from the dynamic model (5.1).

Feature extraction , extract edge pixel map and silhouette pixel map from z_t .

Immune optimization, Optimization $\{s_t^{(n)}\}_n^N$ with CLONGALG:

Initialize antibodies set, $AB_t \leftarrow \{s_t^{(n)}\}_n^N$, $AB_t = \{Ab_t^{(n)}\}_n^N$, evaluate affinity $Af_t^{(n)}$ for each $Ab_t^{(n)}$. it=1 .

- While *Not Converged and it < ITMAXNUM*
  - — Clone
  - — Mutate
  - — Evaluate affinity $Af_t^{'(n)}$ for each $Ab_t^{'(n)}$.
  - — Choose N highest affinity antibodies from $AB_t^{''}$ to compose a new set as memory Ab $AB_t = \{s_t^{(n)}\}_n^N$, $it++$.

Get mature Ab set, $AB_t^* \leftarrow AB_t, AB_t^* = \{Ab_t^{*(n)}\}_n^N$.

Measure ,set $\{s_t^{(n)}, w_t^{(n)}\}_{\text{n}}^N = \{Ab_t^{*(n)}, \exp(Af_t^{(n)}) / \sum_{j=1}^{N} \exp(Af_t^{(j)})\}_{\text{n}}^N$

Estimate, Output a set of particles $\{s_t^{(n)}, w_t^{(n)}\}_n^N$ and

$$\hat{s}_t = E[x_t = s_t \mid z_t] \approx \sum_{j=1}^{J} w_t^{(j)} s_t^{(j)}$$

End.

The Feature extraction step before immune optimization is to increase efficiency. Next,we will describe the immune optimization procedure in detail.

5 Experiments

To evaluate our algorithm for human body motion capture, we use MOCAP [8], the dataset of the synchronized motion capture and image data for a walking human. A total of 530 frames are provided at 60 Hz. The length of body limbs, and radius parameters of 3D ellipsoids and generalized cylinders of body are estimated from

Table 1. Comparison the performance results of anneal particle filter and immune particle filter

Anneal Particle Filter			Immune Particle Filter		
N	Error (pt-pt/frame)	Time (s)	N	Error (pt-pt/frame)	Time (s)
100	49.01	23.14	20	54.45	21.16
200	49.31	50.85	30	49.69	47.92
300	47.10	73.20	40	44.37	75.18
400	47.23	98.40	50	43.85	123.44

motion capture data of the first 50 frames of this dataset. The first 150 frames video data of four cameras is used for testing our algorithm, motion capture data from Vicon system as the ground truth data for quantitative evaluating our algorithms.

Table 1 shows the comparison of anneal particle filter and immune particle filter. N is the number of particles. In immune particle filter, the maximum iteration number of immune optimization is set to 7, clone ratio is 0.4, clone multiplier 0.4 and mutation regulation factor is 0.08. As can be seen in Table 1 the performance of immune particle filter algorithm is better than anneal particle filter.

6 Conclusion

In order to reach a good accuracy with sparse sampling, we propose a modified particle filter based on artificial immune algorithm, and realize this algorithm for body motion capture. Compared to the formal non-population based algorithms, immune particle filter can efficiently eliminate the hypothesizes with low affinity by clone selection, and allocated more computational resources to more probable hypothesizes.

The experiments on MOCAP dataset show that the performance of immune particle filter algorithm is better than anneal particle filter.

References

1. J. MacCormick and M.Isard, Partitioned sampling, articulated objects, and interface-quality hand tracker, ECCV, Vol.2, pp.3-19, 2000.
2. Wu Y., Lin J. and Huang T.S., Capture Natural Hand Articulation, ICCV 2001,, pp. 426-432.
3. V. Pavlovic, J. M. Rehg, T. J. cham, and K. P. Murthy, A dynamic Bayesian network approach to figure tracking using learned dynamic models, ICCV 1999, vol. 1, 94-101.
4. M. Bray, E. Koller-Meier, P. Muller, L. Van Gool, 3D Hand Tracking by Rapid Stochastic Gradient Descent using a Skinining Model, CVMP, pp. 59-68, 2004.
5. M. Bray, E. Koller-Meier, and L. Van Gool, Smart particle filtering for 3D hand tracking, Proc. IEEE Int. Conf. Automatic. Face & Gesture Recognition, pp. 675-680, 2004.
6. J Deutscher, A Blake, I Reid, Articulated body motion capture by annealed particle filtering, IEEE Conference on Computer Vision and Pattern Recognition, CVPR, 2: 1144-1149, 2000.
7. Leandro N. de Castro and Janathan Timmis. Artificial Immune Systems: A New Computational Intelligence Approach. Springer-Verlag, London, UK, 2002.
8. Leonid Sigal, Synchronized Video and MOCAP dataset. Brown University.2004
9. D. Ormoneit, H. Sidenbladh, M. Black, and T. Hastie. Learning and tracking cyclic human motion. In Advances in Neural Information Processing Systems 13, pages 894–900,2001.
10. H. Sidenbladh, M. J. Black, and D. J. Fleet. Stochastic tracking of 3D human figures using2D image motion, In European Conference on Computer Vision, June 2000.
11. A. Balan, L. Sigal and M. Black, A Quantitative Evaluation of Video-based 3D Person Tracking, IEEE Workshop on VS-PETS, 349-356, 2005.

Clustering Protein Interaction Data Through Chaotic Genetic Algorithm

Hongbiao Liu and Juan Liu*

School of Computer, Wuhan University, Wuhan 430079, China
liujuan@whu.edu.cn

Abstract. In this paper, we proposed a Chaotic Genetic Algorithm (CGA) to cluster protein interaction data to find protein complexes. Compared with other computation methods, the main advantage of this method is that it can find as many potential protein complexes as possible. Application on the Yeast genomic data highlights the efficiency of our method.

1 Introduction

Proteomic research is a focus of current study. [1-3] mention that cellular systems depend on multi-protein complexes in which individual proteins assemble into functional modules, so it is meaningful to find the known and forecast still unknown protein complexes within a cell's protein-protein interaction network. Gavin et al. have reported 257 novel protein complexes in the budding yeast through Affinity Purification-Mass Spectrometry this year [3].

Mostly, the common wet experimental methods to distinguish protein complexes include X-ray crystallography [4-6] and multidimensional nuclear magnetic resonance (NMR) [7-8]. But such methods are expensive and time-consuming. Different with wet experimental methods, the computational methods predict protein complexes by theoretical analysis thus can quickly and conveniently supply useful clues of complexes. For examples, King et al. used the Tabu Search algorithm [9]; Bade and Hogue [10] made use of identification of k-cores; Pruzlj [11] employed the graph theory to the study of highly connected subgraphs. Although they used different computational methods, there are some common ideas that will also be used in our work: one is that protein complexes are distinguished from protein-protein interaction (PPI) data, another one is that protein complexes generally correspond to dense subgraphs. Regarding the PPI data as a graph, where the vertices are proteins and the edge between two vertices represents the interaction of them, then above computational methods are in fact to find the dense subgraphs from the graph. Since proteins within the complex are generally interacted with each other, the denser the subgraph, the higher the probability it is. The densest subgraph, usually called as clique, is a complete subgraph. So finding protein complexes from PPI data is equal to finding maximal cliques from the graph. However, finding maximal cliques is NP-hard, usually one can only find near maximal cliques. Most of the existed computational methods have some

* Corresponding author.

T.-D. Wang et al. (Eds.): SEAL 2006, LNCS 4247, pp. 858 – 864, 2006.

drawbacks when solving this problem, such as the suggested near maximal cliques are far from maximal; proteins of different protein complexes are disjoint, which is not true in reality, and so on. So such methods usually can only find small number of potential complexes. In this paper, we will use a new stochastic method, called as Chaotic Genetic Algorithm (CGA), to find as many potential complexes as possible.

The chaos, a general phenomenon in non-linear system, has two special characters: acquiring all kinds of states in a self-rule in a certain range, and being sensitive to tiny change in initial condition. Based on above advantages, some chaos optimization algorithms can prevent local optimization and have a high efficiency. CGA, a combination of chaos optimization and Genetic Algorithm, is the inverse method of Simulated-Anneal Monte Carlo Genetic Algorithm. There are two differences between Simple Genetic Algorithm (SGA) and CGA: first, CGA uses chaotic variables in the initial process to increase the range of the initial populations as large as possible; second, CGA adds chaotic disturbance on the basis of simple genetic operation. In order to prevent good solutions of every generation from being disturbing, CGA puts the part of previous generation with highest fitness values into the next generation directly. Furthermore, we consider both clustering coefficient and effective length (the product of a graph's vertex length and clustering coefficient) to design the fitness function. When used to analyze Saccharomyces cerevisiae data, CGA suggest 149 protein complexes, some of them are still unknown up to now.

The rest of this paper is organized as follows: section 2 describes the CGA in details, section 3 is the application of CGA to the Yeast data, and the last section gives the conclusions and future works..

2 Chaotic Genetic Algorithm

2.1 Chromosome Representation

We use a binary string to describe an operation on the exist edge of a graph. The length of the chromosome is the number of the graph's original edges. Every edge's position in the chromosome is fixed. The '0' in our chromosome means that original edge remains in the new generation while '1' discards. For example, in Fig. 1., there are seven vertex (1-7) and eight edges (a-h). If we only want to discard one edge "d" from the original graph, we use the string "00010000" to means the operation on the graph and to get the partition detailed in figure 1(b) and this chromosome means a partition with two subgraphs {1,2,3} and {4,5,6,7}.

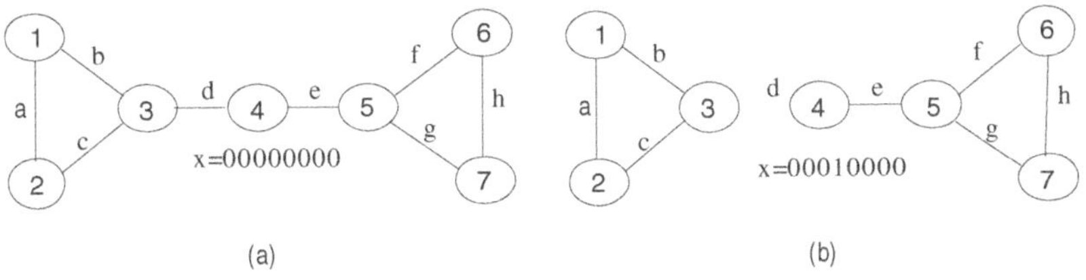

Fig. 1. A simple protein-protein interaction network (a) and its one partition (b)

2.2 Fitness Evaluation

We describe the fitness in the following way:

$$f^{s}(P) = \max g(G_i^{'}) \qquad i = 1,2,..l \tag{1}$$

$$f^{p}(P) = \max (\rho_i^{'}) \qquad i = 1,2,..l \tag{2}$$

$$g(G_i^{'}) = \begin{cases} |V_i^{'}|\rho_i^{'} & \text{if } \rho_i^{'} \geq \rho_T \text{ and } |V_i^{'}| > \lambda \\ 0, & otherwise \end{cases} \tag{3}$$

$$\rho_i^{'} = \begin{cases} \dfrac{|E_i^{'}|}{\dfrac{|V_i^{'}|(|V_i^{'}|-1)}{2}} & \text{if } |V_i^{'}| > \lambda \\ 0, & \text{otherwise} \end{cases} \tag{4}$$

Where P is one partition of the original graph G, and l is the number of connected subgraphs in the partition P, $G_i^{'}$ is a connected subgraph in the partition P, $V_i^{'}$ and $E_i^{'}$ are the vertex set and edge set of $G_i^{'}$, and $|V_i^{'}|$ is the length of $V_i^{'}$, and $|E_i^{'}|$ is the length of $E_i^{'}$, $\rho_i^{'}$ ρ_T and λ are empirical thresholds. In the formula (1), function $f^{s}(P)$ is used to calculate the maximal dot product (denoted as effective length) between vertex number and clustering coefficient. In the formula (2), $f^{p}(P)$ is designed for the maximal clustering coefficient.

At first we only use the formula (4) as the fitness function to evaluate the population, and we find we get many local optimal solutions. Next, we add the factor |V`| in the function. But the adding leads to another problem: it is difficult to tell the difference between a small complete graph with smaller |V`| and a big sparse subgrah with larger |V`|. Next, when we want to use |V`| and ρ to represent the uncompleted subgrah, we find that it is too discrete or too complicated to implement. Finally, we attach above two conditions to fitness function and get the formula (1-2). Furthermore, the clustering coefficient threshold ρ_T and vertex number threshold λ make the function more flexible and practicable. If we increase the value of ρ_T and λ in the program, the process will produce fewer local optimal solutions and converge more quickly. In addition, ρ_T and λ can also be used to filter some random highly dense subgraph and to amend the inaccuracy of experimental data.

2.3 Evolution of CGA

The steps of genetic evolution are listed as following:

Step 1: Set some parameters. The solution range is [0, 2^n], n, the length of the chromosome, is the number of edges of the input graph. The generation size is m=10*n. The maximal number of evolution generations is d=10*m. And the crossover probability is $P_1=P_2=0.9$, and mutation probability is $P_m=0.01$.

Step 2: Set the initial population. We set $\beta^{(0)}=0.2$ and calculate the value of $\beta^{(1)},\ldots,\beta^{(m)}$ according to formula $\beta^{(u+1)}=\gamma*\beta^{(u)}(1-\beta^{(u)})$ Where $\beta^{(u)}$ is a chaotic variable which corresponds to the uth chromosome in the population, u=1,…,m, $0\leq\beta^{(u)}\leq 1$, and the absorber γ is often set as 4.

Step 3: Compute the fitness values of the initial population. We code the first chromosome as the following formula $(x_1)_{10}=2*\beta^{(1)}$. Next, we change the decimal number $(x_1)_{10}$ into binary string which means a partition of the original graph. Then we calculate two fitness of the first chromosome according to formula (1-4). Similarly, we calculate fitness of every chromosome in the population.

Step 4: Produce the new population. We first choose 10% with longest effective lengths from the whole population to the next generation, and then we again choose 10% with largest cluster coefficients to the next generation. The remain 80% of the generation, has to undertake three operations: copy, crossover and mutation and to produce the 80% part of the new generation.

Step 5: Calculate the difference of fitness among the chromosomes in one generation according to the following formula.

$$\overline{f^s(X)} = \frac{1}{m}\sum_{j=1}^{m} f_j^s(x) \tag{5}$$

$$f^s(X)_{\max} = \max\{\ f_j^s(x),\quad j = 1,2,\ldots,\quad m\} \tag{6}$$

$$\left|\overline{f^s(X)} - f^s(X)_{\max}\right| < \varepsilon \tag{7}$$

Where, ε is a threshold. We use the formula (5) to calculate the average fitness of the generation, which is compared with the maximal fitness by the means of formula (6). If the difference between the maximal fitness and average fitness is smaller than the threshold ε as in formula (7), we go to the step 7, otherwise go to following step 6.

Step 6: Add the chaotic disturbance. Among the current generation, we keep the 10% with higher effective length fitness and 10% with higher clustering coefficient fitness, and we add the remaining 80% with chaotic disturbance. We add the chaotic disturbance through the following formula (8-9).

$$\alpha_k = 1 - \left[\frac{k-1}{k}\right]^m \tag{8}$$

$$\delta_k^{(u)'} = (1-\alpha_k)\delta_k^* + \alpha_k \delta_k^{(u)} \tag{9}$$

Where k is the No. of current iteratively computing generation, and m is the number of chromosomes. Where δ_k^* is the optimal chaotic vector of the kth generation and corresponds to the kth generation's optimal solution (x_k^*). $\delta_k^{(u)}$ is the chaotic vector of the uth chromosome in the generation after k times iteratively computing. $\delta_k^{(u)'}$ is the corresponding chaotic vector of $\delta_k^{(u)}$ which is modified by random disturbance. α_k is a self-adjusting variable, set by (8).

According to formula (9), we obtain the new chaotic variable $\delta_k^{'}$ corresponding to variable x_j. After inverse mapping $(x_j^{'})_{10} = 2^n * \delta_k^{'}$, we acquire the new value for the variable x_j. Similarly, we can compute other variable's new value. Then we re-compute the fitness of the generation, if the fitness matches the requirement of the formula (7), we go to next step 7, otherwise go to step 4.

Step 7: Output the solutions and adjust the parameters ρ_T and λ. Different with traditional GAs, we continually decrease the input space to reduce the computing scope to process the large dataset. It is a kind of "Greedy" method. First, through the above steps, if we obtain some solutions which have vertex number greater than λ and the clustering coefficient also greater than ρ_T, we should store the solutions for further process (Which will be explained in the following part.), and at the same time, we delete the edges of the solutions from the input graph and use the decreased graph as the input graph. Next, we decrease the vertex number threshold λ by one and also increase the clustering coefficient threshold by 5%. If the λ is near 3, we should stop our program, otherwise we go to Step 1.

2.4 Process on the Multiple Solutions

In the optimal solution population, some solution have same subgraphs, some have no common subgraph and other have partly identical. If one subgraph is a true subset of another one, we should delete the former smaller one. If one subgraph has no common vertex with another one, we should keep the two. If one subgraph has part vertex similar with the other one, for example, the solution A has part C same as the solution B. If $|C|/|A|>\rho_s$, we delete solution A, where |A| is the number of A's elements; if $|C|/|B|>\rho_s$, we delete the solution B but we do not delete the two solution A and B at the same time if $|C|/|A|>\rho_s$ and $|C|/|B|>\rho_s$. The constant ρ_s is a prune share rate threshold which variables in different requirements. In the next Section, we will discuss the choice of ρ_s.

3 Application on Yeast Genome

The experiment was done on our web server which consisted of two Pentium 4 PCs with 4.8 GHZ CPU and 2G RAM. The Saccharomyces cerevisiae proteomic data [13] has 5321 proteins and 78390 interactions. Since the computation space is very large, we used the greedy method to output the biggest clusters first and then removed its interaction edges from the input space. The parameters are detailed in Table 1. The last two rows of the Table 1 displays the numbers of found clusters before filtering by CGA and RNSC[9]. From Table 1, we see that CGA can find more clusters than RNSC as expected.

During our found clusters, the biggest ones have 66 proteins and the average size is 15.3. Table 2 listed the remain cluster number after filtering with different prune share rate thresholds. The 4240 clusters provide more cluster's clique detail information, while the smallest 149 clusters with lowest sharing rate 0.05 more preferentially partition the input graph into separate, distinct cliques.

We compared the 149 clusters with the MIPS [14] and found that 81 of 149 are matched in the MIPS, which shows that our method can find the known protein complexes. Although some of our results are not matched with MIPS, they are potential protein complexes which need more focus by researchers. More details about the result are listed on the following site: ftp://202.114.70.19/pub/protein/SuppCGAandProteinComplex.pdf

Table 1. CGA and RNSC parameters on Saccharomyces cerevisiae proteomic data [13]

Parameter	Value	Parameter	Value
CGA ε	0.001	CGA max. generation size	100
CGA P_m	0.01	CGA found total clusters (without filtering)	4240
CGA P1,P2	0.8	RNSC found clusters (without filtering)	1811
CGA max. iterate generations	5000		

Table 2. Different prune rate on the results

ρ_s	1.0	0.95	0.75	0.5	0.05
Cluster number	4240	2235	857	435	149

4 Conclusions and Future Works

In this paper, we present a new approach, CGA, to predict of the protein complexes. CGA introduce chaos into GA, thus can try more points in the search space in limited time. The application results show that our method can converge fast and can find more potential complexes than traditional methods.

Although we have successfully applied CGA to the proteomic data, the thresholds in the algorithm seems a little subjective. Moreover, the simple undirected graph model to describe the complicated and intricate relationship among molecular proteins seems a little fragile. We may solve the problems in the near future.

Acknowledgements

This paper is supported by the National Nature Science Foundation of China (60301009), Chenguang Project of Wuhan city (211121009).

References

1. Alberts,B.: The cell as a collection of protein machines: Preparing the next generation of molecular biologists. Cell 92 (1998) 291–294
2. Hartwell,L.H., Hopfield J.J., and Leibler S.,et al.: From molecular to modular cell biology, Nature 402 (1999) C47–C52.
3. Gavin A.C., Aloy, P., and Grandi, P., et al.: Proteome survey reveals modularity of the yeast cell machinery. Nature 440 (2006) 631-6
4. Dziembowski, A. and Seraphin, B.: Recent Developments in the analysis of protein complexes, FEBS Letters 556 (2004) 1-6
5. Bernal,J.D, Fankuchen I., and Perutz, M.F.: An X-Ray study of Chymotrypsin and Haemoglobin, Nature 141 (1938) 523-524
6. Drenth,J.: Principles of protein X-Ray crystallography. Springer Verlag, (1994)
7. Wuthrich,K.: NMR of proteins and nucleic acids. New York: John Wiley and Sons, (1986)
8. Wand, A. and Englander, S.: Protein complexes studied by NMR spectroscopy, Current Opinion in Biotechnology 7 (1996) 403-408
9. King,A.D., Przulj,N. and Jurisica, I.: Protein complex prediction via cost-based clustering, Bioinformatics 20 (2004) 3013-3020
10. Bader,G. & Hogue, C.: An automated method for finding molecular complexes in large protein interaction networks. BMC Bioinformatics 4 (2003)
11. Prˇzulj,N.: Graph theory approaches to protein interaction data analysis. In Jurisica,I., Wigle,D.(eds): Knowledge Discovery in High-Throughput Biological Domains, Interpharm/CRC. (2004)
12. Li Zicai, Zhang Dan and Wang Hong: Simulated optimization method based on chaotic vector. Control and Decision 14 (1999) 382~384
13. von Mering,C., Kraus,R., and Snel,B., et al.: Comparative assessment of large-scale data sets of protein-protein interactions, Nature 417 (2002) 399-403
14. Mewes,H.W., Frishman ., and Guldener,U., et al.: Mips: a database for genomes and protein sequences. Nucleic Acids Research30 (2002) 31-4

Multi-objective Q-bit Coding Genetic Algorithm for Hardware-Software Co-synthesis of Embedded Systems*

Wei Wen-long, Li Bin**, Zou Yi, and Zhuang Zhen-quan

Nature Inspired Computation and Applications Laboratory,
University of Science and Technology of China, Hefei, 230026, China
wwlong@mail.ustc.edu.cn, binli@ustc.edu.cn,
rocky@szonline.net.cn, zqzhuang@ustc.edu.cn

Abstract. One of the key tasks in Hardware-Software Co-design is to optimally allocate, assign, and schedule resources to achieve a good balance among performance, cost, power consumption, etc. So it's a typical multi-objective optimization problem. In this paper, a Multi-objective Q-bit coding genetic algorithm (MoQGA) is proposed to solve HW-SW co-synthesis problem in HW-SW co-design of embedded systems. The algorithm utilizes the Q-bit probability representation to model the promising area of solution space, uses multiple Q-bit models to perform search in a parallel manner, uses modified Q-bit updating strategy and quantum crossover operator to implement the efficient global search, uses an archive to preserve and select pareto optima, uses Timed Task Graph to describe the system functions, introduces multi-PRI scheduling strategy and PE slot-filling strategy to improve the time performance of system. Experimental results show that the proposed algorithm can solve the multi-objective co-synthesis problem effectively and efficiently.

1 Introduction

With the development of IC technology, the traditional design methods that design hardware and software separately and mainly depend on the experience of designers cannot meet the design requirement of modern embedded systems any more. So, a novel method named hardware and software co-design is developed [1]. HW-SW co-design uses automatic and optimized methods to help the designer process rapid prototype development and estimate the performances of the system at a high-level stage, thus to achieve a good balance among performance, cost, power consumption, etc. According to the granularity of the design, the methods of HW-SW co-design can be divided into two kinds[1]: one is called as HW-SW partitioning, which describes the system functions with Control Date Flow Graph(CDFG)[2]and its granularity is BSB(Basic Scheduling Block)level. Another one is called HW-SW co-synthesis, which uses the Task Graph(TG) to describe the system functions and its granularity is task level.

* Supported by the National Natural Science Foundation of China (No.60401015, No. 60572012), and the Nature Science Foundation of Anhui province (No. 050420201).

** Corresponding author.

T.-D. Wang et al. (Eds.): SEAL 2006, LNCS 4247, pp. 865 – 872, 2006.

At present, typical algorithms of HW-SW co-synthesis for distributed heterogeneous multi-processor system can be classified into three kinds: the first one is heuristic multi-processor scheduling, such as [3]. These algorithms schedule the assigned tasks from the standpoint of parallel computation. The second one is linear programming or integer programming [4]. Since denoting the objective function and restriction is so difficult, and the computational complexity of algorithms increases exponentially with the scale of problems, it's hard to use this kind of algorithm to solve large-scale problems. The third one is based on genetic algorithms [5]. This kind of algorithms have good performances due to the high exploration capacity of genetic algorithms, however, they still can be improved. Furthermore, there are several other algorithms [6].

HW-SW co-synthesis can also be treated as a kind of multi-objective optimization problem. In most cases the objectives (performance, cost, power consumption, etc.) are in conflict with each other and there is no single solution that is optimal for all objectives. In this situation, we can only get so called Pareto optimal or non-dominated solutions.

From early 1990s several Multi-Objective Evolutionary Algorithms(MOEAs) were proposed[7]~[11]. Of them, Knowles et. al's PAES [8], T-Ray et. al.'s. Ray-Tai-Seow's algorithm [9], Deb et. al's NSGA-II [10] enjoyed more attention. Carlos A. Coello et al. also proposed a multi-objective particle swarm algorithm (MOPSO[11]).

Q-bit coding evolutionary algorithm was first proposed by Han et al. as GQA[12], and later extended to QIEA[13]. Li et al. proposed a pair-wise quantum crossover to substitute the migration operation in QIEA to improve global search in GAQPR[14]. GAQPR has been used to solve various single objective optimization problems[14]. The experimental results show that the algorithm has a characteristic of keeping good balance between the convergence and diversity. This inspired us to adapt it to multi-objective optimization problems.

In this paper, a Multi-objective Q-bit coding Genetic Algorithm(MoQGA) for HW-SW co-synthesis is proposed. The algorithm is testified on some hypothetical HW-SW co-synthesis problems described with Timed Task Graph (TTG), experimental results show that the proposed algorithm is effective and efficient in solving multi-object HW-SW co-synthesis problem.

The remainder of this paper is organized as follows: section 2 focuses on the model and process for HW-SW co-synthesis. Section 3 proposes a MoQGA for HW-SW co-synthesis. Section 4 provides the experiment result with analysis. Finally, section 5 concludes the paper.

2 Model for HW-SW Co-synthesis

We describe the system functions with Timed Task Graph(TTG)(shown as Fig.1 and Fig.2), which is a directed acyclic graph in which each node is associated with a task. Task graph can be denoted as G = (T, E), where Ti denotes a task. The task set T= {T0,T1,...,Tn} denotes all functions of the system abstractly. The directed edge set E = {(Ti,Tj)|Ti,Tj∈T}denotes data communication between tasks Ti and Tj.

For co-synthesis problem, a task node can be denoted as Ti = {tID, tType, pNum, nNum, sTime, fTime,deadline}, Where tID denotes the task ID, tType denotes the task type, pNum and nNum denote the number of a task's predecessor nodes and successor nodes, sTime and fTime denote the start time and the end time of the task, and deadline means the time by which the task associated with the node must complete its execution.

System resource contains Processor Elements(PEs) and communication resources. A PE can be denoted as P = (C , W ,T), where C denotes the cost, W denotes the power consumption, and T = {t1,t2,...t3}denotes the executing time of each type of task which executes on it.

Communication resources include the system bus that links the PEs and independent linkage. A physical linkage can be denoted as (Ti,Tj) =(Q, C, W), where Q denotes the amount of data transferring on it, C denotes the cost and W denotes the power consumption.

The total cost of system can be denoted as :

$$\text{Cost(S)}= \sum p_i(c)+\sum b_i(c) \ . \tag{1}$$

The first part of equation (1) denotes the cost of PEs and the second part denotes the cost of communications. The power consumption of system can be denoted as fellows:

$$\text{Power(S)}= \sum p_i(w)+\sum b_i(w) \ . \tag{2}$$

The first part of equation (2) denotes the power consumption of PEs and the second part denotes the power consumption of communications.

The HW-SW co-synthesis is a multi-objective optimization problem and can be depicted as follows:

$$\textit{object functions}\begin{cases}\text{Min} & \text{Cost(S)} \\ \text{Min} & \text{Power(S)}\end{cases} . \tag{3}$$

$$\textit{restriction} \quad \text{Time(S)} \le \text{MAX}^{\text{T}}(S) \ . \tag{4}$$

Where Time(S) is the time by which all the functions of system are realized.

3 MoQGA for HW-SW Co-synthesis

3.1 The Procedure of MoQGA

Procedure MoQGA

begin

t←0

1. generate N individuals represented by Q-bit probability amplitude to form the initial population $Q(t)$;
2. sample every individual K times and get K deterministic solutions; if there is any solutions among the K that don't violate any constraint, select a Pareto

optimal solution from such solutions as the individual's current evolution target; if none of the K solutions satisfies all constraints, pick the one with the lowest constraint violation degree value as the current target of the individual;

3. while (not termination-condition) do
   begin
   a)t←t+1;
   b) For each individual in $Q(t-1)$, i.e. q_j^{t-1}, sample it once and get one deterministic solution;
   c) Compare the sampled solution with the current target of q_j^{t-1}, if the sampled solution dominates its current target, replace the current target with the sampled solution; if the sampled solution is dominated by its cur. Age rent target, keep the current target unchanged; otherwise, select one randomly from the two as the individual's next evolutionary target.
   d) update individual according to its new target;
   e) Apply quantum crossover operator to the updated population and archive;
   f) if (re-initialization - condition)
      apply re-initialization strategy;
   end

end

For detailed description of Q-bit Coding, Initialization, Updating strategy of Individuals, and Crossover operator, please refer to literature [14].

Different from [14], the crossover operator in step e) can also be performed on two individuals, one from the population and one from the Archive. The intuition behind is to evolve the individuals towards better solutions to accelerate convergence.

The re-initialization operator in step f) is defined as follows: select some individuals randomly from the population after *m* generations, set them back to the initial state, then the evolution process of the new individuals starts afresh.

3.2 The Archive and Constraint Management

By setting archive, Pareto optimal solutions generated through the evolution process can be stored and selected. For a solution that doesn't break any constraint, add the solution to archive if there is no solution in the archive dominating it and delete the solutions it dominates from the archive.

The maximum size of the archive can be set in advance. When the number of solutions in the archive exceeds pre-set size, a "division" operation is performed. In this paper, liner division method is adopted as follows: for each objective, the minimum and the maximum value are found from the solutions in the archive and used as the lower and upper bound along the dimension respectively, then divide the space into *n* sub-space uniformly. Select the sub-space containing the most solutions and delete one solution from it randomly. The purpose of doing so is to maintain the diversity of the solutions.

The concept violation degree R[10] is adopted to deal with constraints. Suppose there are n constraints: $g_1(x)\le 0, g_2(x)\le 0,\cdots g_n(x)\le 0$, $R_i=0$ if and only if $g_i(A)\le 0$ holds for i=1,2,$\cdots$,n; $R_i=g_i(A)$ if $g_i(A)>0$. The violation degree R of individual A can be calculated using $R = \sum_{i=1}^{n} R_i$. According to the individual's violation degree R, individuals that don't violate any constraint dominate those that violate some constraints; individuals with smaller violation degree value dominate those with greater violation degree value.

3.3 Multi-PRI Scheduling and PE Slot-Filling Strategy

For a certain solution that associated with a scheme of assignment, the scheduling gives an optimal Time(S) of the scheme. A node whose predecessor nodes are all complete scheduled is called a ready node. When two or more ready nodes are assigned to the same PE, the competition happens, which complies with the Multi-PRI shown as follows(ignore the deadline of node):

1. *Instancy* of node, $Instancy = MLFtime - MEFtime$, where MEFtime is the earliest finish time and MLFtime is the latest finish time of node.
2. *Stime* of node, the start time of node.
3. *Ftime* of node, the earliest finish time of node.
4. *Rand* priority.

The above items are priority-ranked. *Instancy* of node well describes the position information in the TG of the node and it's reasonable to be served as the first priority. The priority of *Stime* is to decrease the waiting time of PE. The priority of *Ftime* assures PE available early for another task to use. When the above priorities are same, take the random priority into account.

In order to enhance the efficiency of PE, a PE slot-filling strategy is introduced as follows: in view of communication delay, the task that competes successfully will wait for a moment before execution and the PE is available synchronously. If the idle time can satisfy another ready task, let the ready task executed firstly.

4 Experimental Results and Analysis

To testify the performance of MoQGA, we use it to solve the four nodes and nine nodes tasks introduced in [4] firstly.

Four-node system is shown as in Fig.1. In this system, a task doesn't require all the inputs before starting its execution and it may produce some outputs even before completion. To express this possibility, each input $i_{a,b}$ has a parameter $f_R(i_{a,b})$associated with it which specifies that up to $f_R(i_{a,b})$fraction of the task can proceed with requiring the input $i_{a,b}$. Similarly, each output $o_{a,b}$ has a parameter $f_A(o_{a,b})$associated with it which specifies that the output $o_{a,b}$ becomes available when $f_A(o_{a,b})$fraction of the task is completed. The available PE source of system is shown in Table 1. In nine-node system (Fig.2), a task would require all the inputs before starting its execution and none of the output would be available until the execution is over. The available PE source of system is shown in Table 2. In both four-node system and nine-node system, the kind of communication is point-to-point. The communication volume and the communication delay are both a unit and the cost of communication linkage is a unit too.

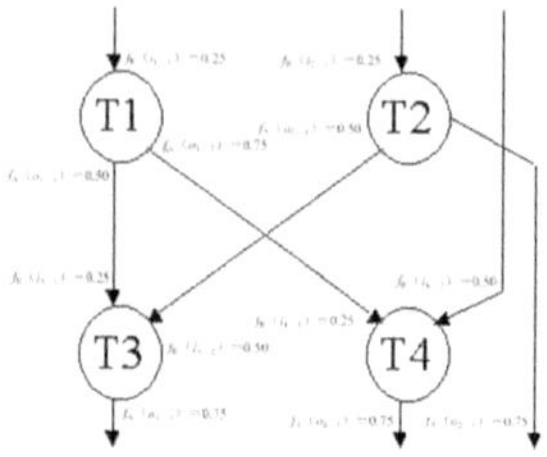

Fig. 1. TG of four-node system

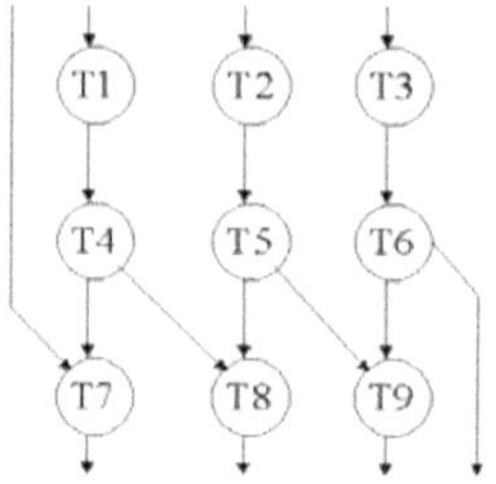

Fig. 2. TG of nine-node system

Table 1. PE sources of four-node system

PE	Cost	Execution time			
		T1	T2	T3	T4
P1	4	1	1	—	3
P2	5	3	1	2	1
P3	2	—	3	1	—

Table 2. PE sources of nine-node system

PE	Cost	Execution time								
		T1	T2	T3	T4	T5	T6	T7	T8	T9
P1	4	2	2	1	1	1	1	3	—	1
P2	5	3	1	1	3	1	2	1	2	1
P3	2	1	1	2	—	3	1	4	1	3

Table 3. Results of multi-object optimization of four and nine-node system(Point—toPoint)

Task	Design		MOGAC[5]			MoQGA		
	No.	Restriction	Cost	Power consumption	Execution time(s)	Cost	Power consumption	Execution time(s)
4 NODES	1	2.5	-	-	-	14	139.705	0.0840
	2	3	-	-	-	13 14	152.140 139.705	0.0930
	3	4	7 15	75.4 64.2	20.1	7 8 10 11 14	314.90 276.90 214.35 141.60 139.705	0.1050
	4	7	5 7 10	44.4 35.1 21.5	16.9	5 11 14	175.00 141.60 139.705	0.1060
9 NODES	1	8	7 12	49.8 40.0	17.2	7 8 9 11 14	431.10 371.33 355.20 323.85 311.13	3.4983
	2	15	5 7 12	48.0 26.8 21.8	22.4	5 8 9 11 14	375.00 371.33 355.20 323.85 311.13	3.6027

The algorithm is implemented in programming language C, and the results are obtained on the computer of 2.0G Pentium Pro CPU with 256MB of main memory. For four-node system, N=20, K=8, tuning pace of rotation angle is 0.08π, crossover adjustment angle is 0.04π, m=200, crossover probability is 0.2, re-initialization probability is 0.5 and the termination number of generations is 100. For nine-node system, N=100, the termination number of generations is 1000, and other parameters are the same as that for four-node system. The results of multi-object optimization of four and nine-node system is shown in Table 3.

For parallel multi-task system, we use the examples introduced by Junwei Hou & Wayne Wolf in [6]. In [6], the authors perform clustering on TG before scheduling (see detail in [6]). The PE source is listed in Table 4. In this experiment, the cost of a communication linkage is 20 and unit transmit delay is 0.8, N=80, the termination number of generations is 2000, and other parameters are the same to four-node system.

The working power-consumption values of P1, P2, P3 are 34.5, 25.0, 44.2 and the idle power-consumption values of P1, P2, P3 are 2.3, 1.25, 4.42. The results are listed in Table 3 and Table 5, and are compared with that of MOGAC[5]. For four and nine nodes system, our results are the same as the optimal results that are obtained by enumerating. From the comparison results, we can conclude that our algorithm can find the optimal solutions for each system and have highly efficiency.

Table 4. PE sources of parallel multi-task system

PE	Cost	Execution time									
		A	B	C	D	E	F	G	H	I	J
P1	100	5	10	5	35	15	30	15	15	7	10
P2	50	12	18	12	85	22	75	25	35	10	28
P3	20	18	40	18	195	80	180	85	47	30	35

Table 5. Results of multi-object optimization of parallel multi-task system(Point–to–Point)

Task	MOGAC[5]			MoQGA		
	Cost	Power consumption	Execution time(s)	Cost	Power consumption	Execution time(s)
Hou1&2(uc)	170	51.8	89.6	170 190 290	50.7583 48.6583 48.3333	14.6558
Hou1&2(c)	170	62.5	9.5	170 200	58.2626 53.1668	5.9850
Hou3&4(uc)	170	48.6	26.3	170 190 210 230 250	50.2105 49.4947 48.8789 48.1447 47.9158	22.0810
Hou3&4(c)	170	43.3	5.1	150 170 200	68.9092 58.5606 50.3686	7.0418

5 Conclusions

The characteristic of Q-bit coding Genetic Algorithm that can keep good balance between convergence and diversity inspired us to adapt it to solve multi-objective HW-SW co-synthesis problems. A archive and associative operators are used to guarantee the quality of the final solution, Timed Task Graph is adopted to describe the system functions, a multi-PRI scheduling strategy and PE slot-filling strategy are introduced to improve the time performance of system. Experiment results show that the algorithm is effective and efficient in solving multi-objective HW-SW co-synthesis problem.

References

1. Ernest R.Codesign of embedded systems: status and trends[J].IEEE Design&Test of Computers,1998,45-54
2. Gupta R K,Micheli G De.System synthesis via hardware-software co-design[R]. Technical Report CSL-TR-92-548,Computer Systems Labroatory, Stanford University,1992-10
3. Yu-Kwong Kwok,Ishfaq Ahmad. Dynamic Critical-Path Scheduling:A Effective Technique for Allocating Task Graphs to Multiprocessors. IEEE Transactions on Parallel and Distributed Systems, May 1996(Vol.7,No.5)
4. S. Prakash and A. Parker. Synthesis of application-specific heterogeneous multi-processor systems. J. Parallel&Distributed Computers,vol.16,pp.338-351,Dec,1992
5. Dick R.P, Jha N.K. MOGAC: a multi-objective genetic algorithm for hardware-software co-synthesis of distributed embedded systems. Computer-Aided Design of Integrated Circuits and Systems, IEEE Transactions on, Volume: 17 , Issue: 10 , Oct. 1998
6. Junwei Hou, Wolf. W. Process Partitioning for Distributed Embedded Systems. IEEE Hardware/Software Co-Design, 1996. (Codes/CASHE '96), Proceedings. Fourth International Workshop on , 18-20 March 1996
7. Srinivas N and Deb Kalyanmoy. Multi-objective optimization using non-dominated sorting in Genetic algorithms. Evolutionary Computation, 1994, 2(3):221-248
8. Joshua D. Knowles and David W. Corne. Approximating the Nondominated Front Using the Pareto Archived Evolution Strategy. Evolutionary Computation, 8(2):149-172, 2000.
9. T.Ray, K.Tai, and C.Seow. An evolutionary algorithm for multi-objective optimization, Eng. Optim., vol.33, no.3, pp.399-424, 2001
10. Kalyanmoy Deb, Amrit Pratap, Sameeer Agarwal and T.Meyarivan. A Fast and Elitist Multi- objective Genetic Algorithm: NSGA- II ,IEEE Transaction On Evolutionary Computation, 2002
11. Carlos A.Coello Coello and Maximino Salazar Lechuga. MOPSO: A Proposal for Multiobjective Particle Swarm Optimization. Evolutionary Computation, CEC'02.Proceedings of the 2002 Congress on Volume2, 12-17 May 2002 Page(s): 1051-1056
12. Han Kuk-Hyun, Kim Jong-Hwan. Genetic Quantum Algorithm and its Application to Combinatorial Optimization Problem[A].Proceeding of the 2000 IEEE Congress on Evolutionary Computation [C].2000,2:1354-1360
13. Kuk-Hyun Han, Jong-Hwan Kim, Quantum-inspired evolutionary algorithm for a class of combinatorial optimization, IEEE Transactions on Evolutionary Computation, Dec. 2002, Vol. 6, Issue: 6, Page(s): 580 –593
14. Li Bin,et al.Genetic Algorithm Based on the Quantum Probability Representation[R]. Yin H, et al.(Eds.),Lecture Notes in Computer Science(LNCS2412),2002-08.500-505

Particle Swarm Optimization for Simultaneous Optimization of Design and Machining Tolerances*

Chi Zhou, Liang Gao, Hai-Bing Gao, and Kun Zan

Department of Industrial & Manufacturing System Engineering
Huazhong Univ. of Sci. & Tech., Wuhan, 430074, China
gaoliang@mail.hust.edu.cn

Abstract. Tolerance assignment is an important issue in product design and manufacturing. However, this problem is commonly formulated as nonlinear, multi-variable and high constrained model. Most of the heuristics for this problem are based on penalty function strategy which unfortunately suffers from inherent drawbacks. To overcome these drawbacks, this paper presented a new powerful tool-Particle Swarm Optimization algorithm (PSO) and meanwhile proposed a sophisticated constraints handling scheme suitable for the optimization mechanism of PSO. An example involving simultaneously assigning both design and machining tolerances based on optimum total machining cost is employed to demonstrate the efficiency and effectiveness of the proposed approach. The experimental result based on the comparison between PSO and GA show that the new PSO model is a powerful tool.

1 Introduction

Tolerance assignment in product design and process planning (machining) affects both the quality and the cost of the overall product cycle. It is a crucial issue to determine how much the tolerance should be relaxed during the assignment process, since a tight tolerance implies a high manufacturing cost and a loose tolerance results in low manufacturing cost. Hence, during tolerance assignment, a balance between a reduction in quality loss and a reduction in manufacturing cost must be considered. Traditionally, in the two stages (product design and process planning) tolerances [1] are often conducted separately. This is probably due to the fact that they deal with different type of tolerances. Product design is concerned with related component tolerances, whereas process designing focus on the process tolerance according to the process specification. However, this separated approach in tolerance design always suffers from several drawbacks. Therefore, we need to develop a simultaneous tolerance design. Singh [2] utilized genetic algorithms and penalty function approach to solve the problem of simultaneous selection of design and manufacturing tolerances based on the minimization of the total manufacturing cost. Gao and Huang [3] utilized a nonlinear programming model for optimal process tolerance simultaneously based on the objective of total manufacturing cost with different weighting factors. Huang,

* This paper is supported by the National Basic Research Program of China (973 Program), No.2004CB719405 and the National Natural Science Foundation of China, No. 50305008.

T.-D. Wang et al. (Eds.): SEAL 2006, LNCS 4247, pp. 873–880, 2006.

Zhong and Xu [4] proposed a robust optimum tolerance design method in a concurrent environment to balance the conflict design targets between manufacturing tolerances and product satisfaction.

Doubtlessly, the tremendous achievement has been obtained in the simultaneous tolerance optimization in the concurrent engineering context. However, this problem is characterized by nonlinear objective, multiple independent variables. Traditional operational research algorithms are successful in locating the optimal solution, but they are usually problem dependent and lack of generality. Some modern heuristic methods are relatively more robust and flexible to solve these complex problems, but they may risk being trapped to a local optimum and are usually slow in convergence and require heavy computational cost. In view of the above problems and the past successful applications of PSO in nonlinear optimization, maybe PSO is a potential remedy to these drawbacks. PSO is a novel population based heuristic, which utilizes the swarm intelligence generated by the cooperation and competition between the particles in a swarm [5][6]. Noorul Hap, et al [7] utilized PSO to achieve the multiple objective of minimum quality loss function and manufacturing cost for the machining tolerance allocation of the over running clutch assembly. The presented method outperforms other methods such as GP and GA, but it considered only two dimensional tolerance allocation of clutch assembly consisting of three components. This paper attempts to solve more complex tolerance assignment problems by PSO with a sophisticated constraints handling strategy.

This paper is organized as follows. In section 2, the problem of simultaneous design was described. The basic PSO algorithm was reviewed and the new sophisticated constraints handling strategy corresponding to PSO was presented in Section 3. Section 4 gave an example and the evaluation of the proposed technique is carried out on the example. Some conclusions and further discussion are offered in Section 5.

2 Simultaneous Design

As mentioned before, design processes are commonly divided into two main stages: product design and process design. Dimensional tolerance analysis is very important in both product and process design. In product design stage, the functional and assembly tolerances should be appropriately distributed among the constituent dimensions, this kind of tolerances are called design tolerances. In the meantime, each design tolerance for the single dimension should be subsequently refined to satisfy the requirement for process plans in machining a part. Such tolerances for the specified machining operation are called manufacturing tolerance. However, the traditional process of design and machining tolerance allocations based on experiences can not guarantee optimum tolerance for minimum production cost. This work aimed at selecting the optimal tolerances sequences to achieve the minimum manufacturing cost considering the two types of tolerances simultaneously by a powerful global optimization tool. This problem is formulated as follows.

2.1 Objective Function

We take the manufacturing cost as the objective function. Generally, the processing of mechanical product is conducted in a series of process plans. Different process consumes different expense because different process is associated with different machining method. Therefore, the cost of manufacture of the product is the summation of all operation cost. In this work, a modified form of the exponential cost [2] function will be adopted. The manufacturing cost of the machining tolerance is formulated as equation (1).

$$c_{ij}(\delta_{ij}) = a_0 e^{-a_1(\delta - a_2)} + a_3 \quad i = 1, \cdots, n \tag{1}$$

The total manufacturing cost of a product will be C, where:

$$C = \sum_{i=1}^{n} \sum_{j=1}^{m_i} c_{ij} \tag{2}$$

Where $c_{ij}(\delta_{ij})$ and δ_{ij} is the manufacturing cost and the tolerance of the jth manufacturing operation associated with the ith dimension respectively. n is the number of the dimensions and m_i is number of operations corresponding to dimension i. The constants a_0, a_1, a_2, a_3 sever as control parameters.

2.2 Constraints

Apart from the constraint of economical manufacturing ranges (process limits), the above objective is subjected to both the design and manufacturing tolerances.

(1) The design tolerances are those on the principal design dimensions (usually assembly dimensions) that relate to the functionality of the components. The principal design usually in turn relies on the other related dimensions which form a dimension chain. This results in a set of constraints on the principal design tolerances that should be suit for the optimal solution of the tolerance assignment. There are many approaches available to formulate the synthesized tolerance. They are different tradeoff between the tolerances and the manufacturing cost. Four commonly used approaches [2] were adopted in this work.

(2) Manufacturing tolerances constraints are equivalent to stock allowance constraints. Stock allowance is associated with the stock removal, the layer to be removed from the surface in the machining process. Due to the tolerances of the dimensions, the stock removal is also not fixed. This gives rise to another kind of tolerances, manufacturing tolerances, which can be formulated as follows:

$$\delta_{ij} + \delta_{i(j-1)} \le \Delta A_{ij} \tag{3}$$

where δ_{ij} and $\delta_{i(j-1)}$ are the machining tolerances of process j and j-1 for part i respectively. ΔA_{ij} is the difference between the nominal and the minimum machining allowances for machining process j.

3 Particle Swarm Optimization

3.1 Background

The investigation and analysis on the biologic colony demonstrated that intelligence generated from complex activities such as cooperation and competition among individuals can provide efficient solutions for specific optimization problems [8]. Inspired by the social behavior of animals such as fish schooling and bird flocking, Kennedy and Eberhart designed the Particle Swarm Optimization (PSO) in 1995 [9].

This method is a kind of evolutionary computing technology based on swarm intelligence. The basic idea of bird flocking can be depicted as follows: In a bird colony, each bird looks for its own food and in the meantime they cooperate with each other by sharing information among them. Therefore, each bird will explore next promising area by its own experience and experience from the others. Due to these attractive characteristics, i.e. memory and cooperation, PSO is widely applied in many research area and real-world engineering fields as a powerful optimization tool.

3.2 Drawbacks of Traditional Constraints Handling Strategy

Although PSO has successfully solved many research problems, the applications are mainly focused on unconstrained optimization problems. Some researchers attempt to solve the constrained problem by optimizing constrained problems indirectly using the traditional penalty function strategy.

Penalty function is an effective auxiliary tool to deal with simple constrained problems and has been the most popular approach because of their simplicity and ease of implementation. Nevertheless, since the penalty function approach is generic, their performance is not always satisfactory. When combined with PSO, the above problem is more severe in that PSO has an inherent mechanism based on memory information. This mechanism can produce high efficiency and effectiveness, but also low the flexibility for constrained optimization simultaneously. It is desirable to design a new constraint handling scheme suit for PSO to effectively solve numerous engineering problems and maintain high efficiency.

3.3 Constraints Handling Strategy for PSO

Taking account of the memory mechanism of PSO and penalty strategy, a new constraint-handling strategy is presented in Figure.1.

The core characteristics of the proposed strategy can be described as follows:

(1) Corresponding to the memory mechanism of PSO, a special notation-Particle has been Feasible (PF) is introduced, which is used to record whether the current particle has ever satisfied all the constraint conditions. This notation preserves historical constrain status for each particle.
(2) Each particle updates its individual best and neighborhood best according to the historical constraint information PF, the current constrain status (Current particle is Feasible, CF) and the objective function with the penalty term.

(3) The algorithm selects the velocity updating strategy according to the historical information PF.
(4) When updating the personal and neighborhood best, the algorithm adopts the static penalty strategy instead of the dynamic and the adaptive ones to guarantee the fairness. The detailed procedure for updating the personal and neighborhood best values based on the above constrain handling strategy is presented in Figure.1.

```
For Each Particle {
  If PF=true Then
    If f(x_i) ≤ f(p_i) and CF= true Then
      p_i = x_i
      If f(p_i) ≤ f(l_i) Then
        p_i = l_i
      End if
    End if
  Else if PF=false Then
    If CF=true Then
      p_i = x_i
      PF=true
      If f(p_i) ≤ f(l_i) Then
        p_i = l_i
      End if
    Else if f(x_i) ≤ f(p_i) Then
      p_i = x_i
    End if
  End if
```

Fig. 1. The proposed constraint handling strategy for PSO

Special attention should be paid that the PSO algorithm based on the proposed constraint handling strategy does not have to guarantee the existence of feasible solutions in the initial population. With the randomized initial velocity, the PSO itself has the ability to explore the feasible space. In addition, the penalty function imposed on the violated particles also direct the search of PSO towards the feasible region. According to the velocity updating formula, each particle will obtain updating information from its neighborhood best particle, so the corresponding particle would return to the feasible solution space immediately.

4 Design Example

To validate the effectiveness of the new proposed strategy and illustrate the application of the concurrent design, the cylinder-piston assembly [2] is described. In this example, the piston diameter is 50.8mm, the cylinder bore diameter is 50.856mm, and the clearance is 0.056±0.025 mm. The ranges of the principal machining tolerances for the piston and cylinder bore were the same as in the [2].

In this problem, we have to consider totally 10 tolerances as follows. (1)The design tolerance parameters: δ_{11d} for the piston and δ_{21d} for the cylinder bore. Four stack-up conditions [2] (worst case, RSS, Spotts' modified method and estimated mean shift criteria) are employed to formulate the corresponding constraints. (2)The machining tolerance parameters are: δ_{ij} where i=1,2 and j=1,2,3,4. Usually, the process tolerance for the final finishing operation is same as the design tolerance, i.e. $\delta_{11d} = \delta_{14}$ and $\delta_{12d} = \delta_{24}$. The machining tolerance constraints are formulated based on Equation 3. The manufacturing decision is the total machining cost and is determined by summing the machining cost-tolerance model as Equation 1 and Equation 2 subjecting to the constraints and ranges of the principal design and machining tolerances. The constant parameters are the same as in [2].

The proposed PSO algorithm with special constraints handling strategy was used to solve this problem. To validate its efficiency, this new approach was compared with GA in [2]. In the optimization process of HPSO, we set the population size *pop-size*=80, the maximum iteration number *itermax*=600. These two parameters are the same as those in GA. The other parameters are set as the common used method. The inertial weight decreases from 0.9 to 0.4 linearly and the accelerated parameters c_1=c_2=2.

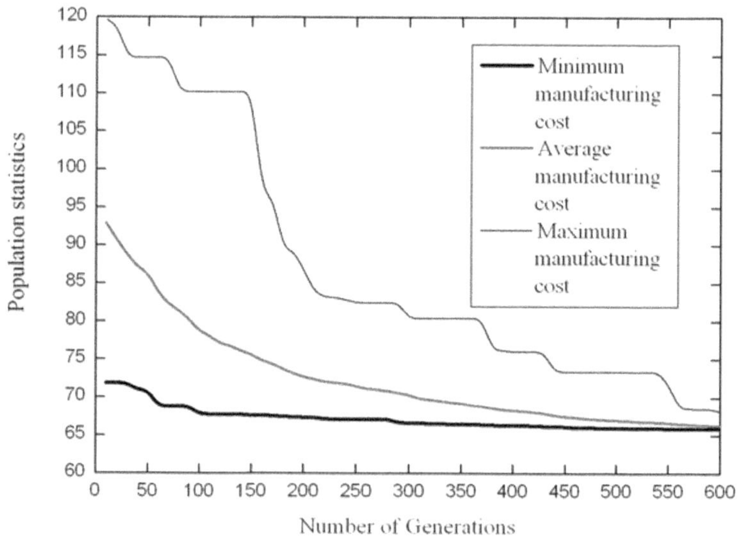

Fig. 2. Variation of the minimum, maximum and average of the manufacturing costs with progress of the algorithm (Greenwood and Chase's unified, or estimated mean shift criteria)

Table 1. Optimal tolerances allocation using GA and PSO

(a) Based on the worst case criteria

GA				PSO					
Piston	Cylinder	Cost	Time (s)	Piston	Cylinder	Min	Ave	Max	Time(s)
0.0162	0.0162	66.85	350	0.0163	0.0163	66.74	66.74	66.74	83
0.0037	0.0038			0.0037	0.0037				
0.0013	0.0012			0.0013	0.0013				
0.0005	0.0005			0.0005	0.0005				

(b) Based on the worst RSS criteria

GA				PSO					
Piston	Cylinder	Cost	Time (s)	Piston	Cylinder	Min	Ave	Max	Time(s)
0.0161	0.0161	65.92	330	0.0161	0.0162	66.82	66.82	66.82	80
0.0039	0.0038			0.0039	0.0038				
0.0011	0.0012			0.0011	0.0012				
0.0007	0.0006			0.0007	0.0006				

(c) Based on the worst Spotts' criteria

GA				PSO					
Piston	Cylinder	Cost	Time (s)	Piston	Cylinder	Min	Ave	Max	Time(s)
0.0160	0.0159	66.23	330	0.0162	0.0162	65.93	65.93	65.93	78
0.0038	0.0038			0.0038	0.0038				
0.0012	0.0012			0.0012	0.0012				
0.0006	0.0005			0.0006	0.0006				

(d) Based on the worst mean shift or Greenwood and Chase's unified criteria

GA				PSO					
Piston	Cylinder	Cost	Time (s)	Piston	Cylinder	Min	Ave	Max	Time(s)
0.0162	0.0151	66.26	350	0.0161	0.0162	65.82	65.82	65.82	82
0.0037	0.0038			0.0039	0.0038				
0.0012	0.0011			0.0011	0.0012				
0.0006	0.0006			0.0006	0.0006				

The optimal tolerance allocated using HPSO and GA based on the above four criteria and the corresponding CPU time are listed in Table 1. The computational results clearly indicate that HPSO outperformed GA in the terms of solution quality as well as computational expense. In addition, HPSO is able to find the optimum in each trial. It is necessary to point out that one important merit of PSO algorithm is the high precision of the solutions. However, due to the limitation of display capacity of the tables, the entire data are rounded.

The statistical results obtained under the Greenwood and Chase's estimated mean shift criteria are demonstrated in Figure.2. Similar curves can be obtained for other cases. Figure.2 reflects the general behavior about convergence of PSO algorithm. Sharply contrast with GA, the PSO algorithm has consistent convergence. The average and worst fitness are not fluctuant as in GA.

5 Conclusion

Tolerance assignment, especially the simultaneous assignment, is very important in product design and machining. However, the optimization task is usually difficult to tackle due to the nonlinear, multi-variable and high constrained characteristics. In view of the memory characteristics of PSO, a new constraints handling strategy suit for PSO is designed. This new strategy can adequately utilize the historical information in PSO algorithm. The application on a cylinder-piston assembly example demonstrates its high efficiency and effectiveness. However, when we attempt to extend the proposed approach to the constrained optimization with large number of complex equality constraints, subtle drawbacks emerged, as the constrained range is so narrow that the equality constraints are hard to satisfy. This problem reveals the new research direction, i.e., the effective equality constraint handling strategy desirable to develop for PSO based nonlinear programming.

References

1. Ngoi, B.K.A., Teck, O.C.: A tolerancing optimization method for product design. Vol. 13. International Journal of Advanced Manufacturing Technology (1997) 290-299
2. Singh, P.K., Jain, P.K., Jain, S.C.: Simultaneous optimal selection of design and manufactur-ing tolerances with different stack-up conditions using genetic algorithms. Vol.41. Interna-tional Journal of Production Research (2003) 2411-2429
3. Gao, Y., Huang, M.: Optimal process tolerance balancing based on process capabilities. Vol.21. International Journal of Advanced Manufacturing Technology (2003) 501–507
4. Huang, M.F., Zhong, Y.R., Xu, Z.G.: Concurrent process tolerance design based on minimum product manufacturing cost and quality loss. Vol.25. International Journal of Advanced Manufacturing Technology (2004) 714-722
5. Kennedy, J., Eberhart, R.C.: Particle swarm optimization. In: Proceedings of IEEE International Conference on Neutral Networks, Perth, Australia (1995) 1942-1948
6. Shi, Y.H., Eberhart, R.C.: A modified particle swarm optimizer. In: Proceedings of IEEE Conference on Evolutionary Computation (1998) 69-73
7. Noorul Hap, A., Sivakumar, K., Saravanan, R., Karthikeyan, K.: Particle swarm optimization (PSO) algorithm for optimal machining allocation of clutch assembly. Vol.27. International Journal of Advanced Manufacturing Technology (2005) 865-869
8. Kennedy, J., Eberhart, R.C., Shi, Y.: Swarm Intelligence. Morgan Kaufman, San Francisco (2001)
9. Kennedy, J., Eberhart, R.C.: Particle swarm optimization. Proceedings of IEEE International Conference on Neutral Networks, Perth, Australia (1995) 1942-1948

A Method to Plan Group Tours with Joining and Forking

Munenobu Nagata[1], Yoshihiro Murata[1], Naoki Shibata[2], Keiichi Yasumoto[1], and Minoru Ito[1]

[1] Nara Institute of Science Technology, Japan
[2] Shiga University, Japan

Abstract. Group sightseeing has some advantages in terms of required budget and so on. Some travel agents provide package tours of group sightseeing, but participants have to follow a predetermined schedule in tour, and thus there may be no plan which perfectly satisfies the tourist's expectation. In this paper, we formalize a problem to find group sightseeing schedules for each user from given users' preferences and time restrictions corresponding to each destination. We also propose a Genetic Algorithm-based algorithm to solve the problem. We implemented and evaluated the method, and confirmed that our algorithm finds efficient routes for group sightseeing.

1 Introduction

Personal navigation system guides its user through a mobile terminal such as mobile phone or PDA. Until now, there has been a lot of researches on personal navigation system. For example, indoor route guidance system[2] and a system which provides sightseeing information through mobile terminal[3] are proposed. These personal navigation systems' primary objective is to guide the user towards single destination or to provide information, and they lack functionality to guide the user through multiple destinations within limited time period, which is common in sightseeing. We have already proposed a personal navigation system P-Tour which guides user through multiple destinations. P-Tour finds a route schedule to tour multiple sightseeing destinations considering user's preferences[9]. The previous P-Tour provides functionality to guide single user only, but it is quite common that a group of members go sightseeing together in order to save traveling cost by riding on the same vehicle. In this paper, we propose an extension for P-Tour which finds a route schedule on group sightseeing. On group sightseeing, the members would prefer (1) visiting each destination with other members, (2) forking from other members and visiting special destinations, (3) visiting each destination taking into account of all members' preferences, and (4) visiting each destination efficiently within limited time. In order to achieve these objectives, we formalized the problem to find routes for each user from user's preferences and restrictions, designed and implemented a GA-based algorithm to solve the problem. Then, we evaluated the method through experiments, and confirmed that our algorithm finds efficient routes for group sightseeing.

T.-D. Wang et al. (Eds.): SEAL 2006, LNCS 4247, pp. 881–888, 2006.

2 Related Works

Genetic algorithm is a combinatorial optimization algorithm inspired from evolution in nature, and it uses crossover, mutation, evaluation and selection operations. There are many applications of genetic algorithm including traditional combinatorial optimization problems such as Knapsack problem[8], Traveling Salesperson Problem [10], Set Coverage Problem[1], and so on. Besides them, genetic algorithm can be used for solving engineering problems such as Job Shop Scheduling Problems[7], and multiple processor scheduling problems[4].

We have also used Genetic Algorithm in our already proposed personal navigation system "P-Tour"[9]. P-Tour can plan schedule around multiple destinations with many restrictions.

There are also problems to gain benefits of whole system (or all users) such as group sightseeing schedule planning problem, explained below.

- Theme park problem[5]: A problem to reduce congestion and improve customer's satisfaction in theme park by adjusting presentation of information and booking of each facilities.
- Delivery scheduling problem[6]: A problem to find the optimal delivery path when there are many parcels with labels to deliver, and many vehicles with limited maximum load. The objective is to minimize the total distance of the path and improve customer's satisfaction.

Since these existing studies do not handle forking and joining of passenger groups, these methods are different from our study.

3 Route Scheduling for Group Sightseeing

In the proposed method, we find a sightseeing schedule in which members of group sightseeing fork and join on the way, shown in Fig.1.

Finding an efficient group sightseeing schedule involves issues which does not exist on scheduling for single person tour.

- **Differences of preferences between members:** Each member may have different preferences for each destination. The system has to decide from: (1) making members visit same destinations by ignoring part of members'

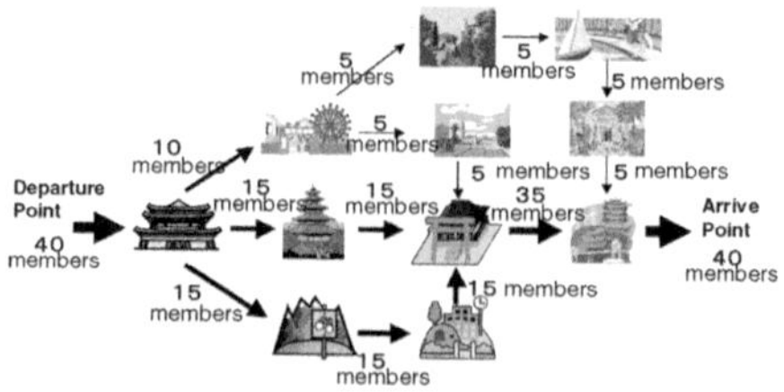

Fig. 1. Concept of group sightseeing schedule

preferences, (2) making compromised route, or (3) making part of members fork from other members.
- **Differences between starting and ending points of the route of each member:** The system has to handle each member's starting point of the route. Besides it, part of members may start sightseeing later than other members. Different time and locations of ending points should also be considered.

4 Definition of the Problem

In this section, we first define the problem to find an efficient schedule for group sightseeing. On group sightseeing, the members usually prefer visiting each destination with other members. But, the members have different conditions regarding to starting location, returning location, and preference of each destination. The objective of the problem is to maximize users' satisfaction which is evaluated by summing up each member's satisfaction value which increases when visiting each destination with other members, and when preferred destinations are included in the route.

4.1 Input

The input of the problem is shown below.

- Map data, given as a directional graph $G = (V, E)$. Each edge is assigned a length. Minimum distance between two given vertices v_1 and v_2 is referred as $dist(v_1, v_2)$.
- Destination data $D = \{d_1, ...\}$, each element is given as tuple of following information.
  1. Name of the destination(e.g. Horyu-ji Temple).
  2. Corresponding vertex $v_h \in V$.
  3. rt_{ij}: The latest arrival time for member u_i at destination d_j. (for example, if rt_{ij} = 12:00, member u_i has to reach destination d_j before 12:00).
  4. dur_{ij}: Restriction of staying time for member u_i at destination d_j.
  5. pre_{ij}: Preference value for member u_i at destination d_j.
- Participant data: each member has following five parameters.
  - $pd_{is} \in D$: Starting point of member u_i.
  - $pd_{ig} \in D$: Returning point of member u_i.
  - pt_{is}, pt_{ig}: Time restriction for member u_i at start / returning points.
  - $speed_i$: Speed of user u_i.

4.2 Notations

The group sightseeing schedule is composed of schedules of each member. The schedule for member u_i is denoted as $s_i = (D_i, Stay_i)$, where D_i is a list of destinations visited by member u_i, defined as $D_i = \langle d'_{i1}, d'_{i2}, ..., d'_{ij}, ..., d'_{i|D_i|} \rangle$, and d'_{ij} is member u_i's j-th destination. $Stay_i$ is the list of stay time for member u_i at each destination, defined as $Stay_i = \langle stay_{i1}, stay_{i2}, ..., stay_{i|D_i|}\rangle$, and $stay_{ij}$ is the stay time for which member u_i stays at destination d_j. The arrival time of member u_i at destination d'_{ij} is denoted as t_{ij}, and calculated as follows : $t_{i(j+1)} = t_{ij} + stay_{ij} + \frac{dist(d'_ij, d'_{i(j+1)})}{speed_i}$. Usually, $stay_{ij}$ is equal to dur_{ij}, but if members have to join at the destination, and some members arrives earlier, other members have to wait at the destination.

4.3 Evaluation Function

Evaluation function f is defined as follows:

$$\begin{aligned} f(S) = \sum_{i=1}^{|U|}\{\alpha \sum_{j=1}^{|D_i|} pre_{ij} \cdot timeok(s_i, d_{ij}) \cdot group(S, i, j) \\ - \sum_{j=1}^{|D_i|-1} (\beta \cdot dist(d_{ij}, d_{i(j+1)})) \\ +\gamma \cdot commonpath(S, i, j, j+1)) - \delta \cdot timegoal(S_i, p_{ig})\} \end{aligned} \tag{1}$$

α, β, γ and δ are constant values.

Function $timeok(s_i, d'_{ij})$ returns 1 iff destination d'_ij is included in route s_i and both restrictions rt_{ij} and dur_{ij} are satisfied.

Function $group(S, i, j)$ returns the number of members who visit destination d_{ij} together on the route S. By this term, evaluation value increases when many members visit a destination together.

Function $commonpath(S, i, j, j+1)$ returns the number of members who move from d_{ij} to $d_{i(j+1)}$ together on the route S. By this term, evaluation value increases when many members move together.

Function $timegoal(s_i, p_{ig})$ returns an absolute value of difference between arrival time and expected arrival time at destination s_i. Evaluation value decreases when they arrive too early or too late.

the size of touring group between destinations.

4.4 Algorithm

In this section, we describe the GA-based scheduling algorithm for group tours.

Fig. 2 shows the routes of two members represented as a list of genes. One chromosome consists of lists of genes for all members. Each gene in the list represents destination, and the traveling order is represented by the path from left side to right side of the list, in the figure. Genes $P_{11}, \ldots, P_{14}$ represent destinations in the route for member u_1, and genes $P_{21}, \ldots, P_{24}$ represent destinations in the route of member u_2.

P_{ij} is genes called normal gene. R_{11}, R_{12}, R_{21}, R_{22}, R_{23} are called reference genes. Reference genes do not represent destinations directly, but points the

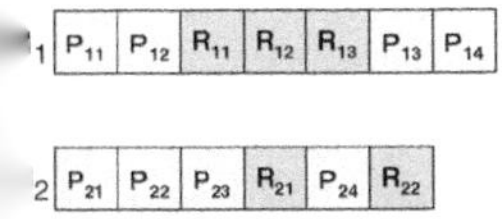

Fig. 2. Chromosome

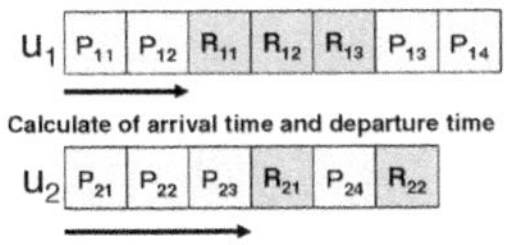

Fig. 3. Calculation of arrival time and departure time

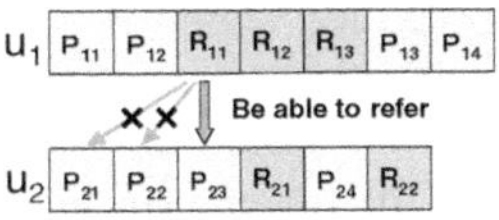

Fig. 4. Search for gene that can be referred

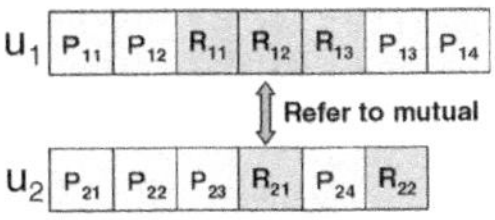

Fig. 5. Mutual reference

destinations visited by another member. Reference genes have three elements: referred member, unique value, hidden gene.

Decoding of reference genes is performed as shown in Fig.3. At first, arrival times and departure times of destinations represented by normal genes are calculated until reference gene appears.

Next, referred gene is searched. Referred gene represents the destination such that the referring member arrives at the destination represented by referred gene before referred member (here, R_{A1} and R_{B1}).

If destination represented by referred gene is included in referring members' route, the hidden gene is activated. Hidden gene represents another destination called hidden destination.

Reference gene may refer to another reference gene, and references between genes can be cyclic(Fig.5). In this case, the unique values of reference genes are compared each other. The hidden gene of the reference gene with the largest unique value is activated.

GA operations are composed of generation of initial population, evaluation, selection, crossover, and mutation. Let M denotes the number of members, N denotes population size, and I denotes the number of generations.

1. Generation of initial population: N chromosomes are randomly generated as initial population. Each chromosome represents candidate solution.
2. Evaluation: The evaluation values are calculated with evaluation function described in section 3.3.
3. Selection: We use elite strategy and tournament selection.
4. Crossover: We use two point crossover. If there are redundant destinations, they are deleted.
5. Mutation: we use random insertion, deletion, exchange, change of reference and conversion.
6. One GA generation is step 2 to 5, and these steps are repeated until a good solution is obtained.

5 Evaluation Experiments

To evaluate the proposed method, we conducted experiments to evaluate the following two points.

– Quality of obtained schedules
– Optimality of obtained schedules

In the experiments, we used the map of Nara prefecture (digital map 25000 issued by Geographical Survey Institute of Japan). We executed the proposed algorithm on an ordinary PC with Pentium M 2.0GHz, 512M Memory, Windows XP pro., Java 1.4.2. We assumed that tourists move by car and their speed is 40 km/h.

Also, we set parameter values from preliminary test as follows: $N = 1000$, $I = 200$, $W = 30$, $L = 20000$, $\alpha = 50$, $\beta = 0.015$, $\gamma = 15$ and $\delta = 10$.

5.1 Quality of Obtained Schedules

We input data of 3 members shown in Table 1, 8 kinds of destination data and evaluated the obtained schedule. Details of the values are as follows: if a user wants to visit one of the destinations, its preference value is set to 5, and if not, it is set to -10. Stay time is 60 minutes for all destinations.

Table 1. Data for 30 destinations

requested by all members	none(∅)	
Yakushi-ji temple(A1) Todai-ji temple(A2) Horyu-ji temple(A3)	NAIST(E1) Taima temple(E2) Kongou shrine(E3) Tamaki shrine(E4) Akisino temple(E5) Suijin tennoryou(E6)	
requested by only u_1	requested by only u_2	requested by only u_3
Kasuga-taisya shrine(B1) Ishigami shrine(B2) Tyougaku temple(B3) Houzan temple(B4)	Zinmu tennoryou(C1) Hase temple(C2) Dansan shrine(C3) Ishibutai kofun(C4)	Syoumu tennoryou(D1) Torinoyama kofun(D2) Keikou tennoryou(D3) Sigi-moutain Nodoka villeage(D4)
requested by u_1, u_2	requested by u_1, u_3	requested by u_2, u_3
Mesuri-moutain kofun(F1) Yashikiyama kofun(F2) Oono temple(F3)	Kamotoba shrine(G1) Oogami shrine(G2) Nukataryou(G3)	Tennnouzan kofun(H1) Ruins of Fujiwarakyu (H2) Suidei kofun(H3)

Next, we evaluate paths obtained by our method. Calculation time for those paths is about 1.3 minutes total. The obtained schedule is shown in Figure 6. Figure 7 is all members' paths overlapped.

Now, we give some explanations regarding to the obtained schedule. Member u_1 departures NAIST(E1) at 8:34, and tours 3 destinations (B3, G3, B1) alone until 11:53. The member sets high preference values for these three destinations. Member u_2 departures Hozanji (B4), joins with u_3, and they tour Tennozan Kofun (H1) together. Member u_2 waits from 9:43 to 11:09 for member u_3 before joining. After that, u_1 joins them at Todaiji(A2), and tours 3 destinations (A2,

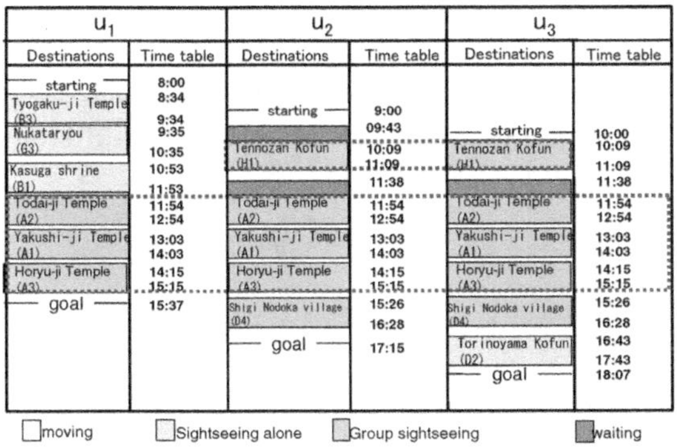

Fig. 6. A time table of obtained schedule

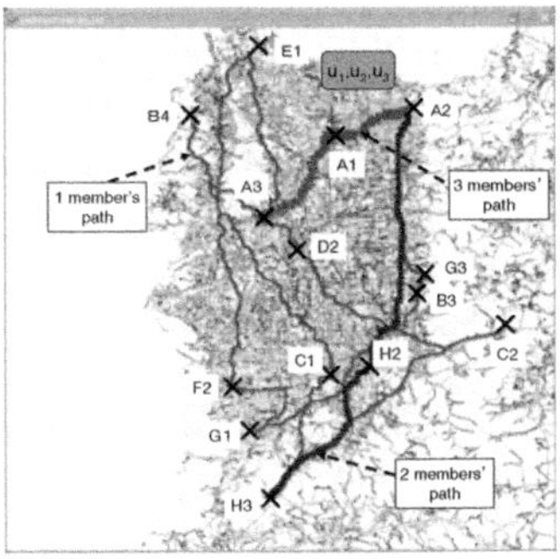

Fig. 7. An obtained path of group

Table 2. Comparison of obtained schedules with the optimal solutions

Number of destinations	Number of combinations (search space)	computation time proposed method	computation time full search	error rate (%)
3	$(21)^3$	about 10(sec.)	about 2(min.)	0%
4	$(142)^3$	about 13(sec.)	about 18(hour)	0%
5	$(12336)^3$	-	-	-

A1, A3) together. All three members set high preference values for these 3 destinations. Also, member u_2 and u_3 have to wait from 11:38 to 11:45 for member u_1 to join. u_1 forks after touring these 3 destinations, and arrives at his final destination. After that, member u_2 and u_3 tour Shigi Nodoka village(D4) together. Then, member u_2 and u_3 splits. Member u_2 goes to his goal, and member u_3 tours Torinoyama Kofun(D2) before reaching his final destination(Hasedera C2). Member u_3 sets a high preference value for Torinoyama Kofun (D2).

As shown above, our method can obtain schedule with joining and forking, and this schedule includes destinations with high preference values for each member.

5.2 Optimality of Obtained Schedules

To evaluate optimality of the obtained schedule by the proposed method, we compared the obtained schedules with the optimal schedules obtained by full search. We set the parameters as follows: the number of members is 3, the number of destination is 3 or 4. The results are shown in Table 2 These results are average values of 10 trials.

In the cases of both 3 and 4 destinations, our method obtains the optimal schedule. Also, our method was a lot faster than the full search algorithm. Since the calculation time of the full search algorithm grows exponentially when the number of destinations increases, we could not observe the case above 4 destinations.

Also, In order to evaluate effect of introducing reference genes, we compared proposed method with and without reference genes.

In the case without reference genes, our method with reference genes achieved 42.5% to 76.5% better fitness values than one without reference genes.

In the case without reference genes, members can join only if their schedules happen to include the same point at the same time. Since members has different starting points, it can find only few schedules with joining.

6 Conclusion

We proposed a method to find a schedule for group tour with different route for each member. Each member of group sets own preference and restrictions. In this paper, we formalized the scheduling problem on group sightseeing. And to solve this problem, we designed scheduling algorithm based on GA. Moreover, we evaluated the method through experiments by using the map data on the northern part of Nara Prefecture, and confirmed that our algorithm finds efficient routes for group sightseeing.

We are planning to improve the search efficiency by introducing the local search method.

References

1. Al-Sultan, K.S., Hussain, M.F., Nizami, J.S., "A Genetic Algorithm for the Set Covering Problem", *Journal of the Oper.Res. Society*, Vol. 47, pp. 702–709, 1996.
2. Butz, A., Baus, J., Krüger, A., Lohse, M., "A Hybrid Indoor Navigation System", *Proc. of IUI2001: International Conf. on Intelligent User Interfaces 2001*, ACM Press, New York, pp. 25–33, 2001.
3. Cheverst, K., Davies, N., Michell, K., Friday, A., "The Design of an Object Model for a Context-Sensitive Tourist Guide", *Computers Graphics Journal*, Vol. 23, No. 6, pp. 883–891, 1999.
4. Hou E.S.H., Ansari N., Ren, H., "A Genetic Algorithm for Multiprocessor Scheduling", *IEEE Transactions on Parallel and Distributed Systems*, Vol. 5, No. 2, pp. 113–120, 1994.
5. Kawamura, H., Kurumatani, K. and Ohuchi, A.: "Modeling of Theme Park Problem with Multiagent for Mass User Support", *Working Note of the IJCAI-03 Workshop on Multiagent for Mass User Support*, pp. 1–7 (2003)
6. Ohta, M., Shinoda, K., Noda, I. and Kunifuji, S., "Usability of Demand-bus in Town Area", *Domestic Conference of Intelligence Transportation Systems* (in Japanese), Vol. 115, pp. 239–245, 2002.
7. Ono, I., Yamamura, M.,Kobayashi, S., "A Genetic Algorithm for Job-shop Scheduling Problems Using Job-based Order Crossover", *Proc. of 1996 IEEE International Conf. on Evolutionary Computation*, pp. 547–552, 1996.
8. Richard, S., "Solving large knapsack problems with a genetic algorithm", *In IEEE International Conf. on Systems, Man and Cybernetics*, Vol. 1, pp. 632–637, 1995.
9. Shiraishi, T., Nagata, M., Shibata, N., Murata, Y., Yasumoto, K., Ito, M., "A Personal Navigation System with a Schedule Planning Facility Based on Multiobjective Criteria", *Proc. of 2nd Int'l. Conf. on Mobile Computing and Ubiquitous Networking*, pp. 104–109, 2005.
10. Whitley, D., Starkweather, T., Fuquay, D., "Scheduling problems and traveling salesmen: The genetic edge recombination operator", *In Proc. of the Third International Conf. on Genetic Algorithms and their Applications*, pp. 133–144, 1989.

A Modified Discrete Binary Ant Colony Optimization and Its Application in Chemical Process Fault Diagnosis

Ling Wang and Jinshou Yu

Research Institution of Automation, East China University of Science & Technology
200237, Shanghai, China
shwl_1212@163.com

Abstract. Considering fault diagnosis is a small sample problem in real chemical process industry, Support Vector Machines (SVM) is adopted as classifier to discriminate chemical process steady faults. To improve fault diagnosis performance, it is essential to reduce the dimensionality of collected data. This paper presents a modified discrete binary ant colony optimization (MDBACO) to optimize discrete combinational problems, and then further combines it with SVM to accomplishing fault feature selection. The tests of optimizing benchmark functions show the developed MDBACO is valid and effective. The fault diagnosis results and comparisons of simulations based on Tennessee Eastman Process (TEP) prove the feature selection method based on MDBACO and SVM can find the essential fault variables quickly and exactly, and greatly increases the fault diagnosis correct rates as irrelevant variables are eliminated properly.

1 Introduction

It is a challenge to apply fault diagnosis to modern complex and large-scaled chemical process as large amounts of variables with noises need be monitored and fault data are deficient in the really productions. To improve the fault diagnosis performance, it is essential to preprocess the sampled data for reducing data dimension. There have been several approaches to preprocess data developed, applied and widely researched in fault diagnosis applications, such as Principal Component Analysis (PCA) [1] and Kernel PCA [2][3], which are well-known methods for feature extraction. But the extracted information by feature extraction methods is not related to the objective of fault diagnosis exclusively. So the number of selected components maybe is still large to contain enough information for diagnosing. Sometimes, even worse, the extracted data are not exactly acceptable for fault diagnosis because the resulting lower dimensional space may contain little of the required faults information, which makes the feature extraction invalid for fault diagnosis.

To make up for this shortage, feature selection method was proposed as an alternative to preprocess the collected data [4], [5], [6]. Feature selection is operated to directly select the essential fault variables and only the selected variables will be retained and used as inputs for fault diagnosis. As the irrelative variables are all removed, the real-time capability and correct rates of fault diagnosis will be greatly improved [7].

In this paper, we use SVM as classifier to diagnose the chemical process steady faults, owing to its remarkable characteristics such as good generalization

T.-D. Wang et al. (Eds.): SEAL 2006, LNCS 4247, pp. 889 – 896, 2006.

performance, the absence of local minimal, and fit for small samples [8], [9], [10]. In order to find the fault feature variables effectively and properly, a modified discrete binary ant colony optimization algorithm is proposed and combined with SVM to realize fault feature selection in the paper.

The reminder of the paper is organized as follows. Section 2 presents the MDBACO algorithm and feature selection method based on MDBACO combined with SVM in detail. The numerical experiments are conducted in Section 3. Section 4 describes the simulations of MDBACO-base feature selection and fault diagnosis. The comparisons with other methods are also presented in this section. Section 5 concludes the results of simulations.

2 Theory

2.1 Modified Discrete Binary Ant Colony Optimization

In the last decade, ACO algorithm [11], [12], [13] has been recognized that not only routing problems as the traveling salesperson problem, but also any other type of combinatorial optimization problems can be encoded as "best path" problems and solved using the ant colony metaphor. In this paper, we adjusted the classical ACO to tackle the discrete binary optimization problems. In the proposed MDBACO algorithm, all bit sequences are represented as graph-ere nodes [14], with the edges between them denoting the choice of the bit, that is, a bit "1" means the corresponding bit is selected by the ant and a bit "0" represents this bit is not selected by the ant. Then the discrete binary optimization problem is transformed into an ant traversal through the graph. The selections of all bits by each ant are collected as a solution and coded as a binary sequence.

The main steps of MDBACO are as follows:

1. Initializing MDBACO algorithm parameters, such as the population of ant colony, a predetermined number of generations and initial pheromone.
2. Each ant traveling from a random beginning bit and visiting all bits to build the solution completely.

   Firstly, a random number *rand* is generated and compared with the exploiting probability parameter P_e. If *rand* is greater than or equal to the parameter P_e, whether the bit j is selected or not by ant i is decided by the pheromone trails, that is, the selection of bit i is decided according to the two probability parameters $\tau_{j,0}$ and $\tau_{j,1}$.

$$Solution_{i,j} = \begin{cases} 0 \ \ if \ \ \tau_{j,1} \le \tau_{j,0} \\ 1 \ \ if \ \ \tau_{j,1} > \tau_{j,0} \end{cases} \tag{1}$$

   If *rand* is less than P_e, whether bit j is chosen is determined by the threshold function:

$$\eta_c(x) = \begin{cases} 0 & c < \eta_0 \\ 1 & c \ge \eta_0 \end{cases} \tag{2}$$

   where c is another random number and η_0 is a constant, for instance, η_0 =0.5.

3. Calculating the fitness of each ant according to its solution after all ant agents completing building solutions.
4. Terminating if termination criteria are reached, else going on the next step.
5. Updating the pheromone intensity of each feature as:

$$\begin{aligned}\tau_{i,0} &= (1-\rho)\tau_{i,0} + \rho \cdot k \cdot Q \\ \tau_{i,1} &= (1-\rho)\tau_{i,1} + \rho \cdot k \cdot Q .\end{aligned} \tag{3}$$

   where ρ is the evaporation rate to avoid unlimited accumulation of pheromone, k and Q are both constants.
6. Updating the global optimal solution and enhancing the pheromones intensities of its feature selections through $\Delta\tau_{i,0}^{k}$ and $\Delta\tau_{i,1}^{k}$ as:

$$\begin{aligned}\tau_{i,0} &= (1-\rho)\tau_{i,0} + \rho \cdot k \cdot Q \text{ if bit } i \text{ not selected by optimal ant} \\ \tau_{i,1} &= (1-\rho)\tau_{i,1} + \rho \cdot k \cdot Q \text{ if bit } i \text{ selected by optimal ant}\end{aligned} \tag{4}$$

7. Going to the step 2.

2.3 Fitness Function

To guide ACO algorithm to search feature correctly, a pre-defined fitness function is applied to evaluate the fitness. To remove the irrelative variables, the fitness function is defined as Eq. (5)

$$f(x_{id}) = f(id) - p \times \frac{m_c}{m_{all}} . \tag{5}$$

where $f(x_{id})$ means the modified fitness function, $f(id)$ represents the correct fault classification rate, m_c is the number of variables chosen by the ant while m_{all} is the dimension of data samples, p is an constant.

2.4 Feature Selection Based on MDBACO

In the proposed MDBACO-based feature selection method, features (i.e., variables of collected data samples) are represented as a binary sequence where a bit "1" denotes the corresponding feature is selected and a bit "0" means its corresponding feature is eliminated. The features selected by ants are taken as the input of SVM for discriminating faults, and then the correct classification rates are supplied to evaluate the fitness of each ant agent.

The process of feature selection based on MDBACO is described as follows:

1. MDBACO algorithm parameters initialization;
2. Building solutions completely according to step 2 of MDBACO algorithm;
3. Using the features selected by each ant to evaluate its fitness by the pre-defined fitness function;
4. Terminating if termination criteria are achieved, or going on the next step;
5. Updating the pheromone intensity and the optimal in accordance with the step 5 and step 6 of MDBACO algorithm;
6. Going to the step 2.

3 Numerical Experiment

In our experimental studies, a set of 6 benchmark optimization functions was employed to validate the feasibility of the proposed MDBACO algorithm and listed as follows:

$$F_1 = 100(x_1^2 - x_2)^2 + (1 - x_1)^2 \qquad -2.048 \le x_i \le 2.048 . \tag{6}$$

$$\begin{aligned} F_2 = {} & [1 + x_1 + x_2 + 1)^2 \cdot (19 - 14x_1 + 3x_1^2 - 14x_2 + 6x_1x_2 + 3x_2^2)] \\ & \cdot [30 + (2x_1 - 3x_2)^2 \cdot (18 - 32x_1 + 12x_1^2 + 48x_2 - 36x_1x_2 + 27x_2^2)] \quad -2 \le x_i \le 2 . \end{aligned} \tag{7}$$

$$F_3 = (x_1^2 + x_2^2)^{0.25}[\sin^2(50(x_1^2 + x_2^2)^{0.1}) + 1.0] \qquad -100 < x_i < 100 . \tag{8}$$

$$F_7 = (x_1^2 + x_2 - 11)^2 + (x_1 + x_2^2 - 7)^2 \qquad -10 \le x_i \le 10 . \tag{9}$$

$$F_5 = 0.5 - \frac{\sin^2 \sqrt{x_1^2 + x_2^2} - 0.5}{(1 + 0.001(x_1^2 + x_2^2))^4} \qquad -100 < x_i < 100 . \tag{10}$$

$$\begin{aligned} F_6 = {} & [-13 + x_1 + ((5 - x_2) \cdot x_2 - 2) \cdot x_2]^2 \\ & + [-29 + x_1 + ((x_2 + 1) \cdot x_2 - 14) \cdot x_2]^2 \quad -10 \le x_i \le 10 . \end{aligned} \tag{11}$$

where F5 has the global maximum, others have the global minimum.

To evaluate the performance of the proposed MDBACO algorithm, classical binary GA and classical discrete particle swarm optimization (DPSO) algorithm were used for comparisons. The parameters set of GA was: the mutation probability P_m=0.05, crossover probability P_c=0.7. The maximum velocity V_{max}, minimum velocity V_{min}, c_1 and c_2 for discrete PSO were set at +8, -8, 2.0 and 2.0, respectively. A setting of exploiting probability P_e=0.1, 0.2 and 0.4 are adopted in MDBACO. All experiments

Table 1. The experimental results of GA, DPSO, MDBACO with Pe=0.1,0.2 and 0.4 on 6 benchmark functions

		F1	F2	F3	F4	F5	F6
Global optimal		0	3	0	0	1	0
GA	V_w	7.49	105.44	0.004	63.09	0.8824	72.89
	L_t	20	20	20	20	20	20
DPSO	V_w	0.25	3.000026	0.087	10	0.9903	0.0013
	L_t	16	8	1	5	12	12
MDBACO	V_w	0	3	0	0.279	1	0.364
(P_e=0.1)	L_t	0	0	0	2	0	3
MDBACO	V_w	0	3	0	0	1	0
(P_e=0.2)	L_t	0	0	0	0	0	0
MDBACO	V_w	0	3	0	2.8E-4	1	0.0072
(P_e=0.4)	L_t	0	0	0	7	0	6

were repeated for 20 runs, a fixed number of maximum generations 2500 was applied to all algorithms. The experimental results (i.e., the worst value V_w and the times of sticking in the local optima L_t,) are list in Table1. From Table 1, MDBACO outperformed the classical GA and discrete PSO algorithm for all the 6 benchmark functions. And MDBACO can escape from the local optimal effectively with the exploiting operator, but a improper setting of exploiting probability also spoil MDBACO search ability seriously. The search performance of all algorithms tested here can be ordered as MDBACO>DPSO>GA.

4 Fault Diagnosis Simulation

4.1 Tennessee Eastman Process

The Tennessee Eastman is a well-known benchmark chemical process, which was firstly introduced by Downs and Vogel [15]. The TEP provides a realistic industrial process for evaluating process control and monitoring methods. Now, the TEP has been widely used for the process monitoring community as a source of data for comparing various approaches [16]. The TEP simulator, coded in Matlab, was used to generate normal data and fault data. The 3 faults researched in this paper are those with stable operating conditions before and after the faults occur, called Fault1, Fault2 and Fault3 below. To make the results comparable, the data used for experiments are given at http://brahms.scs.uiuc.edu.

4.2 Fault Feature Selection

In the fault 1 case, all variables remains steady except the 51-th variable induced a change when the fault occurred. So the fault feature variable of fault1 is just variable 51. Fault2 involves a step changes in variable 1and variable 44, and the other variables are all bothered. The change of variable 1 and variable 44 are so remarkable that any one of them can be taken as fault feature. Variable 45 has a noticeable step change when the fault3 is introduced into the process. And affected by it, other 34 variables deviate significantly from their normal operation behavior and go aback to normal values later by the control of closed loop. In our experiments, 30 and 60 data samples were randomly selected from 3 fault simulations and composed the training and validation data set for running feature selection respectively.

To evaluate and compare the developed algorithm, fault feature selection of three faults were iterated for 20 runs using MDBACO algorithm with P_e=0.2, DPSO algorithm and GA. The maximum iteration was set at 50 times. The other parameters of all algorithms adopted the same settings used in numerical experiment.

The features selected of 3 faults by all algorithms are shown in Fig1-3. From Fig.1-3, the proposed MDBACO selected the fault features of each fault correctly and properly. The performance of DPSO algorithm was better than GA, but both of them did not select the features exactly and were apt to be trapped by local minima and then stagnated. It is obvious that MDBACO algorithm outperforms DPSO and GA for feature selection.

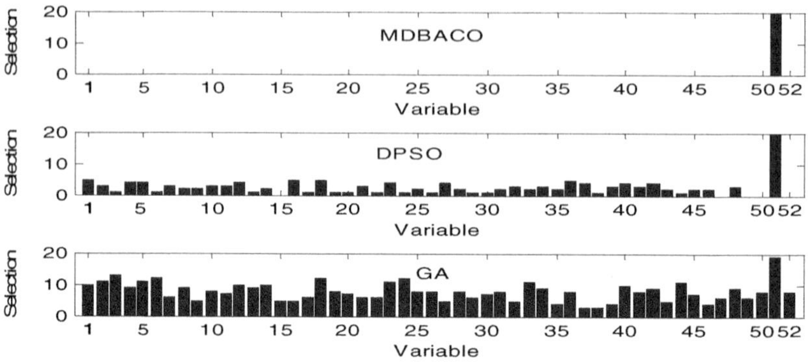

Fig. 1. Fault features of fault 1 selected by MDBACO, DPSO and GA

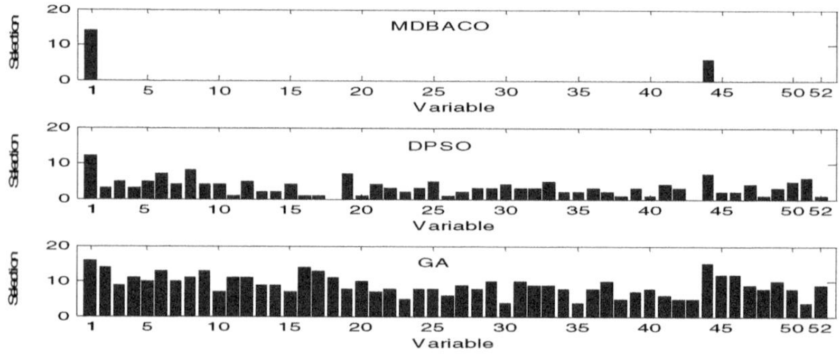

Fig. 2. Fault features of fault 2 selected by MDBACO, DPSO and GA

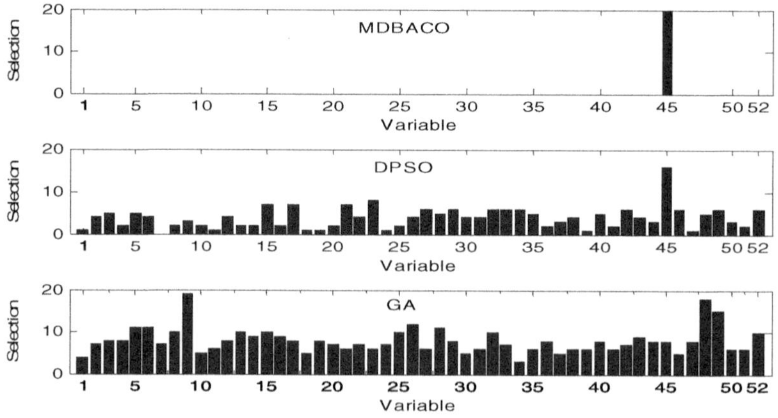

Fig. 3. Fault features of fault 3 selected by MDBACO, DPSO and GA

4.3 Fault Diagnosis Base on SVM with Fault Feature Selection

To estimate the performance of fault diagnosis method based on SVM and MDBACO-based feature selection (MFS), we ran the fault diagnosis simulations with all the test data given by http://brahms.scs.uiuc.edu. Table 2 presents the results of fault diagnosis. To give a comparison, fault diagnosis methods based on SVM with the all collected data, data extracted by PCA and KPCA were tested with the same data set and the results of these three fault diagnosis methods are given the Table 2 as well.

Table 2. The correct diagnosing rates of SVM with all variables, PCA, KPCA and feature selection based on MDBACO

	All variables	PCA	KPCA	MFS
Fault 1	61.3%	68.2%	74.5%	100%
Fault 2	99.4%	99.5%	99.9%	100%
Fault 3	51.8%	59.7%	71.2%	100%

5 Results and Discussion

In this paper, a modified discrete binary ant colony optimization algorithm has been proposed for optimizing the discrete combinational problems. A set of 6 benchmark functions has been used to test the proposed algorithm. The experimental results prove the MDBACO algorithm works better than the classical GA and discrete PSO.

Based on the developed MDBACO algorithm, we further combined it with SVM to select fault features for diagnosing the steady faults in chemical process. The simulation results show the developed MDBACO algorithm can find the essential fault feature exactly and MDBACO-based feature selection method greatly improves the fault diagnosis performance of SVM. Considering SVM is suitable for the limited data sample applications and MDBACO-based feature selection method can select fault feature quickly and properly, all of these excellent characteristics make the SVM combined with MDBACO-based feature selection fault diagnosis method noticeable and attractive in the chemical process fault diagnosis applications.

References

1. Wang, S. W., Xiao, F.: Detection and diagnosis of AHU sensor faults using principal component analysis method. Energy Conversion and Management. 45 (2004) 2667-2686
2. Lee, J. M., Yoo, C. K., Lee, I. B.: Fault detection of batch processes using multiway kernel principal component analysis Computers and Chemical Engineering. 28 (2004) 1837-1847
3. Huang, Y., Gertler, J., McAvoy, T.: Sensor and actuator fault isolation by structured partial PCA with nonlinear extensions. Journal of Process Control. 10 (2000) 459-469
4. Chun, K. Z., Hong, H.: Feature selection using the hybrid of ant colony optimization and mutual information for the forecaster. in: Proceedings of Fourth International Conference on Machine Learning and Cybernetics. (2005) 1728-1732

5. Liu, H., Yu, L.: Toward integrating feature selection algorithms for classification and clustering. IEEE Trans. Knowledge and Data Engineering. 17 (2005) 491-502
6. Kwak, N., Choi, C. H.: Input feature selection for classification problems. IEEE Trans. Neural Networks 13 (2002) 143-159
7. Chiang, Leo H. Pell, Randy J.: Genetic algorithms combined with discriminant analysis for key variable identification. Journal of Process Control. 14 (2004) 143-155
8. Vapnik, V.N.: The Nature of Statistical Learning Theory. Springer, New York. (1995)
9. Fan R. E., Chen, P. H., Lin, C. J. Working set selection using second order information for training SVM. Journal of Machine Learning Research. 6 (2005) 1889-1918
10. Chiang, Leo H., Kotanchek, Mark E., Kordon, Arthur K.: Fault diagnosis based on Fisher discriminant analysis and support vector machines Computers and Chemical Engineering. 28 (2004) 1389-1401
11. Bonabeau, E, Dorigo, M, Theraulaz,G.: Swarm Intelligence from Natural to Artificial System. Oxford University Press, Oxford. (1999)
12. Dorigo, M, Blum, C.: Ant colony optimization theory: A survey. 344 (2005) 243-278
13. Blum, C.: Ant colony optimization: introduction and recent trends. Physics of Life Reviews. 2 (2005) 353-373
14. Gao, H.H., Yang, H.H., Wang. X.Y.: Ant colony optimization based network intrusion feature selection and deltection. in: Proceedings of Fourth International Conference on Machine Learning and Cybernetics. (2005) 3871-3875
15. Downs J. H., Vogel E.F.: A plant-wide industrial process control problem, Comput. Chem. Eng. 17 (1993) 245-255
16. Chiang, L. H., Russell, E. L., Braatz, R.D.: Fault Detection and Diagnosis in Industrial Systems. Springer-Verlag Berlin Heidelberg London (2001)

An Intelligent System for Container Image Recognition Using ART2-Based Self-organizing Supervised Learning Algorithm

Kwang-Baek Kim[1], Young Woon Woo[2], and Hwang-Kyu Yang[3]

[1] Dept. of Computer Engineering, Silla University, Busan, Korea
gbkim@silla.ac.kr
[2] Dept. of Multimedia Engineering, Dong-Eui University, Busan, Korea
ywwoo@deu.ac.kr
[3] Dept. of Multimedia Engineering, Dongseo University, Busan, Korea
hkyang88@hanmail.net

Abstract. This paper proposed an automatic recognition system of shipping container identifiers using fuzzy-based noise removal method and ART2-based self-organizing supervised learning algorithm. Generally, identifiers of a shipping container have a feature that the color of characters is black or white. Considering such a feature, in a container image, all areas excepting areas with black or white colors are regarded as noises, and areas of identifiers and noises are discriminated by using a fuzzy-based noise detection method. Noise areas are replaced with a mean pixel value of the whole image and areas of identifiers are extracted by applying the edge detection by Sobel masking operation and the vertical and horizontal block extraction in turn to the noise-removed image. Extracted areas are binarized by using the iteration binarization algorithm, and individual identifiers are extracted by applying 8-directional contour tracking method. This paper proposed an ART2-based self-organizing supervised learning algorithm for the identifier recognition, which creates nodes of the hidden layer by applying ART2 between the input and the hidden layers and improves the performance of learning by applying generalized delta learning and Delta-bar-Delta algorithm between the hidden and the output layers. Experiments using many images of shipping containers showed that the proposed identifier extraction method and the ART2-based self-organizing supervised learning algorithm are more improved compared with the methods previously proposed.

1 Introduction

Identifiers of shipping containers are given in accordance with the terms of ISO standard, which consist of 4 code groups such as shipping company codes, container serial codes, check digit codes and container type codes [1][2]. And, only the first 11 identifier characters are prescribed in the ISO standard and shipping containers are able to be discriminated by automatically recognizing the first 11 characters. But, other features such as the foreground and background colors,

T.-D. Wang et al. (Eds.): SEAL 2006, LNCS 4247, pp. 897–904, 2006.

the font type and the size of container identifiers, etc., vary from one container to another since the ISO standard doesn't prescribes other features except code type [2][3]. Since identifiers are printed on the surface of containers, shapes of identifiers are often impaired by the environmental factors during the transportation by sea. The damage to a container surface may lead to a distortion of shapes of identifier characters in a container image. So, the variations in the feature of container identifiers and noises make it quite difficult the extraction and recognition of identifiers using simple information like color values [4].

Generally, container identifiers have another feature that the color of characters is black or white. Considering such a feature, in a container image, all areas excepting areas with black or white colors are regarded as noises, and areas of identifiers and noises are discriminated by using a fuzzy-based noise detection method. Noise areas are replaced with a mean pixel value of the whole image area, and areas of identifiers are extracted and binarized by applying the edge detection by Sobel masking operation and the vertical and horizontal block extraction to the conversed image one by one. In the extracted areas, the color of identifiers is converted to black and one of background to white, and individual identifiers are extracted by using 8-directional contour tacking algorithm. This paper proposed an ART2-based self-organizing supervised learning algorithm for the identifier recognition, which creates nodes of the hidden layer by applying ART2 between the input layer and the hidden one and improves performance of learning by applying generalized delta learning and the Delta-bar-Delta algorithm [5]. Experiments using many images of shipping containers showed that the proposed identifier extraction method and the ART2-based supervised learning algorithm is more improved compared with the methods proposed previously.

2 A Proposed Container Identifier Recognition Method

2.1 Extraction of Container Identifier Areas

This paper detects edges of identifiers by applying Sobel masking operation to a grayscale image of the original image and extracts areas of identifiers using information on edges. Sobel masking operation is sensitive to noises so that it detects noises by an external light as edges. To remove an effect of noises in the edge detection, first, this paper detects noise pixels by using a fuzzy method and replaces the pixels with a mean gray value. Next, Applying Sobel masking to the noise-removed image, areas of container identifiers are separated from background areas.

2.2 Fuzzy-Based Noise Detection

To remove noises by an external light, this paper convert an container image to a grayscale one and apply the membership function like Fig. 1 to each pixel of the grayscale image, deciding whether the pixel is a noise or not. In Fig. 1, C and E are categories being likely to belong to an area of identifiers, and D is the

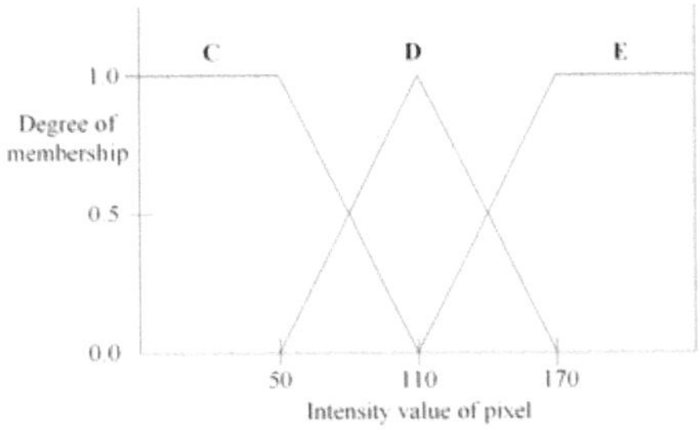

Fig. 1. Membership function(G) for gray-level pixels

Table 1. Criterion to distinguish pixels of noise and non-noise

pixel of non-noise	$u(G) < 0.42$
pixel of noise	$u(G) \geq 0.42$

category being likely to be a noise. The criterion to distinguish pixels of noise and non-noise using the degree of membership in this paper is given in Table 1.

To observe the effectiveness of the fuzzy-based noise detection, results of edge detection by Sobel masking were compared between the original image and the noise-removed image by the proposed method. Fig. 2(a) is the original container image, and Fig. 2(b) is the output image generated by applying only Sobel masking to a grayscale image of Fig. 2(a). Fig. 2(c) is results of edge detection obtained by applying the fuzzy-based noise removal and Sobel masking to Fig.2(a). First, the fuzzy-based noise detection method is applied to a grayscale image of the original image and pixels detected as noises are replaced with a mean gray value. Next, edges of container identifiers are detected by applying Sobel masking to the noise-removed image. As shown in Fig. 2(c), noise removal by the proposed fuzzy method generates more efficient results in the extraction of areas of identifiers.

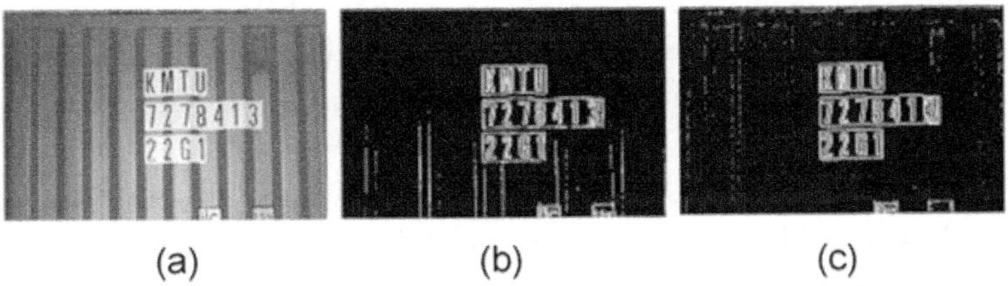

Fig. 2. (a) An original container image, (b) Result of edge detection by only Sobel masking, (c) Result of edge detection by fuzzy-based noise-removal and Sobel masking

2.3 Binarization of Container Identifier Areas

Currently, the iterative binarization algorithm is mainly used in the preprocessing of pattern recognition. The iterative binarization algorithm, first, roughly determines an initial threshold, divides an input image to two pixel groups using the threshold, calculates a mean value for each pixel group, and sets the arithmetic mean of two mean values to a new threshold. And, the algorithm repeats the above processing until there is no variation of threshold value and sets the last value to the threshold value for binarization operation. In the case of a noise-removed container image, since the difference of intensity between the background and the identifiers is great, the iterative algorithm is able to provide a good threshold value.

2.4 Extraction of Individual Identifiers

This paper extracts individual identifiers by applying 8-directional contour tracking method[6] to binarized areas of container identifiers. In the extraction process, the extraction of individual identifiers is successful in the case that the background color is a general color except white one like Fig. 3, and on the other hand, the extraction is failed in the case with white background color as shown in Fig. 4(a).

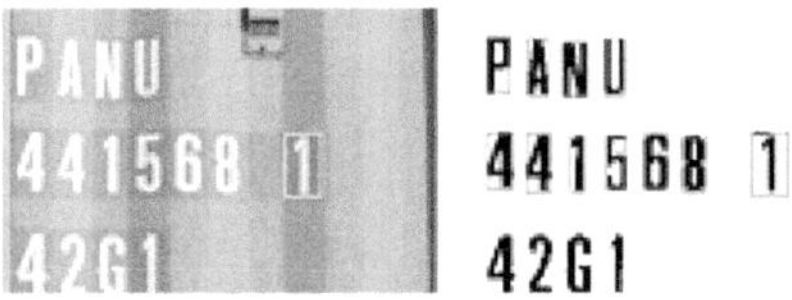

Fig. 3. Identifier area with a general color and successful results of edge extraction

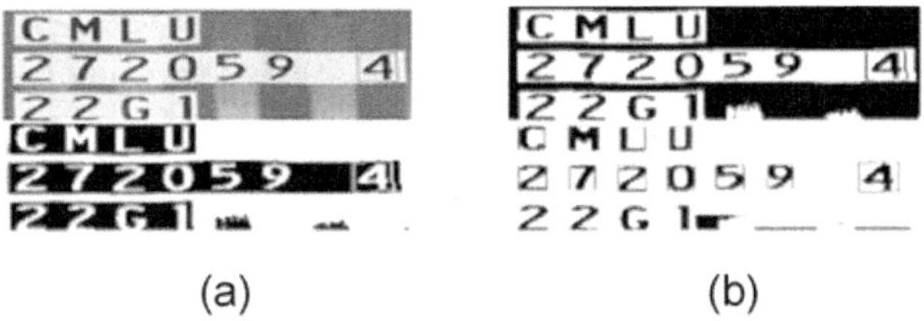

Fig. 4. (a) Identifier area with white color and failed results of edge extraction, (b) Reversed binarized area of Fig. 4(a) and successful result of edge detection

In the binarization process, background pixels of a bright intensity are converted to black and identifier pixels of a dark intensity are converted to white. Since the contour tracking method detects edges of an area with black color, it cannot detect edges of identifiers from target areas with white background. This paper, for identifier areas with white background, reverses a result of binarization process. That is, background pixels are converted to white and identifier pixels to black. Fig. 4(b) shows that the pixel reversal lead to a success of edge detection in an identifier area with white background presented in Fig. 4(a).

The procedure of extracting individual identifiers using 8-directional contour tracking method is as follow: P_i^r and P_i^c are pixels of horizontal and vertical directions being currently scanned in the identifier area, respectively, and P_i^{r+1} and P_i^{c+1} are pixels of the two directions being next scanned in the identifier area. And P_s^r and P_s^c are pixels of horizontal and vertical directions in the first mask of 8-directional contour tracking.

Step 1. Initialize with Eq. (1) in order to apply 8-neighborned contour tracking algorithm to the identifier area, and find the pixel by applying tracking mask as shown in Fig. 5.

$$P_i^{r-1} = P_i^r, \quad P_i^{c-1} = P_i^c \tag{1}$$

Step 2. When a black pixel is found after applying the tracking mask in the current pixel, calculate the value of P_i^r and P_i^c as shown in Eq. (2)

$$P_i^r = \sum_{i=0}^{7} P_i^{r+1}, \quad P_i^c = \sum_{i=0}^{7} P_i^{c+1} \tag{2}$$

Step 3. For the 8 tracking masks, apply Eq. (3) to decide the next tracking mask.

$$P_i^r = P_i^{r+1} \; and \; P_i^c = P_i^{c+1} \text{ then rotates counter-clockwise.} \tag{3}$$

Step 4. If P_i^r and P_i^c return to P_s^r and P_s^c, stop the contour tracking. Otherwise, go to step 1. That is, if $|P_i^r - P_s^r| \leq 1 \; and \; |P_i^c - P_s^c| \leq 1$ then exit the loop, else go to step 1.

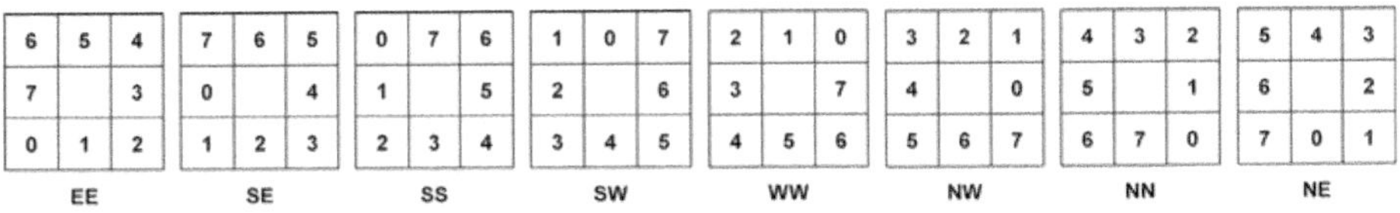

Fig. 5. 8-diectional contour tracking masks

Fig.3 and Fig. 4(b) shows extraction results of individual identifiers by using 8-directional contour tracking method.

2.5 Recognition of Container Identifiers Using ART2-Based Self-organizing Supervised Leaning Algorithm

This paper proposed an ART2-based self-organizing supervised learning algorithm for the recognition of container identifiers. First, a new leaning structure is applied between the input and the middle layers, which applies ART2 algorithm between the two layers, select a node with maximum output value as a winner node, and transmits the selected node to the middle layer. Next, generalized Delta learning algorithm and Delta-bar-Delta algorithm are applied in the learning between the middle and the output layers, improving the performance of learning. The proposed learning algorithm is summarized as follows:

1. The connection structure between the input and the middle layers is like ART2 algorithm and the output layer of ART2 becomes the middle layer of the proposed learning algorithm.

2. Nodes of the middle layer mean individual classes. Therefore, while the proposed algorithm has a fully-connected structure on the whole, it takes the winner node method that compares target vectors and output vectors and back-propagates a representative class and the connection weight.

3. The proposed algorithm performs the supervised learning by applying generalized Delta learning as the learning structure between the middle and the output layers.

4. The proposed algorithm improves the performance of learning by applying Delta-bar-Delta algorithm to generalized Delta learning for the dynamical adjustment of a learning rate. When defining the case that the difference between the target vector and the output vector is less than 0.1 as an accuracy and the opposite case as an inaccuracy, Delta-bar-Delta algorithm is applied restrictively in the case that the number of accuracies is greater than or equal to inaccuracies with respect to total patterns. This prevents no progress or an oscillation of learning keeping almost constant level of error by early premature situation incurred by competition in the learning process.

The detailed description of ART2-based self-organizing supervised learning algorithm is like Fig. 6.

3 Performance Evaluation

The proposed algorithm was implemented by using Microsoft Visual C++ 6.0 on the IBM-compatible Pentium-IV PC for performance evaluation. 79 container images with size of 640x480 were used in the experiments for extraction and recognition of container identifiers. In the extraction of identifier areas, the previously proposed method fails to extract in images containing noises vertically appearing by an external light and the rugged surface shape of containers. On the other hand, the proposed extraction method detects and removes noises by using a fuzzy method, improving the success rate of extraction compared with the previously proposed. The comparison of the success rate of identifier area extraction between the proposed in this paper and the previously proposed is like Table 2.

Table 2. Comparison of the success rate of identifier area extraction

	Previously-proposed method	Proposed method in this paper
Success rate	55/79(69.6%)	72/79(91.1%)

For the experiment of identifier recognition, applying 8-directional contour tracking method to 72 identifier areas extracted by the proposed extraction algorithm, 284 alphabetic characters and 500 numeric characters were extracted. This paper performed the recognition experiments with the FCM-based RBF

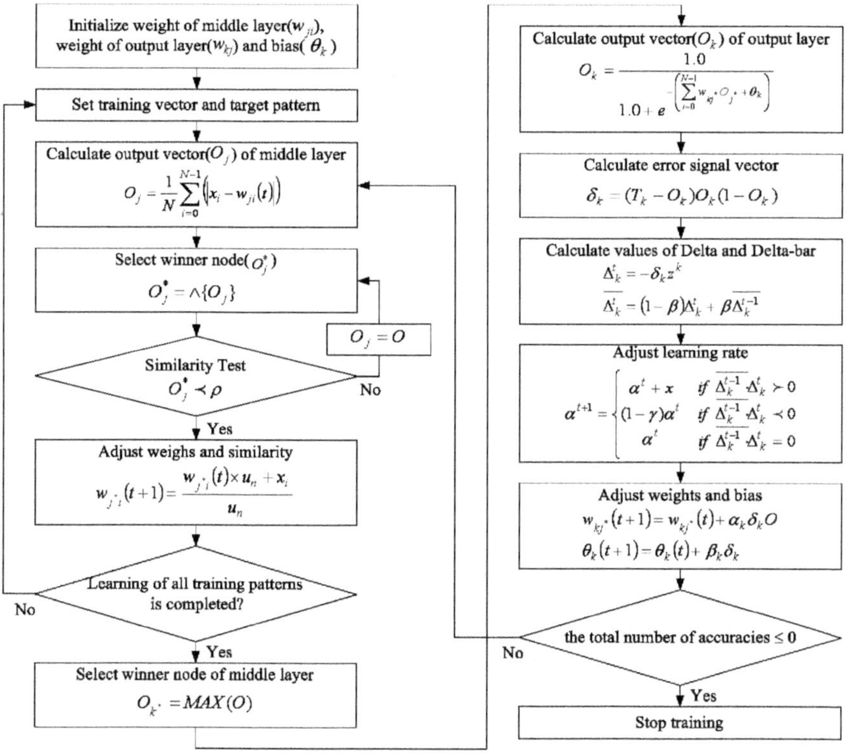

Fig. 6. ART2-based self-organizing supervised learning algorithm

Table 3. Evaluation of recognition performance

	FCM-based RBF network		ART2-base self-organizing supervised learning algorithm	
	# of Epoch	# of success	# of Epoch	# of success
Alphabetic characters(284)	236	240 (84.5%)	221	280 (98.5%)
Numeric characters(500)	161	422 (84.4%)	151	487 (97.4%)

network and the proposed ART2-based self-organizing supervised learning algorithm using extracted identifier characters and compared the recognition performance in Table 3.

In the experiment of identifier recognition, the learning rate and the momentum were set to 0.4 and 0.3 for the two recognition algorithms, respectively. And, for ART2 algorithm generating nodes of the middle layer in the proposed algorithm, vigilance variables of two character types were set to 0.4.

When comparing the number of nodes of the middle layer between the two algorithms, The proposed algorithm creates more nodes than FCM-based RBF network, but via the comparison of the number of Epochs, it is known that the number of iteration of learning in the proposed algorithm is less than

FCM-based RBF network. That is, the proposed algorithm improves the performance of learning. Also, Comparing the success rate of recognition, it is able to be known that the proposed algorithm improves the performance of recognition compared with FCM-based RBF network. Failures of recognition in the proposed algorithm were incurred by the damage of shapes of individual identifiers in original images and the information loss of identifiers in the binarization process.

4 Conclusions

This paper proposed an automatic recognition system of shipping container identifiers using fuzzy-based noise removal method and ART2-based self-organizing supervised learning algorithm. In this paper, after detecting and removing noises from an original image by using a fuzzy method, areas of identifiers are extracted. In detail, the performance of identifier area extraction is improved by removing noises incurring errors using a fuzzy method based on the feature that the color of container identifiers is white or black on the whole. And, individual identifiers are extracted by applying 8-directional contour tracking method to extracted areas of identifiers. This paper proposed an ART2-based self-organizing supervised learning algorithm and applied to the recognition of individual identifiers. Experiments using 79 container images showed that 72 areas of identifiers and 784 individual identifiers were extracted successfully and 767 identifiers among the extracted were recognized by the proposed recognition algorithm. Failures of recognition in the proposed algorithm were incurred by the damage of shapes of individual identifiers in original images and the information loss of identifiers in the binarization process.

A Future work is the development of fuzzy association algorithm that may recover damaged identifiers to improve the performance of extraction and recognition of individual identifiers.

References

1. ISO-6346, Freight Containers-Coding -Identification and Marking, (1995)
2. Kim, K. B.: Recognition of Identifiers from Shipping Container Images using Fuzzy Binarization and Neural Network with Enhanced Learning Algorithm. Applied Computational Intelligence. World Scientific (2004) 215-221
3. Nam, M. Y., Lim, E. K., Heo, N. S., Kim, K. B.: A Study on Character Recognition of Container Image using Brightness Variation and Canny Edge. Proceedings of Korea Multimedia Society 4(1) (2001) 111-115
4. Kim, N. B.: Character Segmentation from Shipping Container Image using Morphological Operation. Journal of Korea Multimedia Society 2(4) (1999) 390-399
5. Vogl, T. P., Mangis, J. K., Zigler, A. K., Zink, W. T., Alkon, D. L.: Accelerating the convergence of the backpropagation method. Biological Cybernetics 59 (1998) 256-264
6. Chen, Y. S., Hsu, W. H.: A systematic approach for designing 2-Subcycle and pseudo 1-Subcycle parallel thinning algorithms. Pattern Recognition 22(3) (1989) 267-282

A Hybrid Estimation of Distribution Algorithm for CDMA Cellular System Design

Jianyong Sun[1], Qingfu Zhang[2], Jin Li[1], and Xin Yao[1]

[1] Cercia, School of Computer Science, University of Birmingham, Edgbaston, Birmingham, B15 2TT
{j.sun, j.li, x.yao}@cs.bham.ac.uk
[2] Department of Computer Science, University of Essex, Wivenhoe Park, CO4 3SQ
qzhang@essex.ac.uk

Abstract. While code division multiple access (CDMA) is becoming a promising cellular communication system, the design for a CDMA cellular system configuration has posed a practical challenge in optimisation. The study in this paper proposes a hybrid estimation of distribution algorithm (HyEDA) to optimize the design of a cellular system configuration. HyEDA is a two-stage hybrid approach built on estimation of distribution algorithms (EDAs), coupled with a K-means clustering method and a simple local search algorithm. Compared with the simulated annealing method on some test instances, HyEDA has demonstrated its superiority in terms of both the overall performance in optimisation and the number of fitness evaluations required.

Keywords: CDMA cellular system configuration design, hybrid estimation of distribution algorithm.

1 Introduction

In the last decade, code division multiple access (CDMA) has become a promising technology for the mobile communication [7]. The rapidly increasing demands for cellular communication raise an important optimisation challenge for cellular service providers. In the design of a cellular communication network, the quality of service, the service coverage and the network cost are three most concerned objectives among many others. The three objectives are largely influenced by certain design parameters, such as the number of based stations (BSs), the locations of BSs, as well as the powers and antennae heights adopted in BSs.

The cellular system configuration problem in this study is taken from [12]: Given a static, spatial distribution of customer demands, find a CDMA cellular system configuration, i.e., the locations of BSs, their corresponding transmitting powers and antennae heights, with the aim of optimizing call quality, service coverage and total cost of system configuration. To simplify this configuration design problem, Melachrinoudis and Rosyidi [12] have transformed this multi-objective optimisation problem into a single objective optimisation problem by aggregating three objectives with weights. In this way, the simulated annealing (SA) method was adopted by [12] to search for optimal solutions.

T.-D. Wang et al. (Eds.): SEAL 2006, LNCS 4247, pp. 905–912, 2006.

In this paper, we shall present a hybrid estimation of distribution algorithm (HyEDA). We apply HyEDA to tackle the problem using the transformed single-objective function given in [12]. HyEDA is a two-stage hybrid evolutionary approach, built on estimation of distribution algorithm [5]. It is incorporated with the well-known K-means clustering and a simple local search algorithm. The first stage aims to find optimal or near-optimal locations of BSs using the hybrid estimation of distribution algorithm (EDA). Given the locations of BSs found in the first stage, the second stage is intended to find optimal or near-optimal corresponding power and antennae height for each BS using a similar simple local search method. Taking this two-stage optimisation is motivated by the intuition that the locations of BSs plays a vital role in achieving good overall performance of a cellular service network in terms of three objectives discussed aforementioned. The effectiveness of this two-stage approach has been further illustrated by our experimental results in this study, in comparison to SA.

The rest of the paper is organized as follows. Section 2 describes the mathematical model of the cellular system configuration design problem proposed by Melachrinoudis and Rosyidi [12]. In Section 3, we shall discuss HyEDA in more detail. Experimental results of HyEDA on several test problems, in comparison to that of the simulated annealing method, are given in Section 4. Section 5 concludes the paper.

2 Mathematical Model

In this section, we would like to describe, in details, some notations for the design problem of a CDMA cellular system configuration, as well as, in brief, the final transformed single-objective mathematical model. The concepts discussed below are generally borrowed from the work done by Melachrinoudis and Rosyidi [12], simply because their formulation is more close to the reality of the problem, thereby suited to be taken for the application.

Given a service area, A, which is a two dimensional geographical region, we could discretise A into a lattice of grid points, each of which, e.g., $g(j, k)$, is identified by their discrete coordinates (j, k), where $1 \leq j \leq M$ and $1 \leq k \leq K$. M and K are the maximum row and column of the lattice respectively, which are set by service engineers. Grid points are used to denote the locations of potential BSs. A cell i is defined as a set of grid points covered by BS_i.

Suppose that for a cellular system, n BSs and the BSs' corresponding powers and antenna heights are going to be configured. The values of power and antenna height can only take discrete values from $P = [p_{min}, p_{max}]$ and $H = [h_{min}, h_{max}]$ with cardinalities $|P|$ and $|H|$, respective. The overall performance of the system configuration is measured by three objectives, i.e., call quality, service coverage and the cost of system.

The call quality can be measured by the bit error rate (BER) at MSs in the process of demodulation. The smaller the BER, the clearer the voice, and the higher the call quality. The BER value of a MS depends on its location within a cell. We denote the maximum (worst) BER value within a cell as a measure of that cell's call quality, e.g., $\max_{(j,k)\in S_i} BER_{jk}$ for cell i. For the whole area covered with all cells, optimizing call quality is transformed into minimizing the maximum BER among all cells. In practice, the deviation value d_{1i}^+ from t_1 (a threshold for the BER) is defined to reflect the call quality of Cell i as follows:

$$d_{1i}^+ = \begin{cases} \max_{(j,k)\in S_i} BER_{jk} - t_1, & \text{if } \max_{(j,k)\in S_i} BER_{jk} \geq t_1; \\ 0, & \text{otherwise.} \end{cases} \quad (1)$$

Similarly, for the measurement of the service coverage for Cell i, the deviation d_{2i}^+ of uncovered service (UCS) from a threshold, t_2 is defined as follows.

$$d_{2i}^+ = \begin{cases} UCS_i - t_2, & \text{if } UCS_i \geq t_2; \\ 0, & \text{otherwise.} \end{cases} \quad (2)$$

The goal regarding service coverage is to minimize the $\max_i d_{2i}^+$.

The total cost of the system configuration, denoted by TCsystem, includes the cost of the base stations construction, their costs of powers and antennae heights. A normalized figure, TC = TCsystem/TCsystem_max, is used, where TCsystem_max is the maximum cost of of the whole system, which is defined as the total cost of n BSs, built at the most expensive locations using the largest possible size, the highest possible antenna. Note that the Tchebycheff distance among grid points is used to calculate the evaluation factors for the system configuration [12].

Eventually, the goal of the cellular system configuration design problem is to minimize the following objective function Z:

$$Z = w_1 \max_i d_{1i}^+ + w_2 \max_i d_{2i}^+ + w_3 TC. \quad (3)$$

with the locations of BSs, their corresponding powers and antenna heights as decision variables. Readers please refer to [12] for details of the considered problem and its formulation.

3 HyEDA

3.1 Algorithm Framework

HyEDA is built mainly on the hybrid estimation of distribution algorithms (EDA) [1][2][9][5]. EDA is a type of evolutionary algorithm (EAs). EAs, such as genetic algorithms (GAs), generate new offspring by using genetic operators like crossover and mutation. Different from GAs, the main characteristic of EDA is that new offspring is generated by sampling from a probabilistic distribution model. EDA extracts global statistical information from visited promising solutions to build the probabilistic distribution model. However, EDA alone cannot efficiently search for optimal or near-optimal solutions of difficult optimisation problems [5]. To improve the efficiency of EDAs, hill climibing algorithms and other techniques can be incorporated [8]. In our previous work, we have hybridized EDA with hill climbing algorithms and other techniques to solve some

$\mathcal{NP}$-hard optimisation problems, such as maximum clique problem [17], and the routing and wavelength assignment problem under shared-risk-link-group constraints [16] arisen in the telecommuncation area, and continuous optimisation problems [15][6].

The general template of the hybrid estimation of distribution algorithm is summarized as follows.

Initialization. Initial the probability model $p_0(x)$, sample *popsize* feasible solutions from it, apply local search algorithm to each solution. The resultant solutions consists of the initial population $P(0)$; Set $t := 0$;

While (stop criteria are not met), do :

1) **Selection** Select *selsize* solutions as the parent set $Q(t)$;
2) **Modeling** Construct a probabilistic model $p_t(x)$ according to $Q(t)$;
3) **Sampling** Sample *cresize* offspring from $p_t(x)$. Apply local search algorithm to each sampled solution;
4) **Replacement** Replace partially the current population with the offspring to constitute a new population $P(t+1)$; set $t := t+1$;

The template consists of five major components, including the fitness definition, the selection, the probability construction, the sampling and the replacement. In the evolutionary process of the proposed first-stage algorithm for the cellular system design problem, the Z value of a solution is defined as its fitness. The truncation selection is adopted. At generation t, N solutions with the smallest Z values are selected to constitute the parent set $Q(t)$ (Step 1). We use the K-means clustering to help the construction of the probability model $p_t(x)$ (Step 2). Details of Step 2 will be described in Section 3.3. The sampling operation creates new offspring using the previously created probability model (step 3). We take the "Accept-Reject" sampling method, which shall be described in Section 3.3. Notably, a local search algorithm is incorporated into the template, in order to tune each solution into local optimum after sampling both in population initialization and in offspring generation. We shall describe the local search algorithm in Section 3.4. To produce $P(t+1)$, the *popsize* solutions with the smallest Z values are selected from the union of the sampled solutions so far and the solutions in $P(t)$.

3.2 Solution Representation and Probability Model Construction

A solution x of the cellular system configuration can be represented as a vector $[(x_1, y_1, p_1, h_1), ..., (x_n, y_n, p_n, h_n)]$, where (x_i, y_i) represents the location of BS i, p_i and h_i are the power and antenna height of BS i, respectively. In the first stage of HyEDA, for each BS, its power and antenna height values are randomly picked from P and H. Those values are fixed over the whole process of optimisation. In the second stage, given its location, (x_i, y_i) is found, we use a simple local search algorithm to optimize (p_i, h_i) with respect to Function Z for each BS.

The construction of the probability model resorts to the K-means clustering. In [4][11], the K-means clustering was used to help the construction of a Gaussian distribution model for continuous optimisation problem. Here, the K-means clustering method is used to cluster a set of grid points with discrete values.

The value of p_{ij} indicates the probability that a BS is located in (i, j). Probability model defines the way of assigning a probability value to each grid at each generation. At generation t, to construct the probability model from the parent set $Q(t)$, i.e., $p_t(x)$ (here, $p_t(x) = \{p_{ij}\}_{i=1}^{M}{}_{j=1}^{K}$), we first cluster the totally $N \times n$ points distributed in the service area into n groups. Then we assign a probability p_{ij} to each grid point (i, j) based on the results of clustering.

Suppose that the coordinates of the cluster centroid are $\{(x_k^*, y_k^*), 1 \leq k \leq n\}$ after clustering, and the corresponding standard deviation of distances between points to the centroid of cluster is $\{\sigma_k, 1 \leq k \leq n\}$. Generally speaking, a larger σ_k means a bigger uncertainty to locate base station k at (x_k^*, y_k^*). Whereas a smaller σ_k means a smaller uncertainty, i.e., we have more confidence to set base station k at the centroid (x_k^*, y_k^*). Let d_l denote the Tchebycheff distance (The Tchebycheff distance between two vectors a and b is defined as $\max_i |a_i - b_i|$) between the grid point (i, j) and the centroid (x_l^*, y_l^*) for $1 \leq l \leq n$, and $k = \arg\min_l d_l$, then p_{ij} is assigned as follows

$$p_{ij} = \begin{cases} \epsilon + \mathcal{N}(\frac{d_k}{d_{\max}^k}; 0, \sigma_k^2) & \text{for } d_k \leq d_{\max}^k; \\ \epsilon, & \text{otherwise.} \end{cases} \tag{4}$$

where ϵ is a positive value to guarantee that all grid point has a probability to be chosen; $\mathcal{N}$ is the normal probability density function (pdf); d_k is the Tchebycheff distance between (i, j) to its nearest centroid (x_k, y_k); and $d_{\max}^k$ is the maximum Tchebycheff distance among the points of cluster k to the centroid. Based on Equation (4), it can be seen that the degree of variation in the values, p_{ij}, assigned to all grid points in a cluster k, would crucially depend on the size of the value of σ_k. If σ_k were smaller, the resultant probability values would vary to a greater extent. Otherwise, they are not much different.

3.3 Sampling Method

EDA employs the same sampling method for generating both an initial population and a number of offspring populations over the generations during evolution. For generating an initial population, probabilities of all $M \times K$ grid points are each set to be $1/(M * K)$. Whereas, in generating a population of offspring, the p_{ij} for each grid point produced in Step 2 is taken.

The sampling process for an individual solution is as follows. To locate the n grid points required for the solution, we select locations based on the probability model one by one. In each step, we employ the well-known roulette wheel selection method to select a location from all available grid points based on their probabilities. A grid point with a higher probability is more likely to be selected. Once the grid point is picked as a BS location, its neighborhoods are not considered as the candidates for the next potential BS location any more. We achieve this through setting the probabilities of its neighborhood to be zero. The neighborhood set of the grid point g_i is defined as $\mathcal{D}(g_i) = \{z : d(z, g_i) \leq d_{\min}\}$, where d_{min} is a user-defined parameter and $d(z, g_i)$ is the Tchebycheff distance between z and g_i.

3.4 Local Search

A local search algorithm has been used in the first stage of HyEDA, to tune each solution after initialization and offspring generation. In the second stage, HyEDA merely takes this local search method to find optimal or near-optimal values of (p, h) for each BS. The only difference of the local method between two stages lies in the definitions of neighborhood.

A solution $z = [(x_1, y_1), \cdots, (x_n, y_n)] = (g_1, \cdots, g_n)$ represents the BS locations assigned in the first stage. For each location $g_i = (x_i, y_i)$, its neighborhood $\mathcal{N}(g_i)$ is defined as the other 8 grid points around g_i, i.e., $\mathcal{N}(g_i) = \{u : d(u, g_i) = 1.\}$. Whereas, for each p_i $(,h_i)$, to be optimized in the second stage, its neighborhood $\mathcal{N}(p_i)$ $(,\mathcal{N}(h_i))$ is defined as all of other $|P| - 1$ $(|H| - 1)$ possible values. The local search algorithm starts from a randomly initialized solution, for each component of the solution, the algorithm searches the component's neighborhood for the "best" component value, which minimize the current objective function value. The best component value updates the current one to form a new solution. The search continues until no better solution can be found.

4 Experimental Results

In [12], a case study for a cellular system configuration was conducted for the service area of the city of Cambridge and its vicinity in Easter Massachusetts, and SA was applied to solve the problem. For that case, the service area is divided into 13×20 grid points, with $n = 11$ base stations supposed to be located for the whole area, and the weights for the objectives set to be $w_i = 1/3, 1 \leq i \leq 3$. We denote this problem as problem 1. To our best knowledge from literature, we are not aware of other algorithms which have been used to address the cellular system configuration design problem, except for the simulated annealing (SA) approach given in [12]. Therefore, our experimental comparison is carried out only between the SA and HyEDA.

To make the comparison between SA and HyEDA more rigourously, apart from the above problem 1, additional four problems, which are newly derived by us based on problem 1, have also been used to compare effectiveness of both algorithms. All 5 problems share the same settings of the service area and the number of BSs required to be located, but have a different scalable setting in resolutions of grid points, which determines how precisely 11 BSs can be located in the area. In problem 2 to 5, the same area are divided into 20×30, 26×40, 39×60, and 65×100 grid points respectively. As a result, finding an optimal locations turns to be even harder progressively from problem 1 to problem 5, simply because of the enlarged search space. In our experiment, the $d_{\min}$ values are set to 2 for the first two problems, 3 for problems 3 and 4, 4 for problem 5. The population size is set to 10 for problems 1 and 2, 20 for the rest problems. The algorithms will terminate if either the algorithm cannot find a better solution in successively 10 generations after 30 generations or a maximum 1,000,000 objective function evaluations are achieved.

Table 1. Comparisons of SA and HyEDA in terms of the mean best objective function values (Mean) and the number of function evaluations (Nfe)

problem	SA		HyEDA	
instance	Mean	Nfe.	Mean	Nfe.
1	0.2414	157,893	0.2302	59,140
2	0.2440	451,120	0.2310	209,053
3	0.2491	618,278	0.2415	380,304
4	0.3167	1,000,000	0.2662	482,288
5	0.3618	1,000,000	0.3058	529,205

For each problem, we run both SA and HyEDA 10 times each. Listed in Table 1 are the average fitness objective value and the average number of fitness evaluations of 10 runs, produced by both methods on each problem. For each of 5 problems, the average fitness value produced by HyEDA is smaller than that achieved by SA, whilst the number of objective funcation evaluations required to achieve the results by HyEDA is less than that in SA. The results have demonstrated that HyEDA outperforms SA in terms of both the solution quality and the number of fitness evaluations that are required. In other words, HyEDA is more likely to find better solutions than SA using a smaller number of fitness evaluations in solving a cellular CDMA system configuration design problem.

5 Conclusion

In this paper, we proposed a two-stage hybrid estimation of distribution algorithm, HyEDA, for solving the CDMA cellular system configuration design problem. HyEDA resorts to the K-means clustering method for probability model construction, and a simple local search algorithm for solution quality improvement. HyEDA has been applied to tackle a problem given in [12], as well as some of its derived and more difficult cases. The experimental results have demonstrated that the proposed algorithm outperforms the simulated annealing approach used in [12], in terms of the quality of the solutions found and the quantity of function evaluations used.

In the future, HyEDA could be improved in several ways. Firstly, we would like to carry out more experiments to thoroughly understand the effects of the components of HyEDA including the two-stage framework, the clustering method, and the Techebycheff distance metric. Secondly, we will try to improve HyEDA itself for its convergence and robustness for solving optimisation problems. Thirdly, we may enhance HyEDA by embedding multiobjective search engines [10], thereby enabling itself to handle the inherently multiobjective optimisation in the CDMA cellular system configuration design. Finally, we shall explore the principle of HyEDA further in solving similar other optimisation problems, such as the terminal assignment [13], the task assignment problem [14] and other optimisation problems raised in telecommunication area.

References

1. S. Baluja, Population-based Incremental Learning: A Method for Integrating Genetic Search Based Function Optimization and Competitive Learning, CMU-CS-94-163, Carnegie Mellon University, Pittersburgh, PA, 1994.
2. H. Mühlenbein and G. Paaß, From Recombination of Gens to the Estimation of Distribution I: Binary Parameters, in Lecture Notes in Computer Science 1411: Parallel Problem Solving from Nature - PPSN IV, pp. 178-187, 1996.
3. V.K. Garg, K. Smolik and J.E. Wilkes, Applications of CDMA in Wireless/Personal Communications, Prentice-Hall, Upper Saddle River, NJ, 1997.
4. M. Pelikan and D.E. Goldberg, Genetic Algorithms, Clustering, and the Breaking of Symmetry, Illinois Genetic Algorithm Laboratory, University of Illinois at Urbana-Champaign, Urbana, IL, 2000.
5. J. Sun, Hybrid Estimation of Distribution Algorithms for Optimization Problems, PhD thesis, University of Essex, 2005.
6. J. Sun, Q. Zhang and E.P.K. Tsang, "DE/EDA: A New Evolutionary Algorithm for Global Optimization," Information Sciences, 169(3), pp. 249–262, 2005.
7. K. I. Kim, "Handbook of CDMA System Design, Engineering and Optimisation," Prentice Hall PTR, 1999.
8. N. Krasnogor, W.Hart and J. Smith (editors), Recent Advances in Memetic Algorithms and Related Search Technologies, to appear, Springer, 2003.
9. P. Larraanaga and J.A. Lozano, Estimation of Distribution Algorithms: A New Tool for Evolutionary Computation. Kluwer Academic Publishers, Norwell, MA, USA, 2001.
10. J. Li, S. Taiwo, Enhancing Financial Decision Making Using Multi-Objective Financial Genetic Programming, to appear in Proceeding of 2006 IEEE Congress on Evolutionary Computation (CEC'06), July 16-21, 2006 Sheraton Vancouver Wall Centre, Vancouver, BC, Canada , 2006
11. Q. Lu and X. Yao, "Clustering and Learning Gaussian Distribution for Continuous Optimisation," IEEE Transcations on Systems, Man, and Cybernetics, Part C: 35(2):195-204, May 2005.
12. E. Melachrinoudis, B. Rosyidi, "Optimizing the Design of a CDMA Cellular System Configuration with Multiple Criteria," Annals of Operations Research, 106, pp. 307-329, 2001.
13. S. Salcedo-Sanz, Y. Xu and X. Yao, "Hybrid Meta-Heuristic Algorithms for Task Assignment in Heterogeneous Computing Systems," Computers and Operations Research, 33(3):820-835, 2006.
14. S. Salcedo-Sanz and X. Yao, "A Hybrid Hopfield Network – Genetic Algorithm Approach for the Terminal Assignment Problem, " IEEE Transactions on System, Man and Cybernetics, Part B: Cybernetics, 34(6): 2343-2353, December, 2004.
15. Q. Zhang, J. Sun, E.P.K. Tsang, J.A. Ford, "Hybrid estimation of distribution algorithm for global optimisation," Engineering Computations, 21(1), pp. 91-107, 2003.
16. Q. Zhang, J. Sun, G. Xiao and E. Tsang, "Evolutionary Algorithm Refining a Heuristic: A Hybrid Method for Shared Path Protection in WDM Networks under SRLG Constraints," to be appeared in IEEE Transactions on System, Man and Cybernetics, 2006.
17. Q. Zhang, J. Sun and E.P.K. Tsang, "Evolutionary Algorithm with the Guided Mutation for the Maximum Clique Problem", IEEE Transactions on Evolutionary Computation, 9(2), pp. 192-200, 2005.

Optimal Motion Generation of Flexible Macro-micro Manipulator Systems Using Estimation of Distribution Algorithm

Yu Zhang, Shude Zhou, Tangwen Yang, and Zengqi Sun

Department of Computer Science and Technology,
Tsinghua University, Beijing 100084, China
Zhang-y-03@mails.tsinghua.edu.cn

Abstract. In this paper, a new approach for motion generation and optimization of the flexible macro-micro manipulator system is proposed based on Estimation of Distribution Algorithm (EDA). The macro-micro manipulator system is a redundant system, of which inverse kinematics remains challenging, with no generic solution to date. Here, the manipulator system configurations, or the optimal joint motions, are generated using the EDA algorithm base on Gaussian probability model. Compared with simple genetic algorithms (SGA), this approach uses fewer parameters and the time for motion optimization is remarkably reduced. The proposed approach shows excellent performance on motion generation and optimization of a flexible macro-micro manipulator system, as demonstrated by the simulation results.

1 Introduction

A flexible macro-micro manipulator comprises a macro flexible manipulator and a micro rigid manipulator. The macro one moves in a large workspace but motion errors are also large due to the link flexibility. Conversely, the small rigid micro manipulator, which is attached to the tip of the macro manipulator, can move fast and precisely, and herein is used to compensate for the motion errors caused by the link deformations and vibrations of the macro manipulator. Usually, the joint numbers of a macro-micro manipulator system are more than the dimensions of its task space. Apparently, it is a redundant robot system. Moreover, flexibility causes large link deformations and vibrations when it moves rapidly. An optimal motion trajectory is thus critical for vibration suppression and trajectory tracking control of a flexible macro-micro manipulator system, as well as the tip error compensation.

Less attention has been paid to motion generation of a flexible macro-micro manipulator system. Usually, the desired joint motion is given a priori for the controller design. Actually, there is no unique closed-form solution to the redundant manipulator systems. The system redundant degree of freedom is used in [1-4] for obstacle and singularity avoidances, control torque optimization, and dexterity improvement.

T.-D. Wang et al. (Eds.): SEAL 2006, LNCS 4247, pp. 913 – 920, 2006.

Cherif et al. [5] used a set of penalty functions to transform the constrained optimization problem into an unconstrained problem, and then formulated a simple optimization problem using neural network. But the expression of the search direction is complex and difficult to derive.

T. Yoshikawa, et al [6] first introduced the compensability concept for the flexible manipulators. The inverse kinematics problem is solved based on this concept. But this algorithm requires computing the inverse or pseudo-inverse Jacobian matrix and a lot of derivation are needed in advance. In our previous work [7], we presented a method to generate optimal motion for a flexible macro-micro manipulator, using genetic algorithms (GA). But the algorithm is time consuming because of the large search space. Moreover, the behavior of the algorithm depends on a large number of parameters such as crossover and mutation operators, crossover and mutation probabilities, population size, and the number of generations, etc.

On the basis of the previous research work, Estimation of Distribution Algorithm (EDA) is employed for the motion generation and optimization of flexible macro-micro manipulator in this paper. EDAs [8] are non-deterministic, stochastic heuristic search strategies within the evolutionary computation approaches, which has recently become the hot research topic. The main characteristic of EDAs compared with other evolutionary search strategies is that the evolution from a generation to the next is done by estimating the probability distribution of the fittest individuals, and afterwards by sampling the induced model. It has widely used in optimization problems. Hu et al. [9] used it to study the biped gait optimization problem.

This paper is organized as follows. In section 2, we first formulate the kinematics and motion constraints of the flexible macro-micro manipulator system and then convert the motion generation problem into the optimization problem. Section 3 the EDA algorithm is proposed and employed for the motion generation and optimization. The simulation results are given in section 4. Finally, conclusions are given in section 5.

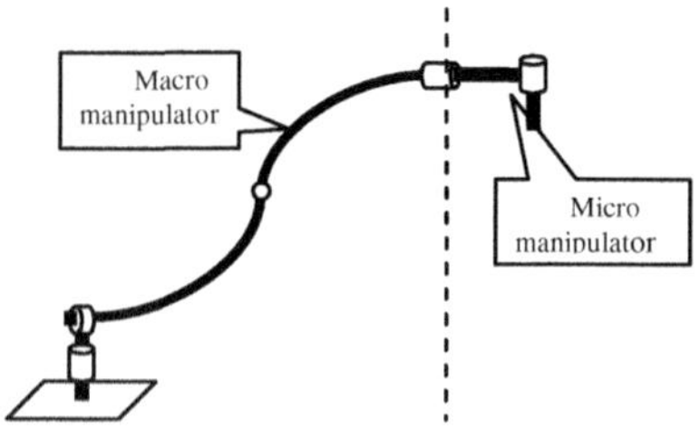

Fig. 1. A flexible macro-micro manipulator system

2 Problem Statement

2.1 Construction of Flexible Macro-micro Manipulator

A flexible macro-micro manipulator system is shown in fig. 1. The macro part uses large, lightweight and slender links; therefore, the deformation and vibration are easily induced. The micro part, which can move fast and precisely, is a small rigid manipulator. Combining the two, wide and precise operation tasks are able to be realized.

2.2 Kinematics Formulation

Let N and n denote the degrees of freedom of the macro manipulator and the micro manipulator, respectively, and k the degrees of freedom of the link deformations of the macro manipulator. An inertial coordinate system $O-xyz$ is defined for the flexible macro-micro manipulator. $\boldsymbol{P}=(p_1, p_2, \cdots p_m)^T$ ($\in R^m$) gives the tip position and orientation of the manipulator system with respect to the inertial coordinate system $O-xyz$. $\boldsymbol{\theta}=(\theta_{\mathrm{M}1}, \theta_{\mathrm{M}2}, \cdots \theta_{\mathrm{M}N}, \theta_{\mathrm{m}1}, \theta_{\mathrm{m}2} \cdots \theta_{\mathrm{m}n})^T$ ($\in R^{N+n}$) is the joint variables vector. $\boldsymbol{\theta}_{\mathrm{M}i}$ ($\in R^N$) and $\boldsymbol{\theta}_{\mathrm{m}i}$ ($\in R^n$) are the joint variables of the macro manipulator and of the micro manipulator, respectively. $\boldsymbol{E}=(e_1, e_2, \cdots e_k)^T$ ($\in R^k$) is the link deformation vector.

$\boldsymbol{P}$ is a nonlinear function of joint variables $\boldsymbol{\theta}$ and deformation $\boldsymbol{E}$, that is

$$\boldsymbol{P}=\boldsymbol{f}(\boldsymbol{\theta},\boldsymbol{E}) \ . \tag{1}$$

Differentiating the above equation, we have

$$\dot{\boldsymbol{P}}=\partial \boldsymbol{f}(\boldsymbol{\theta},\boldsymbol{E})/\partial t=\boldsymbol{J}\cdot\left[\dot{\boldsymbol{\theta}}^T,\dot{\boldsymbol{E}}^T\right]^T , \tag{2}$$

where $\boldsymbol{J}\in R^{m\times(N+n+k)}$ is the Jacobian matrix.

2.3 Compensability of Flexible Macro-micro Manipulator Systems

The compensability of flexible macro-micro manipulator systems is used to measure the capability of the micro manipulator of compensating for the tip motion errors induced by the macro manipulator while its configuration is fixed. The larger the compensability of the micro manipulator is, the less joint adjustment of the micro manipulator is needed for tip error compensation.

Eq. (2) can be further rewritten as

$$\Delta\boldsymbol{P}=\boldsymbol{J}_{\mathrm{M}}\Delta\boldsymbol{\theta}_{\mathrm{M}}+\boldsymbol{J}_{\mathrm{m}}\Delta\boldsymbol{\theta}_{\mathrm{m}}+\boldsymbol{J}_{\mathrm{E}}\Delta\boldsymbol{E} , \tag{3}$$

where the Jacobian matrices $\boldsymbol{J}_{\mathrm{M}}$, $\boldsymbol{J}_{\mathrm{m}}$ and $\boldsymbol{J}_{\mathrm{E}}$ are the functions of the joint variables $\boldsymbol{\theta}$ and the link deformation $\boldsymbol{E}$. $\Delta\boldsymbol{\theta}_{\mathrm{M}}$, $\Delta\boldsymbol{\theta}_{\mathrm{m}}$ and $\Delta\boldsymbol{E}$ are the infinitesimal displacement of $\boldsymbol{\theta}_{\mathrm{M}}$, $\boldsymbol{\theta}_{\mathrm{m}}$ and $\boldsymbol{E}$, respectively.

The motion errors at the tip of the flexible macro-micro manipulator system are compensated by adjusting the motion of the micro manipulator. Therefore, the adjustment of joint variables of macro manipulator $\Delta\boldsymbol{\theta}_{\mathrm{M}}$ is set to be zero. Since the tip errors will be eliminated, namely, $\Delta\boldsymbol{P}=0$, then from Eq. (3) we have

$$\boldsymbol{J}_{\mathrm{m}}\Delta\boldsymbol{\theta}_{\mathrm{m}} + \boldsymbol{J}_{\mathrm{E}}\Delta\boldsymbol{E} = 0 \tag{4}$$

and the compensation motion of the micro manipulator $\Delta\boldsymbol{\theta}_{\mathrm{m}}$ is obtained as

$$\Delta\boldsymbol{\theta}_{\mathrm{m}} = -\boldsymbol{J}_{\mathrm{m}}^{+} \cdot \boldsymbol{J}_{\mathrm{E}}\Delta\boldsymbol{E} \ . \tag{5}$$

In terms of the properties of matrix, we have the relationship of $\Delta\boldsymbol{\theta}_{\mathrm{m}}$ and $\Delta\boldsymbol{E}$,

$$\left\|\Delta\boldsymbol{\theta}_{\mathrm{m}}\right\| = \left\|\boldsymbol{J}_{\mathrm{m}}^{+} \cdot \boldsymbol{J}_{\mathrm{E}}\Delta\boldsymbol{E}\right\| \le \left\|\boldsymbol{J}_{\mathrm{m}}^{+} \cdot \boldsymbol{J}_{\mathrm{E}}\right\| \cdot \left\|\Delta\boldsymbol{E}\right\| \ . \tag{6}$$

Observing the above equation, we can see that, to ensure $\left\|\Delta\boldsymbol{\theta}_{m}\right\|$ be small enough when $\left\|\Delta\boldsymbol{E}\right\|$ is fixed, $\left\|\boldsymbol{J}_{\mathrm{m}}^{+} \cdot \boldsymbol{J}_{\mathrm{E}}\right\|$ should be reduced as much as possible. So, now we define the compensability of the micro manipulator as

$$C = (\left\|\boldsymbol{J}_{\mathrm{m}}^{+} \cdot \boldsymbol{J}_{\mathrm{E}}\right\|)^{-1} \ . \tag{7}$$

Apparently, if C increases, the compensability of the micro manipulator will also increase, and vice versa.

2.4 Optimization Problem Description

Two factors should be considered to generate the joint motions. One is the joint motion limitations of the manipulator system, and the other is the compensability of the micro manipulator. The motion optimal problem can be further described as follows:

1. The generated joint motion θ(k) should be in the allowed span of [-α, α].
2. The joint variation between two subsequent configurations should be as small as possible to reduce the energy consumption. That is

$$fitness1 = \min\left\|\boldsymbol{\theta}(k) - \boldsymbol{\theta}^{*}(k-1)\right\| = \min\left(\sum_{i=1}^{N}\left|\boldsymbol{\theta}_{\mathrm{Mi}}(k) - \boldsymbol{\theta}_{\mathrm{Mi}}^{*}(k-1)\right|^{2} + \sum_{i=1}^{n}\left|\boldsymbol{\theta}_{\mathrm{mi}}(k) - \boldsymbol{\theta}_{\mathrm{m}i}^{*}(k-1)\right|^{2}\right)^{1/2} \ . \tag{8}$$

3. The tip motion errors should be as small as possible, namely

$$fitness2 = \min\left\|\boldsymbol{P} - \boldsymbol{P}_{\mathrm{d}}\right\| = \min(\sum_{i=1}^{m}\left|p_{i} - p_{\mathrm{d}i}\right|^{2})^{1/2} \ , \tag{9}$$

where $\boldsymbol{P}_{\mathrm{d}} = (p_{\mathrm{d}1}, p_{\mathrm{d}2}, \cdots p_{\mathrm{d}m})^{T}$ is the desired tip position and orientation.
4. The compensability of the micro manipulator should be as large as possible.

$$fitness3 = \min C^{-1} = \min\left\|\boldsymbol{J}_{\mathrm{m}}^{+} \cdot \boldsymbol{J}_{\mathrm{E}}\right\| \ . \tag{10}$$

Fitness3 will decrease when compensability increases. Therefore, adding this factor into the objective function for motion generation and optimization will enhance the compensability of the solutions.

We synthesize the objective functions above and get the total fitness function as follows:

$$fitness = \lambda_1 \cdot fitness1 + \lambda_2 \cdot fitness2 + \lambda_3 \cdot fitness3 \ , \tag{11}$$

where, $\lambda_i (i = 1,2,3)$ is the weight for each objective function.

3 EDA Approaches to Optimal Motion Generation

EDAs is a class of novel stochastic optimization algorithms. Compared with the GA algorithms, in EDA algorithms there are no crossover and mutation operators. The new population of individuals is sampled with a probability distribution, which is estimated from a database of the selected individuals from the previous generation. This paper uses UMDA_c, which is one of the EDAs with the Gaussian probability model [10], to solve the motion optimization problem.

3.1 The Algorithm to Motion Generation and Optimization

Let $\boldsymbol{P}_{\mathrm{d}}(t)$ be the desired tip position and orientation, and $\boldsymbol{P}_{\mathrm{d}}(k)$ ($k = 1,2,\cdots Ne$) be the value of $\boldsymbol{P}_{\mathrm{d}}(t)$ at the instant of k. Ne is the number of sampling points. Then the inverse kinematics $\boldsymbol{\theta}^*(k)$ can be solved with the optimal EDA algorithm. One of the main merits of this algorithm is that the previous information about the configuration is used to initialize the mean of the multivariate Gaussian distribution model which is used to get the initial population for the next motion configuration. The algorithm is given as follows,

(1) Generate an initial population with M individuals. Let k=1.
(2) Calculate the fitness value using Eq. (11). Select the best N individuals from the population. If the termination condition is met, $\boldsymbol{\theta}^*(k)$ is generated and go to step (6). Otherwise, go to the next step.
(3) Update the mean μ_k and standard deviation σ_k of the multivariate Gaussian distribution using the best N individuals selected from the previous population.
(4) Generate a population with M individuals by sampling from the current Gaussian distribution model.
(5) Go to step (2).
(6) If $k = Ne$, the algorithm is ended. Else, k=k+1.
(7) The joint variable of the previous configuration is used to initialize mean $\boldsymbol{\mu}_{k+1} = \boldsymbol{\mu}_k$. Set $\sigma_k = [0.1...0.1]^{\mathrm{T}}$.
(8) Go to step (2).

and the flow chart of the algorithm is shown in fig. 2.

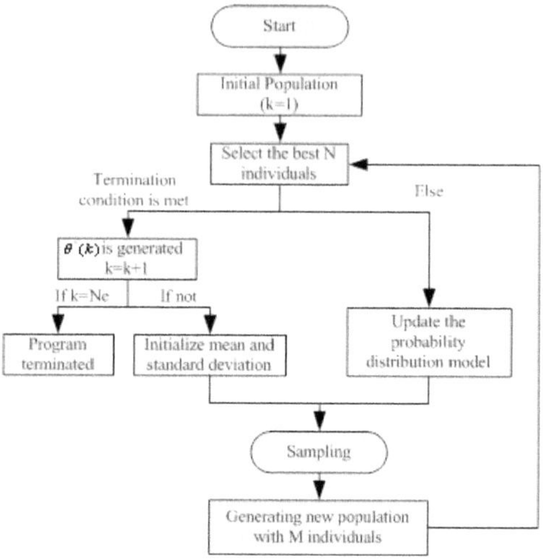

Fig. 2. Flow chart of optimal motion generation

3.2 Generalize the Solutions by Neural Networks

The solutions of inverse kinematics obtained by $UMDA_c$ are discrete, so we need to generalize the solutions by neural networks. The trained neural networks can generate the optimal motion of the manipulator in real time.

The input of neural networks is the desired tip position and orientation $\boldsymbol{P}_{\rm d}(t)$ ($\in R^m$). The output of neural networks is the joint displacement $\theta_{\rm d}(t)$ ($\in R^{N+n}$). The forward neural network which contains one hidden layer is adopted in this paper.

4 Simulation Results

A robot system with 2-DOF flexible macro manipulator and 2-DOF rigid micro manipulator is used for simulation. The task is to track a circle using the tip of the manipulator based on EDA approaches.

The simulation settings of $UMDA_c$ are as follows: population size is set 50, the number of selected individuals used to update probability model is 25, the selection method used is truncation, and the algorithm terminates in the condition that fitness value is less than 10^{-7}. The algorithms run 10 times independently and the average performance is as follows.

$\theta^*(k)(k=1...Ne)$ generated by $UMDA_c$ is shown in Fig. 3 (a). The axis of abscissa denotes the serial number of the sampling points. The axis of vertical denotes joint angle displacement $\theta^*_{\rm M1}(k), \theta^*_{\rm M2}(k), \theta^*_{\rm m1}(k), \theta^*_{\rm m2}(k)$. It is illustrated that the objects (1) and (2) of the solution to inverse kinematics are well satisfied. The joint variables are in the allowed span and the variation of the joint angle is small.

The error between desired tip position and the tip position generated by $UMDA_c$ is shown in Fig. 3 (b). It is indicated that high precision is achieved in motion generation and optimization using $UMDA_c$.

The forward neural network is adopted to generalize the discrete solutions of the inverse kinematics. Fig. 3 (c) illustrates the tip positions calculated by kinematics model of flexible macro-micro manipulator using the joint variables that generated by the trained neural network.

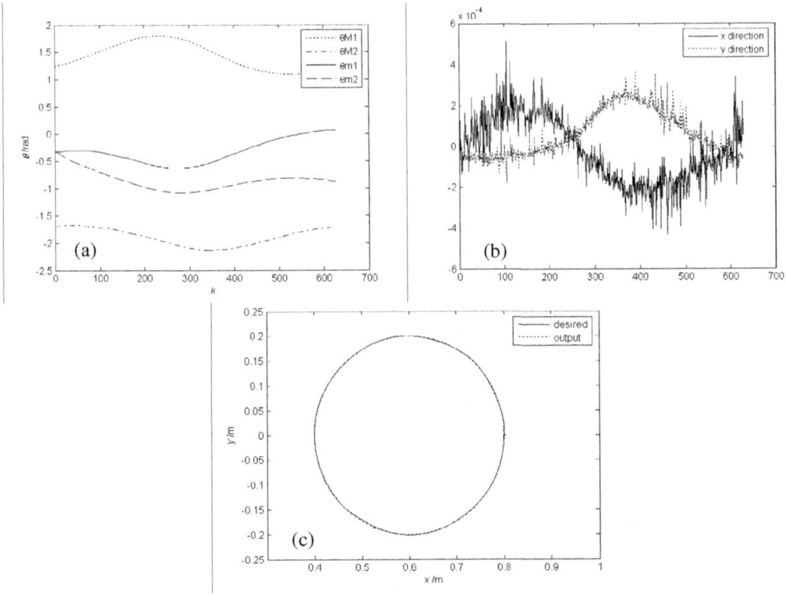

Fig. 3. (a) $\theta^{*}(k)$ acquired by UMDA$_c$; (b) Error of the tip position; (c) Output of the neural network

In order to further verify the performance of the UMDA$_c$, the paper compares it with SGA. The simulation settings of SGA are as follows: population size is 20, crossover probability is set 0.80, mutation probability is 0.05, the selection method used is the proportional selection method, and the algorithm stops evolution when fitness value is less than 10^{-6}. Both algorithms run 10 times independently and the average performance are shown in table 1.

Table 1. The time used in motion generation and optimization using UMDA$_c$ and SGA

	UMDA$_c$	SGA
One point	0.0045 min	0.57 min
One circle (628 points)	2.85 min	360 min

The simulation shows that UMDA$_c$ has better performance than SGA in motion generation and optimization. The time used in motion generation and optimization using UMDA$_c$ and SGA is shown in table 1. The time used to search the optimal solution is remarkably reduced in this simulation.

5 Conclusion

In this paper, a new algorithm of motion generation and optimization for a flexible macro-micro manipulator system is introduced, based on $UMDA_c$. Our method has two significant advantages. One is that it uses fewer parameters to select and adjust than our previous work [7]. Therefore, it is convenient to be applied. The other is that the previous information about the configuration of the manipulator system can be set as the initial mean of the multivariate Gaussian distribution model which is used to get the initial population for the next motion configuration. So the time for search of solutions is remarkably reduced. This approach is easy to implement, and simulation results show its effectiveness.

Acknowledgement. This work is funded by National Key Project for Basic Research of China (G2002cb312205) and the National Science Fund of China (60305008).

References

1. Ait, M., Chevallereau, A.C.: Resolution of Robot Redundancy in the Cartesian Space by Criteria Optimization. IEEE Conference on Robotics and Automation, (1993) 646–651
2. Mayorga, R., Wong, A.K.C., Ma, K.S.: An Efficient Local Approach for the Path Generation of Robot Manipulators. Journal of Robotics Systems, Vol. 7(1). (1990) 23–55
3. Chiacchio, P., Chiaverini, S., Sciavicco, L., Siciliano B.: Closed-loop Inverse Kinematics Schemes for Constrained Redundant Manipulators with Task Space Augmentation and Task Priority Strategy. The International Journal of Robotics Research, Vol. 10. (1991)
4. Seraji, H., Colbaugh, R.: Improved Configuration Control for Redundant Robots. Journal of Robotic Systems, Vol.7(6). (1990) 897–928
5. Ramdane-Cherif, A., Daachi, B., Benallegue, A., Levy, N.: Kinematic Inversion. In: Proceedings of 2002 IEEE/RSJ International Conference on Intelligent Robots and System, (2002)1904–1909
6. Yoshikawa, T., Hosoda, K., Doi, T., Murakami, H.: Quasi-static Trajectory Tracking Control of Flexible Manipulator by Macro-micro Manipulator System. In: Proceedings of 1993 IEEE International Conference on Robotics and Automation, (1993) 210–215
7. Zhang, Y., Sun, Z.Q., Yang, T.W.: Optimal Motion Generation of a Flexible Macro-micro Manipulator System Using Genetic Algorithm and Neural Network. In: Proceedings of 2006 IEEE International Conference on Robotics, Automation and Mechatronics, (2006)
8. Larrañaga, P., Lozano, J.A.: Estimation of Distribution Algorithms: A New Tool for Evolutionary Computation. Kluwer Academic Publishers, (2001)
9. Hu, L.Y., Zhou, C.J., Sun Z.Q.: Biped Gait Optimization Using Estimation of Distribution Algorithm. In: Proceedings of 5th IEEE-RAS International Conference on Humanoid Robots, (2005)
10. Larrañaga, P., Etxeberria, R., Lozano, J.A., Peña, J.M: Optimization in continuous domains by learning and simulation of Gaussian networks. In: Proceedings of the Genetic and Evolutionary Computation Conference, (2000)

Fuzzy Scenarios Clustering-Based Approach with *MV* Model in Optimizing Tactical Allocation

Hsing-Wen Wang

Department of Business Administration, College of Management
National Changhua University of Education
No. 2, Shi-Da Rd., Changhua City, Changhua County, Taiwan 500, R.O.C.
shinwen@cc.ncue.edu.tw

Abstract. A new interactive model for constructing a tactical global assets allocation through integrating fuzzy scenarios clustering- based approaches (*FSCA*) with mean-variance (*MV*) is proposed. This serves as an alternative forecasting rebalance quantitative model to the popular global assets allocation, in which the portfolio is first being observed in contrast with major asset and sub-assets classes which possess upward and downward positive co-movement phenomenon while considering the linkage of cross-market between different time-zones. In addition, fuzzy scenarios clustering would be induced into the *MV* model so as to adjust the weighting of the risk-return structural matrices. It could further enhance the efficient frontier of a portfolio as well as obtaining opportunity of excess return. By means of global major market indices as the empirical evidences, it shows that the new approach can provide a more efficient frontier for a portfolio and there would be less computational cost to solve *MV* model.

Keywords: Fuzzy scenarios clustering- based approach, mean-variance model, tactical global assets allocation, portfolio efficient frontier, computational cost.

1 Introduction

Scenario-based [1], [2], [3], [4] and *MV* [5] are the most popular approaches to the portfolio optimization problems. The scenario-based method is based on the returns originated or expected utility optimization. Especially, the investment management professions have been devoted to the *MV* while expected utility is highly regarded in the academy. That caused the focus on the mean and variance (especially, the downside risk, skewness, kurtosis and the higher moments of portfolio returns). All the efforts are expected to improve the investment performance via *MV* model and tactical global assets allocation [6], [7], [8].

For the assumptions presented based on the theory of the original mean-variance model, as long as its expected return and related coefficient of any portfolio asset can be estimated appropriately, then it is applicable to optimal allocation of local, regional and global arbitrary financial assets, including stock, bond, fixed income securities and exchange etc. However, in fact, the estimation on the related coefficients in portfolios models between assets expected return and risks would be significantly affect

T.-D. Wang et al. (Eds.): SEAL 2006, LNCS 4247, pp. 921 – 928, 2006.

the overall performance of the *MV* relative model, such as multifactor model, single factor model, hedging model, style model or portfolios insurance model, etc [9]. However, most of them are lack of consistency performance because of the change of risk-return structure of portfolios in maturity. Therefore, scenario analysis is necessary to be employed to fine tune the risk-return structure matrix while they are fed to the *MV* models with previous period observations. Besides, the scenario analysis also ought to take the fact into account that the boundary of scenario segmentations is fuzzy-style while clustering the various scenarios, which will affect the performance of model depend on the representative capacity, has not yet been considered. Therefore, in this research, the fuzzy scenarios clustering- based approach is utilized to revise the mean-variance model so as to obtain better portfolio performance.

This paper is based on the assets allocation of major country indices as the empirical cases of presented model. From the result of the empirical study it is discovered that *FSCA* has better profit-making capability.

2 Scenarios-Based Portfolio Selection

Through the probability allocation of incident scenario occurrence, scenario analysis procedure can determine the maximum effect. The classical scenario method can be briefly described as follows: To allow return allocation of portfolio assets *M* there are *S* types of possible return and its corresponding occurrence probability shall be P_s. If final out-come scenario *s* occurs, then asset content will possess the total return r_s^n generated in the portfolio of that asset and the return of the portfolio asset *M* at end of period shall be R_S^M. Therefore, under scenario *s*, the return of that portfolio assets *M* shall be: $R_S^M = \sum_{n=1}^{m} r_s^n M_n$; If combination *S* is considered for all scenarios, then the return of portfolio assets shall be $R_S^M = \sum_{s=1}^{S} \sum_{n=1}^{m} r_s^n M_n$. To maximize the expected utility, that concave function can be best solved by the equation: $Max \quad \sum_{s=1}^{S} P_s(\gamma^{1-\gamma}/1-\gamma)$, subject to: $\gamma \geq 0$ and $0 \leq P \leq 1$, where, γ denotes the parameter of risk aversion.

The main problem of scenario analysis is the grasping of the type of scenario and the acquisition of corresponding probability allocation. In practice, the main factors of the above two items most often will not be measured precisely so easily and therefore there will be many variances. Furthermore, acquisition for the probability of scenario occurrence is time consuming.

3 *MV* Model and Its Extensions

In global assets allocation, mean-variance model is often utilized as the optimization procedure for the first stage so as to determine the optimal proportional allocation of various assets types of asset combination [10]. Generally speaking, execution of global investment can be conducted by portfolio formed by individual investment in individual shares listed in the market or by direct purchase of the main index of that market. If portfolio allocation is based on individual shares, then for setting of portfolio content, if there are n item asset classes, and there are j^{th} assets types then L_j will be formed by n_j individual asset. Then under the designated expected return and through maximization of the risk model, the allocation of weighted equity amount of the m assets of x_{jm} in L_j can be determined. Firstly, the $Index_j$ with higher similarity can be selected to represent L_j, so that $R_{j,(j=1,\ldots,n')}$ shall be the $Index_j$ return and if limit is set as no short sell, then that model can be written as the following formulae: $Min \quad Var[\sum_{j=1}^{n'} R_j x_j]$, subject to: $E[\sum_{j=1}^{n'} R_j x_j] = r$, $\sum_{j=1}^{n'} x_j = 1, x_j \geq 0, j = 1,\ldots,n'$. Out of this, $E[\bullet]$ and $Var[\bullet]$ shall respectively become the expected return and variance(or risk) of the portfolio. If it can be short sell, then delete the constrain $x_j \geq 0$ item in constrains.

From this the optimal ratio x_j of investment in L_j can be determined, and, the *MV* efficient frontier can be obtained by optimization method, the risk model is formed as $[Var[\sum_{j=1}^{n'} R_j x_j(r), r]$. Using global fund as an example, fund manager can select efficient frontier based on the effected optimization principle. After the completion of optimizing allocation of asset types, we proceed to the second stage: Let x_j be the optimal allocation of m_{th} individual asset in asset type L_j so that r_j must satisfy relations $\sum_{j=1}^{n'} r_j x_j = r'$, then the optimal model of asset type content can be written as the equation: $Min \quad Var[\sum_{m=1}^{n_j'} R_{jm} x_{jm}]$, subject to: $E[\sum_{m=1}^{n_j'} R_{jm} x_{jm}] = r$, $\sum_{m=1}^{n_j'} x_{jm} = x_j, m = 1,\ldots,n_j$ [11]. In the following section, the scenarios classification is embedded into the extended *MV* model.

4 Scenarios Classification with an Upward/Downward Tendency

The correlation behavior between cross-country financial asset markets is the most im portant research topic for global assets. One of the important global assets allocation t opics is the expectation to enhance the efficient frontier through cross-country portfoli o content by means of diversifying the system risk of an individual local market [12]. Therefore in traditional tactical global asset allocation method, market with too high a relative correlation will often be eliminated from allocation combination or there will be limit setting on its allocation weight (or percentage) so as to avoid loss of the origi nal intention of diversifying systematic risk.

However, in fact, from this research, there are discoveries from the observation on the practical markets that the correlation behavior between major markets of different countries can be distinguished as the following three types based on scenario analysis: (1). complete dual direction correlation: that means assets price would have the phenomenon of simultaneous rise and fall, which are formed with the fuzzy scenarios clustering- based approaches with *MV* model-type I (*FSCA*-I), and, fuzzy scenarios clustering- based mean variance model-type II (*FSCA*-II); (2). single direction correlation: that means under a majority situation asset price will only have the phenomenon of simultaneous rise or there is only correlation during fall, which are also formed with *FSCA*-I and *FSCA*-II; (3) no correlation relationship: the price change direction has no correlation relationship, which is modeled with *MV* model. In the above three scenarios, we utilize the upward/downward tendency correlation scenarios to extend *MV* model so that through consideration on the characteristics of that scenario to obtain a higher payoff. Besides, the fuzzy scenarios clustering method is helpful to demarcate the scenarios segmentation more precisely for tuning the risk-return structural matrix of allocation.

5 Fuzzy Scenarios Clustering-Based *MV* Model (*FSCA*)

FSCA extracts the implied information to improve the risk-return structure in order to obtain excess return. The following is an explanation based on characteristics of *FSCA* and method of model construction respectively. Consider the leading-price asset S&P500 stock index and the lagging-price asset Nikkei 225 index in our multinational portfolio show as Fig. 1, in which the sliding window method was employed to sample the periodical interval observations of leading-pricing asset for calculating average return to determine the threshold value of corresponding period, so as to classify scenarios of lagging-price asset on previous-period. Assume the trend of asset price on next-period seems clearly estimating by time series model, and leading-price asset (S&P500 index in this case) trending toward rise on previous-period (see Fig. 1: U1→U2), also the linkage between lagging-price asset (Nikkei 225 index in this case) and leading-price asset exist on previous-period, we expect lagging-price asset will trend upward on next-period (see Fig. 1: U2→U3), on the other hand, If leading-price asset trending toward down on previous-period (see Fig. 1: D1→D2), we expect lagging-price asset will trend downward on next-period (see Fig. 1: D2→D3).

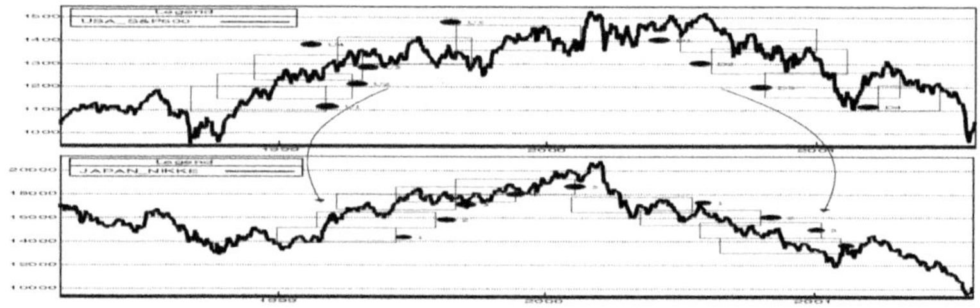

Fig. 1. Sliding window for leading-price asset on previous-period (daily data in the period from 1998 to 2001)

Fuzzy scenarios clustering essentially deal with the task of partitioning a set of scenarios patterns into a number of homogeneous classes (or clusters) with respect to a suitable similarity measure. In our cases, the scenarios patterns belong to any one of the clusters are similar and the scenarios patterns of different scenarios clusters are as dissimilar as possible. In classical cluster analysis, the boundary of different clusters is crisp such that one pattern is assigned to exactly one cluster. In practice, the data are usually not well distributed; therefore the boundary may not be precisely defined. That is, a data point could belong to two or more clusters with different degrees of membership. Therefore, the partitions of a data set can be either hard as in the *k*-means clustering algorithms [13], or fuzzy set as in the fuzzy *c*-means (FCM) algorithm [14]. FCM algorithm partitions a collection of *n* vectors ($X=\{X_1, X_2, \ldots, Xn\} \subset R^P$) into *c* fuzzy groups such that the weighted within-groups sum of squared error objective function is minimized. The objective function for FCM is defined as $J_m = \sum_{i=1}^{c}\sum_{j=1}^{n} u_{ij}^m d^2(\upsilon_i - x_j)$ with constrained conditions, in which, u_{ij}^m is the membership of the jth data point in the ith cluster, v_i is the ith cluster center, and $d(v_i, \chi_j)$ denotes the distance between v_i and χ_j $d^2(\upsilon_i, x_j) = (\upsilon_i - x_j)^T A(\upsilon_i - x_j)$. The distance matrix *A* is chosen depending on the geometric and statistical properties of the assets returns. The parameter *m* in J_m is the fuzzy exponent and $m \in (1,\infty)$. The fuzzy exponent controls the amount of fuzziness in the classification process. In general, the larger *m* is, the fuzzier are the membership assignments of the clustering.

By differentiating J_m with respect to v_i (for fixed U_{ij}, $i = 1, \ldots, c$, $j=1, \ldots, n$) and U_{ij} (for fixed v_i, $i = 1, \ldots, c$), the necessary conditions for J_m to reach its minimum are:

$$\upsilon_i = \left(\sum_{j=1}^{n} u_{ij}^m x_j\right)\left(\sum_{j=1}^{n} u_{ij}^m\right)^{-1}, \qquad i = 1,2,\ldots,c \tag{1}$$

$$u_{ij} = \left[\sum_{k=1}^{c}\left((d_{ij})(d_{kj})^{-1}\right)^{2/(m-1)}\right]^{-1}, \qquad i = 1,2,\ldots,c;\ \ j = 1,2,\ldots,n. \tag{2}$$

Eqs. (1) and (2) are repeated until the objective function is no longer decreasing or within a pre-specified level of accuracy. After the cluster centers and the class memberships are obtained, the fuzzy covariance matrix, F_i, is calculated as: $F_i = (\sum_{j=1}^{n} u_{ij}^m (\upsilon_i - x_j)(\upsilon_i - x_j)^T)(\sum_{j=1}^{n} u_{ij}^m), l \le i \le c.$, in which, the numerator is defined as the fuzzy scatter matrix for i^{th} cluster. The standard deviation for each membership function is then expressed by taking the root of diagonal elements of F_i: $\sigma_{ik} = \sqrt{Diag(F_i)}, l \le k \le p$, where *p* is the dimensions of input vector.

An important question for the FCM algorithm is how to determine the optimal number of scenarios clusters. Since there in no exact solution to this question, some measures of partitioning fuzziness have been used as synthetic indices to point out the most plausible number of clusters in the data set. Another important factor in the FCM algorithm is the fuzzy exponent *m*. The parameter m is selected according to the problem under consideration. When $m \to 1$, the fuzzy *c*-means converges to classical *c*-means. When $m \to \infty$, all scenarios cluster centers tend towards the centroid of the data set. That is, the partition becomes fuzzier with increasing *m*. currently there is no theoretical basis for an optimal choice for the value of *m*. Here, we propose a heuristic strategy incorporated with FCM algorithm to classify upward/downward tendency scenarios into two types of clusters that implied the parameter, *c*, is setting to be 2 and summarize as *Equ*. (3):

$$S_t^s = \begin{cases} FSCA-I,\ \ If\ R_{L_j}(t-1) \le \Gamma_{L_1}^N(t) \ \ , j=2\ldots,n' & (Cluster\ 1:\ \ c_1,\ \ \sigma_1) \\ FSCA-II,\ \ If\ R_{L_j}(t-1) > \Gamma_{L_1}^N(t) \ \ , j=2\ldots,n' & (Cluster\ 2:\ \ c_2,\ \ \sigma_2) \\ MV,\ \ If\ R_{L_j}(t-1) \le \Gamma_{L_1}^N(t) \cup R_{L_j}(t-1) > \Gamma_{L_1}^N(t) \ \ , j=2\ldots,n' & (Universe) \end{cases} \tag{3}$$

where, S_t^s : fuzzy scenarios clusters; *FSCA*-I denotes the fuzzy scenarios clustering-based approaches with *MV* model-type I; *FSCA*-II denotes fuzzy scenarios clustering--based mean variance model-type II; *MV* model denotes original *MV* model that apply previous-period risk matrix to be next-period risk matrix of portfolio; threshold value: $\Gamma_{L_1}^{N}(t)$ is determined by F_i, in which, N is the # of observations for leading-price asset of sampling by moving window on previous-period; L_1 is the leading-price class of portfolio, in which $j = 1$, and if $j = 2,\ldots, n'$, then indicates lagging-price class of portfolio. From *Equ.* (3), if trend estimation of asset seems stable on next-period, *FSCA* model begin to tuning the expected return and risk matrix of portfolio on current-period with *FSCA*-I, *FSCA*-II and *MV* corresponding to upward, downward and fluctuation market. We demonstrate the distribution interval of portfolio returns as Fig. 2.

Under the risk matrix on previous-period, *FSCA* model integrated *MV* model with spirit of scenario approach to take into account the scenarios, for instance, trending upward or downward, from foreign market on previous-period and lagging time-difference domestic market on current-period to modify the weights of asset allocated to take for the next-period. The notable difference is that we add scenarios classification in *MV* approach between original *MV* model and *FSCA* that extend one more dimension besides risk and return dimensions shows as Fig. 2. We classify linkage scenarios into three types according to the FCM algorithms, that are: (1). linkage of bullish market: higher asset returns $R_{t,1}$ at time 1~ t on current-period were sieved out from observations to fit for *FSCA*-I; (2). linkage of bearish market: lower asset returns $R_{t,1}$ at time 1~ t on current-period were sieved out from observations to fit for *FSCA*-II; (3). No exists linkage of asset price: regardless of trending upward, downward or fluctuation, no exists linkage of asset price. The asset returns R_t at time 1~ t on current-period were sieved out from observations to fit for *MV*. Fig. 3 demonstrates the efficient frontier improvement that *FSCA* model plans to achieve in this paper.

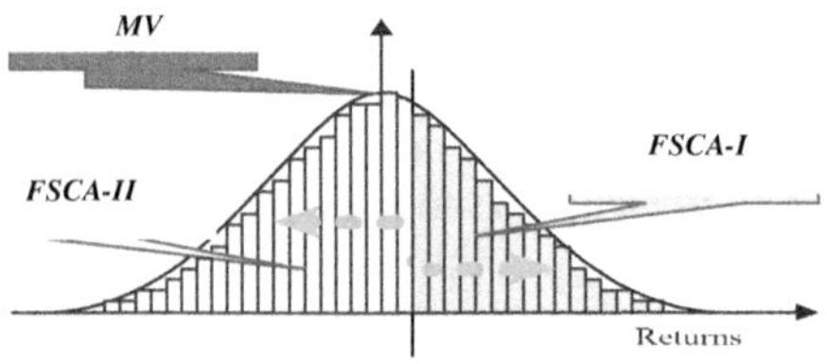

Fig. 2. Distribution allocation of corresponding three types asset returns for *FSCA* model

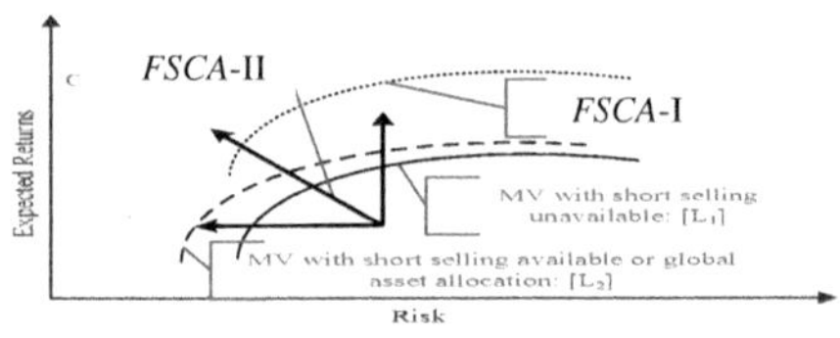

Fig. 3. Improving the efficient frontier of tactical allocation with *FSCA* model

6 Results and Analysis

The summary of results obtained in table 1. It conducts comparison based on the two types of scenarios in *FSCA* model and *MV* model. Under the two types of limitation scenarios of permitted to short sell and not permitted to short sell, the Sharpe ratio is as the guiding index of portfolio performance evaluation. It is obtained from the table that *FSCA* Model-I possesses corresponding high Sharpe ratio; *FSCA* Model-II is in the second place; *MV* model is the least and this conforms to the observation inference and model construction objective. The analytical conclusion is uniform with empirical study result.

Table 1. Comparisons among original Markowitz's *MV* Model, *FSCA* Model-I, & *FSCA* Model-II Model

Panel A.						
$E(R_p)=0.950$	Short sell not available			Short sell available		
	MV	*FSCA-I***	*FSCA-II**	*MV*	*FSCA-I***	*FSCA-II**
$E(R_p)$	1.098370%	1.238475%	1.231364%	1.111204%	1.247085%	1.230121%
Annualized $E(R_p)$	14.006563%	15.810160%	15.782246%	14.180357%	15.901826%	15.743847%
$Var(E(R_p))$	0.143980%	0.162741%	0.132176%	0.144658%	0.167126%	0.132135%
Sharpe Ratio	43.995541%	51.739465%**	56.531782%*	49.618842%	52.417254%**	55.721324%*
Panel B.						
$E(R_p)=1.015$	Short sell not available			Short sell available		
	MV	*FSCA-I***	*FSCA-II**	*MV*	*FSCA-I***	*FSCA-II**
$E(R_p)$	1.098370%	1.238314%	1.231373%	1.111204%	1.247085%	1.230127%
Annualized $E(R_p)$	14.006565%	15.810127%	15.782293%	14.180360%	15.901826%	15.749325%
$Var(E(R_p))$	0.143980%	0.162738%	0.132176%	0.144658%	0.167126%	0.132128%
Sharpe Ratio	43.995579%	51.737182%**	56.532931%*	49.618932%	52.417254%**	55.721941%*
Panel C.						
$E(R_p)=1.080$	Short sell not available			Short sell available		
	MV	*FSCA-I***	*FSCA-II**	*MV*	*FSCA-I***	*FSCA-II**
$E(R_p)$	1.098373%	1.238318%	1.141762%	1.111205%	1.247088%	1.230116%
Annualized $E(R_p)$	14.006602%	15.810129%	13.864215%	14.180362%	15.901899%	15.749300%
$Var(E(R_p))$	0.143981%	0.162738%	0.131274%	0.144658%	0.167126%	0.132128%
Sharpe Ratio	43.996708%	51.737278%**	56.532941%*	49.619007%	52.417476%**	55.721115%*

*: representing corresponding maximum value
**: representing secondary corresponding maximum value

Fig. 4 is the sampling comparison diagram of efficient frontier interval of *FSCA*-I and *FSCA* Model-II. Dotted line is *FSCA*-I. In the diagram, efficient frontier section of *FSCA*-II including *FSCA*-I moves forward further and therefore it can generate a higher performance and its Sharpe ratio is more realistic.

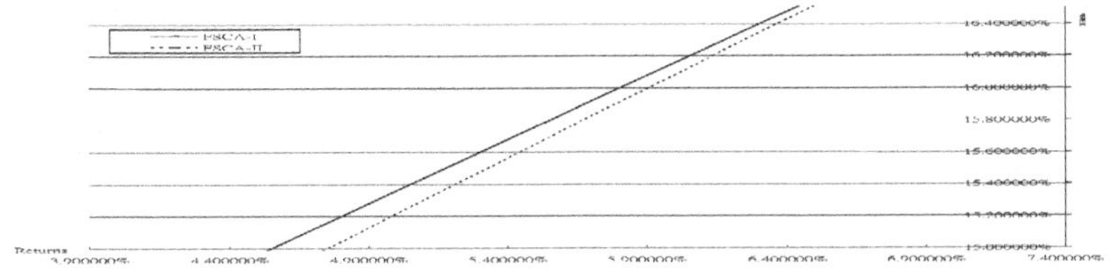

Fig. 4. The efficient frontier area sampling comparison between *FSCA-I* and *FSCA-II*

7 Conclusion

The result obtained from this research can amplify a result proposed by [15] in 2000, describing an "optimal frontier" as opposed to an underlying "efficient frontier". Further more, we obtain a more efficient frontier then original Markowitz's optimal frontier, through integrating the scenario-based approach with *MV* model into global assets allocation.

We also shown how the scenarios-based approach integrated into the portfolio optimization model through: (1) Form a portfolio , then produce major asset and sub-assets;

(2) Determine the pre-period expected return of major assets; (3) Performing the lead-lag analysis via correlation analysis; (4) Incorporate special information from expected return, correlation matrix and variance-covariance matrix with a reasonable level of active risk and Sharpe ratio; (5) Scenario analysis and modeling; (6) Integrating the scenarios clustering and *MV* optimization; (6) Optimizing the assets allocation.

The key to these results is the chosen of major asset and the expected return forecast. That is the difference between non-scenarios forecast, *MV*, with the scenarios forecast, *FSCA*-I & *FSCA-II*. We also have contrasted the results using the scenarios-based approach with *MV* optimization. Form the empirical study of global assets allocation with *MSCI* contents; our conclusion is that when the two approached are put on together, the performance of global assets allocation would be better then original *MV* method.

Acknowledgement

This paper was supported by the National Science Council, Taiwan, R.O.C. under the contract number NSC-95-2516-S-018 -016-.

References

1. Samuelson, P. A. (1969), "Lifetime Portfolio Selection by Dynamic Stochastic Programming," Review of Economic Studies, Vol. 51, No. 3, pp. 239-246.
2. Arrow, K. J. (1970), "Essays in the Theory of Risk Bearing," Amsterdam: North-Holland.
3. Hakansson, N. H. (1970), "Optimal Investment and Consumption Strategies under Risk for a Class of Utility Functions,' Econometrica, Vol. 38, No. 5, pp. 587-607.
4. Merton, R.C. (1990), "Continuous Time Finance," Oxford: Blackwell.
5. Markowitz, H.M. (1959), "Portfolio Selection: Efficient Diversification of Investments," Wiley, New York.
6. Markowitz, H.M. (1988), "Mean Variance Analysis in Portfolio Choice and Capital Markets. Oxford: Basil Blackwell.
7. Sharpe, W. F. (1970)," Portfolio Theory and Capital Markets," New York: McGraw-Hill.
8. Rudd, A. & Rosenberg, B. (1979), "Realistic Portfolio Optimization," TIMS Studies in Management Science: Portfolio Theory, Amsterdam: North Holland Press, No.11, pp. 21-46.
9. Zenios, S. & Kang, P. (1993), "Mean Absolute deviation portfolio optimization for mortgage-backed securities," Annals of Operations Research, Vol. 45, pp. 433-450.
10. Sharpe, W. (1987), "Integrated Asset Allocation,' Financial Analysis.
11. Konno H. & Kobayashi K. (1997), "An Integrated Stock-Bond Portfolio Optimization Model," Journal of Economic Dynamics and Control, pp. 1427-1444.
12. Zhang, L. H. (1998), "Global Asset Allocation with Multirisk Considerations," The Journal of Investing, Vol. 7, No. 3, pp. 7-14.
13. Makhoul J.S. & Gish, Roucos. H. (1985), "Vector Quantization in Speech Coding," Proc IEEE, Vol. 73, No. 11, pp. 1551-1588.
14. Bezdek, J. C. (1987), "Pattern Recognition with Fuzzy Objective Function Algorithms," Plenum Press, New York.
15. Schirripa & Tecotzky (2000), "Schirripa, Felix, Tecotzky, Nan, An Optimal Frontier," The Journal of Portfolio Management, Vol. 26, No. 4, pp. 29-40.

Joint Source-Channel Coding Based on Adaptive Wavelet Pretreatment for Image Transmission over Frequency Selective Fading Channel

Rui Guo and Ji-lin Liu

Department of Information Science and Electronic Engineering Zhejiang University, HangZhou, 310027, China
hbsyrgr@yahoo.com.cn

Abstract. A joint source-channel coding (JSCC) scheme based on adaptive wavelet pretreatment for robust progressive image transmission over wireless fading channels using MIMO-OFDM was proposed. We pointed out that he peak signal to noise ratio (PSNR) of the reconstructed image degrades a little while the complexity decreases sharply when adding pretreatment block before DWT. On the other hand, OFDM can get good performance in the frequency selective fading channel. So in this paper, a JSCC scheme based on adaptive wavelet pretreatment for image transmission over MIMO-OFDM system was analyzed. What`s more, the wavelet pretreatment block is loaded adaptively according to the channel condition. Simulation shows that: after adding the pretreatment block before DWT and post-treatment to the reconstructed image in good channel condition, the complexity of the SPIHT and the channel coder decreases sharply due to the compression ratio, but the PSNR of the receivedimage loses only a little; when the channel condition is getting worse ,the pretreatment block is deleted accordingly.

1 Introduction

JSCC has been proven to be very promising to provide reliable image transmission while maintaining high transmission efficiency [1][2]. An important means to achieve JSCC is to partition source information with different levels of significance so that different levels of error protection may be applied as just required [3]. Among these methods, the most attractive is the SPIHT based JSCC[4]. However, an image contains both low-frequency and high-frequency component. The high-frequency component is corresponding to the detail area of the image, which is not sensitive to human vision. If we can remove the high frequency component using smooth pretreatment before SPIHT code at the transmitter and post-process at the receiver, the data input to the SPIHT encoder and the channel encoder can decrease greatly. Hence the wavelet and SPIHT encoding complexity decrease also. On the other hand, for high bit-rate wireless communications, MIMO-OFDM [5][6] is an attractive technique to be used because of its simplicity in dealing with frequency-selective, time dispersive wireless

T.-D. Wang et al. (Eds.): SEAL 2006, LNCS 4247, pp. 929 – 936, 2006.

fading channels. So in this paper a JSCC scheme based adaptive wavelet pretreatment in connection with MIMO-OFDM is addressed.

2 Wavelet Pretreatment Principle and the Method

2.1 Principle of Wavelet Pretreatment

Suppose the source set is X , human vision spectrum is $F(f)$.In general $F(f)$ is considered as a low-pass filter, $F1$ is the threshold. When use DWT to source X . The result shows as follows [7]:

$$W(x) = L_{F1}(X) + H_{F1}(X) \tag{1}$$

Where $L_{F1}(X)$ is the low-frequency component, which is often used to estimate $W(X)$, denoted as $\overline{W(X)}$. $H_{F1}(X)$ is the high-frequency component, which is not sensitive to human vision and often viewed as error, denoted as $\varepsilon(x)$. If we can decrease the data amount of $H_{F1}(X)$, the complexity of the DWT and SPIHT encoding will reduce too.

2.2 The Method of Wavelet Pretreatment

As we all know, the average operator can smooth an image, decrease the high frequency component . So in this paper we use the smooth arithmetic operators to smooth the image before DWT and SPIHT encoding. Here the smooth arithmetic operator

$$D = \frac{1}{2}\begin{bmatrix} 1 & 1 \\ 1 & 1 \end{bmatrix}$$

At the transmitter, smooth the image using smooth arithmetic operators D before DWT and SPIHT encoding. In practice operation, the first row and column of the image is unchanged, the rest of the image is operated by D , the result is the same as the fellows[8]:

$$F(i,j) = [F(i,j) + F(i-1,j)]/2\,,\ F(i,j) = [F(i,j) + F(i,j-1)]/2$$

Where $0< i <$image Width, $0< j <$image Height

At the receiver, after SPIHT decoding and Inverse DWT, the opposite operation is used to clear the reconstructed image the operators are:

$$D1 = \frac{1}{2}\begin{bmatrix} -1 & 2 \\ -1 & 2 \end{bmatrix},\quad D2 = \frac{1}{2}\begin{bmatrix} -1 & -1 \\ 2 & 2 \end{bmatrix}$$

The result can be explained as:

$$F(i,j) = 2F(i,j) - F(i-1,j)\,,\ F(i,j) = 2F(i,j) - F(i,j-1)$$

3 Performance of the JSCC Based on Adaptive Wavelet Pretreatment over MIMO-OFDM System

3.1 The Structure of the Proposed Image Transmission System

As can be seen from Fig 1, the original image is pretreated before the DWT encoding. After the pretreatment, the smoothed image is then put into DWT and SPIHT encoder, the progressive bit stream is then CRC encoded to gain protection and then put into RS encoder, STBC encoder and at last transmitted after OFDM modulation.

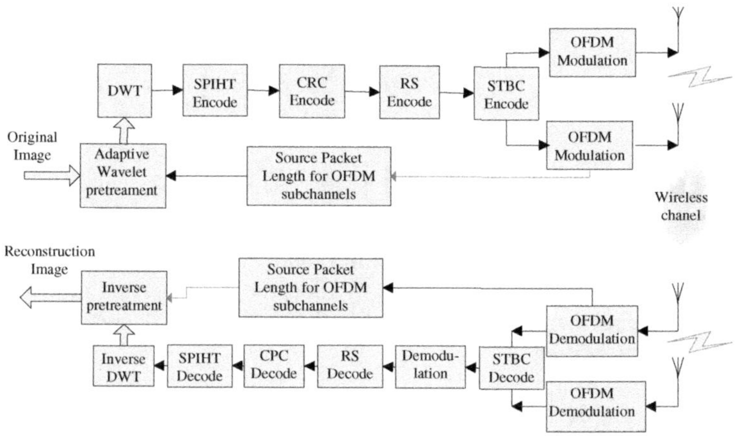

Fig. 1. The structure of the proposed image transmission system

3.2 Performance of JSCC Scheme Based on MIMO-OFDM Systems

Now, the problem considered is as follows. The SPIHT encoded bit stream is to be transmitted over OFDM systems using (M,N) STBC encoding. Suppose the transmission rate is R_S bpp. There are N_S sub channels in the OFDM and each sub channel is modulated by a complex symbol from an M-ary modulation. The SPIHT stream is packed into L source-packets with B_i bits for packet i ($i=1,2..\ L$) , where L is the total number of OFDM blocks decided by R_S :

$$L \leq \frac{R_S HW}{N_S \log_2 M} \tag{2}$$

Where H and W is the dimension of the image, M-ary is the constellation size for sub channels. Here we use the same code scheme as mentioned in [9], each sub channel source packets are CRC-16 outer coded first, CRC is outer code and RS is inner code. The resulted blocks of codeword are of equal size and are subsequently transmitted over the channels as OFDM blocks.

At the receiver, the image is reconstructed only from the decoded bit-stream up to the first packet that is error-detected. RS is used as inner code to combat the burst error. Each OFDM block b_i is a RS codeword over RS(N , K_i) over GF(2^m) , where $N = ((N_S \log_2 M_{ary}) / m$ is the total number of RS code symbols in each OFDM block, and $K_i = B_i / m$ is the number of RS information symbols in OFDM block b_i .We denote the error-correction failure probability of OFDM block $b_i, i = 1,2...L$ as $P_e(b_i)$.

If the first i source-packets are correctly received, the image can be reconstructed to a rate $R = ((\sum_{j=1}^{i}(K_j m - 16)) / HW$ bpp. So the JSCC scheme is becoming the problem to find a source packet allocation scheme A $\{K_1, K_2 ... K_L\}$ to maximize the PSNR of the reconstruction image [9]:

$$PSNR = PSNR(0) P_e(b_1) + PSNR(\sum_{j=1}^{L} K_j m) \prod_{j=1}^{L} [1 - P_e(b_j)]$$
$$\sum_{i=1}^{L-1} PSNR(\sum_{j=1}^{i} K_j m) \times P_e(b_{i+1}) \prod_{j=1}^{i} [1 - P_e(b_j)] \tag{3}$$

Obviously, It is difficult to compute (3) in practice. An alternative criteria is to use the expected number of source bits correctly received as suggested in [10] The result is almost the same as the above function, but the complexity decrease greatly. So the above problem is re-written as:

$$\max_{K_1, K_2 ... K_L} BITS = \sum_{i=1}^{L-1} (\sum_{j=1}^{i} K_j m) P_e(b_{i+1}) \times \prod_{j=1}^{i} (1 - P_e(b_j))$$
$$+ (\sum_{j=1}^{L} K_j m) \prod_{j=1}^{L} (1 - P_e(b_j)) \tag{4}$$

The optimization solution $A^* = \{K_1^* m, ... K_L^* m\}$ is used to pack the SPIHT bit stream and protected by $RS(N, K_i^*)$ over $GF(2^m)$

The decoding failure probability of a OFDM block $RS(N, K)$ over $GF(2^m)$ under the assumption of mutual independence of OFDM sub channels is [9]:

$$P_e(b_i) = 1 - \sum_{v=0}^{(N-K_i)/2} \binom{N}{v} P_s^v (1 - P_s)^{N-v} \tag{5}$$

Where P_S is the error probability of RS code symbol (m bits per RS code symbol), which is:

$$P_s = 1 - (1 - P_{ce})^{m/b} \tag{6}$$

Where b is the number of bits for each sub-channel symbol, and m/b is the number of sub-channel symbols in a RS code symbol. The instantaneous SNR at receiver, denoted as γ_c , is time-varying. Because the transmitter cannot know the γ_c , the JSCC scheme has to be designed for a single SNR, which we call target SNR and denote as γ_T. The target SNR is computed for a given probability P_B such that $P(\gamma_c \le \gamma_T) = P_B$. By specifying a relatively small P_B, the γ_c is larger than γ_T with high probability and the system performance can be guaranteed.

For an OFDM system using (M,N) STBC, the instant SNR γ_c at sub channels with a known average receiver SNR $\overline{\gamma_c}$ is

$$\gamma_c = \frac{\overline{\gamma_c}}{NM}\sum_{i=1}^{NM} C_i^2 \tag{7}$$

Where C_i is the amplitude of complex zero-mean Gaussian random variable with variance 0.5 at each dimension. Let $x = \sum_{i=1}^{NM} C_i^2$, then x is a χ^2 random variable with 2NM-degree freedom. To find the target SNR γ_T such that $P(\gamma_c \le \gamma_T) = P_B$ for a given P_B, we only need to find a real value y such that $P(x \le y) = P_B$, then using (7) we have:

$$\gamma_T = \frac{\overline{\gamma_c}\, y}{NM} \tag{8}$$

After the target SNR γ_T is obtained, the worst-case sub channel symbol error probability of Gray-mapped QPSK is:

$$P_{ce} = bP_b = bQ(\sqrt{\gamma_T}) \tag{9}$$

for a given $\overline{\gamma_T}$ at the receiver and a specified P_B, the target SNR γ_T can be computed using (8), then the BER and sub channel symbol error rate P_{ce} can be given. By(6)and (5), the RS code symbol SER and OFDM block decoding failure probability is computed. Then the optimization solution $A^* = \{K_1^* m, \ldots K_L^* m\}$ is found using (4). At the receiver, the same compute process is used.

3.3 Realization of Adaptive Wavelet Pretreatment

From the above discussion, we find that the channel codewords are of equal length, the length of source packet in each codeword is different. So we can realize the UEP of the SPIHT bits according to the importance to reconstruction. This method is more suitable here because the length of the codewords is limited in block data transmission scheme for OFDM and RS block codes.

SPIHT bit stream is progressive, the bit stream is arranged according to the importance to the image reconstruction. We suppose the importance of bit stream allocated to i sub channel is greater than these allocated to $i+1$ sub channel. Eg, $L(C_1) > L(C_2) \ldots . > L(C_L)$. So the packet scheme A is an indication of the channel condition. For example, When K_1^* is increasing obviously, this means that the channel condition is getting poor. We should increase the high-frequency component, thus weaken the smooth operation. When K_1^* is decreasing, the opposite operation is needed.

In this paper, we use the channel condition indicated by $(K_1^*, K_2^*, ..., K_L^*)$ to choose the wavelet pretreatment module, if the important source packet is increasing to a threshold, the pretreatment module is withdrawn, On the opposite, the smooth process is used to the original image. Here only the length of the most important source packet K_1^* is used for simplicity.

4 Simulation Results

In simulation, a two-ray channel model with delay spread from 0 to 40 μs and Doppler frequency from 10 Hz to 200 Hz is used. Two transmit antennas and two receive antennas are used for STBC encoding. The entire channel bandwidth, 800 kHz, is divided into 128 sub channels. To make the tones orthogonal to each other, the symbol duration is 160 μs . An additional 40 μs guard interval is used to provide protection from ISI due to channel multipath delay spread. QPSK modulation and coherent estimation is used with the assumption of perfect channel information at receivers. The RS code over is GF(2^8) used such that each OFDM block has RS code symbols. The transmit rate used for all the simulations is 0.5bpp, which is approximately equivalent to OFDM blocks. The 256×256 gray image Lena is used. The simulation result is as fellows:

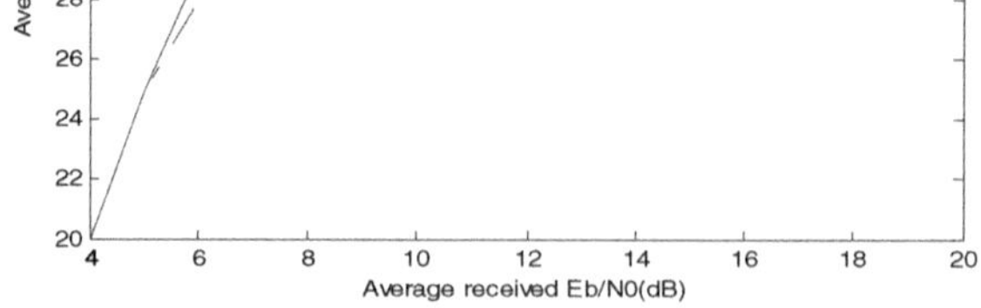

Fig. 2. PSNR comparison between different transmission scheme

Fig 2 shows the PSNR of different transmission scheme. It is clear that: when the channel condition is good, the performance with wavelet pretreatment is almost the same as the scheme without pretreatment; when the channel condition is worse, the pretreatment is removed adaptively to guarantee the PSNR of reconstructed image. Fig 3 and Table 1 give a example when Eb/N0 is 9dB, the PSNR lose is less than 3dB (from

37.33 to 34.72), but the Compression ratio increase greatly (from20.16% to 62.31%) when using the wavelet pretreatment. Fig 4 showed the performance of the JSCC scheme when the channel condition is worse (SNR=4dB). It is clear that the distortion of reconstructed image with pretreatment is too severe to accept. So the pretreatment should be removed adaptively.

(a)original image (b)JSCC without pretreatment (c)JSCC with pretreatment

Fig. 3. Received image of "Lena" when (Eb/N0=9dB)

Table 1. Comparison of the received image form original image and the smoothed image (Eb/N0=9dB)

	Compression ratio after SPIHT encoding (%)	The PSNR of the received image
Without pretreatment	20.16	37.33
With pretreatment	62.31	34.72

(a) JSCC without pretreatment (Compression ratio: 21.54%, PSNR: 24.19) (b) JSCC with pretreatment (Compression ratio: 61.73%, PSNR: 16.35)

Fig. 4. Received image of "Lena" when (Eb/N0=4dB)

5 Conclusion

In this paper, an adaptive joint source-channel code transmission scheme for SPIHT coded image over frequency selective fading channels is proposed. Space–time block coded MIMO-OFDM is used to combat a frequency-selective, time-dispersive wireless channel. An adaptive wavelet pretreatment is used before DWT and SPIHT coder according to the channel condition. When the channel is in good condition, the system with wavelet pretreatment can get almost the same PSNR performance as the system with no wavelet pretreatment, but the compute complexity of the DWT and SPIHT reduce sharply; when the channel condition is poor, the wavelet pretreatment is omitted. What`s more, the change of the most important source packet size is used to indicate the channel condition. Simulation results show that this scheme is robust to different Doppler and multipath delay spread in wireless communications.

References

1. N. Tanabe and N. Farvardin, "Subband image coding using entropy coded quantization over noisy channels," IEEE J. Select. Areas Commun, vol. 10, pp. 926–943, June 1992.
2. H. Zheng and K. J. R. Liu, "The subband modulation: A joint power and rate allocation framework for subband image and video transmission,"IEEE Trans. Circuits Syst. Video Technol., vol. 9, pp. 823–838, Aug.1999.
3. P. G. Sherwood and K. Zeger, "Progressive image coding for noisy channels,"IEEE Signal Processing Lett., vol. 7, pp. 188–191, July 1997.
4. A. Aydın Alatan, Minyi Zhao, and Ali N. Akansu," Unequal Error Protection of SPIHT Encoded Image Bit Streams" IEEE Journal on Selected Areas in Communications, Vol. 18, No. 6, June 2000
5. W. Y. Zou and Y. Wu, "COFDM: An overview," IEEE Trans. Broadcasting, vol. 41, pp. 1–8, Mar. 1995.
6. S.M.Alamouti,A simple transmit diversity technique for wireless communications, IEEE Journal Select Commun.,vol.16,no.8,pp.1451-1458,oct.1998
7. Xiao Zhang, Liu Zhao. A wavelet transform image coding technique with tree character palne. Journal of China Institute of Communications. 1999.20(11):19-24(in Chinese)
8. WANG Yao,ZHU Guang-Xi,LIU Wei .Image Coding Scheme Combined with Wavelet Pretrement. Chinese Journal of Computers Vol 26 No.2 Feb .2003(in Chinese)
9. Jie Song and K. J. Ray Liu,"Robust Progressive Image Transmission Over OFDM Systems Using Space-Time Block Code" IEEE Transactions on Multimedia, Vol. 4, No. 3, September 2002
10. V. Chande and N. Farvardin, "Joint source-channel coding for progressive transmission of embedded source coders," in Proc. Data Compression Conf., 1999, pp. 52–61.

Author Index

Lecture Notes in Computer Science

For information about Vols. 1–4164

please contact your bookseller or Springer

Vol. 4270: H. Zha, Z. Pan, H. Thwaites, A.C. Addison, M. Forte (Eds.), Interactive Technologies and Sociotechnical Systems. XVI, 547 pages. 2006.

Vol. 4265: N. Lavrač, L. Todorovski, K.P. Jantke (Eds.), Discovery Science. XIV, 384 pages. 2006. (Sublibrary LNAI).

Vol. 4264: J.L. Balcázar, P.M. Long, F. Stephan (Eds.), Algorithmic Learning Theory. XIII, 393 pages. 2006. (Sublibrary LNAI).

Vol. 4253: B. Gabrys, R.J. Howlett, L.C. Jain (Eds.), Knowledge-Based Intelligent Information and Engineering Systems, Part III. XXXII, 1301 pages. 2006. (Sublibrary LNAI).

Vol. 4252: B. Gabrys, R.J. Howlett, L.C. Jain (Eds.), Knowledge-Based Intelligent Information and Engineering Systems, Part II. XXXIII, 1335 pages. 2006. (Sublibrary LNAI).

Vol. 4251: B. Gabrys, R.J. Howlett, L.C. Jain (Eds.), Knowledge-Based Intelligent Information and Engineering Systems, Part I. LXVI, 1297 pages. 2006. (Sublibrary LNAI).

Vol. 4249: L. Goubin, M. Matsui (Eds.), Cryptographic Hardware and Embedded Systems - CHES 2006. XII, 462 pages. 2006.

Vol. 4248: S. Staab, V. Svátek (Eds.), Engineering Knowledge in the Age of the Semantic Web. XIV, 400 pages. 2006. (Sublibrary LNAI).

Vol. 4247: T.-D. Wang, X. Li, S.-H. Chen, X. Wang, H. Abbass, H. Iba, G. Chen, X. Yao (Eds.), Simulated Evolution and Learning. XXI, 940 pages. 2006.

Vol. 4243: T. Yakhno, E.J. Neuhold (Eds.), Advances in Information Systems. XIII, 420 pages. 2006.

Vol. 4241: R.R. Beichel, M. Sonka (Eds.), Computer Vision Approaches to Medical Image Analysis. XI, 262 pages. 2006.

Vol. 4239: H.Y. Youn, M. Kim, H. Morikawa (Eds.), Ubiquitous Computing Systems. XVI, 548 pages. 2006.

Vol. 4238: Y.-T. Kim, M. Takano (Eds.), Management of Convergence Networks and Services. XVIII, 605 pages. 2006.

Vol. 4236: L. Breveglieri, I. Koren, D. Naccache, J.-P. Seifert (Eds.), Fault Diagnosis and Tolerance in Cryptography. XIII, 253 pages. 2006.

Vol. 4234: I. King, J. Wang, L. Chan, D. Wang (Eds.), Neural Information Processing, Part III. XXII, 1227 pages. 2006.

Vol. 4233: I. King, J. Wang, L. Chan, D. Wang (Eds.), Neural Information Processing, Part II. XXII, 1203 pages. 2006.

Vol. 4232: I. King, J. Wang, L. Chan, D. Wang (Eds.), Neural Information Processing, Part I. XLVI, 1153 pages. 2006.

Vol. 4229: E. Najm, J.F. Pradat-Peyre, V.V. Donzeau-Gouge (Eds.), Formal Techniques for Networked and Distributed Systems - FORTE 2006. X, 486 pages. 2006.

Vol. 4228: D.E. Lightfoot, C.A. Szyperski (Eds.), Modular Programming Languages. X, 415 pages. 2006.

Vol. 4227: W. Nejdl, K. Tochtermann (Eds.), Innovative Approaches for Learning and Knowledge Sharing. XVII, 721 pages. 2006.

Vol. 4225: J.F. Martínez-Trinidad, J.A. Carrasco Ochoa, J. Kittler (Eds.), Progress in Pattern Recognition, Image Analysis and Applications. XIX, 995 pages. 2006.

Vol. 4224: E. Corchado, H. Yin, V. Botti, C. Fyfe (Eds.), Intelligent Data Engineering and Automated Learning – IDEAL 2006. XXVII, 1447 pages. 2006.

Vol. 4223: L. Wang, L. Jiao, G. Shi, X. Li, J. Liu (Eds.), Fuzzy Systems and Knowledge Discovery. XXVIII, 1335 pages. 2006. (Sublibrary LNAI).

Vol. 4222: L. Jiao, L. Wang, X. Gao, J. Liu, F. Wu (Eds.), Advances in Natural Computation, Part II. XLII, 998 pages. 2006.

Vol. 4221: L. Jiao, L. Wang, X. Gao, J. Liu, F. Wu (Eds.), Advances in Natural Computation, Part I. XLI, 992 pages. 2006.

Vol. 4219: D. Zamboni, C. Kruegel (Eds.), Recent Advances in Intrusion Detection. XII, 331 pages. 2006.

Vol. 4218: S. Graf, W. Zhang (Eds.), Automated Technology for Verification and Analysis. XIV, 540 pages. 2006.

Vol. 4217: P. Cuenca, L. Orozco-Barbosa (Eds.), Personal Wireless Communications. XV, 532 pages. 2006.

Vol. 4216: M.R. Berthold, R. Glen, I. Fischer (Eds.), Computational Life Sciences II. XIII, 269 pages. 2006. (Sublibrary LNBI).

Vol. 4213: J. Fürnkranz, T. Scheffer, M. Spiliopoulou (Eds.), Knowledge Discovery in Databases: PKDD 2006. XXII, 660 pages. 2006. (Sublibrary LNAI).

Vol. 4212: J. Fürnkranz, T. Scheffer, M. Spiliopoulou (Eds.), Machine Learning: ECML 2006. XXIII, 851 pages. 2006. (Sublibrary LNAI).

Vol. 4211: P. Vogt, Y. Sugita, E. Tuci, C. Nehaniv (Eds.), Symbol Grounding and Beyond. VIII, 237 pages. 2006. (Sublibrary LNAI).

Vol. 4210: C. Priami (Ed.), Computational Methods in Systems Biology. X, 323 pages. 2006. (Sublibrary LNBI).

Vol. 4209: F. Crestani, P. Ferragina, M. Sanderson (Eds.), String Processing and Information Retrieval. XIV, 367 pages. 2006.

Vol. 4208: M. Gerndt, D. Kranzlmüller (Eds.), High Performance Computing and Communications. XXII, 938 pages. 2006.

Vol. 4207: Z. Ésik (Ed.), Computer Science Logic. XII, 627 pages. 2006.

Vol. 4206: P. Dourish, A. Friday (Eds.), UbiComp 2006: Ubiquitous Computing. XIX, 526 pages. 2006.

Vol. 4205: G. Bourque, N. El-Mabrouk (Eds.), Comparative Genomics. X, 231 pages. 2006. (Sublibrary LNBI).

Vol. 4204: F. Benhamou (Ed.), Principles and Practice of Constraint Programming - CP 2006. XVIII, 774 pages. 2006.

Vol. 4203: F. Esposito, Z.W. Raś, D. Malerba, G. Semeraro (Eds.), Foundations of Intelligent Systems. XVIII, 767 pages. 2006. (Sublibrary LNAI).

Vol. 4202: E. Asarin, P. Bouyer (Eds.), Formal Modeling and Analysis of Timed Systems. XI, 369 pages. 2006.

Vol. 4201: Y. Sakakibara, S. Kobayashi, K. Sato, T. Nishino, E. Tomita (Eds.), Grammatical Inference: Algorithms and Applications. XII, 359 pages. 2006. (Sublibrary LNAI).

Vol. 4199: O. Nierstrasz, J. Whittle, D. Harel, G. Reggio (Eds.), Model Driven Engineering Languages and Systems. XVI, 798 pages. 2006.

Vol. 4198: O. Nasraoui, O. Zaiane, M. Spiliopoulou, B. Mobasher, B. Masand, P. Yu (Eds.), Advances in Web Minding and Web Usage Analysis. IX, 177 pages. 2006. (Sublibrary LNAI).

Vol. 4197: M. Raubal, H.J. Miller, A.U. Frank, M.F. Goodchild (Eds.), Geographic, Information Science. XIII, 419 pages. 2006.

Vol. 4196: K. Fischer, I.J. Timm, E. André, N. Zhong (Eds.), Multiagent System Technologies. X, 185 pages. 2006. (Sublibrary LNAI).

Vol. 4195: D. Gaiti, G. Pujolle, E. Al-Shaer, K. Calvert, S. Dobson, G. Leduc, O. Martikainen (Eds.), Autonomic Networking. IX, 316 pages. 2006.

Vol. 4194: V.G. Ganzha, E.W. Mayr, E.V. Vorozhtsov (Eds.), Computer Algebra in Scientific Computing. XI, 313 pages. 2006.

Vol. 4193: T.P. Runarsson, H.-G. Beyer, E. Burke, J.J. Merelo-Guervós, L. D. Whitley, X. Yao (Eds.), Parallel Problem Solving from Nature - PPSN IX. XIX, 1061 pages. 2006.

Vol. 4192: B. Mohr, J.L. Träff, J. Worringen, J. Dongarra (Eds.), Recent Advances in Parallel Virtual Machine and Message Passing Interface. XVI, 414 pages. 2006.

Vol. 4191: R. Larsen, M. Nielsen, J. Sporring (Eds.), Medical Image Computing and Computer-Assisted Intervention – MICCAI 2006, Part II. XXXVIII, 981 pages. 2006.

Vol. 4190: R. Larsen, M. Nielsen, J. Sporring (Eds.), Medical Image Computing and Computer-Assisted Intervention – MICCAI 2006, Part I. XXXVVIII, 949 pages. 2006.

Vol. 4189: D. Gollmann, J. Meier, A. Sabelfeld (Eds.), Computer Security – ESORICS 2006. XI, 548 pages. 2006.

Vol. 4188: P. Sojka, I. Kopeček, K. Pala (Eds.), Text, Speech and Dialogue. XIV, 721 pages. 2006. (Sublibrary LNAI).

Vol. 4187: J.J. Alferes, J. Bailey, W. May, U. Schwertel (Eds.), Principles and Practice of Semantic Web Reasoning. XI, 277 pages. 2006.

Vol. 4186: C. Jesshope, C. Egan (Eds.), Advances in Computer Systems Architecture. XIV, 605 pages. 2006.

Vol. 4185: R. Mizoguchi, Z. Shi, F. Giunchiglia (Eds.), The Semantic Web – ASWC 2006. XX, 778 pages. 2006.

Vol. 4184: M. Bravetti, M. Núñez, G. Zavattaro (Eds.), Web Services and Formal Methods. X, 289 pages. 2006.

Vol. 4183: J. Euzenat, J. Domingue (Eds.), Artificial Intelligence: Methodology, Systems, and Applications. XIII, 291 pages. 2006. (Sublibrary LNAI).

Vol. 4182: H.T. Ng, M.-K. Leong, M.-Y. Kan, D. Ji (Eds.), Information Retrieval Technology. XVI, 684 pages. 2006.

Vol. 4180: M. Kohlhase, OMDoc – An Open Markup Format for Mathematical Documents [version 1.2]. XIX, 428 pages. 2006. (Sublibrary LNAI).

Vol. 4179: J. Blanc-Talon, W. Philips, D. Popescu, P. Scheunders (Eds.), Advanced Concepts for Intelligent Vision Systems. XXIV, 1224 pages. 2006.

Vol. 4178: A. Corradini, H. Ehrig, U. Montanari, L. Ribeiro, G. Rozenberg (Eds.), Graph Transformations. XII, 473 pages. 2006.

Vol. 4177: R. Marín, E. Onaindía, A. Bugarín, J. Santos (Eds.), Current Topics in Artificial Intelligence. XV, 482 pages. 2006. (Sublibrary LNAI).

Vol. 4176: S.K. Katsikas, J. Lopez, M. Backes, S. Gritzalis, B. Preneel (Eds.), Information Security. XIV, 548 pages. 2006.

Vol. 4175: P. Bücher, B.M.E. Moret (Eds.), Algorithms in Bioinformatics. XII, 402 pages. 2006. (Sublibrary LNBI).

Vol. 4174: K. Franke, K.-R. Müller, B. Nickolay, R. Schäfer (Eds.), Pattern Recognition. XX, 773 pages. 2006.

Vol. 4173: S. El Yacoubi, B. Chopard, S. Bandini (Eds.), Cellular Automata. XV, 734 pages. 2006.

Vol. 4172: J. Gonzalo, C. Thanos, M. F. Verdejo, R.C. Carrasco (Eds.), Research and Advanced Technology for Digital Libraries. XVII, 569 pages. 2006.

Vol. 4169: H.L. Bodlaender, M.A. Langston (Eds.), Parameterized and Exact Computation. XI, 279 pages. 2006.

Vol. 4168: Y. Azar, T. Erlebach (Eds.), Algorithms – ESA 2006. XVIII, 843 pages. 2006.

Vol. 4167: S. Dolev (Ed.), Distributed Computing. XV, 576 pages. 2006.

Vol. 4166: J. Górski (Ed.), Computer Safety, Reliability, and Security. XIV, 440 pages. 2006.

Vol. 4165: W. Jonker, M. Petković (Eds.), Secure, Data Management. X, 185 pages. 2006.

GPSR Compliance

The European Union's (EU) General Product Safety Regulation (GPSR) is a set of rules that requires consumer products to be safe and our obligations to ensure this.

If you have any concerns about our products, you can contact us on ProductSafety@springernature.com

In case Publisher is established outside the EU, the EU authorized representative is:

Springer Nature Customer Service Center GmbH
Europaplatz 3
69115 Heidelberg, Germany

Batch number: 10844661

Printed by Printforce, the Netherlands